Communications in Computer and Information Science 611

Commenced Publication in 2007
Founding and Former Series Editors:
Alfredo Cuzzocrea, Dominik Ślęzak, and Xiaokang Yang

More information about this series at http://www.springer.com/series/7899

Joao Paulo Carvalho · Marie-Jeanne Lesot
Uzay Kaymak · Susana Vieira
Bernadette Bouchon-Meunier
Ronald R. Yager (Eds.)

Information Processing and Management of Uncertainty in Knowledge-Based Systems

16th International Conference, IPMU 2016
Eindhoven, The Netherlands, June 20–24, 2016
Proceedings, Part II

Editors
Joao Paulo Carvalho
INESC-ID, Instituto Superior Técnico
Universidade de Lisboa
Lisboa
Portugal

Marie-Jeanne Lesot
LIP6
Université Pierre et Marie Curie
Paris
France

Uzay Kaymak
School of Industrial Engineering
Eindhoven University of Technology
Eindhoven
The Netherlands

Susana Vieira
IDMEC, Instituto Superior Técnico
Universidade de Lisboa
Lisboa
Portugal

Bernadette Bouchon-Meunier
LIP6
Université Pierre et Marie Curie, CNRS
Paris
France

Ronald R. Yager
Machine Intelligence Institute
Iona College
New Rochelle, NY
USA

ISSN 1865-0929 ISSN 1865-0937 (electronic)
Communications in Computer and Information Science
ISBN 978-3-319-40580-3 ISBN 978-3-319-40581-0 (eBook)
DOI 10.1007/978-3-319-40581-0

Library of Congress Control Number: 2016941088

Printed on acid-free paper

This Springer imprint is published by Springer Nature
The registered company is Springer International Publishing AG Switzerland

Preface

These are the proceedings of the 16th International Conference on Information Processing and Management of Uncertainty in Knowledge-Based Systems, IPMU 2016. The conference was held during June 20–24, 2016, in Eindhoven, The Netherlands: one of the vibrant hi-tech hot spots of Europe. The IPMU conference is organized every two years with the aim of bringing together scientists working on methods for the management of uncertainty and aggregation of information in intelligent systems.

Since 1986, the IPMU conference has been providing a forum for the exchange of ideas between theoreticians and practitioners working in these areas and related fields. In addition to the many contributed scientific papers, the conference has in the past attracted prominent plenary speakers, including the Nobel Prize winners Kenneth Arrow, Daniel Kahneman, and Ilya Prigogine. Another important feature of the conference is the presentation of the Kampé de Fériet Award for outstanding contributions to the field of uncertainty and management of uncertainty. Past winners of this prestigious award were Lotfi A. Zadeh (1992), Ilya Prigogine (1994), Toshiro Terano (1996), Kenneth Arrow (1998), Richard Jeffrey (2000), Arthur Dempster (2002), Janos Aczel (2004), Daniel Kahneman (2006), Enric Trillas (2008), James Bezdek (2010), Michio Sugeno (2012), and Vladimir N. Vapnik (2014). This year, the recipient was Joseph Y. Halpern from Cornell University, USA.

IPMU 2016 had a rich scientific program. Four invited overview talks (tutorials) were given on the first day, identifying the challenges and discussing the various methods in the field of information processing and the management of uncertainty. Further, the program consisted of five invited plenary talks, 13 special sessions, 127 contributed papers that were authored by researchers from 34 different countries, industry round tables, and discussion panels. The plenary presentations were given by the following distinguished researchers: Chris Dyer (Carnegie Mellon University, USA), Joseph Y. Halpern (Cornell University, USA), Katharina Morik (Technische Universität Dortmund, Germany), Peter P. Wakker (Erasmus University Rotterdam, The Netherlands), and Ronald R. Yager (Iona College, USA). All contributed papers underwent the same review process and were judged by at least two reviewers; 90 % of the papers were reviewed by three or more referees, and some papers by as many as five referees. Furthermore, all papers were scrutinized by the program chairs, meaning that each paper was studied by three to six independent researchers. The review process also respected the usual conflict-of-interest standards, so that all papers received blinded, independent evaluations.

Organizing a conference like IPMU 2016 is not possible without the assistance, dedication, and support of many people and institutions. We want to thank our industry sponsors, the institutional sponsors, and the material sponsors. Our sponsor chair, Paul Grefen, did an excellent job in attracting the interest and support from industry for the success of IPMU 2016. We are also particularly grateful to the organizers of sessions on dedicated topics that took place during the conference—these special sessions have always been a characteristic element of the IPMU conference. Special thanks go to Joao Sousa, who helped evaluate and select the special session proposals. The help of the members of the international Program Committee as well as multiple reviewers was essential in safeguarding the scientific quality of the conference. The local Organizing Committee is very grateful for the efforts of multiple student volunteers who provided practical support during the conference.

Finally, we gratefully acknowledge the technical support of several organizations and institutions, notably the IEEE Computational Intelligence Society, the European Society for Fuzzy Logic and Technology (EUSFLAT), and the Netherlands Research School for Information and Knowledge Systems (SIKS). Last, but not least, our greatest gratitude goes to the authors who submitted their work and presented it at the conference!

April 2016

Rui J. Almeida
Joao Paulo Carvalho
Marie-Jeanne Lesot
Anna M. Wilbik
Bernadette Bouchon-Meunier
Uzay Kaymak
Susana Vieira
Ronald R. Yager

Organization

General Chair

Uzay Kaymak	Technische Universiteit Eindhoven, The Netherlands

Program Chairs

Joao Paulo Carvalho	INESC-ID, Instituto Superior Técnico, Universidade de Lisboa, Portugal
Marie-Jeanne Lesot	LIP6 - Université Pierre et Marie Curie, Paris, France

Executive Directors

Bernadette Bouchon-Meunier	LIP6 - Université Pierre et Marie Curie, CNRS, Paris, France
Ronald R. Yager	Iona College, USA

Finance Chair

Anna Wilbik	Technische Universiteit Eindhoven, The Netherlands

Publicity Chair

Rui Jorge de Almeida	Technische Universiteit Eindhoven, The Netherlands

Special Session Chair

João M.C. Sousa	IDMEC, Instituto Superior Técnico, Universidade de Lisboa, Portugal

Publication Chair

Susana Vieira	IDMEC, Instituto Superior Técnico, Universidade de Lisboa, Portugal

Sponsor Chair

Paul W.P.J. Grefen	Technische Universiteit Eindhoven, The Netherlands

Local Organizing Committee

Uzay Kaymak	Technische Universiteit Eindhoven, The Netherlands
Rui Jorge de Almeida	Technische Universiteit Eindhoven, The Netherlands
Anna Wilbik	Technische Universiteit Eindhoven, The Netherlands
Annemarie van der Aa	Technische Universiteit Eindhoven, The Netherlands
Caro Fuchs	Technische Universiteit Eindhoven, The Netherlands

International Advisory Board

Giulianella Coletti, Italy
Miguel Delgado, Spain
Mario Fedrizzi, Italy
Laurent Foulloy, France
Salvatore Greco, Italy
Julio Gutierrez-Rios, Spain
Eyke Hüllermeier, Germany
Anne Laurent, France
Luis Magdalena, Spain
Christophe Marsala, France
Benedetto Matarazzo, Italy
Manuel Ojeda-Aciego, Spain
Maria Rifqi, France
Lorenza Saitta, Italy
Olivier Strauss, France
Enric Trillas, Spain
Llorenç Valverde, Spain
José Luis Verdegay, Spain
Maria-Amparo Vila, Spain
Lotfi A. Zadeh, USA

Program Committee

Rui Jorge de Almeida
Michał Baczyński
Nalan Bastürk
Gleb Beliakov
Radim Belohlavek
Salem Benferhat
Jim Bezdek
Isabelle Bloch
Ulrich Bodenhofer
Piero Bonissone
Humberto Bustince
Guoqing Chen
Esma Nur Cinicioglu
Carlos Coello Coello
Oscar Cordon
Ana Colubi
Didier Coquin
Inés Couso
Keeley Crockett
Fabio Cuzzolin
Bernard De Baets
Guy De Tré
Sébastien Destercke
Marcin Detyniecki
Antonio Di Nola
Remco Dijkman
Didier Dubois
Fabrizio Durante
Francesc Esteva
Janos Fodor
David Fogel
Sylvie Galichet
Patrick Gallinari
Maria Angeles Gil
Lluis Godo
Fernando Gomide
Gil González Rodríguez
Michel Grabisch
Steve Grossberg
Przemysław Grzegorzewski
Lawrence Hall
Francisco Herrera
Enrique Herrera-Viedma
Ludmilla Himmelspach
Kaoru Hirota
Janusz Kacprzyk
Uzay Kaymak
Cengiz Kahraman
Abraham Kandel
James Keller
Frank Klawonn
Erich Peter Klement
Laszlo Koczy
Vladik Kreinovich
Tomas Kroupa
Rudolf Kruse
Christophe Labreuche
Jérome Lang
Henrik Larsen
Mark Last
Weldon Lodwick
Edwin Lughofer

Jean-Luc Marichal
Trevor Martin
Sebastian Massanet
Mylène Masson
Silvia Massruhá
Gilles Mauris
Gaspar Mayor
Jerry Mendel
Radko Mesiar
Ralf Mikut
Enrique Miranda
Javier Montero
Jacky Montmain
Serafín Moral
Zbigniew Nahorski
Yusuke Nojima
Vilem Novak
Hannu Nurmi
Nikhil Pal
Endre Pap
Simon Parsons
Gabriella Pasi
Witold Pedrycz
Irina Perfilieva
Fred Petry
Vincenzo Piuri
Olivier Pivert
Henri Prade
Anca Ralescu
Dan Ralescu
Mohammed Ramdani
Marek Reformat
Adrien Revault d'Allonnes
Beloslav Riecan
Maria Dolores Ruiz
Thomas Runkler
Enrique Ruspini
Daniel Sanchez
Mika Sato-Ilic
Glen Shafer
Roman Słowinski
Grégory Smits
Joao Sousa
Pilar Sobrevilla
Martin Stepnicka
Umberto Straccia
Michio Sugeno
Eulalia Szmidt
Kay-Chen Tan
Bruno Teheux
Settimo Termini
Konstantin Todorov
Vicenc Torra
I. Burhan Turksen
Bülent Tütmez
Linda van der Gaag
Herman van Dijk
Barbara Vantaggi
Michel Verleysen
Thomas Vetterlein
Susana Vieira
Anna Wilbik
Sławomir Zadrożny
Hans-Jürgen Zimmermann
Jacek Zurada

Additional Reviewers

Cristina Alcalde
Alessandro Antonussi
Louis Aslett
Edurne Barrenechea
Maciej Bartoszuk
Fernando Batista
Benjamin Bedregal
Libor Behounek
Gleb Beliakov
Sarra Ben Hariz
Chiheb-Eddine Ben N'Cir
Magdalena Bendova
Mara José Benítez Caballero
Hanen Borchani
Felix Bou
Reda Boukezzoula
Denis Bouyssou
Yamine Bouzembrak
Christian Braune
Alberto Bugarìn
Ana Burusco
Inma Cabrera
Marta Cardin
Anna Cena
Alireza Chakeri
Mouna Chebbah
Petr Cintula
Maria Eugenia Cornejo Piñero
Miguel Couceiro
Maria Jose Del Jesus
Cyril De Runz
Denisa Diaconescu
Graaliz Dimuro
Alexander Dockhorn
Christoph Doell
Paweł Drygaś
Zied Elouedi
Javier Fernandez

Tommaso Flaminio
María Angeles Galán García
Marek Gagolewski
Juan Gomez Romero
Petra Hodakova
Aoi Honda
Petr Hurtik
Johan Jacquemin
Simon James
Balasubramaniam Jayaram
Ilyes Jenhani
Chee Kau Lim
Frank Klawonn
David Lobo
Nicolas Madrid
Enrico Marchioni
Nicolas Marin
Arnaud Martin
Maria J. Martin-Bautista
Brice Mayag
Jesús Medina
Thuraya Mellah
Jos M. Merigó
Pedro Miranda
Gildas Morvan
Mirko Navara
Tatiane Nogueira Rios
Michael Oberguggenberger
Pere Pardo
Barbara Pękala
Davide Petturiti
Marc Pirlot
Badran Raddaoui
Eloisa Ramírez Poussa
Alejandro Ramos
Jordi Recasens
Ricardo Ribeiro
Juan Vicente Riera
Ricardo Oscar Rodriguez
Jonas Rogger
Christoph Roschger
Daniel Ruiz-Aguilera
Lorenza Saitta
Ahmed Samet
Laura Schnüriger
Karima Sedki
Jose-Maria Serrano
Prakash Shenoy
Andrzej Skowron
Alexander Sostak
Jana Spirkova
Eiichiro Takahagi
Joan Torrens
Gracian Trivino
Esko Turunen
Lev Utkin
Lionel Valet
Llorenc Valverde
Francisco J. Valverde Albacete
Jan Van den Berg
Amanda Vidal Wandelmer
Zdenek Wagner
Gero Walter

Special Session Organizers

Mohamed Anis Bach Tobji	University of Mannouba, Tunisia
Humberto Bustince	Public University of Navarra, Spain
Mouna Chebbah	University of Jendouba, Tunisia
Marek Gagolewski	Polish Academy of Sciences, Warsaw, Poland
Lluis Godo	Artificial Intelligence Research Institute IIIA, Spain
Maria J. Martin-Bautista	University of Granada, Spain
Daniel Sánchez	University of Granada, Spain
Bruno Teheux	University of Luxembourg, Luxembourg
Michał Baczynski	University of Silesia, Poland

Gleb Beliakov	School of Information Technology, Melbourne, Australia
Boutheina Ben Yaghlane	University of Carthage, Tunisia
Didier Coquin	LISTIC - Polytech Annecy-Chambéry, France
Bernard De Baets	Ghent University, Belgium
Sébastien Destercke	Université de Technologie de Compiègne, France
Juan Gómez Romero	University of Granada, Spain
Michel Grabisch	University Paris Sorbonne, Paris, France
Christophe Labreuche	Thales Research and Technology, Palaiseau, France
Sebastia Massanet	University of the Balearic Islands, Spain
Jesús Medina	University of Cádiz, Spain
Javier Montero	Universidad Complutense de Madrid, Spain
Manuel Ojeda Aciego	University of Málaga, Spain
Irina Perfilieva	University of Ostrava, Czech Republic
M. Dolores Ruiz	University of Granada, Spain
Gero Walter	Eindhoven University of Technology, The Netherlands

Contents – Part II

Fuzzy Logic, Formal Concept Analysis and Rough Sets

Graded and Many-Valued Modal Logics

Databases and Information Systems

Computational Aspects of Data Aggregation and Complex Data Fusion

Fuzzy Sets and Fuzzy Logic

Decision Support

Comparison Measures

Contents – Part I

Fuzzy Measures and Integrals

Uncertainty Quantification with Imprecise Probability

Textual Data Processing

Belief Function Theory and Its Applications

Graphical Models

Fuzzy Implication Functions

Fuzzy Logic, Formal Concept Analysis and Rough Sets

(Ir)relevant T-norm Joint Distributions in the Arithmetic of Fuzzy Quantities

Andrea Sgarro[1] and Laura Franzoi[2(✉)]

[1] Department of Mathematics and Geosciences, University of Trieste, Trieste, Italy
sgarro@units.it

[2] Faculty of Mathematics and Computer Science, University of Bucharest, Bucharest, Romania
laura.franzoi@gmail.com

Abstract. *Irrelevance*, a notion put forward in [5,12], is a convenient tool to speed up computations in the arithmetic of *interactive* fuzzy numbers, which were first put forward in the seminal paper [3]. To make our point, below we deal with standard and less standard binary operations for fuzzy quantities whose interactivity is described by a t-norm joint distribution function.

Keywords: Fuzzy arithmetic · Interactive fuzzy numbers · t-norm joint distributions

1 Introduction

In the approach to fuzzy arithmetic which we are taking, an approach which was pioneered in [3] and which is largely followed today, cf. e.g. [1,2,6,8,10], a fuzzy n-tuple $X_1, \ldots, X_n$ is defined by giving its *distribution function*, or simply its *distribution*, $f(\underline{x}) : \mathbb{R}^n \to [0,1]$, $\underline{x} \doteq x_1, \ldots, x_n$, where the equation $f(\underline{x}) = 1$ admits of at least one solution. The *marginal distribution* of any m-tuple thereof, say $X_1, \ldots, X_m$, $m < n$, is defined by taking a supremum:

$$g(x_1, \ldots, x_m) = \sup_{\text{all } x_{m+1}, \ldots, x_n} f(x_1, \ldots, x_n)$$

The advantage of the joint distribution approach is not only its generality, but first and foremost the fact that the computation rules are *the same* as those in the usual arithmetic of crisp numbers, cf. Appendix A. We speak about fuzzy *quantities*, rather than *numbers*, to make it clear that we are not making any special assumption on distribution functions, unless explicitly specified, even if we are quite aware that in practice sensible restrictions are needed, as done below. If $\psi : \mathbb{R}^n \to \mathbb{R}$ is a function, the distribution $Z(z)$ of the fuzzy quantity $Z \doteq \psi(X_1, \ldots, X_n)$ is given by:

$$Z(z) = \sup_{\underline{x}:\, \psi(\underline{x}) = z} f(\underline{x})$$

J.P. Carvalho et al. (Eds.): IPMU 2016, Part II, CCIS 611, pp. 3–11, 2016.
DOI: 10.1007/978-3-319-40581-0_1

with $Z(z) = 0$ when the set on which one takes the supremum is void. A relevant case is when $n = 2$ and so $\psi(x, y)$ is a binary operation $x \circ y$ on fuzzy quantities X and Y of joint distribution $f(x, y)$, $Z = \psi(X, Y) \doteq X \circ Y$.

$$Z(z) = \sup_{x,y:\, x \circ y = z} f(x, y) \doteq \sup_{x,y:\, x \circ y = z} [X, Y](x, y)$$

Let us fix two marginal distributions $X(x)$ and $Y(y)$. A *joint distribution* $f(x, y) = [X, Y](x, y)$ with those marginals can be derived from any t-norm $\top$, cf. [4,11], by just setting $f(x, y) = X(x) \top Y(y)$. Relevant t-norms are e.g.:

$x \top y = x \wedge y$ (*minimum, non-interactive*)
$x \top y = x \times y$ (*product, probabilistic*)
$x \top y = (x + y - 1) \vee 0$ (*Łukasiewicz*)
$x \top 1 = 1 \top x = x$, else $x \top y = 0$ (*drastic*)

The wedge $\wedge$ denotes a minimum, while $\vee$ is a maximum. An example of an important joint distribution with given marginals which is *not* derivable from a t-norm is instead *deterministic equality* $X = Y$; in this case one has to assume *equidistribution*[1] of X and Y, i.e. $X(x) = Y(x)$ for all x, and one sets $f(x, x) \doteq X(x) = Y(x)$, else $f(x, y) = 0$. As for the operands X and Y, Lemma 1 allows one to restrict one's attention to fuzzy quantities of simple types, e.g. increasing or decreasing on the interval which is their support.

In [5,12] the authors have introduced in an explicit and convenient way the notion of *irrelevance* w.r. to a family $\mathcal{F}$ of joint distributions $f(x, y)$ with fixed marginals X and Y: what is meant is that a binary operation $x \circ y$ is given, and that the distribution of $Z = X \circ Y$ remains the same over $\mathcal{F}$: in practice it depends only on the two marginal distributions $X(x)$ and $Y(y)$. In the case of joint distributions derived from a family $\mathcal{N}$ of t-norms, one can equivalently speak of irrelevance w.r. to $\mathcal{N}$, cf. next Section. The *motivation* for this paper is to bring out the convenience of using a tool as is irrelevance to streamline and speed up computations in fuzzy arithmetic. To achieve this goal, we shall spot remarkable cases of irrelevance for relevant binary operations between fuzzy quantities, cf. Theorems 1 and 2 below, but also Lemma 2 and Example 6. Our results on t-norm joint distributions always involve the *drastic* distribution, for which computations are quite easy, cf. Lemma 2; cf. also the examples to follow in Sect. 3. The results of this paper, in particular Corollary 1, generalize results on sums and products already found in literature [1–3,6–8,10]: our monotone binary operations $X \circ Y$ as in the body of the paper do cover as special cases sums and products, in the latter case assuming however "pure" supports (all non-negative or all non-positive), and using in case the identities $xy = (-x) \cdot (-y) = -[(-x) \cdot y]$. As general references for fuzzy set theory, cf. e.g. [4,9].

[1] To avoid misunderstandings, in this paper we carefully distinguish between *equidistribution* $X(x) = Y(x)$ and *deterministic equality* $X = Y$; deterministic equality is a very special *joint distribution* for X and Y, and only when $X = Y$ one of the two symbols X or Y is disposable.

2 Preliminary Lemmas

Lemmas in this section are straightforward; they are quoted explicitly to facilitate referencing, but their easy proofs are omitted or at most hinted at. The fuzzy quantities involved may be quite unruly, unless otherwise specified.

The next lemma explains why one can often operate on separate components, with supports not even necessarily disjoint, e.g. *monotone* components. One needs the distributive property:

$$(a \vee b) \top c = (a \top c) \vee (b \top c)\ , \quad a, b, c \in [0, 1] \tag{1}$$

which holds for any t-norm $\top$, being implied by t-norm monotonicity. First we put forward a notation: by $\langle X_1, X_2 \rangle$ we denote the fuzzy quantity X whose distribution[2] is $X(x) = X_1(x) \vee X_2(x)$, and more generally by $\langle X_i \rangle_{i \in \mathcal{I}}$ we denote the fuzzy quantity X whose distribution is $X(x) = \sup_{i \in \mathcal{I}} X_i(x)$:

$$\langle X_i \rangle_{i \in \mathcal{I}}(x) \ = \ \sup_{i \in \mathcal{I}} X_i(x)$$

Lemma 1. *If the joint distribution of X and Y is derived from a t-norm and $X = \langle X_i \rangle_{i \in \mathcal{I}}$ then*

$$X \circ Y \ = \ \big\langle X_i \circ Y \big\rangle_{i \in \mathcal{I}} \tag{2}$$

where the joint distribution of X_i and Y is derived from the same t-norm, $i \in \mathcal{I}$.

A similar property holds for $Y = \langle Y_j \rangle_{j \in \mathcal{J}}$. Below by $I_{\mathcal{I}}$ we mean a fuzzy quantity whose distribution function is 1 over $\mathcal{I} \subseteq \mathbb{R}$, else is 0; if $\mathcal{I}$ has size 1, $\mathcal{I} = \{a\}$, one re-finds *crisp numbers.* If X is any fuzzy quantity, we set $I_X \doteq I_{K[X]}$ where $K[X] \doteq \{x : X(x) = 1\}$ is the *kernel* of X. Assume *drastic* interactivity and just observe that, whatever the fuzzy quantity X, I_X can be written as the *sup* of crisp a's in the sense of Lemma 1, $a \in K[X]$; observe also that for the fuzzy couple $[X, a]$ only *one* joint distribution has the given marginals:

Lemma 2. *With* drastic interactivity *one has:*

$$Z \doteq X \circ Y \ = \ \big\langle X \circ I_Y\ ,\ I_X \circ Y \big\rangle$$

Moreover, when computing $X \circ I_Y$, or $I_X \circ Y$, the t-norm used is irrelevant.

Example 1. Take X and Y *unimodal*, $X(x) = Y(y) = 1$ only for $x = \alpha$ and $y = \beta$. If the operation is multiplication and $\alpha \times \beta \neq 0$, then $X \times \beta$ has distribution function $X(x)/\beta$, while $Y \times \alpha$ has distribution function $Y(y)/\alpha$. So $Z(z) = \big(X(z)/\beta\big) \vee \big(Y(z)/\alpha\big)$. One might take more general operations as in Theorem 1 below.

---

[2] Observe that *no* joint distribution of $X_1 X_2$ is needed. Note also that $\langle X, Y \rangle$ is *not* the same as $X \vee Y$, just think of two crisp numbers; of course in the case of $X \vee Y$ a joint distribution *is* needed.

Fuzzy quantities[3] usually dealt with in literature are as follows:

Definition 1. *A fuzzy quantity* $X = \langle X_1, I_X, X_2 \rangle$ *with an upper semicontinuous distribution function* $X(x)$ *is a* fuzzy interval (a, b, c, d), $a \le b \le c \le d$, *when* $X_1(x)$ *is (possibly weakly) increasing on the support* $[a, b]$ *of* X_1, I_X *has support* $[b, c] = K[X]$, *and* $X_2(x)$ *is (possibly weakly) decreasing on the support* $[c, d]$ *of* X_2.

Some of the *components* of a fuzzy interval might be lacking: if $b = c$ one finds a *fuzzy number*, or even a crisp number if a, b, c, d are all equal. Actually, all the three components of a fuzzy interval, X_1 increasing, I_X "flat" and X_2 decreasing, are themselves fuzzy intervals, be they "degenerate"; Lemma 1 allows one to deal with them separately, and it is in terms of such simple components that Theorems 1 and 2 are stated: the fuzzy quantities involved might be quite odd, say $X(x) = |\sin \frac{1}{x}|$ for $x > 0$, $X(0) = 0$, which has countably infinite monotone components.

Definition 2. *Given two fuzzy quantities* X *and* Y, *a binary operation* $x \circ y$ *and a family* $\mathcal{F}$ *of* admissible *joint distributions, i.e. of joint distributions with the assigned marginals, there is* irrelevance *with respect to* $\mathcal{F}$ *when the distribution function of* $Z \doteq X \circ Y$ *is the same whatever the admissible joint distribution in* $\mathcal{F}$.

We need two *partially ordered sets.* Norms and distributions can be ordered in an obvious way, by requiring that the corresponding order holds *whatever* the value of their arguments. As for t-norms, as well known [4], there is an *absolute minimum*, the drastic t-norm, and an *absolute maximum*, the non-interactive t-norm; as equally well-known, Łukasiewicz t-norm precedes the probabilistic one. Clearly, two joint distributions $f_1(x, y)$ and $f_2(x, y)$ with fixed marginals obtained by two t-norms such that $\top_1 \le \top_2$ verify the same inequality.

If $f_1(x)$ and $f_2(x)$ are the distribution functions of the fuzzy quantities X_1 and X_2, and if X_2 *dominates* X_1, i.e. if $f_2(x) \ge f_1(x)$ for all x, one soon checks that $X_2 \circ Y$ dominates $X_1 \circ Y$, whatever the binary operation involved, provided only that the two joint distributions are derived from the same t-norm.

If there is irrelevance w.r. to $\mathcal{F} \doteq \{f_1(x, y), f_2(x, y)\}$ with $f_1(x, y) < f_2(x, y)$, one soon checks that there is irrelevance also w.r. to *any* family comprising only joint distributions *intermediate* between the two. In particular, if $f_1(x, y)$ is derived from the drastic t-norm, one can take advantage of the fact that drastic computations are quite easy, cf. above Lemma 2.

[3] The *support* of X is defined as $\{x : X(x) \neq 0\}$; however, with a mild but convenient imprecision, we shall say that a fuzzy quantity of distribution $X(x)$ has *closed* support $[a, b]$ even if $X(a)$ and/or $X(b)$ is 0. Below we might even have $a = -\infty$ and/or $b = +\infty$, but in this case $X(a)$ and/or $X(b)$ are bound to be 0; thus, in the following our supports need *not* be limited.

3 Two Cases of Irrelevance

Below $x \circ y$ is a *commutative* binary operation which is *continuous and strictly increasing* on the supports of the fuzzy operands X and Y.

Theorem 1. Irrelevance of non-interactivity vs. drastic interactivity.
Let X and Y with supports $[a, b]$ and $[\gamma, \delta]$, respectively, be an increasing and a decreasing component as in Definition 1, $X(b) = Y(\gamma) = 1$. Let the joint distribution function of the fuzzy couple $[X, Y]$ be derived from any given t-norm, and take a binary operation $x \circ y$ as above. One has $Z = \langle X \circ \gamma, Y \circ b\rangle$, whatever the t-norm; $X \circ \gamma$ has support $[a \circ \gamma, b \circ \gamma]$ while $Y \circ b$ has support $[b \circ \gamma, b \circ \delta]$.

Proof. Assuming that $x, y, z = x \circ y$ belong to the respective supports, by $z \ominus x$ we denote the single y such that $x \circ y = y \circ x = z$. As soon checked: $x \leq z \ominus y$ iff $z \geq x \circ y$, and $z \ominus [z \ominus y] = y$. Writing $y = z \ominus x$ rather than $x \circ y = z$, one has $Z(z) = \max_x X(x) \top Y(z \ominus x)$, where x is constrained as follows: $\{x : x \leq b\}$ and also $\{x : z \ominus x \geq \gamma\} = \{x : x \leq z \ominus \gamma\}$. The function to be maximized is weakly increasing in x, so being $Y(z \ominus x)$, and therefore the maximum is achieved on the border for $x = b \wedge z \ominus \gamma$. In the case $z \leq b \circ \gamma$, i.e. $b \geq z \ominus \gamma$, one has $Z(z) = X(z \ominus \gamma) \top Y(z \ominus [z \ominus \gamma]) = X(z \ominus \gamma) \top Y(\gamma)$. In the second case, $z \geq b \circ \gamma$, i.e. $b \leq z \ominus \gamma$, one has $Z(z) = X(b) \top Y(z \ominus b)$. Recall that $X(b) = Y(\gamma) = 1$, the neuter element for any t-norm. Since the binary operation is strictly increasing, one soon checks that $X(z \ominus \gamma) = X(z) \circ \gamma$, $Y(z \ominus b) = Y(z) \circ b$. The same result had been found in greater generality for the drastic case, Lemma 1.

Example 2. Let X be linear increasing on its support $[0, 1]$, $Y(y) = e^{2-y}$ for $y \geq 2$ (cf. footnote 3). Take the three operations $x + y$, $x \times y$, $x \diamond y \doteq \sqrt{x + y}$; correspondingly, the supports of $X \circ Y$ are the halflines $z > 2$, $z > 0$, $z > \sqrt{2}$. One has $b \circ \gamma = 3$, 2, $\sqrt{3}$, respectively, and $z \ominus x = z - x$, $\frac{z}{x}$, $z^2 - x$, respectively. In the three cases:
$X + Y = \langle X + 2, Y + 1\rangle$
$X \times Y = \langle X \times 2, Y\rangle$
$X \diamond Y = \langle \sqrt{X + 2}, \sqrt{Y + 1}\rangle$
E.g. in the last case $[X \diamond Y](z)$ is $z^2 - 2$ on $[\sqrt{2}, \sqrt{3}]$ and is e^{3-z^2} for $z \geq \sqrt{3}$.

Unfortunately, unlike Theorem 1, the following Theorem 2 not only requires a certain "shape" of the distribution functions, but is also limited to sums, and to t-norms which do not verify the *positivity* condition $x, y > 0 \Rightarrow x \top y > 0$. In Theorem 2 convexity is required in a weak sense, and so X and/or Y are possibly linear; Example 5 will show that convexity is essential. First, we find it convenient to quote explicitly an obvious lemma for two fuzzy quantities X and Y.

Lemma 3. Linear transformations and sums*:*
If there is irrelevance of $X + Y$ over the family of t-norms $\mathcal{N}$, then there is irrelevance over $\mathcal{N}$ also for $\tilde{X} + \tilde{Y}$, $\tilde{X} \doteq aX + b$, $\tilde{Y} \doteq aY + c$, $a \neq 0$.

Thanks to this lemma, for sums one can restrict the attention to X over $[0,1]$ and Y over $[0,\alpha]$, $\alpha \geq 1$ (so, if $\alpha \neq 1$, Y is the term in the sum with larger support); the sum $Z \doteq X+Y$ has a support equal to or included into $[0,\alpha+1]$. Take any t-norm $x \top y$; one has, with $X(x)$ and $Y(y)$ with those supports:

$$Z(z) \doteq \big[X+Y\big](z) = \max_{x:\, x\in[0,1]\cap[z-\alpha,z]} X(x) \top Y(z-x)$$

In the maximization one conveniently distinguishes three cases:

$$x \in [0,z] \text{ for } z \in [0,1];\ x \in [0,1] \text{ for } z \in [1,\alpha]; \\ x \in [z-\alpha,1] \text{ for } z \in [\alpha,\alpha+1] \tag{3}$$

Theorem 2. *Let X and Y be two increasing components (or two decreasing components) as in Definition 1, and let their distribution functions $X(x)$ and $Y(y)$ be convex-cup on the respective supports. There is* irrelevance *for the sum $Z = X+Y$ w.r.t. to the drastic t-norm and the Łukasiewicz t-norm.*

Proof. For X and Y both increasing, let the supports of X and Y be $[0,1]$ and $[0,\alpha]$, $\alpha \geq 1$, cf. Lemma 3; the support of $Z \doteq X+Y$ is at most $[0,\alpha+1]$. One has to maximize the function $\psi_z(x) \doteq X(x)+Y(z-x)-1$, which is convex, so being the terms in the sum (we "cut" at 0 later, using the obvious interchangeability property $\sup_x(\psi_z(x) \vee 0) = \big(\sup_x \psi_z i(x)\big) \vee 0\big)$. In the case $z \in [0,1]$, use convexity after checking that $\psi_z(0) \leq 0$, $\psi_z(z) \leq 0$. In the other two cases, the function $\psi_z(x)$ being convex-cup in x, one obtains $Y(z-1)$ and $X(z-\alpha) \vee Y(z-1)$, respectively, i.e., $X(z-\alpha)$ being zero on $[1,\alpha]$, the same result found for drastic interactivity, Lemma 2.

Example 3. Let X and $Y = \langle Y_1, Y_2 \rangle$ be two triangles $(0,1,1,1)$ and $(1,2,2,4)$ on $[0,1]$ and $[1,4]$, respectively, with all their three components convex, and the respective *modes* in 1 and 2, i.e. $I_X = \{1\}$, $I_Y = \{2\}$; X is "degenerate" and Y is "skew", the supports of Y_1 and Y_2 being [1,2] and [2,4]. Summing by components and using the obvious associativity of maxima, one has for any t-norm (use Lemma 1 in $\stackrel{1}{=}$, Theorem 1 in $\stackrel{2}{=}$, and the fact that $X+Y_1$ dominates $X+2$): $X+Y = X+\langle Y_1.Y_2\rangle \stackrel{1}{=} \langle X+Y_1, X+Y_2\rangle \stackrel{2}{=} \langle X+Y_1, X+2, Y_2+1\rangle = \langle X+Y_1, Y_2+1\rangle$. Assume the t-norm is intermediate between Łukasiewicz and drastic, e.g. is any convex combination of the two; using Theorem 2 in $\stackrel{3}{=}$ and Lemma 1 in $\stackrel{4}{=}$ one can continue: $X+Y = \langle X+Y_1, Y_2+1\rangle \stackrel{3}{=} \langle X+2, Y_1+1, Y_2+1\rangle \stackrel{1}{=} \langle X+2, Y+1\rangle$ on the support [2,5].

Example 4. Go back to Example 3: *parametric* t-norms [4] yield further cases of joint distributions intermediate between drastic and Łukasiewicz over which there is irrelevance: take Schweizer-Sklar t-norm with any parameter $p \leq 1$, or Yager t-norm with $p \in [0,1]$, or Sugeno-Weber t-norm with $p \in [-1,0]$. For definitions and properties of these t-norms cf. [4].

The following example shows that in Theorem 2 convexity is essential.

Example 5. Take $X(x) = \sqrt{x}$, $Y(y) = y$ on $[0,1]$. Take maxima in two cases left in (3) (the intermediate case vanishes as $\alpha = 1$); $Z \doteq X + Y$ has support at most $[0,2]$. With the following four t-norms, non-interactive, probabilistic, Łukasiewicz and drastic, one obtains, respectively:
$Z_1(z) = \frac{1}{2}\big(\sqrt{4z+1} - 1\big)$, $z \in [0,2]$
$Z_2(z) = \frac{2}{3}z\sqrt{\frac{z}{3}}$ for $z \in [0, \frac{3}{2}]$, and $Z_2 = \sqrt{z-1}$ for $z \in [\frac{3}{2}, 2]$
$Z_3(z) = 0$ for $z \in [0, \frac{3}{4}]$, $Z_3(z) = z - \frac{3}{4}$ for $z \in [\frac{3}{4}, \frac{5}{4}]$, $Z_3(z) = \sqrt{z-1}$ for $z \in [\frac{5}{4}, 2]$
$Z_4(z) = \sqrt{z-1}$ for $z \in [1,2]$, else 0

The four t-norms give four *different* results. As for computations, the case of interactivity is quite familiar, but to enhance self-readability we re-take it in Appendix B. With the probabilistic and the Łukasiewicz t-norms, one just maximizes w.r. to x as in (3) the functions $\psi_x(x) = X(x) \times Y(z-x)$ and $X(x) + Y(z-x) - 1$ with standard analytic methods, using again in the Łukasiewicz case the interchangeability property as in the proof of Theorem 2. For the drastic case cf. Lemma 2 above.

In the following corollary to Theorem 1, we take in ample generality two fuzzy intervals with components X_1 increasing on $[a,b]$, $X_F \doteq I_X$ "flat" on the kernel $[b,c]$, X_2 decreasing on $[c,d]$, Y_1 increasing on $[\alpha,\beta]$, $Y_F \doteq I_Y$ "flat" on the kernel $[\beta,\gamma]$ and Y_2 decreasing on $[\gamma,\delta]$. "Flat" components might be lacking (b=c and/or $\beta = \gamma$), and so we deal also with fuzzy numbers. The two cases ruled out in the corollary are covered directly by Theorem 1.

Corollary 1. *Take a binary operation $x \circ y$ as in Theorem 1 and rule out the two cases $X = X_1, Y = Y_2$ and $X = X_2, Y = Y_1$. One has: $X \circ Y = X_1 \circ Y_1$ on $[a \circ \alpha, b \circ \beta]$, $[X \circ Y](z) = 1$ on $[b \circ \beta, c \circ \gamma]$ and $X \circ Y = X_2 \circ Y_2$ on $[c \circ \gamma, d \circ \gamma]$.*

Proof. Thanks to Lemma 1, one can match separately the various components. As for $X_1 \circ Y_2 = \langle X_1 \circ \gamma, Y_2 \circ b \rangle$ and $X_2 \circ Y_1 = \langle X_2 \circ \beta, Y_1 \circ c \rangle$ their contributions can be ignored by using four times dominance (proceed as in Example 3; this would not hold, had we not ruled out the two cases). Clearly, the support of $I_{X \circ Y}$ is $[b \circ \beta, c \circ \gamma]$, and so we need to deal only with contributions before $b \circ \beta$ or after $c \circ \gamma$. The supports of $X_1 \circ Y_1$ and $X_1 \circ Y_F$ intersect on $[a \circ \beta, b \circ \beta]$, but there one has $X_1 \circ Y_F = X_1 \circ \beta$ and so we can once more use dominance. The same holds for $X_2 \circ Y_2$ and $X_F \circ Y_2$.

The following example exhibits a situation where there is irrelevance for t-norms intermediate between drastic and probabilistic, but *not* w.r. to non-interactivity.

Example 6. Take two degenerate equidistributed fuzzy intervals X and Y both $(0,1,2,2)$ with piecewise linear distribution function, $X(x) = x$ on $[0,1]$. Take the binary operation $x \circ y = -\frac{1}{xy}$; cf. footnote 3. On $[-1, -\frac{1}{4}]$ $Z(z) = 1$ whatever the t-norm; one is left with $X_1 \circ Y_1$ on $]-\infty, -1]$. Assume probabilistic interactivity: one has to maximize in x the function $X(x) \times Y(z \ominus x) = x \cdot \frac{-1}{xz}$ which actually does not depend on x. So $Z(z) = -z^{-1}$ for $z \leq -1$. The same result is obtained with the drastic distribution, Lemma 2, and consequently with any t-norm whose joint distribution is intermediate between drastic and probabilistic. Instead a straightforward computation, cf. Appendix B, shows that under non-interactivity $Z(z) = \sqrt{-z^{-1}}$ for $z \leq -1$.

Appendix

A Montecatini Lemma

With *caveat* 2 below, the following lemma basically states that the arithmetic of fuzzy quantities, however unruly, is the *very same* as for crisp numbers. At an INdAM workshop held in Montecatini, Tuscany, the lemma gave rise to a fruitful discussion; we are tentatively using this odd "attribution" because we were unable to trace back a self-standing and explicit formulation of this result in the literature.

Lemma 4. Montecatini lemma:
$f(x_1, \ldots, x_n) = g(x_1, \ldots, x_n)$ is an identity for crisp numbers iff the two fuzzy quantities $Z_1 \doteq f(X_1, \ldots, X_n)$ and $Z_2 \doteq g(X_1, \ldots, X_n)$ are deterministically equal whatever the joint distribution of $X_1, \ldots, X_n$.

E.g., since $x(y + z) = xy + xz$ for any crisp numbers x, y and z, one has $X(Y + Z) = XY + XZ$ for *any* fuzzy quantities X, Y and Z, whatever their joint distribution. Since $\log xy = \log x + \log y$ for positive x and y one has $\log XY = \log X + \log Y$ for *any* fuzzy quantities X and Y whatever their joint distribution with positive support. To prove the equidistribution of Z_1 and Z_2 just observe that one is taking the supremum of the same function over two sets, $\{\underline{x} : f(\underline{x}) = z\}$ and $\{\underline{x} : g(\underline{x}) = z\}$, which are however equal. As for deterministic equality $Z_1 = Z_2$, one cannot have $f(\underline{x}) \neq g(\underline{x})$, and so the joint distribution of the fuzzy couple $[Z_1, Z_2]$ is zero outside the main diagonal $z_1 = z_2$.

Caveat 1: in more traditional literature on fuzzy arithmetic, where in practice only non-interactivity is used to "glue together" marginal distributions, one comes across statements as are $X \times X \neq X^2$ or $X(Y+Z) \neq XY + XZ$. What is meant is that equidistribution between X_1 and X_2 does not imply equidistribution between $X_1 \times X_2$ and X_1^2 or between $X_1(Y+Z)$ and $X_1Y + X_2Z$. We insist that in our approach a single symbol X is used for two fuzzy numbers only when they are *not only equidistributed but also deterministically equal.* Note that the analogue of Montecatini lemma holds also for *random numbers*, i.e. for random variables as used in probabilistic *distribution calculus*, where one takes much care to distinguish between equidistribution and deterministic equality, the latter being a very special form of joint probability. We recall that non-interactivity of fuzzy quantities is seen as an analogue of probabilistic independence between random variables.

Caveat 2: The lemma is stated in terms of arbitrary fuzzy quantities: if one insists on certain properties, e.g. upper continuity or unimodality, one should of course check whether the result $Z = f(X_1, \ldots, X_n) = g(X_1, \ldots, X_n)$ still verifies those properties, so as to ensure stability.

B Computations for Examples 5 and 6

We shortly cover the very familiar case of *non-interactivity* as in Example 5, when the t-norm is $x \wedge y$. We assume that $X(x)$ and $Y(y)$ are continuous and strictly increasing on [0,1] and $[0, \alpha]$, respectively, $\alpha \geq 1$, cf. Lemma 3; so $Y(z-x)$ is strictly decreasing when seen as a functions of x. For fixed z in the interval $[0, 1+\alpha]$, one soon checks that the equation in x $X(x) = Y(z-x)$ has a single solution $\mu(z)$, $X\big(\mu(z)\big) = Y\big(z - \mu(z)\big)$, and that $\mu(z)$ strictly increases in z from $\mu(0) = 0$ to $\mu(1+\alpha) = 1$. One has three cases, cf. (3); in the first, $x \in [0, z]$, fix $z \in [0, 1]$: on the border $x = 0$ the increasing function $X(x)$ and the decreasing function $Y(z-x)$ take the two values $0 = X(0) < Y(z)$, while on the border $x = z$ they take the two values $X(z) > Y(0) = 0$: the required maximum is found for $x = \mu(z)$. The remaining two cases are dealt in the same way and give the same solution $Z(z) = X\big(\mu(z)\big) = Y\big(z - \mu(z)\big)$, which is found also when $X(x)$ and $Y(y)$ are both strictly decreasing. If the support-intervals are different just use Lemma 3. In Example 5 one has $\alpha = 1$; since $X(x) = \sqrt{x}$, $Y(y) = y$, one gets $\mu(z) = z + \frac{1}{2}\big(1 - \sqrt{4z+1}\big)$, and so $Z(z) = Y\big(z - \mu(z)\big) = \frac{1}{2}(\sqrt{4z+1} - 1)$.

As for Example 6, for fixed z, $z \leq -1$, one has to maximize w.r. to x the minimum of two functions $x \wedge \frac{-1}{xz}$, the first increasing, the second decreasing, over the intersection $\{x : 0 \leq x \leq 1\} \cap \{x : 0 \leq \frac{-1}{xz} \leq 1\}$ i.e. over $\{x : -z^{-1} \leq x \leq 1\}$. The two functions meet at $\mu(z) = \sqrt{-z^{-1}}$, $-z^{-1} \leq \mu(z) \leq 1\}$.

References

1. Carlsson, C., Fullér, R., Majlender, P.: Additions of completely correlated fuzzy numbers. In: IEEE International Conference on Fuzzy Systems, pp. 535–539 (2004)
2. Coroianu, L., Fullér, R.: On multiplication of interactive fuzzy numbers. In: IEEE International Symposium on Intelligent Systems and Informatics, SISY 2013, pp. 181–185 (2013)
3. Dubois, D., Prade, H.: Additions of interactive fuzzy numbers. IEEE Trans. Autom. Control **AC–26**(4), 926–930 (1981)
4. Dubois, D., Prade, H.: Fundamentals of Fuzzy Sets. Kluwer Academic Publishers, Dordrecht (2000)
5. Franzoi, L., Sgarro, A.: (Ir)relevance of interactivity in fuzzy arithmetic. Math. Pannonica **25**(1), 1–11 (2015)
6. Fullér, R., Kersztfalvi, T.: T-norm based addition of fuzzy intervals. Fuzzy Sets Syst. **51**, 155–159 (1992)
7. Fullér, R., Majlender, P.: On interactive fuzzy numbers. Fuzzy Sets Syst. **3**(143), 353–369 (2004)
8. Fullér, R., Majender, P.: On addition of interactive fuzzy numbers. Acta Polytech. Hung. **2**, 59–73 (2005)
9. Klir, G.J., Folger, T.A.: Fuzzy Sets, Uncertainty and Information. Prentice Hall Int., Englewood Cliffs (1988)
10. Mesiar: Triangular-norm-based addition of fuzzy intervals. Fuzzy Sets and Systems. **91**, 231–237 (1997)
11. Nelsen, R.: An Introduction to Copulas. Springer, New York (2006)
12. Sgarro, A., Franzoi, L.: Fuzzy arithmetic for fuzzy n-poles. In: Greco, S., Bouchon-Meunier, B., Coletti, G., Fedrizzi, M., Matarazzo, B., Yager, R.R. (eds.) IPMU 2012. CCIS, vol. 299, pp. 1–8. Springer, Heidelberg (2012)

Knowledge Extraction from L-Fuzzy Hypercontexts

Cristina Alcalde[1(✉)] and Ana Burusco[2]

[1] Dept. Matemática Aplicada, Escuela de Ingeniería de Gipuzkoa, UPV/EHU, Plaza de Europa 1, 20018 San Sebastián, Spain
c.alcalde@ehu.es

[2] Departamento de Automática y Computación, Instituto de Smart Cities, Universidad Pública de Navarra, Campus de Arrosadía, 31006 Pamplona, Spain
burusco@unavarra.es

Abstract. As a generalization of the L-fuzzy contexts, we propose the study of the L-fuzzy hypercontexts where the relation R between the objects X and the attributes Y takes as values other L-fuzzy relations. In this work, we propose the study of these structures using OWA operators in different situations. Finally, the practical case that has motivated this paper is analyzed.

Keywords: L-fuzzy contexts · L-fuzzy concepts · L-fuzzy hypercontexts · OWA operators

1 Introduction

The L-fuzzy concept analysis provides a tool for the extraction of knowledge from tables (L-fuzzy contexts) using L-fuzzy concepts. These L-fuzzy contexts are tuples (L, X, Y, R), with L a complete lattice, X and Y the object and attribute sets and $R \in L^{X \times Y}$ an L-fuzzy relation between the objects and the attributes.

L-fuzzy contexts establishing frameworks that allow us to represent $R(x, y)$, $x \in X$ and $y \in Y$, as a collection of values that has the structure of L-fuzzy context with a set of objects Q_x associated with x and a set of attributes S_y associated with y. These sets Q_x are related to each other, and the same happens with the S_y.

For example, consider a chain of supermarkets that has several establishments in different cities. We want to study the evolution in time Y of the sales of different articles in each of the cities X where they are sold.

Our hypercontexts are the extensions to the fuzzy case of certain multicontexts of Wille [13] verifying some properties with respect to the objects and the attributes. On the other hand, the approach given in this work, using OWA operators, will also be different from the topic discussed by Wille.

The main goal of the paper is the study of the values of R taking into account that for every $x \in X$, we have a set of objects Q_x that can be different in every case and for every $y \in Y$, a different set of attributes S_y.

J.P. Carvalho et al. (Eds.): IPMU 2016, Part II, CCIS 611, pp. 12–22, 2016.
DOI: 10.1007/978-3-319-40581-0_2

Although this problem is related to the construction of formal contexts developed in [9], we extend our study to new structures with values in a complete lattice L. Then, fuzzy logic tools and specifically OWA operators can be applied for obtaining information.

As some of the practical cases are represented in a natural way by a fuzzy relation, we consider the new framework as a good contribution because it increases the range of application of formal and fuzzy concept analysis.

Firstly, we will see some results about L-fuzzy concept analysis and OWA operators [11]:

1.1 Formal Concept Analysis and *L*-Fuzzy Concept Analysis

The Formal Concept Analysis of R. Wille [12] extracts information from a binary table that represents a formal context (X, Y, R) with X and Y finite sets of objects and attributes respectively and $R \subseteq X \times Y$. The hidden information consists of pairs (A, B) with $A \subseteq X$ and $B \subseteq Y$, called formal concepts, verifying $A^* = B$ and $B^* = A$, where $(\cdot)^*$ is the derivation operator that associates the attributes related to the elements of A to every object set A, and the objects related to the attributes of B to every attribute set B. These formal concepts can be interpreted as a group of objects A that shares the attributes of B.

In previous works [6,7] we have defined the L-fuzzy contexts (L, X, Y, R), with L a complete lattice, X and Y sets of objects and attributes respectively and $R \in L^{X \times Y}$ a fuzzy relation between the objects and the attributes. This is an extension of Wille's formal contexts to the fuzzy case when we want to study the relations between the objects and the attributes with values in a complete lattice L, instead of binary values.

In our case, to work with these L-fuzzy contexts, we have defined the derivation operators 1 and 2 given by means of these expressions:

$\forall A \in L^X, \forall B \in L^Y$

$$A_1(y) = \inf_{x \in X} \{\mathcal{I}(A(x), R(x, y))\}$$

$$B_2(x) = \inf_{y \in Y} \{\mathcal{I}(B(y), R(x, y))\}$$

with $\mathcal{I}$ a fuzzy implication operator defined in the lattice $(L, \leq)$.

The information stored in the context is visualized by means of the L-fuzzy concepts that are pairs $(A, A_1) \in L^X \times L^Y$ with $A \in fix(\varphi)$, set of fixed points of the operator φ, being defined from the derivation operators 1 and 2 as $\varphi(A) = (A_1)_2 = A_{12}$. These pairs, whose first and second components are said to be the fuzzy extension and intension respectively, represent a group of objects that share a group of attributes.

Using the usual order relation between fuzzy sets, that is,

$$\forall A, C \in L^X, \quad A \leq C \Longleftrightarrow A(x) \leq C(x) \quad \forall x \in X,$$

we define the set $\mathcal{L} = \{(A, A_1)/A \in fix(\varphi)\}$ with the order relation $\preceq$ defined as:

$$\forall (A, A_1), (C, C_1) \in \mathcal{L}, (A, A_1) \preceq (C, C_1) \text{ if } A \leq C (\text{ or } A_1 \geq C_1)$$

As φ is an order preserving operator, by the theorem of Tarski, the set $fix(\varphi)$ is a complete lattice and then $(\mathcal{L}, \preceq)$ is also a complete lattice that is said to be [6,7] the L-fuzzy concept lattice.

On the other hand, given $A \in L^X$, (or $B \in L^Y$) we can obtain the associated L-fuzzy concept applying twice the derivation operators. In the case of using a residuated implication, as we do in this work, the associated L-fuzzy concept is (A_{12}, A_1) (or (B_2, B_{21})).

Other important papers that generalize the Formal Concepts Analysis using residuated implication operators are due to R. Belohlavek [4,5]. Moreover, extensions of Formal Concept Analysis to the interval-valued case are in [1,8] and to the fuzzy property-oriented and multi-adjoint concept lattices framework in [10].

1.2 OWA Operators

Families of OWA operators were introduced by Yager [11] as a new aggregation technique based on the ordered weighted averaging. This is the definition of these operators:

Definition 1. *A mapping F from $L^n \longrightarrow L$, where $L = [0,1]$ is called an OWA operator of dimension n if associated with F is a weighting $n-$tuple $W = (w_1, w_2 \dots w_n)$ such that $w_i \in [0,1]$ and $\sum_{1 \leq i \leq n} w_i = 1$, where $F(a_1, a_2, \dots a_n) = w_1.b_1 + w_2.b_2 + \dots + w_n.b_n$, with b_i the ith largest element in the collection $a_1, a_2, \dots a_n$.*

There are two particular cases of special interest:

W_* defined by the weighting n-tuple with $w_n = 1$ and $w_j = 0, \forall j \neq n$, and W^* defined by the weighting n-tuple such that $w_1 = 1$ and $w_j = 0, \forall j \neq 1$.

It is proved that $F_*(a_1, a_2, \dots a_n) = \min_j(a_j)$ and $F^*(a_1, a_2, \dots a_n) = \max_j(a_j)$. These operators are said to be *and* and *or*, respectively.

To study the L-fuzzy hypercontexts, we are interested in the use of operators close to *or*. To measure this proximity the orness degree can be used [11].

Furthermore, the use of different weighting vectors provides different results as we will see in the paper.

2 *L*-Fuzzy Hypercontexts

We will begin defining the framework:

Definition 2. *We denote the tuple $(L, X, Y, (Q_x)_{x \in X}, (S_y)_{y \in Y}, R)$ by an L-fuzzy hypercontext, with $L = [0,1]$, X and Y sets of objects and attributes respectively, $(Q_x)_{x \in X}$ and $(S_y)_{y \in Y}$ families of sets associated with the elements of X and Y, and R such that $R(x,y)$ is also a new relation $R_{xy} \in L^{Q_x \times S_y}$, for every $(x,y) \in X \times Y$. This relation R_{xy} defines a new L-fuzzy context (L, Q_x, S_y, R_{xy}).*

Remark 1. The objects Q_x associated with $x \in X$ do not have to be elements of X. Neither for S_y.

The main target of the paper is the study of these hypercontexts when $L = [0,1]$. In some cases, we want to make the study only in respect to the elements of X or Y but, in other cases, we will also be interested in analyzing the results based on the families $(Q_x)_{x \in X}$ and $(S_y)_{y \in Y}$.

We begin with the general case. It is not easy to work with the original R from the point of view of the L-fuzzy concept analysis as it does not represent an L-fuzzy context, the initial idea is to transform this structure in a derived L-fuzzy context $(L, \breve{X}, \breve{Y}, \breve{R})$. Then we can extract the information by means of the construction of its L-fuzzy concepts.

After that, we analyze the particular case where $Q_x = Q, \forall x \in X$ (or $S_y = S, \forall y \in Y$). In this case, the relation values can be studied from the values of an only set Q (or S respectively). Furthermore, if $Card(X)$=1, we have an L-fuzzy sequence studied in [3].

Let us see an example:

Example 1. We want to study the evolution in time Y of the sales of some articles in a supermarket chain in the different cities X in which it works. We define an L-fuzzy hypercontext $(L, X, Y, (Q_x)_{x \in X}, (S_y)_{y \in Y}, R)$ with X and Y sets of objects and attributes respectively. The establishment chain $(Q_x)_{x \in X}$ has different values according to the cities and the family of products $(S_y)_{y \in Y}$ changes over time Y (months). Finally, we define $R_{xy} \in L^{Q_x \times S_y}$, for every (x, y), as a relation that recover the sales of the different products S_y in a month $y \in Y$, in the different establishments Q_x of the city $x \in X$.

R	y_1				y_2				y_3		
x_1	$R_{x_1 y_1}$	$s_{y_{11}}$	$s_{y_{12}}$	$s_{y_{13}}$	$R_{x_1 y_2}$	$s_{y_{21}}$	$s_{y_{22}}$	$s_{y_{23}}$	$R_{x_1 y_3}$	$s_{y_{31}}$	$s_{y_{32}}$
	$q_{x_{11}}$	0.3	1	0.1	$q_{x_{11}}$	0.6	0.9	1	$q_{x_{11}}$	0.4	0.5
	$q_{x_{12}}$	0.7	0.3	0.8	$q_{x_{12}}$	1	0	0.2	$q_{x_{12}}$	0	1
	$q_{x_{13}}$	0.9	0.2	0	$q_{x_{13}}$	0.6	0.8	1	$q_{x_{13}}$	0.9	1
x_2	$R_{x_2 y_1}$	$s_{y_{11}}$	$s_{y_{12}}$	$s_{y_{13}}$	$R_{x_2 y_2}$	$s_{y_{21}}$	$s_{y_{22}}$	$s_{y_{23}}$	$R_{x_2 y_3}$	$s_{y_{31}}$	$s_{y_{32}}$
	$q_{x_{21}}$	0.7	0.8	1	$q_{x_{21}}$	0.3	0.9	0	$q_{x_{21}}$	0.4	0.2
	$q_{x_{22}}$	1	0	0.2	$q_{x_{22}}$	1	0.2	0.8	$q_{x_{22}}$	0	0.6

2.1 General Study

We analyze different ways to transform the L-fuzzy hypercontext in an L-fuzzy context.

(1) In this first case, we are interested in keeping the complete information that we have. So, we give the following definition:

Definition 3. *The L-fuzzy hypercontext* $(L, \breve{X}, \breve{Y}, \breve{R})$ *derived in a natural way from the tuple* $(L, X, Y, (Q_x)_{x \in X}, (S_y)_{y \in Y}, R)$*, is:*

- $\breve{X} = \bigcup \dot{Q}_x$ *with* $\dot{Q}_x = \{x\} \times Q_x$
- $\breve{Y} = \bigcup \dot{S}_y$ *with* $\dot{S}_y = \{y\} \times S_y$
- $\breve{R}((x, q), (y, s)) = R_{xy}(q, s), \forall x \in X, \forall y \in Y, \forall q \in Q_x, \forall s \in S_y.$

Example 2. For our example, this will be our relation $\breve{R}$:

$\breve{R}$	$(y_1, s_{y_{11}})$	$(y_1, s_{y_{12}})$	$(y_1, s_{y_{13}})$	$(y_2, s_{y_{21}})$	$(y_2, s_{y_{22}})$	$(y_2, s_{y_{23}})$	$(y_3, s_{y_{31}})$	$(y_3, s_{y_{32}})$
$(x_1, q_{x_{11}})$	0.3	1	0.1	0.6	0.9	1	0.4	0.5
$(x_1, q_{x_{12}})$	0.7	0.3	0.8	1	0	0.2	0	1
$(x_1, q_{x_{13}})$	0.9	0.2	0	0.6	0.8	1	0.9	1
$(x_2, q_{x_{21}})$	0.7	0.8	1	0.3	0.9	0	0.4	0.2
$(x_2, q_{x_{22}})$	1	0	0.2	1	0.2	0.8	0	0.6

In this case, the derivation operators have the following expression:

Proposition 1. $\forall A \in L^{\breve{X}}, \forall B \in L^{\breve{Y}}, \forall x \in X, \forall y \in Y, \forall q \in Q_x, \forall s \in S_y$

$$\begin{aligned} A_1(y, s) &= \inf_{(x,q) \in \breve{X}} \{\mathcal{I}(A(x, q), \breve{R}((x, q)(y, s))\} \\ &= \inf_{(x,q) \in \breve{X}} \{\mathcal{I}(A(x, q), R_{xy}(q, s)\} \end{aligned}$$

$$\begin{aligned} B_2(x, q) &= \inf_{(y,s) \in \breve{Y}} \{\mathcal{I}(B(y, s)), \breve{R}((x, q)(y, s))\} \\ &= \inf_{(y,s) \in \breve{Y}} \{\mathcal{I}(B(y, s)), R_{xy}(q, s)\} \end{aligned}$$

with $\mathcal{I}$ *an L-fuzzy implication operator defined in* $(L, \leq)$ *and where* A_1 *represents the attributes related to the objects of* A *and* B_2*, to the objects related to the attributes of* B*.*

In this case, every pair (x, q) behaves as an object and every pair (y, s) as an attribute.

This definition does not lose the original information but the size of the obtained context is large.

(2) In some situations, it can be interesting to try to reduce the size of the L-fuzzy context although we lose information related to $Q_x, S_y, x \in X, y \in Y$. We have three possibilities:

(a) Aggregate all the values for every R_{xy}.

Definition 4. *We define the derived L-fuzzy context* $(L, \breve{X}, \breve{Y}, \breve{R})$*:*

- $\breve{X} = X$
- $\breve{Y} = Y$
- $\breve{R} : \breve{X} \times \breve{Y} \longrightarrow L$ *such that* $\forall x \in X, \forall y \in Y$:

$$\breve{R}(x,y) = F_{w_{xy}}(R_{xy}(q_{x_1}, s_{y_1}), R_{xy}(q_{x_1}, s_{y_2}) \ldots R_{xy}(q_{x_{|Q_x|}}, s_{y_{|S_y|}})) = \\ w_1.b_1 + w_2.b_2 + \cdots + w_{|Q_x|.|S_y|}.b_{|Q_x|.|S_y|},$$

where $F_{w_{xy}}$ *is an OWA operator with the associated weighting vector* $w_{xy} = (w_1, w_2, \ldots w_{|Q_x|.|S_y|})$ *and* b_i *the ith largest element of the collection* $R_{xy}(q_{x_1}, s_{y_1}), R_{xy}(q_{x_1}, s_{y_2}) \ldots R_{xy}(q_{x_{|Q_x|}}, s_{y_{|S_y|}})$.

This aggregation allows to establish an study of the elements of X respect to Y. There is not information about the set Q_x neither about S_y.

Let see an example.

Example 3. Suppose that in example 1 we want to give more relevance to the closest to 1 observations, then the use of OWA operators can be a good election.

We can use weights w_{xy} (see Sect. 1.2) such that

$$w_i = \frac{2(n-i+1)}{n(1+n)}, \forall i \in \{1, \ldots, n\}.$$

In this case, we obtain the result applying the definition:

$\breve{R}_{F_{w_{xy}}}$	y_1	y_2	y_3
x_1	0.66	0.84	0.80
x_2	0.80	0.71	0.40

For the extraction of the information, we can now take a set that represents the interest of study and calculate the associated L-fuzzy concept using the Lukasiewicz implication operator.

For instance, if we want to study the second city, then we take $\{x_1/0, x_2/1\}$ and we obtain the following result:

$$(\{x_1/0.86, x_2/1\}, \{y_1/0.8, y_2/0.71, y_3/0.4\})$$

We can interpret this L-fuzzy concept saying that there are important sales in both cities during the first two months.

(b) Aggregate for every $x \in X$ the values of R_{xy} associated with the different Q_x.

Definition 5. *We define the derived L-fuzzy context* $(L, \breve{X}, \breve{Y}, \breve{R})$*:*

- $\breve{X} = X$
- $\breve{Y} = \bigcup \dot{S}_y$ *with* $\dot{S}_y = \{y\} \times S_y, \forall y \in Y$
- $\breve{R} : \breve{X} \times \breve{Y} \longrightarrow L$ *such that* $\forall x \in X, \forall y \in Y, \forall s \in S_y$*:*

$$\breve{R}(x,(y,s)) = F_{w_x}(R_{xy}(q_{x_1}, s), R_{xy}(q_{x_2}, s) \dots R_{xy}(q_{x_{|Q_x|}}, s))$$

where F_{w_x} *is an OWA operator with the associated weighting vector* $w_x = (w_1, w_2, \dots w_{|Q_x|})$ *and* b_i *the ith largest element in the collection* $R_{xy}(q_{x_1}, s)$, $R_{xy}(q_{x_2}, s) \dots R_{xy}(q_{x_{|Q_x|}}, s)$.

In this case, we can analyze the elements of X although there is not information about those of Q_x.

Example 4. We are going to prioritize the membership degrees closest to 1 by means of w. Then, to aggregate the values, we use $w_{x_1} = (3/6, 2/6, 1/6)$ and $w_{x_2} = (2/3, 1/3)$. The result is:

$\breve{R}_{F_{w_x}}$	y_1			y_2			y_3	
	s_{y1_1}	s_{y1_2}	s_{y1_3}	s_{y2_1}	s_{y2_2}	s_{y2_3}	s_{y3_1}	s_{y3_2}
x_1	0.73	0.63	0.43	0.8	0.72	0.87	0.58	0.92
x_2	0.9	0.53	0.73	0.77	0.67	0.53	0.27	0.47

In this example, what is sold in every city x_i in time can be studied although the establishments q_{x_i} information is missing.

For instance, if we want to study the second city, we take $\{x_1/0, x_2/1\}$ and, applying the derivation operator, obtain the following result:

$$\{y_1/(0.9, 0.53, 0.73), y_2/(0.77, 0.67, 0.53), y_3/(0.27, 0.47)\}$$

We can *highlight the sales of* $s_{y_{1_1}}$ *product in all establishments of the second city in the first month.*

We can conclude that *in the second city there are important sales mainly in the first two months* (y_1, y_2).

Moreover, we can also see details of sales of products: there is a larger sale of the first product in the first month followed by the sale, also of the first product, in the second month.

We can establish different nuances that otherwise would not be possible using other OWA operators. For instance, with the minimum ($w_{x_1} = (0, 0, 1)$ and $w_{x_2} = (0, 1)$), and also for the second city, the result is:

$$\{y_1/(0.7, 0, 0.2), y_2/(0.3, 0.2, 0), y_3/(0, 0.2)\}$$

We can *highlight the sales of* $s_{y_{1_1}}$ *product in all establishments of the second city in the first month.*

(c) Aggregate for every $y \in Y$ the values R_{xy} associated with the different S_y.

Definition 6. *We define the derived L-fuzzy context* $(L, \breve{X}, \breve{Y}, \breve{R})$*:*

- $\breve{X} = \bigcup \dot{Q}_x$ *with* $\dot{Q}_x = \{x\} \times Q_x, \forall x \in X$
- $\breve{Y} = Y$
- $\breve{R} : \breve{X} \times \breve{Y} \longrightarrow L$ *such that* $\forall x \in X, y \in Y, q \in Q_x$

$$\breve{R}((x,q),y) = F_{w_y}(R_{xy}(q, s_{y_1}), R_{xy}(q, s_{y_2}) \dots R_{xy}(q, s_{y_{|S_y|}}))$$

where F_{w_y} *is an OWA operator with the associated weighting vector* $w_y = (w_1, w_2, \dots w_{|S_y|})$ *and* b_i *the ith largest element in the collection* $R_{xy}(q, s_{y_1})$, $R_{xy}(q, s_{y_2}) \dots R_{xy}(q, s_{y_{|S_y|}})$.

This aggregation allows the study of the elements of Y (there is not information about the products).

Example 5. We are going to see the evolution of the sales over time (Y). In this case, it is also interesting for us the study of the observations with membership degrees closest to 1 and next to the current instants of time. For the aggregation of the values, we will use $w_{y_1} = (3/6, 2/6, 1/6)$ and $w_{y_2} = (2/3, 1/3)$. The obtained result is:

$\breve{R}_{F_{w_y}}$		y_1	y_2	y_3
x_1	$q_{x_{11}}$	0.62	0.9	0.47
	$q_{x_{12}}$	0.68	0.57	0.67
	$q_{x_{13}}$	0.52	0.87	0.97
x_2	$q_{x_{21}}$	0.88	0.55	0.33
	$q_{x_{22}}$	0.57	0.8	0.4

We can here study the evolution in time of the sales in every supermarket $q_{x_{ij}}$ of every city x_i.

For instance, if we look at the third month (y_3), we can take $\{y_1/0, y_2/0, y_3/1\}$ and obtain the result:

$$\{x_1/(0.47, 0.67, 0.97), x_2/(0.33, 0.4)\}$$

Then, we can conclude that *there are lower sales values in the second city (x_2) and many differences among the establishments of the first one (x_1) with higher sales in the last two.*

Also in this case the results are different if we use other OWA operators. For instance, using the minimum ($w_{y_1} = (0, 0, 1)$ and $w_{y_2} = (0, 1)$) and also for the third month, the result is:

$$\{x_1/(0.4, 0, 0.9), x_2/(0.2, 0)\}$$

Then, we can *highlight the sales of all products of establishment $q_{x_{13}}$ of the city x_1 in the third month.*

2.2 Study of a Particular Case

In this case, when one of the sets (Q_x or S_y) is fixed, we can perform more complete studies. For instance, suppose that $S_y = S, \forall y \in Y$.

For the example:

Example 6. In this case, the articles $S = \{s_1, s_2, s_3\}$ are the same in all the establishments but there are different establishments in the different cities.

R		y_1			y_2			y_3		
		s_1	s_2	s_3	s_1	s_2	s_3	s_1	s_2	s_3
x_1	$q_{x_{11}}$	0.3	1	0.1	0.6	0.9	1	0.4	0.5	0.9
	$q_{x_{12}}$	0.7	0.3	0.8	1	0	0.2	0	1	0.2
	$q_{x_{13}}$	0.9	0.2	0	0.6	0.8	1	0.9	1	0.2
x_2	$q_{x_{21}}$	0.7	0.8	1	0.3	0.9	0	0.4	0.2	1
	$q_{x_{22}}$	1	0	0.2	1	0.2	0.8	0	0.6	0.3

Then, we can apply all the results obtained in the general case and, as the set S is fixed, we can also perform a more individualized study for each one of its values.

Definition 7. *We define the derived L-fuzzy context $(L, \breve{X}, \breve{Y}, \breve{R})$ (Aggregating the values of Y) in this way:*

- $\breve{X} = \bigcup \dot{Q}_x$ *with* $\dot{Q}_x = \{x\} \times Q_x, \forall x \in X$
- $\breve{Y} = S$
- $\breve{R} : \breve{X} \times \breve{Y} \longrightarrow L$ *such that* $\forall x \in X, q \in Q_x, s \in S$:

$$\breve{R}((x, q), s) = F_{w_Y}(R_{xy_1}(q_x, s), R_{xy_2}(q_x, s) \ldots R_{xy_{|Y|}}(q_x, s))$$

where F_{w_Y} is an OWA operator with the associated weighting vector $w_Y = (w_1, w_2, \ldots w_{|Y|})$ and b_i the ith largest element in the collection $R_{xy_1}(q_x, s)$, $R_{xy_2}(q_x, s) \ldots R_{xy_{|Y|}}(q_x, s)$.

Example 7. In our example and using $w_Y = (3/6, 2/6, 1/6)$, we obtain the following table:

We can answer the question about which are the different cities with good sales of the different products over time.

For instance, if we take as the set $\{s_1/0, s_2/1, s_3/0\}$, we obtain the result:

$$\{x_1/(0.88, 0.6, 0.8), x_2/(0.75, 0.37)\}$$

Hence, we can conclude that *there are good sales, mainly of s_2 in all the establishments of x_1.*

However, if we do the same for the product s_1 we have:

$$\{x_1/(0.48, 0.73, 0.85), x_2/(0.53, 0.83)\}$$

and now we have fewer sales only in the last establishment.

$\breve{R}_{F_{w_Y}}$		s_1	s_2	s_3
x_1	$q_{x_{11}}$	0.48	0.88	0.82
	$q_{x_{12}}$	0.73	0.6	0.5
	$q_{x_{13}}$	0.85	0.8	0.57
x_2	$q_{x_{21}}$	0.53	0.75	0.83
	$q_{x_{22}}$	0.83	0.37	0.53

Similar developments can be obtained for $Q_x = Q, \forall x \in X$.

Definition 8. *We define the derived L-fuzzy context* $(L, \breve{X}, \breve{Y}, \breve{R})$ *(Aggregating the values of* X*) in this way:*

- $\breve{X} = Q$
- $\breve{Y} = \bigcup \dot{S}_y$ *with* $\dot{S}_y = \{y\} \times S_y, \forall y \in Y$
- $\breve{R} : \breve{X} \times \breve{Y} \longrightarrow L$ *such that* $\forall x \in X, q \in Q, s \in S_y$:

$$\breve{R}(q,(y,s)) = F_{w_X}(R_{x_1y}(q,s_y), R_{x_2y}(q,s_y) \dots R_{x_{|X|}y}(q,s_y))$$

where F_{w_X} *is an OWA operator with the associated weighting vector* $w_X = (w_1, w_2, \dots w_{|X|})$ *and* b_i *the ith largest element in the collection* $R_{x_1y}(q,s_y)$, $R_{x_2y}(q,s_y) \dots R_{x_{|X|}y}(q,s_y)$.

In this case, we can perform a study in depth of the elements of Q although it has no interest for our example.

3 Conclusions and Future Lines

This paper introduces the study of L-fuzzy hypercontexts by using OWA operators in a complete lattice L with the aim of obtaining the relevant information. These new frameworks allow us to work with L-fuzzy contexts where the set of objects and attributes are variable.

Firstly we have developed a general study in two different situations: the first one without lost of information and the second one, reducing the size of the context and losing the information of the not prioritized set.

Finally, we analyze a particular case where the set of objects or attributes is fixed. Then a more individualized study can be performed.

In all the cases, the use of OWA operators is an interesting tool for obtaining the relevant information and establishing different nuances in our study.

In future works we will use linguistic variables [14] for the representation of the points of interest following the ideas of [2].

Acknowledgements. This paper is partially supported by the Research Group "Intelligent Systems and Energy (SI+E)" of the Basque Government, under Grant IT677-13 and by the Research Group "Artificial Intelligence and Approximate Reasoning" of the Public University of Navarra, under TIN2013-40765-P.

References

1. Alcalde, C., Burusco, A., Fuentes-González, R., Zubia, I.: Treatment of L-fuzzy contexts with absent values. Inf. Sci. **179**(1–2), 1–15 (2009)
2. Alcalde, C., Burusco, A., Fuentes-González, R., Zubia, I.: The use of linguistic variables and fuzzy propositions in the L-fuzzy concept theory. Comput. Math. Appl. **62**, 3111–3122 (2011)
3. Alcalde, C., Burusco, A., Fuentes-González, R.: The study of fuzzy context sequences. Int. J. Comput. Intell. Syst. **6**(3), 518–529 (2013)
4. Bělohlávek, R.: Fuzzy Relational Systems: Foundations and Principles. IFSR International Series on Systems Science and Engineering, vol. 20. Springer, New York (2002)
5. Bělohlávek, R., Galois, F.: Connections, fuzzy concept lattices: from binary relations to conceptual structures. In: Novak, V., Perfileva, I. (eds.) Discovering the World with Fuzzy Logic, pp. 462–494. Physica-Verlag, Heidelberg (2000)
6. Burusco, A., Fuentes-González, R.: The study of the L-fuzzy concept lattice. Math. Soft Comput. **1**(3), 209–218 (1994)
7. Burusco, A., Fuentes-González, R.: Construction of the L-fuzzy concept lattice. Fuzzy Sets Syst. **97**(1), 109–114 (1998)
8. Djouadi, Y., Prade, H.: Interval-valued fuzzy galois connections: algebraic requirements and concept lattice construction. Fundamenta Informaticae **99**(2), 169–186 (2010)
9. Ganter, B., Wille, R.: Formal Concept Analysis: Mathematical Foundations. Springer, Heidelberg (1999)
10. Medina, J., Ojeda-Aciego, M.: Multi-adjoint t-concept lattices. Inf. Sci. **180**(5), 712–725 (2010)
11. Yager, R.R.: On ordered weighted averaging aggregation operators in multi-criteria decision making. IEEE Trans. Syst. Man Cybern. **18**, 183–190 (1988)
12. Wille, R.: Restructuring lattice theory: an approach based on hierarchies of concepts. In: Rival, I. (ed.) Ordered Sets, pp. 445–470. Reidel, Dordrecht (1982)
13. Wille, R.: Conceptual structures of multicontexts. In: Eklund, P.W., Ellis, G., Mann, G. (eds.) Proceedings of the 4th ICCS 1996, Sidney (AU), vol. 1115, pp. 23–39 (1996)
14. Zadeh, L.A.: The concept of a linguistic variable and its application to approximate reasoning-I. Inf. Sci. **8**, 199–249 (1975)

A Semantical Approach to Rough Sets and Dominance-Based Rough Sets

Lynn D'eer[1(✉)], Chris Cornelis[1,2], and Yiyu Yao[3]

[1] Department of Applied Mathematics, Computer Science and Statistics, Ghent University, Ghent, Belgium
lynn.deer@ugent.be

[2] Department of Computer Science and Artificial Intelligence, Research Center on Information and Communications Technology (CITIC-UGR), University of Granada, Granada, Spain
Chris.Cornelis@decsai.ugr.es

[3] Department of Computer Science, University of Regina, Regina, Canada
yyao@cs.uregina.ca

Abstract. There exist two formulations of rough sets: the conceptual and computational one. The conceptual or semantical approach of rough set theory focuses on the meaning and interpretation of concepts, while algorithms to compute those concepts are studied in the computational formulation. However, the research on the former is rather limited. In this paper, we focus on a semantically sound approach of Pawlak's rough set model and covering-based rough set models. Furthermore, we illustrate that the dominance-based rough set model can be rephrased using this semantic approach.

Keywords: Covering-based rough sets · Dominance-based rough sets · Semantics · Pre-order

1 Introduction

Rough set theory has two formulations: a conceptual and a computational one [29]. The former formulation emphasizes the meaning and interpretation of concepts and notions of the theory, while the latter formulation is used to construct procedures and algorithms to compute those notions. A major difference between the formulations is the notion of definability. Nevertheless, both formulations are complementary and they are both fundamental in the research on rough set theory. In addition, it is sometimes necessary to consider both approaches together, for example in the minimal description length principle [8,16] in which it is stated that we need to find a balance between the loss of accuracy (computational) and a more compact description of data models (conceptual) when computing decision reducts.

The research on computational formulations has dominated the rough set theory research field since the seminal paper of Pawlak [10]. For instance, there is a

J.P. Carvalho et al. (Eds.): IPMU 2016, Part II, CCIS 611, pp. 23–35, 2016.
DOI: 10.1007/978-3-319-40581-0_3

broad study on generalized rough set models, in which a binary relation or neighborhood operator is used to describe the indiscernibility relation between objects of the universe [2,3,17,21,24,38]. In addition, several covering-based rough set models are defined in literature [12,19,20,22,23,30–38]. More recently, the classification and comparison of these different rough set models have been discussed [1,14,15,27].

In such contributions on the computational formulations of rough set theory, the notion of definability of subsets of the universe of discourse is determined by the approximation operators. This can be done in different ways: a subset $X \subseteq U$ is definable if X equals its lower approximation, or if X equals its upper approximation, or if its lower and upper approximation are equal.

In comparison, the contribution to the conceptual formulation of rough set theory is limited. Prior to the introduction of rough sets, Pawlak [9] and Marek and Pawlak [6] described a definable set as the extension of a concept, of which its intension is a formula in a descriptive language based on the data. The intension of a concept is an abstract description of the properties characteristic for a concept, while the extension of a concept contains all the objects of the universe of discourse satisfying those properties [29]. Hence, a concept is jointly described by its intension and extension. However, except for a few articles by Marek and Truszczyński [7], Yao and Zhou [25] and Yao [26], this notion of definability, which is semantically superior, has scarcely been discussed.

In this paper, we refocus our attention on the conceptual or semantical approach of rough sets. Given an information or decision table which represents the data, definable subsets of the universe are used as primitive notion. A definable set is an arbitrary union of elementary sets. Such an elementary set is a basic granule which represents an indivisible block of information, obtained from the table. Hence, each elementary and definable set is meaningful. In the original rough set of Pawlak [10], the elementary sets are given by the equivalence classes U/E related to the equivalence relation E, which represents the indiscernibility relation between the objects based on the given data. However, it is not always possible to construct such a partition. For example, when the data is incomplete, or when there is a strict order on the values of certain attribute values. Therefore, we extend the semantical approach of Pawlak's model to covering-based rough set models. The elementary sets are now no longer given by a partition U/E, but by a covering $\mathbb{C}$. As each definable set is the union of elementary sets, the set of definable sets is obtained by closing the set of elementary sets under set union. This results in the Boolean algebra $\mathscr{B}(U/E)$ for the model of Pawlak and in the $\cup$-closed set $\cup^*(\mathbb{C})$ for the covering-based rough set models. As the definability of sets is established, the approximation of undefinable sets by definable sets comes naturally [26]. Therefore, approximation operators are constructed as derived notions in both Pawlak's rough set model and the framework of covering-based rough set models.

To illustrate the semantic approach of covering-based rough sets, we determine the elementary and definable sets in the framework of dominance-based rough sets, introduced by Greco et al. [2,3,18]. In this framework, the indiscernibility relation between objects is given by a dominance relation or pre-order, which is a reflexive and transitive relation. Such an indiscernibility relation is useful, when the data represented in the table are preference-ordered, i.e., based on the different conditional attributes, it is determined whether an object x is preferred over an object y.

Given an object x and a dominance relation $\succeq$, two neighborhoods of x are defined by the objects dominating x and the objects dominated by x. Moreover, each set of neighborhoods results in a covering. Given both coverings, by the semantic approach of covering-based rough sets meaningful lower approximation operators are constructed. It will be shown that these conceptual lower approximation operators are exactly the computational lower approximation operators suggested by Greco et al. [2,3,18].

Furthermore, it is discussed how certain decision rules are obtained from a decision table using dominance-based rough sets. Rule induction [4] is an important technique to extract knowledge from a decision table and can be used for the classification of new objects. For example, to determine whether a new object is preferable to the current objects.

This paper is organized as follows. In Sect. 2, we discuss the semantic approach of Pawlak's rough set model and covering-based rough set models. In Sect. 3, we recall some results on the different characterizations of the lower approximation operator based on a reflexive and transitive neighborhood operator. In Sect. 4, we illustrate that the dominance-based lower approximation operator introduced by Greco et al. is semantically meaningful. Furthermore, we discuss how decision rules are obtained in the dominance-based framework. To end, we state conclusion and future work in Sect. 5.

2 Semantics of Rough Set Models

Rough set analysis is a tool to study data given in an information table. Formally, a complete information table is a tuple $T = (U, At, \{V_a \mid a \in A\}, \{I_a \mid a \in At\})$, where U is a finite non-empty set of objects, At is a finite non-empty set of attributes and for each $a \in A$, V_a represents a non-empty set of values related to the attribute a. Furthermore, the information functions $I_a \colon U \to V_a$ map every object of U to a value in V_a, for each $a \in At$. The table T is called a complete decision table if the set of attributes At consists of the disjoint sets C and $\{d\}$, where C represents the conditional attributes of the table and d represents the decision attribute. In this paper, we use the closed world assumption, i.e., the table contains all objects under consideration [11].

Given such a table T, a basic granule represents an elementary unit of knowledge we can obtain from the table and is formally given by a subset of the universe U. Given a non-empty family of granules $G \subseteq 2^U$, the poset $(G, \subseteq)$ is called a granular structure, where $\subseteq$ is the set-theoretic inclusion relation [28]. By imposing different conditions on the set G, we derive different models (U, G) of granular structures. For example, when G is closed under set intersection, set union and set complement, the model $(U, (G, \cap, \cup, ^c))$ is a Boolean algebra [28].

In the original rough set model of Pawlak [10,11] the basic granules of the information table are given by equivalence classes. Namely, let $A \subseteq C$ be a set of conditional attributes, then we can define an equivalence relation E_A on U as follows:

$$\forall x, y \in U\colon xE_Ay \Leftrightarrow \forall a \in A\colon I_a(x) = I_a(y).$$

The equivalence class of an object x is given by $[x]_{E_A} = \{y \in U \mid xE_Ay\}$. The partition U/E_A can be seen as a family of basic granules and each equivalence class is called an *elementary set*. Moreover, the union of a family of equivalence classes is called a *definable set*, since such a set can be constructed and interpreted from the available data in the information table. The granule structure which is used in rough set theory to represent the definable sets is given by

$$\mathscr{B}(U/E_A) = \{\bigcup F \mid F \subseteq U/E_A\}$$

and is obtained by closing the partition U/E_A under set union. As $\mathscr{B}(U/E_A)$ is closed under set intersection, set union and set complement, it is a Boolean algebra. Therefore, the granule structure $(U, (\mathscr{B}(U/E_A), \cap, \cup, ^{c}))$ which represents the definable sets of the information table in Pawlak's rough set model can be seen as a model of granular structures, as described in [28].

However, not every subset of U is contained in the granular structure $\mathscr{B}(U/E_A)$. Such a set, which we call *undefinable*, must be approximated by subsets in the granular structure. Naturally, the lower approximation of $X \subseteq U$ consists of definable sets which are subsets of X (approximation from below), while the upper approximation of X consists of definable sets which are supersets of X (approximation from above) [10,28].

As in Pawlak's rough set model the definable sets are represented by the Boolean algebra $\mathscr{B}(U/E_A)$, there is a unique greatest set in $\mathscr{B}(U/E_A)$ contained by X, and a unique smallest set in $\mathscr{B}(U/E_A)$ containing X. Therefore, in Pawlak's rough set model the approximations of X are given by

$$\begin{aligned}
\underline{\text{apr}}_A(X) &= \text{the greatest definable set in } \mathscr{B}(U/E_A) \text{ contained by } X\\
&= \bigcup\{Y \in \mathscr{B}(U/E_A) \mid Y \subseteq X\},\\
\overline{\text{apr}}_A(X) &= \text{the smallest definable set in } \mathscr{B}(U/E_A) \text{ containing } X\\
&= \bigcap\{Y \in \mathscr{B}(U/E_A) \mid X \subseteq Y\}.
\end{aligned}$$

Note that for all $X \subseteq U$, its lower and upper approximation are definable, i.e., $\underline{\text{apr}}_A(X) \in \mathscr{B}(U/E_A)$ and $\overline{\text{apr}}_A(X) \in \mathscr{B}(U/E_A)$. Moreover, if X is definable, then

$$\underline{\text{apr}}_A(X) = \overline{\text{apr}}_A(X) = X.$$

Hence, one derives the notion of definability which is used in computational formulations.

Unfortunately, it is not always possible to construct a meaningful equivalence relation between objects based on the attribute values. For example, if the information table is incomplete, or when we have an ordered information table.

While in the former case attribute values are missing which makes it impossible to construct an equivalence relation, the equivalence classes in the latter case will mostly consist of only one object which is unreasonable for applications such as rule induction.

In such cases, the family of basic granules for $A \subseteq C$ is not given by a partition, but by a more general covering $\mathbb{C}_A$. A covering $\mathbb{C}$ is a non-empty family of non-empty subsets of U such that $\bigcup \mathbb{C} = U$. Every set or *patch* in $\mathbb{C}_A$ is called an elementary set and should be constructed using meaningful semantics. In addition, every union of patches of $\mathbb{C}_A$ will be interpretable from the data in the information table. Such a union is called a definable set. The granular structure which represents these definable sets is given by

$$\cup^*(\mathbb{C}_A) = \{\bigcup F \mid F \subseteq \mathbb{C}_A\}$$

and is obtained by closing $\mathbb{C}_A$ under set union. It is $\cup$-closed, contains the empty set and its corresponding model of granular structure is denoted by $(U, (\cup^*(\mathbb{C}_A), \cup))$ [28]. Although it is closed under set union, it is not closed under set intersection and set complement such as in the case of Pawlak's rough set model. As every partition can be seen as a covering of the universe, $(U, (\mathscr{B}(U/E_A), \cap, \cup, ^c))$ is a sub-model of $(U, (\cup^*(\mathbb{C}_A), \cup))$.

Similar to the rough set model of Pawlak, a subset $X \subseteq U$ which is not definable can be approximated by definable sets in $\cup^*(\mathbb{C}_A)$. As $\cup^*(\mathbb{C}_A)$ is closed under set union, there exists a unique greatest definable set in $\cup^*(\mathbb{C}_A)$ contained by X, therefore

$$\underline{\text{apr}}_A(X) = \text{the greatest definable set in } \cup^* (\mathbb{C}_A) \text{ contained by } X \tag{1}$$

$$= \bigcup\{Y \in \cup^*(\mathbb{C}_A) \mid Y \subseteq X\}. \tag{2}$$

Unfortunately, as $\cup^*(\mathbb{C}_A)$ is not closed under set intersection, there does not necessarily exist a unique smallest definable set in $\cup^*(\mathbb{C}_A)$ containing X and thus, $\overline{\text{apr}}_A(X)$ is not necessarily an element in $\cup^*(\mathbb{C}_A)$, but a set of minimal elements in $\cup^*(\mathbb{C}_A)$ [28]:

$$\overline{\text{apr}}_A(X) = \{Y \in \cup^*(\mathbb{C}_A) \mid X \subseteq Y, Y \text{ minimal}\},$$

where Y is minimal if $Y \in \cup^*(\mathbb{C}_A)$, $X \subseteq Y$ and $\forall Z \in \cup^*(\mathbb{C}_A)$ with $X \subseteq Z$, if $Z \subseteq Y$ then $Y = Z$. Note that the upper approximation operator of X is given by the definable sets 'just' above X, as they provide the most accurate information.

Hence, as the upper approximation operator is a subset of $\mathscr{P}(U)$ rather than an element of $\mathscr{P}(U)$, various properties from Pawlak's framework no longer make sense, such as the definability of the upper approximation operator and the duality between the lower and upper approximation operator. Note that for $X \in \cup^*(\mathbb{C}_A)$ it does hold that $\underline{\text{apr}}_A(X) = X$ and $\overline{\text{apr}}_A(X) = \{X\}$.

We illustrate the above approximation operators in the following example:

Example 1. [28] Let $U = \{a, b, c, d, e\}$ and $\mathbb{C} = \{\{a\}, \{b, d\}, \{a, b, c\}, \{b, c, e\}\}$, then

$$\cup^*(\mathbb{C}) = \{\emptyset, \{a\}, \{b, d\}, \{a, b, c\}, \{b, c, e\}, \{a, b, d\}, \{a, b, c, d\}, \{a, b, c, e\}, \{b, c, d, e\}, \{a, b, c, d, e\}\}.$$

The lower and upper approximation of $\{a, b, c\}$ are given by $\underline{\text{apr}}(\{a, b, c\}) = \{a, b, c\}$ and $\overline{\text{apr}}(\{a, b, c\}) = \{\{a, b, c\}\}$ and for $\{b, c, d\}$ they are given by $\underline{\text{apr}}(\{b, c, d\}) = \{b, d\}$ and $\overline{\text{apr}}(\{b, c, d\}) = \{\{a, b, c, d\}, \{b, c, d, e\}\}$.

Next, we discuss different characterizations for the lower approximation operator, when a reflexive and transitive neighborhood operator is used instead of an equivalence relation.

3 Different Characterizations for the Lower Approximation Operator Based on a Reflexive and Transitive Neighborhood Operator

As we saw above, the lower approximation operator of Pawlak's model is given by

$$\underline{\text{apr}}(X) = \bigcup\{Y \in \mathscr{B}(U/E) \mid Y \subseteq X\},$$

where E is an equivalence relation based on the data and X is a subset of the universe U. This characterization is called the *subsystem-based* definition of Pawlak's model [27]. Moreover, there are two equivalent characterizations of the lower approximation operator, called the *element-based* and *granule-based* definition, respectively:

$$\underline{\text{apr}}(X) = \{x \in U \mid [x]_E \subseteq X\},$$
$$\underline{\text{apr}}(X) = \bigcup\{[x]_E \mid x \in U, [x]_E \subseteq X\}.$$

In [24], Yao generalized the element-based and granule-based lower approximation operator of Pawlak's model by using neighborhoods instead of equivalence classes. A neighborhood operator $n \colon U \to \mathscr{P}(U)$ maps an object $x \in U$ to a subset $n(x) \subseteq U$. A neighborhood operator n is called reflexive if $x \in n(x)$ for all $x \in U$ and it is called transitive if $x \in n(y)$ implies $n(x) \subseteq n(y)$ for all $x, y \in U$ [24].

Let n be a neighborhood operator and $X \subseteq U$, then the element-based and granule-based lower approximation of X based on n are defined as follows [24]:

$$\underline{\text{apr}}_{1,n}(X) = \{x \in U \mid n(x) \subseteq X\},$$
$$\underline{\text{apr}}_{2,n}(X) = \bigcup\{n(x) \mid x \in U, n(x) \subseteq X\}$$
$$= \{x \in U \mid \exists y \in U \colon x \in n(y), n(y) \subseteq X\}.$$

In general, the element-based and granule-based definition are no longer equivalent to each other, which was the case in Pawlak's model. However, they are equivalent for a reflexive and transitive neighborhood operator.

Theorem 1. *[24] Let n be a neighborhood operator. The operators $\underline{\mathrm{apr}}_{1,n}$ and $\underline{\mathrm{apr}}_{2,n}$ are equivalent if and only if n is reflexive and transitive.*

Following the discussion in Sect. 2, a meaningful generalization of the subsystem-based definition of Pawlak is provided. The definable sets are now given by the neighborhood system

$$\mathbb{C}_n = \{n(x) \mid x \in U\},$$

which is a covering, instead of the partition U/E. Applying covering $\mathbb{C}_n$ to Eq. (2), we obtain the subsystem-based lower approximation of $X \subseteq U$ based on n:

$$\underline{\mathrm{apr}}_{3,n}(X) = \bigcup \{Y \in \cup^*(\mathbb{C}_n) \mid Y \subseteq X\}.$$

In the following, we discuss that the operator $\underline{\mathrm{apr}}_{3,n}$ is equivalent to the operators $\underline{\mathrm{apr}}_{1,n}$ and $\underline{\mathrm{apr}}_{2,n}$ if n is a reflexive and transitive operator.

Let n be a reflexive and transitive neighborhood operator and let

$$\tau_n = \{X \subseteq U \mid \underline{\mathrm{apr}}_{1,n}(X) = X\}.$$

As n is reflexive, τ_n is a topology [5]. Qin et al. [13] proved the following theorem:

Theorem 2. *[13] If n is a reflexive and transitive operator, then the topology τ_n can be characterized by $\tau_n = \{\underline{\mathrm{apr}}_{1,n}(X) \mid X \subseteq U\}$ and $\underline{\mathrm{apr}}_{1,n}$ is the interior operator of τ_n.*

From this, we derive that for $X \subseteq U$

$$\begin{aligned}\underline{\mathrm{apr}}_{1,n}(X) &= \text{the greatest set in } \tau_n \text{ contained by } X \\ &= \bigcup \{Y \in \tau_n \mid Y \subseteq X\},\end{aligned}$$

since τ_n is a topology and $\underline{\mathrm{apr}}_{1,n}$ is its interior operator.

In the following theorem, we prove that all sets in the topology τ_n are definable and that the operators $\underline{\mathrm{apr}}_{1,n}$ and $\underline{\mathrm{apr}}_{3,n}$ are equivalent.

Theorem 3. *Let n be a reflexive and transitive neighborhood operator, then*

1. *$\tau_n \subseteq \cup^*(\mathbb{C}_n)$,*
2. *$\forall X \subseteq U$: $\underline{\mathrm{apr}}_{1,n}(X) = \underline{\mathrm{apr}}_{3,n}(X)$.*

Proof. 1. Let $Y \in \tau_n$, then $Y = \underline{\mathrm{apr}}_{1,n}(Y)$. By Theorem 1,

$$Y = \underline{\mathrm{apr}}_{2,n}(Y) = \bigcup \{n(y) \mid y \in U, n(y) \subseteq Y\}.$$

Hence, Y is a union of neighborhood operators from $\mathbb{C}_n$, and therefore, $Y \in \cup^*(\mathbb{C}_n)$.

2. Let $X \subseteq U$. From $\tau_n \subseteq \cup^*(\mathbb{C}_n)$, we immediately derive that $\underline{\text{apr}}_{1,n}(X) \subseteq \underline{\text{apr}}_{3,n}(X)$. On the other hand, let $x \in \underline{\text{apr}}_{3,n}(X)$, then there exists a set $Y \in \cup^*(\mathbb{C}_n)$ such that $x \in Y$ and $Y \subseteq X$. As $Y \in \cup^*(\mathbb{C}_n)$, there is a subset $F \subseteq \mathbb{C}_n$ such that
$$Y = \bigcup\{n(y) \in F\}.$$
Therefore, there exists a neighborhood $n(y) \in F$ such that $x \in n(y)$. As n is reflexive and transitive, we have that $\underline{\text{apr}}_{1,n}(n(y)) = n(y)$ [24]. Hence, $n(y) \in \tau_n$ and since $x \in n(y)$ and $n(y) \subseteq Y \subseteq X$, we derive that $x \in \underline{\text{apr}}_{1,n}(X)$.

From the above theorem, we conclude that the operators $\underline{\text{apr}}_{1,n}$ and $\underline{\text{apr}}_{3,n}$ are equivalent for a reflexive and transitive neighborhood operator. While the latter is constructed from a semantical point of view, the former is preferable from a computational point of view.

In the following section, we discuss how the above semantics and different characterizations of the lower approximation operator can be used to obtain decision rules from a dominance-based rough set model.

4 Decision Rules in a Dominance-Based Rough Set Model

The dominance-based rough set model introduced by Greco et al. [2,3,18] extends the rough set model of Pawlak by using a dominance relation instead of an equivalence relation as indiscernibility relation. A dominance relation is reflexive and transitive and is often called a pre-order. It is preferable to choose a dominance relation instead of an equivalence relation when the domains V_a of the attributes in At are preference-ordered, i.e., if there is a natural order on the possible values of an attribute. A real-life example is the overall evaluation of bank clients based on the evaluations of different risk factors.

Formally, an outranking relation $\succeq_a$ is defined for each attribute $a \in At$ based on the natural order on V_a, i.e., an object $x \in U$ dominates an object $y \in U$, or y is dominated by x, with respect to the attribute a if $I_a(x) \succeq_a I_a(y)$. Such a relation $\succeq_a$ is reflexive and transitive. It is assumed that each relation $\succeq_a$ is complete, i.e., that for every pair of objects one object is dominating the other. This way, we also get preference-ordered decision classes D_i, with $D_i = \{x \in U \mid I_d(x) = i\}$, $i \in V_d$. For $i, j \in V_d$, if $i \succeq_d j$, the objects from D_i are strictly preferred to the objects from D_j. E.g., the bank clients with overall evaluation 'good' are preferable to the clients with overall evaluation 'medium'.

As the decision classes are preference-ordered, we obtain the upward and downward union of classes: for $i \in V_d$ we have
$$D_i^{\geq} = \bigcup\{D_j \in U/d \mid j \geq i\}$$
and
$$D_i^{\leq} = \bigcup\{D_j \in U/d \mid j \leq i\}.$$

An object x belongs to $D_i^{\geq}$ if the decision of x is at least i, while x belongs to $D_i^{\leq}$ if the the decision of x is at most i.

Given a set of conditional attributes $A \subseteq C$, we obtain a relation D_A of U based on A as follows: an object $x \in U$ dominates an object $y \in U$ or y is dominated by x with respect to A if and only if $x \succeq_a y$ for all $a \in A$. The relation D_A is a complete pre-order, since all relations $\succeq_a$ are. Given the relation D_A for $A \subseteq C$ and an object $x \in U$, we can define the A-dominating and A-dominated set of x. The former is given by all predecessors of x by D_A, the latter by the successors of x:

$$D_A^p(x) = \{y \in U \mid yD_Ax\}, \tag{3}$$
$$D_A^s(x) = \{y \in U \mid xD_Ay\}. \tag{4}$$

An object y belongs to $D_A^p(x)$ if for all attributes $a \in A$ y dominates x with respect to the attribute a, while y belongs to $D_A^s(x)$ if for all attributes $a \in A$ y is dominated by x with respect to the attribute a. Note that both $D_A^p(x)$ and $D_A^s(x)$ are reflexive and transitive neighborhoods of the object x [24]. Moreover, the sets $\mathbb{C}_A^p = \{D_A^p(x) \mid x \in U\}$ and $\mathbb{C}_A^s = \{D_A^s(x) \mid x \in U\}$ are coverings of the universe U. These coverings are meaningful families of basic granules for $A \subseteq C$ as it is clear that every patch $D_A^p(x)$, respectively $D_A^s(x)$, represents the objects which attributes values on A are bounded from below, respectively from above, by the values of x on A. Moreover, the definable sets are given by the $\cup$-closed sets $\cup^*(\mathbb{C}_A^p)$ and $\cup^*(\mathbb{C}_A^s)$. From Sect. 2, the lower approximation of $X \subseteq U$ using $\cup^*(\mathbb{C}_A^p)$ and $\cup^*(\mathbb{C}_A^s)$ is given, respectively, by

$$\underline{\mathrm{apr}}_{3,D_A^p}(X) = \bigcup\{Y \in \cup^*(\mathbb{C}_A^p) \mid Y \subseteq X\}, \tag{5}$$
$$\underline{\mathrm{apr}}_{3,D_A^s}(X) = \bigcup\{Y \in \cup^*(\mathbb{C}_A^s) \mid Y \subseteq X\}, \tag{6}$$

where we inherit the notation for the lower approximation operators from Sect. 3.

To obtain useful knowledge from the decision table, we want to derive decision rules from the given data. More specifically, Greco et al. obtained certain rules from the following lower approximations of the upward and downward union of classes.

$$\underline{\mathrm{apr}}_{1,D_A^p}(D_i^{\geq}) = \{x \in U \mid D_A^p(x) \subseteq D_i^{\geq}\}, \tag{7}$$
$$\underline{\mathrm{apr}}_{1,D_A^s}(D_i^{\leq}) = \{x \in U \mid D_A^s(x) \subseteq D_i^{\leq}\} \tag{8}$$

By Theorem 3, the lower approximations $\underline{\mathrm{apr}}_{1,D_A^p}(D_i^{\geq})$ and $\underline{\mathrm{apr}}_{1,D_A^s}(D_i^{\leq})$ are equal to $\underline{\mathrm{apr}}_{3,D_A^p}(D_i^{\geq})$ and $\underline{\mathrm{apr}}_{3,D_A^s}(D_i^{\leq})$. Hence, the computational approximation operators used by Greco et al. both have a semantically sound counterpart, provided by the framework from Sect. 2.

The interpretation of the lower approximation of an upward union $D_i^{\geq}$ is the following: an object x certainly belongs to $D_i^{\geq}$, i.e., it belongs to its lower approximation, if for every object y which dominates x with respect to A it holds that the decision of y is at least i. Analogously, x certainly belongs to $D_i^{\leq}$ if every object y which is dominated by x with respect to A has a decision at most i. This way, if the evaluation of an object on A improves, the class assignment of the object does not worsen and vice versa, if the evaluation on A is less good, the class assignment does not improve. Therefore, it is not meaningful to consider $\{x \in U \mid D_A^p(x) \subseteq D_i^{\leq}\}$ and $\{x \in U \mid D_A^s(x) \subseteq D_i^{\geq}\}$ as lower approximations, although it can be done from computational point of view.

To obtain certain decision rules, let $A = \{a_1, a_2, \ldots, a_n\} \subseteq C$ and $i \in V_d$. If the lower approximation $\underline{\text{apr}}_{1,D_A^p}(D_i^{\geq})$ is not empty then we derive the certain decision rule

$$\text{if } I_{a_1}(x) \geq v_1 \wedge I_{a_2}(x) \geq v_2 \wedge \ldots \wedge I_{a_n}(x) \geq v_n, \text{ then } I_d(x) \geq i,$$

where $v_i \in V_{a_i}$. Analogously, if $\underline{\text{apr}}_{1,D_A^s}(D_i^{\leq})$ is not empty, then the following certain decision rule is obtained:

$$\text{if } I_{a_1}(x) \leq v_1 \wedge I_{a_2}(x) \leq v_2 \wedge \ldots \wedge I_{a_n}(x) \leq v_n, \text{ then } I_d(x) \leq i.$$

In the above discussion, we only obtained certain decision rules as we only used the lower approximations of the upward and downward unions of decision classes. However, it is also possible to derive possible rules by using the upper approximations, which we obtain as the dual operators from Eqs. (7) and (8):

$$\overline{\text{apr}}_{1,D_A^p}(D_i^{\geq}) = \{x \in U \mid D_A^s(x) \cap D_i^{\geq} \neq \emptyset\}, \tag{9}$$

$$\overline{\text{apr}}_{1,D_A^s}(D_i^{\leq}) = \{x \in U \mid D_A^p(x) \cap D_i^{\leq} \neq \emptyset\}. \tag{10}$$

Note that these upper approximations are obtained from a computational viewpoint, and not from a conceptual one. Although they provide us with possible

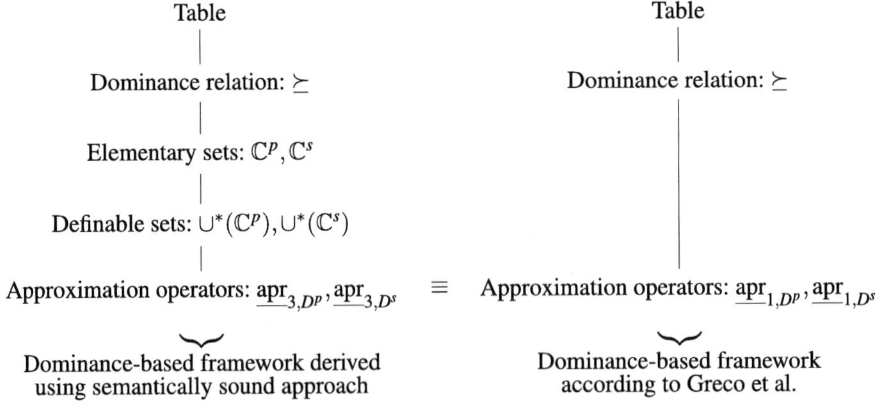

Fig. 1. Comparison between the semantical framework and the dominance-based rough set model

rules which can be used in data analysis, the semantical meaning of these rules is less clear.

To end, we summarize the different steps to obtain the lower approximation operator in the semantically sound approach and the dominance-based rough set approach in Fig. 1. The lower approximation operators from both frameworks are equivalent, but there is no such comparison for the upper approximation operators. By constructing the meaningful coverings $\mathbb{C}^p$ and $\mathbb{C}^s$ via the dominance relation $\succeq$, the dominance-based rough set model can be seen as a special case of the semantically sound framework of rough sets.

5 Conclusion and Future Work

In this paper we have refocussed on the conceptual formulation of rough sets. We discussed a semantical approach of Pawlak's rough set model and covering-based rough set models in which we formalized the elementary and definable sets. Taking the definable sets as primitive notions, meaningful approximation operators are obtained. Unfortunately, since the definable sets in a covering-based rough set model are not closed under set intersection, the upper approximation of a set of the universe in this framework is not a definable set, but it is a set of definable sets.

Furthermore, we have illustrated the semantic approach of covering-based rough sets with the dominance-based rough set model. The obtained conceptual lower approximation operator is in fact equivalent to the known computational lower approximation operator in the dominance-based framework. In addition, we illustrated how to obtain certain decision rules from a preference-ordered decision table.

A future objective is to formalize the elementary and definable sets with respect to a logic language as in [25], by formally describing the intensions and the extensions of the concepts. Moreover, we will study how to obtain a meaningful covering $\mathbb{C}_A$. Furthermore, we want to discuss a semantic approach for covering-based rough sets for an incomplete decision table.

Acknowledgments. Lynn D'eer has been supported by the Ghent University Special Research Fund. This work was partially supported by the Spanish Ministry of Science and Technology under the Project TIN2014-57251-P and the Andalusian Research Plans P10-TIC-6858, P11-TIC-7765 and P12-TIC-2958.

References

1. D'eer, L., Restrepo, M., Cornelis, C., Gómez, J.: Neighborhood operators for covering based rough sets. J. Inf. Sci. **336**, 21–44 (2016)
2. Greco, S., Matarazzo, B., Słowiński, R.: Rough sets theory for multi-criteria decision analysis. Eur. J. Oper. Res. **129**(1), 1–47 (2001)
3. Greco, S., Matarazzo, B., Słowiński, R.: Multicriteria classification by dominance-based rough set approach. In: Kloesgen, W., Zytkow, J. (eds.) Handbook of Data Mining and Knowledge Discovery. Oxford University Press, New York (2002)

4. Grzymala-Busse, J.W.: Rule induction from rough approximations. In: Kacprzyk, J., Pedrycz, W. (eds.) Springer Handbook of Computational Intelligence, pp. 371–385. Springer, Heidelberg (2015)
5. Kondo, M.: On the structure of generalized rough sets. Inf. Sci. **176**, 586–600 (2006)
6. Marek, V.W., Pawlak, Z.: Information storage and retrieval systems: mathematical foundations. Theor. Comput. Sci. **1**, 331–354 (1976)
7. Marek, V.W., Truszczyński, M.: Contributions to the theory of rough sets. Fundamenta Informaticae **39**, 389–409 (1999)
8. Nguyen, H.S.: Approximate boolean reasoning: foundations and applications in data mining. In: Peters, J.F., Skowron, A. (eds.) Transactions on Rough Sets V. LNCS, vol. 4100, pp. 334–506. Springer, Heidelberg (2006)
9. Pawlak, Z.: Mathematical Foundations of Information Retrieval, Research Report CC PAS Report 101. Computation Center, Polish Academy of Sciences (1973)
10. Pawlak, Z.: Rough sets. Int. J. Comput. Inf. Sci. **11**(5), 341–356 (1982)
11. Pawlak, Z.: Rough Sets: Theoretical Aspects of Reasoning About Data. Kluwer Academic Publishers, Boston (1991)
12. Pomykala, J.A.: Approximation operations in approximation space. Bulletin de la Académie Polonaise des Sciences **35**(9–10), 653–662 (1987)
13. Qin, K., Yang, J., Pei, Z.: Generalized rough sets based on reflexive and transitive relations. Inf. Sci. **178**, 4138–4141 (2008)
14. Restrepo, M., Cornelis, C., Gómez, J.: Duality, conjugacy and adjointness of approximation operators in covering based rough sets. Int. J. Approximate Reasoning **55**, 469–485 (2014)
15. Restrepo, M., Cornelis, C., Gómez, J.: Partial order relation for approximation operators in covering-based rough sets. Inf. Sci. **284**, 44–59 (2014)
16. Rissanen, J.: Minimum-Description-Length Principle. Wiley, New York (1985)
17. Słowiński, R., Vanderpooten, D.: A generalized definition of rough approximation based on similarity. IEEE Trans. Knowl. Data Eng. **12**, 331–336 (2000)
18. Słowiński, R., Greco, S., Matarazzo, B.: Rough set based decision support. In: Burke, E.K., Kendall, G. (eds.) Search Methodologies: Introductory Tutorials in Optimization and Decision Support Techniques, pp. 475–527. Springer, New York (2005)
19. Stepaniuk, J., Skowron, A.: Tolerance approximation spaces. Fundamenta Informaticae **27**(2–3), 245–253 (1996)
20. Tsang E., Chen D., Lee J., Yeung D.S.: On the upper approximations of covering generalized rough sets. In: Proceedings of the 3rd International Conference on Machine Learning and Cybernetics, pp. 4200–4203 (2004)
21. Wu, W.Z., Zhang, W.X.: Neighborhood operators systems and approximations. Inf. Sci. **144**, 201–207 (2002)
22. Xu, Z., Wang, Q.: On the properties of covering rough sets model. J. Henan Norm. Univ. (Natural Science) **33**(1), 130–132 (2005)
23. Xu, W., Zhang, W.: Measuring roughness of generalized rough sets induced by a covering. Fuzzy Sets Syst. **158**, 2443–2455 (2007)
24. Yao, Y.: Relational Interpretations of neighborhood operators and rough set approximation operators. Inf. Sci. **101**, 21–47 (1998)
25. Yao, Y., Zhou, B.: A logic language of granular computing. In: Proceedings of the 6th IEEE International Conference on Cognitive Informatics, pp. 178–185 (2007)
26. Yao, Y.: A note on definability and approximations. In: Peters, J.F., Skowron, A., Marek, V.W., Orłowska, E., Słowiński, R., Ziarko, W.P. (eds.) Transactions on Rough Sets VII. LNCS, vol. 4400, pp. 274–282. Springer, Heidelberg (2007)

27. Yao, Y., Yao, B.: Covering based rough sets approximations. Inf. Sci. **200**, 91–107 (2012)
28. Yao, Y., Zhang, N., Miao, D., Xu, F.: Set-theoretic approaches to granular computing. Fundamenta Informaticae **115**(2–3), 247–264 (2012)
29. Yao, Y.: The two sides of the theory of rough sets. Knowl. Based Syst. **80**, 67–77 (2015)
30. Żakowski, W.: Approximations in the space (u, π). Demonstratio Mathematica **16**, 761–769 (1983)
31. Zhu, W., Wang, F.: Reduction and axiomatization of covering generalized rough sets. Inf. Sci. **152**, 217–230 (2003)
32. Zhu, W.: Properties of the first type of covering-based rough sets. In: Sixth IEEE International Conference on Data Mining - Workshops IEEE (2006)
33. Zhu, W.: Properties of the second type of covering-based rough sets. In: Proceedings of the IEEE/WIC/ACM International Conference on Web Intelligence and Intelligent Agent Technology (2006)
34. Zhu, W.: Properties of the third type of covering-based rough sets. In: Sixth IEEE International Conference on Data Mining - Workshops (2006)
35. Zhu, W.: Properties of the fourth type of covering-based rough sets. In: Proceedings of Sixth International Conference on Hybrid Intelligence Systems, vol. 43 (2006)
36. Zhu, W., Wang, F.: A new type of covering rough sets. In: 3rd International IEEE Conference Intelligence Systems (2006)
37. Zhu, W., Wang, F.: On three types of covering based rough sets. IEEE Trans. Knowl. Data Eng. **19**(8), 1131–1143 (2007)
38. Zhu, W.: Relationship between generalized rough sets based on binary relation and covering. Inf. Sci. **179**, 210–225 (2009)

Graded Generalized Hexagon in Fuzzy Natural Logic

Petra Murinová(✉) and Vilém Novák

Institute for Research and Applications of Fuzzy Modeling,
University of Ostrava, 30. dubna 22, 701 03 Ostrava 1, Czech Republic
{petra.murinova,vilem.novak}@osu.cz
http://irafm.osu.cz/

Abstract. In our previous papers, we formally analyzed the generalized Aristotle's square of opposition using tools of fuzzy natural logic. Namely, we introduced general definitions of selected intermediate quantifiers, constructed a generalized square of opposition consisting of them and syntactically analyzed the emerged properties. The main goal of this paper is to extend the generalized square of opposition to graded generalized hexagon.

Keywords: Intermediate quantifiers · Fuzzy natural logic · Evaluative linguistic expressions · Generalized Peterson square · Graded generalized hexagon

1 Introduction

Fuzzy natural logic (FNL) is a formal mathematical theory that consists of three theories: (1) a formal theory of evaluative linguistic expressions (explained in detail in [25]), (2) a formal theory of fuzzy IF-THEN rules and approximate reasoning (presented in [24,27]), and (3) a formal theory of intermediate and generalized fuzzy quantifiers (presented in [16,18,20,26]). This paper is a contribution to (3), namely to extension of the generalized square of opposition to graded generalized hexagon.

Recall that the classical Aristotle's square of opposition [35] consists of the following formulas:

$$^{*}\mathbf{A}: \text{All } B \text{ are } A \qquad (\forall x)(Bx \Rightarrow Ax) \wedge (\exists x)Bx, \tag{1}$$

$$\mathbf{E}: \text{No } B \text{ are } A \qquad (\forall x)(Bx \Rightarrow \neg Ax), \tag{2}$$

$$\mathbf{I}: \text{Some } B \text{ are } A \qquad (\exists x)(Bx \wedge Ax), \tag{3}$$

$$^{*}\mathbf{O}: \text{Some } B \text{ are not } A \qquad (\exists x)(Bx \wedge \neg Ax) \vee \neg(\exists x)Bx. \tag{4}$$

The diagonals correspond to the relation of *contradiction* between the universal affirmative **A** ("All") and the particular negative **O** ("Not all"), as well as between the universal negative **E** ("No") and the particular positive **I** ("Some"). The property of *contrary* holds horizontally at the top between **A**

J.P. Carvalho et al. (Eds.): IPMU 2016, Part II, CCIS 611, pp. 36–47, 2016.
DOI: 10.1007/978-3-319-40581-0_4

("All") and **E** ("No"). The inverse relation of *sub-contraries* holds horizontally at the bottom between **I** ("Some") and **O** "Not all". Finally, the vertical relation between **A** and **I** as well as between **E** and **O** describes the relation of *subaltern* (superaltern).

All the relations can be characterized using the definitions formulated in [19, p. 17] (see also below). Note that the Aristotle's square of opposition works fully with *presupposition* only. In [30,34]), the authors draw a crucial distinction between the "classical" Aristotelian square of opposition and the "modern" duality one based on the concepts of inner and outer negation.

Béziau in [2,3] suggested to extend a square of opposition into a hexagon. This technically means to add two new formulas **U** and **Y**[1] that are defined as disjunction of the two top corners of the square and conjunction of the two bottom corners:

$$\mathbf{U} = \mathbf{A} \vee \mathbf{E} : \text{All or No } B \text{ are } A. \tag{5}$$

$$\mathbf{Y} = \mathbf{I} \wedge \mathbf{O} : \text{Some but Not All } B \text{ are } A. \tag{6}$$

Then we obtain the following Aristotelian hexagon.

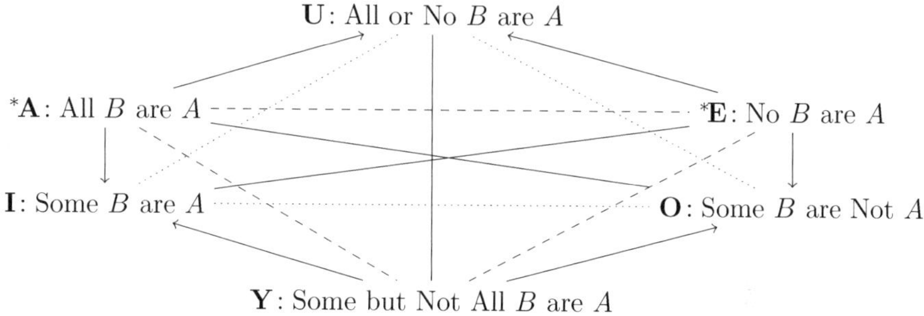

The diagonal lines represent contradictories (denoted by straight lines), the formulas **A** and **E** are contraries (denoted by dashed lines), **A** and **E** entail **U** (denoted by arrow), while **Y** entails both formulas **I** as well as **O**. The formulas **I** and **O** are sub-contraries (denoted by dotted lines). It is interesting to see that the logical hexagon obtains three Aristotle's squares of opposition, namely, **AEIO**, **AYOU** and **EYUI**.

In [32], we can find differences between the Aristotle hexagon and the Duality hexagon. A logical hexagon with many examples and also the cube of opposition was described in [15]. A more complex 3D generalization of the hexagon was proposed by Moretti [17], Pellissier [29] and Smessaert [33]. Applications of the square of opposition in philosophical and mathematical logic, linguistics and psychology were studied in [4–6].

The graded Aristotle square of opposition and also a cube of opposition and its graded version that associates the traditional square of opposition with the dual one were introduced in [13,14]. The structures of opposition in rough set

[1] Blanché in [7] introduced **Y** at first, before completing it with **U** in [8].

theory were analyzed in [10,11]. The gradual hexagon with fuzzy relation and a connection of hexagon and a cube of opposition was described in [12].

In [19], we demonstrated how the generalized[2] Aristotle square of opposition can be represented formally in fuzzy natural logic. In this paper, we first extend this theory and analyze graded Aristotle hexagon as a generalization of the Aristotle classical one. Then we propose a generalization of the Peterson's square of opposition (cf. [31]) to a graded hexagon with intermediate quantifiers (we will call it *graded generalized hexagon*). Let us remark that these results contribute also to development of FNL because the discovered relations among intermediate quantifiers can be applied to formulation of general rules of human reasoning.

The paper is structured as follows: First of all we very briefly remember the basic mathematical formal system. In Sect. 3, we remember the basic definitions of contrary, contradictories, sub-contrary and sub-alterns. Then we prove the main properties of interpretation of three squares of opposition (**AEIO**, **AYOU** and **EYUI**). Finally, we introduce the graded generalized hexagon with five basic intermediate quantifiers.

2 Preliminaries

2.1 The Basic Formal System

The fuzzy natural logic is formulated using tools of Łukasiewicz fuzzy type theory (Ł-FTT) which is a higher-order fuzzy logic. All the details can be found in the papers [19,23,25]. Recall that the basic syntactical objects of Ł-FTT are classical, namely the concepts of type and formula (cf. [1]).

The semantics is defined using the concept of general model in which the type o of truth values is assigned a linearly ordered MV$_{\boldsymbol{\Delta}}$-algebra which is an MV-algebra extended by the delta operation (see [9,28]). In this paper we will consider only models whose algebra of truth values forms the standard Łukasiewicz MV$_{\Delta}$-algebra

$$\mathcal{L} = \langle [0,1], \vee, \wedge, \otimes, \rightarrow, 0, 1, \Delta \rangle. \tag{7}$$

The following special formulas are important in our theory:

$$\Upsilon_{oo} \equiv \lambda z_o \cdot \neg \boldsymbol{\Delta}(\neg z_o), \qquad \text{(nonzero truth value)}$$

$$\hat{\Upsilon}_{oo} \equiv \lambda z_o \cdot \neg \boldsymbol{\Delta}(z_o \vee \neg z_o). \qquad \text{(general truth value)}$$

Thus, $\mathcal{M}(\Upsilon(A_o)) = 1$ iff $\mathcal{M}(A_o) > 0$, and $\mathcal{M}(\hat{\Upsilon}(A_o)) = 1$ iff $\mathcal{M}(A_o) \in (0,1)$ holds in any model $\mathcal{M}$.

The following completeness theorem will be often used below.

Theorem 1 ([23]).

(a) A theory T is consistent iff it has a general model $\mathcal{M}$.
(b) For every theory T and a formula A_o, $T \vdash A_o$ iff $T \models A_o$.

[2] In some papers, the term "generalized Aristotle square" is replaced by "graded on".

2.2 Theories of Evaluative Expressions and Intermediate Quantifiers

The main constituent of FNL is the theory of evaluative linguistic expressions. These are special natural language expressions such as *small, medium, big, very short, more or less deep, quite roughly strong, extremely high*, etc. A formal theory of their semantics was introduced in [25].

Another constituent of FNL is the theory of intermediate quantifiers. These are natural language expressions such as *most, almost all, a few*, etc. Their semantics is formalized using a special theory of Ł-FTT denoted by $T^{\mathrm{IQ}}[\mathcal{S}]$, which extends the theory of evaluative expressions. The detailed structure of $T^{\mathrm{IQ}}[\mathcal{S}]$ and precise definitions can be found in [18,19,26].

Definition 1. *Let $\mathcal{S} \subset$ Types be a set of selected types. Let $T^{IQ}[\mathcal{S}]$ be a theory of intermediate quantifiers. Let $z \in Form_{o\alpha}$, $x \in Form_{\alpha}$ and $A, B \in Form_{o\alpha}$. Then the following special intermediate quantifiers can be introduced:*

$$
\begin{aligned}
\mathbf{A}\text{: } &\textit{All } B \textit{ are } A := (Q^{\forall}_{Bi\boldsymbol{\Delta}}x)(B, A) \equiv (\forall x)(Bx \Rightarrow Ax),\\
\mathbf{E}\text{: } &\textit{No } B \textit{ are } A := (Q^{\forall}_{Bi\boldsymbol{\Delta}}x)(B, \neg A) \equiv (\forall x)(Bx \Rightarrow \neg Ax),\\
\mathbf{P}\text{: } &\textit{Almost all } B \textit{ are } A := (Q^{\forall}_{Bi\ Ex}x)(B, A) \equiv \\
&\quad (\exists z)((\boldsymbol{\Delta}(z \subseteq B)\, \&\, (\forall x)(zx \Rightarrow Ax)) \wedge (Bi\ Ex)((\mu B)z)),\\
\mathbf{B}\text{: } &\textit{Almost all } B \textit{ are not } A := (Q^{\forall}_{Bi\ Ex}x)(B, \neg A) \equiv \\
&\quad (\exists z)((\boldsymbol{\Delta}(z \subseteq B)\, \&\, (\forall x)(zx \Rightarrow \neg Ax)) \wedge (Bi\ Ex)((\mu B)z)),\\
\mathbf{T}\text{: } &\textit{Most } B \textit{ are } A := (Q^{\forall}_{Bi\ Ve}x)(B, A) \equiv \\
&\quad (\exists z)((\boldsymbol{\Delta}(z \subseteq B)\, \&\, (\forall x)(zx \Rightarrow Ax)) \wedge (Bi\ Ve)((\mu B)z)),\\
\mathbf{D}\text{: } &\textit{Most } B \textit{ are not } A := (Q^{\forall}_{Bi\ Ve}x)(B, \neg A) \equiv \\
&\quad (\exists z)((\boldsymbol{\Delta}(z \subseteq B)\, \&\, (\forall x)(zx \Rightarrow \neg Ax)) \wedge (Bi\ Ve)((\mu B)z)),\\
\mathbf{K}\text{: } &\textit{Many } B \textit{ are } A := (Q^{\forall}_{\neg(Sm\ \bar{\boldsymbol{\nu}})}x)(B, A) \equiv \\
&\quad (\exists z)((\boldsymbol{\Delta}(z \subseteq B)\, \&\, (\forall x)(zx \Rightarrow Ax)) \wedge \neg(Sm\ \bar{\boldsymbol{\nu}})((\mu B)z)),\\
\mathbf{G}\text{: } &\textit{Many } B \textit{ are not } A := (Q^{\forall}_{\neg(Sm\ \bar{\boldsymbol{\nu}})}x)(B, \neg A) \equiv \\
&\quad (\exists z)((\boldsymbol{\Delta}(z \subseteq B)\, \&\, (\forall x)(zx \Rightarrow \neg Ax)) \wedge \neg(Sm\ \bar{\boldsymbol{\nu}})((\mu B)z)),\\
\mathbf{I}\text{: } &\textit{Some } B \textit{ are } A := (Q^{\exists}_{Bi\boldsymbol{\Delta}}x)(B, A) \equiv (\exists x)(Bx \wedge Ax),\\
\mathbf{O}\text{: } &\textit{Some } B \textit{ are not } A := (Q^{\exists}_{Bi\boldsymbol{\Delta}}x)(B, \neg A) \equiv (\exists x)(Bx \wedge \neg Ax).
\end{aligned}
$$

3 Graded Aristotle Hexagon

3.1 From Aristotle Square to Graded Square

In this subsection, the main definitions and results that will be used later are summarized.

Definition 2. *Let T be a consistent theory of Ł-FTT and $P_1, P_2 \in Form_o$ be closed formulas of type o.*

(i) *P_1 and P_2 are* contraries *if $T \vdash \neg(P_1 \mathbin{\&} P_2)$. By completeness, this is equivalent to*

$$\mathcal{M}(P_1) \otimes \mathcal{M}(P_2) = 0$$

for every model $\mathcal{M} \models T$[3].

(ii) *P_1 and P_2 are* weak contraries *if $T \vdash \hat{\Upsilon}(P_1 \mathbin{\&} P_2)$. By completeness, this is equivalent to*

$$0 < \mathcal{M}(P_1) \otimes \mathcal{M}(P_2) < 1$$

for every model $\mathcal{M} \models T$.

(iii) *P_1 and P_2 are* sub-contraries *if $T \vdash (P_1 \nabla P_2)$. By completeness, this is equivalent to*

$$\mathcal{M}(P_1) \oplus \mathcal{M}(P_2) = 1$$

for every model $\mathcal{M} \models T$.

(iv) *P_1 and P_2 are* weak sub-contraries *if $T \vdash \Upsilon(P_1 \vee P_2)$. By completeness, this is equivalent to*

$$\mathcal{M}(P_1) \vee \mathcal{M}(P_2) > 0$$

for every model $\mathcal{M} \models T$.

(v) *P_1 and P_2 are* contradictories *if both*

$$T \vdash \neg(\boldsymbol{\Delta} P_1 \mathbin{\&} \boldsymbol{\Delta} P_2) \text{ as well as } T \vdash \boldsymbol{\Delta} P_1 \nabla \boldsymbol{\Delta} P_2.$$

By completeness, this means that both $\mathcal{M}(\boldsymbol{\Delta} P_1) \otimes \mathcal{M}(\boldsymbol{\Delta} P_2) = 0$ as well as $\mathcal{M}(\boldsymbol{\Delta} P_1) \oplus \mathcal{M}(\boldsymbol{\Delta} P_2) = 1$ hold for every model $\mathcal{M} \models T$.

(vi) *The formula P_2 is a* subaltern *of P_1 in T if $T \vdash P_1 \Rightarrow P_2$. By completeness, this means that the inequality*

$$\mathcal{M}(P_1) \leq \mathcal{M}(P_2)$$

holds true in every model $\mathcal{M} \models T$. We will call P_1 a superaltern *of P_2.*

Below we recall the main results which were formally proved in [19]. Recall that we fix the set $\mathcal{S}$ and write T^{IQ} instead of $T^{\mathrm{IQ}}[\mathcal{S}]$. Recall that the graded Aristotle's square of opposition in Ł-FTT works with the following four formulas with presupposition:

***A** :All B are A	$(\forall x)(Bx \Rightarrow Ax) \mathbin{\&} (\exists x)Bx,$	(8)
E :No B are A	$(\forall x)(Bx \Rightarrow \neg Ax),$	(9)
I :Some B are A	$(\exists x)(Bx \wedge Ax),$	(10)
***O** :Some B are not A	$(\exists x)(Bx \wedge \neg Ax) \nabla \neg(\exists x)Bx.$	(11)

[3] Let $\mathcal{M} \models T^{\mathrm{IQ}}$. Then we denote $\mathcal{M}(\top) = 1$ and $\mathcal{M}(\bot) = 0$.

3.2 From Graded Aristotle Square to Graded Aristotle Hexagon

Generalizing the graded Aristotle square to the graded Aristotle hexagon means to define new formulas which will be put to the top and the bottom of the latter:

$$\mathbf{U} := \mathbf{A} \vee \mathbf{E} \quad \text{All or No } B \text{ are } A. \tag{12}$$

$$\mathbf{Y} := \mathbf{I} \,\&\, \mathbf{O} \quad \text{Some but Not All } B \text{ are } A. \tag{13}$$

Lemma 1. *There is no model such that*

(a) $\mathcal{M}(\mathbf{Y}) = 1$ *and* $\mathcal{M}(\mathbf{A}) = 1$.
(b) $\mathcal{M}(\mathbf{Y}) = 1$ *and* $\mathcal{M}(\mathbf{E}) = 1$.

Proof. (a) Let there be a model $\mathcal{M} \models T^{\mathrm{IQ}}$ such that $\mathcal{M}(\mathbf{Y}) = 1$ and $\mathcal{M}(\mathbf{A}) = 1$. Then $\mathcal{M}(\boldsymbol{\Delta}\mathbf{A}) = 1$ and so from the contradictory relation to $\mathbf{O}$ it follows that $\mathcal{M}(\boldsymbol{\Delta}\mathbf{O}) = 0$. Because $\mathbf{A}$ is superaltern of $\mathbf{I}$, $\mathcal{M}(\mathbf{I}) = 1$ and so, $\mathcal{M}(\mathbf{I}\,\&\,\mathbf{O}) = \mathcal{M}(\mathbf{Y}) = 0$ which contradicts the assumption.

(b) Analogously as (a).

Lemma 2. *The following holds in every model* $\mathcal{M} \models T^{IQ}$*:*

(a) $\mathcal{M}(\mathbf{I} \nabla \mathbf{U}) = 1$,
(b) $\mathcal{M}(\mathbf{O} \nabla \mathbf{U}) = 1$.

Proof. (a) Let $\mathcal{M} \models T^{\mathrm{IQ}}$. From the contradictory relation of $\mathbf{E}$ to $\mathbf{I}$ it follows that $\mathcal{M}(\boldsymbol{\Delta}\mathbf{E} \nabla \boldsymbol{\Delta}\mathbf{I}) = 1$. Then

$$1 = \mathcal{M}(\boldsymbol{\Delta}\mathbf{E} \nabla \boldsymbol{\Delta}\mathbf{I}) \leq \mathcal{M}(\mathbf{E} \nabla \mathbf{I}) \leq \mathcal{M}(\mathbf{I} \nabla (\mathbf{E} \vee \mathbf{A})) = \mathcal{M}(\mathbf{I} \nabla \mathbf{U}).$$

(b) Analogously as (a).

Corollary 1. *There is no model of* T^{IQ} *such that*

(a) $\mathcal{M}(\mathbf{U}) = 0$ *and* $\mathcal{M}(\mathbf{I}) = 0$.
(b) $\mathcal{M}(\mathbf{U}) = 0$ *and* $\mathcal{M}(\mathbf{O}) = 0$.

Hence, we conclude the following:

Theorem 2 (Sub-contraries). *The couples of formulas* $\mathbf{I}$ *and* $\mathbf{U}$*, as well as* $\mathbf{O}$ *and* $\mathbf{U}$ *are sub-contraries in* T^{IQ}.

Lemma 3. *Let* $\mathcal{M} \models T^{IQ}$.

(a) If $\mathcal{M}(\hat{\Upsilon}(\mathbf{A})) = 1$ *then* $\mathcal{M}(\mathbf{O}) = 1$.
(b) If $\mathcal{M}(\hat{\Upsilon}(\mathbf{E})) = 1$ *then* $\mathcal{M}(\mathbf{I}) = 1$.

Proof. (a) Let the assumption hold and $\mathcal{M}(\mathbf{O}) < 1$. From the contradictory relation of $\mathbf{A}$ to $\mathbf{O}$ it follows that $\mathcal{M}(\boldsymbol{\Delta}\mathbf{A}) = 1$ and hence $\mathcal{M}(\mathbf{A}) = 1$ which contradicts the assumption.

(b) is proved analogously.

We see from Lemma 3 that in every model of T^{IQ}, if the truth degree of the formulas **A** and **E** is smaller than 1 then the truth degree of the respective formulas **I** and **O** must be equal to 1. We can also immediately see that if $\mathcal{M}(\hat{\Upsilon}(\mathbf{A})) = 1$ and $\mathcal{M}(\hat{\Upsilon}(\mathbf{E})) = 1$ then also $\mathcal{M}(\hat{\Upsilon}(\mathbf{U})) = 1$.

Lemma 4. *Let $\mathcal{M} \models T^{IQ}$. Let $\mathcal{M}(\hat{\Upsilon}(\mathbf{A})) = 1$ and $\mathcal{M}(\hat{\Upsilon}(\mathbf{E})) = 1$. Then the following is true:*

(a) $\mathcal{M}(\mathbf{I}) \otimes \mathcal{M}(\boldsymbol{\Delta}\mathbf{U}) = 0$
(b) $\mathcal{M}(\mathbf{I}) \oplus \mathcal{M}(\boldsymbol{\Delta}\mathbf{U}) = 1$,
(c) $\mathcal{M}(\mathbf{O}) \otimes \mathcal{M}(\boldsymbol{\Delta}\mathbf{U}) = 0$,
(d) $\mathcal{M}(\mathbf{O}) \oplus \mathcal{M}(\boldsymbol{\Delta}\mathbf{U}) = 1$,
(e) $\mathcal{M}(\mathbf{I}) \otimes \mathcal{M}(\mathbf{U}) = a \in (0,1)$,
(f) $\mathcal{M}(\mathbf{O}) \otimes \mathcal{M}(\mathbf{U}) = b \in (0,1)$,
(g) $\mathcal{M}(\mathbf{I}) \oplus \mathcal{M}(\mathbf{U}) = 1$,
(h) $\mathcal{M}(\mathbf{O}) \oplus \mathcal{M}(\mathbf{U}) = 1$.
(i) $\mathcal{M}(\boldsymbol{\Delta}\mathbf{A}) \otimes \mathcal{M}(\boldsymbol{\Delta}\mathbf{Y}) = 0$
(j) $\mathcal{M}(\boldsymbol{\Delta}\mathbf{A}) \oplus \mathcal{M}(\boldsymbol{\Delta}\mathbf{Y}) = 1$,
(k) $\mathcal{M}(\boldsymbol{\Delta}\mathbf{E}) \otimes \mathcal{M}(\boldsymbol{\Delta}\mathbf{Y}) = 0$,
(l) $\mathcal{M}(\boldsymbol{\Delta}\mathbf{E}) \oplus \mathcal{M}(\boldsymbol{\Delta}\mathbf{Y}) = 1$,
(m) $\mathcal{M}(\mathbf{A}) \otimes \mathcal{M}(\mathbf{Y}) = a \in (0,1)$,
(n) $\mathcal{M}(\mathbf{E}) \otimes \mathcal{M}(\mathbf{Y}) = a \in (0,1)$,
(o) $\mathcal{M}(\mathbf{A}) \oplus \mathcal{M}(\mathbf{Y}) = 1$,
(p) $\mathcal{M}(\mathbf{E}) \oplus \mathcal{M}(\mathbf{Y}) = 1$.

Theorem 3. *Let $\mathcal{M} \models T^{IQ}$. Let $\mathcal{M}(\hat{\Upsilon}(\mathbf{A})) = 1$ and $\mathcal{M}(\hat{\Upsilon}(\mathbf{E})) = 1$. Then the following is true:*

(a) The formulas **I** *and* **U** *can be contradictories, sub-contraries, as well as weak contraries.*
(b) The formulas **O** *and* **U** *can be contradictories, sub-contraries, as well as weak contraries.*
(c) The formulas **A** *and* **Y** *can be contradictories, sub-contraries, as well as weak contraries.*
(d) The formulas **E** *and* **Y** *can be contradictories, sub-contraries, as well as weak contraries.*
(e) The formulas **I** *and* **U** *as well as the formulas* **O** *and* **U** *cannot be contraries.*
(f) The formulas **A** *and* **Y** *as well as the formulas* **E** *and* **Y** *cannot be contraries.*

Lemma 5. *Let* **A**, **E** *and* **I**, **O** *be intermediate quantifiers introduced above. Then the following holds in every model $\mathcal{M} \models T^{IQ}$:*

(a) $\mathcal{M}(\mathbf{U}) = 0$ *implies* $\mathcal{M}(\mathbf{Y}) = 1$.
(b) $\mathcal{M}(\mathbf{U}) = 1$ *implies* $\mathcal{M}(\mathbf{Y}) = 0$.

Proof. (a) Let there be a model $\mathcal{M} \models T^{\mathrm{IQ}}$ such that $\mathcal{M}(\mathbf{U}) = \mathcal{M}(\mathbf{A} \vee \mathbf{E}) = 0$. Then $\mathcal{M}(\mathbf{A}) = \mathcal{M}((\forall x)(Bx \Rightarrow Ax)) = 0$ and also $\mathcal{M}(\mathbf{E}) = \mathcal{M}((\forall x)(Bx \Rightarrow \neg Ax)) = 0$. Hence,

$$1 = \mathcal{M}((\exists x)(Bx \,\&\, \neg Ax)) \leq \mathcal{M}((\exists x)(Bx \wedge \neg Ax)) = \mathcal{M}(\mathbf{O})$$

and also

$$1 = \mathcal{M}((\exists x)(Bx \,\&\, Ax)) \leq \mathcal{M}((\exists x)(Bx \wedge Ax)) = \mathcal{M}(\mathbf{I}).$$

We conclude that $\mathcal{M}(\mathbf{I} \,\&\, \mathbf{O}) = \mathcal{M}(\mathbf{Y}) = 1$.

(b) immediately follows from (a).

Lemma 6. *Let $\mathcal{M} \models T^{IQ}$. Let $\mathcal{M}(\hat{\Upsilon}(\mathbf{A})) = 1$ and $\mathcal{M}(\hat{\Upsilon}(\mathbf{E})) = 1$. Then the following is true:*

(a) $\mathcal{M}(\boldsymbol{\Delta}\mathbf{U}) \otimes \mathcal{M}(\boldsymbol{\Delta}\mathbf{Y}) = 0$,
(b) $\mathcal{M}(\boldsymbol{\Delta}\mathbf{U}) \oplus \mathcal{M}(\boldsymbol{\Delta}\mathbf{Y}) = 1$.

Proof. (a) Let $\mathcal{M} \models T^{\mathrm{IQ}}$. Let $\mathcal{M}(\hat{\Upsilon}(\mathbf{A})) = 1$ and $\mathcal{M}(\hat{\Upsilon}(\mathbf{E})) = 1$. Then from the definition of $\hat{\Upsilon}$ it follows that $\mathcal{M}(\boldsymbol{\Delta}(\mathbf{A} \vee \mathbf{E})) = 0$. From Lemma 3 we conclude that $\mathcal{M}(\mathbf{I}\,\&\,\mathbf{O}) = 1$ as well as $\mathcal{M}(\boldsymbol{\Delta}(\mathbf{I}\,\&\,\mathbf{O})) = 1$. Finally, $\mathcal{M}(\boldsymbol{\Delta}\mathbf{U}) \otimes \mathcal{M}(\boldsymbol{\Delta}\mathbf{Y}) = 0$ and $\mathcal{M}(\boldsymbol{\Delta}\mathbf{U}) \oplus \mathcal{M}(\boldsymbol{\Delta}\mathbf{Y}) = 1$ are fulfilled.

As a corollary we immediately obtain the following.

Theorem 4 (Contradictories). *The quantifiers* $\mathbf{U}$ *and* $\mathbf{Y}$ *are contradictories in* T^{IQ}.

Finally, we can demonstrate that in the graded Aristotle hexagon the two extra formulas are perfectly united by means of the four arrows of subalterns.

Theorem 5 (Subalterns). *The following holds true in the theory* T^{IQ}*:*

(a) The formula $\mathbf{U}$ *is subaltern of* $\mathbf{A}$ *and* $\mathbf{E}$*, i.e.,* $T^{IQ} \vdash \mathbf{A} \Rightarrow \mathbf{U}$ *and* $T^{IQ} \vdash \mathbf{E} \Rightarrow \mathbf{U}$.
(a) The formula $\mathbf{Y}$ *is superaltern of* $\mathbf{I}$ *and* $\mathbf{O}$*, i.e.,* $T^{IQ} \vdash \mathbf{Y} \Rightarrow \mathbf{I}$ *and* $T^{IQ} \vdash \mathbf{Y} \Rightarrow \mathbf{O}$.

Corollary 2. *The graded Aristotle hexagon forms three graded squares of opposition, namely* **AEIO**, **AYOU** *and* **EYIU**.

3.3 From Graded Generalized Peterson's Square to Graded Generalized Hexagon

In the papers [19,21], we syntactically analyzed and semantically verified the generalized Peterson square (**5**-square) of opposition in FNL. In this section, we will introduce basic concepts using which the graded generalized hexagon can be formed. We start with the following definitions of new generalized intermediate quantifiers:

$$\mathbf{U}_{Ex\,Bi} := \mathbf{P} \vee \mathbf{B} \quad \text{Almost all } B \text{ are } A \text{ or Almost all } B \text{ are not } A \tag{14}$$

$$\mathbf{U}_{Ve\,Bi} := \mathbf{T} \vee \mathbf{D} \quad \text{Most } B \text{ are } A \text{ or Most } B \text{ are not } A \tag{15}$$

$$\mathbf{Y}_{\neg\,Sm} := \mathbf{K}\,\&\,\mathbf{G} \quad \text{Many } B \text{ are } A \text{ and Many } B \text{ are not } A. \tag{16}$$

We will suppose that the basic fuzzy set used in the definition of the intermediate quantifier is a normal fuzzy set. This is specified by the following definition.

Definition 3. *Let* $B \in Form_{o\alpha}$. *By* $T[B]$ *we denote an extension of the theory* T^{IQ} *such that*

$$T[B] \vdash (\exists x_\alpha)\boldsymbol{\Delta}Bx.$$

Theorem 6 (Contraries, [19]). *The following couples of formulas are contraries in* $T[B]$*:* $\mathbf{B}$ *and* $\mathbf{P}$*, and also* $\mathbf{D}$ *and* $\mathbf{T}$.

Other properties of five basic generalized intermediate quantifiers were proved in [19]. In [22], we analyzed the intermediate quantifier "Many" and its position inside of 5-square of opposition.

Recall that the formula $\mathbf{A}$ is a superaltern of all the formulas $\mathbf{P}, \mathbf{T}, \mathbf{K}, \mathbf{I}$. At the same time, $\mathbf{E}$ is a superaltern of $\mathbf{B}, \mathbf{D}, \mathbf{G}, \mathbf{O}$.

Lemma 7. *The following is provable:*

(a) $T[B] \vdash \mathbf{I} \nabla \mathbf{U}_{Ex\,Bi}$,
(b) $T[B] \vdash \mathbf{I} \nabla \mathbf{U}_{Ve\,Bi}$.

Proof. (a) Because $\mathbf{E}$ is a superaltern of $\mathbf{B}$ then by the properties of the delta operation we obtain $T[B] \vdash \Delta\mathbf{E} \Rightarrow \Delta\mathbf{B}$. Then

$$T[B] \vdash \Delta\mathbf{E} \nabla \Delta\mathbf{I} \Rightarrow \Delta\mathbf{B} \nabla \Delta\mathbf{I}. \tag{17}$$

Furthermore, by the properties of delta we obtain $T[B] \vdash \Delta\mathbf{B} \Rightarrow \mathbf{B}$ as well as, $T[B] \vdash \Delta\mathbf{I} \Rightarrow \mathbf{I}$. Then by properties of Ł-FTT we get

$$T[B] \vdash \Delta\mathbf{B} \nabla \Delta\mathbf{I} \Rightarrow \mathbf{B} \nabla \mathbf{I}. \tag{18}$$

Joining (17) and (18) we obtain

$$T[B] \vdash \Delta\mathbf{E} \nabla \Delta\mathbf{I} \Rightarrow \mathbf{B} \nabla \mathbf{I}. \tag{19}$$

But we know that The quantifiers $\mathbf{E}$ and $\mathbf{I}$ are contradictories and so $T[B] \vdash \Delta\mathbf{E} \nabla \Delta\mathbf{I}$, which gives $T[B] \vdash \mathbf{B} \nabla \mathbf{I}$ and also $T[B] \vdash (\mathbf{B} \vee \mathbf{P}) \nabla \mathbf{I}$.

(b) Analogously as (a).

Theorem 7. *The following couples of formulas are subcontraries in T[B]:* $\mathbf{U}_{Ex\,Bi}$ *and* $\mathbf{I}$*, and also* $\mathbf{U}_{Ve\,Bi}$ *and* $\mathbf{I}$.

Theorem 8. *The following is provable in* $T[B]$*:*

(a) The quantifier $\mathbf{U}$ *is a superaltern of* $\mathbf{U}_{Ex\,Bi}$.
(b) The quantifier $\mathbf{U}_{Ve\,Bi}$ *is a subaltern of* $\mathbf{U}_{Ex\,Bi}$.
(c) The quantifier $\mathbf{Y}_{\neg\,Sm}$ *is a superaltern of* $\mathbf{Y}$.
(d) The quantifiers $\mathbf{P}$ *and* $\mathbf{B}$ *are a superalterns of* $\mathbf{U}_{Ex\,Bi}$.
(e) The quantifiers $\mathbf{T}$ *and* $\mathbf{D}$ *are a superalterns of* $\mathbf{U}_{Ve\,Bi}$.

3.4 Example of the Graded Generalized Hexagon

Let us consider a model $\mathcal{M} \models T[B]$ such that $T^{\mathrm{IQ}} \vdash (\exists x)Bx$ and let $\mathcal{M}(\mathbf{A}) = a > 0$ (e.g., $a = 0.2$). The degrees inside of the generalized Peterson's square follow from the definitions of contraries, contradictories, sub-contraries and subalterns.

The following can be proved: The formulas **I** and **O** are sub-contraries with each of $\mathbf{U}, \mathbf{U}_{Ex\,Bi}, \mathbf{U}_{Ve\,Bi}$. The quantifier **U** is superaltern and the formula **Y** is subaltern of all of the other quantifiers. Furthermore, the quantifiers **A** and **E** are weak contraries with **Y**. The motivation and explanation that the quantifier **A** is not a negation of the formula **O** is explained in [19]. Similarly the quantifiers **P**, **T** and **B**, **D** are weak contraries with **Y**.

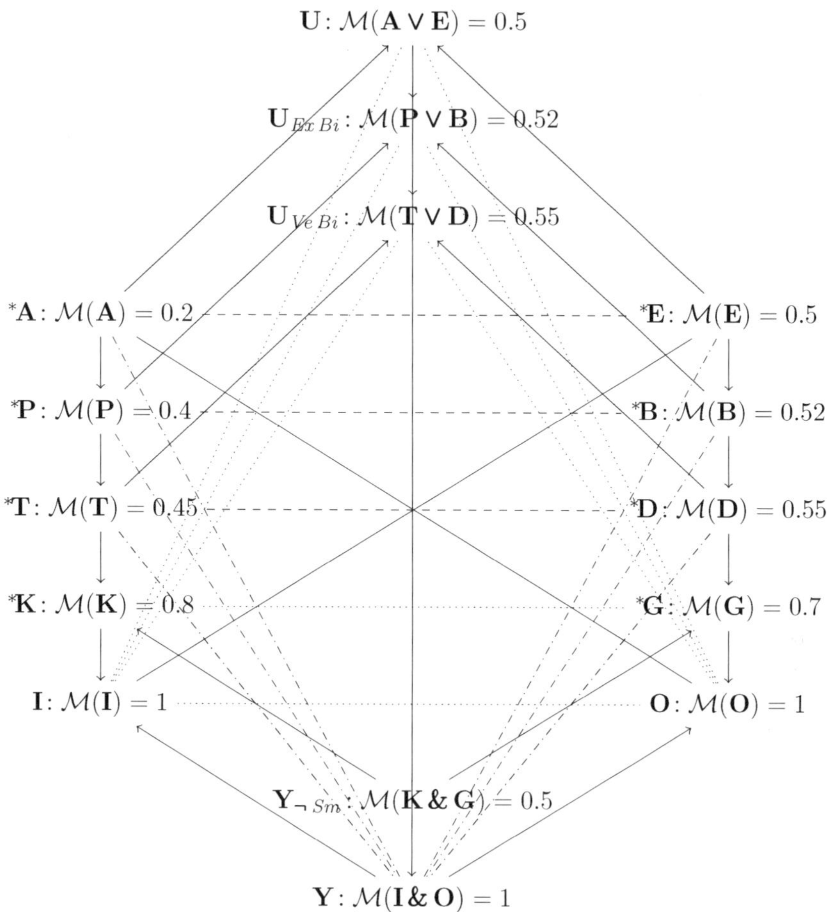

In every classical Aristotle hexagon, three squares **AEIO**, **AYOU** and **EYIU** are formed on the basis of the properties of contraries, contradictories, sub-contraries and sub-alterns. The graded Aristotle square (**AEIO**) contains the same properties as its classical version. It can be seen that the new graded squares (**AYOU** and **EYIU**) prove that the formulas **A** and **Y** as well as the formulas **E** and **Y** are weak contraries.

4 Conclusion

In this paper we extended the theory of graded classical Aristotle square of opposition to the graded Aristotle hexagon. Then we suggested generalization of the Peterson's square of opposition to a graded generalized hexagon, i.e., the hexagon whose vertices contain intermediate quantifiers.

The future work will focus on more detailed analysis of the properties of the graded generalized hexagon, possibly extended by more intermediate quantifiers. Furthermore, we will also study graded cube of opposition. This may open interesting area of study of relations among important classes of properties. We expect that these results will contribute to the development of fuzzy natural logic, namely to formulation of various kinds of general rules used in human reasoning.

References

1. Andrews, P.: An Introduction to Mathematical Logic and Type Theory: To Truth Through Proof. Kluwer, Dordrecht (2002)
2. Béziau, J.: New light on the square of oppositions and its nameless corner. Log. Investig. **10**, 218–233 (2003)
3. Béziau, J.: The power of the hexagon. Logica Universalis **6**(1–2), 1–43 (2012)
4. Béziau, J., Gan-Krzywoszyńska, K.: Handbook of Abstracts of the 2nd World Congress on the Square of Opposition, Corte, Corsica, 17–20 June 2010
5. Béziau, J., Gan-Krzywoszyńska, K.: Handbook of Abstracts of the 3rd World Congress on the Square of Opposition, Beirut, Lebanon, 26–30 June 2010
6. Béziau, J., Gan-Krzywoszyńska, K.: Handbook of Abstracts of the 4th World Congress on the Square of Opposition, Roma, Vatican, pp. 26–30, 5–9 May 2014
7. Blanché, R.: Quantitiy, modality, and other kindred systems of categories. Mind **LXI**(243), 369–375 (1952)
8. Blanché, R.: Sur l'oppostion des concepts. Theoria **19**, 89–130 (1953)
9. Cignoli, R.L.O., D'Ottaviano, I.M.L., Mundici, D.: Algebraic Foundations of Many-Valued Reasoning. Kluwer, Dordrecht (2000)
10. Ciucci, D., Dubois, D., Prade, H.: Oppositions in rough set theory. In: Li, T., Nguyen, H.S., Wang, G., Grzymala-Busse, J., Janicki, R., Hassanien, A.E., Yu, H. (eds.) RSKT 2012. LNCS, vol. 7414, pp. 504–513. Springer, Heidelberg (2012)
11. Ciucci, D., Dubois, D., Prade, H.: The structure of oppositions in rough set theory and formal concept analysis - toward a new bridge between the two settings. In: Beierle, C., Meghini, C. (eds.) FoIKS 2014. LNCS, vol. 8367, pp. 154–173. Springer, Heidelberg (2014)
12. Ciucci, D., Dubois, D., Prade, H.: Structures of opposition induced by relations: the boolean and the gradual cases. Ann. Math. Artif. Intell. **76**(3–4), 351–373 (2015). doi:10.1007/s10472-015-9480-8
13. Dubois, D., Prade, H.: From blanche's hexagonal organization of concepts to formal concepts analysis and possibility theory. Logica Universalis **6**(1–2), 149–169 (2012)
14. Dubois, D., Prade, H.: Gradual structures of oppositions. In: Esteva, F., Magdalena, L., Verdegay, J.L. (eds.) Enric Trillas: Passion for Fuzzy Sets. Studies in Fuzziness and Soft Computing, vol. 322, pp. 79–91. Springer, Switzerland (2015)

15. Dubois, D.: P.H.: from blanches hexagonal organization of concepts to formal concept analysis and possibility theory. Logica Universalis **6**, 149–169 (2012)
16. Dvořák, A., Holčapek, M.: L-fuzzy quantifiers of the type $\langle 1 \rangle$ determined by measures. Fuzzy Sets Syst. **160**, 3425–3452 (2009)
17. Moretti, A.: The geometry of logical oppositions and the opposition of logic to it. In: Bianchi, I., Savaradi, U. (eds.) The Perception and Cognition of Contraries (2009)
18. Murinová, P., Novák, V.: A formal theory of generalized intermediate syllogisms. Fuzzy Sets Syst. **186**, 47–80 (2013)
19. Murinová, P., Novák, V.: Analysis of generalized square of opposition with intermediate quantifiers. Fuzzy Sets Syst. **242**, 89–113 (2014)
20. Murinová, P., Novák, V.: The structure of generalized intermediate syllogisms. Fuzzy Sets Syst. **247**, 18–37 (2014)
21. Murinová, P., Novák, V.: Analysis of the intermediate quantifier "many" in fuzzy natural logic. In: Proceedings of International Conference IFSA-EUSFLAT 2015, pp. 1147–1153 (2015)
22. Murinová, P., Novák, V.: On properties of the intermediate quantifier "many". Fuzzy Sets Syst. (submitted)
23. Novák, V.: On fuzzy type theory. Fuzzy Sets Syst. **149**, 235–273 (2005)
24. Novák, V.: Perception-based logical deduction. In: Reusch, B. (ed.) Computational Intelligence, Theory and Applications. Advances in Soft Computing, vol. 33, pp. 237–250. Springer, Heidelberg (2005)
25. Novák, V.: A comprehensive theory of trichotomous evaluative linguistic expressions. Fuzzy Sets Syst. **159**(22), 2939–2969 (2008)
26. Novák, V.: A formal theory of intermediate quantifiers. Fuzzy Sets Syst. **159**(10), 1229–1246 (2008)
27. Novák, V., Lehmke, S.: Logical structure of fuzzy IF-THEN rules. Fuzzy Sets Syst. **157**, 2003–2029 (2006)
28. Novák, V., Perfilieva, I., Močkoř, J.: Mathematical Principles of Fuzzy Logic. Kluwer, Boston (1999)
29. Pellissier, R.: Setting n-opposition. Logica Universalis **2**(2), 235–263 (2008)
30. Peters, S., Westerståhl, D.: Quantifiers in Language and Logic. Claredon Press, Oxford (2006)
31. Peterson, P.: Intermediate Quantifiers. Logic, linguistics, and Aristotelian Semantics. Ashgate, Aldershot (2000)
32. Smesaert, H.: The classical aristotelian hexagon versus the modern duality hexagon. Logica Universalis **6**, 171–199 (2012)
33. Smessaert, H.: On the 3D-visualisationof logical relations. Logica Universalis **3**(2), 303–332 (2009)
34. Westerståhl, D.: The traditional square of opposition and generalized quantifiers. Stud. Log. **2**, 1–18 (2008)
35. Wikipedia (2004). http://en.wikipedia.org/wiki/aristotle

On a Category of Extensional Fuzzy Rough Approximation L-valued Spaces

Aleksandrs Eļkins[1(✉)], Alexander Šostak[1,2], and Ingrīda Uļjane[1,2]

[1] Department of Mathematics, University of Latvia, Zellu street 25, Riga 1002, Latvia
aleksandrs.elkins@gmail.com

[2] Institute of Mahematics and CS UL Raiņa bulv., Riga 1459, Latvia
sostaks@latnet.lv, ingrida.uljane@lu.lv

Abstract. We establish extensionality of some upper and lower fuzzy rough approximation operators on an L-valued set. Taking as the ground basic properties of these operators, we introduce the concept of an (extensional) fuzzy rough approximation L-valued space. We apply fuzzy functions satisfying certain continuity-type conditions, as morphisms between such spaces, and in the result obtain a category $\mathcal{FRA}\mathbf{SPA}(L)$ of fuzzy rough approximation L-valued spaces. An interpretation of fuzzy rough approximation L-valued spaces as L-fuzzy (di)topological spaces is presented and applied for constructing examples in category $\mathcal{FRA}\mathbf{SPA}(L)$.

Keywords: Fuzzy rough approximation operators · Extensionality · GL-monoid · L-valued set · Fuzzy function · Fuzzy rough approximation L-valued space

1 Introduction and Motivation

The concept of a many-valued, or L-valued set, appears in research of different authors. As the basic source, we refer here to U. Höhle's works [9–12]. Actually, an L-valued set is a pair (X, E) where X is a set and $E : X \times X \to L$ is a fuzzy equivalence, that is a reflexive symmetric transitive fuzzy relation, see Subsect. 2.2. However, specific is the interpretation of the value $E(x, y)$ as the *degree to which elements* $x, y \in X$ *are equal.* Such interpretation provokes, in its turn, the idea to use not usual functions $f : X \to Y$, assigning to each $x \in X$ a unique $y \in Y$, as morphisms between L-valued sets (X, E_X) and (Y, E_Y) but the so called fuzzy functions, that is fuzzy relations $R : X \times Y \to L$, satisfying certain extensionality type conditions [8,27]. The basics of the theory of fuzzy functions in the channel of this work are exposed in Sect. 4.

Returning back to the concept of an L-valued set (X, E) and recalling that $E : X \times X \to L$ is a fuzzy relation, we have the challenge to apply it for the construction of upper and lower rough approximation operators $u : L^X \to L^X$ and $l : L^X \to L^X$, thus approaching the problems of fuzzy rough approximation of fuzzy sets, see Subsect. 3.1. However, aiming to study such structures and to

J.P. Carvalho et al. (Eds.): IPMU 2016, Part II, CCIS 611, pp. 48–60, 2016.
DOI: 10.1007/978-3-319-40581-0_5

stick to fuzzy functions as morphisms between them, we need to assure certain extensionality properties of the approximation operators u and l. This problem is solved, in the "optimal way", in Subsect. 3.2., where we consider extensional fuzzy rough approximation operators $u_E, l_E : L^X \to L^X$. Basic properties of these operators are the subject of Subsect. 3.3

In Sect. 5 we study properties of fuzzy functions on L-valued sets enriched with operators u_E, l_E and apply the obtained results in order to introduce the general category $\mathcal{FRA}\mathbf{SPA}(L)$ of extensional fuzzy rough approximation L-valued spaces as objects and fuzzy functions satisfying certain "continuity-type" properties as morphisms.

In Sect. 6, we briefly touch some issues concerning the interpretation of extensional fuzzy rough approximation L-valued spaces (X, E, u, l) as L-fuzzy ditopological spaces (X, T, K) and the relations between the corresponding categories. Some perspectives for the future work are mentioned in Sect. 7.

2 Prerequisites: The Context of the Work

2.1 Cl-Monoids and Residuated Lattices

Let $(L, \leq, \wedge, \vee)$ denote a complete frame, that is a lattice in which arbitrary suprema (joins) and infima (meets) exist and in which finite meets distribute over arbitrary unions: $\alpha \wedge \{\bigvee \beta_i : i \in I\} = \bigvee_i \{\alpha \wedge \beta_i : i \in I\} \quad \forall \alpha \in L, \ \forall \{\beta_i : i \in I\} \subseteq L$. In particular, the top 1_L and the bottom 0_L elements in L exist and $0_L \neq 1_L$. Following G. Birkhoff [1] by a *cl-monoid* we call a tuple $(L, \leq, \wedge, \vee, *)$ where $(L, \leq, \wedge, \vee)$ is a complete frame and the binary operation $* : L \times L \to L$ satisfies conditions:

(0cl) $*$ is monotone: $\alpha \leq \beta \Longrightarrow \alpha * \gamma \leq \beta * \gamma$ for all $\alpha, \beta, \gamma \in L$;
(1cl) $*$ is commutative: $\alpha * \beta = \beta * \alpha$ for all $\alpha, \beta \in L$;
(2cl) $*$ is associative: $(\alpha * \beta) * \gamma = \alpha * (\beta * \gamma)$ for all $\alpha, \beta, \gamma \in L$;
(3cl) $*$ distributes over arbitrary joins: $\alpha * \left(\bigvee_{i \in I} \beta_i\right) = \bigvee_{i \in I} (\alpha * \beta_i)$ for all $\alpha \in L$, for all $\{\beta_i \mid i \in I\} \subseteq L$,
(4cl) $\alpha * 1_L = \alpha, \quad \alpha * 0_L = 0_L$ for all $\alpha \in L$.

In a cl-monoid a further binary operation $\mapsto$, residuum, is defined: $\alpha \mapsto \beta = \bigvee \{\lambda \in L \mid \lambda * \alpha \leq \beta\}, \ \forall \alpha, \beta, \gamma \in L$. Residuation is connected with operation $*$ by the Galois connection: $\alpha * \beta \leq \gamma \Longleftrightarrow \alpha \leq (\beta \mapsto \gamma)$, see e.g. [5].

Remark 1. A cl-monoid can be defined also as an integral commutative two-sided quantale [25] A cl-monoid $(L, \leq, \wedge, \vee, *)$ interpreted as a tuple $(L, \leq, \wedge, \vee, \mapsto)$ is known also as a residuated lattice, see [18].

2.2 GL-monoids

[10,13] A cl-monoid $(L, \leq, \wedge, \vee, *)$ is called a GL-monoid[1] if it is divisible, that is the following axiom is satisfied:
(GL) If $\alpha \leq \beta, \alpha, \beta \in L$, then there exists $\gamma \in L$ such that $\alpha = \beta * \gamma$.

[1] GL comes as an abbreviation of *Generalized Logic*.

It is known that a GL-monoid has the following important properties, which are not valid in an arbitrary cl-monoid, see [10], [12, pp. 108–109]:

(GL1) $\alpha * (\alpha \mapsto \beta) = \alpha \wedge \beta$;
(GL2) $\alpha * (\beta \wedge \gamma) = (\alpha * \beta) \wedge (\alpha * \gamma)$, $\forall \alpha, \beta, \gamma \in L$;
(GL3) $\alpha * \beta \leq (\alpha * \alpha) \vee (\beta * \beta)$, $\forall \alpha, \beta \in L$.

Important examples of GL-monoids are frames (if $* = \wedge$) and MV-algebras.

A large part of our research can be done if an arbitrary cl-monoid L is used as the range L for fuzzy equivalences and fuzzy sets. However in some cases we additionally need to request divisibility of the cl-monoid. Therefore, in order to make exposition more concise, *we always assume that L denotes a GL-monoid.*

2.3 L-Equivalences and L-valued Sets

A binary fuzzy relation E on X is called an *L-equivalence*, or a *fuzzy equivalence* on X (see [9,11,15,28], et al.) if for all $x, y, z \in X$, the following properties hold:

(1) $E(x,x) = 1$, reflexivity,
(2) $E(x,y) = E(y,x)$, symmetry,
(3) $E(x,y) * E(y,z) \leq E(x,z)$, transitivity.

A pair (X, E), where E is a fuzzy equivalence on X, is called an L-valued set.

2.4 Extensional Fuzzy Sets

A fuzzy set A in an L-valued set (X, E) is called *extensional* if $A(x) * E(x,x') \leq A(x')$ for all $x, x' \in X$, see, e.g., [9,10,14].

3 Fuzzy Rough Approximation Structure of L-valued Sets

3.1 Historical Comments and Statement of the Problem

In 1982 Z. Pawlak introduced the concept of a rough set [19]. Given a set X endowed with a (crisp) equivalence relation $E \subseteq X \times X$ and a subset $A \subseteq X$, a rough set determined by A can be defined as the pair $(A^{\blacktriangledown}, A^{\blacktriangle})$, where $A^{\blacktriangledown} = \{x \in A \mid [x]_E \cap A \neq \emptyset\}$ and $A^{\blacktriangle} = \{x \in A \mid [x]_E \subseteq A\}$; here $[x]_E$ denotes the E-equivalence class of the element $x \in A$.

In 1990 D. Dubois and H. Prade [4] published the first work where the concept of a rough set was extended to the context of fuzzy sets. Later there was done much research where different approaches to the fuzzification of the concept of a rough set were developed. Specifically, an important research work related to fuzzy rough sets is presented in the papers [16,17,20,21,23,26], et al. in which different approaches to the subject of a fuzzy rough set were presented.

When working in the context of L-valued sets and their fuzzy subsets, especially when developing abstract mathematical structures, such as algebraic or topological, the *property of extensionality* of fuzzy sets involved in the research, plays an important role. In particular, aiming to introduce an abstract mathematical concept for a set equipped with some kind of a fuzzy rough approximation structure (u, l), it is natural to request extensionality of these operators. The "classical", Pawlak's definition of a rough set, leads to extensional upper and lower approximation operators $\blacktriangledown, \blacktriangle$, see Remark 3. On the other hand many generalizations of these operators to the fuzzy setting fail to be extensional. The aim of this section to find "the most appropriate" upper and lower extensional fuzzy rough approximation operators u_E and l_E and to establish their basic properties.

3.2 Definition of the Extensional Fuzzy Rough Approximation Operators

Recall (see e.g. [9,10,14]) that the *extensional hull of a fuzzy set* $A \in L^X$ in an L-valued set (X, E) is the smallest extensional fuzzy set $\tilde{A} \in L^X$ that is larger than or equal to A ($\tilde{A} \geq A$.)

We define the *extensional kernel of a fuzzy set* $A \in L^X$ in an L-valued set (X, E) as the largest fuzzy subset $A^0 \in L^X$ that is smaller than or equal to A ($A^0 \leq A$).

Definition 1 [16], see also [26]. *Given an L-valued set (X, E), the upper fuzzy rough approximation operator $u_E : L^X \to L^X$ is defined by:*

$$u_E(A)(x) = \sup_{x'}(E(x,x') * A(x)), \ \forall A \in L^X, \ \forall x \in X,$$

and the lower fuzzy rough approximation operator $l_E : L^X \to L^X$ is defined by:

$$l_E(A)(x) = \inf_{x'}(E(x,x') \mapsto A(x)), \ \forall A \in L^X, \ \forall x \in X.$$

The proof of the following known fact is included for clarification purposes.

Proposition 1. [9,10,14,27]. *The extensional hull of a fuzzy set A in an L-valued set (X, E) equals to its image under upper fuzzy rough approximation operator $u_E : L^X \to L^X$, that is $\tilde{A} = u_E(A)$.*

Proof. By definition of $u_E(A)$, for every $x \in X$, we have $u_E(A)(x) * E(x,x') = \sup_{x'' \in X}(A(x'') * E(x,x'')) * E(x,x') \leq \sup_{x'' \in X} A(x'') * E(x',x'') = u_E(A)(x')$, and hence $u_E(A)$ is extensional. Applying reflexivity of E we see that $u_E(A) \geq A$. Further, by the definition of $u_E(A)$ it is clear, that any extensional fuzzy set containing A will contain also $u_E(A)$, and hence $u_E(A) = \tilde{A}$.

Proposition 2. *Let A be a fuzzy subset of an L-valued set (X, E) and let A^0 be its kernel. Then $A^0 \leq l_E(A)$*

Proof. Since A^0 is extensional, we have $A^0(x) * E(x,x') \leq A^0(x')$ for every x, x' in X, and hence $A^0(x) \leq E(x,x') \mapsto A^0(x') \leq E(x,x') \mapsto A(x')$, $\forall x, x' \in X$. It follows from here that $A^0(x) \leq \inf_{x' \in X}(E(x,x') \mapsto A(x')) = l_E(A)(x)$, $\forall x \in X$, that is $A^0 \leq l(A)$. □

Proposition 3. *$u_E(l_E(A)) = l_E(A)$ for a fuzzy set A in an L-valued set (X, E).*

Proof. From the definition of the operators $u_E, l_E : L^X \to L^X$ we have:

$$u_E(l_E)(A)(x) = \sup_{y \in X} \left(E(x,y) * \inf_{z \in X} \left(E(z,y) \mapsto A(z)\right)\right) \leq$$

$$\sup_{y \in X} \inf_{z \in X} \left(E(x,y) * (E(z,y) \mapsto A(z))\right) \leq \inf_{z \in X} \sup_{y \in X} E(x,y) * (E(z,y) \mapsto A(z))$$

$$\leq \inf_{z \in X} \sup_{y \in Y} \left((E(x,y) \mapsto E(y,z)) \mapsto A(z)\right) \leq \inf_{z \in X} \left(\inf_{y \in Y} (E(x,y) \mapsto E(y,z)) \mapsto A(z)\right)$$

$$\leq \inf_{z \in X} (E(x,z) \mapsto A(z)) = l_E(A)(x).$$

The third and the fourth inequalities in the above series are ensured by the easily established inequalities $a * (b \mapsto c) \leq (a * b \mapsto c)$ and $\bigvee_i (a_i \mapsto b) \leq (\bigwedge_i a_i \mapsto b)$ which hold in every GL-monoid; the last inequality follows from the definition of an L-valued equivalence: the condition $E(x,y) \leq E(x,z) * E(z,y)$ implies that $E(x,z) \leq E(z,y) \mapsto E(y,x), \forall y \in X$.

Thus we have $u_E(l_E(A)) \leq l_E(A)$. Since the inequality $l_E(A) \leq u_E(l_E(A)$ is obvious, we get the requested equality $u_E(l_E(A)) = l_E(A)$. □

Remark 2. In the special case when $L = [0, 1]$ is viewed as a Gödel algebra, that is $* = \wedge$ is the infimum t-norm, the statement of the above theorem is contained in Proposition 9 in [23]. We have reworked the proof of that proposition in order to cover the case of an arbitrary GL-monoid L.

Corollary 1. *For every fuzzy set A in an L-valued set (X, E) its lower fuzzy rough approximation $l_E(A)$ is an extensional fuzzy set.*

From here and Proposition 2 we get the following fundamental for us result:

Theorem 1. *For every fuzzy set A in an L-valued set (X, E) the lower fuzzy rough approximation operator assigns to A its kernel A^0: $l_E(A) = A^0$.*

From this theorem and Proposition 1 we get

Corollary 2. *The equality $l_E(u_E(A)) = l_E(A)$ holds for every fuzzy set A of an L-valued set (X, E).*

In the sequel the pair (u_E, l_E) will be referred to as *the extensional fuzzy rough approximation structure induced by the L-equivalence E.*

Remark 3. Let $L = \{0, 1\} =: 2$ and let $E : X \times X \to \{0, 1\}$ be an equivalence relation. Obviously, in this case E is actually the crisp equivalence relation on X. Then the images of a set $A \in 2^X$ under operators $u_E : 2^X \to 2$ and $l_E : 2^X \to 2$ make the pair $(u_E(A), l_E(A))$ which is actually Pawlak's originally defined rough set $(A^\blacktriangledown, A^\blacktriangle)$ determined by the set A. Indeed, notice first that $u_E(A)(= \tilde{A})$ in this case is just the set of all elements $x \in A$ whose classes $[x]_E$ of E-equivalence have non-empty intersections with A: $[x]_E \cap A \neq \emptyset$, and hence $u_E(A) = A^\blacktriangledown$. On the other hand, $l_E(A)(= A^0)$ is the set of all elements $x \in A$, whose classes of equivalence $[x]_E$ are contained in A: $[x]_E \subseteq A$, and hence $l_E(A) = A^\blacktriangle$.

3.3 Properties of Operators u_E and l_E

The most important for us properties of the operators u_E and l_E are collected in the next proposition:

Proposition 4 (*cf* [6,7,16,26]). *Let (X,E) be an L-valued set and (u_E,l_E) be the extensional fuzzy rough approximation structure of E. Then*

(1u) $u_E(0_X)=0_X$, *where* $0_X : X \to L$ *is the constant mapping* $0_X(x)=0_L$;
(2u) $A \le u_E(A),\ \forall A \in L^X$;
(3u) $u_E(\bigvee_i A_i) = \bigvee_i u_E(A_i),\ \forall\{A_i \mid i \in I\} \subseteq L^X$ *in particular*
(3$'u$) $u_E(A_1 \vee A_2) = u_E(A_1) \vee u(A_2), \forall A_1, A_2 \in L^X$;
(4u) $u_E(u_E(A)) = u_E(A),\ \forall A \in L^X$;
(1l) $l(1_X) = 1_X$ *where* $1_X : X \to L$ *is the constant mapping* $1_X(x)=1_L$;
(2l) $A \ge l_E(A),\ \forall A \in L^X$;
(3l) $l_E(\bigwedge_i A_i) = \bigwedge_i l_E(A_i),\ \forall\{A_i \mid i \in I\} \subseteq L^X$ *in particular*
(3$'l$) $l_E(A_1 \wedge A_2) = l_E(A_1) \wedge u(A_2), \forall A_1, A_2 \in L^X$;
(4l) $l_E(l_E(A)) = l_E(A),\ \forall A \in L^X$.

For completeness we include the proof of this important proposition.

Proof Statement (1u) is obvious. Statement (2u) follows easily taking into account reflexivity of the L-relation E. We prove property (3u) as follows: $u_E(\bigvee_i A_i)(x) = \sup_{x'}(E(x,x') * (\bigvee_i A_i(x'))) = \sup_{x'}(\bigvee_i E(x,x') * A_i(x')) = \bigvee_i(\sup_{x'}(E(x,x') * A_i(x'))) = \bigvee_i(u_E(A_i)(x)) = (\bigvee_i(u_E(A_i))(x)$.
Finally, taking into account transitivity of the L-relation we have: $u_E(u_E(A))(x) = \sup_{x'}(u_E(A)(x') * E(x,x')) = \sup_{x''}\sup_{x'}(A(x'') * E(x,x') * E(x',x'')) \le \sup_{x''} A(x'') * E(x,x'') = u_E(A)(x)$. Since the converse inequality follows from (2u), we get property (4u).

Statement (1l) is obvious. Statement (2l) follows easily taking into account reflexivity of the L-equivalence E. We prove property (3l) as follows: $l_E(\bigwedge_i A_i)(x) = \inf_{x'}(E(x,x') \mapsto \bigwedge_i A_i(x')) = \inf_{x'}\bigwedge_i (E(x,x') \mapsto A_i(x')) = \bigwedge_i \inf_{x'}(E(x,x') \mapsto A_i(x')) = \bigwedge_i l_E(A_i)$.

Finally, taking into account transitivity of the L-equivalence E we have: $l_E(l_E(A))(x) = \inf_{x'}(E(x,x') \mapsto l_E(A)(x')) = \inf_{x'}(E(x,x') \mapsto \inf_{x'} (E(x',x'') \mapsto A(x''))) = \inf_{x'}(\inf_{x''}(E(x,x') * E(x',x'') \mapsto A(x''))) \ge \inf_{x''}(E(x,x'') \mapsto A(x'')) = l_E(A)(x)$. Since the converse inequality follows from (2l), we get (4l). □

4 Fuzzy Functions

The concept of a fuzzy function in the sense, as it is used in this work first appeared in [3], and (independently) in [8]. Further development of the theory of fuzzy functions was conducted in [27] and in the recent research [22]. Here we expose the basics of the theory of fuzzy functions in the form and to the extent, as it is needed in the sequel.

4.1 Fuzzy Functions and the Category $\mathcal{F}\mathbf{SET}(\mathbf{L})$

Let $L = (L, \leq, \wedge, \vee, *)$ be a fixed GL-monoid. Recall that an L-relation (or just a fuzzy relation if the GL-monoid L is fixed) between sets X and Y is a mapping $R : X \times Y \to L$, see e.g. [29], [28]. In case when (X, E_X), (Y, E_Y) are L-valued sets, we specify a special type of fuzzy relations, which "respect" L-equivalences:

Definition 2 [22]. *A double extensional L-fuzzy relation (or a d.e. fuzzy relation for short) between L-valued sets (X, E_X) and (Y, E_Y) is a fuzzy relation $R : X \times Y \to L$ such that*

(1Ff) $R(x, y) * E_Y(y, y') \leq R(x, y'), \ \forall x \in X, \ \forall y, y' \in Y;$
(2Ff) $E_X(x, x') * R(x, y) \leq R(x', y), \ \forall x, x' \in X, \ \forall y \in Y;$

Aiming to distinguish a class of d.e. fuzzy relations that could be interpreted as fuzzy functions, we introduce the degree of functionality for a d.e. fuzzy relation:

Definition 3 [22]. *Given a d.e. fuzzy relation $R : X \times Y \to L$ between L-valued sets (X, E_X) and (Y, E_Y), we define its degree of functionality by* $\phi(R) = \inf_{x \in X, y, y' \in Y} (R(x, y) * R(x, y') \mapsto E_Y(y, y'))$.

Definition 4 [22]. *A fuzzy function from an L-valued set (X, E_X) to an L-valued set (Y, E_Y) is a d.e. fuzzy relation $R : X \times Y \to L$ such that $\phi(R) = 1$.*

One can easily see that a fuzzy function from (X, E_X) to (Y, E_Y) can be defined also as a d.e. fuzzy relation $R : X \times Y \to L$ such that

(3Ff) $R(x, y) * R(x, y') \leq E_Y(y, y'), \ \forall x \in X, \ \forall y, y' \in Y;$

Note, that fuzzy functions, defined as fuzzy relations satisfying properties (1Ff) - (3Ff), were introduced and studied in [3,8,27].

Definition 5. *Let $R : X \times Y \to L$ and $S : Y \times Z \to L$, be fuzzy relations; then their composition is a fuzzy relation $S \circ R : X \times Z \to L$ defined by*

$$(S \circ R)(x, z) = \bigvee_{y \in Y} (R(x, y) * S(y, z)), \ \forall x \in X, \ z \in Z.$$

Proposition 5 [22]. *Let $(X, E_X), (Y, E_Y), (Z, E_Z)$ be L-valued sets and let $R : X \times Y \to L$, $S : Y \times Z \to L$ be d.e. fuzzy relations. Then their composition $S \circ R : X \times Z \to L$ is double extensional.*

Proposition 6 [8,22,27]. *Composition of two fuzzy functions $R : X \times Y \to L$ and $S : Y \times Z \to L$ is a fuzzy function $S \circ R : X \times Z \to L$.*

Given an L-valued set (X, E_X), the fuzzy relation $I : X \times X \to L$ defined by $I(x, x') = 1$ iff $x = x'$ and $I(x, x') = 0$ otherwise, is obviously a fuzzy function $I : X \times X \to L$. Besides $I \circ R = R \circ I = R$ for every fuzzy function R whenever the composition is defined. So, referring to Proposition 6, we come to the following

Proposition 7. *L-valued sets as objects and fuzzy functions as morphisms constitute a category. This category will be denoted $\mathcal{F}\mathbf{SET}(\mathbf{L})$*

4.2 Soundedness Degree of a Fuzzy Function [22,27]

Definition 6. *Given a fuzzy function $R : X \times Y \to L$, we define its degree of soundedness by $\mu(R) = \inf_x \sup_y R(x,y)$. In case $\mu(R) = 1_L$, the fuzzy function R is called sound.*

Remark 4. The intuitive meaning of the value $\mu(R)$ is *to what extent the set X is the domain of the fuzzy function $R : X \times Y \to L$*. We illustrate this with the following *example*: Let X, Y be sets, $X' \subseteq X$ and $f : X' \to Y$ be a function. Interpreting f as the fuzzy function $R_f : X \times Y \to \{0,1\}$ defined by $R_f(x,y) = 1$ iff $y = f(x)$, we have $\mu(R_f) = 1$ iff $X' = X$ and $\mu(R_f) = 0$ otherwise.

Proposition 8 [8,27]. *Let $(X, E_X), (Y, E_Y), (Z, E_Z)$ be L-valued sets, $R : X \times Y \to L$, $S : Y \times Z \to L$ be fuzzy functions and $S \circ R : X \times Z \to L$ be their composition. Then $\mu(S \circ R) \geq \mu(R) * \mu(S)$.*

4.3 Forward and Backward Powerset Operators Induced by Fuzzy Functions

Let (X, E_X) and (Y, E_Y) be L-valued sets and $R : X \times Y \to L$ be a fuzzy relation. Following [22,27] we define the forward and the backward powerset operators induced by $R : X \times Y \to L$ as follows:

Definition 7. *The forward operator $R^{\to} : L^X \to L^Y$ is defined by setting:*

$$R^{\to}(A)(y) = \bigvee_x \left(R(x,y) * A(x)\right), \ \forall A \in L^X, \ \forall y \in Y.$$

The two alternative backward operators $R^{\leftarrow} : L^Y \to L^X$ and $R^{\Leftarrow} : L^Y \to L^X$ are defined by setting for all $B \in L^Y$:

$$R^{\leftarrow}(B)(x) = \bigvee_y R(x,y) * B(y), R^{\Leftarrow}(B)(x) = \bigwedge_y R(x,y) \mapsto B(y), \ \forall x \in X.$$

The fuzzy set $R^{\to}(A)$ is called the image of the fuzzy set A under $R : X \times Y \to L$, and fuzzy sets $R^{\leftarrow}(B)$ and $R^{\Leftarrow}(B)$ are called the preimage and the small preimage of the fuzzy set B under $R : X \times Y \to L$ respectively.[2]

Remark 5. Let E_X and E_Y be $=_X$ and $=_Y$, that is the ordinary equalities on the sets X and Y respectively. If $R = R_f$ is the relation induced by an ordinary function $f : X \to Y$, then $R_f^{\to}$ and $R_f^{\leftarrow}$, $R_f^{\Leftarrow}$ reduce to the definitions of a forward and backward operators $f^{\to} : L^X \to L^Y$ and $f^{\leftarrow} : L^Y \to L^X$, as they were introduced by S.E. Rodabaugh [24].

Basic properties of forward and backward operators are collected in the following proposition, the proof of which can be found in [22,27].

[2] In [22] we show that $R^{\Leftarrow}(B) \leq R^{\leftarrow}(B)$ under some "reasonable" assumptions on R.

Proposition 9. *Let (X, E_X), (Y, E_Y) be L-valued sets and let $R : X \times Y \to L$ be a fuzzy function. Further, let L_E^X and L_E^Y denote the families of extensional subsets of the L-valued sets (X, E_X) and (Y, E_Y) respectively. Then:*

(1) $R^{\to}\left(\bigvee_{i\in I}(A_i)\right) = \bigvee_{i\in I} R^{\to}(A_i), \ \ \forall\{A_i \mid i \in I\} \subseteq L^X$;
(2) $R^{\to}(A_1 \wedge A_2) \leq R^{\to}(A_1) \wedge R^{\to}(A_2), \ \forall A_1, A_2 \in L^X$;
(3) *if R is sound,* $R^{\leftarrow}\left(\bigwedge_{i\in I} B_i\right) = \bigwedge_{i\in I}(R^{\leftarrow} B_i), R^{\Leftarrow}\left(\bigwedge_{i\in I} B_i\right) = \bigwedge_{i\in I}(R^{\Leftarrow} B_i)$;
(4) $R^{\leftarrow}\left(\bigvee_{i\in I} B_i\right) = \bigvee_{i\in I}(R^{\leftarrow} B_i), \ \forall\{B_i : i \in I\} \subseteq L^Y$;
(5) *in case R is sound,* $A \leq R^{\leftarrow}(R^{\to}(A)), \ \forall A \in L^X$;
(6) $R^{\to}(R^{\leftarrow}(B) \leq B, \ \forall B \in L_E^Y$;
(7) $R^{\leftarrow}(\alpha_Y) = \alpha_X$ *and* $R^{\Leftarrow}(\alpha_Y) = \alpha_X$ *whenever R is sound.*

Proposition 10. *Let $R : X \times Y \to L$ and $S : Y \times Z \to L$, be fuzzy relations and let $S \circ R : X \times Z \to L$ be their composition. Then for every $C \in L^Z$ it holds $(S \circ R)^{\leftarrow}(C) = R^{\leftarrow}(S^{\leftarrow}(C))$ and $(S \circ R)^{\Leftarrow}(C) = R^{\Leftarrow}(S^{\Leftarrow}(C))$.*

Proof. Take any $x \in X$. Then $(S \circ R)^{\leftarrow}(C)(x) = \bigvee_{z\in Z}(S \circ R)(x) * C(z) = \bigvee_z \left(\bigvee_y (R(x, y) * S(y, z))\right) * C(z) = \bigvee_y R(x, y) * \bigvee_z (S(y, z) * C(z)) = \bigvee_y R(x, y) * S^{\leftarrow}(C)(y) = R^{\leftarrow}(S^{\leftarrow}(C))(x)$;
$(S \circ R)^{\Leftarrow}(C)(x) = \bigvee_z((S \circ R)(x) \mapsto C(z)) = \bigvee_z \left(\bigvee_y (R(x, y) * S(y, z)) \mapsto C(z)\right) = \bigvee_z \bigvee_y (R(x, y) \mapsto (S(y, z) \mapsto C(z))) = \bigvee_y (R(x, y) \mapsto \bigvee_z (S(y, z) \mapsto C(z))) = \bigvee_y (R(x, y) \mapsto S^{\Leftarrow}(C)(y)) = R^{\Leftarrow}(S^{\Leftarrow}(C))(x)$. □

5 A Category of Extensional Fuzzy Rough Approximation *L*-valued Spaces and Fuzzy Functions

We start with establishing two important properties concerning the behaviour of operators u_E and l_E under fuzzy functions:

Theorem 2. *Let $R : (X, E_X) \to (Y, E_Y)$ be a fuzzy function, and let $u_{E_X}, l_{E_X} : L^X \to L^X$, $u_{E_Y}, l_{E_Y} : L^Y \to L^Y$ be fuzzy rough approximation operators induced by L-valued equivalences E_X and E_Y respectively. Then the following hold:*
(Ru) $R^{\to}(u_{E_X}(A)) \leq u_{E_Y}(R^{\to}(A))$ *for every* $A \in L^X$
(Rl) $l_{E_X}(R^{\Leftarrow} B) \leq R^{\Leftarrow}(l_{E_Y}(B))$ *for every* $B \in L^Y$.

Proof. We prove Property (*Ru*) as follows. Let $A \in L^X$ and $y \in Y$. Then $R^{\to}(u_E(A))(y) = \bigvee_x (R(x, y) * u_E(A)(x)) = \bigvee_x R(x, y) * \bigvee_{x'} (E(x, x') * A(x')) = \bigvee_x \bigvee_{x'} (R(x, y) * E(x, x')) * A(x')) \leq \bigvee_{x'} R(x', y) * A(x') = R^{\to}(A)(y)$, and hence, moreover $R^{\to}(u_E(A))(y) \leq R^{\to} u_E(A)(y)$.

We prove Property (*Rl*) as follows. Let $B \in L^Y$ and $x \in X$. Then $R^{\Leftarrow}(l(B))(x) = \bigwedge_y (R(x, y) \mapsto l(B)(y)) = \bigwedge_y (R(x, y) \mapsto \bigwedge_{y'} (E(y, y') \mapsto B(y'))) = \bigvee_y \bigwedge_{y'} (R(x, y) \mapsto (E(y, y') \mapsto B(y'))) = \bigwedge_y \bigvee_{y'} (R(x, y) * E(y, y') \mapsto B(y')) \geq \bigwedge_y (R(x, y) \mapsto B(y)) \geq R^{\Leftarrow}(B)(x)$, and hence $l_E(R^{\Leftarrow}(B))(x) \leq R^{\Leftarrow}(l_E(B))(x)$.

Remark 6. Property (Ru) in the previous theorem implies that the image $R^{\rightarrow}(A)$ of every $A \in L^X$ is an extensional fuzzy subset in (Y, E_Y). We show that the preimage $R^{\leftarrow}(B)$ of every $B \in L^Y$ is an extensional fuzzy subset in (X, E_X). Indeed, for every $x, x' \in X$ we have $R^{\leftarrow}(B)(x) * E_X(x, x') = \left(\bigvee_y R(x,y) * B(y)\right) * E_X(x,x') = \bigvee_y ((R(x,y) * E_X(x,x')) * B(y) \leq \bigvee_y (R(x',y) * B(y)) = R^{\leftarrow}B(x')$, and hence $R^{\leftarrow}(B)$ is extensional. Specifically, this means that $R^{\leftarrow}(l_E(B)) \leq l_E(R^{\leftarrow}(B))$. The above comments can be summarized as follows:

Proposition 11. *For a fuzzy function $R : (X, E_X) \rightarrow (Y, E_Y)$ the following inequalities hold: (RuR) $R^{\rightarrow} \circ u_{E_X} \leq u_{E_Y} \circ R^{\rightarrow}$; (RlR) $R^{\leftarrow} \circ l_{E_Y} \leq l_{E_X} \circ R^{\leftarrow}$.*

Motivated by the properties of operators u_E, l_E summarized in Proposition 4 and their behaviour under fuzzy functions established in Theorem 2, and anticipating their (di)topological interpretation in Sect. 6, we introduce definitions:

Definition 8. *By a fuzzy rough upper approximation operator on an L-valued set (X, E) we call a mapping $u : L^X \rightarrow L^X_E$ satisfying properties (1u),(2u),(3'u), (4u), see Proposition 4. By a lower rough approximation operator on (X, E) we call a mapping $u : L^X \rightarrow L^X_E$ satisfying properties (1l),(2l),(3'l), (4l). The tuple (X, E, u, l) thus obtained is called a fuzzy rough approximation L-valued space.*

Definition 9. *Given fuzzy rough approximation L-valued spaces (X, E_X, u_X, l_X) and (Y, E_Y, u_Y, l_Y), a fuzzy function $R : (X, E_X, u_Y, l_Y) \rightarrow (X, E_Y, u_Y, l_Y)$ is called continuous if it satisfies properties (Ru) and (Rl) from Theorem 2.*

Let $\mathcal{FRA}\mathbf{SET}(L)$ be the category determined by fuzzy rough approximation L-valued spaces (X, E, u, l) and continuous fuzzy functions $R : (X, E_X, u_X, l_X) \rightarrow (X, E_Y, u_Y, l_Y)$. We introduce a partial order on the set $FRA(X, E)$ of all fuzzy rough approximation structures (u, l) of an L-valued set (X, E) by setting $(u, l) \prec (u', l')$ if and only if $u' \leq u$ and $l' \geq l$. From Theorem 2 it follows that the category $\mathcal{F}\mathbf{SET}(L)$ can be identified with the full subcategory of the category $\mathcal{FRA}\mathbf{SPA}(L)$ whose objects are tuples (X, E, u_E, l_E). Besides, the structure (u_E, l_E) is the largest one in the family $FRA(X, E)$. Specifically, this means that the identity fuzzy function $Id : (X, E, u_E, l_E) \rightarrow (X, E, u, l)$ is continuous for every structure (u, l). The smallest fuzzy rough approximation structure on (X, E) is the pair (u_0, l_0) where $u_0(0_X) = 0_X$ and $u_0(A) = 1_X$ whenever $A \neq 0_X$; $l_0(1_X) = 1_X$ and $l_0(A) = 0_X$ whenever $A \neq 1_X$. Obviously $Id : (X, E, u, l) \rightarrow (X, E, u_0, l_0)$ is continuous for every fuzzy rough approximation structure (u, l).

6 A Topological Outlook on the Fuzzy Rough Approximation L-valued Spaces

Given an L-valued set (X, E), the operators u_E and l_E can be interpreted as closure and interior fuzzy topological operators (Proposition 4). The corresponding fuzzy topology $T_{l_E} = \{A \in L^X \mid l(A) = A\}$ and co-topology

$K_{u_E} = \{A \in L^X \mid A = u(A)\}$ coincide and are equal to the set L_E^X of all extensional fuzzy subsets of X. Besides, it is an Alexandroff fuzzy topology in the sense that it is invariant under taking arbitrary intersections of open fuzzy sets. In a certain sense the family L_E^X is the extensional analogue of the discrete fuzzy topology and is the finest "extensional" fuzzy topology on (X, E).

Generalizing observation in the previous paragraph, let $cl : L^X \to L_E^X$ be an arbitrary closure operator and $int : L^X \to L_E^X$ be an arbitrary (generally unrelated) interior operator, in the sense that they satisfy conditions like (1u), (2u), (3$'u$) (4u) and (1l), (2l), (3$'l$) (4l) respectively in Proposition 4. We define upper an lower approximation operators $u : L^X \to L_E^X$ and $l : L^X \to L_E^X$ by setting $u = cl \circ u_E$ and $l = int \circ l_E$. It is clear that u and l satisfy properties (1u), (2u), (3$'u$) (4u) and (1l), (2l), (3$'l$) (4l) respectively and hence (X, E, u, l) gives an example of a fuzzy rough approximation L-valued space. On the other hand, by setting $T_l = \{A \in L^X \mid l(A) = A\}$ we obtain a fuzzy topology and by setting $K_u = \{A \in L^X \mid u(A) = A\}$ a fuzzy co-topology on X; hence in the result, we get a fuzzy ditopology [2], that is a pair (T_l, K_u) of generally independent families of open and closed extensional fuzzy subsets of (X, E). Since, obviously, $u \geq u_E$ and $l \leq l_E$ the ditopology (T_l, K_u) generated by (u, l) is coarser than the ditopology (T_{l_E}, K_{u_E}). On the other hand every L-fuzzy (di)topological space (X, T, K) can be realized as a fuzzy rough approximation L-valued space (X, E, u, l) by taking the ordinary equality $=_X$ as E and defining $u = cl_K$, $l = int_T$) where int_T and cl_K are the interior and the closure operators determined by T and K respectively.

Remark 7. Note that various aspects of relations of different (fuzzy) rough approximation structures and (fuzzy) topologies, however, as far as we know, without taking any care of the property of extensionality, where considered by different authors, see e.g. [7,16,20,21,26] just to mention a few of them.

7 Conclusion

We establish *extensionality* of fuzzy rough approximation operators u_E and l_E induced by a fuzzy equivalence $E : X \times X \to L$. Basing on this fact we introduce the concept of a *fuzzy rough approximation L-valued space* as a tuple (X, E, u, l) where $u, l : L^X \to L_E^X$ are operators satisfying properties, analogous to the properties of operators u_E, l_E. After brief refreshing information concerning *fuzzy functions* we define the *category of fuzzy rough approximation L-valued spaces* and *fuzzy functions* satisfying certain continuity-type conditions. A *ditopological interpretation* of a fuzzy rough approximation L-valued space is developed. and used for constructing examples of fuzzy rough approximation L-valued spaces.

Concerning perspectives for the future research in the direction of this work, we have in view the following. To study properties of the category $\mathcal{FRA}\mathbf{SPA}(L)$. Specifically, to study the properties of the lattice of fuzzy rough approximation structures (u, l) on an L-valued set (X, E); to investigate operations of products, coproducts, etc., of such spaces; to define initial and final structures in $\mathcal{FRA}$**SPA**(L) etc. In this work we have only briefly touched the relations between fuzzy rough approximation L-valued spaces and (L-)fuzzy ditopological spaces. We think this subject has perspectives for a further study. In particular, it seems interesting to study "inner" conditions on the L-valued equivalence E and/or on operators u and l under which the resulting fuzzy ditopology is a fuzzy topology, that is when u and l are related as the closure and interior operators in an ordinary topological space. Another problem is to elucidate, to what extent the properties of the fuzzy equivalence E are fundamental for the present research. Specifically, whether the results of this work in a certain way can be extended to the so called *local fuzzy equivalences* E [10].

References

1. Birkhoff, G.: Lattice Theory. AMS, Providence (1995)
2. Brown, L.M., Ertürk, R., Dost, Ş.: Ditopological texture spaces and fuzzy topology, I. Basic concepts. Fuzzy Sets Syst. **110**, 227–236 (2000)
3. Demirci, M.: Fuzzy functions and their fundamental properties. Fuzzy Sets Syst. **106**, 239–246 (1999)
4. Dubois, D., Prade, H.: Rough fuzzy sets and fuzzy rough sets. Int. J. Gen. Syst. **17**(2–3), 191–209 (1990)
5. Gierz, G., Hoffman, K.H., Keimel, K., Lawson, J.D., Mislove, M.W., Scott, D.S.: Continuous Lattices and Domains. Cambridge University Press, Cambridge (2003)
6. Eļkins, A., Šostak, A.: On some categories of approximate systems generated by L-relations. In: 3rd Rough Sets Theory Workshop, pp. 14–19. Milan (2011)
7. Han, S.-E., Šostak, A.: M-valued measure of roughness for approximation of L-fuzzy sets and its topological interpretation. In: Merelo, J.J., Rosa, A., Cadenas, J.M., Dourado, A., Madani, K., Filipe, J. (eds.) IJCCI 2014, Part II. SCI, vol. 620, pp. 251–267. Springer, Switzerland (2016)
8. Höhle, U., Porst, H.-E., Šostak, A.: Fuzzy functions: a fuzzy extension of the category SET and some related categories. Appl. Gen. Topol. **1**, 115–127 (2000)
9. Höhle, U.: M-valued sets and sheaves over integral commutative CL-monoids. Chapter 2. In: Rodabaugh, S.E., Klement, E.P., Höhle, U. (eds.) Applications of Category Theory to Fuzzy Sets, pp. 33–73. Kluwer Academic Publishers, Dordrecht (1992)
10. Höhle, U.: Commutative, residuated l-monoids. In: Rodabaugh, S.E., Klement, E.P., Höhle, U. (eds.) Non-classical logics and their applications to Fuzzy Sets, pp. 53–106. Kluwer Academic Publishers, Dordrecht (1995)
11. Höhle, U.: Many-valued equalities, singletons and fuzzy partitions. Soft Comput. **2**, 134–140 (1998)
12. Höhle, U.: Many Valued Topology and its Applications. Kluwer Academic Publisher, Boston (2001)
13. Höhle, U.: Axiomatics for fixed-based fuzzy topologies. Chapter 3. In: Höhle, U., Rodabaugh, S.E. (eds.) Mathematics of Fuzzy Sets: Logic, Topology and Measure Theory. Handbook Series, vol. 3. Kluwer Academic Publishers, Dordrecht (1999)
14. Klawonn, F.: Fuzzy points, fuzzy relations and fuzzy functions. In: Novák, V., Perfilieva, I. (eds.) Discovering the World with Fuzzy Logic, pp. 431–453. Springer, Heidelberg (2000)
15. Klawon, F., Castro, J.L.: Similarity in fuzzy reasoning. Matware Soft Comput. **2**, 197–228 (1995)

16. Järvinen, J., Kortelainen, J.: A unified study between modal-like operators, topologies and fuzzy sets. Fuzzy Sets Syst. **158**, 1217–1225 (2007)
17. Kortelainen, J.: On relationship between modified sets, topological spaces and rough sets. Fuzzy Sets Syst. **61**, 91–95 (1994)
18. Morgan, W., Dilworth, R.P.: Residuated lattices. Trans. Am. Math. Soc. **45**, 335–354 (1939)
19. Pawlak, Z.: Rough sets. Int. J. Comput. Inf. Sci. **11**, 341–356 (1982)
20. Qin, P.: On the topological properties of fuzzy rough sets. Fuzzy Sets Syst. **151**, 601–613 (2005)
21. Qin, P.: Generalized rough sets based on reflexive and transitive relations. Inf. Sci. **178**, 4138–4141 (2008)
22. Perfilieva, I., Šostak, A.: Fuzzy functions: basics of the theory and applications to topology. Fuzzy Sets Syst. (submitted)
23. Radzikowska, A.M., Kerre, E.E.: A comparative study of fuzzy rough sets. Fuzzy Sets Syst. **126**, 137–155 (2002)
24. Rodabaugh, S.E.: Powerset operator based foundation of point-set lattice-theoretic (poslat) fuzzy set theories and topologies. Quaest. Math. **20**(3), 431–462 (1997)
25. Rosenthal, K.I.: Quantales and Their Applications. Pirman Research Notes in Mathematics 234. Longman Scientific & Technical. Wiley, New York (1990)
26. Šostak, A.: Towards the theory of M-approximate systems: fundamentals and examples. Fuzzy Sets Syst. **161**, 2440–2461 (2010)
27. Šostak, A.: Fuzzy functions and an extension of the category L- TOP of Chang-Goguen L-topological spaces. In: Proceedings of the 9th Prague Topological Symposium, pp. 271–294 (2001)
28. Valverde, L.: On the structure of F-indistinguishibility operators. Fuzzy Sets Syst. **17**, 313–328 (1985)
29. Zadeh, L.: Similarity relations and fuzzy orderings. Inf. Sci. **3**, 177–200 (1971)

The Syntax of Many-Valued Relations

Patrik Eklund(✉)

Department of Computing Science, Umeå University, SE-90187 Umeå, Sweden
peklund@cs.umu.se

Abstract. In this paper we show how many-valued relations syntactically can be formulated using powertype constructors. This in turn enables to describe the syntax of generalized relations in the starting point sense where the category sets and relations is isomorphic to the Kleisli category of the powerset monad over the category of sets. We can then generalize to work over monoidal closed categories, and thereby description logic, formal concepts and rough sets can be viewed as depending on that powertype constructor, and within a setting of many-valued λ-calculus. In order to achieve this, we will adopt a three-level arrangement of signatures [4], and demonstrate the benefits of using it. Bivalent and untyped relational adaptations typically appear in terminology and ontology, and we will illuminate this situation concerning classifications in health. Extensions to multivalent and typed nomenclatures provides an enrichment that is beneficial in practical use of health classifications and nomenclatures.

Keywords: Classification · Concept · Functor · Generalized relation · Health · Monad · Nomenclature · Ontology · Signature

1 Introduction

Terms are the foundational cornerstones of logic, and signatures are foundational for the term construction. Type constructors are frequently used to create new sorts from old ones, but type constructors are traditionally adopted *from the outside*. This means that type constructors are not seen as operators in a signature in its own right.

We adopt a three-level arrangement of signatures [4] where the middle level contains type constructors, and the first and third level, respectively, is used to clearly distinguish terms from λ-terms. The conventional definition of λ-terms is informal, and, in fact, not constructive, or at least to say that it hides the underlying formal term construction. Doing so, the conventional definition creates a demand for renaming, which cannot be formally justified, but is an *ad hoc* necessity to avoid ambiguities. Similar hidden phenomena appears, of course, in many branches of computing, even in Turing machines, that hides recursion in a way that makes Church's thesis to be informal only.

We respect Church's view that λ is just an *informal symbol* [1], and we go even further by showing how it can be formalized, when it must be formalized. A fundamental consequence for terms on *signature levels* is then that λ is

J.P. Carvalho et al. (Eds.): IPMU 2016, Part II, CCIS 611, pp. 61–68, 2016.
DOI: 10.1007/978-3-319-40581-0_6

seen not to be a *general abstractor*, but rather that any operator possesses its own capacity to abstract itself, and not to possess any capacity whatsoever to abstract anything else. In an expression like $\lambda x.f$, λ is unique to f, and should be clearly viewed as "f owns its λ". This obviously departs from traditional views and definitions of λ-terms, but on the other hand it disables the appearance of "unwanted" terms. Doing so we, we can also avoid renaming.

A key enabler is then the possibility to use a wide variety of further type constructor, together with the function type constructor producing λ-calculus. The type constructors enable the *transportation* of terms on level one in the three-level arrangement of signatures to become abstracted to λ-terms on the third level. A key issue for the *syntax of relations*, and e.g. for fuzzy description logic, formal concepts and rough sets, is the use of the powertype constructor. This also enables to describe the syntax of generalized relations in the sense where the category of sets and relations is isomorphic to the Kleisli category of the powerset monad over the category of sets.

For many-valuedness to become identifiable within and because of the use of suitable underlying categories, we use monoidal categories, and thereby the notion of many-valued description logic, formal concepts [6] and rough sets [5], can be seen as depending on that powertype constructor. In this paper we focus on description logic, and how it has been adopted within classification and nomenclature in health. This is an unfortunate adoption, since it identifies 'ontology' in web ontology with 'ontology' in health ontology. Doing so, health ontology becomes both bivalent and untyped. Concepts like those in IHTSDO's (International Health Terminology Standards Development Organisation) SNOMED are similarly bivalent as are the alphanumeric codes in WHO's (World Health Organization) ICD, i.e., an encoded disease is assumed either as diagnosed or not. There are considerations of using a 'not (yet) specified' for a disease, as is done in practice, but the severity of a disease as related e.g. to the extent of damage is not encoded in any way. Typical examples appear in the case of cardiac diseases and injuries involving fractures. Fractures with or without dislocation, and in various degrees, are obviously quite different e.g. from subsequent surgery point of view. Having alphanumeric codes rather than only numeric codes in ICD, helps to see the categorization of codes and indeed mostly based on anatomy. In SNOMED, concept identifiers are purely numeric, and unstructurally symbolic in the sense that the sequence of digits building up the identifier is treated as a digitized and purely symbolic name. The identifier of a concept in SNOMED is then attached with a description, which is the textual counterpart for the identifier. SNOMED prefers to speak of 'concept types', but their no formal underlying typing. These types are more like categories, and SNOMED indeed interchangeably uses 'concept type' and 'concept category'. From signature point of view, a 'category' can be seen as a sort, but concepts are indeed only constants as terms (or expressions), so that 'constant of a sort' is the same as 'concept of a type' or 'concept within a category. Needless to say, there is connection whatsoever between 'category' in SNOMED's 'concept category' and 'category' in 'category theory'.

Web ontology is better suited for applications involving natural language, where in fact the depth of applications still is on the level of counting words and finding similarities between sentences. The capacity of web ontology to identify context and meaning is very poor, which typically is seen in tools for identifying plagiarism. For instance, web ontology is unable to clearly understand the distinction between reference and self-reference.

2 Informally Defined Terms in Type Theory

The informal definition of (untyped) λ-terms basically states, firstly, that a variable is a λ-term, secondly, if M is a λ-term, then $\lambda x.M$ is a λ-term, where x is a variable (abstraction), and thirdly, if M and N are λ-terms, then also MN is a λ-term (application).

Traditional λ-calculus looks at terms as becoming abstracted to operators. It is generally seen as a nice trick, but cannot be formally and logically justified. What in fact happens is that an operator is abstracted to another operator, of different arity. Variables cause confusion, and the notion of *free* and *bound* is the root of this inconvenience, trying to make substitution something more than it actually is. Substitution is a morphism in the Kleisli category of a given term monad and over a selected category [4]. Church [1] indeed called "λ" an *improper symbol*, together with "("and")" also being improper symbols. The *proper symbols* are those residing in the signature, or being symbols for variables.

As was pointed out also in [4], Church's ι and o types are not clearly defined. There is a consensus about ι being the 'type of types', but we have to be careful e.g. not say that "ι is a ι", as was noted also by predecessors to modern type theory. The o type for 'propositions' is still not well explained.

3 Levels of Signatures

The syntax of relations, i.e., the powertype, resides as a unary operator on the second level in a three-level arrangement of signatures. Generalized relations, as syntactic objects are therefore λ-terms in a general sense, i.e., generality depending on the semantics of the operator, but in particular, as we shall see, on the underlying category of the term monad producing *types as terms* on that second level for the sake of delivering generalized λ-terms over the third level signature.

In order to explain this in all detail, we adopt the categorically somewhat informal notation $\Sigma = (S, \Omega)$ of a signature. The way it actually needs to be handled in a more strict fashion is explained in [4], which also contains detail on the corresponding formal and fully categorical construction of the term monad with its underlying term functor T_Σ. For a sort $\mathtt{s}$, and a term t of sort $\mathtt{s}$, we may use the notation $t :: \mathtt{s}$. The underlying category is some monoidal biclosed category, but in this treatment we hide detail about this underlying category.

In subsequent papers we will provide detail concerning potential use of signatures like Σ_{ICD}, Σ_{ICF}, and Σ_{SNOMED} in order to enrich their respective classification and nomenclature structure. Making concepts to be semantically explained through their expressions in form of being terms over a signature is far more specific than expecting to have meaning based on textual descriptions of concept identifiers like in SNOMED. Monad compositions like $\mathbf{\Phi} \circ \mathbf{T}_{\Sigma_{SNOMED}}$ [3] point to the underlying functor Φ to structured conglomerates of concepts (as structured terms), where simple hierarchies, like those in SNOMED, can be provided by bivalent powerset functors. Multivalence can be invoked either by using many-valued powerfunctors, or by allowing the term functors to act over underlying categories embracing the required multivalence. In the case of WHO's ICF, the term functor $\mathsf{T}_{\Sigma_{ICF}} : \mathtt{Set}(Q) \to \mathtt{Set}(Q)$ introduced both typing as well as multivalence [2], where Q is the quantale representing ICF's generic scale.

This is where typing comes into play, and multivalence overlooks that completely. Sets X of points must be extended to structures of terms (or expressions). This requires underlying signatures Σ embracing sorts and operators, so that term functors T_Σ provide formal term constructions. The term functor $\mathsf{T}_\Sigma : \mathsf{C} \to \mathsf{C}$ operates typically over monoidal biclosed categories C. If C is $\mathtt{Set}$, we have the traditional bivalent (but typed!) situation, and with the Goguen category $\mathtt{Set}(Q)$, where Q is a quantale (often non-commutative!), we have a full multivalent and typed situation enabled by the signature acting over a selected underlying category [4].

Term functors constructed in this way can be extended to monads, so that substitution can be composed. A substitution in this categorical context is a morphism in the Kleisli category of the related monad. Monad compositions enable to arrive at generalized sets of terms, where the typical example is composing the term functor T_Σ with the powerset functor P in order to obtain the monad $\mathbf{P} \circ \mathbf{T}_\Sigma$. More elaborate generalized set functors Φ can be applied in order to make use of the composition $\mathbf{\Phi} \circ \mathbf{T}_\Sigma$.

On *level one*, we have Σ, and terms over Σ are produced by $\mathsf{T}_\Sigma X$, where X is an *object of variables* in the underlying category. In case of a one-sorted signature over the category of sets and functions, terms are just traditional terms as typically seen in first-order logic.

On *level two*, the level of type constructors, with introduce the single-sorted signature

$$\mathtt{S}_\Sigma = (\{\mathtt{type}\}, \{\mathtt{s} :\to \mathtt{type} \mid \mathtt{s} \in S\} \cup \{\Rrightarrow : \mathtt{type} \times \mathtt{type} \to \mathtt{type}\}),$$

where we then have $\mathsf{T}_{\mathtt{S}_\Sigma}\emptyset$ as the object of all types and constructed types.

On *level three*, the level then includes λ-terms based on the signature $\Sigma' = (S', \Omega')$ where $S' = \mathsf{T}_{\mathtt{S}_\Sigma}\emptyset$ and Ω' is

$$\{\lambda^{\omega}_{i_1,\ldots,i_n} :\to (\mathtt{s}_{i_1} \Rrightarrow \cdots \Rrightarrow (\mathtt{s}_{i_{n-1}} \Rrightarrow (\mathtt{s}_{i_n} \Rrightarrow \mathtt{s})) \mid \omega : \mathtt{s}_1 \times \ldots \times \mathtt{s}_n \to \mathtt{s} \in \Omega\}$$

included with the operator $\mathtt{app}_{\mathtt{s},\mathtt{t}} : (\mathtt{s} \Rrightarrow \mathtt{t}) \times \mathtt{s} \to \mathtt{t}$. In this notation $(i_1, \ldots, i_n)$ is a permutation of $(1, \ldots, n)$. Note also how level one operators are transformed to constant operators on level three. Further, note indeed how "ω owns its abstraction" in $\lambda^{\omega}_{i_1,\ldots,i_n}$. In fact, we could even avoid using the informal symbol 'λ' in this context.

As an example, consider the signature of natural numbers

$$\mathtt{NAT} = (\{\mathtt{nat}\}, \{0 :\to \mathtt{nat}, \mathtt{succ} : \mathtt{nat} \to \mathtt{nat}\}$$

on level one. The 0-ary operator 0 converts to $\lambda_0^0 :\rightarrow \mathtt{nat}$, i.e., as a 0-ary operator on level three. The unary operator succ is (λ-)abstracted to become a 0-ary operator $\lambda_1^{\mathtt{succ}} :\rightarrow (\mathtt{nat} \Rrightarrow \mathtt{nat})$ on level three. Note also that we must not confuse nat on level one with nat on level three, even if for simplicity we use the same notation.

In [4], we pointed out the advantage in avoiding the need of renaming.

On β-reduction we obviously have the following transition from the traditional form to using the three-level signature. Let $[x := t]$ be a substitution, i.e., we have some $\sigma(x) = t$, and choose a $\omega : \mathtt{s}_1 \times \mathtt{s}_2 \rightarrow \mathtt{s}$. Then β-reduction

$$\lambda x.\lambda y.\omega(x,y)\ t \rightarrow_\beta \lambda y.(\omega(x,y)[x := t]) = \lambda y.\omega(t,y)::\mathtt{s}_2 \Rrightarrow \mathtt{s}$$

transforms to

$$(\mu \circ \mathsf{T}\sigma)(\mathsf{app}(\lambda^\omega_{\mathtt{s}_1,\mathtt{s}_2}, x)) \rightarrow_\beta \mathsf{app}(\lambda^\omega_{\mathtt{s}_1,\mathtt{s}_2}, t)::\mathtt{s}_2 \Rrightarrow \mathtt{s}.$$

All these constructions can potentially be used in natural language expressions involving modifiers and quantifiers in expressions like “there are more small balls than large balls in this box”. Obviously, there are no unique solutions to handle this as they are context dependent. Possible encodings of such expression, or related subexpressions, in our three level signatures setting for λ-terms, could view modifiers are closely related to type constructors. Modifiers as operators on level three are then specified using constructed types on level two. We should note that quantifiers are more like abstractors of sentences, and it may therefore be anticipated that the formalization of quantifiers, with quantifier symbols as informal symbols, is similar to the formalization of the way λ acts on expressions.

4 The Syntax of Generalized Relations

In this section we present some background and motivation for generalized relations, and also provide notational considerations needed in this context. We use notation as adopted in [4].

The observation that relations $R \subseteq X \times X$ correspond precisely to functions (in form of substitutions) $\sigma_R : X \rightarrow \mathsf{P}X$, where P is the powerset functor over the category of sets and functions, is the basis for viewing *generalized relations* as morphisms (substitutions) in the Kleisli category over *generalized powerset monads*.

For $\Sigma = (S, \Omega)$ on level one, we now extend $\mathtt{S}_\Sigma$ with further operators beyond just $\Rrightarrow$. Concerning unary operators we may include an $\mathtt{F} : \mathtt{type} \rightarrow \mathtt{type}$, which intuitively is expected to be semantically described by a functor, that is, assuming that the ‘algebra’ of type is the class of objects in some monoidal closed category. Whereas the algebras $\mathfrak{A}(\Sigma)$ of signatures Σ, involving assignments of sorts s to domains $\mathfrak{A}(\mathtt{s})$ of $\mathfrak{A}$, are standard according to universal algebra, the ‘algebra’ of the (Σ-)superseding type signature $\mathtt{S}_\Sigma$ is not immediate since the domain assigned to the sort type clearly cannot be just a set. There are several options for this, and these considerations may go beyond traditional universal algebra. These discussions are outside the scope of this paper.

The 'syntactic functors' view is based on unary type constructors ϕ, ψ : $\mathtt{type} \to \mathtt{type}$, allowing the *composed type constructor* $\psi \circ \phi : \mathtt{type} \to \mathtt{type}$ by $(\psi \circ \phi)\mathtt{s} = \psi(\phi\mathtt{s})$. For unary type constructors $\phi, \psi : \mathtt{type} \to \mathtt{type}$, a *type transformation* τ *from* ϕ *to* ψ, denoted $\tau : \phi \Rightarrow \psi$, if it exists, is assumed, for all $\mathtt{s} \in \mathsf{T}_{\mathsf{S}_\Sigma}\emptyset$, to be given by a unique (constant operator) $\tau_\mathtt{s} :\to (\phi\mathtt{s} \Rrightarrow \psi\mathtt{s})$. Further, we assume that any $\mathtt{f} :\to (\mathtt{s} \Rrightarrow \mathtt{t})$ gives rise to a unique $\phi\mathtt{f} :\to (\phi\mathtt{s} \Rrightarrow \phi\mathtt{t})$.

Various 'syntactic set functors' can be introduced, including the 'powerset' type constructor $\mathsf{P} : \mathtt{type} \to \mathtt{type}$ on level two, intuitively thinking that the 'algebra' of P is the powerset functor, with the underlying monoidal closed category being the category of sets and functions.

From application point of view, it is important now to see how operators like $\omega : \mathtt{s} \to \mathsf{P}\mathtt{t}$ is the underlying syntactic support for enabling typed generalized relations.

5 Description Logic

For transforming description logic into our categorical framework, we use notations in [7]. *Interpretations* $\mathcal{I} = (D^\mathcal{I}, \cdot^\mathcal{I})$, where $\cdot^\mathcal{I}$ *maps every concept description* to a subset of $D^\mathcal{I}$, use D for that universe, which should not be confused with D as used for concept descriptions, e.g., in expressions like $C \bigsqcup D$, where D is not to be understood as the "D in $D^\mathcal{I}$".

With C as a "concept", we have $C^\mathcal{I} \subseteq D^\mathcal{I} \in \mathsf{P}D^\mathcal{I}$. This means that $\mathsf{P}D^\mathcal{I}$ is the actual 'algebra'. *Roles* R are semantically described as relations $R^\mathcal{I} \subseteq D^\mathcal{I} \times D^\mathcal{I}$, i.e., we can equivalently write it as a substitution $R^\mathcal{I} : D^\mathcal{I} \to \mathsf{P}D^\mathcal{I}$. The inverse relation R^{-1} is what is actually used on the semantic side, and, in fact, we have

$$(\exists R : C)^\mathcal{I} = \{a \in D^\mathcal{I} \mid \exists (a,b) \in R^\mathcal{I} : b \in C^\mathcal{I}\} = \mu_{D^\mathcal{I}}(\mathsf{P}R^{-1}C).$$

Note how '$\exists$' in $\exists(a,b) \in R^\mathcal{I} : b \in C^\mathcal{I}$ is different from '$\exists$' in $(\exists R : C)^\mathcal{I}$, where in the latter it appears as an informal symbol providing an abstraction of C. In fact, the "existential quantifier" in $\exists R : C$ is an "R-modality" applied to the powerconcept C. The definition for the semantic expression $(\exists R : C)^\mathcal{I}$ uses the existential quantifier that appears in the assumed underlying set theory.

Concerning the underlying signature and related variables, in [7] the situation is unclear, given the assumption about *the existence of two further disjoint alphabets of symbols*, which are called *individual* and *concept variables*. Logically, variables are not part of any alphabet. Variables are terms, and as such they are terms of a certain type. We should therefore speak of "individual concept" rather than "individual variable", and then use x, y, z as variables for individual concepts, and X, Y, Z as variables for concepts.

Now typing of "concept" and "individual concept" comes into play, and we will need type constructors on level two. As opposed to [7], we say "concept" instead of "individual concept", and "powerconcept" instead of "concept". The underlying signature must be formalized, where concept is a sort in the given underlying signature on level one. On level two, $\mathtt{concept}$ becomes a constant operator, and a type constructor P is then used to produce a new type $\mathsf{P}\mathtt{concept}$, which in their 'algebra' will be understood, respectively, as $D^{\mathcal{I}}$ and $\mathsf{P}D^{\mathcal{I}}$.

Simply typed description logic can now be formally defined in λ-calculus. Let $\Sigma = (S, \Omega)$ be on level one, where $S = \{\mathtt{concept}\}$. Operators in Ω are the constants $\mathtt{c}_1, \ldots, \mathtt{c}_n :\to \mathtt{concept}$. Concepts and powerconcepts must eventually reside in the same signature on level three. Therefore, on level two, we use $\mathtt{S}_\Sigma$, so that $\mathtt{concept} :\to \mathtt{type}$ becomes a constant in $\mathtt{S}_\Sigma$. We then include the type constructor $\mathsf{P} : \mathtt{type} \to \mathtt{type}$ into $\mathtt{S}_\Sigma$, and as the constructed type for "powerconcept". Note that $\mathsf{P}(\mathtt{concept})$ is a term on level two, becoming a sort on level three. A fundamental weakness of traditional description logic is the intertwining of syntax and semantics of the powerconcept. A variable $x \in X_{\mathsf{P}(\mathtt{concept})}$ is a "concept variable" in the sense of [7], and is also a 'term' as an element of $\mathsf{T}_{\Sigma',\mathsf{P}(\mathtt{concept})}(X_\mathtt{s})_{\mathtt{s}\in S'}$. On level three we have $\mathtt{c}_1, \ldots, \mathtt{c}_n \in \mathsf{T}_{\Sigma',\mathtt{concept}}(X_\mathtt{s})_{\mathtt{s}\in S'}$. "Roles" are of the form $\mathtt{r} :\to (\mathsf{P}(\mathtt{concept}) \Rrightarrow \mathsf{P}(\mathsf{P}(\mathtt{concept})))$, which creates the need to include operators $\eta :\to (\mathtt{concept} \Rrightarrow \mathsf{P}(\mathtt{concept}))$ and $\mu :\to (\mathsf{P}(\mathsf{P}(\mathtt{concept})) \Rrightarrow \mathsf{P}(\mathtt{concept}))$ on level three.

A concept

$$\mathtt{c} :\to \mathtt{concept}$$

on level one becomes a "singleton powerconcept"

$$\mathtt{app}_{\mathtt{concept},\mathsf{P}(\mathtt{concept})}(\eta, \mathtt{c})$$

on level three, and the syntactic expression "$\exists r.x$" as a term of type $\mathsf{P}(\mathtt{concept})$ can be defined as

$$\exists r.x = \mathtt{app}_{\mathsf{P}(\mathsf{P}(\mathtt{concept})),\mathsf{P}(\mathtt{concept})}(\mu, \mathtt{app}_{\mathsf{P}(\mathtt{concept}),\mathsf{P}(\mathsf{P}(\mathtt{concept}))}(\mathtt{r}, x)).$$

"Disjunction" and "negation" are added as new type constructors on level two as

$$\sqcup : \mathsf{P}(\mathtt{concept}) \times \mathsf{P}(\mathtt{concept}) \to \mathsf{P}(\mathtt{concept})$$

and

$$\neg : \mathsf{P}(\mathtt{concept}) \to \mathsf{P}(\mathtt{concept}).$$

6 Conclusions

We have shown how syntactic aspects of description logic can be extended to involve generalized relations as compared to just being represented by powerset functors. Double powerset functors, and a range of composed functors can be used, thus representing relations in a more generalized sense. These generalizations are interesting to investigate further over various underlying categories. Viewing many-valued description logic as part of λ-calculus is more general as compared to approaches in [8,9], where fuzzy description logic is basically simply typed description logic with the semantics of P in practice being extended only to the many-valued powerset monad. Our approach reveals the modal nature of description logic more clearly, and shows why it is doubtful to speak about the "existential quantifier" in description logic.

For health applications, the consequence of using our extended relational model is that classifications and nomenclatures [2] can be enriched to become multivalent and typed. Further, this enables a much more elaborate and practically useful approach to structural mapping between classification. Indeed, in the simple bivalent and untyped case, where relations are provided upon unstructured sets, mappings are nothing but mappings of points to points, or points to subsets of points. This is not sufficient in practice. In the case of IHTSDO's approach to nomenclature, SNOMED's logical foundation is all too shallow for practical applications. Ongoing revisions of ICD classification for diseases also indicates adoption of the shallow approach as invited by the use of description logic without extension towards multivalence and typing. The lack of logically multivalent considerations among WHO's reference classifications is particularly unfortunate, as ICD is from time to time criticized of not embracing multiple values, and the ICF classification for functioning in fact is inherently multivalent given its generic scale. In subsequent work, we will expose this situation in more detail.

References

1. Church, A.: A formulation of the simple theory of types. J. Symb. Log. **5**, 56–68 (1940)
2. Eklund, P.: Lative logic accomodating the WHO family of international-classifications, Encyclopedia of E-Health and Telemedicine. In: Cruz-Cunha, M.M., Miranda, I. (eds.) IGI Global (2016). http://www.igi-global.com/book/encyclopedia-health-telemedicine/141916
3. Eklund, P. : Lative Logic for Enriched SNOMED (in preparation)
4. Eklund, P., Galán, M.A., Helgesson, R., Kortelainen, J.: Fuzzy terms. Fuzzy Sets Syst. **256**, 211–235 (2014)
5. Eklund, P., Galán, M.A.: The rough powerset monad. J. Multi. Valued Log. Soft Comput. **13**, 321–334 (2007)
6. Eklund, P., Galán García, M.Á., Kortelainen, J., Ojeda-Aciego, M.: Monadic formal concept analysis. In: Cornelis, C., Kryszkiewicz, M., Ślęzak, D., Ruiz, E.M., Bello, R., Shang, L. (eds.) RSCTC 2014. LNCS, vol. 8536, pp. 201–210. Springer, Heidelberg (2014)
7. Schmidt-Schauß, M., Smolka, G.: Attributive concept descriptions with complements. Artif. Intell. **48**, 1–26 (1991)
8. Straccia, U.: A fuzzy description logic. In: Mostow, J., Rich, C. (eds.) AAAI/IAAI, pp. 594–599. AAAI Press/The MIT Press (1998)
9. Yen, J.: Generalizing term subsumption languages to fuzzy logic. In: IJCAI, pp. 472–477 (1991)

Reduct-Irreducible α-cut Concept Lattices: An Efficient Reduction Procedure to Multi-adjoint Concept Lattices

M. Eugenia Cornejo[1], Jesús Medina[2(✉)], and Eloísa Ramírez-Poussa[2]

[1] Department of Statistic and Operations Research, University of Cádiz, Cádiz, Spain
mariaeugenia.cornejo@uca.es

[2] Department of Mathematics, University of Cádiz, Cádiz, Spain
{jesus.medina,eloisa.ramirez}@uca.es

Abstract. The computation of fuzzy concept lattices is really complex. Hence, looking for mechanisms in order to reduce this complexity is fundamental. This paper presents a new efficient mechanism which combines two procedures. First of all, an attribute reduction is given, which removes the unnecessary attributes, and then a reduction based on a truth degree is applied, which removes the fuzzy attributes with low weight. Different interesting properties and examples of this mechanism are also introduced.

Keywords: Fuzzy sets · Formal Concept Analysis · Concept lattice reduction

1 Introduction

One of the most important mathematical tools for analyzing relational databases and representing conceptual knowledge in a formal way is Formal Concept Analysis (FCA), which is closely related to soft set theory and rough set theory and has been theoretically developed [1,3,12] and in applications [17,18].

Since real databases are usually very large they give rise to complex concept lattices, from which extracting conclusions is a really difficult task. This fact highlights the importance of obtaining new procedures which lets us reduce the concept lattices from both perspectives: attribute reduction and size reduction. These procedures have widely been studied in the classical case [14,15,19], however this development is less advanced in frameworks allowing uncertainties and imprecise data. Besides attribute reduction [7,13], the reduction of the size of a fuzzy concept lattice is also fundamental, but the main information must be preserved. In [6,10] the authors studied a mechanism to reduce the size of concept lattices based on the meet-irreducible elements of the lattice and on a cut value given by the user. This method provides a sublattice of the original concept

Partially supported by the European Regional Development Fund and the Spanish Science Ministry project TIN2012-39353-C04-04.

J.P. Carvalho et al. (Eds.): IPMU 2016, Part II, CCIS 611, pp. 69–80, 2016.
DOI: 10.1007/978-3-319-40581-0_7

lattice, which is called meet-irreducible α-cut concept lattice. Consequently, the most representative knowledge is preserved since the original concepts are not modified and the valuable concepts are not removed. This feature is not shared with other procedures to reduce the size of concept lattices which use hedges [2] or granular computing [11].

This paper introduces a new procedure which combines an attribute reduction and the meet-irreducible α-cut reduction. This mechanism considers a reduct before computing the α-cuts, which provides a more efficient procedure. Moreover, several properties and examples are introduced in order to explain the main features of the new reduction process.

The structure of this paper is the following: Sect. 2 presents a brief summary with basic definitions and results. Section 3 recalls attribute reduction based on the characterizations of the absolutely necessary, relatively necessary and absolutely unnecessary attributes in terms of the meet-irreducible elements of the concept lattice, and the meet-irreducible α-cut concept lattices, respectively. Section 4 introduces the new mechanism combining the attribute reduction with the irreducible α-cut concept lattices, several properties and remarks. Lastly, the paper finishes with the conclusions and prospects for future work.

2 Preliminaries

In the following, we will summarize the essential definitions and results required to make the paper self-contained.

Definition 1. *Given a lattice $(L, \preceq)$, such that $\wedge, \vee$ are the meet and the join operators, and an element $x \in L$ verifying*

1. *If L has a top element $\top$, then $x \neq \top$.*
2. *If $x = y \wedge z$, then $x = y$ or $x = z$, for all $y, z \in L$.*

we call x meet-irreducible ($\wedge$-irreducible) element *of L. Condition (2) is equivalent to*

2′. *If $x < y$ and $x < z$, then $x < y \wedge z$, for all $y, z \in L$.*

Hence, if x is $\wedge$-irreducible, then it cannot be represented as the infimum of strictly greatest elements. A join-irreducible ($\vee$-irreducible) element *of L is defined dually.*

The considered fuzzy concept lattice framework is the one needed to define the multi-adjoint concept lattices [16], in which the adjoint triples are the basic computational operators [8,9].

Definition 2. *Let $(P_1, \leq_1)$, $(P_2, \leq_2)$, $(P_3, \leq_3)$ be posets and $\&\colon P_1 \times P_2 \to P_3$, $\swarrow\colon P_3 \times P_2 \to P_1$, $\nwarrow\colon P_3 \times P_1 \to P_2$ be mappings, then $(\&, \swarrow, \nwarrow)$ is an* adjoint triple *with respect to P_1, P_2, P_3 if:*

$$x \leq_1 z \swarrow y \quad \textit{iff} \quad x \,\&\, y \leq_3 z \quad \textit{iff} \quad y \leq_2 z \nwarrow x \tag{1}$$

where $x \in P_1$, $y \in P_2$ and $z \in P_3$. The condition (1) is also called adjoint property*.*

Now, the notions of multi-adjoint frame and context are included.

Definition 3. *A* multi-adjoint frame $\mathcal{L}$ *is a tuple* $(L_1, L_2, P, \&_1, \dots, \&_n)$ *where* $(L_1, \preceq_1)$ *and* $(L_2, \preceq_2)$ *are complete lattices,* $(P, \leq)$ *is a poset and* $(\&_i, \swarrow^i, \nwarrow_i)$ *is an adjoint triple with respect to* L_1, L_2, P*, for all* $i \in \{1, \dots, n\}$.

Definition 4. *Let* $(L_1, L_2, P, \&_1, \dots, \&_n)$ *be a multi-adjoint frame, a* context *is a tuple* (A, B, R, σ) *such that* A *and* B *are non-empty sets (usually interpreted as attributes and objects, respectively),* R *is a* P*-fuzzy relation* $R\colon A \times B \to P$ *and* $\sigma\colon A \times B \to \{1, \dots, n\}$ *is a mapping which associates any element in* $A \times B$ *with some particular adjoint triple in the frame.*

For each multi-adjoint frame and context, two concept-forming operators ${}^\uparrow\colon L_2^B \longrightarrow L_1^A$ and ${}^\downarrow\colon L_1^A \longrightarrow L_2^B$ can be defined, for all $g \in L_2^B$, $f \in L_1^A$ and $a \in A$, $b \in B$, as

$$g^\uparrow(a) = \inf\{R(a,b) \swarrow^{\sigma(a,b)} g(b) \mid b \in B\} \tag{2}$$

$$f^\downarrow(b) = \inf\{R(a,b) \nwarrow_{\sigma(a,b)} f(a) \mid a \in A\} \tag{3}$$

where L_2^B and L_1^A denote the set of mappings $g\colon B \to L_2$ and $f\colon A \to L_1$, respectively, which form a Galois connection [16].

A *multi-adjoint concept* is a pair $\langle g, f\rangle$, such that $g \in L_2^B$, $f \in L_1^A$ satisfying $g^\uparrow = f$ and $f^\downarrow = g$. The fuzzy subsets of objects g (resp. fuzzy subsets of attributes f) are called *extensions* (resp. *intensions*) of the concepts. The sets of these are denoted as $\mathrm{Ext}(\mathcal{M})$ and $\mathrm{Int}(\mathcal{M})$, respectively.

Definition 5. *Given a multi-adjoint frame* $(L_1, L_2, P, \&_1, \dots, \&_n)$ *and a context* (A, B, R, σ)*. The* multi-adjoint concept lattice *associated with this frame and this context is the set*

$$\mathcal{M} = \{\langle g, f\rangle \mid g \in L_2^B, f \in L_1^A \text{ and } g^\uparrow = f, f^\downarrow = g\}$$

in which the ordering is defined by $\langle g_1, f_1\rangle \preceq \langle g_2, f_2\rangle$ *if and only if* $g_1 \preceq_2 g_2$ *(equivalently* $f_2 \preceq_1 f_1$*).*

Henceforth we will fix a multi-adjoint concept lattice $(\mathcal{M}, \preceq)$ associated with a multi-adjoint frame $(L_1, L_2, P, \&_1, \dots, \&_n)$ and a context (A, B, R, σ).

From the specific family of fuzzy subsets of attributes presented in the following definition, we remind the characterization of the $\wedge$-irreducible elements [5,7].

Definition 6 [7]. *For each* $a \in A$*, the fuzzy subsets of attributes* $\phi_{a,x} \in L_1^A$ *defined, for all* $x \in L_1$*, as*

$$\phi_{a,x}(a') = \begin{cases} x & \text{if } a' = a \\ \bot_1 & \text{if } a' \neq a \end{cases}$$

will be called fuzzy-attributes*, where* $\bot_1$ *is the minimum element in* L_1*. The set of all fuzzy-attributes will be denoted as* $\Phi = \{\phi_{a,x} \mid a \in A, x \in L_1\}$.

Theorem 1 [7]. *The set of* $\wedge$*-irreducible elements of* $\mathcal{M}$, $M_F(A)$, *is formed by the pairs* $\langle \phi^{\downarrow}_{a,x}, \phi^{\downarrow\uparrow}_{a,x} \rangle$ *in* $\mathcal{M}$, *with* $a \in A$ *and* $x \in L_1$, *such that*

$$\phi^{\downarrow}_{a,x} \neq \bigwedge \{\phi^{\downarrow}_{a_i,x_i} \mid \phi_{a_i,x_i} \in \Phi, \phi^{\downarrow}_{a,x} \prec_2 \phi^{\downarrow}_{a_i,x_i}\}$$

and $\phi^{\downarrow}_{a,x} \neq g_{\top_2}$, *where* $\top_2$ *is the maximum element in* L_2 *and* $g_{\top_2}\colon B \to L_2$ *is the fuzzy subset defined as* $g_{\top_2}(b) = \top_2$, *for all* $b \in B$.

3 Reduction Mechanisms

This section will recall the attribute reduction introduced in [7] to multi-adjoint concept lattices, which is based on a classification of the attributes of the context, and the size reduction mechanism based on a α-cut [10], which will be used later to build the new method.

3.1 On the Attribute Classification

First of all, we recall the notions of consistent set, reduct, absolutely necessary attribute, relatively necessary attribute and absolutely unnecessary attribute [7].

Definition 7. *A set of attributes* $Y \subseteq A$ *is a* consistent *set of* (A, B, R, σ) *if the following isomorphism is satisfied:*

$$\mathcal{M}(Y, B, R_Y, \sigma_{Y\times B}) \cong_E \mathcal{M}(A, B, R, \sigma)$$

This is equivalent to say that, for all $\langle g, f \rangle \in \mathcal{M}(A, B, R, \sigma)$, *there exists a concept* $\langle g', f' \rangle \in \mathcal{M}(Y, B, R_Y, \sigma_{Y\times B})$ *such that* $g = g'$.

Moreover, if $\mathcal{M}(Y \setminus \{a\}, B, R_{Y\setminus\{a\}}, \sigma_{Y\setminus\{a\}\times B}) \ncong_E \mathcal{M}(A, B, R, \sigma)$, *for all* $a \in Y$, *then* Y *is called a* reduct *of* (A, B, R, σ).

The core *of* (A, B, R, σ) *is the intersection of all the reducts of* (A, B, R, σ).

Definition 8. *Given a formal context* (A, B, R, σ) *and the set* $\mathcal{Y} = \{Y \subseteq A \mid Y \text{ is a reduct}\}$ *of all reducts of* (A, B, R, σ). *The set of attributes* A *can be divided into the following three parts:*

1. *Absolutely necessary attributes (core attribute)* $C_f = \bigcap_{Y\in\mathcal{Y}} Y$.
2. *Relatively necessary attributes* $K_f = (\bigcup_{Y\in\mathcal{Y}} Y) \setminus (\bigcap_{Y\in\mathcal{Y}} Y)$.
3. *Absolutely unnecessary attributes* $I_f = A \setminus (\bigcup_{Y\in\mathcal{Y}} Y)$.

Taking into account the previous definitions, three classification theorems were introduced in [7].

Theorem 2 [7]. *Given* $a_i \in A$, *we have that* $a_i \in C_f$ *if and only if there exists* $x_i \in L_1$, *such that* $\langle \phi^{\downarrow}_{a_i,x_i}, \phi^{\downarrow\uparrow}_{a_i,x_i} \rangle \in M_F(A)$, *satisfying that* $\langle \phi^{\downarrow}_{a_i,x_i}, \phi^{\downarrow\uparrow}_{a_i,x_i} \rangle \neq \langle \phi^{\downarrow}_{a_j,x_j}, \phi^{\downarrow\uparrow}_{a_j,x_j} \rangle$, *for all* $x_j \in L_1$ *and* $a_j \in A$, *with* $a_j \neq a_i$.

Theorem 3 [7]. *Given $a_i \in A$, we have that $a_i \in K_f$ if and only if $a_i \notin C_f$ and there exists $\langle \phi^{\downarrow}_{a_i,x_i}, \phi^{\downarrow\uparrow}_{a_i,x_i} \rangle \in M_F(A)$ satisfying that E_{a_i,x_i} is not empty and $A \setminus E_{a_i,x_i}$ is a consistent set, where the sets $E_{a_i,x}$ with $a_i \in A$ and $x \in L_1$ are defined as:*

$$E_{a_i,x} = \{a_j \in A \setminus \{a_i\} \mid \text{there exist } x' \in L_1, \text{ satisfying } \phi^{\downarrow}_{a_i,x} = \phi^{\downarrow}_{a_j,x'}\}$$

Theorem 4 [7]. *Given $a_i \in A$, it is absolutely unnecessary, $a_i \in I_f$, if and only if, for each $x_i \in L_1$, we have that $\langle \phi^{\downarrow}_{a_i,x_i}, \phi^{\downarrow\uparrow}_{a_i,x_i} \rangle \notin M_F(A)$, or in the case that $\langle \phi^{\downarrow}_{a_i,x_i}, \phi^{\downarrow\uparrow}_{a_i,x_i} \rangle \in M_F(A)$, then $A \setminus E_{a_i,x_i}$ is not a consistent set.*

From these results, a classification in absolutely necessary, relatively necessary and absolutely unnecessary attributes is given and the reducts are computed as well. Each reduct provides a reduction in the (number of) attributes of the original context, which decreases the computational complexity of the concept lattice.

3.2 Size Reduction by Irreducible α-cuts

In this section, a mechanism to reduce the size of multi-adjoint concept lattices [6,10] is recalled. This reduction procedure only considers the fuzzy-attributes with a truth degree bigger than a value α provided by the user. This threshold represents the least truth-value of a fuzzy-attribute in order to be considered in the computation of the concept lattice. Hence, given a value α, for each attribute a, we only consider the following set of meet-irreducible elements of $(\mathcal{M}, \preceq)$:

$$\widehat{M_F}(A)_\alpha = \{\langle \phi^{\downarrow}_{a,x}, \phi^{\downarrow\uparrow}_{a,x} \rangle \in M_F(A) \mid \alpha \preceq_1 x\}$$

The concepts obtained from the infimum of the elements of $\widehat{M_F}(A)_\alpha$ and the greatest element in $(\mathcal{M}, \preceq)$, that is $\langle g_\top, g_\top^{\uparrow} \rangle$, from a complete lattice.

Definition 9. *Given $\alpha \in L_1$, the set*

$$\widehat{\mathcal{M}}_\alpha = \{\langle g, f \rangle \in \mathcal{M} \mid g = \bigwedge_{i \in I} \phi^{\downarrow}_{a_i,x_i}, \text{ with } \langle \phi^{\downarrow}_{a_i,x_i}, \phi^{\downarrow\uparrow}_{a_i,x_i} \rangle \in \widehat{M_F}(A)_\alpha\} \bigcup \{\langle g_\top, g_\top^{\uparrow} \rangle\}$$

is called meet-irreducible α-cut concept lattice of $\mathcal{M}$, *for short,* irreducible α-cut concept lattice.

In [10], diverse properties of this reduced lattice were introduced. One of the most important ones is that the reduced concept lattice $\widehat{\mathcal{M}}_\alpha$ is a complete sublattice of the original one. Therefore, this mechanism provides a reduction of the original concept lattice without modifying the information given by the concepts.

The following section will merge the philosophies of the two previous mechanisms in order to obtain a more efficient size reduction procedure of fuzzy concept lattices.

4 Reduct-Irreducible α-cut Concept Lattices

An interesting alternative to increase the reduction and decrease the complexity in the computation of the concept lattice is the combination of both reduction mechanisms. First of all, the attribute reduction will be applied and the unnecessary attributes will be removed. Once the different reducts have been obtained, a threshold is considered and the corresponding irreducible α-cuts are computed. This mechanism is formally introduced in the following definition.

Definition 10. *Let (A, B, R, σ) be a context, $(L_1, L_2, P, \&_1, \ldots, \&_n)$ a multi-adjoint frame, $(\mathcal{M}, \preceq)$ its concept lattice, $\alpha \in L_1$ and $(\mathcal{M}^Y, \preceq)$ the concept lattice built from a reduct $Y \subseteq A$. The concept lattice obtained applying the irreducible α-cut CL to $(\mathcal{M}^Y, \preceq)$ is called* r-irreducible α-cut concept lattice *and it is denoted as $\widehat{\mathcal{M}}_\alpha^Y$.*

One of the most important advantages of considering the attribute reduction at the beginning is that we simplify the construction process since unnecessary fuzzy-attributes to compute the concept lattice are removed. Specifically, when the attribute classification satisfies that the set K_f is not empty, then several reducts can be obtained. This fact will determine the subsequent size reduction i.e. depending on the choice of the starting reduct, we will obtain a major or minor reduction of the size. It is interesting to identify sufficient conditions to ensure the reduction is independent of the chosen reduct. This and other properties, together with several examples of this mechanism will be introduced next.

The first property shows that after reducing the number of attributes, if we carry out the r-irreducible α-cut concept lattice to the context with a reduct, we obtain a complete lattice isomorphic to a sublattice of the original one with the same extensions, which is called *sublattice by extensions.*

Proposition 1. *Given a context (A, B, R, σ), a frame $(L_1, L_2, P, \&_1, \ldots, \&_n)$ and the associated multi-adjoint concept lattice $(\mathcal{M}, \preceq)$, we have that for each reduct Y of $(\mathcal{M}, \preceq)$ and $\alpha \in L_1$, the pair $(\widehat{\mathcal{M}}_\alpha^Y, \preceq)$ is a complete sublattice by extensions of $(\mathcal{M}, \preceq)$.*

This property is shown in the following example.

Example 1. Let $\mathcal{L} = (L_1, L_2, L_3, \preceq, \&_G^*)$ be the considered framework, where $L_1 = [0,1]_{10}$, $L_2 = [0,1]_4$ and $L_3 = [0,1]_5$ are the regular partitions of $[0,1]$ in 10, 4 and 5 pieces, respectively, and $\&_G^*$ is discretization of the Gödel conjunctor defined on $L_1 \times L_2$, see [4] for more details. Let (A, B, R, σ) be the fixed context, with $A = \{a_1, a_2, a_3, a_4\}$, $B = \{b_1, b_2, b_3\}$, $R\colon A \times B \to L_3$ given in Table 1, and σ is constantly $\&_G^*$.

The Hasse diagram of the concept lattice $(\mathcal{M}, \preceq)$ is presented in Fig. 1, where it is easy to see that the set of meet-irreducible elements is formed by the concepts C_0, C_5, C_6, C_9 and C_{10}. Below, the fuzzy-attributes associated with the meet-irreducible elements are listed:

Table 1. Relation R of Example 1

R	b_1	b_2	b_3
a_1	0.6	0.8	0.6
a_2	1	1	0
a_3	0.6	0.8	0
a_4	0.2	0	0.2

$$\begin{aligned}
&\langle\phi^{\downarrow}_{a_4,0.3},\phi^{\downarrow\uparrow}_{a_4,0.3}\rangle=\langle\phi^{\downarrow}_{a_4,0.4},\phi^{\downarrow\uparrow}_{a_4,0.4}\rangle=\langle\phi^{\downarrow}_{a_4,0.5},\phi^{\downarrow\uparrow}_{a_4,0.5}\rangle=\langle\phi^{\downarrow}_{a_4,0.6},\phi^{\downarrow\uparrow}_{a_4,0.6}\rangle=\\
&\langle\phi^{\downarrow}_{a_4,0.7},\phi^{\downarrow\uparrow}_{a_4,0.7}\rangle=\langle\phi^{\downarrow}_{a_4,0.8},\phi^{\downarrow\uparrow}_{a_4,0.8}\rangle=\langle\phi^{\downarrow}_{a_4,0.9},\phi^{\downarrow\uparrow}_{a_4,0.9}\rangle=\langle\phi^{\downarrow}_{a_4,1.0},\phi^{\downarrow\uparrow}_{a_4,1.0}\rangle=C_0\\
&\langle\phi^{\downarrow}_{a_2,0.1},\phi^{\downarrow\uparrow}_{a_2,0.1}\rangle=\langle\phi^{\downarrow}_{a_2,0.2},\phi^{\downarrow\uparrow}_{a_2,0.2}\rangle=\langle\phi^{\downarrow}_{a_2,0.3},\phi^{\downarrow\uparrow}_{a_2,0.3}\rangle=\langle\phi^{\downarrow}_{a_2,0.4},\phi^{\downarrow\uparrow}_{a_2,0.4}\rangle=\\
&\langle\phi^{\downarrow}_{a_2,0.5},\phi^{\downarrow\uparrow}_{a_2,0.5}\rangle=\langle\phi^{\downarrow}_{a_2,0.6},\phi^{\downarrow\uparrow}_{a_2,0.6}\rangle=\langle\phi^{\downarrow}_{a_2,0.7},\phi^{\downarrow\uparrow}_{a_2,0.7}\rangle=\langle\phi^{\downarrow}_{a_2,0.8},\phi^{\downarrow\uparrow}_{a_2,0.8}\rangle=\\
&\langle\phi^{\downarrow}_{a_2,0.9},\phi^{\downarrow\uparrow}_{a_2,0.9}\rangle=\langle\phi^{\downarrow}_{a_2,1.0},\phi^{\downarrow\uparrow}_{a_2,1.0}\rangle=\langle\phi^{\downarrow}_{a_3,0.1},\phi^{\downarrow\uparrow}_{a_3,0.1}\rangle=\langle\phi^{\downarrow}_{a_3,0.2},\phi^{\downarrow\uparrow}_{a_3,0.2}\rangle=\\
&\langle\phi^{\downarrow}_{a_3,0.3},\phi^{\downarrow\uparrow}_{a_3,0.3}\rangle=\langle\phi^{\downarrow}_{a_3,0.4},\phi^{\downarrow\uparrow}_{a_3,0.4}\rangle=\langle\phi^{\downarrow}_{a_3,0.5},\phi^{\downarrow\uparrow}_{a_3,0.5}\rangle=\langle\phi^{\downarrow}_{a_3,0.6},\phi^{\downarrow\uparrow}_{a_3,0.6}\rangle=C_5\\
&\langle\phi^{\downarrow}_{a_4,0.1},\phi^{\downarrow\uparrow}_{a_4,0.1}\rangle=\langle\phi^{\downarrow}_{a_4,0.2},\phi^{\downarrow\uparrow}_{a_4,0.2}\rangle=C_6\\
&\langle\phi^{\downarrow}_{a_1,0.9},\phi^{\downarrow\uparrow}_{a_1,0.9}\rangle=\langle\phi^{\downarrow}_{a_1,1.0},\phi^{\downarrow\uparrow}_{a_1,1.0}\rangle=C_9\\
&\langle\phi^{\downarrow}_{a_1,0.7},\phi^{\downarrow\uparrow}_{a_1,0.7}\rangle=\langle\phi^{\downarrow}_{a_1,0.8},\phi^{\downarrow\uparrow}_{a_1,0.8}\rangle=C_{10}
\end{aligned}$$

According to the attribute classification theorems given in Sect. 3.1, we classify the attributes of A as follows:

$$\begin{aligned}C_f&=\{a_1,a_4\}\\ K_f&=\{a_2,a_3\}\end{aligned}$$

From this classification two reducts $Y_1=\{a_1,a_2,a_4\}$ and $Y_2=\{a_1,a_3,a_4\}$ can be obtained. Considering the reduct Y_2 and the value $\alpha=0.7$, we will build the r-irreducible 0.7-cut concept lattice of $(\mathcal{M}^{Y_2},\preceq)$. For that purpose, we need to consider the set of concepts $\langle\phi^{\downarrow^{Y_2}}_{a,x},\phi^{\downarrow^{Y_2}\uparrow_{Y_2}}_{a,x}\rangle$ that belongs to $M_F(Y_2)$ such that $\alpha\preceq x$, with a maximal value x, that is:

$$\widehat{M_F}(Y_2)_{0.7}=\{\langle\phi^{\downarrow^{Y_2}}_{a_4,1.0},\phi^{\downarrow^{Y_2}\uparrow_{Y_2}}_{a_4,1.0}\rangle,\langle\phi^{\downarrow^{Y_2}}_{a_1,1.0},\phi^{\downarrow^{Y_2}\uparrow_{Y_2}}_{a_1,1.0}\rangle,\langle\phi^{\downarrow^{Y_2}}_{a_1,0.8},\phi^{\downarrow^{Y_2}\uparrow_{Y_2}}_{a_1,0.8}\rangle\}$$

Computing the meet-closure of $\widehat{M_F}(Y_2)_{0.7}$ and considering the greatest element of $(\mathcal{M}^{Y_2},\preceq)$, we obtain the lattice $(\widehat{\mathcal{M}}^{Y_2}_{0.7},\preceq)$ as usual, which can be seen in the right side of Fig. 1. Taking into account the notion of complete sublattice and Fig. 1, we can easily verify that $(\widehat{\mathcal{M}}^{Y_2}_{0.7},\preceq)$ is a complete sublattice by extensions of $(\mathcal{M},\preceq)$.

The second property guarantees that when the carrier of the fuzzy subsets of attributes is linear, there always exists a value $\alpha\in L_1$ such that for all $\beta\preceq_1\alpha$, the reduced concept lattices from different reducts are isomorphic.

Proposition 2. *Let (A,B,R,σ) be a context, $(L_1,L_2,P,\&_1,\dots,\&_n)$ be a frame and $(\mathcal{M},\preceq)$ the associated multi-adjoint concept lattice. If $(L_1,\preceq_1)$ is a linear*

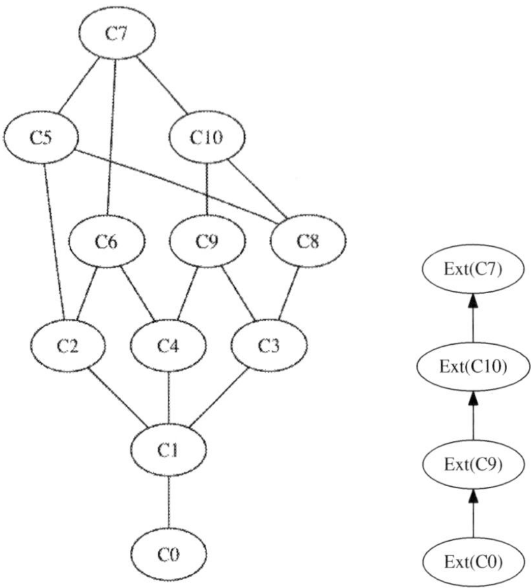

Fig. 1. The Hasse diagram of $(\mathcal{M}, \preceq)$ (left) and $(\widehat{\mathcal{M}}_{0.7}^{Y_2}, \preceq)$ (right).

lattice, then there exists α *greater than the minimum element of* L_1*, such that,* $\widehat{\mathcal{M}}_{\beta}^{Y_1} \cong_E \widehat{\mathcal{M}}_{\beta}^{Y_2}$*, for all* $\beta \preceq_1 \alpha$ *and for all reducts* Y_1, Y_2*.*

Since in Example 1 the considered lattice L_1 is linear, we will continue working with the same framework and context in order to illustrate the previous result.

Example 2. In the environment of Example 1, we can build the concept lattice $(\widehat{\mathcal{M}}_{0.7}^{Y_1}, \preceq)$ considering the following set:

$$\widehat{M_F}(Y_1)_{0.7} = \{\langle \phi_{a_4,1.0}^{\downarrow^{Y_2}}, \phi_{a_4,1.0}^{\downarrow^{Y_2}\uparrow_{Y_2}}\rangle, \langle \phi_{a_1,1.0}^{\downarrow^{Y_2}}, \phi_{a_1,1.0}^{\downarrow^{Y_2}\uparrow_{Y_2}}\rangle, \langle \phi_{a_2,1.0}^{\downarrow^{Y_2}}, \phi_{a_2,1.0}^{\downarrow^{Y_2}\uparrow_{Y_2}}\rangle, \langle \phi_{a_1,0.8}^{\downarrow^{Y_2}}, \phi_{a_1,0.8}^{\downarrow^{Y_2}\uparrow_{Y_2}}\rangle\}$$

The r-irreducible 0.7-cut concept lattice of $(\mathcal{M}^{Y_1}, \preceq)$ is shown in Fig. 2. In the size reduction process when we consider Y_1, we remove the meet-irreducible element C_6. If we observe the concept lattice $(\widehat{\mathcal{M}}_{0.7}^{Y_2}, \preceq)$ in Fig. 1, in this case, we have removed the meet-irreducible elements C_5 and C_6. This is because C_5 is obtained from the fuzzy-attributes $\phi_{a_2,1.0}$ and $\phi_{a_3,0.6}$. Hence, if we regard Y_2 the fuzzy-attribute related to C_5 does not exceed the cut established by $\alpha = 0.7$. Therefore, we conclude that $(\widehat{\mathcal{M}}_{0.7}^{Y_1}, \preceq) \not\cong_E (\widehat{\mathcal{M}}_{0.7}^{Y_2}, \preceq)$.

On the other hand, if a value $\alpha = 0.6$ is fixed, for every value $\beta \preceq_1 \alpha$ the reduced concept lattices from the different reducts, $(\widehat{\mathcal{M}}_{\beta}^{Y_1}, \preceq)$ and $(\widehat{\mathcal{M}}_{\beta}^{Y_2}, \preceq)$, are isomorphic. For instance, when $\beta = 0.5$ we have that:

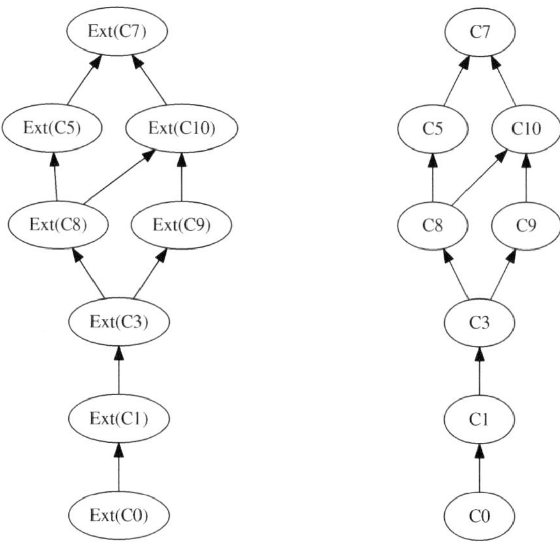

Fig. 2. The Hasse diagram of $(\widehat{\mathcal{M}}_{0.7}^{Y_1}, \preceq)$ (left) and $(\widehat{\mathcal{M}}_{0.5}, \preceq)$ (right).

$$\widehat{M_F}(Y_1)_{0.5} = \{\langle \phi_{a_4,1.0}^{\downarrow^{Y_2}}, \phi_{a_4,1.0}^{\downarrow^{Y_2}\uparrow_{Y_2}}\rangle, \langle \phi_{a_1,1.0}^{\downarrow^{Y_2}}, \phi_{a_1,1.0}^{\downarrow^{Y_2}\uparrow_{Y_2}}\rangle, \langle \phi_{a_2,1.0}^{\downarrow^{Y_2}}, \phi_{a_2,1.0}^{\downarrow^{Y_2}\uparrow_{Y_2}}\rangle, \langle \phi_{a_1,0.8}^{\downarrow^{Y_2}}, \phi_{a_1,0.8}^{\downarrow^{Y_2}\uparrow_{Y_2}}\rangle\}$$
$$\widehat{M_F}(Y_2)_{0.5} = \{\langle \phi_{a_4,1.0}^{\downarrow^{Y_2}}, \phi_{a_4,1.0}^{\downarrow^{Y_2}\uparrow_{Y_2}}\rangle, \langle \phi_{a_1,1.0}^{\downarrow^{Y_2}}, \phi_{a_1,1.0}^{\downarrow^{Y_2}\uparrow_{Y_2}}\rangle, \langle \phi_{a_3,0.6}^{\downarrow^{Y_2}}, \phi_{a_3,0.6}^{\downarrow^{Y_2}\uparrow_{Y_2}}\rangle, \langle \phi_{a_1,0.8}^{\downarrow^{Y_2}}, \phi_{a_1,0.8}^{\downarrow^{Y_2}\uparrow_{Y_2}}\rangle\}$$

Building the concept lattices $(\widehat{\mathcal{M}}_{0.5}^{Y_1}, \preceq)$ and $(\widehat{\mathcal{M}}_{0.5}^{Y_2}, \preceq)$ from the previous sets, we obtain the same Hasse diagram shown in the left side of Fig. 2. That is, these reductions coincide with the reduction given by the concept lattice $(\widehat{\mathcal{M}}_{0.7}^{Y_1}, \preceq)$. Consequently, $(\widehat{\mathcal{M}}_{0.5}^{Y_1}, \preceq) \cong_E (\widehat{\mathcal{M}}_{0.5}^{Y_2}, \preceq)$. As it was previously commented, this fact also arises for all $\beta \preceq_1 0.6$.

The following property establishes that if the r-irreducible α-cut concept lattices obtained from each of the reducts are isomorphic then they also are isomorphic to the meet-irreducible α-cut concept lattice of $\mathcal{M}$.

Proposition 3. *Given a context (A, B, R, σ), a frame $(L_1, L_2, P, \&_1, \dots, \&_n)$ and the associated multi-adjoint concept lattice $(\mathcal{M}, \preceq)$, we obtain that if $\alpha \in L_1$ satisfies $\widehat{\mathcal{M}}_{\alpha}^{Y_1} \cong_E \widehat{\mathcal{M}}_{\alpha}^{Y_2}$, for all reducts Y_1, Y_2, then the reduction obtained from a reduct $Y \subseteq A$ is isomorphic to the one obtained from the original concept lattice, $\widehat{\mathcal{M}}_{\alpha}^{Y} \cong_E \widehat{\mathcal{M}}_{\alpha}$.*

Once again, we return to Example 2 in order to clarify the statement above.

Example 3. Coming back to Example 2, when we compute the concept lattice $(\widehat{\mathcal{M}}_{0.5}, \preceq)$ we obtain the Hasse diagram presented in the right side of Fig. 2. From this figure, we can observe that the r-irreducible 0.5-cut concept lattice of $(\widehat{\mathcal{M}}_{0.5}^{Y_1}, \preceq)$ and $(\widehat{\mathcal{M}}_{0.5}^{Y_2}, \preceq)$ are isomorphic to the reduction of the original one $(\widehat{\mathcal{M}}_{0.5}, \preceq)$.

The last proposition presents how is the reduced concept lattice in the particular case in which the set of relatively necessary attributes is an empty set.

Proposition 4. *Let (A, B, R, σ) be a context, $(L_1, L_2, P, \&_1, \dots, \&_n)$ be a frame and $(\mathcal{M}, \preceq)$ the associated multi-adjoint concept lattice. If $K_f = \varnothing$, then only one reduct exists, which is given by the core set, $Y = C_f$, and the isomorphism $\widehat{\mathcal{M}}_\alpha^Y \cong_E \widehat{\mathcal{M}}_\alpha$ holds, for all $\alpha \in L_1$.*

Finally, a reduction of a concept lattice with only one reduct is introduced.

Example 4. Considering the same framework as in Example 1 and the context (A, B, R, σ), where $A = \{a_1, a_2, a_3\}$, $B = \{b_1, b_2, b_3\}$, $R\colon A \times B \to L_3$ is given in Fig. 3 and σ is constantly $\&_G^*$.

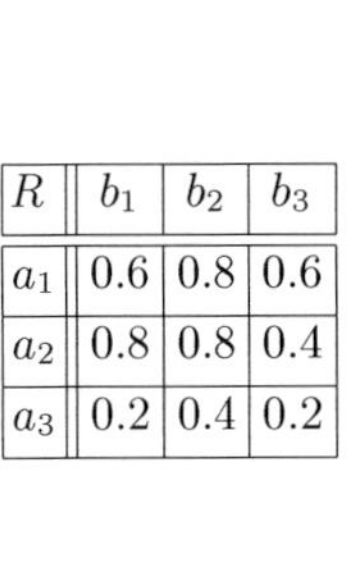

R	b_1	b_2	b_3
a_1	0.6	0.8	0.6
a_2	0.8	0.8	0.4
a_3	0.2	0.4	0.2

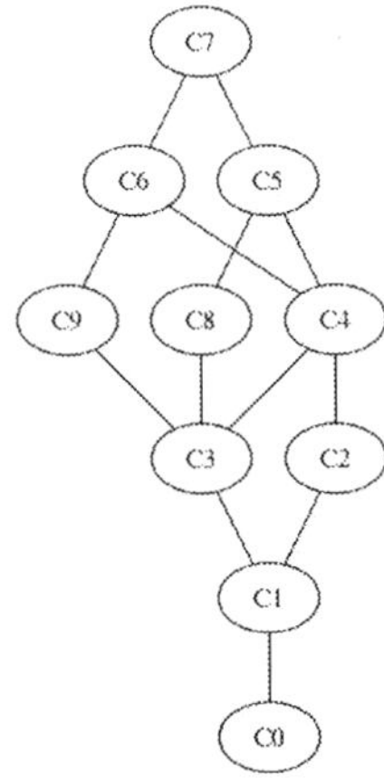

Fig. 3. The definition of the relation R of Example 4 and the Hasse diagram of $(\mathcal{M}, \preceq)$.

From the Hasse diagram of the concept lattice $(\mathcal{M}, \preceq)$ shown in Fig. 3, we can observe that the concepts C_0, C_2, C_5, C_6, C_8 and C_9 are the meet-irreducible elements. In the following, we list the fuzzy-attributes associated with them:

$$\begin{aligned}
\langle \phi_{a_3,0.5}^{\downarrow}, \phi_{a_3,0.5}^{\downarrow\uparrow}\rangle = \langle \phi_{a_3,0.6}^{\downarrow}, \phi_{a_3,0.6}^{\downarrow\uparrow}\rangle = \langle \phi_{a_3,0.7}^{\downarrow}, \phi_{a_3,0.7}^{\downarrow\uparrow}\rangle &= \langle \phi_{a_3,0.8}^{\downarrow}, \phi_{a_3,0.8}^{\downarrow\uparrow}\rangle = \\
\langle \phi_{a_3,0.9}^{\downarrow}, \phi_{a_3,0.9}^{\downarrow\uparrow}\rangle &= \langle \phi_{a_3,1.0}^{\downarrow}, \phi_{a_3,1.0}^{\downarrow\uparrow}\rangle = C_0\\
\langle \phi_{a_3,0.3}^{\downarrow}, \phi_{a_3,0.3}^{\downarrow\uparrow}\rangle &= \langle \phi_{a_3,0.4}^{\downarrow}, \phi_{a_3,0.4}^{\downarrow\uparrow}\rangle = C_2\\
\langle \phi_{a_2,0.5}^{\downarrow}, \phi_{a_2,0.5}^{\downarrow\uparrow}\rangle = \langle \phi_{a_2,0.6}^{\downarrow}, \phi_{a_2,0.6}^{\downarrow\uparrow}\rangle = \langle \phi_{a_2,0.7}^{\downarrow}, \phi_{a_2,0.7}^{\downarrow\uparrow}\rangle &= \langle \phi_{a_2,0.8}^{\downarrow}, \phi_{a_2,0.8}^{\downarrow\uparrow}\rangle = C_5\\
\langle \phi_{a_1,0.7}^{\downarrow}, \phi_{a_1,0.7}^{\downarrow\uparrow}\rangle &= \langle \phi_{a_1,0.8}^{\downarrow}, \phi_{a_1,0.8}^{\downarrow\uparrow}\rangle = C_6\\
\langle \phi_{a_2,0.9}^{\downarrow}, \phi_{a_2,0.9}^{\downarrow\uparrow}\rangle &= \langle \phi_{a_2,1.0}^{\downarrow}, \phi_{a_2,1.0}^{\downarrow\uparrow}\rangle = C_8\\
\langle \phi_{a_1,0.9}^{\downarrow}, \phi_{a_1,0.9}^{\downarrow\uparrow}\rangle &= \langle \phi_{a_1,1.0}^{\downarrow}, \phi_{a_1,1.0}^{\downarrow\uparrow}\rangle = C_9
\end{aligned}$$

Applying the attribute classification theorems, we obtain that $C_f = \{a_1, a_2, a_3\}$. Therefore, the unique reduct that can be considered is $Y = C_f$. In order to obtain the concept lattice $(\widehat{\mathcal{M}}^Y_{0.9}, \preceq)$, we will take into account the following set:

$$\widehat{M_F}(Y)_{0.9} = \{\langle \phi^{\downarrow^Y}_{a_1,1.0}, \phi^{\downarrow^Y \uparrow_Y}_{a_1,1.0}\rangle, \langle \phi^{\downarrow^Y}_{a_2,1.0}, \phi^{\downarrow^Y \uparrow_Y}_{a_2,1.0}\rangle\}$$

The Hasse diagrams corresponding to the reduced concept lattices $(\widehat{\mathcal{M}}^Y_{0.9}, \preceq)$ and $(\widehat{\mathcal{M}}_{0.9}, \preceq)$ are displayed in Fig. 4. Clearly, both diagrams are isomorphic.

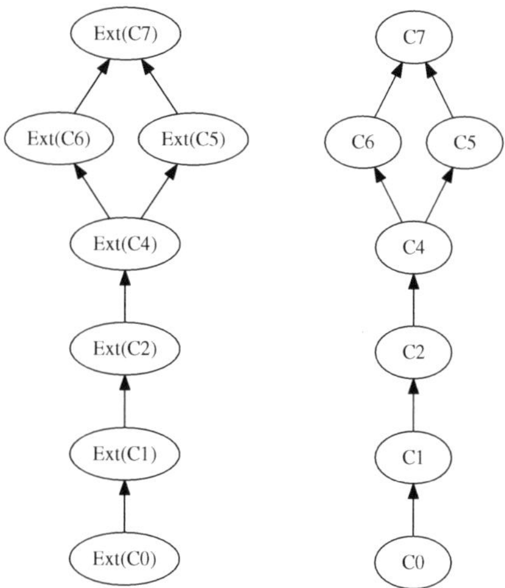

Fig. 4. The Hasse diagram of $(\widehat{\mathcal{M}}^Y_{0.9}, \preceq)$ and $(\widehat{\mathcal{M}}_{0.9}, \preceq)$.

This property is also very interesting since it drastically reduces the complexity of the computation of the concept lattice.

5 Conclusions and Future Work

A new mechanism has been introduced in order to decrease the complexity to obtain a fuzzy concept lattice, which combines two different ones: an attribute reduction is firstly applied and an irreducible α-cut reduction based on a reduct is considered after that. This procedure provides more efficiency, because, for instance, different cuts α may be considered before obtaining the most optimal reduction. Furthermore, several properties and examples have been introduced showing interesting features of the introduced reduction.

Although the presented mechanism is efficient, we need to carry on improving it and, mainly, to apply it to practical examples as well. This last fact will surely need the study of new properties and advances in the future.

References

1. Alcalde, C., Burusco, A., Bustince, H., Jurio, A., Sanz, J.A.: Evolution in time of L-fuzzy context sequences. Inf. Sci. **326**, 202–214 (2016)
2. Bělohlávek, R., Vychodil, V.: Reducing the size of fuzzy concept lattices by hedges. In: The 2005 IEEE International Conference on Fuzzy Systems, pp. 663–668 (2005)
3. Cattaneo, G., Chiaselotti, G., Ciucci, D., Gentile, T.: On the connection of hypergraph theory with formal concept analysis and rough set theory. Inf. Sci. **330**, 342–357 (2016). SI Visual Info Communication
4. Cornejo, M.E., Medina, J., Ramírez-Poussa, E.: A comparative study of adjoint triples. Fuzzy Sets Syst. **211**, 1–14 (2013)
5. Cornejo, M.E., Medina-Moreno, J., Ramírez, E.: On the classification of fuzzy-attributes in multi-adjoint concept lattices. In: Rojas, I., Joya, G., Cabestany, J. (eds.) IWANN 2013, Part II. LNCS, vol. 7903, pp. 266–277. Springer, Heidelberg (2013)
6. Cornejo Piñero, M.E., Medina-Moreno, J., Ramírez-Poussa, E.: Fuzzy-attributes and a method to reduce concept lattices. In: Cornelis, C., Kryszkiewicz, M., Ślęzak, D., Ruiz, E.M., Bello, R., Shang, L. (eds.) RSCTC 2014. LNCS, vol. 8536, pp. 189–200. Springer, Heidelberg (2014)
7. Cornejo, M.E., Medina, J., Ramírez-Poussa, E.: Attribute reduction in multi-adjoint concept lattices. Inf. Sci. **294**, 41–56 (2015)
8. Cornejo, M.E., Medina, J., Ramírez-Poussa, E.: Multi-adjoint algebras versus extended-order algebras. Appl. Math. Inf. Sci. **9**(2L), 365–372 (2015)
9. Cornejo, M.E., Medina, J., Ramírez-Poussa, E.: Multi-adjoint algebras versus non-commutative residuated structures. Int. J. Approximate Reasoning **66**, 119–138 (2015)
10. Cornejo, M.E., Medina, J., Ramírez-Poussa, E.: On the use of irreducible elements for reducing multi-adjoint concept lattices. Know. Based Syst., 1–11 (2015). doi:10.1016/j.knosys.2015.07.003 (in press)
11. Kang, X., Li, D., Wang, S.: Research on domain ontology in different granulations based on concept lattice. Knowl. Based Syst. **27**, 152–161 (2012)
12. Krídlo, O., Krajči, S., Antoni, L.: Formal concept analysis of higher order. Int. J. Gen. Syst. **45**(2), 116–134 (2016)
13. Li, L., Zhang, J.: Attribute reduction in fuzzy concept lattices based on the T-implication. Knowl. Based Syst. **23**(6), 497–503 (2010)
14. Li, M., Wang, G.: Approximate concept construction with three-way decisions and attribute reduction in incomplete contexts. Knowl. Based Syst. **91**, 165–178 (2016). Three-way Decisions and Granular Computing
15. Medina, J.: Relating attribute reduction in formal, object-oriented and property-oriented concept lattices. Comput. Math. Appl. **64**(6), 1992–2002 (2012)
16. Medina, J., Ojeda-Aciego, M., Ruiz-Calviño, J.: Formal concept analysis via multi-adjoint concept lattices. Fuzzy Sets Syst. **160**(2), 130–144 (2009)
17. Rodríguez-Jiménez, J.M., Cordero, P., Enciso, M., Mora, A.: Data mining algorithms to compute mixed concepts with negative attributes: an application to breast cancer data analysis. Mathematical Methods in the Applied Sciences (to appear, 2016)
18. Valverde-Albacete, F.J., González-Calabozo, J.M., Peñas, A., Peláez-Moreno, C.: Supporting scientific knowledge discovery with extended, generalized formal concept analysis. Expert Syst. with Appl. **44**, 198–216 (2016)
19. Wei, L., Qi, J.-J.: Relation between concept lattice reduction and rough set reduction. Knowl. Based Syst. **23**(8), 934–938 (2010)

Towards Galois Connections over Positive Semifields

Francisco J. Valverde-Albacete(✉) and Carmen Peláez-Moreno

Departamento de Teoría de la Señal y de las Comunicaciones,
Universidad Carlos III de Madrid, 28911 Leganés, Spain
{ffva,carmeng}@tsc.uc3m.es

Abstract. In this paper we try to extend the Galois connection construction of $\mathcal{K}$-Formal Concept Analysis to handle semifields which are not idempotent. Important examples of such algebras are the extended non-negative reals and the extended non-negative rationals, but we provide a construction that suggests that such semifields are much more abundant than suspected. This would broaden enormously the scope and applications of $\mathcal{K}$-Formal Concept Analysis.

Keywords: Formal Concept Analysis extensions · $\mathcal{K}$-Formal Concept Analysis · Positive semifields · Galois connection

1 Introduction and Motivation

The most orthodox presentation of Standard FCA is still [1], whose Galois connection is interpreted as between the set of subsets of objects and attributes. But Standard FCA can also be understood in the context of the linear algebra of boolean spaces with sets substituted for characteristic functions, and other extensions, e.g. the $\mathcal{K}$-FCA [2–4], FCA in a fuzzy setting [5], etc., can also be considered in the light of linear algebra over a certain subclass of semirings. In this paper, we will understand a *semiring* [6] to be an algebra $\mathcal{S} = \langle S, \oplus, \otimes, \epsilon, e\rangle$ for which

- the additive structure, $\langle S, \oplus, \epsilon\rangle$, is a commutative monoid,
- the multiplicative structure, $\langle S\backslash\{\epsilon\}, \otimes, e\rangle$, is a monoid,
- multiplication distributes over addition from right and left
- and the zero element is multiplicatively-absorbing i.e. $\forall a \in S,\ \epsilon \otimes a = \epsilon$.

We will only consider commutative semirings, those whose multiplicative structure is commutative.

Specifically, every commutative semiring accepts a canonical preorder, $a \leq b$ if and only if there exists $c \in D$ with $a \oplus c = b$. A *dioid* is a commutative semiring $\mathcal{D}$ where this relation is actually an order. And in fact the mentioned extensions

CPM & FVA have been partially supported by the Spanish Government-MinECo projects TEC2014-53390-P and TEC2014-61729-EXP.

J.P. Carvalho et al. (Eds.): IPMU 2016, Part II, CCIS 611, pp. 81–92, 2016.
DOI: 10.1007/978-3-319-40581-0_8

to FCA all feature subtypes of dioids (e.g. $\mathcal{K}$-FCA or FCA in a fuzzy setting) or dioid-valued entries (e.g. interval-based FCA) in their formal contexts.

One of the most useful extensions to FCA uses $\mathcal{K}$-valued formal contexts where $\mathcal{K}$ is a complete idempotent semifield: this is the basis of $\mathcal{K}$-Formal Concept Analysis [2–4]. An idempotent semiring is one whose addition is idempotent, $u \oplus u = u$ while semifields are semirings whose multiplicative structure is a group. Idempotent semifields like $\overline{\mathbb{R}}_{\max,+}$ and $\overline{\mathbb{R}}_{\min,\times}$ are within this class, but also the semifields of non-negative rationals $\mathbb{Q}_0^+$ and that of completed nonnegative reals $\mathbb{R}_{\geq 0} = \langle [0, \infty], +, *, \bot = 0, e = 1, \top = \infty \rangle$ which is complete and totally ordered in its usual order.

It would be interesting to know whether dioids, in general, are FCA-generating, but for this paper we consider only a proper subclass of dioids: its intersection with the class of semifields, the "positive" semifields.

Regarding this extension we may wonder,

1. whether there are many positive semifields available that generate such extension of FCA.
2. whether the extension is useful, that is whether there are instances of FCA-related problems that are solved with positive semifields,

In this paper we try to address both these concerns. First, we review the theory of dioids and semimodules over them with an emphasis on semifields; next we present and approach to generating semifields and examples of their ubiquity. We also review the construction of Galois Connections from residuated semirings and finally provide a new application of this construction to matrix decomposition.

2 Positive Semifields and Semimodules

2.1 Dioids and Positive Semifields

Complete and Positive Dioids. Recall that a *dioid* is a commutative semiring $\mathcal{D}$ where the canonical preorder relation, $a \preccurlyeq b$ if and only if there exists $c \in D$ with $a \oplus c = b$ is actually an order $\langle D, \preccurlyeq \rangle$. For this order, the additive zero is always the bottom $\bot = \wedge D = \epsilon$.

In a dioid, the canonical orden relation is compatible with both $\oplus$ and $\otimes$ [7, Chap. 1, Prop. 6.1.7]. Dioids are all zero-sum free, that is, they have no non-null additive factors of zero: if $a, b \in D, a \oplus b = \epsilon$ then $a = \epsilon$ and $b = \epsilon$.

A dioid is *complete* if it is complete as an ordered set for the canonical order relation, and the following distributivity properties hold, for all $A \subseteq D, b \in D$,

$$\left(\bigoplus_{a \in A} a\right) \otimes b = \bigoplus_{a \in A} (a \otimes b) \qquad b \otimes \left(\bigoplus_{a \in A} a\right) = \bigoplus_{a \in A} (b \otimes a) \tag{1}$$

In complete dioids, there is already a top element $\top = \oplus_{a \in D} a$.

A semiring is *entire or zero-divisor free* if $a \otimes b = \epsilon$ implies $a = \epsilon$ or $b = \epsilon$. If the dioid is entire, its order properties justifies calling it a *positive dioid* or *information algebra* [7].

Positive Semifields. A semifield, as mentioned in the introduction, is a semiring whose multiplicative structure $\langle K \setminus \{\epsilon\}, \otimes, e, \cdot^{-1}\rangle$ is a group, where $\cdot^{-1} : K \to K$ is the function to calculate the inverse such that $\forall u \in K, u \otimes u^{-1} = e$. Since all semifields are entire, dioids that are at the same time semifields are called *positive semifields*, of which the positive reals or rationals are a paragon.

Example 1 (Semifield of non-negative reals). Since we need the completion property to develop Galois connections, we concentrate on the completed nonnegative reals

$$\mathbb{R}_{\geq 0} = \langle [0, \infty], +, \times, \cdot^{-1}, \bot = 0, e = 1, \top = \infty\rangle$$

which is complete and totally ordered in its usual order $\langle \mathbb{R}_{\geq 0}, \leq\rangle$. Note that the multiplicative $\langle (0, \infty), \times\rangle$ structure must exclude also the infinity, since $0\times\infty = 0$ does not have an inverse. □

Regarding the intrinsic usefulness of positive semifields that are not fields, and apart from the trivial case of $\mathbb{B}$ the booleans, there is not doubt of their usefulness: the best known semifield $\mathbb{R}_{\geq 0}$ is widely used, for instance, in Electrical Network theory, where the series or parallel addition of resistances and conductances is carried out entirely in it.

Pairs of Mutually Inverse Semifields. In fact, this application provides a way forward regarding our first difficulty, to wit the fact that the idempotent semifields used in $\mathcal{K}$-FCA always comes in pairs: $\overline{\mathbb{R}}_{\max,+}$ and $\overline{\mathbb{R}}_{\min,+}$ or $\overline{\mathbb{R}}_{\max,\times}$ and $\overline{\mathbb{R}}_{\min,\times}$. Each member of these pairs appears as the dual of the other member by means of the $\cdot^{-1}$ involution so $\overline{\mathbb{R}}_{\max,+} = (\overline{\mathbb{R}}_{\min,+})^{-1}$ and $\overline{\mathbb{R}}_{\min,+} = (\overline{\mathbb{R}}_{\max,+})^{-1}$.

To settle notation straight, these semifields come in pairs $\left(\overline{\mathcal{K}}, (\overline{\mathcal{K}})^{-1}\right)$ with dual order structures $\overline{\mathcal{K}} = \langle K, \preccurlyeq\rangle$ and $(\overline{\mathcal{K}})^{-1} = \langle K, \succcurlyeq \equiv \preccurlyeq^{\delta}\rangle$, and dual algebraic structures: suppose that $\{\bot, \top\} \subseteq K$, then

$$\overline{\mathcal{K}} = \langle K, \underset{\cdot}{\oplus}, \underset{\cdot}{\otimes}, \cdot^{-1}, \bot, e, \top\rangle \qquad \overline{\mathcal{K}}^{-1} = \langle K, \dot{\oplus}, \dot{\otimes}, \cdot^{-1}, \top, e, \bot\rangle \tag{2}$$

On top of the individual laws as positive semifields, we have the modular laws:

$$(u \underset{\cdot}{\oplus} v) \underset{\cdot}{\otimes} (u \dot{\oplus} v) = u \underset{\cdot}{\otimes} v \qquad (u \underset{\cdot}{\oplus} v) \dot{\otimes} (u \dot{\oplus} v) = u \dot{\otimes} v$$

the analogues of the De Morgan laws:

$$u \underset{\cdot}{\oplus} v = (u^{-1} \dot{\oplus} v^{-1})^{-1} \qquad u \dot{\oplus} v = (u^{-1} \underset{\cdot}{\oplus} v^{-1})^{-1}$$

$$u \underset{\cdot}{\otimes} v = (u^{-1} \dot{\otimes} v^{-1})^{-1} \qquad u \dot{\otimes} v = (u^{-1} \underset{\cdot}{\otimes} v^{-1})^{-1}$$

and the self-dual inequality

$$(u \underset{\cdot}{\otimes} v) \dot{\otimes} w \succcurlyeq u \underset{\cdot}{\otimes} (v \dot{\otimes} w)$$

Note that:

- the notation to "speak" about these semirings tries to follow a convention reminiscent of that of boolean algebra, where the inversion is complement.
- the dot notation is a mnemonic for where do the multiplication of the bottom and top go:

$$\bot \underset{\cdot}{\otimes} \top = \bot \qquad\qquad \bot \overset{\cdot}{\otimes} \top = \top$$

implying that the "lower" addition and multiplication are aligned with the usual order in the semiring while the "upper" addition and multiplication are aligned with its dual.

Example 2 (Dual semifields for the Non-negative Reals). The previous procedure shows that there are some problems with the notation of Example 1, and this led to the definition of the following signatures for this semifield and its inverse in convex analysis [8]:

$$\mathbb{R}_{\geq 0} = \langle [0, \infty], \underset{\cdot}{+}, \underset{\cdot}{\times}, \cdot^{-1}, 0, 1, \infty \rangle \qquad \mathbb{R}_{\geq 0}^{-1} = \langle [0, \infty], \overset{\cdot}{+}, \overset{\cdot}{\times}, \cdot^{-1}, \infty, 1, 0 \rangle \tag{3}$$

Both of these algebras are used, for instance, in (Electrical) Network Analysis: the algebra of complete positive reals to carry out the series sum of resistances, and its dual semifield to carry out parallel summation of resistances. With the convention that $\mathbb{R}_{\geq 0}$ semiring models *resistances*, it is easy to see that the bottom element, $\bot = 0$ models a shortcircuit, that the top element $\top = \infty$ models an open circuit (infinite resistance) and these conventions are swapped in the dual semifield of *conductances*. Interestingly, the required formulae for the multiplication of the extremes:

$$0 \underset{\cdot}{\otimes} \infty = 0 \qquad\qquad 0 \overset{\cdot}{\otimes} \infty = \infty \tag{4}$$

are a no-go for circuit analysis, which suggests that what is actually being operated with are the incomplete versions of these semifields, and the many problems that EE students have in learning how to properly deal with these values may stem from this fact. □

2.2 Semimodules over Positive Semifields

Let $\mathcal{D} = \langle D, +, \times, \epsilon_D, e_D \rangle$ be a commutative semiring. A $\mathcal{D}$-*semimodule* $\mathcal{X} = \langle X, \oplus, \odot, \epsilon_X \rangle$ is a commutative monoid $\langle X, \oplus, \epsilon_X \rangle$ endowed with a scalar action $(\lambda, x) \mapsto \lambda \odot x$ satisfying the following conditions for all $\lambda, \mu \in D$, $x, x' \in X$:

$$\begin{aligned} (\lambda \times \mu) \odot x &= \lambda \odot (\mu \odot x) & \lambda \odot (x \oplus x') &= \lambda \odot x \oplus \lambda \odot x' \\ (\lambda + \mu) \odot x &= \lambda \odot x \oplus \mu \odot x & \lambda \odot \epsilon_X &= \epsilon_X = \epsilon_D \otimes x \\ e_D \odot x &= x & & \end{aligned} \tag{5}$$

Matrices form a $\mathcal{D}$-semimodule $D^{g \times m}$ for given g, m. In this paper, we only use finite-dimensional semimodules where we can identify semimodules with column vectors, e.g. $\mathcal{X} \equiv \mathcal{D}^g$. If $\mathcal{D}$ is commutative, naturally-ordered or complete, then $\mathcal{X}$ is also commutative, naturally-ordered or complete [9]. If $\mathcal{K}$ is a semifield, we may also define an inverse for the semimodule by the coordinate-wise inversion, $(x^{-1})_i = (x_i)^{-1}$.

Similarly, the may define a *matrix conjugate* $(A^{\circledast})_{ij} = A_{ji}^{-1}$. For complete idempotent semifields, the following matrix algebra equations are proven in [10, Ch. 8]:

Proposition 1. *Let $\mathcal{K}$ be an idempotent semifield, and $A \in \mathcal{K}^{m\times n}$. Then:*

1. $A \underset{\cdot}{\otimes}(A^{\circledast} \dot{\otimes} A) = A \dot{\otimes}(A^{\circledast} \underset{\cdot}{\otimes} A) = (A \dot{\otimes} A^{\circledast}) \underset{\cdot}{\otimes} A = (A \underset{\cdot}{\otimes} A^{\circledast}) \dot{\otimes} A = A$ *and* $A^{\circledast} \underset{\cdot}{\otimes}(A \dot{\otimes} A^{\circledast}) = A^{\circledast} \dot{\otimes}(A \underset{\cdot}{\otimes} A^{\circledast}) = (A^{\circledast} \dot{\otimes} A) \underset{\cdot}{\otimes} A^{\circledast} = (A^{\circledast} \underset{\cdot}{\otimes} A) \dot{\otimes} A^{\circledast} = A^{\circledast}$.
2. *Alternating $A - A^{\circledast}$ products of 4 matrices can be shortened as in:*
$$A^{\circledast} \dot{\otimes}(A \underset{\cdot}{\otimes}(A^{\circledast} \dot{\otimes} A)) = A^{\circledast} \dot{\otimes} A = (A^{\circledast} \dot{\otimes} A) \underset{\cdot}{\otimes}(A^{\circledast} \dot{\otimes} A)$$
3. *Alternating $A - A^{\circledast}$ products of 3 matrices and another terminal, arbitrary matrix can be shortened as in:*
$$A^{\circledast} \dot{\otimes}(A \underset{\cdot}{\otimes}(A^{\circledast} \dot{\otimes} M)) = A^{\circledast} \dot{\otimes} M = (A^{\circledast} \dot{\otimes} A) \underset{\cdot}{\otimes}(A^{\circledast} \dot{\otimes} M)$$
4. *The following inequalities apply:*
$$A^{\circledast} \dot{\otimes}(A \underset{\cdot}{\otimes} M) \geq M \qquad\qquad A^{\circledast} \underset{\cdot}{\otimes}(A \dot{\otimes} M) \leq M$$

2.3 Galois Connections over Idempotent Semifields

In this paper we presuppose the setting of [11]. When $\overline{\mathcal{K}}$ is a completed idempotent semifield and $\mathcal{X} \equiv \overline{\mathcal{K}}^g$ and $\mathcal{Y} \equiv \overline{\mathcal{K}}^m$ are idempotent vectors spaces or semimodules, the definition of the Galois connection involves the use of a scalar product $\langle \cdot \mid R \mid \cdot \rangle : X \times Y \to K$ and a scalar $\varphi \in \overline{\mathcal{K}}$ [4]:

$$x^{\uparrow}_{R,\varphi} = \vee\{y \in Y \mid \langle x \mid R \mid y\rangle \leq \varphi\} \qquad y^{\downarrow}_{R,\varphi} = \vee\{x \in X \mid \langle x \mid R \mid y\rangle \leq \varphi\}$$

This definition is quite general and might even be valid for any dioid, but we now want to use it when the semiring has the richer algebraic structure of a complete positive semifield. For simplicity's sake we will consider in this paper that $\varphi = e$. Generalizing it along the lines of [11, Sect. 3.1] is not difficult.

We consider the scalar product $\langle x \mid R \mid y\rangle = x^{\mathrm{T}} \otimes R \otimes y$, where $R \in \overline{\mathcal{K}}^{g\times m}$. Since $x^{\mathrm{T}} \underset{\cdot}{\otimes} R \underset{\cdot}{\otimes} y \leq e \Leftrightarrow y^{\mathrm{T}} \underset{\cdot}{\otimes} R^{\mathrm{T}} \underset{\cdot}{\otimes} x \leq e$, by using residuation we may write:

$$x^{\uparrow}_{R} = (x^{\mathrm{T}} \underset{\cdot}{\otimes} R) \underset{\cdot}{\backslash} e = R^{\circledast} \dot{\otimes} x^{-1} \qquad y^{\downarrow}_{R} = (y^{\mathrm{T}} \underset{\cdot}{\otimes} R^{\mathrm{T}}) \underset{\cdot}{\backslash} e = R^{-1} \dot{\otimes} y^{-1} \qquad (6)$$

involving only transposition, inversion and operation in the dual semifield.

We recall the following proposition:

Proposition 2. *$(\cdot_R^{\uparrow}, \cdot_R^{\downarrow}) : \mathcal{X} \leftharpoondown\rightharpoonup \mathcal{Y}$ is a Galois connection between the semimodules $\mathcal{X} \cong \overline{\mathcal{K}}^g$ and $\mathcal{Y} \cong \overline{\mathcal{K}}^m$: for $x \in X$, $y \in Y$, we have $y \leq x_R^{\uparrow} \Leftrightarrow x \leq y_R^{\downarrow}$.*

Proof. We need only prove in one sense, since the other is similar. If $y \leq x_R^{\uparrow} = R^{\circledast} \dot{\otimes} x^{-1}$, then by inversion, $R^{\mathrm{T}} \underset{\cdot}{\otimes} x \leq y^{-1}$ whence, by residuation $x \leq R^{\mathrm{T}} \underset{\cdot}{\backslash} y^{-1} = R^{-1} \dot{\otimes} y^{-1} = y_R^{\downarrow}$. □

The diagram in Fig. 1 summarizes this Galois connection [4]

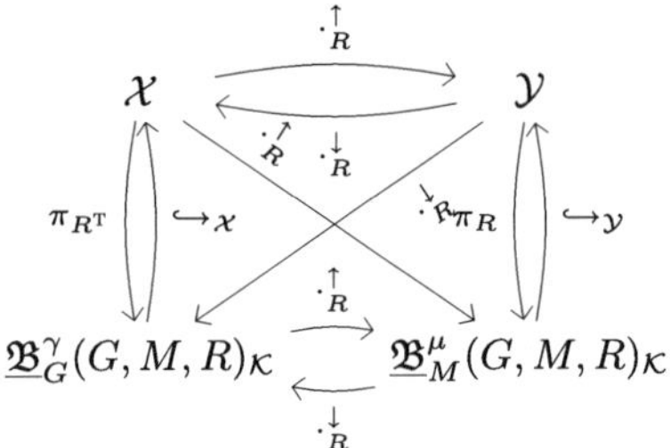

Fig. 1. $(\cdot_R^{\uparrow}, \cdot_R^{\downarrow}) : \mathcal{X} \leftharpoondown\rightharpoonup \mathcal{Y}$, the Galois connection between positive spaces.

This immediately puts at our disposal a number of results which we collect in the following proposition:

Proposition 3. *Consider the Galois connection $(\cdot_R^{\uparrow}, \cdot_R^{\downarrow}) : \mathcal{X} \leftharpoondown\rightharpoonup \mathcal{Y}$. Then:*

1. The polars are antitone, join-inverting functions:

$$(x_1 \underset{\cdot}{\oplus} x_2)_R^{\uparrow} = x_1{}_R^{\uparrow} \dot{\oplus} x_2{}_R^{\uparrow} \qquad (y_1 \underset{\cdot}{\oplus} y_2)_R^{\downarrow} = y_1{}_R^{\downarrow} \dot{\oplus} y_2{}_R^{\downarrow}. \tag{7}$$

2. The compositions of the polars: $\pi_{R^{\mathrm{T}}} : X \to X, \pi_R : Y \to Y$

$$\pi_{R^{\mathrm{T}}}(x) = (x_R^{\uparrow})_R^{\downarrow} = R^{-1} \dot{\otimes} (R^{\mathrm{T}} \underset{\cdot}{\otimes} x) \qquad \pi_R(y) = (y_R^{\downarrow})_R^{\uparrow} = R^{\circledast} \dot{\otimes} (R \underset{\cdot}{\otimes} y)$$

are closures, that is, extensive and idempotent operators.

$$\pi_{R^{\mathrm{T}}}(x) \geq x \qquad \pi_R(y) \geq y$$
$$\pi_{R^{\mathrm{T}}}(\pi_{R^{\mathrm{T}}}(x)) = \pi_{R^{\mathrm{T}}}(x) \qquad \pi_R(\pi_R(y)) = \pi_R(y)$$

3. The polars are mutual pseudo-inverses:

$$(\cdot)_R^{\uparrow} \circ (\cdot)_R^{\downarrow} \circ (\cdot)_R^{\uparrow} = (\cdot)_R^{\uparrow} \qquad (\cdot)_R^{\downarrow} \circ (\cdot)_R^{\uparrow} \circ (\cdot)_R^{\downarrow} = (\cdot)_R^{\downarrow}$$

One of the advantages of working in idempotent semimodules is that we can strengthen Statement 1 in Proposition 3 to reveal that the polars are idempotent semimodule morphisms:

Proposition 4. *The polar of intents of the Galois connection transforms a $\overline{\mathcal{K}}$-semimodule of extents into a $\overline{\mathcal{K}}^{-1}$-semimodule of intents, and dually for the polar of the extents.*

Proof. For linearity, consider $x_1{}^{\uparrow}_{R} = R^{\circledast} \dot{\otimes} x_1^{-1}$ and $x_2{}^{\uparrow}_{R} = R^{\circledast} \dot{\otimes} x_2^{-1}$.

$$
\begin{aligned}
(\lambda_1 \underset{\cdot}{\otimes} x_1 \underset{\cdot}{\oplus} \lambda_2 \underset{\cdot}{\otimes} x_2)^{\uparrow}_{R} &= R^{\circledast} \dot{\otimes} (\lambda_1 \underset{\cdot}{\otimes} x_1 \underset{\cdot}{\oplus} \lambda_2 \underset{\cdot}{\otimes} x_2)^{-1} = \\
&= R^{\circledast} \dot{\otimes} (\lambda_1^{-1} \dot{\otimes} x_1^{-1} \dot{\oplus} \lambda_2^{-1} \dot{\otimes} x_2^{-1}) = \\
&= (\lambda_1^{-1} \dot{\otimes} R^{\circledast} \dot{\otimes} x_1^{-1}) \dot{\oplus} (\lambda_2^{-1} \dot{\otimes} R^{\circledast} \dot{\otimes} x_1^{-1}) = \\
&= (\lambda_1^{-1} \dot{\otimes} x_1{}^{\uparrow}_{R}) \dot{\oplus} (\lambda_2^{-1} \dot{\otimes} x_2{}^{\uparrow}_{R}).
\end{aligned}
$$

For the polar of extents the proof is similar. □

Note that this is the $\mathcal{K}$-FCA analogue of the fact that the polars are join-inverting. But the novelty is that the scalings for one semimodule and the other are inverted. We need two more results from [11]:

Lemma 1. *Let I_G and I_M be the identity matrices of dimension $g \times g$ and $m \times m$ in $\overline{\mathcal{K}}$. Then the object- and attribute-concepts of the Galois connection are:*

$$
\gamma_R(I_G) = (R^{-1} \dot{\otimes} R^{\mathrm{T}}, R^{\circledast}) \qquad \mu_R(I_M) = (R^{-1}, R^{\circledast} \dot{\otimes} R)
$$

taken as pairs of co-indexed vectors.

Corollary 1. *Consider the Galois connection $(\cdot^{\uparrow}_{R}, \cdot^{\downarrow}_{R}) : \widetilde{\mathcal{X}}^{\gamma} \rightleftharpoons \widetilde{\mathcal{Y}}^{\mu}$. Then, its system of extents is $\overline{\mathcal{K}}^{-1}$-generated by the attribute-extents. Dually, its system of intents is $\overline{\mathcal{K}}^{-1}$-generated by the object-intents.*

3 A Construction for Positive Semifields

There is a non-countable number of semifields obtainable from $\mathbb{R}_{\geq 0}$. Their discovery is probably due to Maslov, but we present here the generalized procedure introduced by Pap and collaborators that include Maslov results. The application to positive semifields is our own statement:

Construction 1 (Pap's dioids and semifields). Let $\mathbb{R}_{\geq 0}$ be the semiring of non-negative reals, and consider a strictly monotone *generator function* g on an interval $[a, b] \subseteq [-\infty, \infty]$ with values in $[0, \infty]$. Since g is strictly increasing it admits an inverse g^{-1}, so set

1. the pseudo-addition, $u \oplus v = g^{-1}(g(u) + (g(v))$
2. the pseudo-multiplication, $u \otimes v = g^{-1}(g(u) \underset{\cdot}{\times} (g(v))$
3. neutral element, $e = g^{-1}(1)$
4. inverse, $x^* = g^{-1}(\frac{1}{g(x)})$,

Then,

1. if g is strictly increasing such that $g(a) = 0$ and $g(b) = \infty$, then a complete positive semifield whose order is aligned with that of $\mathbb{R}_0^+$ is:

$$\mathcal{K}_g = \langle [a, b], \underset{\cdot}{\oplus}, \underset{\cdot}{\otimes}, \cdot^*, \bot = a, e, \top = b\rangle.$$

2. if g is strictly decreasing such that $g(a) = \infty$ and $g(b) = 0$, then a complete positive semifield whose order is aligned with that of $(\mathbb{R}_{\geq 0})^{-1}$ is

$$(\mathcal{K}_g)^{-1} = \langle [a, b], \dot{\oplus}, \dot{\otimes}, \cdot^*, \bot^{-1} = b, e, \top^{-1} = a\rangle.$$

Proof. See [12,13] for the basic dioid, and [7, p. 44] for the inverse operation and the fact that it is a semifield, hence a positive semifield. □

Our use of Construction 1 is to generate different kind of semifields by providing different generator functions:

Construction 2 (Multiplicative-cost semifields [14]). Consider a free parameter $\lambda \in [-\infty, 0) \bigcup (0, \infty]$ and the function $g(x) = x^\lambda$ in $[a, b] = [0, \infty]$ in Construction 1. For the operations we obtain:

$$u \oplus_\lambda v = \left(u^\lambda \underset{\cdot}{+} v^\lambda\right)^{\frac{1}{\lambda}} \quad u \otimes_\lambda v = \left(u^\lambda \underset{\cdot}{\times} v^\lambda\right)^{\frac{1}{\lambda}} = u \underset{\cdot}{\times} v \quad u^{\circledast} = \left(\frac{1}{x^\lambda}\right)^{\frac{1}{\lambda}} = x^{-1} \tag{8}$$

where the basic operations are to be interpreted in $\mathbb{R}_{\geq 0}$. Now,

- if $\lambda \in (0, \infty]$ then $g(x) = x^\lambda$ is strictly monotone increasing whence $\bot_\lambda = 0$, $e_\lambda = 1$, and $\top_\lambda = \infty$, and the complete positive semifield generated, order-aligned with $\mathbb{R}_{\geq 0}$, is:

$$\mathbb{R}_{\geq 0,\lambda} = \langle [0, \infty], \underset{\cdot}{\oplus}_\lambda, \underset{\cdot}{\times}, \cdot^{-1}, \bot_\lambda = 0, e, \top_\lambda = \infty\rangle \tag{9}$$

- if $\lambda \in [-\infty, 0)$ then $g(x) = x^\lambda$ is strictly monotone decreasing whence $\bot_\lambda = \infty$, $e_\lambda = 1$, and $\top_\lambda = 0$, and the complete positive semifield generated, order-aligned with $(\mathbb{R}_{\geq 0})^{-1}$, or dually aligned with $\mathbb{R}_{\geq 0}$, is:

$$\mathbb{R}_{\geq 0,-\lambda} = \mathbb{R}_{\geq 0,\lambda}^{-1} = \langle [0, \infty], \dot{\oplus}_\lambda, \dot{\times}, \cdot^{-1}, \bot_\lambda^{-1} = \infty, e, \top_\lambda^{-1} = 0\rangle \tag{10}$$

Proof By instantiation of the basic case. □

In particular, consider the cases:

Proposition 5. *In the previous Construction 2, if $\lambda \in \{\pm 1\}$ then*

$$\mathbb{R}_{\geq 0,1} = \mathbb{R}_{\geq 0} \qquad (\mathbb{R}_{\geq 0,1})^{-1} = (\mathbb{R}_{\geq 0})^{-1} \tag{11}$$

and

$$\lim_{\lambda \to \infty} \mathbb{R}_{\geq 0,\lambda} = \overline{\mathbb{R}}_{\max,\times} \qquad \lim_{\lambda \to -\infty} \mathbb{R}_{\geq 0,\lambda}^{-1} = \overline{\mathbb{R}}_{\min,\times} \tag{12}$$

Proof. The proof of (11) by inspection. For (12) see [14]. □

This suggests the following corollary:

Corollary 2. $\mathbb{R}_{\geq 0,\lambda}$ *and* $\mathbb{R}_{\geq 0,\lambda}^{-1}$ *are inverse semifields.*

4 The Idempotent Singular Value Decomposition

This is our answer to the second question presented in the introduction as to the usefulness of the Galois Connection in the setting of semifields.

The singular value decomposition (SVD) is a well-known decomposition scheme for real- or complex-valued rectangular matrices [15].

Theorem 1. *Given a matrix $M \in \mathcal{M}_{m\times n}(\mathcal{K})$ where $\mathcal{K}$ is a field, there is a factorization $M = U\Sigma V^*$ where $\cdot^*$ stands for conjugation, given in term of three matrices*

- *$U \in \mathcal{M}_{n\times n}(\mathcal{K})$ is a unitary matrix of* left singular vectors.
- *$\Sigma \in \mathcal{M}_{m\times n}(\mathcal{K})$ is a diagonal matrix of non-negative real values called the* singular values.
- *$V \in \mathcal{M}_{n\times n}(\mathcal{K})$ is a unitary matrix of* right singular vectors.

Often the singular values are listed in descending order, and the left and right singular eigenvalues are re-ordered accordingly.

Note that M can also be written using outer products as:

$$M = \sum_{i=1}^{\min(m,n)} \sigma_i u_i v_i^* \tag{13}$$

hence, since the SVD is a costly procedure, it is also interesting to find $k < \min(m,n)$ such that using the k greatest singular values we may approximate:

$$M \approx \sum_{i=1}^{k} \sigma_i u_i v_i^* . \tag{14}$$

Note that, in particular, singular vectors of the null eigenvalue never contribute to the reconstruction so they may be discarded. This approximation procedure is particularly useful in applications like Latent Semantic Analysis [16].

Equation (13) suggests that the triples (σ_i, u_i, v_i) of a singular value and related left and right singular vectors have a special status in the theory. Indeed, [17] already suggested the use of formal concepts for this purpose. We are going next to introduce the idempotent Singular Value Decomposition (iSVD) for matrices over an idempotent semifield, in several flavours.

A full singular value decomposition is particularly easy in idempotent semifields. Recall from property 1 of Proposition 1 that we have no less than four decompositions of $A \in \mathcal{M}_{m\times n}(\mathcal{K})$ in terms of two other matrices:

$$A \underset{\cdot}{\otimes} (A^{\circledast} \overset{\cdot}{\otimes} A) = A \overset{\cdot}{\otimes} (A^{\circledast} \underset{\cdot}{\otimes} A) = (A \overset{\cdot}{\otimes} A^{\circledast}) \underset{\cdot}{\otimes} A = (A \underset{\cdot}{\otimes} A^{\circledast}) \overset{\cdot}{\otimes} A = A$$

Why these are SVDs is the answer produced by $\mathcal{K}$-Formal Concept Analysis.

Proposition 6. *The σ-concepts (u, v) of a matrix $M \in \mathcal{M}_{m\times n}(\mathcal{K})$ over a completed idempotent semifield $\mathcal{K}$ are (u, σ, v) triples in a singular value decomposition of M in the linear algebra of $\mathcal{K}$.*

Proof. Consider the massaged SVD equation in the dual semiring:

$$M = (U^{\mathrm{T}})^{\circledast} \dot{\otimes} \Sigma \dot{\otimes} V^{\circledast} \tag{15}$$

where the columns of U are left singular vectors, those of V are right singular vectors and Σ is the (dual) diagonal matrix of singular values, whose off-diagonal entries are $\top$. We relax the equation $M \leq (U^{\mathrm{T}})^{\circledast} \dot{\otimes} \Sigma \dot{\otimes} V^{\circledast}$ to bring residuation in the dual semimodule into the picture and then use the residuation equalities to find:

$$(U^{\mathrm{T}})^{\circledast} \dot{\backslash} M \dot{/} V^{\circledast} \leq \Sigma \qquad \Leftrightarrow U^{\mathrm{T}} \underset{\cdot}{\otimes} M \underset{\cdot}{\otimes} V \leq \Sigma \tag{16}$$

These are actually $m \times n$ inequations, but those with $i \neq j$ for $i \in \{1, \ldots, m\}, j \in \{1, \ldots, n\}$, are trivial $u_i^{\mathrm{T}} \underset{\cdot}{\otimes} M \underset{\cdot}{\otimes} v_j \leq \top$, so we concentrate on the ones involving the triples $(u_k, \sigma_k, v_k), k \in \{1, \ldots, \min(m, n)\}$ where $\sigma_k = \Sigma_{kk}$ and u_k, v_k, are, respectively the k-th columns of U, V so $u_k^{\mathrm{T}} \underset{\cdot}{\otimes} M \underset{\cdot}{\otimes} v_k \leq \sigma_k$

Now consider the definition of the σ-polars

$$u^{\uparrow}_{R,\sigma} = \{v \in K^n \mid u^{\mathrm{T}} \underset{\cdot}{\otimes} M \underset{\cdot}{\otimes} v \leq \sigma\} \quad u^{\uparrow}_{R,\sigma} = \{u \in K^m \mid u^{\mathrm{T}} \underset{\cdot}{\otimes} M \underset{\cdot}{\otimes} v \leq \sigma\} \tag{17}$$

So for a σ-concept (u, v) clearly $M \leq (u^{\mathrm{T}})^{\circledast} \dot{\otimes} \sigma \dot{\otimes} v^{\circledast}$ whence

$$M \leq \bigwedge_{(u,v) \in \underline{\mathfrak{B}}^{\sigma}(G,M,R)_{\mathcal{K}}} (u^{\mathrm{T}})^{\circledast} \dot{\otimes} \sigma \dot{\otimes} v^{\circledast} \tag{18}$$

□

Note that the number of SVD triples in (18) is enormous, hence the description is not practical. We can, of course, ignore in this description, all triples with the null singular value $\sigma = \bot$. But, more importantly, we can ignore collinear concepts, whereby we mean $\lambda \underset{\cdot}{\otimes} (u, v) = (\lambda \underset{\cdot}{\otimes} u, \lambda^{-1} \underset{\cdot}{\otimes} v)$:

Lemma 2. *If (u_1, v_1) and (u_2, v_2) are collinear, one of them can be ignored in the reconstruction of the matrix.*

Proof. Mere algebra on the concepts as per Proposition 3. □

Another step would be to prove that a linear combination of concepts does not add to the individual generating concepts, but we can actually do much more than that.

Proposition 7. *Let $M \in \mathcal{M}_{m \times n}(\mathcal{K})$ be a matrix over a completed idempotent semifield $\mathcal{K}$. Then it can be synthesized from the join- and meet-irreducibles of the concept lattice $\underline{\mathfrak{B}}^{e}(G, M, R)_{\mathcal{K}}$ as:*

$$M = \left(M \underset{\cdot}{\otimes} M^{\circledast}\right) \dot{\otimes} M \qquad M = M \dot{\otimes} \left(M^{\circledast} \underset{\cdot}{\otimes} M\right) \tag{19}$$

Proof. The proof is easy: from Lemma 1 and Corollary 1 select the triples with $\sigma = e$ and write for the join-irreducible and meet irreducible concepts, respectively:

$$M = \left(M^{-1} \dot{\otimes} M^{\mathrm{T}}\right)^{-1} \dot{\otimes} (M^{\circledast})^{\circledast} \qquad M = (M^{-1})^{-1} \dot{\otimes} \left(M^{\circledast} \dot{\otimes} M\right)^{\circledast} \tag{20}$$

Then simplify algebraically. □

Note how the last result accounts for two of the previously introduced decompositions: the other two are their duals in the inverse semifield. Note also, how scalar multiples of any of the join- or meet-irreducibles may be further ignored.

5 Discussion and Further Work

We have presented two different contributions in this paper. First an analysis of the possibilities of positive semifields to generate Galois Connections, possibly by residuation. In this respect, we have mixed results:

- On the one hand, we have described Pap's construction which, instantiated with a suitable function, is able to generate dual pairs of idempotent semifields (e.g. max-times and min-times), among a plethora of other, commuative, complete non-idempotent semifields.
- On the other hand, commutative complete dioids are already complete residuated lattices, which make them good candidates to support Galois connections by residuation, although we have not been able to provide closed expressions for the polars.

Note that the Pap semifields are a side result of the lengthier process of defining a g-calculus [12,13]. Also, in [18, Sect. 6.2,7] functions like g above a called *transforms*. This would seem to imply that we would have a non-standard calculus associated to semifields, as well as a non-standard algebra, e.g. concept lattices, in this algebraic setting.

The second contribution was a new application of Concept Lattices to decompose nonnegative matrices (NMF) by providing an analogue of Singular Value Decomposition. Since the procedure was developed in the completed idempotent semifield notation, this means there are different instantiations for semifields with different carrier sets, hence this is a *generic procedure*.

References

1. Ganter, B., Wille, R.: Formal Concept Analysis: Mathematical Foundations. Springer, Heidelberg (1999)
2. Valverde-Albacete, F.J., Peláez-Moreno, C.: Galois connections between semimodules and applications in data mining. In: Kuznetsov, S.O., Schmidt, S. (eds.) ICFCA 2007. LNCS (LNAI), vol. 4390, pp. 181–196. Springer, Heidelberg (2007)

3. Valverde-Albacete, F.J., Peláez-Moreno, C.: Further Galois connections between semimodules over idempotent semirings. In: Diatta, J., Eklund, P. (eds.) Proceedings of the 4th Conference on Concept Lattices and Applications (CLA 2007), pp. 199–212, Montpellier (2007)
4. Valverde-Albacete, F.J., Peláez-Moreno, C.: Extending conceptualisation modes for generalised formal concept analysis. Inf. Sci. **181**, 1888–1909 (2011)
5. Bělohlávek, R.: Fuzzy Relational Systems. Foundations and Principles. IFSR International Series on Systems Science and Engineering, vol. 20. Kluwer Academic, New York (2002)
6. Golan, J.S.: Semirings and Their Applications. Kluwer Academic, Dordrecht (1999)
7. Gondran, M., Minoux, M.: Graphs, Dioids and Semirings. New Models and Algorithms. Operations Research Computer Science Interfaces Series. Springer, New York (2008)
8. Moreau, J.J.: Inf-convolution, sous-additivité, convexité des fonctions numériques. J. Math. Pures Appl. **49**(9), 109–154 (1970). (in French)
9. Golan, J.S.: Semirings and Affine Equations over Them: Theory and Applications (Mathematics and Its Applications). Kluwer Academic, Dordrecht (2003)
10. Cuninghame-Green, R.: Minimax Algebra. Lecture notes in Economics and Mathematical Systems, vol. 166. Springer, Heidelberg (1979)
11. Valverde-Albacete, F.J., Peláez-Moreno, C.: The linear algebra in formal concept analysis over idempotent semifields. In: Baixeries, J., Sacarea, C., Ojeda-Aciego, M. (eds.) ICFCA 2015. LNCS, vol. 9113, pp. 97–113. Springer, Heidelberg (2015)
12. Pap, E., Ralević, N.: Pseudo-Laplace transform. Nonlinear Anal. Theor. Methods Appl. **33**, 533–550 (1998)
13. Pap, E.: g-calculus. Zbornik Radova Prirodno-Matematichkog Fakulteta, Serija za Matematiku. Rev. Res. Fac. Sci. Math. Ser. **23**, 145–156 (1993)
14. Mesiar, R., Pap, E.: Idempotent integral as limit of g-integrals. Fuzzy Sets Syst. **102**, 385–392 (1999)
15. Golub, G.H., Van Loan, C.F.: Matrix Computations. 3rd edn. JHU Press (2012)
16. Landauer, T.K., McNamara, D.S., Dennis, S., Kintsch, W.: Handbook of Latent Semantic Analysis. Lawrence Erlbaum Associates, Mahwah (2007)
17. Belohlavek, R., Vychodil, V.: Formal concepts as optimal factors in boolean factor analysis: implications and experiments. In: Proceedings of the 5th International Conference on Concept Lattices and their Applications (CLA 2007), 24–26 October 2007, Montpellier, France (2007)
18. Grabisch, M., Marichal, J.L., Mesiar, R., Pap, E.: Aggregation Functions. Cambridge University Press, Cambridge (2009)

Graded and Many-Valued Modal Logics

From Kripke to Neighborhood Semantics for Modal Fuzzy Logics

Petr Cintula[1], Carles Noguera[2], and Jonas Rogger[3(✉)]

[1] Institute of Computer Science, Czech Academy of Sciences, Pod Vodárenskou věží 2, 182 07 Prague, Czech Republic
cintula@cs.cas.cz

[2] Institute of Information Theory and Automation, Czech Academy of Sciences, Pod Vodárenskou věží 4, 182 08 Prague, Czech Republic
noguera@utia.cas.cz

[3] Mathematical Institute, University of Bern, Bern, Switzerland
jonas.rogger@math.unibe.ch

Abstract. The majority of works on modal fuzzy logics consider Kripke-style possible worlds semantics as the principal semantics despite its well known axiomatizability issues when considering fuzzy accessibility relations. The present work offers the first (two) steps towards exploring a more general semantical picture, namely a fuzzified version of the classical neighborhood semantics. First we prove the fuzzy version of the classical relationship between Kripke and neighborhood semantics. Second, for any axiomatic extension of MTL (one of the main fuzzy logics), we define its modal expansion by a $\Box$-like modality, and, in the presence of some additional conditions, we prove that the resulting logic can be axiomatized by adding the (E)-rule to the corresponding Hilbert-style calculus of the starting logic.

1 Introduction

The study of propositional systems expanded with modal operators has recently been gaining momentum in the field of Mathematical Fuzzy Logic (MFL). Several works such as [3–5,13,14,19] follow the steps of the initial developments in [10, 12]. In this line of research, modal fuzzy logics are endowed with a Kripke-style semantics which generalizes the classical one by allowing a fuzzy scale for either (or for both) the truth-values of propositions at each possible world and for the degree of accessibility from one world to another. However, axiomatizing such semantics over a given algebra (or class of algebras) of truth-values is in general a difficult problem. Also, conversely, proof systems with natural syntactic conditions may fail to be complete with any such Kripke-style semantics.

In classical modal logic, the neighborhood semantics [15,18] is seen as a more general framework (compared to Kripke frames) that allows us to prove completeness for non-normal modal logics, where the Kripke-style semantics would not work. Similarly, the goal of the present paper is to propose a form of neighborhood semantics (as already considered in [16,17] in a slightly different setting) for

J.P. Carvalho et al. (Eds.): IPMU 2016, Part II, CCIS 611, pp. 95–107, 2016.
DOI: 10.1007/978-3-319-40581-0_9

modal fuzzy logics that provides a relational semantics for a bigger class of logics. After recalling the main notions from the classical case in Sect. 2.1, standard notions from MFL in Sect. 2.2 and fuzzy class theory in Sect. 2.3, we introduce neighborhood frames for fuzzy logics in Sect. 3. Then we show their relationship with the Kripke-style semantics in Sect. 4 and we obtain an axiomatization of the logic given by all neighborhood frames in Sect. 5.

2 Preliminaries

In this section, we briefly introduce the classical neighborhood semantics and the basics on fuzzy sets, MTL logic and its algebraic counterpart, along with the different notations we use throughout the paper.

First, we fix a language $\mathcal{L}_\Box$, consisting of a fixed countably infinite set *Var* of (propositional) variables, denoted $p, q, \ldots$, binary connectives $\wedge$, $\vee$, $\&$, and $\rightarrow$, constants $\overline{0}$, $\overline{1}$, and the unary operator $\Box$. The set of *formulas* $Fm_\Box$, with arbitrary members denoted $\varphi, \psi, \chi, \ldots$ is defined inductively as usual, as are *subformulas* of formulas. We call formulas of the form $\Box\varphi$ *box-formulas* and fix the *length* of a formula φ to be the number of symbols occurring in φ. We also define $\neg\varphi = \varphi \rightarrow \overline{0}$ and $\varphi \leftrightarrow \psi = (\varphi \rightarrow \psi) \wedge (\psi \rightarrow \varphi)$.

Note that in the setting defined in Sect. 3, the diamond modality $\Diamond$ is not definable from the box $\Box$ and would need to be introduced as another primitive unary connective. This is due to the fact that in MTL, as in many other logics weaker than classical, double negation elimination fails. However in this paper, for the sake of simplicity, we focus on modal fuzzy logics only with $\Box$.

2.1 Classical Neighborhood Semantics

Introduced independently by Scott [18] and Montague [15], neighborhood semantics is a kind of possible worlds semantics for modal logics, similar in spirit to the well-known Kripke semantics, but resulting in a weaker logic. A good overview of these semantics is [8].

A *neighborhood model*, or shortly SM-*model*, is a triple $\mathfrak{M} = \langle W, N, V\rangle$, where W is a non-empty set of *worlds* while N is a function $N\colon W \rightarrow 2^{2^W}$ ($2 = \{0, 1\}$, denoting the domain of two-element Boolean algebra) that assigns to each world x a set of subsets of W, called the *neighborhood* of $x \in W$. V is an *evaluation* $V\colon Var \times W \rightarrow 2$ that is extended to all formulas inductively as in classical propositional logic (where & and $\wedge$ coincide and both denote classical conjunction), while for a box-formula:

$$V(\Box\varphi, x) = 1 \qquad \text{iff} \qquad [\![\varphi]\!]_\mathfrak{M} \in N(x)\ ,$$

where $[\![\varphi]\!]_\mathfrak{M} = \{y \in W \mid V(\varphi, y) = 1\}$, the set of worlds where "φ is true".

We say that a formula $\varphi \in Fm_\Box$ is *valid in an* SM-*model* $\mathfrak{M} = \langle W, N, V\rangle$ if $V(\varphi, x) = 1$ for all $x \in W$ (which we can equivalently formulate as $[\![\varphi]\!]_\mathfrak{M} = W$), written $\mathfrak{M} \models_{\mathrm{SM}} \varphi$. For a set of formulas $\Gamma \subseteq Fm_\Box$, we use the shorthand notation $\mathfrak{M} \models_{\mathrm{SM}} \Gamma$, if for all $\varphi \in \Gamma$, $\mathfrak{M} \models_{\mathrm{SM}} \varphi$. Furthermore, a formula $\varphi \in Fm_{\Box}$ is called an SM-*consequence* of a set of formulas $\Gamma \subseteq Fm_{\Box}$, if for all SM-models $\mathfrak{M}$, such that $\mathfrak{M} \models_{\mathrm{SM}} \Gamma$, also $\mathfrak{M} \models_{\mathrm{SM}} \varphi$, written $\Gamma \models_{\mathrm{SM}} \varphi$. Note that we consider the so-called *global* consequence relations; the reformulations of all our definitions to the *local* variant is straightforward. If $\emptyset \models_{\mathrm{SM}} \varphi$, we write $\models_{\mathrm{SM}} \varphi$ and say that φ is SM-*valid*. Clearly, SM-valid formulas are the same for global and local consequences.

A *Kripke model*, or shortly K-*model*, is a triple $\mathcal{M} = \langle W, R, V\rangle$, where W is a non-empty set of worlds, R is an *accessibility relation* on W, i.e. a binary relation $R \subseteq W^2$, and V is an evaluation, i.e., a mapping $V\colon Var \times W \to 2$. For convenience, we write Rxy instead of $\langle x, y\rangle \in R$ and denote the R-image of $x \in W$ as $R[x]$, i.e., the set $\{y \in W \mid Rxy\}$. An evaluation V is extended to formulas $\varphi \in Fm_{\Box}$ inductively as in classical propositional logic (where again & and $\wedge$ coincide and both denote classical conjunction) while for a box-formula:

$$V(\Box\varphi, x) = 1 \qquad \text{iff} \qquad V(\varphi, y) = 1, \text{ for all } y \in R[x] \ .$$

Note that we can equivalently write this condition as: $R[x] \subseteq [\![\varphi]\!]_{\mathcal{M}}$, where $[\![\varphi]\!]_{\mathcal{M}}$ is defined as in the case of SM-semantics. K-validity and K-consequence are defined analogously to SM-validity and SM-consequence.

It is not hard to see, that given any K-model $\mathcal{M} = \langle W, R, V\rangle$, we obtain an SM-model $\mathfrak{M} = \langle W, N_R, V\rangle$ by setting for all $x \in W$,

$$N_R(x) = \{X \in 2^W \mid R[x] \subseteq X\} \ ,$$

and the truth values of all formulas are preserved in all worlds.

Conversely, given any SM-model $\mathfrak{M} = \langle W, N, V\rangle$, we can define a K-model $\mathcal{M} = \langle W, R_N, V\rangle$ by setting for all $x, y \in W$,

$$R_N xy \qquad \text{iff} \qquad y \in X, \text{ for each } X \in N(x) \ .$$

Note that this entails that $R_N[x] = \bigcap N(x) = \bigcap_{X \in N(x)} X$. However, in order to preserve the truth of all formulas at each world, we need the original SM-model $\mathfrak{M}$ to satisfy the following two additional conditions for each $x \in W$:

- $N(x)$ contains its core, i.e. the set $(\bigcap_{X \in N(x)} X) \in N(x)$,
- $N(x)$ is closed under taking supersets, i.e. if $X \in N(x)$ and $X \subseteq Y$, then $Y \in N(x)$.

In this case, $\mathfrak{M}$ is called *augmented*. The following results about these transitions can be found for example in [8].

Theorem 1.

(a) Let $\mathcal{M} = \langle W, R, V\rangle$ be a K*-model. Then $R_{N_R} = R$, $\mathfrak{M} = \langle \widehat{W}, N_R, \widehat{V}\rangle$, defined by $\widehat{W} = W$ and $\widehat{V} = V$, is an augmented* SM*-model, and for all $\varphi \in Fm_{\Box}$ and all $x \in W$, $\widehat{V}(\varphi, x) = V(\varphi, x)$.*

(b) Let $\mathfrak{M} = \langle W, N, V\rangle$ *be an augmented* SM*-model. Then* $N_{R_N} = N$, $\mathcal{M} = \langle \widehat{W}, R_N, \widehat{V}\rangle$*, defined by* $\widehat{W} = W$ *and* $\widehat{V} = V$*, is a* K*-model, and for all* $\varphi \in Fm_\Box$ *and all* $x \in W$, $\widehat{V}(\varphi, x) = V(\varphi, x)$.

If we define a semantical consequence relation $\models_{\mathrm{ASM}}$ by considering only augmented instead of all SM-models, the following corollary is immediate.

Corollary 1. *For any subset* $\Gamma \subseteq Fm_\Box$ *and formula* $\varphi \in Fm_\Box$,

$$\Gamma \models_{\mathrm{ASM}} \varphi \qquad \textit{iff} \qquad \Gamma \models_{\mathrm{K}} \varphi\,.$$

Furthermore, let CPC denote any Hilbert-style axiomatization of classical propositional logic, and let (E) denote the following rule:

$$\frac{\varphi \leftrightarrow \psi}{\Box\varphi \leftrightarrow \Box\psi}$$

Given $\Gamma \cup \{\varphi\} \subseteq Fm_\Box$, we denote the fact that there is a proof of φ from Γ in SM = CPC + (E) by $\Gamma \vdash_{\mathrm{SM}} \varphi$. We then have the following completeness result:

Theorem 2. *Let* $\Gamma \cup \{\varphi\} \subseteq Fm_\Box$*, then*

$$\Gamma \vdash_{\mathrm{SM}} \varphi \qquad \textit{iff} \qquad \Gamma \models_{\mathrm{SM}} \varphi\,.$$

2.2 MTL Logic and MTL-Algebras

An MTL-*algebra* (introduced in [9]) is a prelinear commutative bounded integral residuated lattice. That is, using the algebraic language $\mathcal{L} = \mathcal{L}_\Box \setminus \{\Box\}$, the algebraic structure

$$\boldsymbol{A} = \langle A, \wedge, \vee, \&, \rightarrow, \overline{0}, \overline{1}\rangle$$

is an MTL-*algebra* if the following conditions are satisfied:

- $\langle A, \wedge, \vee, \overline{0}, \overline{1}\rangle$ is a bounded lattice,
- $\langle A, \&, \overline{1}\rangle$ is a commutative monoid,
- & and $\rightarrow$ form a residuated pair, i.e. $a \mathbin{\&} b \leq c$ iff $a \leq b \rightarrow c$, for all $a, b, c \in A$,
- $(a \rightarrow b) \vee (b \rightarrow a) = \overline{1}$ is satisfied for all $a, b \in A$.

The class of all MTL-algebras forms a variety, i.e. an equational class of algebras. An MTL-algebra $\boldsymbol{A}$ is called an MTL-*chain* if it is linearly ordered and we will call it *complete*, if $\bigvee B$ and $\bigwedge B$ exist in A, for any subset $B \subseteq A$. The set of formulas built from propositional variables in *Var* and connectives in $\mathcal{L}$ is denoted by *Fm*. Fixing an MTL-algebra $\boldsymbol{A}$, we define an $\boldsymbol{A}$-*evaluation* as a function $e\colon Var \rightarrow A$ that extends to $e\colon Fm \rightarrow A$ by interpreting the connectives in $\mathcal{L}$ by the corresponding operations of $\boldsymbol{A}$, i.e. a homomorphism from the algebra of formulas to $\boldsymbol{A}$. For a subset $\Gamma \cup \{\varphi\} \subseteq Fm$, we say that φ is an $\boldsymbol{A}$-*consequence* of Γ, written $\Gamma \models_{\boldsymbol{A}} \varphi$, if $e(\varphi) = \overline{1}$ for all $\boldsymbol{A}$-evaluations e such that $e[\Gamma] = \{e(\psi) \mid \psi \in \Gamma\} \subseteq \{\overline{1}\}$. Furthermore, we say that φ is an MTL-*consequence* of Γ, written $\Gamma \models_{\mathrm{MTL}} \varphi$, if $\Gamma \models_{\boldsymbol{A}} \varphi$, for all MTL-algebras $\boldsymbol{A}$.

MTL is the logic given by the consequence relation $\models_{\mathrm{MTL}}$. It was axiomatized (see [9,11]) by a Hilbert-style calculus with *modus ponens* rule (MP) as the only inference rule.

MTL is not only sound and complete with respect to the variety of all MTL-algebra, but also with respect to the class of all standard MTL-algebras, while an MTL-algebra $\langle A, \wedge, \vee, \&, \rightarrow, \overline{0}, \overline{1}\rangle$ is called *standard*, if $A = [0,1]$, the real unit interval, $\wedge$ and $\vee$ are the minimum and maximum, respectively, and $\&$ is a left-continuous t-norm (i.e. an associative, commutative binary operation with 1 as neutral element).

The logic $\mathrm{MTL}_\triangle$ is the expansion of MTL with an extra unary operator $\triangle$, axiomatized in [9]. The additional operation $\triangle$ is interpreted on any chain as:

$$\triangle(a) \quad = \quad \begin{cases} \overline{1} & \text{if } a = \overline{1}\,, \\ \overline{0} & \text{otherwise}\,. \end{cases}$$

By $\boldsymbol{A}_\triangle$ we denote the $\mathrm{MTL}_\triangle$-chain resulting from adding $\triangle$ to the MTL-chain $\boldsymbol{A}$ and say that $\boldsymbol{A}_\triangle$ is the $\triangle$-*expansion* of $\boldsymbol{A}$.

2.3 Fuzzy Sets and Their Notation

In order to formulate neighborhood semantics over MTL-algebras, rather than the two-element Boolean algebra, we need to talk about fuzzy subsets of worlds and fuzzy sets of fuzzy subsets. To do this efficiently, we introduce a convenient notation inspired by the syntax of fuzzy class theory (FCT), see e.g. in [1].

Given a complete $\mathrm{MTL}_\triangle$-chain $\boldsymbol{A}_\triangle = \langle A, \wedge, \vee, \&, \rightarrow, \triangle, \overline{0}, \overline{1}\rangle$, and a (classical) set of worlds W, a *fuzzy subset* X of W is a function $X : W \rightarrow A$. Intuitively, a world $x \in W$ is a member of X to the degree $X(x) \in A$ and thus we also write '$x \in X$' to denote the value $X(x)$ in A. A fuzzy set $\mathcal{X}$ of fuzzy subsets of W is a function $\mathcal{X}\colon A^W \rightarrow A$ and we also write '$X \in \mathcal{X}$' for the value $\mathcal{X}(X)$ in A. We usually use lower case letters $x, y, z, \ldots$ to denote members of W, upper case letters $X, Y, Z, \ldots$ to denote members of A^W and upper case calligraphic letters $\mathcal{X}, \mathcal{Y}, \mathcal{Z}, \ldots$ to denote members of A^{A^W}.

Obviously, as e.g. $x \in X$ and $X \in \mathcal{X}$ represent values in A, we can use the operation symbols in $\mathcal{L} \cup \{\triangle\}$, representing operations of $\boldsymbol{A}_\triangle$, to form what we call *meta-formulas*, e.g. $(x \in X) \rightarrow (X \in \mathcal{X})$, which themselves represent values in A. Furthermore, we use quantifier symbols $\forall$ and $\exists$ to represent infima and suprema over $\boldsymbol{A}_\triangle$; e.g., the meta-formula $(\forall X)(\exists Y)(\forall x)((x \in X) \leftrightarrow (x \in Y))$ represents the following value in A:

$$\bigwedge_{X \in A^W} \bigvee_{Y \in A^W} \bigwedge_{x \in W} ((x \in X) \leftrightarrow (x \in Y))\,.$$

Given a meta-formula $\varphi(x)$ (e.g. $\varphi(x) = (\exists X)(x \in X)$), we define a fuzzy set $X = \{x \in W \mid \varphi(x)\}$ to which each element y belongs exactly to the same degree as is the value of $\varphi(y)$, i.e.

$$\varphi(y) \quad = \quad (y \in \{x \in W \mid \varphi(x)\})\,.$$

Using the same idea, we introduce fuzzy sets of fuzzy subsets of W by comprehension terms $\{X \in A^W \mid \varphi(X)\}$, e.g. for $\varphi(X) = (\forall x)(x \in X)$. Finally, we write '$X \subseteq Y$' to denote the value in A represented by $(\forall x)((x \in X) \to (x \in Y))$.

3 Neighborhood and Kripke Semantics for Modal MTL

Let us fix an MTL-chain $\boldsymbol{A}$. We define an $\boldsymbol{A}$*-neighborhood frame* (short: SM($\boldsymbol{A}$)-frame) to be a pair $\langle W, N\rangle$ such that W is a non-empty (classical) set of worlds while N is a function $N\colon W \to A^{A^W}$ that assigns to each world $x \in W$ a fuzzy set of fuzzy subsets of W, called the $\boldsymbol{A}$*-neighborhood* of $x \in W$.

We define an $\boldsymbol{A}$*-neighborhood model* (short: SM($\boldsymbol{A}$)-model) to be a triple $\langle W, N, V\rangle$, where $\langle W, N\rangle$ is an SM($\boldsymbol{A}$)-frame and V is an evaluation $V\colon Var \times W \to A$ that is extended to formulas $\varphi \in Fm_\Box$ inductively as follows: the non-modal connectives are interpreted locally at each world as the corresponding operations of $\boldsymbol{A}$ while for a box-formula,

$$V(\Box\varphi, x) \quad = \quad (\llbracket\varphi\rrbracket_{\mathfrak{M}} \in N(x))\ ,$$

where for any formula $\varphi \in Fm_\Box$, $\llbracket\varphi\rrbracket_{\mathfrak{M}}$ denotes the fuzzy subset of W to which y belongs to the degree $V(\varphi, y)$, i.e., the fuzzy subset $\{y \in W \mid V(\varphi, y)\}$.

Furthermore, if $\boldsymbol{A}$ is a complete MTL-chain, we define an $\boldsymbol{A}$*-Kripke frame* (short: K($\boldsymbol{A}$)-frame) to be a pair $\langle W, R\rangle$ such that W is a non-empty (classical) set of worlds while R is a function $R\colon W \times W \to A$. For any $x \in W$ we define $R[x] = \{y \in W \mid Rxy\}$, i.e., the fuzzy subset of W to which y belongs to the degree Rxy.

An $\boldsymbol{A}$*-Kripke model* (short: K($\boldsymbol{A}$)-model) is a triple $\mathcal{M} = \langle W, R, V\rangle$, where $\langle W, R\rangle$ is a K($\boldsymbol{A}$)-frame and V is an evaluation $V\colon Var \times W \to A$ that extends to formulas $\varphi \in Fm_\Box$ inductively as follows: the non-modal connectives are interpreted locally at each world as the corresponding operations of $\boldsymbol{A}$ while for a box-formula,

$$V(\Box\varphi, x) \;=\; \bigwedge\{Rxy \to V(\varphi, y) \mid y \in W\} \;=\; (R[x] \subseteq \llbracket\varphi\rrbracket_{\mathcal{M}}).$$

Given an SM($\boldsymbol{A}$)-model $\mathfrak{M} = \langle W, N, V\rangle$, a formula $\varphi \in Fm_\Box$ is *valid in* $\mathfrak{M}$, if $V(\varphi, x) = \overline{1}$ for all $x \in W$, written $\mathfrak{M} \models_{\mathrm{SM}(\boldsymbol{A})} \varphi$. For a subset $\Gamma \cup \{\varphi\} \subseteq Fm_\Box$, we say that φ is an SM($\boldsymbol{A}$)*-consequence* of Γ, written $\Gamma \models_{\mathrm{SM}(\boldsymbol{A})} \varphi$, if for all SM($\boldsymbol{A}$)-models $\mathfrak{M}$ such that $\mathfrak{M} \models_{\mathrm{SM}(\boldsymbol{A})} \Gamma$, also $\mathfrak{M} \models_{\mathrm{SM}(\boldsymbol{A})} \varphi$. The notion of validity in a K($\boldsymbol{A}$)-model and K($\boldsymbol{A}$)*-consequence* ($\Gamma \models_{\mathrm{K}(\boldsymbol{A})} \varphi$) are defined analogously.

An SM($\boldsymbol{A}$)-frame $\langle W, N\rangle$ is $\boldsymbol{A}$*-augmented* if the following meta-formulas is valid (i.e., the meta-formula represents $\overline{1}$ in $\boldsymbol{A}_\triangle$, the $\triangle$-expansion $\boldsymbol{A}$) for each $x \in W$:

$$(\exists X)\triangle(\forall Y)(X \subseteq Y \leftrightarrow Y \in N(x))\ .$$

Note that validity of this formula, for a given x, means that there is a fuzzy subset C_x of W such that $(C_x \subseteq Y) = (Y \in N(x))$. Note that this implies that $(C_x \in N(x)) = \overline{1}$ (because for each fuzzy subset X we have $X \subseteq X = \overline{1}$ and

$\overline{1} \to a = a$ is satisfied in any MTL-algebra) and if there would be two such fuzzy sets C_x and C'_x, we would have $C_x = C'_x$ (just consider $X = C_x$ and $Y = C'_x$, and vice-versa). This fuzzy subset, for a given x, is called the *core of* $N(x)$. Clearly, an SM($\boldsymbol{A}$)-frame $\langle W, N\rangle$ is $\boldsymbol{A}$-augmented iff for each element x, $N(x)$ has a core.

4 Relating Neighborhood and Kripke Semantics

For this section, let $\boldsymbol{A}$ be a complete MTL-chain. We show that also in the fuzzy setting, there is, analogously to the classical case, a close relationship between fuzzy neighborhood semantics and fuzzy Kripke semantics. While the (fuzzy) neighborhood function N allows more freedom, it becomes equivalent to the more restricted (fuzzy) binary relation R when it is required to be ($\boldsymbol{A}$-)augmented.

Similarly to the classical case, given a K($\boldsymbol{A}$)-frame $\langle W, R\rangle$, we define an SM($\boldsymbol{A}$)-frame $\langle W, N_R\rangle$ as follows. For all $x \in W$, let

$$N_R(x) \quad = \quad \{X \in A^W \mid (\forall y)(Rxy \to y \in X)\}\,.$$

Note that $(\forall y)(Rxy \to y \in X) = (R[x] \subseteq X)$. On the other hand, given an SM($\boldsymbol{A}$)-frame $\langle W, N\rangle$, we define a K($\boldsymbol{A}$)-frame $\langle W, R_N\rangle$ as follows:

$$R_N[x] \quad = \quad \{y \in W \mid (\forall X)(X \in N(x) \to y \in X)\}\,.$$

Similarly to the classical case, our goal is to prove that for a K($\boldsymbol{A}$)-frame $\langle W, R\rangle$, we have $R_{N_R} = R$, and if an SM($\boldsymbol{A}$)-frame $\langle W, N\rangle$ is $\boldsymbol{A}$-augmented, then $N_{R_N} = N$. These proofs follow the same ideas as in the classical case (see e.g. [8]), but obviously an adaptation to deal with fuzzy sets of fuzzy subsets of W is needed.

Lemma 1. *Let $\langle W, N\rangle$ be an $\boldsymbol{A}$-augmented* SM($\boldsymbol{A}$)*-frame, $x \in W$, and let C_x be the core of $N(x)$. Then $C_x = R_N[x]$.*

Proof. We prove that $C_x = R_N[x]$ by showing that for all $y \in W$, $(y \in R_N[x]) \leq (y \in C_x)$ and $(y \in C_x) \leq (y \in R_N[x])$. First note that because C_x is the core of $N(x)$, it is the case that $(C_x \in N(x)) = \overline{1}$. Fixing a $y \in W$, it follows that

$$\begin{aligned}(y \in R_N[x]) &= (\forall Y)(Y \in N(x) \to y \in Y)\\ &\leq (C_x \in N(x) \to y \in C_x)\\ &= (y \in C_x)\end{aligned}$$

as '$(\forall Y)$' denotes $\bigwedge$ and its instantiation by C_x is greater. The last equality is justified by the fact that the equation $\overline{1} \to a = a$ is satisfied in any MTL-algebra.

For the other inequality, note first that for all $y \in W$ and all $Y \in A^W$,

$$(C_x \subseteq Y) \;=\; (\forall z)(z \in C_x \to z \in Y) \;\leq\; (y \in C_x \to y \in Y)\,.$$

By residuation and commutativity of the & operation, it follows that

$$(y \in C_x \;\&\; C_x \subseteq Y) \;=\; (C_x \subseteq Y \;\&\; y \in C_x) \;\leq\; (y \in Y)\,,$$

for all $y \in W$, and thus by residuation again

$$(y \in C_x) \ \leq \ (C_x \subseteq Y \to y \in Y) .$$

From this, the fact that C_x is the core of $N(x)$, and by the definition of R_N, we can complete the proof by the following chain of (in)equalities

$$\begin{aligned}(y \in C_x) &= (\forall Y)(y \in C_x) \\ &\leq (\forall Y)(C_x \subseteq Y \to y \in Y) \\ &= (\forall Y)(Y \in N(x) \to y \in Y) = (y \in R_N[x]) .\end{aligned}$$

□

Lemma 2. *If $\langle W, R\rangle$ is a* K($\boldsymbol{A}$)*-frame, then the* SM($\boldsymbol{A}$)*-frame $\langle W, N_R\rangle$ is $\boldsymbol{A}$-augmented.*

Proof. We prove that for each world $x \in W$, $R[x]$ is the core of $N_R(x)$ and so $\langle W, N_R\rangle$ is $\boldsymbol{A}$-augmented. For this, we recall the observation after the definition of N_R and note that for all $Y \in A^W$, we have

$$(R[x] \subseteq Y) \ = \ (Y \in N_R(x)) .$$

Thus, we obtain that the meta-formula

$$\triangle(\forall Y)(X \subseteq Y \leftrightarrow Y \in N_R(x))$$

is satisfied for $X = R[x]$. □

Theorem 3. *Let $\langle W, N\rangle$ be an* SM($\boldsymbol{A}$)*-frame. Then $\langle W, N\rangle$ is $\boldsymbol{A}$-augmented iff $N_{R_N} = N$.*

Proof. For the direction from left to right, let $\langle W, N\rangle$ be an $\boldsymbol{A}$-augmented $\boldsymbol{A}$-neighborhood frame. Then notice for all $x \in W$:

$$\begin{aligned}N_{R_N}(x) &= \{Y \in A^W \mid (\forall y)(R_N xy \to y \in Y)\} && (1)\\ &= \{Y \in A^W \mid R_N[x] \subseteq Y\} && (2)\\ &= \{Y \in A^W \mid C_x \subseteq Y\} && (3)\\ &= \{Y \in A^W \mid Y \in N(x)\} && (4)\\ &= N(x) . && (5)\end{aligned}$$

While the first two and the last equalities are just notational facts, step (2) to (3) is justified by Lemma 1, and we get from (3) to (4) because $\langle W, N\rangle$ is $\boldsymbol{A}$-augmented.

The right to left direction is an easy consequence of Lemma 2. □

Theorem 4. *If $\langle W, R\rangle$ is a* K($\boldsymbol{A}$)*-frame, then $R_{N_R} = R$.*

Proof. Let $\langle W, R\rangle$ be a $\mathrm{K}(\boldsymbol{A})$-frame and fix an $x \in W$, then

$$\begin{aligned}
R_{N_R}[x] &= \{y \in W \mid (\forall Y)(Y \in N_R(x) \to y \in Y)\} \\
&= \{y \in W \mid (\forall Y)(Y \in \{Z \in A^W \mid (\forall y)(Rxy \to y \in Z)\} \to y \in Y)\} \\
&= \{y \in W \mid (\forall Y)(Y \in \{Z \in A^W \mid R[x] \subseteq Z\} \to y \in Y)\} \\
&= \{y \in W \mid (\forall Y)(R[x] \subseteq Y \to y \in Y)\}\,.
\end{aligned}$$

It then remains to be shown that for all $y \in W$, $(\forall Y)(R[x] \subseteq Y \to y \in Y) = (y \in R[x])$. For this, note first that for all $y \in W$ and all $Y \in A^W$,

$$(R[x] \subseteq Y) \;=\; (\forall z)(z \in R[x] \to z \in Y) \;\leq\; (y \in R[x] \to y \in Y)\,.$$

By residuation and commutativity of the & operation, we obtain $(y \in R[x]) \leq (R[x] \subseteq Y \to y \in Y)$ for all $y \in W$ and $Y \in A^W$, and thus also for all $y \in W$,

$$(y \in R[x]) \;\leq\; (\forall Y)(R[x] \subseteq Y \to y \in Y)\,.$$

On the other hand, by instantiation,

$$(\forall Y)(R[x] \subseteq Y \to y \in Y) \;\leq\; (R[x] \subseteq R[x] \to y \in R[x]) \;=\; (y \in R[x])\,,$$

and thus $R_{N_R}[x] = \{y \in W \mid (\forall Y)(R[x] \subseteq Y \to y \in Y)\} = R[x]$. □

Having established a tight connection between $\boldsymbol{A}$-neighborhood and $\boldsymbol{A}$-Kripke frames, the extension of this connection to the level of models does not come as a surprise.

Theorem 5.

(a) Given a $\mathrm{K}(\boldsymbol{A})$*-model* $\mathcal{M} = \langle W, R, V\rangle$*, define the* $\mathrm{SM}(\boldsymbol{A})$*-model* $\mathfrak{M} = \langle \widehat{W}, \widehat{N}, \widehat{V}\rangle$ *with* $\widehat{W} = W$*,* $\widehat{N} = N_R$*, and* $\widehat{V} = V$*. Then for all* $\varphi \in Fm_\Box$ *and all* $x \in W$*:*

$$\widehat{V}(\varphi, x) = V(\varphi, x)\,.$$

(b) Given an $\boldsymbol{A}$*-augmented* $\mathrm{SM}(\boldsymbol{A})$*-model* $\mathfrak{M} = \langle W, N, V\rangle$*, define the* $\mathrm{K}(\boldsymbol{A})$*-model* $\mathcal{M} = \langle \widehat{W}, \widehat{R}, \widehat{V}\rangle$ *with* $\widehat{W} = W$*,* $\widehat{R} = R_N$*, and* $\widehat{V} = V$*. Then for all* $\varphi \in Fm_\Box$ *and all* $x \in W$*:*

$$\widehat{V}(\varphi, x) = V(\varphi, x)\,.$$

Proof. We proceed by induction over the complexity of a formula $\varphi \in Fm_\Box$. For (a) and (b), the case where $\varphi \in Var$ or φ is a constant follows by the definition of $\widehat{V}$ while the case where φ is not a box-formula follows trivially from the induction hypothesis (as only box-formulas depend on R or N). Let $\varphi = \Box\psi$ for some $\psi \in Fm_\Box$.

For (a), note that by the induction hypothesis, for any $x \in \widehat{W} = W$,

$$
\begin{aligned}
\widehat{V}(\Box\psi, x) &= (\llbracket\psi\rrbracket_{\mathfrak{M}} \in N_R(x)) \\
&= (\llbracket\psi\rrbracket_{\mathfrak{M}} \in \{Y \in A^W \mid (\forall y)(Rxy \to y \in Y)\}) \\
&= (\forall y)(Rxy \to y \in \llbracket\psi\rrbracket_{\mathfrak{M}}) \\
&= \bigwedge\{Rxy \to \widehat{V}(\psi, y) \mid y \in \widehat{W}\} \\
&= \bigwedge\{Rxy \to V(\psi, y) \mid y \in W\} \\
&= V(\Box\psi, x)\ .
\end{aligned}
$$

For (b), on the other hand, by the fact that $\mathfrak{M}$ is $\boldsymbol{A}$-augmented and thus, for any $x \in \widehat{W} = W$, $R_N[x]$ is the core of $N(x)$ by Lemma 1, we can use the induction hypothesis to conclude the proof by the following chain of equalities:

$$
\begin{aligned}
V(\Box\psi, x) &= (\llbracket\psi\rrbracket_{\mathfrak{M}} \in N(x)) \\
&= (R_N[x] \subseteq \llbracket\psi\rrbracket_{\mathfrak{M}}) \\
&= (\forall y)(y \in R_N[x] \to y \in \llbracket\psi\rrbracket_{\mathfrak{M}}) \\
&= \bigwedge\{R_N xy \to V(\psi, y) \mid y \in W\} \\
&= \bigwedge\{R_N xy \to \widehat{V}(\psi, y) \mid y \in \widehat{W}\} \\
&= \widehat{V}(\Box\psi, x)\ .
\end{aligned}
$$

□

Corollary 2. *For all subsets $\Gamma \cup \{\varphi\} \subseteq Fm_\Box$,*[1]

$\Gamma \models_{\mathrm{K}(\boldsymbol{A})} \varphi$ *iff* $\mathfrak{M} \models_{\mathrm{SM}(\boldsymbol{A})} \varphi$ *for all* $\boldsymbol{A}$*-augmented* $\mathrm{SM}(\boldsymbol{A})$*-models* $\mathfrak{M}$ *such that* $\mathfrak{M} \models_{\mathrm{SM}(\boldsymbol{A})} \Gamma$.

Proof. For the contraposition of the right-to-left direction, let us assume $\Gamma \not\models_{\mathrm{K}(\boldsymbol{A})} \varphi$, that is, there is a $\mathrm{K}(\boldsymbol{A})$-model $\mathcal{M} = \langle W, R, V\rangle$ such that $V[\Gamma, W] = \{V(\psi, x) \mid \psi \in \Gamma, x \in W\} \subseteq \{\overline{1}\}$ and $V(\varphi, y) < \overline{1}$, for some world $y \in W$. Define the $\mathrm{SM}(\boldsymbol{A})$-model $\mathfrak{M} = \langle \widehat{W}, \widehat{N}, \widehat{V}\rangle$ by $\widehat{W} = W, \widehat{N} = N_R$, and $\widehat{V} = V$ and notice that by Lemma 2, $\mathfrak{M}$ is $\boldsymbol{A}$-augmented, and that for all $x \in W$ and all $\psi \in Fm_\Box$, $\widehat{V}(\psi, x) = V(\psi, x)$, by Theorem 5(a). It therefore follows that $\widehat{V}[\Gamma, \widehat{W}] = V[\Gamma, W] \subseteq \{\overline{1}\}$ and $\widehat{V}(\varphi, y) = V(\varphi, y) < \overline{1}$ and thus the right-hand side of the claim is false.

For the contraposition of the left-to-right direction, let us assume that there is an $\boldsymbol{A}$-augmented $\mathrm{SM}(\boldsymbol{A})$-model $\mathfrak{M} = \langle W, N, V\rangle$, such that $V[\Gamma, W] \subseteq \{\overline{1}\}$ and $V(\varphi, y) < \overline{1}$, for some world $y \in W$. Define the $\mathrm{K}(\boldsymbol{A})$-model $\mathcal{M} = \langle \widehat{W}, \widehat{R}, \widehat{V}\rangle$ by $\widehat{W} = W, \widehat{R} = R_N$, and $\widehat{V} = V$ and notice that for all $x \in W$ and all $\psi \in Fm_\Box$, $\widehat{V}(\psi, x) = V(\psi, x)$, by Theorem 5(b). It therefore follows that $\widehat{V}[\Gamma, \widehat{W}] = V[\Gamma, W] \subseteq \{\overline{1}\}$ and $\widehat{V}(\varphi, y) = V(\varphi, y) < \overline{1}$ and thus $\Gamma \not\models_{\mathrm{K}(\boldsymbol{A})} \varphi$. □

[1] It is easy to check that this result also holds for the local consequence.

5 An Axiomatization of SM($\boldsymbol{A}$)

A logic L is an axiomatic extension of MTL if it is obtained by adding new axiom schemes. It is well known that MTL, and hence all its axiomatic extensions, is algebraizable in the sense of Blok and Pigozzi (see [2,7]). Given an axiomatic extension L, we call L-*algebras* the members of its equivalent algebraic semantics, which form a subvariety of MTL-algebras. It is also well known that all axiomatic extensions of MTL are *semilinear*, that is, each member of the equivalent algebraic semantics is representable as a subdirect product of chains. Prominent axiomatic extensions of MTL include Gödel-Dummett logic G, Łukasiewicz logic Ł, Product logic Π, and Hájek's basic fuzzy logic HL (see [7,12]).

We say that an axiomatic extension L of MTL is *strongly complete* with respect to an L-chain $\boldsymbol{C}$ if for every $\Gamma \cup \{\varphi\} \subseteq Fm$ we have: $\Gamma \vdash_{\mathrm{L}} \varphi$ iff $\Gamma \models_{\boldsymbol{C}} \varphi$. We say that L is *finitely strongly complete* with respect to $\boldsymbol{C}$ if the same property holds for each *finite* set $\Gamma \cup \{\varphi\} \subseteq Fm$. The four prominent axiomatic extensions of MTL mentioned above are finitely strongly complete (G is even strongly complete) with respect to the respective 'standard' chains, i.e., algebras defined over the real unit interval $[0, 1]$ ordered in the usual way (see [7,12]).

For each axiomatic extension L of MTL, we define the modal logic LSM in the language $\mathcal{L}_\Box$ as the logic axiomatized by any axiomatic system of L and the additional rule (E) (see Sect. 2.1). Clearly, LSM remains an algebraizable logic, since (E) ensures the congruence law for the new connective $\Box$, hence it has an equivalent algebraic semantics. If we assume that L is (finitely) strongly complete with respect to an L-chain $\boldsymbol{C}$ we are able to prove (finite) strong completeness of LSM with respect to a semantics of SM($\boldsymbol{C}$)-models.

Theorem 6. *Let* L *be an axiomatic extension of* MTL *and let* $\boldsymbol{C}$ *be an* L-*chain. If* L *is (finitely) strongly complete with respect to* $\boldsymbol{C}$, *then for each (finite)* $\Gamma \cup \{\varphi\} \subseteq Fm_\Box$ *we have:*

$$\Gamma \vdash_{\mathrm{LSM}} \varphi \qquad \textit{iff} \qquad \Gamma \models_{\mathrm{SM}(\boldsymbol{C})} \varphi\,.$$

Proof. For the left-to-right directions, we only need to check the soundness of the rule (E). Let us assume that for some SM($\boldsymbol{C}$)-model $\mathfrak{M} = \langle W, N, V\rangle$ and some formulas $\psi, \chi \in Fm_\Box$, $\mathfrak{M} \models_{\mathrm{SM}(\boldsymbol{C})} \psi \leftrightarrow \chi$, then

$$\begin{aligned}
\mathfrak{M} \models_{\mathrm{SM}(\boldsymbol{C})} \psi \leftrightarrow \chi \quad &\Rightarrow \quad V(\psi, x) = V(\chi, x), \text{ for all } x \in W,\\
&\Rightarrow \quad [\![\psi]\!]_{\mathfrak{M}} = [\![\chi]\!]_{\mathfrak{M}}\\
&\Rightarrow \quad ([\![\psi]\!]_{\mathfrak{M}} \in N(x)) = ([\![\chi]\!]_{\mathfrak{M}} \in N(x)), \text{ for all } x \in W,\\
&\Rightarrow \quad V(\Box\psi, x) = V(\Box\chi, x), \text{ for all } x \in W,\\
&\Rightarrow \quad \mathfrak{M} \models_{\mathrm{SM}(\boldsymbol{C})} \Box\psi \leftrightarrow \Box\chi.
\end{aligned}$$

For the reverse implication in the finite strong completeness case, assume that $\Gamma \nvdash_{\mathrm{LSM}} \varphi$ for a finite set $\Gamma \cup \{\varphi\} \subseteq Fm_\Box$. By the algebraizability of LSM, there is an LSM-algebra $\boldsymbol{A}$ and an evaluation $e\colon Fm_\Box \to A$ such that $e[\Gamma] \subseteq \{\overline{1}^{\boldsymbol{A}}\}$ and $e(\varphi) \neq \overline{1}^{\boldsymbol{A}}$. The $\mathcal{L}$-reduct of $\boldsymbol{A}$ is an L-algebra, and because L is a semilinear logic, it is representable as a subdirect product of L-chains $\{\boldsymbol{A}_i \mid i \in I\}$ (see e.g. [6, Proposition 2.14]). Let X be the finite set of the subformulas of $\Gamma \cup \{\varphi\}$. By the completeness assumption for L, we have that each $\boldsymbol{A}_i$ is partially embeddable into $\boldsymbol{C}$ [6, Theorem 3.8]. For each $i \in I$, $\pi_i[e[X]]$ is a finite subset of A_i; let g_i be a corresponding partial embedding into $\boldsymbol{C}$. For each $i \in I$, we take an arbitrary $\boldsymbol{C}$-evaluation e_i such that $e_i(\psi) = (g_i \circ \pi_i \circ e)(\psi)$ for each $\psi \in X$.

Now we have all the ingredients to build an SM($\boldsymbol{C}$)-model $\mathfrak{M} = \langle W, N, V\rangle$. Let $W = I$ and define for all $p \in \mathit{Var}$ and all $j \in W$, $V(p, j) = e_j(p)$ and

$$(\langle a_i\rangle_{i\in W} \in N(j)) = \begin{cases} e_j(\Box\psi) & \text{if there is } \psi \in X \text{ s.t. } \langle a_i\rangle_{i\in W} = \langle e_i(\psi)\rangle_{i\in W} \\ \overline{0}^{\boldsymbol{C}} & \text{otherwise.} \end{cases}$$

Then for each formula $\psi \in X$ and each world $j \in W$, we have $V(\psi, j) = e_j(\psi)$ (by an easy induction where the step for $\Box$ follows from the following chain of equalities: $V(\Box\psi, j) = ([\![\psi]\!]_{\mathfrak{M}} \in N(j)) = (\langle e_i(\psi)\rangle_{i\in W} \in N(j)) = e_j(\Box\psi)$). Therefore $\mathfrak{M}$ is a model of Γ (because for each $\psi \in \Gamma$ we have $e(\psi) = 1^{\boldsymbol{A}}$ and so for each $j \in W$: $V(\psi, j) = e_j(\psi) = (g_i \circ \pi_i \circ e)(\psi) = \overline{1}^{\boldsymbol{C}}$) but it is not a model of φ (because $e(\varphi) \neq \overline{1}^{\boldsymbol{A}}$, there has to be a $j \in W$ such that $(\pi_j \circ e)(\varphi) \neq \overline{1}^{\boldsymbol{A}_j}$ and so $V(\varphi, j) = e_j(\varphi) = (g_j \circ \pi_j \circ e)(\varphi) \neq \overline{1}^{\boldsymbol{C}}$).

The proof of the right-to-left direction for the strong completeness case is very similar; in this case we even know that the projections of the countable subalgebras generated by the values of the subformulas of $\Gamma \cup \{\varphi\}$ by the evaluation e are totally embeddable into $\boldsymbol{C}$ (by [6, Theorem 3.5]). $\Box$

Acknowledgments. The work of the P. Cintula and C. Noguera is supported by the joint project of Austrian Science Fund (FWF) I1897-N25 and Czech Science Foundation (GACR) GF15-34650L. P. Cintula also acknowledges institutional support RVO: 67985807. Furthermore, J. Rogger is supported by the Swiss National Science Foundation grant 200021_146748. The authors would also like to thank the anonymous referees for their helpful comments and remarks.

References

1. Běhounek, L., Cintula, P.: Fuzzy class theory. Fuzzy Sets Syst. **154**(1), 34–55 (2005)
2. Blok, W.J., Pigozzi, D.L.: Algebraizable Logics, vol. 396. Memoirs of the American Mathematical Society, Providence (1989)
3. Bou, F., Esteva, F., Godo, L., Rodríguez, R.O.: On the minimum many-valued modal logic over a finite residuated lattice. J. Logic Comput. **21**(5), 739–790 (2011)
4. Caicedo, X., Metcalfe, G., Rodríguez, R.O., Rogger, J.: Decidability of order-based modal logics. J. Comput. Syst. Sci. (to appear)
5. Caicedo, X., Rodríguez, R.O.: Standard Gödel Modal Logics. Studia Logica **94**(2), 189–214 (2010)
6. Cintula, P., Esteva, F., Gispert, J., Godo, L., Montagna, F., Noguera, C.: Distinguished algebraic semantics for t-norm based fuzzy logics: methods and algebraic equivalencies. Ann. Pure Appl. Logic **160**(1), 53–81 (2009)

7. Cintula, P., Hájek, P., Noguera, C. (eds.): Handbook of Mathematical Fuzzy Logic. Studies in Logic, Mathematical Logic and Foundations, vol. 37, 38. College Publications, London (2011)
8. Chellas, B.F.: Modal Logic: An Introduction. Cambridge University Press, Cambridge (1980)
9. Esteva, F., Godo, L.: Monoidal t-norm based logic: towards a logic for left- continuous t-norms. Fuzzy Sets Syst. **124**(3), 271–288 (2001)
10. Fitting, M.: Many-valued modal logics. Fundamenta Informaticae **15**, 235–254 (1992)
11. Jenei, S., Montagna, F.: A proof of standard completeness for Esteva and Godo's logic MTL. Studia Logica **70**(2), 183–192 (2002)
12. Hájek, P.: Metamathematics of Fuzzy Logic, Trends in Logic, vol. 4. Kluwer, Dordrecht (1998)
13. Kroupa, T., Teheux, B.: Modal extension of Łukasiewicz logic for modelling coalitional power. J. Logic Comput. (2015)
14. Marti, M., Metcalfe, G.: A Hennessy-Milner Property for Many-Valued Modal Logics. Adv. Modal Logic **10**, 407–420 (2014)
15. Montague, R.: Universal grammar. Theoria **36**(3), 373–398 (1970)
16. Rodríguez, R.O., Godo, L.: Modal uncertainty logics with fuzzy neighborhood semantics. In: IJCAI-13 Workshop on Weighted Logics for Artificial Intelligence (WL4AI-2013), pp. 79–86 (2013)
17. Rodríguez, R.O., Godo, L.: On the fuzzy modal logics of belief KD45(A) and Prob(Ln): axiomatization and neighbourhood semantics. In: IJCAI-15 Workshop on Weighted Logics for Artificial Intelligence (WL4AI-2015), pp. 64–71 (2015)
18. Scott, D.: Advice on modal logic. In: Lambert, K. (ed.) Philosophical Problems in Logic. Synthese Library, vol. 29, pp. 14–173. Springer, Netherlands (1970)
19. Vidal, A.: On modal expansions of t-norm based logics with rational constants, PhD dissertation, University of Barcelona (2015)

Łukasiewicz Public Announcement Logic

Leonardo Cabrer[1], Umberto Rivieccio[2(✉)], and Ricardo Oscar Rodriguez[3]

[1] Institute of Computer Languages, Technische Universität Wien, Wien, Austria
leonardo.cabrer@logic.at

[2] Department of Informatics and Applied Mathematics, UFRN, Natal, Brazil
urivieccio@dimap.ufrn.br

[3] Departamento de Computación, FCEyN-UBA, Buenos Aires, Argentina
ricardo@dc.uba.ar

Abstract. In this work we lay a theoretical framework for developing dynamic epistemic logics in a many-valued setting. We consider in particular the logic of Public Announcements, which is one of the simplest and best-known dynamic epistemic systems in the literature. We show how to develop a Public Announcement Logic based on finite-valued Łukasiewicz modal logic. We define our logic through a relational semantics based on many-valued Kripke models, and also introduce an alternative but equivalent algebra-based semantics using MV-algebras endowed with modal operators. We provide a Hilbert-style calculus for our logic and prove completeness with respect to both semantics.

Keywords: Łukasiewicz modal logic · MV-algebras · Public Announcements Logic · Epistemic logics

1 Introduction

Dynamic epistemic logics (DEL) are formal systems designed to model the change brought about by *epistemic actions*, that is, actions that affect the cognitive state of (a group of) reasoning agents, rather than the facts of the world themselves. A prominent and simple example of an epistemic action is the *public announcement* of a certain proposition α. In this scenario one considers how α becoming publicly known affects the beliefs of a group of agents. In the tradition of modal logic, DEL formally represent the beliefs of agents as algebraic or relational (Kripke-style) models. An epistemic action such as a public announcement induces a change on such models, which is accordingly modelled through an algebraic or frame-theoretic construction.

For example, viewing the set of beliefs of an agent as a Kripke model (a set of worlds that the agent considers possible, plus an accessibility relation and a valuation), the public announcement of a proposition α causes certain worlds (those where α is not true) to be no longer plausible, that is, precisely, to be no longer "possible worlds" from the agent's perspective. These worlds are therefore "deleted" and a new Kripke model is created, whose underlying set of worlds is essentially just the extension of α (all worlds which satisfied α

J.P. Carvalho et al. (Eds.): IPMU 2016, Part II, CCIS 611, pp. 108–122, 2016.
DOI: 10.1007/978-3-319-40581-0_10

in the original model) and where the relation and the valuation are restricted in a straightforward way. This construction is known as an *epistemic update.* One then uses this new model to define the semantics of sentences of type $\langle\alpha\rangle\varphi$, whose intended meaning is: the proposition α has been announced, and after that φ holds. Now if φ itself contained an epistemic modal operator, then we would for example have a sentence like $\langle\alpha\rangle\Box\psi$, whose intended meaning is: the proposition α has been announced, and after that the agent knows (believes, etc.) that ψ is the case.

The system of DEL that was introduced to deal with public announcements is called precisely Public Announcement Logic (PAL): see e.g. [1,2]. This is essentially a language expansion of classical modal logic with so-called *dynamic* modal operators $\langle\alpha\rangle$ and $[\alpha]$, whose meaning and semantics is the one introduced above. A series of more recent papers [7,9,10] considers systems of PAL built on a non-classical propositional base. The motivation for the introduction of non-classical PAL is along the following lines.

On the one hand, PAL and related systems based on classical propositional logic have been used to provide a formal solution to epistemic problems such as the Muddy Children Puzzle and the Byzantine Generals Problem. It can be shown, however, that at least for some of these scenarios the full power of classical inference principles may not be needed: for example [7] provides a constructive solution to the Muddy Children Puzzle using intuitionistic logic.

Secondly, there are reasoning contexts where the strength of classical logic makes it unsuitable, a prominent example being reasoning in the presence of inconsistent information. The papers [9,10] provide a framework for building a logic of public announcements that may be applied in such contexts.

Lastly, from a theoretical point of view, one may ask which structural conditions make it possible to extend a given modal epistemic system to a dynamic setting. The above-mentioned papers [6,7,9,10] point at certain conditions that seem to be sufficient, at least in the cases that have been studied so far, and the present paper, as we will argue, provides further evidence and tools in this direction. However, the more general problem of formulating mathematically precise conditions that can be proven to be necessary and sufficient is still open, and constitutes, in our opinion, an intriguing direction for future research.

In the present paper we take the generalization of PAL to non-classical settings proposed in [7,9,10] one step further: namely, we show how to define a logic of public announcements having as base n-valued Łukasiewicz modal logic, n being an arbitrary positive integer. In doing so, we believe we are providing a tool that can be useful for reasoning in all those contexts where graded properties and predicates are involved, which are the main scope of many-valued and fuzzy logics. On the other hand, as mentioned above, we also hope that our study will shed further light on the mathematical nature of epistemic updates performed on relational structures and on algebras, thus helping to define a most general context in which these constructions can be performed.

The epistemic update constructions and the logical methods used in the present paper are essentially those of [6,7], that we extend using insights of

[9,10]. These methods are applied to the family of finite-valued Łukasiewicz modal systems developed in [5], relying on on the duality for finitely-generated modal MV-algebras of [8].

The paper is organized as follows. Section 2 contains an introduction to Public Announcement Logic and to Łukasiewicz modal logic, which are the two main ingredients that are going to be combined in our treatment. In Sect. 3 we officially introduce Łukasiewicz Public Announcement Logic (Ł$_n$PAL). We define a relational semantics based on many-valued Kripke models and present a Hilbert-style calculus for Ł$_n$PAL. Section 4 is an intermezzo on the equivalence between algebraic and relational semantics of Łukasiewicz modal logic: this result is needed in order to introduce and develop our algebraic semantics for Ł$_n$PAL. In Sect. 5 we define the mechanism of epistemic updates in the algebraic setting of modal $\mathcal{MV}$-algebras (Subsect. 5.1). This will allow us to provide an algebraic semantics for Ł$_n$PAL (Subsect. 5.2) that is alternative but equivalent to the relational one, and to prove soundness (and completeness) for our Hilbert-style calculus (Subsect. 5.3). Lastly, Sect. 6 contains concluding remarks and suggestions for future work.

2 Preliminaries

2.1 The Logic of Public Announcements

The logic of public announcements [1,2] is a dynamic logic that models the epistemic change on the cognitive state of a group of agents resulting from a given fact (expressed by some proposition, that we will denote by α) becoming publicly known.

From a syntactic point of view, PAL is a language expansion of (classical) modal logic where, besides the so-called "static" modal operators $\Box$ and $\Diamond$ (whose intended interpretation is epistemic), we have "dynamic" operators $\langle\alpha\rangle$ and $[\alpha]$ for each formula α in the language. The intended meaning of a formula of type $\langle\alpha\rangle\varphi$ is: the proposition α has been announced, and after the announcement φ is the case. The other dynamic operator, which in the classical and Łukasiewicz case is the dual of $\langle\alpha\rangle$, has the following interpretation: $[\alpha]\varphi$ means that if the proposition α has been announced, then after the announcement φ is the case.

The formulas of PAL are built from a countable set of propositional letters Var through the following inductive rule:

$$\varphi ::= p \in Var \mid \neg\varphi \mid \varphi \to \varphi \mid \Box\varphi \mid [\varphi]\varphi$$

In this paper we focus on this language, because all other connectives are term-definable from these in both classical and Łukasiewicz logic. In particular we have $\Diamond\varphi := \neg\Box\neg\varphi$ and $\langle\alpha\rangle\varphi := \neg[\alpha]\neg\varphi$.

As mentioned above, the underlying (static) modal logic is usually taken to be a system modeling the knowledge of an agent, for example modal logic S5, whose semantics is provided in the standard way by relational models $\langle W, R, v\rangle$ with R an equivalence relation. In order to provide a semantics for formulas involving dynamic modalities, one needs to introduce the epistemic update construction. Let $M = \langle W, R, v \rangle$ be a model, where $v \colon Fm \times W \to \{0,1\}$ is the valuation map, and let α be a formula in the above-defined language. We define a new model $M^\alpha = \langle W^\alpha, R^\alpha, v^\alpha \rangle$ where $W^\alpha := \{w \in W : M, w \models \alpha\}$, $R^\alpha := R \cap (W^\alpha \times W^\alpha)$ and $v^\alpha \colon Fm \times W^\alpha \to \{0,1\}$ is the restriction of the map v to W^α. Notice that, although M and M^α are different models, the set W^α can be embedded into W in the obvious way. This justifies the abuse of language of the following definitions:

$$M, w \models [\alpha]\varphi \qquad \text{iff} \qquad M, w \models \alpha \quad \text{implies} \quad M^\alpha, w \models \varphi.$$

Dually, one sets

$$M, w \models \langle\alpha\rangle\varphi \qquad \text{iff} \qquad M, w \models \alpha \quad \text{and} \quad M^\alpha, w \models \varphi.$$

Given these definitions, the notion of modal consequence (usually one focuses on the so-called *local* one) is introduced in the standard way.

Classical PAL admits a simple Hilbert-style axiomatization. One expands the set of axioms and rules for modal logic (S5) with the following schemes:

1. $\langle\alpha\rangle p \leftrightarrow (\alpha \wedge p)$
2. $\langle\alpha\rangle\neg\varphi \leftrightarrow (\alpha \wedge \neg\langle\alpha\rangle\psi)$
3. $\langle\alpha\rangle(\varphi \vee \psi) \leftrightarrow (\langle\alpha\rangle\varphi \vee \langle\alpha\rangle\psi)$
4. $\langle\alpha\rangle\Diamond\varphi \leftrightarrow (\alpha \wedge \Diamond(\alpha \wedge \langle\alpha\rangle\varphi))$

where $p \in Var$ and $\alpha, \varphi, \psi \in Fm$, together with the monotonicity rule:

$$\text{from} \quad \emptyset \vdash \varphi \to \psi \quad \text{derive} \quad \emptyset \vdash \langle\alpha\rangle\varphi \to \langle\alpha\rangle\psi.$$

The restriction that $p \in Var$ in the first axiom reflects the important fact that the consequence relation of PAL is not substitution-invariant. Taken together, the above axioms suggest that any PAL-formula having the dynamic operator as main connective can be proven to be inter-derivable in the calculus to one where the dynamic operator has been pushed inside the scope of some other propositional or static modal operator. This is indeed the case, as we will see, and is the key to a completeness proof that relies on reducing any PAL-formula to a formula that does not contain any dynamic operator. In fact, also when moving from classical to a non-classical version of PAL, it is sufficient to check soundness of the set of proposed axioms with respect to the intended semantics, and completeness can be proven using the same reduction strategy that works for the classical case (see [7,9,10]). This is also the approach that we will take in our treatment of Łukasiewicz PAL.

2.2 Łukasiewicz n-valued Modal Logic

The main ingredients that we need to introduce our Łukasiewicz PAL are: (1) a "static" Łukasiewicz modal logic base to build upon, defined in terms of a (Hilbert-style) calculus which is sound and complete with respect to a relational semantics; (2) a suitable adaptation of the techniques for dealing with non-classical dynamic epistemic logics developed in [6,7,9,10]; (3) in order to be able to successfully apply these techniques, it is also desirable to have at hand a workable duality theory connecting the relational and the algebraic semantics of (the static fragment of) the logic. As mentioned earlier on, the first item is provided by the work of Hansoul and Teheux [5], while the third is due to Teheux [8].

In this section we recall the main definitions and facts that we shall need about the modal extension of Łukasiewicz logic introduced in [5], which is going to be the *static* base of our Łukasiewicz PAL. We begin by recalling the essentials of Łukasiewicz (non-modal) logic, and then turn to its modal counterpart.

Łukasiewicz logic can be defined as the logic of $\mathcal{MV}$-algebras (built in the language $\langle \rightarrow, \neg, 1 \rangle$ of type $\langle 2, 1, 0 \rangle$ and defined as in [3, Definition 4.2.1]). We use the following abbreviations: $0 := \neg 1$, $x \oplus y := \neg x \rightarrow y$, $x \odot y := \neg(x \rightarrow \neg y)$, $x \vee y := (x \rightarrow y) \rightarrow y$, $x \wedge y := \neg(\neg x \vee \neg y)$. We also abbreviate $x^m := x \odot \ldots \odot x$ (m times) and $m x := x \oplus \ldots \oplus x$ (m times). As the notation suggests, the $\{\wedge, \vee, 0, 1\}$-reduct of any $\mathcal{MV}$-algebra is a bounded (distributive) lattice. As often happens in many-valued logics, algebras whose lattice order is total play a key role within the variety of $\mathcal{MV}$-algebras. It is in particular well-known that, for every positive integer n, there is up to isomorphism only one totally ordered n-element $\mathcal{MV}$-algebra. We denote it by $\mathbf{Ł}_n$, and by $\mathcal{MV}_n$ the variety generated by $\mathbf{Ł}_n$.

The logics that we shall focus on have as non-modal base *n-valued Lukasiewicz logic* . This is the logic defined by the the logical matrix $\langle \mathbf{Ł}_n, \{1\} \rangle$ in the standard way. For formulas $\Gamma \cup \{\varphi\} \subseteq Fm$, we set $\Gamma \models_{\mathbf{Ł}_n} \varphi$ iff, for every $\mathcal{MV}$-algebra homomorphism $h \colon \mathbf{Fm} \rightarrow \mathbf{Ł}_n$, it holds that $h(\varphi) = 1$ whenever $h[\Gamma] \subseteq \{1\}$. Logically, we think of $\rightarrow$ as an implication and $\neg$ as a negation, while $\oplus, \vee$ are two different types of disjunction and $\odot, \wedge$ are two different conjunctions.

The above consequence relation is extended in [5] to the language $\langle \rightarrow, \neg, \Box, 1 \rangle$ which is augmented with a unary (necessity-type) modal operator $\Box$ through a many-valued generalization of Kripke semantics. The notion of *Kripke frame* is defined in the usual way: a frame is a structure $\mathcal{F} = \langle W, R \rangle$ with W a non-empty set of 'worlds' and $R \subseteq W \times W$ an accessibility relation. An *n-valued Kripke model* is a structure $\mathcal{M} = \langle W, R, v \rangle$ such that $\langle W, R \rangle$ is a frame and $v \colon \mathbf{Fm} \times W \rightarrow \mathbf{Ł}_n$ is a valuation map satisfying the following requirements: for all $\varphi, \psi \in Fm$ and any $w \in W$,

- $v(\neg\varphi, w) = \neg^{\mathbf{Ł}_n} v(\varphi, w)$
- $v(\varphi \rightarrow \psi, w) = v(\varphi, w) \rightarrow^{\mathbf{Ł}_n} v(\psi, w)$
- $v(\Box\varphi, w) = \bigwedge^{\mathbf{Ł}_n} \{v(\varphi, w') : wRw'\}$ where $\bigwedge^{\mathbf{Ł}_n}$ is the lattice meet in $\mathbf{Ł}_n$.

Note that arbitrary meets always exist in the finite algebra $\mathbf{Ł}_n$, and in fact the meet is a minimum because $\mathbf{Ł}_n$ is a chain. A dual possibility operator $\Diamond$ can be defined by $\Diamond\varphi := \neg\Box\neg\varphi$, and it is easy to check that, for any valuation v, any $\varphi \in Fm$ and all $w \in W$, we have $v(\Diamond\varphi, w) = \bigvee^{\mathbf{Ł}_n} \{v(\varphi, w') : wRw'\}$, where $\bigvee^{\mathbf{Ł}_n}$ is the lattice join (or maximum) in $\mathbf{Ł}_n$.

As usual, we say that an n-valued model $\mathcal{M} = \langle W, R, v \rangle$ *satisfies* a formula φ at a world $w \in W$, denoted $\mathcal{M}, w \models \varphi$, when $v(\varphi, w) = 1^{\mathbf{Ł}_n}$ (we omit the parameters n and $\mathbf{Ł}_n$ when they are clear from the context). We can then define a *(local) modal consequence* relation by setting $\Gamma \models^l_n \varphi$ iff, for every n-valued model $\mathcal{M}$ and every $w \in W$, $\mathcal{M}, w \models \gamma$ for all $\gamma \in \Gamma$ implies $\mathcal{M}, w \models \varphi$.

In this paper (as in [5]) we focus on the local logic. The local consequence relation over a given $\mathbf{Ł}_n$ is axiomatized by the following Hilbert-style calculus [5, Theorem 6.2].

The set of *axioms* is the least set of formulas Σ that is closed under *modus ponens* (if $\varphi, \varphi \to \psi \in \Sigma$, then $\psi \in \Sigma$), under substitutions, necessitation (if $\varphi \in \Sigma$, then $\Box\varphi \in \Sigma$) and that contains:

- an axiomatic base for Łukasiewicz n-valued logic (see [4]),
- $\Box(\varphi \to \psi) \to (\Box\varphi \to \Box\psi)$
- $\Box(\varphi \oplus \varphi) \leftrightarrow (\Box\varphi \oplus \Box\varphi)$
- $\Box(\varphi \odot \varphi) \leftrightarrow (\Box\varphi \odot \Box\varphi)$.

The only *inference rule* that can be used without restrictions is modus ponens.

Notice that the set of axioms Σ of the above calculus includes all propositional Łukasiewcz tautologies, and that if φ is a propositional tautology, then, by necessitation, $\Box\varphi \in \Sigma$. However, it is *not* true that if we have derived φ from a set of formulas Γ in the whole calculus, then we can derive $\Box\varphi$ from Γ. This reflects the fact that the calculus is designed to capture the notion of *local* modal consequence relation, not the global one. If we add necessitation as a rule that can be applied without restrictions, then indeed we obtain a calculus for the global modal consequence relation.

Writing $\Gamma \vdash^l_n \varphi$ when there is a proof (in the standard sense) of φ from Γ using the axioms and rule of the above calculus, we can state completeness as follows:

$$\Gamma \vdash^l_n \varphi \quad \text{if and only if} \quad \Gamma \models^l_n \varphi.$$

3 Łukasiewicz Public Announcement Logic

The language of Łukasiewicz Public Announcement Logic is the same as classical PAL. For the single-agent case[1], formulas are built from a countable set of propositional letters Var through the following inductive rule:

$$\varphi ::= p \in Var \mid \neg\varphi \mid \varphi \to \varphi \mid \Box\varphi \mid [\varphi]\varphi$$

Following the usual conventions, we set $\Diamond\varphi := \neg\Box\neg\varphi$ and $\langle\alpha\rangle\varphi := \neg[\alpha]\neg\varphi$.

For a given n, we can define the semantics of the static modal fragment of n-valued Łukasiewicz Public Announcement Logic (abbreviated Ł$_n$PAL) using

[1] The multi-agent case is a straightforward generalization of the single-agent one: one just needs to index the static modal operator by the agents. At this point we do not deal with more complicated operators such as those for common knowledge; these may provide interesting lines for future research.

n-valued Kripke models as described in the preceding section. In order to provide a semantics for formulas of type $[\alpha]\varphi$, we need to define a notion of *epistemic update on n-valued Kripke models.*

Let $\mathcal{M} = \langle W, R, v \rangle$ be an n-valued model, and let α be a formula. Mimicking the classical case, we set

$$W^\alpha := \{w \in W : v(\alpha, w) = 1^{\mathbf{Ł}_n}\}.$$

Notice that $W^\alpha = W^{\alpha \odot \alpha}$, because $v(\alpha, w) = 1^{\mathbf{Ł}_n}$ iff $v(\alpha \odot \alpha, w) = 1^{\mathbf{Ł}_n}$. In fact, writing α^m instead of $\alpha \odot \ldots \odot \alpha$ (m times), we have $W^\alpha = W^{\alpha^m}$ for any $m \geq 1$. This means that there is no difference between announcing α and announcing α^m, because we are only looking at formulas that are "absolutely true". This fact will play an important role in our definition of epistemic updates in the n-valued setting.

The accessibility relation can be restricted just as in the classical case, that is we set $R^\alpha := R \cap (W^\alpha \times W^\alpha)$. Similarly, we define $v^\alpha \colon Fm \times W^\alpha \to \mathbf{Ł}_n$ to be the restriction of the map v to W^α. In this way we obtain the *updated model* $\mathcal{M}^\alpha = \langle W^\alpha, R^\alpha, v^\alpha \rangle$, which allows us to define

$$M, w \models [\alpha]\varphi \qquad \text{iff} \qquad M, w \models \alpha \quad \text{implies} \quad M^\alpha, w \models \varphi.$$

Given that we are in a many-valued setting, the reader might wonder whether the above definition is sufficient to determine the semantic value of arbitrary formulas involving dynamic operators. This may not evident at this point, but will be easily checked by looking at the algebraic semantics that we are going to define in Sect. 5.2 (which is, as we will show, equivalent to the relational one via duality). This is indeed one of the main reasons why we find it useful to present a double perspective (relational as well as algebraic) on Ł$_n$PAL.

Having provided a notion of satisfaction for arbitrary formulas, the definition of consequence in Ł$_n$PAL is just a reformulation of the one we have stated above for the static fragment. We define the *(local) modal consequence* relation by setting $\Gamma \models^l_{\text{Ł}_n\text{PAL}} \varphi$ iff, for every n-valued model $\mathcal{M}$ and every $w \in W$, $\mathcal{M}, w \models \gamma$ for all $\gamma \in \Gamma$ implies $\mathcal{M}, w \models \varphi$.

We now introduce an axiomatization that we will later on prove to be sound and complete with respect to the above-defined semantics. We abbreviate $\langle\alpha\rangle\varphi := \neg[\alpha]\neg\varphi$. The axioms of Ł$_n$PAL are all axioms and rules of $\mathcal{MMV}_n$ [5, Definition 3.1] plus the following:

Interaction with 1	$[\alpha]1 \leftrightarrow 1$
Interaction with $\rightarrow$	$[\alpha](\varphi \rightarrow \psi) \leftrightarrow (\langle\alpha\rangle\varphi \rightarrow \langle\alpha\rangle\psi)$
Interaction with $\neg$	$[\alpha]\neg\varphi \leftrightarrow (\alpha^n \rightarrow \neg[\alpha]\varphi)$
Interaction with $\Box$	$[\alpha]\Box\varphi \leftrightarrow (\alpha^n \rightarrow \Box[\alpha]\varphi)$
Preservation of facts	$[\alpha]p \leftrightarrow (\alpha^n \rightarrow p)$

where φ, ψ, α are arbitrary formulas, while p is a propositional variable. We further require the set of theorems to be closed under $[\alpha]$-monotonicity:

$$\vdash \varphi \rightarrow \psi \quad \Rightarrow \quad \vdash [\alpha]\varphi \rightarrow [\alpha]\psi.$$

The only rule of our calculus is modus ponens: $\varphi,\ \varphi \to \psi \vdash \psi$.

This defines the calculus $\vdash^{l}_{Ł_n\mathrm{PAL}}$ for Ł$_n$PAL.

As the reader will have noticed, the shape of the axioms defining the interaction of the dynamic operator with the other connectives resembles that of the classical (and intuitionistic: see [7]) case, and their role in our completeness proof is analogous. The main difference worth mentioning is the presence of the formula α^n, which is a shorthand for $\alpha \odot \ldots \odot \alpha$ (n times; recall that $\alpha \odot \alpha$ is itself a shorthand for $\neg(\alpha \to \neg\alpha)$). The parameter n obviously distinguishes one finite-valued Łukasiewicz public announcement logic from another. However, for the purpose of axiomatization, all that matters is that the n appearing in the axioms be big enough. In fact we might use any $m \geq n$ for providing a complete axiomatization of Ł$_n$PAL, for α^m is semantically equivalent in Ł$_n$PAL to α^n for each $m \geq n$. The role of these exponents will be further clarified when we look at the algebraic semantics of Ł$_n$PAL.

Let us also highlight that the static modal logic we build on is a many-valued analogue of the minimal normal modal logic K (rather than, say, of modal logic S5). This choice was made just for the sake of generality, for we will show that, once we have axiomatized the minimal Ł$_n$PAL, it is easy to deal with extensions obtained by adding axioms to its static fragment.

As mentioned earlier, completeness of Ł$_n$PAL can be proved using the same strategy as the classical case. For this, we will need to check the soundness of the above-introduced axioms, which is most easily done with respect to the algebraic semantics of Ł$_n$PAL. Then, assuming we know that the algebraic and the relational semantics are equivalent, we will have completed the proof. This equivalence can indeed be easily obtained exploiting Teheux's duality for modal $\mathcal{MV}$-algebras [8]. It is therefore convenient to recall the main results of this duality before we introduce our algebraic semantics for Ł$_n$PAL and go on to develop the epistemic update construction on algebraic models.

4 On Duality for Modal $\mathcal{MV}$-algebras

The Hilbert-style calculus for Łukasiewicz n-valued modal logic of Subsect. 2.2 not only enjoys completeness with respect to the relational semantics of n-valued Kripke models, but can also be endowed with an algebraic semantics provided by the class of *modal n-valued $\mathcal{MV}$-algebras* ($\mathcal{MMV}_n$-algebras). These play a key role in our semantic approach to Ł$_n$PAL, therefore in this section we take a closed look at them and in particular at the duality relating $\mathcal{MMV}_n$-algebras to n-valued Kripke frames.

An *$\mathcal{MMV}$-algebra* is a structure $\langle \mathbf{A}, \to, \neg, \Box, 1\rangle$ such that the reduct $\langle \mathbf{A}, \to, \neg, 1\rangle$ is an $\mathcal{MV}$-algebra and the following equations are satisfied:

(MO1) $\Box 1 = 1$
(MO2) $\Box(x \to y) \to (\Box x \to \Box y) = 1$
(MO3) $\Box(x \oplus x) = \Box x \oplus \Box x$ and $\Box(x \odot x) = \Box x \odot \Box x$
(MO4) $\Box(x \oplus x^m) = \Box x \oplus (\Box x)^m$ for every natural number m.

A dual operator $\Diamond$ can be defined by $\neg\Box\neg$ as for classical modal logic. An $\mathcal{MMV}_n$*-algebra* is an $\mathcal{MMV}$-algebra whose $\Box$-free reduct belongs to $\mathcal{MV}_n$. In this case one can prove that (MO4) follows from (MO1)-(MO3), which also explains why we have omitted the corresponding axiom in the Hilbert-style calculus for n-valued modal logic of Subsect. 2.2.

An important notion, both from a logical and a duality point of view, is that of filter. Given an algebra $\mathbf{A}$ having an $\mathcal{MV}$-algebra reduct (thus also any algebra in $\mathcal{MMV}$ or $\mathcal{MMV}_n$), a *filter* of $\mathbf{A}$ is defined as a non-empty set $F \subseteq A$ which is an up-set w.r.t. the lattice order of $\mathbf{A}$ and is moreover closed under the $\odot$ operation, i.e. $a, b \in F$ implies $a \odot b \in F$. Every $\mathcal{MV}$-algebra $\mathbf{A}$ has a least filter which is the singleton $\{1^{\mathbf{A}}\}$.

An *algebraic model* of modal n-valued Łukasiewicz logic is a pair $\langle \mathbf{A}, v\rangle$ where $\mathbf{A} \in \mathcal{MMV}$ and $v\colon \mathbf{Fm} \to \mathbf{A}$ is a homomorphism of $\mathcal{MMV}$-algebras. Through this notion we can define, as with the relational semantics, a local consequence relation. We set $\Gamma \models_{\mathcal{MMV}_n} \varphi$ when, for any algebraic model $\langle \mathbf{A}, v\rangle$ and any filter $F \subseteq A$, we have $v(\varphi) \in F$ whenever $v(\gamma) \in F$ for all $\gamma \in \Gamma$.

The Hilbert-style calculus of [5] introduced in Subsect. 2.2 is complete, for each modal n-valued Łukasiewicz system, w.r.t. to the corresponding above-defined local consequence; this result can be proved either directly or, via duality (see below), using completeness w.r.t. to n-valued Kripke models. For details and proofs about the duality for n-valued Łukasiewicz modal logics, we refer the reader to [5,8]. Here we just recall the main bits that are needed for our treatment of Ł$_n$PAL.

If we have an algebraic model for our logic $\langle \mathbf{A}, v\rangle$, where $\mathbf{A} \in \mathcal{MMV}_n$ and $v\colon \mathbf{Fm} \to \mathbf{A}$, we can use duality to turn it into a n-valued Kripke model as follows. We construct the *canonical frame* $\mathbf{A}_+ = \langle \mathcal{MV}(\mathbf{A}, \mathbf{Ł}_n), R\rangle$, where $\mathcal{MV}(\mathbf{A}, \mathbf{Ł}_n)$ is the set of $\mathcal{MV}$-algebra homomorphisms (i.e., not necessarily $\Box$-preserving) from $\mathbf{A}$ to $\mathbf{Ł}_n$ and the accessibility relation $R_\Box \subseteq \mathcal{MV}(\mathbf{A}, \mathbf{Ł}_n) \times \mathcal{MV}(\mathbf{A}, \mathbf{Ł}_n)$ is defined, for all $h, h' \in \mathcal{MV}(\mathbf{A}, \mathbf{Ł}_n)$, by

$$\langle h, h'\rangle \in R_\Box \quad \text{iff} \quad \forall a \in A \;\; h(\Box a) = 1^{\mathbf{Ł}_n} \;\text{ implies }\; h'(a) = 1^{\mathbf{Ł}_n}.$$

Denoting by $[\mathbf{0}, \mathbf{1}]$ the $\mathcal{MV}$-algebra having as universe the real interval $[0, 1]$, it is easy to show that $\mathcal{MV}(\mathbf{A}, \mathbf{Ł}_n) \cong \mathcal{MV}(\mathbf{A}, [\mathbf{0}, \mathbf{1}])$ for any $\mathbf{A} \in \mathcal{MMV}_n$. This explains why we can rephrase [5, Definition 5.2] replacing the algebra $[\mathbf{0}, \mathbf{1}]$ by $\mathbf{Ł}_n$. Given an algebraic model $\langle \mathbf{A}, v\rangle$, the *canonical model* $\langle \mathbf{A}_+, v_+\rangle$ is obtained by defining the valuation $v_+\colon \mathbf{Fm} \times \mathcal{MV}(\mathbf{A}, \mathbf{Ł}_n) \to \mathbf{Ł}_n$ as

$$v_+\langle h, \varphi\rangle = (h \cdot v)(\varphi)$$

for all $h \in \mathcal{MV}(\mathbf{A}, \mathbf{Ł}_n)$ and $\varphi \in Fm$. It can be checked that v_+ indeed respects all connectives of the logic and is therefore a modal valuation. It is also easy to check that, for any formula φ, we have $v(\varphi) = 1^{\mathbf{A}}$ if and only if $v_+\langle h, \varphi\rangle = 1^{\mathbf{Ł}_n}$ for all $h \in \mathcal{MV}(\mathbf{A}, \mathbf{Ł}_n)$, that is, if and only if φ is valid in the model $\langle \mathbf{A}_+, v_+\rangle$. Conversely, any n-valued Kripke model $\mathcal{M} = \langle \mathcal{F}, v\rangle$, where $\mathcal{F} = \langle W, R\rangle$ and $v\colon Fm \times W \to \mathbf{Ł}_n$, can be turned into an algebraic one in the following way.

We have the $\mathcal{MV}_n$-algebra $\mathcal{F}^+ = \mathbf{Ł}_n^W$ whose elements are all maps $f\colon W \to \mathbf{Ł}_n$, which can be endowed with a modal operator $\Box_R$ defined by

$$\Box_R(f)(w) := \bigwedge^{\mathbf{Ł}_n}\{f(u) : u \in W \text{ and } wRu\}. \tag{1}$$

We have then a $\mathcal{MMV}_n$-algebra $\mathcal{F}^+ = \langle \mathbf{Ł}_n^W, \Box_R\rangle$, on which we define a valuation v^+, for all $\varphi \in Fm$ and $w \in W$, as

$$v^+(\varphi)(w) := v(\varphi, w).$$

A formula φ is valid in $\mathcal{M}$ if and only if $v^+(\varphi)$ is the constant map $1^{\mathbf{Ł}_n}$.

Analogously to the case of modal Boolean algebras, it can be shown that every n-valued frame $\mathcal{F}$ is embeddable into its double dual $(\mathcal{F}^+)_+$. Likewise, an arbitrary $\mathcal{MMV}_n$-algebra $\mathbf{A}$ need not be isomorphic to its double dual $(\mathbf{A}_+)^+$ but it will be embeddable in it. This is sufficient to prove that the algebra-based and the frame-based semantics are equivalent.

We prove the equivalence by contraposition. First, suppose $\Gamma \not\models_{\mathcal{MMV}_n} \varphi$, which means that there is an algebraic model $\langle \mathbf{A}, v\rangle$ and a filter $F \subseteq A$ with $v(\varphi) \notin F$ and $v(\gamma) \in F$ for all $\gamma \in \Gamma$. Now F can be extended to a maximal filter $F' \supseteq F$ with $v(\varphi) \notin F'$, and F' corresponds to an $\mathcal{MV}$-algebra homomorphism $h_{F'} \in \mathcal{MV}(\mathbf{A}, \mathbf{Ł}_n)$ such that $h_{F'}(\gamma) = 1^{\mathbf{Ł}_n}$ for all $\gamma \in \Gamma$ and $h_{F'}(\varphi) \neq 1^{\mathbf{Ł}_n}$. Thus we have $\langle \mathbf{A}_+, v_+\rangle, h_{F'} \models \gamma$ for all $\gamma \in \Gamma$ while $\langle \mathbf{A}_+, v_+\rangle, h_{F'} \not\models \varphi$. Hence, $\Gamma \not\models_n^l \varphi$.

Conversely, assume $\Gamma \not\models_n^l \varphi$, which means that there is an n-valued model $\mathcal{M} = \langle \mathcal{F}, v\rangle$ with $\mathcal{F} = \langle W, R\rangle$ and a point $w \in W$ such that $\mathcal{M}, w \models \gamma$ for all $\gamma \in \Gamma$ while $\mathcal{M}, w \not\models \varphi$. Thus, in the dual algebra $\mathcal{F}^+ = \langle \mathbf{Ł}_n^W, \Box_R\rangle$ we have elements a_γ for each $\gamma \in \Gamma$ and an element a_φ such that $a_\gamma(w) = 1^{\mathbf{Ł}_n} \neq a_\varphi(w)$. This means that $a_\gamma \not\leq a_\varphi$ for each $\gamma \in \Gamma$, which implies that the filter F generated by all a_γ does not contain a_φ. Thus, the algebraic model $\langle \mathcal{F}^+, v^+\rangle$ together with the filter F is a counter-model to $\Gamma \models_{\mathcal{MMV}_n} \varphi$, as was required to prove.

5 Algebraic Models and Completeness

5.1 Epistemic Updates on Modal $\mathcal{MV}$-algebras

Given the n-valued Kripke model $\mathcal{M} = \langle \mathcal{F}, v\rangle$ where $\mathcal{F} = \langle W, R\rangle$, we have an $\mathcal{MV}_n$-algebra $\mathbf{Ł}_n^W$ whose elements are all maps $f\colon W \to \mathbf{Ł}_n$. Following Teheux, we endow this algebra with a modal operator $\Box_R$ defined as in (1) and we have an $\mathcal{MMV}_n$-algebra $\langle \mathbf{Ł}_n^W, \Box_R\rangle$. Now let $\alpha \in Fm$, and consider the updated model $\mathcal{M}^\alpha = \langle W^\alpha, R^\alpha, v^\alpha\rangle$ defined as in Sect. 3. To this model corresponds, via duality, the $\mathcal{MMV}_n$-algebra $\langle \mathbf{Ł}_n^{W^\alpha}, \Box_{R^\alpha}\rangle$, whose elements are maps $g\colon W^\alpha \to \mathbf{Ł}_n$, which are precisely the restrictions of the maps in $\mathbf{Ł}_n^W$ to W^α. We are going to see that the algebra $\langle \mathbf{Ł}_n^{W^\alpha}, \Box_{R^\alpha}\rangle$ is isomorphic to a *pseudo-quotient* of $\mathcal{MMV}_n$-algebra $\langle \mathbf{Ł}_n^W, \Box_R\rangle$, defined as below.

Let $\mathbf{A} = \langle \mathbf{Ł}_n^W, \Box_R \rangle$ be the algebra dual to $\langle W, R \rangle$, and let $a \in A$ be the element corresponding to the map $v(\alpha, \cdot) \colon W \to \mathbf{Ł}_n$. We define the following equivalence relation: for all $b, c \in \mathbf{A}$,

$$b \equiv_a c \quad \text{iff} \quad b \odot a^n = c \odot a^n.$$

Notice that, since the non-modal reduct of $\mathbf{A}$ is in the variety generated by $\mathbf{Ł}_n$, the equation $a^n = a^{n+1}$ holds in $\mathbf{A}$. This means that a^n is an idempotent element of $\mathbf{A}$, which implies that in fact we have:

$$b \equiv_a c \quad \text{iff} \quad b \odot a^n = c \odot a^n \quad \text{iff} \quad b \wedge a^n = c \wedge a^n.$$

Thanks to the idempotency of a^n, the above-defined relation is a congruence of the $\mathcal{MV}$-algebra reduct of $\mathbf{A}$. In fact, the possibility of finding such an idempotent element one of the main technical reasons for working with finitely-generated algebras: at the moment we are not aware of a definition of pseudo-quotient that would work for general $\mathcal{MV}$-algebras.
Although $\equiv_a$ need not be compatible with the modal operator $\Box$, we can define, for each $[b]_{\equiv_a} \in \mathbf{A}/{\equiv_a}$,

$$\Box^a [b]_{\equiv_a} := [\Box(a^n \to b)]_{\equiv_a}$$

and we obtain an $\mathcal{MMV}_n$-algebra $\langle \mathbf{A}/{\equiv_a}, \Box^a \rangle$, which we call the *pseudo-quotient* and denote by $\mathbf{A}^a$ (see Proposition 1 below). As usual, the operator dual to $\Box^a$ is defined by $\Diamond^a [b]_{\equiv_a} := \neg \Box^a \neg [b]_{\equiv_a}$ which gives

$$\Diamond^a [b]_{\equiv_a} = [\Diamond(b \wedge a^n)]_{\equiv_a}$$

Lemma 1. *The algebra* $\langle \mathbf{Ł}_n^{W^\alpha}, \Box_{R^\alpha} \rangle$ *is isomorphic to the pseudo-quotient* $\mathbf{A}^a$.

Proof. Let $\eta \colon \mathbf{A}^a \to \mathbf{Ł}_n^{W^\alpha}$ be defined by $\eta[b] = b \restriction W^\alpha$ (from now on we shall write $[b]$ instead of $[b]_{\equiv_a}$ to simplify the notation). Notice that if $[b] = [c]$, i.e. $b \wedge a^n = c \wedge a^n$, then, for all $w \in W^\alpha$, we have $a(w) = 1$ and so $a^n(w) = 1$. Hence, $b(w) = (b \wedge a^n)(w) = (c \wedge a^n)(w) = c(w)$. That is, if $[b] = [c]$, then $\eta[b] = \eta[c]$. Thus the map η is well-defined. To see that it is injective, notice that, for any $w \in W$, the element $a^n(w)$ is an idempotent of $\mathbf{Ł}_n$, that is, $a^n(w) \in \{0^{\mathbf{Ł}_n}, 1^{\mathbf{Ł}_n}\}$. Now, assuming $[b] \neq [c]$, we have $(b \wedge a^n)(w) \neq (c \wedge a^n)(w)$ for some $w \in W$. If $w \notin W^\alpha$, one would have $1^{\mathbf{Ł}_n} > a(w)$ and so necessarily $1^{\mathbf{Ł}_n} > a(w) \geq a^n(w) = 0^{\mathbf{Ł}_n}$. So we must have $w \in W^\alpha$, which means that $a^n(w) = 1^{\mathbf{Ł}_n}$ and so $b(w) = (b \wedge a^n)(w) \neq (c \wedge a^n)(w) = c(w)$, i.e. $\eta[b] \neq \eta[c]$. Surjectivity of η is straightforward: if $b \in \mathbf{Ł}_n^{W^\alpha}$, then b is the restriction of some $b' \in \mathbf{Ł}_n{}^W$, and so $\eta[b'] = b$.
Hence η is a bijection between the universe of $\langle \mathbf{Ł}_n^{W^\alpha}, \Box_{R^\alpha} \rangle$ and that of $\mathbf{A}^a$. It is also obvious that η is an $\mathcal{MV}$-algebra homomorphism. Furthermore, it is easy to show that, for all $w \in W^\alpha$,

$$\eta(\Box_R^a[b])(w) = \eta[\Box_R(a^n \to b)](w) = \Box_{R^\alpha}(b)(w) = (\Box_{R^\alpha}(\eta[b]))(w).$$

The equality

$$\bigwedge^{\mathbf{Ł}_n}\{(a^n \to b)(u) : u \in W \text{ and } wRu\} = \bigwedge^{\mathbf{Ł}_n}\{(a^n \to b)(u) : u \in W^\alpha \text{ and } wR^\alpha u\}$$

holds because, when computing the infimum, any world outside W^α can be disregarded. In fact, if $u \notin W^\alpha$, then $a^n(u) = 0^{\mathbf{Ł}_n}$ as we have seen earlier, and so $(a^n \to b)(u) = 1^{\mathbf{Ł}_n}$.
We thus conclude that η is an $\mathcal{MMV}_n$-algebra isomorphism.

In light of Lemma 1, we shall adopt the above-defined pseudo-quotient as our official construction for epistemic updates on $\mathcal{MMV}_n$-algebras.

Lemma 2. *Let $u \in A$ be an idempotent element of an $\mathcal{MV}$-algebra $\mathbf{A}$. Then, for all $a, b \in A$,*

(i) $u \odot a = u \wedge a$ *and* $u \oplus a = u \vee a$
(ii) $u \to (a \odot b) = (u \to a) \odot (u \to b)$
(iii) $u \to (a \oplus b) = (u \to a) \oplus (u \to b)$

Proof. It suffices to check that the above items hold in any $\mathcal{MV}$-chain, for it is well-known that the variety of $\mathcal{MV}$-algebras is generated (as a quasivariety) by its chains. It is also well-known that in an $\mathcal{MV}$-chain the only idempotents are the top and the bottom element, for which the above statements follow trivially.

Proposition 1. *For any $\mathbf{A} \in \mathcal{MMV}_n$ and any $a \in A$, we have $\mathbf{A}^a = \langle \mathbf{A}/{\equiv_a}, \Box^a \rangle \in \mathcal{MMV}_n$.*

Proof. That $\equiv_a$ is a congruence follows from, e.g., [3, Proposition 1.2.6], so we have $\mathbf{A}/{\equiv_a} \in \mathcal{MV}$, and if $\mathbf{A} \in \mathcal{MV}_n$ then $\mathbf{A}/{\equiv_a} \in \mathcal{MV}_n$. It remains to check that $\Box^a$ is a modal operator satisfying equations (MO1)–(MO3) of [8, Definition 3.1]. The proof of these facts follows straighforwardly from Lemma 2 and the fact that a^n is idempotent.

As mentioned in Sect. 3, we take as static base the minimal n-valued modal Łukasiewicz logic. It is however easy to see that the pseudo-quotient construction introduced above can be applied also to $\mathcal{MMV}_n$-algebras satisfying additional equations, which correspond to axiomatic extensions of the basic logic. All that needs to be checked is that, if $\mathbf{A} \in \mathcal{MMV}_n$ satisfies some extra equation δ, then the pseudo-quotient algebra $\mathbf{A}^a$ will also satisfy δ. For equations in the pure language of $\mathcal{MV}$-algebras, this is straightforward, for the non-modal reduct of $\mathbf{A}^a$ is actually just a quotient of the corresponding reduct of $\mathbf{A}$. On the other hand, if δ contains some modal operator, then it may not be preserved. A simple example is the equation $\Diamond 1 = 1$. On the contrary, it is easy to check e.g. that the equation $\Box x \to x = 1$ (which corresponds, also in the setting of n-valued frames, to reflexivity of the relation) is preserved.

In general, reasoning on frames, it is not hard to see that a sufficient condition for an equation to be preserved is that it corresponds to some property that is preserved by the update, which consists essentially in deleting some worlds from a model. This is of course not a characterization, and indeed providing such a characterization might prove an interesting topic for future research. It is, however, sufficient to establish that it is possible to define an n-valued analogue of, e.g., modal logic S5 by adding appropriate axioms/equations, and to take this logic as the static base for Ł$_n$PAL.

5.2 Algebraic Semantics for Ł$_n$PAL

We are now going to use the notion of algebraic model for modal Łukasiewicz logic introduced in Sect. 4 to provide an alterative but equivalent semantics for Ł$_n$PAL. This semantics, which is based on $\mathcal{MMV}_n$-algebras, is in certain respects easier to handle than the relational one, and will allow us to give us a simple(r) proof of completeness.

Recall that an *algebraic model for Łukasiewicz n-valued modal logic* is a pair $\langle \mathbf{A}, v \rangle$ where $\mathbf{A} \in \mathcal{MMV}_n$ (for we focus here on finitely-generated $\mathcal{MMV}$-algebras) and $v\colon \mathbf{Fm} \to \mathbf{A}$ is a homomorphism of $\mathcal{MMV}$-algebras (thus, $\Box$-preserving, too). To a formula α that is being announced corresponds an element $v(\alpha) = a \in \mathbf{A}$, which we can use to build the pseudo-quotient algebra $\mathbf{A}^a$ as shown earlier. Notice that each equivalence class $[b] \in \mathbf{A}^a$ has a minimum element (w.r.t. the lattice order of $\mathbf{A}$), namely $b \wedge a^n$. This means (cfr. [7, Fact 6.1]) that we can define an injective map $\iota\colon \mathbf{A}^a \to \mathbf{A}$ given by $\iota[b] := b \wedge a^n$. Via ι we can view $\mathbf{A}^a$ as a *pseudo-subalgebra* of $\mathbf{A}$, and we can also extend the algebraic semantics for n-valued modal logic to formulas with dynamic operators, as shown in the following definition.

An *algebraic model for* Ł$_n$*PAL* is a pair $\langle \mathbf{A}, v \rangle$ where $\mathbf{A} \in \mathcal{MMV}_n$ and $v\colon \mathbf{Fm} \to \mathbf{A}$ is a homomorphism of $\mathcal{MMV}$-algebras. The map v is extended to formulas containing dynamic operators as follows:

$$v([\alpha]\varphi) := v(\alpha^n) \to^{\mathbf{A}} (\iota \cdot v^\alpha)(\varphi).$$

where $v^\alpha\colon \mathbf{Fm} \to \mathbf{A}^{v(\alpha)}$ is the unique $\mathcal{MMV}_n$-homomorphism extending the map $v^\alpha\colon Var \to \mathbf{A}^{v(\alpha)}$ defined by $v^\alpha(p) := [v(p)]_{\equiv_{v(\alpha)}}$ for each $p \in Var$. We do not take the dynamic diamond $\langle\alpha\rangle$ as primitive, but if we did, we would define, analogously to classical and intuitionistic PAL,

$$v(\langle\alpha\rangle\varphi) := v(\alpha^n) \wedge^{\mathbf{A}} (\iota \cdot v^\alpha)(\varphi).$$

At this point we can use algebraic models to introduce a notion of (local) consequence relation. We set $\Gamma \models^l_{\text{Ł}_n\text{PAL}} \varphi$ if and only if for every algebraic model $\langle \mathbf{A}, v \rangle$ and every filter $F \subseteq A$, we have that $[\![\gamma]\!] \in F$ for all $\gamma \in \Gamma$ implies $[\![\varphi]\!] \in F$.

5.3 Soundness and Completeness

As mentioned earlier, in the case of public announcement logics the less straightforward part of the completeness proof consists in proving soundness. For this we find it more convenient to work with algebraic rather than relational models, and we will need to establish a few technical lemmas. We won't include proofs because of space constraints, but these are analogous to the corresponding ones in [7,10].

Lemma 3. *Let* $\mathbf{A}$ *be an* $\mathcal{MMV}_n$*-algebra and* $\mathbf{A}^a$ *the pseudo-quotient relative to some* $a \in \mathbf{A}$*. Then the map* $\iota\colon \mathbf{A}^a \to \mathbf{A}$ *defined by* $\iota[b] := a^n \wedge^{\mathbf{A}} b$ *for all* $b \in \mathbf{A}$ *satisfies the following:*

(i) $\iota(\neg^{\mathbf{A}^a}[b]) = a^n \wedge^{\mathbf{A}} \neg^{\mathbf{A}}\iota[b]$.
(ii) $\iota(\Box^a[b]) = a^n \wedge^{\mathbf{A}} \Box^{\mathbf{A}}(a^n \to^{\mathbf{A}} \iota[b])$.
(iii) $\iota([b] \to^{\mathbf{A}^a} [c]) = a^n \wedge^{\mathbf{A}} (\iota[b] \to^{\mathbf{A}^a} \iota[c])$.

Lemma 4. *Let* $(\mathbf{A}, v)$ *be an algebraic model,* $\alpha, \varphi, \psi \in Fm$ *and* $p \in Var$*. Then,*

(i) $v(\langle\alpha\rangle\varphi) = v(\alpha^n) \wedge^{\mathbf{A}} (\iota \cdot v^\alpha)(\varphi)$
(ii) $v([\alpha]\,1) = 1^{\mathbf{A}}$
(iii) $v([\alpha]\,p) = v(\alpha^n) \to^{\mathbf{A}} v(p)$
(iv) $v([\alpha]\neg\varphi) = v(\alpha^n) \to^{\mathbf{A}} \neg^{\mathbf{A}} v([\alpha]\varphi)$
(v) $v([\alpha](\varphi \to \psi)) = v(\langle\alpha\rangle\varphi) \to^{\mathbf{A}} v(\langle\alpha\rangle\psi)$
(vi) $v([\alpha]\Box\varphi) = v(\alpha^n) \to^{\mathbf{A}} \Box^{\mathbf{A}} v([\alpha]\varphi)$
(vii) if $v(\varphi \to \psi) = 1^{\mathbf{A}}$*, then* $v([\alpha]\varphi \to [\alpha]\psi) = 1^{\mathbf{A}}$.

Notice that the first item of Lemma 4 actually shows that the dynamic operators $[\alpha]$ and $\langle\alpha\rangle$ are dual to one another, which justifies our choice of focusing on one only.

Theorem 1. *The calculus* $\vdash^l_{Ł_n PAL}$ *is sound and complete w.r.t. the consequence of* $Ł_n PAL$*.*

6 Conclusions and Future Work

The present paper is part of an ongoing enterprise that aims, on the applied logic side, at extending dynamic epistemic logics outside the boundaries of classical reasoning and, on the theoretical side, at better understanding the mechanism itself of epistemic updates. Many issues are still open on both sides. It has been shown in [6,7] that certain epistemic reasoning contexts can be alternatively, and perhaps more appropriately be handled using intuitionistic logic instead of the classical one. In the case of $Ł_n$PAL this is an almost trivial exercise, for we can simply recover classical logic by restricting semantic valuations to $\{0,1\}$-valued ones. More interesting will be the study of specific scenarios for which we can argue that classical logic would be unsuitable altogether; for this enterprise we hope to have at least provided a mathematically sound framework which can serve as a starting point.

The framework itself can and most likely needs to be improved in many directions. An obvious extension is to consider a more general *product update* construction such as those of [1,6], which would allow us to consider more complex epistemic actions than public announcements. This might prove a relatively straightforward task, but also one that may lead to interesting applications. Another potentially promising line of research could result from dropping the assumption that propositions are announced with the highest possible truth degree, which entails, as mentioned earlier, that announcing α is equivalent to announcing α^m for any m. This fact played a central rôle in this paper from a technical point of view; in a many-valued setting, however, a very natural thing to do would be to allow for *graded* announcements such as "α has at least true degree k/n". This choice, which is supported by a strong semantic intuition, would give rise to the novel notion of graded epistemic action. A technically more challenging issue, even in the simplest public announcement setting, is whether and how it is possible to apply our methods to infinite-valued Łukasiewicz modal logic.

References

1. Baltag, A., Moss, L., Solecki, A.: The logic of public announcements, common knowledge, and private suspicions, CWI technical report SEN-R9922 (1999)
2. van Benthem, J.: Logical Dynamics of Information and Interaction. Cambridge University Press, Cambridge (2011)
3. Cignoli, R., D'Ottaviano, I.M.L., Mundici, D.: Algebraic Foundations of Many-Valued Reasoning. Trends in Logic—Studia Logica Library, vol. 7. Kluwer Academic Publishers, Dordrecht (2000)
4. Grigolia, R.S.: Algebraic analysis of Łukasiewicz-Tarski's n-valued logical systems. In: Wójciki, R., Malinowski, G. (eds.) Selected Papers on Łukasiewicz Sentential Calculi, Ossolineum, Wronclaw, pp. 81–92 (1977)
5. Hansoul, G., Teheux, B.: Extending Łukasiewicz logics with a modality: algebraic approach to relational semantics. Studia Logica **101**(3), 505–545 (2013)
6. Kurz, A., Palmigiano, A.: Epistemic updates on algebras. Logical Methods Comput. Sci. **9**(4:17), 1–28 (2013)
7. Ma, M., Palmigiano, A., Sadrzadeh, M.: Algebraic semantics and model completeness for Intuitionistic Public Announcement Logic. Ann. Pure Appl. Logic **165**, 963–995 (2014)
8. Teheux, B.: A duality for the algebras of a Łukasiewicz $n+1$-valued modal system. Studia Logica **87**(1), 13–36 (2008)
9. Rivieccio, U.: Algebraic semantics for bilattice public announcement logic. In: Indrzejczak, A., Kaczmarek, J., Zawidzki, M. (eds.) Proceedings of Trends in Logic XIII, Lodz, Poland, 2–5 July 2014, pp. 199–215. Lodz University Press (2014)
10. Rivieccio, U.: Bilattice public announcement logic. In: Goré, R., Kooi, B., Kurucz, A. (eds.) Advances in Modal Logic, vol. 10. College Publications, pp. 459–477 (2014)

Possibilistic Semantics for a Modal $KD45$ Extension of Gödel Fuzzy Logic

Félix Bou[1], Francesc Esteva[1], Lluís Godo[1(✉)], and Ricardo Oscar Rodriguez[2]

[1] Artificial Intelligence Research Institute, IIIA - CSIC, Bellaterra, Spain
{fbou,esteva,godo}@iiia.csic.es

[2] Departamento de Computación, FCEyN - UBA, Buenos Aires, Argentina
ricardo@dc.uba.ar

Abstract. In this paper we provide a simplified semantics for the logic $KD45(\mathbf{G})$, i.e. the many-valued Gödel counterpart of the classical modal logic $KD45$. More precisely, we characterize $KD45(\mathbf{G})$ as the set of valid formulae of the class of possibilistic Gödel Kripke Frames $\langle W, \pi \rangle$, where W is a non-empty set of worlds and $\pi : W \to [0,1]$ is a normalized possibility distribution on W.

1 Introduction

Possibilistic logic [7,8] is a well-known uncertainty logic to reasoning with graded beliefs on classical propositions by means of necessity and possiblity measures. These measures are defined in terms of possibility distributions. A (normalized) possibility distribution is a mapping $\pi : \Omega \to [0,1]$, with $\sup_{w \in \Omega} \pi(w) = 1$, on the set Ω of classical interpretations of a given propositional language that ranks interpretations according to its plausibility level: $\pi(w) = 0$ means that w is rejected, $\pi(w) = 1$ means that w is fully plausible, while $\pi(w) < \pi(w')$ means that w' is more plausible than w. A possibility distribution π induces a pair of dual possibility and necessity measures on propositions, defined respectively as:

$$\Pi(\varphi) = \sup\{\pi(w) \mid w \in \Omega, w(\varphi) = 1\}$$
$$N(\varphi) = \inf\{1 - \pi(w) \mid w \in \Omega, w(\varphi) = 0\}.$$

From a logical point of view, possibilistic logic can be seen as a sort of graded extension of the non-nested fragment of the well-known modal logic of belief $KD45$ [9], in fact, $\{0,1\}$-valued possibility and necessity measures over classical propositions can be taken as equivalent semantics for the modal operators of the logic $KD45$ [1].

When trying to extend the possibilistic belief model beyond the classical framework of Boolean propositions to many-valued propositions, one has to come up with appropriate extensions of the notion of necessity and possibility measures for them (see e.g. [6]). In the particular context of Gödel fuzzy logic [11], natural generalizations that we will consider in this paper are the following. A possibility distribution $\pi : \Omega \to [0,1]$ on the set Ω of Gödel propositional interpretations induces the following generalized possibility and necessity measures over Gödel logic propositions:

J.P. Carvalho et al. (Eds.): IPMU 2016, Part II, CCIS 611, pp. 123–135, 2016.
DOI: 10.1007/978-3-319-40581-0_11

$$\Pi(\varphi) = \sup_{w\in\Omega}\{\min(\pi(w), w(\varphi))\}$$
$$N(\varphi) = \inf_{w\in\Omega}\{\pi(w) \Rightarrow w(\varphi)\},$$

where $\Rightarrow$ is Gödel implication, that is, for each $x, y \in [0,1]$, $x \Rightarrow y = 1$ if $x \leq y$, $x \Rightarrow y = y$, otherwise.[1] These expressions agree with the ones commonly used in many-valued modal Kripke frames (W, R) to respectively evaluate modal formulas $\Diamond\varphi$ and $\Box\varphi$ (see for example [2] and references therein) when the $[0,1]$-valued accessibility relation $R : W \times W \to [0,1]$ is defined by a possibility distribution $\pi : W \to [0,1]$ as $R(w, w') = \pi(w')$, for any $w, w' \in W$.

Actually, modal extensions of Gödel fuzzy logic have been studied by Caicedo and Rodriguez [5], providing sound and complete axiomatizations for different classes of $[0,1]$-valued Kripke models. These structures are triples $M = (W, R, e)$, where W is a set of worlds, $R = W \times W \to [0,1]$ is a many-valued accessibility relation and $e : W \times Var \to [0,1]$ is such that, for every $w \in W$, $e(w,\cdot)$ is a Gödel $[0,1]$-valued evaluation of propositional variables (more details in next section) that extends to modal formulas as follows:

$$e(w, \Diamond\varphi) = \sup_{w'\in W}\{\min(R(w, w'), e(w', \varphi))\}$$
$$e(w, \Box\varphi) = \inf_{w'\in W}\{R(w, w') \Rightarrow e(w', \varphi)\}.$$

We will denote by $\mathcal{KD}45(G)$ the class of $[0,1]$-models $M = (W, R, e)$ where R satisfies the following many-valued counterpart of the classical properties:

- *Seriality*: $\forall w \in W$, $\sup_{w'\in W} R(w, w') = 1$.
- *Transitivity*: $\forall w, w', w'' \in W$, $\min(R(w, w'), R(w', w'')) \leq R(w, w'')$.
- *Euclidean*: $\forall w, w', w'' \in W$, $\min(R(w, w'), R(w, w'')) \leq R(w', w'')$.

In this setting, the class $\Pi\mathcal{G}$ of *possibilistic Kripke models* (W, π, e), where $\pi : W \to [0,1]$ is a normalized possibility distribution on the set of worlds W, can be considered as the subclass of $\mathcal{KD}45(G)$ models (W, R, e) where R is such that $R(w, w') = \pi(w')$. Since $\Pi\mathcal{G} \subsetneq \mathcal{KD}45(G)$, it follows that the set $Val(\mathcal{KD}45(G))$ of valid formulas in the class of $\mathcal{KD}45(G)$ is a subset of the set $Val(\Pi\mathcal{G})$ of valid formulas in the class $\Pi\mathcal{G}$, i.e. $Val(\mathcal{KD}45(G)) \subseteq Val(\Pi\mathcal{G})$.

In the classical case (where truth-evaluations, accessibility relations and possibility distributions are $\{0,1\}$-valued) it is well known that (see e.g. [13]) that the semantics provided by the class of Kripke frames with serial, transitive and euclidean accessibility relations is equivalent to the class of Kripke frames with semi-universal accessibility relations (that is, relations of the form $R = W \times E$, where $\emptyset \neq E \subseteq W$). But the latter models are nothing else than $\{0,1\}$-valued possibilistic models, given by the characteristic functions of the E's.

However, over Gödel logic, the question of whether the semantics provided by the class of $[0,1]$-valued serial, transitive and euclidean Kripke frames is equivalent to the possibilistic semantics, that is, whether $Val(\Pi\mathcal{G}) = Val(\mathcal{KD}45(G))$ also holds, is not known. In this paper we positively solve this problem. Indeed

[1] Strictly speaking, the possibility measure is indeed a generalization of the classical one, but the necessity measure is not, since $x \Rightarrow 0 \neq 1 - x$.

we show that Caicedo-Rodriguez's Gödel modal logic $KD45(\mathbf{G})$ [5] properly captures the above possibilistic semantics. In this way, we extend the results obtained in [6] for the non-nested fragment of the modal language. We also note that this problem has already been solved for logics over finite and linearly ordered residuated lattices (MTL chains), thus in particular for finite-valued Gödel logics, but with a language expanded with truth-constants and with Baaz-Monteiro operator Δ, see [3,12].

After this introduction, in the next section we first summarize the main results by Caicedo-Rodriguez on Gödel modal logic $KD45(\mathbf{G})$ and its semantics given by $[0,1]$-valued serial, transitive and euclidean Kripke models. Then we consider our many-valued possibilistic Kripke semantics, and prove in the last section that it is equivalent to the relational one. We conclude with some open questions that we leave as future research. We also include an appendix with several technical proofs.

2 Gödel Kripke Frames

In their paper [5] Caicedo and Rodriguez consider a modal logic over Gödel logic. The language $\mathcal{L}_{\Box\Diamond}(Var)$ of propositional *bi-modal logic* is built from a countable set Var of propositional variables, connectives symbols $\vee, \wedge, \rightarrow, \bot$, and the modal operator symbols $\Box$ and $\Diamond$. We will simply write $\mathcal{L}_{\Box\Diamond}$ assuming Var is known and fixed. Then, the modal semantics is defined as follows.

Definition 1. A *Gödel-Kripke frame (GK-frame)* will be a structure $\mathcal{F} = \langle W, R\rangle$ where W is a non-empty set of objects that we call *worlds of* $\mathcal{F}$, and $R : W \times W \rightarrow [0,1]$. *A $\mathcal{F}$-Kripke Gödel model is a pair $M = \langle \mathcal{F}, e\rangle$ where $\mathcal{F}$ is a GK-frame and $e : W \times Var \rightarrow [0,1]$ provides in each world an evaluation of variables. e is inductively extended to arbitrary formulas in the following way:*

$$e(w, \varphi \wedge \psi) = \min(e(w,\varphi), e(w,\psi)) \quad e(w, \varphi \vee \psi) = \max(e(w,\varphi), e(w,\psi))$$
$$e(w, \varphi \rightarrow \psi) = e(w,\varphi) \Rightarrow e(w,\psi) \qquad e(w,\bot) = 0$$
$$e(w, \Box\varphi) = \inf_{w' \in W}\{R(w,w') \Rightarrow e(w',\varphi)\}$$
$$e(w, \Diamond\varphi) = \sup_{w' \in W}\{\min(R(w,w'), e(w',\varphi))\}.$$

Truth, validity and entailment are defined as usual: given a GK-model $M = (W, R, e)$, we write $(M, w) \models \varphi$ when $e(w, \varphi) = 1$, and $M \models \varphi$ if $(M, w) \models \varphi$ for every $w \in W$; given a class of GK-models $\mathcal{N}$, and a set of formulas T, we write $T \models_{\mathcal{N}} \varphi$ if, for every model $M = (W, R, e)$ and $w \in W$, $(M, w) \models \varphi$ whenever $(M, w) \models \psi$ for every $\psi \in T$.

In [5] it is shown that the set $Val(\mathcal{K}) = \{\varphi \mid \models_{\mathcal{K}} \varphi\}$ of valid formulas in $\mathcal{K}$, the class of all GK-frames, is axiomatized by adding the following additional axioms and rule to those of Gödel fuzzy logic G (see e.g. [11]):

$(K_\Box)\ \Box(\varphi \to \psi) \to (\Box\varphi \to \Box\psi)$ $(K_\Diamond)\ \Diamond(\varphi \vee \psi) \to (\Diamond\varphi \vee \Diamond\psi)$
$(F_\Box)\ \Box\top$ $(P)\ \Box(\varphi \to \psi) \to (\Diamond\varphi \to \Diamond\psi)$
$(FS2)\ (\Diamond\varphi \to \Box\psi) \to \Box(\varphi \to \psi)$ (Nec) from φ infer $\Box\varphi$.

The resulting logic will be denoted $K(G)$. Moreover, in [5] it is also shown that the set $Val(\mathcal{KD}45(G))$ of valid formulas in the subclass of GK-models $\mathcal{KD}45(G)$ is axiomatized by adding the following additional axioms:

$(D)\ \Diamond\top$
$(4_\Box)\ \Box\varphi \to \Box\Box\varphi$ $(4_\Diamond)\ \Diamond\Diamond\varphi \to \Diamond\varphi$
$(5_\Box)\ \Diamond\Box\varphi \to \Box\varphi$ $(5_\Diamond)\ \Diamond\varphi \to \Box\Diamond\varphi$.

The logic obtained by adding these axioms to $K(G)$ will be denoted $KD45(\mathbf{G})$.

3 More About $KD45(\mathbf{G})$

Let $\vdash_G$ denote deduction in Gödel fuzzy logic G. Let $\mathcal{L}(X)$ denote the set of formulas built by means of the connectives $\wedge, \to$, and $\bot$, from a given subset of variables $X \subseteq Var$. For simplicity, the extension of a valuation $v : X \to [0,1]$ to $\mathcal{L}(X)$ according to Gödel logic interpretation of the connectives will be denoted v as well. It is well known that G is complete for validity with respect to these valuations. We will need the fact that it is actually sound and complete in the following stronger sense, see [4].

Proposition 1. *(i) If* $T \cup \{\varphi\} \subseteq \mathcal{L}(X)$*, then* $T \vdash_G \varphi$ *iff* $\inf v(T) \leq v(\varphi)$ *for any valuation* $v : X \to [0,1]$.
(ii) If T *is countable, and* $T \nvdash_G \varphi_{i_1} \vee .. \vee \varphi_{i_n}$ *for each finite subset of a countable family* $\{\varphi_i\}_{i\in I}$ *there is an evaluation* $v : \mathcal{L}(X) \to [0,1]$ *such that* $v(\theta) = 1$ *for all* $\theta \in T$ *and* $v(\varphi_i) < 1$ *for all* $i \in I$.

The following are some theorems of $K(G)$, see [5]:

$T1.\ \neg\Diamond\theta \leftrightarrow \Box\neg\theta$
$T2.\ \neg\neg\Box\theta \to \Box\neg\neg\theta$
$T3.\ \Diamond\neg\neg\varphi \to \neg\neg\Diamond\varphi$
$T4.\ (\Box\varphi \to \Diamond\psi) \vee \Box((\varphi \to \psi) \to \psi)$
$T5.\ \Diamond(\varphi \to \psi) \to (\Box\varphi \to \Diamond\psi)$.

The first one is an axiom in Fitting's systems in [10], the next two were introduced in [5], the fourth one will be useful in our completeness proof and is the only one depending on prelinearity. The last is known as the first connecting axiom given by Fischer Servi.

Next we show that in $KD45(\mathbf{G})$ iterated modalities can be simplified. This is in accordance with our intended possibilistic semantics for $KD45(\mathbf{G})$ that will be formally introduced in next section.

Proposition 2. *The logic* $KD45(\mathbf{G})$ *proves the following schemes:*

$(F_{\Box\Diamond})\ \ \Diamond\Box\top \leftrightarrow \Box\Diamond\top \leftrightarrow \neg\bot$
$(U_\Diamond)\ \ \Diamond\Diamond\varphi \leftrightarrow \Diamond\varphi \leftrightarrow \Box\Diamond\varphi$
$(U_\Box)\ \ \Box\Box\varphi \leftrightarrow \Box\varphi \leftrightarrow \Diamond\Box\varphi$.

Proof. It is easy to prove $F_{\Box\Diamond}$ using axioms $F_\Box$ and D. The details are left to reader. For schemes $U_\Diamond$ and $U_\Box$, axioms $4_\Box$, $4_\Diamond$, $5_\Box$ and $5_\Diamond$ give one direction of them. The opposite directions are obtained as follows:

Proof 1.

$\Diamond\varphi \to \Box\Diamond\varphi$	axiom $5_\Diamond$
$\Box(\varphi \to \Diamond\varphi)$	by MP and $FS2$
$\Diamond\varphi \to \Diamond\Diamond\varphi$	by MP and P

Proof 2.

$\Diamond\Box\varphi \to \Box\varphi$	axiom $5_\Box$
$\Box(\Box\varphi \to \varphi)$	by MP and $FS2$
$\Box\Box\varphi \to \Box\varphi$	by MP and K

Proof 3.

$\Diamond\varphi \to \Diamond\varphi$	prop. taut.
$\Diamond(\Diamond\varphi \to \Diamond\varphi)$	by D
$\Box\Diamond\varphi \to \Diamond\Diamond\varphi$	by MP and $T5$
$\Box\Diamond\varphi \to \Diamond\varphi$	by $4_\Diamond$

Proof 4.

$\Box\varphi \to \Box\varphi$	prop. taut.
$\Diamond(\Box\varphi \to \Box\varphi)$	by D
$\Box\Box\varphi \to \Diamond\Box\varphi$	by MP and $T5$
$\Box\varphi \to \Diamond\Box\varphi$	by $4_\Box$

■

From now on we will use $ThKD45(\mathbf{G})$ to denote the set of theorems of $KD45(\mathbf{G})$. We close this section with the following observation: deductions in $KD45(\mathbf{G})$ can be reduced to derivations in pure propositional Gödel logic G.

Lemma 1. *For any theory T and formula φ in $\mathcal{L}_{\Box\Diamond}$, it holds that $T \vdash_{KD45(\mathbf{G})} \varphi$ iff $T \cup ThKD45(\mathbf{G}) \vdash_G \varphi$.*

4 Possibilistic Semantics and Completeness

In this section we will show that $KD45(\mathbf{G})$ is also complete with respect to the class of possibilistic Gödel frames.

Definition 2. *A* possibilistic Gödel frame *(ΠG-frame) will be a structure $\langle W, \pi\rangle$ where W* is a non-empty set of worlds, and *$\pi : W \to [0,1]$ is a normalized possibility distribution over W, that is, such that* $\sup_{w\in W} \pi(w) = 1$.

A possibilistic Gödel model *is a triple $\langle W, \pi, e\rangle$ where $\langle W, \pi\rangle$ is a ΠG-frame frame and $e : W \times Var \to [0,1]$ provides an evaluation of variables in each world. For each $w \in W$, $e(w, \cdot)$ extends to arbitrary formulas in the usual way for the propositional connectives and for modal operators in the following way:*

$$e(w, \Box\varphi) := \inf_{w'\in W}\{\pi(w') \Rightarrow e(w', \varphi)\}$$
$$e(w, \Diamond\varphi) := \sup_{w'\in W}\{\min(\pi(w'), e(w', \varphi))\}.$$

Observe that the evaluation of formulas beginning with a modal operator is in fact independent from the current world. As we already mentioned in the introduction, it is clear that a possibilistic frame $\langle W, \pi\rangle$ is equivalent to the GK-frame $\langle W, R_\pi\rangle$ where $R_\pi = W \times \pi$.

In the rest of the paper we provide a completeness proof of the logic $KD45(\mathbf{G})$ with respect of the class $\Pi\mathcal{G}$ of possibilistic Gödel models, in fact we are going to prove weak completeness for deductions from finite theories.

In what follows, for any formula φ we denote by $Sub(\varphi) \subseteq \mathcal{L}_{\Box\Diamond}$ the set of subformulas of φ and containing the formula $\bot$. Moreover, let $X := \{\Box\theta, \Diamond\theta :$

$\theta \in \mathcal{L}_{\Box\Diamond}\}$ be the set of formulas in $\mathcal{L}_{\Box\Diamond}$ beginning with a modal operator; then $\mathcal{L}_{\Box\Diamond}(Var) = \mathcal{L}(Var \cup X)$. That is, any formula in $\mathcal{L}_{\Box\Diamond}(Var)$ may be seen as a propositional Gödel formula built from the extended set of propositional variables $Var \cup X$. In addition, for a given formula φ, let $\sim_\varphi$ be equivalence relation in $[0,1]^{Var\cup X} \times [0,1]^{Var\cup X}$ defined as follows:

$$u \sim_\varphi w \text{ iff } \forall \psi \in Sub(\varphi) : u(\Box\psi) = w(\Box\psi) \text{ and } u(\Diamond\psi) = w(\Diamond\psi).$$

Now, assume that a formula φ is not a theorem of $KD45(\mathbf{G})$. Hence by completeness of Gödel calculus and Lemma 1, there exists a Gödel valuation v such that $v(ThKD45(\mathbf{G})) = 1$ and $v(\varphi) < 1$. Following the usual canonical model construction, once fixed the valuation v, we define next a canonical $\Pi\mathcal{G}$-model M^v_φ in which we will show φ is not valid.

The *canonical model* $M^v_\varphi = (W^v, \pi^\varphi, e^\varphi)$ is defined as follows:

- W^v is the set $\{u \in [0,1]^{Var\cup X} \mid u \sim_\varphi v \text{ and } u(ThKD45(\mathbf{G})) = 1\}$.
- $\pi^\varphi(u) = \min_{\psi\in Sub(\varphi)}\{\min(v(\Box\psi) \to u(\psi), u(\psi) \to v(\Diamond\psi))\}$.
- $e^\varphi(u,p) = u(p)$ for any $p \in Var$.

In this context, we call the elements of $\Delta_\varphi = \{\Box\theta, \Diamond\theta : \theta \in Sub(\varphi)\}$, the *fixed points* of the Canonical Model.

Note that having $\nu(ThKD45(\mathbf{G})) = 1$ does not guarantee that ν belongs to the canonical model because it may not take the appropriated values for the fixed points, i.e. it may be that $v \not\sim_\varphi \nu$. However, the next lemma shows how, in certain conditions, to transform such an evaluation into another belonging to the canonical model.

Lemma 2. *Let $u \in W^v$ and let $\nu : Var \cup X \mapsto [0,1]$ be a Gödel valuation. Define $\alpha = \max\{u(\lambda) : \nu(\lambda) < 1$ and $\lambda \in \Delta_\varphi\}$ and $\beta = \min\{u(\lambda) : \nu(\lambda) = 1$ and $\lambda \in \Delta_\varphi\}$. If ν satisfies the following conditions:*

a. $\nu(ThKD45(\mathbf{G})) = 1$.
b. *for any $\psi, \phi \in \{\lambda : u(\lambda) \leq \alpha$ and $\lambda \in \Delta_\varphi\}$, $\nu(\psi) < \nu(\phi)$ iff $u(\psi) < u(\phi)$.*
c. *$\nu(\lambda) = 1$ for every $\lambda \in \Delta_\varphi$ such that $u(\lambda) > \alpha$,*

then, there exists a Gödel valuation $w \in W^v$ such that, for any formulas ψ, ϕ:

1. *$\nu(\psi) = 1$ implies $w(\psi) \geq \delta$.*
2. *$\nu(\psi) < 1$ implies $w(\psi) < \delta$.*
3. *$1 \neq \nu(\psi) \leq \nu(\phi)$ implies $w(\psi) \leq w(\phi)$.*
4. *$\nu(\psi) < \nu(\phi)$ implies $w(\psi) < w(\phi)$.*
5. *$\nu(\psi) = \nu(\phi) = 1$ and $u(\psi) \leq u(\phi)$ imply $w(\psi) \leq w(\phi)$.*
6. *$\nu(\psi) = \nu(\phi) = 1$ and $u(\psi) < u(\phi)$ imply $w(\psi) < w(\phi)$.*

Proof. See Appendix.

Completeness will follow from the next truth-lemma.

Lemma 3 (Truth-lemma). *$e^{\varphi}(u, \psi) = u(\psi)$ for any $\psi \in Sub(\varphi)$ and any $u \in W^v$.*

Proof. For simplicity, we write W for W^v. We prove the identity by induction on the complexity of the formulas in $Sub(\varphi)$, considered now as elements of $\mathcal{L}_{\Box\Diamond}(Var)$. For $\bot$ and the propositional variables in $Sub(\varphi)$ the equation holds by definition. The only non trivial inductive steps are: $e^{\varphi}(u, \Box\psi) = u(\Box\psi)$ and $e^{\varphi}(u, \Diamond\psi) = u(\Diamond\psi)$ for $\Box\psi, \Diamond\psi \in Sub(\varphi)$. By the inductive hypothesis we may assume that $e^{\varphi}(u', \psi) = u'(\psi)$ for every $u' \in W$; thus we must prove

$$\inf_{u' \in W} \{\pi^{\varphi}(u') \Rightarrow u'(\varphi)\} = u(\Box\varphi) \tag{1}$$

$$\sup_{v' \in W} \{\min(\pi^{\varphi}(u'), u'(\varphi))\} = u(\Diamond\varphi). \tag{2}$$

By definition, $\pi^{\varphi}(u') \leq (v(\Box\psi) \Rightarrow u'(\psi))$ and $\pi^{\varphi}(u') \leq (u'(\psi) \Rightarrow v(\Diamond\psi))$ for any $\psi \in Sub(\varphi)$ and $u' \in W$; therefore, $u(\Box\psi) = v(\Box\psi) \leq (\pi^{\varphi}(u') \Rightarrow u'(\psi))$ and $\min(\pi^{\varphi}(u'), u'(\psi)) \leq v(\Diamond\psi) = u(\Diamond\psi)$. Taking infimum over u' in the first inequality and the supremum in the second we get

$$u(\Box\psi) \leq \inf_{u' \in W} \{\pi^{\varphi}(u') \Rightarrow u'(\psi)\}, \quad \sup_{u' \in W} \{\min(\pi^{\varphi}(u'), u'(\psi))\} \leq u(\Diamond\psi).$$

Hence, if $u(\Box\psi) = 1$ and $u(\Diamond\psi) = 0$ we obtain (1) and (2), respectively. Therefore, it only remains to prove the next two claims for $\Box\psi, \Diamond\psi \in Sub(\varphi)$.

Claim 1. *If $u(\Box\psi) = \alpha < 1$, for every $\varepsilon > 0$, there exists a valuation $w \in W$ such that $\pi^{\varphi}(w) > w(\psi)$ and $w(\psi) < \alpha + \varepsilon$, and thus $(\pi^{\varphi}(w) \Rightarrow w(\psi)) < \alpha + \varepsilon$.*

Claim 2. *If $u(\Diamond\psi) = \alpha > 0$ then, for any $\varepsilon > 0$, there exists a valuation $w' \in W$ such that $w'(\psi) = 1$ and $\pi^{\varphi}(w') \geq \alpha - \varepsilon$, and thus $\min(w'(\psi), \pi^{\varphi}(w')) \geq \alpha - \varepsilon$.*

The proof of these two claims are rather involved and they can be found in the appendix. ∎

Theorem 1 (Finite strong completeness). *For any finite theory T and formula φ in $\mathcal{L}_{\Box\Diamond}$, $T \models_{\Pi\mathcal{G}} \varphi$ implies $T \vdash_{KD45(\mathbf{G})} \varphi$.*

Proof. One direction is soundness, and it is easy to check that the axioms are valid in the class $\Pi\mathcal{G}$ of models. As for the other direction, assume $T = \emptyset$ and $\not\vdash_{KD45(\mathbf{G})} \varphi$. Then $ThKD45(\mathbf{G}) \not\vdash_{G} \varphi$ by Lemma 1, and thus there is, by Proposition 1, a Gödel valuation $v : Var \cup X \to [0, 1]$ such that $v(\varphi) < v(ThKD45(\mathbf{G})) = 1$. Then v is a world of the canonical model M_v^{φ} and by Lemma 3, $e^{\varphi}(v, \varphi) = v(\varphi) < 1$. Thus $\not\models_{\Pi\mathcal{G}} \varphi$. This proof can be easily generalized when T is a non empty and finite. ∎

5 Conclusions

In this paper we have studied the logic over Gödel fuzzy logic arising from many-valued Gödel Kripke models with possibilistic semantics, and have shown that it actually corresponds to a simplified semantics for the logic $KD45(\mathbf{G})$, the extension of Caicedo and Rodriguez's bi-modal Gödel logic with many-valued versions of the well-known modal axioms D, 4 and 5. The truth-value of a formula $\Diamond\varphi$ in a possibilistic Kripke model is indeed a proper generalization of the possibility measure of φ when φ is a classical proposition, however the semantics of $\Box\varphi$ is not. This is due to the fact that the negation in Gödel logic is not involutive.

Therefore, a first open problem we leave for further research is to consider to extension of the logic $KD45(\mathbf{G})$ with an involutive negation and investigate its possibililistic semantics. A second open problem is to investigate the logic arising from *non-normalized* possibilistic Gödel frames. In the classical case, one can show that this corresponds to the modal logic $K45$, that is, without the axiom D, see e.g. [13]. However, over Gödel logic this seems to be not as straightforward as in the classical case.

Acknowledgments. The authors are grateful to the anonymous reviewers for their helpful comments. They acknowledge partial support by the H2020-MSCA-RISE-2015 project SYSMICS, the Spanish MINECO/FEDER project RASO (TIN2015-71799-C2-1-P) and the Argentinean project PIP CONICET 11220150100412CO.

Appendix

Proof of Lemma 2

Proof. First of all, notice that if ν satisfies the condition **b**, then necessarily $\alpha < \beta$. Let $B = \{\nu(\lambda) : \lambda \in \Delta_\varphi, \nu(\lambda) < 1\} \cup \{0\} = \{b_0 = 0 < b_1 < \ldots b_N\}$. Obviously, $b_N < 1$. For each $0 \leq i \leq N$, pick $\lambda_i \in \Delta_\varphi$ such that $\nu(\lambda_i) = b_i$. Define now a continuous strictly function $g : [0,1] \mapsto [0,\delta) \cup \{1\}$ such that

$$g(1) = 1$$
$$g(b_i) = v(\lambda_i) \text{ for every } 0 \leq i \leq N$$
$$g[(b_N, 1)] = (\alpha, \delta).$$

Notice that $\alpha = g(b_N)$. In addition, define another continuous strictly increasing function $h : [0,1] \mapsto [\delta, 1]$ such that

$$h(0) = \delta$$
$$h[(0,\beta)] = (\delta, \beta)$$
$$h(x) = x, \text{ for } x \in [\beta, 1].$$

Then we define the valuation $w : Var \cup X \to [0,1]$ as follows:

$$w(p) = \begin{cases} g(\nu(p)), \text{ if } \nu(p) < 1, \\ h(u(p)), \text{ if } \nu(p) = 1. \end{cases}$$

First of all, let us show by induction that this extends to any propositional formula, that is,

$$w(\varphi) = \begin{cases} g(\nu(\varphi)), \text{ if } \nu(\varphi) < 1, \\ h(u(\varphi)), \text{ if } \nu(p) = 1. \end{cases}$$

Note that, since g and h are strictly increasing mappings, $g \circ \nu$ and $h \circ u$ are valuations as well. So, in the induction steps below we only need to check that everything is fine when both are used at the same time when evaluating a compound formula. The base case holds by definition.

- Assume $\psi = \psi_1 \wedge \psi_2$. We only check the case when $v(\psi) < 1$ and $v(\psi_1) < 1$ and $v(\psi_2) = 1$. Then $w(\psi) = \min(w(\psi_1), w(\psi_2)) = \min(g(\nu(\psi_1)), h(u(\psi_2))) = g(\nu(\psi_1))$, since $g(\nu(\psi_1)) < \delta \leq h(u(\psi_2)$. But, $g(\nu(\psi_1)) = \min(g(\nu(\psi_1)), 1) = \min(g(\nu(\psi_1)), g(\nu(\psi_2))) = g(\nu(\psi_1 \wedge \psi_2)) = g(\nu(\psi))$.
- Assume $\psi = \psi_1 \to \psi_2$, and consider two subcases:
  (1) $v(\psi_1) < 1$ and $v(\psi_2) = 1$. Then $v(\psi_1 \to \psi_2) = 1$ and $w(\psi) = w(\psi_1) \Rightarrow w(\psi_2) = g(\nu(\psi_1)) \Rightarrow h(u(\psi_2)) = 1 = h(u(\psi_1)) \Rightarrow h(u(\psi_2)) = h(u(\psi_1 \to \psi_2)) = h(u(\psi))$.
  (2) $v(\psi_1) = 1$ and $v(\psi_2) < 1$. Then $v(\psi_1 \to \psi_2) = v(\psi_2) < 1$ and $w(\psi) = w(\psi_1) \Rightarrow w(\psi_2) = h(u(\psi_1)) \Rightarrow g(\nu(\psi_2)) = g(\nu(\psi_2)) = g(\nu(\psi_1)) \Rightarrow g(\nu(\psi_2)) = g(\nu(\psi_1 \to \psi_2)) = g(\nu(\psi))$.

Properties 1 – 6 in the statement of Lemma 2 now directly follow from the above.

Finally, we prove that $w \in W^v$. By definition of w it is clear that $w \sim_\varphi v$. It remains to check that w validates all the axioms. The axioms of $\mathcal{G}$ are evaluated to 1 by any Gödel valuation. As for the specific axioms of $KD45(\mathbf{G})$, it is an immediate consequence of Property 3 because it implies that if $\nu(\psi \to \phi) = 1$ then $w(\psi \to \phi) = 1$. ■

Claim 1 from Lemma 3. *If* $u(\Box\psi) = \alpha < 1$, *for every* $\varepsilon > 0$, *there exists a valuation* $w \in W$ *such that* $\pi^\varphi(w) > w(\psi)$ *and* $w(\psi) < \alpha + \varepsilon$, *and thus* $\pi^\varphi(w) \Rightarrow w(\psi) = w(\psi) < \alpha + \varepsilon$.

Proof. By definition of Gödel's implication $\Rightarrow$ in [0,1], to grant the required conditions on w it is enough to find $w \in W$ such that $\alpha \leq w(\psi)$ and, for any $\theta \in Sub(\varphi)$, $u(\Box\theta) \leq w(\theta) \leq u(\Diamond\theta) \leq \alpha$. This is achieved in two stages:

- first producing a valuation $\nu \in W$ satisfying $\nu(\psi) < 1$ and preserving the relative ordering conditions the values $w(\theta)$ must satisfy, conditions which may be coded by a theory $\Gamma_{\psi,u}$;
- and then moving the values $\nu(\theta)$, for $\theta \in Sub(\varphi)$, to the correct valuation w by composing ν with an increasing bijection of [0,1].

Assume $u(\Box\psi) = \alpha < 1$, and define (all formulas involved ranging in $\mathcal{L}_{\Box\Diamond}(Var)$)

$$
\begin{aligned}
\Gamma_{\psi,u} = \; & \{\lambda : \lambda \in \Delta_\varphi \text{ and } u(\lambda) > \alpha\} \\
& \cup \{\lambda \to \theta : \lambda \in \Delta_\varphi \text{ and } u(\lambda) \leq u(\Box\theta)\} \\
& \cup \{(\theta \to \lambda) \to \lambda : \lambda \in \Delta_\varphi \text{ and } u(\lambda) < u(\Box\theta) < 1\} \\
& \cup \{\theta \to \lambda : \lambda \in \Delta_\varphi \text{ and } u(\Diamond\theta) \leq u(\lambda)\} \\
& \cup \{(\lambda \to \theta) \to \theta : \lambda \in \Delta_\varphi \text{ and } u(\Diamond\theta) < u(\lambda) < 1\}
\end{aligned}
$$

Then we have $u(\Box\xi) > \alpha$ for each $\xi \in \Gamma_{\psi,u}$. Indeed, first recall that, by $U_\Box$ and $U_\Diamond$ of Proposition 2, for any $\lambda \in \Delta_\varphi$ we have $u(\lambda) = u(\Box\lambda) = u(\Diamond\lambda)$. We analyse case by case. For the first set of formulas, it is clear by construction. For the second set, we have $u(\Box(\lambda \to \theta)) \geq u(\Diamond\lambda \to \Box\theta) = u(\Diamond\lambda) \Rightarrow u(\Box\theta) = u(\lambda) \Rightarrow u(\Box\theta) = 1$, by $FS2$. For the third, by FS2 and P, we have $u(\Box((\theta \to \lambda) \to \lambda)) \geq u(\Diamond(\theta \to \lambda) \to \Box\lambda) \geq u((\Box\theta \to \Diamond\lambda) \to \Box\lambda) = 1$, since $u(\Box\lambda) = u(\Diamond\lambda) = u(\Box\theta \to \Diamond\lambda) < 1$. The fourth and fifth cases are very similar to the second and third ones respectively.

This implies

$$\Gamma_{\psi,u} \nvdash_{KD45(\mathbf{G})} \psi,$$

otherwise there would exist $\xi_1, \ldots, \xi_k \in \Gamma_{\psi,u}$ such that $\xi_1, \ldots, \xi_k \vdash_{KD45(\mathbf{G})} \psi$. In such a case, we would have $\Box\xi_1, \ldots, \Box\xi_k \vdash_{KD45(\mathbf{G})} \Box\psi$ by Nec and $K_\Box$. Then $\Box\xi_1, \ldots, \Box\xi_k, ThKD45(\mathbf{G}) \vdash_G \Box\psi$ by Lemma 1 and thus by Proposition 1 (i), and recalling that $u(ThKD45(\mathbf{G})) = 1$,

$$\alpha < \inf u(\{\Box\xi_1, \ldots, \Box\xi_k\} \cup ThKD45(\mathbf{G})) \leq u(\Box\psi) = \alpha,$$

a contradiction. Therefore, by Proposition 1 (ii) there exists a valuation $\nu : Var \cup X \mapsto [0,1]$ such that $\nu(\Gamma_{\psi,u} \cup ThKD45(\mathbf{G})) = 1$ and $\nu(\psi) < 1$. This implies the following relations between u and ν, that we list for further use. Given $\lambda \in \Delta_\varphi, \theta \in \mathcal{L}_{\Box\Diamond}(Var)$, we have :

#1. If $u(\lambda) > \alpha$ then $\nu(\lambda) = 1$ (since then $\lambda \in \Gamma_{\psi,u}$).

#2. If $u(\lambda) \leq u(\Box\theta)$ then $\nu(\lambda) \leq \nu(\theta)$ (since then $\lambda \to \theta \in \Gamma_{\psi,u}$). In particular, if $\lambda_1, \lambda_2 \in \Delta_\varphi$ and $u(\lambda_1) \leq u(\Box\lambda_2) = u(\lambda_2)$ then $\nu(\lambda_1) \leq \nu(\lambda_2)$. Furthermore,if $\Box\theta \in \Delta_\varphi$ then from $u(\Box\theta) = u(\Box\theta)$ by #2, $\nu(\Box\theta) \leq \nu(\theta)$. That means, taking $\theta = \psi$, $\nu(\Box\psi) \leq \nu(\psi) < 1$.

#3. If $u(\lambda) < u(\Box\theta) < 1$ then $\nu(\lambda) < \nu(\theta)$ or $\nu(\lambda) = 1$ (since then $(\theta \to \lambda) \to \lambda) \in \Gamma_{\psi,u}$). In particular, if $\lambda_1, \lambda_2 \in \Delta_\varphi$, $u(\lambda_1) < u(\lambda_2)$ and $u(\lambda_2) \leq u(\Box\psi) = \alpha$ then $\nu(\lambda_1) < \nu(\psi) < 1$ and thus $\nu(\lambda_1) < \nu(\lambda_2)$. This means that ν preserves in a strict sense the order values by u of the formulas $\lambda \in \Delta_\varphi$ such that $u(\lambda) \leq \alpha$.

#4. If $u(\Diamond\theta) \leq u(\lambda)$ then $\nu(\theta) \leq \nu(\lambda)$ (because $\theta \to \lambda \in \Gamma_{\psi,u}$). In particular, if $\Diamond\theta \in \Delta_\varphi$ then $\nu(\theta) \leq \nu(\Diamond\theta)$.

#5. If $u(\Diamond\theta) < u(\lambda) < 1$ then $\nu(\theta) < \nu(\lambda)$ or $\nu(\theta) = 1$. In particular, if $\lambda_1, \lambda_2 \in \Delta_\varphi$ and $u(\lambda_1) < u(\lambda_2) \leq \alpha = u(\Box\psi)$ then $\nu(\lambda_1) < \nu(\lambda_2)$. Furthermore, if $u(\lambda_2) > 0$ then $\nu(\lambda_2) > 0$ (making $\lambda_1 := \Diamond\bot$ since $u(\bot) = u(\Diamond\bot) = 0$).

According to the properties #1, #2 and #3, it is clear that ν satisfies the conditions of Lemma 2. Consequently, for all $\epsilon > 0$ (such that $\alpha + \varepsilon < \beta$), taking $\delta = \alpha + \varepsilon$ in Lemma 2, there exists a valuation $w \in W^v$ such that $w(\psi) < \alpha + \varepsilon = \delta$. Then in order to finish our proof, it remains to show that:

$$\pi^\varphi(w) = \inf_{\lambda \in sub(\varphi)} \min(v(\Box\lambda) \Rightarrow w(\lambda), w(\lambda) \Rightarrow v(\Diamond\lambda)) > w(\varphi) \tag{3}$$

To do so, we will prove that, for any $\lambda \in sub(\varphi)$, both implications in (3) are greater than δ.[2] First we prove it for the first implication by cases:

[2] Remember that $u \sim_\varphi v \sim_\varphi w$.

- If $v(\Box\lambda) \leq \alpha < 1$ then $\min(v(\Box\lambda) \Rightarrow w(\lambda), w(\lambda) \Rightarrow v(\Diamond\lambda)) = 1$. Indeed, first of all, by #2, from $u(\Box\lambda) = v(\Box\lambda) \leq \alpha = u(\Box\psi)$ it follows $\nu(\Box\lambda) \leq \nu(\psi) < 1$. Now, since $u(\Box\lambda) \leq u(\Box\lambda)$, by #2, we have $1 \neq \nu(\Box\lambda) \leq \nu(\lambda)$, and by 3 of Lemma 2 we have $v(\Box\lambda) = w(\Box\lambda) \leq w(\lambda)$. Then $v(\Box\lambda) \Rightarrow w(\lambda) = 1$.
- If $v(\Box\lambda) > \alpha$ then by #1 and #2, $1 = \nu(\Box\lambda) \leq \nu(\lambda)$. Therefore, by 1 of Lemma 2, $w(\lambda) > \delta$ which implies $v(\Box\lambda) \Rightarrow w(\lambda) > \delta$.

For the second implication we also consider two cases:

- If $v(\Diamond\lambda) = u(\Diamond\lambda) > \delta$ then it is obvious that $w(\lambda) \Rightarrow v(\Diamond\lambda) > \delta$.
- If $u(\Diamond\lambda) < \delta$, by definition of δ and taking into account that $\Diamond\lambda \in \Delta_\varphi$, then $u(\Diamond\lambda) < \alpha$. Now from $u(\Diamond\lambda) = u(\Diamond\lambda)$ we obtain by #4, that $\nu(\lambda) \leq \nu(\Diamond\lambda) < 1$. Then by Lemma 2 we have $w(\lambda) \leq w(\Diamond\lambda) = v(\Diamond\lambda)$ and thus $w(\lambda) \Rightarrow v(\Diamond\lambda) = 1$. ■

Claim 2 from Lemma 3. *If* $u(\Diamond\psi) = \alpha > 0$ *then, for any* $\varepsilon > 0$, *there exists a valuation* $w' \in W$ *such that* $w'(\psi) = 1$ *and* $\pi^\varphi(w') \geq \alpha - \varepsilon$, *and thus* $\min(w'(\psi), \pi^\varphi(w')) \geq \alpha - \varepsilon$.

Proof. Assume $u(\Diamond\psi) = \alpha > 0$ and define $\Gamma_{\psi,u}$ in the same way that it was defined in the proof of Claim 1. Then we consider two cases:

- **If** $u(\Diamond\psi) = 1$, let $U_{\psi,u} = \{\lambda :\ \lambda \in \Delta_\varphi \text{ and } u(\lambda) < 1\}$. We claim that

$$\psi, \Gamma_{\psi,u} \not\vdash_{KD45(\mathbf{G})} \bigvee U_{\psi,u}\ ,$$

otherwise we would have $\theta_1, \ldots, \theta_n \in \Gamma_{\psi,u}$ such that $\vdash_{KD45(\mathbf{G})} \psi \to ((\theta_1 \wedge \ldots \wedge \theta_n) \to \bigvee U_{\psi,u})$, and then we would also have $\vdash_{KD45(\mathbf{G})} \Diamond\psi \to \Diamond((\theta_1 \wedge \ldots \wedge \theta_n) \to \bigvee U_{\psi,u})$, that would imply in turn that $\vdash_{KD45(\mathbf{G})} \Diamond\psi \to ((\Box\theta_1 \wedge \ldots \wedge \Box\theta_n) \to \Diamond \bigvee U_{\psi,u})$. In that case, taking the evaluation u it would yield: $1 = u(\Diamond\psi) \leq u(\Box\theta_1 \wedge \ldots \wedge \Box\theta_n) \Rightarrow u(\Diamond \bigvee U_{\psi,u})$, a contradiction, since $u(\Box\theta_1 \wedge \ldots \wedge \Box\theta_n) = 1$ and $u(\Diamond \bigvee U_{\psi,u})) < 1$.[3]

Therefore, there is a Gödel valuation ν' (not necessarily in W) such that $\nu'(\psi) = \nu'(\Gamma_{\psi,u}) = \nu'(ThKD45(\mathbf{G})) = 1$ and $\nu'(\bigvee U_{\psi,u}) < 1$. By #2 and #3, it follows that for any $\lambda_1, \lambda_2 \in \Delta_\varphi$ such that $u(\lambda_1), u(\lambda_2) \leq \alpha$, we have $u(\lambda_1) < u(\lambda_2) \leq \alpha$ iff $\nu'(\lambda_1) < \nu'(\lambda_2)$. Thus, ν' satisfies the conditions of Lemma 2 because it is strictly increasing in Δ_φ (i.e. it satisfies condition **b** of Lemma 2), and $\nu'(ThKD45(\mathbf{G})) = 1$. Therefore, there exists a valuation $w' \in W$ such that $w'(\psi) = 1$.

It remains to show that $\pi^\varphi(w') = 1$. Indeed, by construction, it holds that $u(\Box\theta) \leq w'(\theta) \leq u(\Diamond\theta)$, and hence $\min(u(\Box\theta) \Rightarrow w'(\theta), w'(\theta) \Rightarrow u(\Diamond\theta)) = 1$.

- **If** $1 > u(\Diamond\psi) = \alpha > 0$, then we let $U_{\psi,u} =\ (\Diamond\psi \to \psi) \to \psi$. We claim that

[3] Note that, in this case, the first member of the union defining $\Gamma_{\psi,u}$ is the empty set.

$$\Box\top, \Gamma_{\psi,u} \nvdash_{KD45(\mathbf{G})} U_{\psi,u} ,$$

otherwise there would exist $\theta_1, \ldots, \theta_n \in \Gamma_{\psi,u}$ such that $\vdash_{KD45(\mathbf{G})} \Box\top \to ((\theta_1 \wedge \ldots \wedge \theta_n) \to U_{\psi,u})$, and then we would have $\vdash_{KD45(\mathbf{G})} \Diamond\Box\top \to \Diamond((\theta_1 \wedge \ldots \wedge \theta_n) \to U_{\psi,u})$, which would imply $\vdash_{KD45(\mathbf{G})} \Box\top \to ((\Box\theta_1 \wedge \ldots \wedge \Box\theta_n) \to \Diamond U_{\psi,u})$. In that case, evaluating with u it would yield $1 = u(\Box\top) \leq u(\Box\theta_1 \wedge \ldots \wedge \Box\theta_n) \Rightarrow u(\Diamond U_{\psi,u}))$, contradiction, since $u(\Box\theta_1 \wedge \ldots \wedge \Box\theta_n) > \alpha$ and $u(\Diamond U_{\psi,u})) \leq \alpha$ (because $u(\Diamond((\Diamond\psi \to \psi) \to \psi)) \leq u(\Box(\Diamond\psi \to \psi) \to \Diamond\psi) \leq u(\Diamond\psi) \leq \alpha$).

Therefore, there is an evaluation ν' such that $\nu'(ThKD45(\mathbf{G})) = \nu'(\Gamma_{\psi,u}) = 1$ and $\nu'(\bigvee U_{\psi,u}) < 1$. Hence, we can conclude that the three pre-conditions **a**, **b** and **c** required in Lemma 2 are satisfied. In addition, the following condition is also satisfied:

d. $\nu'(\Diamond\psi) = \nu'(\psi)$.

At this point, we can now do a proof dual to the one for Claim 1. Again, by Lemma 2 for $\delta = \frac{\beta-\alpha}{2}$, we obtain from ν' an evaluation $w' \in W^v$ such that $w'(\psi) = \alpha$. It only remains then to show that $\pi^{\varphi}(w) > \alpha$. But in this case, the proof is the same than the one given for Eq. (3) using w' instead of w. This finishes the proof. ■

References

1. Banerjee, M., Dubois, D.: A simple logic for reasoning about incomplete knowledge. Int. J. Approximate Reasoning **55**(2), 639–653 (2014)
2. Bou, F., Esteva, F., Godo, L., Rodríguez, R.: On the minimum many-valued modal logic over a finite residuated lattice. J. Logic Comput. **21**(5), 739–790 (2011)
3. Bou, F., Esteva, F., Godo, L.: On possibilistic modal logics defined over MTL-chains. In: Montagna, F. (ed.) Petr Hájek on Mathematical Fuzzy Logic. Outstanding Contributions to Logic, vol. 6, pp. 225–244. Springer, Switzerland (2015)
4. Caicedo, X., Rodríguez, R.: Standard Gödel modal logics. Studia Logica **94**(2), 189–214 (2010)
5. Caicedo, X., Rodríguez, R.: Bi-modal Gödel modal logics. J. Logic Comput. **25–1**, 37–55 (2015)
6. Dellunde, P., Godo, L., Marchioni, E.: Extending possibilistic logic over Gödel logic. Int. J. Approx. Reasoning **52**(1), 63–75 (2011)
7. Dubois, D., Lang, J., Prade, H.: Possibilistic logic, In: Gabbay, et al. (ed.) Handbook of Logic in Artificial Intelligence and Logic Programming, Non monotonic Reasoning and Uncertain Reasoning, vol. 3, pp. 439–513. Oxford UP (1994)
8. Dubois, D., Prade, H.: Possibilistic logic: a retrospective and prospective view. Fuzzy Sets Syst. **144**, 3–23 (2004)
9. Dubois, D., Prade, H., Schockaert, S.: Reasoning about uncertainty and explicit ignorance in generalized possibilistic logic. Proc. ECAI **2014**, 261–266 (2014)
10. Fitting, M.: Many valued modal logics. Fundamenta Informaticae **15**, 325–254 (1991)
11. Hájek, P.: Metamathematics of Fuzzy Logic. Trends in Logic, vol. 4. Kluwer, Dordrecht (1998)

12. Hájek, P., Harmancová, D., Esteva, F., Garcia, P., Godo, L.: On modal logics for qualitative possibility in a fuzzy setting. In: Proceeding of the 94 Uncertainty in Artificial Intelligence Conference (UAI 1994), pp. 278–285. Morgan Kaufmann (1994)
13. Pietruszczak, A.: Simplified Kripke style semantics for modal logics K45, KB4 and KD45. Bull. Sect. Logic **38**(3/4), 163–171 (2009)

A Calculus for Rational Łukasiewicz Logic and Related Systems

Paolo Baldi(✉)

Institute of Computer Languages, TU Wien, Favoritenstrasse 9-11, 1040 Wien, Austria
baldi@logic.at

Abstract. We introduce hypersequent calculi for Rational Łukasiewicz logic and for the logic $KZ(\pi)$, an extension of Kleene-Zadeh logic, motivated by game semantic investigations.

1 Introduction

Fuzzy Logic is nowadays a vast research area, which offers many different methods and tools to handle vagueness for computational purposes. In particular, in the area of so-called Fuzzy Logic in narrow sense or Mathematical Fuzzy Logic [4] many axiomatic systems have been so far introduced and investigated, in order to characterize valid inferences under vagueness. One of the most prominent such system is *Łukasiewicz logic* Ł. This logic is an important example of a *t*-norm based logic, together with Gödel and Product logic, see e.g. [4,9]. In its intended or *standard* semantics, truth values are taken over the real interval $[0,1]$ and the (strong) conjunction and implication connectives are interpreted by the well known Łukasiewicz t-norm $x * y = \max(0, x + y - 1)$ and its residuum $x \to y = min(1, 1 - x + y)$, respectively.

In this paper we focus on *Rational Łukasiewicz logic* RŁ, an expansion of Ł with a family of unary connectives $\{\delta_n\}_{n\in\mathbb{N}}$, standing for *division* operators. In other words, the intended evaluation v over the real interval $[0,1]$ of a formula $\delta_n\alpha$ is defined by $v(\delta_n\alpha) = v(\alpha)/n$ where/stands for the usual division.

The name of the logic hints at the fact that constants corresponding to all the rational numbers in $[0,1]$ are definable in RŁ. Not surprisingly, therefore, RŁ has been shown in [3] to be a conservative extension of the so-called Rational Pavelka logic [9]. The logic has also a nice functional representation, in analogy to the famous McNaughton theorem for Łukasiewicz [11]: formulas in RŁ correspond to continuous piecewise linear functions with rational coefficients over $[0,1]$, see [2,3].

RŁ has been systematically investigated in [3], where a Hilbert system and a corresponding algebraic semantics DMV (divisible MV algebras) have been introduced.

In this work we present a hypersequent calculus HRŁ for RŁ, which extends the calculus for Łukasiewicz logic introduced in [12]. In Sect. 2 we introduce the calculus and show its soundness and completeness with respect to the standard semantics. In Sect. 3 we then move to consider the logic $KZ(\pi)$, an expansion

J.P. Carvalho et al. (Eds.): IPMU 2016, Part II, CCIS 611, pp. 136–147, 2016.
DOI: 10.1007/978-3-319-40581-0_12

of Kleene-Zadeh logic (KZ in the following), introduced in [7] by considerations of game-semantic nature. A hypersequent calculus for $KZ(\pi)$ is easily obtained from suitable restrictions on $HR\text{Ł}$. Indeed, the logic $KZ(\pi)$ can be seen as a proper fragment of $R\text{Ł}$. In Sect. 4 we conclude by pointing to future work. In particular, we suggest that our calculi may provide a useful framework for a proof-theoretic investigation of fuzzy logics extended with so-called fuzzy quantifiers [8], such as *many, few, about half*, etc.

2 The Hypersequent Calculus *HRŁ*

In this section we introduce a calculus for the logic $R\text{Ł}$, i.e. the expansion of Łukasiewicz logic with division operators. Recall that in a language for propositional Łukasiewicz logic only the constant (or 0-ary connective) $\bot$ the connective $\to$ are needed, other connectives $\neg, \cdot, \oplus, \wedge, \vee, \top$ being definable in terms of $\to, \bot$. In the following, we call *atomic formula* a propositional variable or the constant $\bot$. As usual, formulas are built recursively from atomic formulas. Any evaluation v on the standard semantics assigns truth values in $[0,1]$ to propositional variables, the value 0 to $\bot$, and is extended truth functionally by letting $v(\alpha \to \beta) = \min(1, 1 - v(\alpha) + v(\beta))$. For the remaining connectives, evaluations v are determined as follows:

$$\begin{array}{ll} v(\top) = 1, & v(\neg\alpha) = 1 - v(\alpha) \\ v(\alpha \cdot \beta) = \min(0, v(\alpha) + v(\beta) - 1), & v(\alpha \oplus \beta) = \min(1, v(\alpha) + v(\beta)) \\ v(\alpha \wedge \beta) = \min(v(\alpha), v(\beta)), & v(\alpha \vee \beta) = \max(v(\alpha), v(\beta)) \end{array}$$

Notation. In what follows, given an integer n, we denote by α^n a multiset of α's and by $n\alpha$ the formula $\alpha \oplus \cdots \oplus \alpha$. More precisely, we let

$$\alpha^1 = \alpha \quad \alpha^n = \alpha, \alpha^{n-1} \quad \text{and} \quad 1\alpha = \alpha \quad n\alpha = \alpha \oplus (n-1)\alpha.$$

The language of $R\text{Ł}$ is obtained extending that of Ł with the set of unary connectives $\{\delta_n\}_{n\in\mathbb{N}}$. (Standard) Evaluations for $R\text{Ł}$ are defined extending those for Ł with the condition:

$$v(\delta_n\alpha) = \frac{v(\alpha)}{n}$$

for any δ_n. Clearly $v(\delta_m(\delta_n\alpha)) = v(\delta_{mn}\alpha)$ and $v(\delta_1\alpha) = v(\alpha)$. Hence we will identify in the following any formula of the kind $\delta_m(\delta_n\alpha)$ with $\delta_{mn}\alpha$ and $\delta_1\alpha$ with α. Note that for any rational number n/m in $[0,1]$, a corresponding constant $n(\delta_m\top)$ is definable in $R\text{Ł}$ and clearly satisfies $v(n(\delta_m\top)) = n/m$ for any evaluation v.

A Hilbert-style axiomatization of the logic $R\text{Ł}$ has been introduced in [3]. It is obtained by adding the following axioms to the Hilbert system for Łukasiewicz logic

$$\begin{array}{ll} (\delta 1a) & n(\delta_n\varphi) \to \varphi \\ (\delta 1b) & \varphi \to n(\delta_n\varphi) \\ (\delta 2) & \neg\delta_n\varphi \oplus \neg(n-1)(\delta_n\varphi) \end{array}$$

The axiomatic system is shown in [3] to be complete w.r.t. the standard semantics over $[0, 1]$, via algebraic methods. More precisely, a corresponding general algebraic semantics, the variety of *divisible MV algebras* (*DMV* algebras) is introduced and shown to be generated by its members on the real interval $[0, 1]$. In what follows we introduce a Gentzen-style calculus for the logic RŁ that is based on *hypersequents*. We exhibit a direct proof of the completeness of the calculus w.r.t the standard semantics. First, we recall the notion of sequent and hypersequent (see e.g. [1,13]).

Definition 1. *A* hypersequent *is a non-empty finite multiset* $S_1 \mid \ldots \mid S_n$ *where each* $S_i, i = 1, \ldots, n$ *is a sequent, called a* component *of the hypersequent. A* (multiple-conclusioned) sequent *is in turn an object of the form* $\Gamma \Rightarrow \Pi$*, where* Γ, Π *are multisets of formulas.*

Our hypersequent calculus for RŁ is an extension of the hypersequent calculus for Łukasiewiz logic introduced in [12]. In Table 1 we recall the calculus HŁ for Ł , with some unessential modifications. We include also rules for the connectives $\oplus, \neg, \top$, although they are not necessary, being derivable from the rules for $\rightarrow, \bot$.

Table 1. Hypersequent calculus HŁ for Łukasiewicz logic

$$\frac{}{\Rightarrow}\ (emp) \qquad \frac{}{\alpha \Rightarrow \alpha}\ (id) \qquad \frac{}{\bot \Rightarrow \alpha}\ (\bot)$$

$$\frac{G \mid \Gamma_1, \Gamma_2 \Rightarrow \Delta_1, \Delta_2}{G \mid \Gamma_1 \Rightarrow \Delta_1 \mid \Gamma_2 \Rightarrow \Delta_2}\ (split) \qquad \frac{G \mid \Gamma_1 \Rightarrow \Delta_1 \quad G \mid \Gamma_2 \Rightarrow \Delta_2}{G \mid \Gamma_1, \Gamma_2 \Rightarrow \Delta_1, \Delta_2}\ (mix) \qquad \frac{}{\Rightarrow \top}\ (\top)$$

$$\frac{G \mid H \mid H}{G \mid H}\ (ec) \qquad \frac{G}{G \mid H}\ (ew) \qquad \frac{G \mid \Gamma \Rightarrow \Delta}{G \mid \Gamma, \Pi \Rightarrow \Delta}\ (wl)$$

$$\frac{G \mid \Gamma, \beta \Rightarrow \alpha, \Delta \mid \Gamma \Rightarrow \Delta}{G \mid \Gamma, \alpha \rightarrow \beta \Rightarrow \Delta}\ (\rightarrow l) \qquad \frac{G \mid \Gamma \Rightarrow \Delta \quad G \mid \Gamma, \alpha \Rightarrow \beta, \Delta}{G \mid \Gamma \Rightarrow \alpha \rightarrow \beta, \Delta}\ (\rightarrow r)$$

$$\frac{G \mid \Gamma, \alpha, \beta \Rightarrow \bot, \Delta}{G \mid \Gamma, \alpha \oplus \beta \Rightarrow \Delta}\ (\oplus l) \qquad \frac{G \mid \Gamma, \bot \Rightarrow \alpha, \beta, \Delta \quad G \mid \Gamma \Rightarrow \Delta}{G \mid \Gamma \Rightarrow \alpha \oplus \beta, \Delta}\ (\oplus r)$$

$$\frac{G \mid \Gamma, \bot \Rightarrow \alpha, \Delta}{G \mid \Gamma, \neg\alpha \Rightarrow \Delta}\ (\neg l) \qquad \frac{G \mid \Gamma, \alpha \Rightarrow \bot, \Delta}{G \mid \Gamma \Rightarrow \neg\alpha, \Delta}\ (\neg r)$$

We are now ready to introduce the calculus for RŁ.

Definition 2. *The calculus HRŁ is obtained by adding to the calculus for Łukasiewicz logic in Table 1 the rules in Table 2.*

Hypersequents are usually interpreted as particular formulas in a logic: for instance, the symbol $\mid$ is generally interpreted as a disjunction $\vee$ and $\Rightarrow$ as an implication $\rightarrow$. This is not the case of HŁ, where hypersequents are directly

Table 2. Additional rules for $HR\L$

$$\frac{G \mid \Gamma, (\delta_n \alpha)^n \Rightarrow \Delta, \bot^{n-1}}{G \mid \Gamma, \alpha \Rightarrow \Delta}\ (\delta \uparrow l) \qquad \frac{G \mid \Gamma, \bot^{n-1} \Rightarrow (\delta_n \alpha)^n, \Delta}{G \mid \Gamma \Rightarrow \alpha, \Delta}\ (\delta \uparrow r)$$

$$\frac{G \mid \Gamma, \alpha, \bot^{n-1} \Rightarrow \Delta}{G \mid \Gamma, (\delta_n \alpha)^n \Rightarrow \Delta}\ (\delta \downarrow l) \qquad \frac{G \mid \Gamma \Rightarrow \alpha, \bot^{n-1}, \Delta}{G \mid \Gamma \Rightarrow (\delta_n \alpha)^n, \Delta}\ (\delta \downarrow r)$$

$$\frac{G \mid \Gamma, \delta_n \beta, \delta_n \top \Rightarrow \delta_n \alpha, \Delta \mid \Gamma, \delta_n \top \Rightarrow \Delta}{G \mid \Gamma, \delta_n(\alpha \to \beta) \Rightarrow \Delta}\ (\delta \to l) \qquad \frac{G \mid \Gamma, \delta_n \alpha \Rightarrow \delta_n \top, \delta_n \beta, \Delta \quad \Gamma \Rightarrow \delta_n \top, \Delta}{G \mid \Gamma \Rightarrow \delta_n(\alpha \to \beta), \Delta}\ (\delta \to r)$$

interpreted over the standard semantics of the logic. The evaluation of a multiset Γ of formulas is defined in [12,13] for Ł as:

$$v(\Gamma) = 1 + \sum_{\alpha \in \Gamma} (v(\alpha) - 1).$$

We will adopt the same notion for the evaluation of a multiset of formulas in the logic $R\L$. The validity of a hypersequent is then defined as follows.

Definition 3. *Let $G = \Gamma_1 \Rightarrow \Delta_1 \mid \dots \mid \Gamma_n \Rightarrow \Delta_n$ be a hypersequent in $HR\L$. We say that G is valid and denote it by $\models_{R\L} G$ iff for any valuation v there is a component $\Gamma_i \Rightarrow \Delta_i$ such that $v(\Gamma_i) \leq v(\Delta_i)$ ($i \in \{1, \dots, n\}$).*

As usual, we denote by $\vdash_{HR\L} G$ the fact that a hypersequent G is derivable in $HR\L$. Note that the rules for $(\delta \to)$ allow for a form of deep inference: they do not necessarily operate on the most external connective, i.e. on δ_n, but inside the formula. As an example to illustrate the functioning of the calculus, we show how to derive the axiom $(\delta 2)$:

$$\cfrac{\cfrac{\cfrac{\cfrac{\cfrac{\bot^n \Rightarrow \bot^n}{\varphi, \bot^n \Rightarrow \bot^n}\,(wl)}{\bot, (\delta_n \varphi)^n \Rightarrow \bot^n}\,(\delta \downarrow l)}{\bot, \delta_n \varphi, (n-1)(\delta_n \varphi) \Rightarrow \bot, \bot}\,(\oplus l) \times (n-2)}{\bot \Rightarrow \neg \delta_n \varphi, \neg (n-1)(\delta_n \varphi)}\,(\neg r) \times 2 \qquad \cfrac{}{\Rightarrow}\,(emp)}{\Rightarrow \neg \delta_n \varphi \oplus \neg (n-1)(\delta_n \varphi)}\,(\oplus r)$$

where $\bot^n \Rightarrow \bot^n$ is clearly derivable by repeated applications of (mix) with the axiom $\bot \Rightarrow \bot$.

Lemma 1. *The rules for $HR\L$ in Table 2 are sound and invertible.*

Proof. We consider only two rules, the others being similar. First, we consider the rule $(\delta \to r)$. Assume that the premises hold. The case where the context G is valid is trivial. W.l.o.g. let us assume thus:

$$(*) \qquad v(\Gamma) + (\frac{v(\alpha)}{n} - 1) \leq (\frac{1}{n} - 1) + (\frac{v(\beta)}{n} - 1) + v(\Delta)$$

and

$$(**) \qquad v(\Gamma) \leq v(\Delta) + (\frac{1}{n} - 1).$$

In case $v(\alpha) \leq v(\beta)$, we have $v(\alpha \to \beta) = 1$, hence the conclusion just amounts to $(**)$. In case $v(\beta) \leq v(\alpha)$, we have $v(\alpha \to \beta) = 1 - v(\alpha) + v(\beta)$, hence the conclusion holds iff

$$v(\Gamma) \leq (\frac{1 - v(\alpha) + v(\beta)}{n} - 1) + v(\Delta)$$

which follows from $(*)$ by easy computations. For invertibility, we assume that

$$(***) \qquad v(\Gamma) \leq (\frac{v(\alpha \to \beta)}{n} - 1) + v(\Delta).$$

In case $v(\alpha) \leq v(\beta)$ this amounts to $(**)$ i.e. the right premise. Combining $v(\alpha) \leq v(\beta)$ and $(**)$ we can easily obtain $(*)$, i.e. the left premise. In case $v(\beta) \leq v(\alpha)$ we obtain the left premise $(*)$ by easy computations. From $(***)$ we easily get

$$v(\Gamma) \leq \frac{1}{n} + \frac{v(\beta) - v(\alpha)}{n} - 1 + v(\Delta) \leq \frac{1}{n} - 1 + v(\Delta)$$

i.e. the right premise $(**)$.

Let us consider now the rule $(\delta \uparrow l)$ and assume its premises hold, i.e.

$$v(\Gamma) + n(\frac{v(\alpha)}{n} - 1) \leq v(\Delta) - (n - 1).$$

This is clearly equivalent to

$$v(\Gamma) + v(\alpha) - 1 \leq v(\Delta).$$

that is, the conclusion of $(\delta \uparrow l)$. The same reasoning gives also the invertibility of $(\delta \uparrow l)$.

In what follows we call δ-*atomic* any formula of kind $\delta_n \alpha$, with α atomic formula.[1] We call a hypersequent δ-atomic if it only contains δ-atomic formulas. Towards the completeness theorem, we show first two useful technical lemmas.

Lemma 2. *(i) The following rules are derivable in HRŁ and invertible.*

$$\frac{G \,|\, \Gamma^m, \bot^{(m-1)n} \Rightarrow \Delta^m}{G \,|\, \Gamma, (\delta_m \top)^n \Rightarrow \Delta} \,(div\ l) \qquad \frac{G \,|\, \Gamma^m \Rightarrow \bot^{(m-1)n}, \Delta^m}{G \,|\, \Gamma \Rightarrow (\delta_m \top)^n, \Delta} \,(div\ r)$$

(ii) The sequent $\delta_m \alpha \Rightarrow \delta_m \top$ is derivable in HRŁ.

[1] A formula of the kind $\delta_n(\delta_m \alpha)$ with α atomic is considered δ-atomic as well. Recall that we identify such formulas with $\delta_{mn}\alpha$.

Proof. (i). The rule $(div\ l)$ is derivable as follows:

$$\dfrac{\dfrac{\dfrac{\dfrac{G \mid \Gamma^m, \bot^{(m-1)n} \Rightarrow \Delta^m}{G \mid \Gamma^m, \top^n, \bot^{(m-1)n} \Rightarrow \Delta^m}\ (wl)}{G \mid \Gamma^m, ((\delta_m \top)^m)^n \Rightarrow \Delta^m}\ (\delta\downarrow l) \times n}{G \mid \Gamma, (\delta_m \top)^n \Rightarrow \Delta \mid \ldots \mid \Gamma, (\delta_m \top)^n \Rightarrow \Delta}\ (split) \times m}{G \mid \Gamma, (\delta_m \top)^n \Rightarrow \Delta}\ (ec) \times m$$

For invertibility, note that $(\delta \downarrow l)$ and (ec) are invertible in general. The applications of $(split)$ and (wl) above can be easily shown to be invertible as well. The rule $(div\ r)$ is derivable in a similar way, using repeated (mix) with the sequent $\Rightarrow \top$ instead of (wl) and $(\delta \downarrow r)$ instead of $(\delta \downarrow l)$.

(ii). A derivation of $\delta_m \alpha \Rightarrow \delta_m \top$ is obtained as follows:

$$\dfrac{\dfrac{\dfrac{\bot^{m-1} \Rightarrow \bot^{m-1}}{\alpha, \bot^{m-1} \Rightarrow \bot^{m-1}}\ (wl)}{(\delta_m \alpha)^m \Rightarrow \bot^{m-1}}\ (\delta\downarrow l)}{\delta_m \alpha \Rightarrow \delta_m \top}\ (div\ r)$$

where $\bot^{m-1} \Rightarrow \bot^{m-1}$ is derivable by repeated applications of (mix) with the axiom $\bot \Rightarrow \bot$.

Lemma 3. *If* $\vdash_{HR Ł} G \mid \Gamma, \delta_n \top \Rightarrow \Delta$ *then* $\vdash_{HR Ł} G \mid \Gamma, \delta_n \alpha \Rightarrow \Delta$.

Proof. We reason by induction on the length of the derivation of $G \mid \Gamma, \delta_n \top \Rightarrow \Delta$. For the base case, if we have an axiom of the form $\delta_n \top \Rightarrow \delta_n \top$, we replace it by the derivable sequent $\delta_n \alpha \Rightarrow \delta_n \top$ (see Lemma 2). In the remaining cases, the lemma just follows by a suitable application of the induction hypothesis on the last applied rule in a derivation of $G \mid \Gamma, \delta_n \top \Rightarrow \Delta$.

We are now ready for the completeness theorem, which follows the basic structure of the argument for Łukasiewicz logic as presented e.g. in [5,12,13].

Theorem 1. *[Completeness] Let* G *be a hypersequent in* $HRŁ$. *If* $\models_{RŁ} G$, *then* $\vdash_{HRŁ} G$

Proof. By the invertibility of logical rules (see Lemma 1), it is sufficient to show the claim only for δ-atomic hypersequents. We reason by induction on the number k of different propositional variables occurring on the left hand side of the components of G. In case $k = 0$, there can only be $\bot, \delta_n \top$ on the left hand side of any component. By applying (mix) backwards with $\bot \Rightarrow \bot$, $\delta_n \top \Rightarrow \delta_n \top$, we remove any simultaneous occurrence of $\bot$ and $\delta_n \top$ on both sides of a sequent. We then apply the rules $(div\ l)$ and $(div\ r)$ backwards (see Lemma 2) to obtain a hypersequent G_1 where no occurrences of $\delta_n \top$ appear. It is clear that G_1 is valid iff G is valid and moreover, if G_1 is derivable, G is derivable as well. To conclude the base case, we are now left to show that if G_1 is valid, it is provable. Note that any component of G_1 can only be of the form $(\bot)^n \Rightarrow \Delta$ for a certain n. If there is a component such that $n \geq |\Delta|$, then the whole hypersequent is derivable by $(\bot)$, (mix),(wl) and (ew). Assume this is not the case and consider an evaluation which assigns the value 0 to any propositional variable. It can be easily shown that this evaluation would falsify the whole hypersequent G_1, thus contradicting our assumption that G_1 is valid.

We consider now the case where $k > 0$ and we pick an atomic variable q occurring on the left of at least one sequent in G. By suitable backwards application of the rules $(\delta \uparrow r)$ and $(\delta \uparrow l)$, we can obtain a hypersequent where all occurrences of q are of the form $\delta_n q$, for the same integer n. W.l.o.g. we assume $n \geq 2$ (in case $n = 1$ the proof proceeds as the one in [13]). We remove any occurrence of $\delta_n q$ on both sides of each sequent, applying backwards repeatedly (mix) with the axiom $\delta_n q \Rightarrow \delta_n q$. We obtain thus a valid hypersequent, whose components contain $\delta_n q$ either only on the right or on the left. We multiply the components of this hypersequent applying (ec) and $(split)$ backwards, so to obtain

$$G' = G_0 \mid \{\Gamma_i, (\delta_n q)^\lambda \Rightarrow \Delta_i \mid \Pi_j \Rightarrow (\delta_n q)^\lambda, \Sigma_j\}_{i\in I, j\in J}$$

where I, J are finite sets of indices and $\lambda \in \mathbb{N}$. Clearly we still have $\models_{RL} G'$. Let us consider now the hypersequent

$$H = G_0 \mid \{\Gamma_i, \Pi_j \Rightarrow \Sigma_j, \Delta_i \mid \Gamma_i, (\delta_n \top)^\lambda \Rightarrow \Delta_i \mid \Pi_j \Rightarrow (\delta_n q)^\lambda, \Sigma_j\}_{i\in I, j\in J}$$

which contains fewer distinct variables on the left than G'. We claim that, if H is derivable, G' is derivable as well. Indeed, from a derivation of H, by suitable applications of (mix) with $\delta_n q \Rightarrow \delta_n q$ and $(split)$, we can obtain a derivation of

$$G_0 \mid \{\Gamma_i, (\delta_n q)^\lambda \Rightarrow \Delta_i \mid \Gamma_i, (\delta_n \top)^\lambda \Rightarrow \Delta_i \mid \Pi_j \Rightarrow (\delta_n q)^\lambda, \Sigma_j\}_{i\in I, j\in J}$$

Applying Lemma 3 to the latter hypersequent and (ec), we obtain our desired derivation of G'. It suffices now to show that H is valid, as in this case we obtain $\vdash_{HRL} H$ by the induction hypothesis. For a contradiction, let us suppose that there exists a valuation v such that $v(\Gamma) > v(\Delta)$ for all components $\Gamma \Rightarrow \Delta \in H$. We let

$$x = \max(\{v(\Delta_i) - v(\Gamma_i)\}_{i\in I} \cup \{-\lambda\})$$

$$y = \min(\{v(\Pi_j) - v(\Sigma_j)\}_{j\in J} \cup \{0\}).$$

Assume $x \geq y$. We would have either $v(\Gamma_i) + v(\Pi_j) \leq v(\Sigma_j) + v(\Delta_i)$ or $-\lambda \geq v(\Pi_j) - v(\Sigma_j)$ or $v(\Delta_i) - v(\Gamma_i) \geq 0$. In any of these cases, we can easily obtain a contradiction with the assumption that the valuation v does not satisfy H. Hence we have $x < y$. We claim that there is a $w \in [0,1]$ such that $x < \lambda(\frac{w}{n} - 1) < y$. Towards this aim, let us first show the two following facts:

$$(a) \quad x < \lambda(\frac{1}{n} - 1) \quad (b) \quad \lambda(\frac{v(q)}{n} - 1) < y$$

Let us start from (a). In case $x = -\lambda$ we get $-\lambda < \lambda(\frac{1}{n} - 1)$ which clearly holds. Assume instead that $x = v(\Delta_i) - v(\Gamma_i)$ for some $i \in I$. We have $v(\Delta_i) - v(\Gamma_i) < \lambda(\frac{1}{n} - 1)$ as otherwise v would satisfy the component $\Gamma_i, (\delta_n \top)^\lambda \Rightarrow \Delta_i$ of H and this would contradict our assumption. Let us now consider the inequation (b) and assume it does not hold. In case $y = 0$, we would have $\lambda(\frac{v(q)}{n} - 1) \geq 0$, which is clearly a contradiction. Otherwise, there would be an index $j \in J$ such that $y = v(\Pi_j) - v(\Sigma_j) \leq \lambda(\frac{v(q)}{n} - 1)$. Hence we would have $v(\Pi_j) \leq v(\Sigma_j, (\delta_n q)^\lambda)$, which contradicts the assumption that v does not satisfy the hypersequent H. Recall now that $x < y$. If either $x < \lambda(\frac{v(q)}{n} - 1) < y$ or $x < \lambda(\frac{1}{n} - 1) < y$ we are done. Otherwise we have

$$\lambda(\frac{v(q)}{n} - 1) < x < y < \lambda(\frac{1}{n} - 1)$$

Also in this latter case we can find a $w \in [0,1]$ (actually in $(v(q), 1)$) such that $x < \lambda(\frac{w}{n} - 1) < y$. We define now a new valuation $v'(q)$ which differs from v only for letting $v'(q) = w$. We have thus $x < \lambda(\frac{v'(q)}{n} - 1) < y$. Hence $v'(\Delta_i) - v'(\Gamma_i) < \lambda(\frac{v'(q)}{n} - 1)$ and $\lambda(\frac{v'(q)}{n} - 1) < v'(\Pi_j) - v'(\Sigma_j)$, i.e.

$$v'(\Gamma_i, (\delta_n q)^\lambda) > v'(\Delta_i) \qquad v'(\Pi_j) > v'(\Sigma_j, (\delta_n q)^\lambda)$$

for any $i \in I, j \in J$. This means that G' is not valid, which contradicts our initial assumption.

In Theorem 1 we have directly shown the completeness of the hypersequent calculus with respect to the *standard* semantics over the real interval $[0, 1]$. Notice that our calculus does not include the (cut) rule

$$\frac{G \mid \Gamma, \alpha \Rightarrow \Delta \quad \Sigma \Rightarrow \alpha, \Pi}{G \mid \Gamma, \Sigma \Rightarrow \Pi, \Delta}\ (cut)$$

which can be easily proved to be sound with respect to the standard semantics. The completeness of our (cut)-free calculus shows thus that the (cut) rule is actually *admissible* for $HR\text{Ł}$. This means that the addition of (cut) to the calculus $HR\text{Ł}$ would not change the set of derivable formulas.

3 A Hypersequent Calculus for the Logic $KZ(\pi)$

In this section we introduce a calculus for the logic $KZ(\pi)$, which we obtain as a restriction of the calculus $HR\text{Ł}$. $KZ(\pi)$ was introduced in [7] in the context of an investigation into Hintikkas's game semantics for fuzzy logic. Hintikka-games [10] are essentially two-person zero-sum games. The players, say Myself and You, in each move stepwise reduce a complex logic formula, until atomic formulas are reached. A state of the game is fully determined by the formula at stake and by an attribution of roles (attacker and defender) to the two players. For propositional (classical) logic, the rules for decomposing complex into atomic formulas are as follows:

($R_\wedge$) If I assert (i.e. defend) $\alpha \wedge \beta$ then You attack by pointing either to the left or to the right subformula. As corresponding defense, I then have to assert either α or β, according to Your choice.

($R_\vee$) If I assert $\alpha \vee \beta$ then I have to assert either α or β at My own choice.

($R_\neg$) If I assert $\neg\alpha$ then You have to assert α. In other words, our roles are switched: the game continues with You as defender and Me as attacker (of α).

Once the players reach an atomic formula, the game ends. We say that I win if in a final state I assert an atomic formula α and $v(\alpha) = 1$ (my payoff is 1). Similarly, I lose if $v(\alpha) = 0$ (my payoff is 0). In case in a final state You assert an atomic formula α, the winning conditions and related payoffs are reverted.

The game-theoretical framework just sketched provides an alternative characterization for truth in classical logic: the truth of a formula corresponds to the existence of a winning strategy (i.e. ending with payoff 1) for Myself in the corresponding game.

Hintikka games were not originally meant to deal with many-valued logic: atomic formulas are indeed interpreted only as either true or false. Nevertheless, it is possible to drop this requirement and admit evaluations over $[0, 1]$, while retaining the basic game-theoretical framework. As shown in [7], this results in a game-theoretic semantics for the $\{\wedge, \vee, \neg, \bot\}$-fragment of Łukasiewicz logic, i.e. the so-called Kleene Zadeh logic (KZ). More precisely, a formula α in KZ evaluates to $w \in [0, 1]$ under a certain evaluation iff the corresponding Hintikka game has a payoff value w for Myself. Moreover, it is shown in [7] that any additional rule in a Hintikka game, involving only choices between two players and role switches, (such as $(R_\wedge), (R_\vee), (R_\neg)$) always corresponds to a definable connective in KZ. Hence a different kind of game rule is needed to go beyond the logic KZ. The logic $KZ(\pi)$ is obtained in [7] expanding KZ with a new binary connective π, characterized by the following *random choice* rule:

(R_π) If the current formula is $\alpha\pi\beta$ then a uniformly random choice determines whether the game continues with α or with β.

The corresponding truth function for this connective is obtained as

$$v(\alpha\pi\beta) = (v(\alpha) + v(\beta))/2.$$

This truth function matches the corresponding game semantics, provided that we consider *expected* payoff instead of payoff. More precisely, it is shown in [7] that a formula of $KZ(\pi)$ has a value w under a given evaluation iff the expected payoff in the corresponding Hintikka game for Myself is w. Note that the logic $KZ(\pi)$ is a proper extension of KZ, but it is incomparable with Ł: indeed, the connective π is not definable from the connectives of Ł, nor can $\cdot, \oplus, \rightarrow$ be defined from $\pi, \wedge, \vee, \neg, \bot$. The addition to $KZ(\pi)$ of a further unary connective, standing for a doubling of the truth values, is needed to capture the whole Ł while retaining the Hintikka-style game-semantics.

We can see $KZ(\pi)$, however, as a fragment of the logic RŁ, which we considered in Sect. 2. The fragment is generated by the atomic formulas and the connectives $\wedge, \vee, \neg, \pi$ where $x\pi y := \delta_2 x \oplus \delta_2 y$. Note that, in turn the unary connective δ_2 is definable in $KZ(\pi)$ by letting $\delta_2\alpha = \alpha\pi\bot$. By these simple observations we can thus obtain a hypersequent calculus for $KZ(\pi)$ as a fragment of that for RŁ. We present the calculus explicitly in Table 3. Note that only δ-formulas of the kind $\delta_{2^m}\alpha$ can occur in a proof of a hypersequent in the language of $KZ(\pi)$[2].

Table 3. Calculus $HKZ(\pi)$ for $KZ(\pi)$

$$\frac{G \mid \Gamma \Rightarrow \Delta}{G \mid \Gamma, \Pi \Rightarrow \Delta}\ (wl)$$

$$\overline{\Rightarrow}\ (emp) \qquad \overline{\alpha \Rightarrow \alpha}\ (id)$$

$$\overline{(\delta_{2^m})\bot \Rightarrow \alpha}\ (\bot l) \qquad \overline{\Rightarrow \top}\ (\top)$$

$$\frac{G \mid \Gamma_1, \Gamma_2 \Rightarrow \Delta_1, \Delta_2}{G \mid \Gamma_1 \Rightarrow \Delta_1 \mid \Gamma_2 \Rightarrow \Delta_2}\ (split) \qquad \frac{G \mid \Gamma_1 \Rightarrow \Delta_1 \quad G \mid \Gamma_2 \Rightarrow \Delta_2}{G \mid \Gamma_1, \Gamma_2 \Rightarrow \Delta_1, \Delta_2}\ (mix)$$

$$\frac{G \mid H \mid H}{G \mid H}\ (ec) \qquad \frac{G}{G \mid H}\ (ew)$$

$$\frac{G \mid \Gamma, (\delta_{2^m})\ \alpha \Rightarrow \Delta \quad G \mid \Gamma, (\delta_{2^m})\ \beta \Rightarrow \Delta}{G \mid \Gamma, (\delta_{2^m})\ \alpha \vee \beta \Rightarrow \Delta}\ (\vee l) \qquad \frac{G \mid \Gamma \Rightarrow (\delta_{2^m})\ \alpha, \Delta \mid \Gamma \Rightarrow (\delta_{2^m})\ \beta, \Delta}{G \mid \Gamma \Rightarrow (\delta_{2^m})\ \alpha \vee \beta, \Delta}\ (\vee r)$$

$$\frac{G \mid \Gamma, (\delta_{2^m})\ \alpha \Rightarrow \Delta \mid \Gamma, (\delta_{2^m})\ \beta \Rightarrow \Delta}{G \mid \Gamma, (\delta_{2^m})\ \alpha \wedge \beta \Rightarrow \Delta}\ (\wedge l) \qquad \frac{G \mid \Gamma \Rightarrow (\delta_{2^m})\ \alpha, \Delta \quad G \mid \Gamma \Rightarrow (\delta_{2^m})\ \beta, \Delta}{G \mid \Gamma \Rightarrow (\delta_{2^m})\ \alpha \wedge \beta, \Delta}\ (\wedge r)$$

$$\frac{G \mid \Gamma, (\delta_{2^{m+1}})\ \alpha, (\delta_{2^{m+1}})\ \beta \Rightarrow \bot, \Delta}{G \mid \Gamma, (\delta_{2^m})\ \alpha\pi\beta \Rightarrow \Delta}\ (\pi l) \qquad \frac{G \mid \Gamma, \bot \Rightarrow (\delta_{2^{m+1}})\ \alpha, (\delta_{2^{m+1}})\ \beta, \Delta}{G \mid \Gamma \Rightarrow (\delta_{2^m})\ \alpha\pi\beta, \Delta}\ (\pi r)$$

$$\frac{G \mid \Gamma, \bot, (\delta_{2^m})\ \top \Rightarrow (\delta_{2^m})\ \alpha, \Delta}{G \mid \Gamma, (\delta_{2^m})\ (\neg\alpha) \Rightarrow \Delta}\ (\neg l) \qquad \frac{G \mid \Gamma, (\delta_{2^m})\ \alpha \Rightarrow \bot, (\delta_{2^m})\ \top, \Delta}{G \mid \Gamma \Rightarrow (\delta_{2^m})\ (\neg\alpha), \Delta}\ (\neg r)$$

$$\frac{G \mid \Gamma, \delta_2\alpha, \delta_2\alpha \Rightarrow \Delta, \bot}{G \mid \Gamma, \alpha \Rightarrow \Delta}\ (\delta \uparrow l) \qquad \frac{G \mid \Gamma, \bot \Rightarrow \delta_2\alpha, \delta_2\alpha, \Delta}{G \mid \Gamma \Rightarrow \alpha, \Delta}\ (\delta \uparrow r)$$

$$\frac{G \mid \Gamma, \alpha, \bot \Rightarrow \Delta}{G \mid \Gamma, \delta_2\alpha, \delta_2\alpha \Rightarrow \Delta}\ (\delta \downarrow l) \qquad \frac{G \mid \Gamma \Rightarrow \alpha, \bot, \Delta}{G \mid \Gamma \Rightarrow \delta_2\alpha, \delta_2\alpha, \Delta}\ (\delta \downarrow r)$$

Lemma 4. *The logical rules and $(\delta \uparrow), (\delta \downarrow)$ in Table 3 are sound and invertible for RŁ.*

Proof. By simple arithmetic computation, as for the proof of Lemma 1. Notice that the rules $(\delta \uparrow)$ and $(\delta \downarrow)$ are just particular cases of the corresponding ones in Table 1. Similarly, the rules for $(\neg)$ and (π) are just special cases of the rules $(\delta \rightarrow)$ and $(\oplus)$, respectively.

[2] As for RŁ, we identify $\delta_{2^m}\delta_{2^n}\alpha$ with $\delta_{2^{m+n}}\alpha$ and $\delta_{2^0}\alpha$, i.e. $\delta_1\alpha$, with α.

Being the logical rules invertible, the completeness proof in Theorem 1 can be adapted to the fragment $KZ(\pi)$.

Theorem 2. *Let G be a hypersequent in the language of $KZ(\pi)$. If $\models_{KZ(\pi)} G$, then $\vdash_{HKZ(\pi)} G$*

Proof. Using the invertibility of the logical rules, we can reduce G to an atomic hypersequent. Applying the rules $(\delta \uparrow l)$ and $(\delta \uparrow r)$ backwards we obtain a valid hypersequent which contains only δ-atomic formulas of the kind $(\delta_{2^m})\ \alpha$, for a given m. The rest of the proof proceeds as in Theorem 1 (note that Lemmas 2 and 3 apply to $HKZ(\pi)$ as well).

4 Conclusions and Future Work

Variants of $KZ(\pi)$, with similar game theoretical motivations, can also be defined as fragments of $R\L$. First, in the definition of the game rule (R_π) in page 9 we can drop the requirement that the formula is chosen according to a random uniform distribution. A generalized connective π_r for any rational number r in $[0,1]$ can be introduced via the game rule:

(R_{π_r}) If the current formula is $\alpha\pi_r\beta$ then the game continues with α with probability r and with β with probability $1-r$.

Let $r = m/n$, for m, n natural numbers. The corresponding truth function for π_r is $v(\alpha\pi_r\beta) = (m/n)v(\alpha) + (1-m/n)v(\beta)$. The connective π_r is clearly definable in $R\L$ as $\alpha\pi_r\beta := m(\delta_n\alpha) \oplus (n-m)\delta_n\beta$.

In a different direction, we can also consider π-like connectives of arbitrary arity[3], i.e. connectives of kind π_n, arising from the following game rule:

(R_{π_n}) If the current formula is $\pi_n(\alpha_1, \ldots, \alpha_n)$ then a uniform random choice determines whether the game continues with one of the $\alpha_1, \ldots, \alpha_n$.

The corresponding truth function is clearly the average of the truth values $v(\alpha_1), \ldots, v(\alpha_n)$, i.e.

$$v(\pi_n(\alpha_1, \ldots, \alpha_n)) = \sum_{i=1,\ldots,n} \frac{v(\alpha_i)}{n}.$$

The connective is definable in $R\L$ by letting $\pi_n(\alpha_1, \ldots, \alpha_n) = \delta_n\alpha_1 \oplus \cdots \oplus \delta_n\alpha_n$. The connective π_n is strictly related to the *random witness quantifier*, introduced by game semantics means in [6], as an extension of first-order Łukasiewicz logic. The random witness quantifier is determined by the following game rule

(R_Π) If the current formula is $\Pi x F(x)$ then an element c from the domain D is chosen randomly and the game continues with $F(c)$.

[3] Note that π is not associative in general.

As we might expect, for a finite domain D, the corresponding truth function for Πx is defined as:

$$v(\Pi x F(x)) = \sum_{d \in D} \frac{v(F(d))}{|D|}$$

This coincides with the truth function of the connective π^n for $n = |D|$. It is thus possible to investigate the properties of the quantifier $\Pi(x)$ in a finite domain by means of the corresponding connective π^n, which is in turn definable in $R\text{Ł}$. As shown in [6, 7], the mechanism of random choice provides a guiding principle for the characterization and systematic introduction of families of so-called *fuzzy quantifiers*, i.e. expressions such as "few", "many", "about half". Many such quantifiers are indeed definable over an extension of first-order Łukasiewicz logic with $\Pi(x)$. Our calculus $HR\text{Ł}$ can thus provide a natural framework where a proof-theoretical study of these quantifiers can be further developed.

We leave also as a topic of future research the closer investigation of the connection between the calculi $HR\text{Ł}, HKZ(\pi)$ and the game semantics of the corresponding logic, along the lines of [5].

References

1. Avron, A.: A constructive analysis of RM. J. Symbolic Logic **52**(4), 939–951 (1987)
2. Baaz, M., Veith, H.: Interpolation in fuzzy logic. Arch. Math. Log. **38**(7), 461–489 (1999)
3. Gerla, B.: Rational Lukasiewicz logic and DMV-algebras. Neural Netw. World **11**, 579–584 (2001)
4. Běhounek, L., Cintula, P., Hájek, P.: Introduction to mathematical fuzzy logic. In: Cintula, P., et al. (eds.) Handbook of Mathematical Fuzzy Logic, vol. 1, pp. 1–102. College Publications, London (2011)
5. Fermüller, C.G., Metcalfe, G.: Giles's game and the proof theory of Łukasiewicz logic. Studia Logica **92**(1), 27–61 (2009)
6. Fermüller, C.G., Roschger, C.: Randomized game semantics for semi-fuzzy quantifiers. Logic J. IGPL **22**(3), 413–439 (2014). Special Issue on Non-Classical Modal and Predicate Logics
7. Fermüller, C.G.: Hintikka-style semantic games for fuzzy logics. In: Beierle, C., Meghini, C. (eds.) FoIKS 2014. LNCS, vol. 8367, pp. 193–210. Springer, Heidelberg (2014)
8. Glöckner, I.: Fuzzy Quantifiers: A Computational Theory. Studies in Fuzziness and Soft Computing, vol. 193. Springer, Heidelberg (2006)
9. Hájek, P.: Metamathematics of Fuzzy Logic. Kluwer, Dordrecht (1998)
10. Hintikk, J., Sandu, G.: Game-theoretical semantics. In: van Benthem, J., et al. (eds.) Handbook of Logic and Language, pp. 279–282. Elsevier, Amsterdam (1997)
11. McNaughton, R.: A theorem about infinite-valued sentential logic. J. Symbolic Logic **16**(1), 1–13 (1951)
12. Metcalfe, G., Olivetti, N., Gabbay, D.M.: Sequent and hypersequent calculi for abelian and Łukasiewicz logics. ACM Trans. Comput. Logic **6**(3), 578–613 (2005)
13. Metcalfe, G., Olivetti, N., Gabbay, D.M.: Proof Theory for Fuzzy Logics. Applied Logic Series, vol. 36. Springer, Heidelberg (2008)

Negation of Graded Beliefs

Bénédicte Legastelois[1(✉)], Marie-Jeanne Lesot[1], and Adrien Revault d'Allonnes[2]

[1] Sorbonne Universités, UPMC Univ Paris 06, CNRS, LIP6 UMR 7606, 4 Place Jussieu, 75005 Paris, France
{Benedicte.Legastelois,Marie-Jeanne.Lesot}@lip6.fr
[2] Université Paris 8 - EA 4383 - LIASD, 93526 Saint-Denis, France
Adrien.Revault-d_Allonnes@paris8.fr

Abstract. Negation is a key element in the construction of logical systems and plays a central role in reasoning and information manipulation tools. This paper considers the issue of negating graded beliefs, in the framework of a graded doxastic logic. It studies three interpretations of negation for these high level pieces of information, where negation is transferred to the three components of graded beliefs: the formula about which a belief is expressed, the belief modality and the belief level. The paper discusses the choice of appropriate formal frameworks for each of them, considering modal, fuzzy and many-valued logics; it characterises their use and underlines their relations, in particular regarding their effects on the belief degrees.

Keywords: Belief reasoning · Modal logic · Weighted logic · Doxastic logic

1 Introduction

One of the authors of this paper firmly believes unicorns exist, the other two do not. One does *not firmly* believe they exist, the other firmly does *not believe* they exist. Most people can tell the difference between these two points of view and can even tell them apart from someone who firmly believes unicorns do *not exist.* All three assertions are distinct negations of the original belief, which serves to show how tricky it is to negate a high level concept like graded beliefs.

The very notion of graded belief is a complex one, for which several interpretations can be considered: first, degrees of beliefs can be interpreted in terms of certainty. They can for instance be related to the -subjective- certainty an agent associates with the fact about which he expresses a belief [1]. Another type of uncertainty arises when an agent reasons about the beliefs of another agent, as revealed by the latter [2]: this uncertainty can be modelled in a possibilistic framework, as offered by Generalised Possibilistic Logic [8,9] a graded version of Meta-Epistemic Logic [2].

This paper takes a different point of view: interpreting degrees of beliefs in terms of belief strength. Graded beliefs are used to express partial beliefs, i.e. beliefs that are more or less true, but all attached to the same level of certainty.

J.P. Carvalho et al. (Eds.): IPMU 2016, Part II, CCIS 611, pp. 148–160, 2016.
DOI: 10.1007/978-3-319-40581-0_13

For this general type of graded beliefs, the paper offers a discussion of the negation issue, proposing to model the three interpretations of the negation given in the introduction and, in particular, their cognitive underpinnings. It focuses on choosing appropriate formal paradigms to model all three, specifically looking at modal, many-valued and fuzzy logics: the former with its doxastic variant for belief representation, the latter two weighted logics to help represent and manipulate gradual concepts, with distinct truth degrees, beyond the classic binary case. In these high level logics, negation models notions which cannot be addressed in classical logics.

After setting a formal representation of the tackled issues in Sect. 2, the paper reviews the principles of negation in these logics in Sect. 3. The three interpretations of negation for graded beliefs are then discussed in turn in the following three sections. Conclusion and future works are presented in Sect. 7.

2 Formalising the Graded Belief Negation Issue

In order to formalise the issue of graded belief negation, this section first introduces the notation proposed to represent graded beliefs and then expresses the three interpretations of their negation outlined in the introduction.

2.1 Notations for Graded Beliefs

A graded belief extends the notion of belief by introducing a measure of the extent to which something is believed. We propose to denote this $B(\varphi, \alpha)$, read 'φ is believed to a degree α': this notation simultaneously represents (i) the considered formula, φ, about which a modal assertion is expressed; (ii) the type of non factual modality, belief, B; (iii) the degree to which belief is partially held, α.

Using this notation, a sentence such as 'I firmly believe that unicorns exist' can be represented as $B(\varphi, \alpha)$ for which φ is the formula 'unicorns exist' and α represents the belief degree 'firmly', e.g. 0.8 if a numerical transposition were chosen, the choice of a correspondence between a numerical value and a linguistic term being out of the scope of this paper, 0.8 is given here as a mere example.

This notation highlights the relation between graded beliefs and the doxastic variant of modal logic [12], which aims at representing and reasoning about beliefs, as well as to weighted logics, such as many-valued or fuzzy logics, which manipulate degrees, respectively with ordinal and numerical values.

The negation of a graded belief is linguistically expressed as '*it does not hold that* φ is believed to a degree α'. We formally denote this $\mathcal{N}(B(\varphi, \alpha))$ where $\mathcal{N}$ is a general negation operator whose meaning is the topic of this paper: we propose a special notation to distinguish it from the classical negation, so as not to confuse it with $\neg B(\varphi, \alpha)$ and avoid implicitly transposing classical results known for $\neg$ to the case of graded belief negation.

2.2 Objects of the Negation of Graded Beliefs

The challenge of negating graded beliefs comes from the fact that several interpretations can be considered, depending on which of their components is seen as the negated object: formally, negation can be considered as transferred to each of them, leading to examine the following formulae, illustrated below:

global negation	transferred on	modality	formula	degree
$\mathcal{N}(B(\varphi, \alpha))$		$\mathcal{N}(B)(\varphi, \beta)$	$B(\mathcal{N}(\varphi), \gamma)$	$B(\varphi, \mathcal{N}(\alpha))$

The purpose of this work is to discuss them, questioning the choice of appropriate formal frameworks and definitions for the three types of negation operators $\mathcal{N}$. Note that, in the general notation above, different degrees, α, β, γ and $\mathcal{N}(\alpha)$, are used to avoid implicitly imposing *a priori* constraints. Establishing their relations to one another, including the possibility of their being equal, is an integral part of the problem of negation interpretation: the three object transpositions above show a simplified view of the problem. The three components of graded beliefs are actually intertwined and their combination and mutual influence matters.

First, note that, when considering the transfer to the believed formula φ, i.e. $B(\mathcal{N}(\varphi), \gamma)$, the general negation $\mathcal{N}$ applies to a logical formula and is therefore naturally interpreted as the logical negation, i.e. $\mathcal{N} \equiv \neg$. The question is then whether the statement $\mathcal{N}(B(\varphi, \alpha))$, read '*it does not hold that* φ is believed to a degree α', allows to draw some conclusion about the belief in $\neg\varphi$, formally written $B(\neg\varphi, \gamma)$: if *it does not hold that* 'I firmly believe that unicorns exist', do I believe that unicorns do *not exist* and, if so, how much?

Second, transferring the negation to the belief degree means that 'I believe φ' does hold, but to a degree other than α: using the running unicorn example, the question is to specify the meaning of the level for which it holds that 'I *not firmly* believe that unicorns exist'. The general negation $\mathcal{N}$ then applies to the belief degrees and is denoted N, some suitable negation operator for degrees, to be defined. This interpretation questions the relationship between the global negation $\mathcal{N}(B(\varphi, \alpha))$ and $B(\varphi, N(\alpha))$.

Third, the modality itself comes into question, considering that the statement '*it does not hold that* φ is believed to a degree α' gives some information about a 'non-belief' of φ, written $\mathcal{N}(B)(\varphi, \alpha)$, to indicate that the negation applies to the modality. Depending on how it is taken, this interpretation may lead to the introduction and manipulation of a second, opposite modality, on top of belief, something along the lines of disbelief.

3 Literature Review of Negation Principles

Negation is a central component of reasoning and information manipulation tools and constitutes an essential part of any logical system: its specification, together with, e.g., that of implication and the definition of an inference rule, suffices to define fully such a logical system. If, in classical logic, it establishes simple relations from true to false formulae, for systems with higher expressive power, it offers more complex, and richer, options and behaviours.

Since graded beliefs are related to both the formal frameworks of modal and weighted logics, the interpretation of their negation is related to the negation they respectively define: after reviewing the reference case of classical logic, this section discusses the richness of negation meaning when applied to high level notions e.g. representing knowledge or beliefs in the case of modal logics, and to truth degrees for many-valued and fuzzy logics.

3.1 Negation in Classical Logic

In classical logic, the manipulation of negation relies on the laws of excluded middle (EM) and non-contradiction (NC) which respectively state that $\varphi \vee \neg\varphi$ is a tautology and that $\varphi \wedge \neg\varphi$ is a contradiction. (EM) imposes that any formula is either true or its negation is; (NC) imposes that they cannot be true simultaneously. Together they mean that exactly one among φ and $\neg\varphi$ is true.

This principle can also be expressed in terms of truth values: denoting by $\mathcal{F}$ the set of all well-formed formulae and $d : \mathcal{F} \to \mathbb{B}$, the function that computes the truth value of any formula, the negation principle can be expressed as $d(\neg\varphi) = n(d(\varphi))$, where $n{:}\{0,1\} \to \{0,1\}$ is the function $n(x) = 1 - x$.

From an informational point of view, $\neg\varphi$ is usually taken as the opposite piece of information with respect to φ: if, e.g., φ is 'unicorns exist', $\neg\varphi$ usually means 'unicorns do not exist'.

3.2 Negation in Modal Logic

Modal logics [3,7] manipulate non factual pieces of information, such as knowledge and beliefs, through the modal operator $\Box$. Their combination with negation then raises the issue of the relations between the formulae $F_1 = \Box\varphi$, $F_2 = \neg\Box\varphi$, $F_3 = \Box\neg\varphi$ and, applying negation both before and after the modal operator, $F_4 = \neg\Box\neg\varphi = \Diamond\varphi$. The mere existence of these four variants underlines the richness of meaning negation can express in modal logics.

The negation behaviours expressed in classical logic still hold in modal logics. In particular, applying (EM) to the modal formula $\Box\varphi$ implies that $\Box\varphi \vee \neg\Box\varphi$ is a tautology, which establishes a relation between F_1 and F_2, and similarly between F_3 and F_4.

Relations between F_1 and F_4 are established by axioms (D) and (CD), reciprocal of each other, which state that:

$$(D) \quad \vdash \Box\varphi \to \neg\Box\neg\varphi \qquad (CD) \quad \vdash \neg\Box\neg\varphi \to \Box\varphi \tag{1}$$

Since φ is any formula, taking it to be a negated one and using the double negation property $\neg\neg\varphi = \varphi$ makes it possible to establish relations between F_3 and F_2, when similarly applying (D) and (CD) to $\Box\neg\varphi$.

Among the variants of modal logics, doxastic logic [12] interprets the modal operator as a belief operator, offering a formalism to represent and reason with beliefs. The doxastic reading of axiom (D) expresses that if φ is believed, then its contrary is not: it conveys the impossibility of believing both a formula and its contrary. (D) is thus considered as modelling one facet of the assumed rationality of the agent whose beliefs are represented. It is known as the consistency axiom and is included in the usual axiomatic definition of doxastic logic, viz. $KD45$.

Its reciprocal, (CD), conveys a similar, if complementary, consistency: in a doxastic reading, it would state that if a formula is not believed, then its contrary is. Its premise applies to an absence of belief and its conclusion to a belief. Rewriting the implication with negation and disjunction, it, therefore, imposes that an agent either believe a formula or its contrary, leading to a modal form of (EM), stating that $\Box\neg\varphi \vee \Box\varphi$ is a tautology. This constraint, which requires he take a stance on his belief in φ, excludes modelling a neutral, perplexed or undetermined frame of mind, i.e. the absence of belief. It is considered as too restrictive and generally not included in the axiomatic definition of doxastic logic.

These axioms illustrate the fact that processing non factual pieces of information leads to discussions and allows the expression of negations with specific behaviours, as opposed to what classical logic offers.

3.3 Negation in Many-Valued Logic

Graded logics, including many-valued [18] and fuzzy [20] logics, respectively reviewed in this subsection and the next, consider factual pieces of information, like classical logic does, but extend the set of admissible truth values beyond the binary case $\{0, 1\}$. They thus manipulate truth degrees, with major consequences on the behaviour of the negation: they relax the law of non-contradiction, allowing both φ and $\neg\varphi$ to be partially true together. The negation principles are thus redefined to set how far a formula and its negation can simultaneously hold.

Formally, for many-valued logics [18] the set of admissible truth values is the totally ordered set $\mathcal{L}_M = \{\tau_0, \ldots, \tau_{M-1}\}$, where M is a predefined positive integer and $\forall\alpha, \beta \in \{0, \ldots, M-1\}, (\alpha \leq \beta \Leftrightarrow \tau_\alpha \leq \tau_\beta)$. $\mathcal{L}_M$ represents intermediary truth values between 'false', written τ_0, and 'true', τ_{M-1}, at various levels of granularity, depending on the total number of levels, M.

As in classical logic, the negation principle can be written $d(\neg\varphi) = n(d(\varphi))$. The truth negation function n is modified, compared to the classic negation (see Sect. 3.1), to process truth degrees and no longer just binary truth values. The usual definition is:

$$\begin{aligned} n : \mathcal{L}_M &\to \mathcal{L}_M \\ \tau_i &\mapsto \tau_{M-1-i} \end{aligned} \tag{2}$$

This extension preserves compatibility with classical logic: the negation of 'false', $n(\tau_0) = \tau_{M-1}$, is still 'true', and vice-versa. For intermediary truth degrees, this negation operator can be seen as computing the symmetrical value with respect to the scale's middle value $\tau_{\frac{M-1}{2}}$. In the case where the number of truth levels M is odd, i.e. when this middle value indeed belongs to the scale, it is its own negation $n(\tau_{\frac{M-1}{2}}) = \tau_{\frac{M-1}{2}}$. This means that a formula whose truth degree is $\tau_{\frac{M-1}{2}}$ has the same truth degree as its negation. This specific behaviour of the negation, allowing a formula and its negation to be somewhat true together, can be interpreted in terms of a generalised law of non-contradiction.

Many-valued logic is further motivated by its capacity to provide linguistic representations of truth degrees, using a correspondence between the discrete ordered scale $\mathcal{L}_M$ and a set of linguistic labels, e.g. based on adverbs qualifying the truth degree: for instance, $\mathcal{L}_5$ is in bijection with the set {'false', 'rather false', 'neither', 'rather true', 'true'}. Taken thus, the negation operation expressed in Eq. (2) can be interpreted in terms of the linguistic notion of antonymy [17].

3.4 Negation in Fuzzy Logic

Fuzzy logic [20] is an infinitely many-valued logic, for which truth values are not defined on a discrete scale but on the real interval $[0,1]$. As detailed below, two levels account for the richness of its negation behaviour: fuzzy logic offers many negation operators on truth degree, as well as two negation types for predicates.

As in classical and many-valued logics, the negation principle can be expressed as $d(\neg\varphi) = n(d(\varphi))$, where the truth negation function n is adapted to manipulate values in $[0,1]$. Such a function is called a *fuzzy negator* [19] and is defined by three properties: (i) domain and co-domains: n is a function $n : [0,1] \to [0,1]$; (ii) monotonicity: n is non-increasing; (iii) boundary conditions: $n(0) = 1$ and $n(1) = 0$.

Note that the general definition does not impose involutivity, which makes a major difference with the classic case when processing double negations: the truth degree of $\neg\neg\varphi$ can be different from φ's.

However, the most usual definition of a fuzzy negator [20],

$$\begin{aligned} n : [0,1] &\to [0,1] \\ x &\mapsto 1 - x \end{aligned} \tag{3}$$

is both involutive and strict (i.e. strictly decreasing and continuous). Note that it is a straightforward generalisation of the negation operator defined for classic logic (see Sect. 3.1) from the domain $\{0,1\}$ to $[0,1]$; the preservation of the cases when the truth value is 0 or 1 are guaranteed by the boundary conditions. It also generalises the many-valued negation operator (see Eq. (2)) when the degrees τ_i are mapped to a discretisation of the interval $[0,1]$, $\tau_i \to i/(M-1)$.

Other fuzzy negators include (see e.g. [19]) $n(x) = \sqrt{1-x^2}$ or $n(x) = 1 - x^2$ which is strict but not involutive. The Gödel operator, defined by $n(x) = 0$ if $x > 0$ and $n(0) = 1$, and its dual, defined by $n(x) = 1$ if $x < 1$ and $n(1) = 0$, are neither continuous nor involutive. These two examples are drastic choices: their use restricts the truth values to the Boolean $\{0,1\}$. The variety of functions satisfying the general definition of fuzzy negators allows to define complex negation stances, richer than the behaviours the classic case can express.

Further, fuzzy logic also offers two types of negations for predicates defined as sub-intervals of a fuzzy partition over a numerical universe: complements and antonyms (see [16], for instance). Formally, for a predicate with membership function $A : [a^-, a^+] \to [0,1]$, its complement, based on a fuzzy negator, and its antonym are respectively defined by the membership functions $\overline{A}(x) = n(A(x))$ and $\widehat{A}(x) = A(a^+ - a^- - x)$. They usually differ one from another. Observe that, if the predicate is a modality of a Ruspini partition, its antonym can be interpreted as the symmetrical modality with respect to the central one, establishing a relationship with the principle of the many-valued negation operator defined in Eq. (2). It is then also related to the already mentioned linguistic definition of antonymy [17]: fuzzy predicates thus allow, for instance, to represent and distinguish between the three notions 'hot', 'not hot' and 'cold'.

4 From $\mathcal{N}(B(\varphi, \alpha))$ to $B(\neg\varphi, \beta)$: Negating Formulae

This paper addresses the issue of negation in the case of graded beliefs, at the crossroads of doxastic and weighted logics: this issue can thus be discussed as extensions of the manipulation rules reviewed in the previous section.

As described in Sect. 2, the general form of negation, written $\mathcal{N}(B(\varphi, \alpha))$, reads '*it does not hold that* φ is believed to a degree α'. A first interpretation of this negation considers its transposition to the formula about which the belief is expressed, leading to examine the relation between $\mathcal{N}(B(\varphi, \alpha))$ and $B(\neg\varphi, \beta)$.

After discussing the general principles underlying this interpretation, this section reads it as a graded variant of the consistency axiom (D).

4.1 General Principles

Establishing a relation between $\mathcal{N}(B(\varphi, \alpha))$ and $B(\neg\varphi, \beta)$ raises the question of drawing conclusions about a belief in $\neg\varphi$ from the negation of a belief in φ to a degree α: in the running example, this interpretation aims at establishing a relation between the facts '*it does not hold that* I firmly believe that unicorns exist' and 'I $\sharp$ believe that unicorns *do not exist*', where $\sharp$ represents an appropriate modulating adverb, yet to be determined.

Let us underline that, even if this interpretation is expressed as a shifting of the negation to the formula, the associated belief degree, β, will likely also be impacted, allowing for a discussion: imposing it *a priori* to be equal to any value, in particular to α, would limit the expressiveness of the considered interpretation.

4.2 Graded Variant of the Consistency Axiom

We propose to study the transfer of the global negation to the formula in the framework of doxastic logic, respectively interpreting $\mathcal{N}$ and $\neg$ as an outer and an inner negation with respect to the belief operator: setting the degrees aside, for a moment, this introduces a correspondence between the formulae $\mathcal{N}(B(\varphi, \alpha))$ and $B(\neg\varphi, \beta)$, on the one hand, and $\neg\Box\varphi$ and $\Box\neg\varphi$, on the other.

This interpretation of $\mathcal{N}$ requires that it satisfies the basic properties of a negation operator and, in particular, involutivity. From its very definition, this is indeed the case, as can be informally illustrated by its proposed reading: $\mathcal{N}(\mathcal{N}(B(\varphi, \alpha))$ is read 'it does not hold that *it does not hold that* φ is believed to a degree α'. It can be argued that this awkward expression is expected to be equivalent to 'φ is believed to a degree α'.

In doxastic logic, a relation is established between formulae $\neg\Box\varphi$ and $\Box\neg\varphi$, denoted F_2 and F_3 in Sect. 3.2, as the contrapositive of axiom (D), according to which $\vdash \Box\neg\varphi \rightarrow \neg\Box\varphi$. This implication is equivalently obtained when applying (D) to formula $\neg\varphi$.

Studying this negation for graded beliefs further leads to question a graded equivalent for axiom (D). There exist several graded extensions of modal logics, mainly specified from a semantic point of view: they consider enriched Kripke frame definitions [4–6, 14] or introduce counting functions in the semantic definition of the modal operator [10, 11, 13, 15]. Using a relative counting approach to introduce a weighted modality $\Box_\alpha$, we established in previous work [15] a graded variant of (D) stating that $\Box_\alpha\varphi \rightarrow \neg\Box_\beta\neg\varphi$ is a tautology, for any $\beta > 1-\alpha$, on the same semantic frame hypotheses as (D).

This weighted extension makes a relevant candidate for the desired relation between the global belief negation $\mathcal{N}(B(\varphi,\alpha))$ and the belief in the negated formula $B(\neg\varphi,\beta)$: applying (D_α) to $\neg\varphi$ and still considering $\mathcal{N}$ as the outer and $\neg$ as the inner negation w.r.t. the modal operator: $B(\neg\varphi,\alpha) \rightarrow \mathcal{N}(B(\varphi,\beta))$ for all $\alpha,\beta \in [0,1]$ such that $\beta > 1-\alpha$. This formula can be read read 'if $\neg\varphi$ is believed at degree α, then it does not hold that φ is believed at degree β'. As a consequence, φ and $\neg\varphi$ can both be partially believed together, so long as their respective degrees satisfy the inequality constraint.

5 From $\mathcal{N}(B(\varphi,\alpha))$ to $B(\varphi,N(\alpha))$: Negating Degrees

Considering now that the negation bears not on the formula but on the degree, a new set of possibilities arises. Indeed, interpreting '*it does not hold that* φ is believed to a degree α' as 'φ is believed to a degree $N(\alpha)$', where N remains to be defined, offers new interpretations, both on the choice of operator and on the meaning these impose on the ensuing doxastic reading. At a fundamental level, this question depends on the interpretation of the degrees.

After detailing this interpretation, this section proposes to read the degrees as partial membership to a belief set, in the fuzzy set formalism, and studies the relevance of fuzzy negators, from a doxastic point of view.

5.1 General Principle

Note, first, that this section focuses on the case where the belief degrees are represented as numerical values, more specifically in the interval $[0,1]$. A discretisation of this range could be considered, but it makes the interpretation harder to grasp. Indeed, it blurs the difference with the case where linguistic labels are used to express belief degrees but these hold no meaning when interpreted alone, apart from the modality. These will be considered when the negation is applied to the modality, as discussed in Sect. 6.

A literal interpretation of the negation of the degree considers a negative statement '*it does not hold that* φ is believed to a degree α' as meaning 'φ is believed to a degree which is *not* α', suggesting that any value other than α is suitable: it may lead to define $N(\alpha)$ as a set and not a value, e.g. $N(\alpha) = [0,1]\backslash\{\alpha\}$. This interpretation, although acceptable to the letter, leads to a highly imprecise and uninformative understanding of the statement: we do not examine it further and propose to establish N as a function from $[0,1]$ to $[0,1]$.

5.2 Belief Degree as Fuzzy Membership to a Belief Set

Considering the universe of all well-formed formulae $\mathcal{F}$, the set of gradual beliefs $\mathcal{B}$ of an agent can be defined as a fuzzy subset of $\mathcal{F}$: each formula more or less belongs to $\mathcal{B}$ and the belief degrees are interpreted as membership degrees.

The boundary case, where the degree is $B(\varphi, 1)$, that is, a formula φ representing a fully believed fact, believed without any restriction, is interpreted as totally belonging to the belief set $\mathcal{B}$. From a defuzzified point of view, it can be seen as equivalent to $\Box\varphi$ in doxastic logic. A formula with a lower membership degree to $\mathcal{B}$ is believed less: defuzzifying the belief set through an α-cut, with $\alpha < 1$, can then be interpreted as enlarging the set of beliefs taken into account for reasoning, to extend beyond the full-fledged ones. The other boundary case appears to be somewhat harder to interpret. $B(\varphi, 0)$, means that φ does not belong to the belief set at all and represents a formula about which no belief holds. This interpretation raises the question of introducing an additional modality, to represent 'disbelief', and is discussed in the next section.

5.3 Relevance of Fuzzy Negators

Considering belief degrees as fuzzy membership degrees to a belief set suggests to use a fuzzy negator for the negation operator N on degrees, as reviewed in Sect. 3.4. This subsection examines their properties, regarding boundary conditions, monotonicity and involutivity, and their effect on a doxastic reading.

Boundary Conditions. Using the running example, the relevance of fuzzy negator boundary conditions questions the relevance of equating 'I believe to a degree $N(1)$ that unicorns exist' to 'I believe to a degree 0 that unicorns exist' – and vice-versa– and therefore depends on the interpretation of these degrees. Considering $B(\varphi, 1)$ as equivalent to $\Box\varphi$, as suggested above, the boundary conditions can be interpreted as preserving the compatibility with the classic case of the modal expression of (EM), which states that $\Box\varphi \vee \neg\Box\varphi$ is a tautology.

Monotonicity. From a doxastic point of view, the monotonicity of fuzzy negators implies comparing beliefs –or rather their levels– and their negations.

Let us extend the running example slightly and suppose we hold the following graded beliefs: B_1: 'I believe to a degree $N(0.8)$ that unicorns exist' and B_2: 'I believe to a degree $N(0.6)$ that leprechauns exist'. Negating both these beliefs with a fuzzy, and therefore non-decreasing, negator entails that $\mathcal{N}(B_1)$ is believed, at most, as much as $\mathcal{N}(B_2)$.

Additionally, any increasing operator would contradict the scale's structure, e.g. creating situations where $N(\alpha) < N(1) = 0$. The different fuzzy negators, be they symmetric, drastic or if they offer some form of compromise, allow ways of modelling different belief stances, none of which contradict monotonicity.

Involutivity. Finally, a property that not all fuzzy negators share but which is held by the most usual ones is involutivity. Choosing to believe $\mathcal{N}(\mathcal{N}(B_1))$ as much as B_1 can be understood as refusing to gain or lose any belief by *not not accepting 0.8* for a belief level. Even if the expression is somewhat farfetched, the underlying notion seems desirable.

6 From $\mathcal{N}(B(\varphi, \alpha))$ to $\mathcal{N}(B)(\varphi, \beta)$: Negating Modalities

The third and final interpretation of negated graded beliefs transfers negation to the modality, where the negative statement '*it does not hold that* φ is believed to a degree α' is seen as providing information about a disbelief, therefore requiring an additional modality, besides belief. At a more intuitive, yet nuanced, level, in the case where belief degrees are linguistically expressed by adverbs, we propose to consider several modalities, intrinsically combining the modality with the level, e.g. distinguishing between the modalities 'weakly believe' and 'firmly believe'. This section considers these two cases in turn.

The difference with the transfer of negation to the formula should be underlined: the latter discusses how far one can believe a fact and its contrary simultaneously, with different degrees. This transfer considers a single fact and questions the links between believing and disbelieving it.

6.1 Belief and Disbelief

The introduction of a disbelief modality proposes to interpret '*it does not hold that* φ is believed to a degree α' as 'φ is *disbelieved* to a degree β', transferring the global negation to the modal operator, and simultaneously allowing an effect on the associated degrees, so as to avoid limiting the expressiveness of the considered negation interpretation *a priori*.

Note that the β coefficient describes a degree of disbelief, and not a belief degree. This induces a major change as opposed to the previously discussed interpretations, where degrees are measured on a single scale and all have the same meaning. As, nevertheless, the belief and disbelief modalities are related to one another, the associated degrees open the way to a signed scale rather than two independent ones. This interpretation was mentioned in Sect. 5.2 where understanding a 0-belief as utter disbelief was suggested. It then begs the question of at which level, i.e. around which α, does belief become disbelief.

The fuzzy interpretation discussed in the previous section can be considered as an extreme case, where the open interval (0,1] represents various belief degrees

Table 1. Discussed interpretations of graded belief negation

Framework	Negation expression	Degree constraint
Graded modal	$B(\neg\varphi, \alpha) \rightarrow \mathcal{N}(B(\varphi, \beta))$	$\alpha, \beta \in [0, 1]$ and $\beta > 1 - \alpha$
Fuzzy	$\mathcal{N}(B(\varphi, \alpha)) = B(\varphi, N(\alpha))$	e.g. $N(\alpha) = 1 - \alpha$
Bimodal	$\mathcal{N}(B(\varphi, \alpha)) = Db(\varphi, f(\alpha))$	f isotone function
Multimodal	$\mathcal{N}(B_t(\varphi)) = B_{n(t)}(\varphi)$	$t \in \mathcal{L}_M$, $n(\tau_i) = \tau_{M-1-i}$

and a single value, 0, is considered for disbelief. Such an interpretation can be considered as bridging the gap between the interpretation of graded belief negation in terms of degree and in terms of modalities.

A natural value for changing from belief to disbelief considering a signed scale is probably 0.5, the scale's middle value: (0.5, 1] then represent intermediate degrees of belief and [0, 0.5) intermediate degrees of disbelief and 0.5 a neutral value. However, one could choose another value, to emphasise different attitudes with regards to belief, e.g. to have more positive values than negative ones, even if both sides are uncountable.

6.2 Finite Set of Belief Modalities

When considering the transfer of negation to the modality, a variant of special interest is the case where a belief level is considered as integrated in the modality, e.g. when 'firmly believe' and 'weakly believe' are considered to be two separate, yet comparable, modalities, instead of a single, continuous, belief modality nuanced with a degree: this interpretation is equivalent to considering a finite set of modalities with more than the two levels, belief and disbelief, discussed in the previous subsection. One can for instance consider five levels of belief {'low', 'weak', 'moderate', 'strong' and 'high'} or more, with more linguistic labels.

This choice naturally leads to exploit a formalisation in many-valued logic, introducing several modalities formally defined as B_t, $t \in \mathcal{L}_M = \{\tau_0, \ldots, \tau_{M-1}\}$. Negation then requires to define $\mathcal{N}(B_t)$, $t \in \mathcal{L}_M$, which can be translated to $B_{n(t)}$ with the many-valued negation operator n (see Eq. (2)). The negation can then be considered as being applied to the adverb, which represents the degree-modality combination, and therefore interpreted in terms of linguistic antonyms.

7 Conclusion and Future Works

This paper proposed a discussion of the problem of negating graded beliefs. In the formal frameworks of modal, fuzzy and many-valued logics, it examined three main interpretations of this negation, depending on what it bears on: the formula about which a modal assertion is expressed; the modality, so as to express belief; or the degree, which represents the level of belief. Table 1 gathers the given interpretations and the associated frameworks. Although the objects of the negation have been considered separately, the results show them to be closely related to one another, in particular in their influence on the associated belief degree. This suggests that the core component of graded beliefs is the level and, thus, their manipulation revolves around the construction of the rules governing its evolution.

Beyond negation, graded belief manipulation, in order to allow the combination of available beliefs into new ones, requires the definition of reasoning tools, in particular regarding their conjunction and disjunction. Future work will aim at extending the principles established for the negation operation to other connectives, leading to a general formal framework to reason about graded beliefs.

References

1. Bacchus, F., Grove, A.J., Halpern, J.Y., Koller, D.: From statistical knowledge bases to degrees of belief. Artif. Intell. **87**(1), 75–143 (1996)
2. Banerjee, M., Dubois, D.: A simple modal logic for reasoning about revealed beliefs. In: Sossai, C., Chemello, G. (eds.) ECSQARU 2009. LNCS, vol. 5590, pp. 805–816. Springer, Heidelberg (2009)
3. Blackburn, P., De Rijke, M., Venema, Y.: Modal Logic. Cambridge University Press, Cambridge (2001)
4. Bou, F., Esteva, F., Godo, L., Rodrıguez, R.: Characterizing fuzzy modal semantics by fuzzy multimodal systems with crisp accessibility relations. In: Proceedings of the Joint 2009 IFSA World Congress and EUSFLAT Conference, pp. 1541–1546 (2009)
5. Boutilier, C.: Modal logics for qualitative possibility theory. Int. J. Approximate Reasoning **10**(2), 173–201 (1994)
6. Fariñas del Cerro, L., Herzig, A.: A modal analysis of possibility theory. In: Jorrand, P., Kelemen, J. (eds.) Fundamentals of Artificial Intelligence Research. LNCS, vol. 535, pp. 11–18. Springer, Heidelberg (1991)
7. Chellas, B.F.: Modal Logic: An Introduction, vol. 316. Cambridge University Press, Cambridge (1980)
8. Dubois, D., Prade, H.: Generalized possibilistic logic. In: Benferhat, S., Grant, J. (eds.) SUM 2011. LNCS, vol. 6929, pp. 428–432. Springer, Heidelberg (2011)
9. Dubois, D., Prade, H., Schockaert, S.: Rules and meta-rules in the framework of possibility theory and possibilistic logic. Scientia Iranica **18**, 566–573 (2011)
10. Fattorosi-Barnaba, M., Cerrato, C.: Graded modalities I. Studia Logica **47**, 99–110 (1988)
11. Fine, K.: In so many possible worlds. Notre Dame J. Formal Logic **13**(4), 516–520 (1972)
12. Hintikka, J.: Knowledge and Belief: An Introduction to the Logic of the Two Notions, vol. 181. Cornell University Press, Ithaca (1962)
13. van der Hoek, W., Meyer, J.J.C.: Graded modalities in epistemic logic. In: Nerode, A., Taitslin, M. (eds.) Tver 1992. LNCS, vol. 620, pp. 503–514. Springer, Heidelberg (1992)
14. Laverny, N., Lang, J.: From knowledge-based programs to graded belief-based programs. part I: on-line reasoning. In: Proceedings of ECAI. IOS Press (2004)
15. Legastelois, B., Lesot, M.J., Revault d'Allonnes, A.: Typology of axioms for a weighted modal logic. In: Proceedings of the IJCAI Workshop on Weighted Logics for Artificial Intelligence (WL4AI), pp. 40–48 (2015)

16. Moyse, G., Lesot, M.J., Bouchon-Meunier, B.: Oppositions in fuzzy linguistic summaries. In: Proceedings of the IEEE International Conference on Fuzzy Systems. IEEE (2015)
17. Muehleisen, V.L.: Antonymy and semantic range in English. Ph.D. thesis, Northwestern University (1997)
18. Rescher, N.: Many-valued logic. In: Rescher, N. (ed.) Topics in Philosophical Logic. Synthese Library, vol. 17, pp. 54–125. Springer, Dordrecht (1968)
19. Weber, S.: A general concept of fuzzy connectives, negations and implications based on t-norms and t-conorms. Fuzzy Sets Syst. **11**(1), 103–113 (1983)
20. Zadeh, L.: Fuzzy sets. Inf. Control **8**, 338–353 (1965)

Comparing Some Substructural Strategies Dealing with Vagueness

Pablo Cobreros[1], Paul Egré[2], David Ripley[3], and Robert van Rooij[4(✉)]

[1] Department of Philosophy, University of Navarra, Pamplona, Spain
[2] CNRS/ENS, PSL/SCAS, Institut Jean-Nicod, Paris, France
[3] Department of Philosophy, University of Connecticut, Storrs, USA
[4] Institute for Logic, Language and Communication, Amsterdam, The Netherlands
r.a.m.vanrooij@uva.nl

1 Introduction

It is well-known that in combination with further premises that look less controversial, the tolerance principle – the constraint that if Pa holds, and a and b are similar in P-relevant respects, Pb holds as well – leads to contradiction, namely to the sorites paradox. According to many influential views of the sorites paradox (e.g. Williamson 1994), we therefore ought to reject the principle of tolerance as unsound.

There are reasons to think of such a view as too drastic and as missing out on the role that such a principle plays in categorization and in ordinary judgmental and inferential practice. Taking a different perspective, the tolerance principle ought not to be discarded that fast, even when viewed normatively. Instead, it corresponds to what might be called a *soft constraint*, or a *default*, namely a rule that we can use legitimately in reasoning, but that must be used with care.

One family of approaches represents the tolerance principle by a certain conditional sentence, of the form: $Pa \wedge a \sim_P b \rightarrow Pb$, and bestows special properties to the conditional to turn it into a *soft constraint*. One natural option is to use fuzzy logic, where $v_{\mathcal{M}}(A)$ can be anywhere in $[0, 1]$ and $v_{\mathcal{M}}(A \rightarrow B) = Min\{1, 1 - v_{\mathcal{M}}(A) + v_{\mathcal{M}}(B)\}$. One can demand that the tolerance conditional may never have a value below $1 - \varepsilon$ for some small ε. Given an appropriate sorites sequence, it will be possible to have: $v_{\mathcal{M}}(Pa_1 \rightarrow Pa_2) = 1-\varepsilon$, $v_{\mathcal{M}}(Pa_2 \rightarrow Pa_3) = 1 - \varepsilon$, without having $v_{\mathcal{M}}(Pa_1 \rightarrow Pa_3) = 1 - \varepsilon$.

A different option is to treat the tolerance conditional as expressing a *defeasible rule* (like when '$\rightarrow$' expresses a counterfactual conditional). Say that $Pa \wedge a \sim_P b \rightarrow Pb$ is true provided Pb is true in all 'optimal' $(Pa \wedge a \sim_P b)$-worlds. Call a world $(Pa \wedge a \sim_P b)$-optimal if a is P-similar to b but is not close to a borderline case of P. From $Pa \wedge a \sim_P b, Pa \wedge a \sim_P b \rightarrow Pb$, it need not follow that Pb, since a world may satisfy $Pa \wedge a \sim_P b$ without being

Except for the last section, this paper is an abridged version of a longer paper entitled "The Tolerance Principle: Nontransitive Reasoning or Nonmonotonic Reasoning?". We are indebted to two anonymous reviewers for helpful comments.

J.P. Carvalho et al. (Eds.): IPMU 2016, Part II, CCIS 611, pp. 161–172, 2016.
DOI: 10.1007/978-3-319-40581-0_14

$(Pa \wedge a \sim_P b)$-optimal, precisely when b is a borderline case of P. This conditional fails to satisfy modus ponens, and it is also nontransitive. But moreover it is nonmonotonic, since a world that is $Pa \wedge a \sim_P b$-optimal need not be $Pa \wedge Pc \wedge a \sim_P b$-optimal. On that view, the tolerance conditional represents a defeasible rule, usable only if the main premise correspond to an optimal world.

Both the fuzzy approach and the nonmonotonic approach have some appeal. On the definition of fuzzy validity that allows tolerance to be a sound premise, the sorites paradox is solved by saying that modus ponens is not a valid rule any more (where $\Gamma \models A$ iff $\forall \mathcal{M} : Min\{v_{\mathcal{M}}(\gamma) : \gamma \in \Gamma\} \leq v_{\mathcal{M}}(A)$). On the nonmonotonic approach, the sorites paradox is solved by saying that modus ponens is a defeasible rule: the sorites argument has sound premises, but is not undefeasibly valid. Despite that, both approaches suffer an important limitation, which concerns their treatment of the tolerance principle in terms of a special conditional connective. As is well known, a sorites argument can be stated using only conjunction and negation, by saying that it is not the case that there are two cases a and b that are very similar, but are such that Pa and not Pb. But a nonmonotonic treatment of the conditional does not tell us how to address that alternative version of the sorites. Similarly, for the fuzzy case.

In this paper, we are interested in accounts of vagueness that, instead of relying on a special conditional connective in a way that leaves intuitively desirable properties of a conditional connective in place, and in a way suitable to deal with the sorites argument in its conjunctive form as well as its conditional one. We will focus on two structural approaches which mirror the nontransitive and nonmonotone conditional to some extent, but shift those properties up one level, namely to the consequence relation. The first is the *nontransitive* treatment of logical consequence favored in our past work, on which the principle of tolerance comes out as valid in rule form, but cannot be iterated without risk (soft consequence as *permissive consequence*, see Cobreros et al. (2014) for an overview). The second is the *nonmonotone* treatment of logical consequence, on which the principle of tolerance too can come out as valid, but in a way that is sensitive to context and to the addition of further premises (soft consequence as defeasible consequence).

2 Validating Tolerance Using Non-standard Entailment

We have argued in Cobreros et al. (2012, 2014) that the tolerance principle should be adopted both in rule form and in sentential form. We were able to do so by (i) interpreting the language in standard three valued models, (ii) adding similarity relations (one for each predicate P) to the language and interpreting that in a specific way, and (iii) by formulating a new consequence relation, $\models^{st}_{\sim}$. In this section we first rehearse the details of our approach in more detail. We first concentrate on (i) and (iii). We then consider two broader variants of our initial strategy, which both rely on a more general notion of pragmatic entailment which turn out to be non-monotonic.

2.1 The Logic *st*

Let $\mathcal{M} = \langle \mathcal{D}, \mathcal{I} \rangle$, with $\mathcal{I}$ a total function from atomic sentences to $\{0, \frac{1}{2}, 1\}$. This model extends to formulas according to the strong Kleene valuation scheme:

- $\mathcal{V}_{\mathcal{M}}(\phi) = \mathcal{I}_{\mathcal{M}}(\phi)$, if ϕ is atomic
- $\mathcal{V}_{\mathcal{M}}(\neg\phi) = 1 - \mathcal{V}_{\mathcal{M}}(\phi)$
- $\mathcal{V}_{\mathcal{M}}(\phi \wedge \psi) = min\{\mathcal{V}_{\mathcal{M}}(\phi), \mathcal{V}_{\mathcal{M}}(\psi)\}$
- $\mathcal{V}_{\mathcal{M}}(\phi \vee \psi) = max\{\mathcal{V}_{\mathcal{M}}(\phi), \mathcal{V}_{\mathcal{M}}(\psi)\}$
- $\mathcal{V}_{\mathcal{M}}(\forall x\phi) = min\{\mathcal{V}_{\mathcal{M}}([^{x}/_{\underline{d}}]\phi) : d \in D\}$[1]

We say that ϕ is **strictly true** in $\mathcal{M}$ iff $\mathcal{V}_{\mathcal{M}}(\phi) = 1$, and that ϕ is **tolerantly true** iff $\mathcal{V}_{\mathcal{M}}(\phi) \geq \frac{1}{2}$. In terms of this semantics we can define some well-known logics: Kleene's $K3$ and Priest's LP. Both logics understand entailment as preservation of truth in all models, the difference is that while for $K3$ *truth* means *strict truth*, for LP means *tolerant truth*:

$$\begin{gathered}\Gamma \models_{K3} \Delta \text{ just in case for all } \mathcal{M}: \\ \text{if } \forall A \in \Gamma : \mathcal{V}_{\mathcal{M}}(A) = 1, \text{ then } \exists B \in \Delta : \mathcal{V}_{\mathcal{M}}(B) = 1 \\ \Gamma \models_{LP} \Delta \text{ just in case for all } \mathcal{M}: \\ \text{if } \forall A \in \Gamma : \mathcal{V}_{\mathcal{M}}(A) > 0, \text{ then } \exists B \in \Delta : \mathcal{V}_{\mathcal{M}}(B) > 0\end{gathered}$$

A fundamental idea in Cobreros et al. (2012) was to define entailment from *s*trict to *t*olerant:

$$\begin{gathered}\Gamma \models^{st} \Delta \text{ just in case for all } \mathcal{M}: \\ \text{if } \forall A \in \Gamma : \mathcal{V}_{\mathcal{M}}(A) = 1, \text{ then } \exists B \in \Delta : \mathcal{V}_{\mathcal{M}}(B) > 0\end{gathered}$$

Thus, although we don't give up the idea that entailment is truth-preserving, we allow the standard of assertion of the conclusions to be weaker than the standard of assertion of the premises. A surprising feature of this logic is that although the semantics makes use of three truth-values, the consequence relation is exactly the familiar consequence relation of classical logic for the standard language fragment. This in contrast with $K3$ and LP, which give up many classically valid arguments.

Now, despite its classicality, this new semantics makes room for tolerance without falling prey to the sorites paradox. In order to account for tolerance, we extend the language with similarity relations, $\sim_P$, one for each predicate P.[2] One interpretation is the following:

- $\mathcal{V}_{\mathcal{M}}(a \sim_P b) = 1$ iff $|\mathcal{V}_{\mathcal{M}}(Pa) - \mathcal{V}_{\mathcal{M}}(Pb)| < 1$, 0 otherwise

[1] We assume here for convenience that each $d \in D$ has a name $\underline{d}$.

[2] See Halpern (2008) and van Rooij (2010) for more on the link between vagueness and nontransitive similarity.

The resulting logic $st^{\sim}$ is a conservative extension of classical logic, in the sense that any classically valid argument in the old vocabulary remains valid. In addition, the tolerance formula $(\forall x, y((Px \wedge x \sim_P y) \rightarrow Py))$ becomes valid, as does the tolerance argument: $Pa, a \sim_P b \models^{st}_{\sim} Pb$. The endorsement of tolerance does not lead to paradox, however, since tolerance in $st^{\sim}$ leads to non-transitivity: $Pa, a \sim_P b \models^{st}_{\sim} Pb$ and $Pb, b \sim_P c \models^{st}_{\sim} Pc$ BUT $Pa, a \sim_P b, b \sim_P c \not\models^{st}_{\sim} Pc$.

We felt, and still feel, that this is a very intuitive and appealing treatment of the sorites paradox. The treatment however, comes with the limitation that we should make a distinction about the diagnosis of the sorites paradox depending on its formulation. If we look at the sorites as a step-by-step argument based on the validity of tolerance inferences then, though each tolerance inference is valid, validity breaks when we try to chain these inferences. If we consider the sorites argument with the tolerance formula $(\forall x, y((Px \wedge x \sim_P y) \rightarrow Py))$ as an explicit premise, then although that formula is valid, the resulting argument is valid but unsound (the tolerance formula is valid but cannot automatically be appealed to as a premise; it is not suppressible). In short, although the tolerance formula is valid, according to the logic $st^{\sim}$ we are not in a position to draw on it as a premise without further ado. However, there are contexts in which we would like to assert the tolerance formula, in much the same way there are contexts in which we would like to assert a contradiction; we would like assert, from time to time, *in a tolerant sense*

A number of recent psycholinguistic studies (e.g., Alxatib and Pelletier 2011; Ripley 2011; Egré et al. 2013) show that naive speakers find a logical contradiction like 'John is tall and John is not tall' acceptable in cases where John is a borderline tall man. This seems to show that we need to take account of tolerant truth, since tolerant truth exhibits this exact behavior. However, just relying on the notion of tolerant truth would mean that the assertion 'John is tall' would be acceptable in the same situation. The same experimental evidence shows, however, that this is not the case: 'John is tall' is taken to be acceptable only if John is *really* tall. In terms of our three-valued models this could be modeled by saying that the assertion 'John is tall' is acceptable only if John is strictly tall. Similarly, Serchuk et al. (2011) found that classical tautologies like '$Tj \vee \neg Tj$' are not automatically accepted if John is borderline tall. So making use of tolerant *and* of strict truth (which exhibits this latter behavior) seems required.

The conclusion we draw from the previous discussion (cf. Cobreros et al. 2012) is that we should interpret a sentence strictly if possible, and tolerantly otherwise. This interpretation strategy is in line with Grice's (1967) strategy to account for scalar implicatures. Unfortunately, this interpretation strategy taken at face value gives rise to trouble for more complex sentences. Alxatib et al. (2013) show that we wrongly predict that a sentence like 'Adam is tall and not tall, *or* John is rich' not only entails, but even means that John is strictly rich, although it should not entail this and intuitively should mean that *either* Adam is borderline tall *or* John is strictly rich. In Cobreros et al. (2015)

we responded by providing a more sophisticated pragmatic interpretation rule to strengthen the meaning of a sentence.

2.2 Pragmatic Interpretation

To account for this pragmatic strengthening we make use of truth-makers. We propose that the pragmatic interpretation of ϕ makes one exact truth-maker of ϕ as true as possible. To determine what the truth-makers of a sentence are, we follow van Fraassen (1969). We start with a set of basic *state of affairs*, SOA. It is assumed that for every element $\mathbf{p}$ of SOA there is also its complement $\overline{\mathbf{p}} \in SOA$ for which it holds that $\overline{\overline{\mathbf{p}}} = \mathbf{p}$. For simplicity we assume a close correspondence between atomic sentences of the language and the SOAs: with each literal (atomic sentence or its negation) of the language there corresponds exactly one SOA: the state of affairs that makes this literal true. The set of *facts*, $\mathcal{F}$, is just $\wp(SOA) - \{\emptyset\}$, so any non-empty subset of SOA is thought of as a fact. If $\mathbf{p}, \mathbf{q} \in SOA$, then $\{\mathbf{p}\}$ and $\{\mathbf{q}\}$ are atomic facts, and $\{\mathbf{p}, \mathbf{q}\}$ is a conjunctive fact. A fact is what makes a sentence true. But, of course, a sentence might have more than one truth-maker. Atomic sentence p is not only made true by atomic fact $\{\mathbf{p}\}$, but also by conjunctive fact $\{\mathbf{p}, \mathbf{q}\}$. The former one is a *more minimal* truth-maker than the latter. More interestingly, disjunctive sentences might have several minimal truth-makers. The disjunction $p \vee q$, for instance, has two minimal truth-makers: $\{\mathbf{p}\}$ and $\{\mathbf{q}\}$. What we are after, however, is the notion of the *exact* truth-makers for ϕ. We say that the disjunction $p \vee (p \wedge q)$ – although it has only $\{\mathbf{p}\}$ as its minimal truth-maker – has *two* exact truth-makers: $\{\mathbf{p}\}$ and $\{\mathbf{p}, \mathbf{q}\}$. We can give the following simultaneous recursive definition of the set of exact truth- and falsity-makers of ϕ, $T(\phi)$ and $F(\phi)$, respectively:

$$
\begin{array}{ll}
T(p) = \{\{\mathbf{p}\}\} & F(p) = \{\{\overline{\mathbf{p}}\}\} \text{ for atomic } p. \\
T(\neg\phi) = F(\phi) & F(\neg\phi) = T(\phi). \\
T(\phi \wedge \psi) = T(\phi) \otimes T(\psi) & F(\phi \wedge \psi) \ = F(\phi) \cup F(\psi). \\
= \{X \cup Y | X \in T(\phi), Y \in T(\psi)\} & \\
T(\phi \vee \psi) \ = T(\phi) \cup T(\phi) & F(\phi \vee \psi) \ = F(\phi) \otimes F(\psi).
\end{array}
$$

Notice that according to these rules, $T(p) = \{\{\mathbf{p}\}\}$, $T(\neg p) = \{\{\overline{\mathbf{p}}\}\}$, $T(p \vee q) = \{\{\mathbf{p}\}, \{\mathbf{q}\}\}$ and $T(p \wedge q) = \{\{\mathbf{p}, \mathbf{q}\}\}$. We analyse conditionals like $\phi \rightarrow \psi$ as material implication, that is $p \rightarrow q \equiv \neg p \vee q$, and thus $T(p \rightarrow q) = \{\{\overline{\mathbf{p}}\}, \{\mathbf{q}\}\}$.

To account for quantifiers, we assume that for each n-place predicate P the model contains facts like $\mathbf{Pd_1}, , \cdots \mathbf{d_n}$, with each $\mathbf{d_i} \in D$ an individual. We assume for simplicity that each $\mathbf{d} \in D$ has a unique name $\underline{d}$ in the language.

- $T(P\underline{d_1}, \cdots, \underline{d_n}) \ = \ \{\{\mathbf{Pd_1}, \cdots, \mathbf{d_n}\}\}$ $F(P\underline{d_1}, \cdots, \underline{d_n}) \ = \ \{\{\overline{\mathbf{Pd_1}, \cdots, \mathbf{d_n}}\}\}$
- $T(\forall x \phi) \ = \ \bigotimes_{d \in D} T(\phi[^x/_{\underline{d}}])$ $F(\forall x \phi) \ = \ \bigcup_{d \in D} F(\phi[^x/_{\underline{d}}])$.
- $T(\exists x \phi) \ = \ \bigcup_{d \in D} T(\phi[^x/_{\underline{d}}])$ $F(\exists x \phi) \ = \ \bigotimes_{d \in D} F(\phi[^x/_{\underline{d}}])$.

Observe that $T(\forall x Px) = T(P\underline{a}) \otimes T(P\underline{b}) = \{\{\mathbf{Pa}, \mathbf{Pb}\}\}$, if $D = \{a, b\}$. Similarly, $T(\exists x Px) = T(P\underline{a}) \cup T(P\underline{b}) = \{\{\mathbf{Pa}\}, \{\mathbf{Pb}\}\}$. Notice that facts might not only be incomplete (neither verify nor falsify a sentence), they might also be inconsistent and both verify and falsify a sentence. Indeed, we have not ruled out facts like $\{\mathbf{Tj}, \overline{\mathbf{Tj}}\}$. Such inconsistent facts are crucial for us to model the meaning of vague sentences, expressing in this case that John is borderline tall.

$T(\phi)$ can be thought of as a fine-grained *semantic* interpretation of ϕ. It can be used to determine its standard truth-conditional meaning as given by possible worlds semantics, if a world is taken to be a maximally consistent conjunctive fact. In that case the standard truth-conditional meaning of ϕ, $[\![\phi]\!]$, can be recovered as the set of worlds in which ϕ has a truth-maker:

- $[\![\phi]\!] \quad \stackrel{def}{=} \quad \{w \in W \mid \exists f \in T(\phi) : f \subseteq w\}.$

For our purposes, we retain the insistence that worlds be maximal, that $\mathbf{p} \in w$ or $\overline{\mathbf{p}} \in w$ for each atomic fact $\mathbf{p}$ and world w. But we allow for worlds to be inconsistent, for some worlds to contain both $\mathbf{p}$ and $\overline{\mathbf{p}}$, for some atomic facts $\mathbf{p}$. This allows us to capture the difference between strict and tolerant satisfaction at a world, connecting this truth-maker semantics to our three-valued *st*-models. For each atomic sentence p and *st*-model, or world, w we define $\mathcal{V}_w(p) = 1$ iff $\mathbf{p} \in w$ and $\overline{\mathbf{p}} \notin w$; $\mathcal{V}_w(p) = 0$ iff $\mathbf{p} \notin w$ and $\overline{\mathbf{p}} \in w$, and $\mathcal{V}_w(p) = \frac{1}{2}$ otherwise.

But we did not introduce truth-makers just to recover notions we already had. Our purpose in introducing truth-makers is to define a notion of pragmatic meaning in terms of which we can strengthen the semantic meaning of a sentence. We have suggested above that although we allow for inconsistencies, we can still pragmatically infer that $\neg p$ is not true from the fact that 'p' is said by a reasoning analogue to those involving scalar implicatures. In linguistic pragmatics it is not uncommon to use *minimal models* (e.g. van Rooij and Schulz 2004) to account for scalar pragmatic implicatures. For us, a minimal model, or world, is one that is minimally inconsistent: it doesn't contain more inconsistencies than required. To model this, we will make use of the following definition, with $v <_f w \quad \text{iff}_{df}$ $\{x \in SOA : x \in f \,\&\, \overline{x} \in v\} \subsetneq \{x \in SOA : x \in f \,\&\, \overline{x} \in w\}$):

- $PRAG(\phi) \quad \stackrel{def}{=} \quad \{w \in W \mid \exists f \in T(\phi) : f \subseteq w \,\&\, \neg\exists v \supseteq f : v <_f w\}.$

PRAG gets many predictions correct: (i) 'John is tall' is pragmatically interpreted to mean that John is strictly tall, (ii) 'John is not tall' is predicted to mean that John is not even tolerantly tall, (iii) 'John is tall and John is not tall' means that John is borderline tall, and (iv) 'John is tall or not tall' means that John is not borderline tall. All these predictions are in accordance with recent experimental results reported by Alxatib and Pelletier (2011), Ripley (2011), Serchuk et al. (2011) and Egre et al. (2013). Furthermore, 'John is tall and not tall, and Mary is rich' is pragmatically interpreted to mean that John is borderline tall and Mary strictly rich, which seems intuitively correct. Finally, 'John is tall and not tall, or Mary is rich' is correctly interpreted as saying that John is borderline tall, or Mary is strictly rich.

Let us go back now to the tolerance principle. How *PRAG* interprets it depends partly on the way we interpret similarity statements. We might use similarity statements to constrain our models at least in the following ways:

(i) $\mathcal{V}_{\mathcal{M}}(a \sim_P b) = 1$ iff $|\mathcal{V}_{\mathcal{M}}(Pa) - \mathcal{V}_{\mathcal{M}}(Pb)| < 1$, 0 otherwise, or
(ii) $\mathcal{V}_{\mathcal{M}}(a \sim_P b) = 1 - |\mathcal{V}_{\mathcal{M}}(Pa) - \mathcal{V}_{\mathcal{M}}(Pb)|$.

Notice that they are incompatible with each other, at least if there are any a, b with $|\mathcal{V}_{\mathcal{M}}(Pa) - \mathcal{V}_{\mathcal{M}}(Pb)| = \frac{1}{2}$.

Option (i) is the one we mentioned in Sect. 2.1, as an example constraint. According to this option that $\mathcal{V}_w(Pa \wedge a \sim_P b) = 1$ does not force it to be the case that $\mathcal{V}_w(Pb) = 1$. In Sect. 2.1 this was used to account for the nontransitivity of the logic *st*. For one of the logics discussed below, however, we will adopt option (ii).

Making use of facts and truth-makers, this means that we should assume that $T(a \sim_P b) = \{\{\mathbf{Pa}, \mathbf{Pb}\}, \{\overline{\mathbf{Pa}}, \overline{\mathbf{Pb}}\}\}$ and $F(a \sim_P b) = \{\{\mathbf{Pa}, \overline{\mathbf{Pb}}\}, \{\overline{\mathbf{Pa}}, \mathbf{Pb}\}\}$. Notice that if $\mathcal{V}_w(Pa) = 1$ and $\mathcal{V}_w(a \sim_P b) = 1$, it follows that $\mathcal{V}_w(Pb) = 1$. To see this using facts, observe that from $\mathcal{V}_w(Pa) = 1$, it follows that $\mathbf{Pa} \in w$ and $\overline{\mathbf{Pa}} \notin w$. Because $a \sim_P b$ is true and not false, it follows that $\mathbf{Pb} \in w$ and $\overline{\mathbf{Pb}} \notin w$, and thus that $\mathcal{V}_w(Pb) = 1$.[3]

2.3 Tolerance and Inference Relations

Our pragmatic interpretation rule can be included in the definition of logical consequence to try to overcome the limitations we pointed out above about the assertability of the tolerance formula in $st^{\sim}$. Consider, in particular, the following notion of pragmatic consequence, $\models^{prt}$, that goes from *pr*agmatically strongest to *t*olerant (see Cobreros et al. 2015):

- $\phi \models^{prt} \psi \quad \text{iff} \quad PRAG(\phi) \subseteq [\![\psi]\!]^t$.

Thus, for inference we take into account what is (pragmatically) meant by the premise. The fact that we look at what was meant by the premise means that, even though $\phi \wedge \neg\phi \models^{prt} \phi$, it does not hold that $\phi \wedge \neg\phi \models^{prt} \psi$. Thus, explosion is not valid. In this sense, *prt*-entailment is a type of paraconsistent entailment relation. Notice, moreover, that if we extend the language with our similarity relation, our new consequence relation $\models^{prt}_{\sim}$ validates the tolerance formula just as much as $\models^{st}_{\sim}$ did.

How should we extend $\models^{prt}$ so as to allow for multiple premises? This is somewhat tricky. The first thought that comes to mind is the following:

- $\Gamma \models^{prt} \phi \quad \text{iff}_{df} \quad \bigcap_{\gamma \in \Gamma} PRAG(\gamma) \subseteq [\![\phi]\!]^t$

One shortcoming of that definition, however, is that the resulting notion of consequence fails the adjunction property (the property that $\phi, \psi \models \chi$ provided $\phi \wedge \psi \models \chi$). In order to avoid that problem, we introduce the following variant on that definition. We restrict attention to finite sets of premisses, and say that

[3] To be sure, we could make use of truth-makers as well to implement the analysis of similarity relation in (i). To do so, however, is somewhat involved, and we won't go into that here.

a finite set of premisses Γ entails ϕ provided the pragmatic meaning of the conjunction of the premisses entails the tolerant meaning of the conclusion. We note $\bigwedge \Gamma$ the conjunction of the premisses in Γ, and call Prt the corresponding notion of validity, to distinguish it from the former:

- $\Gamma \models^{Prt} \phi \quad \text{iff}_{df} \quad PRAG(\bigwedge \Gamma) \subseteq [\![\phi]\!]^t$

Notice that if we extend the language with our similarity relation, our new consequence relation $\models^{Prt}_{\sim}$ validates the tolerance formula just as much as $\models^{st}_{\sim}$ did. The fact that we look at what was meant by the premisses means that, even though $\phi, \neg\phi \models^{Prt} \phi$, it does not hold that $\phi, \neg\phi \models^{Prt} \psi$. Thus, explosion is not valid. In this sense, *Prt*-entailment is a type of paraconsistent entailment relation, just like LP.

Distinctive about $\models^{Prt}$ are the following three properties: (i) conjunction elimination is valid, which implies that $p \wedge \neg p \models^{Prt} p$; (ii) $\models^{Prt}_{\sim}$ is nontransitive, if based on similarity relation (i); and (iii) it is nonmonotonic. As for (ii), even if both Pa and $a \sim_P b$ have value 1, it is required that Pb have at least value $\frac{1}{2}$, but not that it have value 1. As for (iii), in contrast with $\models^{st}_{\sim}$, the notion $\models^{Prt}_{\sim}$ is *nonmonotonic* in the sense that if $\phi_1 \models^{Prt}_{\sim} \chi$, it might still be the case that $\phi_1, \phi_2 \not\models^{Prt}_{\sim} \chi$. In particular, $Pa,\ a\ \sim_P\ b \models^{Prt}_{\sim}\ Pb$, but $Pa, a \sim_P b, \neg Pa \not\models^{Prt}_{\sim}$ Pb.

There are reasons for which one might be unhappy with $\models^{Prt}$, however. We argued in Sects. 1 and 2 that nontransitivity and nonmonotonicity might be desirable features to account for vagueness. One might wonder, however, whether we need both of these properties. Second, if pragmatic interpretation captures what is *meant* by the speaker, one might wonder whether either conjunct can be inferred from the premisse $Pa \wedge \neg Pa$. With this sentence the speaker wants to impart that a is borderline tall. But if a conjunct like 'Pa' is asserted alone, it is pragmatically interpreted to mean that a is strictly tall, and thus that a is not borderline tall. If we want a consequence relation capturing what can be asserted on the basis of antecedent assertions, the inference from $Pa \wedge \neg Pa$ to Pa should *not* be valid according to such a relation (see Alxatib and Pelletier 2011).

To account for the latter type of consequence relation we therefore define the following inference relation (from *Pr*agmatic to *Pr*agmatic interpretation), again restricting ourselves to finite sets of premisses:

- $\Gamma \models^{PrPr} \phi \quad \text{iff}_{df} \quad PRAG(\bigwedge \Gamma) \subseteq PRAG(\phi)$

Thus, for inference we take into account what is (pragmatically) meant by the premisses and by the conclusion. It follows that $\models^{PrPr}$ does not satisfy conjunction elimination: in particular, $p \wedge \neg p \not\models^{PrPr} p$. Even though $\phi \wedge \psi \not\models^{PrPr} \phi$ for arbitrary ϕ and ψ, still $p \wedge q \models^{PrPr} p$. Important for the analysis of vagueness is that $(p \wedge \neg p) \vee q \not\models^{PrPr} q$. Similarly, $\phi \wedge \neg\phi \not\models^{PrPr} \psi$, that is, explosion is not valid. In this sense, *PrPr*-entailment is again a type of paraconsistent entailment relation, just like Prt. Likewise, the tolerance inference is valid: $Pa, a \sim_P b \models_{\sim}^{PrPr} Pb$, on either of the two interpretations of the similarity relations discussed in the previous section. Again, $PrPr$ is *nonmonotonic* for even as $\phi_1 \models^{Prt} \chi$, it can happen that $\phi_1, \phi_2 \not\models^{PrPr} \chi$. This is already clear from the fact that even though $p \models^{PrPr} p$, we have that $p, \neg p \not\models^{PrPr} p$. And in context of soritical reasoning, $Pa, a \sim_P b \models_{\sim}^{PrPr} Pb$, but $Pa, a \sim_P b, \neg Pa \not\models_{\sim}^{PrPr} Pb$.

3 Comparisons

With the machinery introduced in Sect. 2 we have come to define three different consequence relations: $\models^{st}$, $\models^{Prt}$ and $\models^{PrPr}$. In Sect. 2.2 we presented a way to capture similarity relations, in order to be able to express tolerance. In these logics tolerance is *internalized* since, in fact, the inference from 'Pa' and 'a $\sim_P$ b' to 'Pb' is valid. In this section we review how these logics deal with sorites arguments.

The logic st is based on the idea that premises and conclusions of an argument need not be subject to equal standards of satisfaction. If the premises of an argument are true to some strict standard, it suffices for validity if the conclusion is true to some less strict standard. Intuitively, this will lead to breaches of transitivity and this is precisely what happens, according to this logic, in sorites arguments.

The logic Prt combines two features: pragmatic interpretation for the premises and tolerance for the conclusion. If the premises of an argument are classically satisfiable, the argument is Prt-valid just in case it is st-valid. When the premises are not classically satisfiable, pragmatic interpretation enters the scene. The logic is nontransitive, as we would expect by its affinities with st. It is also nonmonotonic as pragmatic interpretation of the premises will change the range of models.

The logic $PrPr$, like the previous ones, depends on the existence of different standards of satisfaction. This time however, the driving idea for $PrPr$ is that the validity of an argument should be evaluated in connection with those models that provide the highest standards of satisfaction compatible with the statements contained in the premises and in the conclusion. Intuitively, that the set of models vary with what is in the premises or in the conclusion will lead to breaches of monotonicity and this is precisely what happens, according to this logic, in sorites arguments.

Hence, we have outlined three distinct consequence relations: st, Prt, and $PrPr$. Of these, the first is monotonic but nontransitive, the second is both nonmonotonic and nontransitive, and the third is nonmonotonic but transitive. All validate tolerance, in both its theorem and argument forms. So each gives a different approach to capturing tolerance, and each allows the soft status of tolerance to be recognized by failing to obey the full budget of usual structural rules. Which consequence relation, then, do we recommend?

None, or all, depending on how our recommendation will be understood. None of these consequence relations captures what we take to be the full story in play with vague predicates; each is only one window on the underlying phenomena. In particular, we have offered a nonstandard (but reasonably familiar) model-theoretic semantics, on which the models involve two distinct notions of satisfaction, plus a pragmatic story involving van-Fraassen-style truthmakers and a particular bias towards stronger interpretations. No single consequence relation will capture the full texture of this story, but each can reflect something important about it.

For example, consider the data reported by Alxatib and Pelletier (2011), where some participants accept the claim that a borderline case is both tall and not tall, while rejecting the claim that he is tall and rejecting the claim that he is not tall. That is, there are situations in which these participants accept $Ta \wedge \neg Ta$ but reject both Ta and $\neg Ta$. This suggests that a consequence relation in which $Ta \wedge \neg Ta$ does not entail either Ta or $\neg Ta$ may capture some patterns in these participants' judgments; indeed, the relation $PrPr$ invalidates these arguments.

Importantly, however, we do not hold up $PrPr$ for this use simply because its pattern of entailments tracks (some) speakers' behavior. We have in fact offered a model of the pragmatic processes underlying these speaker judgments, and it is $PrPr$ that is sensitive to the outputs of this model both in its premises and in its conclusions. Since speaker judgments are sensitive to pragmatic processes, we predict that speakers will judge in ways that accord with $PrPr$-validity.

On the other hand, it is hard not to think that the classical tradition is on to something, in taking conjunctions to entail their conjuncts, and in other ways besides. $PrPr$, of course, does not reflect this; there are cases in which conjunctions are correctly assertible without either of their conjuncts being so. But since *st* validates every classically-valid argument, we have a story available about just what it is that classical logic gets right: any classically-valid argument whose premises are strictly satisfied must have some conclusion that is tolerantly satisfied.

Again, though, we do not hold up *st* for this use simply because it agrees with classical logic. Rather, *st* is fully semantic, involving only the notions of strict and tolerant satisfaction; the pragmatic part of our apparatus does not enter into it. Classical logic, of course, is much more plausible as a logic for semantics than as a logic for pragmatics;[4] the unified picture we have given respects that.

The logic *Prt*, finally, gives a picture about an interesting mixed phenomenon: when the premises of a *Prt*-valid argument are correctly assertible, then some conclusion must be at least tolerantly satisfied.

We have given the raw materials to define nine different (two-sided) consequence relations. None of these is itself the full story; they all reveal different interactions between strict satisfaction, tolerant satisfaction, and the pragmatic processes we have outlined. Some, like $K3$, fail to validate tolerance in any form; it is part of our theory, then, that tolerance (the formula) is not always strictly satisfied, and that the tolerance argument is not guaranteed to preserve strict satisfaction. Overall, then, our approach not only explains *how* tolerance can be valid without the sorites wreaking disaster, it also gives a detailed picture of *the ways in which* tolerance is valid.

[4] The point goes back at least to Grice (1967).

4 Outlook

To finish this paper, we will suggest that by making use of our rule of pragmatic interpretation *PRAG*, and the consequence relation $\models^{PrPr}$ we can account for a pragmatic solution to the sorites paradox, and solve an expressive limitation problem for the theory of transparent truth.

According to this radical pragmatic solution to the problem (Gaifman 2010; Pagin 2010, 2011; Rayo 2010 and van Rooij 2010, arguably all based on Wittgenstein 1953), we can appropriately use a predicate P in a context if and only if it helps to clearly demarcate the set of objects that have property P from those that do not (the *gap-hypothesis*). This solution seems natural: the division of the set of all relevant objects into those that do have property P and those that do not is (i) easy to make and (ii) worth making. In those circumstances, the tolerance principle $(\forall x, y((Px \wedge x \sim_P y) \rightarrow Py))$ will not give rise to inconsistency, but serves its purpose quite well. Only in exceptional situations — i.e., when we are confronted with long sequences of pairwise indistinguishable objects — do things go wrong.

Unfortunately, there is a major problem with this approach: the gap hypothesis seems too strong: Even if there is no clear demarcation between the bigger and the smaller persons of the domain, certainly the tallest person can be called 'tall'. Thus, the gap-hypothesis doesn't seem to allow for such exceptional situations. Fortunately, it does, if we make use of our nonmonotonic consequence relation $\models^{PrPr}$. Notice first that the following holds: $Px, \neg Py \models^{PrPr} x \not\sim_P y$. This basically says that if we have a sequence going from truth value 1 to 0, you expect this to be due to a gap (a pair x_i, x_j such that $x_i \not\sim_P x_j$). Similarly, it holds that if you explicitly say that $x \sim_P y$ (and do not say much more) then you expect that Px and Py have the same truth value. Still, this expectation can be cancelled if it is explicitly said that another individual in the transitive closure of the similarity relation (of course, the similarity relation is only transitively closed with respect to strict truth) doesn't have property P. This cancellation holds if we use our nonmonotonic consequence relation $\models^{PrPr}$. Thus, $Px_1, \neg Px_n,\ x_1 \sim_P x_2, \cdots, x_{n-1} \sim_P x_n \not\models^{PrPr} Px_n$.

Consider, finally, the extension of the language with a transparent truth predicate (and the possibility of self-reference). One can express within the language the truth-conditions of sentences of that language. Unfortunately, this normally also immediately gives rise to the Liar paradox. In Cobreros et al. (2013) it is shown, however, that this problem can be solved, making use of the non-transitive consequence-relation $\models^{st}$. There, however, we had to put limits on expressibility: paradox reappears if a sentence can say that it is, e.g. *strictly true*.

Is there perhaps a way of communicating that a sentence is *only true* or *only false* without explicitly saying it? Priest (2006) proposes that to communicate that a sentence is 'only true' or 'only false', a speaker either makes use of an independent speech act of denial (or rejection), or relies on a conversational implicature. In this paper we can suggest the latter: the pragmatic reasoning from the assertion that ϕ is true, to the conclusion that ϕ is *only* true (and not also false) can be seen as an implicature, and immediately follows from the pragmatic interpretation of ϕ as here proposed.

References

Alxatib, S., Pelletier, F.R.: The psychology of vagueness: borderline cases and contradictions. Mind Language. **26**, 287–326 (2011)

Alxatib, S., Pagin, P., Sauerland, U.: Acceptable contradictions: pragmatics or semantics? a reply to Cobreros et al. J. Philos. Logic **42**, 619–634 (2013)

Cobreros, P., Egré, P., Ripley, D., van Rooij, R.: Tolerant, classical, strict. J. Philos. Logic **41**, 347–385 (2012)

Cobreros, P., Egré, P., Ripley, D., van Rooij, R.: Reaching transparent truth. Mind **122**, 841–866 (2013)

Cobreros, P., Egrá, P., Ripley, D., van Rooij, R.: Vagueness, truth and permissive consequence. In: Achourioti, T., Galinon, H., Fernández, J.M., Fujimoto, K. (eds.) Unifying the Philosophy of Truth. Logic, Epistemology, and the Unity of Science, vol. 36, pp. 409–430. Springer, Netherlands (2015)

Cobreros, P., Egré, P., Ripley, D., van Rooij, R.: Pragmatic interpretations of vague expressions. J. Philos. Logic **44**, 375–393 (2015)

van Fraassen, B.: Facts and tautological entailment. J. Philos. **66**, 477–487 (1969)

Gaifman, H.: Vagueness, tolerance and contextual logic. Synthese **174**, 5–46 (2010)

Grice, H.P.: Logic and conversation, William James Lectures, Harvard University, Reprinted in Studies in the Way of Words, 1989. Harvard University Press, Cambridge (1967)

Halpern, J.: Vagueness and intransitivity. Rev. symbolic Logic **1**(4), 530–547 (2008)

Pagin, P.: Vagueness and central gaps. In: Dietz, R., Moruzzi, S. (eds.) Cuts and Clouds, pp. 254–272. Oxford University Press, Oxford (2010)

Pagin, P.: Vagueness and domain restriction. In: Egré, P., Klinedinst, N. (eds.) Vagueness and Language Use, pp. 283–307. Palgrave MacMillan, Basingstoke (2011)

Priest, G.: Minimally inconsistent LP. Studia Logica **50**, 321–331 (1991)

Priest, G.: In: Contradiction, 2nd edn. Oxford University Press, Oxford (2006)

Rayo, A.: A metasemantic account of vagueness. In: Dietz, R., Moruzzi, S. (eds.) Cuts and Clouds: Vagueness, its Nature and its Logic, pp. 23–45. Oxford University Press, Oxford (2010)

Ripley, D.: Contradictions at the borders. In: Nouwen, R., van Rooij, R., Sauerland, U., Schmitz, H.C. (eds.) ViC 2009. LNCS (LNAI), vol. 6517, pp. 169–188. Springer, Heidelberg (2011)

van Rooij, R.: Vagueness and linguistics. In: Ronzitti, G. (ed.) Vagueness: A Guide. Logic, Epistemology, and the Unity of Science, vol. 19, pp. 123–170. Springer, Netherlands (2011)

van Rooij, R., Schulz, K.: Exhaustive interpretation of complex sentences. J. Logic, Lang. Inf. **13**, 491–519 (2004)

Serchuk, P., Hargreaves, I., Zach, R.: Vagueness, logic and use: four experimental studies on vagueness. Mind Lang. **26**, 540–573 (2011)

Wittgenstein, L.: Philosophische Untersuchungen/Philosophical Investigations. Blackwell, Oxford (1953). Translated by Anscombe, G.E.M

Imperfect Databases

An Incremental Algorithm for Repairing Training Sets with Missing Values

Bas van Stein(✉) and Wojtek Kowalczyk

Leiden Institute of Advanced Computer Science, Leiden University,
Niels Bohrweg 1, Leiden, The Netherlands
{b.van.stein,w.j.kowalczyk}@liacs.leidenuniv.nl

Abstract. Real-life datasets that occur in domains such as industrial process control, medical diagnosis, marketing, risk management, often contain missing values. This poses a challenge for many classification and regression algorithms which require complete training sets. In this paper we present a new approach for "repairing" such incomplete datasets by constructing a sequence of regression models that iteratively replace all missing values. Additionally, our approach uses the target attribute to estimate the values of missing data. The accuracy of our method, Incremental Attribute Regression Imputation, IARI, is compared with the accuracy of several popular and state of the art imputation methods, by applying them to five publicly available benchmark datasets. The results demonstrate the superiority of our approach.

Keywords: Missing data · Imputation · Regression · Classification · Random forest

1 Introduction

In industrial processes and many other real-world applications, data points are collected to gain insight into the process and to make important decisions. Understanding and making predictions for these processes are vital for their optimization. Missing values in the collected data cause additional problems in building predictive models and applying them to fresh data. Unfortunately, missing values are very common and occur in many processes, for example, sensors that collect data from a production line may fail; a physician that examines a patient might skip some tests; questionnaires used in market surveys often contain unanswered questions, etc. This problem leads to the following questions:

1. How to build high quality models for classification and regression, when some values in the training set are missing?
2. How to apply trained models to records with missing values?

In this paper we address only the first question, leaving the answers to the second one for further research.

J.P. Carvalho et al. (Eds.): IPMU 2016, Part II, CCIS 611, pp. 175–186, 2016.
DOI: 10.1007/978-3-319-40581-0_15

There are several methods developed for tackling this problem, see e.g., [4,5,11,15,16]. The most common method, *imputation*, reconstructs the missing values with help of various estimates such as means, medians, or simple regression models which predict the missing values. In this paper we present a more sophisticated approach, Incremental Attribute Regression Imputation, IARI, which prioritizes all attributes with missing values and then iteratively "repairs" each of them, one by one, using values of all attributes that have no missing values or are already repaired, as predictors. Additionally, the target variable is also used as a predictor in the repair process. Repairing an attribute is achieved by constructing a regression model and applying it for estimation of missing values. We use here the Random Forest algorithm, [3,6], due to its accuracy, robustness, and versatility: it can be used to model both numerical and categorical variables. Obviously, after repairing all attributes with missing values a final model for the original target variable is trained on the repaired training set.

We tested our algorithm on five datasets: *Digits*,*Page Blocks*,*Concrete*, and *CoverType* from the UCI Machine Learning Repository, [2], and *Housing 16H* from mldata.org [1], first removing some values at random, and then reconstructing them with help of IARI and several common imputation algorithms. Finally we compared the quality of these imputation methods by measuring the accuracy of regression and classification models trained on the reconstructed datasets. The results demonstrate that in most cases, no matter how many attributes were spoiled and how severely, the IARI outperformed other imputation methods both in terms of the accuracy of the final models and the accuracy of imputation. On the other hand, the IARI algorithm is computationally very demanding–it builds as many Random Forests as the number of attributes that should be repaired. Fortunately, due to the parallel nature of the Random Forest algorithm, the runtime of the IARI algorithm can be dramatically reduced by running it on a system with multiple cores or CPUs.

The paper is organized as follows. After introducing various types of missing data and providing an overview of the relevant research on imputation methods, we present the IARI algorithm. Next, we describe in more detail an experimental framework and results of our experiments. Finally, we draw some conclusions and make recommendations for further research.

2 Missing Data Types

There are three categories of missing data [6,8,10,11,13]: *Missing Completely at Random* (MCAR), *Missing at Random* (MAR), and *Missing Not at Random* (MNAR). In many cases, data are MNAR, meaning that the probability that a value of a variable is missing somehow depends on the actual (observed or not) values of this or other variables. A value of a variable is MAR if the probability of being missing does not depend on the (unobserved) value of this variable. And a value of a variable is MCAR if the probability of being missing does not depend on (observed or unobserved) values of this or other variables. In real world scenarios one often cannot determine if the missing data are MCAR, MAR or MNAR because the mechanism behind missingness is not known. In such situations domain expertise is of vital importance and it can guide the choice of a strategy for handling missing values.

3 Relevant Research

There are many ways of dealing with missing data when building a regression or classification model. Some of the most popular methods are:

Complete Case Analysis **(CCA):** This method simply ignores all records that have missing values and selects only records with no missing values [5,7]. When the percentage of complete records is relatively high and the data are missing at random or completely at random, this method does not affect model accuracy. However, if the amount of missing data is large the prediction accuracy will be low (not enough complete cases) and when the data are missing not at random then this method generates bias.
Missing Indicator Variable **(MIV):** This method uses a dummy variable as an indicator for missing values [7]. For every variable that might be missing, a dummy variable is introduced, where the value of this dummy variable is 1 when the input variable is missing and 0 when the input variable is not missing. While this method is more efficient than the Complete Case Analysis, it can also create bias in the final model.
Predictive Value Imputation **(PVI):** PVI replaces missing values by some estimates of their values [9]. In many cases the unconditional mean is used (the mean value of all non-missing values of the attribute) or a conditional mean (the mean of a specific group of records where the record with a missing value belongs to). The problem with this method is that the predictive values are always derived from the complete cases and that might introduce some bias. However, some additional mechanisms can be added to PVI which lower this bias. For example, PVI might use the conditional mean over the K nearest neighbors of a record with a missing value, and then the bias can be limited by first imputing the dataset with unconditional mean and then using the K nearest neighbors on the completed dataset to predict the values of the originally missing data. By counting the number of missing data in the neighbors, one can create a weighted average that incorporates the uncertainty of the measurements. There are several other methods to do single-value predictive imputation like *hot-deck imputation*, *cold-deck imputation* and *last observation carried forward*, where the dataset is sorted on specific variables and when a missing value is encountered, the value is replaced by the value of its predecessor.
Regression Imputation **(RI):** Regression Imputation [9] is a PVI variant where we use regression models (Support Vector Machines, Random Forests, etc.) to estimate the imputed value. One way is to build the models to estimate the missing values using the complete cases. However, it is usually better to also incorporate the non-complete cases by first imputing the missing values with a more simple imputation method (like the unconditional mean). In the first case (using only complete cases), there might be too few complete cases to generate

good models, in the latter case there is a danger of bias by training the model with imputed (wrong) data.
Multiple Imputation **(MI):** This is a general imputation framework by Rubin et al. [4,13–15]. The idea is to generate multiple versions of imputed (completed) datasets, which result in multiple models. Each model is then combined into a final predictor. The framework uses a single value imputation algorithm of choice and a random component that represents the uncertainty of the imputation. By creating multiple imputed datasets, the distribution of the imputed values will reflect the distribution of the already known values and therefore reduce bias. This method allows any non-deterministic imputation algorithm to be used. After imputing the dataset several times, creating several copies, a model is being built for each complete dataset. The results of each model are combined using Rubin's Rules [4]. The combined result leads to less biased and more accurate predictions. One of the major advantages of MI is that it can be used with almost any imputation algorithm. Because of this, we do not add MI in our comparison because each of the imputation algorithms can be wrapped with Multiple Imputation.

Most of the above methods can also be used for handling missing data at prediction time. The CCA method is here an obvious exception, but imputation or using dummy variables are valid ways to deal with missing values at prediction time. It should also be mentioned that in addition to the classical "off-line" scenario, where the training set is fixed and is not changing over time, some researchers were considering an "on-line" scenario, where the model is continuously updated while processing a stream of data [18].

In this paper we propose a novel strategy that uses regression models in an attribute wise algorithm to impute missing values in the training stage using the target attribute as one of the predictors. We compare our model strategy with commonly used imputation methods and an imputation method that also uses regression models: *Regression Imputation.*

4 IARI: Incremental Attribute Regression Imputation

There are two ideas behind our method for incremental repair of training sets. First, attributes with missing values are repaired one by one, according to the priority of the attribute. The attribute with the highest priority is repaired first, the attribute with the lowest priority is repaired last. Second, the data used for repairing an attribute include all attributes that are already repaired and additionally the target attribute of the original dataset. The choice of the repair algorithm is arbitrary, in principle any regression algorithm can be used here. In our experiments we used Random Forest [3], due to its superior accuracy, speed and robustness. Random Forest requires little to no tuning, which is very important when numerous models have to be developed without human assistance. Additionally, the Random Forest algorithm provides a heuristic for ranking attributes according to their importance. The IARI algorithm uses this heuristic for ordering the attributes.

It might seem counter-intuitive to include the target attribute in the set of predictors to impute an input attribute–it resembles a circular process. However, our goal is to repair a training set with help of any data we have. When the training set is fixed, a final model is trained and it can be applied to fresh data that were not used in the training process, so there is no circularity here. Moreover, results of our experiments demonstrate that including the target variable in the imputation process substantially increases the accuracy of the final model which is validated on data that were not used in the imputation process.

The IARI algorithm consists of two steps: initialization and main loop. During the initialization all attributes are split into two groups: those that contain no missing values (REPAIRED), and all others (TO_BE_REPAIRED). We assume here that the target attribute, y, contains no missing values so it falls into the REPAIRED group. Additionally, the set of attributes with missing values is ordered according to their importance. This is achieved in three steps. First, the training set is repaired with help of a simple imputation method which replaces missing values of continuous attributes by their mean values and missing values of discrete variables are replaced by their most frequent values. Second, a Random Forest model is built on the repaired training set to predict values of y. Finally, the model is applied to randomized out-of-bag samples to measure the importance of all attributes, as described in [6].

When the initialization step is finished, the algorithm enters the main loop which repairs attributes with missing values, one by one, in the order of their importance (from most to least important). To repair an attribute x, IARI creates a temporary training set which contains all attributes that are already repaired (including y) as predictors and x as the target. All records where the value of x is missing are removed from this training set and, depending on the type of x, a classification or regression variant of the Random Forest algorithm is used to model x. Finally, the model is used to impute all missing values of x and x is moved from the TO_BE_REPAIRED to the REPAIRED set.

The pseudo-code of a generic version of the *IARI* algorithm is provided below.

5 Experimental Setup

To compare the existing algorithms with our approach we used five, very different, datasets from various Machine Learning Repositories: *Digits*, *Cover Type*, *House 16H*, *Page Blocks*, and *Concrete Compressive Strength*. For a complete overview of these datasets, see the public IARI repository, [17].

5.1 Parameters

In our experiments we used a popular implementation of the Random Forest algorithm that comes with the *Scikit-learn* Python package, [12]. The key learning parameter, the number of estimators, was set to 100, and the remaining parameters had default values.

For each dataset we run several experiments with 75 % of the attributes containing missing values and 25 % of the attributes (randomly chosen) containing no missing values. The amount of missing values in the attributes with

Algorithm 1. Incremental Attribute Regression Imputation

Given: A training set X with input attributes $x_1, \ldots, x_n$, a target attribute y, and a classification or regression algorithm ALG

Initialization:
for all *attributes* $x_i \in X$ **do**
  $Nmissing[i] = Count\_missing(x_i)$
  $Importance[i] = ImportanceMeasure(X, x_i, y)$
end for
$REPAIRED = y \cup \{$All attributes x_i where $Nmissing[i] = 0\}$
$TO_BE_REPAIRED = \{$All attributes x_i where $Nmissing[i] > 0\}$
while $TO\_BE\_REPAIRED \; ! = \emptyset$ **do**
  $Repair\_Attribute = SELECT\_X_i(TO\_BE\_REPAIRED, Importance)$
  $Repair\_Target = Delete\_Missing\_Values(Repair\_Attribute)$
  $Model = ALG.train(REPAIRED, Repair\_Target)$
  for all *records* $A_j \in Repair\_Attribute$ **do**
    if $is\_missing(A_j)$ **then**
      $A_j = ALG.predict(REPAIRED[j])$
    end if
  end for
  $REPAIRED = REPAIRED \cup Repair\_Attribute$
  $TO\_BE\_REPAIRED = TO\_BE\_REPAIRED \setminus Repair\_Attribute$
end while
return $REPAIRED$

missing data, was set to 10, 20, 30, 40, 50, 60 percent and for each setup we run 20 experiments using different random seeds. In each experiment, the complete dataset was split in a training (80 %) and a test set (20 %). The deletion of values, repairing the training set and final modeling was performed on the training set. The test set was used to estimate the accuracy of the final model. When removing values from the training set we used two strategies: "missing at random", *MAR*, where values were removed uniformly at random, and "missing not at random", *MNAR*, where only values bigger than the median value of the attribute, were removed uniformly at random.

5.2 Performance Indicators

We measured two aspects of the quality of the imputation. First, we estimated, with help of cross-validation, the accuracy of the final model that was trained on the repaired dataset. The accuracy was measured either by the ratio of correctly classified cases (in case of classification) or by the *coefficient of determination*, R^2, (in case of regression):

$$R^2 = 1 - \frac{\sum_i (p_i - y_i)^2}{\sum_i (y_i - \bar{y})^2}$$

where y_i denotes the target value and p_i the predicted value.

This score indicates how well the model fits the test data. The maximal value of R^2 is 1, meaning the perfect fit; values smaller than 1 reflect the error. Furthermore, the R^2 and accuracy scores of each dataset are measured on final models that were developed with three algorithms: Random Forests, Support Vector Machines and Gradient Boosted Decision Trees. This is to demonstrate that the value of R^2 (or the accuracy score) depends on the regressor or classifier that is being used in the final modeling, and that it not always reflects the quality of the imputation itself.

Second, we measured the quality of the approximation of the imputed values. As all the imputed variables were numeric, we used the *Root Mean Squared Error, RMSE*, to measure the difference between the observed and imputed values:

$$\text{RMSE} = \sqrt{\frac{\sum \left(v_{\text{observed}} - v_{\text{imputed}}\right)^2}{n}}$$

To make the comparison of results over various datasets meaningful, we standardized attributes of all training sets by centering them around 0 and dividing by their standard deviations. As the last indicator of algorithm's performance we measured the execution time. For bigger datasets the cpu time might be an issue to consider.

6 Results

For each dataset, we performed 12 experiments: one for each of the percentage levels of missing values (from 10 to 60) combined with the type of missingness (MAR or MNAR). Each experiment was repeated 20 times (with different random seeds) and the results were averaged. Additionally, for each reconstructed training set, we run three algorithms, Random Forests, Support Vector Machines and Gradient Boosted Decision Trees, to build the final models.

The results of our experiments, the accuracy of the final model (R^2 or the ratio of correctly classified cases) and the accuracy of imputation ($RMSE$), are presented in the following subsection. Each row contains averaged results of 20 runs of the same experiment with different random seeds. The amount of missing values and the type of missing values (MAR or MNAR) are shown as well. For the sake of space we report only results for the percentage of missing values 20 %, 40 %, and 60 % for the MAR model, and various percentages for the MNAR model where we used the missing percentages 20 %, 40 %, and 60 % as upper bounds for the percentage of missing values per attribute, but were not always able to delete that many values of the attribute due to the restriction of deleting only values bigger than the median. Let us note, that it may happen that the fraction of records with a value of an attribute bigger than its median might be arbitrarily small, e.g., when an attribute is almost constant. Moreover, in the results presented below, we show the average number of missing values taken over all attributes with missing values.

For the first three datasets (Cover Type, Digits and Houses 16H) we show the results from the Random Forest final model; for the remaining data sets and final model options we do not show the results due to space limitations. A complete overview of the results, together with software and data used in our experiments, can be found in the public IARI repository, [17].

Each table contains several columns. The first two columns contain information about the percentage of missing values and the type of missingness. The next column, *Ref*, contains the accuracy of the model trained on the original complete dataset: either R^2 for regression problems or classification accuracy for classification problems. The following columns contain results of various imputation methods: Imputation by Mean, Imputation by Median, Imputation by Most Frequent, Predictive Value Imputation using 2-Nearest Neighbour over a dataset imputed by the Mean, Regression Imputation using Random Forests and last but not least, our own algorithm: IARI. Entries in boldface are significantly better than all other entries with the same settings. The significance is tested using the *t-test*, with significance level $p = 0.05$. The absence of a bold entry in the row means that none of the results were significantly better than the others.

6.1 Cover Type Dataset Results

In Tables 1 and 2 the accuracy of the model (Accuracy Score) and the quality of imputation ($RMSE$) are shown for the imputation algorithms on 40.000 instances of the Cover Type dataset.

From our test results we can observe that the maximum average number of MNAR values we can delete from each attribute is around the 12 %. Which implies that approximately 88 % of the dataset is filled with values below or equal the median of each attribute (probably 0). In Table 3 the execution time for each algorithm is shown for the case of 50 % values MAR, which is representative for all the tests on this dataset. Our approach is not the fastest, Replace by Median, Replace by Mean and Replace by Most Frequent are almost instant while PVI, RI and IARI are more complex and take some time. The execution time is mostly dependent on the size of the dataset and on the number of attributes, and not so much on the number of missing values.

Table 1. Model accuracy score on the cover type dataset with 40000 instances using random forests

Miss.%	Type	Ref.	Mean	Median	Freq.	PVI NN	RI	IARI
6	MNAR	0.911	0.871	0.864	0.860	0.868	0.868	0.881
10	MNAR	0.911	0.815	0.809	0.803	0.806	0.805	0.839
12	MNAR	0.911	0.670	0.678	0.656	0.657	0.663	0.693
20	MAR	0.911	0.874	0.887	0.886	0.883	0.880	0.899
40	MAR	0.911	0.834	0.859	0.858	0.845	0.845	0.878
60	MAR	0.911	0.776	0.824	0.822	0.787	0.799	0.847

Table 2. Imputation quality (RMSE) of each imputation algorithm on the cover type dataset with 40000 instances

Miss.%	Type	Mean	Median	Freq.	PVI NN	RI	IARI
6	MNAR	0.786	0.795	0.813	0.786	0.776	0.760
10	MNAR	0.848	0.852	0.867	0.847	0.838	0.791
12	MNAR	0.894	0.884	0.889	0.894	0.894	0.877
20	MAR	0.380	0.389	0.414	0.370	0.330	0.266
40	MAR	0.540	0.552	0.588	0.533	0.496	0.422
60	MAR	0.661	0.676	0.718	0.658	0.630	0.564

Table 3. Execution time of imputation algorithms on the cover type dataset with values 50 % MAR in seconds.

Mean	Median	Freq.	PVI NN	RI	IARI
0.03	0.11	0.48	61.47	381.75	119.12

6.2 Digits Dataset Results

In the table below (Table 4) the *accuracy* of the models created using the different imputed datasets as training set are shown.

In Table 5 the $RMSE$ values for every imputation algorithm are presented for all the combinations of missing data percentage and missing data type. Our IARI approach outperforms the other imputation algorithms in most of the MAR cases with respect to the accuracy. In the MNAR cases our algorithm works well but imputation by Mean sometimes has a slightly better accuracy for the Random Forest model.

Table 4. Model accuracy score on the digits dataset using random forests

Miss.%	Type	Ref.	Mean	Median	Freq.	PVI NN	RI	IARI
16	MNAR	0.972	0.969	0.967	0.953	0.967	0.967	0.968
25	MNAR	0.972	0.967	0.951	0.904	0.954	0.961	0.947
27	MNAR	0.972	0.950	0.923	0.815	0.934	0.932	0.944
20	MAR	0.972	0.964	0.963	0.961	0.967	0.968	0.970
40	MAR	0.972	0.954	0.957	0.952	0.959	0.960	0.962
60	MAR	0.972	0.944	0.943	0.934	0.944	0.948	0.953

Table 5. Imputation quality (RMSE) of each imputation algorithm on the digits dataset

Miss.%	Type	Mean	Median	Freq.	PVI NN	RI	IARI
16	MNAR	0.608	0.649	0.752	0.566	0.565	0.479
25	MNAR	0.858	0.903	1.037	0.841	0.829	0.646
27	MNAR	0.974	0.994	1.103	0.960	0.963	0.850
20	MAR	0.380	0.399	0.511	0.299	0.316	0.231
40	MAR	0.537	0.564	0.721	0.470	0.472	0.345
60	MAR	0.658	0.692	0.883	0.619	0.608	0.451

6.3 Houses 16H Dataset Results

In Table 6 the R^2 scores are shown and in Table 7 the $RMSE$ results are shown for the imputation algorithms on the Houses 16 H dataset.

Table 6. Model accuracy score (R^2) on the houses dataset using random forests

Miss.%	Type	Ref.	Mean	Median	Freq.	PVI NN	RI	IARI
20	MNAR	0.636	0.604	0.598	0.580	0.606	0.603	0.617
40	MNAR	0.636	0.534	0.491	0.485	0.511	0.520	0.531
49	MNAR	0.636	−0.277	−0.287	−0.545	−0.405	−0.171	−0.450
20	MAR	0.636	0.604	0.599	0.586	0.610	0.598	0.620
40	MAR	0.636	0.544	0.533	0.511	0.552	0.521	0.590
60	MAR	0.636	0.423	0.402	0.375	0.458	0.414	0.536

Table 7. Imputation quality (RMSE) of each imputation algorithm on the houses dataset

Miss.%	Type	Mean	Median	Freq.	PVI NN	RI	IARI
20	MNAR	0.486	0.517	0.709	0.485	0.428	0.342
40	MNAR	0.785	0.801	1.007	0.787	0.753	0.587
49	MNAR	0.956	0.925	1.134	0.955	0.954	0.927
20	MAR	0.386	0.395	0.542	0.370	0.328	0.280
40	MAR	0.545	0.558	0.764	0.535	0.487	0.412
60	MAR	0.669	0.685	0.925	0.665	0.625	0.531

7 Conclusion

We presented a novel algorithm, IARI, for imputing missing values into training sets. IARI can handle both regression and classification problems. The key advantage of IARI over other imputation methods is the superior accuracy of the final models which are trained on the repaired training sets, and more accurate reconstruction of missing values. On the other hand, IARI requires much more computing resources than its alternatives: 2-3 orders of magnitude. Fortunately, the main algorithm behind IARI, Random Forest, can be efficiently distributed along multiple nodes, significantly reducing the real (wall clock) computation time.

In principle, IARI is a generic algorithm which can be configured in various ways by changing the measure of importance of attributes, ordering of attributes, and the base algorithm used for imputation. Also the initialization step, where only attributes with no missing values are used as a starting set of predictors, can be modified: sometimes adding to this set several attributes with just a few missing values and removing incomplete records from it, lead to better results.

During our experiments with IARI, we noticed that sometimes a simple imputation method may lead to better results than those obtained with IARI. This happens in case of the *Digits* dataset, where values were removed "not at random", see the *IARI repository* [17]. As expected, the quality of IARI approximations of missing values was always significantly better than those obtained by imputing means, but surprisingly, the opposite holds for the quality of the corresponding final models. This is probably caused by the nature of the classification problem and the fact that the Random Forest is not suitable for image classification. Almost in all other cases the IARI algorithm outperforms other imputation methods: both in terms of the accuracy of imputation and the accuracy of the final model.

In most real world cases it is difficult to determine how well a certain imputation algorithm will work. The quality of imputation depends a lot on the dataset and the reason of why values are missing. However, when we know little about a dataset, the IARI algorithm is probably the best choice.

References

1. PASCAL Machine Learning Benchmarks Repository - mldata.org. http://mldata.org/repository/data
2. Bache, K., Lichman, M.: UCI machine learning repository (2013). http://archive.ics.uci.edu/ml
3. Breiman, L.: Random forests. Mach. Learn. **45**(1), 5–32 (2001)
4. Carpenter, J.R., Kenward, M.G.: Multiple Imputation and its Application, 1st edn. Wiley, New York (2013)
5. Greenland, S., Finkle, W.: A critical look at methods for handling missing covariates in epidemiologic regression analyses. Am. J. Epidemiol. **142**(12), 1255–1264 (1995)
6. Hastie, T., Tibshirani, R., Friedman, J.: The Elements of Statistical Learning: Data Mining, Inference and Prediction, 2nd edn. Springer, Verlag (2009)

7. Henry, A.J., Hevelone, N.D., Lipsitz, S., Nguyen, L.L.: Comparative methods for handling missing data in large databases. J. Vasc. Surg. **58**(5), 1353–1359 (2013)
8. Howell, D.C.: The Analysis of Missing Data. Sage, London (2007)
9. Junninen, H., Niska, H., Tuppurainen, K., Ruuskanen, J., Kolehmainen, M.: Methods for imputation of missing values in air quality data sets. Atmos. Environ. **38**(18), 2895–2907 (2004)
10. Lakshminarayan, K., Harp, S.A., Samad, T.: Imputation of missing data in industrial databases. Appl. Intell. **11**, 259–275 (1999)
11. Little, R.J.A., Rubin, D.B.: Statistical Analysis with Missing Data, 2nd edn. Wiley, New Year (2002)
12. Pedregosa, F., Varoquaux, G., Gramfort, A., Michel, V., Thirion, B., Grisel, O., Blondel, M., Prettenhofer, P., Weiss, R., Dubourg, V., Vanderplas, J., Passos, A., Cournapeau, D., Brucher, M., Perrot, M., Duchesnay, E.: Scikit-learn: machine learning in Python. J. Mach. Learn. Res. **12**, 2825–2830 (2011)
13. Rubin, D.B.: Inference and missing data. Biometrika **63**(3), 581–592 (1976)
14. Rubin, D.B., Schenker, N.: Multiple imputation in healthcare databases: an overview and some applications. Stat. Med. **10**, 585–598 (1991)
15. Rubin, D.B.: Multiple Imputation for Nonresponse in Surveys, vol. 81. Wiley, New York (2004)
16. Seaman, S.R., White, I.R.: Review of inverse probability weighting for dealing with missing data. Stat. Methods Med. Res. **22**(3), 278–295 (2013)
17. Stein, B. van: A Python implementation of the incremental attributeregression imputation (2015). http://basvanstein.github.io/IARI/
18. Žliobaitė, I., Hollmén, J.: Optimizing regression models for data streams with missing values. Mach. Learn. **99**(1), 47–73 (2015)

Analysis and Visualization of Missing Value Patterns

Bas van Stein(✉), Wojtek Kowalczyk, and Thomas Bäck

Leiden Institute of Advanced Computer Science, Leiden University, Niels Bohrweg 1, Leiden, The Netherlands
{b.van.stein,w.j.kowalczyk,t.h.w.baeck}@liacs.leidenuniv.nl

Abstract. Missing values in datasets form a very relevant and often overlooked problem in many fields. Most algorithms are not able to handle missing values for training a predictive model or analyzing a dataset. For this reason, records with missing values are either rejected or repaired. However, both repairing and rejecting affects the dataset and the final results, creating bias and uncertainty. Therefore, knowledge about the nature of missing values and the underlying mechanisms behind them are of vital importance. To gain more in-depth insight into the underlying structures and patterns of missing values, the concept of *Monotone Mixture Patterns* is introduced and used to analyze the patterns of missing values in datasets. Several visualization methods are proposed to present the "patterns of missingness" in an informative way. Finally, an algorithm to generate missing values in datasets is provided to form the basis of a benchmarking tool. This algorithm can generate a large variety of missing value patterns for testing and comparing different algorithms that handle missing values.

Keywords: Patterns · Imputation · Missing values · Monotone

1 Introduction

In the current era of data science and big data, missing values in datasets become an increasingly important problem. Many state of the art algorithms for building predictive models for supervised learning, unsupervised learning and other algorithms that use datasets, are most of the time not able to handle missing values, or perform significantly worse if missing values are present. To address this issue, several algorithms have already been proposed, ranging from various forms of single value imputation, multiple imputation to expectation maximization [2,9,10,12]. However, as one might expect, performance of these algorithms strongly depends on the dataset to which they are applied – an algorithm that performs very well on one dataset can fail on another dataset. Therefore, gaining more insights into the patterns of the missing values is an important factor for selecting algorithms that are appropriate for a given dataset.

A theory about missing value patterns and mechanisms [3–6,8] already exists, but the existing theory is insufficient to gain a clear understanding of each possible pattern of missing values because it only defines a small set of possibilities.

J.P. Carvalho et al. (Eds.): IPMU 2016, Part II, CCIS 611, pp. 187–198, 2016.
DOI: 10.1007/978-3-319-40581-0_16

In this paper, an extension to the current theory is proposed, covering all patterns of missing values occurring in a dataset, ranging from a completely *Univariate Pattern* to a completely *Arbitrary Pattern*. Using this new concept, a greedy algorithm that analyzes datasets is proposed, together with various visualization techniques that provide a clear overview of the patterns of missing values occurring in the dataset. Besides analyzing missing values, it is also interesting that the same techniques can be used to analyze the patterns of occurrence of a specific value. For example, the patterns in sparse datasets (where 0 would be the unique value to analyze, or even more interesting where all non-zero values are analyzed).

In Sect. 2, an overview of types of missing data, patterns and missing value mechanisms is given as well as an extension of the current theory. In Sect. 3, a greedy algorithm for analyzing missing value patterns is proposed and several visualization methods are presented in Sect. 4. In Sect. 5 the quality of the greedy algorithm is tested, and in Sect. 6 a generator of missing values is proposed as a basis of a benchmarking tool for algorithms handling missing values.

2 Missing Data

In the following two sections an overview of common definitions of several mechanisms behind missing data is given, and several new concepts of "patterns of missingness" are introduced.

2.1 Missing Data Types

Rubin [8] defined three major classes of missing values: *Missing Completely At Random* (MCAR), *Missing At Random* (MAR), and *Missing Not At Random* (MNAR). Informally, we say that values in a dataset are Missing Completely At Random (MCAR), if the probability distribution of "being missing" is completely independent on the observed or missing data. When the probability distribution of "being missing" somehow depends on the observed (non-missing) values, then we talk about the Missing At Random (MAR) scenario. Finally, when "being missing" depends on the actual, unobserved values, then we talk about the Missing Not At Random (MNAR) scenario.

To illustrate these three definitions, let us consider data of patients collected in a hospital. When a doctor decides not to measure patient's body temperature because she can already see that the temperature is too high, then we have the MNAR scenario - the decision of not measuring the parameter depends on its actual value. On the other hand, if the temperature is systematically measured, but from time to time the data registration process malfunctions (independently on the measured values), then we have the MCAR scenario. Finally, if the doctor has a habit of not measuring the temperature of patients with high blood pressure (and blood pressure is always registered), then we have a MAR scenario.

Formally, the three scenario's can be summarized in the following definition, [3]:

Definition 1. *Let y denote a target attribute y, X a matrix of input attributes with missing values, X_{obs} the observed entries in X, $Z = (y, X)$, $Z_{obs} = (y, X_{obs})$. Additionally, let R denote an indicator matrix with ijth entry 1 if x_{ij} is missing and 0 otherwise.*

We say that the data is Missing Completely At Random if:

$$Pr(R|Z, \theta) = Pr(R|\theta).$$

We say that the data is Missing At Random if:

$$Pr(R|Z, \theta) = Pr(R|Z_{obs}, \theta).$$

We say that the data is Missing Not At Random if it is not Missing at Random. (Here we assume that probability distributions are parametrized by parameters θ.)

2.2 Patterns of Missing Values

In addition to some general probabilistic mechanisms behind missing values, one can also look at the shape of missing values in the data table. In the literature [6], three definitions of missing value patterns exist; namely *Univariate*, *Monotone* and *Arbitrary* pattern.

A *univariate* pattern (Fig. 1a) of missing values means that one or several attributes (columns) contain missing values in exactly the same records and no other values are missing. When attributes can be organized in several groups $G_1, ..., G_k$, such that each group forms a univariate pattern and records with missing values in G_i have also missing values in G_{i-1}, for $i = k, ..., 2$, then we have a *monotone* pattern (Fig. 1b). An *arbitrary* pattern, is anything else. Obviously, in order to visualize patterns of missing values, one has to permute columns and rows of the data matrix to create "rectangular regions of missingness".

It is very important to understand the patterns of missing values in a dataset, because they might provide insight into why values are missing and relations of attributes that are missing in groups. As an example: a camera system fails to recognize the bar-code of a certain product, which in turn makes it impossible for the next two sensors to save their measurements to the database, resulting in two missing values. Additionally, identifying important patterns of missing values in the data can lead to using a different strategy for handling these missing values. However, in reality there are many more patterns that are now falling under the category *arbitrary pattern*, but that are not arbitrary at all. Consider a dataset with a Monotone pattern of missing values, now remove one value from a column that does not contain any missing values yet. The dataset with the extra removed value falls under the arbitrary category, while in reality the dataset is almost completely falling into the category of a monotone pattern. Another example, imagine that a survey takes place led by two volunteers, both volunteers ask the same ten questions to a hundred different people, but volunteer a asks the questions in order, and volunteer b asks the questions in reverse order. Due to time limitations, people start to drop out after the sixth question. The combined dataset seems to have an arbitrary pattern of missing values, while if we look more closely we can identify two partitions of the data with both one monotone pattern of missing values.

To fill the gap between the definitions of missing value patterns, we introduce the concept of *Mixtures of Monotone patterns*. This requires a more precise definition of the *Univariate* and *Monotone* patterns.

Let us consider a dataset D of size $N \times k$, with N the number of records in D and k the number of attributes in D, and a missing indicator matrix I. Here, $I_{ij} = 1$ if D_{ij} is missing, and 0 otherwise.

Definition 2 (Univariate Pattern). *Missing values in D form a* univariate pattern *if and only if there exists a set of attributes A such that:*

$$\forall x \in D : \{a : x_a \text{ is missing}\} = A \text{ or } \emptyset$$

So for all records in D, the record has either missing values in all attributes in the attribute set A or the record has no missing values.

Definition 3 (Monotone Pattern). *A dataset D has missing values in a* monotone pattern *if and only if there exists an ordering of all attributes A, $a_1 \dots a_k$ such that:*

$$\forall i \in \{1, ..., N\}, \forall j \in \{1, ..., k\} : I_{i,j} = 0 \Rightarrow I_{i,j+1} = 0, \dots, I_{i,k} = 0$$

Note that Definition 3 is a generalization of Definition 2, in other words, a univariate pattern is a special case of a monotone pattern. A monotone pattern can also be seen as a collection of record groups, where each group of records has a univariate pattern of missing values. For example, a dataset has twenty attributes and forty records, five of the records have attribute one and five missing this is denoted as: $(1, 5)$. Ten records have attributes one, five and nineteen missing: $(1, 5, 19)$, and twenty records have only attribute five missing: (5). The remaining five records are complete. The complete dataset has a monotone pattern of missing values, which can be denotes as set $p = \{(5), (1, 5), (1, 5, 19)\}$. Each element in p stands for a univariate pattern that holds within a subset of the complete dataset.

This way we can further generalize into a *Mixture of Monotone patterns* (Fig. 1c).

Definition 4 (k-Monotone Mixture Pattern). *A dataset D has missing values in a* k-monotone mixture pattern *if and only if here is a partitioning of D, $S = S_0 \dots S_{k-1}$ of size k such that $S_0 \cup S_1 \cdots \cup S_{k-1} = D$ and $\forall S_i, S_j \in S, i \neq j$: $S_i \cap S_j = \emptyset$ and $\forall S_i \in S$: S_i has values missing in a monotone pattern.*

A univariate pattern can be seen as a rectangular area of missingness, a monotone pattern can be seen as a stack of adjacent rectangular regions and a monotone mixture pattern can be seen as a union of disjoint monotone patterns.

Note that **any** dataset has values missing in a *k-Monotone Mixture Pattern* where $k \leq N$. When there exists a 1–*Monotone Mixture Pattern*, the pattern is completely monotone, when there exists only a high k mixture pattern, the pattern is close to arbitrary. In this manner, a transition between completely monotone and arbitrary patterns can be identified.

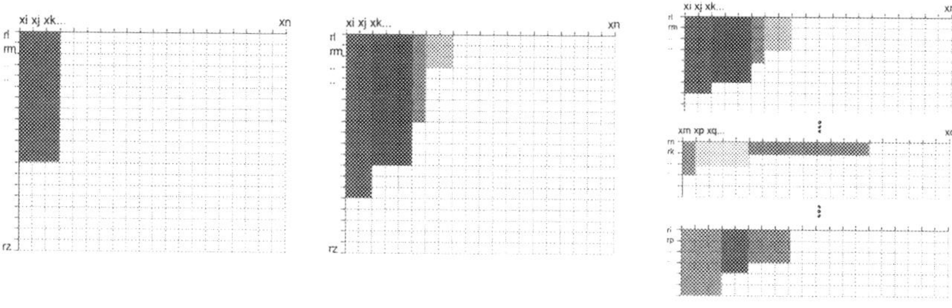

(a) Missing values in a univariate pattern. (b) Missing values in a monotone pattern. (c) Missing values in a k-monotone mixture pattern.

Fig. 1. Dataset of records (y-axis) and attributes (x-axis) with missing values denoted by the colored bars (Color figure online).

3 Heuristic Search for Monotone Mixture Patterns

It is possible to analyze a dataset and identify the existing monotone mixture patterns using our novel *MMP-Finder* (Algorithm 1). In this algorithm, first a dictionary with all existing monotone patterns of missing values is build and sorted by the number of rows per pattern. Then, mixtures of monotone patterns are constructed by adding the next monotone pattern to an already existing mixture, or by defining a new mixture. The MMP-Finder uses a greedy approach to construct the mixtures of monotone patterns.

The complexity of the proposed greedy approach is $O(n + m^2)$, where n is the number of records and m is the number of unique sets of missing attributes. Of course $m \leq n$, since every record can have a unique set of attributes missing, but usually m is much smaller than n.

Using the MMP-Finder, all identified monotone patterns in the partitions of dataset X are returned, together with the number of records and record indexes that belong to each monotone pattern. Notice that the solution returned is not a *unique solution*, it is possible that a specific univariate pattern can belong to multiple monotone patterns. For example, two monotone pattern sets are defined: $\{(1,5),(1,5,8)\}$ and $\{(4,5),(4,5,9)\}$, the next univariate pattern that occurs is (5), this pattern might belong to the first monotone pattern set, or the second. The proposed algorithm handles these choices in a greedy manner, the univariate patterns are handled in an order depending on the coverage, the pattern that covers most records is handled first, the pattern that covers the least records is handled last. This way it is very likely to identify "the biggest" monotone pattern. Since the missingness mechanism is usually not known, it is impossible to find the "correct" monotone patterns.

Once the monotone patterns and their support are known, it is easier to verify why certain attributes contain missing values, and whether there are relations between the various attributes inside the monotone patterns. This can not only

Algorithm 1. MMP-Finder

```
Given: A training set X with input attributes x_0, ..., x_n containing missing values, a target
  attribute y
  # Create a dictionary with all unique missing attribute combinations
  CombinationsMissing = dict()
  RecordsPerCombination = dict()
  for all records x_i ∈ X do
      combi = GetMissingAttributes(x_i)
      if combi ∉ CombinationsMissing then
          CombinationsMissing[combi] = 0
          RecordsPerCombination[combi] = ∅
      end if
      CombinationsMissing[combi]+ = 1
      RecordsPerCombination[combi].append(x_i)
  end for
  # Sort the combinations by size
  sortedComb = sort(CombinationsMissing)
  Mixtures = []
  MixtureRecords = []
  # Construct the mixtures
  for all comb ∈ sortedComb do
      for all M ∈ Mixtures do
          # If the combination is a sub or super set for all combinations in M add it to M
          if ∀c ∈ M : comb ⊆ c ∨ comb ⊇ c then
              Mixtures[M].append(comb)
              MixtureRecords_M.append(RecordsPerCombination[comb])
              Added = True
              Break
          end if
      end for
      if Added == False then
          # If the combination does not belong to an existing Mixture, add a new Mixture
          Mixtures.append([comb])
          MixtureRecords.append([RecordsPerCombination[comb]])
      end if
  end for

      return Mixtures
```

provide valuable insight, but also help in choosing a good imputation or modeling algorithm.

4 Analysis of Existing Datasets

Fourteen datasets with missing values from the *UCI machine learning repository* [1] were analyzed using Algorithm 1. The output of the algorithm is shown in Table 1, and a visualization of the result is provided in Figs. 2 and 3. The visualization and textual summaries are generated directly from the output of the MMP-finder algorithm.

In Fig. 2, the monotone mixture patterns found in the *Wiki* dataset can be observed using two kinds of visualization techniques. In Fig. 2a each record in the

Table 1. Textual summaries for datasets with missing values

dataset	#Mixtures	Ratio of each mixture	Miss.%	Category
Post-oper.	1	[1.0]	0.0333333333333	Monotone
Wisconsin	1	[1.0]	0.0228898426323	Monotone
Dermato.	1	[1.0]	0.0218579234973	Monotone
Cleveland	2	[0.667 0.33]	0.019801980198	Two monotone patterns
Adult	2	[0.776 0.224]	0.0741165390443	Two monotone patterns
census	7	[0.948 0.030 0.006 0.001 ...]	0.5268	Mostly monotone
automobile	4	[0.826 0.087 0.043 0.043]	0.224390243902	Mostly monotone
hepatitis	7	[0.707 0.093 0.093 0.053 ...]	0.483870967742	70 % mono., 30 % rand.
Mammogr.	5	[0.550 0.244 0.160 0.038 ...]	0.136316337149	Monotone mixture
Bands	18	[0.39 0.259 0.086 0.086 ...]	0.322820037106	60 % two patterns
Wiki	116	[0.503 0.091 0.030 0.016 ...]	0.807228915663	50 % mono., 50 % rand.
Marketing	39	[0.353 0.128 0.112 0.111 ...]	0.235405315245	Random
Horse-colic	82	[0.221 0.061 0.058 0.044 ...]	0.98097826087	Random

dataset that contains missing values is labeled with a color and a position on the x axis. This way it is easy to observe where several monotone mixture patterns are located in the dataset and if there are specific regions in the dataset where these patterns occur. Additionally, the horizontal length of each bar depends on the number of attributes that are missing. For each mixture, the longest pattern (pattern with most attributes missing) has a length of one, and each other univariate pattern belonging to the same monotone pattern, has a length proportional to the ratio of missing attributes over the longest pattern.

This visualization technique to present the various monotone patterns in a dataset can be useful in understanding the underlying missing data mechanisms. For example, in the Wiki dataset, the two monotone patterns that cover most of the records are located in a very specific order in the dataset, which might be relevant information regarding the missing data mechanism. Even more specific, the three most occurring *univariate patterns* occur exactly after each other in the dataset. In Fig. 2b, the same patterns can be observed, but now in a histogram plot. The distribution of records belonging to each univariate pattern can be observed and the largest monotone patterns in terms of the number of records and in terms of the number of univariate patterns can be identified easily. Using this visualization technique it is easy to observe the different distributions in between the patterns. In Fig. 3, a visualization of all the datasets with natural missing values is shown using the first visualization technique.

5 Quality Measurements

As already mentioned, the monotone mixture patterns found are not one unique solution, patterns can sometimes be combined in multiple ways. Because of these different possible monotone mixture patterns, some quality measurements are proposed to see which of the quality measurements are optimized by the MMP-finder (Algorithm 1). The assumption is that finding the largest monotone

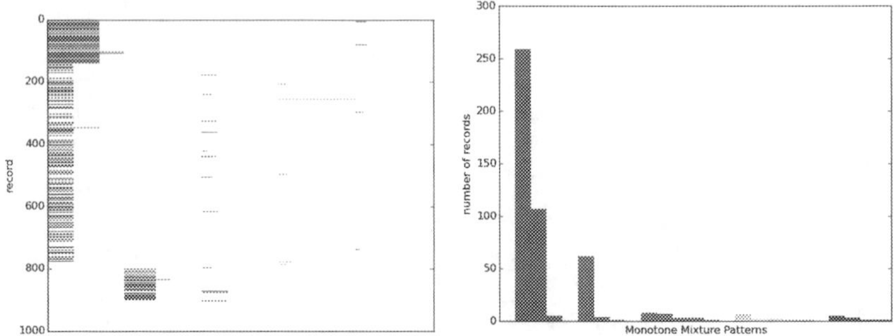

(a) Visualization of missing values per record. Each column (color) represents a mixture of monotone patterns, the length of each bar is proportionate to the number of missing attributes versus the maximum number of missing attributes in its mixture.

(b) Visualization by the number of records affected per pattern. Each color is a mixture of monotone patterns, each bar is a monotone pattern.

Fig. 2. First five monotone mixture patterns for the *Wiki* dataset (Color figure online).

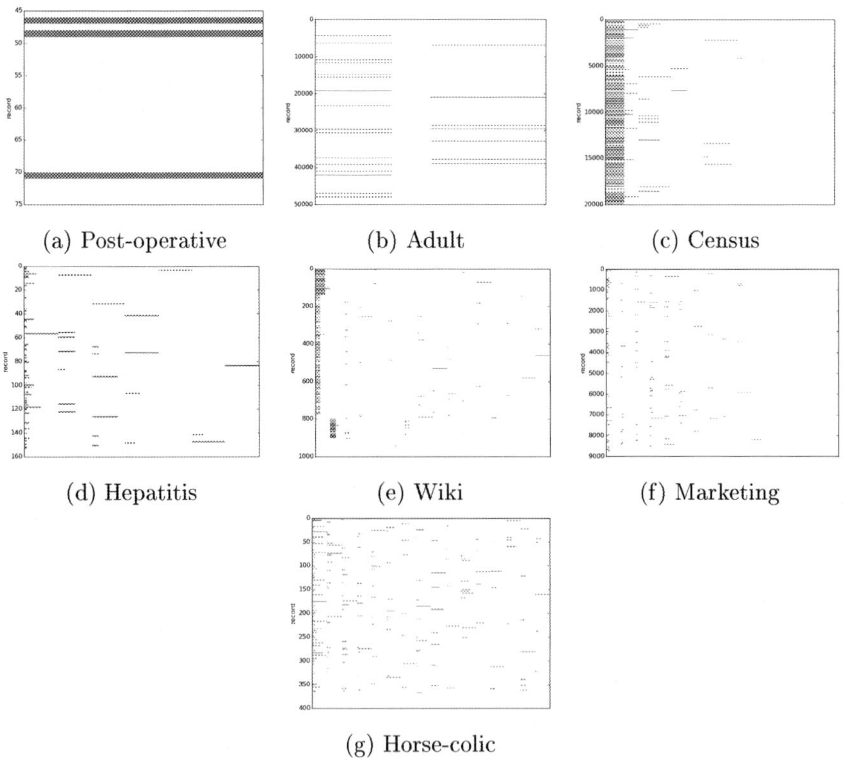

(a) Post-operative (b) Adult (c) Census

(d) Hepatitis (e) Wiki (f) Marketing

(g) Horse-colic

Fig. 3. Visualization of datasets with natural occurring missing values.

patterns (in terms of records they cover) is of major importance. One very large pattern with multiple tiny patterns is generally preferred over two middle sized patterns with a few smaller patterns. On the other hand, finding the minimal set of monotone patterns might be an objective as well, which can conflict with the previous objective. Yet a third objective would be to maximize the size in terms of univariate patterns covered by one monotone pattern.

Therefore, three measurements are proposed: the first two are *Skewness of Monotone Pattern Coverage* (SMPC) and *Skewness of Monotone Pattern Lengths* (SMPL), which are both basically the sum of squares, also called the *Surprise number* [7] in Mining Data Streams, of records covered per monotone pattern and pattern lengths. The third quality measure is the *Monotone Mixture Size* (MMS), which is the k in *k-Monotone Mixture Patterns* (Eq. 1). In the equations below, s denotes one possible solution of a k-monotone mixture pattern, M_s is the collection of monotone patterns belonging to s, $m_{records}$ is the number of records covered by one monotone pattern m and $m_{upatterns}$ is the number of univariate patterns belonging to monotone pattern m.

$$MMS(s) = |s| \tag{1}$$

The *Skewness of Monotone Pattern Coverage* (Eq. 2) is the sum of all squared monotone pattern record counts:

$$SMPC(s) = (\sum_{\forall m \in M_s} |m_{records}|^2) \tag{2}$$

The *Skewness of Monotone Pattern Lengths* (Eq. 3) is the sum of all squared monotone pattern sizes, where the size of a monotone pattern is the number of univariate patterns that belong to this monotone pattern.

$$SMPL(s) = (\sum_{\forall m \in M_s} |m_{upatterns}|^2) \tag{3}$$

For both skewness measurements, a higher value is better, for MMS less is better. A higher *SMPC* means a lower number of monotone patterns that cover most of the records with missing values. A higher *SMPL* means a lower number of monotone patterns that cover most of the univariate patterns. In both cases it implicitly means a more unbalanced distribution, which we assume is preferred. Note that both the total number of records with missing values and the total number of univariate patterns is always the same per dataset. To visualize these measurements in a convenient manner and compare the measurement with the optimal possible solution for each of the measurement, the relative performance (RP) is calculated like:

$$f_{max} = \max_{\forall s \in S}(f(s)) \tag{4}$$

$$f_{min} = \min_{\forall s \in S}(f(s)) \tag{5}$$

$$RP_f(s) = \frac{f(s) - f_{min}}{f_{max} - f_{min}} \tag{6}$$

Where f is a placeholder for the measurements ($SMPC$, $SMPL$ and MMS) and S is the set of all possible solutions, and $s \in S$ is one particular possible solution. Here a relative performance of 1.0 means the maximum possible for this measurement, and 0.0 meaning the lowest possible, so for MMS, a value of 0.0 is optimal, while for the other two measures a value of 1.0 is optimal.

Table 2. Performance Greedy algorithm versus optimal

Dataset	MSS	SMPL	SMPC	Comb.
Cleveland	**0.00**	**1.00**	**1.00**	2
Adult	**0.00**	**1.00**	**1.00**	2
Census	**0.00**	**1.00**	**1.00**	6
Automobile	**0.00**	**1.00**	**1.00**	1
Mammogr.	**0.00**	**1.00**	**1.00**	4
Bands	0.75	0.50	0.97	3683
Wiki	0.40	0.59	0.04	>100000*
Marketing	0.50	0.78	0.91	>100000*
Horse-colic	0.00	0.77	**1.00**	>100000*

$*$ The optimal solutions are calculated using a backtracking algorithm that was limited to find a maximum of 100.000 possible solutions.

It is shown in Table 2 that the proposed greedy algorithm performs well in most quality measurements for the tested datasets, only the number of mixtures found is not always minimal. The last column (Comb.) is the number of all possible combinations, limited at 100.000 for time complexity reasons.

6 Missing Value Generator

Using the definitions of Subsect. 2.2, a missing data generator is constructed to remove values from a given dataset using user defined parameters to cover the complete spectrum from *univariate* to completely *arbitrary* missing value patterns.

Using the *k-monotone mixture pattern* concept, it is possible to generate missing values in any dataset that obeys the number of monotone patterns defined by the user. For this the user should be able to define; the distribution of the monotone patterns (one pattern may be applicable to most of the data while a second only affects a small portion of the data), and the number of attributes that contain missing values.

It is also possible to analyze datasets that already contain missing values using Algorithm 1, and use the return values of this algorithm to specify a "similar" dataset for testing purposes.

6.1 Open Source

Both the missing value generator and the analysis and visualization methods are publicly available on Github [11]. The missing value generator uses a wide set of parameters, k as the number of monotone mixture patterns, a variety of parameters to set the various distributions of patterns and records with missing values and parameters to set the number of missing values and patterns.

Using this broad set of parameters, there are many possible patterns of missing values that can be generated, ranging from completely univariate (where we set $k = 1$), to completely arbitrary (with $k = N$ and each distribution set to "random"). The generator can be used as a benchmarking tool to compare the performance of algorithms on different distributions and patterns of missing values.

The missing data mechanism used by the algorithm is "MCAR" at the moment, but this can be extended by "MAR" and "MNAR" mechanisms. However since these are open to interpretation and there are many ways to accomplish these mechanisms, this it is left for further research.

7 Conclusions

A new concept to analyze and visualize missing value patterns in datasets is proposed and it is shown that with a greedy method called MMP-finder, a dataset with missing values can be analyzed. Using the concept of *k-Monotone Mixture Patterns*, a better in-depth understanding of the underlying patterns of missing values or unique values can be obtained.

Furthermore, a missing value generator is developed and made publicly available to make it possible to test and compare different algorithms on a wide set of both generated and natural datasets. The proposed missing value generator can be used as a tool to generate benchmark datasets for algorithms that handle certain kinds of missing values, giving insight in what kind of situations an algorithm might work well and in what situations certain algorithms would not perform well.

For future work, it would be interesting to incorporate various mechanisms of missing values into the generator to also cover *MNAR* situations. Another interesting extension to the proposed analysis algorithm would be to include expert knowledge and assumptions to come up with the most likely partition of monotone patterns.

Finally, another challenge would be developing efficient heuristics aimed at optimizing the proposed objective functions (MSS, SMPL and SMPC).

References

1. Bache, K., Lichman, M.: UCI machine learning repository (2013). http://archive.ics.uci.edu/ml
2. Carpenter, J.R., Kenward, M.G.: Multiple Imputation and its Application, 1st edn. Wiley, New York (2013)

3. Hastie, T., Tibshirani, R., Friedman, J.: The Elements of Statistical Learning: Data Mining, Inference and Prediction, 2nd edn. Springer, New York (2009)
4. Howell, D.C.: The Analysis of Missing Data. Sage, London (2007)
5. Lakshminarayan, K., Harp, S.A., Samad, T.: Imputation of missing data in industrial databases. Appl. Intell. **11**, 259–275 (1999)
6. Little, R.J., Rubin, D.B.: Statistical Analysis with Missing Data, 2nd edn. Wiley, New York (2002)
7. Rajaraman, A., Ullman, J.D., Ullman, J.D., Ullman, J.D.: Mining of Massive Datasets, vol. 77. University Press Cambridge, Cambridge (2012)
8. Rubin, D.B.: Inference and missing data. Biometrika **63**(3), 581 (1976)
9. Saar-tsechansky, M., Provost, F.: Handling missing values when applying classification models. J. Mach. Learn. Res. **8**, 1625–1657 (2007)
10. Seaman, S.R., White, I.R.: Review of inverse probability weighting for dealing with missing data. Stat. Methods Med. Res. **22**(3), 278–295 (2013)
11. van Stein, B.: Missing data visualisation (2015). https://github.com/Basvanstein/MissingDataVis
12. Williams, D., Carin, L.: Incomplete-data classification using logistic regression. In: Proceedings of the 22nd International Conference on Machine learning, pp. 972–979. ACM (2005)

Efficient Skyline Maintenance over Frequently Updated Evidential Databases

Sayda Elmi[1,2], Mohamed Anis Bach Tobji[2], Allel Hadjali[1(✉)], and Boutheina Ben Yaghlane[3]

[1] LIAS, Université de Poitiers, ENSMA, Poitiers, France
{saida.elmi,allel.hadjali}@ensma.fr
[2] LARODEC, Université de Tunis, ISG, Tunis, Tunisia
anis.bach@isg.rnu.tn
[3] LARODEC - IHEC Carthage, Université de Carthage, Carthage, Tunisia
boutheina.yaghlane@ihec.rnu.tn

Abstract. In many recent applications, data are intrinsically uncertain, noisy and error-prone. That is why, uncertain database management has attracted the attention of several researchers. Data uncertainty can be modeled in the evidence theory setting. On the other hand, skyline analysis is a powerful tool in a wide spectrum of real applications involving multi-criteria optimal decision making. It relies on Pareto dominance relationship. However, the skyline maintenance is not an easy task when the queried database is updated. This paper addresses the problem of the maintenance of the skyline objects of frequently updated evidential databases. In particular, we propose algorithms for maintaining evidential skyline in the case of object insertion or deletion. Extensive experiments are conducted to demonstrate the efficiency and scalability of our proposal.

Keywords: Evidential database · Skyline · *b*-Skyline · Database update

1 Introduction

Efficient processing of uncertain data, which differs from handling certain data in traditional databases, has become increasingly important. This kind of processing is a crucial requirement in different and several domains such as object identification [4], sensor networks [12] and medical databases [7]. Uncertain data in those applications are generally caused by factors like data randomness and incompleteness, limitations of measuring equipment, and delay or loss in data transfer. To deal with uncertain data, several models were proposed: probabilistic databases [1,14], possibilistic databases [6] and evidential databases [3,9]. The advantage of the evidential databases model is twofold: (i) it allows modelling both uncertainty and imprecision in data; and (ii) in some cases, it can be a generalization of probabilistic and possibilistic models.

On the other hand, the present decade has seen a revival of interest in preference queries which aim at retrieving the tuples from a database which are dominated by no other tuple. Skyline queries [5] are a popular example of preference

J.P. Carvalho et al. (Eds.): IPMU 2016, Part II, CCIS 611, pp. 199–210, 2016.
DOI: 10.1007/978-3-319-40581-0_17

queries and are very useful in multi-criteria decision making applications. They rely on Pareto dominance relationship. Substantial research work has addressed the problem of skyline analysis on uncertain data from different perspectives and within various communities, including databases like [10].

Though the skyline queries can control selection, unfortunately, there exist not much works that can handle skyline queries under database updates. The maintenance of the skyline set is then very useful. It allows users to get informed about the new interesting objects. Let us mention the work done in [13,15] where the authors introduce an optimal skyline deletion maintenance for certain data. Xia et al. [16] present efficient update algorithms for compressed skyline Cubes. Closely relating to the maintenance of skyline results, let us point out the study related to the progressive skyline query evaluation and maintenance done in [18]. However, up to our knowledge, there is no work about the skyline maintenance issue in the uncertain/evidential databases context.

In [9], the authors introduced a method for extracting skyline objects from an evidential database. When evidential data are updated, the skyline set could be computed, again from the overall updated database. It is the trivial maintenance of the skyline set. In this paper, the aim is to incrementally maintain the skyline set, without starting from the scratch. Our objective is to reduce the computation cost of the maintenance by using the skyline set already computed. We address then the following major challenges:

- We propose efficient methods to maintain the skyline results in the evidential database context when an object is inserted or deleted.
- We perform an extensive experimental evaluation to demonstrate the scalability of the algorithms proposed for the evidential skyline maintenance.

The rest of the paper is organized as follows. Section 2 contains a reminder about skyline notions and provides some basic concepts about evidential theory, evidential databases and skyline. In Sect. 3, we formally propose a new approach for the incremental maintenance of evidential skyline. Our experimental evaluation is reported in Sect. 4. Finally, Sect. 5 concludes the paper.

2 Preliminaries

In this section, we first present a reminder about skyline operator. Then, we provide some notions about evidential theory as well as evidential databases. We conclude with a recall on b-dominant evidential skyline initially introduced in [9].

2.1 Skyline Set

We present here a basic formal model of Pareto dominance. Let $\mathcal{O}$ be a collection of objects defined on a set of d attributes $\mathcal{A} = \{a_1, a_2, \ldots, a_d\}$.

Definition 1. *(Pareto Dominance) Given two objects $o_i, o_j \in \mathcal{O}$, o_i dominates o_j (in the sense of Pareto), denoted by $o_i \succ o_j$, if and only if o_i is as good or better than o_j on all attributes and strictly better in at least one attribute, i.e., $\forall a_k \in \mathcal{A} : o_i.a_k \leq o_j.a_k \wedge \exists a_\ell \in \mathcal{A} : o_i.a_\ell < o_j.a_\ell$ where $o_i.a_r$ and $o_j.a_r$ stand for the rth attribute of o_i and o_j, respectively.*

The skyline includes those objects that are dominated by no other object. Without loss of generality, we assume through all the paper that the smaller the value the better.

Definition 2. *(Skyline) The skyline of $\mathcal{O}$, denoted by $Sky_{\mathcal{O}}$, includes objects of $\mathcal{O}$ that are dominated by no other object, i.e., $Sky_{\mathcal{O}} = \{o_i \in \mathcal{O} \mid \nexists\, o_j \in \mathcal{O}, o_j \succ o_i\}$.*

2.2 Evidential Databases

Evidence Theory. The theory of evidence is a generalization of the Bayesian theory of subjective probabilities [8]. Let Θ be a finite and exhaustive set whose elements are mutually exclusive, Θ is called a frame of discernment. A basic belief assignment (*bba*), also called a mass function is a mapping $m : 2^\Theta \longrightarrow [0,1]$ such that $m(\emptyset) = 0$ and $\sum_{A \subseteq \Theta} m(A) = 1$.

An element A of 2^Θ is called a focal element whenever $m(A) > 0$. The mass $m(A)$ represents the level of confidence in the truth allocated to the subset A. The belief of A, reflects the total weight of evidence (confidence) in A, denoted by $bel(A)$, and defined as the sum of the masses assigned to every subset B of A, i.e., $bel(A) = \sum_{B \subseteq A} m(B)$.

The plausibility of A, denoted by $pl(A)$, is defined as the sum of the masses assigned to every subset B of Θ that intersects A, i.e., $pl(A) = \sum_{B \cap A \neq \emptyset} m(B)$.

Table 1. Evidential database example.

Property	Price	Distance
p_1	$\langle\{150, 160, 180\}, 0.1\rangle$, $\langle\{190, 200\}, 0.9\rangle$	$\langle 90, 0.3\rangle$, $\langle 100, 0.7\rangle$
p_2	$\langle 100, 0.7\rangle$ $\langle\Theta, 0.3\rangle$	$\langle\{70, 80\}, 0.8\rangle$, $\langle 80, 0.2\rangle$
p_3	70	Θ

Definition 3. *Let x, y two bba whose mass functions are $m_x, m_y : 2^\Theta \longrightarrow [0,1]$, respectively. Then, for $A, B \subseteq \Theta$, $bel(x \leq y) = \sum_{A \subseteq \Theta}(m_x(A) \sum_{B \subseteq \Theta, A \leq^\forall B} m_y(B)))$, Where $A \leq^\forall B$ means that $a \leq b$ for all $a \in A$ and $b \in B$.*

Example 1. Assume we have the following information about the price of properties p_1 and p_2 denoted by x and y, respectively: $x = \langle\{10\}, 0.3\rangle, \langle\{10, 20\}, 0.7\rangle$ and $y = \langle\{10\}, 1\rangle$. It is easy to check that $bel(x \leq y) = 0.3 * 1 = 0.3$

Evidential Databases. The paradigm of evidential databases [2,11], aims at handling imprecise, missing and uncertain data. An evidential database, is a collection of objects $\mathcal{O}$ defined on a set of d attributes $\mathcal{A} = \{a_1, a_2, \ldots, a_d\}$ where each attribute a_k has a domain Θ_{a_k}, and the relation between the ith object and the kth attribute is expressed by a *bba*. That is: $m_{ik} = \{A \subseteq \Theta_{a_k}, m_{ik}(A) > 0\}$, where $m_{ik} : 2^{\Theta_{a_k}} \longrightarrow [0,1]$, with $m_{ik}(\emptyset) = 0$ and $\sum_{A \subseteq \Theta_{a_k}} m_{ik}(A) = 1$.

Example 2. Table 1 represents an evidential database describing properties over their price and distance to the user office. Assume that the data about the properties are pervaded with some imperfection modelled thanks to evidence theory. A *bba* defined on each property and on each attribute, may have one or more focal elements. For example, the *bba* defined on object 1 and Attribute *price*, (i.e., the price of the property p_1) includes two focal elements $\langle\{150, 160, 180\}, 0.1\rangle$ and $\langle\{190, 200\}, 0.9\rangle$. That is, we believe that the price of p_1 is either 150, 160 or 180 with mass function 0.1 or one of the values 190 or 200 with mass 0.9. However, we do not know how much each single element is credible. Note that evidential databases can store various kind of data imperfection: Probabilistic data: focal elements are singletons ($p_{1.Distance}$), Possibilistic data: focal elements are nested ($p_{2.Distance}$), Partial ignorance: $0 < m(\Theta_{a_k}) < 1$, ($p_{2.Price}$), Perfect data: focal element is a singleton and its mass is equal to one ($p_{3.Price}$), Total ignorance: $m(\Theta_{a_k}) = 1$, ($p_{3.Distance}$).

2.3 Evidential Skyline

Given a set of objects $\mathcal{O} = \{o_1, o_2, \ldots, o_n\}$ defined on a set of attributes $\mathcal{A} = \{a_1, a_2, \ldots, a_d\}$, with $o_i.a_k$ denoting the *bba* (the set of focal elements) of object o_i w.r.t. attribute a_k. The belief that an object o_i is better than or equal to another object o_j w.r.t. an attribute a_k is given by [3]:
$bel(o_i.a_k \le o_j.a_k) = \sum_{A \subseteq Dom(a_k)}(m_{ik}(A) \sum_{B \subseteq Dom(a_k), A \le^{\forall} B} m_{jk}(B))$, where $Dom(a_k)$ stands for the domain of attribute a_k, and $A \le^{\forall} B$ stands for $a \le b$ for all $(a, b) \in A \times B$. Given two objects o_i and o_j in $\mathcal{O} : o_i \neq o_j$, the belief that o_i dominates o_j is defined in [9] as follows: $bel(o_i \succ o_j) = \prod_{a_k \in \mathcal{A}} bel(o_i.a_k \le o_j.a_k)$. The notions of b-dominance and b-skyline were introduced in [9] where b is a threshold defined by the user.

Definition 4. *(b-dominance) Given two objects $o_i, o_j \in \mathcal{O}$ and a belief threshold b, o_i b-dominates o_j, denoted by $o_i \succ_b o_j$ if and only if $bel(o_i \succ o_j) \ge b$.*

Definition 5. *(b-skyline) The skyline of $\mathcal{O}$, denoted by b-$Sky_{\mathcal{O}}$, comprises those objects in $\mathcal{O}$ that are b-dominated by no other object, i.e., $b - Sky_{\mathcal{O}} = \{o_i \in \mathcal{O} \mid \nexists\ o_j \in \mathcal{O}, o_j \succ_b o_i\}$.*

3 Incremental Maintenance of Evidential Skyline

In this section, we discuss the maintenance problem of the b-skyline after an insertion or a deletion occurs in the evidential database EDB.

3.1 Object Insertion

Because the b-dominance relationship is intransitive as proved in [9], a naive approach to see if a new object has an impact on the b-skyline, is to compare all objects in $\mathcal{O}$ against this inserted object. This way results in a very costly procedure. To avoid the full scan of the database, we could prune the search space in some cases.

We present in this section an approach for optimizing b-skyline updating by using the notions of "*IdealPoint*" and "*HeaderPoint*" which keep up a concise summary of the already visited regions of the objects' space. This summary allows a fraction of objects in $\mathcal{O}$ to be pruned from b-skyline processing phases required in the naive approach, thus reducing the overall cost of expensive dominance checks.

An *IdealPoint* and a *HeaderPoint* summarize the region of data explored in earlier iterations. They enable a newly inserted object to be compared against this summary rather than to perform multiple comparisons against the whole objects in the b-skyline. Our goal is to leverage these two points to determine whether a newly inserted object should be considered as a candidate skyline object or be pruned in advance from the expensive skyline processing phases. The *IdealPoint* is a virtual object having the most interesting values across all attributes.

Definition 6. *(IdealPoint) Let b-$sky_{\mathcal{O}} = \{S_1, S_2, \ldots, S_n\}$ be the set of objects being in the b-skyline. An IdealPoint IP of b-$sky_{\mathcal{O}}$ is a certain object defined such as: $<MIN(val.a_1), MIN(val.a_2), \ldots, MIN(val.a_d)>, \forall S_i \in b\text{-}sky_{\mathcal{O}}$ where $MIN(val.a_k)$ is the function which returns the minimum value defined on the attribute a_k and $val.a_k$ is the set of distinct values defined on a_k.*

Table 2. 0.4-Skyline.

Skyline objects S_i	Distance (m)	Price (K)
S_1	$\langle\{50, 51\}0.7\rangle, \langle\{55, 56\}0.3\rangle$	$\underline{20}$
S_2	$\{40, 41\}$	$\langle\{30\}0.9\rangle, \langle\{31\}0.1\rangle$
S_3	$\{\underline{10}, 11\}$	$\{60\}$

Example 3. Suppose we have the 0.4-skyline objects presented in Table 2. The most interesting values defined on distance and price attributes, are returned by the MIN() function and appear in an underlined form in Table 2. Thus, *IdealPoint* is $IP(10, 20)$.

Let $\mathcal{O}^* \subseteq b\text{-sky}_{\mathcal{O}}$ be the set of skyline points having the most interesting values in one or more attributes, i.e., $\mathcal{O}^* = \{S \in b\text{-sky}_{\mathcal{O}} / \exists item \in S.a_k, item = \text{MIN}(val.a_k)\}$ where $val.a_k$ is the set of distinct values in attribute a_k occurring in the skyline and $item$ is a single proposition in the bba "$S.a_k$".

Example 4. In Table 2, $\mathcal{O}^*$ comprises the skyline objects S_1 and S_3 since they have the interesting values 10 and 20 defined respectively on distance and price.

Definition 7. *(HeaderPoint) Let $\mathcal{O}^*$ be the set of skyline points having the most interesting values. A HeaderPoint HP of b-sky$_{\mathcal{O}}$ = $\{S_1, S_2, ..., S_n\}$ is a certain virtual object defined such as:*
$\forall S_i \in \mathcal{O}^*$, such that, $<MAX(val.a_1), MAX(val.a_2), \ldots, MAX(val.a_d)>$.

Example 5. One can check that in Table 2, $\mathcal{O}^* = \{S_1, S_3\}$. The maximum values defined on distance and price for S_1 and S_3 are 56 and 60, respectively. Then the Header Point is $HP(56, 60)$.

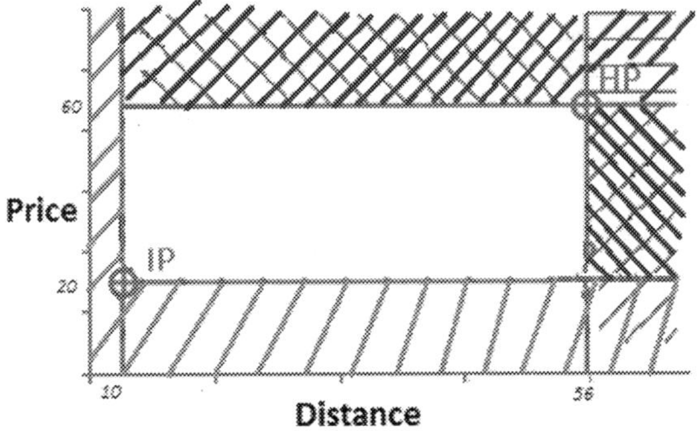

Fig. 1. Ideal and header points.

Our goal with the Header Point and the Ideal Point is to determine whether the newly inserted object should be considered as a candidate skyline object or be pruned in advance from the expensive skyline processing phases. As shown in Fig. 1, if the newly inserted point denoted by P^+ is in the hatched area, i.e., is strictly better than IP in at least one dimension, then it is directly added to the b-skyline. In this case, P^+ should be compared to all skyline's points because it could dominate one or several of them. We also propose another pruning strategy by adapting the notion of Header Point (HP) to the evidential database context. If P^+ is in the crosshatch area, then it cannot be in the b-skyline because it has a value attribute which is worse than the HP. In other words, if a newly inserted object is not better than the Ideal Point in at least one dimension and it is worse than the Header Point, then, the new object cannot be a skyline object and can be discarded.

If an evidential object is inserted, we have to refer to Definition 3 in order to compare the new object against the Header Point and the Ideal Point. Let $P^+.a_k$ be the new object value defined on attribute a_k. Let also b be the threshold introduced by the user. Note that b is already considered to compute the original set of skyline points b-sky$_{\mathcal{O}}$.

Property 1. If $\exists a_k \in \mathcal{A}$ such that $bel(P^+.a_k < IP.a_k) \geq b$, then P^+ is added to the b-skyline.

Property 2. If Property 1 is not satisfied and $\exists a_k \in \mathcal{A}$ such that $bel(P^+.a_k \leq HP.a_k) < b$ then P^+ cannot be in the b-skyline.

Proof. Suppose $\forall a_i \in \mathcal{A}$, $bel(P^+.a_i \leq HP.a_i) = 1$ and $\exists a_k \in \mathcal{A}$ such that $bel(P^+.a_k \leq HP.a_k) = x < b$.

$We have bel(P^+ \succ HP) = \prod_{i=1}^{d} bel(P^+.a_k \leq HP.a_k) = 1 * 1 * \cdots * 1 * x = x < b$. Thus P^+ does not b-dominates HP.

Algorithm 1. Incremental maintenance after Insertion (MAI)

Input: $b\text{-}Sky_{\mathcal{O}}$, P^+: the object to insert; HP$(a_1, a_2, \ldots, a_d)$: Header Point, IP$(a_1, a_2, \ldots, a_d)$: Ideal Point

begin
- $var1 \leftarrow true$;
- **while** a_k *in* $\mathcal{A}$ *and* $var1$ **do**
  - **if** $bel(P^+.a_k < IP.a_k) \geq b$ **then**
    - $b\text{-}Sky_{\mathcal{O}'} \leftarrow P^+$;
    - Compare P^+ to all point in $b\text{-}Sky_{\mathcal{O}}$;
    - $var1 \leftarrow false$;
  - **else**
  - **if** $bel(HP.a_k < P^+.a_k) \geq b$ **then**
    - P^+ cannot be in the b-skyline;
    - $var1 \leftarrow false$;
  - **else**
  - Execute b-Sky$_{\mathcal{O}}$();

A naive approach for skyline insertion maintenance is to recompute from scratch the b-sky$_{\mathcal{O}}$ considering the newly inserted object (Baseline Maintenance Algorithm after Insertion denoted by BMAI). Clearly, this approach may result in a high computational cost since we recompute the b-dominance relationship between all objects in $\{\mathcal{O}\} \cup \{P^+\}$. Algorithm 1 shows the algorithmic description of the proposed method in order to decrease the check space.

As shown in Algorithm 1, we compare the *bbas* of the inserted object defined on each attribute against all values of the Ideal Point and the Header Point defined on the set of attributes $\mathcal{A}$. If it does exist an attribute $a_k \in \mathcal{A}$ such that $bel(P^+.a_k < IP.a_k) \geq b$, then the newly inserted object is added to the b-skyline and is compared to all skyline points since it may dominate some of them. If Property 1 is not satisfied and it does exist an attribute $a_k \in \mathcal{A}$ such that $bel(\text{HP}.a_k < \text{P}^+.a_k) \geq b$, then the newly inserted object is directly discarded from the b-skyline since the Header Point b-dominates the new object P^+ in at

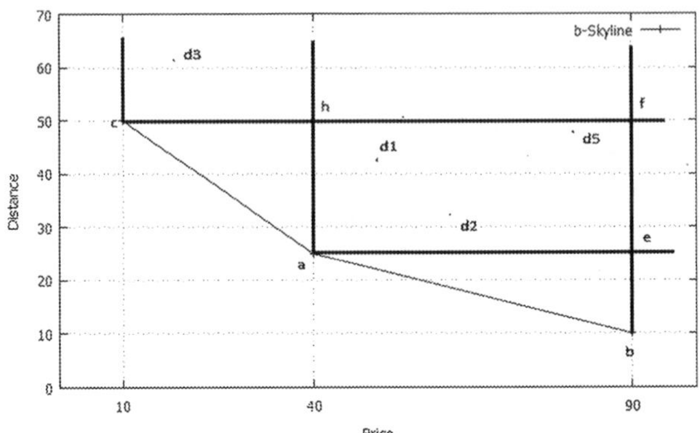

Fig. 2. Exclusive dominance region example.

least one dimension. Else, because of the well-known non transitivity of the *b*-dominance relationship, we have to re-compute the *b*-skyline from scratch which is represented by the function "*b*-Sky$_{\mathcal{O}}$()".

3.2 Object Deletion

In this section, we study the impact of skyline object deletion on the set of skyline points. Two simple approaches are discussed here. The most straightforward method for skyline deletion maintenance is to recompute from scratch (represented by the Baseline Deletion Algorithm BMAD) the *b*-Sky$_{\mathcal{O}}$. Clearly, this approach is overly simplistic and may result in a high computational cost because a considerable portion of evidential objects is not affected by the deleted point at all. This BMAD computation can be easily optimized for the purpose of deletion maintenance.

For a given skyline object $S_i \in b\text{-}Sky_{\mathcal{O}}$, we define its *b*-dominance region designed by *b*-DR(S_i) as the whole objects space that is dominated by S_i, and its exclusive *b*-dominance region, designed by EDR(S_i), which contains the objects space that is only dominated by S_i. For instance, in Fig. 2, the *b*-DR(a) can be represented as rectangle $ahfe$. However, objects d_1 and d_2 are exclusively *b*-dominated by object a. Thus, the exclusive *b*-dominance region of skyline object a is defined as follows: EDR(a) $= \{d_1, d_2\}$. As a result, both d_1 and d_2 are promoted in the skyline after object a is deleted. EDR(S_i) presents the smallest region that may contain the new skyline objects after deletion of S_i. Intuitively, EDR(S_i) contains those points that must be added to the new skyline after S_i is deleted, since those points are exclusively dominated by S_i.

Let $b\text{-}Sky_{\mathcal{O}}$ denotes the original skyline, $b\text{-}Sky_{\mathcal{O}}$'($S_i$) denotes the new skyline after the deletion of skyline object S_i, i.e., $S_i \in b\text{-}Sky_{\mathcal{O}}$. Let ΔS denotes the skyline objects that are expected to be added in the new skyline $b\text{-}Sky_{\mathcal{O}}$', then $b\text{-}Sky'_{\mathcal{O}}$ - $b\text{-}Sky_{\mathcal{O}}$ $= \Delta S$. ΔS is exactly the exclusive dominance region

Table 3. b-Dominance region

Skyline Objects S_i	b-DR(S_i)
$S_1 = o_2$	$\{o_1, o_3\}$
$S_2 = o_4$	$\{o_3\}$
$S_3 = o_6$	$\{o_5\}$

of deleted object. The key issue now is to compute the EDR of the skyline object S_i in order to find the exact b-$Sky_{\mathcal{O}}$' with optimal I/O performances.

Example 6. Suppose we have an evidential database that contains a set of objects $\mathcal{O} = \{o_1, o_2, \ldots, o_6\}$ and b-$Sky_{\mathcal{O}} = \{o_2, o_4, o_6\}$. Table 3 gives the objects that are dominated by each object in b-$Sky_{\mathcal{O}}$, i.e., the b-dominance region of skyline objects. Suppose o_2 is deleted, only objects that are exclusively dominated by o_2 are promoted as new skyline objects. However, o_3 can not be promoted to the b-skyline because it is dominated by one or more other objects. Table 4 shows that o_2 exclusively b-dominates o_1. It is then promoted to be in the new skyline b-$Sky_{\mathcal{O}}$'. As a result, b-$Sky_{\mathcal{O}}$'$= \{o_1, o_4, o_6\}$

Table 4. Exclusive b-dominance region of skyline points.

Object Skyline S_i	EDR(S_i)
o_2	$\{o_1\}$
o_4	$\emptyset$
o_6	$\{o_5\}$

To compute the b-skyline over an evidential database, we refer to the algorithm proposed in [9] called b-Sky$_{\mathcal{O}}$. In b-Sky$_{\mathcal{O}}$ algorithm, while computing the b-dominance degrees between objects in $\mathcal{O}$, one can automatically save the exclusive b-dominance region of each skyline point EDR(S_i).

Algorithm 2. Incremental maintenance after Deletion (MAD)

Input: $\mathcal{O}$: evidential database, b-$Sky_{\mathcal{O}}$: evidential skyline, S_i: the skyline point to delete;

Output: b-$Sky_{\mathcal{O}}$': the new evidential skyline;

```
begin
    b-Sky'_O ← b-Sky_O ∪ EDR(S_i);
    return b-Sky'_O;
```

As it can be seen, once information about exclusive b-dominance region of each skyline object is available, we can detect via Algorithm 2, which object can be directly promoted as a new skyline object.

4 Experimental Evaluation

In this section, we present an extensive experimental evaluation of our approaches. More specifically, we focus on the scalability of our proposed methods for maintaining the evidential skyline. For comparison purpose, we also implemented the baseline algorithms for maintenance after insertion and after deletion referred to as BMAI and BMAD. The generation of evidential data sets is controlled by the parameters in Table 5. In each experimental setup, we investigate the effect of one parameter, while we set the remaining ones to their default values. K means a thousand of evidential objects.

Table 5. Parameters and examined values.

Parameter	Symbol	Values	Default
Skyline size	S	1K, 3K, 5K, 7K, 10K	10K
Number of objects in $\mathcal{O}$	n	10K, 30K, 50K, 70K, 100K	10K
Number of attributes	d	2, 4, 6, 8	4
Number of focal elements/attribute	f	3, 5, 7, 9	5

Figure 3 depicts the execution time of the implemented algorithms MAI and BMAI w.r.t. S, d and f. Overall, MAI outperforms BMAI. More specifically, MAI is faster than BMAI since it can detect immediately whether an inserted object is better or not than another existing in the b-Sky$_{\mathcal{O}}$. A simple comparison between the attributes values of P^+ and the certain points IP and HD, makes the algorithm MAI more efficient than BMAI. This later aims at comparing P^+ against all objects in the skyline set in order to check if the newly inserted object is in the evidential skyline or not. However, MAI decreases the research space by discarding points which are worst than the Header Point on the one hand, and inserting P^+ in the b-skyline if it is better than the Ideal Point on the other hand. As expected, Fig. 3a shows that the performance of the algorithm BMAI deteriorates with the increase of the skyline size S. This is because when S increases the number of dominance checks becomes larger. Observe that MAI is one order of magnitude faster than BMAI since it can quickly identify if an object can be in the skyline or not with a simple operation of comparison with both Header Point and Ideal Point. As shown in Fig. 3b and c, MAI is not affected by increasing d and f as it makes a simple check. However, BMAI does not scale with d and f. Figure 3 depicts also the execution time of the implemented algorithms MAD and BMAD w.r.t. n, d and f. Figure 3d shows that the execution time of the algorithm MAD slightly increases with the increase of n as we have to check more and more objects in $\mathcal{O}$, but it outperforms BMAD. Observe that BMAD has a high computational cost if we increase all parameters. It is not the case for MAD.

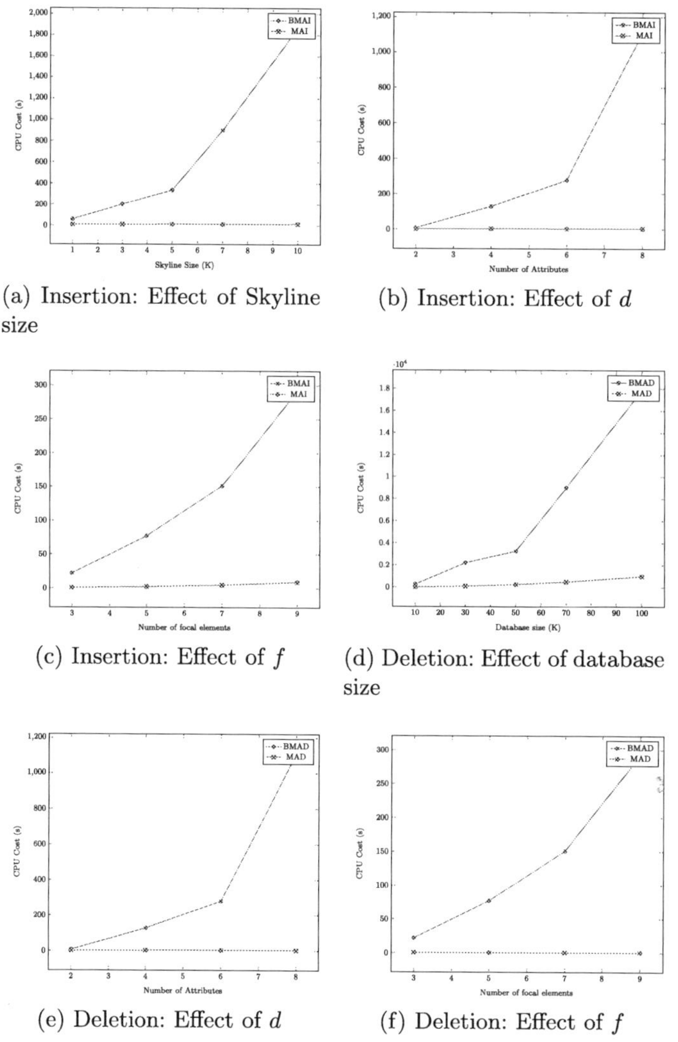

(a) Insertion: Effect of Skyline size

(b) Insertion: Effect of d

(c) Insertion: Effect of f

(d) Deletion: Effect of database size

(e) Deletion: Effect of d

(f) Deletion: Effect of f

Fig. 3. Elapsed time for maintenance operations.

5 Conclusion

In this paper, we have addressed the important problem of efficiently maintaining an evidential skyline result set in response to an object insertion or deletion.

Our solutions guarantees I/O optimality and can be easily implemented. Experimental results show that our methods outperform the naive methods. An interesting future direction is to introduce the notion of confidence level to b-skyline computing and updating in the spirit of [17].

References

1. Aggarwal, C.C., Yu, P.S.: A survey of uncertain data algorithms and applications. IEEE Trans. Knowl. Data Eng. **21**(5), 609–623 (2009)
2. Tobji, M.A.B., Yaghlane, B.B.: Maintaining evidential frequent itemsets in case of data deletion. In: Hüllermeier, E., Kruse, R., Hoffmann, F. (eds.) IPMU 2010. CCIS, vol. 80, pp. 218–227. Springer, Heidelberg (2010)
3. Bell, D.A., Guan, J.W., Lee, S.K.: Generalized union and project operations for pooling uncertain and imprecise information. Data Knowl. Eng. **18**(2), 89–117 (1996)
4. Bohm, C., Pryakhin, A., Schubert, M.: The gauss-tree : Efficient object identification in databases of probabilistic feature vectors. In: ICDE 2006, p. 9. IEEE (2006)
5. Borzsonyi, S., Kossmann, D., Stocker, K.: The skyline operator. In: IN ICDE, pp. 421–430 (2001)
6. Bosc, P., Pivert, O.: Modeling, querying uncertain relational databases : a survey of approaches based on the possible worlds semantics. J. Fuzziness Knowl. Based Syst. **18**(5), 565–603 (2010)
7. Chen, L., Li, X., Yang, Y., Kurniawati, H., Sheng, Q.Z., Hu, H.-Y., Huang, N.: Personal health indexing based on medical examinations: a data mining approach. Decis. Support Syst. **81**, 54–65 (2016)
8. Dempster, A.P.: A generalization of bayesian inference. J. Roy. Stat. Soc. **30**, 205–247 (1968)
9. Elmi, S., Benouaret, K., Hadjali, A., Bach Tobji, M.A., Ben Yaghlane, B.: Computing skyline from evidential data. In: Straccia, U., Calì, A. (eds.) SUM 2014. LNCS, vol. 8720, pp. 148–161. Springer, Heidelberg (2014)
10. Jiang, B., Pei, J., Lin, X., Yuan, Y.: Probabilistic skylines on uncertain data: model and bounding-pruning-refining methods. J. Intell. Inf. Syst. **38**(1), 1–39 (2012)
11. Lee, S.K.: Imprecise, uncertain information in databases : An evidential approach. In: Proceedings on Data Engineering, pp. 614–621 (1992)
12. Li, M., Liu, Y.: Underground coal mine monitoring with wireless sensor networks. ACM Trans. Sen. Netw. **5**(2), 1–29 (2009)
13. Papadias, D., Tao, Y., Fu, G., Seeger, B.: Progressive skyline computation in database systems. ACM Trans. Database Syst. **30**(1), 41–82 (2005)
14. Pei, J., Jiang, B., Lin, X., Yuan, Y.: Probabilistic skylines on uncertain data. In: VLDB, pp. 15–26 (2007)
15. Wu, P., Agrawal, D., Egecioglu, Ö., El Abbadi, A. Deltasky.: Optimal maintenance of skyline deletions without exclusive dominance region generation. In: ICDE, pp. 486–495 (2007)
16. Xia, T., Zhang, D.: Refreshing the sky : the compressed skycube with efficient support for frequent updates. In: SIGMOD, pp. 491–502. ACM (2006)
17. Yong, H., Jin, H.K., Seung, W. H.: Skyline ranking for uncertain data with maybe confidence. In: ICDE, pp. 572–579 (2008)
18. Zhang, Z., Cheng, R., Papadias, D., Tung, A.K.: Minimizing the communication cost for continuous skyline maintenance. In: SIGMOD, pp. 495–508 (2009)

Multiple Criteria Decision Methods

Prediction Model with Interval Data -Toward Practical Applications-

Michihiro Amagasa[1] and Kiyoshi Nagata[2(✉)]

[1] Faculty of Business Administration, Hokkai Gakuen University, 1-40, Asahimachi, Toyohira-ku, Sapporo 062-8605, Japan
amagasa@ba.hokkai-s-u.ac.jp

[2] Faculty of Business Administration, Daito Bunka University, 1-9-1, Takashimadaira, Itabashi-ku, Tokyo 175-8571, Japan
nagata@ic.daito.ac.jp

Abstract. The regression model is one of typical model for predicting some values by analyzing existing numerical data collected in various ways. If data are not crisp numbers, they are usually transformed into numerical crisp values by means of some methods such as quantification method. In companies' decision making process, collected and referred data usually have uncertainty which sometimes play an important roles for business performance. One of authors have proposed a model deriving predicting values with uncertainty by handling data with uncertainties. In this paper, we review some method for this purpose, then describe the model by applying it to some test cases.

Keywords: Prediction model · Interval data · Application test

1 Introduction

In various stages of decision making, properly predicted data from available information are valuable, so many people not only in research area but also in practical area are investigating and proposing several methods or methodologies with certain model.

One of most typical model is the regression model based on relationships between objective value and set of explanatory values which are collected beforehand. For instance, the price of a production or a service are determined from several factors, such as price of raw materials, selling expenses, consumer demand, etc. The price also has high correlation coefficient with customer value of product or service. Bradley T. Gale proposed a scenario where price satisfaction carries 40 % of the weight and non-price attributes 60 % in the customer-value equation, and showed a figure representing the relationship between relative performance overall score and relative price for luxury cars based on data [8, pp. 218–219]. In that figure, the relative price is generically expressed in linguistic values such as "Higher", "Lower", etc., but these values are translated into numerical values in order to plot corresponding points on the performance-price plane.

J.P. Carvalho et al. (Eds.): IPMU 2016, Part II, CCIS 611, pp. 213–224, 2016.
DOI: 10.1007/978-3-319-40581-0_18

In the real world, it is usually happened that the collected data are expressed in linguistic values, then these values are transformed into numerical ones for applying well-known and authorized stochastic methods such as regression analysis. When we estimate the price of a product, we should consider certain amount of customer related factors which usually have uncertainties. Inoue et al. proposed a sale price prediction model by fuzzy regression, [5], and one of authors, Michihiro Amagasa, also proposed a method to handle data with uncertainty in the model of regression analysis as an extension of their model, [2]. He has aimed at constructing a model for a sale price prediction and proposed a regression model whose input and output are both expressed in fuzzy sets. In the paper, outline of the model and necessarily formulas are described, however there are no comparative studies nor even test examples presented. In this paper, we try to study the model by considering it in some particular conditions, and give some examples.

The rest of the paper is organized as follows: In the next section, we will review some existing research especially on fuzzy regression model. The following section is dedicated to the details of the Amagasa's proposed model. Examples to see how the proposed model works are coming up in the following section with some discussions. The last section is the conclusion and the future works.

2 Existing Research on Regression Model

There are several researches and papers on fuzzy regression analysis, and there are many systems in various fields in which fuzzy regression model are involved. Here we refer some papers on essential models of fuzzy regression.

2.1 Original Model by Tanaka et al.

Among some of papers in this field, the most referred works are by Tanaka et al. (e.g. [11,12]), in which the following type of regression model was proposed,

$$Y = A_0 + \sum_{i=1}^{n} A_i x_i, \tag{1}$$

where Y, A_i $(i = 0, \ldots, n)$ are symmetrical triangular fuzzy numbers (STFN) and x_i $(i = 0, \ldots, n)$ are crisp numbers. This formula looks like ordinary linear regression model which is usually solved by the least square method, but in the fuzzy case it can be solved by linear programming analysis so that minimizing the total value of the range of intervals of A_i times the sum of input data under several linear constraints.

Tanaka mentioned that this type of regression analysis is based on interval regression analysis and error terms in ordinary regression analysis are absorbed in intervals of objective fuzzy set, [13].

2.2 On Review by Shapiro

Arnold F. Shapiro published an article reviewing studies on the fuzzy regression and discussing issues related to Tanaka's approach as the original work in possibilistic approach, [14].

In the article, he pointed out followings as problematic situations, and the reason why fuzzy regression model was required.

- Number of observations is inadequate, i.e., data set is not large enough
- Difficulty of verifying that the error is normally distributed
- Vagueness in the relationship between the independent and dependent variables
- Ambiguity associated with the event or degree
- Inappropriate assumption of linearity

On the other hand, he summarized some researches pointing out problems of Tanaka's model. Followings are some of identified problems in the paper.

- It is unclear what the relation is to a least squares concept, or that any measure of best fit by residuals is present. (by Diamond, [3])
- The original Tanaka model was extremely sensitive to the outliers. (by Peters, [9])
- There is no proper interpretation about the fuzzy regression interval. (by Wand and Tsaur, [15])
- The fuzzy linear regression tends to have multi-collinearity when more independent variables are collected. (by Kim et al., [6])
- etc.

There are many techniques to detect outliers of data set, and some of them could be applied for fuzzy case. So the extreme sensitivity of fuzzy regression model to outliers could be solved by removing them before hand.

Wang and Tsaur proposed a question on interpretation of fuzzy regression interval, then they tried to give a proper interpretation in the paper [15].

The least squares concept can be applied for fuzzy regression model, and Diamond implemented the fuzzy least squares regression (FLSR) model using distinct measures, [3]. FLSR is actually the other type of important model for fuzzy regression, and Shapiro also referred FLSR in his paper. We will refer this type of fuzzy regression models in the next subsection.

2.3 Fuzzy Least Square Regression (FLSR)

Fuzzy regression model is classified into two classes according to type of approach. One is by possibilistic approach introduced by Tanaka et al. and the other is the least square approaches. The concept of FLSR model is similar to that of ordinary regression model, that is to minimize the sum of distances between given data and estimated values.

D'Urso adopts the least square approach, because he insists that "the possibilistic approach (in which linear programming methods are utilized to estimate the regression coefficients) presents some difficulties in the estimation procedures", [4]. He handles several types of input-output data, such as crisp-fuzzy, fuzzy-crisp, and fuzzy-fuzzy, with not only type1 fuzzy set but also type2 fuzzy set. Here we describe the case of both input and output are type1 fuzzy set.

Let X_j $(j = 1, \ldots, k)$, Y be explanatory variables and the objective variable respectively. As these variables are type1 fuzzy sets, they are expressed as $X_j = (x_j, w_j, z_j)$ and $Y = (y, p, q)$ with centers x_j and y, left spread w_j and p, and right spread z_j and q, respectively. When we have a dataset of n number of data, $\{(Y_i; X_{i1}, \cdots, X_{ik})\}_{i=1,\ldots,n}$ with $X_{ij} = (x_{ij}, w_{ij}, z_{ij})$ and $Y_i = (y_i, p_i, q_i)$, consider

$$\begin{cases} \boldsymbol{y} = (\boldsymbol{X}\boldsymbol{a} + \boldsymbol{W}\boldsymbol{r} + \boldsymbol{Z}\boldsymbol{s}) + \boldsymbol{\varepsilon} = \boldsymbol{y}^* + \boldsymbol{\varepsilon} \\ \boldsymbol{p} = (b\boldsymbol{y}^* + d\boldsymbol{1}) + \boldsymbol{\lambda} = \boldsymbol{p}^* + \boldsymbol{\lambda} \\ \boldsymbol{q} = (g\boldsymbol{y}^* + h\boldsymbol{1}) + \boldsymbol{\rho} = \boldsymbol{q}^* + \boldsymbol{\rho} \end{cases}, \tag{2}$$

where b, d, g, h are numbers,

$$\boldsymbol{y} = \begin{pmatrix} y_1 \\ \vdots \\ \vdots \\ y_n \end{pmatrix}, \boldsymbol{a} = \begin{pmatrix} a_0 \\ a_1 \\ \vdots \\ a_k \end{pmatrix}, \boldsymbol{r} = \begin{pmatrix} r_0 \\ r_1 \\ \vdots \\ r_k \end{pmatrix}, \boldsymbol{s} = \begin{pmatrix} s_0 \\ s_1 \\ \vdots \\ s_k \end{pmatrix}, \boldsymbol{\varepsilon} = \begin{pmatrix} \varepsilon_1 \\ \vdots \\ \vdots \\ \varepsilon_n \end{pmatrix},$$

and

$$\boldsymbol{X} = \begin{pmatrix} 1 & x_{11} & \cdots & x_{1k} \\ \vdots & \vdots & \cdots & \vdots \\ \vdots & \vdots & \cdots & \vdots \\ 1 & x_{n1} & \cdots & x_{nk} \end{pmatrix}, \boldsymbol{W} = \begin{pmatrix} 1 & w_{11} & \cdots & w_{1k} \\ \vdots & \vdots & \cdots & \vdots \\ \vdots & \vdots & \cdots & \vdots \\ 1 & w_{n1} & \cdots & w_{nk} \end{pmatrix}, \boldsymbol{Z} = \begin{pmatrix} 1 & z_{11} & \cdots & z_{1k} \\ \vdots & \vdots & \cdots & \vdots \\ \vdots & \vdots & \cdots & \vdots \\ 1 & z_{n1} & \cdots & z_{nk} \end{pmatrix}.$$

The problem is to find numbers b, d, g, h and vectors $\boldsymbol{a}, \boldsymbol{r}, \boldsymbol{s}$ minimizing

$$\Delta(\boldsymbol{a}, \boldsymbol{r}, \boldsymbol{s}, b, d, g, h) = \pi_c |\boldsymbol{y} - \boldsymbol{y}^*|^2 + \pi_p |\boldsymbol{p} - \boldsymbol{p}^*|^2 + \pi_q |\boldsymbol{q} - \boldsymbol{q}^*|^2, \tag{3}$$

with certain weights π_c, π_p, π_q. Here $|\boldsymbol{v}|$ means the norm of the vector $\boldsymbol{v}$.

D'Urso gives practical examples for some input-output cases, one of which is on price and its related factors such as cooking and cellar from Italian specialized book on the Roman restaurant.

2.4 Dual Models for Possibilistic Regression

Guo and Tanaka have investigated the possibilistic regression model, and dual possibilistic regression models with crisp input and fuzzy output are analyzed in [7]. Although they consider not only linear model but also non-linear model, here we refer only linear model.

Let $Y = (y, p) = (y, p, p)$ be a symmetric fuzzy type output from crisp input values for variables x_j $(j = 1, \ldots, k)$. Then consider a model,

$$Y = A_1 x_1 + A_2 x_2 + \cdots + A_k x_k, \tag{4}$$

with symmetric fuzzy coefficients $A_j = (a_j, r_j)$ $(j = 1, \ldots, k)$. Once we have the model, the value of Y is obtained by calculate $(\sum_{j=1}^{k} a_j c_j, \sum_{j=1}^{k} r_j |c_j|)$ with given explicit values $c_1, \ldots, c_k$ for each variable. When we have a dataset of n number of data, $\{(Y_i; x_{i1}, \cdots, x_{ik})\}_{i=1,\ldots,n}$ with crisp x_{ij} and symmetric fuzzy $Y_i = (y_i, p_i)$, we consider the upper regression model and the lower regression model.

In the upper regression model, try to find fuzzy coefficients $A_j^* = (a_j^*, r_j^*)$ such that

$$\begin{aligned} &\text{Minimizing: } J(\boldsymbol{r}^*) = \textstyle\sum_{j=1}^{k} r_j^* \left(\sum_{i=1}^{n} |x_{ij}| \right), \text{under the condition} \\ &Y_i \subseteq Y_i^* = A_1^* x_{i1} + \cdots + A_k^* x_{ik} \qquad (i = 1, \ldots, n). \end{aligned} \tag{5}$$

Since the shapes of fuzzy set are supposed to be similar, the inclusion relation is dependent only on the value of spread and the condition above can be expressed by the following equations.

$$\begin{cases} y_i - p_i \geq \sum_{j=1}^{k} a_j^* x_{ij} - \sum_{j=1}^{k} r_j^* |x_{ij}| \\ y_i + p_i \leq \sum_{j=1}^{k} a_j^* x_{ij} + \sum_{j=1}^{k} r_j^* |x_{ij}| \\ r_j^* \geq 0 \end{cases}. \tag{6}$$

In the lower regression model, try to find fuzzy coefficients $A_{j*} = (a_{j*}, r_{j*})$ such that

$$\begin{aligned} &\text{Maximizing: } J(\boldsymbol{r}_*) = \textstyle\sum_{j=1}^{k} r_{j*} \left(\sum_{i=1}^{n} |x_{ij}| \right), \text{under the condition} \\ &Y_i \supseteq Y_{i*} = A_{1*} x_{i1} + \cdots + A_{k*} x_{ik} \qquad (i = 1, \ldots, n). \end{aligned} \tag{7}$$

Then the condition above can be expressed by the following equations.

$$\begin{cases} y_i - p_i \leq \sum_{j=1}^{k} a_{j*} x_{ij} - \sum_{j=1}^{k} r_{j*} |x_{ij}| \\ y_i + p_i \geq \sum_{j=1}^{k} a_{j*} x_{ij} + \sum_{j=1}^{k} r_{j*} |x_{ij}| \\ r_{j*} \geq 0 \end{cases}. \tag{8}$$

Guo and Tanaka also showed two theorems, one is on the existence of upper regression model and the other is on that of lower regression model.

Theorem 1. *(by Guo and Tanaka, [7])*

1. *There always exists an optimal solution in the upper regression model (5) under (6) .*
2. *There exists an optimal solution in the lower regression model (7) under (8) if and only if there exist $a_{1*}^{(0)}, a_{2*}^{(0)}, \ldots, a_{k*}^{(0)}$ satisfying*

$$y_i - p_i \leq \sum_{j=1}^{k} a_{j*}^{(0)} x_{ij} \leq y_i + p_i \qquad (i = 1, \ldots, n). \tag{9}$$

Proof.

1. This is trivial as a sufficiently large r_j^* satisfies each condition of (6).
2. If there exist $A_{j*} = (a_{j*}, r_{j*})$ $(j = 1, \dots, k)$ satisfying (8), then

$$y_i - p_i \leq \sum_{j=1}^{k} a_{j*}x_{ij} - \sum_{j=1}^{k} r_{j*}|x_{ij}| \leq \sum_{j=1}^{k} a_{j*}x_{ij},$$
$$y_i + p_i \geq \sum_{j=1}^{k} a_{j*}x_{ij} + \sum_{j=1}^{k} r_{j*}|x_{ij}| \geq \sum_{j=1}^{k} a_{j*}x_{ij}.$$

Conversely, for $a_{j*}^{(0)}$ satisfying (9), put $A_{j*}^{(0)} = (a_{j*}^{(0)}, 0)$ and they satisfy the condition (8). □

Remark: Original expression of the condition (9) is $Y_i \cap Y_{i*}^{(0)} \neq \phi$ (the empty set) for some $A_{i*}^{(0)} = (a_{i*}^{(0)}, r_{i*}^{(0)})$ with $\sum_{j=1}^{k} r_{j*}^{(0)}|x_{ij}| = 0$ for $i = 1, \dots, n$.

3 Details of Our Model

Souhir Charfeddine et al. proposed a system as original Tanaka's extension using h-cut with crisp input and L-R fuzzy output, [10]. Michihiro Amagasa already proposed a possibilistic model with fuzzy input and fuzzy output using h-cut, [2]. In the paper, he constructed the system with L-R fuzzy number for both input and output. Here we first describe the system in much more general case with proofs, then review it as the limited case. We also give some remarks some of which are related to the value of h.

3.1 Situation and Necessary Settings

We consider two functions L and R satisfying the following conditions for the left and right spreading parts of a fuzzy number, respectively .

(1) L and R are defined in the interval $[0, +\infty)$, and monotone decreasing,
(2) $L(0) = 1$ and $R(0) = 1$,
(3) $L(1) = 0$ and $R(1) = 0$.

Then the member-ship function of a L-R fuzzy number $Y = (y, p, q)$ is defined as

$$\mu_Y(u) = \begin{cases} L(\frac{y-u}{p}) & \text{if } u \leq y \\ R(\frac{u-y}{q}) & \text{if } y \leq u \\ 0 & \text{otherwise} \end{cases}. \tag{10}$$

The inverse functions $L^{-1}(h)$ and $R^{-1}(h)$ are defined for any non-negative value h by putting $L^{-1}(0) = 1$ and $R^{-1}(0) = 1$. And $\mu_Y(u) = h$ implies

$$u = \begin{cases} y - pL^{-1}(h) & \text{if } u \leq y \\ y + qR^{-1}(h) & \text{if } y < u \end{cases}. \tag{11}$$

We consider the same situation as in the Subsect. 2.3, i.e., there are dataset of L-R fuzzy numbers $\{(Y_i; X_{i1}, \cdots, X_{ik})\}_{i=1,\dots,n}$ with $X_{ij} = (x_{ij}, w_{ij}, z_{ij})$ and $Y_i = (y_i, p_i, q_i)$.

3.2 Proposed Model

We consider the possibilistic fuzzy regression model

$$Y = A_1X_1 + A_2X_2 + \cdots + A_kX_k, \tag{12}$$

with L-R fuzzy variables $Y = (y,p,q)$ and $X_j = (x_j,w_j,z_j)$ and L-R fuzzy coefficients $A_j = (a_j,r_j,s_j)$ $(j = 1,\ldots,k)$.

When the support of fuzzy number Y above h-cut line is denoted by $[Y]_h$, we have

$$\begin{aligned}
&[Y]_h = [y - pL^{-1}(h), y + qR^{-1}(h)],\\
&[X_j]_h = [x_j - w_jL^{-1}(h), x_j + z_jR^{-1}(h)],\\
&[A_j]_h = [a_j - r_jL^{-1}(h), a_j + s_jR^{-1}(h)].
\end{aligned}$$

Applying commonly known multiplication and summation of L-R fuzzy numbers, for example see [1], we have

$$\begin{aligned}
[\textstyle\sum_{j=1}^k A_jX_j]_h = [&\textstyle\sum_{j=1}^k (a_j - r_jL^{-1}(h))(x_j - w_jL^{-1}(h)),\\
&\textstyle\sum_{j=1}^k (a_j + s_jR^{-1}(h))(x_j + z_jR^{-1}(h))]_h,
\end{aligned}$$

and the range of the interval, J, is calculated by subtract the left end value from the right end value. Then

$$\begin{aligned}
J = &\textstyle\sum_{j=1}^k \{(z_jR^{-1}(h) + w_jL^{-1}(h))a_j\\
&+(x_j + z_jR^{-1}(h))R^{-1}(h)s_j + (x_j - w_jL^{-1}(h))L^{-1}(h)r_j\}.
\end{aligned}$$

In the same way as Guo and Tanaka did, we consider upper and lower models, and describe the inclusion relation of the support of Y_i and that of obtained fuzzy number in the regression model for a given dataset.

Now let ZW_j, XZ_j, XW_j be as follows,

$$\begin{cases}
ZW_j = (\sum_{i=1}^n z_{ij})R^{-1}(h) + (\sum_{i=1}^n w_{ij})L^{-1}(h)\\
XZ_j = ((\sum_{i=1}^n x_{ij}) + (\sum_{i=1}^n z_{ij})R^{-1}(h))R^{-1}(h)\\
XW_j = ((\sum_{i=1}^n x_{ij}) - (\sum_{i=1}^n w_{ij})L^{-1}(h))L^{-1}(h)
\end{cases}. \tag{13}$$

Then our upper model Y^* is constructed if we have $A_j^* = (a_j^*, r_j^*, s_j^*)$ such that

$$\begin{aligned}
&\text{Minimizing: } J(\mathbb{A}^*) = \textstyle\sum_{j=1}^k (ZW_ja_j^* + XZ_js_j^* + XW_jr_j^*),\\
&\text{where } \mathbb{A}^* = (A_1^*,\ldots,A_k^*),
\end{aligned} \tag{14}$$

under the condition that for all i

$$\begin{cases}
y_i - p_iL^{-1}(h) \geq \sum_{j=1}^k (a_j^* - r_j^*L^{-1}(h))(x_{ij} - w_{ij}L^{-1}(h))\\
y_i + q_iR^{-1}(h) \leq \sum_{j=1}^k (a_j^* + s_j^*R^{-1}(h))(x_{ij} + z_{ij}R^{-1}(h))\\
r_j^*,\ s_j^* \geq 0
\end{cases}. \tag{15}$$

The lower model Y_* is similarly constructed if we have $A_{j*} = (a_{j*}, r_{j*}, s_{j*})$ such that

$$\begin{aligned}
&\text{Maximizing: } J(\mathbb{A}_*) = \textstyle\sum_{j=1}^k (ZW_ja_{j*} + XZ_js_{j*} + XW_jr_{j*}),\\
&\text{where } \mathbb{A}_* = (A_{1*},\ldots,A_{k*}),
\end{aligned} \tag{16}$$

under the condition that for all i

$$\begin{cases} y_i - p_i L^{-1}(h) \leq \sum_{j=1}^{k}(a_{j*} - r_{j*}L^{-1}(h))(x_{ij} - w_{ij}L^{-1}(h)) \\ y_i + q_i R^{-1}(h) \geq \sum_{j=1}^{k}(a_{j*} + s_{j*}R^{-1}(h))(x_{ij} + z_{ij}R^{-1}(h)) \\ r_{j*},\ s_{j*} \geq 0 \end{cases}. \tag{17}$$

3.3 Some Properties of the Model

First we show the similar theorem to the Theorem 1 on the existence of models under assumption on the given dataset.

Theorem 2. *When $x_{ij} - w_{ij}L^{-1}(h) > 0$ $(i = 1, \ldots, n,\ j = 1, \ldots, k)$, then*

1. *There always exists an optimal solution in the upper regression model (14) under (15) .*
2. *There exists an optimal solution in the lower regression model (16) under (17) if and only if there exist $a_{1*}^{(0)}, a_{2*}^{(0)}, \ldots, a_{k*}^{(0)}$ satisfying*

$$\begin{cases} y_i - p_i L^{-1}(h) \leq \sum_{j=1}^{k}(x_{ij} - w_{ij}L^{-1}(h))a_{j*}^{(0)} \\ y_i + q_i R^{-1}(h) \geq \sum_{j=1}^{k}(x_{ij} + z_{ij}R^{-1}(h))a_{j*}^{(0)} \end{cases}. \tag{18}$$

Proof.

1. If $x_{ij} - w_{ij}L^{-1}(h) \geq 0$ in (15), then $x_{ij} > 0$ from $w_{ij} \geq 0$ and $0 \leq L^{-1}(h) \leq 1$. Therefore $x_{ij} + z_{ij}R^{-1}(h)$ are also non-negative, and sufficiently large r_j^* and s_j^* satisfy the condition.
2. If there exist $A_{j*} = (a_{j*}, r_{j*}, s_{j*})$ $(j = 1, \ldots, k)$ satisfying (17), then we have the condition (18). Conversely, for $a_{j*}^{(0)}$ satisfying (18), put $A_{j*}^{(0)} = (a_{j*}^{(0)}, 0, 0)$ and they satisfy the condition (17).

□

Remark 1: The assumption in the Theorem 2 are reasonable condition especially when we consider fuzzy numbers. If the data for independent variables are given in linguistic values, they are usually transformed into fuzzy numbers satisfying the assumption. Even if given data has negative center value, they sometimes can be transformed in the positive one by translating them by a certain constant value.

Remark 2: The condition (18) represents the inclusion relation between Y_i and the resulted fuzzy number Y_{i*}. When the condition hold, then area between h-cut horizontal line and the base-line ($h = 0$) of the Y_{i*} is included in the area of Y_i.

Remark 2.1: In case of $h = 1$, $L^{-1}(1) = R^{-1}(1) = 0$ and (18) is reduced to

$$y_i = \sum_{j=1}^{k} x_{ij}a_{j*}^{(0)},$$

which means that the line segment of Y_{i*} is in the area of Y_i.

Remark 2.2: In case of $h = 0$, $L^{-1}(0) = R^{-1}(0) = 1$ and (18) is reduced to

$$\begin{cases} y_i - p_i \leq \sum_{j=1}^{k}(x_{ij} - w_{ij})a_{j*}^{(0)} \leq \sum_{j=1}^{k} x_{ij}a_{j*}^{(0)} \\ y_i + q_i \geq \sum_{j=1}^{k}(x_{ij} + z_{ij})a_{j*}^{(0)} \geq \sum_{j=1}^{k} x_{ij}a_{j*}^{(0)} \end{cases}.$$

which means that $Y_{i*} \cap Y_i \neq \phi$.

In Amagasa's paper [2], although a multi-objective variables model is constructed, only the case that left spread and right spread are equal and $L = R$ is considered. In that case, ZW_j, XZ_j, XW_j in (13) become as follows,

$$\begin{cases} Z_j = ZW_j = 2(\sum_{i=1}^{n} z_{ij})L^{-1}(h) \\ X_j^{+} = XZ_j = ((\sum_{i=1}^{n} x_{ij}) + (\sum_{i=1}^{n} z_{ij})L^{-1}(h))L^{-1}(h) \\ X_j^{-} = XW_j = ((\sum_{i=1}^{n} x_{ij}) - (\sum_{i=1}^{n} z_{ij})L^{-1}(h))L^{-1}(h) \end{cases},$$

and the objective functions becomes

$$\begin{aligned} J(\mathbb{A}) &= \sum_{j=1}^{k}(Z_j a_j + X_j^{+} r_j + X_j^{-} r_j) \\ &= L^{-1}(h) \sum_{j=1}^{k}((\sum_{i=1}^{n} z_{ij})a_j + (X_j^{+} + X_j^{-})r_j). \\ &= 2L^{-1}(h) \sum_{j=1}^{k}((\sum_{i=1}^{n} z_{ij})a_j + (\sum_{i=1}^{n} x_{ij})r_j) \end{aligned}$$

Then the upper model Y^* for $A_j^* = (a_j^*, r_j^*)$ is

$$\begin{aligned} &\text{Minimizing: } J'(\mathbb{A}^*) = \sum_{j=1}^{k}((\sum_{i=1}^{n} z_{ij})a_j^* + (\sum_{i=1}^{n} x_{ij})r_j^*), \\ &\text{where } \mathbb{A}^* = (A_1^*, \ldots, A_k^*), \end{aligned} \tag{19}$$

under the condition that for all i

$$\begin{cases} y_i - p_i L^{-1}(h) \geq \sum_{j=1}^{k}(a_j^* - r_j^* L^{-1}(h))(x_{ij} - z_{ij}L^{-1}(h)) \\ y_i + p_i L^{-1}(h) \leq \sum_{j=1}^{k}(a_j^* + r_j^* L^{-1}(h))(x_{ij} + z_{ij}L^{-1}(h)) \\ r_j^* \geq 0 \end{cases}. \tag{20}$$

The lower model Y_* is

$$\begin{aligned} &\text{Maximizing: } J'(\mathbb{A}_*) = \sum_{j=1}^{k}((\sum_{i=1}^{n} z_{ij})a_{j*} + (\sum_{i=1}^{n} x_{ij})r_{j*}), \\ &\text{where } \mathbb{A}_* = (A_{1*}, \ldots, A_{k*}), \end{aligned} \tag{21}$$

under the condition that for all i

$$\begin{cases} y_i - p_i L^{-1}(h) \leq \sum_{j=1}^{k}(a_{j*} - r_{j*} L^{-1}(h))(x_{ij} - z_{ij}L^{-1}(h)) \\ y_i + p_i L^{-1}(h) \geq \sum_{j=1}^{k}(a_{j*} + r_{j*} L^{-1}(h))(x_{ij} + z_{ij}L^{-1}(h)) \\ r_{j*} \geq 0 \end{cases}. \tag{22}$$

4 Test Examples

Here we apply Amagasa's model for some STFN type dataset. First dataset is very small only with two independent variables, one is crisp and the other is fuzzy, shown in the Table 1. Using EXCEL's solver to solve the LP problem of the upper and lower model for $h = 0.63$, we have

$$Y^* = (2.7435, 2.7641)X_1 + (0.6745, 0.0000)X_2,$$
$$Y_* = (2.5194, 0.0000)X_1 + (0.6932, 0.0260)X_2.$$

The spread value of the coefficient of X_1 in Y^* seemed to be too large. But when substituting $X_1 = (1.0, 0.2)$ and $X_2 = (8.0, 0.5)$ as a test data, the resulted values $Y^* = (8.2153, 3.65)$ and $Y_* = (8.0665, 1.0587)$ might be acceptable.

For comparison, calculate three regression formulas Y_c(center values), Y_l(left end) and Y_r(right end) of each dataset of $[Y_i]_{0.63}$, here we call a "simple FLSR model".

$$Y_c = 0.71x_2 + 2.559, \quad Y_l = 0.684x_2 + 2.125, \quad Y_r = 0.733x_2 + 2.984,$$

and the corresponding value for the test data is $(8.2153, 0.7446, 0.7446)$.

Table 1. Crisp and fuzzy input 1st

	(y, p)	(x_1, z_1)	(x_2, z_2)
1	(3.5, 1.5)	(1.0, 0.0)	(2.0, 0.5)
2	(4.5, 2.0)	(1.0, 0.0)	(2.0, 0.5)
3	(7.0, 2.5)	(1.0, 0.0)	(6.5, 0.5)
4	(9.5, 2.0)	(1.0, 0.0)	(9.5, 1.0)
5	(11.0, 3.0)	(1.0, 0.0)	(12.0, 1.0)

Next we consider 6 variable model with 8 data, as in Table 2, and performed the calculation for $h = 0.7$ and 0.8. The ordinary regression formula for Y_c calculated from values for x_i $(i = 1, \ldots, 6)$ is alway the same as

$$Y_c = 0.162x_1 - 0.056x_2 + 0.104x_3 + 0.353x_4 + 1.232x_5 + 0.005x_6 - 0.394.$$

We show the resulted formulas of our proposed model and of the simple FLSR model for each h, and give corresponding values by substituting $X_1 = (1, 0), X_2 = (8, 0.5), X_3 = (1.5, 0.5), X_4 = (3, 1), X_5 = (4, 0.3), X_6 = (5, 0.5)$ as test data.

In case that $h = 0.7$

$$\begin{aligned}
Y^* &= (0.2898, 0)X_1 + (0, 0.0774)X_2 + (0.023, 0)X_3 \\
&\quad +(0.2633, 0.1434)X_4 + (1.1619, 0)X_5 + (0, 0.2111)X_6 \\
Y_* &= (0.4691, 0)X_1 + (0.1082, 0)X_2 + (0, 0)X_3 \\
&\quad +(0.3712, 0)X_4 + (0.8873, 0.0894)X_5 + (0, 0.0262)X_6 \\
Y_l &= 0.862x_1 + 0.144x_2 - 0.061x_3 + 0.477x_4 + 0.851x_5 + 0.04x_6 - 1.433, \\
Y_r &= -0.261x_1 - 0.167x_2 + 0.162x_3 + 0.271x_4 + 1.431x_5 + 0.002x_6 + 0.508.
\end{aligned}$$

The resulted values are $Y^* = (5.7879, 2.7279)$, $Y_* = (6.0013, 1.18)$, and (5.488, 0.2016, 0.4218) from the simple FLSR.

In case that $h = 0.8$

$$
\begin{aligned}
Y^* &= (0.2421, 0)X_1 + (0, 0.1004)X_2 + (0.0416, 0)X_3 \\
&\quad +(0.273, 0.261)X_4 + (1.1621, 0)X_5 + (0, 0.0645)X_6 \\
Y_* &= (0.2903, 0)X_1 + (0.0449, 0)X_2 + (0.0409, 0)X_3 \\
&\quad +(0.3606, 0)X_4 + (1.0115, 0.0884)X_5 + (0, 0)X_6 \\
Y_l &= 0.7019x_1 + 0.0924x_2 - 0.022x_3 + 0.4384x_4 + 0.9556x_5 + 0.0489x_6 - 1.2015, \\
Y_r &= -0.512x_1 - 0.248x_2 + 0.2477x_3 + 0.2495x_4 + 1.5756x_5 - 0.0672x_6 + 0.6497.
\end{aligned}
$$

The resulted values are $Y^* = (5.7856, 2.5513)$, $Y_* = (5.8399, 1.0607)$, and (5.488, 0.0556, *-0.112*) from the simple FLSR.

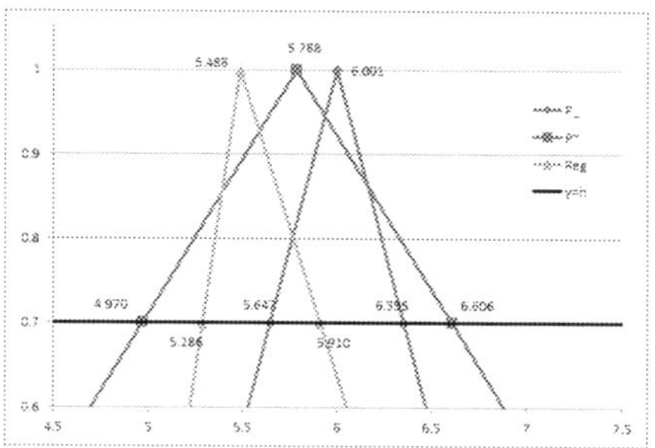

Fig. 1. Fuzzy sets for $h = 0.7$

Fig. 2. Fuzzy sets for $h = 0.8$

We can see that the resulted fuzzy sets of our model have higher ambiguities than that of the simple FLSR model, but they preserve at least the shape of fuzzy set. In case of $h = 0.8$, the right end value of the simple FLSR model is minus, so the shape of the corresponding set in the Fig. 2 is extremely sharp.

Table 2. Crisp and fuzzy input with six variables

	(y, p)	(x_1, z_1)	(x_2, z_2)	(x_3, z_3)	(x_4, z_4)	(x_5, z_5)	(x_6, z_6)
1	(3.5, 1.5)	(1.0, 0.0)	(2.0, 0.5)	(2.0, 0.3)	(4.0, 0.5)	(1.8, 0.0)	(3.0, 0.5)
2	(4.5, 2.0)	(1.0, 0.0)	(2.0, 0.5)	(1.0, 0.5)	(3.5, 0.5)	(2.8, 0.3)	(6.0, 0.5)
3	(7.0, 2.5)	(1.0, 0.0)	(6.5, 0.5)	(2.0, 0.5)	(5.0, 0.5)	(4.6, 0.3)	(4.0, 0.5)
4	(9.5, 2.0)	(1.0, 0.0)	(9.5, 1.0)	(1.0, 0.2)	(6.0, 1.0)	(6.5, 1.0)	(2.0, 0.0)
5	(10.0, 3.0)	(1.0, 0.0)	(10.0, 1.0)	(3.0, 0.5)	(4.5, 0.0)	(7.2, 0.5)	(6.0, 0.0)
6	(8.0, 2.0)	(2.0, 0.0)	(3.0, 1.0)	(1.0, 0.5)	(5.6, 1.0)	(5.0, 0.3)	(3.0, 0.3)
7	(6.0, 1.5)	(2.5, 0.0)	(5.0, 0.5)	(5.0, 1.0)	(2.3, 0.5)	(4.0, 0.5)	(2.0, 0.5)
8	(7.0, 1.5)	(1.5, 0.0)	(4.0, 0.5)	(6.0, 0.5)	(8.6, 0.5)	(3.0, 1.0)	(4.0, 0.5)

5 Conclusion and Discussion

We have reviewed some regression models of several types of input-output. Our purposes is to see the possibility of application of our proposed model to real problems, but it is still on the way. We could see that the model is feasible, and the resulted fuzzy values are reasonable as long as judging from the examples.

It is not easy to see advantage or disadvantage of our model just from the calculated values in the previous section. We need a dataset conforming to reality and some standard to see what kind of values are effective for decision making in a certain business performance.

References

1. Alim, A., Johora, F.T., Babu, S., Sultana, A.: Operations, elementary,on L-R fuzzy number. Adv. Pure Math. **5**, 131–136 (2015)
2. Amagasa, M.: Formulation of a sale price prediction model based on fuzzy regression analysis. In: Klatte, D., Lüthi, H.J. (eds.) Operations Research Proceedings 2011. Operations Research Proceedings, pp. 567–572. Springer, Heidelberg (2012)
3. Diamond, P.: Fuzzy least squares. Inf. Sci. **46**(3), 141–157 (1988)
4. D'Urso, P.: Linear regression analysis for fuzzy/crisp input and fuzzy/crisp output data. Comput. Stat. Data Anal. **42**, 47–72 (2003)
5. Inoue, Y., Uematsu, Y., Amagasa, M., Tomizawa, G.: The method of setting forecast selling price model by fuzzy regression analysis. In: Proceedings of Japan industrial Management Association, Autumn Congress, pp. 194–195 (1991)
6. Kim, K.J., Moskowitz, H., Koksalan, M.: Fuzzy versus statistical linear regression. Eur. J. Oper. Res. **92**(2), 417–434 (1996)
7. Guo, P., Tanaka, H.: Dual models for possibilistic regression analysis. Comput. Stat. Data Anal. **51**(1), 252–266 (2006)
8. Gale, B.T.: Managing Customer Value. Free Press, New York (1994)
9. Peters, G.: Fuzzy linear regression with fuzzy intervals. Fuzzy Sets Syst. **63**(1), 45–55 (1994)
10. Souhir, C., Karim, Z., Felix, M.-C.: Fuzzy regression analysis using trapezoidal fuzzy numbers. In: Proceedings of the Joint 4th Conference of the European Society for Fuzzy Logic and Technology and 11th Rencontres Francophones sur la logique Floue et ses Applications, pp. 1213–1218 (2005)
11. Tanaka, H., Uejima, S., Asai, K.: Linear regression analysis with fuzzy model. IEEE Trans. Syst. Man Cybern. SMC-12 **6**, 903–907 (1982)
12. Tanaka, H.: Fuzzy analysis by possibility linear models. Fuzzy Sets Syst. **24**, 363–375 (1987)
13. Tanaka, H., Entani, T., Sugihara, K.: Approach to Interval Evaluation in Decision Making. J. Jpn. Soc. Fuzzy Theory Intell. Inf. **17**(4), 406–412 (2005)
14. Shapiro, A.F.: Fuzzy Regression Models. ARC2005: Actuarial Research Clearing House 2006 (2005)
15. Wang, H.F., Tsaur, R.C.: Insight of a fuzzy regression model. Fuzzy Sets Syst. **112**(3), 355–369 (2000)

β-Robustness Approach for Fuzzy Multi-objective Problems

Oumayma Bahri[1,2(✉)], Nahla Ben Amor[1], and El-Ghazali Talbi[2]

[1] LARODEC Laboratory, ISG Tunis, Le Bardo, Tunisia
oumayma.b@gmail.com, nahla.benamor@gmx.fr
[2] INRIA Laboratory, CRISTAL-CNRS, Lille, France
el-ghazali.talbi@lifl.fr

Abstract. The paper addresses the robustness of multi-objective optimization problems with fuzzy data, expressed via triangular fuzzy numbers. To this end, we introduced a new robustness approach able to deal with fuzziness in the multi-objective context. The proposed approach is composed of two main contributions: First, new concepts of β-robustness are proposed to analyze fuzziness propagation to the multiple objectives. Second, an extension of our previously proposed evolutionary algorithms is suggested for integrating robustness. These proposals are illustrated on a multi-objective vehicle routing problem with fuzzy customer demands. The experimental results on different instances show the efficiency of the proposed approach.

Keywords: Robustness · Multi-objective optimization · Triangular fuzzy numbers · Evolutionary algorithms · VRP

1 Introduction

Multi-objective problems under uncertainty have gained more and more attention in recent years since they closely reflect the reality of many practical applications. These problems are characterized by the necessity of optimizing simultaneously different objectives, while considering that some input data are uncertain and without knowing what their full effects will be. Indeed, input uncertainty in a multi-objective problem can affect the set of objectives and/or constraints to be satisfied and thereby may lead to great deviations on the quality of solutions. To handle such a problem, specific optimization algorithms that consider uncertainty propagation along the decision making process should be used. Then, it is important to examine the performance of these algorithms and to analyse the influence of unavoidable uncertainties on the generated solutions. To this end, robustness analysis becomes necessary since it is intuitively connected to the idea that in the presence of uncertain inputs, the outputs should be relatively insensitive. Moreover, the goal of robustness in multi-objective optimization is to achieve a set of solutions that are not only optimal but also safe, reliable and robust. Several and various approaches exist in the literature to assess sensitivity and robustness of achieved results. Some of them use a combination of

J.P. Carvalho et al. (Eds.): IPMU 2016, Part II, CCIS 611, pp. 225–237, 2016.
DOI: 10.1007/978-3-319-40581-0_19

Monte-Carlo simulation and optimization to generate robust solutions [16]. Some others consider min-max criteria to produce solutions having the best possible performance in the worst-case [1,7]. However rather than gambling on the possible performance, it may be more efficient to give a lower limit on the overall acceptable performance and to state the confidence in being able to achieve that level. This is the main idea of β-robust approach [3] to take into account the subjective aspect of robustness through a target level specified by the decision maker. Unfortunately, all these approaches are limited to a single-objective only and often fail to consider the robustness requirement for real-world applications. Indeed, there are very few approaches regarding the robustness in multi-objective setting [2,5,6]. In this paper, we first propose to generalize the standard β-robustness concepts to our fuzzy multi-objective context. Then, we suggest to extend two fuzzy evolutionary algorithms from our previous work, for integrating robustness. Otherwise, the new robustness concepts are included into the search process of our algorithms in order to ensure the convergence towards robust optimal solutions. The remainder of this paper is organized as follows: Sect. 2 recalls some known concepts on which our approach is based. Section 3 defines the new β-robustness concepts after briefly describing the problem and summarizing our previous achievements. Section 4 presents the algorithmic refinements for integrating robustness and finally Sect. 5 reports the obtained results.

2 Background

We define here a multi-objective optimization problem (MOP) in both deterministic and uncertain cases and discuss some existing robustness approaches.

2.1 Deterministic vs. Uncertain MOP

A deterministic MOP usually involves the simultaneous optimization of two or more conflicting objectives and implies obtaining not only a single solution but a set of solutions called *Pareto optimal set* [17]. Formally, a MOP defined in the sense of minimization of all objectives, consists of solving the following mathematical program:

$$\min \; F(x) = (f_1(x), \ldots, f_n(x)) \; s.t. \; x \in X \tag{1}$$

where $F(x)$ is the vector of n ($n \geq 2$) objective functions to be minimized and $x = (x_1, \ldots, x_k)$ is the vector of decision variables from the feasible decision space $X \subseteq \mathbb{R}^n$. In the objective space, F can be defined as a cost function by assigning an objective vector $y = (y_1, \ldots, y_n)$ which represents the quality of solutions.

$$F : X \rightarrow Y \subseteq \mathbb{R}^n, \; F(x) = y = (y_1, \ldots, y_n) \tag{2}$$

While deterministic MOPs have been extensively discussed and reviewed in the literature [17], only few studies exist today for dealing with uncertain MOPs.

An uncertain MOP is characterized by the necessity of optimizing simultaneously several objectives in the presence of some uncertain data. Uncertainty in such problems may be associated with the key elements of decision making such as preference parameters, decision variables, constraints and/or objectives. In our study, we focus on the most critical and difficult case where uncertainty is assumed to occur in the objective functions. Formally, an uncertain MOP can be defined as follows:

$$\min\ F(x,\xi) = (f_1(x,\xi), f_2(x,\xi), \ldots, f_n(x,\xi))\ s.t.\ \ x \in \mathbb{R}^n, \xi \in U \tag{3}$$

where F is the vector of uncertain objectives that may depend on uncertainty scenarios U and $\xi = (\xi_1, \xi_2, \ldots, \xi_q)$ is the vector of uncertain variables. It should be noted that ignoring uncertainty propagation in the optimization process can lead to very poor decisions with often misleading simulation results. It becomes therefore necessary to estimate the influence of possible uncertainties on outputs. So far, very little research works investigate the combination of uncertain multi-objective optimization and output sensitivity. In fact, almost all the existing approaches for handling a MOP with uncertain objectives are limited to transform the problem into a mono-objective one by aggregating the set of objectives or also to reduce it to a deterministic MOP by using expectation values. Only few approaches have been developed for dealing with an uncertain MOP as-is without erasing any of its multi-objective or uncertain characteristics [10]. Besides, the robustness is not well handled by these approaches. Thus, a need for special methods able to deal with uncertainty in multi-objective problems while respecting their flexibility and robustness, is evident.

2.2 Robustness and Some Related Approaches

Robustness is a very important and essential concept for many optimization problems in diverse areas. It can be described as the ability of a resolution system to remain unaffected despite potential perturbations due to uncertain parameters. The approaches to cope with robustness in decision making are multiple and varied. For instance, in [16] a robust approximation approach combined with the Monte Carlo method is introduced. In [1,7], min-max regret approaches are applied to seek solutions having the best possible performance in the worst case. Yet, these robust min-max approaches may be deemed as extremely conservative when the worst case is not crucial and so an overall acceptable performance is preferred. Consequently, other approaches are introduced to take into account the subjective aspect of robustness through a target level specified by the decision maker. Among them, we cite the popular β-robustness approach [3] to which we are interested in this work. The goal of this approach is to maximise the likelihood that a solutions's actual performance is not worse than a given threshold. In fact, many studies have already used the β-robustness to assess simulation results. For example, in [3] the authors developed constraint models to achieve β-robust solutions that minimize the risk of costs exceeding the fixed threshold. In [14] the authors proposed a non-linear programming model to seek β-robust solutions minimizing the probability of achieving poor system performance and in [13] they introduced concepts of necessary and possible β-robustness. However, all these approaches can only deal with specific mono-objective problems for which they have been developed. In other words, until now, only few studies are devoted to robustness in multi-objective optimization where the aim is to find the set of robust non-dominated solutions. For example, in [9], a robust multi-objective genetic algorithm is proposed to achieve the trade-off between robustness and optimality. In [5], robustness is defined by either adding constraints to fix a predefined limit of variation or by replacing objectives with their mean functions. In [2], a multi-objective approach based on the degree of robustness of solutions is presented. In [20], an hypervolume-based robustness approach is considered and recently in [6] the concept of min-max robustness is extended to multi-objective optimization. Unlike these approaches, the approach we propose readily generalizes the standard β-robustness concepts to the multi-objective fuzzy context. Specifically, our approach retains the basic idea of β-robustness and extends it for evaluating any multi-objective problem with fuzzy data, especially with fuzzy-valued objective functions. To the best of our knowledge, there was no similar work done in the multi-objective optimization domain that involves β-robustness concepts in the context of fuzzy objectives.

3 Proposed Approach

This section presents the new β-robustness concepts for fuzzy MOPs and briefly describes the treated problem.

3.1 Problem Description

In the following, we focus on MOPs with fuzzy data in which fuzziness can be associated with the linguistic vagueness or ambiguity of information due to limited knowledge. Indeed, fuzzy sets are frequently used in many real-world applications since they offer a natural and efficient way to express different aspects of uncertainty [19]. First, we assume that uncertain input data are modeled using the most popular and simplest shape of fuzzy sets, namely, triangular fuzzy numbers or TFNs. A TFN denoted $A = [\underline{a}, \widehat{a}, \overline{a}]$ with an interval of possible values $[\underline{a}, \overline{a}]$ and a modal value $\widehat{a}$, is represented by a piecewise linear membership function μ_A which assigns a value within $[0, 1]$ to each element in A (see Fig. 1). In practical use of TFNs, the two dual measures: *Possibility* Π and *Necessity* N can be applied to express the degree of plausibility of any subset of A. Formally, the degree that A is less than a real number r may be derived from measures $\Pi(A \leq r) = sup_{A \leq r}\, \mu_A(x)$ and $N(A \leq r) = 1 - \Pi(A > r) = 1 - sup_{A>r}\, \mu_A(x)$, which are given by:

$$\Pi(A \leq r) = \begin{cases} 1 & \text{if } \widehat{a} < r \\ \frac{r-\underline{a}}{\widehat{a}-\underline{a}} & \text{if } \underline{a} < r < \widehat{a}, \\ 0 & \text{if } \underline{a} \geq r. \end{cases} \quad N(A \leq r) = \begin{cases} 1 & \text{if } \overline{a} < r \\ \frac{r-\widehat{a}}{\overline{a}-\widehat{a}} & \text{if } \widehat{a} \leq r \leq \overline{a} \\ 0 & \text{if } \widehat{a} \geq r. \end{cases} \tag{4}$$

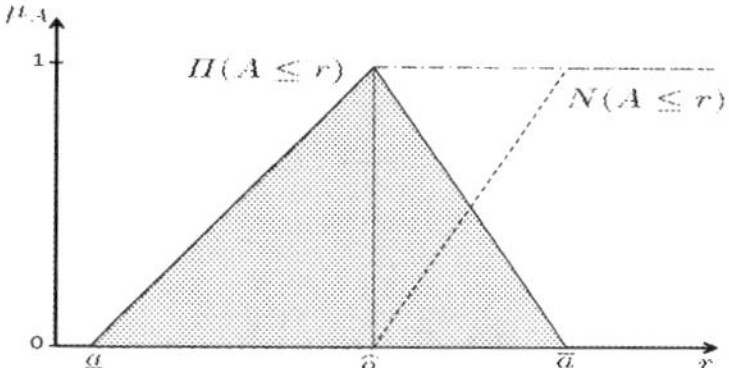

Fig. 1. Triangular Fuzzy number

As illustrated in Fig. 1, for any value of $r \in \mathbb{R}$, $N(A \leq r) \leq \Pi(A \leq r)$. This means that if N is positive, we are somewhat certain that Π matches the requirement. For this reason, the necessity degree is of great importance than the possibility degree [4]. Passing now to the semantic of problem outputs, we suppose that fuzziness of input data is propagated through the optimization process. This propagation has an impact on the objective functions that will be disrupted by the used fuzzy shape. Consequently, the problem outcomes or solutions will be also affected by fuzziness and obtained as vectors of triangular fuzzy values. Therefore, a fuzzy MOP with fuzzy-valued objectives can be formulated as:

$$F : X \rightarrow Y \subseteq (\mathbb{R} \times \mathbb{R} \times \mathbb{R})^n,$$

$$F(x) = y = \begin{pmatrix} y_1 = [\underline{y_1}, \widehat{y_1}, \overline{y_1}] \\ \dots \\ y_n = [\underline{y_n}, \widehat{y_n}, \overline{y_n}] \end{pmatrix} \tag{5}$$

To deal with such a problem, we have previously proposed a framework composed of two main stages:

- A fuzzy Pareto dominance for ranking the generated fuzzy-valued solutions since the standard dominance can only be applied for deterministic case [11].
- An extension of two Pareto-based evolutionary algorithms in order to enable them working in fuzzy space. The extended algorithms, denoted E-SPEA2 and E-NSGAII, use the fuzzy Pareto dominance as a fitness assignment strategy [12]. As well, the source code of these algorithms is shared via: http://www.oumaymabahri.com/sourcecodes.

While our framework can be efficiently applied for solving any MOP with fuzzy-valued objectives, there are still some limitations that need to be addressed. In fact, for performance assessment, we have approximated the generated fuzzy solutions into crisp ones and then simply applied classical multi-objective indicators (i.e., Hypervolume and epsilon indictors). However, fuzziness of solutions must not be ignored because if the input data or domain parameters are highly ambiguous, how can the optimizer simply state that the outputs are robust? It may be feasible only for simplicity or other practical reasons as long as the algorithm performance will not be affected. Moreover, approximating the generated fuzzy solutions to exact values may be criticised since it reduces the information provided and so affects their reliability. To this end, sensitivity analysis should be performed in order to draw further conclusions about the robustness of solutions.

3.2 New β-Robustness concepts for fuzzy MOPs

We present here a generic robustness approach able to determine robust solutions for any fuzzy MOP, in which fuzziness is modeled via triangular fuzzy numbers. To illustrate the new approach, we consider a well-studied combinatorial problem: the MO-VRPTW-UD (Multi-objective Vehicle Routing Problem with Time Windows and Uncertain Demands). The main idea of this problem is to find optimal routes for a fleet of identical vehicles which serve a set of customers whose exact demands are known only upon arrival at the customer's location. Besides, every route starts and ends at a central depot and each customer has a time window for its delivery. A detailed description of the problem is given in [18]. Without loss of generality, we assume that we have to minimize two objectives, namely the *total traveled distance* and *total tardiness time.* Moreover, the uncertain customer demands are assumed to be triangular fuzzy numbers.

Example 1. Figure 2 is an illustration of the MO-VRPTW-UD, with a central depot, 3 vehicles ($V1, V2, V3$) and a set of 8 customers represented by nodes. Each customer has a triangular fuzzy demand; the fuzzy demand of a customer $i = 1, \ldots, 8$ is $dm_i = [\underline{dm_i}, \widehat{dm_i}, \overline{dm_i}])$.

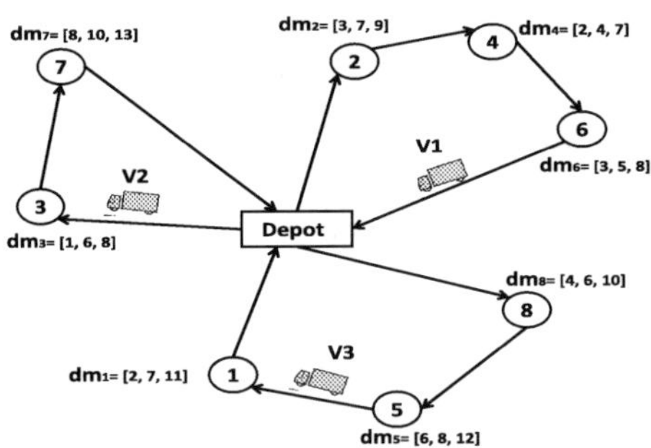

Fig. 2. Example of MO-VRPTW-UD

As the sequence of customers and service time for every vehicle depend primely on the amount of demands to be delivered, the transportation cost will be clearly disrupted by the fuzziness of demands. In consequence, both objective functions are affected by the used fuzzy shape and obtained as triangular fuzzy vectors. The challenge is so to achieve a set of robust optimal routes with respect to the fuzzy-valued objectives. Then as we want to generalize the β-robustness concepts to fuzzy multi-objective context, our aim becomes to obtain robust routes that yield actual cost no worse than a given quality threshold. We start by defining the β-robust solutions of the MO-VRPTW-UD problem:

Let $D = [\underline{D}, \widehat{D}, \overline{D}]$ and $T = [\underline{T}, \widehat{T}, \overline{T}]$ denote respectively triangular fuzzy variables for the traveled distance and the tardiness time, we assume that:

Definition 1. *β-Robust Routes are routes with a certain confidence that the total traveled distance D and tardiness time T will be less than a given threshold of desired optimal cost.*

Notice that, in real-life problems, a quality threshold is often provided by one or more experts opinions. However, as we are treating synthetic problems, such threshold is not always available. Then, knowing that the objective functions in a MOP are usually considered to be independent from each other (i.e., they depend only on the decision variable), we suggest to define a threshold to each objective as follows:

Let D^* be a traveled distance threshold, T^* be a tardiness time threshold and let $A = [\underline{a}, \widehat{a}, \overline{a}]$ and $B = [\underline{b}, \widehat{b}, \overline{b}]$ be the best known solutions for D and T respectively, then we have:

$$D^* = \widehat{a} + TF \times (\overline{a} - \widehat{a}), \quad T^* = \widehat{b} + TF \times (\overline{b} - \widehat{b}) \tag{6}$$

where TF is a given tightness factor of best possible performance selected by a parametric analysis which consists of examining the variability and closeness of triangular fuzzy values to the peaked ones (i.e. the most plausible values).

The issue now is so to maximise the confidence that the traveled distance D and tardiness time T of each route will "for sure" be less than the fixed thresholds D^* and T^*, respectively. Thus, we propose to maximize the degrees of necessity that each objective value lies within its confidence level. In possibilistic setting, these degrees are provided by the necessity measure N which gives a pessimistic view in decision making (see Eq. 5). In what follows, we assume that the necessary β-robustness, denoted β_N^i for every objective i, is equivalent to the necessity degree N and so should ideally be close to 1 (i.e., the higher β_N^i is the better). Subsequently, suppose we have computed the β_N degrees for both objectives: $\beta_N^1 = N(D \leq D^*)$ and $\beta_N^2 = N(T \leq T^*)$, the question is how to evaluate the obtained values. More generally, we need to check if a solution has a good or insufficient level of necessary robustness based on its different β_N values. We suggest hence to use the t-norm operator min [4] which is classically applied to interpret the conjunction. Based on the min operator, we propose to aggregate the set of β_N degrees of each solution:

$$(\beta_N^1 \cap \beta_N^2) = \min\ (\beta_N^1, \beta_N^2) \tag{7}$$

This aggregation allows us to make a decision based on the most pessimistic value given by the minimum necessary robustness. Thereafter, to avoid achieving a solution with low β_N values, we propose to enhance them with an interval of desired robustness level $[R, 1]$ where R is the lowest value fixed by the decision maker. The necessary robustness of our problem may be defined as follows:

Definition 2. *A route with traveled distance D and tardiness time T is said to be necessarily β_N-robust w.r.t. thresholds D^* and T^* respectively, iff:*

$$\beta_N^1 = N(D \leq D^*), \beta_N^2 = N(T \leq T^*) \text{ and } \min\ (\beta_N^1, \beta_N^2) \in [R, 1]$$

Example 2. We consider an example of β_N-robustness evaluation with a set of 4 solutions $\{x_1, x_2, x_3, x_4\}$, their corresponding β_N degrees: $\beta_N(x_1) = (0.30, 0.60)$, $\beta_N(x_2) = (0.55, 0.80)$, $\beta_N(x_3) = (0.95, 0.15)$ and $\beta_N(x_4) = (0.75, 1.00)$ and a confidence parameter $R = 0.5$ to suppose that the attitude of decision maker is neither pessimistic nor optimistic. We clearly remark that solutions x_1 and x_3 are not robust because the minimum of their β_N values is low than the confidence interval $[0.5, 1]$: $\min \beta_N(x_1) = 0.3$ and $\min \beta_N(x_3) = 0.15$. On the other hand, solutions x_2 and x_4 are judged as necessary robust since they reach the desired level of robustness (i.e., all their β_N values are within $[0.5, 1]$).

Furthermore, as necessity and possibility are inter-definable and their dual expression $N(A) = 1 - \Pi(\neg A)$ induces $N(A) \leq \Pi(A)$, we conclude that by maximising the necessary robustness of any solution, we are also maximising its possible robustness denoted β_Π^i for every objective $i = (1, \ldots, n)$. Obviously, we have:

Definition 3. *A route with traveled distance D and tardiness time T is said to be possibly β-robust w.r.t. thresholds D^* and T^* respectively, iff:*

$$\beta_\Pi^1 = \Pi(D \leq D^*) = 1 - N(D > D^*) \text{ and } \beta_\Pi^2 = \Pi(T \leq T^*) = 1 - N(T > T^*).$$

Finally, β_N and β_Π can be seen as lower and upper bounds of the degree that a solution is β−robust and so we may deduce that:

Definition 4. *If a route is necessarily and possibly β-robust w.r.t. the same thresholds, then we have:*

$$\forall i = \{1, \ldots, n\}, \quad \beta_N^i \leq \beta^i \leq \beta_\Pi^i$$

4 Algorithmic Refinements

To develop the new β-robustness approach, we suggest to extend our previously proposed evolutionary algorithms E-SPEA2 and E-NSGAII [12]. The idea is how to integrate the robustness concepts as evaluation criteria into the search process of these algorithms. As described above, our aim is to solve the MO-VRP-FD while maximizing the necessary β-robustness of routes. Notice that the discussion to follow focuses only on the necessary robustness β_N since the possible robustness β_P can always be deduced from the dual relationship. In order to enable the algorithms achieving β–robust solutions, we suggest firstly to replace the fitness function of each solution by the degree of necessary robustness, specifically with the minimum of its β_N values (Eq. 7). Yet, the initial random population usually generates poor solutions that may yield a β_N value closer or equal to zero for any reasonable threshold and thereby will prevent the algorithm from converging more quickly. Therefore to avoid returning zero-fitness values, we follow the method of "adaptive" threshold [13] which consists in a set of successive smaller thresholds with linearly decreasing approximations. For the MO-VRPTW-UD problem, we begin the initial population with two first thresholds D_0^* and T_0^* obtained as most pessimistic values of the best traveled-distance and tardiness-time. Then, populations and thresholds are generated iteratively with more demanding values until reaching D^* and T^* which remain fixed in the last generations. These latter thresholds are used to define the necessary β_N robustness. Secondly in order to perform selection, the quality of each solution should be evaluated according to a Pareto dominance relationship. At this level, we may simply use the fuzzy Pareto dominance from our previous work [11] for comparing or ranking the generated triangular fuzzy solutions. Hence, this dominance is not enough to evaluate their quality and cannot discriminate between the robust ones. Thus, we suggest to extend it for integrating the β-robustness criterion. We present here the refinements of three mono-objective dominance relations: *Total β-robust dominance*, *Strong β-robust dominance* and *Weak β-robust dominance.*

Definition 5. *Total β-robust dominance*
Let $y = [\underline{y}, \widehat{y}, \overline{y}]$ and $y' = [\underline{y}', \widehat{y}', \overline{y}']$ two triangular fuzzy solutions. y totally and robustly dominates y' (denoted by $y \prec_T^\beta y'$) iff:

$$(\overline{y}<\underline{y}')\, and\, (\beta_N(y) \geq \beta_N(y'))$$

Definition 6. *Strong β-robust dominance*
Let $y = [\underline{y}, \widehat{y}, \overline{y}]$ and $y' = [\underline{y}', \widehat{y}', \overline{y}']$ be two triangular fuzzy solutions. y strongly and robustly dominates y' (denoted by $y \prec_S^\beta y'$) iff:

$$(\overline{y} \geq \underline{y}') \wedge (\widehat{y} \leq \underline{y}') \wedge (\overline{y} \leq \widehat{y}') \wedge (\beta_N(y) \geq \beta_N(y'))$$

Definition 7. *Weak β-robust dominance*
Let $y = [\underline{y}, \widehat{y}, \overline{y}] \subseteq \mathbb{R}$ and $y' = [\underline{y}', \widehat{y}', \overline{y}'] \subseteq \mathbb{R}$ be two triangular fuzzy solutions. y weakly and robustly dominates y' (denoted by $y \prec_w y'$) iff:

$$[(\underline{y}<\underline{y}') \wedge (\overline{y}<\overline{y}')] \wedge [\beta_N(y) \geq \beta_N(y')] \wedge$$

$$[((\widehat{y} \leq \underline{y}') \wedge (\overline{y}>\widehat{y}')) \vee ((\widehat{y}>\underline{y}') \wedge (\overline{y} \leq \widehat{y}')) \vee ((\widehat{y}>\underline{y}') \wedge (\overline{y}>\widehat{y}'))].$$

Afterwards, the Pareto dominance between vectors of fuzzy solutions is determined based on the type of mono-objective dominance found for all objectives. For instance, a β-robust Pareto dominance holds if there is at least a total or strong β-robust dominance in one of objective and a weak β-robust dominance in another. All these dominance relations are then integrated into the search process of each algorithm, especially into the fitness assignment strategy. Finally, to implement the new robust evolutionary algorithms denoted R-SPEA2 and R-NSGAII, we have followed and extended the repository of classical algorithms (i.e., SPEA2 and NSGAII) of platform ParadisEO-MOEO [8].

5 Experimental Results

For the experimental study, we have conducted some tests to examine the robustness of obtained solutions in solving the MO-VRP-FD problem. In fact, we have used 56 fuzzy instances sampled uniformly at random from the crisp Solomon's benchmark [15]. More precisely, we have generated for each Solomon's instance its fuzzy sampled version in which the exact demand values are replaced by triangular fuzzy values in the following manner: First, the kernel value $(\widehat{dm})$ for each triangular fuzzy demand dm is kept the same as the crisp demand value dm_i of the current instance. Then, the lower $(\underline{dm})$ and upper $(\overline{dm})$ bounds of this triangular fuzzy demand are uniformly sampled at random in the intervals $[50\,\%dm, 95\,\%dm]$ and $[105\,\%dm, 150\,\%dm]$, respectively. All the 56 sampled fuzzy instances were then tested on the refined algorithms R-SPEA2 and R-NSGAII. In parametric analysis, we have taken the following values: a randomly initialized population with size = 100, crossover rate = 0.8, mutation rate = 0.05, tightness factor $TF = 0.75$ and number of generations = 1000 from which the last 100 use the threshold values D^* and T^*. Both algorithms have been executed 30 times on each fuzzy instance, i.e., $2 \times 56 \times 30 = 3360$ runs. For empirical assessment, we follow the method of fuzzy semantics from [13] and use Monte-Carlo simulations. In our setting, fuzzy routes are taken as a-priori solutions found when the customer demands are not exactly known. Each route consists of a precise sequence of visits to customers with minimum traveled-distance and tardiness-time. This sequence is necessary to evaluate the behaviour of solutions on a family of N crisp instances interpreted as possible a-posteriori realizations. To this end, we have generated, for each fuzzy instance, 10 deterministic samples by simulating exact demands at random according to different probability distributions which are coherent with the triangular fuzzy demands. We have then used the sequences of customers provided by the resolution of fuzzy instances in order to obtain a simulation of real traveled distance and tardiness time that may be under or above the thresholds D^* and T^*. Finally, we have computed the proportion n of those values among the N which are below the thresholds. This gives us an empirical measure of the real robustness n-rob, i.e., a good degree of β_N should correspond to a high n. Table 1 presents the robustness results for six fuzzy instances (labeled as Fuzz-C101, Fuzz-C201, etc.) and it shows for each instance, the thresholds D^* and T^*, the β_N values of the best solution across 30 runs, the robustness of simulated proportion n and the runtime in seconds. Notice that, the lowest desired level of robustness is already fixed to $R = 0.25$. As we can see, the necessary β_N robustness is >0 in all the cases, so the possible robustness β_Π is 1. Then observing the minimum of β_N degrees, we can deduce that almost all solutions provide a good level of robustness $(\min(\beta_N^1, \beta_N^1) \in [0.25, 1])$. Besides, the simulated real robustness values $n-$rob are always 1 or ideally close to 1 even when β_N is low. This means that the traveled-distance and tardiness-time values for all simulations are below the fixed thresholds. It should be noticed that we achieved the same robustness results for other not shown instances, since they are completely similar and exhibit the same trend. In that sense, we may conclude that the robustness we are looking for in our solutions is satisfiable. Figure 3 illustrates the results by box plots, such that each box presents the $\min(\beta_N)$ values of 30 runs of each algorithm tested on one fuzzy instance. Clearly, for the six illustrated instances,

Table 1. Robustness results of R-SPEA2 and R-NSGAII

Instances	Algorithms	D^*	T^*	β_N^1	β_N^2	$\min(\beta_N^1, \beta_N^2)$	Runtime	$n-$rob
Fuzz-C101	R-SPEA2	2548.00	272894.25	0.647	0.553	0.553	7.4 s	0.996
	R-NSGAII	2781.00	281100.50	0.489	0.502	0.489	8.5 s	0.892
Fuzz-C201	R-SPEA2	2958.75	286533.50	0.469	0.704	0.469	7.5 s	0.954
	R-NSGAII	3101.75	301994.00	0.233	0.641	0.233	9.1 s	0.912
Fuzz-R101	R-SPEA2	2504.25	219522.00	0.732	0.815	0.772	6.6 s	0.989
	R-NSGAII	2822.50	220531.00	0.471	0.723	0.471	7.4 s	0.869
Fuzz-R201	R-SPEA2	3018.50	214915.00	0.623	0.913	0.623	7.7 s	0.994
	R-NSGAII	3214.50	250975.00	0.554	0.736	0.554	8.9 s	0.979
Fuzz-RC101	R-SPEA2	3747.00	295513.50	0.389	0.642	0.389	8.3 s	0.934
	R-NSGAII	3998.00	309812.00	0.215	0.571	0.215	9.0 s	0.887
Fuzz-RC201	R-SPEA2	3879.00	318017.50	0.361	0.617	0.361	7.9 s	0.997
	R-NSGAII	4086.00	325519.00	0.327	0.496	0.327	9.5 s	0.985

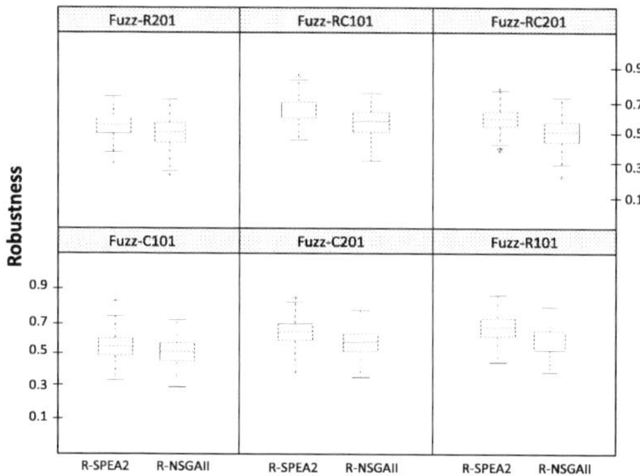

Fig. 3. Comparison of R-SPEA2 and R-NSGAII

the boxes of R-SPEA2 are less sensitive to variations and higher than those of R-NSGAII. Consequently, R-SPEA2 provides better robust solutions than the R-NSGAII.

6 Conclusion

In this paper, we have presented a new approach of β-robustness that allows us to achieve robust optimal solutions for any fuzzy multi-objective problem. We have also described an extension of two evolutionary algorithms for integrating robustness. Moreover, we have applied our approach on a vehicle routing problem and empirically assessed the actual robustness of obtained solutions using Monte-Carlo simulations. As future work, we intend to extend multi-objective performance indicators (i.e., Hypervolume indicator) to the robust fuzzy context.

It would also be interesting to validate the proposed approach for different fuzzy multi-objective problems.

References

1. Bazgan, C., Aissi, H.: Min-max and min-max regret versions of combinatorial optimization problems: a survey. Eur. J. Oper. Res. **197**, 427–438 (2009)
2. Barrico, C., Antunes, C.H.: Robustness analysis in multi-objective optimization using a degree of robustness concept. In: IEEE congree on CEC, pp. 6778–6783 (2006)
3. Daniels, R.L., Carrillo, J.E.: β-robust scheduling for single-machine systems with uncertain processing times. IIE Trans. **29**, 977–985 (1997)
4. Dubois, D., Prade, H.: A class of fuzzy measures based on triangular norms. Int. J. General Syst. **8**, 43–61 (1982)
5. Deb, K., Gupta, H.: Introducing robustness in multiple-objective optimization. Evol. Comput. **14**, 463–494 (2006)
6. Ehrgott, M., Ide, J., Schöbel, A.: Minmax robustness for multi-objective optimization problems. Eur. J. Oper. Res. **239**(1), 17–31 (2014)
7. Kasperski, A., Kule, M.: Choosing robust solutions in discrete optimization problems with fuzzy costs. Fuzzy Sets Syst. **160**, 667–682 (2009)
8. Liefooghe, A., Basseur, M., Jourdan, L., Talbi, E.-G.: ParadisEO-MOEO: a framework for evolutionary multi-objective optimization. In: Obayashi, S., Deb, K., Poloni, C., Hiroyasu, T., Murata, T. (eds.) EMO 2007. LNCS, vol. 4403, pp. 386–400. Springer, Heidelberg (2007)
9. Li, M., Azarm, S.: A multi-objective genetic algorithm for robust design optimization. In: 7th conference on GECCO, pp. 771–778. ACM (2005)
10. Limbourg, P.: Multi-objective optimization of problems with epistemic uncertainty. In: Coello Coello, C.A., Hernández Aguirre, A., Zitzler, E. (eds.) EMO 2005. LNCS, vol. 3410, pp. 413–427. Springer, Heidelberg (2005)
11. Bahri, O., Ben Amor, N., El-Ghazali, T.: New pareto approach for ranking triangular fuzzy numbers. In: Laurent, A., Strauss, O., Bouchon-Meunier, B., Yager, R.R. (eds.) IPMU 2014, Part II. CCIS, vol. 443, pp. 264–273. Springer, Heidelberg (2014)
12. Oumayma, B., Nahla, B.A., Talbi, E.-G.: Optimization algorithms for multi-objective problems with fuzzy data. In IEEE International Symposium on MCDM, pp. 194–201 (2014)
13. Palacios, J.J., González-Rodríguez, I., Vela, C.R., Puente Peinador, J.: β-Robust solutions for the fuzzy open shop scheduling. In: Laurent, A., Strauss, O., Bouchon-Meunier, B., Yager, R.R. (eds.) IPMU 2014, Part I. CCIS, vol. 442, pp. 447–456. Springer, Heidelberg (2014)
14. Pishevar, A., Tavakkoi, R.: β-Robust parallel machine scheduling with uncertain durations. Ind. Bus. Manage. **2**(3), 69–74 (2014)
15. Solomon, M.M.: Algorithms for the vehicle routing and scheduling problem with time window constraints. Oper. Res. **35**(2), 254–265 (1987)
16. Solano-Charris, E.L., Prins, C., Santos, A.C.: Heuristic approaches for the robust vehicle routing problem. In: Fouilhoux, P., Gouveia, L.E.N., Mahjoub, A.R., Paschos, V.T. (eds.) ISCO 2014. LNCS, vol. 8596, pp. 384–395. Springer, Heidelberg (2014)
17. Talbi, E.-G.: Metaheuristics: From Design to Implementation. Wiley, Lille (2009)

18. Toth, P., Vigo, D.: The vehicle routing problem. SIAM Monogr. Discrete Math. Appl. (2002)
19. Zadeh, L.A.: Fuzzy sets as a basis for a theory of possibility. Fuzzy Sets Syst. **1**, 3–28 (1978)
20. Zitzler, E., Bader, J.: Robustness in hypervolume-based multiobjective search. Technical report, Computer Engineering and Networks Laboratory (2010)

Construction of an Outranking Relation Based on Semantic Criteria with ELECTRE-III

Miriam Martínez-García(✉), Aida Valls, and Antonio Moreno

ITAKA (Intelligent Technologies for Advanced Knowledge Acquisition) Research Group, Departament d'Enginyeria Informàtica i Matemàtiques, Universitat Rovira i Virgili, Tarragona, Catalonia, Spain
{miriam.martinez, aida.valls, antonio.moreno}@urv.cat

Abstract. ELECTRE-III is a well-known multi-criteria decision analysis method that ranks a set of alternatives in terms of a set of heterogeneous evaluation criteria. It is based on constructing and exploiting a pairwise outranking relation between alternatives, which are defined with numerical and ordinal values. However, nowadays it is very common the use of descriptive linguistic tags, which is information that requires a qualitative treatment rather than a numerical one. In this paper we propose to store the user preferences about a set of tags in an ontological structure and to use this knowledge to construct the outranking relation by means of a semantic analysis of the tags associated to the alternatives. Uncertainty is handled by means of fuzzy concordance and discordance functions. The method is illustrated with a case study related to the recommendation of touristic activities.

Keywords: MCDA · Ontology · Semantic profile · Outranking relations

1 Introduction

The problem of ranking a set of alternatives has been thoroughly studied in *Multi-Criteria Decision Aid* (MCDA). Some well-known ranking methods, like the ELECTRE (*ELimination and Choice Expressing REality*) family [2], construct a preference structure from a pairwise comparison of a set of possible alternatives, which is based on two voting-inspired ideas: *concordance* (or "the choice of the majority") and *discordance* (or "the respect to minorities"). These *outranking* techniques have been widely used in many fields [4].

An important advantage of these methods is that they can work directly with purely ordinal scales, without requiring their transformation into abstract ones with an arbitrary range. A second advantage is that indifference and preference thresholds can be used to model uncertain knowledge. However, one of the main shortcomings of ELECTRE is that alternatives can only be defined in terms of numerical and ordinal criteria. Nowadays it is becoming increasingly common to find decisional situations in which alternatives may also include non-numerical information, represented in the form of *semantic criteria*, which may take as values the concepts of a given domain ontology. For example, the description of a touristic destination may include numerical

J.P. Carvalho et al. (Eds.): IPMU 2016, Part II, CCIS 611, pp. 238–249, 2016.
DOI: 10.1007/978-3-319-40581-0_20

criteria (e.g. number of inhabitants, average temperature) but also semantic ones (e.g. sports that may be practiced in the city, cultural attractions that may be visited).

Ontologies are knowledge structures that commonly store the main concepts of a domain, the taxonomic and non-taxonomic relationships between them, and their attributes [8]. A *semantic user profile* [10] may contain the degree of preference of the user with respect to some domain concepts. This information may be exploited to compare and rank a set of alternatives. The main aim of this paper is to show how ELECTRE-III, a well-known outranking method of the ELECTRE family, may be enhanced to deal with alternatives defined on semantic criteria. The contribution of this paper is twofold. First, we propose a mechanism by which the preference scores missing in a semantic user profile may be estimated from the known ones, using the taxonomical relationships of the ontology. Second, we present a method that constructs an outranking relation on a set of alternatives defined on semantic multi-valued criteria using a new definition of the concordance and discordance fuzzy functions. This relation is later exploited to rank the alternatives and show the best ones to the user so that she can take the final decision.

The rest of the paper is organised as follows. The next section explains how the classical ELECTRE-III method builds an outranking relation that is used to rank the alternatives in a partial pre-order structure. Section 3 introduces *multi-valued semantic criteria*, which are those that can take as values the concepts defined in a reference ontology. The following section proposes a way to represent a *semantic user profile* in an ontological structure and to complete the missing preferential information by taking into account the available one. These preferences are used in Sect. 5 to propose an extension of ELECTRE-III that builds an outranking relation on alternatives defined on semantic criteria. This method is illustrated with a case study in which it is employed to order tourist attractions. The paper finishes with some conclusions and lines of future work.

2 The ELECTRE-III Ranking Method

The ELECTRE-III ranking method considers the following input data [2]:

- Alternatives A: they are the potential actions or solutions for the decision problem. $A = \{a, b, c, d, \ldots\}$ is the finite set of n alternatives.
- Criteria G: they are the numerical or ordinal indicators on which the alternatives are evaluated based on the goals of the decision maker. $G = \{g_{1,}g_2, \ldots, g_m\}$ is the finite set of m criteria.
- Weights W: they indicate the relevance of each criterion on the final decision. W is the addition of the m weights in vector W, i.e.$\sum_{j=1}^{m} w_j = \mathrm{W}$.

ELECTRE-III constructs the outranking relation taking into account the uncertainty and imprecision associated to the pairwise comparison of the alternatives using pseudo-criteria. For this reason, each criterion is associated with the following two discrimination thresholds:

- Indifference threshold $q_j(a)$: given two alternatives a and b, it is the difference on criterion g_j below which the decision maker is indifferent between both options.
- Preference threshold $p_j(a)$: given two alternatives a and b, it is the difference on criterion g_j above which the decision maker shows a clear strict preference in favour of aψover b.

ELECTRE-III also includes the veto rule, which is the right of giving essential reasons for rejecting the outranking relation. This is introduced as another threshold:

- Veto threshold $v_j(a)$: given two alternatives a and b, a discordant difference larger than the veto in favour of bψwith respect to aψinψcriterion g_j will require the negation of the outranking relation aSb (thus, if there is a criterion in which b is much better than a, it will not be possible to claim that a is at least as good as b).

The ELECTRE-III ranking procedure has two steps:

Step 1 (Construction of the outranking relation): The outranking relation S is built for each pair of alternatives $(a, b) \in AxA$ by comparing their performance on the set of criteria G. The alternative a outranks alternative b if, taking into account the decision maker's preferences, a is at least as good as b and there is no strong argument against this claim. Two indices are applied to evaluate this relation: concordance and discordance. For each criterion $g_j \in G$, the *partial concordance* is calculated as:

$$c_j(a,b) = \begin{cases} 1 & \textit{if } g_j(a) \geq g_j(b) - q_j(b) \\ 0 & \textit{if } g_j(a) \leq g_j(b) - p_j(b) \\ \frac{g_j(a) - g_j(b) + p_j(b)}{p_j(b) - q_j(b)} & \textit{otherwise.} \end{cases} \tag{1}$$

Once the partial concordances have been measured, an overall *concordance index* is computed for each pair a, b as follows:

$$c(a,b) = \frac{1}{W} \sum\nolimits_{j=1}^{m} w_j c_j(a,b) \tag{2}$$

On the other hand, the *partial discordance* index is defined as:

$$d_j(a,b) = \begin{cases} 1 & \textit{if } g_j(a) \leq g_j(b) - v_j(a) \\ 0 & \textit{if } g_j(a) \geq g_j(b) - p_j(a) \\ \frac{g_j(b) - g_j(a) - p_j(a)}{v_j(a) - p_j(a)} & \textit{otherwise} \end{cases} \tag{3}$$

Finally, the degree of credibility of the outranking relation aSb, $\rho(a,b)$, is calculated using the global concordance and the partial discordance indices of the set $J(a,b)$ of criteria for which the discordance is larger than the overall concordance.

$$\rho(a,b) = \begin{cases} c(a,b) & \textit{if } \forall_j d_j(a,b) \leq c(a,b), \\ c(a,b).\prod_{j \in J(a,b)} \frac{1 - d_j(a,b)}{1 - c(a,b)} & \textit{otherwise.} \end{cases} \tag{4}$$

Step 2 (Distillation): The outranking relation is exploited to build a partial pre-order among the alternatives in A. It is an iterative process that selects at each step a subset of alternatives, taking into account the credibility values of the outranking relation. This procedure yields two complete pre-orders (descending and ascending distillation chains), which are intersected to generate the final partial pre-order.

3 Semantic Criteria

When structuring information in data matrices (as in ELECTRE), a *semantic variable* is defined as an unbounded categorical variable whose linguistic values can be interpreted based on background knowledge; thus, values are *concepts* rather than simple modalities [3]. Moreover, semantic criteria are usually multi-valued, so an alternative has a list of concepts (*tags*) in each criterion, rather than a single value.

The semantic interpretation of the tags requires a structured representation of the domain concepts. In the Artificial Intelligence literature, the most successful knowledge representation models are *ontologies*. They offer a formal and explicit description of a shared conceptualization [8], in which semantic interrelations are modelled as links between concepts. Ontologies enable the formal articulation of domain knowledge at a high level of expressiveness and the implementation of automatic reasoning procedures. Each semantic criterion may be associated to a different ontology depending on its domain. Specialized search engines such as Swoogle [1] permit to find domain ontologies.

As an example, Table 1 presents a small dataset with different touristic activities. The first column shows the name of the activities. The second one shows a semantic criterion with a list of tags that describe it. The third one corresponds to a numerical criterion that gives the price of the activity, and the last column is another semantic criterion which includes tags that describe the best weather conditions to perform each activity. It can be seen that each semantic criterion may take a different number of tags on each alternative. Due to the different meaning of the two semantic criteria, a different ontology would be used for each one. For example, in [5] a weather ontology was constructed, and in [6] some Tourism ontologies are presented.

Table 1. Example of a dataset with semantic and numerical criteria

Activity name	Touristic description	Cost	Best weather
Montsant Mountain	Paragliding, ClimbingWall, Rappelling	80 €	NoPrecipitation, PartlyCloudy
Tarragona Beach	BeachPicnic, FamilyBeaches, Sunbathing	40 €	HighSun, LightAir
Archeological Museum	UniqueBuilding, HumanHeritage, Ruins, CultureRoutes, HistoricBuilding	50 €	LightPrecipitation, NeutralState, OverCast
Adventure and Journey	HorseRiding, Car4 × 4, PaintBall, ShoppingArea	60 €	ModerateSun, OptimumHumidity

4 User Profiling with Ontologies

Ontologies are mainly used to represent the concepts and relations in a certain domain. However, they can also be employed [10] to store information about the preferences of users on different concepts, i.e. a *semantic user profile*.

The semantic user profile model proposed in this work consists on adding a numerical property to each of the most specific concepts of the ontology (i.e. the leaves of the ontological tree), called *Tag Interest Score* (*TIS*). This value represents the degree of interest of the decision maker in the corresponding concept, with a satisfaction score in [0,1]. An example is shown in Fig. 1, where we can see that some of the leaves of the ontology have a score, like *TIS*(Boating) = 0.7, *TIS* (Rowing) = 0.4, *TIS*(WaterSkiing) = 0.8 and *TIS* (Fishing) = 0.6. Notice that the leaves of the tree may appear at different levels.

Initially, we may know the scores of some leaf concepts, which may have been obtained explicitly from the user or elicited implicitly with learning algorithms that analyse the interaction of the user with the system [6]. Provided that ontologies usually have hundreds of leaves, the system can estimate unknown scores from the ones available in the ontology. The procedure proposed to estimate the missing score for a concept *c* is the following:

1. Following the taxonomic relations we find concepts that are semantically similar to *c*. A set of leaf concepts is built by following the taxonomical relations in the ontology using *relatives*(*c*,*l*), where *l* is the number of is-a levels we want to

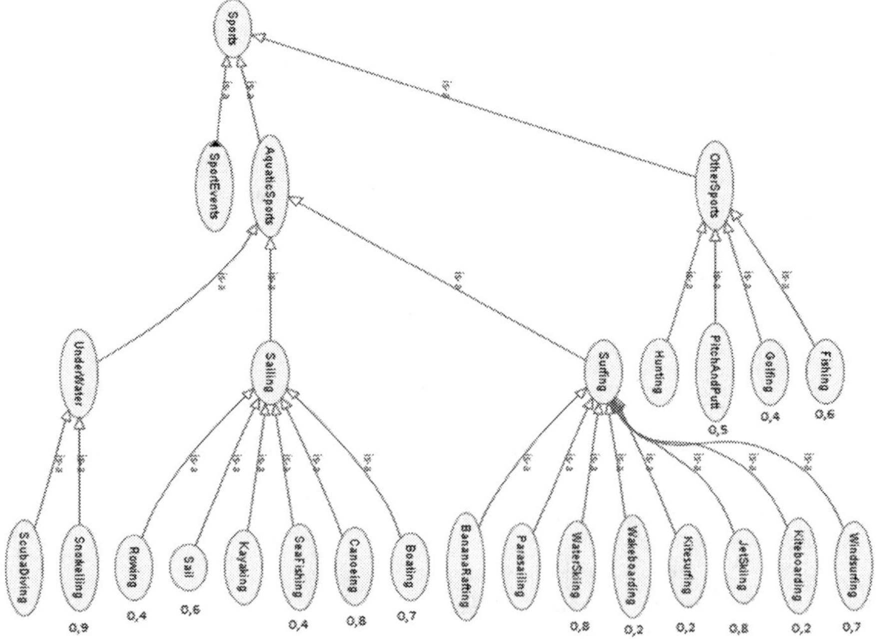

Fig. 1. Ontology portion dealing with Sports concepts

explore. The function *leaves(t)* returns the set of leaves (basic concepts) of the ontology subtree whose root is *t*.

$$relatives(c,l) = \begin{cases} leaves(father(c)) & l = 1 \\ leaves(father(c), l-1) & l > 1 \end{cases} \tag{5}$$

We propose to start with $l = 1$ (i.e. leaf concepts descending from the father of *c*). If the number of elements with a known score is below a given threshold, we can move to $l = 2$ (i.e. including also leaf concepts descending from the grandparent of *c*). If the number of known scores is still low, we continue increasing *l*.

2. The value of *TIS(c)* is calculated using some averaging operator on the known scores given to the relatives of *c*. In particular, the *WOWA* (*Weighted Ordered Weighted Average*) operator with two weighting factors is proposed [9]. The classical *OWA* weights allow the definition of different aggregation policies [11]. With conjunctive parameters the resulting score is penalized when similar concepts have low scores (pessimistic approach), whereas with disjunctive parameters the score is based only on the highest scores of the similar concepts (optimistic approach). A neutral configuration is also possible, which leads to the classic arithmetic average. In addition, with WOWA we can give a different importance to each of the values that are aggregated in terms of the number of levels used to find it (so that for instance concepts obtained in level 1 are given more relevance than concepts found in level 2). For a concept c_k found at level *i*, the corresponding weight w_k is calculated with the following expression, in which *#concepts(p)* is the number of new concepts that would be discovered in level *p*:

$$w_k = \frac{1}{i \cdot V}, \text{ where } V = \sum\nolimits_{l=1...max} \frac{\#concepts(l)}{l} \tag{6}$$

Considering Fig. 1, if we want to estimate the score of the concept '*Kayaking*', first we construct a set of similar concepts: *relatives*(*Kayaking*,1) = {*Rowing*, *Sail*, *SeaFishing*, *Canoeing*, *Boating*} which are all descendants of '*Sailing*'. As all of them have a known score we can proceed to the calculation of *TIS*(*Kayaking*). Using an optimistic policy (*e.g.* averaging the 2 highest scores) the result is 0.75, using a pessimist policy (*e.g.* averaging the 2 lowest scores) the result is 0.4, and with an average of all the scores we get 0.58. If we wanted to calculate *TIS*(*ScubaDiving*) all the leaves under '*AquaticSports*' should be considered in the WOWA calculation, because there is no scoring information in other '*Underwater*' sports.

5 Concordance and Discordance Indices for Semantic Criteria

In Sect. 2 it was shown how ELECTRE-III calculates the concordance and discordance indices in the case of numerical criteria. In this section we propose a novel way to calculate those indices when *semantic* criteria are considered. Those indices would be used to construct an outranking relation between the set A of alternatives. Each semantic criterion is defined as a pseudo-criterion, with two discriminant thresholds as well as the veto threshold. This procedure follows the same principles than the classic ELECTRE-III method. We define new concordance and discordance indices as fuzzy functions based on thresholds in the sense of [7], but in terms of the pairwise comparison of Tag Interest Scores.

First, we define how to measure the strength of the assertion aSb in terms of one semantic variable:

Definition 1. *Semantic Win Rate* $SWR_j(a,b)$: it is a numerical value in [0,1] that indicates the degree of performance of alternative a with respect to b on the semantic criterion g_j. It is based on the two sets of tags $g_j(a) = \{t_{1,a}, t_{2,a}, t_{3,a}, \ldots, t_{|g_j(a)|,a}\}$ and $g_j(b) = \{t_{1,b}, t_{2,b}, t_{3,b}, \ldots, t_{|g_{j(b)}|,b}\}$ and it is calculated as follows:

$$SWR_j(a,b) = \frac{\sum_{t_{i,a} \in g_j(a)} \sum_{t_{k,b} \in g_j(b)} f(t_{i,a}, t_{k,b})}{|g_j(a)| \cdot |g_j(b)|}, \tag{7}$$

where

$$f(x,y) = \begin{cases} 1 & \text{if } TIS(x) \geq TIS(y) \\ 0 & \text{if } TIS(x) < TIS(y) \end{cases}$$

Thus, $SWR_j(a,b)$ is the percentage of pairwise comparisons between the tags of a and b for the semantic criterion g_j for which the user has a higher (or equal) preference for the a-tag than for the b-tag. Using this value, the partial concordance and partial discordance indices are defined as follows:

Definition 2. *Partial concordance and discordance indices for semantic criteria*

$$c_j(a,b) = \begin{cases} 1 & \text{if } SWR_j(a,b) \geq \mu_j \\ 0 & \text{if } SWR_j(a,b) \leq \mu_j - p_j \\ \frac{SWR_j(a,b) - (\mu_j - p_j)}{p_j} & \text{otherwise} \end{cases} \tag{8}$$

$$d_j(a,b) = \begin{cases} 1 & \text{if } SWR_j(a,b) \leq \mu_j - v_j \\ 0 & \text{if } SWR_j(a,b) \geq \mu_j - p_j \\ \frac{(\mu_j - p_j) - SWR_j(a,b)}{v_j - p_j} & \text{otherwise} \end{cases} \tag{9}$$

As $SWR_j(a,b)$ is a percentage that represents the comparison of the performance of a over b, the thresholds are not parameterised and they have this meaning:

- μ_j replaces the q_j indifference threshold. It is the minimum value for the strength of $SWR_j(a,b)$ to consider a maximum concordance with aSb.
- p_j indicates the maximum difference between $SWR_j(a,b)$ and μ_j that still shows some preference of a with regards to b, thus still supporting the relation aSb to a certain degree.

- v_j is the veto threshold, which shows the minimum negative difference between $SWR_j(a, b)$ and μ_j that requires the full discordance with the outranking relation.

Figure 2 shows how these thresholds delimit the fuzzy functions for concordance and discordance in semantic criteria. This new partial concordance and discordance indices can be used in the standard ELECTRE-III procedure explained in Sect. 2 to calculate the overall concordance (2) and the credibility (4). After that, the standard distillation procedure is applied to obtain the partial pre-order among the alternatives.

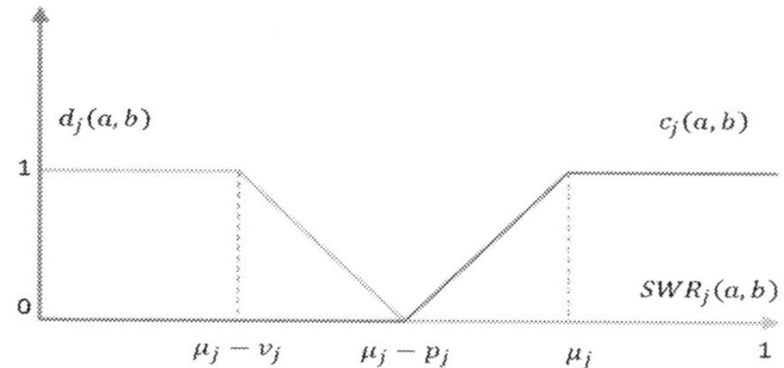

Fig. 2. Fuzzy relations for concordance and discordance in semantic criteria

6 Case Study

This new procedure to construct concordance and discordance indices for semantic criteria is illustrated in this section with a case study, in which alternatives are tourist attractions in the city of Tarragona. As shown in Table 2, there are 20 alternatives, defined on 2 criteria: a multi-valued semantic one (*Touristic Tags*) and a numerical one (*Cost*). We have used a Tourist ontology developed in the Scientific and Technological Park for Tourism and Leisure [6], in which 343 concepts are structured in a 5-level hierarchy. The identifier of each alternative shows its stronger focus: C-*Culture*, E-*Event*, S-*Sport* and L-*Leisure*. We consider the case of a very sportive tourist, who has a mild interest in events and leisure activities but is not keen on cultural activities (except *UrbanLandscape*). It is assumed that all the preference scores of the ontology leaves have been calculated using the procedure described on Sect. 4, from some basic initial information on the user's preferences.

Test 1: This test shows the influence of the discordance index in the construction of the outranking relation. ELECTRE-III has been executed with or without discordance (i.e. without veto power) in each criterion. The parameters are the following: *Cost* (num, min, $q = 0$, $p = 10$, $v = 20$) and *Touristic Tags* (semantic, max, $\mu = 0.7$, $p = 0.1$, $v = 0.4$).

In this test both criteria have the same weight. The results are displayed in Fig. 3 (a. partial pre-order with veto in both criteria, b. without veto in *Cost*, c. without veto in *Touristic Tags*, d. without any veto). In the first case we consider both the semantic information and the cost, so the best options are the cheapest sports (S5 and S3) and some cheap events and leisure activities. When the *Cost* veto is not considered the

Table 2. List of alternatives

Id	Touristic tags	Cost
C1	CultureRoutes, Cathedral, Palace, Tower	30 €
C2	UrbanLandscape, CultureRoutes, HistoricBuilding, Tower, Baroque, Castle	30 €
C3	UniqueBuilding, Ruins, HumanHeritage, CultureRoutes, HistoricBuilding	30 €
C4	WineFairs, CultureRoutes, Ruins, Amphitheater	15 €
C5	BookFairs, TraditionalCelebrations, HistoricBuilding	5 €
E1	TraditionalCelebrations, MusicFestivals, DanceFestivals, GastronomyFestivals	20 €
E2	WineFairs, MusicFestivals, ChampagneFestivals, BookFairs, DanceFestivals	20 €
E3	ChampagneFestivals, ArtsAndCraftsEvents, MusicFestivals, GastronomyFestivals	40 €
E4	BeachPicnic, DanceFestivals, BigGroupsAtmosphere, TapasCuisine, TraditionalCuisine	30 €
E5	TapasCuisine, ArtsAndCraftsEvents, BookFairs, TraditionalCelebrations, WineFairs	5 €
S1	Canoeing, Kayaking, BananaRafting, Windsurfing, WaterSkiing, Wakeboarding, ScubaDiving	80 €
S2	Snorkelling, Rappelling, ZipLine, BananaRafting, Kayaking	60 €
S3	HorseRiding, Car4 × 4, PaintBall, ShoppingArea	30 €
S4	SafariPark, HorseRiding, Car4 × 4, PaintBall	40 €
S5	Paragliding, ClimbingWall, Rappelling	10 €
L1	BeachPicnic, FamilyBeaches, Pizzeria, SafariPark	40 €
L2	TapasCuisine, ShoppingCenter, SpaResorts, Vegetarian, LocalMarket	20 €
L3	WineRoutes, TapasCuisine, WineFestivals, BookFairs	10 €
L4	Car4 × 4, PaintBall, HorseRiding, Pizzeria, SafariPark	40 €
L5	Bars, Discos, ShoppingArea, BeachPicnic, TraditionalCuisine, WineRoutes, SpaResorts	20 €

sportive activities are promoted, so S1 and S2 are able to outrank more options (like C5, C4 and E3). In the third case, in which there is no veto for *Touristic Tags*, the price takes more importance and activities like S3 and S1 go down in the ranking. Finally, when no discordance is used (fourth case), the result is very similar to the first one, because there are only two criteria. Notice that in all cases we identify some incomparability relations between activities that have good performance in one criterion but bad in the other (for instance between C5 -cheap but not in the interests of the user- and L1 –more interesting but much more expensive-). In the first positions, S3 (sport, 30 €) is better than E5 (event, 5 €) when there is no veto on cost, but the relation is reversed when there is no veto on the semantic criterion. These results show that the formulation of concordance and discordance indices for semantic data leads to plausible results when applied in the ELECTRE-III distillation procedure.

Test 2: This test studies the influence of veto power (i.e. discordance) when there is a strong difference on the criteria weights (0.9 vs 0.1). The same previous 4 cases (with

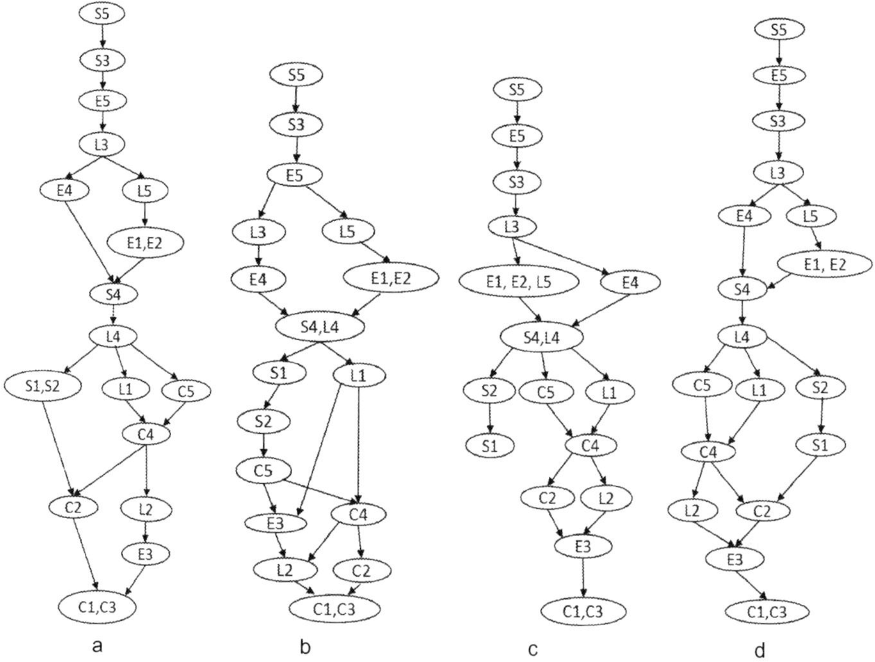

Fig. 3. Partial pre-orders obtained in test 1.

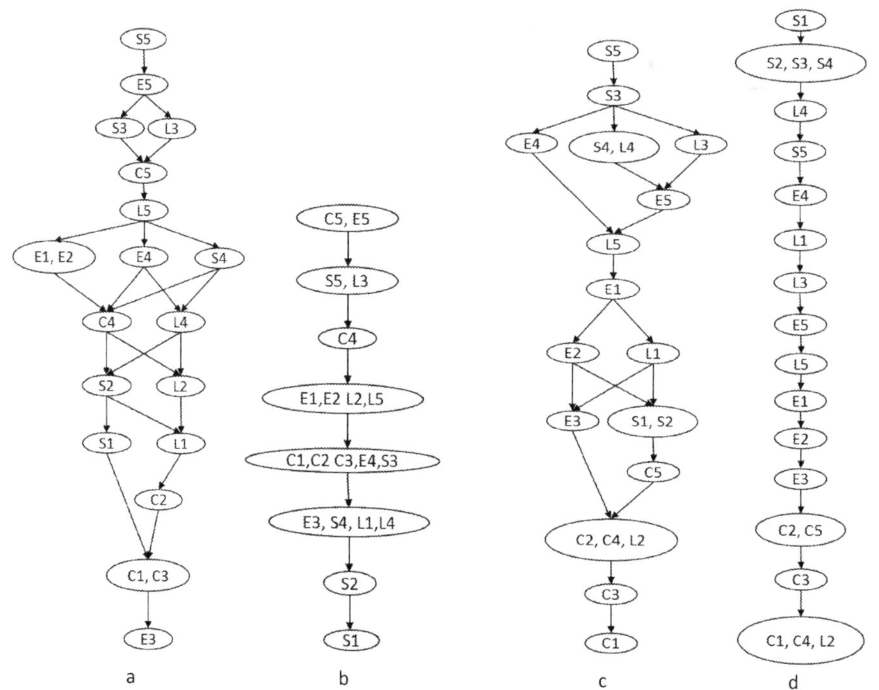

Fig. 4. Partial pre-orders obtained in test 2

and without veto) have been studied, using the same parameters for the thresholds as in Test 1. The partial pre-orders are displayed in Fig. 4.

In the two figures on the left (4a and 4b) $w_{cost} = 0.9$ and $w_{tags} = 0.1$, whereas in the two figures on the right (4c and 4d) $w_{cost} = 0.1$ and $w_{tags} = 0.9$.

When cost is more important, using or not its veto power (the discordance) leads to the same result, which is the first partial pre-order in Fig. 4 (a). The right of veto of the semantic criterion is able to place S5 (10 €) at the top, S3 (30 €) in the third position and most cultural activities in the lowest ones, despite the extremely high importance of cost. However, when there is no veto in the semantic criterion (Fig. 4b) the ranking is mainly based on the cost and touristic tags are almost neglected, due to its low weight: C5 and E5(5 €) are the first options, S5 and L3 (10 €) the second ones, etc. This test shows that the formulation of semantic discordance and concordance proposed in the paper has the expected effect in the construction of the partial pre-orders.

Similarly, when importance is given to the semantic data, the ranking with cost veto (Fig. 4c) places S5 as the best alternative, S3 is the second one and (S4, E4, L3, L4) appear in the third place, because they are quite cheap and fit with the user's preferences. However, when there is no veto on cost (Fig. 4d) the ranking depends mainly on the scores of the touristic tags, which are based on the Semantic Win Rate, because the cost has an extremely low weight.

Therefore, we can see that the veto power of the less relevant criterion is able to influence the result, both in the case on numerical data (classic procedure) and semantic values (new proposal).

7 Conclusions and Future Work

Semantic information is nowadays frequent in some datasets and requires new analysis methods. In this paper, an extension of the ELECTRE-III multi-criteria decision making method is proposed. In particular, the procedure for constructing a fuzzy outranking relation is modified with the definition of new concordance and discordance indices. Those indices are calculated with a fuzzy function that depends on three thresholds (two for concordance and one for discordance) similar to those of classic ELECTRE-III. However, they are based on Semantic Win Rate that is a new measure that permits to compare the lists of tags of a pair of alternatives using the user's preferences about the tags. The illustrative example shows that the behaviour of the method is similar for numeric and semantic data, which was the aim of the proposal. An analysis with real data is our next step, this will help us to evaluate the robustness the decision with regards to the new parameters based on the TIS scores. The influence of user's preference scores variations will be also studied.

The user's preference about different tags is easily stored inside an ontology by adding a new property to the classes, called Tag Interest Score (TIS). This score is only required for the most specific concepts of the ontology (the leaves), so the user do not need to know the structure of the ontology. As a manual initialization of big ontologies may be not feasible (*cold start* problem), the paper proposes the WOWA operator to estimate unknown scores in terms of the existing ones. It remains an open research line the definition of appropriate learning methods to help the user in the initialization and updating of the scores that are stored in the ontology.

The thorough study of other weighting policies for WOWA is our next step. Semantic similarity measures between concepts could be used in this stage.

Acknowledgements. This work has been funded by Spanish research project SHADE (TIN-2012-34369): *Semantic and Hierarchical Attributes in Decision Making*. The first author is a Martí-Franquès grant holder at the Universitat Rovira i Virgili (Tarragona, Spain). URV also supports this work through grant 2014PFR-URV-B2-60.

References

1. Ding, L., Finin, T., Joshi, A., Pan, R., Cost, R.S., Peng, Y., Reddivari, P., Doshi, V.C., Sachs, J.: Swoogle: a search and metadata engine for the semantic web. In: Proceedings of the 13th ACM Conference on Information and Knowledge Management, New York, USA, pp. 652–659 (2004)
2. Figueira, J.R., Greco, S., Roy, B., Słowiński, R.: An overview of ELECTRE methods and their recent extensions. J. Multi-Criteria Decis. Anal. **20**(1–2), 61–85 (2013)
3. Gibert, K., Valls, A., Batet, M.: Introducing semantic variables in mixed distance measures: impact on hierarchical clustering. Knowl. Inf. Syst. **40**(3), 559–593 (2014)
4. Govindan, K., Jepsen, M.B.: ELECTRE: a comprehensive literature review on methodologies and applications. Eur. J. Oper. Res. **250**, 1–29 (2015)
5. Kofler, M.J., Reinisch, C., Kastne, W.: An ontological weather representation for improving energy-efficiency in interconnected smart home systems. In: Proceedings of Applied Simulation and Modelling/Artificial Intelligence and Soft Computing (ASC 2012), Napoli, Italy, pp. 256–263 (2012)
6. Moreno, A., Valls, A., Isern, D., Marin, L., Borràs, J.: SigTur/E-destination: ontology–based personalized recommendation of tourism and leisure activities. Eng. Appl. Artif. Intell. **26** (1), 633–651 (2013)
7. Perny, P., Roy, B.: The use of fuzzy outranking relations in preference modelling. Fuzzy Sets Syst. **49**(1), 33–53 (1992)
8. Staab, S., Studer, R. (eds.): Handbook on Ontologies. Springer, New York (2013)
9. Torra, V.: The WOWA operator: a review. In: Yager, R.R., Kacprzyk, J., Beliakov, G. (eds.) Recent Developments in the Ordered Weighted Averaging Operators: Theory and Practice. STUDFUZZ, vol. 265, pp. 17–28. Springer, Heidelberg (2011)
10. Valls, A., Moreno, A., Borràs, J.: Preference representation with ontologies. In: Doumpos, M., Grigoroudis, E. (eds.) Multicriteria Decision Aid and Artificial Intelligence, pp. 77–99. Wiley, Chichester (2013)
11. Yager, R.R.: Quantifier guided aggregation using OWA operators. Int. J. Intell. Syst. **11**(1), 49–73 (1996)

Argumentation and Belief Revision

Argumentation Framework Based on Evidence Theory

Ahmed Samet[1(✉)], Badran Raddaoui[2], Tien-Tuan Dao[1], and Allel Hadjali[2]

[1] Sorbonne University, Université de technologie de Compiègne, CNRS, UMR 7338 Biomechanics and Bioengineering, Compiègne, France
{ahmed.samet,tien-tuan.dao}@utc.fr

[2] LIAS - ENSMA/University of Poitiers, Poitiers, France
{badran.raddaoui,allel.hadjali}@ensma.fr

Abstract. In many fields of automated information processing it becomes crucial to consider imprecise, uncertain or inconsistent pieces of information. Therefore, integrating uncertainty factors in argumentation theory is of paramount importance. Recently, several argumentation based approaches have emerged to model uncertain data with probabilities. In this paper, we propose a new argumentation system called *evidential argumentation framework* that takes into account imprecision and uncertainty modeled by means of evidence theory. Indeed, evidence theory brings new semantics since arguments represent expert opinions with several weighted alternatives. Then, the evidential argumentation framework is studied in the light of both Smets and Demspter-Shafer interpretations of evidence theory. For each interpretation, we generalize Dung's standard semantics with illustrative examples. We also investigate several preference criteria for pairwise comparison of extensions in order to select the ones that represent potential solutions to a given decision making problem.

Keywords: Argumentation theory · Evidence argumentation framework · Pignistic scenario graph · Belief scenario graph

1 Introduction

Argumentation has long been a major topic in Artificial Intelligence (see e.g., [1,2] and for more recent accounts e.g., [3]) that has concerned a large variety of application domains for more than a decade, like e.g., medicine [4], law [5,6], negotiation [7], decision making [8], multiagent systems [9,10], semantic web [11], and databases [12], etc. Argumentation is basically concerned with the exchange of interacting arguments. This set of arguments may come either from a dialogue between several agents but also from the available (and possibly contradictory) pieces of information at the disposal of one unique agent. Usually, the interaction between arguments takes the form of a conflict, called *attack*. Two main families of computational models for argumentation have been studied in the literature: namely, the abstract and the logic-based argumentation frameworks. Following

J.P. Carvalho et al. (Eds.): IPMU 2016, Part II, CCIS 611, pp. 253–264, 2016.
DOI: 10.1007/978-3-319-40581-0_21

the seminal work of [13], the first family is based on graph-oriented representations and focuses mainly on the interaction between arguments without taking the possible internal structure of the involved arguments into account. Different acceptability semantics for abstract argumentation frameworks have also been proposed that highlight different aspects of argumentation. Basically, each of these semantics corresponds to some properties which certify whether a set of arguments can be profitably used to support a point of view in a discussion. On the contrary, the logic-based approaches (e.g., [14–17]) exploit the logical internal structure of arguments and adopt inconsistency as a pivotal paradigm: any pair of conflicting arguments must be contradictory.

Although Dung's frameworks are widely approved tools for abstract argumentation, their abstractness make expressing notions such as support or uncertainty very difficult. So far a plethora of works have been introduced in order to model the uncertainty of an argument with a probability [18–20]. However, weighting arguments in this manner has several drawbacks, due mainly to uncertainty representation. To illustrate this point, let us consider, for example the following three doctors that prescribe medication to a patient. To cure the patient headache, the first doctor prescribes several paracetamol-family drugs (e.g. P_1 and P_2) with weights to express his/her preferences. The second doctor favours all biological medical products and therefore (s)he prescribes an andrographis drug. Finally, a third doctor is hesitating whether to give the paracetamol-based product P_1 or the andrographis one without a further complementary analysis on the patient. In this argumentation context, uncertainty is ubiquitous in all opinions. In addition, doctors are confronting arguments about the convenience and the applicability of drug alternatives (drug brands) rather a type of medication. This kind of example is difficult to handle with classical probability-based approaches.

The aim of this paper is to extend uncertainty consideration in argumentation theory. In fact, uncertainty is modelled thanks to evidence theory rather with classical probabilities to solve problems as the one described above. A new framework, called *evidential argumentation framework*, is introduced to model experts' opinion over alternatives. The latter is studied and interpreted in the light of two evidence theory interpretations: the non probabilistic model of Smets [21] and the probabilistic one of Dempster-Shafer [22,23]. Therefore, several acceptability semantics are generalized based on scenario graphs derived from the evidential argumentation framework.

The roadmap of this paper is organized as follows. The basic foundations of abstract argumentation theory are detailed in Sect. 2. In Sect. 3, the evidential argumentation framework is motivated and introduced. Then, our approach is studied and interpreted in the light of two evidence theory interpretations. Moreover, acceptability semantics are generalized for evidential argumentation framework. We also introduce various preference relations at the semantics level in order to determine what are desirable outcomes of the argumentation framework. Finally, we conclude and sketch potential issues for the future work.

2 Abstract Argumentation Framework: Brief Overview

In this section, we briefly outline the notion of abstract argumentation framework and various semantics studied in the literature.

An abstract argumentation framework [13], AF for short, is a pair $\langle \mathcal{A}, \mathcal{R} \rangle$, where $\mathcal{A}$ is a finite set, whose elements are called *arguments*, and $\mathcal{R} \subseteq \mathcal{A} \times \mathcal{A}$ is a binary relation over $\mathcal{A}$, whose elements are referred to as attacks. An argument is an abstract entity whose role is entirely determined by its relationships with other arguments.

An AF can simply be represented as a directed graph, called *attack graph*, where nodes are the arguments and edges represent the attack relation. Throughout the paper examples are using this graph representation.

Given two arguments A and B, we say that A *attacks* B iff there is $(A, B) \in \mathcal{R}$. Moreover, a set $S \subseteq \mathcal{A}$ *attacks* an argument $B \in \mathcal{A}$ iff there is $A \in S$ s.t. A attacks B. A set $S \subseteq \mathcal{A}$ of arguments is said to be *conflict-free* if there are no arguments $A, B \in S$ s.t. A attacks B. An argument A is *defended* by a set $S \subseteq \mathcal{A}$ iff $\forall B \in \mathcal{A}$ s.t. B attacks A, there is $C \in S$ s.t. C attacks B.

Using the notions of conflict-freeness and defense, we can define a number of argumentation semantics, each embodying a particular rationality criterion, in order to identify reasonable sets of arguments, called *extensions*.

Definition 1 (Acceptability semantics). *Given an argumentation framework $\mathcal{F} = \langle \mathcal{A}, \mathcal{R} \rangle$. A set $S \subset \mathcal{A}$ of arguments is said to be:*

- admissible *iff S is conflict-free and all its arguments are defended by S*
- *a* stable *extension iff S is conflict-free and S attacks each argument in $\mathcal{A} \setminus S$*
- *a* complete *extension iff S is admissible and S contains all and only the arguments it defends*
- *a* grounded *extension iff S is a minimal (w.r.t. set inclusion) complete set of arguments*
- *a* preferred *extension iff S is a maximal (w.r.t. set inclusion) admissible set of arguments*
- *an* ideal *extension iff S is admissible and S is contained in every preferred set of arguments.*

Example 1. Consider the AF $\mathcal{F} = \langle \mathcal{A}, \mathcal{R} \rangle$ such that $\mathcal{A} = \{A, B, C, D, E\}$ and $\mathcal{R} = \{(A, B), (C, B), (C, D), (D, C), (D, E)\}$. The graph representation of $\mathcal{F}$ is indicated on Fig. 1. This argumentation framework has two preferred extensions: $\mathcal{E}_1 = \{A, C, E\}$ and $\mathcal{E}_2 = \{A, D\}$; these are also the unique stable extensions. Moreover, $\mathcal{F}$ possesses a unique ideal extension $\{A\}$.

3 Evidence Theory Based Argumentation Framework

One of the abstract argumentation frameworks shortcomings is the insufficient handling of the levels of uncertainty, an aspect which typically occurs in domains, where diverging opinions are raised. In this section, we describe a new framework

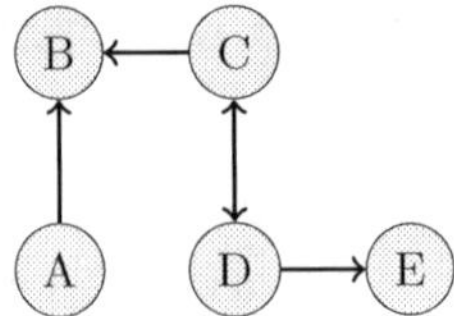

Fig. 1. Attack graph of the AF from Example 1

of abstract argumentation based on evidence theory in which uncertainty has two dimensions. In other words, we intend to model an argumentation system that handles imprecise opinions as arguments. Indeed, each argument (i.e., opinion) is described over the powerset of alternatives (hypotheses). In the sequel, we denote by Θ_A the *frame of discernment* of the alternatives of the argument A.

Definition 2 (Evidential Argumentation Framework). *An evidential argumentation framework is a tuple* $\mathcal{V} = \langle \mathcal{A}, \mathcal{R}, m \rangle$ *where* $\langle \mathcal{A}, \mathcal{R} \rangle$ *is an AF,* $m = \{m_A, A \in \mathcal{A}\}$*, and* m_A *(basic belief assignment (bba)) is a mapping from elements of the powerset* 2^{Θ_A} *onto* $[0, 1]$ *such that:*

$$\begin{cases} m_A(\emptyset) = 0 \\ \sum\limits_{a \in \Theta_A} m_A(a) = 1 \end{cases}$$

In Definition 2, a corresponds to a hypothesis of the argument A and is called an *alternative* of A. Clearly, an evidential argumentation framework differs from classical argumentation models since it deals with arguments' alternatives rather with arguments. In this study, the arguments are supposed to be independent.

Example 2. Let us consider the evidential argumentation graph shown in Fig. 2. Five arguments are considered $\mathcal{A} = \{A, B, C, D, E\}$. Each argument highlights the diagnostic of a doctor and the intended prescription. Each prescription contains either one of two alternatives from a drug family. In other words, an argument represents a drug type family having several brands. A drug brand of a single family has its own properties, effects and prescriptions. Therefore, a mass is given to each drug brand depending on the treated patient. For example, $A =$ "*Patient has hypertension so prescribe drug family A with a higher preference to* A_1 *of all other same family products*". In addition, a doctor can hesitate between same drug family alternatives. For example argument $E =$ "*Patient has hypertension so prescribe one of the medication of the drug family E without any preference to any of them.*". Here, we assume that C and D attack each other because we should only give one treatment and so giving one precludes the other, and we assume that A and C attack B because they provide a counterargument to all A family medications.

Several interpretations exist for evidence theory such as [21,22,24]. In the remainder, we build our contributions following the Transferable Belief Model

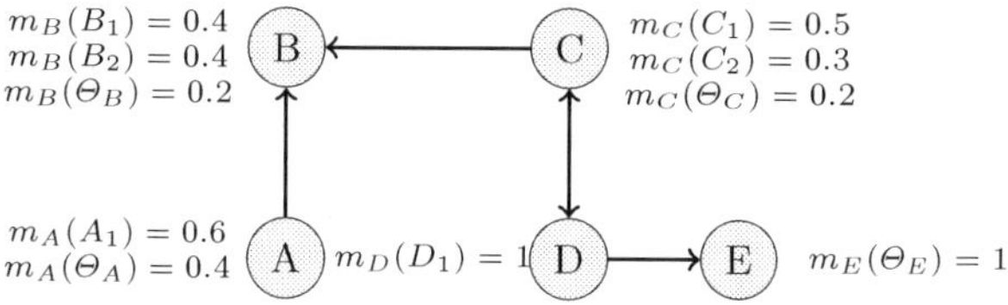

Fig. 2. Example of an evidential argumentation graph

(TBM), that was introduced by Smets [21] and the probabilistic interpretation of Dempster-Shafer [22,23]. The TBM model is a non-probabilistic interpretation of the theory, that aims at representing quantified beliefs based on two levels: *(i)* a credal level where beliefs are entertained and quantified by belief functions; *(ii)* a pignistic level where beliefs can be used to make decisions and are quantified by probability functions. In the following subsection, we provide the argumentation framework based on the TBM interpretation.

3.1 Pignistic Argumentation Framework

In the following, we analyse the evidential argumentation framework using Smets TBM interpretation. Even if TBM is not limited to normalized bbas (i.e., bba with a null mass over the empty-set), we restrict our study to bbas without conflict.

Definition 3. *Let $\mathcal{V} = \langle \mathcal{A}, \mathcal{R}, m \rangle$ be an evidential argumentation framework. A pignistic scenario graph is a tuple $G_p = \langle \mathcal{A}, \mathcal{R}, \mathcal{P} \rangle$ such that $\mathcal{P} = \{P_A, A \in \mathcal{A}\}$ where P_A is a pignistic probability defined as:*

$$P_A(a) = \sum_{x \subseteq \Theta_A} \frac{|a \cap x|}{|x|} \times m_A(x) \qquad \forall a \in \Theta_A$$

where m_A is the bba of the argument $A \in \mathcal{A}$ and $|.|$ is the cardinality operator.

Obviously, the value assigned by $P_A(a)$ represents the probability that the alternative a actually occurs. This probability also considers the absolute ignorance (i.e., Θ_A) when probabilities are built. Comparatively to the state-of-the-art works, the pignistic scenario graph analyses the attacks of the hypotheses of arguments rather than arguments. This could be seen as an extension of previous works [13]. Indeed, to recover the method of [13], a *certain bba*[1] must be constructed over single hypothesis. In addition, it is important to note that the number of pignistic scenario graphs N that could be retrieved from an evidential argumentation framework is computed as $N = \prod |\Theta_A|$, $A \in \mathcal{A}$. Indeed, the number of scenarios depends on the arguments' frame of discernment size. However, a simple heuristic could be applied to drop alternatives with a low

[1] A certain bba expresses the total certainty. It is defined as follows: $m(A) = 1$ and $m(B) = 0$ for all $B \neq A$ and $B \subseteq \Theta$, where A is a singleton event of Θ.

pignistic probability. This approach consists of dropping alternatives with a pignistic probability lower than 0.5. Thus, the number of retrieved scenario graphs is drastically reduced.

Example 3. Fig. 3 shows an example of pignistic scenario graph built from the evidential argumentation framework presented in Example 2.

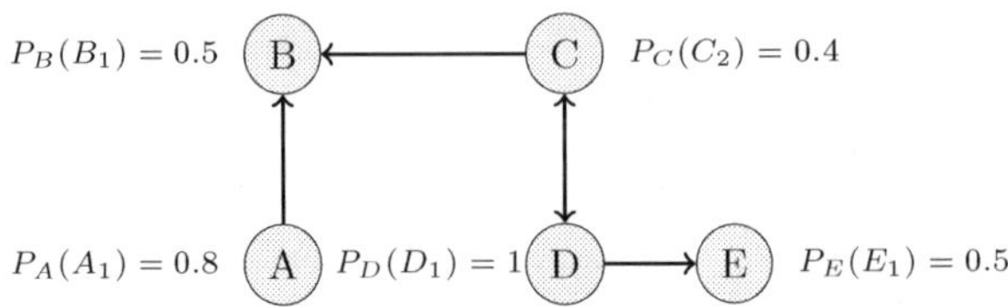

Fig. 3. A pignistic scenario graph from Example 2

Note that some of the provided constraints of [25] in probabilistic argumentation framework could be recovered in our pignistic argumentation framework. The pignistic probability function may take different aspects of the structure of the argument graph into account. More formally,

- P is **coherent** w.r.t. G_p if for every $a \in \Theta_A$, $\sum_{a \subseteq \Theta_A} P_A(a) = 1$.
- P is **semi-founded** w.r.t. G_p if $P_A(a) \geq 0.5$ for every unattacked $a \in \Theta_A$.
- P is **founded** w.r.t. G_p if $P_A(a) = 1$ for every unattacked $a \in \Theta_A$.
- P is **semi-optimistic** w.r.t. G_p if $P_A(a) \geq 1 - \sum_{(a,b) \in \mathcal{R}} P_B(b)$ for every $a \in \Theta_A$ that has at least one attacker.
- P is **optimistic** w.r.t. G_p if $P_A(a) \geq 1 - \sum_{(a,b) \in \mathcal{R}} P_B(b)$ for every $a \in \Theta_A$.
- P is **justifiable** w.r.t. G_p if P is coherent and optimistic.
- P is **ternary** w.r.t. G_p if $P_A(a) \in \{0, 0.5, 1\}$ for every $a \in \Theta_A$.
- P is **rational** w.r.t. G_p if for every $a \in \Theta_A$, $b \in \Theta_B$, if $(A, B) \in \mathcal{R}$ then $P_A(a) > 0.5$ implies $P_B(b) \leq 0.5$.
- P is **neutral** w.r.t. G_p if $P_A(a) = 0.5$ for every $a \in \Theta_A$.
- P is **involutary** w.r.t. G_p if for every $a \in \Theta_A$, $b \in \Theta_B$, if $(A, B) \in \mathcal{R}$ then $P_A(a) = 1 - P_B(b)$.
- P is **maximal** w.r.t. G_p if $P_A(a) = 1$ for every $a \in \Theta_A$.
- P is **minimal** w.r.t. G_p if $P_A(a) = 0$ for every $a \in \Theta_A$.

Definition 4. *Let $G_p = \langle \mathcal{A}, \mathcal{R}, \mathcal{P} \rangle$ be a pignistic scenario graph such that $a \in \Theta_A$ and $b \in \Theta_B$. Then, a is stronger than b, denoted by $a \succ_s b$, if and only if*

$$P_A(a) > P_B(b)$$

In order to characterize a particular argumentation semantics for a pignistic argumentation framework, we define a version of the semantics introduced in Definition 1.

Definition 5. *Let $G_p = \langle \mathcal{A}, \mathcal{R}, \mathcal{P} \rangle$ be a pignistic scenario graph s.t. $a \in \Theta_A$, $b \in \Theta_B$ and $S \subseteq \Theta_{\mathcal{A}}$. We say that:*

- *a* defeats *b if and only if a attacks b, b does not attack a, and $a \succ_s b$.*
- *a is* p-acceptable *with respect to S, if $\forall b$ which defeats a there exists $c \in S$ such that c defeats b.*
- *a set S of arguments is p-conflict-free if there are no arguments $a, b \in S$ such that a attacks b.*
- *S is* p-admissible *iff S is conflict-free and all its arguments are acceptable w.r.t. S.*
- *S is a* p-stable *extension iff S is conflict-free and S attacks each argument in $\Theta_{\mathcal{A}} \setminus S$.*
- *S is a* p-complete *extension iff S is admissible and S contains all and only the arguments it defends.*
- *S is a* p-grounded *extension iff S is a minimal (w.r.t. set inclusion) complete set of arguments.*
- *S is a* p-preferred *extension iff S is a maximal (w.r.t. set inclusion) admissible set of arguments.*
- *S is a* p-ideal *extension iff S is admissible and S is contained in every preferred set of arguments.*

Let $\mathtt{Ext}_x(G_p)$ denote the set of extensions of the pignistic scenario graph G_p under semantics x where $x \in \{a, s, c, g, p, i\}$ and a (resp. s, c, g, p, i) stands for p-admissible (resp. p-stable, p-complete, p-grounded, p-preferred and p-ideal). When the semantics are not important, or when it is clear from the context to which semantics we refer to, we use the notation $\mathtt{Ext}(\mathcal{F})$ for short.

Example 4. Let us consider the pignistic scenario graph depicted in Fig. 3. We have two p-preferred extensions $\mathcal{E}_1 = \{A_1, C_2, E_1\}$ and $\mathcal{E}_2 = \{A_1, D_1\}$.

By applying a basic argumentation semantics to a pignistic scenario graph, one can infer different extensions which represent potential solutions to a given decision making problem. Since an extension has different arguments that argue for a particular decision, some criteria for selecting a suitable decision are worth defining. In other words, it is desirable to compare extensions based on the arguments which support a decision with level of certainty. There exist several different approaches to induce a preference relation over extensions. The first comparison criterion is based on the cardinality of the set of arguments.

Definition 6. *Let $G_p = \langle \mathcal{A}, \mathcal{R}, P \rangle$ be a pignistic scenario graph and $\mathcal{E}_1, \mathcal{E}_2 \in \mathtt{Ext}(G_p)$. Then, $\mathcal{E}_1$ is cardinality-preferred to $\mathcal{E}_2$, denoted by $\mathcal{E}_1 \succ_c \mathcal{E}_2$, iff $|\mathcal{E}_1| > |\mathcal{E}_2|$.*

In certain applications, counting is not the best method of defining an order between extensions. Therefore, a more cautious preference relation can be defined based on probability of the arguments induced in a given extension. For the next preference relation, we need to provide a new metric. To do so, let $\mathcal{E} = \{a_1, \ldots, a_n\} \in \texttt{Ext}(G_p)$ be a given extension, the weight of $\mathcal{E}$ is $W(\mathcal{E}) = \sum_{i=1}^{n} P_A(a_i)$. Now, let us define an ordering for pairwise comparison of extensions as follows.

Definition 7. *Let $G_p = \langle \mathcal{A}, \mathcal{R}, \mathcal{P} \rangle$ be a pignistic scenario graph. Let $\mathcal{E}_1, \mathcal{E}_2 \in \texttt{Ext}(G_p)$. Then, $\mathcal{E}_1$ is probability-preferred to $\mathcal{E}_2$, denoted by $\mathcal{E}_1 \succ_p \mathcal{E}_2$, iff $W(\mathcal{E}_1) > W(\mathcal{E}_2)$.*

Example 5. Let us consider again the pignistic scenario graph depicted in Fig. 3. We have $W(\mathcal{E}_1) = 1.7$ and $W(\mathcal{E}_2) = 1.8$, then $\mathcal{E}_2 \succ_p \mathcal{E}_1$.

Notice that several extensions can be obtained through each pignistic scenario graph. Now, in order to select the outcome of the original argumentation framework, it will be of interest to compare different extensions from the different pignistic scenario graphs. To do this, let us consider the following definition.

Definition 8. *Let $\mathcal{V} = \langle \mathcal{A}, \mathcal{R}, m \rangle$ be an evidential argumentation framework and $G_p^1, \ldots, G_p^n$ the set of pignistic scenario graphs obtained from $\mathcal{V}$. Let $\mathcal{W}(G_p^i) = \sum_{j=1}^{m} W(\mathcal{E}_j)$ such that $\{\mathcal{E}_1, \ldots, \mathcal{E}_m\}$ are the extensions over the pignistic scenario graph G_p^i. Then, the outcome of $\mathcal{V}$, denoted by $\texttt{Ext}(\mathcal{V})$, is defined as:*

$$\texttt{Ext}(\mathcal{V}) = \{\mathcal{E} \mid \mathcal{E} \in \texttt{Ext}(G_p),\ G_p = argmax_{i=1}^{n} \mathcal{W}(G_p^i)\}$$

The pignistic scenario graph represents the TBM-based approach to handle the evidential argumentation framework. Another interpretation can be obtained based on Dempster-Shafer works. In fact, a lower and an upper bound on the degree of belief of a single argument's alternative could be assigned.

3.2 Belief Argumentation Framework

In the following, we intend to analyse the evidential argumentation framework in the light of Dempster-Shafer interpretation. In this context, each alternative's pertinence is bounded by an upper and a lower bound.

Definition 9. *Let $\mathcal{V} = \langle \mathcal{A}, \mathcal{R}, m \rangle$ be an evidential argumentation framework. A belief scenario graph is a tuple $G_B = \langle \mathcal{A}, \mathcal{R}, \mathcal{B}el, \mathcal{P}l \rangle$ such that for all $A \in \mathcal{A}$, $Bel_A : \Theta_A \to [0,1]$ and $Pl_A : \Theta_A \to [0,1]$ are, respectively, the belief and the plausibility functions over the hypotheses of A where:*

$$Bel_A(a) = \sum_{\emptyset \neq x \subseteq a} m_A(x) \tag{1}$$

$$Pl_A(a) = \sum_{a \cap x \neq \emptyset} m_A(x) \tag{2}$$

Example 6. Fig. 4 shows an example of a belief scenario graph built from the evidential argumentation framework of Example 2. Each argument is labelled by a belief and a plausibility functions of a single alternative.

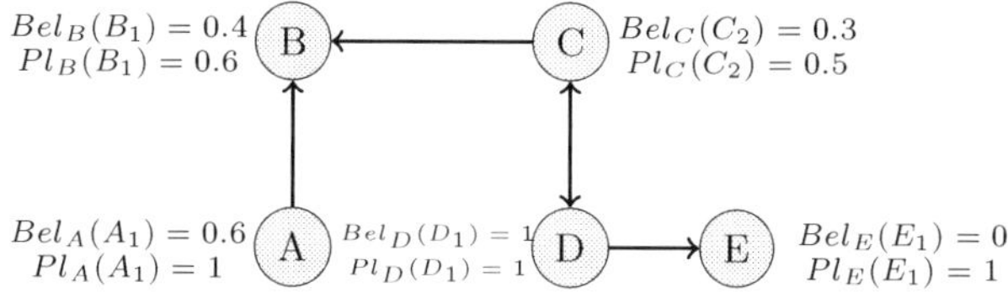

Fig. 4. Belief scenario attack graph built from Example 2

Now, we consider some constraints on the belief scenario graph which may take different aspects of the structure of the argument graph into account. For two arguments $a \in \Theta_A$, $b \in \Theta_B$ we have:

- An argument a is **believable** over b, if $Bel_A(a) \geq Bel_B(b)$ for every, such that $(a, b) \in \mathcal{R}$.
- Bel (resp. Pl) is **semi-founded**, if $Bel_A(a) \geq 0.5$ (resp. $Pl_A(a) \geq 0.5$) for every unattacked a.
- Bel (resp. Pl) is **founded**, if $Bel_A(a) = Pl_B(a) = 1$ for every unattacked a.
- An argument a is **plausible** over b, if $Pl_A(a) \geq Pl_B(b)$ for every such as $(a, b) \in \mathcal{R}$.
- An argument a is **preferred** over b, if $Bel_A(a) \geq Pl_B(b)$ for every such as $(a, b) \in \mathcal{R}$.
- An argument a is **weak** over b, if $Pl_A(a) \leq Bel_B(b)$ for every such as $(a, b) \in \mathcal{R}$.
- Bel_A (resp. Pl) is **maximal** w.r.t. G_B if $Bel_A(a) = 1$ (resp. $Pl_A(a) = 1$).
- Bel_A (resp. Pl)is **minimal** w.r.t. G_B if $Pl_A(a) = 0$ (resp. $Pl_A(a) = 0$).

Proposition 1. *If a is preferred, then a is believable and plausible.*

Definition 10. *Let $G_B = \langle \mathcal{A}, \mathcal{R}, \mathcal{B}el, \mathcal{P}l \rangle$ be a belief scenario graph and $a \in \Theta_A$. The strength of a is defined as $(Bel_A(a), Pl_A(a))$.*

The use of a dual value for dealing with the strength of an argument, as it is done in Definition 10, was explored in the context of possibilistic theory in [26]. The strength of an argument allows us to compare pairs of arguments. Informally, an argument is all the better as it uses more certain knowledge and refers to an important goal. This can be formally captured by a Pareto-based comparison criterion[2].

Definition 11. *Let $G_B = \langle \mathcal{A}, \mathcal{R}, \mathcal{B}el, \mathcal{P}l \rangle$ be a belief scenario graph and $a \in \Theta_A$ and $b \in \Theta_B$. Then, a is stronger than b, denoted $a \succ_b b$, if and only if:*

$$(Bel_A(a), Pl_A(a)) \geq_{pareto} (Bel_B(b), Pl_B(b))$$

[2] Let $x_1, x_2, x_1^{'}, x_2^{'}$ be four alternatives. Then $(x_1, x_2) \geq_{pareto} (x_1^{'}, x_2^{'})$ iff $\forall i \in [1, 2]$, $x_i \geq x_i^{'}$ and $\exists\, j$, such that $x_j > x_j^{'}$.

Definition 12. *Let $G_B = \langle \mathcal{A}, \mathcal{R}, \mathcal{B}el, \mathcal{P}l\rangle$ be a belief scenario graph s.t. $a \in \Theta_A$, $b \in \Theta_B$ and $S \subseteq \Theta_{\mathcal{A}}$. We say that:*

- *a* defeats *b if and only if a attacks b, b does not attack a, and $a \succ_b b$.*
- *a is* b-acceptable *with respect to S, if $\forall b$ which defeats a there exists $c \in S$ such that c defeats b.*
- *a set S of arguments is b-conflict-free if there are no arguments $a, b \in S$ such that a attacks b.*
- *S is* b-admissible *iff S is conflict-free and all its arguments are acceptable w.r.t. S.*
- *S is a* b-stable *extension iff S is conflict-free and S attacks each argument in $\Theta_{\mathcal{A}} \setminus S$.*
- *S is a* b-complete *extension iff S is admissible and S contains all and only the arguments it defends.*
- *S is a* b-grounded *extension iff S is a minimal (w.r.t. set inclusion) complete set of arguments.*
- *S is a* b-preferred *extension iff S is a maximal (w.r.t. set inclusion) admissible set of arguments.*
- *S is a* b-ideal *extension iff S is admissible and S is contained in every preferred set of arguments.*

Proposition 2. *Let $G_B = \langle \mathcal{A}, \mathcal{R}, \mathcal{B}el, \mathcal{P}l\rangle$ such that $\mathcal{F} = \langle \mathcal{A}, \mathcal{R}\rangle$ be a belief scenario graph and $\mathcal{E} \in \mathtt{Ext}(G_B)$. Then, $\succ_b$ is a partial order relation over arguments in $\mathcal{E}$.*

Notice that the condition in Definition 11 follows the principle of Pareto optimality according to which an argument is preferred if it is better or equal to another in all attributes and strictly better in at least one attribute. The set of best arguments is represented by the Pareto frontier which contains arguments which are not dominated by any other arguments. A way for computing the Pareto frontier is by means of the skyline operator [27]. It is important to observe that the Pareto relation can be used for defining the acceptability of arguments. This means that a belief argumentation framework can be instantiated as $G_B = \langle \mathcal{F}, Bel, Pl, \succ_b\rangle$. In this case, any basic argumentation semantics applying to G_B could use $>_{pareto}$ for defining the acceptability of arguments from $\mathcal{A}$.

Now in order to compare extensions, of a given belief scenario graph, we can consider the following ordering criteria.

Definition 13. *Let $G_B = \langle \mathcal{A}, \mathcal{R}, \mathcal{B}el, \mathcal{P}l\rangle$ and $\mathcal{E}_1, \mathcal{E}_2 \in \mathtt{Ext}(G_B)$, Then:*

1. *$\mathcal{E}_1 \succ_{B1} \mathcal{E}_2$ if $|\mathcal{E}_1| > |\mathcal{E}_2|$.*
2. *$\mathcal{E}_1 \succ_{B2} \mathcal{E}_2$ if $\forall a \in \mathcal{E}_1, \forall b \in \mathcal{E}2$, $a \succ_b b$.*
3. *$\mathcal{E}_1 \succ_{B3} \mathcal{E}_2$ if the number of arguments in $\mathcal{E}_1$ non attacked by $\mathcal{E}_2$ is greater than the number of arguments in $\mathcal{E}_2$ non attacked by $\mathcal{E}_1$.*

Note that the first relation is a basic ordering based on the size of extensions. The last two ones give a more fine-grained ordering since they are based on the degree of certainty of arguments ($\succ_{B2}$) and the number of non-attacked arguments ($\succ_{B3}$).

Example 7. Consider the belief scenario graph from Example 6. Then, $\texttt{Ext}_a(G_B) = \{\mathcal{E}_1, \mathcal{E}_2, \mathcal{E}_3\}$ where $\mathcal{E}_1 = \{A_1\}$ and $\mathcal{E}_2 = \{D_1\}$, $\mathcal{E}_3 = \{A_1, D_1\}$. Therefore, we have $\mathcal{E}_3 \succ_{B1} \mathcal{E}_1$, $\mathcal{E}_3 \succ_{B2} \mathcal{E}_2$, and $\mathcal{E}_3 \succ_{B3} \mathcal{E}_2$.

Last, Definition 8 can be naturally extended in order to compute the outcome of the original argumentation framework. This can be done by considering all the belief scenario graphs obtained from the original evidential argumentation framework.

4 Conclusion

In this paper, we have presented a new argumentation framework that handles uncertainty based on evidence theory. The evidential argumentation framework allows to model arguments expressed as opinions and preferences over several alternatives. From this argumentation framework, two families of scenario graphs are distinguished. Each one relies on a specific interpretation of the evidence theory. Moreover, new acceptability semantics are provided on the pignistic and the belief scenario graphs to select acceptable arguments. We have also introduced several criteria for pairwise comparison of extensions and a method for selecting only the best extensions given the winners of pairwise duels.

There is still work needed on the topic. First to propose other criteria to compare and rationalize extensions and to explore the notion of skyline in argumentation theory. Second, we plan to study aggregation methods for evidential abstract argumentation by taking as input a profile of evidential argumentation frameworks, and give as result an argumentation framework that represents the beliefs of the group. Finally, an ambitious research agenda would be to study the computational complexity of our framework and practical algorithms to compute extensions.

References

1. Bench-Capon, T.J.M., Dunne, P.E.: Argumentation in artificial intelligence. Artif. Intell. (Spec. Issue Argumentation) **171**(10–15), 619–641 (2007)
2. Rahwan, I., Simari, G.R. (eds.): Argumentation in Artificial Intelligence. Springer, New York (2009)
3. Verheij, B., Szeider, S., Woltran, S. (eds.): Computational Models of Argument - Proceedings of COMMA 2012, Vienna, Austria, 10–12 September. Frontiers in Artificial Intelligence and Applications, vol. 245. IOS Press (2012)
4. Das, S.K., Fox, J., Krause, P.: A unified framework for hypothetical and practical reasoning (1): theoretical foundations. In: Proceedings of the International Conference on Formal and Applied Practical Reasoning (FAPR), pp. 58–72 (1996)
5. Prakken, H.: An argumentation framework in default logic. Ann. Math. Artif. Intell. **9**(1–2), 93–132 (1993)
6. Prakken, H., Sartor, G.: A dialectical model of assessing conflicting arguments in legal reasoning. Artif. Intell. Law **4**(3–4), 331–368 (1996)
7. Parsons, S., Sierra, C., Jennings, N.R.: Agents that reason and negotiate by arguing. J. Logic Comput. **8**(3), 261–292 (1998)

8. Ferguson, G., Allen, J.F., Miller, B.W.: Trains-95: towards a mixed-initiative planning assistant. In: Proceedings of the Third International Conference on Artificial Intelligence Planning Systems (AIPS 1996), pp. 70–77 (1996)
9. Parsons, S., Wooldridge, M., Amgoud, L.: Properties and complexity of some formal inter-agent dialogues. J. Logic Comput. **13**(3), 347–376 (2003)
10. McBurney, P., Parsons, S., Rahwan, I.: Argumentation in Multi-Agent Systems. In: 6th International Workshop, Valencia (2012)
11. Schneider, J., Groza, T., Passant, A.: A review of argumentation for the social semantic web. Seman. Web **4**(2), 159–218 (2013)
12. Cabrio, E., Cojan, J., Villata, S., Gandon, F.: Argumentation-based inconsistencies detection for question-answering over dbpedia. In: International Semantic Web Conference (2013)
13. Dung, P.M.: On the acceptability of arguments and its fundamental role in non-monotonic reasoning, logic programming and n-person games. Artif. Intell. **77**(2), 321–358 (1995)
14. Besnard, P., Hunter, A.: A logic-based theory of deductive arguments. Artif. Intell. **128**(1–2), 203–235 (2001)
15. García, A.J., Simari, G.R.: Defeasible logic programming: an argumentative approach. Theor. Pract. Logic Program. **4**(1–2), 95–138 (2004)
16. Besnard, P., Grégoire, É., Piette, C., Raddaoui, B.: Mus-based generation of arguments and counter-arguments. In: Proceedings of the 11th IEEE International Conference on Information Reuse and Integration (IRI 2010), pp. 239–244 (2010)
17. Besnard, P., Grégoire, É., Raddaoui, B.: A conditional logic-based argumentation framework. In: Liu, W., Subrahmanian, V.S., Wijsen, J. (eds.) SUM 2013. LNCS, vol. 8078, pp. 44–56. Springer, Heidelberg (2013)
18. Hunter, A.: Probabilistic qualification of attack in abstract argumentation. Int. J. Approximate Reasoning **55**(2), 607–638 (2014)
19. Hunter, A.: A probabilistic approach to modelling uncertain logical arguments. Int. J. Approximate Reasoning **54**(1), 47–81 (2013)
20. Hadoux, E., Beynier, A., Maudet, N., Weng, P., Hunter, A.: Optimization of probabilistic argumentation with markov decision models. In: Twenty-Fourth International Joint Conference on Artificial Intelligence, pp. 2004–2010 (2015)
21. Smets, P., Kennes, R.: The transferable belief model. Artif. Intell. **66**(2), 191–234 (1994)
22. Dempster, A.: Upper and lower probabilities induced by multivalued mapping. AMS-38 (1967)
23. Shafer, G.: A Mathematical Theory of Evidence. Princeton University Press, Princeton (1976)
24. Gärdenfors, P.: Probabilistic reasoning and evidentiary value. In: Evidentiary Value: Philosophical, Judicial, and Psychological Aspects of a Theory: Essays Dedicated to Sören Halldén on His Sixtieth Birthday. C.W.K. Gleerups (1983)
25. Hunter, A., Thimm, M.: Probabilistic argumentation with incomplete information. In: European Conference on Artificial Intelligence, pp. 1033–1034 (2014)
26. Amgoud, L., Prade, H.: Using arguments for making and explaining decisions. Artif. Intell. **173**(3–4), 413–436 (2009)
27. Börzsönyi, S., Kossmann, D., Stocker, K.: The skyline operator. In: Proceedings of the 17th International Conference on Data Engineering, Heidelberg, Germany, 2–6 April 2001, pp. 421–430 (2001)

Constrained Value-Based Argumentation Framework

Karima Sedki[1,2(✉)] and Safa Yahi[3]

[1] LIMICS (INSERM UMRS 1142), Université Paris 13, Sorbonne Paris Cité, 93017 Bobigny, France
karima.sedki@univ-paris13.fr

[2] UPMC Université Paris 6, Sorbonne Universités, Paris, France

[3] Domaine Universitaire de Saint Jérôme, Avenue Escadrille Normandie Niemen, 13397 Marseille Cedex 20, France

Abstract. Value-based argumentation framework (*VAF*) is an extension of Dung argumentation framework where arguments promote specific values. In *VAF*, an argument a defeats b only if the value promoted by b is not preferred than the value promoted by a according to some total ordering on values given by a specific audience. However, despite the interesting idea of considering the preference relation between arguments' values, *VAF* does not offer a way to express further requirements, like "no arguments promoting *expensive* value" or "if we accept arguments promoting *expensive* value, then we accept arguments promoting *healthy* value". This paper extends *VAF* by incorporating some constraints, expressed as propositional formulas on either the arguments' values or on the arguments. We propose two inference relations for defining some acceptability semantics in such constrained value-based argumentation framework (*CVAF*). The first inference relation is more prudent than the second one since it derives less arguments.

1 Introduction

Dung argumentation framework (AF) has been extended by considering different aspects: preferences [1,9,11], values [2], weights for attacks [4,8], etc. Extensions of AF where every argument promotes a specific value and audiences constitute a preference relation between the arguments' values are called value-based argumentation framework (*VAF*). In *VAF*, an argument a successfully attacks (defeats) b according to a given audience only if the value promoted by b is not preferred than the value of a according to that audience. Despite the interesting idea of considering preferences between the values of arguments, *VAF* does not offer a way to express further requirements. For example, assume that a group of friends want to choose a good restaurant. Using *VAF*, the friends can give their preferences like "restaurants proposing gourmet recipes are preferred to those proposing healthy recipes", but they cannot express further requirements such as "no expensive restaurant" or "if we accept expensive restaurant, then we accept restaurant proposing *healthy* recipes".

J.P. Carvalho et al. (Eds.): IPMU 2016, Part II, CCIS 611, pp. 265–278, 2016.
DOI: 10.1007/978-3-319-40581-0_22

A quite natural approach for handling constraints in argumentation framework has been proposed in [5]. But to the best of our knowledge, there is no work incorporating constraints in *VAF*. So, our proposition consists in extending *VAF* by incorporating additional information, expressed as propositional formulas on either the arguments' values or on the arguments. Such information represent constraints about the set of arguments to be admissible. Considering constraints is interesting and can concern many real applications. For example, in medical decision making where doctors have to take a decision concerning a given patient (e.g., what is the appropriate drug to prescribe?), there is not only need to consider their preferences about a set of appropriate treatments such as "low side-effects treatments are preferred to high side-effects treatments" but additional constraints are also needed such as "reimbursable by social security treatments", "if hospitalization is required then we do not want lengthy duration treatments".

Our aim consists to point out some semantics and evaluate the set of arguments on the basis on the preferences and the additional constraints. Thus, we propose two inference relations. The first inference relation aims to handle the preferences as in *VAF* and the constraints as in constrained argumentation framework (*CAF* for short) presented in [5]. More precisely, for handling preferences, the process considers Dung *AF* with removing unsuccessful attacks on the basis of the preferences represented in a specific audience. For handling the constraints, the process corresponds to consider Dung *AF* with the condition of satisfying the constraints. Namely, a given set of arguments is admissible if and only if it is admissible for *VAF*, admissible for Dung *AF* and satisfies the constraints. In the second inference relation, a given set of arguments is admissible if and only if it is admissible for *VAF* and satisfies the constraints. We show that the second inference relation is less prudent than the first one since it allows to derive more admissible sets of arguments.

The remainder of the paper is organized as follows: Sect. 2 recalls some basics on Dung *AF* and *VAF*. We present *CVAF* in Sect. 3. Section 4 concludes the paper.

2 Argumentation Framework: Preliminaries

This section briefly recalls Dung's AF [7] and VAF [2].

2.1 Dung's Argumentation Framework

An argumentation framework [7] is defined on a set of arguments and a set of attacks between them. An attack expresses conflicts between arguments. Each argumentation system can be represented with a directed graph where nodes are the arguments, and the edges represent the attacks between them.

Definition 1 (Dung's argumentation framework). *An argumentation framework is a pair* $\mathrm{AF} = \langle \mathcal{A}, \mathcal{R} \rangle$ *where* $\mathcal{A}$ *is a finite set of arguments and* $\mathcal{R}$ *is a binary attack relation defined on* $\mathcal{A} \times \mathcal{A}$. *Given two arguments* a *and* b, $a\ \mathcal{R}\ b$ *or* $(a, b) \in \mathcal{R}$ *means* a *attacks* b.

The output of an AF is a set of sets of acceptable arguments, called extensions which are also called acceptability semantics (for example, preferred).

Definition 2. *Let* $AF = \langle \mathcal{A}, \mathcal{R} \rangle$ *be an argumentation framework and* $\mathcal{S} \subseteq \mathcal{A}$,

- $\mathcal{S}$ *is conflict-free of* AF *iff there are no arguments* $a,\ b \in \mathcal{S}$ *s.t.* $a\ \mathcal{R}\ b$.
- $a \in \mathcal{A}$ *is acceptable w.r.t.* $\mathcal{S}$ *iff* $\forall b \in \mathcal{A}$ *s.t.* $b\ \mathcal{R}\ a$, $\exists\ c \in \mathcal{S}$ *s.t.* $c\ \mathcal{R}\ b$.
- $\mathcal{S}$ *is an admissible set iff it is conflict-free and each argument in* $\mathcal{S}$ *is acceptable w.r.t.* $\mathcal{S}$.
- $\mathcal{S}$ *is a preferred extension iff it is maximal (for set inclusion) among admissible sets.*

2.2 Value-Based Argumentation Framework

VAF [2] is an extension of Dung *AF* where arguments promote specific values like, economy, health, etc. Compared to Dung's framework, where all attacks are always successful, in *VAF*, we must distinguish attack from defeat. For example, if an argument promoting the value *health* attacks an argument promoting the value *economy* then the attack succeeds only if *economy* is not preferred to *health*. The concept of audience is introduced in *VAF* where the values constitute a preference relation (transitive, irreflexive and asymmetric). For example, given two values *health* and *economy*, we can have an audience where *health* preferred to *economy*. In the remainder of the paper, an audience is called preference-audience.

Definition 3 (Value-based Argumentation Framework [2]**).** *A Value-based Argumentation Framework is a 5-tuple* $VAF = \langle \mathcal{A}, \mathcal{R},\ V,\ val,\ \mathcal{P} \rangle$, *where* $\mathcal{A}$ *is a set of arguments,* $\mathcal{R}$ *is an irreflexive binary attack relation defined on* $\mathcal{A} \times \mathcal{A}$, V *is a non empty set of values, val is a function which maps elements of* $\mathcal{A}$ *to elements of* V *and* $\mathcal{P}$ *is the set of preference-audiences.*

Definition 4. *An AF specific to a preference-audience* α *is a 5-tuple* $AVAF_\alpha = \langle \mathcal{A}, \mathcal{R},\ V,\ val,\ \succ_\alpha \rangle$ *where* $\mathcal{A}, \mathcal{R},\ V$ *and val are as a VAF and* $\succ_\alpha$ *is a (transitive, irreflexive and asymmetric) preference relation on* V *for the audience* α.

An *AVAF* is a *VAF* where the attack relation of the *VAF* is replaced by a defeat relation for that preference-audience (i.e., the attacks which are unsuccessful for that preference-audience are removed). An *AVAF* can be treated as a Dung-style *AF* since for Dung, all attacks are always successful. Thus, the defeat relation is relative to a preference-audience. We use $\mathcal{D}_\alpha$ to denote the defeat relation w.r.t. the preference-audience α and $val(a) \succ_\alpha val(b)$ means that the value promoted by a is preferred to the value of b w.r.t. the preference-audience α. The preference relation for the set of audiences should constitute a total order on V. For example, if $V = \{v_1,\ v_2,\ v_3\}$ and $\mathcal{P} = \{v_2 \succ v_1,\ v_2 \succ v_3\}$ then the preference relation for $\mathcal{P}$ is not a total order on V. So, another audience ($v_3 \succ v_1$ or $v_1 \succ v_3$) should be added to $\mathcal{P}$ in order to obtain a total order on V. Some details about the notion of an audience can be found in [3].

Definition 5. *Let* $\langle \mathcal{A}, \mathcal{R}, V, val, \succ_\alpha \rangle$ *be an AVAF and let a, b be two arguments in* $\mathcal{A}$. $a\ \mathcal{D}_\alpha\ b$ *iff* $a\ \mathcal{R}\ b$ *and not* $val(b) \succ_\alpha val(a)$.

Definition 5 states that an argument a defeats another argument b for the preference-audience α iff it attacks it and the value of b is not preferred to its value for that audience. Note that the attack succeeds if both arguments promote the same value, or if no preference between the values has been defined. If V contains a single value, then the *VAF* becomes a standard *AF*.

Bench-Capon [2] introduces the notions of acceptability, conflict freeness, admissible sets and preferred extensions for *VAF* which are relative to a preference-audience.

Definition 6. *Let* $\langle \mathcal{A}, \mathcal{R}, V, val, \succ_\alpha \rangle$ *be an AVAF,* α *be a preference-audience and* $\mathcal{S}$ *be a subset of arguments. An argument a is acceptable for* α *w.r.t.* $\mathcal{S}$ *iff* $\forall b \in \mathcal{A}$ *s.t.* $b\ \mathcal{D}_\alpha\ a$, $\exists\ c \in \mathcal{S}$ *s.t.* $c\ \mathcal{D}_\alpha\ b$.

Definition 7. *Let* $\langle \mathcal{A}, \mathcal{R}, V, val, \succ_\alpha \rangle$ *be an AVAF,* α *be a preference-audience and* $\mathcal{S}$ *be a subset of arguments.*

- *A set* $\mathcal{S}$ *is conflict-free for the preference-audience* α *iff for all* $a, b \in \mathcal{S}$, *it is not the case that* $a\mathcal{D}_\alpha b$.
- $\mathcal{S}$ *is an admissible set for* α *iff it is conflict-free for* α *and* $\forall a \in \mathcal{S}$, *a is acceptable for* α *w.r.t.* $\mathcal{S}$.
- $\mathcal{S}$ *is a preferred extension for* α *iff it is maximal (for set inclusion) among admissible sets for the preference-audience* α.

Definition 8 (Status of arguments in *VAF*). *Let* $\langle \mathcal{A}, \mathcal{R}, V, val, \mathcal{P} \rangle$ *be a VAF and* $\varepsilon_1, \ldots, \varepsilon_n$ *its preferred extensions for the set of all preference-audiences. Let* $a \in \mathcal{A}$.

- *a is objectively accepted iff* $\forall i \in \{1 \ldots n\}, a \in \varepsilon_i$.
- *a is subjectively accepted iff* $\exists i \in \{1 \ldots n\}$ *s.t.* $a \in \varepsilon_i$.
- *a is rejected iff* $\forall i \in \{1 \ldots n\}, a \notin \varepsilon_i$.

From Definition 8, it follows that an argument is objectively accepted iff it is in every preferred extension for every preference-audience, subjectively accepted iff it is in preferred extension for some preference-audiences and rejected iff it does not belong to preferred extension of any preference-audience.

Note that in [2], different notions such as admissible sets and preferred extensions are relative to a preference-audience. For our contribution, we need to introduce the following definitions concerning admissible sets and preferred extensions for *VAF*.

Definition 9. *Let* $VAF = \langle \mathcal{A}, \mathcal{R}, V, val, \mathcal{P} \rangle$, $AVAF_{\alpha_1}, \ldots, AVAF_{\alpha_n}$ *be n AF specific to the given preference-audience* $\alpha_1, \ldots, \alpha_n$ *and* $\mathcal{S} \subseteq \mathcal{A}$. *Let* $Adm(AVAF_{\alpha_1}), \ldots, Adm(AVAF_{\alpha_n})$ *be the set of admissible sets for* $AVAF_{\alpha_1}, \ldots, AVAF_{\alpha_n}$. $\mathcal{S}$ *is an admissible set for VAF iff* $\exists i \in \{1 \ldots n\}$ *s.t.* $\mathcal{S} \in Adm(AVAF_{\alpha_i})$.

Definition 10. *Let VAF=$\langle \mathcal{A}, \mathcal{R}, V, val, \mathcal{P}\rangle$, $AVAF_{\alpha_1}$,..., $AVAF_{\alpha_n}$ be n AF specific to the given preference-audience α_1,...,α_n and $\mathcal{S} \subseteq \mathcal{A}$. Let Pref($AVAF_{\alpha_1}$),..., Pref($AVAF_{\alpha_n}$) be the set of preferred extensions for $AVAF_{\alpha_1}$,..., $AVAF_{\alpha_n}$. $\mathcal{S}$ is a preferred extension for VAF iff $\exists i \in \{1 \ldots n\}$ s.t. $\mathcal{S} \in$ Pref($AVAF_{\alpha_i}$).*

In *VAF*, the status assigned to each argument is relative to preferred extensions of the audiences (i.e. the union of preferred extensions of each audience is considered). For example, a given argument is subjectively accepted in *VAF* iff it is in the preferred extension for at least one preference-audience. However, preferred extensions are maximal (for set inclusion) among admissible sets. This means that if *VAF* has at least one subjectively accepted argument, then there is at least one audience having this argument in its preferred extension and then there is at least one admissible set to which this argument belongs, as stated in Definitions 9 and 10.

Example 1. In Fig. 1, we have an example of *VAF* with $\mathcal{A}$=$\{a, b, c, d\}$, $\mathcal{R}$ = $\{(a, b), (b, c), (c, b), (c, d)\}$. V=$\{v_1, v_2, v_3\}$, $val(a)$=v_1, $val(b)$=v_2, $val(c)$=v_3 and $val(d)$=v_3. Assume that we have the following preference-audiences: $v_2 \succ v_1$ and $v_3 \succ v_2$. For the preference-audience $v_2 \succ v_1$: $\{a, b, d\}$, $\{a, c\}$ are the preferred extensions. For this preference-audience, the attack from a to b is unsuccessful (i.e., the attack is removed for that preference-audience) because the value of a is not preferred to the value of b. For $v_3 \succ v_2$: $\{a, c\}$ is the preferred extension. For this preference-audience, b attacks c but it does not defeat it because its value is not preferred to the value of b, so the attack from b to c is removed for that preference-audience. Thus, the argument a is objectively accepted, while b, c and d are subjectively accepted.

Fig. 1. An example of *VAF*

3 Constrained Value-Based Argumentation Framework

Our aim in this section is to extend *VAF* by incorporating constraints defined on either the arguments' values or on the arguments. More precisely, instead of considering only preferences between the arguments' values as in *VAF*, we propose to consider additional information. Let us consider the following notations to be used in the remainder of the paper. Let $PROP_{PS}$ be the language of propositional formulas defined on a set PS of propositional symbols, the classical logic connectives $\neg$, $\wedge$, $\vee$, $\rightarrow$, $\Leftrightarrow$ and the symbols $\top$, $\bot$ which denote tautology and contradiction respectively. The symbol $\models$ denotes classical inference relation. $PROP_{\mathcal{A}}$ is the language of propositional formulas defined on a set $\mathcal{A}$ of arguments, the classical logic connectives and the symbols $\top$, $\bot$. $PROP_V$ is the language of propositional formulas defined on a set V of values, the classical logic connectives and the symbols $\top$, $\bot$.

Definition 11. *A Constrained Value-based Argumentation Framework is a 6-tuple* $CVAF$=$\langle \mathcal{A}, \mathcal{R}, V, val, \mathcal{P}, \mathcal{C}$=$\mathcal{C}_\mathcal{A} \cup \mathcal{C}_V \rangle$*, where* $\mathcal{A}$ *is a set of arguments,* $\mathcal{R}$ *is an irreflexive binary attack relation defined on* $\mathcal{A} \times \mathcal{A}$*,* V *is a non emptyset of values, val is a function which maps elements of* $\mathcal{A}$ *to elements of* V*,* $\mathcal{P}$ *is the set of preference-audiences and* $\mathcal{C}$ *is a set of constraints such that* $\mathcal{C}_\mathcal{A}$ *are constraints defined as propositional formulas from* $PROP_\mathcal{A}$ *and* $\mathcal{C}_V$ *are constraints defined as propositional formulas from* $PROP_V$*.*

In *CVAF*, an audience can be seen as a user in group decision making. Thus, the term preference (resp. constraint) can be interchangeably used with preference-audience (resp. constraint-audience). Namely, the element $\mathcal{P}$ of *CVAF* contains preference-audiences defined over the arguments' values (as for *VAF*) and the element $\mathcal{C}$ contains all constraints where some of them are defined as propositionnal formulas over the arguments' values ($\mathcal{C}_V$) and others are defined as propositionnal formulas over the arguments ($\mathcal{C}_\mathcal{A}$). Note that it is also possible to have constraints only from $\mathcal{C}_\mathcal{A}$ (resp. $\mathcal{C}_V$). For simplicity, we sometimes use $\mathcal{C}$ instead of $\mathcal{C}$=$\mathcal{C}_\mathcal{A} \cup \mathcal{C}_V$.

Example 2. Let us consider the *CVAF* obtained by adding two constraints c_V defined on $PROP_V$ and $c_\mathcal{A}$ defined on $PROP_\mathcal{A}$ to the *VAF* of Fig. 1 (Example 1). We have $CVAF$=$\langle \mathcal{A}, \mathcal{R}, V, val, \mathcal{P}, \mathcal{C}_\mathcal{A} \cup \mathcal{C}_V \rangle$ with $\mathcal{A} = \{a, b, c, d\}$, $\mathcal{R}$=$\{(a, b), (b, c), (c, b), (c, d)\}$. V=$\{v_1, v_2, v_3\}$, $val(a)$=v_1, $val(b)$=v_2, $val(c)$=$val(d)$=v_3, $\mathcal{P}$=$\{v_2 \succ v_1, v_3 \succ v_2\}$. $\mathcal{C}_V$=$\{c_V\}$ s.t. c_V=$\neg v_3$ and $\mathcal{C}_\mathcal{A}$=$\{c_\mathcal{A}\}$ s.t. $c_\mathcal{A}$=a.

Contrary to the approach proposed in [5] where the additional constraint is defined on the arguments, for *CVAF*, constraints are defined either on the arguments or on the arguments' values. Thus, we introduce the following useful definitions for handling the constraints.

Definition 12. *Let* $CVAF$=$\langle \mathcal{A}, \mathcal{R}, V, val, \mathcal{P}, \mathcal{C} \rangle$ *be a constrained value-based argumentation framework. We define* Arg: $V \rightarrow 2^\mathcal{A}$ *a function that associates for each value* v *in* V*, a set of arguments* $\mathcal{S} \subseteq \mathcal{A}$*, denoted by* $Arg(v)$=$\mathcal{S}$ *s.t. for each* $a \in \mathcal{S}$*, we have that* $val(a) = v$*.*

Definition 12 states that the function *Arg* groups the arguments promoting the same value in the same subset. More precisely, given a set of arguments $\mathcal{A}$ and the set of values V=$\{v_1, v_2,.., v_m\}$ of *CVAF*, the result of *Arg* is $\{\mathcal{S}_1, \mathcal{S}_2,..., \mathcal{S}_m\}$ s.t. $\mathcal{S}_1$ (resp. $\mathcal{S}_2,..., \mathcal{S}_m$) $\subseteq \mathcal{A}$ and $Arg(v_1)$=$\mathcal{S}_1$, $Arg(v_2) = \mathcal{S}_2$,..., $Arg(v_m) = \mathcal{S}_m$.

Example 3. Le us consider the *CVAF* of Example 2. We have $Arg(v_1) = \{a\}$, $Arg(v_2) = \{b\}$ and $Arg(v_3) = \{c, d\}$.

Let us introduce the following definition. Each subset $\mathcal{S}$ of arguments corresponds to an interpretation over $\mathcal{A}$ given by the completion [5] of $\mathcal{S}$.

Definition 13. (Completion). *Let CVAF=$\langle \mathcal{A}, \mathcal{R}, V, val, \mathcal{P}, \mathcal{C}\rangle$ be a constrained value-based argumentation framework. Let $\mathcal{S} \subseteq \mathcal{A}$. The completion of $\mathcal{S}$ denoted by $\hat{\mathcal{S}}$ is obtained as follows: $\hat{\mathcal{S}} = \{a \mid a \in \mathcal{S}\} \cup \{\neg a \mid a \in \mathcal{A}\setminus \mathcal{S}\}$.*

In the following, we define how a given set of arguments $\mathcal{S}$ satisfies a constraint c defined on $PROP_{\mathcal{A}}$. This definition is inspired from the one given in [5].

Definition 14. *Let CVAF=$\langle \mathcal{A}, \mathcal{R}, V, val, \mathcal{P}, \mathcal{C}\rangle$ be a constrained value-based argumentation framework, c be a constraint defined on $PROP_{\mathcal{A}}$ and $\mathcal{S} \subseteq \mathcal{A}$ be a subset of arguments. $\mathcal{S}$ satisfies c iff the completion of $\mathcal{S}$ is a model of c (denoted by $\hat{\mathcal{S}} \models c$).*

Example 4. Let us consider the *CVAF* obtained by adding a constraint $c_{\mathcal{A}}=\neg b \vee \neg c$ to the *VAF* of Fig. 1. Let E_1 ={a, d} be a set of arguments of *CVAF*. Applying Definition 13, we have $\hat{E_1}=\{a, d, \neg b, \neg c\}$ is a completion of E_1. Following Definition 14, $\hat{E_1} \models c_{\mathcal{A}}$. So, E_1 satisfies $c_{\mathcal{A}}$. However, E_2 ={c, b} does not satisfy $c_{\mathcal{A}}$.

In the following, we define situations where a given set of arguments $\mathcal{S}$ satisfies a constraint c_V defined on $PROP_V$.

Definition 15. *Let CVAF=$\langle \mathcal{A}, \mathcal{R}, V, val, \mathcal{P}, \mathcal{C}\rangle$ be a constrained value-based argumentation framework, c, c' be two constraints from $PROP_V$ and $\mathcal{S} \subseteq \mathcal{A}$ be a subset of arguments. The following items give the definition of satisfaction of a constraint c by $\mathcal{S}$:*

1. *$\forall\ \mathcal{S} \subseteq \mathcal{A}$, $\mathcal{S}$ satisfies $\top$.*
2. *$\forall\ \mathcal{S} \subseteq \mathcal{A}$, $\mathcal{S}$ does not satisfy $\bot$.*
3. *$\mathcal{S}$ satisfies v_i iff $\exists a \in \mathcal{S}$ s.t. $val(a) = v_i$ (for a propositional atom v_i).*
4. *$\mathcal{S}$ satisfies $\neg c$ iff $\mathcal{S}$ does not satisfy c.*
5. *$\mathcal{S}$ satisfies $c \wedge c'$ iff $\mathcal{S}$ satisfies c and $\mathcal{S}$ satisfies c'.*
6. *$\mathcal{S}$ satisfies $c \vee c'$ iff $\mathcal{S}$ satisfies c or $\mathcal{S}$ satisfies c'.*

Example 5. Let us consider the *CVAF* of Example 2 where $c_V=\neg v_3$. Let $\mathcal{S}_1 = \{a, b\}$ and $\mathcal{S}_2 = \{c, d\}$. From Definition 15, we have $\mathcal{S}_1$ satisfies c_V, $\mathcal{S}_2$ does not satisfy c_V.

3.1 From Constraints over Arguments' Values to Constraints over Arguments

The following definition introduces a function, which associates with each constraint c_V from $PROP_V$, its corresponding constraint $c_{\mathcal{A}}$ from $PROP_{\mathcal{A}}$. The possibility to give constraints over the values in *CVAF* seems natural since arguments promote values. However, transforming constraints defined on $PROP_V$ to constraints defined on $PROP_{\mathcal{A}}$ is important from computational point of view, since it allows us to reuse existing results and algorithms developed in argumentation theory (e.g. [5,6,10]).

Definition 16. *Let* $CVAF$*=*$\langle\mathcal{A},\mathcal{R},\ V,\ val,\ \mathcal{P},\ \mathcal{C}=\mathcal{C}_{\mathcal{A}}\cup\mathcal{C}_V\rangle$ *be a constrained value-based argumentation framework,* c_V*,* c'_V *be two constraints in* $\mathcal{C}_V$*. Let* $F_{\mathcal{A}}$ *be a function which associates with each constraint* c_V*, its represented constraint* $c_{\mathcal{A}}$*. Then, we have:*

1. $F_{\mathcal{A}}(\top)=\top$.
2. $F_{\mathcal{A}}(\bot)=\bot$.
3. *If* $c_V = v$ *then* $F_{\mathcal{A}}(c_V) = \bigvee a_i$ *s.t.* $a_i \in Arg(v)$.
4. $F_{\mathcal{A}}(\neg(c_V)) = \neg F_{\mathcal{A}}(c_V)$.
5. $F_{\mathcal{A}}(c_V \vee c'_V)=F_{\mathcal{A}}(c_V) \vee F_{\mathcal{A}}(c'_V)$.
6. $F_{\mathcal{A}}(c_V \wedge c'_V)=F_{\mathcal{A}}(c_V) \wedge F_{\mathcal{A}}(c'_V)$.

From Item 1 (resp. 2) of Definition 16, it follows that having a constraint $c_V = \top$ (resp. $c_V = \bot$) from $PROP_V$, then its corresponding constraint from $PROP_{\mathcal{A}}$ is $\top$ (resp. $\bot$). Item 3 of Definition 16 states that having a constraint $c_V = v$ from $PROP_V$, then its corresponding constraint from $PROP_{\mathcal{A}}$ is obtained by replacing v by $\bigvee a_i$ s.t. $a_i \in Arg(v)$. The rest of items give the way to transform any constraint from $PROP_V$ defined using the usual connective (negation, disjunction and conjunction) to its corresponding constraint from $PROP_{\mathcal{A}}$.

Example 6. In Example 2, the considering $CVAF$ has two constraints: c_V=$\neg v_3$ and $c_{\mathcal{A}}$=a. From Definition 16, there is a constraint $c'_{\mathcal{A}}$ defined on $PROP_{\mathcal{A}}$ that corresponds to c_V. We have, $Arg(v_3) = \{c, d\}$, then, applying Item 3 of Definition 16, $F_{\mathcal{A}}(c'_V) = c'_{\mathcal{A}}$=$\neg(c \vee d)$.

Proposition 1. *Let* $CVAF$*=*$\langle\mathcal{A},\mathcal{R},\ V,\ val,\ \mathcal{P},\ \mathcal{C}=\mathcal{C}_{\mathcal{A}}\cup\mathcal{C}_V\rangle$ *be a constrained value-based argumentation framework. Let* $\mathcal{S} \subseteq \mathcal{A}$ *and* c_V *be a constraint in* $\mathcal{C}_V$*. Then, it holds that* $\mathcal{S}$ *satisfies* c_V *if and only if the completion* $\hat{\mathcal{S}}$ *is a model of* $F_{\mathcal{A}}(c_V)$ *(denoted by* $\hat{\mathcal{S}} \models F_{\mathcal{A}}(c_V)$*) where* $F_{\mathcal{A}}(c_V)$ *is obtained from Definition 16.*

Proof. Let us reason by induction of the structure of c_V.

- **Case 1:** Let $c_V \equiv v_i$ (v_i is a propositional atom from $PROP_V$). Recall that from Definition 16, we have $F_{\mathcal{A}}(c_V) = \bigvee a_j$ s.t. $a_j \in Arg(v_i)$.
  - Suppose that $\mathcal{S}$ satisfies c_V. Following Definition 15, this means that $\exists a_k \in \mathcal{S}$ s.t. $val(a_k) = v_i$. This means that $a_k \in Arg(v_i)$. Besides that, $a_k \in \hat{\mathcal{S}}$ (i.e. completion of $\mathcal{S}$). Thus, $\hat{\mathcal{S}} \models \bigvee a_j$ s.t. $a_j \in Arg(v_i)$ which means that $\hat{\mathcal{S}} \models F_{\mathcal{A}}(c_V)$.
  - Assume that $\mathcal{S}$ satisfies $F_{\mathcal{A}}(c_V)$, thus, $\hat{\mathcal{S}} \models \bigvee a_j$ s.t. $a_j \in Arg(v_i)$. This means that $\exists a_k \in \hat{\mathcal{S}}$ with $a_k \in Arg(v_i)$. Thus, $\exists a_k \in \mathcal{S}$ with $val(a_k) = v_i$, which means that $\mathcal{S}$ satisfies c_V.
- **Case 2:** Let $c_V \equiv \neg(c'_V)$. Suppose that the proposition is satisfied with respect to c'_V and show that it is satisfied for c_V too. By definition, $\mathcal{S}$ satisfies $\neg(c'_V)$ means that $\mathcal{S}$ does not satisfy c'_V. Thus, following the induction hypothesis we have $\mathcal{S}$ does not satisfy $F_{\mathcal{A}}(c'_V)$. Which means that $\hat{\mathcal{S}} \not\models F_{\mathcal{A}}(c'_V)$. Thus, we have $\hat{\mathcal{S}} \models \neg F_{\mathcal{A}}(c'_V)$. However from Definition 16, we have $\neg F_{\mathcal{A}}(c'_V) = F_{\mathcal{A}}(\neg(c'_V))$. Thus, $\hat{\mathcal{S}} \models F_{\mathcal{A}}(\neg c'_V)$.

- **Case 3:** Suppose that the proposition is verified for both c'_V and c''_V and show that it is verified for $c_V \equiv c'_V \wedge c''_V$ too. $\mathcal{S}$ satisfies both c'_V and c''_V means that $\mathcal{S}$ satisfies c'_V and satisfies c''_V (according to Definition 15). Following the induction hypothesis, this is equivalent to say that $\mathcal{S}$ satisfies $F_{\mathcal{A}}(c'_V)$ and $\mathcal{S}$ satisfies $F_{\mathcal{A}}(c''_V)$. Thus, we have $\hat{\mathcal{S}} \models F_{\mathcal{A}}(c'_V)$ and $\hat{\mathcal{S}} \models F_{\mathcal{A}}(c''_V)$. Equivalently, $\hat{\mathcal{S}} \models F_{\mathcal{A}}(c'_V) \wedge F_{\mathcal{A}}(c''_V)$. But from Definition 16, $F_{\mathcal{A}}(c'_V) \wedge F_{\mathcal{A}}(c''_V)$ is nothing that $F_{\mathcal{A}}(c'_V \wedge c''_V)$. Finally, $\hat{\mathcal{S}} \models F_{\mathcal{A}}(c'_V \wedge c''_V)$ which means that $\mathcal{S}$ satisfies $F_{\mathcal{A}}(c_V)$.
- **Case 4:** Suppose that the proposition is verified for c'_V and c''_V and show that it is verified for $c_V \equiv c'_V \vee c''_V$ too. $\mathcal{S}$ satisfies $c'_V \vee c''_V$ means that $\mathcal{S}$ satisfies c'_V or satisfies c''_V (according to Definition 15). Following the induction hypothesis, this is equivalent to say that $\mathcal{S}$ satisfies $F_{\mathcal{A}}(c'_V)$ or $\mathcal{S}$ satisfies $F_{\mathcal{A}}(c''_V)$. Stated otherwise, $\hat{\mathcal{S}} \models F_{\mathcal{A}}(c'_V)$ or $\hat{\mathcal{S}} \models F_{\mathcal{A}}(c''_V)$. Equivalently, $\hat{\mathcal{S}} \models F_{\mathcal{A}}(c'_V) \vee F_{\mathcal{A}}(c''_V)$. But $F_{\mathcal{A}}(c'_V) \vee F_{\mathcal{A}}(c''_V)$ is nothing that $F_{\mathcal{A}}(c'_V \vee c''_V)$ (according to Definition 16). Finally, $\hat{\mathcal{S}} \models F_{\mathcal{A}}(c'_V \vee c''_V)$ which means that $\mathcal{S}$ satisfies $F_{\mathcal{A}}(c_V)$.

3.2 Acceptability Semantics in *CVAF*

To define some acceptability semantics and evaluating the set of arguments in *CVAF*, we propose two inference relations. The first one is more prudent than the second one since it derives less acceptable arguments. Each inference relation requires the satisfaction of the preferences and the additional constraints. So, let us firstly introduce the following observation regarding the satisfaction of the constraints.

Observation 1. *To verify if a given set of arguments $\mathcal{S}$ satisfies a constraint c, we have the following cases. If c is defined on $PROP_{\mathcal{A}}$, then Definition 14 is applied. If c is defined on $PROP_V$ then either Definition 15 is applied without transforming c into its corresponding constraint over arguments or Definition 16 is firstly applied to transform each c to its corresponding constraint over arguments and then Definition 14 is applied.*

First Inference Relation. For handling constraints in *CVAF*, this relation consists to generalize the approach proposed in [5]. The preferences are handled as in *VAF*. More precisely, for handling preferences, the process corresponds to consider Dung *AF* with removing unsuccessful attacks on the basis of the preferences represented in a specific audience. For handling the constraints, the process corresponds to consider Dung *AF* with the condition of satisfying the constraints. Namely, as defined in the following, considering both preferences and constraints, a given set of arguments $\mathcal{S}$ is admissible for *CVAF* iff it is admissible for at least one preference-audience (i.e. $\mathcal{S}$ is admissible for *VAF* as stated in Definition 9), admissible for *AF* and it satisfies each constraint.

Definition 17. *Let CVAF=$\langle \mathcal{A}, \mathcal{R}, V, val, \mathcal{P}, \mathcal{C}\rangle$ be a constrained value-based argumentation framework. Let $c \in \mathcal{C}$ be a constraint and $\mathcal{S}$ be a subset of arguments. $\mathcal{S}$ is admissible for CVAF iff*

1. $\mathcal{S}$ is admissible for VAF=$\langle \mathcal{A}, \mathcal{R}, V, val, \mathcal{P}\rangle$, and
2. $\mathcal{S}$ is admissible for AF=$\langle \mathcal{A}, \mathcal{R}\rangle$, and
3. $\mathcal{S}$ satisfies c (see Observation 1).

Definition 17 states that $\mathcal{S}$ is admissible for *CVAF* iff $\mathcal{S}$ is admissible for *VAF* (condition of Item 1) and $\mathcal{S}$ is admissible for $\langle \mathcal{A}, \mathcal{R}, C\rangle$ (conditions of Item 2 and Item 3) which corresponds to constrained argumentation framework (*CAF*) defined in [5] and recalled in Definition 18. More precisely, in this inference relation, we consider each constraint-audience as a *CAF*. Note that in *CAF* defined in [5], additional constraints are defined over the arguments while in *CVAF*, constraints are defined either on the arguments' values or on the arguments. However, it is shown above that each constraint defined over arguments' values has its corresponding constraint over the arguments. Let us define the *CAF* proposed in [5].

Definition 18. *A constrained argumentation framework is a triple CAF = $\langle \mathcal{A}, \mathcal{R}, C\rangle$ where $\mathcal{A}$ is a set of arguments, $\mathcal{R}$ is a binary attack relation defined on $\mathcal{A}\times\mathcal{A}$ and $\mathcal{C}$ is a propositional formula from $PROP_{\mathcal{A}}$.*

Definition 18 states that *CAF* extends Dung *AF* in order to take account for constraints over arguments defined as propositionnal formulas from $PROP_{\mathcal{A}}$.

Let us now introduce the following property.

Property 1. Let *CVAF*=$\langle \mathcal{A}, \mathcal{R}, V, val, \mathcal{P}, \mathcal{C}\rangle$ be a constrained value-based argumentation framework and c be a constraint in $\mathcal{C}$. In case of applying the first inference relation, it holds that $\langle \mathcal{A}, \mathcal{R}, \mathcal{C}\rangle$ is a *CAF* defined in Definition 18.

Proof. It is mentioned above that for handling the constraints in *CVAF*, the first inference relation is the generalization of the approach defined for *CAF* in [5] where a given set of arguments $\mathcal{S}$ is admissible for *CAF* if it is admissible for *AF* and satisfies the constraints. These conditions are the same to those of Item 2 and Item 3 of Definition 17. Thus, Definition 17 can be rewritten, where $\mathcal{S}$ is admissible for CVAF iff $\mathcal{S}$ is admissible for *VAF*=$\langle \mathcal{A}, \mathcal{R}, V, val, \mathcal{P}\rangle$, and $\mathcal{S}$ is admissible for $\langle \mathcal{A}, \mathcal{R}, \mathcal{C}\rangle$. In addition, *CVAF* can contain constraints that are defined on $PROP_{\mathcal{A}}$ which is the case for *CAF*. If the constraints of *CVAF* are defined on $PROP_V$, then from Definition 16, each constraint defined on $PROP_V$ has its corresponding constraint defined on $PROP_{\mathcal{A}}$. Thus, given *CVAF*=$\langle \mathcal{A}, \mathcal{R}, V, val, \mathcal{P}, \mathcal{C}\rangle$, then $\langle \mathcal{A}, \mathcal{R}, \mathcal{C}\rangle$ is a *CAF* defined in [5].

Considering the first inference relation, we can observe that each *CVAF* is composed of two formalisms *VAF* and *CAF* where each preference-audience is an AVAF defined in Definition 4 and each constraint-audience is considered as a *CAF*.

Second Inference Relation. The idea of the second inference relation consists to consider a given set of arguments $\mathcal{S}$ be admissible for *CVAF* iff it is admissible for *VAF* and it satisfies the constraint. Namely, this inference relation does not require additional condition about is admissibility for Dung AF as it is the case in the first inference relation. So, this relation allows to derive more acceptable arguments than the first one.

Definition 19. *Let* $CVAF{=}\langle\mathcal{A},\mathcal{R},V,val,\mathcal{P},\mathcal{C}\rangle$ *be a constrained value-based argumentation framework. Let* $c\in\mathcal{C}$ *be a constraint and* $\mathcal{S}$ *be a subset of arguments.* $\mathcal{S}$ *is admissible for* $CVAF$ *iff*

1. $\mathcal{S}$ *is admissible for* $VAF{=}\langle\mathcal{A},\mathcal{R},V,val,\mathcal{P}\rangle$, *and*
2. $\mathcal{S}$ *satisfies* c *(see Observation 1).*

Definition 19 states that a given set of arguments $\mathcal{S}$ is admissible for $CVAF$ iff $\mathcal{S}$ is admissible for VAF and $\mathcal{S}$ satisfies each constraint.

The following definitions are available either for applying the first or the second inference relation.

Definition 20. *Let* $CVAF{=}\langle\mathcal{A},\mathcal{R},V,val,\mathcal{P},\mathcal{C}\rangle$ *be a constrained value-based argumentation framework. A preferred extension of* $CVAF$ *is a subset* S *of* $\mathcal{A}$ *s.t.* S *is maximal w.r.t.* $\subseteq$ *among the set of admissible sets for* $CVAF$*. The set of preferred extensions of* $CVAF$ *is denoted by* $Pref_{CVAF}$.

Definition 21 (Status of arguments in *CVAF*). *Let* $CVAF{=}\langle\mathcal{A},\mathcal{R},V,val,\mathcal{P},\mathcal{C}\rangle$ *be a constrained value-based argumentation framework,* $\varepsilon_1,\ldots,\varepsilon_m$ *be its preferred extensions. Let* $a\in\mathcal{A}$.

- a *is objectively accepted iff* $\forall i\in\{1\ldots m\}, a\in\varepsilon_i$.
- a *is subjectively accepted iff* $\exists i\in\{1\ldots m\}$ *s.t.* $a\in\varepsilon_i$.
- a *is rejected iff* $\nexists i\in\{1\ldots m\}$ *s.t.* $a\in\varepsilon_i$.

From Definition 21, it follows that an argument is objectively accepted iff it is in every preferred extension for $CVAF$, subjectively accepted iff it is in some preferred extensions for $CVAF$, rejected iff it does not belong to any preferred extension for $CVAF$.

Property 2. Let $CVAF{=}\langle\mathcal{A},\mathcal{R},V,val,\mathcal{P},\mathcal{C}\rangle$ be a constrained value-based argumentation framework, $Adm1_{CVAF}$ (resp. $Adm2_{CVAF}$) be the set of admissible sets of $CVAF$ by applying the first (resp. second inference relation). Then, it holds that $Adm1_{CVAF}\subseteq Adm2_{CVAF}$.

Proof. It can be given directly from Definitions 17 and 19 where defining admissible sets of $CVAF$ by applying the first inference relation requires the two conditions of the second inference relation with another condition (admissibility for AF). Namely, all admissible sets for $CVAF$ obtained by applying the first relation are also admissible for $CVAF$ when applying the second relation but the converse is not true.

Example 7. Assume that a group of friends want to choose a good restaurant. They perceived a restaurant and one of them ask the question: "Are we going to choose this restaurant?". On the basis of the reputation of the restaurant and some displayed information, each member of the group can give its arguments in order to take a decision.

- **Argument a:** This restaurant is expensive because the cook is a star chef.
- **Argument b:** The restaurant is expensive but the reason is that recipes are prepared in the restaurant and the products are fresh and local.
- **Argument c:** Of course, the ingredients are freach and local, so the proposed recipes are certainly healthy.
- **Argument d:** Our goal is to make us pleasure and we do not interest neither for the price nor for healthy recipes.

Assume that we have the following information concerning the group.

- **Group Preferences:** p_1=*Pleasure*≻*Expensive* and p_2=*Healthy*≻*Pleasure*.
- **Group constraints:** c=*Pleasure*∧¬*Expensive*.

The corresponding *CVAF* to the group dialogue, the preferences and the constraints is *CVAF*=$\langle \mathcal{A}, \mathcal{R}, V, val, \mathcal{P}, \mathcal{C} \rangle$ with $\mathcal{A} = \{a, b, c, d\}$, $\mathcal{R}$=$\{(a, b), (b, a), (c, a), (d, a), (d, b)$, (d, c)}. V={*Healthy*, *Expensive*, *Pleasure*}, $val(a)$ =$val(b)$=*Expensive*, $val(c)$=*Healthy* and $val(d)$=*Pleasure*. $\mathcal{P}$={*Pleasure*≻*Expensive*, *Healthy*≻*Pleasure*}. $\mathcal{C}$={c} s.t. c=*Pleasure*∧¬*Expensive*. The graph of *CVAF* is given in Fig. 2.

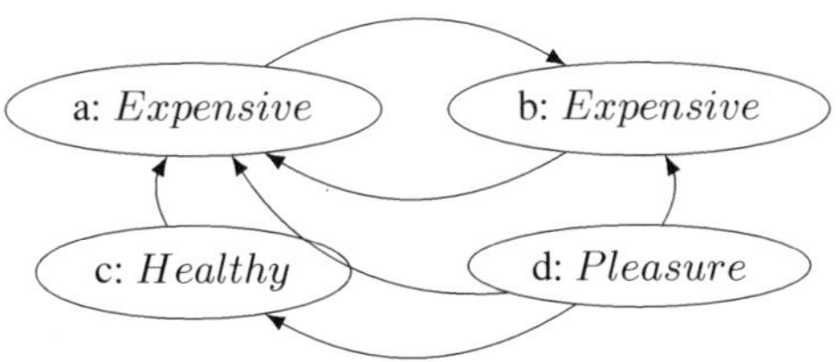

Fig. 2. The graph of *VAF*

The *AF* specific to p_1 is $AVAF_{p_1}$=$\langle \mathcal{A}, \mathcal{R}, V, val$, *Pleasure*≻*Expensive*$\rangle$. The *AF* specific to p_2 is $AVAF_{p_2}$=$\langle \mathcal{A}, \mathcal{R}, V, val$, *Healthy*≻*Pleasure*$\rangle$. Applying Definition 7, admissible sets for $AVAF_{p_1}$ are: {d}. Admissible sets for $AVAF_{p_2}$ are: {c}, {d}, {c, d}. Thus, from Definition 9, admissible sets for *VAF* are: {c}, {d}, {c, d}. Applying Definition 2, admissible sets for $\langle \mathcal{A}, \mathcal{R} \rangle$ are: {d}. We have {d} and {c, d} satisfy c while {c} does not satisfy it.

In the following, we define admissible sets, preferred extensions and the status of each argument for *CVAF* by applying the two proposed inference relations.

Applying the first inference relation: From Definition 17, we have Adm_{CVAF} ={d} since it is the unique set which is admissible for *VAF* and for *AF* and satisfies the constraint. Applying Definition 20, we have $Pref_{CVAF}$ ={d}. From Definition 21, the arguments d is objectively accepted (under the preferred extensions) while a, b and c are rejected.

Applying the second inference relation: From Definition 19, we have Adm_{CVAF} = {d}, {c, d} since they are admissible for *VAF* and each one satisfies the constraint. From Definition 20, we have $Pref_{CVAF}$ ={c, d}. Thus, from

Definition 21, the arguments c and d are objectively accepted, while a and b are rejected (under the preferred extensions).

We can observe that the result obtained by applying the second inference relation is more coherent with the preferences and the constraints expressed by the group. For example, the argument c is rejected by applying the first inference relation although $val(c)=Healthy$ and $Healthy$ is the first choice of the preference-audience p_2 and it is not excluded in the constraint. However, c is objectively accepted by applying the second inference relation which is the expected result.

4 Conclusion

In this paper, we have presented a generalization of VAF by incorporating additional constraints defined either on the arguments' values or on the arguments. We have proposed two inference relations for defining some acceptability semantics in the new framework called, constrained value-based argumentation framework. The first inference relation is more prudent than the second one since it requires more conditions than the second one for defining admissible sets of arguments. We have also shown that each constraint over arguments' values has its corresponding constraint over arguments. $CVAF$ can be used in many applications. It can also be used for multiple criteria decision making problems where values can be considered as criteria, preferences as a relative importance between the criteria and constraints as a requirements about the criteria with the possibility to give arguments. As future work, we would study different points including: (i) other properties of the two inference relations, (ii) determining other acceptability semantics for $CVAF$ and (iii) applying $CVAF$ for medical decision making.

References

1. Amgoud, L., Vesic, S.: Rich preference-based argumentation frameworks. Int. J. Approximate Reasoning **55**, 585–606 (2014)
2. Bench-Capon, T.J.M., Atkinson, K., Chorley, A.: Persuasion and value in legal argument. J. Logic Comput. **15**(6), 1075–1097 (2005)
3. Bench-Capon, T.J.M., Doutre, S., Dunne, P.E.: Audiences in argumentation frameworks. Artif. Intell. **171**(1), 42–71 (2007)
4. Cayrol, C., Lagasquie-Schiex, M.-C.: From preferences over arguments to preferences over attacks inabstract argumentation: a comparative study. In: Proceedings of the IEEE International Conference on Tools with Artificial Intelligence, pp. 1–8 (2013)
5. Coste-Marquis, S., Devred, C., Marquis, P.: Constrained argumentation frameworks. In: Proceedings of the Tenth International Conference on Principles of Knowledge Representation and Reasoning, pp. 112–122 (2006)
6. Devred, C., Doutre, S., Lefèvre, C., Nicolas, P.: Dialectical proofs for constrained argumentation. In: Computational Models of Argument: Proceedings of COMMA 2010, Desenzano del Garda, Italy, pp. 159–170, 8–10 September 2010 (2010)

7. Dung, P.M.: On the acceptability of arguments and its fundamental role in non-monotonic reasoning, logic programming and n-person games. Artif. Intell. **77**(2), 321–358 (1995)
8. Dunne, P.E., Hunter, A., McBurney, P., Parsons, S., Wooldridge, M.: Weighted argument systems: basic definitions, algorithms, and complexity results. Artif. Intell. **175**(2), 457–486 (2011)
9. Modgil, S., Prakken, H.: A general account of argumentation with preferences. Artif. Intell. **195**, 361–397 (2013)
10. Nofal, S., Atkinson, K., Dunne, P.E.: Algorithms for decision problems in argument systems under preferred semantics. Artif. Intell. **207**, 23–51 (2014)
11. Simari, G.R., Loui, R.P.: A mathematical treatment of defeasible reasoning and its implementation. Artif. Intell. **53**(2–3), 125–157 (1992)

Belief Revision and the EM Algorithm

Inés Couso[1(✉)] and Didier Dubois[2]

[1] Universidad Oviedo, Gijon, Spain
couso@uniovi.es
[2] IRIT, CNRS and University of Toulouse, Toulouse, France

Abstract. This paper provides a natural interpretation of the EM algorithm as a succession of revision steps that try to find a probability distribution in a parametric family of models in agreement with frequentist observations over a partition of a domain. Each step of the algorithm corresponds to a revision operation that respects a form of minimal change. In particular, the so-called expectation step actually applies Jeffrey's revision rule to the current best parametric model so as to respect the frequencies in the available data. We also indicate that in the presence of incomplete data, one must be careful in the definition of the likelihood function in the maximization step, which may differ according to whether one is interested by the precise modeling of the underlying random phenomenon together with the imperfect observation process, or by the modeling of the underlying random phenomenon alone, despite imprecision.

Keywords: Maximum likelihood · Belief revision · Incomplete information · Expectation-Maximization

1 Introduction

The EM (Expectation-Maximization) algorithm is an iterative technique aiming to find a parameterized model achieving a local maximum of the likelihood function when there is no closed-form solution for the maximum likelihood estimator. Another case where EM is repeatedly used is when there are missing data (unsupervised or semi-supervised learning). In order to do so, a latent (unobserved) variable (artificial, in the first case, meaningful in the second case) is used, whose distribution depends on the same parameter as the observed one. The procedure starts with the assessment of an initial value (or vector of values) for the parameter. Each iteration alternates two steps, the "expectation" (E) step and the "maximization" (M) step. The expectation step postulates an empirical distribution for the unobserved variable that agrees with the observed data. During the maximization step, the maximum likelihood estimator based on the joint empirical distribution of both the latent and the observed variable is determined. The process iterates until some stability is reached. The procedure is known to provide an increasing sequence of values for the likelihood function. It converges to a local maximum when some additional conditions are satisfied [12].

J.P. Carvalho et al. (Eds.): IPMU 2016, Part II, CCIS 611, pp. 279–290, 2016.
DOI: 10.1007/978-3-319-40581-0_23

In the following we study the EM algorithm for likelihood-based estimation, where an observed random variable Y goes along with a latent variable X with range $\mathcal{X}$, and where, based on a sequence of precise observations $y_1, \ldots, y_N$, a likelihood function of the form $\prod_{i=1}^N P_Y(y_i;\theta)$, understood as a likelihood function on the joint space $\prod_{i=1}^N P_{(X,Y)}(\mathcal{X}\times\{y_i\};\theta)$, is maximized. The EM algorithm proceeds based on an alternating optimisation scheme, where at each step, a fictitious precise data set pertaining to (X, Y) and agreeing with the observed result on Y is generated in agreement with the optimal probabilistic model obtained at the previous step from the previous fictitious data set pertaining to (X, Y) and agreeing with the observed result on Y.

The aim of the paper is to better understand the nature of the solution provided by the EM algorithm on the range of (X, Y). We provide an interpretation of the EM algorithm in terms of a sequence of revision steps. More specifically, the E step consists in determining the sample that minimises Kullback divergence with respect to the parametrical distribution postulated during the M step of the last iteration, while respecting the constraints imposed by the data. We show that it corresponds to a natural use of Jeffrey's rule of revision, that comes down to an imputation of sample values for the latent variable. This result enables a better understanding of what the EM algorithm actually aims to. To the best of our knowledge the relationship between the EM algorithm and Jeffrey's rule has not been previously pointed out.

Moreover, we provide an example-based preliminary discussion on cases of incompletely informed data where the EM algorithm should not be used without caution, either because the collection of postulated parametrized distributions is large enough in order to contain all the joint distributions in agreement with the empirical one, or because, in case of overlapping incomplete observations, the definition of the proper likelihood function is a delicate issue.

The paper is organized as follows: Sect. 2 proposes an original introduction to the EM algorithm where the basic steps are formally justified. In Sect. 3, we recall Jeffrey's rule of revision, the properties it satisfies and its connection with the minimization of divergence. We then reinterpret the EM algorithm as a succession of revision steps. Finally, in Sect. 4, we give some examples of anomalies due to an inefficient or incautious usage of the EM algorithm.

2 Introduction to the EM Algorithm

Let X be a random variable, namely a mapping from a sample space $(\Omega, \mathcal{A}, P)$ to the range of X. For simplicity, we assume that $\mathcal{X}$ is finite, and P_X, the probability function attached to X depends on a parameter θ, i.e. $P_X(\cdot;\theta)$ is a model of the random process driving X. We suppose that instead of observing X, another random quantity Y is observed, also driven by parameter θ. Y incompletely informs about the realization of X, in the sense that if $Y = b \in \mathcal{Y} = \{b_1, \ldots, b_n\}$, we only know that $X \in \Gamma(b) \subseteq \mathcal{X}$, for some multimapping Γ [3]. Dempster et al. [4] give a version of the EM algorithm when the observations y_i are viewed as incomplete perceptions of a latent variable X, assuming that the observations bear on a partition of the whole state space. So, the range of Y is of the form $\{\{A_1\}, \ldots, \{A_r\}\}$, where the A_i's form a partition of $\mathcal{X}$.

Let us consider a sequence of N iid copies of $Z = (X, Y)$. We will use the nomenclature $\mathbf{z} = ((x_1, y_1), \ldots, (x_N, y_N)) \in (\mathcal{X} \times \mathcal{Y})^N$ to represent a specific sample of the vector (X, Y). Thus, $\mathbf{y} = (y_1, \ldots, y_N)$ will denote the observed sample (an observation of the vector $\mathbf{Y} = (Y_1, \ldots, Y_n)$), and $\mathbf{x} = (x_1, \ldots, x_N)$ will denote an arbitrary artificial sample from $\mathcal{X}$ for the latent variable X, that we shall vary in $\mathcal{X}^N$. Let us also use the nomenclature $L^{\mathbf{y}}(\theta) = \log \mathbf{p}(\mathbf{y}; \theta)$ for the log-likelihood function, where $\mathbf{p}(\mathbf{y}; \theta) = \prod_{i=1}^{N} p(y_i; \theta)$ denotes the probability of observing $\mathbf{y} \in \mathcal{Y}^N$, assuming that the value of the parameter is θ. The final goal of EM is to find a value of the parameter θ that is a (maybe local) maximum of $L^{\mathbf{y}}(\theta)$.

We are interested in modelling the likelihood function associated to the result of the random process driving the random variable X *despite* imprecision. Namely, behind the measurement report $(y_1, \ldots, y_N)$ there exists a sequence of precise outcomes for X, $(x_1^*, \ldots, x_N^*)$ that would have been observed, had the measurement device been accurate (had Γ been a one-to-one function).

2.1 From the Likelihood Function to the EM Criterion

Let $\mathcal{P}^{\mathcal{X}^N}$ be the set of all probability measures $\mathbf{P}$ we can define on the measurable space $(\mathcal{X}^N, \wp(\mathcal{X}^N))$.When the optimisation of the log-likelihood $L^{\mathbf{y}}(\theta) = \log \sum_{\mathbf{x} \in \mathcal{X}^N} \mathbf{p}(\mathbf{x}, \mathbf{y}; \theta)$ is too difficult, a trick is to optimize a lower bound $F(\mathbf{P}, \theta)$ of it that is simpler to optimize. This is allowed by the introduction of arbitrary latent or fake variables[1] and the use of Jensen inequality. Haas [8] proposes the simple following derivation of the functional F:

$$
\begin{aligned}
L^{\mathbf{y}}(\theta) &= \log \sum_{\mathbf{x} \in \mathcal{X}^N} \mathbf{p}(\mathbf{x}, \mathbf{y}; \theta) = \log \sum_{\mathbf{x} \in \mathcal{X}^N} \frac{\mathbf{p}(\mathbf{x})\mathbf{p}(\mathbf{x}, \mathbf{y}; \theta)}{\mathbf{p}(\mathbf{x})} \\
&\geq \sum_{\mathbf{x} \in \mathcal{X}^N} \mathbf{p}(\mathbf{x}) \log \left[\frac{\mathbf{p}(\mathbf{x}, \mathbf{y}; \theta)}{\mathbf{p}(\mathbf{x})} \right] \quad \text{(Jensen's inequality)} \\
&= \sum_{\mathbf{x} \in \mathcal{X}^N} \mathbf{p}(\mathbf{x}) \log \left[\frac{\mathbf{p}(\mathbf{x}|\mathbf{y}; \theta)\mathbf{p}(\mathbf{y}; \theta)}{\mathbf{p}(\mathbf{x})} \right] \\
&= \sum_{\mathbf{x} \in \mathcal{X}^N} \mathbf{p}(\mathbf{x}) \log \mathbf{p}(\mathbf{y}; \theta) + \sum_{\mathbf{x} \in \mathcal{X}^N} \mathbf{p}(\mathbf{x}) \log \left[\frac{\mathbf{p}(\mathbf{x}|\mathbf{y}; \theta)}{\mathbf{p}(\mathbf{x})} \right] \\
&= L^{\mathbf{y}}(\theta) - D(\mathbf{P}, \mathbf{P}(\cdot|\mathbf{y}; \theta)) = F(\mathbf{P}, \theta).
\end{aligned}
$$

where $D(\mathbf{P}, \mathbf{P}') = \sum_{\mathbf{x} \in \mathcal{X}^N} \mathbf{p}(\mathbf{x}) \log[\frac{\mathbf{p}(\mathbf{x})}{\mathbf{p}'(\mathbf{x})}]$ is the Kullback-Leibler divergence of $\mathbf{P}'$ from $\mathbf{P}$, and $\mathbf{p}$ is the mass function associated to $\mathbf{P}$.[2]

[1] In some cases, they are not artificial, and are naturally present in the problem.

[2] In the expression in line 2 of the above derivation, $F(\mathbf{P}, \theta)$ could be, with some abuse of notation, written $-D(\mathbf{P}, \mathbf{P}(\cdot, \mathbf{y}; \theta))$ as it is a kind of divergence from $\mathbf{P}(\cdot, \mathbf{y}; \theta))$. However the sum on $\mathcal{X}^N$ of the latter quantities is not 1 (it is $\mathbf{p}(\mathbf{y}; \theta)$) and this pseudo-divergence can be negative.

Some authors use the nomenclature $\ell(\theta|\theta^{(n-1)}) = F(\mathbf{P}(\cdot|\mathbf{y};\theta^{(n-1)});\theta)$. According to the definition of F, the properties of logarithms, we can alternatively express $\ell(\theta|\theta^{(n-1)})$ as follows:

$$\ell(\theta|\theta^{(n-1)}) = \log(\mathbf{p}(\mathbf{y};\theta)) + \sum_{\mathbf{x}\in\mathcal{X}^N} \mathbf{p}(\mathbf{x}|\mathbf{y};\theta^{(n-1)}) \log \frac{\mathbf{p}(\mathbf{x}|\mathbf{y};\theta)}{\mathbf{p}(\mathbf{x}|\mathbf{y};\theta^{(n-1)})}.$$

Moreover, taking into account the fact that $\mathbf{p}(\cdot|\mathbf{y};\theta^{(n-1)}) : \mathcal{X}^N \to [0,1]$ is a mass function (the sum of the masses is equal to 1), $\ell(\theta|\theta^{(n-1)})$ also reads

$$\sum_{\mathbf{x}\in\mathcal{X}^N} \mathbf{p}(\mathbf{x}|\mathbf{y};\theta^{(n-1)}) \log \frac{\mathbf{p}(\mathbf{x}|\mathbf{y};\theta)\mathbf{p}(\mathbf{y};\theta)}{\mathbf{p}(\mathbf{x}|\mathbf{y};\theta^{(n-1)})} = \sum_{\mathbf{x}\in\mathcal{X}^N} \mathbf{p}(\mathbf{x}|\mathbf{y};\theta^{(n-1)}) \log \frac{\mathbf{p}(\mathbf{x},\mathbf{y};\theta)}{\mathbf{p}(\mathbf{x}|\mathbf{y};\theta^{(n-1)})}. \quad (1)$$

since $\mathbf{p}(\mathbf{x},\mathbf{y};\theta) = \mathbf{p}(\mathbf{x}|\mathbf{y};\theta)\mathbf{p}(\mathbf{y};\theta)$. We can therefore express $\ell(\theta;\theta^{(n-1)})$ as the sum of an entropy and a term that takes the form of an expectation:

$$\ell(\theta|\theta^{(n-1)}) = H(\mathbf{P}(\cdot|\mathbf{y};\theta^{(n-1)}) + E_{\cdot|\mathbf{y};\theta^{(n-1)}}[\log \mathbf{p}(\mathbf{X},\mathbf{y};\theta)]. \quad (2)$$

The last term represents indeed the expectation of a function of the random variable $\mathbf{X}$ taking the value $\log \mathbf{p}(\mathbf{x},\mathbf{y};\theta)$ with probability $\mathbf{p}(\mathbf{x}|\mathbf{y};\theta^{(n-1)})$ for every $\mathbf{x}\in\mathcal{X}^N$.

The main structure of the EM algorithm is then as follows. We first provide an initial value for the parameter, $\theta^{(0)} \in \Theta$. Each iteration of the algorithm, $n \geq 1$ consists of two steps, respectively called "expectation" (E) and "maximization" (M). According to [13], they can be described as follows:

- *Expectation step*: We compute the expectation $E_{\cdot|\mathbf{y};\theta^{(n-1)}}[\log \mathbf{p}(\mathbf{X},\mathbf{y};\theta)]$.
- *Maximization step*: We maximize $\ell(\theta|\theta^{(n-1)})$ wrt θ. According to Eq. (1), this is equivalent to minimizing the divergence $D(\mathbf{P}(\cdot|\mathbf{y};\theta^{(n-1)});\mathbf{P}(\cdot|\mathbf{y};\theta))$.

2.2 The EM Algorithm as a Succession of Revision Steps

Computing $E_{\cdot|\mathbf{y};\theta^{(n-1)}}[\log \mathbf{p}(\mathbf{X},\mathbf{y};\theta)]$ requires the determination of the conditional distribution $\mathbf{P}(\cdot|\mathbf{y};\theta^{(n-1)})$. The algorithm can then be alternatively described as follows:

- *"Expectation" step*: We compute the first argument of F as the probability measure determined by the mass function $\mathbf{p}(\cdot|\mathbf{y};\theta^{(n-1)}) : \mathcal{X}^N \to [0,1]$. In other words, we find the value of the first argument of the function F in order to fulfill the equality $F(\mathbf{P},\theta^{(n-1)}) = L^{\mathbf{y}}(\theta^{(n-1)})$.
- *Maximization step*: We determine $\theta^{(n)} = \arg\max_{\theta\in\Theta} F(\mathbf{P}(\cdot|\mathbf{y};\theta^{(n-1)}),\theta)$.

Note that in this presentation, the E-step is no longer, strictly speaking, computing an expectation, as it yields a mass function on $\mathcal{X}^N$. In this case, the computation of the expectation proper takes place when determining $F(\mathbf{P}(\cdot|\mathbf{y};\theta^{(n-1)}),\theta)$.

With these two steps, it is easy to guarantee that the sequence $(L^{\mathbf{y}}(\theta^{(n)}))_{n\in\mathbb{N}}$ is increasing. Namely as noticed above, we have that

$F(\mathbf{P}(\cdot|\mathbf{y};\theta^{(n-1)}),\theta^{(n-1)}) = L^{\mathbf{y}}(\theta^{(n-1)})$, for an arbitrary n. Now, since $\theta^{(n)} = \arg\max_{\theta\in\Theta} F(\mathbf{P}(\cdot|\mathbf{y};\theta^{(n-1)}),\theta)$, we have that $F(\mathbf{P}(\cdot|\mathbf{y};\theta^{(n-1)}),\theta^{(n)}) \geq F(\mathbf{P}(\cdot|\mathbf{y};\theta^{(n-1)}),\theta^{(n-1)}) = L^{\mathbf{y}}(\theta^{(n-1)})$. Taking into account the non-negativity of Kullback-Leibler's divergence (due to Jensen's inequality), we can deduce that $L^{\mathbf{y}}(\theta^{(n)}) \geq F(\mathbf{P}(\cdot|\mathbf{y};\theta^{(n-1)}),\theta^{(n)})$, and therefore that $L^{\mathbf{y}}(\theta^{(n)}) \geq L^{\mathbf{y}}(\theta^{(n-1)})$.

Some authors also describe the EM algorithm as a maximization-maximization procedure, since both steps refer to the maximization of the function F:

- *Expectation step*: We maximize $F(\mathbf{P},\theta^{(n-1)})$ with respect to $\mathbf{P}$; we get $\mathbf{P} = \mathbf{P}(\cdot|\mathbf{y};\theta^{(n-1)})$.
- *Maximization step*: maximize $F(\mathbf{P}(\cdot|\mathbf{y};\theta^{(n-1)}),\theta)$ with respect to θ; we get $\theta = \theta^{(n)}$.

3 The EM Algorithm from a Belief Revision Perspective

In this section, we shall prove that the E-step is an example of application of Jeffrey's revision rule governed by the minimal change principle. As the M-step also implements a form of minimal change, we thus show that the EM algorithm tries to iteratively find a statistical model that is as close as possible to a distribution of latent variables that is compatible with the observed incomplete data, oscillating from one distribution to the other.

3.1 Jeffrey's Revision Rule

In probability theory, there is a natural method for revising a prior probability P on a set S of mutually exclusive alternatives, in the presence of new probabilistic information I: a distribution $\rho_1,\ldots,\rho_r$ on elements of a partition $\{A_1,\ldots,A_r\}$ of S. The coefficients ρ_i sum to 1 and act as constraints on the posterior probability of elements A_i of the partition. Such an updating rule is proposed by Jeffrey [11]. Jeffrey's rule provides an effective means to revise a prior probability distribution P to a posterior P', given input I. Some axioms guide the revision process:

$$P'(A_i) = \rho_i. \tag{3}$$

This axiom clearly expresses that P' should respect the input information which is of the same nature as the prior probability, with priority given to the input.

Jeffrey's method also relies on the assumption that, while the probability on a prescribed subalgebra of events is enforced by the input information, the probability of any event $B \subseteq S$ conditional to any uncertain event A_i in this subalgebra is the same in the original and the revised distributions. Namely,

$$\forall A_i, \forall B, P(B|A_i) = P'(B|A_i). \tag{4}$$

The underlying interpretation of minimal change implied by the constraint of Eq. (4) is that the revised probability measure P' must preserve the conditional probability degree of any event B knowing event A_i has occurred. Jeffrey's rule of conditioning yields the unique distribution that satisfies (3) and (4) and takes the following form:

$$P'(B) = \sum_{i=1}^{r} \rho_i \cdot \frac{P(B \cap A_i)}{P(A_i)}. \tag{5}$$

Jeffrey's rule respects the probability kinematics principle, whose objective is to minimize change, usually in the sense of an informational distance between probability distributions [1]: The posterior probability P' minimizes the Kullback-Leibler divergence $D(P, P') = \sum_{s \in S} p'(s) \log[\frac{p'(s)}{p(s)}]$ with respect to the original distribution under the probabilistic constraints (3) defined by the input I (as explained in [16]).

3.2 The EM Algorithm from the Standpoint of Joint Distributions: E-step

Since we have assumed that $\mathbf{z}$ represents a sequence of N i.i.d. copies of (X, Y), we can decompose the probability mass $\mathbf{p}(\cdot|\mathbf{y}; \theta) : \mathcal{X}^N \to [0, 1]$ into a product of N mass functions, each one determining a distribution on $\mathcal{X}$. Let us now denote by n_{kj} the number of times that the pair (a_k, b_j) appears in the sample $\mathbf{z}$. Now, in order to denote the product mass function, we will use the nomenclature

$$\mathbf{p}(\mathbf{x}|\mathbf{y}; \theta) = \prod_{i=1}^{N} p(x_i|y_i; \theta) = \prod_{k=1}^{m} \prod_{j=1}^{r} p(a_k|b_j; \theta)^{n_{kj}}, \tag{6}$$

where $p(\cdot|b_j; \theta)$ denotes the mass function associated to the j-th marginal distribution:

$$p(a_k|b_j; \theta) = \frac{p^{\theta}_{kj}}{p^{\theta}_{.j}}, \ \forall j = 1, \ldots, r.$$

At the expectation step of the nth iteration of the EM algorithm, we compute the conditional probabilities $p(\cdot|b_j; \theta^{(n-1)}), \forall j = 1, \ldots, r$. If we consider the joint probability that results from combining those conditional probabilities with the marginal distribution on $(\mathcal{Y}, \wp(\mathcal{Y}))$ determined by the empirical distribution associated to the observed sample $\mathbf{y}$, $(\frac{n_{.1}}{N}, \ldots, \frac{n_{.r}}{N})$, where $n_{.j} = \sum_{k=1}^{m} n_{kj}$ is the number of times b_j appears in the observed sample, we will get the following joint mass distribution on $(\mathcal{X} \times \mathcal{Y}, \wp(\mathcal{X}) \times \wp(\mathcal{Y}))$:

$$\hat{p}^{(n-1)}(a_k, b_j) := \frac{n_{.j}}{N} \cdot p(a_k|b_j; \theta^{(n-1)}) = \frac{n_{.j}}{N} \cdot \frac{p^{\theta^{(n-1)}}_{kj}}{p^{\theta^{(n-1)}}_{.j}} \tag{7}$$

The E-step thus leads to a joint probability measure $\hat{P}^{(n-1)}$, on $\mathcal{X} \times \mathcal{Y}$ that, if the terms $n_{.j} \cdot \frac{p^{\theta^{(n-1)}}_{kj}}{p^{\theta^{(n-1)}}_{.j}}$ are integers, corresponds to an artificial sample $\mathbf{z}^{(n-1)} \in$

$(\mathcal{X}\times\mathcal{Y})^N$ involving the latent variable X, that is in agreement with the observed sample $\mathbf{y}$. Let us denote by $\mathcal{P}_\mathbf{y}$, the set of such joint probability measures on $(\mathcal{X}\times\mathcal{Y}, \wp(\mathcal{X})\times\wp(\mathcal{Y}))$ whose marginal distribution on $\mathcal{Y}$ coincides with the empirical distribution $(p_{.1},\ldots,p_{.r}) = (\frac{n_{.1}}{N},\ldots,\frac{n_{.r}}{N})$, associated to the sample $\mathbf{y}$.

Proposition 1. *The result $\hat{p}^{(n-1)}$ of the E-step is the posterior probability distribution generated by Jeffrey's rule of conditioning where the input information is given by the observed sample probabilities.*

Proof: Compare Eqs. (5) and (7). In the above Eq. (7), let $S = \mathcal{X}\times\mathcal{Y}$, the prior probability P is the parametric one with mass function $p(a_k, b_j; \theta^{(n-1)})$, the input comes from the observable sample $\mathbf{y}$, in the sense that $A_j = \mathcal{X}\times\{b_j\}$, with probabilities $\rho_j = \frac{n_{.j}}{N}$.

According to the result provided in [16] by P.M. Williams, if we consider the collection, $\mathcal{P}_\mathbf{y}$, of joint probability measures on $(\mathcal{X}\times\mathcal{Y}, \wp(\mathcal{X})\times\wp(\mathcal{Y}))$ whose marginal distribution on $\mathcal{Y}$ coincides with the empirical distribution associated to the sample $\mathbf{y}$, $(p_{.1},\ldots,p_{.r}) = (\frac{n_{.1}}{N},\ldots,\frac{n_{.r}}{N})$, the above joint probability measure, $\hat{P}^{(n)}$, is, among all of them, the one that minimizes Kullback-Leibler's divergence with respect to the joint distribution $p(\cdot,\cdot;\theta^{(n-1)}) : \mathcal{X}\times\mathcal{Y}\to\mathbb{R}$ obtained in the maximization step of the previous iteration.

3.3 The EM Algorithm from the Standpoint of Joint Distributions: M-step

Next we will check that the M step aims at looking for the Maximum Likelihood Estimate (MLE) of θ, given the joint empirical distribution proposed in Eq. (7). The criterion to be optimised at the nth M- step is

$$F(\mathbf{P}(\cdot|\mathbf{y};\theta^{(n-1)}),\theta) = L^{\mathbf{y}}(\theta) - D\left(\mathbf{P}(\cdot|\mathbf{y};\theta^{(n-1)}), \mathbf{P}(\cdot|\mathbf{y};\theta)\right).$$

Let us also notice that:

$$\begin{aligned} D(\mathbf{P}(\cdot|\mathbf{y};\theta^{(n-1)}), \mathbf{P}(\cdot|\mathbf{y};\theta)) &= \sum_{i=1}^N D\left(P(\cdot|y_i;\theta^{(n-1)}), P(\cdot|y_i;\theta)\right) \\ &= \sum_{i=1}^N\sum_{k=1}^m p(a_k|y_i;\theta^{(n-1)})\cdot\log\frac{p(a_k|y_i;\theta^{(n-1)})}{p(a_k|y_i;\theta)}. \end{aligned}$$

On the other hand, due to the properties of the logarithmic function, we can write $L^{\mathbf{y}}(\theta) = \sum_{i=1}^N \log p(y_i;\theta)$. Moreover, taking into account the fact that $p(\cdot|y_i;\theta^{(n-1)}) : \mathcal{X}\to[0,1]$ is a mass function ($\sum_{k=1}^m p(a_k|y_i;\theta^{(n-1)}) = 1$), we can equivalently write:

$$L^{\mathbf{y}}(\theta) = \sum_{i=1}^N \log p(y_i;\theta) = \sum_{i=1}^N\sum_{k=1}^m p(a_k|y_i;\theta^{(n-1)})\log p(y_i;\theta).$$

Again, taking into account the properties of logarithm and also the commutativity of the sum, we can write:

$$F(\mathbf{P}(\cdot|\mathbf{y};\theta^{(n-1)}),\theta) = -\sum_{i=1}^{N}\sum_{k=1}^{m} p(a_k|y_i;\theta^{(n-1)}) \cdot \log \frac{p(a_k|y_i;\theta^{(n-1)})}{(p(a_k|y_i;\theta)\cdot p(y_i;\theta))}$$
$$= -\sum_{i=1}^{N}\sum_{k=1}^{m} p(a_k|y_i;\theta^{(n-1)}) \cdot \log \frac{p(a_k|y_i;\theta^{(n-1)})}{p(a_k,y_i;\theta)}$$
$$= \sum_{i=1}^{N} H(P(\cdot|y_i;\theta^{(n-1)})) + \sum_{i=1}^{N}\sum_{k=1}^{m} p(a_k|y_i;\theta^{(n-1)}) \cdot \log p(a_k,y_i;\theta),$$

where H stands for Shannon entropy. For each $j = 1,\ldots,r$, recall that $n_{.j}$ is the number of occurrences of $b_j \in \mathcal{Y}$ in the observed sample $\mathbf{y} = (y_1,\ldots,y_N)$. Then we can rewrite the above expression of $F(\mathbf{P}(\cdot|\mathbf{y};\theta^{(n-1)}),\theta)$ as follows:

$$-\sum_{j=1}^{r} n_{.j} H(P(\cdot|b_j;\theta^{(n-1)})) + \sum_{j=1}^{r}\sum_{k=1}^{m} n_{.j} \cdot p(a_k|b_j;\theta^{(n-1)}) \cdot \log p(a_k,b_j;\theta).$$

And due to the properties of logarithm, we can rewrite $F(\mathbf{P}(\cdot|\mathbf{y};\theta^{(n-1)}),\theta)$ as:

$$-\sum_{j=1}^{r} n_{.j} H(P(\cdot|b_j;\theta^{(n-1)})) + \log\left(\prod_{j=1}^{r}\prod_{k=1}^{m} p(a_k,b_j;\theta)^{n_{.j}p(a_k|b_j;\theta^{(n-1)})}\right). \quad (8)$$

According to the nomenclature established in (Eq. (7)), the above exponent $n_{.j}p(a_k|b_j;\theta^{(n-1)})$ coincides with $N\cdot\hat{p}^{(n)}(a_k,b_j)$. Such an exponent can be seen as the number of occurrences of (a_k,b_j) in an artificial sample inducing the empirical distribution determined by $\hat{p}^{(n)}$ (the joint distribution characterised by the mass function displayed in Eq. (7)). Moreover the entropy term in (8) does not depend on θ. Therefore, maximizing the above expression with respect to θ is equivalent to finding the maximum likelihood estimator associated to such an artificial sample on $\mathcal{X}\times\mathcal{Y}$.

In a nutshell, the M step at iteration n actually finds the MLE associated to the fake sample (the joint distribution) determined by Eq. (7). If the algorithm stops at iteration n^*, we have determined the collection of maximum likelihood estimators associated to all the joint artificial samples on $\mathcal{X}\times\mathcal{Y}$ constructed for the first n^* iterations (the n^* samples inducing the empirical distributions determined by the collection of joint mass functions $\{\hat{p}^{(n)} : n = 1,\ldots,n^*\}$). Let the reader notice that, for a specific iteration n, the exponent $n_{.j}\cdot p(a_k|b_j;\theta^{(n-1)}) = N\cdot\hat{p}^{(n)}(a_k,b_j)$ may not be an integer necessarily, and therefore such an empirical joint distribution is not necessarily in total correspondence with some feasible joint sample. In some papers, the fake sample is interpreted as a probability distribution over possible imputations (see e.g., the short paper by Do and Batzoglou [5]), over which the expectation is then taken. This makes the fact that the fake sample could be unobservable much less problematic from an interpretation standpoint.

4 Some Difficulties with the EM Algorithm for Handling Incomplete Information

In this section we show that when the set $\mathcal{P}_\theta$ of parameterized joint distributions includes the set $\mathcal{P}_{\mathbf{y}}$ of joint distributions whose marginals on $\mathcal{Y}$ agree with the empirical distribution induced by $\mathbf{y}$, the EM algorithm cannot be properly used. Moreover in the case of overlapping pieces of incomplete data, a non-careful definition of the likelihood function leads to anomalous results.

Case of Imprecise Data Forming a Partition. As above, we consider the situation where each observation $y = b_i$ on $\mathcal{Y}$ is interpreted as a report providing an element A_i of a partition of $\mathcal{X}$. The maximum likelihood estimator of θ based on the observed sample $\mathbf{y}$ will be the value of the argument for which the likelihood of $\mathbf{y}$ is maximal among all the maximum likelihood estimators associated to all the joint empirical distributions compatible with $\mathbf{y}$. If the probabilities of elements of $\mathcal{X}$ and $\mathcal{Y}$ are not related to each other via enough constraints, there will generally be several MLE distributions on $\mathcal{X}$ in agreement with the observed sample on $\mathcal{Y}$. Moreover, the collection of n^* joint distributions determined by the n^* E-steps of the algorithm are, in general, just a fraction of this collection of compatible joint distributions.

Example 1. Consider the random experiment that consists of rolling a dice. We do not know whether the dice is fair or not. Suppose we only get reports on whether the outcomes are even or odd. Let X be the random variable denoting the actual outcome of the dice roll (from $a_1 = 1$ to $a_6 = 6$) and let Y be a binary variable taking the values b_1 (odd) and b_2 (even). Let the 6-dimensional vector $\theta = (p_1, \ldots, p_6)$ represent the actual (unknown) probability distribution of Z, with $p_6 = 1 - \sum_{i=1}^{5} p_i$. Let $\pi = p_2 + p_4 + p_6$ and $1 - \pi = p_1 + p_3 + p_5$ respectively denote the probabilities of getting an even or an odd number. Based on a sample of $n_{.1}$ occurrences of b_1 and $n_{.2}$ occurrences of b_2 in a sample of $N = n_{.1} + n_{.2}$ trials, the maximum likelihood estimator of π would be $\hat{\pi} = \frac{n_{.2}}{N}$. Also, we can easily check that any vector $(\hat{p}_1, \ldots, \hat{p}_6)$ satisfying the constraints $\hat{p}_2 + \hat{p}_4 + \hat{p}_6 = \frac{n_{.2}}{N}$ is a maximum likelihood estimator of θ given the observed sample. Now, let us suppose that we use the EM algorithm in order to find such an MLE. We first initialize the vector θ, by means of selecting some $\theta^{(0)} = (p_1^{(0)}, \ldots, p_6^{(0)})$. Then, we have to apply the E-step, that is, Jeffrey's rule with $\rho_1 = \frac{n_{.1}}{N}, A_1 = \{1,3,5\}, \rho_2 = \frac{n_{.2}}{N}, A_2 = \{2,4,6\}$. We get $(p_1^{(1)}, \ldots, p_6^{(1)})$, where:

$$p_i^{(1)} = \frac{n_{.1}}{N} \frac{p_i^{(0)}}{p_1^{(0)} + p_3^{(0)} + p_5^{(0)}}, i = 1,3,5; \quad p_i^{(1)} = \frac{n_{.2}}{N} \frac{p_i^{(0)}}{p_4^{(0)} + p_6^{(0)} + p_6^{(0)}}, i = 2,4,6.$$

For instance, if we take the starting point $\left(p_1^{(0)}, \ldots, p_6^{(0)}\right) = (\frac{1}{6}, \ldots, \frac{1}{6})$ then, we will get $p_i^{(1)} = \frac{n_{.1}}{3N}, i = 1,3,5$ and $p_i^{(1)} = \frac{n_{.2}}{3N}, i = 2,4,6$. Such a vector is also the maximum likelihood estimator of θ based on a fake sample with equal numbers of 1, 3, 5's and equal numbers of 2, 4, 6's. This vector is thus the optimum of the first M step based on this fake sample. A different postulated initial vector would be identified with a different imputation $\left(p_1^{(1)}, \ldots, p_6^{(1)}\right)$.

The previous example illustrates a case where the MLE based on the observed sample $\mathbf{y}$ is not unique, and an MLE is reached after the first iteration of the EM algorithm. Whatever the starting point $\theta^{(0)}$, the estimate based on the subsequent iteration of the algorithm, $\theta^{(1)}$ is an MLE of θ based on $\mathbf{y}$, which completely depends on $\theta^{(0)}$. Using EM in this situation sounds questionable. In cases where the probabilities on $\mathcal{X}$ are tightly constrained, the MLE for $\mathbf{y}$ can be unique and is asymptotically reached after several iterations of the EM algorithm, independently of the initial choice of the parameter (see the first example in the paper by Dempster et al. [4]).

Anomalies when Imprecise Observations Overlap. When the elements of the range of the observed variable $\mathcal{Y}$ correspond to elements of a partition of $\mathcal{X}$, the likelihood function of Y takes the form $\prod_{j=1}^r P(X \in A_j)^{n_{.j}}$, with $\sum_{j=1}^r n_{.j} = N$. Suppose now that the images $\{A_1, \ldots, A_r\}$ of Γ do not form a partition of $\mathcal{X}$. In other words, if $x \in X$ there may be several $A_i's$ enclosing outcome x_j. Maximizing the product $\prod_{j=1}^r P(X \in A_j)^{n_{.j}}$ instead of $L^{\mathbf{y}}(\theta) = \prod_{j=1}^r P(Y = \{A_j\})^{n_{.j}}$ leads to counter-intuitive results, as we show in the following example.

Example 2. Suppose that a dice is tossed, as in the previous example. Suppose we are told either that the result has been less than or equal to 3 or that it has been greater than or equal to 3. Then $A_1 = \{1, 2, 3\}$ and $A_2 = \{3, 4, 5, 6\}$. Let us denote both responses by y_1 and y_2, respectively. After each toss, when the actual result (X) is 3, the reporter says y_1 or y_2 but we do not know how it is chosen. Let us take a sample of N tosses of the dice and let us assume that we have been told us n_1 times "less than or equal to 3" and $n_2 = N - n_1$ times "greater than or equal to 3". Suppose we take as a likelihood function

$$h(\theta) = P(Z \in A_1)^{n_1} \cdot P(Z \in A_2)^{n_2} = (p_1 + p_2 + p_3)^{n_1} \cdot (p_3 + p_4 + p_5 + p_6)^{n_2},$$

where $\theta = (p_1, \ldots, p_6) \in [0, 1]^6$ such that $\sum_{i=1}^6 p_i = 1$. We can easily observe that it reaches its maximum $(h(\theta) = 1)$ for any vector θ satisfying the constraint $p_3 = 1$. But such a prediction of θ would not be a reasonable estimate for θ. Worse, the EM algorithm applied to this case would also stop after the first iteration and fail to reach this maximum, for the same reason as in the previous example.

The difficulty comes from the fact that, with overlapping pieces of data, the function $h(\theta)$ is arguably not a likelihood function. Edwards ([6], p. 9) defines a likelihood function as follows:

> Let $P(R|\theta)$ be the probability of obtaining results R given the hypothesis θ, according to the probability model ... The likelihood of the hypothesis θ given data R, and a specific model, is proportional to $P(R|\theta)$, the constant of proportionality being arbitrary.

Edwards mentions that "this probability is defined for any member of the set of possible results given any one hypothesis ... As such its mathematical properties are well-known. A fundamental axiom is that if R_1 and R_2 are two of the possible results, mutually exclusive, then $P(R_1 \text{ or } R_2|\theta) = P(R_1|\theta) + P(R_2|\theta)$".

The key point in our problem with overlapping imprecise observations is what we understand by "a result". Actually, an imprecise result taking the form of a subset A_i of $\mathcal{X}$ should be modelled by a singleton $R_i = \{A_i\}$ of the power set of $\mathcal{X}$ in order to satisfy the requirements of Edwards. In other words, if the possible observable results are $\{\{A_1\}, \ldots, \{A_r\}\}$ then $\sum_{i=1}^{r} P(\{A_i\}|\theta) = 1$. In our case, a result is not an event A_i, it is an elementary event $\{A_i\}$ (a report carrying imprecise information about X). Only elementary events can be observed. For instance, when tossing a die, you cannot observe the event "odd". What you see is 1, 3 or 5. But some source may report "{odd}". So, a likelihood function is proportional to $P(\{A_i\}|\theta)$ where R is an elementary event. For instance, $P(\mathcal{X}|\theta) = 1$ cannot be viewed as the likelihood of θ given the sure event.

In order to properly apply the EM algorithm to find the distribution of X, in the case of overlapping observations A_i, we have to introduce a parametric model describing which A_i is chosen by the reporter when the outcome of X is x_j, say a conditional probability $P_\theta(\{A_i\} \mid x_j)$ and let the likelihood function $L^y(\theta)$ account for it, e.g. $P(\{A_i\}|\theta) = \sum_{i=1,m} P_\theta(\{A_i\} \mid x_j)P(x_j \mid \theta)$. Generally, $P_\theta(\{A_i\} \mid x_j) > 0$ only if $x_j \in A_i$. For instance, the superset assumption [10] considers $P_\theta(\{A_i\} \mid x_j)$ to be constant over all supersets of x_j.

In the above example, suppose we model the measurement device by assuming $P(\Gamma = \{1,2,3\}|Z = 3) = \alpha$. If m denotes the mass function associated to the imprecise observations, we have that $m(A_1) = P(Y = y_1) = p_1 + p_2 + \alpha\, p_3$, $m(A_2) = (1-\alpha)p_3 + p_4 + p_5 + p_6$.

Notice that in this case $P(X \in A)$ does not coincide with $m(A) = P(Y = \{A\})$. It would, only under the special situations where $\alpha = 1$ or $p_3 = 0$. Moreover, the difficulty due to the inclusion $\mathcal{P}_\theta \supseteq \mathcal{P}_{\mathbf{y}}$, making the EM algorithm inefficient, remains.

5 Conclusion

What our results show is that the EM algorithm oscillates between the set $\mathcal{P}_\theta$ of parameterized joint distributions and the set $\mathcal{P}_{\mathbf{y}}$ of joint distributions whose marginals on $\mathcal{Y}$ agree with the empirical distribution induced by $\mathbf{y}$: In the initial step a probability measure in $\mathcal{P}_\theta$ is chosen, and then it is updated in the E-step into a probability measure $\hat{P}^{(0)}$ in $\mathcal{P}_{\mathbf{y}}$ using Jeffrey's rule of revision, thus producing an artificial sample on $\mathcal{X} \times \mathcal{Y}$; on this basis, an MLE estimate $\theta^{(1)}$, hence a probability measure $P(\cdot;\theta^{(1)}) \in \mathcal{P}_\theta$ is computed in the M-step, based on the artificial sample underlying $\hat{P}^{(0)}$, and so on, until convergence of the $\theta^{(n)}$ sequence. At each stage n, the log-likelihood function $L^{\mathbf{y}}(\theta)$ increases. However, we have shown cases of incomplete information management where this method does not seem to work properly. In future works we shall propose a more systematic analysis of situations when the EM algorithm stops at the first iteration under the partition assumption, and explore alternative ways of posing the problem of maximum likelihood estimation under incomplete overlapping data [2,7,9,10]. Another issue is to investigate the cogency of the fake sample found by the EM algorithm viewed as an imputation method [14,15].

Acknowledgements. This work is partially supported by ANR-11-LABX-0040-CIMI (Centre International de Mathématiques et d'Informatique) within the program ANR-11-IDEX-0002-02, while Inés Couso was a visiting expert scientist at CIMI, Toulouse, and by TIN2014-56967-R (Spanish Ministry of Science and Innovation) and FC-15-GRUPIN14-073 (Regional Ministry of the Principality of Asturias).

References

1. Domotor, Z.: Probability kinematics - conditional and entropy principles. Synthese **63**, 74–115 (1985)
2. Couso, I., Dubois, D.: Statistical reasoning with set-valued information: ontic vs. epistemic views. Int. J. Approximate Reasoning **55**(7), 1502–1518 (2014)
3. Dempster, A.P.: Upper and lower probabilities induced by a multivalued mapping. Ann. Math. Stat. **38**, 325–339 (1967)
4. Dempster, A.P., Laird, N.M., Rubin, D.B.: Maximum likelihood from incomplete data via the EM algorithm (with discussion). J. Roy. Stat. Soc. B **39**, 1–38 (1977)
5. Do, C.B., Batzoglou, S.: What is the expectation maximization algorithm? Nat. Biotechnol. **26**(8), 897–899 (2009)
6. Edwards, A.W.F.: Likelihood. Cambridge University Press, Cambridge (1972)
7. Guillaume, R., Dubois, D.: Robust parameter estimation of density functions under fuzzy interval observations. In: 9th ISIPTA Symposium, Pescara, Italy, pp. 147–156 (2015)
8. Haas, S.E.: The Expectation-Maximization and Alternating Minimization Algorithms. MIT, Cambridge (2002)
9. Hüllermeier, E.: Learning from imprecise and fuzzy observations: data disambiguation through generalized loss minimization. Int. J. Approximate Reasoning **55**(7), 1519–1534 (2014)
10. Hüllermeier, E., Cheng, W.: Superset learning based on generalized loss minimization. In: Appice, A., Rodrigues, P.P., Santos Costa, V., Gama, J., Jorge, A., Soares, C. (eds.) ECML PKDD 2015. LNCS, vol. 9285, pp. 260–275. Springer, Heidelberg (2015)
11. Jeffrey, R.C.: The Logic of Decision. McGraw-Hill, New York (1965), 2nd edn.: University of Chicago Press, Chicago, (1983). Paperback Edition (1990)
12. McLachlan, G., Krishnan, T.: The EM Algorithm and Extensions. Wiley, New York (2007)
13. Russell, S.: The EM algorithm, Lecture notes CS 281, Computer Science Department University of California, Berkeley (1998)
14. Schafer, J.L.: Multiple imputation: a primer. Stat. Methods Med. Res. **8**, 3–15 (1999)
15. Ting, A., Lin, H.: Comparison of multiple imputation with EM algorithm and MCMC method for quality of life missing data. Qual. Quant. **44**, 277–287 (2010)
16. Williams, P.: Bayesian conditionalization and the principle of minimum information. Brit. J. Philos. Sci. **31**, 131–144 (1980)

Causal Belief Inference in Multiply Connected Networks

Oumaima Boussarsar(✉), Imen Boukhris, and Zied Elouedi

LARODEC, Institut Supérieur de Gestion de Tunis,
Université de Tunis, Tunis, Tunisia
oumaima.boussarsar@hotmail.fr, imen.boukhris@hotmail.com,
zied.elouedi@gmx.fr

Abstract. The belief function theory is an efficient tool to represent causal knowledge under uncertainty. Therefore, causal belief inference process is important to evaluate the impact of an observation or an intervention on the system. However, existing algorithms only deal with the propagation of observational data in belief networks. This paper addresses propagation algorithms of causal knowledge in multiply connected causal belief networks. To handle this propagation, we have first to transform the initial network into a tree structure. Therefore, we propose some modifications to construct a new structure by exploiting independence relations in the initial network. This structure is called hybrid binary join tree composed of conditional distributions and non conditional ones. Then, we develop a causal belief propagation algorithm using the belief graph mutilation or the graph augmentation methods.

Keywords: Belief function theory · Causality · Causal belief networks · Hybrid binary join tree · Propagation process · Interventions

1 Introduction

Causality is an important concept involved in many fields. It amounts to determine what truly causes what and what it matters. It allows to describe, interpret and analyze information and events and it plays an important role in the expression of our perception of our environment. Besides, it enables to identify if two events are related in a causal way or not, using interventions which are external actions that force a target variable to have a specific value. While conditioning is used to compute the effect of observations, the "do" operator [12] is used as a tool to represent interventions on causal networks. Thus, this operator is used to compute the impact of external action [5].

Bayesian networks [11] are successful graphical models. However, probability distribution does not naturally distinguish between equiprobability and ignorance situations. To tackle this problem, some researchers [2,3,17,23] have proposed alternative networks where uncertainty is quantified with the belief function theory [13]. This latter is well established as a general framework for

J.P. Carvalho et al. (Eds.): IPMU 2016, Part II, CCIS 611, pp. 291–302, 2016.
DOI: 10.1007/978-3-319-40581-0_24

reasoning with uncertainty, and has well understood connections to other frameworks such as probability, possibility and imprecise probability theories. A causal belief network [5] is a graphical structure. It is used to represent causal relations between nodes under the belief function framework. Two different graphical approaches to represent interventions in causal belief networks are provided namely, the mutilated and the augmented based approaches [5].

A network is said to be multiply connected if there are several paths between two variables in the network. Since, the adaptation of propagation algorithms in singly connected graphs (i.e., where there is at most one directed path between any two nodes) [6, 7] is not appropriate, the idea in this case is to transform the initial graph (a multiply connected graph) into a tree structure (a singly connected graph) such as join trees (JT) [16], binary join trees (BJT) [15], modified binary join (MBJT) [1] trees and junction trees [10] to remove the loops in order to avoid infinite traffic messages.

In this paper, we explain how we can propagate interventions in multiply connected networks. Since, there are some problems when propagating beliefs in the modified binary join tree [8], we present first a new structure called hybrid binary join tree based on conditional nodes and non-conditional ones. We propose some improvements in cases representing some ambiguities. This structure uses the two rules proposed by Smets: the disjunctive rule of combination (DRC) and the generalized Bayesian theorem (GBT) between nodes having a conditional node and extension and marginalisation between non-conditional nodes. The second contribution consists of presenting an algorithm of propagation of interventions in multiply connected networks using the mutilated and the augmented approaches on this data structure. We explain that this algorithm allows the propagation of several observations and interventions.

The rest of the paper is organized as follows: Sect. 2 presents the basic concepts of the belief function theory. In Sect. 3, we recall belief networks. The causal belief propagation algorithm in hybrid binary join tree is described in Sect. 4 as well as the changes made to the construction of the tree structure. Section 5 concludes the paper.

2 Belief Function Theory

The belief function theory [13, 20] represents an appropriate framework for experts to express their beliefs in a flexible way and to reason under uncertainty.

Let Θ the finite non empty set including n elementary events representing the solutions of a given problem. These events are assumed to be exhaustive and mutually exclusive. The set Θ is called the frame of discernment. Beliefs are expressed on subsets belonging to the powerset of Θ denoted 2^{Θ}. The basic belief assignment (*bba*), denoted by m^{Θ} or m, is a mapping from 2^{Θ} to [0,1] such that: $\sum_{A \subseteq \Theta} m(\mathrm{A}){=}1$.

$m(A)$ is a basic belief mass (*bbm*) assigned to the event A. It represents the part of belief exactly committed to the event A of Θ. Subsets of Θ such that $m(\mathrm{A}) > 0$ are called focal elements. The mass function respecting the constraint m(∅) = 0 is called normalized. A *bba* is said to be certain if the whole mass is allocated to a unique singleton of Θ and Bayesian when all focal elements are singletons. If the *bba* has Θ as a unique focal element, it is called vacuous and it represents the case of total ignorance.

The belief function theory offers interesting tools for aggregating basic belief assignments. Two *bbas* provided by two distinct and independent sources m_1 and m_2 may be combined to give one resulting mass using the Dempster rule of combination (several explanations for the origin and the unicity of this rule are given in [9,18]).

$$m_1 \oplus m_2(A) = K \cdot \sum_{B \cap C = A} m_1(B)m_2(C), \forall B, C \subseteq \Theta \tag{1}$$

where $K^{-1} = 1 - \sum_{B \cap C = \emptyset} m_1(B)m_2(C)$.
Dempster's rule of conditioning allows to update the knowledge of an expert following the new information. $m(A|B)$ denotes the degree of belief of A in the context of B with A, B $\subseteq \Theta$.

$$m(A|B) = \begin{cases} K. \sum_{C \subseteq \bar{B}} m(A \cup C) & \text{if } A \subseteq B, A \neq \emptyset \\ 0 \text{ if } A \not\subseteq B \end{cases} \tag{2}$$

where K^{-1}=1 − $m(\emptyset)$.
Vacuous extension is useful when new variables are added to the referential. It allows to express the marginal mass function m^{Θ} defined on Θ over the frame $\Theta \times \Omega$ as follows:

$$m^{\Theta \uparrow \Theta\Omega}(B) = m^{\Theta}(A) \;\; if \;\; B = A \times \Omega \tag{3}$$

$$such \;\; that \;\; A \subseteq \Theta, B \subseteq \Theta \times \Omega$$

Given the product space $\Theta \times \Omega$ and a mass distribution defined on this product space. Marginalization allows to map over a subset of the product space by dropping the extra coordinates.

$$m^{\Theta\Omega \downarrow \Theta}(A) = \sum_{C \subseteq \Theta \times \Omega, C^{\downarrow \Theta} = A} m^{\Theta\Omega}(C), A \subseteq \Theta \tag{4}$$

The ballooning extension [4,14] is useful when, after conditioning, an expert would reconstruct the initial distribution. The ballooning extension transforms a mass belief function $m(A|\omega)$ defined on Θ given $\omega \in \Omega$ into a new mass function over $\Theta \times \Omega$. To get rid of conditioning, we have to compute the ballooning extension defined as:

$$m_{\omega}^{\Theta \Uparrow \Theta\Omega}(C) = \begin{cases} m^{\Theta}(A|\omega) \;\; if \;\; C = (A \times \omega \cup \Theta \times \bar{\omega}) \\ 0 \;\; otherwise \end{cases} \tag{5}$$

$$such \;\; that \;\; A = (C \cap \omega^{\Theta\Omega})^{\downarrow \Theta}$$

3 Causal Belief Networks

Causal belief networks [5] are graphical models representing imperfect causal knowledge under an uncertain environment where the uncertainty is quantified by mass distributions. They represent an efficient model due to their capacity to handle causal relationships between variables. These graphical models represent an alternative and an extension to causal Bayesian networks [21,22] that offer interesting tools to handle interventions. They allow the detection of causal relationships under the belief function framework. Moreover, they offer a flexible way to define conditional beliefs without the need to specify all a priori distributions using vacuous *bbas*. They are defined on two levels:

- Qualitative level: represented by a directed acyclic graph (DAG) G=(N,E), where nodes N represent different random variables and edges E encode the causal links among variables. Parents of the node A denoted by $PA(A)$ are their immediate causes and children of A denoted by $CH(A)$ are considered as their effects. $PA_j(A)$ is a single parent of A. A subset from the set of the parents of A is denoted by $Pa(A)$ and a subset from $PA_j(A)$ is denoted by $Pa_j(A)$. The conditional nodes are denoted by C.
- Quantitative level: represented by the set of mass distributions associated to each node in the graph. Conditional distributions can be defined for each subset of each variable A_i ($sub_{ik} \subseteq \Theta_{A_i}$) in the context of its parents such that: $\sum_{sub_{ik} \subseteq \Theta_{A_i}} m^{A_i}(sub_{ik}|Pa(A_i)) = 1$

In order to predict the effects of external actions on the system, the construction of the belief causal network must be different from belief network and the conditioning on an observation should be distinguished from a conditioning on an external action. Handling interventions and computing their effects on the system can be done by making changes on the structure of the belief causal network [5]: after acting on a variable, we assume that its initial causes are no more responsible of its state. Accordingly, arcs linking the variable of interest to its parents should be deleted. The resulting graph is a mutilated causal belief network. Another alternative is to add a new fictitious variable "DO" as a parent node of the variable concerned by an intervention. This added variable will totally control its status. The resulting graph is called an augmented causal belief network.

4 Causal Inference in Hybrid Binary Join Tree

The inference process is used to evaluate the influence of the various observations and interventions of the system using equivalent local computations. To ensure the belief causal inference, we have to compute the effect of observations and interventions with two different equivalent ways, i.e., mutilating the graph or using the augmented graph method.

In order to propagate beliefs efficiently, the computation of joint belief functions on the product space is involved using the disjunctive rule of combination (DRC) and the generalized Bayesian theorem (GBT) [19].

Let us consider $pl^{\Omega}(c|a_i)$ and $a_i \in a$ where $a \subseteq \Theta$ and $c \subseteq \Omega$. Θ or Ω may well represent the frame of discernment of one variable or multi-variables.

The message sent from the parent node A to its child C using the DRC is defined as follows:

$$pl^{\Omega}(c) = \sum_{a \subseteq \Theta} m^{\Theta}(a)(1 - \prod_{a_i \in \Theta} (1 - pl^{\Omega}(c|a_i))) \tag{6}$$

The message sent from the child node to its parent node A using the GBT is defined as follows:

$$pl^{\Theta}(a) = \sum_{c \subseteq \Omega} m^{C}(c)(1 - \prod_{a_i \in \Theta} (1 - pl^{\Omega}(c|a_i))) \tag{7}$$

4.1 Hybrid Binary Join Tree

Since in binary join tree [15], computation time is minimized while a lot of memory space is required to store intermediate results, authors in [1] proposed a refinement of the binary join tree, called modified binary join tree which exploits the structure of the original belief network by transforming it into a binary join tree and then doing some modifications by manipulating conditional relations between variables instead of joint distributions.

The construction of the modified binary join tree follows three steps [1]:

- arrange the subsets the hypergraph in a binary join tree where each node has no more than three neighbors
- attach singleton subsets to the binary join tree and always ensure that the tree is binary
- replace circle containing the list of these variables by a rectangle containing the conditional relations between variables

As explained in [8], the construction of the modified binary join tree have some limitations:

- Limitation 1: sometimes it is not possible to distinguish between the child and the parent nodes
- Limitation 2: conditional nodes on the MBJT represented by rectangle nodes do not correspond to the ones initially represented on the original network.
- Limitation 3: conditional distributions initially defined per edge are represented conditional distributions defined for all parents on the MBJT.

To tackle these problems, we first propose in this paper a new solution to make the propagation process efficient so called hybrid binary join tree. It is a combination of the two structure "the modified binary join tree" exploiting the independence relations and "the binary join tree".

- Solution for limitation 1: do not replace the node containing the list of variables by a conditional one. Replace the circle containing the list of variables (non-conditional nodes) with a rectangle nodes representing the a priori conditional distribution in the case where the parent and child nodes are present in the initial network.
- Solution for limitation 2: do not replace the node containing the list of variables by a conditional one.
- Solution for limitation 3: combine the distributions of the parent nodes in a fictional node using Dempster's rule of combination [7].

The propagation process will be based on the two rules DRC and GBT (Eqs. 6 and 7) between non-conditional nodes having between them a conditional node. Otherwise, extension and marginalisation (Eqs. 3 and 4) will be used between nodes containing a list of variables.

Example 1. *Let us consider the multiply connected belief network depicted in Figure* 1 *originally constituted by 8 nodes* $U=\{A,T,L,E,S,B,D,X\}$ *representing the variables of the problem (note that we use the same network proposed by* [1] *to explain the limitations and the proposed modifications).*

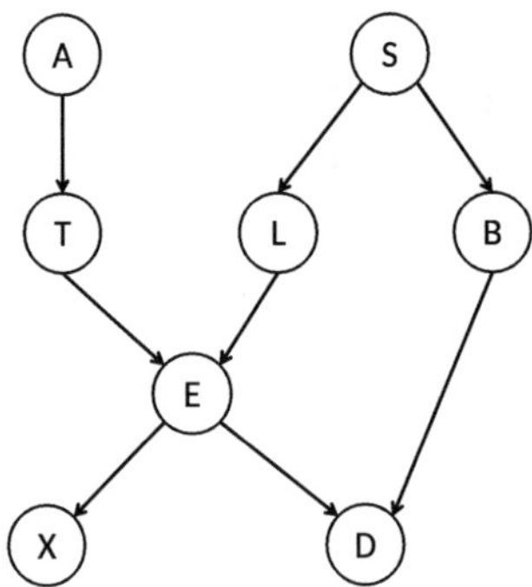

Fig. 1. The multiply connected belief network.

In the modified binary join tree represented in Figure 2, *we cannot deduce which is the parent or the child nodes between* $\{T,E\}$ and $\{L,E\}$ *given the conditional node* $\{E|T,L\}$ *to perform the propagation. Besides, the node* $\{L,B|S\}$ *is not a conditional node in the initial network (we only have conditional mass distributions* $\{L|S\}$ *and* $\{B|S\}$*). Accordingly, we cannot found the mass distribution of the conditional node* $\{L,B|S\}$. *Moreover, for* $\{D|E,B\}$, *initially the conditional distributions are defined per edge. So, the mass distribution represented in the modified binary join tree is not the same. The suggested modifications are represented in the hybrid binary join tree depicted in Figure* 3.

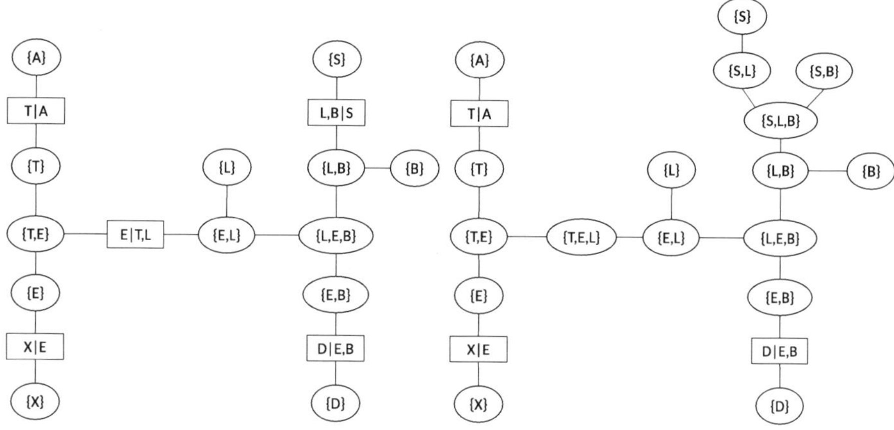

Fig. 2. The corresponding modified binary join tree.

Fig. 3. Hybrid binary join tree.

4.2 Propagation Algorithm in Hybrid Binary Join Tree

To ensure the propagation using DRC and GBT rules, some elements of the message passing scheme have to be used. When receiving a message, each node updates three local vectors, the vector $m(x_1, \cdots, x_n)$ (the mass distribution), the vector $\pi(x_1, \cdots, x_n)$ (the vector concerning messages received by its parents) and the vector $\lambda(x_1, \cdots, x_n)$ (the vector concerning messages received by its children). Each node sends and receives messages from its neighbors, the π-message (the message sent from a parent node to a child node) and the λ-message (the message sent from a child node to a parent node). A post-order (in direction of pivot) and a pre-order (from pivot to other nodes) will be defined to propagate information.

In this section, we propose new propagation algorithms for multiply connected causal belief networks where beliefs are quantified by conditional masses. To propagate beliefs in the causal multiply connected network, we have to transform it into an hybrid binary join tree. Then, we will perform the propagating algorithm which comprises three steps (see Fig. 4):

1. the initialization phase in which all vectors are instantiated
2. the collect of information algorithm will be performed in which the nodes send messages towards the pivot node
3. the distribution of information algorithm will be applied where messages are sent from the pivot to other nodes

In initialization step, the algorithm initializes all fields π, λ and m.

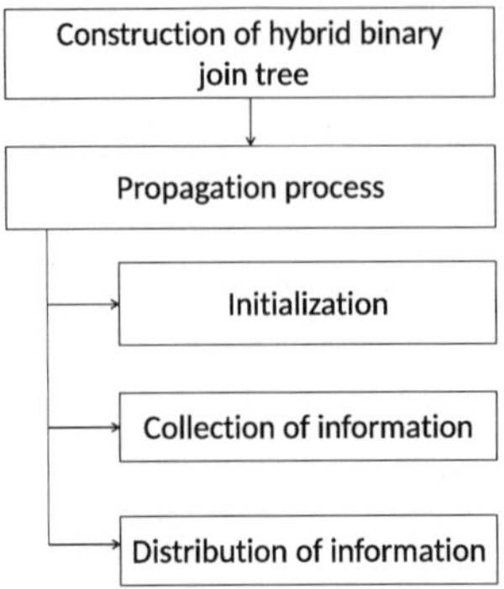

Fig. 4. Propagation process

```
Algorithm. Initialization
For i from 1 to length(U)
|  m^X ← m_0^X
|  π_X ← m^X
|  λ_X ← vacuous bba
End for
```

To collect information, if the adjacent node is a conditional node, a message is sent to the next node. Otherwise, there will be an extension to the product space and a combination.

```
Algorithm. Collect of information
For each A ∈ post-order
|  If adj_node in C
|  |  If adj_node in PA(adj_node)
|  |  |  Send a π message
|  |  |  else
|  |  |  send a λ message
|  |  End if
|  |  else
|  |  Extend to the product space.
|  |  Combine the mass distributions.
|  End if
End for
```

To distribute information, if the adjacent node is a conditional node, a message is sent to the next node. Otherwise, there will be an extension to the product space and a combination.

Algorithm. distribution of information

```
For each A ∈ pre-order
  If adj_node in C
    If adj_node in PA(adj_node)
      Send a π message
      else
      send a λ message
    End if
    else
    Extend to the product space.
    Combine the mass distributions.
  End if
End for
```

4.3 Inference in the Mutilated Modified Binary Join Tree

The external action $do(a_{ij})$ imposes a specific value a_{ij} to the variable A_i. The conditional distribution of the target variable becomes a certain *bba* which is defined as follows:

$$m^{A_i}_{G_{mut}}(sub_{ik}) = \begin{cases} 1 & \text{if } sub_{ik} = \{a_{ij}\} \\ 0 & \text{otherwise} \end{cases} \tag{8}$$

Propagation in this graph consists of three main steps: the mutilation step, the construction of the hybrid binary join tree and the propagation process.

Algorithm. Propagation using the mutilated based approach

```
  Cutting all edges pointing to the node concerned by the intervention.
  Construction of the hybrid binary join tree.
  Choose a pivot.
  Initialization.
  Collect of information.
  Distribution of information.
```

Note that using this approach, we had to construct a new hybrid binary join tree for new interventions.

4.4 Inference in the Augmented Modified Binary Join Tree

After adding the "DO" node, the conditional distribution of the node concerned by the intervention A given parent nodes must be updated. The parents set of the variable A is transformed to $PA' = PA \cup \{DO\}$. The DO node takes values in $do(x)$, $x \in \{\Theta_A \cup \{nothing\}\}$. do(*nothing*) represents the case when no interventions are made. When the DO node is taking the value a_i. This means that a certain action succeeds to put its target variable A at a precise value which makes it completely independent of its original causes. Let PA(A) be the parents of the variable A except the DO node, the conditional distribution of A is defined as follows:

$$m(a_k|PA(A), do(x)) = \begin{cases} 1 \text{ if } x = a_i \\ 0 \text{ if } x \neq a_i \\ m(a_k|PA(A), do(x)) \;\; x = nothing \end{cases} \tag{9}$$

Propagation using this approach consists of the augmentation of the graph. The conditional distribution of the node concerned by the intervention becomes certain. Then, the propagation process described above is handled.

Algorithm. Propagation using the augmented based approach

Add the node DO as a parent of the node concerned by the intervention.
Updating its conditional mass distribution.
Construction of the MBJT.
Choose a pivot.
Initialization.
Collect of information.
Distribution of information.

The use of the augmented approach allows to represent the effect of interventions and also observations. The main advantage of this approach in multiply connected networks, is that the reconstruction of the tree structure after each evidence can be avoided. Indeed, it is possible to add the DO node before or after the construction of the hybrid binary join tree. The two methods are equivalent since the nodes resulting from the augmented hybrid binary join tree and the nodes resulting from hybrid binary join tree obtained from the augmented initial network are the same.

Proposition 1. *Let G be the initial network. G_{aug} represents the augmented causal belief network where interventions are initially represented and $G^{'}$ its*

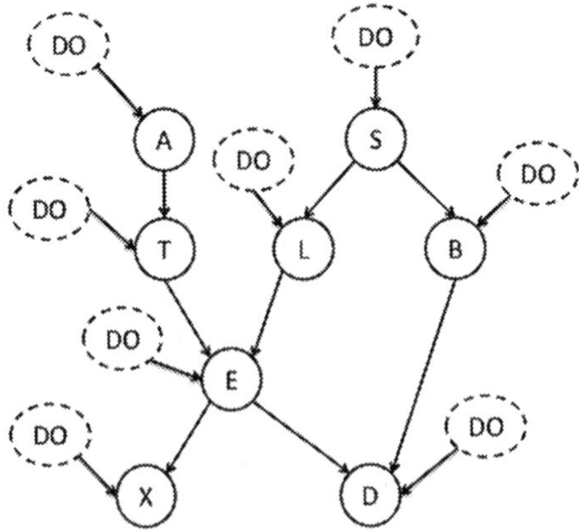

Fig. 5. An augmented multiply connected network.

corresponding hybrid binary join tree. Let G'' be the hybrid binary join tree constructed from the G and then augmented by adding DO nodes. These two structures (G' and G'') are equivalent.

By adding the DO node as a parent for each variable, the structure of the initial network graph is stable regardless the number of observations and interventions. There is no need to build another hybrid binary join tree. The conditional distribution given the DO node is obtained using Eq. 9. In the case where a node in not concerned by an intervention, it takes the value nothing by avoiding the regeneration of the hybrid binary join tree for each evidence (see Fig. 5).

5 Conclusion

In this paper, we explained how we can propagate observations and interventions in causal multiply connected networks using the hybrid binary join tree. We proposed a propagation algorithm on this structure allowing the inference of several observations and interventions in the augmented and mutilated graphs. As future work, we intend to treat inference of non standard interventions in singly and multiply connected networks. Inference in causal belief networks can be used in several applications like those allowing the intrusion detection and or ensuring system reliability.

References

1. Ben Yaghlane, B., Mellouli, K.: Inference in directed evidential networks based on the transferable belief model. Int. J. Approximate Reasoning **48**, 399–418 (2008)
2. Ben Yaghlane, B., Smets, P., Mellouli, K.: Directed evidential networks with conditional belief functions. In: Nielsen, T.D., Zhang, N.L. (eds.) ECSQARU 2003. LNCS (LNAI), vol. 2711, pp. 291–305. Springer, Heidelberg (2003)
3. Boukhris, I., Benferhat, S., Elouedi, Z.: Representing belief function knowledge with graphical models. In: Xiong, H., Lee, W.B. (eds.) KSEM 2011. LNCS, vol. 7091, pp. 233–245. Springer, Heidelberg (2011)
4. Boukhris, I., Elouedi, Z., Benferhat, S.: Analyzing belief networks with conditional beliefs. In International Conference on Intelligent Systems Design and Applications, pp. 959–964. IEEE Computer Society (2011)
5. Boukhris, I., Elouedi, Z., Benferhat, S.: Dealing with external actions in causal belief networks. Int. J. Approximate Reasoning **54**(8), 978–999 (2013)
6. Boussarsar, O., Boukhris, I., Elouedi, Z.: A direct propagation method in singly connected causal belief networks with conditional distributions for all causes. In: Aranda-Corral, G.A., Calmet, J., Martín-Mateos, F.J. (eds.) AISC 2014. LNCS, vol. 8884, pp. 64–75. Springer, Heidelberg (2014)
7. Boussarsar, O., Boukhris, I., Elouedi, Z.: Inference in hybrid causal belief networks. In: IEEE International Symposium on Computational Intelligence and Informatics, pp. 285–290 (2014)
8. Boussarsar, O., Boukhris, I., Elouedi, Z.: Inference in junction trees using the belief function theory. In: International Conference on Intelligent Systems Design and Applications (2015)

9. Dubois, D., Prade, H.: On the unicity of Dempster rule of combination. Int. J. Intell. Syst. **1**, 133–142 (1986)
10. Jensen, F.V., Lauritzen, S.L., Olesen, K.G.: Bayesian updating in causal probabilistic networks by local computation. Comput. Stat. Q. **4**, 269–282 (1990)
11. Pearl, J.: Graphical models for probabilistic and causal reasonning. In: Gabbay, D.M., Smets, P. (eds.) The Handbook of Defeasible Resonning and Uncertainty Management Systems, pp. 367–389 (1998)
12. Pearl, J.: Causality: Models, Reasonning and Inference. Cambridge University Press, New York (2000)
13. Shafer, G.: A Mathematical Theory of Evidence. Princeton Univ. Press, Princeton (1976)
14. Shafer, G.: Belief functions and parametric models. J. Roy. Stat. Soc. B **44**, 322–352 (1982)
15. Shenoy, P.: Binary join trees for computing marginals in the Shenoy-Shafer architecture. Int. J. Approximate Reasoning **17**(2), 239–263 (1997)
16. Shenoy, P., Shafer, G.: Axioms for probability and belief functions propagation. In: Uncertainty in Artificial Intelligence (UAI), pp. 159–198 (1990)
17. Simon, C., Weber, P., Evsukoff, A.: Bayesian networks inference algorithm to implement Dempster-Shafer theory in reliability analysis. Reliab. Eng. Syst. Saf. **93**, 950–963 (2008)
18. Smets, P.: The combination of evidence in the transferable belief model. IEEE Pattern Anal. Mach. Intell. **12**, 447–458 (1990)
19. Smets, P.: Belief functions: the disjunctive rule of combination and the generalized Bayesian theorem. Int. J. Approximate Reasoning **9**(1), 1–35 (1993)
20. Smets, P.: The application of the transferable belief model to diagnostic problems. Int. J. Intell. Syst. **13**(2–3), 127–157 (1998)
21. Spirtes, P., Glymour, C., Scheines, R.: Causation, Prediction and Search, vol. 1. MIT Press, Cambridge (2001)
22. Verma, T., Pearl, P.: Equivalence and synthesis of causal models. In: Uncertainty in Artificial Intelligence, pp. 255–270 (1990)
23. Xu, H., Smets, P.: Evidential reasoning with conditional belief functions. In: Uncertainty in Artificial Intelligence, pp. 598–606 (1994)

Databases and Information Systems

Indexing Possibilistic Numerical Data: The Interval B$^+$-tree Approach

Guy De Tré(✉), Robin De Mol, and Antoon Bronselaer

Department of Telecommunications and Information Processing, Ghent University, Sint-Pietersnieuwstraat 41, 9000 Ghent, Belgium
{Guy.DeTre,Robin.DeMol,Antoon.Bronselaer}@UGent.be

Abstract. When record sets become large, indexing becomes a required technique for speeding up querying. This holds for regular databases, but also for 'fuzzy' databases. In this paper we propose a novel indexing technique, supporting the querying of imperfect numerical data. A possibility based relational database setting is considered. Our approach is based on a novel adaptation of a B$^+$-tree, which is currently still one of the most efficient indexing techniques for databases. The leaf nodes of a B$^+$-tree are enriched with extra data and an extra tree pointer so that interval data can be stored and handled with them, hence the name Interval B$^+$-tree (IBPT). An IBPT allows to index possibility distributions using a single index structure, offering almost the same benefits as a B$^+$-tree. We illustrate how an IBPT index can be used to index fuzzy sets and demonstrate its benefits for supporting 'fuzzy' querying of 'fuzzy' databases. More specifically, we focus on the handling of elementary query criteria that use the so-called compatibility operator IS, which checks whether stored imperfect data are compatible with user preferences (or not).

Keywords: Indexing · Possibilistic databases · B$^+$-tree

1 Introduction

Query processing can be time-consuming. Regular database management systems cope with this problem by providing indexing mechanisms. In its simplest form, an index can be seen as an ordered file of attribute values where each value has an associated disk block address, denoting in which disk block a database record with that value is stored. Among the most popular indexing mechanisms for handling numerical data is the B$^+$-tree [1].

However, when considering the management of imperfect data in a database system, indexing mechanisms must be adapted in order to efficiently cope with the data format in which the imperfect data are represented. A commonly used technique for representing imperfect data is to model the uncertainty that is caused by the imperfections, by means of a possibility distribution [11]. This approach is known as the possibility based database modelling approach and was originally presented in [8]. A possibility distribution can be represented by

J.P. Carvalho et al. (Eds.): IPMU 2016, Part II, CCIS 611, pp. 305–316, 2016.
DOI: 10.1007/978-3-319-40581-0_25

a fuzzy set, which on its turn is characterized by a membership function [11]. For practical reasons, membership functions for possibility distributions over numerical data are often approximated by trapezoidal shape functions.

Hence, indexes for imperfect numerical data should support the efficient querying of 'trapezoidal' data (instead of numerical data). The research presented in this paper, aims to contribute in the study of such indexing techniques.

Although very relevant and indispensable for real-world databases, not many researchers have studied indexes for imperfect data. In [3], Bosc and Galibourg proposed an indexing principle that aims to preselect and efficiently retrieve only those records that potentially satisfy the query condition under consideration. This principle forms the basis for the work presented in this paper. Boss and Helmer studied the use of superimposed coding to annotate the support and core of a possibility distribution and use these annotations for index construction [4]. In [9,10], Yazici and Cibiceli proposed to use a single, multidimensional index for indexing similarity-based fuzzy data. This technique is only applicable for attribute domains that consist of a finite number of fuzzy values that are represented by linguistic labels. A technique, based on so-called G-trees, has been presented by Liu et al. [7]. This technique, called 1GT, is suitable for indexing convex possibility distributions, of which the support is mapped to a two-dimensional data point, with a single G-tree. Finally, in [2] Barranco et al. presented a B^+-tree based indexing technique, called 2BPT, for fuzzy numerical data where two indexes are constructed: one for the lower bounds of the supports of the trapezoidal possibility distributions, and one for their upper bounds. The efficiency of 2BPT is similar to the efficiency of GT, however 2BPT is more stable and much easier to implement as it is based on B^+-trees [2].

In this paper we propose the Interval B^+-tree (IBPT) indexing technique, which is an alternative to 2BPT using only one IBPT. The idea is to extend the concept of a B^+-tree, so that it is able to store and handle numerical intervals. This new tree structure is called IBPT. An IBPT can then be used to store and index the lower bound and upper bound points of the support of the possibility distributions representing the values of a given 'fuzzy' database attribute. This leads to lesser storage needs (only one IBPT is needed) and a faster preselection for query condition checking (only one IBPT has to be traversed).

The remainder of the paper has the following structure. In Sect. 2 some preliminaries on possibilistic data modelling and B^+-trees are described. The novel IBPT indexing technique is presented in Sect. 3. In Sect. 4 an illustrative example, demonstrating the practical use of IBPT indexing is presented and discussed. Finally, in Sect. 5 some conclusions and directions for further work are given.

2 Prelimaries

In this section the basic concepts of the possibility based database modelling approach and B^+-trees are briefly described.

2.1 Possibility Based Database Modelling

Possibility based database modelling has been introduced for relational databases [5] by Prade and Testemale in [8] and is based on possibility theory [6,11]. Consider a relation R with schema $R(A_1 : T_1, \ldots, A_n : T_n)$ of a relational database. R is characterized by a finite attribute set $\{A_1 : T_1, \ldots, A_n : T_n\}$, where each attribute $A_i : T_i$, $i = 1, \ldots, n$ has a name A_i and associated data type T_i implying that A_i can take values from the domain (set of allowed values) dom_{T_i} of T_i. Hence, for every tuple $t(A_1 : v_1, \ldots, A_n : v_n)$ of R it holds that $v_i \in dom_{T_i}$. In traditional databases, all attribute values in tuples should be crisp or null, the latter denoting that the actual value is unknown or undefined. This approach does not support the explicit handling of imperfect data.

The basic idea in possibility based database modelling is to model the uncertainty that characterizes imperfect data by means of a possibility distribution. A possibility distribution reflecting the available knowledge about which candidate values are plausible (or less plausible) to be the actual value of an attribute $A : T$, is a mapping $\pi_A : dom_T \rightarrow [0, 1]$. Extreme forms of partial knowledge like complete knowledge (i.e. $\exists v' \in dom_T : \pi_A(v') = 1$ and $\forall v \neq v' \in dom_T : \pi_A(v) = 0$) and complete ignorance (i.e. $\forall v \in dom_T : \pi_A(v) = 1$) can be reflected. Each possibility distribution π_A is normalized, which implies that $\exists v \in dom_T : \pi_A(v) = 1$. A possibility distribution π_A can be modelled by means of a fuzzy set F with membership function $\mu_F : dom_T \rightarrow [0, 1]$. In such a case, $\forall v \in dom_T : \pi_A(v) = \mu_F(v)$. Every normalized membership function can be used but, for practical reasons, often trapezoidal membership functions are used for numerical domains. Trapezoidal membership functions can be fully characterized by a quadruple consisting of the values values a (lower bound of the support), b (lower bound of the core), c (upper bound of the core) and d (upper bound of the support). This is illustrated in Fig. 1.

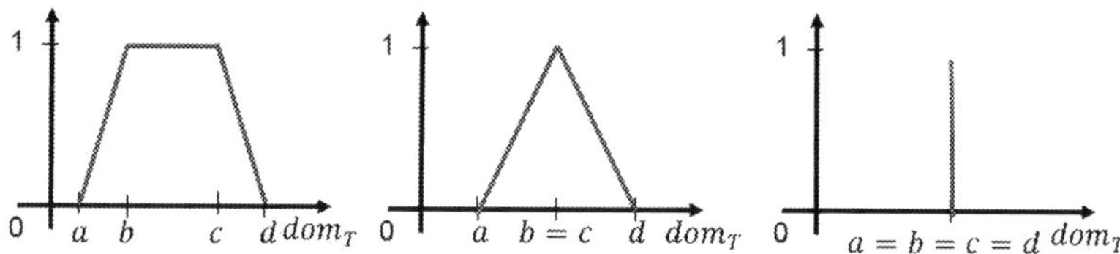

Fig. 1. Examples of trapezoidal membership functions, characterized by a quadruple (a, b, c, d).

If uncertainty is reflected in the data, this uncertainty propagates to the query results and should therefore not be neglected. In a possibility based approach this uncertainty is often reflected by a pair consisting of a possibility degree and necessity degree. These degrees respectively denote to which extent it is plausible and to which extent it is certain that the (stored value in the) tuple satisfies a given query criterion. A basic facility for querying imperfect data is the so-called compatibility operator IS. In its general form

$$A\ IS\ P \tag{1}$$

the operator checks whether the stored value for attribute A is possibly and necessarily compatible with the preferences P (for A) as specified by the user. For the sake of generality, it is hereby assumed that the user preferences P are expressed by means of a (trapezoidal) membership function $\mu_P : dom_T \rightarrow [0, 1]$. For each $v \in dom_T$, $\mu_P(v)$ then expresses to which extent v corresponds to the user's preferences regarding the values of A. The user preferences are hence defined by all $v \in dom_T : \mu_P(v) \neq 0$. Consider a tuple $t(A_1 : v_1, \ldots, A_n : v_n)$ of R and let the attribute value of $A : T$ in t be shortly denoted by $t[A]$. In case of imperfect data, $t[A]$ will be a possibility distribution, which will be denoted by $\pi_{t[A]}$. With these considerations, the definition of the evaluation of $A\ IS\ P$ given in [6] can be written as

$$\Pi(A\ IS\ P)(t) = \sup_{v \in dom_T} \min(\pi_{t[A]}(v), \mu_P(v)) \tag{2}$$

and

$$\begin{aligned} N(A\ IS\ P)(t) &= \inf_{v \in dom_T} \max(1 - \pi_{t[A]}(v), \mu_P(v)) \\ &= 1 - \sup_{v \in dom_T} \min(\pi_{t[A]}(v), 1 - \mu_P(v)) \end{aligned} \tag{3}$$

where $\Pi(A\ IS\ P)(t)$ is the possibility degree and $N(A\ IS\ P)(t)$ is the necessity degree. The index proposed in this paper aims to support the evaluation of queries that contain a criterion that uses the compatibility operator IS.

2.2 B$^+$-Trees

B$^+$-trees [1] are considered to be among the most efficient techniques used for indexing numerical data in traditional databases. A B$^+$-tree is a tree structure where each node corresponds to a disk block (the unit of data transfer in hard disk drive I/O operations). The tree is balanced, which implies that all leaf nodes are at the same level. Because data pointers are only stored in leaf nodes, all data can be accessed with the same number of disk accesses (to traverse the tree from the root to a leaf node and one extra to load the disk block that contains the data). All nodes in a B$^+$-tree are kept between half full and full (except for the root node if this is the only node in the tree).

As depicted in Fig. 2, each internal node of a B$^+$-tree of order p can contain at most p tree pointers P_i, $i = 1, \ldots, q$, $q \leq p$ and $p - 1$ search values K_i, $i = 1, \ldots, q - 1$, $q \leq p$. The search values in the node are sorted such that $K_1 < K_2 < \cdots < K_{q-1}$. Each tree pointer P_i refers to the root node of a subtree S_i. If $1 < i < q$, then each search value K in S_i satisfies $K_{i-1} \leq K < K_i$. Each search value K in S_1 satisfies $K < K_1$, and each search value K in S_q satisfies $K_{q-1} \leq K$. Every internal node should be kept at least half full, i.e. it should contain at least $\lceil (p/2) \rceil$ tree pointers and $\lceil (p/2) \rceil - 1$ search values.

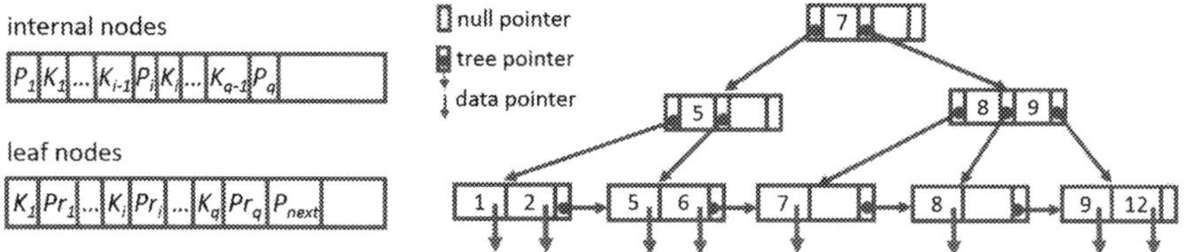

Fig. 2. The structure of internal and leaf nodes (left) and an example of a B$^+$-tree of orders $p = 3$ and $p_{leaf} = 2$ (right).

Leaf nodes are structured in a different way because these are the only nodes that contain data pointers that refer to the disk blocks of stored data (containing records with the search value). Usually leaf nodes also have a different order p_{leaf}. As depicted in Fig. 2, each leaf node can contain at most p_{leaf} search values K_i and p_{leaf} data pointers Pr_i, $i = 1, \ldots, q$, $q \leq p_{leaf}$. It also contains a tree pointer P_{next} that refers to the next leaf node in the tree. This supports efficient sequential traversal of the indexed data (in ascending order of the search values). The search values in each node are sorted such that $K_1 < K_2 < \cdots < K_q$. Every leaf node, except if it is the root node, should be kept at least half full, i.e. it should contain at least $\lceil (p_{leaf}/2) \rceil$ data pointers and $\lceil (p_{leaf}/2) \rceil$ search values.

In the right hand side of Fig. 2, an example of a B$^+$-tree is shown. The values in this tree are inserted in the order 8, 5, 1, 7, 2, 12, 9, and 6. B$^+$-trees are dynamically constructed. Inserting data can result in new nodes, which can eventually propagate to a tree growth (adding one level). Deleting data can result in the removal of nodes, which can propagate to a tree shrinkage (removing one level). Efficient algorithms for constructing (node insertion and removal) and searching B$^+$-trees are described in [1] and other works on data structuring.

3 Interval B$^+$-trees

In this section we present our proposal for indexing database attributes that contain imperfect numerical data. Considering the possibility based setting as described in Sect. 2, we hereby assume that imperfect data are modelled by trapezoidal possibility distributions. As illustrated in Fig. 1, each such a trapezoidal distribution is fully specified by a quadruple (a, b, c, d).

3.1 Background

Our goal is to support the processing of the compatibility operator $A\ IS\ P$ (cf. Eqs. (2)-(3)), where P is modelled by a trapezoidal fuzzy set. Without an index, for each tuple t of R, the stored possibility distribution $\pi_{t[A]}$ has to be compared with the given fuzzy set P. If R contains a lot of tuples, this can be time-consuming. This is why Bosc and Galibourg proposed an indexing principle to preselect only those possibility distributions $\pi_{t[A]}$ that overlap with the user preferences P [3]. Indeed, for possibility distributions that do not overlap with P, it holds that $\Pi(A\ IS\ P) = N(A\ IS\ P) = 0$.

The preselection principle of Bosc and Galibour forms the basis of the 2BPT indexing technique proposed by Barranco et al. [2]. In 2BPT indexing, two B$^+$-trees are used. Assume that the possibility distribution $\pi_{t[A]}$ for the attribute A in tuple t is specified by the quadruple $(a_{t[A]}, b_{t[A]}, c_{t[A]}, d_{t[A]})$, then one B$^+$-tree of 2BPT is used to index the lower bounds $a_{t[A]}$ of the support of the possibility distributions $\pi_{t[A]}$, and the other B$^+$-tree is used to index the upper bounds $d_{t[A]}$ of this support. Let the fuzzy set P be specified by the quadruple (a_P, b_P, c_P, d_P). With 2BPT, preselection boils down to using the first B$^+$-tree to find the set of those possibility distributions that start before the end of P, i.e. for which $a_{t[A]} \leq d_P$. Subsequently, using the second B$^+$-tree to find the set of those possibility distributions that end after the start of P, i.e. for which $d_{t[A]} \geq a_P$, and finally, taking the intersection of both sets [2].

3.2 Index Structure

With IBPT indexing, we aim to further improve the indexing of imperfect data by using only one tree structure instead of two. The interval B$^+$-tree is used to index the upper bounds $d_{t[A]}$ of the supports of the possibility distributions $\pi_{t[A]}$, but is constructed in such a way that extra data about the lower bounds $a_{t[A]}$ are stored in the leaf nodes. We propose to define the nodes of an IBPT as depicted in Fig. 3.

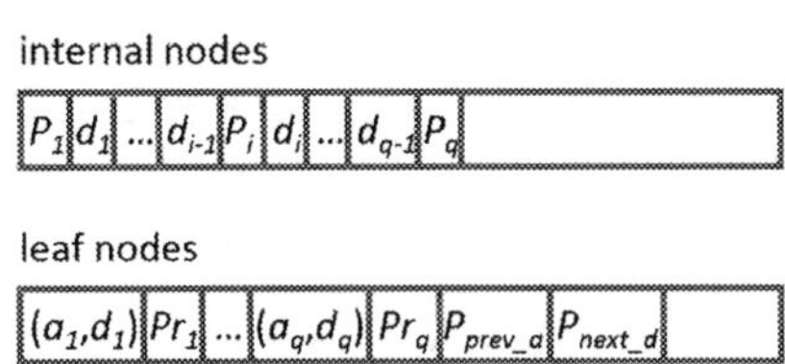

Fig. 3. The structure of internal and leaf nodes of an IBPT.

Internal nodes are structured like the internal nodes of a regular B$^+$-tree. Each internal node of order p can contain at most p tree pointers P_i, $i = 1, \ldots, q$, $q \leq p$ and $p - 1$ search values d_i, $i = 1, \ldots, q-1$, $q \leq p$. Each search value corresponds to an upper bound of the support of a possibility distribution $\pi_{t[A]}$ that is stored as value for attribute A in a tuple t of R. The search values in an internal node are sorted such that $d_1 < d_2 < \cdots < d_{q-1}$. Like with a regular B$^+$-tree, each tree pointer P_i refers to the root node of a subtree S_i. If $1 < i < q$, then each search value d in S_i satisfies $d_{i-1} \leq d < d_i$. Each search value d in S_1 satisfies $d < d_1$, and each search value d in S_q satisfies $d_{q_1} \leq d$. Every internal node should be kept at least half full.

Each leaf node of order p_{leaf} can contain at most p_{leaf} data values $[a_i, (d_i)]$ and p_{leaf} data pointers Pr_i. Each data value $[a_i, (d_i)]$ is composed of the lower bound a_i and upper bound d_i of the support of a possibility distribution $\pi_{t[A]}$. (For notational clarity, we put parentheses around the upper bound values, herewith denoting that these are the (search) values based on which the tree has been constructed.) Hence, information about the lower bound and the upper bound of each support is stored in the leaf nodes of the IBPT. The data values in each node are sorted such that $d_1 < d_2 < \cdots < d_q$. Each leaf node also contains two tree pointers P_{prev_a}, P_{next_d}. The P_{next_d} pointer is used to construct a linked list of leaf nodes that supports to traverse the data values in ascending order of the upper bound values d_i. Because the IBPT is constructed like a B$^+$-tree, using the upper bound values d_i, the P_{next_d} pointer will always refer to the next leaf in the tree (except for the last node, which contains a null pointer). The P_{prev_a} pointer is new and introduced to set up a second linked list of leaf nodes that supports the processing of all data values with a smaller lower bound value than a given data value. For that purpose, the smallest lower bound value a_{min} of all data values $[a_i, (d_i)]$ in each leaf node is selected and leaf nodes are connected with a linked list in descending order of their smallest lower bound value a_{min}. Every leaf node, except if it is the root node, should be kept at least half full.

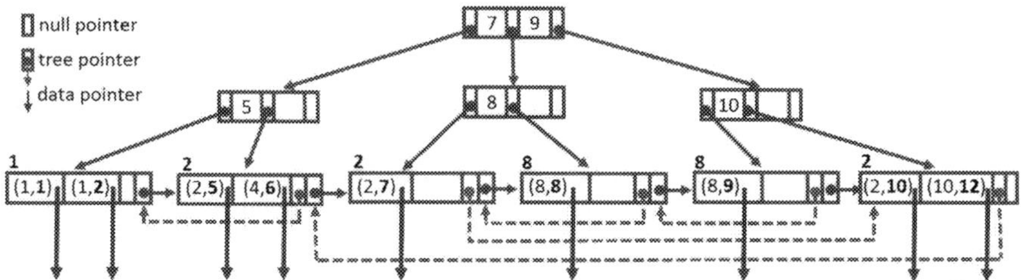

Fig. 4. An example of a IBPT index structure.

An example of a IBPT index structure is given in Fig. 4. The IBPT is used to index possibility distributions with (upper bound of) support $[8, (8)]$, $[2, (5)]$, $[1, (1)]$, $[2, (7)]$, $[1, (2)]$, $[10, (12)]$, $[8, (9)]$, $[4, (6)]$, and $[2, (10)]$ (inserted in that order). The P_{prev_a} tree pointers are depicted using dotted arrows. For the sake of clarity, the upper bound values in the leaf nodes are presented in boldface and the smallest lower bound value of each leaf node is shown in boldface, left on top of the leaf node.

3.3 Preselection

Preselection aims to support the querying of imperfect data by preselecting only those database tuples, for which it is known that they could satisfy the query condition. Or, as seen from the opposite point of view, by discarding those tuples from further query processing for which it is known that they certainly can not satisfy the query condition.

When supporting the processing of query conditions that are based on the compatibility operator IS, preselection can in its simplest form be guided by interval comparison. Consider a query condition $A\ IS\ P$ where A is an attribute of a relation R and P is a fuzzy set, denoting the user's preferences with respect to the values of attribute A, as described in Sect. 2. Assume that the possibility distribution $\pi_{t[A]}$, representing the available knowledge about the potential values of A in tuple t, and the membership function μ_P of P have a trapezoidal shape that is respectively characterized by the quadruples $(a_{t[A]}, b_{t[A]}, c_{t[A]}, d_{t[A]})$ and (a_P, b_P, c_P, d_P). Under this assumption it is certain that tuples for which the support of the value $\pi_{t[A]}$ of attribute A does not overlap with the support of μ_P will not satisfy $A\ IS\ P$, because for these tuples t Eqs. (2)-(3) will yield a necessity $N(A\ IS\ P)(t) = 0$ and a possibility $\Pi(A\ IS\ P)(t) = 0$.

Not overlapping translates to the conditions:

$$d_{t[A]} < a_P \tag{4}$$

expressing that the end point $d_{t[A]}$ of the support of $\pi_{t[A]}$ is located before the start point a_P of the support of μ_P, or

$$d_P < a_{t[A]} \tag{5}$$

expressing that the start point $a_{t[A]}$ of the support of $\pi_{t[A]}$ is located before the end point d_P of the support of μ_P. This is depicted in Fig. 5.

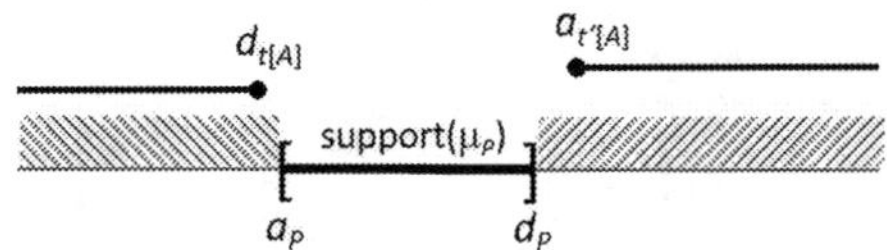

Fig. 5. Preselection conditions for processing $A\ IS\ P$.

When using an IBPT index structure, preselection of candidate values that potentially satisfy the query condition can be done with a three step procedure.

- **Step 1. Find the first candidate value.** Search the IBPT for a_P (the lower bound of the support of μ_P). After searching through the internal nodes, a tree pointer to the leaf node that should contain a value $[a_{t[A]}, (a_P)]$ will be obtained.
  - If a_P is in the index, i.e. if a tuple t with value $(a_{t[A]}, b_{t[A]}, c_{t[A]}, a_P)$ exists, then the first candidate value is the value $[a_{t[A]}, (a_P)]$ in the leaf node. (If multiple tuples with value $(a_{t[A]}, b_{t[A]}, c_{t[A]}, a_P)$ exist, then the tree will store a reference to a list of data pointers –each referring to one of those tuples– instead of a single data pointer.) Because the data values $[a, (d)]$ in each leaf node are sorted increasingly based on their value for d, it is certain that all values located left from this first candidate value have an upper bound for which Eq. (4) holds and hence do not belong to the preselection.

  - If a_P is not in the index, i.e. if a tuple t with value $(a_{t[A]}, b_{t[A]}, c_{t[A]}, a_P)$ does not exist in the database, then the first candidate value is the first value $[a, (d)]$ of the next leaf node which is obtained by using the P_{next_d} tree pointer. All the values located in leave nodes that are located in the linked list before this node have an upper bound for which Eq. (4) holds and hence do not belong to the preselection.
- **Step 2. Traverse the leaf nodes using the P_{next_d} tree pointers.** From the previous step a candidate value $[a, (d)]$ is obtained. This candidate value is the starting point for a traversal of the values in the leaf nodes. The values are traversed in increasing order of their end point (d). This traversal is straightforward, because by construction of the IBPT, all the values in the leaf nodes are already sorted in the proposed traversal order and the P_{next_d} tree pointer determines the next leaf node to traverse.
During traversal, each value $[a, (d)]$ is considered to be a candidate and will be checked using Eq. (5). If the candidate value does not satisfy the condition, i.e. if $d_P \geq a$, then we know that $\pi_{t[A]}$ (partially) overlaps with μ_P. So, the candidate value should be put in the preselection and we can move to the next candidate value. Traversal stops when a candidate value is found for which Eq. (5) holds, or when the last value in the list is checked. If Eq. (5) is satisfied for a candidate value $[a, (d)]$, i.e. if $d_P < a$, then we know that $\pi_{t[A]}$ starts after μ_P and hence can not overlap with it. So, the candidate value should not be put in the preselection.
When the traversal stops, we consider the last value $[a, (d)]$ that has been checked and continue with the third step.
- **Step 3. Traverse the leaf nodes using the P_{prev_a} tree pointers.** In the previous step we started with the construction of the preselection. The traversal based on the P_{next_d} tree pointers stops when the last data value is reached or when Eq. (5) is satisfied for a candidate value. This last stop condition gives no guarantee that all values that should be preselected are found. Indeed, it is still possible that there exist data values with a larger upper bound (d) for which Eq. (5) is not satisfied. This will be illustrated in the example in the next section. To find such values, we traverse the leaf nodes backwards using the P_{prev_a} tree pointers, starting with the leaf node of the last value that has been checked in step 2.

  Before we start the traversal, all the remaining data values $[a, (d)]$ in the leaf node have to be checked for potential membership to the preselection. If Eq. (5) does not hold, i.e. if $d_P \geq a$, then the value $[a, (d)]$ should be in the preselection (unless Eq. (4) holds). If it is not already there, it has to be added. If we find a data value for which Eq. (4) holds, i.e. for which $d < a_P$, or if P_{prev_a} is a null pointer, then the traversal stops. Else we move on to the leaf node that is referred to by P_{prev_a} and check all data values in that node for missing preselection values using Eqs. (5) and (4). Step 3 ends if the traversal stops.

After finishing the three step procedure described above, the preselection for processing the compatibility operator IS is determined. The data pointers associated with the data values in the preselection can then be used to efficiently load the preselected data from the storage for further query processing.

4 Illustrative Example

Reconsider the IBPT index structure that is presented in Fig. 4. This index is obtained after subsequently inserting the upper bounds of the supports [8, (8)], [2, (5)], [1, (1)], [2, (7)], [1, (2)], [10, (12)], [8, (9)], [4, (6)], and [2, (10)]. Assume that these are the supports of the (trapezoidal) possibility distributions used as values for the attribute $A : T$ in the different tuples t that are stored in the database. Moreover, assume that the (trapezoidal) membership function μ_P expressing the user's preferences in a query condition $A\ IS\ P$ has the interval [3, 7] as support. This information in depicted in Fig. 6. The supports in this figure are depicted bottom-up in increasing order of their upper bound values (the lowest support has the lowest upper bound value, whereas the highest support has the highest upper bound value).

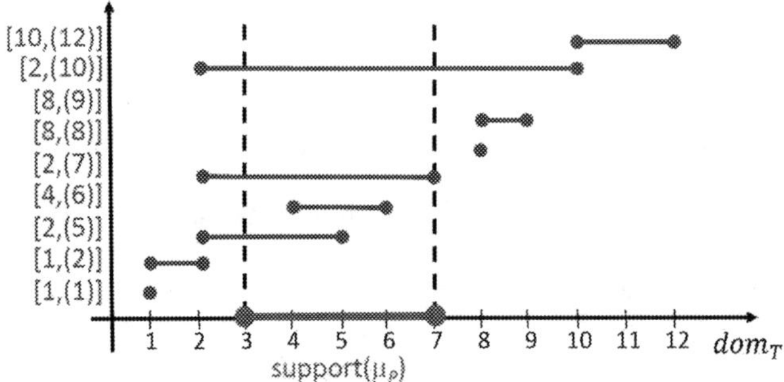

Fig. 6. The supports of the possibility distributions $\pi_{t[A]}$ stored in a database.

When applying preselection, the preselection should consist of those intervals that (partially) overlap with the interval [3, 7]. As can be seen in Fig. 6 (consider the vertical dotted lines) the preselection should be the set

$$S = \{[2, (5)], [4, (6)], [2, (7)], [2, (10)]\}.$$

Applying our three step procedure, presented in Sect. 3, yields the following. Assume that $S = \emptyset$.

- **Step 1.** Search the IBPT of Fig. 4 for the lower bound value (3) of the support of μ_P. Traversing the IBPT leads to the most left leaf node. The value (3) is not in the index, hence the first candidate value is the value [2, (5)], being the first value in the leaf node that is referred to by the P_{next_d} tree pointer of the most left leaf node.

- **Step 2.** Check the candidate value $[2,(5)]$. It holds that $2 \leq 7$, so $S = \{[2,(5)]\}$. Move to the next candidate value, which is $[4,(6)]$. It holds that $4 \leq 7$, so $S = \{[2,(5)],[4,(6)]\}$. Move to the next candidate value, which is $[2,(7)]$. It holds that $2 \leq 7$, so $S = \{[2,(5)],[4,(6)],[2,(7)]\}$. Move to the next candidate value, which is $[8,(8)]$. Now $8 > 7$, hence Step 2 ends.
- **Step 3.** Search for missing values in S. Reconsider the leaf node containing the last data value $[8,(8)]$ that has been checked. There are no subsequent data values in this node (cf. Fig. 4), hence we follow the P_{prev_a} tree pointer and move on to the leaf node containing the data value $[2,(7)]$. This value is already in S, so we can move on to the next leaf node, which is the node containing the data values $[2,(10)]$ and $[10,(12)]$. For $[2,(10)]$ it holds that $2 \leq 7$, so this value should be in S, but it is not. A missing data value is found and should be added to the preselection, so $S = \{[2,(5)],[4,(6)],[2,(7)],[2,(10)]\}$. Check the value $[10,(12)]$. This value should not be in S. Move on to the next leaf node, which is the node containing the data values $[2,(5)]$ and $[4,(6)]$. Both data values are already in S, so we can move on to the next leaf node, which is the node containing the values data $[1,(1)]$ and $[1,(2)]$. Because $2 < 3$ and $1 < 3$, Eq. 4 holds for at least one data value in the leaf node and we can stop the traversal. Hence S is found and equal to $\{[2,(5)],[4,(6)],[2,(7)],[2,(10)]\}$.

5 Conclusions and Future Work

In this paper, we proposed a novel indexing technique for possibilistic numerical data: Interval B$^+$-Trees (IBPT). The main advantage of an IBPT is that it allows to index supports of fuzzy sets using a single tree structure, while still offering the benefits of a regular B$^+$-tree. This should result in smaller index storage needs and faster preselection of candidate results in the processing of 'fuzzy' queries (because only one B$^+$-tree is used and should be traversed). We demonstrated the use of IBPT indexing with an illustrative example. More extensive experiments with real databases are still needed. Also, comparative studies with existing indexing techniques, taking into consideration the number of performed operations, are required. This is subject to ongoing and future work.

References

1. Aho, A.V., Hopcroft, J.E., Ullman, J.D.: Data Structures and Algorithms. Addison-Wesley, Boston (1983)
2. Barranco, C.D., Campana, J.R., Medina, J.M.: A B$^+$-tree based indexing technique for fuzzy numerical data. Fuzzy Sets Syst. **159**, 1431–1449 (2008)
3. Bosc, P., Galibourg, M.: Indexing principles for a fuzzy data base. Inf. Syst. **14**(6), 493–499 (1989)
4. Boss, B., Helmer, S.: Indexing a fuzzy database using the technique of superimposed coding - cost models and measurements. Reihe Informatik **2**, 24 (1996)
5. Codd, E.F.: A relational model of data for large shared data banks. Commun. ACM **13**(6), 377–387 (1970)

6. Dubois, D., Prade, H.: Possibility Theory: An Approach to Computerized Processing of Uncertainty. Plenum Press, New York (1988)
7. Liu, C., Ouksel, A., Sistla, P., Wu, J., Yu, C., Rishe, N.: Performance evaluation of G-tree and its application in fuzzy databases. In: Proceeding of the 5th International Conference on Information and Knowledge Management CIKM 1996, pp. 235–242. ACM, New York (1996)
8. Prade, H., Testemale, C.: Generalizing database relational algebra for the treatment of incomplete or uncertain information and vague queries. Inf. Sci. **34**(2), 115–143 (1984)
9. Yazici, A., Cibiceli, D.: An index structure for fuzzy databases. In: Proceeding of the 5th IEEE International Conference on Fuzzy Systems, vol. 2, pp. 1375–1381, New Orleans (1996)
10. Yazici, A., Cibiceli, D.: An access structure for similarity-based fuzzy databases. Inf. Sci. **115**(1–4), 137–163 (1999)
11. Zadeh, L.A.: Fuzzy sets as a basis for a theory of possibility. Fuzzy Sets Syst. **1**, 3–28 (1978)

Ordinal Assessment of Data Consistency Based on Regular Expressions

Antoon Bronselaer(✉), Joachim Nielandt, Robin De Mol, and Guy De Tré

Department of Telecommunication and Information Processing, Ghent University, Sint-Pietersnieuwstraat 41, 9000 Ghent, Belgium
{antoon.bronselaer,joachim.nielandt,robin.demol,guy.detre}@ugent.be

Abstract. In this paper, a novel assessment method for measurement of consistency of individual, text-valued attributes is proposed. The first novelty of this method is that it allows to express a broad range of well-known consistency measurements in a simple, elegant and standardized way. This property is obtained by relying on the standardized framework of regular expressions to support measurement. The key advantage of using such a highly standardized expression syntax, is that knowledge about consistency becomes portable, exchangeable and easy to access. The second novelty of the method, is that it examines the advantages of using a finite and ordinal scale for expression of measurement. These advantages include a high degree of interpretation and efficient calculations both in terms of time and space complexity.

Keywords: Data quality · Regular expressions · Ordinal measurement

1 Introduction

Throughout the past decades, data has become one of the most important assets of a modern organization. Understanding data leads to valuable insights, aids in making strategic decisions and creates a lead with respect to competitors. The quality of data has therefore become an important factor in its practical usefulness. As a result, the field of data quality has become more and more mature and the main principles of data quality research have been established. Perhaps the most accepted principle is that data quality is a *multi-dimensional* problem ([2,23]), meaning that quality of data has many different aspects also known as *dimensions*. Many such dimensions have been proposed and investigated, but the most commonly studied are correctness, completeness, consistency and the three time-related dimensions being timeliness, currency and volatility [3,22]. Within this paper, the focus lies on the dimension of *consistency*.

According to [3], consistency is a dimension of data quality that "*captures the violation of semantic rules defined over data items*". Basically, a piece of data is consistent if it satisfies all of the known rules that apply to it. Within the scope of the relational database model, consistency can be partially enforced through integrity constraints such as functional and inclusion dependencies, unique constraints and check constraints. In more recent storage systems such as NoSQL

J.P. Carvalho et al. (Eds.): IPMU 2016, Part II, CCIS 611, pp. 317–328, 2016.
DOI: 10.1007/978-3-319-40581-0_26

database, such support for consistency is omitted for the sake of higher throughput and better performance. To cope with this decreased support of consistency verification, several authors have investigated the measurement of consistency independently from the type of storage system [1,4,7].

As a running example throughout the paper, the scope of Belgian Social Security Numbers (SSNs) is considered. These are 11-digits numbers that must satisfy a set of rules in order to be a valid SSN. An example[1] of such an SSN is 18.01.01 – 022.42. The rules for a valid Belgian SSN will be introduced gradually throughout the paper and will be used to illustrate the measurement of consistency proposed in this paper. To explain the novelty of our approach, we revise two important issues of current measurement techniques for consistency.

First, based on the three functional forms for quality expression described by Pipino et al. [21], there is a perseverance in expressing measurement of consistency in the unit interval $[0,1]$. This was originally advocated by Pipino et al. [21] and has been widely adopted since then. However, this choice has never been well motivated and is not the result of a formal, measure-theoretic treatment of data quality [19]. To make things worse, it is often assumed that numbers in the unit interval expressing quality of data are by default interval-scaled. However, what do we know about two pieces of data that are assigned a consistency score of respectively 0.73 and 0.86? How do we interpret the difference between those two numbers? In this paper, a first small step towards a more interpretation-oriented method for measuring consistency is envisioned.

Second, there is currently a great dispersion in the knowledge about data consistency. In order to know the rules that define the consistency of a particular type of data (e.g., a Belgian SSN), a human agent has to look for those rules on the web or another source. If available, the rules are usually described in words, mathematical formulae or in the best case in pseudo-code. There is no common agreement on how to express rules for consistency, making it hard to access and share such knowledge. An interesting contribution with respect to this problem was made by Fürber et al., who proposed the Data Quality Management (DQM) ontology as a standard vocabulary to express data quality rules [13]. The authors of the current paper believe that this effort deserves to be further examined by proposing a more concrete syntax for expression of consistency rules.

In this paper, a method for measurement of consistency is proposed that overcomes the two problems mentioned above. In our approach, an ordinal scale is adopted to express quality. Each level in this scale has a clear interpretation in terms of predicates on which measurement is based. In order to standardize measurement of consistency, the basic predicates underlying measurement are defined entirely in terms of regular expressions. Hereby, two types of predicates are distinguished. On the one hand, a predicate can verify whether or not data matches a given regular expression pattern. On the other hand, a predicate can also verify additional assertions in terms of captured groups of a matched pattern. It is pointed out that these predicates allow a broad coverage of existing consistency measurements, but describe these measurements in a standardized

[1] This example is fictional.

manner. In addition, it is shown that our approach has appealing properties in terms of computational complexity.

The remainder of this paper is structured as follows. In Sect. 2, the most relevant contributions that have been made in the field of data quality are summarized and reviewed with respect to the current paper. In Sect. 3, some preliminary notations are introduced. Next, in Sect. 4, a novel method for measurement of consistency is proposed and two kinds of predicates are defined in terms of standard regular expressions. In Sect. 5, a discussion regarding the practical usefulness of the proposed framework is given. Section 6 provides some insights in future research and finally, the main contributions of this paper are summarized in Sect. 7.

2 Related Work

Throughout the past decades, a consensus has grown that data quality is a multi-dimensional problem [22–24]. Based on this observation, several authors defined a broad range of different data quality dimensions [2,3]. When it comes to *measurement* of data quality, it has been argued that a distinction can be made between objective and subjective measurement [21]. This distinction is investigated in depth in [10–12] and led to the proposal of an axiomatic definition of data quality measurement, which was further refined in [14]. Apart from the developments with respect to quality measurement in general, there have been several contributions that proposed techniques for measurement of specific dimensions. In that respect, completeness, accuracy and currency of data have been investigated in terms of the above mentioned axiomatic definition of data quality [15–17]. Interestingly, when it comes to measurement of data consistency, there is a great dispersion in the nature of existing techniques. It has been motivated in [13] that such heterogeneity is cumbersome and for that reason, the Data Quality Management (DQM) ontology is proposed as a standard vocabulary to express data quality rules. Despite this important effort to standardize the definitions of quality measurement, it is argued here that the exercise of standardization should be done at a more fundamental level. The current paper aims to fill this gap by proposing a standardized way of measuring consistency of data and in case of textual attributes, this is obtained by relying on regular expressions.

3 Notations and Preliminaries

In the remainder of this paper, the relational database model is assumed [5]. With $\mathcal{A}$ a countable set of attributes, a *(relational) schema* $\mathcal{R}$ is defined by a non-empty and finite subset of $\mathcal{A}$. The domain of an attribute $a \in \mathcal{A}$ is denoted by $\mathrm{dom}(a)$. A *relation* R over the schema $\mathcal{R}$ is defined by a subset of the crossproduct of all attribute domains, i.e., $R \subseteq \mathrm{dom}(a_1) \times \ldots \times \mathrm{dom}(a_m)$. An element of the relation R with schema $\mathcal{R}$ is called a tuple t over $\mathcal{R}$. The projection of R with schema $\mathcal{R}$ over a set of attributes $A \subseteq \mathcal{R}$ is denoted as $R[A]$. It is defined by a relation with schema A that contains the tuples from R projected over attributes in A. In the case where R and A are given by singletons $\{t\}$ and $\{a\}$, the notation $t[a]$ is adopted.

When proposing measurements of consistency for textual attributes, the concept of regular expressions will be adopted. A regular expression Σ is basically a sequence of characters that forms a pattern to which textual data can be matched. At the end of the previous century, first attempts were made by IEEE to standardize regular expressions and in this paper, the current POSIX standard is assumed [18]. Basically, a pattern can contain both literals and metacharacters, which are characters with a special meaning and role within the pattern. The key role of metacharacters is to denote patterns in a compact way. The most important metacharacters are summarized in Table 1.

Table 1. An overview of the most important meta-character constructions in regular expressions.

Metacharacter construction	Meaning
.	Matches any character
[ ]	Matches any character within the specified range
[ˆ]	Matches any character not within the specified range
( )	Defines a capture group
$\backslash d$	Matches any digit
$\backslash w$	Matches any word character
$\backslash s$	Matches any whitespace
|	Matches the preceding or the following
$*$	The preceding must occur zero or more times
$+$	The preceding must occur one or more times
?	The preceding must occur zero or one time
$\{n\}$	The preceding must occur exactly n times
$\{n, m\}$	The preceding must occur between n and m times

Within this paper, special attention is given to the concept of "capture groups". In an arbitrary pattern Σ, each pair of round brackets defines a specific part of the pattern that is known as a capture group or group for short. This name refers to the ability that each group can be "captured" in each piece of data matches the pattern. More specifically, if a string s matches the pattern Σ, then each group in Σ uniquely identifies a substring of s. As such, the definition of groups within a pattern fits a string s into a data structure that is induced by the groups of Σ and the different parts of this structure can be accessed easily. This powerful concept will allow us to define additional constraints on data that are known to match a pattern Σ in the remainder of this paper.

4 Ordinal Assessment of Consistency

Consider a relation R with schema $\mathcal{R}$ and consider attributes $A \subseteq \mathcal{R}$ for which consistency is to be measured. As mentioned in the previous, data is consistent if it satisfies a set of semantic rules. This is formalized by assuming a set of predicates, where each predicate verifies the satisfaction of one rule. A predicate is thus is defined by a function $p : \mathrm{dom}(A) \rightarrow \mathbb{B}$. The entire set of semantic rules then corresponds to a set $P = \{p_1, ..., p_n\}$ where each predicate p_i models one semantic rule. In order to translate predicates into a measurement of quality, we assume a total order relation $\prec$ on the set P such that $p_i \prec p_j$ expresses that predicate p_i needs to be evaluated before p_j. The subset of P that contains the predicates that should be evaluated first under $\prec$ is then denoted by $P_{(i)}$. In order to express quality, a finite ordinal scale $\mathbb{S} = \{s_1, ..., s_k\}$ is considered such that:

$$s_1 < s_2 < ... < s_k. \tag{1}$$

The smallest element of $\mathbb{S}$ is referred to as $\mathbb{0}$ and the largest element is referred to as $\mathbb{1}$. With these notations at hand, the definition of a quality function can be introduced as follows.

Definition 1 (Quality Function). *Consider a relation R with schema $\mathcal{R}$ and $A \subseteq \mathcal{R}$ for which the predicates P are given. A quality function Q for attributes A on the scale $\mathbb{S}$ is defined by a function:*

$$Q : \mathrm{dom}(A) \rightarrow \mathbb{S} \tag{2}$$

that satisfies the boundary constraints:

$$\forall t \in \mathrm{dom}(A) : (\forall p \in P : p(t) = F) \Rightarrow Q(t) = \mathbb{0} \tag{3}$$

$$\forall t \in \mathrm{dom}(A) : (\forall p \in P : p(t) = T) \Rightarrow Q(t) = \mathbb{1} \tag{4}$$

and is monotonic w.r.t. the evaluation order $\prec$ on P, which means that for any $t \in \mathrm{dom}(A)$ and $t' \in \mathrm{dom}(A)$ we have that:

$$Q(t) \geq Q(t') \Leftrightarrow \max\left\{i \mid \forall p \in P_{(i)} : p(t)\right\} \geq \max\left\{j \mid \forall p \in P_{(j)} : p(t')\right\} \tag{5}$$

Informally, Definition 1 dictates that quality of $t \in \mathrm{dom}(a)$ is determined by evaluating predicates in a specific order (i.e., indicated by $\prec$). The more predicates are true, the higher the appreciation[2]. The main advantage of assessing consistency in this way, is a clear interpretation of a measurement. Each $s_i \in \mathbb{S}$ implies a set of predicates that have succeeded and a set of predicates for which at least one has failed. A special kind of quality function is obtained when (i) there are as much predicates as there are non-zero levels in the scale (i.e., $|P| = |\mathbb{S}| + 1$) and (ii) level s_i is obtained if and only if all predicates in $P_{(i)}$ are true and $p_{(i+1)}$

[2] Such a paradigm is also used to determine the normal form of a relational database [5,6].

is false. A quality function that meets these requirements is called a *complete* quality function here. In what follows, the case of complete quality functions is assumed. In order to apply a quality function in the case of consistency, two kinds of measurement should be distinguished.

The first kind of measurement considers attributes separately and quality functions are defined for singleton sets of the type $\{a\}$ with $a \in \mathcal{R}$. The need for this kind of measurement comes from the fact that a database system usually provides a very small set of very generic data structures such as integer numbers, dates and character strings. When an attribute a is assigned with such a generic data structure, the domain of this data structure is usually a superset of the set of all valid values for a. A very generic way of limiting such a domain is the following. For an attribute a, consider k nested sets $N_1 \subset N_2 \subset ... \subset N_k \subseteq \text{dom}(a)$ and consider the predicates:

$$\forall i \in \{1, ..., k\} : p_i\left(t[a]\right) \triangleq t[a] \in N_{k-i+1}. \tag{6}$$

With this construction, for any $v \in \text{dom}(a)$ we have that $Q(v)$ is calculated by finding the smallest N_i that contains v. It is noted that this system of measurement bears a very close relationship with the way a possibility measure can be defined in the qualitative case [9]. The second kind of measurement has a broader scope and measures directly on entire groups of attributes. Rather than considering single attributes independently of each other, it is recognized that there may exist *dependencies* between attributes. These dependencies can be used to verify whether the co-occurrence of multiple attribute values within the same tuple is in adherence to the rules of consistency.

In the remainder of this paper, we will focus on the first kind of measurement and it will be shown how the nested sets N_i can be denoted in a very generic yet compact manner if the domain is that of textual character strings. We rely hereby on the concept of regular expressions and more specifically on (capture) groups. Consider an attribute a such that $\text{dom}(a)$ is the set of all character strings. For a pattern Σ, a Σ-predicate p is characterized by:

$$p(v) \triangleq v \models \Sigma. \tag{7}$$

A Σ-predicate is thus a predicate that verifies whether a value v matches the pattern Σ. In the scope of the running example of Belgian SSNs, we may consider the pattern Σ as follows:

$$(\backslash d\{2\})\ \backslash.?\ (\backslash d\{2\})\ \backslash.?\ (\backslash d\{2\})\ \backslash-?\ (\backslash d\{3\})\ \backslash.?\ (\backslash d\{2\}). \tag{8}$$

A value v matches this pattern if it starts with two digits (i.e., $\backslash d\{2\}$), optionally followed by a dot character. Next, there must again be two digits, again optionally followed by a dot character etcetera. As mentioned in Sect. 3 and as can be seen in the above pattern, groups can be used to identify specific parts of data that match Σ. If such groups are defined with a pattern Σ, they allow for an easy definition of additional predicates on the data as explained in the following.

In the context of a pattern Σ, a *group expression* λ is generally defined as a formula in which variables may occur that refer to groups defined in Σ. These variables are denoted by the symbol @ followed by the index of the group it refers to[3]. In order to allow flexibility in the way group expressions are to be interpreted, each variable that references to a group, can be assigned a data type. The default data type of a variable is `text`. Other data types that are allowed are currently the basic numerical data types `float`, `int` and `long`. For simplicity, it is assumed in this paper that all variables within the same group expression λ have the same data type. If the data type of variables within λ must be denoted explicitly, the notation $[\lambda] :: < \text{type} >$ will be used. If we wish to explicitly mention the pattern Σ in context of which λ must be put, we shall denote this as $[\lambda \mid \Sigma]$.

For a value $v \in \text{dom}(a)$, λ is *resolved* by first replacing all variables with the text that matches the referenced group and then evaluate the formula. In case of data type `text`, evaluation simply maintains the textual representation of λ. In case of numerical data types, λ is evaluated in an analytical manner. Let us illustrate this idea on the running example of Belgian SSNs. Consider v to be 18.01.01 – 022.42 and consider Σ as defined above. Note that Σ defines five groups. When applied to v, the first group comprises "18" while the fifth and last group comprises "42". Consider now the group expression $\lambda = @1 + @3 + @5$. This group expression concatenates groups with indices 1, 3 and 5, interleaved by a '+' character. If this group expression is resolved on v under the assumption that variables are of data type `text`, then we get the string "18 + 01 + 42". If we resolve λ as an `int`, then we get the integer number 61 as the sum of 18, 1 and 42. Let us now assume that, for an attribute a and a pattern Σ, g group expressions are defined. In general, a predicate on these g group expressions can be defined by:

$$\theta : \mathcal{X}_1 \times \ldots \times \mathcal{X}_g \rightarrow \mathbb{B} \tag{9}$$

where $\mathcal{X}_i$ is the domain of resolution of λ_i. In many practical cases, predicate functions with $0 < g \leq 2$ appear to be sufficient. In the example of Belgian SSNs, there are two constraints that need to be verified besides the satisfaction of pattern Σ as defined above. First, in Belgian SSNs, the first six digits constitute the birth date of the person in question. As such, for a value v, the first six digits must at least constitute a valid date in order for v to be an SSN. This constraint can be verified by the following predicate:

$$[@1/@2/@3] :: \texttt{text}\ \text{MATCHES}\ \texttt{yy/MM/dd} \tag{10}$$

The above predicate verifies that, after resolution, the resolved text constitutes a valid date under the date pattern `yy/MM/dd`. Second, it must be verified whether the control number is valid and this check is somewhat more complex. The reason for this is that the structure of the Belgian SSN uses a two-digit year notation. As a result, the calculation of the control number is different for people born in

[3] The assignment of indices to groups is standardized and is a based on the order in which opening brackets appear in the pattern.

or after the year 2000. Therefore, the check of the control number consists of two subtests that are combined with a Boolean disjunction. These subtests are:

$$[97 - (@1@2@3@4\,\%97)] :: \text{int} = [@5] :: \text{int} \tag{11}$$

and

$$[97 - (2@1@2@3@4\,\%97)] :: \text{int} = [@5] :: \text{int} \tag{12}$$

The first subtest takes the first four groups from Σ and concatenates them. Because the data type of resolution is `int`, the expression is analytically resolved into an integer by applying modulo 97 on the number formed by the first four groups. This result is then subtracted from 97. The predicate then checks if this result is equal to the number given by the fifth group. The second subtest does the same thing, but prefixes the digit formed by the first four groups with the digit 2. The control number predicate is then the Boolean disjunction of these two predicates. With respect to the data fragment that we consider, the first four groups constitute the number 180101022. We have that:

$$97 - 180101022\,\%97 = 42 \tag{13}$$

$$97 - 2180101022\,\%97 = 71 \tag{14}$$

Because the fifth group of the data fragment equals 42, the first subtest passes and thus the predicate on the control number evaluates to true. In the following section, the advantages of the described approach are discussed.

5 Discussion

In the previous section we have introduced quality functions and we have proposed to define such functions for text-based attributes by using regular expressions. Thereby, two kinds of predicates came apparent: Σ-predicates that verify the pattern Σ and θ-predicates that verify additional constraints in terms of groups of a pattern Σ. In order to verify the usefulness of these concepts, an implementation in Java was made. In this implementation, group expressions can be defined and resolved either as text, (4 or 8 bytes) integer numbers or floating point (8 bytes) numbers. Analytical resolution obeys the standard rules of precedence and the most important mathematical functions are defined. In addition, custom mappings can be defined. Next to Σ and θ-predicates, it is also possible to define a predicate as a Boolean function (conjunction, disjunction, negation, implication) on other predicates. This implementation allows us to illustrate some important advantages of the proposed framework.

Interpretation. The use of the ordinal scale $\mathbb{S}$ results in a framework of measurement with a high degree of interpretation. For a given $s \in \mathbb{S}$, there is an immediate feedback on the *kind* of degradation of data. In turn, this enables concrete actions to be undertaken to *improve* both current and future data.

Standardization. An important advantage of the approach is that the definition of predicates relies on POSIX standard regular expressions and basic (mathematical) operators to define predicates on group expressions. Quality functions for consistency are therefore defined in terms of strongly standardized concepts. As a result, whereas knowledge about consistency is very dispersed and represented heterogeneously at the time of writing, our framework opens the path to an open repository in which knowledge about consistency measurement can be easily accessed, shared and extended. In such a repository, attribute names are standardized by means of Uniform Resource Indicators (URIs) and measurement of consistency becomes as easy as asserting that data should match a specific standardized attribute in the open repository.

Coverage. Despite its simple structure, the current implementation of the proposed framework covers a broad range of the functionalities required to verify data consistency. Currently, properties like gender, email addresses, social security numbers, zip codes, bank accounts, various EAN codes and many more are supported. It has been verified that both basic checksum algorithms (e.g., the Luhn algorithm [20]) as well as more sophisticated, state-of-the-art algorithms (e.g., the Damm algorithm [8]) are covered. In addition, the current implementation can easily be extended to cover more data type resolutions and suitable predicate operators.

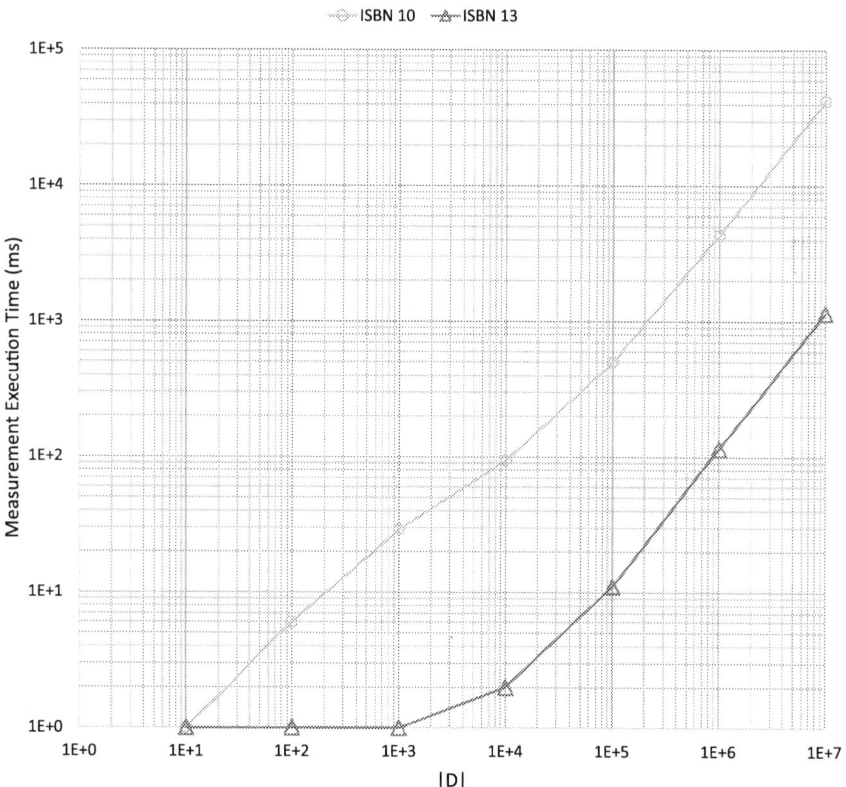

Fig. 1. Execution time (ms) of consistency measurement for ISBN numbers for different dataset sizes.

Computational Complexity. Finally, the proposed approach has some appealing features with respect to computational complexity. First, because $\mathbb{S}$ is finite, measurements of consistency for an attribute can be efficiently stored and represented as a histogram with a linear space complexity in terms of $|\mathbb{S}|$. Such a histogram representation can be updated *incrementally* and allows for very efficient calculation of many different aggregations.

Second, the actual measurement of consistency is efficient in terms of time complexity. Figure 1 shows the execution time in milliseconds of measuring the consistency of ISBNs for different sizes of data. Hereby, both the ISBN-10 and ISBN-13 structure are mixed in one dataset and must therefore both be checked. However, about 95 % of the samples have the ISBN-10 structure. For 10M samples of data it takes about 42 seconds to measure consistency in terms of the ISBN-10 structure. For the ISBN-13 structure, it takes only 1 second. This difference can be explained by noting that in case of the ISBN-13 structure, most data does not adhere to the basic pattern Σ and results in an "early abort" of the measurement. For the ISBN-10 structure, all predicates are verified and this results in additional computations. However, it can be seen that execution time increases linearly in terms of the size of the data. In combination with the incremental properties discussed above, this yields very efficient computations.

6 Future Work

In this paper, a novel approach to measurement of consistency has been proposed. Although some advantages of this approach have been pointed out, some promising paths of research are left open here. First, efforts should be taken to further implement and manage an open repository that contains quality functions for measurement of consistency. This way, standardized definitions of consistency are made possible and can be shared between researchers and end users. Second, group expressions as introduced here seem to provide an interesting way to describe dependencies between data. For example, in the case of Belgian SSNs, the 9th digit must be even for females and odd for males. As a result, the gender of a person can be *inferred* from its SSN. It is very easy to denote this inference in terms of a group expression. If such inferences are at hand, this opens the path to an automated data repair algorithm.

7 Conclusion

In this paper, quality functions are introduced as an ordinal-scaled measurement technique for consistency of data. In the case of text-based attributes, an approach to define such quality functions in terms of regular expressions has been described. It has been shown that the proposed approach enables a highly interpretable, standardized and efficient way to measure consistency. Whereas knowledge about consistency rules is nowadays very disperse and heterogeneously structured, the proposed approach clears the path to an open repository in which this knowledge becomes easy to access, share and extend.

References

1. Ballou, D., Pazer, H.: Modeling completeness versus consistency tradeoffs in information decision systems. IEEE Trans. Knowl. Data Eng. **15**(1), 240–243 (2003)
2. Batini, C., Cappiello, C., Francalanci, C., Maurino, A.: Methodologies for data quality assessment and improvement. ACM Comput. Surv. **41**(3), 16–52 (2009)
3. Batini, C., Scannapieco, M.: Data Quality: Concepts, Methodologies and Techniques. Data-Centric Systems and Applications. Springer, Heidelberg (2006)
4. Clark, P.G., Grzymala-Busse, J.W., Rzasa, W.: Consistency of incomplete data. Inf. Sci. **322**, 197–222 (2015)
5. Codd, E.F.: A relational model of data for large shared data banks. Commun. ACM **13**(6), 377–387 (1970)
6. Codd, E.F.: Recent investigations in relational data base systems. In: IFIP Congress, pp. 1017–1021 (1974)
7. Cong, G., Wenfei, F., Geerts, F., Jia, X., Ma, S.: Improving data quality: consistency and accuracy. In: Proceedings of the VLDB Conference, pp. 315–326 (2007)
8. Damm, M.: Total anti-symmetrische Quasigruppen. Ph.D. thesis, Philipps-Universität Marburg (2004)
9. Dubois, D., Prade, H.: Practical methods for constructing possibility distributions. Int. J. Intell. Syst. **31**, 215–239 (2015)
10. Even, A., Shankaranarayanan, G.: Value-driven data quality assessment. In: Proceedings of the International Conference on Information Quality, pp. 265–279 (2005)
11. Even, A., Shankaranarayanan, G.: Understanding impartial versus utility-driven quality assessment in large data-sets. In: Proceedings of the International Conference on Information Quality, pp. 265–279 (2007)
12. Even, A., Shankaranarayanan, G.: Utility-driven assessment of data quality. DATA BASE Adv. Inf. Syst. **38**(2), 75–93 (2007)
13. Fürber, C., Hepp, M.: Towards a vocabulary for data quality management in semantic web architectures. In: Proceedings of the 1st International Workshop on Linked Web Data Management (LWDM2011), pp. 265–279 (2011)
14. Heinrich, B., Kaiser, M., Klier, M.: How to measure data quality? A metric based approach. In: Proceedings of the International Conference on Information Systems, pp. 1–15 (2007)
15. Heinrich, B., Kaiser, M., Klier, M.: Does the EU insurance mediation directive help to improve data quality? A metric-based analysis. In: European Conference on Information Systems, pp. 1871–1882 (2008)
16. Heinrich, B., Klier, M.: Metric-based data quality assessment - developing and evaluation a probability-based currency metric. Decis. Supp. Syst. **72**, 82–96 (2015)
17. Heinrich, B., Klier, M., Kaiser, M.: A procedure to develop metrics for currency and its application in CRM. ACM J. Data Inf. Qual. **1**(1), 5:1–5:28 (2009)
18. IEEE: ISO/IEC/IEEE 9945: 2009 information technology portable operating system interface (posix) base specifications, issue 7 (2009)
19. Krantz, D., Luce, D., Suppes, P., Tversky, A.: Foundations of Measurement: Additive and Polynomial Representations, vol. I. Academic Press, New York (1971)
20. Luhn, H.P.: Computer for verifying numbers , US Patent 2,950,048 (1960)
21. Pipino, L., Lee, Y., Wang, R.: Data quality assessment. Commun. ACM **45**(4), 211–218 (2002)

22. Redman, T.: Data Quality for the Information Age. Artech-House, Boston (1996)
23. Wang, R., Storey, V., Firth, C.: A framework for analysis of data quality research. IEEE Trans. Knowl. Data Eng. **7**(4), 623–640 (1995)
24. Wang, R., Strong, D.: Beyond accuracy: what data quality means to data consumers. J. Manage. Inf. Syst. **12**(4), 5–34 (1996)

A Fuzzy Approach to the Characterization of Database Query Answers

Aurélien Moreau(✉), Olivier Pivert, and Grégory Smits

Irisa – University of Rennes 1, Technopole Anticipa, 22305 Lannion Cedex, France
{aurelien.moreau,pivert}@enssat.fr, gregory.smits@univ-rennes1.fr

Abstract. This paper describes an approach helping users to better understand the results of their queries. These results are structured with a clustering algorithm and described using a personal vocabulary. The goal is to find what the elements of a cluster have in common that also differentiates them from the elements of the other clusters. The data considered for characterizing each cluster of answers are not limited to attributes used in the query, revealing unexpected correlations to the user. The originality of this work resides in the definition and use of fuzzy-set-based characterizations and their properties.

Keywords: Databases · Cooperative answering · Clustering · Fuzzy logic

1 Introduction

Providing users with additional information when answering their queries is one of the objectives of *cooperative query answering*, along with taking into account personal preferences, or helping users formulate their queries correctly for instance in the field of databases [3]. Many approaches aiming at helping users harness databases have been proposed in the past. Helping users explore databases is a form of cooperative answering, along with handling failing queries [6], which can be dealt with by relaxing the selection conditions; or queries yielding a plethoric answer set which on the opposite may need more conditions to filter the answers; or ranking the answers to return only the *top-k* ones.

Several approaches consider clustering to tackle the many answer problem such as [9]. In this case the authors offer the users to refine their results by presenting the most representative answers. However they do not provide any additional information regarding the formed clusters beyond the attributes used by the user. Providing end users with a mechanism to understand the answer set and eventually narrow it down according to unexpected criteria is one of our objectives. For instance, if one looks for possible prices for houses to let obeying some (possibly fuzzy) specifications, and that two clusters of prices are found, one may discover, e.g., that this is due to two categories of houses having, or not, some additional valuable equipment such as a swimming pool. This latter issue, which constitutes the topic of the present paper, has been

J.P. Carvalho et al. (Eds.): IPMU 2016, Part II, CCIS 611, pp. 329–340, 2016.
DOI: 10.1007/978-3-319-40581-0_27

previously dealt with in [2]. The authors used a possibilistic representation of the data to determine which elements were most present and responsible of peaks. Here, we propose an alternative approach that first uses a clustering algorithm to detect groups of answers (a group corresponds to elements that have similar values on the attributes from the projection clause of the query) — this is the description step, that makes use of a fuzzy vocabulary specified by the user. Then we look for common properties between the elements of each cluster (that are not possessed by elements from other clusters) for the other attributes – this is the characterization step. This paper is a refinement of the approach presented in [10], in which the characterization was based on a Boolean definition, resulting in a lack of robustness and flexibility in the case of overlapping clusters. In this paper, the characterization is now based on fuzzy set theory, increasing the chance to explain a result since a characterization is now assigned a specificity degree in [0, 1] (whereas it had to be fully specific in the Boolean version of the approach).

The remainder of the paper is structured as follows. Related work is discussed in Sect. 2. Section 3 provides a refresher on fuzzy sets and fuzzy partitions. In Sect. 4, we describe the principle of the approach and present experimental results in Sect. 5. Finally, Sect. 6 recalls the main contributions and outlines perspectives for future work.

2 Related Work

Fuzzy approaches to answer explanations have been previously proposed in [1,2]. In [1], the answers to a fuzzy query are ranked according to an overall aggregation function and additional information (positive and negative) is provided about the different results. A bridge between formal concept analysis and bipartite graph analysis is established in [4]. The authors introduce new operators and discuss topics that could benefit from this parallel such as community detection.

Case-based reasoning is at the heart of [2], as the authors study the similarities between situations and their resulting outcomes. To do so, queries with a single output attribute are considered and the result is presented in the form of (1) a possibility distribution reflecting the values taken by this attribute, and (2) a function giving the number of cases supporting a particular outcome attribute value. The fact that a single attribute is considered makes it relatively easy to detect clusters of answers (they correspond to distinct peaks of the distribution) by looking at the associated curve. However, the authors do not give any detail about how this detection process could be generalized and automated (which we do by using a clustering technique). To find explanations for a given distribution, they propose to look for attribute values that are shared by elements in one peak and different in the others, through the use of fuzzy sets, membership functions and similarity measures. The authors point out that the explanations found may not always be meaningful with sets containing values that are too different. Our use of a vocabulary helps the user understand which ranges of values are considered. Also the authors do not make clear how to compute "joint ranges" to find explanations based on several attributes (in the case no single attribute can explain a peak).

In [11], explanations based on causality and provenance are defined. The objective of the authors is different from ours insofar as they do not provide any insight regarding the structure of the results of the queries but rather illustrate causality with "intervention", *i.e.* removing tuples from the database and assessing how the results are modified. A close research direction deals with "why not" answers in [5], looking for explanations for missing elements in an answer set. Causality and provenance are here the keys to figuring out which tuples and which conditions prevented some tuples from being part of the result.

3 Fuzzy Vocabulary

Fuzzy set theory was introduced by Zadeh [13] for modeling classes or sets whose boundaries are not clear-cut. For such objects, the transition between full membership and full mismatch is gradual rather than crisp. Typical examples of such fuzzy classes are those described using adjectives of the natural language, such as *young, cheap, fast*, etc. Formally, a fuzzy set F on a referential U is characterized by a membership function $\mu_F : U \rightarrow [0,1]$ where $\mu_F(u)$ denotes the grade of membership of u in F. In particular, $\mu_F(u) = 1$ reflects full membership of u in F, while $\mu_F(u) = 0$ expresses absolute non-membership. When $0 < \mu_F(u) < 1$, one speaks of partial membership.

In the approach we propose, it is assumed that the user specifies a vocabulary defined by means of fuzzy partitions. Let R be a relation defined on a set $\mathcal{A}$ of q categorical or numerical attributes $\{A_1, A_2, \ldots, A_q\}$. A fuzzy vocabulary on R is defined by means of fuzzy partitions of the q domains. A partition $\mathcal{P}_i$ associated with the domain of attribute A_i is composed of m_i fuzzy predicates $\{P_{i,1}, P_{i,2}, \ldots, P_{i,m_i}\}$, such that for all $x \in domain(A_i)$:

$$\sum_{j=1}^{m_i} \mu_{P_{ij}}(x) = 1$$

where $\mu_{P_{ij}}(x)$ denotes the degree of membership of x to the fuzzy set P_{ij}.

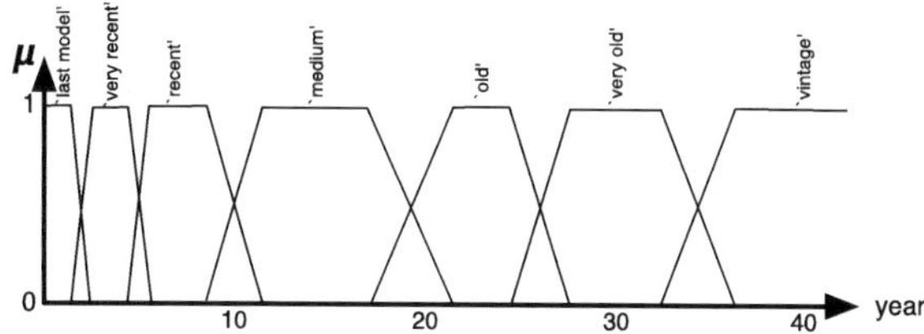

Fig. 1. A partition over the domain of the attribute *year*

More precisely, we consider partitions for numerical attributes (Fig. 1) composed of fuzzy sets, where a set, say P_i, can only overlap with its predecessor

P_{i-1} or/and its successor P_{i+1} (when they exist). For categorical attributes, we simply impose that for each value of the domain the sum of the satisfaction degrees on all elements of a partition is equal to 1. These partitions are specified by an expert during the database design step and represent "common sense partitions" of the domains. Each $\mathcal{P}_i$ is associated with a set of lingiustic labels $\{L_1^i, L_2^i, \ldots, L_{m_i}^i\}$.

As an example, let us consider a database containing ads about second hand cars and a view named *secondHandCars* of schema (*id*, *model*, *description*, *year*, *mileage*, *price*, *make*, *length*, *height*, *nbseats*, *consumption*, *acceleration*, *co2emission*) as the result of a join-query over the database. A common sense partition and labelling of the domain of the attribute *year* is illustrated in Fig. 1.

4 Principle of the Approach

Let us denote by R the relation concerned by the selection-projection query Q considered (note that R that may be the result of a join operation on multiple relations). $\mathcal{A}$ being the set of attributes of R, let us denote by $\mathcal{A}_\pi$ the subset of $\mathcal{A}$ made of the attributes onto which R is projected (i.e., the attributes of the resulting relation), by $\mathcal{A}_\sigma$ the subset of $\mathcal{A}$ concerned by the selection condition, and let us denote $\mathcal{A}_\omega = \mathcal{A} \backslash (\mathcal{A}_\pi \cup \mathcal{A}_\sigma)$. Let us consider a set of clusters of answers, formed based on the attributes from $\mathcal{A}_\pi$ (with a clustering algorithm). The two main steps are:

1. **description** of the clusters: projecting them on the vocabulary defined on the domains of the attributes from $\mathcal{A}_\pi$ (Subsect. 4.1);
2. **characterization** of each cluster in terms of the vocabulary defined on the domains of the attributes from $\mathcal{A}_\omega$ (Subsect. 4.2).

Step 1 is about using a fuzzy vocabulary to describe each one of these clusters. Step 2 aims at providing one or several characterizations for each of these clusters. A characterization is considered as additional information as it concerns attributes that do not appear in the query. Descriptions and characterizations both appear in the form of a conjunction of modalities (*i.e.* fuzzy labels) from the vocabulary, the only difference being in the origin of the attributes considered. The objective is to find properties that will permit to describe the clusters with attributes used to produce them (from $\mathcal{A}_\pi$) and then characterize them with attributes not involved in the query (from $\mathcal{A}_\omega$).

A characterization is related to the set of clusters built in step 1. It is made of a set of linguistic descriptions, one for each cluster. Let us denote by $\mathcal{C} = \{C_1, \ldots, C_n\}$ the set of clusters obtained.

Definition 1. *A characterization (*resp. *description)* E_{C_i} *attached to a cluster* C_i *is a conjunction of couples (attribute, fuzzy set of labels) of the form*

$$E_{C_i} = \{(A_j, F_{i,j}) \mid A_j \in \mathcal{A}_\omega \text{ (resp. } \mathcal{A}_\pi) \textit{ and } F_{i,j} \textit{ is a fuzzy set of linguistic labels from the partition of the domain of } A_j\}.$$

Example 1. Let us consider a query looking for the year and mileage of second-hand cars. Thus $\mathcal{A}_\pi = \{year, mileage\}$ and $\mathcal{A}_\omega = \{price, consumption, make, ...\}$.

The following descriptions and characterizations may be obtained:

- Cluster 1 is described by:
  "*(year is recent (0.8) or medium (0.2))* and *(mileage is small (1))*";
  A characterization could be:
  "*(consumption is medium (1))* and *(price is expensive (0.7) or medium (0.3))*".
- Cluster 2 is described by:
  "*(year is old (0.6) or very old (0.4))* and *(mileage is high (1))*";
  A characterization could be:
  "*(consumption is high (0.8) or medium (0.2))* and *(price is low (0.7) or very low (0.3))*".◇

Table 1. Correspondance between modalities and clusters: Example 1

	Year	*Mileage*	*Price*	*Consumption*	...
C_1	*{0.8/recent 0.2/medium}*	*small*	*{0.7/expensive 0.3/medium}*	*medium*	...
C_2	*{0.6/old 0.4/very old}*	*high*	*{0.7/low 0.3/very low}*	*0.8/high 0.2/medium*	...

4.1 Description Step

Once the clusters are formed, they are projected on the vocabulary in order to provide the user with a description of the clusters using terms of the natural language. The projection of C_i on the partition of an attribute $A_j \in \mathcal{A}_\pi$ is represented by a fuzzy set of labels $F_{i,j} = \{\mu_{L_k^j}(C_i)/L_k^j \mid L_k^j \in \mathcal{P}_j\}$ where

$$\mu_{L_k^j}(C_i) = \frac{\sum_{x \in C_i} \mu_{L_k^j}(x)}{|C_i|} \tag{1}$$

and $\mu_{L_k^j}(x)$ is the degree of membership of x to L_k^j. It is assumed that the only labels that appear in $F_{i,j}$ are such that $\mu_{L_k^j}(C_i) > 0$. Note that the fuzzy set $F_{i,j}$ is not normalized in general, but this does not matter here. The degree associated with each label is related to the number of points verifying it and to their membership degrees, hence making descriptions representative of each cluster.

4.2 Characterization Step

The first step to discovering characterizations (in the sense of Definition 1) consists in filling a table associating each cluster with its projection on the attributes of $\mathcal{A}_\omega$ (cf. Formula 1, considering this time that $A_j \in \mathcal{A}_\omega$). For every A_j ($j \in [1, |\mathcal{A}_\omega|]$) in $\mathcal{A}_\omega$, we indicate which modality L_k^j, $k \in [1, |\mathcal{P}_j|]$ (or fuzzy set of modalities) is satisfied by each cluster and to which degree $\mu_{L_k^j}(C_i)$.

To be informative, a characterization should satisfy two properties: specificity and minimality.

Property 1. Specificity: the specificity degree $\mu_{spec}(E_C)$ determines how representative a characterization E_C is for a given cluster C, and not so for the other clusters.

Since the cluster projections are fuzzy sets of labels, the notion of specificity must itself be viewed as a gradual concept. Being specific for a cluster characterization E means that there does not exist any other cluster with the same characterization, i.e., with fuzzy sets that are not disjoint from those of E for every attribute. It is then necessary to define the extent to which two such fuzzy sets are disjoint. Let us first consider a characterization involving a single attribute. Let E_1 and E_2 be the respective projections of the clusters C_1 and C_2 onto an attribute A_j of $\mathcal{A}_\omega$, whose associated fuzzy partition is denoted by $\mathcal{P}_j$. One may define:

$$\mu_{disjoint}(E_1, E_2) = 1 - \max_{L \in \mathcal{P}_j} \min(\mu_{L_k^j}(C_1), \mu_{L_k^j}(C_2)), \tag{2}$$

which corresponds to the fuzzy interpretation of the constraint $\nexists L \in \mathcal{P}_j$ such that both C_1 and C_2 are L. When several attributes – let us denote by $\mathcal{A}$ this set of attributes – are involved, two characterizations are globally disjoint if they are so on at least one attribute and we get:

$$\mu_{disjoint}(E_1, E_2) = \max_{A_j \in \mathcal{A}} (1 - \max_{L \in \mathcal{P}_j} \min(\mu_{L_k^j}(C_1), \mu_{L_k^j}(C_2))). \tag{3}$$

Finally, the specificity degree attached to a candidate characterization associated with a given cluster C may be defined as:

$$\mu_{spec}(E_C) = \min_{C' \neq C} \mu_{disjoint}(E_C, E_{C'}), \tag{4}$$

where $E_{C'}$ denotes the projection of C' onto the attributes present in E_C.

Property 2. Minimality: viewing a characterization as a conjunction of fuzzy sets of predicates, one says that E_C is a minimal characterization of the cluster C iff $\nexists E'_C \subset E_C$ so that E'_C characterizes C with a specificity degree equal or greater than that of E_C.

Formally, we use the inclusion in the sense of Zadeh ($F_1 \subseteq F_2$ iff $\forall x \in U, \mu_{F_1}(x) \leq \mu_{F_2}(x)$ where U denotes the universe on which fuzzy sets F_1 and F_2 are defined) and we get:

$$E_C \text{ is minimal iff } \nexists E'_C \text{ such that } \forall A_j \in \mathcal{A}_\omega,\ E'_C[A_j] \subseteq E_C[A_j] \text{ and } \mu_{spec}(E'_C) \geq \mu_{spec}(E_C) \tag{5}$$

where $E_C[A_j]$ denotes the fuzzy set related to attribute A_j in E_C.

Here is a crisp example for the sake of clarity. If we consider houses to let, and identify a subset of answers whose characterization is $E =$ (price is expensive (1)) $\wedge$ (swimming pool = yes (1)) $\wedge$ (garden is big (1)), there should not exist a characterization *e.g.* $E' =$ (price is expensive (1)) $\wedge$ (swimming pool = yes (1)) also characterizing this cluster only *i.e.* so that $\mu_{spec}(E') \geqslant \mu_{spec}(E)$.

4.3 Characterization Algorithms

Given the definition of specificity, a characterization involving every attribute from $\mathcal{A}_\omega$ will have the highest specificity degree possible, denoted $maxSpec$. (*Elements of proof:* adding attributes to characterizations will add more terms to the aggregate $\max_{A_j \in \mathcal{A}}$, thus potentially raising the specificity degree).

The first step of the characterization process is to determine for each cluster the maximal specificity degree $maxSpec$ one may expect for its characterizations. Clusters whose maximal specificity degree is greater than a predefined threshold λ are said to be fully characterizable. For the others, two strategies may be envisaged: to accept a less demanding specificity threshold, or to try to find specific characterizations on subsets (of points) of the clusters concerned. Hereafter, we investigate the second option and propose a solution based on the notion of cluster focusing. With this method, one expects to be able to generate specific enough characterizations of an interesting subset of a non fully characterizable cluster. Our goal being to characterize a set of items gathered particularly according to their closeness to each other, it appears obvious to focus on the most central points of the cluster concerned. It is nevertheless worth noticing that the central points of a cluster built on the attributes from $\mathcal{A}_\pi$ do not necessarily form a compact and characterizable set on the attributes from $\mathcal{A}_\omega$.

Thus, Algorithm 1 is applied on each cluster to determine its maximal specificity degree, and if necessary to determine the largest subset of central points for which a characterization of a high enough specificity degree may be found.

This focusing step is done with the *clusterFocus* function, which requires three parameters: the cluster $originalC_i$, a focusing step α and the number of focusing steps *focus-factor*. It returns a limited part of the cluster, $(100 - \alpha)\%$ of $originalC_i$. The new $maxSpec$ value for this cluster is then computed (line 9). For this calculation, all clusters are considered in their entirety (whether some have already been focused or not) except for the current one.

If it is still not characterizable, this step can be repeated until the cluster is reduced to its medoid/centroid (line 6), always computing the new size of the cluster focusing based on the original cluster C_i (line 8). In other words, clusters are automatically truncated to provide users with the best characterizations possible *i.e.* with a specificity degree higher than λ. When displaying characterizations, users will be informed whether or not said characterizations concern a full or a focused cluster.

```
   Input: n clusters C ; |A_ω| attributes/values for each cluster ; specificity
          threshold λ ; focusing step α
   Output: one maxSpec for each cluster
1  begin
2      foreach cluster C_i do
3          compute maxSpec;
4          focus-factor ← 0;
5          originalC_i ← C_i;
6          while maxSpec < λ ∧ |C_i| > 1 do
7              focus-factor ← focus-factor + 1;
8              C'_i ← clusterFocus(originalC_i, focus-factor, α);
9              compute maxSpec for C'_i;
10             C_i ← C'_i;
11         end
12     end
13     characterize each cluster (focusing) with Algorithm 2;
14 end
```

Algorithm 1. Cluster Characterizer

Once the maximal specificity degree has been computed for each cluster, (either complete or truncated), Algorithm 2 is applied to determine for each cluster all the possible characterizations of a minimal size and a maximal specificity.

This algorithm takes as input the number of clusters, the *maxSpec* value for each of them computed with Algorithm 1 as well as the data from Table 1. For each cluster C_i (line 2), we look for characterizations (line 5) composed first of a single fuzzy set of labels (for one attribute only), then with two of them, then three, etc., and check whether candidate characterizations are specific and minimal. If so, they are added to the set of characterizations (line 9).

Remark 1. Some attributes from $\mathcal{A}_\sigma$ may also be added to $\mathcal{A}_\omega$: those concerned by inequality conditions $(<, \leqslant, \geqslant, >, \neq)$ as results may have several satisfying values for these attributes, and participate in the characterization process.

5 Experimentation

Concerning the choice of the clustering algorithm, we need an algorithm that does not imply to know in advance the number of clusters to obtain. We used the

```
Input: n clusters C; |A_ω| attributes/values for each cluster; one maxSpec for
       each cluster;
Output: a set of characterizations for each cluster;
begin
    foreach cluster C_i do
        Charact(C_i) ← ∅;
        if maxSpec ⩾ λ then
            for j ← 1 to |A_ω| do
                for every characterization E of size j that is not a superset of
                any element of Charact(C_i) of specificity maxSpec do
                    if μ_spec(E) ⩾ λ then
                        if E is minimal then
                            Charact(C_i) ← Charact(C_i) ⋃ E
                        end
                    end
                end
            end
        end
    end
end
```

Algorithm 2. Characterizations Finder

l-cmed-select algorithm, a crisp variant of the *l-fcmed-select* technique proposed in [8], which belongs to the framework of incremental clustering and combines relational clustering and medoid-based methods. *l-fcmed-select* is an extension of the linearised fuzzy c-medoids clustering algorithm [7], *l-fcmed*. The *l-cmed-select* algorithm possesses the following main characteristics: (i) it does not require a precise number of clusters to operate, simply an over-estimation; (ii) it exploits a linear approximation scheme to update the cluster medoid, looking for the new medoid in the vicinity of its previous position. This approximation alleviates computational costs. The distance measure used is $dist(x, y) = |x-y|/\max(x, y)$.

5.1 Illustrative Examples

To test our approach, we performed a preliminary experimentation with a real dataset of second hand cars ads extracted from LeBonCoin.fr. The attributes considered were *price*, *mileage*, *year*, *option level*, *consumption*, *horse power*, *brand* and *model*. The first two $\mathcal{A}_\pi = \{price, mileage\}$ were the ones according to which the groups of data were formed, while the others $\mathcal{A}_\omega = \{year, horse\text{-}power, ...\}$ were used to find characterizations for each cluster, both specific and minimal. Several examples are presented illustrating different situations.

Querying for the prices and mileage of cars of make 'Audi', from 2010 onwards and costing less than 15,000€ (Query 1), the clusters obtained are presented in Fig. 2a. We empirically chose $\lambda = 0.7$ and got:

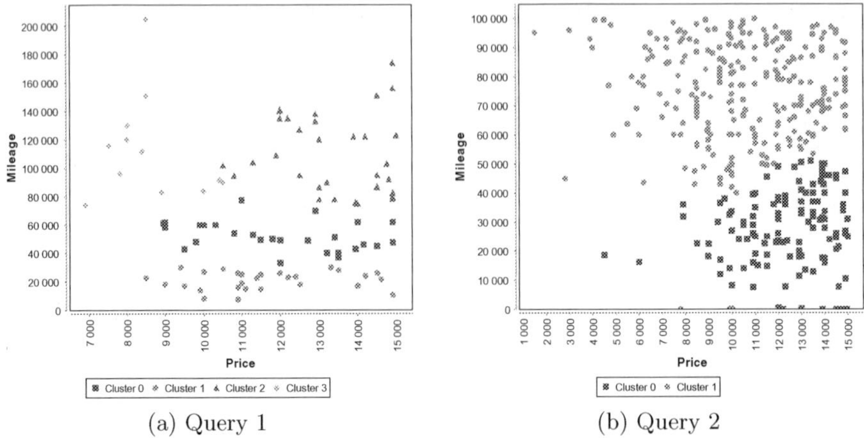

(a) Query 1 (b) Query 2

Fig. 2. Full clusters of second hand cars over the attributes *price* and *mileage* (Color figure online)

- Cluster 1: description: *(price is medium (0.69) or expensive (0.31))* and *(mileage is very low (0.68) or low (0.32)*; characterization: specificity 0.83, *(year is recent (0.15) or very recent (0.85))*;
- Cluster 2: description: *(price is expensive (0.77) or medium (0.23))* and *(mileage is medium (0.85) or high (0.15)*; characterization: specificity 0.71, *(option level is high (0.70) or medium (0.13) or low (0.13))* and *(consumption is high (0.76) or low (0.12) or medium (0.11)*;
- Cluster 3: description: *(price is medium (1))* and *(mileage is medium (0.78) or high (0.22)*; characterization: specificity 0.75, *(year is recent (0.83) or very recent (0.17))* and *(option level is medium (0.5) or low (0.28) or high (0.22))*;

but no characterizations for cluster 0. After a double focusing (62 %), we got:

- Cluster 0 (62 %): specificity 0.71, *(year is recent (0.87) or very recent (0.13))* and *(consumption is low (0.33) or medium (0.33) or high (0.3)*.

We then considered cars of make 'BMW', 'Seat' or 'Volkswagen' costing less than 15,000€ with a mileage inferior to 100,000 km (Query 2). The clusters are presented in Fig. 2b.

- Cluster 0: description *(price is expensive (0.58) or medium (0.41))* and *(mileage is low (0.62) or very low (0.38))*; characterization: specificity 0.74, *year is very recent (0.65) or recent (0.27)*;
- Cluster 1: description *(price is medium (0.64) or expensive (0.29))* and *(mileage is medium (0.73) or low (0.26))*; characterization: specificity 0.74, *year is recent (0.63) or medium (0.3)*.

Two characterizations were found for the entire clusters, however since they were not very well separated, descriptions and characterizations have many labels in common, albeit with different degrees. Labels whose degree is inferior to 0.1 are omitted for the sake of readability, which explains why the sum of the description or characterization degrees is not always equal to 1.

5.2 Performances

To assess the efficiency of this approach, we used a synthetic dataset with randomly-generated values on a Macbook Pro with a 3 GHz Intel Core i7 processor and 16 GB RAM. We checked the impact of two parameters on the processing time: the cardinality of the dataset and the number of attributes in $\mathcal{A}_\omega$. $|\mathcal{A}_\pi|$ was set to 3 for both experimentations. In the first experiment (Fig. 3a), $|\mathcal{A}_\omega|$ was set to 10. The clustering part processing times are acceptable under 10,000 tuples of data and those for the explanation process (description and characterization) are below one second for answer sets of up to 10,000 tuples. Let us emphasize that the clustering step is performed on a *set of answers*, not on a base relation, and one may consider that 10,000 already corresponds to a rather large answer set. The number of tuples raises the computation times of Table 1, which has to be updated for every focusing. However the rest of the characterization process is not impacted by the number of tuples considered. In the second experiment, we set the number of tuples to 10,000. The results (Fig. 3b) show that the processing times remain low as long as $|\mathcal{A}_\omega|$ is under 15. The complexity of Algorithm 2 is exponential in the number of attributes $|\mathcal{A}_\omega|$, and follows the growth of $2^{|\mathcal{A}_\omega|}$.

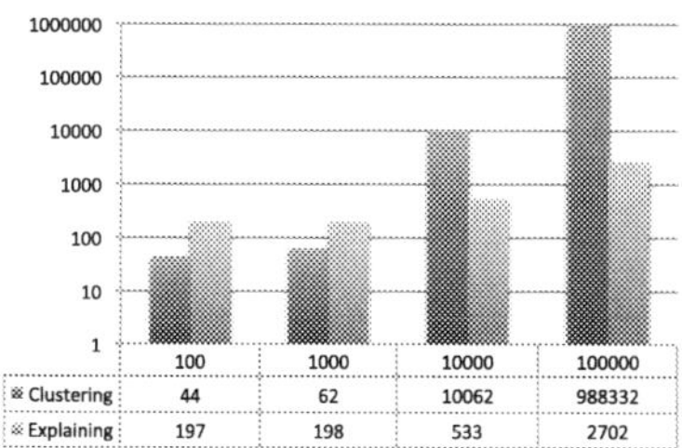

(a) Overall processing time (in ms, log scale) depending on the number of tuples

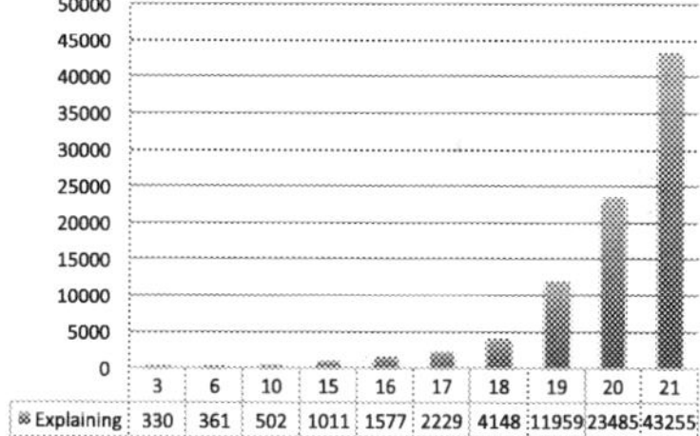

(b) Processing time (in ms) depending on the number of attributes in $\mathcal{A}_\omega$ for the explanation part

Fig. 3. Experimentations (Color figure online)

6 Conclusion

In this paper, we have presented an approach aimed to characterize subsets of answers to database queries, using two steps: (i) description: the clusters obtained beforehand are described in terms of a fuzzy vocabulary; (ii) characterization: other attributes (not involved in the clustering process) are used to highlight the particular properties of each cluster.

Preliminary experimental results show that the approach is indeed effective in finding characterizations including in cases where the approach described in [10] would fail because of its rigidity. The use of fuzzy sets to characterize clusters offers flexibility when dealing with clusters with mixed borders, and cluster focusing limits the impact of borderline elements. Perspectives include considering cluster focusing based on typicality (in the sense of [12]) instead of distance only, in order to better identify these borderline elements.

Acknowledgments. This work has been partially funded by the French DGE (Direction Générale des Entreprises) under the project ODIN (Open Data INtelligence).

References

1. Amgoud, L., Prade, H., Serrut, M.: Flexible querying with argued answers. In: Proceedings of the 14th IEEE International Conference on Fuzzy Systems (FUZZ-IEEE 2005), pp. 573–578, Reno, Nevada, USA (2005)
2. de Calmès, M., Dubois, D., Hüllermeier, E., Prade, H., Sedes, F.: Flexibility and fuzzy case-based evaluation in querying: an illustration in an experimental setting. Int. J. Uncertainty, Fuzziness Knowl. Based Syst. **11**(1), 43–66 (2003)
3. Gaasterland, T., Godfrey, P., Minker, J.: An overview of cooperative answering. J. Intell. Inf. Syst. **1**(2), 123–157 (1992)
4. Gaume, B., Navarro, E., Prade, H.: Clustering bipartite graphs in terms of approximate formal concepts and sub-contexts. Int. J. Comput. Intell. Syst. **6**(6), 1125–1142 (2013)
5. Herschel, M.: Wondering why data are missing from query results? Ask conseil why-not. In: He, Q., Iyengar, A., Nejdl, W., Pei, J., Rastogi, R. (eds.) CIKM, pp. 2213–2218. ACM (2013)
6. Koudas, N., Li, C., Tung, A.K.H., Vernica, R.: Relaxing join and selection queries. In: Proceedings of the 32nd International Conference on Very Large Data Bases, pp. 199–210 (2006). http://dl.acm.org/citation.cfm?id=1182635.1164146
7. Krishnapuram, R., Joshi, A., Nasraoui, O., Yi, L.: Low-complexity fuzzy relational clustering algorithms for web mining. IEEE T. Fuzzy Syst. **9**(4), 595–607 (2001)
8. Lesot, M.-J., Revault d'Allonnes, A.: Credit-card fraud profiling using a hybrid incremental clustering methodology. In: Hüllermeier, E., Link, S., Fober, T., Seeger, B. (eds.) SUM 2012. LNCS, vol. 7520, pp. 325–336. Springer, Heidelberg (2012)
9. Liu, B., Jagadish, H.V.: DataLens: making a good first impression. In: Proceedings of the ACM SIGMOD International Conference on Management of Data, pp. 1115–1118 (2009)
10. Moreau, A., Pivert, O., Smits, G.: A clustering-based approach to the explanation of database query answers. In: Andreasen, T., et al. (eds.) FQAS 2015. AISC, vol. 400, pp. 307–319. Springer, Switzerland (2015)
11. Roy, S., Suciu, D.: A formal approach to finding explanations for database queries. In: Proceedings of the 2014 ACM SIGMOD International Conference on Management of Data, SIGMOD 2014, pp. 1579–1590. ACM, New York (2014)
12. Smits, G., Pivert, O.: Linguistic and graphical explanation of a cluster-based data structure. In: Beierle, C., Dekhtyar, A. (eds.) SUM 2015. LNCS, vol. 9310, pp. 186–200. Springer, Heidelberg (2015)
13. Zadeh, L.: Fuzzy sets. Inf. Control **8**, 338–353 (1965)

Making the Skyline Larger: A Fuzzy-Neighborhood-Based Approach

Djamal Belkasmi[1,2], Allel Hadjali[2(✉)], and Hamid Azzoune[3]

[1] DIF-FS/UMBB, Boumerdes, Algeria
[2] LIAS/ENSMA, Poitiers, France
{djamal.belkasmi,allel.hadjali}@ensma.fr
[3] LRIA/USTHB, Algiers, Algeria
azzoune@yahoo.fr

Abstract. Skyline queries have gained much attention in the last decade and are proved to be valuable for multi-criteria ranking. They are based on the concept of Pareto dominance. In many real-life applications, the skyline returns only a small number of non-dominated objects which could be insufficient for the user. In this paper, we discuss an approach to enriching the small skyline with particular points that could serve the decision makers' needs. The idea consists in identifying the most interesting non-skyline points belonging to the fuzzy neighborhood of a skyline point and then adding them to the classical skyline. To do so, a particular fuzzy closeness relation is introduced. The relaxed skyline obtained which include the classical skyline, is a discriminated set. Furthermore, an efficient algorithm to compute the relaxed skyline is proposed. Extensive experiments are conducted to demonstrate the effectiveness of our approach and the performance of the proposed algorithm.

Keywords: Fuzzy sets · Databases · Skyline queries · Closeness · Relaxation

1 Introduction

In recent years, preference queries have received a great attention by many database researchers. Skyline queries [1] are specific example of SQL extensions that allow users to express preference in queries. Based on Pareto dominance relationship, skyline queries select all non-dominated objects based on a multi-criteria comparison. This means that, given a set D of d-dimensional points, a skyline query returns, the skyline S, set of points of D that are not dominated by any other point of D. A point p dominates another point q iff p is better than or equal to q in all dimensions and strictly better than q in at least one dimension. One can see that skyline points are incomparable. Several research studies have been conducted to develop efficient algorithms and introduce multiple variants of skyline queries [2–5]. However, querying a d-dimensional data sets using a skyline operator may lead to two possible scenarios: (i) a large number of skyline points

J.P. Carvalho et al. (Eds.): IPMU 2016, Part II, CCIS 611, pp. 341–354, 2016.
DOI: 10.1007/978-3-319-40581-0_28

returned, which could be less informative for users, (ii) a small number of skyline points returned, which could be insufficient for users. To solve the two above problems, various approaches have been proposed to refine the skyline, therefore reducing its size [6–13], but only very few works exist to relax the skyline in order to increase the number of skyline results [10,14–17]. Goncalves and Tineo [15] propose a flexible dominance relationship using fuzzy comparison operators. This increases the skyline with points that are only weakly dominated by any other point. In [10], Hadjali et al. discuss some ideas of relaxing the skyline. In [14], and taking as starting point the study in [10], we develop an approach, called $\mathcal{MP2R}$ ($\mathcal{M}$uch $\mathcal{P}$referred $\mathcal{R}$elation for $\mathcal{R}$elaxation), for skyline relaxation. This approach relies on a novel fuzzy dominance relationship *Much Preferred (MP)* which makes more demanding the dominance between the points of D.

In this paper, we investigate another way of relaxing the skyline S. The idea is to consider that a non-skyline point p still belongs to a *fuzzily extended skyline* S_{FE} if p is *close* to a skyline point q. We then develop an approach, called $\mathcal{C2R}$ ($\mathcal{C}$loseness $\mathcal{R}$elation for $\mathcal{R}$elaxation), to enlarging the small skyline with points that are closest to skyline points (keep in mind that those points are ruled out from the skyline when applying the classical Pareto dominance). The approach makes use of a particular appropriate fuzzy *"Closeness (C)"* relation. Each element in the relaxed skyline obtained S_{FE} is then associated with a degree ($\in [0,1]$) expressing the extent to which it belongs to S_{FE}. In summary, the main contributions made are as follows:

- We provide the definition and semantic basis for a relaxed variant of skyline S_{FE}.
- We develop and implement an algorithm to compute S_{FE} efficiently.
- We conduct a set of experiments to study and analyze the relevance and effectiveness of S_{FE}.
- Finally, we present a comparative study between S_{FE} and S_{Relax} (i.e., the relaxed skyline obtained by the $\mathcal{MP2R}$ approach of [14]).

The paper is structured as follows: Sect. 2 provides some necessary background on skyline queries and on MP2R-based approach to skyline relaxation. In Sect. 3, we introduce a new approach for skyline relaxation based on fuzzy closeness relationship. An algorithm to efficiently compute S_{FE} is presented and discussed. Section 4 is devoted to the experimental study. Finally, Sect. 5 concludes the paper and draws some lines for future works.

2 Background

In this section, we recall some notions on skyline queries. Then, we present our MP2R-based approach for Skyline relaxation.

2.1 Skyline Queries

The notion of skyline queries was pioneered in [1]. Subsequently, the interest in this area has exploded: [1] has garnered over 1800 citations (Google Scholar,

January 2016). Skyline queries are a specific, yet relevant, example of preference queries. They rely on Pareto dominance principle which can be defined as follows:

Definition 1. *Let D be a set of d-dimensional data points and u_i and u_j two points of D. u_i is said to dominate in Pareto sense u_j (denoted $u_i \succ u_j$) iff u_i is better than or equal to u_j in all dimensions and better than u_j in at least one dimension. Formally, we write*

$$u_i \succ u_j \Leftrightarrow (\forall k \in \{1,..,d\}, u_i[k] \geq u_j[k]) \wedge (\exists l \in \{1,..,d\}, u_i[l] > u_j[l]) \quad (1)$$

where each tuple $u_i = (u_i[1], u_i[2], u_i[3], ..., u_i[d])$ with $u_i[k]$ stands for the value of the tuple u_i for the attribute A_k.

In (1), without loss of generality, we assume that the largest value, the better.

Definition 2. *The skyline of D, denoted by S, is the set of points which are not dominated by any other point.*

$$u \in S \Leftrightarrow \nexists u' \in D, u' \succ u \quad (2)$$

Skyline queries compute the set of Pareto-optimal tuples in a relation, i.e., those tuples that are not dominated by any other tuple in the same relation.

Example 1. To illustrate the concept of the skyline, let us consider a database containing information on candidates as shown in Table 1. The list of candidates includes the following informations: Code, Age, Management experience (man_exp in years), Technical experience (tec_exp in years) and distance work to Home (dist_wh in Km). Ideally, personnel manager is looking for a candidate with the largest management and technical experience (Max man_exp and Max tec_exp), ignoring other informations. Applying the traditional skyline will returns the following candidates: M_5, M_8. As can be seen, such results are the most interesting candidates (see Fig. 1).

2.2 MP2R-based Approach for Skyline Relaxation

In [14] we have proposed an approach to relax skyline called $\mathcal{MP2R}$. Its relies on a new dominance relationship that allows enlarging the skyline with the most interesting points among those ruled out when computing the initial skyline S. This new dominance relationship uses a fuzzy relation, named *"Much Preferred (MP)"* to compare two tuples u and u'. So, u is an element of S_{relax} if there is no tuple $u' \in U$ such that u' is *much preferred* to u (denoted $MP(u', u)$) in all skyline attributes. Formally, we write:

$$u \in S_{relax} \Leftrightarrow \nexists u' \in U, \forall i \in \{1, ..., d\}, MP_i(u'_i, u_i) \quad (3)$$

where, MP_i is a fuzzy preference relation defined on the domain $\mathbb{D}_i$ of the attribute A_i and $MP_i(u'_i, u_i)$ expresses the extent to which the value u'_i is *much preferred* to the value u_i. Each element u of S_{relax} is associated with a degree ($\in [0, 1]$). The semantics of this relation is represented by the trapezoidal function $(\gamma_{i1}, \gamma_{i2}, \infty, \infty)$, and denoted $MP_i^{(\gamma_{i1}, \gamma_{i2})}$, see Fig. 2. Figure 3 shows the relaxed version, S_{relax}, of the skyline S of the Example 1.

Table 1. List of candidates

Code	Age	man_exp	tec_exp	dist_wh
M1	32	5	10	35
M2	41	7	5	19
M3	37	5	12	45
M4	36	4	11	39
M5	40	8	10	18
M6	30	4	6	27
M7	31	3	4	56
M8	36	6	13	12
M9	33	6	6	95
M10	40	7	9	20

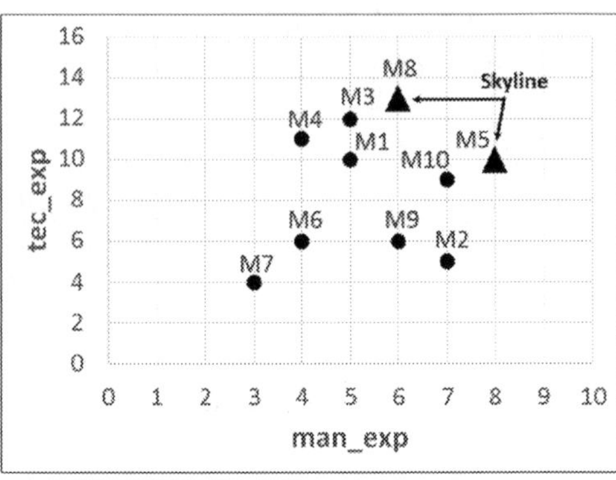

Fig. 1. Skyline of candidates

3 $\mathcal{C2R}$: An Efficient Approach to Enlarging the Skyline

Let $\mathbb{D} = (\mathbb{D}_1, \mathbb{D}_2, ..., \mathbb{D}_d)$ a d-dimensional space where $\mathbb{D}_i$ is the domain attribute of A_i and $R(A_1, A_2, ..., A_d)$ a relation defined in $\mathbb{D}$. We assume the existence of a total order relationship on each domain $\mathbb{D}_i$. $U = (u_1, u_2, ..., u_n)$ is a set of n tuples belonging to a relation R. Let S be the skyline of U and S_{FE} the relaxed skyline of U computed by $\mathcal{C2R}$ approach.

3.1 Principe of the Approach

Our approach relies on the idea of identifying interesting points that are in the neighborhood of skyline points and adding them to the skyline S. Let u be a tuple of $U - S$, and u' a tuple of S. Then, $u \in S_{FE}$ if u is *close* to u'. We write:

$$u \in S_{FE} \Leftrightarrow \exists u' \in S,\ such\, that\ \forall i \in \{1, ..., d\}, (u_i, u'_i) \in C_i \quad (4)$$

where, C_i is a reflexive, symmetrical approximate indifference (or equality) relation defined on the domain $\mathbb{D}_i$ of the attribute A_i and $C_i(u_i, u'_i)$ expresses the extent to which the value u_i is close to the value u'_i. Since C_i is of a gradual

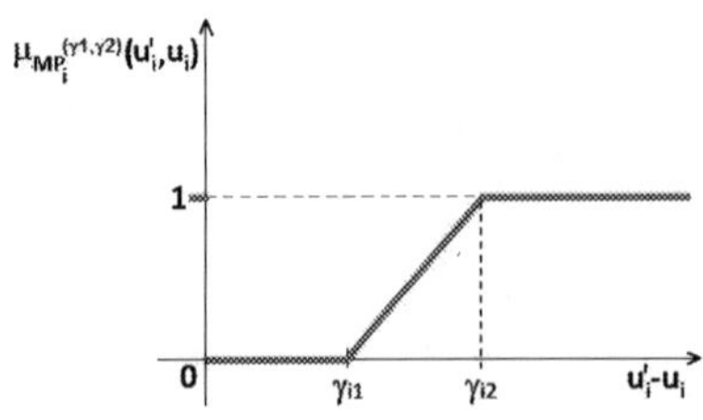

Fig. 2. μ_{MP_i} function

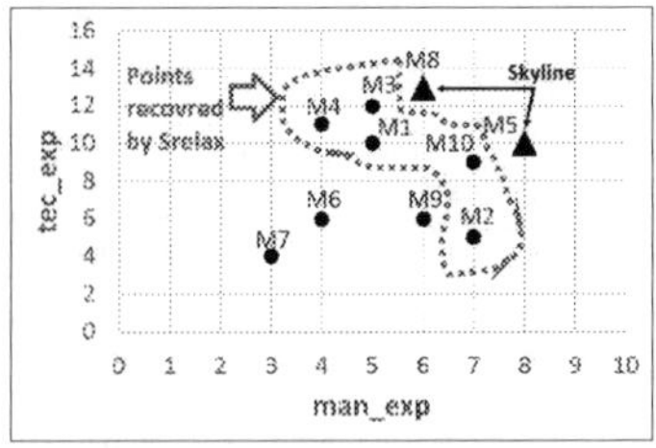

Fig. 3. S_{relax}

nature, each element u of S_{FE} is associated with a degree ($\in [0,1]$) expressing the extent to which u belongs to S_{FE}. In fuzzy set terms, we write:

$$\mu_{S_{FE}}(u) = \max_{u' \in S} \min_{i} \mu_{C_i}(u_i, u'_i) \tag{5}$$

As for C_i relation on $\mathbb{D}_i$, its semantics can be provided by the formulas (6) (see also Fig. 4). In terms of t.m.f., C_i writes $(0, 0, \gamma_{i1}, \gamma_{i2})$, and denoted $C_i^{(\gamma_{i1},\gamma_{i2})}$. It is easy to check that $C_i^{(0,0)}$ corresponds to the classical equality "=".

$$\mu_{C_i^{(\gamma_{i1},\gamma_{i2})}}(u_i, u'_i) = \begin{cases} 0 & \text{if } |u_i - u'_i| \geq \gamma_{i2} \\ 1 & \text{if } |u_i - u'_i| \leq \gamma_{i1} \\ \frac{(\gamma_{i2} - |u_i - u'_i|)}{\gamma_{i2} - \gamma_{i1}} & \text{else} \end{cases} \tag{6}$$

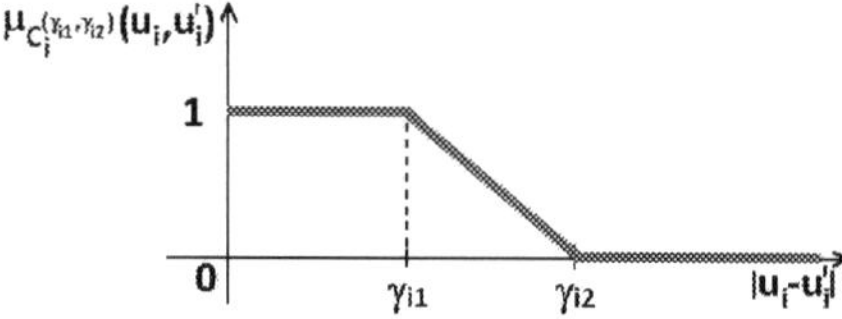

Fig. 4. The membership function $\mu_{C_i^{(\gamma_{i1},\gamma_{i2})}}$

Let $\gamma = ((\gamma_{11}, \gamma_{12}), \cdots, (\gamma_{d1}, \gamma_{d2}))$ be a vector of pairs of parameters where $C_i^{(\gamma_{i1},\gamma_{i2})}$ denotes the C_i relation defined on the attribute A_i and $S_{FE}^{(\gamma)}$ denotes the extended skyline computed on the basis of the vector γ. One can easily check that the classical Skyline S is equal to $S_{FE}^{(\mathbf{0})}$, where $\mathbf{0} = ((0,0), \cdots, (0,0))$.

Definition 3. *Let γ and γ' be two vectors of parameters. We say that $\gamma \geq \gamma'$ if and only if $\forall i \in \{1, \cdots, d\}$, $(\gamma_{i1}, \gamma_{i2}) \geq (\gamma'_{i1}, \gamma'_{i2})$ (i.e., $\gamma_{i1} \geq \gamma'_{i1} \wedge \gamma_{i2} \geq \gamma'_{i2}$).*

Proposition 1. *Let γ and γ' be two vectors of parameters. The following property holds: $\gamma \leq \gamma' \Rightarrow S_{FE}^{(\gamma)} \subseteq S_{FE}^{(\gamma')}$.*

Proof. Let $\gamma \leq \gamma'$, one can deduce that $\forall i, C_i^{\gamma} \subseteq C_i^{\gamma'}$. Let $u \in S_{FE}^{(\gamma)}$
$\Rightarrow \exists u' \in S, \forall i \in \{1, \cdots, d\}, (u_i, u'_i) \in C_i^{(\gamma_{i1},\gamma_{i2})}$
$\Rightarrow \exists u' \in S, \forall i \in \{1, \cdots, d\}, \mu_{C_i^{(\gamma_{i1},\gamma_{i2})}}(u_i, u'_i) > 0$
$\Rightarrow \exists u' \in S, \forall i \in \{1, \cdots, d\}, \mu_{C_i^{(\gamma'_{i1},\gamma'_{i2})}}(u_i, u'_i) > \mu_{C_i^{(\gamma_{i1},\gamma_{i2})}}(u_i, u'_i) > 0$
$\Rightarrow \exists u' \in S, \forall i \in \{1, \cdots, d\}, (u_i, u'_i) \in C_i^{(\gamma'_{i1},\gamma'_{i2})} \Rightarrow u \in S_{FE}^{(\gamma')}$
So we have $S_{FE}^{(\gamma)} \subseteq S_{FE}^{(\gamma')}$ □

Lemma 1. *Let $\gamma = ((0, \gamma_{12}), \cdots, (0, \gamma_{d2}))$ and $\gamma' = ((\gamma'_{11}, \gamma'_{12}), \cdots, (\gamma'_{d1}, \gamma'_{d2}))$, the following holds: $S_{FE}^{(0)} \subseteq S_{FE}^{(\gamma)} \subseteq S_{FE}^{(\gamma')}$*

Table 2. Degrees of the elements of S_{FE}

Mat	M5	M8	M3	M10	M1	M2	M4	M6	M7	M9
$\mu_{S_{FE}}$	1	1	0.66	0.66	0.28	0	0	0	0	0

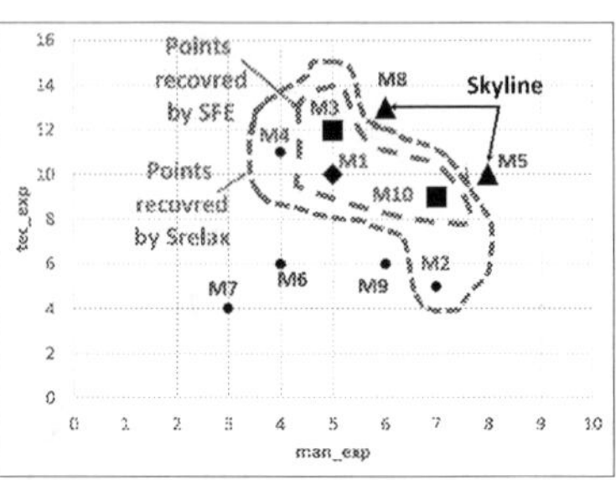

Fig. 5. Points retrieved by S_{FE}

Example 2. Let us come back to the skyline calculated in Example 1. Assume that the fuzzy *"Closeness"* relations corresponding to the skyline attributes (man_exp and tec_exp) are respectively given by:

$$\mu_{C^{(1/2,2)}_{man\_exp}}(u,u') = \begin{cases} 1 & \text{if } |u-u'| \leq 1/2 \\ 0 & \text{if } |u-u'| \geq 2 \\ (-2|u-u'|+4)/3 & \text{else} \end{cases} \tag{7}$$

$$\mu_{C^{(1/2,4)}_{tec\_exp}}(u,u') = \begin{cases} 1 & \text{if } |u-u'| \leq 1/2 \\ 0 & \text{if } |u-u'| \geq 4 \\ (-2|u-u'|+8)/7 & \text{else} \end{cases} \tag{8}$$

Now, applying our approach to relax the skyline $S = \{M_5, M_8\}$ found in Example 1, leads to the following $S_{FE} = \{(M_5, 1), (M_8, 1), (M_3, 0.66), (M_{10}, 0.66), (M_1, 0.28)\}$, see Table 2. One can note that some candidates that were not in S are now elements of S_{FE} (such M_3, M_{10} and M_1) see Fig. 5. As can be seen, S_{FE} is larger than S and $S_{FE} \subseteq S_{relax}$. Let us now take a glance at the content of S_{FE}, one can observe that (i) the skyline elements of S are still elements of S_{FE} with a degree equal to 1; (ii) Appearance of new elements recovered by our approach whose degrees are less than 1 (such as M_3). Interestingly, the user can select from S_{FE}: (i) the Top-k elements (k is a user-defined parameter), or (ii) the subset of elements, denoted $(S_{FE})_\sigma$, with a degrees higher than a threshold σ provided by the user. In the context of Example 2, it is easy to check that $Top-5 = \{(M_5, 1), (M_8, 1), (M_3, 0.66), (M_{10}, 0.66), (M_1, 0.28)\}$ and $(S_{FE})_{0.66} = \{(M_5, 1), (M_8, 1), (M_3, 0.66), (M_{10}, 0.66)\}$.

3.2 S_{FE} Computation

To compute S_{FE}, we proceed in two steps (see Fig. 6). Firstly we compute the skyline S using a slightly modified version of **BNL** algorithm [14], then we execute our FES algorithm to relax the skyline S (see Algorithm 1).

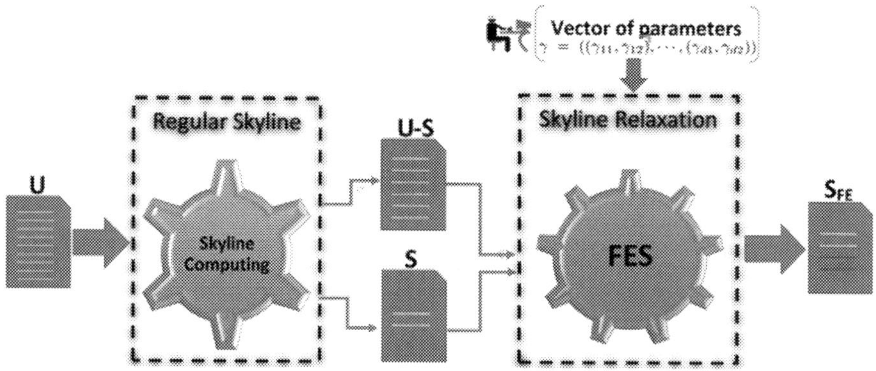

Fig. 6. Enlarging skyline process

Algorithm 1: FES

Input: Set of n tuples $U - S$; a Skyline of m tuples S; γ a vector of parameters;
Output: The relaxed skyline S_{FE};

begin
  $S_{FE} = S$;
  for $i = 1$ *to* n **do**
    $V_{max} = 0$;
    for $j = 1$ *to* m **do**
      $V_{min} = 1$;
      for $k = 1$ *to* d **do**
        $V_{min} = MIN(V_{min}, \mu_{C_k}(u_i, u_j))$;
        if $V_{min} = 0$ **then**
          break;
      $V_{max} = MAX(V_{max}, V_{min})$;
      $\mu_{S_{FE}}(u_i) = V_{max}$;
      if $V_{max} = 1$ **then**
        break;
    if $\mu_{S_{FE}}(u_i) > 0$ **then**
      $S_{FE} = S_{FE} \cup \{u_i\}$;
  rank S_{FE} in decreasing order w.r.t. $\mu_{S_{FE}}(u_i)$;
  - to **return** top-K;
  - or to **return** u_i satisfying $\mu_{S_{FE}}(u_i) \geq \sigma$, ($\sigma$ is a user-defined threshold).

4 Experimental Study

The goal of this study is to demonstrate the effectiveness of the approach proposed and its ability to relax small skylines with the most interesting tuples. We also compare the results obtained with those computed by the $\mathcal{MP2R}$ approach.

4.1 Experimental Environment

We use a Linux OS, on a machine with an i7 processor, a RAM of 8 GB and a 250 GB of disk. Algorithms were implemented with Java. Dataset benchmark is generated using method described in [1] following three distribution schema (correlated, anti-correlated and independent). For each dataset, we consider different sizes (5 K to 750 K). Each tuple contains an integer identifier (4 bytes), 12 decimal fields (96 bytes) with values belonging to the interval [0,1], and a string field with length of 10 characters. Therefore, the size of one tuple is 110 bytes.

4.2 Experimental Results

We vary a collection of parameters that could impact the results. This collection includes the dataset size [D] (5 K, 10 K, 50 K, 100 K, 250 K, 500 K, 750 K), dataset distribution schema [DIS] (independent, correlated, anti-correlated), the number of skyline dimensions [d] (2, 4, 6, 8, 10, 12) and the relaxation thresholds [$\gamma = (\gamma_{i1}, \gamma_{i2})$, i$\in \{1, \ldots, d\}$] where ($\gamma_{i1}, \gamma_{i2} \in$[0,1] and $\gamma_{i1} \leq \gamma_{i2}$). The default values of these parameters are D = 5 K; DIS = "Correlated"; d = 2; γ=((0,0.25),(0,0.25)). In our experiment, we consider that the less the value, the better. Also, we address the issue of comparison between S_{FE} and S_{relax} in terms of Data distribution scheme [DIS], Number of skyline dimension [d], Data size [D] and Variation of the values of $(\gamma_{i1}, \gamma_{i2})$.

–S_{FE} **vs** S_{relax} **w.r.t [DIS]**. Figure 7 shows that the particularity of correlated data minimize the seize of S_{FE} and S_{relax}. We observe also that $\mathcal{C2R}$ approach retrieves fewer tuples than $\mathcal{MP2R}$ because it is more demanding when relaxation processes. We note that the execution time of $\mathcal{C2R}$, for the three distributions, is largely low compared with the time of $\mathcal{MP2R}$ approach.

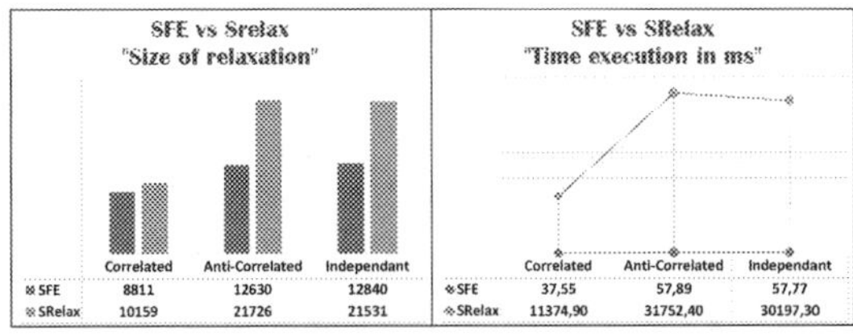

Fig. 7. S_{FE} vs S_{relax} w.r.t [DIS]. (Color figure online)

–S_{FE} **vs** S_{relax} **w.r.t [d]**. When dimensionality increases (from 2 to 12) the size of S_{FE} and S_{relax} increases proportionally (see Fig. 8). We also note that S_{FE} outperforms S_{relax} in terms on computing time.

–S_{FE} **vs** S_{relax} **w.r.t [D]**. The analysis of Fig. 9 shows that the size of S_{FE} and S_{relax} are proportional to the size of the dataset. While in terms of execution time, the computation of S_{FE} is extremely faster.

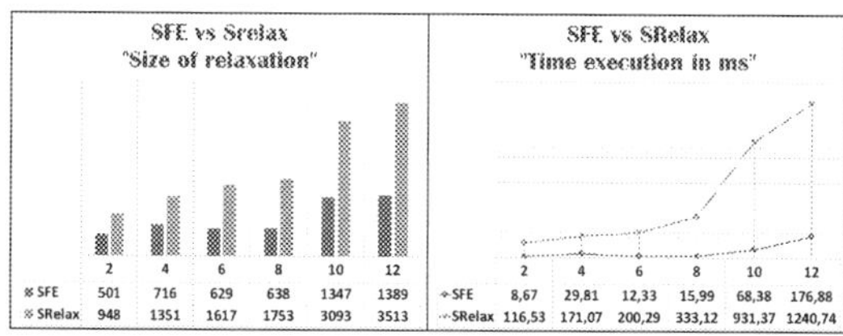

Fig. 8. S_{FE} vs S_{relax} w.r.t [d] (Color figure online)

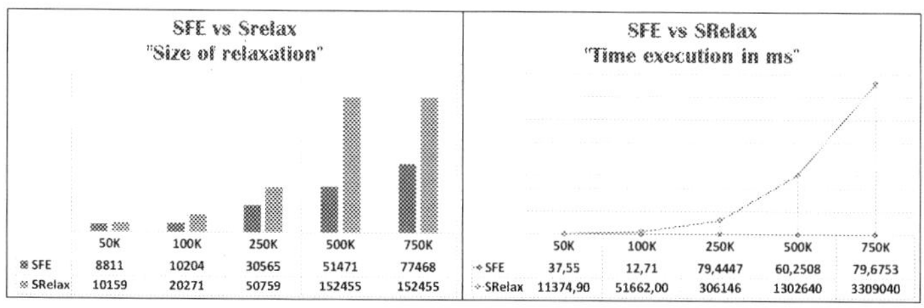

Fig. 9. S_{FE} vs S_{relax} w.r.t [D] (Color figure online)

As can be seen, this first part of the experimental study shows that $\mathcal{C2R}$ approach is better and more optimal than $\mathcal{MP2R}$ approach.

Variation of $(\gamma_{i_1}, \gamma_{i_2})$ **values.** Now, we show the influence of the variation of $(\gamma_{i_1}, \gamma_{i_2})$ values on the size and the computation time of S_{FE} and S_{relax}.The idea is to vary both thresholds. For the sake of simplicity, and since the data are normalized, we will apply the same values of (γ_1, γ_2) for all skyline dimensions. Note that the size of the skyline is equal to 1 and we will analyze the variation of the number of tuples whose degree $\mu_{S_{FE}}(u) > 0$. The following scenarios are worth to be discussed:

Scenario 1: In this scenario, we fix γ_{i1} and vary γ_{i2} to increase the relaxation zone. We observe the following cases:

- $\gamma_{i1} = 0$ and $\gamma_{i2} \in \{0; 0.25; 0.5; 0.75; 1\}$ (see Fig. 10)
- $\gamma_{i1} = 0.25$ and $\gamma_{i2} \in \{0.25; 0.5; 0.75; 1\}$ (see Fig. 11)
- $\gamma_{i1} = 0.5$ and $\gamma_{i2} \in \{0.5; 0.75; 1\}$ (see Fig. 12)
- $\gamma_{i1} = 0.75$ and $\gamma_{i2} \in \{0.75; 1\}$ (see Fig. 13)

The analysis of Fig. 10 shows that the size of S_{FE} and S_{relax} increases when the value of γ_{i2} increases. We also note that there are no tuples whose degrees of relaxation is equal to 1 (this is due to the value of $\gamma_{i1} = 0$). In Figs. 11, 12 and 13 we note that the value of γ_{i2} controls the size of relaxation (by S_{FE} or S_{relax}).

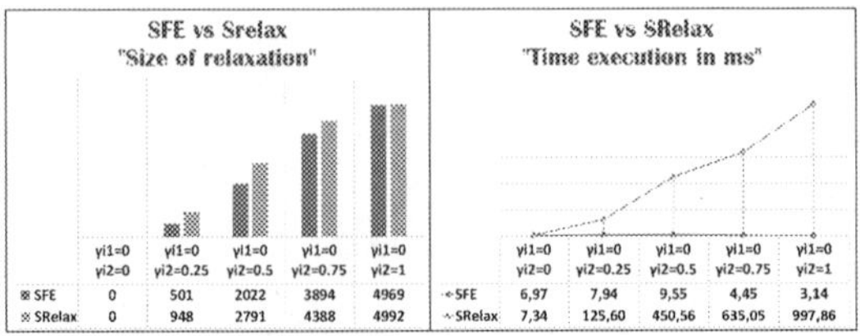

Fig. 10. Scenario 1: Fix γ_{i1} and vary γ_{i2} (case1) (Color figure online)

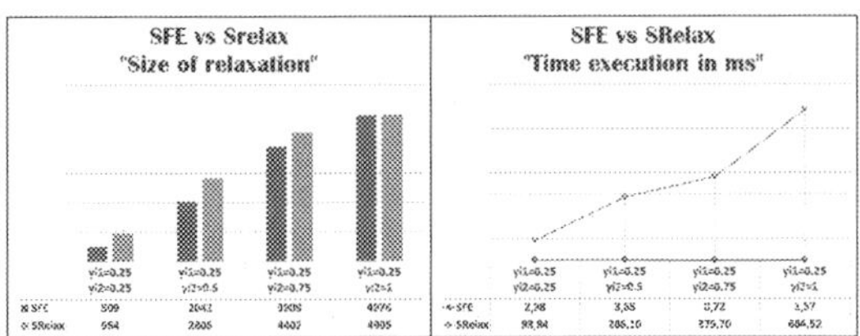

Fig. 11. Scenario 1: Fix γ_{i1} and vary γ_{i2} (case2) (Color figure online)

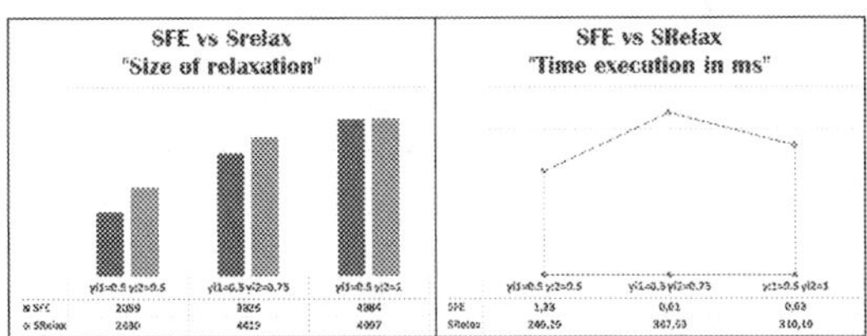

Fig. 12. Scenario 1: Fix γ_{i1} and vary γ_{i2} (case3) (Color figure online)

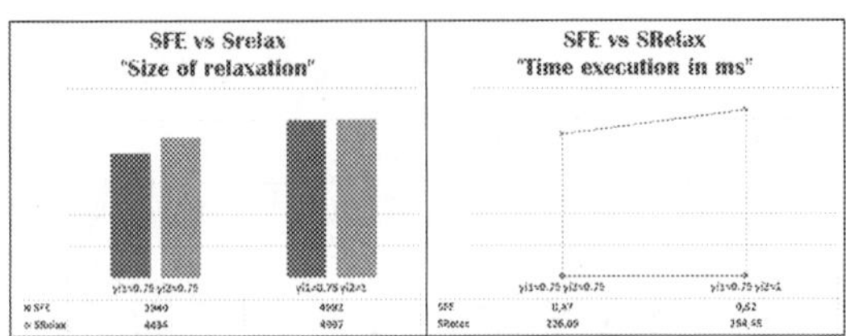

Fig. 13. Scenario 1: Fix γ_{i1} and vary γ_{i2} (case4) (Color figure online)

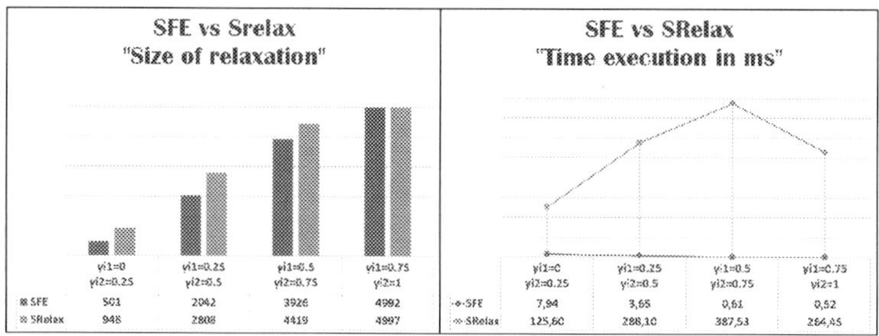

Fig. 14. Varying γ_{i1} and γ_{i2} (Color figure online)

However, we observe the appearance of retrieved tuples with degrees equal 1. It noted that, whatever the value of γ_{i1} and γ_{i2}, the $\mathcal{C2R}$ is more efficient than $\mathcal{MP2R}$ in terms of computation time.

Scenario 2: In this scenario, we vary both thresholds. The obtained results are shown in Fig. 14. The analysis of these curves shows that the relaxation process becomes more permissive when thresholds move away from the origin. Nevertheless, S_{FE} is always more selective than S_{relax} on the number of tuples retrieved (i.e., $|S_{FE}| < |S_{relax}|$)[1] and more efficient in terms of computation time. The Fig. 15 illustrates the distribution of tuples recovered by S_{FE} according to their degrees of relaxation.

Scenario 3: In the previous scenarios, the vector $\gamma = (\gamma_{i_1}, \gamma_{i_2})$ is similar when computing S_{relax} and S_{FE}. Here we will show the impact of using different vectors γ and γ' respectively for S_{FE} and S_{relax}. Table 3 summarizes the results obtained. One can observe that $|S_{FE}| < |S_{relax}|$ if $\gamma \leqslant \gamma'$, $|S_{FE}| > |S_{relax}|$ otherwise (Fig. 16).

Table 3. Impact of the vector γ and γ'.

	#Tuples		
	S_{FE}	S_{relax}	**Conclusion**
Cas1: $\gamma < \gamma'$	2043	4998	$\lvert S_{FE}\rvert < \lvert S_{relax}\rvert$
Cas2: $\gamma > \gamma'$	4985	4403	$\lvert S_{FE}\rvert > \lvert S_{relax}\rvert$
Cas3: $\gamma = \gamma'$	2042	2808	$\lvert S_{FE}\rvert < \lvert S_{relax}\rvert$

[1] Even the relation $S_{FE} \subseteq S_{relax}$ holds in this context.

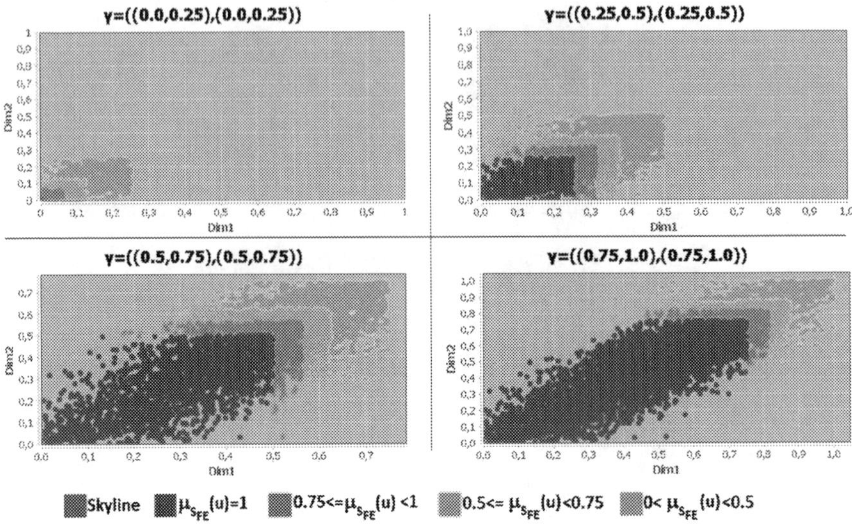

Fig. 15. Distribution of tuples recovered by S_{FE} (Color figure online)

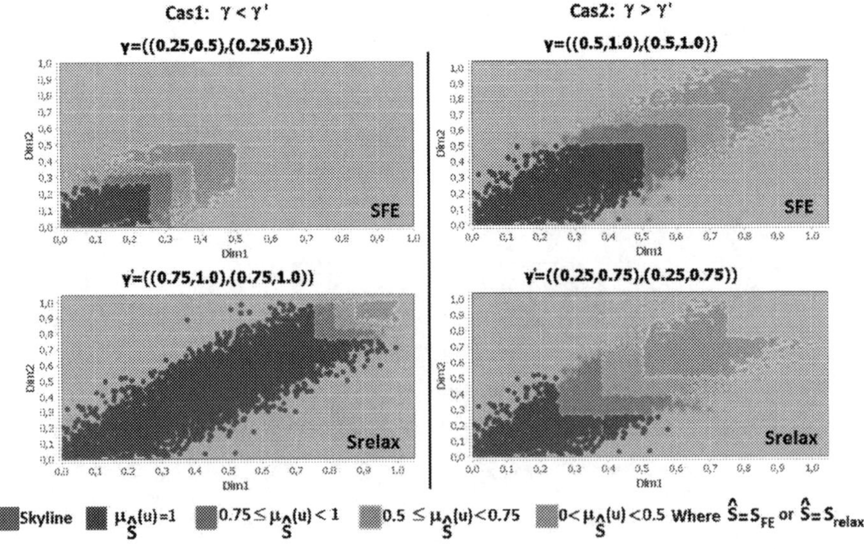

Fig. 16. Distribution of tuples recovered by S_{FE} and S_{relax} (Color figure online)

5 Conclusion

In this paper, we addressed the problem of skyline relaxation, especially less skylines. We propose a new approach for relaxing the skyline, called $\mathcal{C2R}$. This approach is based on a particular fuzzy *Closeness* relation whose semantics is a user-defined. In addition, a new algorithm called **FES** to compute the relaxed skyline is proposed. The experimental study we done has shown that, on the one hand, and in some cases, the $\mathcal{C2R}$ approach is more restrictive than $\mathcal{MP2R}$ approach when relaxing classic skyline and, on the other hand, the computation cost of $\mathcal{C2R}$ is more acceptable. Furthermore, $\mathcal{C2R}$ like $\mathcal{MP2R}$ involves various parameters, which can be used to control the size and the quality of the relaxed skyline. As for future work, we will consider the $\mathcal{C2R}$ approach using a relative fuzzy closeness relation. Then, we will investigate the issue of skyline relaxation in the categorical attributes context.

References

1. Börzsönyi, S., Kossmann, D., Stocker., K.: The skyline operator. In: ICDE, pp. 421–430 (2001)
2. Hadjali, A., Pivert, O., Prade, H.: Possibilistic contextual skylines with incomplete preferences. In: SoCPaR, pp. 57–62 (2010)
3. Khalefa, M.E., Mokbel, M.F., Levandoski, J.J.: Skyline query processing for incomplete data. In: IEEE ICDE, pp. 556–565 (2008)
4. Pei, J., Jiang, B., Lin, X., Yuan, Y.: Probabilistic skylines on uncertain data. In: VLDB, pp. 15–26 (2007)
5. Yiu, M.L., Mamoulis, N.: Efficient processing of top-k dominating queries on multi-dimensional data. In: VLDB, pp. 483–494 (2007)
6. Abbaci, K., Hadjali, A., Lietard, L., Rocacher, D.: A linguistic quantifier-based approach for skyline refinement. In: IFSA/NAFIPS, pp. 321–326 (2013)
7. Chan, C.Y., Jagadish, H.V., Tan, K., Tung, A.K.H., Zhang, Z.: Finding k-dominant skylines in high dimensional space. In: ACM SIGMOD, pp. 503–514 (2006)
8. Chan, C.-Y., Jagadish, H.V., Tan, K.-L., Tung, A.K.H., Zhang, Z.: On high dimensional skylines. In: Ioannidis, Y., Scholl, M.H., Schmidt, J.W., Matthes, F., Hatzopoulos, M., Böhm, K., Kemper, A., Grust, T., Böhm, C. (eds.) EDBT 2006. LNCS, vol. 3896, pp. 478–495. Springer, Heidelberg (2006)
9. Endres, M., Kießling, W.: Skyline snippets. In: Christiansen, H., De Tré, G., Yazici, A., Zadrozny, S., Andreasen, T., Larsen, H.L. (eds.) FQAS 2011. LNCS, vol. 7022, pp. 246–257. Springer, Heidelberg (2011)
10. Hadjali, A., Pivert, O., Prade, H.: On different types of fuzzy skylines. In: Kryszkiewicz, M., Rybinski, H., Skowron, A., Raś, Z.W. (eds.) ISMIS 2011. LNCS, vol. 6804, pp. 581–591. Springer, Heidelberg (2011)
11. Hüllermeier, E., Vladimirskiy, I., Prados Suárez, B., Stauch, E.: Supporting case-based retrieval by similarity skylines: basic concepts and extensions. In: Althoff, K.-D., Bergmann, R., Minor, M., Hanft, A. (eds.) ECCBR 2008. LNCS (LNAI), vol. 5239, pp. 240–254. Springer, Heidelberg (2008)
12. Lin, X., Yuan, Y., Zhang, Q., Zhang, Y.: Selecting stars: the k most representative skyline operator. In: ICDE, pp. 86–95 (2007)

13. Papadias, D., Tao, Y., Fu, G., Seeger, B.: An optimal and progressive algorithm for skyline queries. In: ACM SIGMOD, pp. 467–478 (2003)
14. Belkasmi, D., Hadjali, A.: $\mathcal{MP}\in\mathcal{R}$: a human-centric skyline relaxation approach. In: Proceedings of FQAS, Cracow, Poland, 26–28 October, pp. 227–241 (2015)
15. Goncalves, M., Tineo, L.J.: Fuzzy dominance skyline queries. In: Wagner, R., Revell, N., Pernul, G. (eds.) DEXA 2007. LNCS, vol. 4653, pp. 469–478. Springer, Heidelberg (2007)
16. Balke, W., Güntzer, U., Siberski, W.: Restricting skyline sizes using weak pareto dominance. Inform. Forsch. Entwickl. **21**(3–4), 165–178 (2007)
17. Jin, W., Han, J., Ester, M.: Mining thick skylines over large databases. In: Boulicaut, J.-F., Esposito, F., Giannotti, F., Pedreschi, D. (eds.) PKDD 2004. LNCS (LNAI), vol. 3202, pp. 255–266. Springer, Heidelberg (2004)

Describing Rough Approximations by Indiscernibility Relations in Information Tables with Incomplete Information

Michinori Nakata[1(✉)] and Hiroshi Sakai[2]

[1] Faculty of Management and Information Science, Josai International University, 1 Gumyo, Togane, Chiba 283-8555, Japan
nakatam@ieee.org

[2] Department of Mathematics and Computer Aided Sciences, Faculty of Engineering, Kyushu Institute of Technology, Tobata, Kitakyushu 804-8550, Japan
sakai@mns.kyutech.ac.jp

Abstract. Rough approximations, a pair of lower and upper approximations, and rule induction are described by directly using indiscernibility relations in information tables containing incomplete information. A set of values is used to express incomplete information. The indiscernibility relations are constructed from viewpoints of both certainty and possibility. First, rough approximations and rule induction are described in information tables with complete information. Second, they are addressed in three cases under incomplete information. One is that a set of objects is approximated by objects with incomplete information. Another is that a set of objects with incomplete information is approximated by objects with complete information. The other is the most general case where a set of objects with incomplete information is approximated by objects with incomplete information. Consequently, we obtain four approximations: certain lower, certain upper, possible lower, and possible upper approximations. Using these approximations, rough approximations are expressed by interval sets. The rough approximations have the complementarity property linked with lower and upper approximations, as is valid under complete information. Last, rule induction are addressed in information tables with incomplete information. Rough approximations under incomplete information do not give sufficient information on rules that an object supports. This is resolved by introducing formulae dealing with pairs of an object and a rule that it supports. The pairs are classified into certain and consistent, possible and consistent, certain and inconsistent, and possible and inconsistent pairs.

Keywords: Rough sets · Incomplete information · Indiscernibility relation · Lower and upper approximations · Rule induction

1 Introduction

Incomplete information systems consist of objects whose attribute values are described by a set of values. When a set of values is obtained as an attribute

J.P. Carvalho et al. (Eds.): IPMU 2016, Part II, CCIS 611, pp. 355–366, 2016.
DOI: 10.1007/978-3-319-40581-0_29

value, one element of the set is the actual one, but we cannot know it without additional information. Such a situation frequently appears in our daily life. For example, when we obtain the incomplete information "John's age is round 20," "round 20" is expressed by $\{19, 20, 21\}$. The actual age is in $\{19, 20, 21\}$, but it is unknown which is actual.

The framework of rough sets, constructed by Pawlak [10], is used as an effective tool for data science including various fields such as data analysis, pattern recognition, machine learning, data mining, and so on. The rough sets are based on indiscernibility of objects whose characteristic values are indistinguishable. The fundamental framework is given by rough approximations that consist of lower and upper approximations. The original rough approximations are usually derived from interrelationships, inclusion and intersection, between equivalence classes. The equivalence classes are obtained from indiscernibility relations in an information table containing only complete information.

Some extensions are imposed on the original rough approximations to deal with incomplete information. Kryszkiewicz constructed a discernibility relation by giving indiscernibility of a missing value with any value under an assumption [2]. Some authors propose indiscernibility relations under different assumptions from Kryszkiewicz [1,3,11]. This approach creates poor results of rough approximations [6,11], because it considers only the possibility that a missing value may be equal to another value. In addition, the approach does not give the same rough approximations as the method based on Lipski's one under possible world semantics [6].

A missing value has two possibilities. One possibility is that it may be equal to another value. The other is that it may not be equal to the value. It is unknown which possibility is true without additional information. From this standpoint, Nakata and Sakai have developed an approach based on possible equivalence classes [7]. The number of possible equivalence classes exponentially increases, as the number of missing values does. They avoid the computational complexity by using minimum and maximum possible equivalence classes. Their approach gives the same rough approximations as the work based on Lipski. However, the approach is limited in the case of obtaining possible equivalence classes.

To remove the limitation, we show an approach directly using indiscernibility relations, but not equivalence classes obtained from the indiscernibility relations. The approach is applicable to various types of information. Nakata and Sakai develop the approach for possibilistic information and give successful results [8,9]. In this paper, we apply the approach to incomplete information expressed by a set of values. We formulate rough approximations and rule induction from the viewpoint of both certainty and possibility, as Lipski did in incomplete databases, to deal with incomplete information that includes present but unknown type of missing values as special cases.

The paper is organized as follows. In Sect. 2, an approach based on indiscernibility relations is briefly addressed in the case of information tables with complete information, called complete information systems. In Sect. 3, we develop the approach in the case of information tables with incomplete information, called incomplete information systems. The approach is described from the viewpoint of both certainty and possibility. In Sect. 4, conclusions are addressed.

2 Rough Sets by Indiscernibility Relations in Complete Information Systems

A data set is represented as a table, called an information table, where each row and each column represent an object and an attribute, respectively. A mathematical model of an information table with complete information is called a complete information system. The complete information system is a triplet expressed by $(U, AT, \{D(a_i) \mid a_i \in AT\})$. U is a non-empty finite set of objects called the universe, AT is a non-empty finite set of attributes such that $a_i : U \rightarrow D(a_i)$ for every $a_i \in AT$ where $D(a_i)$ is the domain of attribute a_i. Binary relation R_{a_i} for indiscernibility of objects on attribute $a_i \in AT$, which is called the indiscernibility relation for a_i, is:

$$R_{a_i} = \{(o, o') \in U \times U \mid a_i(o) = a_i(o')\}, \tag{1}$$

where $a_i(o)$ is the value for attribute a_i of object o. From the indiscernibility relation, indiscernible class $[o]_{a_i}$ for object o is obtained:

$$[o]_{a_i} = \{o' \mid (o, o') \in R_{a_i}\}. \tag{2}$$

The condition $a_i(o) = a_i(o')$ in formula (1) makes $[o]_{a_i}$ an equivalence class. The condition can be replaced by another condition. For example, $a_i(o)$ and $a_i(o')$ are similar. In this case, $[o]_{a_i}$ is not always an equivalence class. Family $\mathcal{E}_{a_i}$ of indiscernible classes on a_i is:

$$\mathcal{E}_{a_i} = \{[o]_{a_i} \mid o \in U\}. \tag{3}$$

When $[o]_{a_i}$ is an equivalence class, U is uniquely partitioned by a_i; namely, this is the classification induced by a_i.

Using indiscernibility relation R_{a_i}, lower approximation $\underline{apr}_{a_i}(\mathcal{O})$ and upper approximation $\overline{apr}_{a_i}(\mathcal{O})$ for a_i of set $\mathcal{O}$ of objects are:

$$\underline{apr}_{a_i}(\mathcal{O}) = \{o \mid \forall o' \in U \ (o, o') \notin R_{a_i} \vee o' \in \mathcal{O}\}, \tag{4}$$

$$\overline{apr}_{a_i}(\mathcal{O}) = \{o \mid \exists o' \in U \ (o, o') \in R_{a_i} \wedge o' \in \mathcal{O}\}. \tag{5}$$

When we focus on object o, o is an element of the lower approximation of $\mathcal{O}$, if all objects that are indiscernible with o are included in $\mathcal{O}$. On the other hand, the object is an element of the upper approximation of $\mathcal{O}$, if some objects that are indiscernible with o are in $\mathcal{O}$. Thus, if $o \in \underline{apr}_{a_i}(\mathcal{O})$, then $o \in \overline{apr}_{a_i}(\mathcal{O})$; namely, $\underline{apr}_{a_i}(\mathcal{O}) \subseteq \overline{apr}_{a_i}(\mathcal{O})$. It is well known that the lower and upper approximations are linked with each other, which is called complementarity property:

$$\underline{apr}_{a_i}(\mathcal{O}) = U - \overline{apr}_{a_i}(U - \mathcal{O}). \tag{6}$$

From this formula, $o \in \underline{apr}_{a_i}(\mathcal{O})$ if and only if $o \notin \overline{apr}_{a_i}(U-\mathcal{O})$ and $o \in \overline{apr}_{a_i}(\mathcal{O})$ if and only if $o \notin \underline{apr}_{a_i}(U-\mathcal{O})$.

When objects are characterized by values of attributes, a set of objects being approximated have some structures. In the case where $a_i(o) = a_i(o')$ is used in formula (1), the set of objects is partitioned by equivalence classes obtained from the values of attribute a_i being equal. Under this consideration, lower approximation $\underline{apr}_{a_i}(\mathcal{O}/a_j)$ and upper approximation $\overline{apr}_{a_i}(\mathcal{O}/a_j)$ for a_i are:

$$\underline{apr}_{a_i}(\mathcal{O}/a_j) = \{o \mid \exists o'' \in \mathcal{O}\ \forall o' \in U\ (o,o') \notin R_{a_i} \vee (o',o'') \in R_{a_j} \wedge o' \in \mathcal{O}\}, \tag{7}$$

$$\overline{apr}_{a_i}(\mathcal{O}/a_j) = \{o \mid \exists o'' \in \mathcal{O}\ \exists o' \in U\ (o,o') \in R_{a_i} \wedge (o',o'') \in R_{a_j} \wedge o' \in \mathcal{O}\} \tag{8}$$

In the two approximations, if $o \in \underline{apr}_{a_i}(\mathcal{O}/a_j)$, then $o \in \overline{apr}_{a_i}(\mathcal{O}/a_j)$; namely inclusion relation $\underline{apr}_{a_i}(\mathcal{O}/a_j) \subseteq \overline{apr}_{a_i}(\mathcal{O}/a_j)$ holds. On the other hand, the complementarity property does not hold. If $o \in \underline{apr}_{a_i}(\mathcal{O}/a_j)$, then $o \notin \overline{apr}_{a_i}((U-\mathcal{O})/a_j)$ and if $o \in \overline{apr}_{a_i}(\mathcal{O}/a_j)$, then $o \notin \underline{apr}_{a_i}((U-\mathcal{O})/a_j)$.

We induce rules that hold between attributes from lower and upper approximations. From the lower approximation, when $o \in \underline{apr}_{a_i}(\mathcal{O}/a_j)$, $\exists E_{a_j=v} \in \mathcal{E}_{a_j}\ [o]_{a_i=u} \subseteq E_{a_j=v}$, where $E_{a_j=v}$ is the indiscernible class characterized by value v of a_j and $[o]_{a_i=u}$ is the indiscernible class including o characterized by value u for a_i:

$$E_{a_j=v} = \{o \mid a_j(o) = v \wedge v \in D(a_j)\},$$
$$[o]_{a_i=u} = \{o' \mid a_i(o') = a_i(o) \wedge a_i(o) = u \wedge u \in D(a_i)\}.$$

All objects in $[o]_{a_i=u}$ supports the rule denoted by $a_i = u \rightarrow a_j = v$ where o has u and v of a_i and a_j; namely , $a_i(o) = u$ and $a_j(o) = v$, respectively. Thus, o consistently supports $a_i = u \rightarrow a_j = v$. This is denoted by $(o, a_i = u \rightarrow a_j = v)$. From the upper approximation, when $o \in \overline{apr}_{a_i}(\mathcal{O}/a_j)$, $\exists E_{a_j=v} \in \mathcal{E}_{a_j}\ [o]_{a_i=u} \cap E_{a_j=v} \neq \emptyset$ if o has u of a_i. From $\underline{apr}_{a_i}(\mathcal{O}/a_j) \subseteq \overline{apr}_{a_i}(\mathcal{O}/a_j)$, $o \in (\overline{apr}_{a_i}(\mathcal{O}/a_j) - \underline{apr}_{a_i}(\mathcal{O}/a_j))$ inconsistently supports a rule denoted by $a_i = u \rightarrow a_j = v$ where o has u of a_i, but o does not always have v of a_j, although this is also expressed by $(o, a_i = u \rightarrow a_j = v)$. All objects included in $[o]_{a_i=u}$ do not support $a_i = u \rightarrow a_j = v$. The consistency degree, called accuracy, is evaluated by $|[o]_{a_i=u} \cap E_{a_j=v}|/|[o]_{a_i=u}|$. Clearly, this degree is equal to 1, if $o \in \underline{apr}_{a_i}(\mathcal{O}/a_j)$.

For formulae on sets A and B of attributes,

$$R_A = \cap_{a_i \in A} R_{a_i}, \tag{9}$$

$$[o]_A = \{o' \mid (o,o') \in R_A\} = \cap_{a_i \in A}[o]_{a_i}, \tag{10}$$

$$\underline{apr}_A(\mathcal{O}) = \{o \mid \forall o' \in U\ (o,o') \notin R_A \vee o' \in \mathcal{O}\}, \tag{11}$$

$$\overline{apr}_A(\mathcal{O}) = \{o \mid \exists o' \in U\ (o,o') \in R_A \wedge o' \in \mathcal{O}\}, \tag{12}$$

$$\underline{apr}_A(\mathcal{O}/B) = \{o \mid \exists o'' \in \mathcal{O}\ \forall o' \in U\ (o,o') \notin R_A \vee (o',o'') \in R_B \wedge o' \in \mathcal{O}\}, \quad (13)$$

$$\overline{apr}_A(\mathcal{O}/B) = \{o \mid \exists o'' \in \mathcal{O}\ \exists o' \in U\ (o,o') \in R_A \wedge (o',o'') \in R_B \wedge o' \in \mathcal{O}\}. \quad (14)$$

3 Rough Sets by Indiscernibility Relations in Incomplete Information Systems

In incomplete information systems, $a_i : U \rightarrow s_{a_i}$ for every $a_i \in AT$ where s_{a_i} is the set of all subsets over domain $D(a_i)$ of attribute a_i. $v \in a_i(o)$ is a possible value that may be the actual one as the value of attribute a_i in object o. The possible value is the actual one if $|a_i(o)| = 1$.

The indiscernibility relation for a_i in an incomplete information system is expressed by using two relations CR_{a_i} and PR_{a_i}. CR_{a_i} is a certain indiscernibility relation and PR_{a_i} is a possible one:

$$CR_{a_i} = \{(o,o') \mid o = o' \vee a_i(o) = a_i(o') \text{ with } |a_i(o)| = |a_i(o')| = 1\}, \quad (15)$$

$$PR_{a_i} = \{(o,o') \mid o = o' \vee u = v \wedge u \in a_i(o) \wedge v \in a_i(o')\}. \quad (16)$$

The certain indiscernibility relation is reflexive, symmetric, and transitive, but the possible one is not transitive although it is reflexive and symmetric. We have three patterns. One case is that a pair of objects are not in both certain and possible indiscernibility relations, which means that they are discernible. Another is that they are not in the certain indiscernibility relation, but in the possible one, which means that they are discernible and indiscernible. The other is that they are in both certain and possible indiscernibility relations, which means that they are indiscernible.

Example 1. Let information table T be obtained as follows:

T

U	a_1	a_2
1	$\{x\}$	$\{a,c\}$
2	$\{x,y\}$	$\{a,b\}$
3	$\{y\}$	$\{b\}$
4	$\{y\}$	$\{b\}$
5	$\{w\}$	$\{c\}$
6	$\{w,z\}$	$\{c\}$

In information table T, $U = \{o_1, o_2, o_3, o_4, o_5, o_6\}$, where domains $D(a_1)$ and $D(a_2)$ of attributes a_1 and a_2 are $\{w,x,y,z\}$ and $\{a,b,c\}$, respectively. Using formulae (15) and (16), certain and possible indiscernibility relations for a_1 in T are:

$$CR_{a_i} = \{(o_1,o_1),(o_2,o_2),(o_3,o_3),(o_3,o_4),(o_4,o_3),(o_4,o_4),(o_5,o_5),(o_6,o_6)\},$$

$$PR_{a_i} = \{(o_1,o_1),(o_1,o_2),(o_2,o_1),(o_2,o_2),(o_2,o_3),(o_2,o_4),(o_3,o_2),(o_3,o_3), (o_3,o_4),(o_4,o_2),(o_4,o_3),(o_4,o_4),(o_5,o_5),(o_5,o_6),(o_6,o_5),(o_6,o_6)\}.$$

Lipski showed that certain and possible answers, not the actual answer, are obtained in query processing under incomplete information [4,5]. This is true for rough approximations. We cannot definitely obtain whether or not an object belongs to rough approximations, but we can know whether or not the object certainly or possibly belongs to rough approximations. Therefore, we show certain rough approximations (resp. possible rough approximations) whose object certainly (resp. possibly) belongs to the actual rough approximations.

Let $\mathcal{O}$ be a set of objects. Certain lower approximation $C\underline{apr}_{a_i}(\mathcal{O})$ and possible one $P\underline{apr}_{a_i}(\mathcal{O})$ are:

$$C\underline{apr}_{a_i}(\mathcal{O}) = \{o \mid \forall o' \in U\ (o,o') \notin PR_{a_i} \vee o' \in \mathcal{O}\}, \tag{17}$$
$$P\underline{apr}_{a_i}(\mathcal{O}) = \{o \mid \forall o' \in U\ (o,o') \notin CR_{a_i} \vee o' \in \mathcal{O}\}. \tag{18}$$

Proposition 1. $C\underline{apr}_{a_i}(\mathcal{O}) \subseteq P\underline{apr}_{a_i}(\mathcal{O})$.

Similarly, Certain upper approximation $C\overline{apr}_{a_i}(\mathcal{O})$ and possible one $P\overline{apr}_{a_i}(\mathcal{O})$ are:

$$C\overline{apr}_{a_i}(\mathcal{O}) = \{o \mid \exists o' \in U\ (o,o') \in CR_{a_i} \wedge o' \in \mathcal{O}\}, \tag{19}$$
$$P\overline{apr}_{a_i}(\mathcal{O}) = \{o \mid \exists o' \in U\ (o,o') \in PR_{a_i} \wedge o' \in \mathcal{O}\}. \tag{20}$$

Proposition 2. $C\overline{apr}_{a_i}(\mathcal{O}) \subseteq P\overline{apr}_{a_i}(\mathcal{O})$.

Proposition 3. $C\underline{apr}_{a_i}(\mathcal{O}) \subseteq C\overline{apr}_{a_i}(\mathcal{O})$ and $P\underline{apr}_{a_i}(\mathcal{O}) \subseteq P\overline{apr}_{a_i}(\mathcal{O})$.

Proposition 4. $C\underline{apr}_{a_i}(\mathcal{O}) \subseteq P\underline{apr}_{a_i}(\mathcal{O}) \subseteq \mathcal{O} \subseteq C\overline{apr}_{a_i}(\mathcal{O}) \subseteq P\overline{apr}_{a_i}(\mathcal{O})$.

Four approximations are linked with each other.

Proposition 5. $P\underline{apr}_{a_i}(\mathcal{O}) = U - C\overline{apr}_{a_i}(U - \mathcal{O})$ and $C\underline{apr}_{a_i}(\mathcal{O}) = U - P\overline{apr}_{a_i}(U - \mathcal{O})$.

Using four approximations denoted by formulae (17)–(20), lower and upper approximations are expressed by interval sets as follows:

$$\underline{apr}_{a_i}(\mathcal{O}) = [C\underline{apr}_{a_i}(\mathcal{O}), P\underline{apr}_{a_i}(\mathcal{O})], \tag{21}$$
$$\overline{apr}_{a_i}(\mathcal{O}) = [C\overline{apr}_{a_i}(\mathcal{O}), P\overline{apr}_{a_i}(\mathcal{O})]. \tag{22}$$

Certain and possible approximations are the lower and upper bounds of the actual approximation. The lower and upper approximations depend on each other; namely, the complementarity property linked with them holds, as is so in complete information systems.

Proposition 6.

$$\underline{apr}_{a_i}(\mathcal{O}) = U - \overline{apr}_{a_i}(U - \mathcal{O}).$$

Example 2. Let us go back to Example 1. Let set $\mathcal{O}$ of objects be $\{o_2, o_3, o_4\}$. Using formulae (17)–(20),

$$\begin{aligned} C\underline{apr}_{a_1}(\mathcal{O}) &= \{o_3, o_4\}, \\ P\underline{apr}_{a_1}(\mathcal{O}) &= \{o_2, o_3, o_4\}, \\ C\overline{apr}_{a_1}(\mathcal{O}) &= \{o_2, o_3, o_4\}, \\ P\overline{apr}_{a_1}(\mathcal{O}) &= \{o_1, o_2, o_3, o_4\}. \end{aligned}$$

Thus, using formulae (21)–(22),

$$\begin{aligned} \underline{apr}_{a_1}(\mathcal{O}) &= [\{o_3, o_4\}, \{o_2, o_3, o_4\}], \\ \overline{apr}_{a_1}(\mathcal{O}) &= [\{o_2, o_3, o_4\}, \{o_1, o_2, o_3, o_4\}]. \end{aligned}$$

Subsequently, we describe the case where a set of objects characterized by incomplete information is approximated by objects with complete information. Let objects in U have complete information for a_i and $\mathcal{O}$ be characterized by a_j with incomplete information. Four approximations are:

$$C\underline{apr}_{a_i}(\mathcal{O}/a_j) = \{o \mid \exists o'' \in \mathcal{O}\ \forall o' \in [o]_{a_i} (o', o'') \in CR_{a_j} \wedge o' \in \mathcal{O}\}, \quad (23)$$
$$P\underline{apr}_{a_i}(\mathcal{O}/a_j) = \{o \mid \exists o'' \in \mathcal{O}\ \forall o' \in [o]_{a_i} (o', o'') \in PR_{a_j} \wedge o' \in \mathcal{O}\}, \quad (24)$$
$$C\underline{apr}_{a_i}(\mathcal{O}/a_j) = \{o \mid \exists o'' \in \mathcal{O}\ \exists o' \in [o]_{a_i} (o', o'') \in CR_{a_j} \wedge o' \in \mathcal{O}\}, \quad (25)$$
$$P\underline{apr}_{a_i}(\mathcal{O}/a_j) = \{o \mid \exists o'' \in \mathcal{O}\ \exists o' \in [o]_{a_i} (o', o'') \in PR_{a_j} \wedge o' \in \mathcal{O}\}. \quad (26)$$

Combining the above two cases, we can obtain four approximations in the case where both objects used to approximate and objects approximated are characterized by attributes with incomplete information. Certain lower approximation $C\underline{apr}_{a_i}(\mathcal{O}/a_j)$ and possible one $P\underline{apr}_{a_i}(\mathcal{O}/a_j)$ are:

$$\begin{aligned} &C\underline{apr}_{a_i}(\mathcal{O}/a_j) \\ &\quad = \{o \mid \exists o'' \in \mathcal{O}\ \forall o' \in U\ (o, o') \notin PR_{a_i} \vee (o', o'') \in CR_{a_j} \wedge o' \in \mathcal{O}\}, \quad (27) \\ &P\underline{apr}_{a_i}(\mathcal{O}/a_j) \\ &\quad = \{o \mid \exists o'' \in \mathcal{O}\ \forall o' \in U\ (o, o') \notin CR_{a_i} \vee (o', o'') \in PR_{a_j} \wedge o' \in \mathcal{O}\}. \quad (28) \end{aligned}$$

Proposition 7. $C\underline{apr}_{a_i}(\mathcal{O}/a_j) \subseteq P\underline{apr}_{a_i}(\mathcal{O}/a_j)$.

Similarly, certain upper approximation $C\overline{apr}_{a_i}(\mathcal{O}/a_j)$ and possible one $P\overline{apr}_{a_i}(\mathcal{O}/a_j)$ are:

$$\begin{aligned} &C\overline{apr}_{a_i}(\mathcal{O}/a_j) \\ &\quad = \{o \mid \exists o'' \in \mathcal{O}\ \exists o' \in U\ (o, o') \in CR_{a_i} \wedge (o', o'') \in CR_{a_j} \wedge o' \in \mathcal{O}\}, \quad (29) \\ &P\overline{apr}_{a_i}(\mathcal{O}/a_j) \\ &\quad = \{o \mid \exists o'' \in \mathcal{O}\ \exists o' \in U\ (o, o') \in PR_{a_i} \wedge (o', o'') \in PR_{a_j} \wedge o' \in \mathcal{O}\}. \quad (30) \end{aligned}$$

Proposition 8. $C\overline{apr}_{a_i}(\mathcal{O}/a_j) \subseteq P\overline{apr}_{a_i}(\mathcal{O}/a_j)$.

Proposition 9. $C\underline{apr}_{a_i}(\mathcal{O}/a_j) \subseteq C\overline{apr}_{a_i}(\mathcal{O}/a_j)$ and $P\underline{apr}_{a_i}(\mathcal{O}/a_j) \subseteq P\overline{apr}_{a_i}(\mathcal{O}/a_j)$.

Proposition 10. $C\underline{apr}_{a_i}(\mathcal{O}/a_j) \subseteq P\underline{apr}_{a_i}(\mathcal{O}/a_j) \subseteq \mathcal{O} \subseteq C\overline{apr}_{a_i}(\mathcal{O}/a_j) \subseteq P\overline{apr}_{a_i}(\mathcal{O}/a_j)$.

Lower and upper approximations are:

$$\underline{apr}_{a_i}(\mathcal{O}/a_j) = [C\underline{apr}_{a_i}(\mathcal{O}/a_j), P\underline{apr}_{a_i}(\mathcal{O}/a_j)], \quad (31)$$

$$\overline{apr}_{a_i}(\mathcal{O}/a_j) = [C\overline{apr}_{a_i}(\mathcal{O}/a_j), P\overline{apr}_{a_i}(\mathcal{O}/a_j)]. \quad (32)$$

Example 3. Let us go back to information table T in Example 1. Let $\mathcal{O}$ be $\{o_2, o_3, o_4\}$ that is characterized by values of attribute a_2. Using formulae (27)–(32),

$$\underline{apr}_{a_1}(\mathcal{O}/a_2) = [\{\emptyset\}, \{o_2, o_3, o_4\}],$$
$$\overline{apr}_{a_1}(\mathcal{O}/a_2) = [\{o_2, o_3, o_4\}, \{o_1, o_2, o_3, o_4\}].$$

An object that belongs to certain rough approximations does not certainly support a rule. For example, $C\underline{apr}_{a_1}(U/a_2) = \{o_5, o_6\}$ in T of Example 1. o_5 certainly supports rule $a_1 = w \rightarrow a_2 = c$, but o_6 does not certainly supports rule $a_1 = w \rightarrow a_2 = c$, because $a_1 = \{w, z\}$. To clarify how an object supports a rule, we derive certain and possible indiscernibility relations $CR_{a_i=u}$ and $PR_{a_i=u}$ where all pairs of objects are characterized by value u of attribute a_i.

$$CR_{a_i=u} = \{(o, o') \mid a_i(o) = a_i(o') = u\}, \quad (33)$$

$$PR_{a_i=u} = \{(o, o') \mid u \in a_i(o) \wedge u \in a_i(o')\}. \quad (34)$$

Example 4. In T of Example 1, using formulae (33) and (34),

$$CR_{a_1=x} = \{(o_1, o_1)\},$$
$$PR_{a_1=x} = \{(o_1, o_1), (o_1, o_2), (o_2, o_1), (o_2, o_2)\},$$
$$CR_{a_1=y} = \{(o_3, o_3), (o_3, o_4), (o_4, o_3), (o_4, o_4), \},$$
$$PR_{a_1=y} = \{(o_2, o_2), (o_2, o_3), (o_2, o_4), (o_3, o_2), (o_3, o_3), (o_3, o_4), (o_4, o_2), (o_4, o_3), (o_4, o_4), \},$$
$$CR_{a_1=z} = \emptyset,$$
$$PR_{a_1=z} = \{(o_6, o_6)\},$$
$$CR_{a_1=w} = \{(o_5, o_5)\},$$
$$PR_{a_1=w} = \{(o_5, o_5), (o_5, o_6), (o_6, o_5), (o_6, o_6)\}.$$

Using indiscernibility relations $CR_{a_i=u}$ and $PR_{a_i=u}$ that are characterized by a value of an attribute, we obtain four sets of pairs of an object and a rule that it supports: certain lower, possible lower, certain upper, and possible upper sets, which correspond to the above four approximations.

Certain lower set $C\underline{r}_{a_i}(\mathcal{O}/a_j)$, which corresponds to $C\underline{apr}_{a_i}(\mathcal{O}/a_j)$, is:

$$\begin{aligned} C\underline{r}_{a_i}(\mathcal{O}/a_j) = \{&(o, a_i = u \rightarrow a_j = v) \mid \exists o' \in U\ (o, o') \in CR_{a_i=u} \wedge \\ &(\exists o'' \in \mathcal{O}\ \forall o' \in U\ (o, o') \notin PR_{a_i=u} \vee (o', o'') \in CR_{a_j=v} \wedge o' \in \mathcal{O})\}. \end{aligned} \tag{35}$$

Possible lower set $P\underline{r}_{a_i}(\mathcal{O}/a_j)$, which corresponds to $P\underline{apr}_{a_i}(\mathcal{O}/a_j)$, is:

$$\begin{aligned} P\underline{r}_{a_i}(\mathcal{O}/a_j) = \{&(o, a_i = u \rightarrow a_j = v) \mid \exists o' \in U\ (o, o') \in PR_{a_i=u} \wedge \\ &(\exists o'' \in \mathcal{O}\ \forall o' \in U\ (o, o') \notin CR_{a_i=u} \vee (o', o'') \in PR_{a_j=v} \wedge o' \in \mathcal{O})\}. \end{aligned} \tag{36}$$

Proposition 11. $C\underline{r}_{a_i}(\mathcal{O}/a_j) \subseteq P\underline{r}_{a_i}(\mathcal{O}/a_j)$.

This proposition shows that the possible lower set includes the certain lower set.

Certain upper set $C\overline{r}_{a_i}(\mathcal{O}/a_j)$, which corresponds to $C\overline{apr}_{a_i}(\mathcal{O}/a_j)$, is:

$$\begin{aligned} C\overline{r}_{a_i}(\mathcal{O}/a_j) = \{&(o, a_i = u \rightarrow a_j = v) \mid \\ &\exists o'' \in \mathcal{O}\ \exists o' \in U\ (o, o') \in CR_{a_i=u} \wedge (o', o'') \in CR_{a_j=v} \wedge o' \in \mathcal{O}\}. \end{aligned} \tag{37}$$

Possible upper set $P\overline{r}_{a_i}(\mathcal{O}/a_j)$, which corresponds to $P\overline{apr}_{a_i}(\mathcal{O}/a_j)$, is:

$$\begin{aligned} P\overline{r}_{a_i}(\mathcal{O}/a_j) = \{&(o, a_i = u \rightarrow a_j = v) \mid \\ &\exists o'' \in \mathcal{O}\ \exists o' \in U\ (o, o') \in PR_{a_i=u} \wedge (o', o'') \in PR_{a_j=v} \wedge o' \in \mathcal{O}\}. \end{aligned} \tag{38}$$

Proposition 12. $C\overline{r}_{a_i}(\mathcal{O}/a_j) \subseteq P\overline{r}_{a_i}(\mathcal{O}/a_j)$.

This proposition shows that the possible upper set includes the certain upper set.

Proposition 13. $C\underline{r}_{a_i}(\mathcal{O}/a_j) \subseteq C\overline{r}_{a_i}(\mathcal{O}/a_j)$ and $P\underline{r}_{a_i}(\mathcal{O}/a_j) \subseteq P\overline{r}_{a_i}(\mathcal{O}/a_j)$.

This proposition shows that the certain upper set includes the certain lower set and the possible upper set includes the possible lower set.

Using the above four sets, sets $\underline{r}_{a_i}(\mathcal{O}/a_j)$ and $\overline{r}_{a_i}(\mathcal{O}/a_j)$, which correspond to $C\underline{apr}_{a_i}(\mathcal{O}/a_j)$ and $C\overline{apr}_{a_i}(\mathcal{O}/a_j)$, are also expressed by interval sets:

$$\underline{r}_{a_i}(\mathcal{O}/a_j) = [C\underline{r}_{a_i}(\mathcal{O}/a_j), P\underline{r}_{a_i}(\mathcal{O}/a_j)], \tag{39}$$

$$\overline{r}_{a_i}(\mathcal{O}/a_j) = [C\overline{r}_{a_i}(\mathcal{O}/a_j), P\overline{r}_{a_i}(\mathcal{O}/a_j)]. \tag{40}$$

Pairs of an object and a rule are classified into four cases: certain and consistent, certain and inconsistent, possible and consistent, and possible inconsistent pairs. Objects that appear in certain set $C\underline{r}_{a_i}(\mathcal{O}/a_j)$ certainly support rules with consistency. From Proposition 13, $C\underline{r}_{a_i}(\mathcal{O}/a_j) \subseteq C\overline{r}_{a_i}(\mathcal{O}/a_j)$. So, objects that appear in $(C\overline{r}_{a_i}(\mathcal{O}/a_j) - C\underline{r}_{a_i}(\mathcal{O}/a_j))$ certainly support rules with inconsistency. From Proposition 11, $C\underline{r}_{a_i}(\mathcal{O}/a_j) \subseteq P\underline{r}_{a_i}(\mathcal{O}/a_j)$. So, objects that appear in $(P\underline{r}_{a_i}(\mathcal{O}/a_j) - C\underline{r}_{a_i}(\mathcal{O}/a_j))$ possibly support rules with consistency. From Propositions 12 and 13, $C\overline{r}_{a_i}(\mathcal{O}/a_j) \subseteq P\overline{r}_{a_i}(\mathcal{O}/a_j)$ and $P\underline{r}_{a_i}(\mathcal{O}/a_j) \subseteq P\overline{r}_{a_i}(\mathcal{O}/a_j)$. So, objects that appear in $(P\overline{r}_{a_i}(\mathcal{O}/a_j) - P\underline{r}_{a_i}(\mathcal{O}/a_j) - C\overline{r}_{a_i}(\mathcal{O}/a_j))$ possibly support rules with inconsistency.

Example 5. Let us go back to Example 1. Using formulae (35)–(38) and then gathering objects supporting the same rule in a set,

$$\begin{aligned}
C\underline{r}_{a_1}(U/a_2) &= \{(\{o_5\}, a_1 = w \rightarrow a_2 = c)\},\\
C\overline{r}_{a_1}(U/a_2) &= \{(\{o_3, o_4\}, a_1 = y \rightarrow a_2 = b), (\{o_5\}, a_1 = w \rightarrow a_2 = c)\},\\
P\underline{r}_{a_1}(U/a_2) &= \{(\{o_1\}, a_1 = x \rightarrow a_2 = c), (\{o_1, o_2\}, a_1 = x \rightarrow a_2 = a),\\
&\quad (\{o_2, o_3, o_4\}, a_1 = y \rightarrow a_2 = b), (\{o_5, o_6\}, a_1 = w \rightarrow a_2 = c),\\
&\quad (\{o_6\}, a_1 = z \rightarrow a_2 = c)\},\\
P\overline{r}_{a_1}(U/a_2) &= \{(\{o_1, o_2\}, a_1 = x \rightarrow a_2 = a), (\{o_1, o_2\}, a_1 = x \rightarrow a_2 = b),\\
&\quad (\{o_1, o_2\}, a_1 = x \rightarrow a_2 = c), (\{o_2, o_3, o_4\}, a_1 = y \rightarrow a_2 = a),\\
&\quad (\{o_2, o_3, o_4\}, a_1 = y \rightarrow a_2 = b), (\{o_5, o_6\}, a_1 = w \rightarrow a_2 = c),\\
&\quad (\{o_6\}, a_1 = z \rightarrow a_2 = c)\}.
\end{aligned}$$

Using these formulae, the certain set of pairs of objects and a rule with consistency, which are in $C\underline{r}_{a_1}(U/a_2)$, is:

$$\{(\{o_5\}, a_1 = w \rightarrow a_2 = c)\}.$$

The certain set of pairs with inconsistency, which are in $(C\overline{r}_{a_1}(U/a_2) - C\underline{r}_{a_1}(U/a_2))$, is:

$$\{(\{o_3, o_4\}, a_1 = y \rightarrow a_2 = b)\}.$$

The possible set of pairs with consistency, which are in $(P\underline{r}_{a_1}(U/a_2) - C\underline{r}_{a_1}(U/a_2))$, is:

$$\begin{aligned}
&\{(\{o_1\}, a_1 = x \rightarrow a_2 = c), (\{o_1, o_2\}, a_1 = x \rightarrow a_2 = a),\\
&(\{o_2, o_3, o_4\}, a_1 = y \rightarrow a_2 = b), (\{o_6\}, a_1 = w \rightarrow a_2 = c),\\
&(\{o_6\}, a_1 = z \rightarrow a_2 = c)\}.
\end{aligned}$$

A possible set of pairs with inconsistency, which are in $(P\overline{r}_{a_1}(U/a_2) - P\underline{r}_{a_1}(U/a_2) - C\overline{r}_{a_1}(U/a_2))$, is:

$$\begin{aligned}
&\{(\{o_1, o_2\}, a_1 = x \rightarrow a_2 = b), (\{o_2\}, a_1 = x \rightarrow a_2 = c),\\
&(\{o_2, o_3, o_4\}, a_1 = y \rightarrow a_2 = a)\}.
\end{aligned}$$

4 Conclusions

We have described an approach based on rough sets in an incomplete information system. An attribute value is expressed by a set of values in the incomplete information system. The approach is based on directly using indiscernibility relations from the viewpoint of both certainty and possibility. First, We have shown rough approximations for the case where only objects used to approximate are characterized by attributes with incomplete information. Second, we have shown the case where only objects in a set approximated have incomplete information. Finally, rough approximations have been shown in the case where both objects used to approximate and objects approximated are characterized by attributes with incomplete information.

We have four approximations: certain lower, possible lower, certain upper, and possible upper ones. These are linked with each other. Lower and upper approximations consists of a pair of certain and possible lower ones and a pair of certain and possible upper ones, respectively. This is essential in incomplete information systems. As a result, the complementarity property linked with lower and upper approximations holds, as is valid under complete information.

Objects that belongs to certain rough approximations do not always support certain rule. To clarify how rules an object supports, we have introduced expressions where we deal with pairs of an object and a rule that it supports. By using the expressions, we can obtain four sets of pairs of an object and a rule that it supports: certain and consistent, possible and consistent, certain and inconsistent, and possible and inconsistent sets. In other words, pairs of an object and a rule are classified into four types.

Our approach is applicable to the case where equivalence classes are not obtained, because we directly use indiscernibility relations.

References

1. Grzymała-Busse, J.W.: Data with missing attribute values: generalization of indiscernibility relation and rule induction. In: Peters, J.F., Skowron, A., Grzymała-Busse, J.W., Kostek, B., Swiniarski, R.W., Szczuka, M.S. (eds.) Transactions on Rough Sets I. LNCS, vol. 3100, pp. 78–95. Springer, Heidelberg (2004)
2. Kryszkiewicz, M.: Rules in incomplete information systems. Inf. Sci. **113**, 271–292 (1999)
3. Leung, Y., Li, D.: Maximum consistent techniques for rule acquisition in incomplete information systems. Inf. Sci. **153**, 85–106 (2003)
4. Lipski, W.: On semantics issues connected with incomplete information databases. ACM Trans. Database Syst. **4**, 262–296 (1979)
5. Lipski, W.: On databases with incomplete information. J. ACM **28**, 41–70 (1981)
6. Nakata, M., Sakai, H.: Applying rough sets to information tables containing missing values. In: Proceedings of 39th International Symposium on Multiple-Valued Logic, pp. 286–291. IEEE Press (2009)
7. Nakata, M., Sakai, H.: Twofold rough approximations under incomplete information. Int. J. Gen. Syst. **42**, 546–571 (2013)

8. Nakata, M., Sakai, H.: An approach based on rough sets to possibilistic information. Commun. Comput. Inf. Sci. **444**, 61–70 (2014)
9. Nakata, M., Sakai, H.: Formulating rough approximations in information tables with possibilistic information. In: Proceedings of ICGCRSIS 2015: International Conference on Granular Computing, Rough Sets and Intelligent Systems, Copenhagen, Denmark, pp. 1898–1903, 11–12 June 2015
10. Pawlak, Z.: Theoretical Aspects of Reasoning About Data. Kluwer Academic Publishers, Dordrecht (1991)
11. Stefanowski, J., Tsoukiàs, A.: Incomplete information tables and rough classification. Comput. Intell. **17**, 545–566 (2001)

A Possibilistic Treatment of Data Quality Measurement

Antoon Bronselaer(✉) and Guy De Tré

Department of Telecommunication and Information Processing, Ghent University, Sint-Pietersnieuwstraat 41, 9000 Gent, Belgium
{antoon.bronselaer,guy.detre}@ugent.be

Abstract. The ever growing capabilities of data storage systems have created the need to assess the quality of data in an efficient manner. In this paper, we consider a framework of data quality measurement that relies on basic predicates formulated on the data. It is then motivated that in some cases, the evaluation of predicates is hindered due to a lack of information. As a result, the truth value of a predicate can not be determined with complete certainty. In this paper, it is first shown how such uncertainty about the evaluation of predicates can be modelled. Such uncertainty can then be propagated throughout the measurement process. This establishes a possibilistic measurement of data quality.

Keywords: Data quality measurement · Uncertainty modelling · Possibility theory

1 Introduction

The ever growing amount of data that is collected and made available nowadays, has caused a setting in which modern organisations increasingly rely on information drawn from data they posses to make important and strategic decisions. This phenomenon is sometimes referred to as "Business Intelligence". Needless to say, if data become the cornerstone for decisions, the quality of data strongly influences the accuracy of these decisions and thus gains importance. As a result, methods and techniques for the assessment (and improvement) of data quality have drawn much attention of researchers in the past decade [1].

In this paper, a recently proposed framework is adopted in which quality of data is measured on an ordinal scale. In the very essence, the measurement of data considers a set of predicates that models the requirements for data to be of the best possible quality. The level of quality for given data is then obtained as an appreciation on an ordinal scale of all true predicates for that data. In the scope of this framework, predicates can only be evaluated if all information is at hand to do this. Otherwise, the evaluation of a predicate is hindered and there is uncertainty about its truth value. Such a lack of information can have many different causes. A first cause is when the data for which we want to measure quality, is not known completely. In this case, rather than an exact value, we

J.P. Carvalho et al. (Eds.): IPMU 2016, Part II, CCIS 611, pp. 367–378, 2016.
DOI: 10.1007/978-3-319-40581-0_30

are given for example an interval or a possibility distribution. A second cause is when the agent that evaluates the predicate is not entirely trustworthy. The uncertainty about the truth value of p is then caused by the lack of trust we have in the evaluation. Finally, lack of information also occurs when a predicate is not formulated precisely. Consider for example the predicate "X IS high". Hereby, the term 'high' is difficult to quantify in a precise manner and as a result, there exists an implicit uncertainty about the truth value of the predicate. An important application where such imprecise predicates naturally arise, is the measurement and assessment of *relationship cardinality*. In a database, an object can be in a one-to-one, a one-to-many or a many-to-many relationship with another object [4]. It is hereby almost impossible to precisely quantify the term 'many' as it may vary from case to case (i.e., from object to object). However, it is often possible to model the term 'many' as a fuzzy set that describes the reasonable ranges of acceptable cardinalities. This fuzzy set then serves as the basis for an imprecise predicate.

The two main contributions of this paper are the following. First, it is shown how the uncertainty about the truth value of predicates can be modelled by means of possibility theory. Hereby, a novel approach is described for the calculation of possibilities when uncertainty comes from imprecision in the predicate formulation. It is shown that this approach better respects the connection between the given information and the derived uncertainty. Second, it is shown how this uncertainty on the level of predicates can be propagated throughout the process of measurement. This leads to a model for measurement of data quality on the ordinal scale under circumstances of uncertainty.

The remainder of this paper is structured as follows. In Sect. 2, a concise overview of related work from different fields is presented. In Sect. 3, some preliminary concepts regarding possibility theory are presented. In Sect. 4, we introduce the basic concept of a quality function based on predicates. It is then shown how to treat uncertainty caused by an imprecise formulation of these predicates and the propagation of this uncertainty throughout the measurement process is studied. The properties of the approach are investigated. Section 5 highlights the main results of the paper.

2 Related Work

The field of data quality assessment and measurement has received many contributions in the last decades. Virtually all of these contributions adhere to a multi-dimensional model for data quality [1]. Numerous such dimensions exist but the most important are accuracy, consistency, completeness and timeliness. For these dimensions, several measurement techniques have been proposed [7–9]. The theoretical and technical maturity created by all these contributions has led recently to more application-oriented research. In [12], a tool is proposed to visualise all discrepancies and quality-degrading issues in a tabular-structured dataset. In [2], the connection between degraded quality and outcome of data mining techniques is studied. In [3], data quality is studied in the context of big data. In order to inform data consumers better on the quality of the data, an ontological approach for the binding between quality measurements and data is studied in [6]. A similar goal is envisioned in [16], where it is studied how quality demands can be incorporated in an SQL-like query language.

Apart from these contributions, a specific line of research has focussed on the aspect of uncertainty in data quality assessment. The quality of data in XML documents in the presence of uncertainty is investigated in [13]. Hereby, four measures of quality are proposed: uncertainty density, answer decisiveness, adapted precision and adapted recall. These measures allow to assess the quality of an answer that is posed to an XML database under uncertainty. From a different perspective, the various contributions on automated linguistic summaries of data can also be regarded as data assessment techniques. Yager has initially established the basic framework for summaries of the type "Q objects are S" where Q is a fuzzy set that models a quantity and S is a fuzzy set that models a linguistic constraint on the objects under consideration [19]. In addition, Yager proposed a measure to validate the degree of truth of a summary. This idea was further elaborated in later research and several other types of linguistic summaries were proposed [10,11]. Raschia et al. have proposed SaintEtiq, which is an incremental (i.e., online) approach to the construction of linguistic summaries [15].

3 Preliminaries

Possibility theory is an uncertainty theory that allows to quantify the confidence about the occurrence of uncertain events. It is differentiated from probability theory by the fact that uncertainty is caused by *incomplete information* rather than randomness in the outcome of experiments. In the finite case, possibilistic information can be represented as a possibility distribution $\pi : X \rightarrow [0,1]$ that is normalized in the sense that $\sup_{x\in X} \pi(x) = 1$. A widely studied application of possibility theory [5,14,18] is the case where the universe X is the set of Boolean values $\mathbb{B} = \{T, F\}$. In that case, possibility distributions are called *possibilistic truth values* (PTVs). For a Boolean proposition p, the possibilistic truth value $\widetilde{p}$ represents the uncertainty about the truth value of p, where uncertainty is caused by a lack of information about p. In the following, we shall denote a PTV in the couple notation $(\widetilde{p}(T), \widetilde{p}(F))$. The natural order of PTVs is given by:

$$\widetilde{p}_1 \geq \widetilde{p}_2 \Leftrightarrow \begin{cases} \widetilde{p}_1(F) \leq \widetilde{p}_2(F) \textbf{ if } & \widetilde{p}_1(T) = \widetilde{p}_2(T) = 1 \\ \widetilde{p}_1(T) \geq \widetilde{p}_2(T) \textbf{ else.} \end{cases} \tag{1}$$

If we wish to omit the distribution $\widetilde{p}$ for notational purposes, we will denote:

$$\mathrm{Pos}(p = T) = \widetilde{p}(T) \tag{2}$$

$$\mathrm{Pos}(p = F) = \widetilde{p}(F) \tag{3}$$

Several researchers have investigated the extensions of Boolean connectives to the epistemological case of PTVs [5,18]. The extension of the negation operator is defined by:

$$\widetilde{\neg}\, \widetilde{p} = (\mathrm{Pos}(p = F), \mathrm{Pos}(p = T))\,. \tag{4}$$

The extensions of Boolean conjunction and disjunction of propositions is based on the principle of t-independence [5] where t is an appropriate triangular norm [17]. For two propositions p_1 and p_2 that are t-independent, the possibilistic extension of $\wedge$ is defined by:

$$\text{Pos}\left(p_1 \wedge p_2 = T\right) = t\left(\text{Pos}\left(p_1 = T\right), \text{Pos}\left(p_2 = T\right)\right) \tag{5}$$

$$\text{Pos}\left(p_1 \wedge p_2 = F\right) = \max\left(\text{Pos}\left(p_1 = F\right), \text{Pos}\left(p_2 = F\right)\right) \tag{6}$$

and the possibilistic extension of $\vee$ is defined by:

$$\text{Pos}\left(p_1 \vee p_2 = T\right) = \max\left(\text{Pos}\left(p_1 = T\right), \text{Pos}\left(p_2 = T\right)\right) \tag{7}$$

$$\text{Pos}\left(p_1 \vee p_2 = F\right) = t\left(\text{Pos}\left(p_1 = F\right), \text{Pos}\left(p_2 = F\right)\right) \tag{8}$$

The corresponding PTVs are denoted as $\widetilde{p}_1 \,\widetilde{\wedge}_t\, \widetilde{p}_2$ and $\widetilde{p}_1 \,\widetilde{\vee}_t\, \widetilde{p}_2$.

4 Quality Measurement Under Uncertainty

4.1 Quality Functions

Consider a universe of discourse U in which quality measurement is required and suppose a set of predicates $P = \{p_1, ..., p_n\}$ where each predicate is given by a function $p : U \rightarrow \mathbb{B}$. Basically, the set P models the sufficient constraints for data in U to be of the best possible quality. The quality of $u \in U$ is obtained by evaluating all $p \in P$ and by expressing an appreciation of those evaluations on a scale $\mathbb{S} = \{s_1, ..., s_k\}$ that is at least ordinal. For such a scale, it will be assumed that $s_1 < s_2 < ... < s_k$. The smallest and largest elements of $\mathbb{S}$ will be denoted as $\mathbb{0}$ and $\mathbb{1}$ respectively. For simplicity, it is assumed here that the appreciation of evaluations is based on a total order relation $\prec$ on P where $p_i \prec p_j$ expresses that p_i needs to be evaluated before p_j. Such an evaluation order indicates there is a logical order in which predicates must be evaluated and naturally arises in many applications such as database normalization and consistency verification. Let us denote the set of the i predicates that must be evaluated first under $\prec$ by $P_{(i)}$ and the ith predicate as $p_{(i)}$. We can then provide the following definition.

Definition 1 (Quality Function). *Consider a universe of discourse U and a set of predicates P on U. A quality function Q for U on the scale $\mathbb{S}$ is defined by a function $Q : U \rightarrow \mathbb{S}$ that satisfies the boundary constraints:*

$$\forall u \in U : (\forall p \in P : p(u) = F) \Rightarrow Q(u) = \mathbb{0} \tag{9}$$

$$\forall u \in U : (\forall p \in P : p(u) = T) \Rightarrow Q(u) = \mathbb{1} \tag{10}$$

and is monotonic w.r.t. the evaluation order $\prec$ on P, which means that for any u and u' we have that:

$$Q(u) \geq Q(u') \Leftrightarrow \max\left\{i \mid \forall p \in P_{(i)} : p(u)\right\} \geq \max\left\{j \mid \forall p \in P_{(j)} : p(u')\right\} \tag{11}$$

A quality function basically evaluates predicates in the order specified by $\prec$ and the more predicates are true, the higher the quality on the scale $\mathbb{S}$. A *complete* quality function is a quality function for which $|P| = |\mathbb{S}| + 1$ and:

$$\forall u \in U : Q(u) = s_i \Leftrightarrow p_{(1)}(u) \wedge ... \wedge p_{(i-1)}(u) \wedge \neg p_{(i)}(u). \tag{12}$$

With a complete quality function, each additional predicate that evaluates to true increases the quality with one level. If none of the predicates are true, the lowest level of quality (i.e., $\mathbb{0} = s_1$) is assigned. Unless explicitly stated otherwise, a quality function is assumed to be complete.

4.2 Imprecise Predicates

As mentioned in the introduction of this paper, it may naturally occur that the predicates in P can not be formulated in a precise manner. As a result, evaluation of predicates for observed values $u \in U$ is hindered by a lack of information and uncertainty about the evaluation arises. For the sake of simplicity, we restrict ourselves to predicates of the form "u IS L" where L is a fuzzy set on U with membership function $\widetilde{L}$.

When a predicate p is formulated imprecisely, its evaluation to a truth value deals with uncertainty that can be expressed as a PTV $\widetilde{p}(u)$. Before we treat the general case of how $\widetilde{p}(u)$ can be calculated, two special cases can be considered: binary and ternary evaluation. In the case of *binary* evaluation, there is no uncertainty about the predicate evaluation and we have that $\widetilde{p}(u) \in \{(1,0),(0,1)\}$. It is then either completely certain that the predicate evaluates to true or completely certain that the predicate evaluates to false. A strongly related case is that of *ternary* evaluation, where there is either no uncertainty or complete uncertainty about the predicate evaluation. This means that $\widetilde{p}(u) \in \{(1,0),(1,1),(0,1)\}$. As such, it is either known with complete certainty that the predicate evaluates to true or false, or nothing about the predicate is known at all. Ternary evaluation is a useful extension of binary evaluation to account for missing data (i.e., NULL values) and has a strong affinity with Kleene's three-valued logic.

In the general case where we consider an imprecise predicate of the form "u IS L", a state of *partial* information occurs. To infer $\widetilde{p}(u)$ for an evaluation, there exist some approaches but it is shown in the following that these approaches cope with some counter intuitive issues. Therefore, a novel way of inferring $\widetilde{p}(u)$ is proposed here.

A first option to calculate $\widetilde{p}(u)$ is to model the operator "IS" as a binary relation on U. Typically, an equivalence relation is used to indicate which values in U are compatible with each other. When using such a binary relation, the Extension Principle (EP) of Zadeh [20] allows to obtain $\widetilde{p}(u)$. For an equivalence relation $\mathcal{E}$, we obtain that:

$$\text{Pos}(\text{“}u \text{ IS } L\text{”} = T) = \text{Pos}(u \equiv_{\mathcal{E}} L) = \sup_{v \in U, (v,u) \in \mathcal{E}} \widetilde{L}(v) \tag{13}$$

and

$$\text{Pos}(\text{“}u \text{ IS } L\text{”} = F) = \text{Pos}(\neg (u \equiv_{\mathcal{E}} L)) = \sup_{v \in U, (v,u) \notin \mathcal{E}} \widetilde{L}(v). \tag{14}$$

The above given formulae assume that u is a proper value from U. In the case where u is *unknown* (i.e., a NULL-value), the previous reasoning can be generalized by modelling a NULL-value with a possibility distribution π_{NULL} that satisfies $\forall v \in U : \pi_{\text{NULL}}(v) = 1$. The formulae for uncertainty inference are then given by:

$$\text{Pos}(\text{“}\pi_{\text{NULL}} \text{ IS } L\text{”} = T) = \sup_{(v_1,v_2)\in\mathcal{E}} \min\left(\widetilde{L}(v_1), \pi_{\text{NULL}}(v_2)\right) \tag{15}$$

$$\text{Pos}(\text{“}\pi_{\text{NULL}} \text{ IS } L\text{”} = F) = \sup_{(v_1,v_2)\notin\mathcal{E}} \min\left(\widetilde{L}(v_1), \pi_{\text{NULL}}(v_2)\right). \tag{16}$$

By normalization of $\widetilde{L}$ and reflexivity of $\mathcal{E}$, we get that Pos(“π_{NULL} IS L” $= T) = 1$. In addition, if $\mathcal{E} \neq U \times U$ and thus induces at least two equivalence classes on U, we have that Pos(“π_{NULL} IS L” $= F) = 1$. As such, we get that the uncertainty about the proposition “u IS L” is given by $(1, 1)$ in the case where $u =$ NULL. This indicates that we have no knowledge at all about the truth value of the proposition, which makes perfect sense as we have no knowledge at all about the value of u.

The usage of the EP to infer uncertainty has some disadvantages. It relies on the assumption that there exists a binary relation that properly models the concept of compatibility. It can be seen that this assumption might not always hold. In case of interval and ratio scales a distance-based technique can be used, but this confronts us with some parameters that must be chosen. In case of ordinal and nominal scales, the construction of an equivalence relation is even more difficult. In the special case where the equality relation is adopted, more objections for the EP approach are found. For example, we have that:

$$\left|\text{core}\left(\widetilde{L}\right)\right| > 1 \Rightarrow (\text{Pos}(\text{“}u \text{ IS } L\text{”} = F) = 1). \tag{17}$$

In other words, if the core of $\widetilde{L}$ contains more than one value, it is always completely possible that proposition “u IS L” is false. As such, $\widetilde{L}(u) = 1$ implies that $\widetilde{p} = (1, 1)$. This means that if a value u is completely compatible with a linguistic concept L, there is complete uncertainty about the corresponding predicate p, which is a counter intuitive result.

A second option to derive the uncertainty about the proposition “u IS L” is to let the possibilities for T and F be proportional to the membership degree $\widetilde{L}(u)$ and the complement thereof [18]. This approach requires an explicit step to ensure a normalized distribution. The calculation of $\widetilde{p}(u)$ following this approach yields:

$$\text{Pos}(p(u) = T) = \frac{\widetilde{L}(u)}{\max\left(\widetilde{L}(u), 1 - \widetilde{L}(u)\right)} \tag{18}$$

$$\text{Pos}(p(u) = F) = \frac{1 - \widetilde{L}(u)}{\max\left(\widetilde{L}(u), 1 - \widetilde{L}(u)\right)} \tag{19}$$

In addition, if $u = \text{NULL}$ then $\widetilde{p}(u) \triangleq (1,1)$. The advantage of this approach over usage of the EP, is that it does not require any input in the form of a binary relation and is thus always directly applicable. However, just as was the case with application of the EP, this method allows to infer that $\widetilde{p}(u) = (1,1)$ when u is a known value from U. More specifically, this holds if $\widetilde{L}(u) = 0.5$. Again, the indication of complete uncertainty in a case where u is a known value, is a counter-intuitive result. Indeed, if $u \in U$ and $\widetilde{L}(u) = 0.5$, we most definitely have information about the proposition "u IS L" and this information should be reflected in $\widetilde{p}(u)$.

To solve the problem that complete uncertainty is inferred while information is present, a novel method of uncertainty derivation is proposed here. This method relies on a separate measurement for the possibilities of T and F[1]. In our approach, we treat $\widetilde{L}$ as a source of information that supports the evaluation of the predicate "u IS L". We then use this information to determine which truth value is most plausible. Based on the outcome of this, one of the truth values (i.e., T or F) is considered to be *completely* possible. Next, we use the information given by $\widetilde{L}$ to calculate the certainty we have for this truth value. It is shown in the following that we can implement this process in two different ways, depending on how we perceive information from $\widetilde{L}$. More specifically, if we employ a pessimistic view on information, then failure of the predicate is considered to be unacceptable. In this case, a decreasing membership of u to $\widetilde{L}$ indicates a decreasing possibility that "u IS L" is true. In other words, if u only partially belongs to the fuzzy set $\widetilde{L}$, then it is not completely possible that the corresponding predicate evaluates to true. Under this pessimistic perception of information, the uncertainty about "u IS L" is given by:

$$\widetilde{p^-}(u) = \begin{cases} (1,1) & \textbf{if} \quad u = \text{NULL} \\ (1,0) & \textbf{if} \quad \widetilde{L}(u) = 1 \\ \left(\widetilde{L}(u), 1\right) & \textbf{otherwise} \end{cases} \tag{20}$$

According to this formula, there is complete uncertainty about the truth value of the predicate "u IS L" if u is unknown. If $\widetilde{L}(u) = 1$, then the truth value of the predicate "u IS L" is known to be T. In any other case, it is considered completely possible that the truth value of the predicate is F and to some extent possible that the truth value is T. This reflects the following decision rule $u \notin \text{core}\left(\widetilde{L}\right) \Rightarrow p(u) = F$. In the notation $\widetilde{p^-}(u)$, the minus sign indicates the pessimistic and suspicious treatment of information which is reflected by the fact that under partial information, it will remain completely possible that the predicate is falsified. Opposed to this, a more confident and optimistic treatment of information can be embraced. In this case, a decreasing membership of u to $\widetilde{L}$ indicates an increasing possibility that the proposition "u IS L" is false. In other words, if u only partially belongs to the fuzzy set $\widetilde{L}$, then it is to some

[1] In his formal derivation of the theory of PTVs, De Cooman observed that it is essential that information about $p = T$ and $p = F$ can be separated [5].

extent possible that the corresponding predicate evaluates to false. Under this optimistic perception of information, the uncertainty about "u IS L" is given by:

$$\widetilde{p^{+}}(u) = \begin{cases} (1,1) & \textbf{if} \quad u = \text{NULL} \\ (0,1) & \textbf{if} \quad \widetilde{L}(u) = 0 \\ \left(1, 1 - \widetilde{L}(u)\right) & \textbf{otherwise} \end{cases} \tag{21}$$

According to this formula, there is again complete uncertainty about the truth value of the proposition "u IS L" if u is unknown. If $\widetilde{L}(u) = 0$ then the truth value of the proposition "u IS L" is known to be F. In any other case, it is considered completely possible that the truth value of the proposition is T and to some extent possible that the truth value is F. This reflects the following decision rule $u \in \text{supp}\left(\widetilde{L}\right) \Rightarrow p(u) = T$. In the notation $\widetilde{p^{+}}(u)$, the plus sign indicates an optimistic and confident treatment of information.

Figure 1 illustrates the derivation of uncertainty in the case where $U = \mathbb{R}$ and under the assumption of respectively a strong and a weak predicate. It can be seen that the membership degree $\widetilde{L}(u)$ influences the possibility of only one truth value. The other possibility depends on whether or not u is an element of resp. the core and the support of $\widetilde{L}$. There are some interesting dualities between the pessimistic and optimistic treatment of information. These are discussed in the following.

Property 1 (Dominance). For a predicate "u IS L" it holds that $\widetilde{p^{-}}(u) \leq \widetilde{p^{+}}(u)$.

Proof. The proof follows immediately from (20) and (21). □

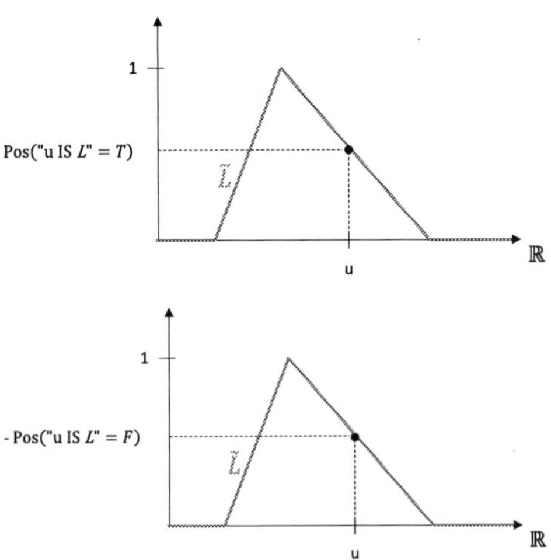

Fig. 1. Uncertainty inference for a predicate "u IS L" for pessimistic (upper panel) and optimistic (lower panel) treatment of information.

Property 2 (Complete Uncertainty). For a predicate "u IS L" we have that:

$$u = \text{NULL} \Leftrightarrow \widetilde{p^-}(u) = (1,1) \tag{22}$$

$$u = \text{NULL} \Leftrightarrow \widetilde{p^+}(u) = (1,1). \tag{23}$$

Proof. The proof follows immediately from (20) and (21). □

Property 2 states that whenever u is a proper value of U, we have that $\widetilde{p}(u) \neq (1,1)$. The sole cause of complete uncertainty is when there is no information at all. This property establishes a certain consistency between information and uncertainty. Property 2 differentiates our approach from both usage of the EP and normalized membership degrees. Some additional properties are the following.

Property 3 (Boundary Values). For a predicate "u IS L" we have that:

$$\widetilde{L}(u) = 1 \Leftrightarrow \widetilde{p^-}(u) = \widetilde{p^+}(u) = (1,0) \tag{24}$$

$$\widetilde{L}(u) = 0 \Leftrightarrow \widetilde{p^-}(u) = \widetilde{p^+}(u) = (0,1). \tag{25}$$

Proof. The proof follows immediately from (20) and (21). □

Property 4 (Trivalence). For a "u IS L" we have that:

$$\text{supp}(\widetilde{L}) = \text{core}(\widetilde{L}) \Rightarrow \widetilde{p^-}(u) \in \{(1,0),(1,1),(0,1)\} \tag{26}$$

$$\text{supp}(\widetilde{L}) = \text{core}(\widetilde{L}) \Rightarrow \widetilde{p^+}(u) \in \{(1,0),(1,1),(0,1)\}. \tag{27}$$

Proof. The proof follows immediately from (20) and (21) on the one hand and the definition of ternary predicates on the other hand. □

It is noted that the reasoning about predicates of the type "u IS L" can be easily generalized. In the binary case, predicates of the type "u_1 AND u_2 ARE L" can be considered where L is a linguistic term modelled by a *binary fuzzy relation* $\widetilde{L} \in \mathcal{F}(U_1) \times \mathcal{F}(U_2)$. Further generalization to the n-ary case is then straightforward.

4.3 A Possibilistic Measurement of Quality

Let us now investigate how uncertainty about predicate evaluation affects measurement of quality on the scale $\mathbb{S}$. As mentioned before, our main interest lies with complete quality functions. Equation (12) shows that for such functions, the calculation of the level of quality can be written as a Boolean function of the predicates under consideration. By using the possibilistic extensions of operators described in Sect. 3, the propagation of uncertainty about predicates can be modelled as follows. For $u \in U$ and $s_i \in \mathbb{S}$, let us consider the Boolean predicate "$Q(u) = s_i$" and let us denote this predicate with the shorthand notation p_{s_i}. Taking into account the definitions of possibilistic extension (Sect. 3), application of Eq. (12) implies that for any $i \in \{1,...,n+1\}$, with n the number of predicates, we have that:

$$\widetilde{p}_{s_i}(u) = \widetilde{p}_{(1)}(u) \mathbin{\widetilde{\wedge}_t} \dots \mathbin{\widetilde{\wedge}_t} \widetilde{p}_{(i-1)}(u) \mathbin{\widetilde{\wedge}_t} \widetilde{\neg}\, \widetilde{p}_{(i)}(u). \tag{28}$$

As a boundary case, we have that $p_{(n+1)}$ is a contradiction and is therefore falsified for any $u \in U$ by definition. Based on this, a possibilistic inference of uncertainty is established under the assumption of mutual t-independence of predicates in P. From the definition of the extended operators, it can be seen that $\text{Pos}\left(p_{s_i}(u) = T\right)$ is given by:

$$t\left(\text{Pos}\left(p_{(1)}(u) = T\right), ..., \text{Pos}\left(p_{(i-1)}(u) = T\right), \text{Pos}\left(p_{(i)}(u) = F\right)\right) \tag{29}$$

and that $\text{Pos}\left(p_{s_i}(u) = F\right)$ is given by:

$$\max\left(\text{Pos}\left(p_{(1)}(u) = F\right), ..., \text{Pos}\left(p_{(i-1)}(u) = F\right), \text{Pos}\left(p_{(i)}(u) = T\right)\right) \tag{30}$$

It is now shown that the derivation of uncertainty from imprecisely formulated predicates implies the following interesting property.

Property 5. For a complete quality function Q based on a set of predicates P of the type "u IS L", we have that:

$$u = \text{NULL} \Leftrightarrow \forall s_i \in \mathbb{S} : \text{Pos}\left(Q(u) = s_i\right) = \text{Pos}\left(Q(u) \neq s_i\right) = 1 \tag{31}$$

Proof. On the one hand, if u = NULL, then for all predicates $p_{(i)}$ we have that $\widetilde{p}_{(i)}(u) = (1, 1)$. By construction of the possibilistic inference, we have that $\widetilde{p}_{s_i}(u) = (1, 1)$ for all $s_i \in \mathbb{S}$. On the other hand, if for all $s_i \in \mathbb{S}$ we have that $\widetilde{p}_{s_i}(u) = (1, 1)$ then by construction of possibilistic inference all $\widetilde{p}_{(i)}(u)$ must equal $(1, 1)$ which means that u = NULL. □

Property 5 shows that, only if u is a NULL value, the outcome of quality measurement is completely uncertain. This illustrates that the interpretation attached to the concept 'no information' is maintained on the level of quality measurement. Although this is a nice result, a stronger result can be shown with respect to the possibilistic inference of uncertainty.

Theorem 1. *Let Q be a complete quality function based on a set of predicates P of the type "u IS L". If $u \neq$ NULL there exists one and only one $s \in \mathbb{S}$ for which:*

$$\text{Pos}\left(Q(u) = s\right) = 1. \tag{32}$$

Proof. Consider $P = \{p_1, ..., p_n\}$ and let m be the smallest number in $\{1, ..., n+1\}$ such that $\text{Pos}\left(p_{(m)}(u) = T\right) < 1$ and

$$\forall j < m : \text{Pos}\left(p_{(j)}(u) = T\right) = 1. \tag{33}$$

Because $u \neq$ NULL we have that $\forall j < m : \text{Pos}\left(p_{(j)}(u) = F\right) < 1$. It can now be shown in three steps that the possibility of the predicate "$Q(u) = s$" is 1 if $s = s_m$ and is smaller than 1 for any other element of $\mathbb{S}$.

Step 1. For the levels of quality s_j where $j > m$, we have that $m \leq j - 1$ and $\text{Pos}\left(p_{(m)}(u) = T\right) < 1$. From the definition of a t-norm it follows that:

$$\text{Pos}\left(p_{s_j}(u) = T\right) < 1. \tag{34}$$

Step 2. For the levels of quality s_j where $j < m$, we have for any j that $\text{Pos}\left(p_{(j)}(u) = F\right) < 1$. It thus follows from the definition of a t-norm that:

$$\text{Pos}\left(p_{s_j}(u) = T\right) < 1. \tag{35}$$

Step 3. For the level of quality s_m, we have by construction of m and by the law of normalization that $\text{Pos}\left(p_{s_m}(u) = T\right) = 1$. □

Theorem 1 implies that, if $u \in U$ is known, propagation of uncertainty on the level of predicates with an imprecise formulation, induces a unimodal possibility distribution on $\mathbb{S}$. This possibility distribution $\pi : \mathbb{S} \to [0, 1]$ is characterized by:

$$\pi(s) = \text{Pos}\left(p_s(u) = T\right). \tag{36}$$

This distribution represents a possibilistic measurement of quality in the presence of imprecise formulated predicates. Note that the unimodality of the distribution comes from the optimistic or pessimistic treatment of information. However, it is clear that the propagation schema modelled by Eq. (28) can also be applied if uncertainty about predicate evaluation is caused by something else than an imprecise formulation. More specifically, a possibilistic measurement of quality can also be achieved in cases where data is incomplete (i.e., u is a possibility distribution) or when evaluation is not entirely confident (i.e., the assessor that evaluates p is not entirely trustworthy). Having this said, there is a rich set of applications in which possibilistic measurement of quality can play a central role. Examples thereof include peer assessment, community-driven assessment and vague data mining tasks like sentiment analysis.

5 Conclusion

In this paper, data quality measurement is studied in the presence of uncertainty. Hereby, a framework of measurement is considered where quality of data is expressed as an appreciation of the results of predicate tests on an ordinal scale. It is then argued that in some cases, these predicates can not be formulated in a precise manner. An example thereof is the assessment of relationship cardinality between entities in a database, where the concepts 'one-to-many' and 'many-to-many' are hard to quantify precisely. When such imprecise predicates occur, evaluation of these predicates is hindered by a lack of information and uncertainty about the evaluation can be modelled by means of possibility theory. For that purpose, a novel approach to model the uncertainty about predicate evaluation is presented here. It is shown that, by usage of this approach, uncertainty propagates nicely throughout the process of quality measurement and eventually, a possibility distribution on the ordinal scale of measurement is obtained.

References

1. Batini, C., Cappiello, C., Francalanci, C., Maurino, A.: Methodologies for data quality assessment and improvement. ACM Comput. Surv. **41**(3), 16–52 (2009)
2. Blake, R., Mangiameli, P.: The effects and interactions of data quality and problem complexity on classification. J. Data Inf. Qual. **2**(2), 8:1–8:28 (2011)
3. Caballero, I., Serrano, M., Piattini, M.: A data quality in use model for big data. In: Indulska, M., Purao, S. (eds.) ER Workshops 2014. LNCS, vol. 8823, pp. 65–74. Springer, Heidelberg (2014)
4. Chen, P.: The entity-relationship model - toward a unified view of data. ACM Trans. Database Syst. **1**(1), 9–36 (1976)
5. De Cooman, G.: Towards a possibilistic logic. In: Ruan, D. (ed.) Fuzzy Set Theory and Advanced Mathematical Applications. International Series in Intelligent Technologies, vol. 4, pp. 89–133. Springer, New York (1995)
6. Debattista, J., Lange, C., Auer, S.: DaQ, an ontology for dataset quality information. In: Linked Data on the Web (LDOW), pp. 1–8 (2014)
7. Even, A., Shankaranarayanan, G.: Utility-driven assessment of data quality. Database Adv. Inf. Syst. **38**(2), 75–93 (2007)
8. Heinrich, B., Klier, M.: Metric-based data quality assessment - developing and evaluation a probability-based currency metric. Decis. Support Syst. **72**, 82–96 (2015)
9. Heinrich, B., Klier, M., Kaiser, M.: A procedure to develop metrics for currency and its application in CRM. ACM J. Data Inf. Qual. **1**(1), 5:1–5:28 (2009)
10. Kacprzyk, J., Yager, R., Zadrozny, S.: A fuzzy logic based approach to linguistic summaries. Int. J. Appl. Math. Comput. Sci. **10**, 813–834 (2000)
11. Kacprzyk, J., Zadrozny, S.: Computing with words in intelligent database querying: standalone and internet-based applications. Inf. Sci. **134**(14), 71–109 (2001)
12. Kandel, S., Parikh, R., Paepcke, A., Hellerstein, J.M., Heer, J.: Profiler: integrated statistical analysis and visualization for data quality assessment. In: International Working Conference on Advanced Visual Interfaces, pp. 547–554 (2012)
13. de Keijzer, A., van Keulen, M.: Quality measures in uncertain data management. In: Prade, H., Subrahmanian, V.S. (eds.) SUM 2007. LNCS (LNAI), vol. 4772, pp. 104–115. Springer, Heidelberg (2007)
14. Prade, H.: Possibility sets, fuzzy sets and their relation to lukasiewicz logic. In: Proceedings of the International Symposium on Multiple-Valued Logic, pp. 223–227 (1982)
15. Raschia, G., Mouaddib, N.: SAINTETIQ: a fuzzy set-based approach to database summarization. Fuzzy Sets Syst. **129**, 137–162 (2002)
16. Sampaio, S.D.F.M., Dong, C., Sampaio, P.: DQ(2)S - a framework for data quality-aware information management. Expert Syst. Appl. **42**(21), 8304–8326 (2015)
17. Schweizer, B., Sklar, A.: Probabilistic Metric Spaces. Elsevier, New York (1983)
18. Van Schooten, A.: Ontwerp en implementatie van een model voor de representatie en manipulatie van onzekerheid en imprecisie in databanken en expert systemen. Ph.D. thesis, Ghent University (1988)
19. Yager, R.: A new approach to the summarization of data. Inf. Sci. **28**, 69–86 (1982)
20. Zadeh, L.: Fuzzy sets. Inf. Control **8**, 338–353 (1965)

Computing Theoretically-Sound Upper Bounds to Expected Support for Frequent Pattern Mining Problems over Uncertain Big Data

Alfredo Cuzzocrea[1] and Carson K. Leung[2(✉)]

[1] DIA Department, University of Trieste and ICAR-CNR, Trieste, TS, Italy
alfredo.cuzzocrea@dia.units.it

[2] Department of Computer Science, University of Manitoba, Winnipeg, MB, Canada
kleung@cs.umanitoba.ca

Abstract. *Frequent pattern mining* aims to discover implicit, previously unknown, and potentially useful knowledge in the form of sets of frequently co-occurring items, events, or objects. To mine frequent patterns from *probabilistic datasets of uncertain data*, where each item in a transaction is usually associated with an existential probability expressing the likelihood of its presence in that transaction, the UF-growth algorithm captures important information about uncertain data in a UF-tree structure so that *expected support* can be computed for each pattern. A pattern is considered frequent if its expected support meets or exceeds the user-specified threshold. However, a challenge is that the UF-tree can be large. To handle this challenge, several algorithms use smaller trees such that *upper bounds to expected support* can be computed. In this paper, we examine these upper bounds, and determine which ones provide tighter upper bounds to expected support for frequent pattern mining of uncertain big data.

Keywords: Uncertainty · Data analysis · Big data · Data science · Data mining

1 Introduction

Uncertain big data (e.g., [21,33,34,38]) are becoming more and more popular in modern applications [23] (e.g., social computing [20,22], data warehousing and OLAP [10]) because (big) data in real-life scenarios are typically *imprecise and uncertain* (e.g., [14,17,19]). *Mining uncertain big data* (e.g., [6,40]) is problematic due to the fact that models, techniques, and algorithms running on such data must consider uncertainty as a fundamental characteristic of big data while this challenging property is not foreseen by classical large-scale data mining approaches. As a consequence, mining uncertain big data is a first-class problem to deal with, and several interesting initiatives that focus the attention on this problem are appearing recently in active literature [12,35,45].

Among the wide class of data mining tasks [4,16,42,43], *frequent pattern mining* [2] is a very popular problem that has attracted the attention of a large

J.P. Carvalho et al. (Eds.): IPMU 2016, Part II, CCIS 611, pp. 379–392, 2016.
DOI: 10.1007/978-3-319-40581-0_31

community of data miners. Frequent pattern mining aims to discover implicit, previously unknown, and potentially useful knowledge in the form of sets of frequently co-occurring items, events, or objects (i.e., frequent patterns). It also serves as building blocks for various other data mining tasks such as stream mining [8,9,25,26] (which mines data that come at a high velocity), constrained mining [13], and social network mining [27,41]. Many existing algorithms mine frequent patterns from high volumes of precise data, in which users definitely know whether an item is present in, or absent from, a transaction in databases of precise data. However, there are situations in which users are uncertain about the presence or absence of items (e.g., a physician may suspect, but may not guarantee, that a fevered patient got a flu or Zika virus) in a *probabilistic dataset of uncertain data.* In such dataset, each item x_i in a transaction t_j is associated with an *existential probability* $P(x_i, t_j)$ expressing the likelihood of the presence of x_i in t_j.

To mine frequent patterns from high varieties of (high-value) uncertain data, various algorithms [1,3,44] have been proposed, including UF-growth [30]. The UF-growth algorithm first constructs a UF-tree structure with the goal of capturing important contents on uncertain data, from which frequent patterns can then be mined recursively. A pattern X is considered *frequent* if its expected support $expSup(X)$ in the entire uncertain dataset meets or exceeds the user-specified minimum support threshold *minsup* [24]. Here, $expSup(X)$ over all n transactions in the uncertain dataset can be computed in terms of the sum of $expSup(X, t_j)$ over every transaction t_j containing X, as follows:

$$expSup(X) = \sum_{j=1}^{n} expSup(X, t_j) \tag{1}$$

while $expSup(X, t_j)$ can be computed in terms of the product of the existential probability $P(x_i, t_j)$ of every independent item x_i within the pattern $X = \{x_1, \ldots, x_k\}$, as follows:

$$expSup(X, t_j) = \prod_{i=1}^{k} P(x_i, t_j) \tag{2}$$

In order to accurately compute the expected support of each pattern, paths in the corresponding UF-tree are shared only if tree nodes on the paths have the same item and the same existential probability. Due to this restrictive path sharing requirement, the UF-tree may be quite large.

A way to solve this large tree-size issue is to explore alternative mining approaches (e.g., UH-Mine algorithm [1] that uses hyper-structures, as well as sampling-based or vertical mining approaches [5]). Another way is to make the tree compact by capturing less information about uncertain data but sufficient for computing *upper bounds to the expected support* of patterns. Over the past few years, different computations on the upper bounds to expected support have been proposed. Many of them are reported to lead to more compact tree structures for capturing uncertain data than the UF-tree. These, in turn, shorten the tree traversal time during the mining process, and thus help reduce the overall runtime. In addition, another benefit of using these upper bounds is that they are guaranteed *not* to generate any false negatives. Indeed, if an upper bound to expected support of a pattern X is less than *minsup*, then X is guaranteed to be infrequent. Moreover, these upper bounds are reported to be so tight that not too many false positives are generated-and-tested. Then, interesting questions to ask include the following: Among these upper bounds, which one is tighter? Which one leads to shorter runtime or mining time? In this paper, we examine these upper bounds, and re-formulate them so that we can compare them and determine which ones provide tighter upper bounds to expected support of patterns when mining frequent patterns from a high variety of high volumes of high-value uncertain data that may come at a high velocity (i.e., uncertain "streaming" big data). Our *key contributions* of this paper include our computation of theoretically-sound upper bounds to expected support for frequent pattern mining problems over uncertain big data.

The remainder of this paper is organized as follows. In Sect. 2, we provide a formal unifying model for computing upper bounds to expected support, as to obtain a (formal) model to be used thought the paper. The section also contains relevant related work for our research. Section 3 reports a theoretical analysis on the bounds. In Sect. 4, we provide an experimental assessment and evaluation of our methods for computing these upper bounds, according to several experimental parameters. Finally, Sect. 5 presents conclusions and proposes future work of our research.

2 Computing Upper Bounds: A Unifying Model from the State-of-the-Art Analysis

In this section, we re-formulate upper bounds to expected support (as provided by the state-of-the-art analysis) via using a common expression or notion so that we can introduce a unifying model for easily comparing among the various proposals available in literature. This section also serves as analysis of related work that is relevant to our research.

To approximate an upper bound to expected support of a pattern X, CUF-growth [31] introduces the concept of *transaction cap* (*TC*), which is defined as the product of the two highest existential probabilities in the entire transaction $t_j=\{y_1,\ldots,y_{r-1},y_r,\ldots,y_h\} \supseteq \{x_1,\ldots,x_k\}=X$ (where $x_k = y_r$), as follows:

$$TC(X,t_j) = \begin{cases} P(y_1,t_j) & \text{if } h=1 \\ TM_1(t_j) \times TM_2(t_j) & \text{if } h \geq 2 \end{cases} \tag{3}$$

where (i) $TM_1(t_j) = \max_{i\in[1,h]} P(y_i,t_j)$ is the *transaction maximum*, which is defined as the highest existential probability in t_j; and (ii) $TM_2(t_j) = \max_{i\in[1,h]\wedge(i\neq g)} P(y_i,t_j)$ is the second highest existential probability in t_j for $y_g = \text{argmax}_{i\in[1,h]} P(y_i,t_j)$ (i.e., $TM_1(t_j) = P(y_g,t_j)$).

While this transaction cap serves as a good upper bound to 2-itemsets, it may not be too tight for k-itemsets (where $k \geq 3$). To tighten the upper bound to expected support for 3^+-itemsets, *CUF*-growth* [31] extends the concept of transaction cap as to use the product of the three highest existential probabilities in t_j, as follows:

$$CUF^*(X,t_j) = \begin{cases} TC(X,t_j) & \text{if } k \leq 2 \\ TM_1(t_j) \times TM_2(t_j) \times [TM_3(t_j)]^{k-2} & \text{if } k \geq 3 \end{cases} \tag{4}$$

where $TM_3(t_j) = \max_{i \in [1,h] \wedge (i \neq g) \wedge (i \neq s)} P(y_i, t_j)$ is the third highest existential probability in t_j for $y_s = \text{argmax}_{i \in [1,h] \wedge (i \neq g)} P(y_i, t_j)$ (i.e., $TM_2(t_j) = P(y_s, t_j)$).

On the one hand, the transaction cap can be easily pre-computed. On the other hand, it may not involve any items in X. To tighten the upper bound, *item cap* (*IC*) [37] involves at least one item in X. Specifically, the item cap is defined as the product of $P(x_k, t_j)$ and the highest existential probability $TM_1(t_j)$ in t_j, as follows:

$$IC(X,t_j) = \begin{cases} P(y_1,t_j) & \text{if } h = 1 \\ P(x_k,t_j) \times TM_1(t_j) & \text{if } h \geq 2 \end{cases} \tag{5}$$

For the special case where $TM_1(t_j) = P(x_k, t_j)$, *DISC-growth* [37] avoids multiplying $TM_1(t_j)$ twice. Instead, it multiplies $P(x_k, t_j)$ by the second highest existential probability $TM_2(t_j)$ in t_j, as follows:

$$DISC(X,t_j) = \begin{cases} P(y_1,t_j) & \text{if } h = 1 \\ P(x_k,t_j) \times TM_1(t_j) & \text{if } h \geq 2 \wedge x_k \neq y_g \\ P(x_k,t_j) \times TM_2(t_j) & \text{if } h \geq 2 \wedge x_k = y_g \end{cases} \tag{6}$$

To deal with 3^+-itemsets, *DISC*-growth* [37] uses the self-product of $TM_2(t_j)$. For special cases where (i) $TM_1(t_j) = P(x_k, t_j)$ or (ii) $TM_2(t_j) = P(x_k, t_j)$, DISC*-growth uses the self-product of the third highest existential probability $TM_3(t_j)$ in t_j, as follows:

$$DISC^*(X,t_j) = \\ \begin{cases} DISC(X,t_j) & \text{if } k \leq 2 \\ P(x_k,t_j) \times TM_1(t_j) \times [TM_2(t_j)]^{k-2} & \text{if } k \geq 3 \wedge x_k \neq y_g \wedge x_k \neq y_s \\ P(x_k,t_j) \times TM_1(t_j) \times [TM_3(t_j)]^{k-2} & \text{if } k \geq 3 \wedge x_k = y_s \\ P(x_k,t_j) \times TM_2(t_j) \times [TM_3(t_j)]^{k-2} & \text{if } k \geq 3 \wedge x_k = y_g \end{cases} \tag{7}$$

Recall from Eq. (2) that the expected support of X can be computed as the product of $P(x_k, t_j)$ and existential probabilities of proper prefix of x_k. Hence, it is more logical to approximate an upper bound to expected support of X by involving $P(x_k, t_j)$ and existential probabilities of proper prefix of x_k. This leads to the concept of *prefixed item cap* (*PIC*) [29], which is defined as the product of $P(x_k, t_j)$ and the highest existential probability $PM_1(y_r, t_j)$ among items in the proper prefix of $x_k = y_r$, as follows:

$$PIC(X,t_j) = \begin{cases} P(y_1,t_j) & \text{if } h = 1 \\ P(x_k,t_j) \times PM_1(y_r,t_j) & \text{if } h \geq 2 \end{cases} \tag{8}$$

where (i) $PM_1(y_r, t_j) = \max_{i \in [1,r-1]} P(y_i, t_j)$ is the *prefixed maximum*, which is defined as the highest existential probability in $\{y_1, \ldots, y_{r-1}\} \subset t_j$.

PUF-growth [32] makes use of the above prefixed item cap to approximate a tight upper bound to expected support of 2-itemsets. To handle 3^+-itemsets, *PUF*-growth* [36] multiplies $PIC(X, t_j)$ with self-product of the second highest existential probability $PM_2(y_r, t_j)$ in $\{y_1, ..., y_{r-1}\} \subset t_j$, as follows:

$$PUF^*(X, t_j) = \begin{cases} PIC(X, t_j) & \text{if } k \leq 2 \\ P(x_k, t_j) \times PM_1(y_r, t_j) \times [PM_2(y_r, t_j)]^{k-2} & \text{if } k \geq 3 \end{cases} \quad (9)$$

where $PM_2(y_r, t_j) = \max_{i \in [1,r-1] \wedge (i \neq g)} P(y_i, t_j)$ is the second highest existential probability in $\{y_1, \ldots, y_{r-1}\} \subset t_j$ for $y_g = \text{argmax}_{i \in [1,h]} P(y_i, t_j)$ (i.e., $PM_1(y_r, t_j) = P(y_g, t_j)$).

Alternatively, the *BLIMP-growth* algorithm [28] multiplies $PIC(X, t_j)$ with existential probabilities of the first $(k-2)$ items in the proper prefix $\{y_1, ..., y_{r-1}\} \subset t_j$, as follows:

$$BLIMP(X, t_j) = \begin{cases} PIC(X, t_j) & \text{if } k \leq 2 \\ P(x_k, t_j) \times PM_1(y_r, t_j) \times \prod_{i=1}^{k-2} P(y_i, t_j) & \text{if } k \geq 3 \end{cases} \quad (10)$$

3 Theoretical Analysis and Results

After re-formulating upper bounds to expected support of patterns in Sect. 2, let us analyze and evaluate these bounds by taking advantages from the unifying model introduced above. When dealing with singletons (1-itemsets), we do not need to use upper bounds because we could scan the entire uncertain dataset of n transactions and accurately obtain the expected support of each pattern $\{x\}$ by summing existential probabilities of $\{x\}$ in every transaction t_j containing $\{x\}$:

$$expSup(\{x\}) = \sum_{j=1}^{n} P(x, t_j) \quad (11)$$

For any 2-itemset X, the upper bound computing models of Sect. 2 specialize as follows:

$$CUF^*(X, t_j) = TC(X, t_j) \quad (12)$$

$$DISC^*(X, t_j) = DISC(X, t_j) \quad (13)$$

$$PUF^*(X, t_j) = BLIMP(X, t_j) = PIC(X, t_j) \quad (14)$$

Among these groups of upper bounds of Eqs. (12)–(14), *PIC* involves the item having the maximum existential probability $PM_1(y_r, t_j)$ in the proper prefix of y_r, whereas *IC* (used by DISC-growth) involves the item having the maximum existential probability $TM_1(t_j)$ in the proper prefix of y_r as well as its suffix. So, as $PM_1(y_r, t_j) \leq TM_1(t_j)$, we derive the following theoretical result:

$$PIC(X, t_j) \leq IC(X, t_j) \quad (15)$$

Moreover, *IC* also uses $P(x_k, t_j)$, whereas *TC* uses $TM_2(t_j)$—which may not even involve any items in X—when $x_k \neq y_g$. So, as $P(x_k, t_j) \leq TM_2(t_j)$, we get the following result:

$$IC(X, t_j) \leq TC(X, t_j) \tag{16}$$

Hence, it is generally that

$$PIC(X, t_j) \leq IC(X, t_j) \leq TC(X, t_j) \tag{17}$$

i.e., *PIC* generally provides the tightest upper bounds to expected support when mining frequent 2-itemsets from high volumes of high-value uncertain data.

When mining 3^+-itemsets, the following property holds:

$$CUF^*(X, t_j) \leq TC(X, t_j) \tag{18}$$

This is due to the extra multiplication term $[TM_3(t_j)]^{k-2}$ in *CUF** such that $0 < [TM_3(t_j)]^{k-2} \leq 1$. Hence, *CUF** provides tighter upper bounds to expected support than TC when mining frequent 3^+-itemsets from high volumes of high-value uncertain data. Similar comments, due to the same reason, apply to *DISC** (when compared with *DISC*), as well as *PUF** and *BLIMP* (when both compared with *PIC*):

$$DISC^*(X, t_j) \leq DISC(X, t_j) \tag{19}$$

$$PUF^*(X, t_j) \leq PIC(X, t_j) \tag{20}$$

$$BLIMP(X, t_j) \leq PIC(X, t_j) \tag{21}$$

After analyzing the intra-group relationships among the aforementioned algorithms, let us analyze the inter-group relationships among *CUF**, *DISC**, *PUF**, and *BLIMP* when they mine 3^+-itemsets. If $x_k = y_g$, then the following property holds:

$$DISC^*(X, t_j) = CUF^*(X, t_j) \tag{22}$$

because $P(x_k, t_j) = P(y_g, t_j) = TM_1(t_j)$. The same property also holds when $x_k = y_s$ because $P(x_k, t_j) = P(y_s, t_j) = TM_2(t_j)$. Hence, when x_k is associated with the highest or the second highest existential probability in t_j, both *DISC** and *CUF** provide the same upper bounds to expected support when mining frequent 3^+-itemsets. Moreover, if $x_k \neq y_g$ and $x_k \neq y_s$, then the following property holds:

$$PUF^*(X, t_j) \leq DISC^*(X, t_j) \tag{23}$$

because both $PM_1(y_r, t_j) \leq TM_1(t_j)$ and $PM_2(y_r, t_j) \leq TM_2(t_j)$. Hence, when x_k does not associated with the highest or the second highest existential probability in t_j, *PUF** provides tighter upper bounds to expected support than *DISC**.

Furthermore, if $P(x_{k-1}, t_j) = PM_1(y_r, t_j)$ and $P(x_i, t_j) = P(y_i, t_j)$ for $i \in [1, k-2]$, then we obtain:

$$BLIMP(X, t_j) = expSup(X, t_j) \tag{24}$$

Hence, when X is the first k items in t_j such that $P(x_{k-1}, t_j)$ happens to be the highest existential probability in the proper prefix $\{y_1, ..., y_{r-1}\} \subset t_j$, *BLIMP* provides upper bounds that are so tight that they are indeed the expected support.

Note that all the aforementioned algorithms do not generate any false negatives. With tighter upper bounds to expected support, fewer false positives are produced. Hence, shorter runtime is needed to verify whether or not a pattern is true positive (i.e., frequent) or false positive (i.e., potentially frequent w.r.t. upper bounds but truly infrequent w.r.t. *minsup*).

In terms of memory consumption, the aforementioned frequent pattern mining algorithms are all tree-based. The number of nodes in the corresponding tree is small. With appropriate item ordering, the number of tree nodes for uncertain big data mining is identical to that of the FP-tree [18] for mining precise data. Note that each node in the FP-tree captures an item x and its actual support, respectively. Conversely, when mining 2-itemsets, each tree node captures x and its TC for CUF-growth. Similarly, each tree node captures x and *DISC* for DISC-growth; and each tree node captures x and *PIC* for PUF-growth. When mining 3^+-itemsets, each tree node captures an additional information such as $TM_3(t_j)$ for the CUF*-growth algorithm, $TM_2(t_j)$ or $TM_3(t_j)$ for the DISC*-growth algorithm, $PM_2(y_r, t_j)$ for PUF*-growth, as well as $P(y_i, t_j)$ for BLIMP-growth, respectively.

It should be noted, as these theoretical results allow us to find tight upper bounds to expected support for frequent pattern mining problems over uncertain big data, they also introduce the nice amenity of effectively lowering the overall algorithm runtime efficiently. This will be completely demonstrated in our experimental assessment and analysis in Sect. 4.

4 Experimental Assessment and Evaluation

In this section, we evaluate several performance aspects on the optimization opportunities offered by the six different upper bounds to expected support described in Sect. 3. As regards the data layer of our experimental campaign, we considered the following well-known datasets: (i) IBM synthetic dataset, and (ii) mushroom dataset from the UC Irvine Machine Learning Depository. In particular, these datasets have been artificially made uncertain via a simple sampling-based routine that injects the existential probabilities as associated to the values of a pre-determined sub-set of attributes of the input dataset. As regards metrics, we focused on the following experimental benchmarks: (i) memory consumption, (ii) accuracy, and (iii) runtime. The final goal of our experimental campaign is to provide a comparative analysis and confirm our analytical findings provided in Sect. 3.

4.1 Memory Consumption Analysis

First, we analytically evaluate the memory consumption of the different approximations. Among them, we observe the following main behaviors that are relevant to our research:

- CUF-growth (which uses TC) requires the least amount of memory space because they are solely dependent on transaction t_j. In other words, only a single value (TC) is needed for each transaction t_j.
- CUF*-growth (which uses CUF^*) requires slightly more memory space because two values—both TC and $TM_3(t_j)$—are needed for each transaction t_j in order to compute the CUF^* value for patterns of different cardinality k. Both CUF-growth and CUF*-growth do not need to store existential probabilities of any items in transaction t_j.
- DISC-growth and PUF-growth each requires a total of h values for each transaction t_j. Specifically, for each transaction $t_j = \{y_1, y_2, \ldots, y_r \ldots, y_h\}$ with h items, a single value (IC or PIC) is needed for each item y_i in t_j.
- DISC*-growth, as an extension to DISC-growth, needs to store an additional value—namely, $TM_2(t_j)$ or $TM_3(t_j)$ depending on whether $x_k = y_g$ or y_s—for each item x_k ($= y_r$) in transaction t_j.
- PUF*-growth, as an extension to PUF-growth, needs to store an additional value—namely, $PM_2(y_r, t_j)$—for each item y_r in transaction t_j. Both DISC*-growth and PUF*-growth require the most amount of memory space because each of them requires a total of $2h$ values for each transaction t_j.

4.2 Accuracy Analysis

We measure the accuracy of the different frequent pattern mining algorithms when the derived theoretical upper bounds are applied. In this experiment series, we compare the tightness of the upper bounds as approximated expected support. From Sect. 3, Eqs. (12)–(14) are confirmed by results shown in Fig. 1. Note the following:

- CUF*-growth and CUF-growth lead to the same number of false positives for 2-itemsets (i.e., cardinality = 2).
- DISC*-growth and DISC-growth, as well as PUF*-growth and PUF-growth, also lead to the same number of false positives for 2-itemsets.

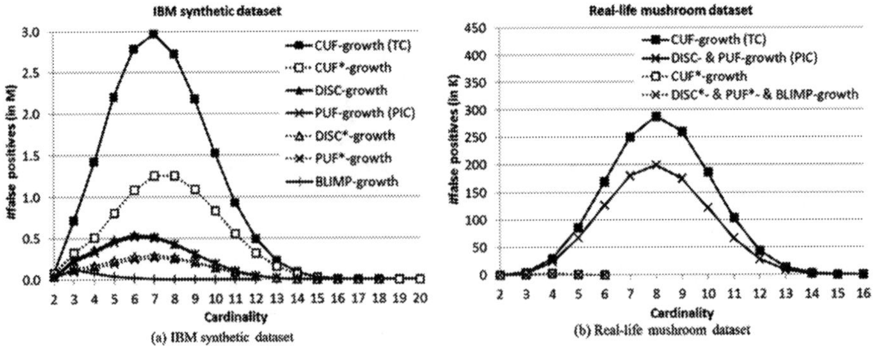

Fig. 1. Experimental results on accuracy analysis

Among these three groups of upper bounds, we also observe the following:

- PUF-growth involves the item having the maximum existential probability $PM_1(y_r, t_j)$ in the proper prefix of y_r.
- DISC-grwoth involves the item having the maximum existential probability $TM_1(t_j)$ in the proper prefix of y_r as well as its suffix.

As a consequence, since $PM_1(y_r, t_j) \leq TM_1(t_j)$, we can also experimentally illustrate Eq. (15). Moreover, IC also uses $P(x_k, t_j)$, whereas TC uses $TM_2(t_j)$—which may not even involve any items in X—when $x_k \neq y_g$. So, as $P(x_k, t_j) \leq TM_2(t_j)$, we can also experimentally illustrate Eq. (16). These two experimental evidences support the observation that PUF-growth generally provides the tightest upper bounds to expected support when mining frequent 2-itemsets from high volumes of high-value uncertain data.

When mining 3^+-itemsets, following the analysis provided in Sect. 3, we further observe the following results, which are also confirmed by our experimental evaluation (see Fig. 1):

- $DISC^*(X, t_j) \leq TC(X, t_j)$ due to the extra multiplication term $[TM_3(t_j)]^{k-2}$ in CUF*-growth such that $0 < [TM_3(t_j)]^{k-2} \leq 1$. Hence, CUF*-growth provides tighter upper bounds to expected support than CUF-growth when mining frequent 3^+-itemsets from high volumes of high-value uncertain data.
- $DISC^*(X, t_j) \leq IC(X, t_j)$ and $PUF^*(X, t_j) \leq PIC(X, t_j)$ due to the same reason, i.e., the extra multiplication terms—which are in the range (0,1]—in DISC*-growth and PUF*-growth.

After analyzing the intra-group relationships between the aforementioned upper bounds, let us analyze the inter-group relationships among the four extensions when they mine k-itemsets, and simultaneously checking it on the experimental results shown Fig. 1 (which further confirm our theoretical analysis provided in Sect. 3):

- If $x_k{=}y_g$, then $DISC^*(X, t_j) = CUF^*(X, t_j)$ because $P(x_k, t_j) = P(y_g, t_j) = TM_1(t_j)$.
- If $x_k{=}y_s$, then $DISC^*(X, t_j) = CUF^*(X, t_j)$ because $P(x_k, t_j) = P(y_s, t_j) = TM_2(t_j)$.
- If $x_k{\neq}y_g$ and $x_k{\neq}y_s$, then $PUF^*(X, t_j) \leq DISC^*(X, t_j)$ because both $PM_1(y_r, t_j){\leq}TM_1(t_j)$ and $PM_2(y_r, t_j){\leq}TM_2(t_j)$.

Hence, it follows that, when x_k is associated with the highest or the second highest existential probability in t_j, both DISC*-growth and CUF*-growth provide the same upper bounds to expected support when mining frequent 3^+-itemsets. Moreover, when x_k is not associated with the highest or the second highest existential probability in t_j, PUF*-growth provides tighter upper bounds to expected support than DISC*-growth.

The evaluation above shows the tightness of our upper bounds to expected support. Note that all these bounds do not lead to any false negatives but only false positives. The tighter the bound, the lower is the number of false positives.

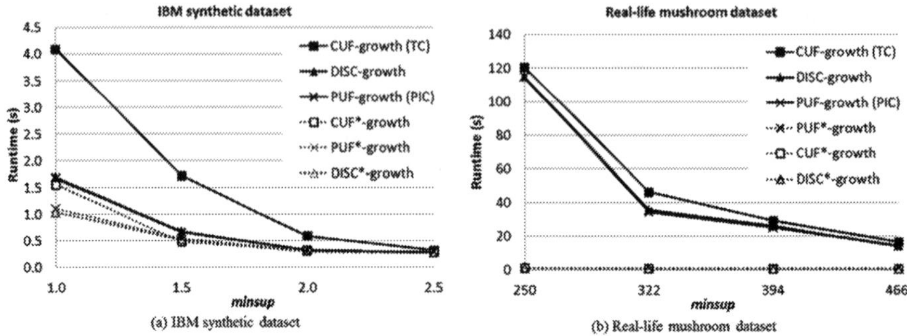

Fig. 2. Experimental results on runtime analysis

Our experimental results shown in Fig. 1 clearly support our analytical results. Specifically, CUF-growth leads to the highest numbers of false positives, whereas PUF*-growth leads to the lowest numbers (with DISC*-growth led to a close second lowest numbers) of false positives in IBM synthetic dataset and mushroom real-life dataset. Also, it is interesting to note that the tightness of the upper bound to expected support provided by the following extensions: CUF*-growth, DISC*-growth, PUF*-growth and BLIMP-growth. In fact, they do not generate any false positives beyond cardinality 6 for the mushroom dataset, as shown in Fig. 1.

4.3 Runtime Analysis

Recall that knowledge discovery and data mining algorithms use the aforementioned caps *TC*, *IC* and *PIC* to approximate expected support (see Sect. 2). The related algorithms find patterns with upper bounds to expected support meeting or exceeding the user-specified threshold *minsup*. This results in a collection of all potentially frequent 2^+-itemsets that include both true positive (i.e., truly frequent patterns) and false positive (i.e., potentially frequent with respect to upper bounds but truly infrequent with respect to *minsup*). With tighter upper bounds to expected support, fewer false positives are produced. Hence, shorter runtimes result. Figure 2 shows overall runtime of the various alternatives using the proposed upper bounds. From the analysis Fig. 2, the following observations can be derived:

- Due to its highest number of false positives generated, CUF-growth introduces the longest runtime.
- As all three extensions (CUF*-growth, PUF*-growth and DISC*-growth) produce fewer false positives than the counterparts (CUF-growth, PUF-growth and DISC-growth), runtimes for the former are also shorter.
- As usual, when *minsup* increases, runtime decreases.
- Recall that $PUF^*(X, t_j) \leq DISC^*(X, t_j)$ if $x_k = y_g$ and $x_k = y_s$. For the cases where $x_k = y_g$ or $x_k = y_s$, it is possible (but not guaranteed) that $PUF^*(X, t_j)$

$\leq DISC^*(X, t_j)$. However, for some other cases (e.g., for short transactions in the IBM synthetic dataset or short frequent patterns mined from the real-life mushroom dataset), DISC*-growth beats PUF*-growth.

4.4 Comparative Analysis

After evaluating the seven approximations as upper bounds to expected support, we observe the following:

- CUF-growth requires the least amount of memory space (with a single value per transaction), and CUF*-growth requires the second least amount of memory space (with two values per transaction);
- DISC*-growth and PUF*-growth both produce fewest false positives due to the tightness of their bounds;
- DISC*-growth takes the shortest runtime, where PUF*-growth and CUF*-growth take just slightly longer than DISC*-growth.

Hence, our recommendation is as follows: *If memory is an issue, it is better to use CUF*-growth due to its small memory requirements, few false positives and short runtimes. Otherwise, it is better to use DISC*-growth or PUF*-growth because their relatively low memory requirements* ($2h$ *values for* h *items in a transaction*) *while they produce fewer false positives and run faster than others.*

5 Conclusions and Future Work

In this paper, we have examined the concepts of transaction cap *TC*, item cap *IC* and prefixed item cap *PIC* by viewing them as tight upper bounds to expected support of frequent k-itemsets when mining uncertain big data. Among these upper bounds, *PIC* provides the tightest upper bounds when mining frequent 2-itemsets, and thus produces the fewest false positives and the fastest running. When mining frequent 3^+-itemsets, the concepts of *TC*, *IC*, and *PIC* have been extended to become *CUF**, *DISC**, *PUF**, and *BLIMP*. Our experimental results confirm our analytical findings that any of these four extensions could provide tighter upper bounds to expected support of frequent 3^+-itemsets than the other three extensions on different mining parameters and/or distributions of uncertain data.

Future work is mainly oriented towards (i) studying optimization alternatives particularly targeted to distributed environments (e.g., *fragmentation techniques* [11,15], which could allow us to improve the efficiency of our framework, and (ii) extending the proposed framework according to modern *big data analytics* predicates [7,21,39].

Acknowledgements. This project is partially supported by NSERC (Canada) and University of Manitoba.

References

1. Aggarwal, C.C., Li, Y., Wang, J., Wang, J.: Frequent pattern mining with uncertain data. In: ACM KDD 2009, pp. 29–37 (2009)
2. Agrawal, R., Srikant, R.: Fast algorithms for mining association rules. In: VLDB 1994, pp. 487–499 (1994)
3. Ahmed, A.U., Ahmed, C.F., Samiullah, M., Adnan, N., Leung, C.K.: Mining interesting patterns from uncertain databases. Inf. Sci. **354**, 60–85 (2016)
4. Aryadinata, Y.S., Lin, Y., Barcellos, C., Laurent, A., Libourel, T.: Mining epidemiological dengue fever data from Brazil: a gradual pattern based geographical information system. In: Laurent, A., Strauss, O., Bouchon-Meunier, B., Yager, R.R. (eds.) IPMU 2014, Part II. CCIS, vol. 443, pp. 414–423. Springer, Heidelberg (2014)
5. Calders, T., Garboni, C., Goethals, B.: Efficient pattern mining of uncertain data with sampling. In: Zaki, M.J., Yu, J.X., Ravindran, B., Pudi, V. (eds.) PAKDD 2010, Part I. LNCS (LNAI), vol. 6118, pp. 480–487. Springer, Heidelberg (2010)
6. Chen, L., Liu, C., Zhang, C.: Mining probabilistic representative frequent patterns from uncertain data. In: SIAM SDM 2013, pp. 73–81 (2013)
7. Cuzzocrea, A.: Analytics over big data: exploring the convergence of data warehousing, OLAP and data-intensive cloud infrastructures. In: IEEE COMPSAC 2013, pp. 481–483 (2013)
8. Cuzzocrea, A.: Approximate OLAP query processing over uncertain and imprecise multidimensional data streams. In: Decker, H., Lhotská, L., Link, S., Basl, J., Tjoa, A.M. (eds.) DEXA 2013, Part II. LNCS, vol. 8056, pp. 156–173. Springer, Heidelberg (2013)
9. Cuzzocrea, A.: Retrieving accurate estimates to OLAP queries over uncertain and imprecise multidimensional data streams. In: Cushing, J.B., French, J., Bowers, S. (eds.) SSDBM 2011. LNCS, vol. 6809, pp. 575–576. Springer, Heidelberg (2011)
10. Cuzzocrea, A., Bellatreche, L., Song, I.-Y.: Data warehousing and OLAP over big data: current challenges and future research directions. In: ACM DOLAP 2013, pp. 67–70 (2013)
11. Cuzzocrea, A., Darmont, J., Mahboubi, H.: Fragmenting very large XML data warehouses via K-means clustering algorithm. Int. J. Bus. Intell. Data Min. **4**(3/4), 301–328 (2009)
12. Cuzzocrea, A., Leung, C.K.: Upper bounds to expected support for frequent itemset mining of uncertain big data. In: ACM SAC 2015, pp. 919–921 (2015)
13. Cuzzocrea, A., Leung, C.K., MacKinnon, R.K.: Mining constrained frequent itemsets from distributed uncertain data. Future Gener. Comput. Syst. **37**, 117–126 (2014)
14. Cuzzocrea, A., Saccà, D., Ullman, J.D.: Big data: a research agenda. In: IDEAS 2013, pp. 198–203 (2013)
15. Daenen, J., Neven, F., Tan, T.: Gumbo: guarded fragment queries over big data. In: EDBT 2015, pp. 521–524 (2015)
16. Diaz-Valenzuela, I., Martin-Bautista, M.J., Vila, M.-A.: A fuzzy semisupervised clustering method: application to the classification of scientific publications. In: Laurent, A., Strauss, O., Bouchon-Meunier, B., Yager, R.R. (eds.) IPMU 2014, Part I. CCIS, vol. 442, pp. 179–188. Springer, Heidelberg (2014)
17. Fan, W., Bifet, A.: Mining big data: current status, and forecast to the future. ACM SIGKDD Explor. **14**(2), 1–5 (2012)

18. Han, J., Pei, J., Yin, Y.: Mining frequent patterns without candidate generation. In: ACM SIGMOD 2000, pp. 1–12 (2000)
19. Hodáková, P., Perfilieva, I., Hurtík, P.: F-transform and its extension as tool for big data processing. In: Laurent, A., Strauss, O., Bouchon-Meunier, B., Yager, R.R. (eds.) IPMU 2014, Part III. CCIS, vol. 444, pp. 374–383. Springer, Heidelberg (2014)
20. Jiang, F., Kawagoe, K., Leung, C.K.: Big social network mining for "following" patterns. In: C3S2E 2015, pp. 28–37 (2015)
21. Jiang, F., Leung, C.K.: A data analytic algorithm for managing, querying, and processing uncertain big data in cloud environments. Algorithms **8**(4), 1175–1194 (2015)
22. Jiang, F., Leung, C.K., Liu, D.: Efficiency improvements in social network communication via MapReduce. In: IEEE DSDIS 2015, pp. 161–168 (2015)
23. Leung, C.K.: Big data mining applications and services. In: BigDAS 2015, pp. 1–8 (2015)
24. Leung, C.K.: Uncertain frequent pattern mining. In: Aggarwal, C.C., Han, J. (eds.) Frequent Pattern Mining, pp. 417–453. Springer, Switzerland (2014)
25. Leung, C.K., Cuzzocrea, A.: Frequent subgraph mining from streams of uncertain data. In: C3S2E 2015, pp. 18–27 (2015)
26. Leung, C.K.-S., Cuzzocrea, A., Jiang, F.: Discovering frequent patterns from uncertain data streams with time-fading and landmark models. In: Hameurlain, A., Küng, J., Wagner, R., Cuzzocrea, A., Dayal, U. (eds.) TLDKS VIII. LNCS, vol. 7790, pp. 174–196. Springer, Heidelberg (2013)
27. Leung, C.K., Jiang, F., Pazdor, A.G.M., Peddle, A.M.: Parallel social network mining for interesting 'following' patterns. Concurrency Computat. Pract. Exper. (2016). doi:10.1002/cpe.3773
28. Leung, C.K.-S., MacKinnon, R.K.: BLIMP: a compact tree structure for uncertain frequent pattern mining. In: Bellatreche, L., Mohania, M.K. (eds.) DaWaK 2014. LNCS, vol. 8646, pp. 115–123. Springer, Heidelberg (2014)
29. Leung, C.K., MacKinnon, R.K., Tanbeer, S.K.: Tightening upper bounds to expected support for uncertain frequent pattern mining. Procedia Comput. Sci. **35**, 328–337 (2014)
30. Leung, C.K.-S., Mateo, M.A.F., Brajczuk, D.A.: A tree-based approach for frequent pattern mining from uncertain data. In: Washio, T., Suzuki, E., Ting, K.M., Inokuchi, A. (eds.) PAKDD 2008. LNCS (LNAI), vol. 5012, pp. 653–661. Springer, Heidelberg (2008)
31. Leung, C.K.-S., Tanbeer, S.K.: Fast tree-based mining of frequent itemsets from uncertain data. In: Lee, S., Peng, Z., Zhou, X., Moon, Y.-S., Unland, R., Yoo, J. (eds.) DASFAA 2012, Part I. LNCS, vol. 7238, pp. 272–287. Springer, Heidelberg (2012)
32. Leung, C.K.-S., Tanbeer, S.K.: PUF-Tree: a compact tree structure for frequent pattern mining of uncertain data. In: Pei, J., Tseng, V.S., Cao, L., Motoda, H., Xu, G. (eds.) PAKDD 2013, Part I. LNCS (LNAI), vol. 7818, pp. 13–25. Springer, Heidelberg (2013)
33. Li, X., Wang, Y., Li, X., Wang, X., Yu, J.: GDPS: an efficient approach for skyline queries over distributed uncertain data. Big Data Res. **1**, 23–36 (2014)
34. Liu, C., Chen, L., Zhang, C.: Summarizing probabilistic frequent patterns: a fast approach. In: ACM KDD 2013, pp. 527–535 (2013)
35. Liu, Y.-H.: Mining time-interval univariate uncertain sequential patterns. Data Knowl. Eng. **100**, 54–77 (2015)

36. MacKinnon, R.K., Leung, C.K.-S., Tanbeer, S.K.: A scalable data analytics algorithm for mining frequent patterns from uncertain data. In: Peng, W.-C., Wang, H., Bailey, J., Tseng, V.S., Ho, T.B., Zhou, Z.-H., Chen, A.L.P. (eds.) PAKDD 2014 Workshops. LNCS (LNAI), vol. 8643, pp. 404–416. Springer, Heidelberg (2014)
37. MacKinnon, R.K., Strauss, T.D., Leung, C.K.: DISC: efficient uncertain frequent pattern mining with tightened upper bounds. In: IEEE ICDM 2014 Workshops, pp. 1038–1045 (2014)
38. Nguyen, H.T.H., Cao, J.: Trustworthy answers for top-k queries on uncertain big data in decision making. Inf. Sci. **318**, 73–90 (2015)
39. Pei, J.: Some new progress in analyzing and mining uncertain and probabilistic data for big data analytics. In: Ciucci, D., Inuiguchi, M., Yao, Y., Ślęzak, D., Wang, G. (eds.) RSFDGrC 2013. LNCS (LNAI), vol. 8170, pp. 38–45. Springer, Heidelberg (2013)
40. Petry, F.E.: Data mining approaches for geo-spatial big data: uncertainty issues. Int. J. Organ. Collective Intell. **3**(1), 52–71 (2012)
41. Rahman, Q.M., Fariha, A., Mandal, A., Ahmed, C.F., Leung, C.K.: A sliding window-based algorithm for detecting leaders from social network action streams. In: IEEE/WIC/ACM WI-IAT 2015, vol. 1, pp. 133–136 (2015)
42. Saati, S., Hatami-Marbini, A., Tavana, M., Agrell, P.J.: A fuzzy data envelopment analysis for clustering operating units with imprecise data. Int. J. Uncertainty Fuzziness Knowl. Based Syst. **21**(1), 29–54 (2013)
43. Samet, A., Lefèvre, E., Ben Yahia, S.: Classification with evidential associative rules. In: Laurent, A., Strauss, O., Bouchon-Meunier, B., Yager, R.R. (eds.) IPMU 2014, Part I. CCIS, vol. 442, pp. 25–35. Springer, Heidelberg (2014)
44. Tong, Y., Chen, L., Cheng, Y., Yu, P.S.: Mining frequent itemsets over uncertain databases. PVLDB **5**(11), 1650–1661 (2012)
45. Xu, J., Li, N., Mao, X.-J., Yang, Y.-B.: Efficient probabilistic frequent itemset mining in big sparse uncertain data. In: Pham, D.-N., Park, S.-B. (eds.) PRICAI 2014. LNCS (LNAI), vol. 8862, pp. 235–247. Springer, Heidelberg (2014)

In-Database Feature Selection Using Rough Set Theory

Frank Beer(✉) and Ulrich Bühler

University of Applied Sciences Fulda, Leipziger Straße 123, 36037 Fulda, Germany
{frank.beer,u.buehler}@informatik.hs-fulda.de

Abstract. Despite their traditional roles, database systems increasingly became attractive as scalable analytical platforms using extensible SQL over the last decade. This methodology is termed in-database processing and provides several advantages over traditional mining attempts. In this work we bring Variable Precision Rough Sets to the domain of databases as a common framework to unlock hidden knowledge from data. Our derived model is built upon pure relational operations and thus very efficient. We further demonstrate its applicability for feature selection by introducing two in-database algorithms. Our experiments indicate, the model scales and is comparable to existing approaches in terms of performance but superior when applied to real-life applications.

Keywords: Concept approximation · Feature selection · In-database analytics · Knowledge discovery in databases · Relational algebra · Relational database systems · Rough Set Theory · SQL · Variable Precision Rough Sets

1 Introduction

In the 1990s and early 2000s, traditional mining frameworks relied on a client-only or client-server architecture. These frameworks produce predictive models or pursue other kinds of mining tasks by either loading data from flat files or external repositories. While the former requires inefficient file operations, the latter suffers from enduring data transmissions. These characteristics are crucial particularly when processing enormous datasets and have been critized in literature due to their poor performance and their inability to meet agile exploratory analysis today [1,2]. By that time, Rough Set Theory (RST) broadly emerged as a sophisticated instrument of extracting hidden knowledge from data. Therefore a lot of rough set-based software systems and libraries are based on those traditional architectures (e.g. [3–5]) and thus suffer from time-consuming operations or high communication costs.

In this work, we try to overcome those historical drawbacks of classic rough set-based systems by computing Variable Precision Rough Sets (VPRS) in-database. In-database processing took off more recently as a flexible paradigm in data science (e.g. [6,7]). It orchestrates conventional database systems to build

J.P. Carvalho et al. (Eds.): IPMU 2016, Part II, CCIS 611, pp. 393–407, 2016.
DOI: 10.1007/978-3-319-40581-0_32

up reliable mining algorithms in a data-centric fashion using SQL and statistical extensions. Thus it provides essential benefits, because hidden knowledge is stored in relational repositories predominantly given through transactional data or warehouses. As such, data transports are minimized and processing time can be reduced to a large extend. Exploiting this paradigm, we redefine the concept approximation of VPRS as an extension of an earlier work [8] and derive an equivalent model based on common database operations, which enriches most conventional databases. Unlike classic rough sets, it is tolerant to minor data irregularities and thus valueable for real-life scenarios. Further motivation comes from the following three points: (i) We observed that the availability of RST algorithms for in-database analytics is very limited and (ii) those existing approaches are either inefficient or unable to cope with uncertainty. (iii) Leveraging efficient relational database infrastructures (parallelism, algorithms, data structures and statistics) provide promising opportunities for the practical application of RST particularly dimensionality reduction, pattern extraction or classification. In this work, we concentrate on Feature Selection (FS) as a key application in RST and introduce new relational expressions to compute core and reducts. Based on that, we present two filter algorithms utilizing efficient in-database processing.

The remainder is organized as follows: First we introduce central concepts of RST and VPRS (Sect. 2). In Sect. 3 we review related models and systems. Section 4 restructures VPRS for in-database processing and presents two FS algorithms. Then we compare close related approaches to our model (Sect. 5) and conclude in Sect. 6.

2 Rough Set Preliminaries

RST is a mathematical framework to analyze uncertain data proposed by Z. Pawlak [9,10]. An extension to that theory are VPRS coined by W. Ziarko [11]. In this section, we introduce rudiments of both including the basic data structures, the concept approximation and the definition of core and reduct.

2.1 Information Systems and Indiscernibility Relation

In RST, data is represented in a table-like data structure, i.e. the Information System (IS). It consists of objects (rows) $U = \{x_1, ..., x_n\}$ and attributes (columns) $A = \{a_1, ..., a_m\}, n, m \in \mathbb{N}$. Thus, it can be expressed in the tuple $\langle U, A\rangle$, where each $a \in A$ poses a mapping from U to a's value range V_a, i.e. $a : U \rightarrow V_a$. An extension to an IS is the Decision System (DS), which even holds features $d_i \in D$ with $d_i : U \rightarrow V_{d_i}, 1 \leq i \leq p \in \mathbb{N}$, representing a context-specific decision made by an expert or teacher. It is denoted by $\langle U, A, D\rangle$.

To discern objects inside an IS or DS, RST supplies the indiscernibility relation w.r.t. the feature subset $B \subseteq A$. Formally, it is denoted by $IND(B) = \{(x, y) \in U^2 \mid \forall a \in B : a(x) = a(y)\}$ in a given $\langle U, A\rangle$. Its resulting partition $U/IND(B)$ produces disjoint equivalence classes K_j over U w.r.t. B. Out of convenience we write $U/B = \{K_1, ..., K_q\}, 1 \leq j \leq q \in \mathbb{N}$.

2.2 Variable Precision Rough Sets

In RST, we approximate concepts based on equivalence classes and standard subset inclusion. These tools are relaxed in VPRS by applying a majority inclusion. As such, a controlled degree of overlapping is permitted in order to address minor irregularities in the data that would, in turn, fall out or be considered inconsistent in the classic rough set model. The substantial part of VPRS relies on the relative inclusion measure

$$c(X,Y) = \begin{cases} 1 - \frac{|X \cap Y|}{|X|} & \text{, if } X \neq \emptyset \\ 0 & \text{, otherwise} \end{cases} \tag{1}$$

where X and Y are two sets. To obtain the majority inclusion, we define the bound $0 \leq \beta < 0.5$ such that $c(X,Y) \leq \beta$. In this regard, X is said to be included in Y w.r.t. the permitted overlapping β and we write $X \subseteq_\beta Y$. Combining this relaxation and the indiscernibility relation, a given target concept can be described in terms of VPRS: Let $\langle U, A \rangle, B \subseteq A, \beta \in [0, 0.5)$ and the concept $X \subseteq U$, then for any fixed β the β-lower approximation of X is defined by

$$\underline{X}_B^\beta = \bigcup \{K \in U/B \mid c(K,X) \leq \beta\} \tag{2}$$

whereas the β-upper approximation can be specified by

$$\overline{X}_B^\beta = \bigcup \{K \in U/B \mid c(K,X) < 1 - \beta\}. \tag{3}$$

Both expressions form the β-approximation of X w.r.t. the knowledge of B and precision β. For $\underline{X}_B^\beta \neq \overline{X}_B^\beta$ we get an indication that information in B is insufficient to express the knowledge of X properly. Objects causing this uncertainty can be determined by the β-boundary approximation

$$\overline{\underline{X}}_B^\beta = \bigcup \{K \in U/B \mid \beta < c(K,X) < 1 - \beta\}. \tag{4}$$

In some domains of interest, there might be multiple concepts to address, which can be modelled in a DS easily. Consequently, RST and VPRS provide further definitions to extract all approximations. Given $\langle U, A, D \rangle$, $B \subseteq A, E \subseteq D$ and $\beta \in [0, 0.5)$, all concepts induced by the partition U/E can be evaluated using the β-positive region

$$POS_B^\beta(E) = \bigcup_{X \in U/E} \underline{X}_B^\beta \tag{5}$$

and the β-boundary region

$$BND_B^\beta(E) = \bigcup_{X \in U/E} \overline{\underline{X}}_B^\beta. \tag{6}$$

Denote, (5) holds those kind of samples which we can classify certainly to be in the concepts $X \in U/E$ employing knowledge B with β-precision, while (6) contains inconsistent objects. With both β-regions we get a complete view on the quality of B to describe E properly. Ultimately, the tolerance of this model can be mitigated. In case of $\beta = 0$, VPRS is identical to RST.

2.3 Core and Reducts

In many applications, the question arises whether we require all condition attributes to predict or describe a decision well. Thus, there may be attributes which do not contribute much while others are essential. Using the β-positive region, VPRS provide us with a quality measure to seek for such attribute subsets. Let $\langle U, A, D\rangle$ be a DS with $B \subseteq A, E \subseteq D$. The attribute set $R \subseteq B$ is called β-reduct w.r.t. E if

$$|POS_R^\beta(E)| = |POS_B^\beta(E)| \tag{7}$$

and $\forall a \in R:$

$$|POS_{R\setminus\{a\}}^\beta(E)| \neq |POS_R^\beta(E)|. \tag{8}$$

By definition, R yields the same predictive power compared to B and any attribute removal changes the classification of E. In that sense, a reduct is minimal. Intersecting all β-reducts reveals the β-core holding most essential information. Particularly it consists of those attributes a that are indispensable, i.e.

$$|POS_{B\setminus\{a\}}^\beta(E)| \neq |POS_B^\beta(E)|. \tag{9}$$

Note, the β-reduct definition can be further relaxed by modifying (7) with "$\geq$" instead of "$=$", which is possible in VPRS permitting to extract much more compact reducts. However, this change does not jointly work with (9).

3 Related Work

One of the first systems combining RST and database systems is presented in [3]. To compute the concept approximation, SQL commands are embedded inside external programming logic for a final row-by-row processing, which causes enduring network i/o particularly for enormous datasets. To manage this shortcoming of many mining frameworks today, Nguyen suggests to perform simple in-database aggregations rather than transmitting huge volumes of data over the wire [1]. Thus, latency can be decreased to a large extend but a client-server architecture is implied. In contrast, the work in [12] and [13] fully leverage database operations for FS, but without the explicit computation of the concept approximation. They share an efficient algorithm to extract core attributes and present distinctive approaches to retrieve a final set of significant attributes. In both, inconsistent data is considered noise and handled differently. While in [12] a noise measure is developed on predefined thresholds, such records are simply eliminated in [13]. Another work extracting value reducts is proposed in [14] using database operations. Similar to the two previous approaches, this method runs on consistent data only and thus preprocessing is an obligatory step. Other approaches exploiting database technology rely on procedural structures rather than pure and efficient database operations [15–17]. More recently, a new database system emerged known as Infobright [18]. Inspired by RST, it focuses on fast data processing and estimated query results towards ad-hoc querying.

Most existing approaches such as [3,15–17] leverage database technology only partially using procedural structures or external programming. Only few RST methods fully exploit database operations [12–14] but do not rely on the concept approximation, which limits their methods to consistent data only. In contrast, our proposed model is compliant with RST and VPRS. Thus it can handle uncertainty and irregularities in the data using efficient relational operations.

4 In-Database Feature Selection

This section redefines fundamental concepts of RST and VPRS. We obtain a new model based on relational algebra and demonstrate its applicability for FS.

4.1 Basic Database Notation

Being aware of their differences, a DS and a database table share common structures, i.e. rows and columns. Therefore we suppose a data table R with columns $A \cup D$ and underlying schema $\langle a_1, ..., a_m, d_1, ..., d_p \rangle$, where $A = \{a_1, ..., a_m\}, m \in \mathbb{N}$ and $D = \{d_1, ..., d_p\}, p \in \mathbb{N}$. For short, we write $R_{\langle a_1,...,a_m,d_1,...,d_p \rangle}$ or $R_{\langle A,D \rangle}$ intuitively. Concerning basic database operations, we draw on conventional projection π, selection σ, grouping $\mathcal{G}$, join $\bowtie$ and renaming operation ρ as follows: $\pi_B(R_{\langle A \rangle})$ allows to project tuples $t \in R$ to a specified feature subset $B \subseteq A$ while removing duplicates, whereas a projection without duplicate elimination is indicated by $\pi^+_B(R_{\langle A \rangle})$. The selection $\sigma_\phi(R_{\langle A \rangle})$ only holds those tuples $t \in R$ fulfilling condition ϕ with retaining feature schema $\langle A \rangle$. $\mathcal{G}_{\{f_1,...,f_r\},G,B}(R_{\langle A \rangle})$ groups tuples of R according to $G \subseteq A$ while applying the aggregation functions $f_1, ..., f_r$. The output features are $B \cup \{f_1, ..., f_r\}$ with $B \subseteq G, r \in \mathbb{N}_0$. For $B = G = \emptyset$, $\mathcal{G}$ returns only one row performing $f_1, ..., f_r$ on all given rows. Out of simplicity, we write $\mathcal{G}_{\{f_1,...,f_r\}}(R_{\langle A \rangle})$. The natural join $S_{\langle G \rangle} \bowtie T_{\langle H \rangle}$ assembles both tables S and T to a new relation R such that $s.b = t.b, \forall s \in S, t \in T$ and attributes $b \in G \cap H$. Note, R consists of all attributes in G and H, where overlapping attributes are shown only once. Finally, the operation $\rho_{\langle b_1,...,b_m \rangle}(R_{\langle a_1,...,a_m \rangle})$ renames attributes a_i in table R to its new name $b_i, 1 \leq i \leq m \in \mathbb{N}$.

To compute the different levels of the indiscernibility relation, our model makes use of $\mathcal{G}$. For our purpose we simply count the number of members in each elementary class of a given table $R_{\langle A \rangle}$, i.e. the cardinality expressed by the aggregate `count`, and include it as new feature c. Consolidated, we make use of the following notation $\mathcal{I}^G_{B \cup \{c\}}(R) := \rho_{\langle c,b_1,...,b_m \rangle}(\mathcal{G}_{\{\texttt{count}\},G,B}(R_{\langle A \rangle}))$ with $B = \{b_1, ..., b_m\}, B \subseteq G \subseteq A$ and refer to $\mathcal{I}^G_{B \cup \{c\}}(R)$ as our compressed multiset representation using schema $\langle c, b_1, ..., b_m \rangle$.

4.2 VPRS Using Relational Operations

We redefine the β-approximation using Propositions 1 and 2. Thus, we can bring VPRS immediately to the domain of databases through Theorems 1 and 2 using relational operations from Sect. 4.1. Furthermore we examine the runtime of our proposed model utilizing established database algorithms in Theorem 3.

Proposition 1. *Let $\langle U, A\rangle$ and $B \subseteq A$. For any $X \subseteq U$ and a fixed $\beta \in [0, 0.5)$, the β-approximation of X can be described by*

$$\bigcup\{K \in U/B \mid \exists H \in X/B : \phi\} \text{ with} \\ \phi : \begin{cases} c(K,H) \leq \beta, & \text{for } \underline{X}_B^\beta \\ c(K,H) < 1-\beta, & \text{for } \overline{X}_B^\beta \\ \beta < c(K,H) < 1-\beta, & \text{for } \overline{\underline{X}}_B^\beta. \end{cases} \tag{10}$$

Proof. We have to compare classes $K \in U/B$ which have elements in $X \subseteq U$ and $H \in X/B$. Because of $X \subseteq U$ we obtain for $K \cap X \neq \emptyset$: $K \cap X = H$ and thus $c(K, X) = 1 - \frac{|K \cap X|}{|K|} = 1 - \frac{|H|}{|K|} = 1 - \frac{|K \cap H|}{|K|} = c(K, H)$. It follows: $c(K, X) = c(K, H) \leq \beta$ which is proposed by $\underline{X}_B^\beta$. Likewise, we can show $c(K, X) < 1 - \beta$ is equivalent to $c(K, H) < 1 - \beta$ which holds $\overline{X}_B^\beta$. From those two justifications we conclude $\overline{\underline{X}}_B^\beta$. □

Proposition 2. *Let $\langle U, A, D\rangle, B \subseteq A$ and $E \subseteq D$. For any fixed $\beta \in [0, 0.5)$, the β-regions $POS_B^\beta(E)$ and $BND_B^\beta(E)$ can be described by*

$$\bigcup\{K \in U/B \mid \exists H \in U/(B \cup E) : \phi\} \text{ with} \\ \phi : \begin{cases} c(K,H) \leq \beta, & \text{for } POS_B^\beta(E) \\ \beta < c(K,H) < 1-\beta, & \text{for } BND_B^\beta(E). \end{cases} \tag{11}$$

Proof. Using the equality $\{H \in X/B \mid X \in U/E\} = U/(B \cup E)$, we conclude Proposition 2 directly from Proposition 1. □

Theorem 1. *Let $T_{\langle A\rangle}$, $B \subseteq A, \beta \in [0, 0.5)$ and let the target concept $C_{\langle A\rangle}$ be a subset of T. We can compute the β-lower ($\mathcal{L}_B^\beta(T, C)$), β-upper ($\mathcal{U}_B^\beta(T, C)$) and β-boundary approximation ($\mathcal{B}_B^\beta(T, C)$) of C using the relational operations*

$$\pi^+_{c_t,b_1,\ldots,b_m}(\sigma_\phi(\mathcal{I}^B_{B\cup\{c_t\}}(T) \bowtie \mathcal{I}^B_{B\cup\{c_p\}}(C))) \text{ with} \\ \phi : \begin{cases} 1 - \frac{c_p}{c_t} \leq \beta, & \text{for } \underline{X}_B^\beta \\ 1 - \frac{c_p}{c_t} < 1-\beta, & \text{for } \overline{X}_B^\beta \\ \beta < 1 - \frac{c_p}{c_t} < 1-\beta, & \text{for } \overline{\underline{X}}_B^\beta. \end{cases} \tag{12}$$

Theorem 2. *Let $T_{\langle A,D\rangle}$, $B \subseteq A$, $E \subseteq D$ and $\beta \in [0, 0.5)$. The β-positive region ($\mathcal{L}_{B,E}^\beta(T)$) and β-boundary region ($\mathcal{B}_{B,E}^\beta(T)$) can be computed by*

$$\pi_{c_t,b_1,\ldots,b_m}(\sigma_\phi(\mathcal{I}^B_{B\cup\{c_t\}}(T) \bowtie \mathcal{I}^{B\cup E}_{B\cup\{c_p\}}(T))) \text{ with} \\ \phi : \begin{cases} 1 - \frac{c_p}{c_t} \leq \beta, & \text{for } POS_B^\beta(E) \\ \beta < 1 - \frac{c_p}{c_t} < 1-\beta, & \text{for } BND_B^\beta(E). \end{cases} \tag{13}$$

Theorem 3. *Our VPRS model based on extended relational algebra (see Theorems 1 and 2) can be computed in $\mathcal{O}(nm)$, where n is the number of tuples and m the number of attributes.*

Proof. The grouping ($\mathcal{G}$) and projection (π) can be implemented using hash aggregation, which requires nm time for either operation. Therefore the comparison ($\bowtie$) of both partitions utilizing the hash join results in $4nm$. At most, the selection (σ) requires a sequential scan followed by the final projection (π). Thus six subsequent scans need to be performed overall, which is $\mathcal{O}(nm)$. □

4.3 Feature Selection

One of the main challenges in machine learning and related subjects is the task of finding one or multiple subsets of all condition attributes with same or similar expressiveness compared to all conditions, i.e. FS. Particularly in RST, FS plays a central role, because it provides built-in concepts to address this subject (see Sect. 2.3). However, finding such subsets is not trivial and computational expensive [19]. The integration of FS algorithms with databases is most promising for this subject, because of the efficient operations and data structures provided.

Only few RST methods exist fully exploiting database operations for FS [12,13]. Despite their efficiency, they do not rely on the concept approximation and therefore have a crucial drawback, i.e. their limited applicability to inconsistent data. This fact leads to two key findings: These methods are less favorable for (i) mining in continuous environments and (ii) reduce FS quality in terms of rough sets. For (i) there are real-life scenarios where the removal of inconsistencies is an inpractical task or simply not an option. Particularly in dynamic or near real-time systems producing decision rules as data becomes available, FS is an essential prerequisite. In these scenarios, inconsistent entities are of great value and should be kept in the mining process. They constitute an exceptional source for the conflict resolution of potentially new concept descriptors, existing rules or the fusion of them. No less critical is finding (ii). We argue that imperfection in data can give rise to core attributes which would not be detectable in cleaned environments[1]. Thus we generally miss the chance to identify essential features by ignoring the nature of inconsistency. Denote, in RST a special treatment for ambiguous data is not required, because it features built-in capabilities to handle uncertainty by definition. So, inconsistent data can remain in place. Our model preserves these traditional bounds. Thus data movement or removal is not a requirement and proper reducts can be obtained. In what follows, we demonstrate the applicability of our model by presenting two FS approaches relying on attribute dependency as a heuristic function to maximize where inconsistencies are permitted. All intensive calculations are lifted by the database engine and huge data transports can be avoided. In order to comprehend the details, we figure out the concept of core and reduct in terms of database operations in advance.

The core is the intersection of all available reducts in a given dataset. Thus, it contains most relevant attributes to address classification problems, i.e. all

[1] Let $\langle U, A, D\rangle$ be inconsistent. Thus, we have $K, K' \in U/A : K \subseteq_\beta X$ and $K' \cap X \neq \emptyset$ with $K' \nsubseteq_\beta X, X \in U/D$. For indispensable attributes $a \in A$, we consider $U/(A\backslash\{a\})$ and may get $K^* \in U/(A\backslash\{a\})$ with $K^* = K \cup K'$ and $|POS^{\beta}_{A\backslash\{a\}}(D)| \neq |POS^{\beta}_{A}(D)|$. Hence a is in the β-core. This case is not covered when $\langle U, A, D\rangle$ is cleaned up-front.

indispensable attributes. These attributes can be computed in a straightforward fashion and therefore their involvement in the search for reducts is most valuable. According to (9) and in line with our proposed model (see Theorem 2), Corollary 1 determines indispensable attributes using relational expressions.

Corollary 1. *Let $T_{\langle A,D\rangle}, B \subseteq A, E \subseteq D$ and $\beta \in [0, 0.5)$. An attribute $a \in B$ is a β-core attribute w.r.t. to the classification of E if*

$$\mathcal{G}_{\{\mathtt{sum}(c_t)\}}(\mathcal{L}^{\beta}_{B\setminus\{a\},E}(T)) \neq \mathcal{G}_{\{\mathtt{sum}(c_t)\}}(\mathcal{L}^{\beta}_{B,E}(T)). \tag{14}$$

Likewise, from the reduct definition (see (7) and (8)) and Theorem 2, we can transfer the reduct properties to the relational domain in addition. Corollary 2 states the two required properties.

Corollary 2. *Let $T_{\langle A,D\rangle}, B \subseteq A, E \subseteq D, \beta \in [0, 0.5)$ and $R \subseteq B$. R is a β-reduct w.r.t. E if*

$$\mathcal{G}_{\{\mathtt{sum}(c_t)\}}(\mathcal{L}^{\beta}_{R,E}(T)) = \mathcal{G}_{\{\mathtt{sum}(c_t)\}}(\mathcal{L}^{\beta}_{B,E}(T)) \tag{15}$$

and $\forall a \in R$:

$$\mathcal{G}_{\{\mathtt{sum}(c_t)\}}(\mathcal{L}^{\beta}_{R\setminus\{a\},E}(T)) \neq \mathcal{G}_{\{\mathtt{sum}(c_t)\}}(\mathcal{L}^{\beta}_{R,E}(T)). \tag{16}$$

Denote, the final aggregations $\mathcal{G}$ in Corollaries 1 and 2 imply the cardinality of all tuples in the designated β-region. The next Corollary 3 provides insight to the runtime of both corollaries.

Corollary 3. *For a given data table with n tuples and m attributes, the costs for Corollary 1 are $2nm$ and two additional scans for the aggreations. Hence, we obtain $\mathcal{O}(nm)$. Consequently, the entire core computation is $\mathcal{O}(nm^2)$ inspecting all m attributes. Corollary 2 requires $4nm$ for (15) and $4nm^2$ for (16) under the strong assumption that a reduct consists of all condition attributes. Therefore checking an attribute set to be a valid β-reduct takes $\mathcal{O}(nm^2)$.*

In RST, one of the most prominent FS algorithm is QuickReduct introduced in [20]. This forward selection greedy hill climber starts off with an empty attribute set and constantly appends the attribute with highest significance. The algorithm ends if either a reduct is found or a local optimum is reached. For illustration purpose, we reimplemented QuickReduct using our relational model based on VPRS. The pseudo code is given in Algorithm 1 (QuickReductDB). Its runtime can be observed as follows: Let n be the number of tuples and m the feature size of a given data table. At most, the for-loop has $\frac{1}{2}m(m+1)$ iterations (see line 5 to 9) whereas the stop criterion (see line 11) is evaluated m times. Thus, using our model (see (13)), we get a worst-case runtime of $\mathcal{O}(nm^3)$. In practice, however, QuickReductDB performs much better. This examination is supported by experiments obtained from the original algorithm, where the average runtime appeared to be fairly linear [21].

Algorithm 1. QuickReductDB

Input: $T_{\langle A,D\rangle}, B \subseteq A, E \subseteq D, \beta \in [0, 0.5)$
Output: $R \subseteq B$
Procedure:

BEGIN
  $R \leftarrow \{\}$
  LOOP
    $S \leftarrow R$
    FOR $a \in B \setminus R$ **LOOP**
      IF $\mathcal{G}_{\{\mathtt{sum}(c_t)\}}(\mathcal{L}^{\beta}_{R\cup\{a\},E}(T)) > \mathcal{G}_{\{\mathtt{sum}(c_t)\}}(\mathcal{L}^{\beta}_{S,E}(T))$ **THEN**
        $S \leftarrow R \cup \{a\}$
      END IF
    END LOOP
    $R \leftarrow S$
    EXIT WHEN $\mathcal{G}_{\{\mathtt{sum}(c_t)\}}(\mathcal{L}^{\beta}_{R,E}(T)) = \mathcal{G}_{\{\mathtt{sum}(c_t)\}}(\mathcal{L}^{\beta}_{B,E}(T))$
  END LOOP
END

QuickReduct or our port QuickReductDB generally suffers from its best-first heuristic which may result in a local optimum. The next algorithm we present tries to overcome such situations and allows to extract multiple reducts. The basic idea of Algorithm 2 (BacktrackReductsDB) is to pursue a forward selection approach utilizing the same climbing engine as QuickReduct. In contrast, it does not start with an empty configuration, but uses the core as an ideal initialization. Additionally, it relaxes the strict best-first strategy by expanding its heuristic with all best solutions in the current round. This means, we constantly append one or multiple succeeding attribute subsets to a queue Q for further observation. This enables us to explore various alternative paths, which QuickReduct simply ignores and we overcome potential local optima by backtracking to previously identified branches with same validity than the current path seeked. As a result, we may extract several reducts R. Obviously, these can be valuable in a number of applications including co-training, ensemble learning or rule induction to name a few. However, the benefits of BacktrackReductsDB come at a price, i.e. complexity. The main steps consist of the core computation (see line 2), the reduct examination (see line 3) and the main loop (see line 7 to 20), where most actions are performed. Hence, the dominant factor concerning time is the number of iterations required. This number, though, is unknown in advance and highly dependent to the underlying data. Therefore we make a rather theoretical assessment towards the upper bound: Let us assume we constantly find $k > 1$ attribute subsets per iteration. This means we append k new candidates to Q per round while processing one only. Inquiring this statement, we obtain an overall runtime of $\mathcal{O}(nmk^m)$ for the forward selection including backtracking, where n is the number of tuples, m the feature size and constant k. One can verify this exponential upper bound is pessimistic especially when considering real-world datasets. In fact, experiments on various inputs including some from [22] and internal datasets indicate that the number of expected alternative paths per round is way below 2.0 on average. This exposes that alternatives are rarely explored compared to the amount of ordinary iterations with a single best path only. In turn, that leads to a non-exhaustive search in practice while collecting multiple reducts. The average costs of BacktrackReductsDB are rather comparable to the worst-case runtime of QuickReductDB.

Denote, both presented algorithms may incorporate with variable precision (i.e. $\beta > 0$) which is beneficial for certain domains of interest. However, its usage needs to be treated with caution, because the β-positive region is not necessarily monotonic.

Algorithm 2. BacktrackReductsDB

Input: $T_{\langle A,D\rangle}, B \subseteq A, E \subseteq D, \beta \in [0, 0.5)$
Output: $R \subseteq \mathcal{P}(B)$
Procedure:

BEGIN
  $R, Q \leftarrow \emptyset, C \leftarrow$ find core (according to (14))
  IF C is reduct (see (15) and (16))
    $R \leftarrow C$
    RETURN
  END IF
  LOOP
    $p_{max} \leftarrow \max(\mathcal{G}_{\{\mathtt{sum}(c_t)\}}(\mathcal{L}^{\beta}_{C\cup\{a\},E}(T)), \forall a \in B \setminus C)$
    FOR $a \in B \setminus C$ **LOOP**
      IF $\mathcal{G}_{\{\mathtt{sum}(c_t)\}}(\mathcal{L}^{\beta}_{C\cup\{a\},E}(T)) = p_{max}$ **THEN**
        IF $p_{max} < \mathcal{G}_{\{\mathtt{sum}(c_t)\}}(\mathcal{L}^{\beta}_{B,E}(T))$ **THEN**
          $Q \leftarrow Q \cup (C \cup \{a\})$
        ELSIF $C \cup \{a\}$ fulfills (16) **THEN**
          $R \leftarrow R \cup (C \cup \{a\})$
        END IF
      END IF
    END LOOP
    EXIT WHEN $Q = \emptyset$
    $C \leftarrow$ first element of Q, $Q \leftarrow Q \setminus C$
  END LOOP
END

4.4 Implementation Details

To highlight implementation details for our model, we use common SQL notation. Starting with the β-approximation for a single concept, let $T_{\langle A,D\rangle}$ be given by conditions A, decision $D = \{d\}$ and the considered target $C = \sigma_{d=1}(T)$. Using $B \subseteq A$ with $B = \{b_1, ..., b_m\}$ and $\beta \in [0, 0.5)$, Theorem 1 can be computed by the following SQL statement. The *WHERE* clause specifies the β-approximation.

```
SELECT T.* FROM (
        SELECT COUNT(*) AS ct, b1,...,bm FROM T
        GROUP BY b1,...,bm
) AS T JOIN (
        SELECT COUNT(*) AS cp, b1,...,bm FROM T WHERE d=1
        GROUP BY b1,...,bm
) AS C ON T.b1=C.b1 AND ... AND T.bm=C.bm
```

$$WHERE \begin{cases} 1 - CAST(cp\ AS\ DECIMAL)/ct <= \beta, & for\ \mathcal{L}_B^\beta(T, C) \\ 1 - CAST(cp\ AS\ DECIMAL)/ct < 1 - \beta, & for\ \mathcal{U}_B^\beta(T, C) \\ \beta < 1 - CAST(cp\ AS\ DECIMAL)/ct\ AND & \\ 1 - CAST(cp\ AS\ DECIMAL)/ct < 1 - \beta, & for\ \mathcal{B}_B^\beta(T, C) \end{cases}$$

For Theorem 2 we consider $T_{\langle A,D \rangle}, \beta \in [0, 0.5), B \subseteq A, E \subseteq D$ with $B = \{b_1, ..., b_m\}$ and $E = \{e_1, ..., e_p\}$. It can be implemented by the next SQL command. Note, the *WHERE* clause depends on whether to compute $\mathcal{L}_{B,E}^\beta(T)$ or $\mathcal{B}_{B,E}^\beta(T)$.

SELECT DISTINCT T. FROM (*
SELECT COUNT(*) AS ct, $b_1,...,b_m$ *FROM T*
GROUP BY $b_1,...,b_m$
) AS T JOIN (
SELECT COUNT(*) AS cp, $b_1,...,b_m$ *FROM T*
GROUP BY $b_1,...,b_m,\ e_1,...,e_p$
) AS C ON $T.b_1 = C.b_1$ *AND ... AND* $T.b_m = C.b_m$

$$WHERE \begin{cases} 1 - CAST(cp\ AS\ DECIMAL)/ct <= \beta, & for\ \mathcal{L}_{B,E}^\beta(T) \\ \beta < 1 - CAST(cp\ AS\ DECIMAL)/ct\ AND & \\ 1 - CAST(cp\ AS\ DECIMAL)/ct < 1 - \beta, & for\ \mathcal{B}_{B,E}^\beta(T) \end{cases}$$

Finally, the aggregation $\mathcal{G}_{\{\mathtt{sum}(c_t)\}}(Q)$ can be implemented by

SELECT SUM(ct) FROM (Q)

where Q is replaced by either one of the queries above.

5 Comparative Study

For comparison reasons, this evaluation concentrates on the time analysis of the positive region, a measure frequently used in FS. We compare our model (see Sect. 4) against two related approaches, i.e. RSDM [3] and RSMDS [13]. Additionally we state the responses of the classic RST framework RSES [5] and the R package "RoughSets" (RSR) [23] for reference.

All evaluated approaches except RSMDS provided an implementation of the positive region. Therefore the corresponding metrics of RSMDS were examined instead. For RSDM, we redeveloped its algorithm in two versions to verify the potentials: an in-database cursor implementation and a mixed variant consisting of a client program implemented in Java supplied with data of a conventional database query. The measures of RSES and RSR were based on latest software but processing relied on a single core only. The configuration[2] of our experiment was based on a standalone server environment, whereas six machine learning datasets ranging from 148 K to 11 M records with varying dimensions, data

[2] OS: Microsoft Windows 2012 R2 (Standard edition x64); DBs: Microsoft SQL Server 2014 (Developer edition 12.0.2, x64), Oracle 12c (Enterprise edition 12.1.0.2, x64); Misc: JDK 1.8.0.51, latest JDBC, R 3.2.0 (x64), RSESLib 3.0.4, RSR 1.3.0; Memory: 48 GByte; CPU: 32x2.6 GHz Intel Xeon E312xx (Sandy Bridge); HDD: 500 GByte.

types and distribution were chosen from [22]: HIGGS [24], KDD99m[3], PAMAP2 [25], Poker Hand [26], Covertype [27], NSLKDD [28]. For RSDM, RSMDS and our model, we ran our experiment inside the two established database systems SQL Server (MSSQL) and Oracle (ORA) providing us with parallelism up to 32 cores. Denote, all compared models and systems were tuned according to documentation and best knowledge to ensure optimal outcomes.

Using the best query plan and most optimal fetch size for the mixed variant of RSDM revealed very high network i/o waits degrading performance especially for HIGGS and PAMAP2. A similar weak performance was encountered for the in-database version. In neither case the in-database implementation could outperform the mixed variant of RSDM. On average the combination of Java and MSSQL was 5 times faster and the interaction of Java and ORA completed 3 times earlier than the corresponding in-database cursor. The direct comparison to our model turned out that on average our model using a very efficient hash-based query plan was 10 times faster on both database engines. Despite that huge difference in latency, RSDM generally benefits from coarse-grained datasets because its underlying query compresses the data already at the database end. Consequently fewer records are passed and processed at the application level. In our experiment, this effect took place on KDD99m where the amount of records could be reduced from 4 M to 1 M. Under these circumstances our model was only 4.6 times faster on ORA and 2.5 times faster utilizing MSSQL. The comparison to RSES and RSR showed that our model was at least 13 times faster benefitting from parallelism (see details in Fig. 1(a)). Furthermore, we analyzed the runtime of RSMDS using a similar hash-based plan to our model. Disregarding preprocessing, the partial terms to compute core and reduct attributes outperformed our model because both expressions require fewer operations than our implementation, i.e. the join operator. The core query rests upon one projection and an aggregation, while its reduct metric takes twice as much operations and time on average. On both databases, RSMDS was 30 % faster on average. When we considered preprocessing, which is generally required for RSMDS, our model performed much better. It was 26 % faster on HIGGS, PAMAP2 and KDD99m on both database engines. For Poker Hand, Covertype and NSLKDD it was only 10 % quicker on average. However, this superior is cushioned by the fact that preprocessing for RSMDS is only required once in a FS task. Further details about this particular comparison are stated in Fig. 1(b).

Performance-wise, this evaluation showed our VPRS implementation is able to compete with close related concepts and is faster than conventional approaches. In the selected environment, our model is more than 10 times faster compared to RSDM. On inconsistents dataset our model outruns RSMDS by 18 %, while both approaches show similar responses computing core attributes. In cleaned environments, however, RSMDS takes advantage of its elaborated queries, which is 30 % faster to find reducts and 60 % faster for core attributes.

[3] KDD99m is a modification of the original KDD99 dataset available in [22]. In contrast, it holds one additional attribute resulting in evenly sized equivalence classes.

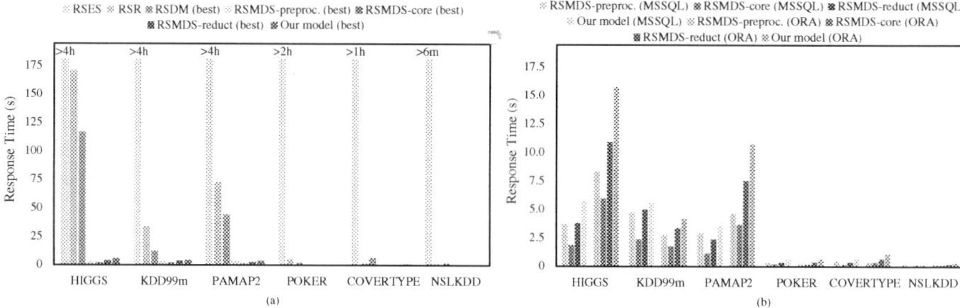

Fig. 1. The experimental results on six different datasets measuring the positive region and equivalent heuristics for FS. (a) shows an overview of the best measures obtained from each evaluated approach, where (b) compares the two best models in detail. (Color figure online)

6 Conclusion and Future Work

Although Rough Set Theory has proven to be well suited for a wide range of mining problems in the past, few implementations exist fully supporting its concept approximation for in-database analytics. In this work, we made an attempt to bring Variable Precision Rough Sets to the domain of databases exploiting the efficient and well-established algorithms provided by these systems. Furthermore, we demonstrated the applicability of our derived model for feature selection by introducing two algorithms utilizing efficient in-database processing. From a quantitative perspective, our experiments revealed that the proposed model is faster or comparable to existing approaches. Particularly in mining real-life scenarios, we outlined the model's qualitative advantages due to variable precision and its ability to cope with uncertainty. As part of our current research in the field of intrusion detection, the proposed model is central. We will use the presented algorithms and concept approximation to induce reliable attack signatures from incoming network traffic. These rules will be computed in near real-time and build a sound base to detect reoccuring cyberattacks in ambiguous environments. Another objective is to bring our idea to distributed architectures in order to obtain a scale-out rather than a scale-up approach. Additionally our model can be employed to develop new rough set algorithms, while existing approaches based on the concept approximation can be ported easily leveraging reliable in-database capabilities.

References

1. Nguyen, H.S.: Approximate boolean reasoning: foundations and applications in data mining. In: Peters, J.F., Skowron, A. (eds.) Transactions on Rough Sets V. LNCS, vol. 4100, pp. 334–506. Springer, Heidelberg (2006)

2. Shreya, P., Fard, A., Gupta, V., Martinez, J., LeFevre, J., Xu, V., Hsu, M., Roy, I.: Large-scale predictive analytics in vertica: fast data transfer, distributed model creation, and in-database prediction. In: Proceedings of the 2015 ACM SIGMOD International Conference on Management of Data, pp. 1657–1668 (2015)
3. Fernandez-Baizán, M.C., Menasalvas Ruiz, E., Peña Sánchez, J.M.: Integrating rdms and data mining capabilities using rough sets. In: Proceedings of the 6th International Conference on IPMU, pp. 1439–1445 (1996)
4. Ohrn, A., Komorowski, J.: ROSETTA - a rough set toolkit for analysis of data. In: Proceedings of the 3rd International Joint Conference on Information Sciences, pp. 403–407 (1997)
5. Bazan, J., Szczuka, M.S.: The rough set exploration system. In: Peters, J.F., Skowron, A. (eds.) Transactions on Rough Sets III. LNCS, vol. 3400, pp. 37–56. Springer, Heidelberg (2005)
6. Tileston, T.: Have your cake & eat it too! accelerate data mining combining SAS & teradata. In: Teradata Partners 2005 "Experience the Possibilities" (2005)
7. Hellerstein, J.M., Re, C., Schoppmann, F., Wang, D.Z., Fratkin, E., Gorajek, A., Ng, K.S., Welton, C., Feng, X., Li, K., Kumar, A.: The MADlib analytics library or MAD skills, the SQL. Proc. VLDB Endowment **5**(12), 1700–1711 (2012)
8. Beer, F., Bühler, U.: An in-database rough set toolkit. In: Proceedings of the LWA 2015 Workshops: KDML, FGWM, IR and FGDB, pp. 146–157 (2015)
9. Pawlak, Z.: Rough Sets. Int. J. Comput. Inf. Sci. **11**(5), 341–356 (1982)
10. Pawlak, Z.: Rough Sets - Theoretical Aspects of Reasoning about Data. Springer, Netherlands (1991)
11. Ziarko, W.: Variable precision rough set model. J. Comput. Syst. Sci. **46**(1), 39–59 (1993)
12. Kumar, A.: New techniques for data reduction in a database system for knowledge discovery applications. JIIS **10**(1), 31–48 (1998)
13. Hu, X.T., Lin, T.Y., Han, J.: A new rough sets model based on database systems. In: Wang, G., Liu, Q., Yao, Y., Skowron, A. (eds.) Rough Sets, Fuzzy Sets, Data Mining, and Granular Computing. LNCS, vol. 2639, pp. 114–121. Springer, Heidelberg (2003)
14. Vaithyanathan, K., Lin, T.Y.: High frequency rough set model based on database systems. In: Annual Meeting of the NAFIPS, pp. 1–6 (2008)
15. Sun, H.Q., Xiong, Z., Wang, Y.: Research on integrating ordbms and rough set theory. In: Tsumoto, S., Słowiński, R., Komorowski, J., Grzymała-Busse, J.W. (eds.) RSCTC 2004. LNCS (LNAI), vol. 3066, pp. 169–175. Springer, Heidelberg (2004)
16. Chan, C.-C.: Learning rules from very large databases using rough multisets. In: Peters, J.F., Skowron, A., Grzymała-Busse, J.W., Kostek, B., Swiniarski, R.W., Szczuka, M.S. (eds.) Transactions on Rough Sets I. LNCS, vol. 3100, pp. 59–77. Springer, Heidelberg (2004)
17. Naouali, S., Missaoui, R.: Flexible query answering in data cubes. In: Tjoa, A.M., Trujillo, J. (eds.) DaWaK 2005. LNCS, vol. 3589, pp. 221–232. Springer, Heidelberg (2005)
18. Slezak, D., Wroblewski, J., Eastwood, V., Synak, P.: Brighthouse: An Analytic Data Warehouse for Ad-hoc Queries. Proc. of the VLDB Endowment **1**, 1337–1345 (2008)
19. Skowron, A., Rauszer, C.: The discernibility matrices and functions in information systems. Intelligent Decision Support. Theory and Decision Library, pp. 331–362. Springer, Netherlands (1992)

20. Shen, Q., Chouchoulas, A.: A modular approach to generating fuzzy rules with reduced attributes for the monitoring of complex systems. Eng. Appl. Artif. Intell. **13**(3), 263–278 (2000)
21. Jensen, R., Shen, Q.: Computational Intelligence and Feature Selection: Rough and Fuzzy Approaches. Wiley, Hoboken (2008)
22. Bache, K., Lichman, M.: UCI Machine Learning Repository. University of California, Irvine, January 2016. http://archive.ics.uci.edu/ml
23. Riza, L.S., Januszb, A., Bergmeira, C., Cornelisa, C., Herreraa, F., ŚleZak, D., Benítez, J.: Implementing algorithms of rough set theory and fuzzy rough set theory in the R package roughsets. Inf. Sci. **287**, 68–89 (2014)
24. Baldi, P., Sadowski, P., Whiteson, D.: Searching for exotic particles in high-energy physics with deep learning. Nat. Commun. 5 (2014)
25. Reiss, A., Stricker, D.: Introducing a new benchmarked dataset for activity monitoring. In: Proceedings of the 16th ISWC, pp. 108–109 (2012)
26. Cattral, R., Oppacher, F., Deugo, D.: Evolutionary data mining with automatic rule generalization. In: Recent Advances in Computers, Computing and Communications, pp. 296–300 (2002)
27. Blackard, J.A., Dean, D.J.: Comparative accuracies of neural networks and discriminant analysis in predicting forest cover types from cartographic variables. In: Proceedings of the 2nd Sourthern Forestry GIS Conference, pp. 189–199 (1998)
28. NSL-KDD: Data Set for Network-based Intrusion Detection Systems, January 2016. http://nsl.cs.unb.ca/NSL-KDD

Computational Aspects of Data Aggregation and Complex Data Fusion

Linear Optimization for Ecological Indices Based on Aggregation Functions

Gleb Beliakov[1], Andrew Geschke[1], Simon James[1(✉)], and Dale Nimmo[2]

[1] Deakin University, Geelong, Australia
{gleb,sjames}@deakin.edu.au, geschke.andrew@gmail.com
[2] Charles Sturt University, Albury, Australia
dnimmo@csu.edu.au

Abstract. We consider an optimization problem in ecology where our objective is to maximize biodiversity with respect to different land-use allocations. As it turns out, the main problem can be framed as learning the weights of a weighted arithmetic mean where the objective is the geometric mean of its outputs. We propose methods for approximating solutions to this and similar problems, which are non-linear by nature, using linear and bilevel techniques.

Keywords: Aggregation functions · Linear programming · Weight learning · Ecology · Biodiversity

1 Introduction

We consider the problem of distributing a human population across a finite land area in such a way that negative impact to local flora and fauna is minimized. A simple version of the problem, optimization of abundance for a single species, is one that is easily solved with a linear programming approach, however ecologists are usually more interested in how the land-use allocations affect *biodiversity*. A number of quantitative indices exist for biodiversity, which incorporate both the number of species present (the richness) along with how evenly distributed the species are. As has been observed in [1], many of these can be expressed in terms of common aggregation functions. For instance, the geometric mean of species abundances is being increasingly used as a proxy for biodiversity [2], providing an average abundance that is more sensitive to smaller values (rare species). Whilst the optimization of these more complicated indices is non-linear in nature, we will show that close approximations can be achieved using mostly linear techniques and capitalizing on the ability to express quasi-arithmetic means in terms of generating functions. We illustrate the techniques using bird occupancy data from surveys conducted in Melbourne, Australia, and include details of our implementations as an appendix.

The article will be set out as follows: In Sect. 2, we outline our notation along with the necessary underlying concepts from the field of aggregation functions. In Sect. 3, we introduce the ecological context and go through the associated

J.P. Carvalho et al. (Eds.): IPMU 2016, Part II, CCIS 611, pp. 411–422, 2016.
DOI: 10.1007/978-3-319-40581-0_33

problems with our proposed optimization solutions. We then provide an example in Sect. 4, before concluding in Sect. 5.

2 Preliminaries

We will give an overview of the preliminary concepts of aggregation functions as relevant to the problem of land-use allocation. We consider an input dataset consisting of an $m \times n$ matrix where the entries x_{ij} denote the predicted abundances of the i-th species for the j-th land-use type. In practice, such abundances are measured by reporting rates calculated after conducting surveys. In our case, the n land-use types correspond with increasing densities of human population but these need not be numeric or even ordered. With respect to a total human population P and available land area A, the values w_j denote the percentage allocation to each land-type, so that $\sum_{j=1}^{n} w_j = 1$. These values will correspond with the weights of our aggregation functions.

Aggregation functions are employed in various contexts for summarizing data. Overviews of the important families, properties and definitions can be found in [3–6].

Definition 1. *An aggregation function $f : [a, b]^n \to [a, b]$ is a function monotone in each argument and satisfying the boundary conditions $f(a, \ldots, a) = a$ and $f(b, \ldots, b) = b$ (with $a < b$).*

Of particular interest to us is the weighted arithmetic mean, perhaps the most commonly employed aggregator across various contexts. It is expressed,

$$WAM(x_1, \ldots, x_n) = \sum_{j=1}^{n} w_j x_j. \tag{1}$$

In our case, for a given species i, the aggregated value $WAM(x_{i1}, \ldots, x_{in})$ denotes its abundance per unit of area.

Another aggregation function important in ecology is the geometric mean. For an input vector $\mathbf{x}$, the geometric mean is given by,

$$G(x_1, \ldots, x_n) = \left(\prod_{j=1}^{n} x_j \right)^{\frac{1}{n}}. \tag{2}$$

In ecology, the geometric mean of species abundance is often used to give a measure of abundance that is more sensitive to rare species. So if we have $\mathbf{s} = (s_1, s_2, \ldots, s_m)$ denoting the set of species abundances for each of the m species, $G(\mathbf{s})$ is a proxy measure for biodiversity. We note that $G(\mathbf{s}) \leq AM(\mathbf{s})$ where AM is the weighted arithmetic mean with equal weights, and that the values will be closer the more even the species abundances are.

The geometric mean can also be obtained as a special case of the quasi-arithmetic mean, which generalizes[1] the WAM. Specifically, we have

$$G(x_1, \ldots, x_n) = g^{-1}\left(\sum_{j=1}^{n} w_j g(x_j)\right), \tag{3}$$

where $g(t) = \ln t$ is the generating function, its inverse is $g^{-1}(t) = \exp(t)$ and in our context we have $w_j = 1/n$ for all j.

3 Finding Optimum Land-Use Allocations with Respect to Species Diversity

The process of urbanization is a major contributor to biodiversity loss [7], with the expansion of cities leading to habitat loss, climatic changes in temperature as well as other disruptions to local species dynamics. However while some species respond negatively to increases in human population density, other species (pigeons for example) can actually benefit. In planning for the development of cities and towns, two theories of conservation have arisen in ecology literature [8]: *land-sharing*, whereby the human population is spread as evenly as possible over a given area; and *land-sparing*, which fits the human population to small areas of high density so that the remaining area can be reserved to preserve flora and fauna.

The way individual species respond to changes in human population density can be considered in terms of response curves (see Fig. 1).

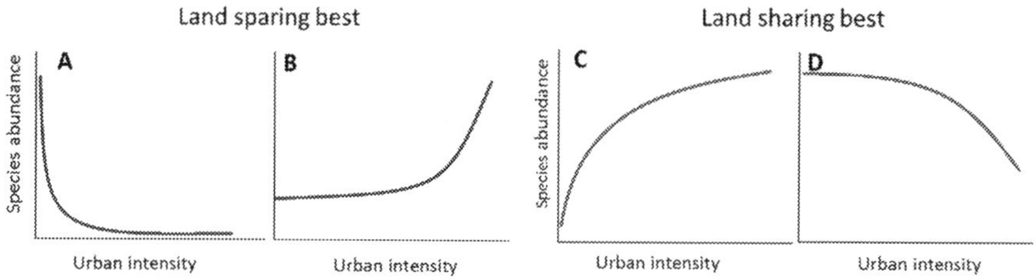

Fig. 1. Examples of response curves for species that benefit most from either a land-sparing or land-sharing approach to urban development. Species that respond to urban density according to curves like A and B are best suited to land-sparing, since it aims to segment a city into either very high or very low density living (where the abundances for these species are highest). On the other hand, response curves like C and D relate to species who would be better off with a land-sharing approach, since they have high abundance for mid-range urban intensity.

[1] More information about such generalizations can be found in any of [3–6], however we will restrict ourselves to the relevant cases to our problem.

We consider different ranges of population density to constitute a 'land-type', with each species having a predicted abundance. For example, if we considered 5 levels of density, for each species we would have data of the following form.

Land type (j)	1	2	3	4	5
Human population density (per 25ha)	0	50	100	250	500
Species abundance (per 25ha)	20	14	12	10	7

In the following subsections, we will present methods for finding the best allocation of each land-type subject to area and human population constraints where we are interested in optimizing either (1) the total abundance of all species, (2) the geometric mean of species abundances, and (3) the biodiversity as calculated using Shannon's diversity index. We focus on linear methods over more general approaches for two reasons. Firstly, although the dataset we use here is relatively small, both the number of species and the number of land-types can potentially be very large in practice and we want the method to be scalable. Secondly, we have a number of constraints that are more difficult to incorporate in more general optimisation models[2], such as the land area and population.

3.1 Optimization of Total Abundance

In previous works we have used linear optimization to learn the weights of various aggregation functions from data [9–14]. In those cases, we considered a set of input and output pairs with the aim of minimizing differences between observed and predicted outputs. Our aim here is to find the best allocation of land types. The percentage allocations which correspond with our aggregation weights w_j are our decision variables. We denote by d_j the population density for the j-th land-type. For any given species i, we have

$$\begin{aligned} \underset{\mathbf{w}}{\text{Maximize}} \quad & \sum_{j=1}^{n} w_j x_{ij}, \\ \text{s.t.} \quad & A\sum_{j=1}^{n} w_j d_j \geq P, \\ & \sum_{j=1}^{n} w_j = 1, w_j \geq 0, j = 1, \ldots, n. \end{aligned} \tag{4}$$

In order to maximize the total abundance, we note that we can simply sum the total abundances across all species for each land-type, so that the objective equation becomes,

[2] However since the constraints are linear, quadratic programming formulations would also be fine.

$$\underset{\mathbf{w}}{\text{Maximize}} \sum_{j=1}^{n} w_j \left(\sum_{i=1}^{m} x_{ij} \right). \tag{5}$$

3.2 Maximizing the Geometric Mean of Species Abundances

As discussed previously, we are often more interested in maximizing the geometric mean of abundances, which is more sensitive to rare species. This is so that the impression of abundance is not inflated by having a very common species. Our objective becomes,

$$\underset{\mathbf{w}}{\text{Maximize}} \prod_{j=1}^{m} \left(\sum_{j=1}^{n} w_j x_{ij} \right)^{\frac{1}{n}}. \tag{6}$$

We can ignore the $1/n$ power since the product and geometric mean will have the same maximum. This is still a non-linear objective, however we can use Eq. (3) and consider maximizing the sum of the logarithms of each species. In terms of the decision variables we have,

$$\underset{\mathbf{w}}{\text{Maximize}} \sum_{i=1}^{m} \ln \left(\sum_{j=1}^{n} w_j x_{ij} \right), \tag{7}$$

and whilst this representation remains non-linear, we can find an approximate solution to any desired precision by taking advantage of the fact that the log function is concave and hence can be expressed as the maximum value with respect to a set of bounding linear equations.

We transform the log function and write it as,

$$\ln t = \lim_{K \to \infty} \min \left(f_1(t),\, f_2(t),\, f_3(t), \ldots, f_K(t) \right),$$

where $f_k(t)$ denote the tangent lines of $\ln t$ across its domain, with $f_k(t) = \alpha_k t + \beta_k, \alpha_k = \frac{d}{dt}(\ln t_k) = 1/t_k, \beta_k = \ln t_k - \alpha_k t_k$ where t_k are the points at which log is evaluated. In other words, the logarithm is expressed in terms of the minimum of its K affine functions. Figure 2 helps demonstrate this visually.

Equation (7) hence becomes piecewise linear and the objective can be reduced to a linear program if the constraints are also linear. For each species i and each of our tangent functions given by $f_k(t) = \alpha_k t + \beta_k$, we introduce constraints of the form, $-\alpha_k s_i + y_i \leq \beta_k$, where s_i is the abundance of the i-th species, i.e.

$$-\alpha_k (w_1 x_{i1} + w_2 x_{i2} + \ldots + w_n x_{in}) + y_i \leq \beta_k.$$

The variables y_i now become decision variables in the optimization formulation. We optimize for the maximum sum of these values, however each y_i is bounded from above by the tangent lines described by the K constraints.

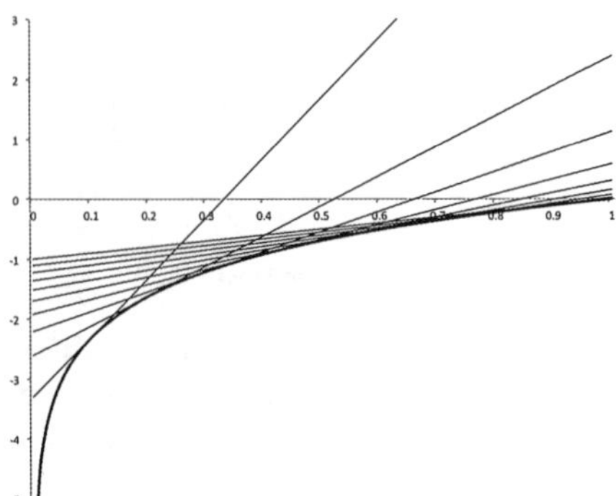

Fig. 2. Example of the natural log function being bound from above by 10 approximating affine functions, equispaced over the interval [0, 1].

Summarizing, we have the following linear programming formulation for this problem.

$$\underset{\mathbf{w},\mathbf{y}}{\text{Maximize}} \sum_{i=1}^{m} y_i$$

$$\text{s.t.} \quad A\sum_{j}^{n} w_j d_j \geq P.$$

$$-\alpha_k A(w_1 x_{i1} + w_2 x_{i2} + \ldots + w_n x_{in}) + y_i \leq \beta_k, k = 1, \ldots, K, i = 1, \ldots, m$$

$$\sum_{j=1}^{n} w_j = 1, \quad w_j \geq 0, j = 1, \ldots, n. \tag{8}$$

3.3 Maximization of Shannon's Diversity

Whilst the approach of the previous section could be adapted to the optimization of any convex function of abundance values that can be expressed as the sum of generating functions, there are a number of ecological indices that are not based on species abundances but rather on proportional abundance, i.e. the values p_i such that,

$$p_i = \frac{\sum\limits_{j=1}^{n} w_j x_{ij}}{\sum\limits_{i=1}^{m}\sum\limits_{j=1}^{n} w_j x_{ij}}.$$

Shannon's diversity index is one such example, expressed in terms of the p_i,

$$\sum_{i=1}^{m} -p_i \ln p_i. \tag{9}$$

For a given number of species (m) present in a given community, it reaches a maximum $\ln m$ when all species have equal abundance, and approaches zero if a single species dominates, i.e. if $p_i = 1$ for any i.

The use of p_i values makes it impossible to express this in terms of linear multiples of the constraints as we did previously, however we can take a different approach that results in a bi-level optimization problem. As we will see, it remains feasible for finding practical solutions with real datasets.

We introduce a variable M which indicates the total abundance, i.e.

$$M = \sum_{i=1}^{m} w_j \left(\sum_{j=1}^{n} x_{ij} \right).$$

With M known, we can therefore use this to scale our variables so that they are equivalent to proportions. We then have the capacity to solve the optimization, *provided* we know M. As before, we create affine functions from the curve, $-t \ln t$ and maximize such that the y_i values are bounded by these lines. In this case, accuracy bounds pose less of a problem since we know that all p_i are less than 1. The constraints will now be of the form,

$$-\alpha_k (w_1 \frac{x_{i1}}{M} + w_2 \frac{x_{i2}}{M} + \ldots + w_n \frac{x_{in}}{M}) + y_i \leq \beta_k.$$

We then can find the M that gives the best result for Shannon's diversity, which we implement as a bilevel problem. We have,

$$\begin{aligned}
&\text{Maximize}_{M} \; Z \\
&Z = \max_{\mathbf{w},\mathbf{y}} \sum_{i=1}^{m} y_i \\
&\text{s.t.} \; A \sum_{j}^{n} w_j d_j \geq P \\
-\frac{\alpha_k A}{M}&(w_1 x_{i1} + w_2 x_{i2} + \ldots + w_n x_{in}) + y_i \leq \beta_k, k = 1, \ldots, K, i = 1, \ldots, m \\
&\sum_{j=1}^{n} w_j = 1, \quad w_j \geq 0, j = 1, \ldots, n. \qquad (10)
\end{aligned}$$

4 Example: Bird Surveys Data

Survey data reporting presence or absence of bird species across 28 landscapes in the wider Melbourne area was collected over a period of four months (May to August) in 2015. All landscapes were one hectare in area and the human population densities were determined from census data. The report rates for each species were the result of four separate observation rounds. An example of the fitted response curves (showing the probability of a species occurring in a land-type with that human density) across the 28 sites for three bird species are shown in Fig. 3.

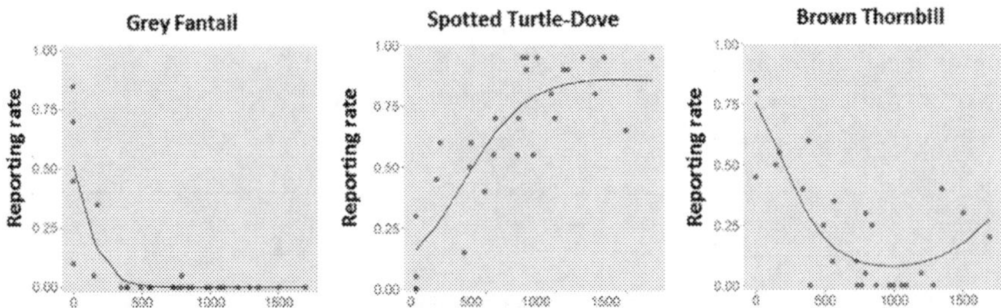

Fig. 3. Data collected for three bird species at sites with increasing human population density. While the *grey fantail* virtually only lives in reserves (i.e. all vegetation and no housing), the *spotted turtle dove* benefits from high urban population density and the *brown thornbill* was present across the range.

The dataset we will use to illustrate the methods proposed here includes that relating to 21 native species, with response rates calculated at densities from 0 to 1600 at intervals of 100 ($n = 17$ land types). We consider allocating the optimum allocation of a human population of 2.744 million, i.e. the current population of the Melbourne residential area (outside the central business district). The area under consideration is 964 km^2.

Table 1. Summary results from applying the methods for optimizing total abundance, the geometric mean, and Shannon's diversity index respectively.

Objective	Densities					
	w_1	w_2	w_3	Total	Geometric	Shannon's
	0	1000	1600	Abundance	Mean	Diversity
Abundance	0.5551	0	0.4449	17445	549.4	2.7425
Geometric mean	0.5210	0.0911	0.3879	17068	564.3	2.7443
Shannon's diversity	0.5247	0.0813	0.3941	17108	564.0	2.7443

* weights and Shannon's diversity rounded to 4 dp, geometric mean rounded to 1 dp

The results are shown in Table 1. These do not vary greatly based on the objective used, however we do note slight changes. Obviously all three measures will be somewhat correlated, with each essentially capturing some overall measure of how many individuals are present. We have only displayed three weights because in all models the remaining densities were all given zero allocation, regardless of the method. In terms of the ecological interpretation, we see that the *land-sparing* approach to biodiversity conservation is preferred overall for this particular set of species, allocating areas of high population density as well as reserves for wildlife, with only small amounts of land at medium population density (Table 2).

Table 2. Summary results from applying the methods for optimizing total abundance, the geometric mean, and Shannon's diversity index respectively with a smaller subset of the data (only 4 species).

Objective	Densities							
	w_1	w_2	w_3	w_4	w_5	Total	Geom.	Shan.
	0	600	800	1000	1600	Abund.	Mean	Div.
Abundance	0.1102	0	0.8898	0	0	5736	990.0	1.1106
Geometric mean	0.3582	0	0	0.5251	0.1167	5534	1079.6	1.2156
Shannon's diversity	0.3499	0.3284	0	0	0.3217	5244	1054.3	1.2289

* weights and Shannon's diversity rounded to 4 dp, geometric mean rounded to 1 dp

To help give some insight into the difference between these approaches, we have also optimized for a smaller set of species. We used the four species with response curves shown in Fig. 4. We note that in this case, we have four very different response curves including quite common species, e.g. the *Australian magpie*, and rare species such as the *eastern rosella*.

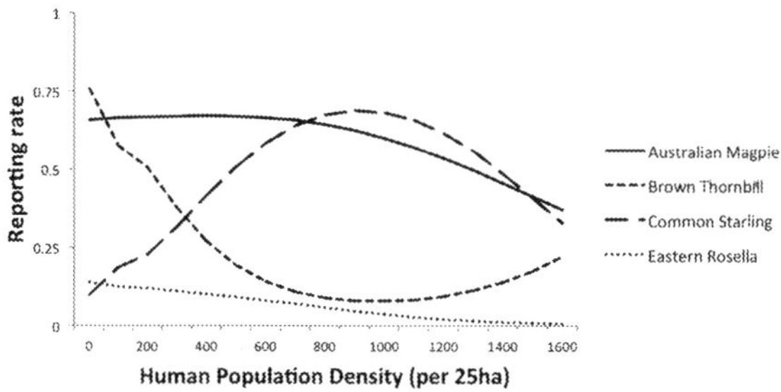

Fig. 4. Four species and their response curves. The *Australian magpie* is present at high levels across all population densities while the *brown thornbill* and the *eastern rosella* both generally decrease as the human population becomes more and more dense.

In Fig. 5 we can observe how each of the species change individually with respect to the different optimization objectives. When the overall abundance is maximized, low abundance in the *brown thornbill's* population is compensated for by high abundance with the *Australian magpie* and *common starling*[3].

[3] As a side note the *common starling* is an introduced species and was not included in the previous example of 21 native bird species.

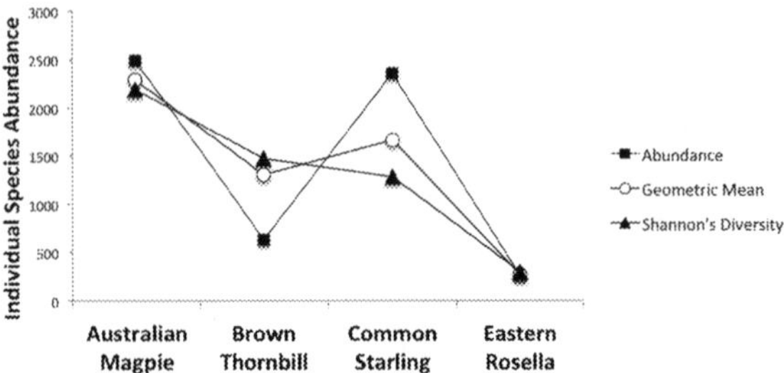

Fig. 5. Comparison of individual species' abundances obtained with each of the optimization objectives.

Both the geometric mean and Shannon's diversity approaches obtain a more evenly spaced distribution for these three species, however we note that in all cases we are not able to raise the abundance of the *eastern rosella*.

5 Conclusion

We have focused on an optimization problem that arises in ecology where land management decisions can have an impact on the local biodiversity. In this context, the optimization problems take on a similar form to what we have in learning parameters for aggregation functions, however with objective functions whose arguments are expressible in terms of weighted arithmetic means of our data. We have shown that these non-linear objectives can be approximated with linear techniques, the advantages of which are that they are quickly solvable, are guaranteed to reach a global optimum and are scalable in terms of time and computation complexity. In our main problem involving the geometric mean of species abundances, we represented its additive generating function as the maximum of bounding affine functions. We have made all R algorithms available at our website.

Appendix: Implementations

We have implemented all three approaches to optimization as functions in an R library available at our website[4].

Optimizing total abundance: `eco.opti()`.

Description of inputs

`species.data` - matrix of species abundances per unit of land area, i.e. with x_{ij} denoting the i-th species and its abundance for land type j;

[4] http://aggregationfunctions.wordpress.com.

`densities` - vector of densities per unit of land area, i.e. with d_j denoting the human population density for a given land-type j;
`tot.pop` - the total population required to be fit into the given land area;
`tot.land` - the total area over which we need to distribute the population.

Additional optional constraints

`w.min / w.max` - vectors denoting minimum or maximum bounds on the land-types, e.g. if we want to ensure that at least 20 % of the land is populated at minimum density we incorporate the constraint $w_1 \geq 0.2$ (assuming w_1 is the land-type with minimum density), or alternatively we may wish to limit high density housing to at most 40 % of the land area etc.;
`spec.min / spec.max` - vector placing minimum or maximum bounds for a particular species, for example, if we want to make sure that a rare species is above a given threshold γ_i, the linear constraint $\sum_{j=1}^{n} w_j x_{ij} \geq \gamma_i$ is added for that species.

The function also gives as output a number of ecological indices such as the individual species abundances, and the Simpson and Shannon diversity indices.

Maximizing the geometric mean of species abundances: `eco.opti.gm()`

Description of inputs

In addition to all inputs and constraints used with `eco.opti()`, this function has two additional optional parameters to control the precision.

`fprec` - a positive integer giving the number of tangent functions to be defined. The default setting is 100 linear segments, and so gains in accuracy can be achieved with settings of 500, 1000 etc., however obviously at the cost of computation time;
`max.x` - a real number giving the maximum value for the domain over which the tangent functions are calculated, the default setting is 10000, and so depending on the scale given it could be necessary to increase this value (or decrease it for finer accuracy) or the optimization will be the same as it would be for maximizing abundance.

For the number of tangent lines K, optimizing over 5 species with 100 linear segments will require $5 \times 100 = 500$ additional constraints, use of 1000 linear segments will require 5000 additional constraints and so on. We need to be careful when reducing the precision, since the log function's gradient changes more drastically for values closer to zero than it does for large values. Rather than taking equal step sizes in calculating our tangent lines, they were distributed using $t_k = \exp(-k \cdot \max(s)/K)$ where $\max(s)$ is the `max.x` parameter above.

We note also that by default the setting for `spec.min` will be 1 for all species. It could be adjusted to a fractional value if desired.

Maximizing Shannon's diversity index: `eco.opti.sh()`

This function uses the same inputs as with the previous two. The program first solves a maximum and minimum problem using `eco.opti()` in order to find the feasible bounds to search for M. Note that $-t\ln t$ is concave for $t \in (0,1]$.

Another biodiversity index used as an objective and included in the code made available online is Simpson's diversity index $1/(\sum_{i=1}^{m} p_i^2)$. This is performed in a similar manner, however now we are minimizing for a convex function $y = t^2$ rather than maximizing for a concave function and so we need to make the appropriate changes when using the linear framework above.

References

1. Tuomisto, H.: An updated consumer's guide to evenness and related indices. Oikos **121**, 1203–1218 (2012)
2. Kelly, L.T., Bennett, A.F., Clarke, M.F., McCarthy, M.A.: Optimal fire histories for biodiversity conservation. Conserv. Biol. **29**, 473–481 (2015)
3. Beliakov, G., Bustince, H., Calvo, T.: A Practical Guide to Averaging Functions. Springer, New York (2015)
4. Beliakov, G., Pradera, A., Calvo, T.: Aggregation Functions: A Guide for Practitioners. Springer, Heidelberg (2007)
5. Grabisch, M., Marichal, J.-L., Mesiar, R., Pap, E.: Aggregation Functions. Cambridge University Press, Cambridge (2009)
6. Torra, Y., Narukawa, V.: Modeling Decisions. Information Fusion and Aggregation Operators. Springer, Heidelberg (2007)
7. Grimm, N.B., Faeth, S.H., Golubiewski, N.E., Redman, C.L., Wu, J., Bai, X., Briggs, J.M.: Global change and the ecology of cities. Science **319**, 756–760 (2008)
8. Lin, B.B., Fuller, R.A.: FORUM: sharing or sparing? how should we grow the world's cities? J. Appl. Ecol. **50**, 1161–1168 (2013)
9. Beliakov, G., James, S.: Citation-based journal ranks: the use of fuzzy measures. Fuzzy Sets Syst. **167**, 101–119 (2011)
10. Beliakov, G., James, S.: Using linear programming for weights identification of generalized Bonferroni means in R. In: Torra, V., Narukawa, Y., López, B., Villaret, M. (eds.) MDAI 2012. LNCS, vol. 7647, pp. 35–44. Springer, Heidelberg (2012)
11. Beliakov, G., James, S., Gómez, D., Rodríguez, J.T., Montero, J.: Learning stable weights for data of varying dimension. In: Proceedings of the 8th International Summer School on Aggregation Operators, University of Silesia, Katowice, Poland (2015)
12. Beliakov, G., James, S., Gómez, D., Rodríguez, J.T., Montero, J.: Approaches to learning strictly-stable weights for data with missing values. Fuzzy Sets and Systems (2016, submitted)
13. Beliakov, G., James, S., Nimmo, D.: Learning aggregation weights from 3-tuple comparison sets. In: Proceedings of IFSA 2014, Edmonton, Canada. pp. 1–6 (2013)
14. Beliakov, G., James, S., Nimmo, D.: Using aggregation functions to model human judgements of species diversity. Inf. Sci. **306**, 21–33 (2015)

A Qualitative Approach to Set Achievable Goals During the Design Phase of Complex Systems

Diadie Sow(✉), Abdelhak Imoussaten, Pierre Couturier, and Jacky Montmain

Centre de Recherche LGI2P/Ecole des mines d'Alès, Site EERIE, Parc scientifique G. Besse, 30035 Nîmes cedex 1, France
Diadie.Sow@mines-ales.fr

Abstract. The problem addressed in this paper is "how to set ambitious targets when improving or designing a product while these targets remain within the reach of the manufacturer". Thus, improvements to be focused on are those which both have a significant positive impact on product performance and correspond to operational changes properly under control by the manufacturer. While some approaches in the literature have already addressed each of the two issues of the improvement problem, few deal with both of them at the same time. In this paper we investigate a qualitative approach that conciliates both points of view as an optimization problem. The notion of interaction between any two objectives to be simultaneously satisfied is central in our framework. An illustrative example related to the design phase of autonomous robot is provided.

Keywords: Multi criteria decision support · Criteria interaction · Conceptual design · Performance management · Qualitative reasoning

1 Introduction

Competition in industry is becoming increasingly intense; therefore, in order to satisfy fluctuating demand and customers' increasing expectations, deal with the competition and remain or become market leader; industries must focus on searching for sustainable advantages. The survival of a company is heavily dependent on its capacity to identify new customer requirements and develop new products [1]. However, dealing with new products or improving existing ones in today's technology–driven market presents significant risks. Many constraints must be taken into account into the design or improvement phases [2, 3]. The number of components, functions, and interactions within complex products/systems is becoming so substantial that responding to the requirements of customers, and moreover at lower cost is anything but obvious [4]. To remain successful, manufacturers must adapt to an unprecedented rate of change in their processes and practices. Such a challenge requires large forecasting capacities. Industrials must be able to produce challenging but achievable goals. The corresponding optimization problem to be solved for the industrial manufacturer is how to design new products or improve products according to customers' requirements at the

J.P. Carvalho et al. (Eds.): IPMU 2016, Part II, CCIS 611, pp. 423–434, 2016.
DOI: 10.1007/978-3-319-40581-0_34

limit of what is technically feasible as he is aware of his available enterprise-level skills.

Defining achievable targets is a matter of situation awareness to relevantly manage the balance between strategic ambition and manufacturing realism. Thus, improvements to be made as a priority, i.e. targets that must be raised considerably, are those which allow both significant positive impacts on product/system performance but also correspond to actions that are derived from the expertise of the manufacturer. Industrials are thus supposed to relevantly make a prediction about the positive impact one improvement or another one might have on the global performance of the product/system but simultaneously be aware of the actions they can actually undertake with regard to their skill [5]. Forecasting the performances of a new product is far from obvious: it may be confusing to associate preferential interactions between goals and behavioral influences actions may have on these goals. This is further complicated by the qualitative characteristic of the knowledge used to do this forecasting: impacts of changes in the system/product configuration upon the expected performances cannot generally be properly quantified. Defining achievable targets is thus a risky and not deterministic process. The two aspects of the problem that this paper addresses when, improving a product or a system are:

- Assessing which performances should be improved first to maximize customer expectations [6];
- Identifying the configurations of the system that should best fulfill these expected goals [7].
- Sequential management of these two steps may lead to non-convergent situations: expected goals do not necessarily match feasible ones. Identifying achievable goals necessitates a conjoint management of these two steps.
- Moreover, the problem is all the more complicated because knowledge regarding actions-and-goals relationships is mostly imprecise, especially in the preliminary stage of product design [8]. Two challenges are hereafter considered when improving a complex system: at the strategic level, which changes in system outputs would bring actual improvements that would best fulfill customers' expectations, and at the operational level, how system configuration adjustments have to be carried out in order to achieve these targets. This paper is based on this semantic distinction to organize approaches in the literature and finally propose a unified framework.

These two problems have been extensively studied in the literature. Nevertheless few works integrate both of them. The first point of view focuses on defining strategic targets without actual feasibility considerations [6, 9], which may lead to focus on unachievable goals. Conversly, the second point of view focuses on the capacity to achieve specific goals while not taking into account the commitment to do it [10–12], which may lead to focus on meaningless goals for the strategic development of the industrial manufacturer. Some attempts seem nevertheless to be aware of the necessity to deal with both aspects when improving and designing systems [13, 14]. The MAUT-like formalism in [13] is attractive, but hides the difficulty of establishing the behavioral model. In [15], we proposed to identify the coalition of criteria to be improved first using the index value defined in [6] while being the least difficult to be reached; both identifications are merged in a unique optimization problem, unlike to [7] where they are sequentially processed. This sequencing might lead to divergent cases where strategic targets never met feasible ones. The model in [15] avoids this misleading process, but is limited to deal with quantitative knowledge. Yet the difficulty is still compounded by the only imprecise or incomplete data and knowledge that are generally available in the conceptual phase of design. In this paper, we propose to reformulate the problem defined in [15] in a qualitative framework. Adequate qualitative operators as Sugeno integral and median operator will replace quantitative ones: Choquet integral and weighted average. Section 2 poses the problematic and the unified approach with the related optimization problem and proposes redefining the two sub problems in a qualitative framework. Section 3 proposes an illustration with a case study and discusses the results. Finally Sect. 4 gives some conclusions and perspectives.

2 Modeling Policy Aspirations and Capacity to Act

2.1 Problem Characterization and Notations

In order to study complex system, we should characterize it by a set of parameters. For example, an autonomous robot can be characterized by several parameters: type of rolling base, size, engine type, embedded energy capacity etc. Each configuration of these parameters is an instantiation of a robot. To check if there is an improvement when changing the configuration, the decision maker has to consider the objectives that have to be achieved by the robot in terms of autonomy, reliability, mobility, cost etc. The possible configurations perform different outputs that more or less satisfy the objectives. In a general context, let a system be characterized by its parameters $\gamma_1, \gamma_2, \ldots \gamma_p$. Let Γ be the set of all possible values of the vector $(\gamma_1, \gamma_2, \ldots \gamma_p)$. A system is then defined by a configuration $\gamma \in \Gamma$. Improving a system is to make it evolve from a configuration $\gamma \in \Gamma$ to a configuration $\gamma' \in \Gamma$ which gives a better satisfaction of the objectives that have been fixed for the system under cost constraints (money, risk, time etc.). Let us denote $N = \{1, 2, ..n\}$, the set of criteria. The system performance is evaluated by its elementary performance measures $p_1, p_2, \ldots p_n$. An action a_j is associated with each change of parameter γ_j . The set of actions is denoted by A. As there are operational constraints over the set of actions, some of them cannot be performed together: they are said to be mutually exclusive. We define an action plan denoted *ap* as a subset of non-exclusive actions in A. We are searching among the available action plans *ap*, the ones that better improve the customer's satisfaction and that are compliant with the capacity of the manufacturer to apply them. The greater the number of individual performances to be improved, the more difficult it is to find the proper action plan.

Optimistically, one can seek for an action plan that improves all objectives associated to the system. But, it can be more relevant for the decision maker to improve a subset of criteria I of N that leads to reasonable satisfaction degree when the improvement is written-down to its achievability. For $I \subseteq N$, let $S(I)$ denote the degree to which improving criteria of I seems achievable considering the available actions; and

$W(I)$ the expected degree of satisfaction the improvement on criteria in I should provide. In [16], we proposed an optimization problem (1) in a multi-criteria aggregation framework to identify the set of criteria I^* that maximizes the expected degree of satisfaction under achievability constraints:

$$\begin{cases} \max_{I \subseteq \mathrm{N}} W(I) \\ \textit{subject to}\ W(I) \leq S(I) \end{cases} \quad (1)$$

The problem can be seen as a multi-criteria decision making problem subject to feasibility. However, feasibility has little to do with explicit operational constraints. *S* (*I*) rather assesses how much confidence there is that the actions the manufacturer intends to commit will give satisfaction. The idea is to make comparable the degree of confidence in achieving one goal and the expected degree of satisfaction this goal would provide.

In [15], *S*(*I*) and *W*(*I*) were defined in a quantitative setting. In this paper we propose to redefine these quantities in a qualitative setting. Indeed, the system's response to configuration changes and the decision-maker expectations are generally only imprecisely and incompletely known during the conceptual design phase. Expected benefits and achievability of targets are assessed through adequate operators in this qualitative framework.

2.2 Qualitative Characterization of the Capacity to Achieve a Subset of Criteria

The qualitative approach involved in this section has been originally proposed in [7, 16] and is summarized in the following. It models the impacts that actions have on the system's performances. It covers three sub problems:

- how to characterize the impact of one parameter change on a given performance?
- how to merge the impact of several parameters' changes on a given performance?
- how to assess the overall impact a configuration may have on a set of criteria?

Let $S_i(a_j)$ denote the degree to which the action a_j may support performance p_i. As soon as $S_i(a_j) > 0$, the action a_j contributes to the satisfaction of the criterion i. Let $D_i(a_j)$ be the degree to which the action a_j may distract the performance p_i. As soon as $D_i(a_j) > 0$, the action a_j harms the criterion i. The qualitative action-performance relationship has then to be extended to action plans. The major difficulty is that for a given action plan *ap* and a criterion i, several actions in *ap* may affect p_i positively and several other actions in *ap* may affect p_i negatively. What is then the resulting effect of *ap* on the performance p_i? To answer this question, standard multi-criteria approaches cannot be applied since both positive and negative impacts on performance level are considered here.

Indeed, the estimation of the merged impact of an action plan naturally depends on the system behavior, as well as on the designer's/operator's decisional behavior: a pessimistic attitude (whereby a risk aversion position will focus attention on the lowest positive merged impacts of the action plan) vs. an optimistic attitude (whereby risk acceptance will focus attention on the most highly positive merged impacts). In both cases, we suppose that the merged negative impacts cannot be under rated. For example, from an optimistic point of view, the merged effect can be defined as (Considering $D_i(ap) = \min_{a\in A_i^D} D_i(a)$, should be too permissive.): $S_i(ap) = \max_{a\in A_i^S} S_i(a)$ and $D_i(ap) = \max_{a\in A_i^D(ap)} D_i(a)$ While in a pessimistic approach: $S_i(ap) = \min_{a\in A_i^S(ap)} S_i(a)$ and $D_i(ap) = \max_{a\in A_i^D(ap)} D_i(a)$

Where $A_i^S(ap) = \{\text{a} \in \text{ap} : S_i(a) > 0\}$ and $A_i^D(ap) = \{\text{a} \in \text{ap} : D_i(a) > 0\}$.

Then, the degree to which an action plan may affect a subset of criteria is yet to be defined. Let $I \subseteq N$ be a subset of criteria and *ap* an action plan. The resulting degree to which *ap* should contribute to the improvement of I while not deteriorating the other criteria in $N\backslash I$ can be assessed, from a pessimistic point of view by: $s_I(ap) = \begin{cases} \min_{i\in I} S_i(ap)\, if\ \forall j \in N\backslash I\left[S_j(ap) > D_j(ap)\, or\, A_j^D(ap) = \emptyset\right] \\ 0\, otherwise \end{cases}$, (respectively $\begin{cases} \max_{i\in I} S_i(ap)\, if\ \forall j \in N\backslash I\left[S_j(ap) > D_j(ap)\, or\, A_j^D(ap) = \emptyset\right] \\ 0\, otherwise \end{cases}$ from an optimistic point of view).

More precisely $S_I(ap) > 0$ if any criterion in I is improved by *ap*, whereas no criterion in $N\backslash I$ is distracted[1].

Finally, $\max_{ap} S_I(ap)$ characterizes the highest degree to which criteria in I can be expected to be improved considering all possible action plans. Hence, the quantity $s(I) = \max_{ap} S_I(ap)$ is considered here as the assessment of the capacity to achieve improvement of I.

2.3 Qualitative Characterization of the Commitment to Achieve a Subset of Criteria

In this subsection, we are interested in determining the criteria that should be improved first to satisfy as much as possible the decision-maker expectations. For this purpose, we build a value function w_p defined on the set of all the criteria subsets I of N, and that estimates the expected overall satisfaction the improvement of criteria in I would provide, knowing that the initial performances are given by the vector of performances $p^0 = \left(p_1^0, p_2^0, \ldots, p_n^0\right)$. In our qualitative multi-criteria framework, the overall satisfaction related to any vector of partial performances $(p_1, p_2, \ldots, p_n)$ is modeled as the qualitative aggregation of the p_i's performance on the ith criterion by the Sugeno integral. This operator both allows modeling the importance of criteria but also the preferential interactions among them. Let us start by recalling some notations concerning the Sugeno integral.

[1] In fact we believe that only the pessimistic point of view is a reasonable attitude to compute $s_I(ap)$ since a conjunctive aggregation is necessary in order to guarantee that all criteria of I are expected to be improved.

Let L be an ordered qualitative scale with 1_L its highest value and 0_L its lowest value. A function μ: $2^N \rightarrow L$ is called a fuzzy measure if it satisfies the following conditions:

1. $\mu(\emptyset) = 0_L$
2. $\mu(N) = 1_L$
3. μ is monotonic non decreasing for inclusion, i.e., for any $A, B \subseteq N$ $A \subseteq B \Rightarrow \mu(A) \leq \mu(B)$.

The monotonicity of μ means that the weight of a subset of criteria cannot decrease when new criteria are added to it (see [9, 10] for further details).

Let μ be a fuzzy measure on N taking values in the scale L. Let consider a function f: $N \rightarrow L$, then the Sugeno integral of f with respect to μ, denoted $S_\mu(f)$ is given by:

$$S_\mu(f) = \max_{1 \leq i \leq n} \left(\min(f_{\sigma(i)}, \mu(A_{\sigma(i)})\right),$$

where σ is a permutation on N such that $f_{\sigma(1)} \leq f_{\sigma(2)} \leq \cdots f_{\sigma(n)}$ and $A_{\sigma(i)} = \{\sigma(i), \sigma(i+1), \ldots \sigma(n)\}$.

This integral is monotone, and obviously presents a compromise behavior. It searches the importance exceeding a certain level, and then performs a compromise deal between the selected values. Another point of view is to see this combination as the disjunction of conjunctions. All the criteria, their interactions and their importance, are taken into account to assess the aggregated score of a given vector of elementary ratings [17–19].

Let us consider that the overall performance of the system is evaluated from criteria performances using a Sugeno integral operator with respect to the fuzzy measure $\mu : S_\mu(p) = \max_{1 \leq i \leq n}\left(\min(p_{\sigma(i)}, \mu(A_{\sigma(i)})\right)$ where p_i is the ith performance with a value in the same ordinal scale $L = \{0_L, \ldots, 1_L\}$ as the capacity μ. By considering the set of criteria I and the initial vector of performances p^0, several vectors of performances p^k are possible improvements on I. If $p^0 = (p_1^0, p_2^0, \ldots, p_n^0)$, then any $p^k = (p_1^k, p_2^k, \ldots, p_n^k)$ such that $\forall i \in I, p_i^k > p_i^0$ is a possible improvement. Let us note $Improvt_{p^0}$ the set of all these vectors such that p^k is obtained from p^{k-1} by increasing each p_i^{k-1} one step up (steps are the levels of L) for all the criteria in I. In other words, similarly to [6], only improvements obtained by progressing from p^0 to 1_I along the diagonal of the $|I|$-cube are considered (where $|I|$ denotes the cardinal of I) in order to be sure to improve all criteria in I. To each expected improvement p^k is associated its overall performance $S_\mu(p^k)$. Then, the expected overall performance in this qualitative framework of performance can be assessed by the median of the $S_\mu(p^k)'s$. The median provides how worth it is improving criteria on I (the median is an associative qualitative compensatory aggregation operator [21]):

$$w_{p^0}(I) = med\left(S_\mu(p^0), S_\mu(p^1), \ldots S_\mu(p^{max})\right), \text{where } p^{max} = \left(1_I, p^0_{N \backslash I}\right).$$

Note that the cardinal number of $Improvt_{p^0}$ may be odd or even. Since in the last case the median consists of finding a value between the two middle points, such a choice depends on the decision maker's attitude. A pessimistic attitude would take first middle point as the median, while an optimistic one would take the second point.

2.4 Optimization Problem in the Qualitative Setting

Note that by construction, $I \subseteq I' \Rightarrow s_I(ap) \geq s_{I'}(ap), \forall ap$, and therefore $s(I) \geq s(I')$. In the other hand, $I \subseteq I' \Rightarrow w_{p^0}(I) \leq w_{p^0}(I')$ for a given initial performance vector p^0. Hence these two functions are respectively non-increasing and non-decreasing with respect to I (with respect to the inclusion relationship), and so coincide at an optimal given I^* solution of the optimization problem (1). Hence, the problem of searching the subset of criteria that are both profitable and achievable is equivalent to the optimization problem (1) with s(I) and $w_{p^0}(I)$ as arguments. However precautions are necessary to properly process the inequality constraint of problem (1).

In the case where these functions take their values in different ordered qualitative scales, a third ordered qualitative scale can be introduced where values have the semantic of satisfaction degree or of possibility degree in order to draw the comparison of the inequality in (1).

Solving this problem may be extremely hard with $2^{|AP|}$ action plans and $2^{|N|}$ subsets of criteria to search. The first thorny exponential problem comes from the computation of $s(I)$: heuristics have already been introduced in a branch and bound algorithm in [7] to compute $s(I)$ in a reasonable time. Further heuristics are needed to solve (1) while computing the minimum of $s(I)'$s values; they will be addressed in our future works. We give here just an overview of two heuristics. The first heuristic will be based on the inverse monotonicity of $s(I)$ and $w_p(I)\left(I \subseteq I' \Rightarrow s(I) \geq s(I') \text{ and}, I \subseteq I' \Rightarrow w_{p^0}(I) \leq w_{p^0}(I')\right)$; a second heuristic will introduce lower and upper bounds for $s(I)$. In the following study case $|N|$ and $|AP|$ are small enough to perform exhaustively all the required computations.

3 Design Application

The following example has no validation ambition but aims at illustrating the proposed approach. The robotic challenge Robafis is organized annually by the French association of Systems Engineering AFIS to promote Systems Engineering practice in engineers' schools[2]. The scope of the challenge is for instance (Robafis_2013) to build an autonomous mobile robot able to compete with other robots and using some provided and imposed materials. Each robot is limited to a 0.3^3m^3 cube and has to achieve the following mission as quickly as possible: to grasp and transport some various colored spheres between several stock devices spread over a plan playground. Some dark lines are drawn on the ground to guide the robot between stock devices. The autonomous robot was broken down into four sub-systems: a gripper device, sensors equipment, a rolling base, a control device. The programmable control device type is

[2] (http://www.robafis.fr/RobAFIS/Bienvenue.html).

Table 1. Solution principles

Gripper device	Rolling base	Sensors equipment
G1: Fork (taking the sphere from below)	R1: Four wheels rolling base	S1: two color sensors for following the dark lines, one light sensor for recognizing the sphere color
G2: Lateral gripper (pinching laterally the sphere)	R2: Two wheel drive and one free wheel rolling base	S2: one color sensor for following the dark line, one color sensor for recognizing the sphere color
G3: Grapnel (taking over the sphere)	R3: rolling base with tracks	

also imposed to the competitors. The physical alternatives (the possible robot configurations) depend on the design options and on the skills of each competitor. The preference model depends on the strategy followed by each competitor. The solution principles are resumed in Table 1. There are three principles for designing the gripper device, three for the rolling base and two for the sensors equipment. Thus, there are 18 actions plans to be compared, corresponding to the $3 \times 3 \times 2$ admissible configurations.

In this example an action is the choice of one element of the configuration, and an action plan is a configuration.

3.1 Data of the Application

Four criteria are considered to decide between configurations: the sparsity of used components (Cr1), the robot speed capacity (Cr2), the reliability of the robot (Cr3), and the maintainability of the robot (Cr4). Taking into account the competition rules, Table 2 provides an example of competitors' preferences between the criteria modeled by a qualitative capacity function $\mu : 2^N \rightarrow L$ where N = {1,2,3,4}, and $\mathrm{L} = \{0, \mathrm{a}, \mathrm{b}, \mathrm{c}, \mathrm{d}, \mathrm{e} = 1_L\}$; 0 means: not important at all and e means very important. Table 3 presents, for each of the four criteria, the positive impact (denoted by "+") or the negative impact (denoted by "−") of each elementary technical choice on the satisfaction level of each criteria. Furthermore a confidence degree in the fulfillment of such an impact is defined on the same qualitative scale L.

It must be emphasized here that the approach differs from just aggregating sub-criteria of multiple options and choosing the best one since actions may have positive or negative impacts on the criteria (it is a bipolar problem) and that Table 3

Table 2. Competitors' preferences: the qualitative capacity function values $\mu(I)$

I	$\mu(I)$	I	$\mu(I)$	I	$\mu(I)$	I	$\mu(I)$
∅	0	{C4}	a	{C1,C3}	b	{C1,C3,C4}	d
{C1}	a	{C1,C4}	b	(C2,C3)	d	{C2,C3,C4}	d
{C2}	a	{C2, C4}	c	(C1,C2,C3)	e	{C1,C2,C3,C4}	e
{C1,C2}	b	{C1,C2,C4}	d	{C3}	b	{C3, C4}	b

Table 3. Impacts of configuration

	Number of pieces	Speed	Reliability	Maintainability
	Cr1	Cr2	Cr3	Cr4
G1: fork	+e	−a	−b	+e
G2: gripper	−b	−a	+d	−a
G3: graspel	−d	−a	+c	−d
R1: 4wheels	+e	−c	−d	+c
R2: 3 wheels	−c	+d	+c	−b
R3: trackers	−d	+c	−c	−d
S1: 2 + 1	−c	+e	+c	−c
S2: 1 + 1	+b	−d	−c	+c

results from an analysis of a behavior model of the robot and describes the confidence the designers have in the impacts the actions may have on each partial performance.

3.2 Computation of w_{p^i}

Let's consider some given initial performance vectors: $p^0 = (0\,0\,0\,0)$, $p^{0_1} = (0\,a\,0\,a)$, $p^{0_2} = (0\,d\,b\;\;c)$. The Sugeno integral for each of these initial performance vectors are respectively: $S_\mu(p^0) = 0, S_\mu(p^{0_1}) = a, S_\mu(p^{0_2}) = d$ with respect to the capacity defined in Table 2. Hence, the values of the $w_{p^{0_i}}$ for each initial performance vectors are given in the Table 4:

Table 4. $w_{p^{0_i}}(I)$ values for different initial performances p^{0_i}

w_{p^i}/I	{1}	{1,2}	{1,3}	{1,4}	{1,2,3}	{1,2,4}	{1,3,4}	{1,2,3,4}
$w_{p^0}(I)$	a	b	b	b	c	c	c	c
$w_{p^{0_1}}(I)$	a	b	b	b	c	c	c	c
$w_{p^{0_2}}(I)$	c	c	d	c	d	c	d	d
w_{p^i}/I	{2}	{2,3}	{2,4}	{2,3,4}	{3}	{3,4}	{4}	
$w_{p^0}(I)$	a	c	c	c	b	b	a	
$w_{p^{0_1}}(I)$	a	c	c	c	b	b	a	
$w_{p^{0_2}}(I)$	c	d	c	d	d	d	c	

It can be easily checked in Table 4 that $I \subseteq I' \Rightarrow w_{p^{0_i}}(I) \le w_{p^{0_i}}(I')$ and $w_{p^{0_i}}$ depends on the initial performance: the higher the initial global performance value $S_\mu(p^{0_i})$, the higher $w_{p^{0_i}}(I)$. As the configuration to be selected is the first design choice, we assume that the initial performance vector is the null vector (p^0). Then, with such an initial vector, the Sugeno integral properties, the median operator and the ordinal scale L imply that w_{p^0} cannot exceed the value c.

3.3 Computation of *s*(*I*)

The values of the $s(I)'s$ for each subset of criteria I are computed with an optimistic attitude (see Sect. 2.2) and are given in Table 5:

Table 5. *s*(*I*) values

s / I	{1}	{2}	{1,2}	{3}	{1,3}	{2,3}	{1,2,3}	{4}
s(*I*)	e	e	e	d	d	d	d	e
s / I	{1,4}	{2,4}	{1,2,4}	{3,4}	{1,3,4}	{2,3,4}	{1,2,3,4}	
s(*I*)	e	e	e	c	c	c	c	

Considering the initial performance p^0, *s*(*I*) takes its values on $\{c,d,e\}$, The only valid coalitions I^* that satisfy the inequality constraint of (1) are the coalition $\{2,4\}$ and $\{1,2,4\}$ which happen to maximize, in this particular case both $s(I)$ and $w_{p_1^0}(I)$. Thus, according to the competitors' preference model defined in Table 4, the best strategy to follow is to build a robot with high speed and easily maintainable. Considering Table 3, the three configurations that improve I^* and do not degrade the performance on N/I^* are: a1 = $\{$G1,R1,S2$\}$, a2 = $\{$G1,R2,S2$\}$ and a3 = $\{$G1,R3,S2$\}$. The three rolling bases are acceptable at this stage of the design. A sensibility analysis has to be undergone in order to consolidate this configurations choice. Such analysis should be included in a wider risk analysis that should be the scope of further research work.

4 Conclusion

Dealing with new products or improving existing ones in today's technology-driven markets present important risks. In the case of complex industrial systems, the number of constraints and goals rapidly becomes inextricable. Manufacturers are faced with an unprecedented rate of changes in their processes and practices. This challenge requires large forecasting capacities to produce challenging but achievable goals. In [15], we have proposed a model in a quantitative framework of performance to better manage the balance between strategic ambition and manufacturing realism. Nevertheless, the forecasts are yet compounded by the only imprecise or incomplete knowledge generally available in the conceptual phase of design. In this paper, we have proposed to reformulate the problem defined in [15] in a qualitative framework. This qualitative approach allows lifting the constraint encountered in [15]. It allows qualitative reasoning with a gradient like approach, which seems more consistent with the available knowledge during design phases. The case study of a design autonomous robot shows the necessity of considering simultaneously know-how and ambition in design project to set achievable goals. Identifying achievable goals necessitates a conjoint management of feasibility and ambition in their definition contrarily to [7] where they were sequentially managed. In future works, we will propose another point of view to this conjoint control issue. The idea is to integrate the capacity *s*(*I*) in the computation of w_I in such a way that improvements are all the less credible than they require high level of know-how whereas improvements are all considered equally probable in the models of [7, 15].

References

1. Shen, X.X., Tan, K.C., Xie, M.: An integrated approach to innovative product development using Kano's model and QFD. Eur. J. Innov. Manage. **3**(2), 91–99 (2000)
2. Baykasoglu, A., Oztas, A., Ozbay, E.: Prediction and multi-objective optimization of high strength concrete parameters via soft computing approaches. Expert Syst. Appl. **36**, 6145–6155 (2002)
3. Ng, K.W.: A critical analysis of current engineering design methodologies from a decision making perspective. In: Pham, D.T., Eldukhri, E.E., Soroka, A.J. (eds.) Intelligent Production Machines and Systems. Elsevier Science Ltd., Oxford (2006)
4. Moulianitis, V.C., Aspragathos, N.A., Dentsoras, A.J.: A model for concept evaluation in design—an application to mechatronics design of robot grippers. Mechatronics **14**(6), 599–622 (2004)
5. Hauser, J.R., Clausing, D.: The house of quality. Harv. Bus. Rev. **66**(3), 63–73 (1988)
6. Labreuche, C.: Determination of the Criteria to be Improved First in Order to Improve as Much as Possible the Overall Evaluation. IPMU, Perugia (2004)
7. Montmain, J., Labreuche, C., Imoussaten, A., Trousset, F.: Multi-criteria improvement of complex systems. Inf. Sci. **291**, 61–84 (2015)
8. Antonsson, E.K., Otto, K.N.: Imprecision in engineering design. ASME J. Mech. Des. **117B**, 25–32 (1995)
9. Keeney, R., Raiffa, H.: Decisions with Multiple Objectives: Preferences and Value Tradeoffs. Wiley, New York (1976)
10. Giorgini, P., Mylopoulos, J., Nicchiarelli, E., Sebastiani, R.: Reasoning with goal models. In: Spaccapietra, S., March, S.T., Kambayashi, Y. (eds.) ER 2002. LNCS, vol. 2503, pp. 167–181. Springer, Heidelberg (2002)
11. Felix, R.: Relationships between goals in multiple attribute decision making. Fuzzy Sets Syst. **67**, 47–52 (1994)
12. Montmain, J., Clivillé, V., Berrah, L., Mauris, G.: Preference and causal fuzzy models for manager's decision aiding in industrial performance improvement. In: FUZZ-IEEE 2010, Barcelona, Spain (2010)
13. Clivillé, V., Corrente, S., Greco, S., Rizzon, B.: Feasible Optimization. In: MCDA 2015, Annecy (2015). http://www.polytech.univsavoie.fr/fileadmin/polytech_autres_sites/sites/mcda2015/doc/session/MCDA81_session5_1.pdf
14. Fleurey, F., Dehlen, V., Bencomo, N., Morin, B., Jézéquel, J.-M.: Modeling and validating dynamic adaptation. In: Chaudron, M.R. (ed.) MODELS 2008. LNCS, vol. 5421, pp. 97–108. Springer, Heidelberg (2009)
15. Sow, D., Imoussaten, A., Couturier, P., Montmain, J.: Between Policy Aspirations and Capacity to Act: How Designing Achievable Goals. LFA, Poitiers (2015)
16. Imoussaten, A.: Modélisation et pilotage de la phase de délibération dans une décision collective-vers le management d'activités à risques, thèse de Doctorat de l'Ecole des mines de Paris (2011)
17. De Campos, L., Lamata, M., Moral, S.: A unified approach to define fuzzy integrals. Fuzzy Sets Syst. **39**(1), 75–90 (1991)

18. Dubois, D., Marichal, J.-L., Prade, H., Roubens, M., Sabbadin, R.: The use of the discrete sugeno integral in decision-making: a survey. Int. J. Uncertain. Fuzziness and Knowl. Based Syst. **9**(5), 539–561 (2001)
19. Grabisch, M.: The symmetric sugeno integral. Fuzzy Sets Syst. **139**(3), 473–490 (2003)
20. Dubois, D., Grabish, M., Prade, H., Smets, P.: Using the transferable belief model and a qualitative possibility theory approach on an illustrative example: the assessment of the value of a candidate. Int. J. Intell. Syst. **16**, 1245–1272 (2001)
21. Arif, M.: AWFO (un opérateur d'agrégation) pour la reconnaissance des formes. In: 15e congrès francophone de Reconnaissance des Formes et Intelligence Artificielle (2006)
22. Mouelhi, O., Couturier, P., Redarce, T.: A hybrid search algorithm for multi-criteria optimization and evaluation in mechatronic products design. Advanced intelligent Mechatronics, Singapore (2009)

Unbalanced OWA Operators for Atanassov Intuitionistic Fuzzy Sets

Laura De Miguel[1,2(✉)], Edurne Barrenechea[1,2], Miguel Pagola[1,2], Aranzazu Jurio[1,2], Jose Sanz[1,2], Mikel Elkano[1,2], and Humberto Bustince[1,2]

[1] Departamento de Automatica y Computacion, Universidad Publica de Navarra, Campus Arrosadia s/n, 31006 Pamplona, Spain
{laura.demiguel,edurne.barrenechea,miguel.pagola,aranzazu.jurio, joseantonio.sanz,mikel.elkano,bustince}@unavarra.es

[2] Institute of Smart Cities, Universidad Publica de Navarra, Campus Arrosadia s/n, 31006 Pamplona, Spain

Abstract. In this work we introduce a new class of OWA operators for Atanassov intuitionistic fuzzy sets which distinguishes between the weights for the membership degree and the weights for the nonmembership degree; we call these operators Unbalanced Atanassov Intuitionistic OWA operators. We also study under which conditions these operators are aggregation functions with respect to the Atanassov intuitionistic admissible linear orders. Finally, we apply these aggregation functions in an illustrative example of a decision making problem.

Keywords: Atanassov Intuitionistic Fuzzy Set · OWA operators · Unbalanced OWA operators

1 Introduction

Aggregation functions have shown to be a useful tool in problems where information should be fused. Although a partial order is used in some generalizations of aggregation function on other sets (see, for example [1]), some particular classes of these functions such as OWA operators and Choquet or Sugeno integrals require all the elements being comparable. Consequently a linear order is needed. However, these orders are not trivially generated in the extensions of fuzzy sets where more than one value is used to define the membership degree. This is the case, for instance of Interval-Valued Fuzzy Sets (IVFSs) [2] or Atanassov Intuitionistic Fuzzy Sets (AIFSs) [3].

Although some constructions of linear orders on AIFSs have already been studied [4], more works generalizing different notions using linear orders are indispensable for its use in applications. In particular, we aim to define on AIFSs a new class of OWA operators which may apply different weight vectors for the membership and nonmembership degree. We denote these operators Unbalanced Atanassov Intuitionistic OWA operators (UAIOWAs). Taking into account that OWA operators are a particular class of aggregation functions frequently used

J.P. Carvalho et al. (Eds.): IPMU 2016, Part II, CCIS 611, pp. 435–444, 2016.
DOI: 10.1007/978-3-319-40581-0_35

in applications, our final goal is to study when $UAIOWAs$ satisfy the properties demanded to the aggregation functions. Finally, we introduce an illustrative example on a decision making problem where the Unbalanced Atanassov intuitionistic OWA operators are a suitable option to solve the problem.

The structure of the work is as follows: In Sect. 2 we introduce some well-known concepts which are necessary for the development of this work. The notion of Unbalanced Atanassov intuitionistic OWA operators is introduced in Sect. 3 where we study when these operators are aggregation functions. Section 4 shows an example where Unbalanced Atanassov intuitionistic OWA operators are applied to a decision making problem. We close the study with some conclusions and open problems for future research.

2 Preliminaries

This section is devoted to briefly introduce several well-known basic concepts and to fix the notation used in this work. We first recall the notion of aggregation function on a poset which becomes crucial in the development of this work. For more information see [1,5].

Definition 1. *Given a poset $(P, \preceq)$ with bottom and top, 0_P and 1_P respectively, an aggregation function M on P with respect to the order $\preceq$ is a mapping $M : P^n \to P$ satisfying:*

- $M(0_P, \ldots, 0_P) = 0_P, \quad M(1_P, \ldots, 1_P) = 1_P,$
- $M(x_1, \ldots, x_n) \preceq M(y_1, \ldots, y_n)$ *whenever* $(x_1, \ldots, x_n) \preceq (y_1, \ldots, y_n),$

where $(x_1, \ldots, x_n) \preceq (y_1, \ldots, y_n)$ if and only if $x_i \preceq y_i$ for all $i \in \{1, \ldots, n\}$.

A particular instance of aggregation functions frequently used in many applications are OWA operators given by Yager [6].

Definition 2. [6] *Let w be a weight vector, i.e., $w = (w_1, \ldots, w_n) \in [0,1]^n$ with $w_1 + \ldots + w_n = 1$. The Ordered Weighted Averaging operator associated with w, OWA_w, is a mapping $OWA_w : [0,1]^n \longrightarrow [0,1]$ defined by*

$$OWA_w(x_1, \ldots, x_n) = \sum_{i=1}^{n} w_i x_{(i)},$$

where $x_{(i)}$, $i = 1, \ldots, n$, denotes the $i-th$ greatest component of the input $(x_1, \ldots, x_n)$.

In this work, we focus on Atanassov intuitionistic fuzzy sets which were presented in 1986 by Atanassov.

Definition 3. [3] *An Atanassov intuitionistic fuzzy set A over the universe $X \neq \emptyset$ is defined as*

$$A = \{(x, \mu_A(x), \nu_A(x)) \mid x \in X\},$$

where $\mu_A(x), \nu_A(x) \in [0,1]$ *are respectively, the membership and nonmembership degree of the element* x *to* A *and they satisfy* $\mu_A(x) + \nu_A(x) \leq 1$.

We call $(\mu_A(x), \nu_A(x))$ *Atanassov Intuitionistic Fuzzy pair (AIF-pair) and we denote by* $\mathcal{L}([0,1])$ *the set of all possible AIF-pairs, i.e.*

$$\mathcal{L}([0,1]) = \{(\mu, \nu) \mid \mu, \nu \in [0,1] \text{ and } \mu + \nu \leq 1\}.$$

For the sake of simplicity, when the Atanassov intuitionistic fuzzy set and the element of the referential could not be misunderstood we denote the AIF-pair (μ, ν).

In [3] a partial order on AIF-pairs is introduced. This order is certainly enough for defining some aggregation functions but for a suitable definition of OWA operators on these sets a linear order is required. In this way, some recent studies define and construct admissible orders for the different generalizations of fuzzy sets [7,8].

In the following we introduce a construction method of an Atanassov intuitionistic admissible order [4], namely, a linear order which refines the partial order introduced in [3] by Atanassov. That is, a linear order that satisfies that for all $(\mu_1, \nu_1), (\mu_2, \nu_2) \in \mathcal{L}([0,1])$ such that $\mu_1 \leq \mu_2$ and $\nu_1 \geq \nu_2$ then $(\mu_1, \nu_1) \leq (\mu_2, \nu_2)$.

Proposition 1. *Let* M_1, M_2 *be two aggregation functions* $M_1, M_2 : [0,1]^2 \to [0,1]$ *such that for all* $(\mu_1, \nu_1), (\mu_2, \nu_2) \in \mathcal{L}([0,1])$ *the equalities* $M_1(\mu_1, 1-\nu_1) = M_1(\mu_2, 1-\nu_2)$ *and* $M_2(\mu_1, 1-\nu_1) = M_2(\mu_2, 1-\nu_2)$ *hold simultaneously if and only if* $\mu_1 = \mu_2$ *and* $\nu_1 = \nu_2$.

The relation $\leq_{M_1,M_2}$ *on* $\mathcal{L}([0,1])$ *given by* $(\mu_1, \nu_1) \leq_{M_1,M_2} (\mu_2, \nu_2)$ *if and only if*

(i) $M_1(\mu_1, 1-\nu_1) < M_1(\mu_2, 1-\nu_2)$ *or*
(ii) $M_1(\mu_1, 1-\nu_1) = M_1(\mu_2, 1-\nu_2)$ *and* $M_2(\mu_1, 1-\nu_1) \leq M_2(\mu_2, 1-\nu_2)$

is an admissible order on $\mathcal{L}([0,1])$.

Notice that taking

- $M_1(\mu, 1-\nu) = \mu$ and $M_2(\mu, 1-\nu) = 1-\nu$ we recover the intuitionistic lexicographic 1 order on $\mathcal{L}([0,1])$. We denote it by $\leq_{ilex1}$ and it is given by:

$$(\mu_1, \nu_1) \leq_{ilex1} (\mu_2, \nu_2) \text{ if and only if } \mu_1 < \mu_2 \text{ or } (\mu_1 = \mu_2 \text{ and } \nu_1 \geq \mu_2). \quad (1)$$

- $M_1(\mu, 1-\nu) = 1-\nu$ and $M_2(\mu, 1-\nu) = \mu$ we recover the intuitionistic lexicographic 2 order on $\mathcal{L}([0,1])$. We denote it by $\leq_{ilex2}$ and it is given by:

$$(\mu_1, \nu_1) \leq_{ilex2} (\mu_2, \nu_2) \text{ if and only if } \nu_1 > \nu_2 \text{ or } (\nu_1 = \nu_2 \text{ and } \mu_1 \leq \mu_2). \quad (2)$$

3 Unbalanced Atanassov Intuitionistic OWA Operators

In the literature we can find several constructions of OWA operators on the intuitionistic field. In the following, we introduce the first construction of OWA operators on AIFSs considering the partial order introduced by Atanassov on [9]. It is worth mentioning we do not use the original notation on [9] but we rewrite the method following the notation introduced in Sect. 2.

Definition 4. *The OWA aggregation of intuitionistic fuzzy set associated with $\tilde{w}$ a weight vector ($\tilde{w} = (w_1, \dots, w_n)$ in $[0,1]^n$) such that $w_1 + \dots + w_n = 1$ is given by*

$$UAIOWA_{[\tilde{w},\tilde{v},\leq]}((\mu_1,\nu_1),\dots,(\mu_n,\nu_n)) = \left(\sum_{i=1}^{n} w_i\mu_{(i)}, \sum_{i=1}^{n} w_{n-i+1}\nu_{(i)}\right), \quad (3)$$

where $\mu_{(n)} \leq \dots \leq \mu_{(1)}$ and $\nu_{(n)} \leq \dots \leq \nu_{(1)}$.

More recently, some other works about similar aggregation functions but using linear orders are presented. For instance, in [10] where Xu and Yager order is presented, some geometric operators are defined. These operators are IFWG, IFOWG and IFHG. However, in all these operators the same weight vector is considered for both membership and nonmembership degree. In this way, the novelty of the concept of Unbalanced Atanassov intuitionistic OWA operators lies on the use of two different weight vectors $\tilde{w}$ and $\tilde{v}$.

Definition 5. *An Unbalanced Atanassov Intuitionistic OWA(UAIOWA) operator associated with an admissible order $\leq$ on $\mathcal{L}([0,1])$ and $\tilde{w}, \tilde{v}$ weight vectors ($\tilde{w} = (w_1, \dots, w_n), \tilde{v} = (v_1, \dots, v_n)$ in $[0,1]^n$ such that $w_1 + \dots + w_n = 1$ and $v_1 + \dots + v_n = 1$) is a mapping $UAIOWA_{[\tilde{w},\tilde{v},\leq]} : (\mathcal{L}([0,1]))^n \longrightarrow [0,1]^2$ given by*

$$UAIOWA_{[\tilde{w},\tilde{v},\leq]}((\mu_1,\nu_1),\dots,(\mu_n,\nu_n)) = \left(\sum_{i=1}^{n} w_i\mu_{(i)}, \sum_{i=1}^{n} v_i\nu_{(i)}\right), \quad (4)$$

where $(\mu_{(n)},\nu_{(n)}) \leq \dots \leq (\mu_{(1)},\nu_{(1)})$.

OWA operators on fuzzy sets are particular instances of aggregation functions. In the following we study under which conditions $UAIOWA$ operators are also particular examples of these functions. Since the boundary conditions $UAIOWA_{[\tilde{w},\tilde{v},\leq]}((1,0),\dots,(1,0)) = (1,0)$ and $UAIOWA_{[\tilde{w},\tilde{v},\leq]}((0,1),\dots,(0,1)) = (0,1)$ are trivially satisfied, we only need to study the monotonicity with respect to the considered order $\leq$ and when they are well defined, i.e., the codomain is $\mathcal{L}([0,1])$. That is, we have to study when the image of n AIF-pairs satisfies

$$\sum_{i=1}^{n} w_i\mu(i) + \sum_{i=1}^{n} v_i\nu_{(i)} \leq 1.$$

Proposition 2. *Let $\leq$ be the order $\leq_{ilex1}$ or $\leq_{ilex2}$ on $\mathcal{L}([0,1])$ (generated as in Eq. (1) or (2), respectively) and $\tilde{w}, \tilde{v} \in (0,1]^n$. Then $UAIOWA_{[\tilde{w},\tilde{v},\leq]}$ operator satisfies monotonicity.*

Proof. Straight by the monotonicity of the OWA operators when the space considered is the unit interval.

Notice that in Proposition 2, $w_i, v_i \neq 0$ for all $i = 1, ..., n$ is imposed. This fact is crucial as it can be seen in the following example.

Example 1. Let $\leq_{ilex2}$ be the order generated as in Eq. (2), $w = (0.5, 0.5)$ and $v = (0, 1)$ Then

$UAIOWA_{[w,v,\leq_{ilex2}]}((0.9, 0.1), (0, 1)) = (0.9 \cdot 0.5 + 0 \cdot 0.5, 0.1 \cdot 0 + 1 \cdot 1) = (0.45, 1)$.

Similarly, $UAIOWA_{[\tilde{w},\tilde{v},\leq_{ilex2}]}((0.8, 0), (0, 1)) = (0.8 \cdot 0.5 + 0 \cdot 0.5, 0 \cdot 0 + 1 \cdot 1) = (0.4, 1)$. Since $(0.9, 0.1) \leq_{ilex2} (0.8, 0)$ but $(0.45, 1) \geq_{ilex2} (0.4, 1)$, then $UAIOWA$ is not monotonic.

Finally, we study when the image of the operators are always intuitionistic pairs, namely,

$$\left(\sum_{i=1}^{n} w_i \mu_{(i)} + \sum_{i=1}^{n} v_i \nu_{(i)} \leq 1\right). \quad (5)$$

It is a simple calculation to see that in the more restrictive case, when $\nu_{(i)} = 1 - \mu_{(i)}$, the equation is reduced to

$$\sum_{i=1}^{n} w_i \mu_{(i)} \leq \sum_{i=1}^{n} v_i \mu_{(i)}. \quad (6)$$

Lemma 1. *Let $\tilde{w}, \tilde{v} \in [0, 1]^n$ be two weight vectors. Then the following statements are equivalent.*

(i) $\sum_{j=1}^{i} w_j \leq \sum_{j=1}^{i} v_j$ *for all* $i = 1, \ldots, n$.

(ii) $\sum_{i=1}^{n} w_i t_i \leq \sum_{i=1}^{n} v_i t_i$ *for all* $t_i \in [0, 1]$ *such that* $t_1 \geq t_2 \geq \ldots \geq t_n \geq 0$.

Proof. We first prove (i.) implies (ii.). As

$$\begin{array}{ll} w_1 \leq v_1 \text{ then for all } a_1 \geq 0 & a_1 w_1 \leq a_1 v_1 \\ w_1 + w_2 \leq v_1 + v_2 \text{ then for all } a_2 \geq 0 & a_2(w_1 + w_2) \leq a_2(v_1 + v_2) \\ \vdots & \vdots \\ w_1 + \ldots + w_n \leq v_1 + \ldots + v_n \text{ then for all } a_n \geq 0 & a_n(w_1 + \ldots + w_n) \leq a_n(v_1 + \ldots + v_n). \end{array} \quad (7)$$

If we sum

$(a_1 + \ldots + a_n)w_1 + (a_2 + \ldots + a_n)w_2 + \ldots + a_n w_n \leq (a_1 + \ldots + a_n)v_1 + (a_2 + \ldots + a_n)v_2 + \ldots + a_n v_n$ for all $a_1, \ldots, a_n \geq 0$.

Taking $t_1 = (a_1 + \ldots + a_n)$, $t_2 = (a_2 + \ldots + a_n), \ldots, t_n = a_n$ it satisfies (ii.).

Let us see that (ii.) implies (i.). But this is trivial taking $t_1 = t_2 = \ldots = t_i = 1$ and $t_{i+1} = t_{i+2} = \ldots = t_n = 0$.

Finally, we have the following characterization of $UAIOWA$ operators.

Theorem 1. *Let $\tilde{w}, \tilde{v} \in (0, 1]^n$ be weight vectors. Then the following statements are equivalent:*

(i) $UAIOWA$ operator associated with $\tilde{w}, \tilde{v}$ and the order $\leq_{ilex1}$ is an aggregation function.

(ii) $\sum_{i=1}^{n} w_i t_i \leq \sum_{i=1}^{n} v_i t_i$ *for all* $t_i \in [0,1]$ *such that* $t_1 \geq t_2 \geq \ldots \geq t_n \geq 0$.

Proof. Notice that since the boundary conditions and the monotonicity holds true, $UIOWA$ is an aggregation operator if the codomain of the function is $\mathcal{L}([0,1])$, namely, the image of n AIF-pairs is always an AIF-pair.

Let us show that (i.) implies (ii.). Suppose $UAIOWA$ is an aggregation function. Then it satisfies Eq. (6) for all $\mu_i \in [0,1]$, $i = 1, \ldots, n$. Due to $\leq_{ilex1}$, $\mu_{(1)} \geq ... \geq \mu_{(n)}$, i.e., they are ordered in a decreasing way. Taking $t_i = \mu_{(i)}$ it satisfies (ii.).

Finally, let us show that (ii.) implies (i.). First of all (ii.) can be rewritten as

$$\sum_{i=1}^{n}(w_i - v_i)t_i \leq 0, \text{ for all } t_i \in [0,1] \text{ such that } t_1 \geq t_2 \geq \ldots \geq t_n \geq 0. \quad (8)$$

Let (μ_i, ν_i) for $i = 1, \ldots, n$, be n intuitionistic pairs. The expression of $UAIOWA$ associated with $\tilde{w}, \tilde{v}$ and the order $\leq_{ilex1}$ is

$$UAIOWA_{[\tilde{w},\tilde{v},\leq_{ilex1}]}((\mu_1,\nu_1),\ldots,(\mu_n,\nu_n)) = \left(\sum_{i=1}^{n} w_i\mu_{(i)}, \sum_{i=1}^{n} v_i\nu_{(i)}\right),$$

where $\mu_{(1)} \geq \mu_{(2)} \geq \ldots \geq \mu_{(n)}$ due to the order $\leq_{ilex1}$ used.

Considering that $\mu_{(i)} + \nu_{(i)} \leq 1$ and $v_1 + v_2 + \ldots + v_n = 1$ then

$$\sum_{i=1}^{n} w_i\mu_{(i)} + \sum_{i=1}^{n} v_i\nu_{(i)} \leq \sum_{i=1}^{n} w_i\mu_{(i)} + \sum_{i=1}^{n} v_i(1-\mu_{(i)}) = 1 + \sum_{i=1}^{n}(w_i - v_i)\mu_{(i)} \leq 1,$$

where the last inequation is due to Eq. (8).

Corollary 1. *Let* $\tilde{w}, \tilde{v} \in (0,1]^n$ *be weight vectors. Then the following statements are equivalent:*

(i) $UAIOWA$ *operator associated with* $\tilde{w}, \tilde{v}$ *and the order* $\leq_{ilex1}$ *is an aggregation function.*

(ii) $\sum_{j=1}^{i} w_j \leq \sum_{j=1}^{i} v_j$ *for all* $i = 1, \ldots, n$.

Proof. Straight by Lemma 1 and Theorem 1.

Lemma 2. *Let be* $\tilde{w}, \tilde{v} \in [0,1]^n$ *two weight vectors. Then the following statements are equivalent:*

(i) $\sum_{j=i}^{n} w_j \geq \sum_{j=i}^{n} v_j$ *for all* $i = 1, \ldots, n$.

(ii) $\sum_{i=1}^{n} w_i t_i \geq \sum_{i=1}^{n} v_i t_i$ *for all* $t_i \in [0,1]$ *such that* $t_n \geq t_{n-1} \geq \ldots \geq t_1 \geq 0$.

Proof. Similar to Lemma 1.

Theorem 2. *Let $\tilde{w}, \tilde{v} \in (0,1]^n$ be weight vectors. Then the following statements are equivalent:*

(i) $UAIOWA$ operator associated with $\tilde{w}, \tilde{v}$ and the order $\leq_{ilex2}$ is an aggregation function.

(ii) $\sum_{i=1}^{n} w_i t_i \geq \sum_{i=1}^{n} v_i t_i$ for all $t_i \in [0,1]$ such that $t_n \geq t_{n-1} \geq \ldots \geq t_1 \geq 0$.

Proof. Similar to Theorem 1.

Corollary 2. *Let $\tilde{w}, \tilde{v} \in (0,1]^n$ be weight vectors. Then the following statements are equivalent*

(i) $UAIOWA$ operator associated with $\tilde{w}, \tilde{v}$ and the order $\leq_{ilex2}$ is an aggregation function.

(ii) $\sum_{j=1}^{i} w_j \leq \sum_{j=1}^{i} v_j$ for all $i = 1, \ldots, n$

Proof. Straight by Lemma 2 and Theorem 2.

Remark 1.
It can be seen that:

$$\sum_{j=1}^{i} w_j \leq \sum_{j=1}^{i} v_j \qquad \text{for } i = 1, ..., n-1 \text{ (the condition for } i = n \text{ is trivial)}$$

if and only if

$$1 + \sum_{j=1}^{i} w_j \leq 1 + \sum_{j=1}^{i} v_j \qquad \text{for } i = 1, ..., n-1$$

if and only if

$$1 - \sum_{j=1}^{i} v_j \leq 1 - \sum_{j=1}^{i} w_j \qquad \text{for } i = 1, ..., n-1$$

if and only if

$$\sum_{j=i+1}^{n} v_j \leq \sum_{j=i+1}^{n} w_j \qquad \text{for } i = 1, ..., n-1.$$

Consequently, taking into account that $\tilde{w}$ and $\tilde{v}$ are weight vectors, i.e., $\sum_{i=1}^{n} w_i = 1$ and $\sum_{i=1}^{n} v_i = 1$, the condition that the weight vector must satisfy to be aggregation functions is the same for the orders $\leq_{ilex1}, \leq_{ilex2}$.

Notice that the weight vectors $w_i = v_i$ for all $i = 1, \ldots, n$ satisfy the condition required. In this way, the UAIOWA operators obviously increase the expressiveness and add more flexibility to the aggregation result than other operators. In fact, we believe a deep study on the optimization of the weight vectors could be useful in the improvement of the applications.

Example 2. Take $\tilde{w} = (0.2, 0.3, 0.5)$, $\tilde{v} = (0.4, 0.25, 0.35)$ and $\leq_{ilex2}$. Given the AIF-pairs $(0.3, 0.7)$, $(0.4, 0.2)$ and $(0.1, 0.8)$

$$UAIOWA_{[\tilde{w},\tilde{v},\leq_{ilex2}]}((0.3,0.7),(0.4,0.2),(0.1,0.8)) =$$
$$(0.2 \cdot 0.4 + 0.3 \cdot 0.3 + 0.5 \cdot 0.1, 0.4 \cdot 0.2 + 0.25 \cdot 0.7 + 0.35 \cdot 0.8) = (0.22, 0.535)$$

which satisfies $0.22 + 0.535 \leq 1$.

4 Illustrative Example: Application to a Decision Making Problem

In this section we make use of $UAIOWA$ operators in a decision making problem where information represented by Atanassov intuitionistic fuzzy set needs to be fused. However, we never intended to introduce a real application, but rather to show how it can be used.

We recall that a decision making problem consists on finding which is the best alternative in a set of n elements, $\mathcal{X} = \{x_1, \ldots, x_n\}$. In particular, in this problem we consider a set of four companies where some money can be invested. We ask a set of 50 experts who give their opinion in the following way:

- If they believe investing in the company is a good option, they vote *in favour* of the company.
- If they believe investing in the company is not a good option, they vote *against* the company.
- If they are not sure they vote *abstain*.

In this way, after all the votes we have the results given in Table 1.

We can construct the Atanassov intuitionistic fuzzy set, considering the universe $X = \{\text{Company 1}, \text{Company 2}, \text{Company 3}, \text{Company 4}\}$ and generating the membership and nonmembership degrees dividing the values in favour and against the company by the number of experts.

Table 1. Opinions of the experts with respect of the 4 companies

	Favour	Against	Abstain
Company 1	15	10	25
Company 2	28	14	8
Company 3	30	10	10
Company 4	8	13	29

For instance, with the information of Table 1 the following AIFS is generated:

$$A_1 = \{(C_1, (0.3, 0.2)), (C_2, (0.56, 0.28)), (C_3, (0.6, 0.2)), (C_4, (0.16, 0.26))\}.$$

Due to the international nature of the companies, the cultural differences may have a negative effect on the result. To avoid this situation, we repeat this process in three different countries: Spain, China and Brasil.

The results of the three countries Spain, China and Brasil are summarized in the AIFSs A_1, A_2, A_3, respectively.

$$A_2 = \{(C_1, (0.46, 0.42)), (C_2, (0.4, 0.6)), (C_3, (0.2, 0.5)), (C_4, (0.75, 0.2))\},$$

$$A_3 = \{(C_1, (0.12, 0.34)), (C_2, (0.26, 0.58)), (C_3, (0.7, 0.3)), (C_4, (0.44, 0.26))\}.$$

In the process of choice, the first step to be taken is the fusion of the three AIFSs. In this example, such fusion is carried out using an UAIOWA operator. Since the aim of this section is purely illustrative and for the sake of simplicity and clarity, the order and the weights vectors in this example are set arbitrarily. Nevertheless, in real applications some kind of optimization algorithm should be used to fine-tune them. In the present example the considered order is $\leq_{ilex1}$, and the weight vectors are set to $\tilde{w} = (0.2, 0.25, 0.55)$ and $\tilde{v} = (0.25, 0.35, 0.4)$ (which satisfy the condition of Corollary 1).

The results are

$$UAIOWA_{[\tilde{w},\tilde{v} \leq_{ilex1}]}((0.3, 0.2), (0.46, 0.42), (0.12, 0.34)) = (0.233, 0.311),$$

$$UAIOWA_{[\tilde{w},\tilde{v} \leq_{ilex1}]}((0.56, 0.28), (0.4, 0.6), (0.26, 0.58)) = (0.355, 0.512),$$

$$UAIOWA_{[\tilde{w},\tilde{v} \leq_{ilex1}]}((0.6, 0.2), (0.2, 0.5), (0.7, 0.3)) = (0.4, 0.345),$$

$$UAIOWA_{[\tilde{w},\tilde{v} \leq_{ilex1}]}((0.16, 0.26), (0.75, 0.2), (0.44, 0.26)) = (0.348, 0.245),$$

which generate the AIFS $\tilde{A}$, given by:

$$\tilde{A} = \{(C_1, (0.233, 0.311)), (C_2, (0.355, 0.512)), (C_3, (0.4, 0.345)), (C_4, (0.348, 0.245))\}.$$

Notice that $\tilde{A}$ summarizes the information of the experts of the three countries about the companies. Moreover, since the result is an Atanassov intuitionistic fuzzy set, a linear order is required to take a decision. In this example, we take $\leq_{ilex1}$ since it is the order used for the $UAIOWA$. The ranking of the alternatives is:

Company 3 *better than* Company 2 *better than* Company 4 *better than* Company 1.

Consequently, the best alternative in this illustrative example is to invest in the third company.

5 Conclusion

In the last years there has been an increasing interest in the study of linear orders for the different extensions of Fuzzy Sets. These orders let us define some theoretical notions which could not be trivially generalized. In particular, in this work, we have defined a new class of aggregation functions slightly different from OWA operators which makes use of different weight vectors for the membership and nonmembership degrees. However, the number of linear orders for Atanassov intuitionistic fuzzy sets in which these operators satisfy the monotonicity are really scarce. We let for future work the study of the linear orders in these sets generated by aggregation functions which satisfy the monotonicity.

In the context of the applications, we have only introduced an illustrative example where the parameters are set arbitrarily. Thereby, a deep study of algorithms which fine-tunes the parameters (order and weight vectors) in real applications is left to future researchs.

Acknowledgments. The work has been supported by projects TIN2013-40765-P of the Spanish Ministry of Science and the Research Services of the Universidad Publica de Navarra.

References

1. Komorníková, M., Mesiar, R.: Aggregation functions on bounded partially ordered sets and their classification. Fuzzy Sets Syst. **175**(1), 48–56 (2011)
2. Grattan-Guinness, I.: Fuzzy membership mapped onto intervals and many-valued quantities. Math.l Logic Q. **22**(1), 149–160 (1976)
3. Atanassov, K.T.: Intuitionistic fuzzy sets. Fuzzy Sets Syst. **20**(1), 87–96 (1986)
4. De Miguel, L., Bustince, H., Fernandez, J., Jurio, A., Paternain, D., El problema del orden en conjuntos Atanassov intuicionistas. In: Actas de la XVI Conferencia de la Asociación Española para la Inteligencia Artificial, CAEPIA 2015, Albacete 9-12 Noviembre 2015
5. Beliakov, G., Bustince, H., Calvo, T.: A Practical Guide to Averaging Functions. Studies In Fuzziness and Soft Computing. Springer, Heidelberg (2016)
6. Yager, R.R.: On ordered weighted averaging aggregation operators in multicriteria decision making. IEEE Trans. Syst. Man Cybern. **18**(1), 183–190 (1988)
7. Bustince, H., Fernandez, J., Kolesárová, A., Mesiar, R.: Generation of linear orders for intervals by means of aggregation functions. Fuzzy Sets Syst. **220**, 69–77 (2013)
8. De Miguel, L., Bustince, H., Fernandez, J., Induráin, E., Kolesárová, A., Mesiar, R.: Construction of admissible linear orders for interval-valued Atanassov intuitionistic fuzzy sets with an application to decision making. Inf. Fusion **27**, 189–197 (2016)
9. Yager, R.R.: OWA aggregation of intuitionistic fuzzy sets. Int. J. General Syst. **38**(6), 617–641 (2009)
10. Xu, Z., Yager, R.R.: Some geometric aggregation operators based on intuitionistic fuzzy sets. IInt. J. General Syst. **35**(4), 417–433 (2006)

Fuzzy K-Minpen Clustering and K-nearest-minpen Classification Procedures Incorporating Generic Distance-Based Penalty Minimizers

Anna Cena[1] and Marek Gagolewski[1,2(✉)]

[1] Systems Research Institute, Polish Academy of Sciences, ul. Newelska 6, 01-447 Warsaw, Poland

[2] Faculty of Mathematics and Information Science, Warsaw University of Technology, ul. Koszykowa 75, 00-662 Warsaw, Poland

{cena,gagolews}@ibspan.waw.pl

Abstract. We discuss a generalization of the fuzzy (weighted) k-means clustering procedure and point out its relationships with data aggregation in spaces equipped with arbitrary dissimilarity measures. In the proposed setting, a data set partitioning is performed based on the notion of points' proximity to generic distance-based penalty minimizers. Moreover, a new data classification algorithm, resembling the k-nearest neighbors scheme but less computationally and memory demanding, is introduced. Rich examples in complex data domains indicate the usability of the methods and aggregation theory in general.

Keywords: Fuzzy k-means algorithm · Clustering · Classification · Fusion functions · Penalty minimizers

1 Introduction

Aggregation theory [14] till very recently was mainly focused on various methods to determine individual objects that are representative to a set of n numeric values, typically in the $[0,1]$ interval. Such functions are very useful, e.g., in univariate statistics, decision making, and fuzzy control. Yet, we are currently observing a growing interest in aggregation of more complex data types [2,12] – in particular, on diverse ordering structures and spaces equipped with dissimilarity measures. In the latter case, a *fusion function* F to aggregate n objects from a set $\mathcal{X}$, is usually required at least to:

- return an object of the same type as the type of the inputs,
- for all $\mathbf{x} \in \mathcal{X}$ fulfill $\mathsf{F}(\mathbf{x}, \ldots, \mathbf{x}) = \mathbf{x}$.

In other words, it is an *idempotent* function like $\mathsf{F} : \mathcal{X}^n \to \mathcal{X}$. Most commonly, such fusion functions are defined as minimizers of some penalties, compare [6],

J.P. Carvalho et al. (Eds.): IPMU 2016, Part II, CCIS 611, pp. 445–456, 2016.
DOI: 10.1007/978-3-319-40581-0_36

given via a proper aggregation of dissimilarity degrees between each input observation and the representative object being sought.

In this contribution we study the role of the mentioned penalty minimizers in partitional clustering and object proximity-based classification procedures. First of all, let us note that Leisch in [18] and Bock in [4] – among others – observed that the k-means [19] procedure that aims to cluster a set of n objects in $\mathbb{R}^d$, $d \geq 2$, into a fixed number $k > 1$ of disjoint and nonempty clusters does not necessarily have to rely on the squared Euclidean metric as a basis for measuring the within-cluster sums of distances. These lead to a common framework bracketing the classical k-means, k-medians, and – by extension – the k-medoids algorithms together. What is more, Yu and Yang in [22], see also the references therein, made similar observations concerning the relative arbitrariness of each observation's cluster membership degrees in the weighted k-means (also known as fuzzy c-means) procedure, yet, this time, they were only considering the squared Euclidean metric.

In this paper, we consider a natural generalization of the Yu and Yang [22] fuzzy k-means procedure to spaces equipped with arbitrary dissimilarity measures. As it shall turn out, fusion functions defined as minimizers of some distance-based penalties play a key role in its definition. We explore various examples of clustering of complex data types, including rankings, strings, intervals, and fuzzy numbers.

Additionally, we introduce a new classification algorithm that extends the ideas behind the nearest centroid classifier [13,20] and incorporates the fuzzy k-means procedure in its data pre-processing part. Such a method, similarly as the k-nearest neighbor algorithm, works in arbitrary spaces equipped with a dissimilarity measure, but is less time and memory demanding. Moreover, it creates a natural and nicely interpretable data *model* that can be used to understand the form of observations in each data class.

This contribution is arranged in the following manner. Sect. 2 introduces the generalized fuzzy k-means procedure and discusses the role of distance-based penalty minimizers in its formulation. In Sect. 3 we present various examples of clustering of data of different complexity and discuss the importance of a proper choice of the algorithm's free parameters. Sect. 4 introduces the new penalty minimizers-based classification method, which in turn is illustrated in Sect. 5. Lastly, in Sect. 6 we conclude the paper and indicate noteworthy future research topics.

2 Fuzzy K-Minpen Clustering

Let $(\mathcal{X}, \mathfrak{d})$ be a space equipped with a pairwise dissimilarity measure, i.e., $\mathfrak{d} : \mathcal{X} \times \mathcal{X} \to [0, \infty]$ such that for any $\mathbf{x}, \mathbf{y} \in \mathcal{X}$ it at least holds $\mathfrak{d}(\mathbf{x}, \mathbf{x}) = 0$ and $\mathfrak{d}(\mathbf{x}, \mathbf{y}) = \mathfrak{d}(\mathbf{y}, \mathbf{x})$.

Let us consider the following generalization of the fuzzy (weighted) k-means [3] clustering task. For some fixed $k > 1$, $\mathbf{x}^{(1)}, \ldots, \mathbf{x}^{(n)} \in \mathcal{X}$, we seek:

$$\mathop{\arg\min}_{\boldsymbol{\mu}^{(1)},\ldots,\boldsymbol{\mu}^{(k)}\in\mathcal{X}'} \sum_{j=1}^{k}\sum_{i=1}^{n} \psi\left(\frac{\vartheta(\mathfrak{d}(\mathbf{x}^{(i)},\boldsymbol{\mu}^{(j)}))}{\sum_{l=1}^{k}\vartheta(\mathfrak{d}(\mathbf{x}^{(i)},\boldsymbol{\mu}^{(l)}))}\right)\varphi\left(\mathfrak{d}(\mathbf{x}^{(i)},\boldsymbol{\mu}^{(j)})\right), \tag{1}$$

where $\mathcal{X}' \subseteq \mathcal{X}$, $\vartheta : [0,\infty] \to [0,\infty]$ is a strictly decreasing, and $\varphi : [0,\infty] \to [0,\infty]$, $\psi : [0,1] \to [0,\infty]$ are strictly increasing continuous functions such that $\varphi(0) = \psi(0) = 0$. In the original fuzzy k-means setting, we have $\mathcal{X} = \mathcal{X}' = \mathbb{R}^d$, $\vartheta(d) = d^{-2/(m-1)}$ for some $m > 1$ called *fuzzifier*, $\varphi(d) = d^2$, and $\psi(d) = d^m$.

Note that already in [22] it was suggested that an arbitrary increasing function φ can be used in the fuzzy k-means algorithm. Nevertheless, the authors were focused only on the Euclidean space, because they observed that otherwise the results were difficult to study analytically (of course, as we indicate in the section to follow, this does not contradict the fact that in other settings a clustering task may end up with a practically meaningful solution). Moreover, it is worth pointing out that they assumed $\psi(d) = d^m$ for some $m > 1$ and that ϑ is functionally dependent on φ.

A typical choice of $\mathfrak{d}$ is some relatively easily computable metric. By definition, $\varphi \circ \mathfrak{d}$ is then a dissimilarity measure. The choice of φ, ϑ, and ψ is quite arbitrary, however, as we shall see further on, obviously not all the choices lead to a unique solution. By changing the generator function, we may control – among others – the level of *fuzziness* of the obtained partition as well as the method's sensitivity to outliers.

The optimization problem (1) aims to find the centers of k clusters. Let us make a few observations. For a fixed k and a fixed set of input observations, let $\boldsymbol{\mu}^{*(1)}, \ldots, \boldsymbol{\mu}^{*(k)} \in \mathcal{X}'$ be the solution to (1) (assuming it exists and is unique). Assuming that $\sum_{l=1}^{k}\vartheta(\mathfrak{d}(\mathbf{x}^{(i)},\boldsymbol{\mu}^{*(l)})) > 0$, denote with $w_{i,j}$ the following quantity:

$$w_{i,j} = \frac{\vartheta(\mathfrak{d}(\mathbf{x}^{(i)},\boldsymbol{\mu}^{*(j)}))}{\sum_{l=1}^{k}\vartheta(\mathfrak{d}(\mathbf{x}^{(i)},\boldsymbol{\mu}^{*(l)}))}. \tag{2}$$

As each $w_{i,j} \geq 0$ and $\sum_{j=1}^{k} w_{i,j} = 1$, we have that $(w_{i,1}, \ldots, w_{i,k})$ is a weighting vector. Thus, $w_{i,j}$ can be interpreted as the degree of belongingness of $\mathbf{x}^{(i)}$ to the j-th cluster.

Remark 1. In fact, the w_j functions given for any $\mathbf{x} \in \mathcal{X}$ and $j = 1, \ldots, k$ by $\mathsf{w}_j(\mathbf{x}) = \frac{\vartheta(\mathfrak{d}(\mathbf{x},\boldsymbol{\mu}^{*(j)}))}{\sum_{l=1}^{k}\vartheta(\mathfrak{d}(\mathbf{x},\boldsymbol{\mu}^{*(l)}))}$ define a *fuzzy pseudopartition* of whole data domain, $\mathcal{X}$, into k subsets. As usual in unsupervised learning methods, its quality is strongly related to the representativeness of a given sample $\mathbf{x}^{(1)}, \ldots, \mathbf{x}^{(n)}$.

We see that the closer $\mathbf{x}^{(i)}$ is to some $\boldsymbol{\mu}^{*(j)}$, the higher is its impact in the computation of $\boldsymbol{\mu}^{*(j)}$. The strength of this impact may be moderated via the ψ function. Yet, please note that $\psi(w_{i,j})$ does not possess the same appealing interpretation as the sole $w_{i,j}$.

In the current context, it is easily seen that for each j:

$$\boldsymbol{\mu}^{*(j)} = \arg\min_{\boldsymbol{\mu}\in\mathcal{X}'} \varphi^{-1}\left(\sum_{i=1}^{n} v_i \varphi\left(\mathfrak{d}(\mathbf{x}^{(i)}, \boldsymbol{\mu})\right)\right), \tag{3}$$

where $v_i = \psi(w_{i,j})/\sum_{l=1}^{n}\psi(w_{l,j})$. Recall that our aforementioned assumptions on φ guarantee that φ^{-1} exists. In other words, each $\boldsymbol{\mu}^{*(j)}$ is a minimizer of a distance-based penalty function given as a weighted quasi-arithmetic mean. Hence the name of the discussed procedure: k-penalty minimizers (k-minpen for brevity). Each $\boldsymbol{\mu}^{*(j)}$ is of course generated by an idempotent fusion function.

Remark 2. A common choice of $\mathcal{X}'$ is either $\mathcal{X}$ or $\{\mathbf{x}^{(1)}, \ldots, \mathbf{x}^{(n)}\}$. In the latter case we obtain a cluster center that in fact is a kind of a *set exemplar.* In particular, in the case $\varphi(d) = d$ such a penalty minimizer is known as the *medoid* in the literature, while its unconstrained version – the *1-median.* In complex data domains (character strings, rankings, etc.), set exemplars are much easier to compute, as exact algorithms to determine, e.g., the 1-median, may not exist or be computationally too demanding.

Remark 3. An interesting special case is when ϑ, φ, ψ are power functions. Let $\vartheta(d) = d^{-\alpha}$ for some $\alpha > 0$, $\psi(d) = d^m$, $m > 0$, $\varphi(d) = d^p$, $p > 0$, $p \neq \alpha m$. Then the objective function may be rewritten as $\sum_{j=1}^{k}\sum_{i=1}^{n}\mathfrak{d}(\mathbf{x}^{(i)}, \boldsymbol{\mu}^{(j)})^{-\alpha m+p}/\left(\sum_{l=1}^{k}\mathfrak{d}(\mathbf{x}^{(i)}, \boldsymbol{\mu}^{(l)})^{-\alpha}\right)^m$. Assume that for every $i = 1, \ldots, n$, $\min_j \mathfrak{d}(\mathbf{x}^{(i)}, \boldsymbol{\mu}^{(j)})$ is unique. Let $\vee$ and $\wedge$ denote the maximum and minimum operators, respectively. Then we may observe that as $\alpha \to 0^+$, we have $\sum_{l=1}^{k}\mathfrak{d}(\mathbf{x}^{(i)}, \boldsymbol{\mu}^{(l)})^{-\alpha} = \left(\bigvee_{l=1}^{k} 1/\mathfrak{d}(\mathbf{x}^{(i)}, \boldsymbol{\mu}^{(l)})\right)^{\alpha}$. In such a case the objective function coincides with $\sum_{i=1}^{n}\bigwedge_{j=1}^{k}\mathfrak{d}(\mathbf{x}^{(i)}, \boldsymbol{\mu}^{(j)})^p$, which is in fact equivalent to the criterion used in the generalized k-means algorithm, compare [4].

Remark 4. Investigation carried out in this paper is concerned with a possible generalization of the fuzzy k-means through the modifications of the "centroid" part of Eq. (1). Interestingly, a different focus can be found in [17], where the emphasis is put on the properties and possible generalizations of the "fuzzifier" part, i.e., the weights w_{ij}. Some examples of such a fuzzifier were also studied in [21].

Of course, one should ask how to approximate the solution to Eq. (1) for practical use. If $\mathcal{X} = \mathbb{R}^d$, then a nonlinear solver like, e.g., the BFGS (Broyden-Fletcher-Goldfarb-Shanno) or CMA-ES (Covariance Matrix Adaptation-Evolution Strategy) algorithm, can be used for that purpose. In discrete spaces, e.g., genetic algorithms may be used. We can also rely on the following well-known heuristic:

1. Initialize $\boldsymbol{\mu}^{(1)}, \ldots, \boldsymbol{\mu}^{(k)}$, e.g., randomly;
2. Based on current $\boldsymbol{\mu}^{(1)}, \ldots, \boldsymbol{\mu}^{(k)}$, compute the new weights according to Eq. (2);

3. Based on current weights, compute the new $\boldsymbol{\mu}^{(1)}, \ldots, \boldsymbol{\mu}^{(k)}$ using the penalty-based fusion function from Eq. (3);
4. Go to Step 2. until some convergence criterion is met.

As usual with any iterative optimization procedure, the above should be repeated a few times and the output leading to the least overall objective criterion should be chosen as an approximation to the solution.

3 Clustering Examples

3.1 Aggregation of Multidimensional Real Vectors

Let us consider the most typical clustering case, i.e., $\mathcal{X}' = \mathcal{X} = \mathbb{R}^d$. It is well known that if $\mathfrak{d}$ is the Euclidean metric and $\varphi(d) = d^2$, then the penalty minimizer is the weighted centroid (componentwise weighted arithmetic mean, see the classical k-means and fuzzy k means algorithm). Moreover, if $\mathfrak{d}$ is the Manhattan distance and $\varphi(d) = d$, then the penalty minimizer is the componentwise weighted median (compare the k-medians algorithm), see [4,18].

Many different instances of the k-minpen algorithm can easily be generated. For example, let $\mathfrak{d}$ be the Euclidean metric but this time $\varphi(d) = d$. The solution to Eq. (3) is called the weighted geometric (1-)median. In general, it cannot be expressed with an analytic formula. However, e.g., the Weiszfeld (Vázsonyi) procedure, compare [12], may be applied in this case. Note that the 1-median is more robust to outlying observations as opposed to the centroid. Therefore, in some cases, one may obtain more reliable results, especially if data follow a skewed distribution.

3.2 Aggregation of Intervals

Let $\mathcal{X} = \mathcal{I}(\mathbb{R})$ be the space of all closed subintervals in the real line. Among popular interval metrics, compare [2,12], we have the *Moore interval metric*, given by: $\mathfrak{d}_{\mathrm{M}}\left([\underline{x}, \overline{x}], [\underline{y}, \overline{y}]\right) = |\underline{x} - \underline{y}| \vee |\overline{x} - \overline{y}|$. Interestingly, if an object in $\mathcal{I}(\mathbb{R})$ is interpreted as a point in $\mathbb{R}^2$, the Moore metric is exactly the Chebyshev distance, $\mathfrak{d}_\infty$. On the other hand, please note that each interval can be represented as a pair $(x \pm r)$, where $x = (\underline{x} + \overline{x})/2$ is its *midpoint* and $r = (\overline{x} - \underline{x})/2$ is its *halfwidth*. In such a setting, this metric is the Manhattan one: it holds $\mathfrak{d}_{\mathrm{M}}(x \pm r_x, y \pm r_y) = |x - y| + |r_x - r_y|$. The $\mathfrak{d}_{\mathrm{M}}$-based weighted 1-median of $\mathbf{x} \in \mathcal{I}(\mathbb{R})^n$ is equal to the componentwise weighted median of the inputs' midpoints and halfwidths, see [8, Theorem 1].

Moreover, if we assume that $\mathfrak{d}_{\mathrm{M}_2}(x \pm r_x, y \pm r_y) = \sqrt{|x - y|^2 + |r_x - r_y|^2}$, then the $\mathfrak{d}_{\mathrm{M}_2}$-based weighted centroid of $\mathbf{x} \in \mathcal{I}(\mathbb{R})^n$ is equal to the componentwise weighted arithmetic mean of midpoints and halfwidths, see [8, Theorem 3].

Let us consider a dataset representing the average low and high temperatures (in °C) recorded in 860 cities all over the World, which we shall treat as interval data. Data were web scraped on January 12, 2016 from the English Wikipedia

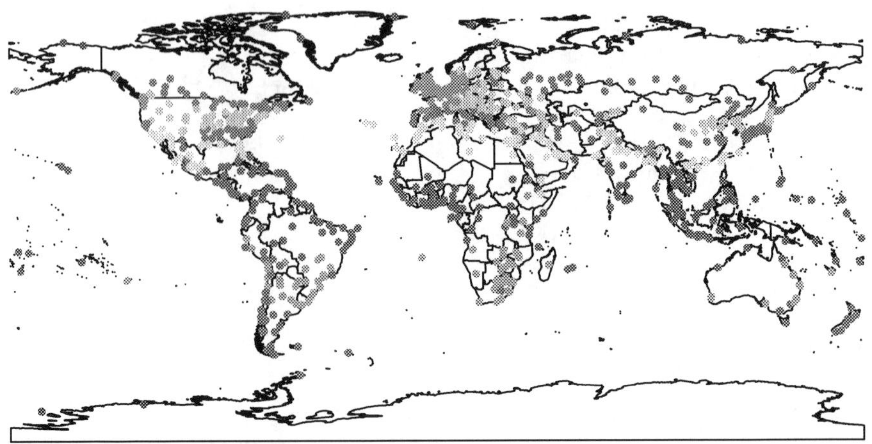

Fig. 1. The partitioning obtained with the k-minpen algorithm on interval temperature data. The color version of this figure is available at http://gagolewski.rexamine.com/resources/data/wikiweather/.

(http://en.wikipedia.org/) and are available for download at http://gagolewski.rexamine.com/resources/data/wikiweather/.

Let us partition the cities into $k = 9$ clusters using the k-minpen algorithm based on the temperatures recorded in January and July. In such a setting, $\mathcal{X} = \mathcal{X}' = \mathcal{I}(\mathbb{R}) \times \mathcal{I}(\mathbb{R})$. During the computations we rely on the midpoint±halfwidth representation, i.e., $\mathbf{x}^{(i)} = \{(x_1^{(i)} \pm r_1^{(i)}), (x_2^{(i)} \pm r_2^{(i)})\} \in \mathcal{X}$, $\vartheta(d) = d^{-4}$, $\psi(d) = d$, $\mathfrak{d} = \mathfrak{d}_{M_2}$, $\varphi(d) = d^2$, i.e., a weighted centroid-based approach. Figure 1 gives a graphical representation of the corresponding partitioning. Moreover, Table 1 relates each clusters' centroids with the well-known Köppen climate classification. We see that the two data division schemes agree with each other quite nicely.

Nevertheless, we observe that the choice of $\mathfrak{d}$, φ, ψ, and ϑ strongly influences the resulting partitionings. For instance, let us compare the aforementioned setting with the one based on $\mathfrak{d} = \mathfrak{d}_M$, $\varphi(d) = d$, $\vartheta(d) = d^{-1}$, and $\psi(d) = d^2$. The Fowlkes-Mallows (FM-)index, a measure of the agreement between two clusterings, is equal to ca. 0.652. On the other hand, if $\mathfrak{d} = \mathfrak{d}_M$, $\varphi(d) = d$, $\vartheta(d) = d^{-2}$, and $\psi(d) = d$, then the FM-index equals to ca. 0.599 (with FM = 0.765 for the two $\mathfrak{d}_M$-based variants).

3.3 Aggregation of Fuzzy Numbers

Let $\mathbb{F}(\mathbb{R})$ denote the space of all fuzzy numbers and let:

$$\mathfrak{d}_2(A, B) = \sqrt{\int_0^1 (A_L(\alpha) - B_L(\alpha))^2 d\alpha + \int_0^1 (A_U(\alpha) - B_U(\alpha))^2 d\alpha}$$

Table 1. The cluster centers obtained with the k-minpen algorithm related to the Köppen climate classification.

	January		July		
No	Ave. low°C	Ave. high°C	Ave. low°C	Ave. high°C	Köppen climate classification
1	−27.86	−20.85	0.42	6.90	Polar and alpine climates
2	−14.50	−5.77	13.76	24.46	Taiga
3	0.19	6.24	13.22	22.51	Oceanic climate
4	−4.96	2.79	16.88	27.81	Warm summer continental
5	2.05	11.10	20.95	31.14	Mediterranean climates
6	9.25	20.08	23.64	33.45	Hot desert climate
7	20.72	29.87	22.47	30.46	Tropical wet and dry
8	17.30	27.46	9.92	21.35	Cold desert climate
9	10.51	21.69	3.42	13.01	Humid subtropical climate

be the Euclidean metric on $\mathbb{F}(\mathbb{R})$, where, e.g., A_L and A_R denote the lower and upper α-cut bounds, respectively. Note that in a very similar manner, arbitrary weighted Minkowski metrics may be introduced, see [15].

The most commonly used subclass of $\mathbb{F}(\mathbb{R})$ is formed by the so-called *trapezoidal fuzzy numbers*, i.e., fuzzy numbers with linear sides, compare [1]. A membership function of a trapezoidal fuzzy number T is given by:

$$\mu_T(x) = \begin{cases} 0 & \text{if } x < t_1 \text{ or } x > t_4, \\ \frac{x-t_1}{t_2-t_1} & \text{if } t_1 \le x < t_2, \\ \frac{t_4-x}{t_4-t_3} & \text{if } t_3 < x \le t_4, \\ 1 & \text{if } t_2 \le x \le t_3, \end{cases}$$

where $t_1 \le t_2 \le t_3 \le t_4$. Since the membership function of a trapezoidal fuzzy number T is completely defined by these four real numbers, we denote it usually as $T = \mathrm{T}(t_1, t_2, t_3, t_4)$. It is easy to show that the α-cuts have the following lower and upper bounds: $T_L(\alpha) = t_1 + (t_2 - t_1)\alpha$, $T_U(\alpha) = t_4 - (t_4 - t_3)\alpha$. The set of all trapezoidal fuzzy numbers is denoted by $\mathbb{F}^T(\mathbb{R})$.

Ban et al. in [1] derived an algorithm for determining a $\mathfrak{d}_2$-based centroid of n such fuzzy sets, i.e., $\mathbf{t}^{(i)} = \mathrm{T}(t_1^{(i)}, \ldots, t_4^{(i)})$ for $i = 1, \ldots, n$. It turns out that it can be expressed as a trapezoidal fuzzy number of the form $\mathrm{T}(\bar{t}_1, \bar{t}_2, \bar{t}_3, \bar{t}_4)$, where $\bar{t}_i = \frac{1}{n}\sum_{j=1}^{n} t_i^{(j)}$. The weighted version can be obtained by replacing the arithmetic mean with its weighted version.

Let us consider climate data from Sect. 3.2, in the case of 135 European cities (the dataset size has been reduced due to missing observations). This time, we shall represent record low, average low, average high, and record high temperatures (in °C) as trapezoidal fuzzy numbers. Again, we rely on data from January and July. Moreover, additionally we consider the average precipitation observed during these two months. Therefore, $\mathcal{X} = \mathcal{X}' = \mathbb{F}^T(\mathbb{R}) \times \mathbb{F}^T(\mathbb{R}) \times \mathbb{R}^2$ and for all $\mathbf{x} = (\mathbf{t}_1, \mathbf{t}_2, p_1, p_2) \in \mathcal{X}$, the dissimilarity measure is given as a fusion of metrics on two different spaces: $\mathfrak{d}(\mathbf{x}^{(i)}, \mathbf{x}^{(j)}) = \lambda_1 \mathfrak{d}_2^2(\mathbf{t}_1^{(i)}, \mathbf{t}_1^{(j)}) + \lambda_2 \mathfrak{d}_2^2(\mathbf{t}_2^{(i)}, \mathbf{t}_2^{(j)}) +$

Table 2. Exemplary cities clustered into $k = 3$ groups with the k-minpen algorithm.

Amsterdam	Oslo	Gijón
Luxembourg	Paris	Gibraltar
Bordeaux	Warsaw	Tunis
Oxford	Saint Petersburg	Tehran
Milan	Tbilisi	Thessaloniki
Linz	Skopje	Ankara
Sarajevo	Samara	Florence
...	...	...

$\lambda_3|p_1^{(i)} - p_1^{(j)}|^2 + \lambda_4|p_2^{(i)} - p_2^{(j)}|^2$. In Table 2 we present an exemplary clustering of a rectangular World map region including Europe into $k = 3$ city groups, $\lambda_1 = \lambda_2 = 0.9, \lambda_3 = \lambda_4 = 0.2$, $\varphi(d) = d$, $\vartheta(d) = d^{-2}$, $\psi(d) = d$. Note that in this case we can rely on various other metrics on the $\mathbb{F}^T(\mathbb{R})$ space, including the weighted ones.

3.4 Aggregation of Rankings

Let us assume that $(\mathcal{X}, \mathfrak{d})$ is the space of all possible full rankings of length $d \geq 2$, where $\mathfrak{d}$ is the Kendall rank (bubble sort) distance, i.e.,

$$\mathfrak{d}(\mathbf{x}, \mathbf{y}) = |\{(k, l) : (r_{\mathbf{x}}(k) - r_{\mathbf{x}}(l))(r_{\mathbf{y}}(k) - r_{\mathbf{y}}(l)) < 0\}|,$$

where, e.g., $r_{\mathbf{x}}(k)$ denotes the rank of k in $\mathbf{x}$. For instance, $r_{(1,3,4,2,5)}(2) = 4$ and $\mathfrak{d}((1,2,3,4,5),(1,5,2,3,4)) = 3$.

The 1-median with respect to the Kendall distance is called Kemeny optimal aggregation. Unfortunately, the problem of computing such a fusion function is known to be NP-complete even when $n = 4$ [11]. However, we can employ an approximate weighted 1-median in the process of computing the fuzzy k-minpen algorithm. Firstly, let us recall that the Borda count method computes the averaged elements' ranks, $\mathbf{r} = \left(\sum_{i=1}^n r_{\mathbf{x}^{(i)}}(1)/n, \ldots, \sum_{i=1}^n r_{\mathbf{x}^{(i)}}(d)/n\right)$, and then returns as the result the ordering permutation of $\mathbf{r}$. It has been shown in [9] that the Borda count may serve as a 5-approximation to the Kemeny-optimal aggregation problem. In our case, the weighted 1-median can be based on the weighted Borda count. Additionally, we suggest to employ the following heuristic aiming at improving the approximate result. For some $l \geq 1$:

1. Let $\boldsymbol{\mu}$ be the weighted Borda count computed on the input data set;
2. Generate l candidate rankings by applying on $\boldsymbol{\mu}$ a single bubble sort swap at a random position;
3. Set as the new $\boldsymbol{\mu}$ the best (in terms of the 1-median objective function) ranking among the candidate ones;
4. Go to Step 2 until there are no further improvements in terms of the objective function;

As usual, such a procedure should be repeated a few times.

Let us consider a set of 100 vectors, each representing the number of citations to the publications written by an author listed in the Elsevier Scopus database (available at http://cena.rexamine.com/research). By computing the h-index (H), the number of publications (N), and the total number of citations ($\sum$) of each agent, we obtain 3 rankings $\boldsymbol{\mu}^{(1)}, \boldsymbol{\mu}^{(2)}, \boldsymbol{\mu}^{(3)}$ of length 100.

A benchmark set, consisting of 90 rankings and 3 equal-sized classes was created in such a way that $\mathbf{x}^{(j)}$ is a version of some $\boldsymbol{\mu}^{(i)}$ disturbed by consecutively applying l bubble sort swaps at random positions. The total Kendall distance between the true $\boldsymbol{\mu}^{(i)}$ and the cluster centers computed with the fuzzy k-minpen algorithm (averaged over 35 runs) equals to ca. 0.2 in the $l = 50$ case and 5.6 in the $l = 100$ case ($\varphi(d) = d, \vartheta(d) = d^{-2}, \psi(d) = d$).

3.5 Aggregation of Strings

Numeric Strings. Let $\mathcal{X} = \{(x_1, \ldots, x_d) \in \bigcup_{d \geq 1} \mathbb{R}^d : x_1 \geq x_2 \geq \cdots \geq x_d\}$ denote the space of non-increasingly ordered numeric lists of arbitrary length, e.g., citation sequences. The nature of such data may be situated "somewhere between" multidimensional real data and the character string domain. On the one hand, observations are real numbers but, on the other, their number is not established a priori. In [7], the authors considered the dissimilarity measure on $\mathcal{X}$ given for some $p, r > 0$ by $\mathfrak{d}_{p,r}(\mathbf{x}, \mathbf{y}) = \mathfrak{d}_2^2(\tilde{\mathbf{x}}, \tilde{\mathbf{y}}) + p|n_x^r - n_y^r|$, where $\tilde{\mathbf{x}} = (x_1, x_2, \ldots, x_{n_x}, 0, 0, \ldots)$ and $\mathfrak{d}_2$ denotes the Euclidean distance on $\mathbb{R}^\infty$. Then, a fuzzy k-means-like algorithm was used to cluster real-world bibliometric and StackExchange datasets.

Character Strings. For some finite Σ called alphabet, let $\mathcal{X} = \bigcup_{d \geq 1} \Sigma^d$ denote the set of all character strings over Σ. For instance, if $\Sigma = \{\mathtt{a}, \mathtt{c}, \mathtt{t}, \mathtt{g}\}$, then $\mathcal{X}$ is the set of all DNA sequences. In such a case, the clustering procedure may be based on various metrics and other dissimilarity measures, including the longest common subsequence, Levenshtein, q-gram, or Dinu rank distance, see, e.g., [5]. In particular, the Dinu rank distance-based 1-median can be computed in polynomial time exactly, see [10]. In other cases, we may rely on approximate or set exemplar-based approaches, compare [12].

4 K-nearest-minpen Classification

Let us now assume that the domain $\mathcal{X}$ is divided into $c \geq 2$ classes and together with each input item $\mathbf{x}^{(i)} \in \mathcal{X}$ we also observe its true class label $y_i \in \{1, \ldots, c\}$. The aim of a classification task, compare, e.g., [16], is to construct a model (decision rule) that, based on $\{(\mathbf{x}^{(i)}, y^{(i)})\}_{i=1,\ldots,n}$, can be used to classify any observation in $\mathcal{X}$.

Let us recall that in the current context we deal with a space equipped with a dissimilarity measure only. It turns out that there are only a few methods in the literature for such a type of supervised learning task. Among them we find:

- the k-nearest neighbors algorithm, compare [16], which classifies $\mathbf{x} \in \mathcal{X}$ based on the mode of the class labels of the $k \geq 1$ closest to $\mathbf{x}$ observations (in terms of $\mathfrak{d}$) in the input data set,
- the nearest centroid algorithm [13,20], which assigns to $\mathbf{x} \in \mathcal{X}$ the class label $\arg\min_{j=1,\ldots,c} \mathfrak{d}(\mathbf{x}, \boldsymbol{\mu}^{(j)})$, where $\boldsymbol{\mu}^{(j)}$ is the result of applying a $\mathfrak{d}$-based penalty minimizer on the set $\{\mathbf{x}^{(i)} : y^{(i)} = j, i = 1, \ldots, n\}$ (originally: the centroid with respect to the Euclidean distance).

Unfortunately, the first method requires access to the whole data set, which – in the big data era – makes the algorithm time-consuming if a large number of data points is to be classified (proximity search data structures may be used but due to the curse of dimensionality they are only advantageous in low dimensional spaces). On the other hand, the second method is way faster (as soon as the pre-processing stage is already performed), require significantly less storage, and provides a data analyst with an interpretable model.

Based on the aforementioned methods, we propose the following algorithm called *k-nearest-minpen*. Fix $k, k_1, \ldots, k_c \geq 1$, and φ, ϑ, ψ.

Pre-processing stage.

1. For all $j = 1, \ldots, c$:
   1.1. Determine the cluster centers $\boldsymbol{\mu}^{(j,1)}, \ldots, \boldsymbol{\mu}^{(j,k_j)}$ of the set $\{\mathbf{x}^{(i)} : y^{(i)} = j, i = 1, \ldots, n\}$ using the fuzzy k_j-minpen clustering algorithm;

Classification stage. Classify $\mathbf{x} \in \mathcal{X}$ as follows:

1. Determine $\boldsymbol{\mu}^{(j_1,i_1)}, \ldots, \boldsymbol{\mu}^{(j_k,i_k)}$ – the k-nearest (w.r.t. $\mathfrak{d}$) to $\mathbf{x}$ cluster centers;
2. Return $\arg\max_{j=1,\ldots,c} |\{j_i : j_i = j, i = 1, \ldots, k\}|$, i.e., the mode of $\{j_1, \ldots, j_k\}$ as result;

Note that for $k = k_1 = \cdots = k_c = 1$ we get the generalized nearest centroid algorithm. On the other hand, for $k_j = |\{y^{(i)} : y^{(i)} = j, i = 1, \ldots, n\}|$, by idempotence of the $\mathfrak{d}$-based penalty minimizer, we approach exactly the k-nearest neighbor scheme. Nevertheless, for a typical use case we recommend the use of $k = 1$ as well as $k_1 = \cdots = k_c$ with k_i being merely a fraction of the average number of observations in each clusters.

5 Classification Examples

Table 3 gives classification accuracy over 5 exemplary data sets from the UCI Machine Learning Repository: `iris` (150 observations in $\mathbb{R}^4$, 3 classes), `wine` (178 observations in $\mathbb{R}^{13}$, 3 classes), `seeds` (210 observations in $\mathbb{R}^7$, 3 classes), `ecoli` (327 observations in $\mathbb{R}^6$, 5 classes; underrepresented classes 3,4,7 removed), and `glass` (175 observations in $\mathbb{R}^9$, 3 classes; classes 3,5,6 removed). Reported quality measures are computed by repeated random sub-sampling validation and were averaged over 1000 random data splits into two parts: 80 % used as a training, whereas the remaining 20 % as a test sample. Attributes were standardized before performing the experiments. We set $\mathfrak{d}$ – Euclidean distance, $\varphi(d) = d^2$, $m = 2$, $\vartheta(d) = d^m$, and $\psi(d) = d^{-2/(m-1)}$. Recall that $(k; k_i) = (1; 1)$ gives the nearest centroid classifier. Not surprisingly, we observe that for $k = 1$, our algorithm has – on average – a slightly better performance than the nearest centroids method.

Table 3. Classification accuracy (in % of correct label prediction).

	k-nearest neighbors (k)				k-nearest minpen ($k; k_i$)					
dataset	1	3	5	7	1;1	1;3	1;5	3;5	1;7	3;7
`iris`	94.32	94.44	94.96	95.32	85.94	95.73	95.17	92.96	95.29	94.09
`wine`	95.12	95.42	96.09	96.37	96.89	96.19	96.17	96.28	95.57	95.76
`seeds`	92.82	91.91	92.64	91.87	92.20	91.19	91.45	90.18	91.57	91.13
`ecoli`	81.55	84.91	86.78	87.74	86.09	82.47	79.71	84.54	78.34	84.13
`glass`	80.49	79.00	76.01	74.69	67.51	76.67	77.49	62.99	76.61	67.62
average	88.86	89.14	89.30	89.20	85.73	88.45	88.00	85.39	87.48	86.55

6 Conclusion

We introduced two data analysis techniques: the fuzzy k-minpen clustering and the k-nearest-minpen classification algorithms. Both of them employ generic distance-based penalty minimizers and thus serve as an application of complex data aggregation in machine learning. Let us stress that the discussed methods are applicable in arbitrary spaces equipped with dissimilarity measures.

Future work on the introduced clustering method shall involve including a generalized fuzzifier term, as in [17, 21], and the study of its effects on the clustering quality.

Acknowledgments. This study was supported by the National Science Center, Poland, research project 2014/13/D/HS4/01700.

References

1. Ban, A.I., Coroianu, L., Grzegorzewski, P.: Trapezoidal approximation and aggregation. Fuzzy Sets Syst. **177**(1), 45–59 (2011)
2. Beliakov, G., Bustince, H., Calvo, T.: A Practical Guide to Averaging Functions. Studies in Fuzziness and Soft Computing. Springer, Heidelberg (2016)
3. Bezdek, J.C.: Pattern Recognition with Fuzzy Objective Function Algorithms. Advanced Applications in Pattern Recognition. Springer, Heidelberg (1981)
4. Bock, H.H.: Origins and extensions of the k-means algorithm in cluster analysis. Electron. J. Hist. Probab. Stat. **4**(2), 1–18 (2008)
5. Boytsov, L.: Indexing methods for approximate dictionary searching: comparative analyses. ACM J. Exp. Algorithmics **16**, 1–86 (2011)
6. Calvo, T., Beliakov, G.: Aggregation functions based on penalties. Fuzzy Sets Syst. **161**, 1420–1436 (2010)
7. Cena, A., Gagolewski, M.: Aggregation and soft clustering of informetric data. In: Baczynski, M., De Baets, B., Mesiar, R. (eds.) Proceeding 8th International Summer School on Aggregation Operators (AGOP 2015), pp. 79–84. University of Silesia, Katowice (2015)
8. Chavent, M., Saracco, J.: Central tendency and dispersion measures for intervals and hypercubes. Commun. Stat. Theor. Methods **37**, 1471–1482 (2008)

9. Coppersmith, D., Fleischer, L., Rudra, A.: Ordering by weighted number of wins gives a good ranking for weighted tournaments. In: Proceeding 17th Annual ACM-SIAM Symposium Discrete Algorithms (SODA 2006), pp. 776–782. ACM (2006)
10. Dinu, L.P., Manea, F.: An efficient approach for the rank aggregation problem. Theor. Comput. Sci. **359**(1–3), 455–461 (2006)
11. Dwork, C., Kumar, R., Naor, M., Sivakumar, D.: Rank aggregation methods for the web. In: Proceedings of the 10th International Conference on World Wide Web, pp. 613–622. ACM (2001)
12. Gagolewski, M.: Data Fusion: Theory, Methods, and Applications. Institute of Computer Science, Polish Academy of Sciences, Warsaw (2015)
13. Golub, T., et al.: Molecular classification of cancer: class discovery and class prediction by gene expression monitoring. Science **286**, 531–537 (1999)
14. Grabisch, M., Marichal, J.L., Mesiar, R., Pap, E.: Aggregation Functions. Cambridge University Press, Cambridge (2009)
15. Grzegorzewski, P.: Metrics and orders in space of fuzzy numbers. Fuzzy Sets Syst. **97**, 83–94 (1998)
16. Hastie, T., Tibshirani, R., Friedman, J.: The Elements of Statistical Learning: Data Mining, Inference, and Prediction. Springer Series in Statistics. Springer, NewYork (2013)
17. Klawonn, F., Höppner, F.: What is fuzzy about fuzzy clustering? Understanding and improving the concept of the fuzzifier. In: Berthold, M., Lenz, H.-J., Bradley, E., Kruse, R., Borgelt, C. (eds.) IDA 2003. LNCS, vol. 2810, pp. 254–264. Springer, Heidelberg (2003)
18. Leisch, F.: A toolbox for K-centroids cluster analysis. Computat. Stat. Data Anal. **51**(2), 526–544 (2006)
19. MacQueen, J.B.: Some methods for classification and analysis of multivariate observations. In: Proceeding Fifth Berkeley Symposium on Mathematical Statistics and Probability, vol. 1, pp. 281–297. University of California Press, Berkeley (1967)
20. Tibshirani, R., Hastie, T., Narasimhan, B., Chu, G.: Diagnosis of multiple cancer types by shrunken centroids of gene expression. Proc. Nat. Acad. Sci. **99**(10), 6567–6572 (2002)
21. Winkler, R., Klawonn, F., Kruse, R.: Fuzzy clustering with polynomial fuzzifier in connection with M-estimators. Appl. Comput. Math. **10**, 146–163 (2011)
22. Yu, J., Yang, M.S.: Optimality test for generalized FCM and its application to parameter selection. IEEE Trans. Fuzzy Syst. **13**(1), 164–176 (2005)

Fuzzy Sets and Fuzzy Logic

Adjoint Fuzzy Partition and Generalized Sampling Theorem

Irina Perfilieva[1(✉)], Michal Holčapek[1], and Vladik Kreinovich[2]

[1] Centre of Excellence IT4Innovations, Institute for Research and Applications of Fuzzy Modeling, University of Ostrava, 30. Dubna 22, 701 03 Ostrava 1, Czech Republic
{Irina.Perfilieva,Michal.Holcapek}@osu.cz
[2] Department of Computer Science, University of Texas at El Paso, 500 W. University, El Paso, TX 79968, USA
vladik@utep.edu

Abstract. A new notion of adjoint fuzzy partition is introduced and the reconstruction of a function from its F-transform components is analyzed. An analogy with the Nyquist-Shannon-Kotelnikov sampling theorem is discussed.

Keywords: F-transform · Adjoint fuzzy partition · Sampling theorem · Nyquist-Shannon-Kotelnikov reconstruction

1 Introduction

We analyze the problem of whether a function can be reconstructed from a countable set of its F-transform components. We prove that if a function fulfills the same conditions as in the Nyquist-Shannon-Kotelnikov theorem (also known as a sampling theorem), see [4,6,12], then the above mentioned reconstruction is possible and moreover, the sampling theorem is its particular case.

Our inspiration came from the following analogy: similar to the F-transform components, signal samples can be computed on the basis of the partition generated by Dirac's delta function δ. On the other hand, the reconstruction is performed with the help of another partition generated by the function sinc. We analyzed the interconnection between δ and sinc and extracted a principal characteristic that we call *adjointness*. If partitions are generated by adjoint functions, they are called *adjoint* as well. Adjoint fuzzy partitions are used in the direct and newly defined inverse F-transform so that their mutually inverse correspondence is guaranteed for functions that fulfill the same conditions as in the standard sampling theorem.

The F-transform is very useful in many applications such as image and signal processing, image compression, time series prediction, etc.; see, e.g., [2,5,8,9]. The initially proposed inverse F-transform [8] is lossy; i.e., except for constant functions, it produces a result that is different from an original object. This fact motivated us to modify the definition of the inverse F-transform to extend

J.P. Carvalho et al. (Eds.): IPMU 2016, Part II, CCIS 611, pp. 459–469, 2016.
DOI: 10.1007/978-3-319-40581-0_37

the space of original functions, for which direct and inverse F-transforms are mutually inverse.

In the proposed contribution[1], we give a short overview of the F-transform theory and its evolution. We discuss various fuzzy partitions and extend the notion of the inverse F-transform. We introduce a notion of an adjoint fuzzy partition and discuss its properties. Finally, we prove the main theoretical result about reconstruction from a countable set of F-transform components.

2 Preliminaries: Nyquist-Shannon-Kotelnikov Reconstruction

In this section, we provide a short review of the background of the sample-based reconstruction of a band-limited signal.

We assume that a digital signal is identified with a function varying in time, which is assumed to have a Fourier transform that is zero outside some bounded interval (in other words, a signal is *band-limited* to a given *bandwidth*). The sampling theorem (also known as Nyquist-Shannon-Kotelnikov theorem, see [4, 6, 12]) characterizes what is sufficient for full reconstruction of a signal from a set of its samples.

Theorem 1 (Sampling Theorem). *Let $x \in L_2(\mathbb{R})$ be continuous and band-limited, i.e., $\widehat{x}(\omega) = 0$ for $|\omega| > \Omega$ where $\widehat{x}$ is the Fourier transform of x and Ω is some positive constant. Then, x can be determined by its values at a discrete set of points:*

$$x(t) = \sum_{k=-\infty}^{\infty} x\left(\frac{k\pi}{\Omega}\right) \cdot \frac{\sin(\Omega t - k\pi)}{\Omega t - k\pi}. \tag{1}$$

We will be using the following notation: $h = \frac{\pi}{\Omega}$, $t_k = \frac{k\pi}{\Omega} = k \cdot h$ and the corresponding reconstruction formula:

$$x(t) = \sum_{k=-\infty}^{\infty} x(t_k) \cdot \operatorname{sinc}\left(\frac{t}{h} - k\right), \tag{2}$$

where

$$\operatorname{sinc}(t) \stackrel{\text{def}}{=} \frac{\sin(\pi t)}{\pi t}.$$

3 The F-Transform: Short Overview and Evolution

The F-transform (originally, *fuzzy transform*) is a particular integral transform whose peculiarity consists in using a *fuzzy partition* of a universe of discourse (usually, $\mathbb{R}$). We observe that the F-transform method was motivated by the

[1] The extended version of this contribution together with the application to the problem of function "de-noising" was submitted to [11].

ideas and techniques of fuzzy logic (see, e.g., [15]) and especially by the Takagi-Sugeno models [14]. In addition, the idea of a fuzzy partition was derived from observing a collection of antecedents in a fuzzy rule based system. The direct F-transform components are possible consequents in the Takagi-Sugeno model with singletons.

The F-transform has two phases: direct and inverse (see details in [8]). The direct F-transform is applied to functions from $L_2(\mathbb{R})$ and maps them linearly onto sequences (originally finite) of numeric/functional components. The inverse F-transform smoothly approximates the original function.

Let us remark that almost all fuzzy approximation models, including Takagi-Sugeno models [14], are based on linear-like combinations of fuzzy sets with numeric or functional coefficients. The principal difference between them and the inverse F-transform is in the computation of coefficients. In the F-transform case, these coefficients are weighted orthogonal projections on subdomains, such that the best approximation in a local sense is guaranteed. In Takagi-Sugeno models, the coefficients guarantee that the corresponding approximating function is a best approximation on a whole domain in the sense of the L_2 metric. Similar models have been considered in [1,7].

3.1 Fuzzy Partition

The notion of a fuzzy partition does not have a nonambiguous meaning in fuzzy literature. We will not go into full detail but concentrate on an evolution of this notion in connection with the F-transform (see [3,10,13]).

A *fuzzy partition with the Ruspini condition* was introduced in [8] as a collection of bell-shaped fuzzy sets $A_1, \ldots, A_n$ on the real interval $[a, b]$ with continuous membership functions, such that for all $x \in [a, b]$,

$$\sum_{k=1}^{n} A_k(x) = 1.$$

This partition can be characterized as a "partition-of-unity".

In [10], a generalized fuzzy partition without the Ruspini condition was proposed with the purpose of obtaining a better approximation by the inverse F-transform.

Below, in Definition 1, we introduce a particular case of a generalized fuzzy partition that is determined by a generating function. We say that function $a : \mathbb{R} \to [0, 1]$ is a *generating function of a fuzzy partition* (a *generating function*, for short), if it is non-negative, continuous, even, bell-shaped and moreover, it vanishes outside $[-1, 1]$ and fulfills $\int_{-1}^{1} a(t)\,dt = 1$. Below, we give the example of a generating function, which we call the *raised cosine*:

$$a^{cos}(t) = \begin{cases} \frac{1}{2}(1 + \cos(\pi t)), & -1 \le t \le 1, \\ 0, & \text{otherwise}. \end{cases} \tag{3}$$

Generating function a produces infinitely many *rescaled* functions $a_H : \mathbb{R} \to [0, 1]$ such that

$$a_H(t) \stackrel{\text{def}}{=} a\left(\frac{t}{H}\right),$$

where H is a positive number called a *scale factor.*

Definition 1. *Let $a : \mathbb{R} \to [0,1]$ be a generating function of a fuzzy partition, i.e., a is non-negative, continuous, even, bell-shaped, vanishes outside $[-1,1]$ and fulfills $\int_{-1}^{1} a(t)\,dt = 1$. Let $h > 0$, $t_k = t_0 + k \cdot h$, $k \in \mathbb{Z}$, be uniformly distributed nodes*[2] *in $\mathbb{R}$. Let $H > \frac{h}{2}$ and a_H be an H-rescaled version of a. With each node t_k, we correspond the translation $a_k(t) = a_H(t_k - t)$. We say that the set $\{a_k,\ k \in \mathbb{Z}\}$ establishes an (h,H)-uniform fuzzy partition of $\mathbb{R}$. Functions a_k are called basic functions.*

By the condition $H > \frac{h}{2}$, each point from $\mathbb{R}$ is "*covered*" by at least one basic function - by this we mean that the value of this function at this point is greater than zero. By the condition $h > 0$, each point from $\mathbb{R}$ is covered by at most a finite number of basic functions.

It is easy to see that (substituting $s = \frac{t}{H}$)

$$\int_{-\infty}^{\infty} a_H(t)\,dt = \int_{-H}^{H} a_H(t)\,dt = \int_{-H}^{H} a\left(\frac{t}{H}\right)\,dt = H \cdot \int_{-1}^{1} a(s)\,ds = H. \quad (4)$$

If $h = H$, then an (h,H)-uniform fuzzy partition is called an *h-uniform fuzzy partition.*

The following lemma will be used in the sequel.

Lemma 1. *Let $a : \mathbb{R} \to [0,1]$ be a generating function so that it is continuous, even, bell-shaped, vanishes outside $[-1,1]$ and fulfills $\int_{-1}^{1} a(t)\,dt = 1$. Then, the following is valid:*

$$\frac{1}{2} \le \|a\|^2 \le 1, \quad (5)$$

where $\|a\|$ is the norm in $L_2([-1,1])$.

In particular, if $a = a^{cos}$, then $\|a^{cos}\|^2 = \frac{3}{4}$.

3.2 Direct and Inverse F-Transform

In this section, we review formal notions of the direct and inverse F-transforms as introduced in [8] and extend the latter.

Assume that $x \in L_2(\mathbb{R})$ and $\{a_k,\ k \in \mathbb{Z}\}$ is an (h,H)-uniform fuzzy partition of $\mathbb{R}$, where $a_k(t) = a_H(t_k - t)$, a_H is the H-rescaled generating function a, and $t_k = k \cdot h$, $k \in \mathbb{Z}$, are nodes. The sequence $F[x] = \{X_k,\ k \in \mathbb{Z}\}$, where

$$X_k = \frac{\int_{-\infty}^{\infty} a_k(s) \cdot x(s)\,ds}{\int_{-\infty}^{\infty} a_k(s)\,ds}, \quad (6)$$

[2] For simplicity of representation, we assume that $t_0 = 0$.

is called the *(direct) F-transform* of x with respect to $\{a_k,\ k \in \mathbb{Z}\}$. Real numbers $X_k, k \in Z$, are called the *F-transform components* of x. Due to the assumption of uniformity of the partition and by (4), the representation (6) of X_k can be simplified as follows:

$$X_k = \frac{\int_{-\infty}^{\infty} a_H(t_k - s) \cdot x(s)\, ds}{\int_{-\infty}^{\infty} a_H(t_k - s)\, ds} = \frac{1}{H} \int_{-\infty}^{\infty} a_H(t_k - s) \cdot x(s)\, ds. \tag{7}$$

It is easy to see that if $x, y \in L_2(\mathbb{R})$, $\alpha \in \mathbb{R}$, then

$$\begin{aligned} F[x + y] &= F[x] + F[y], \\ F[\alpha x] &= \alpha F[x]. \end{aligned} \tag{8}$$

The basic idea of the F-transform is to "capture" a local behavior of an original function and characterize it by a certain value. It follows from (6) that the F-transform can be effectively computed for a rather wide class of functions. In particular, all continuous functions on compact domains can be originals of the F-transform.

Let $\mathbf{x} = (X_k,\ k \in \mathbb{Z})$ be an arbitrary sequence of reals and $\{a_k,\ k \in \mathbb{Z}\}$ be an (h, H)-uniform fuzzy partition of $\mathbb{R}$ with the H-rescaled generating function a. The following *inversion formula*

$$\hat{\mathbf{x}}^F(t) = \frac{\sum_{k=-\infty}^{\infty} X_k \cdot a_k(t)}{\sum_{k=-\infty}^{\infty} a_k(t)},\ t \in \mathbb{R}, \tag{9}$$

converts the sequence $\mathbf{x}$ into the real valued function $\hat{\mathbf{x}}^F$. Because the parameter h in an (h, H)-uniform fuzzy partition $\{a_k,\ k \in \mathbb{Z}\}$ of $\mathbb{R}$ is greater than zero, both sums in (9) contain only a finite number of non-zero summands. Because $H > \frac{h}{2}$, each point from $\mathbb{R}$ is covered by at least one basic function, so that the denominator in (9) is always non-zero. Therefore, the expression in (9) is well defined.

We say that the function $\hat{\mathbf{x}}^F$ is the *inverse F-transform of the sequence* $\mathbf{x} = (X_k,\ k \in \mathbb{Z})$ with respect to the fuzzy partition $\{a_k,\ k \in \mathbb{Z}\}$. If the sequence $\mathbf{x}$ consists of the F-transform components of some function x with respect to $\{a_k,\ k \in \mathbb{Z}\}$, then $\hat{\mathbf{x}}^F$ is simply called the *inverse F-transform* of x.

The inverse F-transform $\hat{\mathbf{x}}^F$ of a continuous function x can approximate x with an arbitrary precision. The desired quality of approximation can be achieved by a special choice of a partition. This fact can be easily proved using the technique introduced in [8].

4 Reconstruction from the F-Transform Components

The F-transform is the result of a linear correspondence between a set of functions from $L_2(\mathbb{R})$ and a set of sequences of reals. In general, the inversion formula does not define the inverse correspondence. In [8], it has been shown that the inverse F-transform can approximate a continuous function with an arbitrary precision. In the later publications [1,7], other smooth approximations for functions from $L_2(\mathbb{R})$ by the inverse F-transforms were proposed.

Below, we show even more; namely, the original function can be reconstructed from its F-transform components. Of course, this result can be established for a narrower than $L_2(\mathbb{R})$ class of functions. Our motivation stems from the Nyquist-Shannon-Kotelnikov reconstruction theorem discussed above.

4.1 Adjoint Partition

If a fuzzy partition is fixed, then both direct and inverse F-transforms are uniquely determined by this partition. If we require the inverse F-transform to be coincident with the original function, we shall change its main parameter – the fuzzy partition.

Definition 2. *Let $\{a_k,\, k \in \mathbb{Z}\}$ be an (h, H)-uniform fuzzy partition of $\mathbb{R}$, where $a_k(t) = a_H(t_k - t)$, a_H is the H-rescaled generating function a and $t_k = k \cdot h$, $k \in \mathbb{Z}$, are uniformly distributed nodes. We say that the set of functions $\{b_k,\, k \in \mathbb{Z}\}$, establishes an adjoint (h, H)-*uniform partition *of $\mathbb{R}$ (with respect to $\{a_k,\, k \in \mathbb{Z}\}$), if $b_k(t) = b_H(t - t_k)$ are translations of the continuous function $b_H : \mathbb{R} \to \mathbb{R}$ with the same nodes t_k, $k \in \mathbb{Z}$, and b_H is determined by*

$$\widehat{a_H} \cdot \widehat{b_H} = \mathbf{1}_{[-\Omega,\Omega]}, \tag{10}$$

where $\Omega > 0$ is some positive constant, $\mathbf{1}_{[-\Omega,\Omega]}$ is a characteristic function of $[-\Omega, \Omega]$ and $\widehat{a_H}$, $\widehat{b_H}$ are the Fourier transforms of a_H and b_H, respectively.

The lemma given below gives a necessary and sufficient condition on an $(h, 1)$-uniform fuzzy partition that guarantees the existence of the adjoint one.

Lemma 2. *Let $\{a_k,\, k \in \mathbb{Z}\}$, be an $(h, 1)$-uniform fuzzy partition of $\mathbb{R}$ with generating function $a : \mathbb{R} \to [0, 1]$, such that $a_k(t) = a(t - t_k)$ and $t_k = k \cdot h$, $k \in \mathbb{Z}$, are nodes. Then, the adjoint partition $\{b_k,\, k \in \mathbb{Z}\}$ exists if and only if there exists $\Omega > 0$ such that for all $\omega \in [-\Omega, \Omega]$,*

$$\widehat{a}(\omega) \neq 0. \tag{11}$$

Moreover, the adjoint partition $\{b_k,\, k \in \mathbb{Z}\}$ is determined by h-translations of function $b : \mathbb{R} \to \mathbb{R}$ such that

$$b(t) = \frac{1}{2\pi} \int_{-\Omega}^{\Omega} \frac{e^{i\omega t}}{\widehat{a}(\omega)}\, d\omega. \tag{12}$$

Remark 1. Let $\{a_k,\, k \in \mathbb{Z}\}$ be an (h, H)-uniform fuzzy partition of $\mathbb{R}$, where $a_k(t) = a_H(t_k - t)$ and a_H is the H-rescaled generating function a. Let $\{b_k,\, k \in \mathbb{Z}\}$, where $b_k(t) = b_H(t - t_k)$ be the adjoint (h, H)-uniform partition of $\mathbb{R}$ with respect to $\{a_k,\, k \in \mathbb{Z}\}$.

In Remark 1, we discuss some particular properties of functions b_k, $k \in \mathbb{Z}$.

(i) The function b_H is a rescaled version of a certain function $b : \mathbb{R} \to \mathbb{R}$ in both vertical and horizontal directions. Specifically,

$$b_H(t) = \frac{1}{H^2} \cdot b\left(\frac{t}{H}\right), \tag{13}$$

where b is determined as follows:

$$\widehat{a} \cdot \widehat{b} = \mathbf{1}_{[-H\Omega, H\Omega]}. \tag{14}$$

Indeed, equality (13) easily follows from (14) and the scaling property of the Fourier transform applied to the function a:

$$\widehat{a_H}(\omega) = H\widehat{a}(H\omega).$$

(ii) The explicit representation of a particular function b_k, $k \in \mathbb{Z}$ as a translation and rescaling of the function b is as follows:

$$b_k(t) = b_H(t - t_k) = \frac{1}{H^2} \cdot b\left(\frac{t - t_k}{H}\right). \tag{15}$$

This representation justifies the name "partition", assigned to the set $\{b_k,\ k \in \mathbb{Z}\}$. Moreover, as we see in Lemma 3 below, the generating function b fulfills the extended Ruspini condition (16).

We call b a *generating function of the adjoint (h, H)-uniform partition* $\{b_k,\ k \in \mathbb{Z}\}$,[3], which corresponds to the (h, H)-uniform fuzzy partition $\{a_k,\ k \in \mathbb{Z}\}$, determined by a. If $h = H$, we simply call both partitions as h-uniform.

As the following result shows, the set of translations (without rescaling) of a generating function of an adjoint H-uniform partition establishes the Ruspini partition. This is an additional argument in favor of using the word "partition" in the notion of adjoint partition.

Lemma 3. *Let $a : \mathbb{R} \to [0, 1]$ be a generating function such that for all $\omega \in [-\Omega, \Omega]$, $\widehat{a}(\omega) \neq 0$, where $\Omega > 0$ is some positive constant. Let $H = \frac{\pi}{\Omega}$ and $\{a_k,\ k \in \mathbb{Z}\}$, be an H-uniform fuzzy partition such that $a_k(t) = a_H(t - t_k)$, a_H is the H-rescaled generating function a and $t_k = k \cdot H$, $k \in \mathbb{Z}$. Let $\{b_k,\ k \in \mathbb{Z}\}$, where $b_k(t) = b_H(t - t_k)$, be the adjoint H-uniform partition of $\mathbb{R}$ with respect to $\{a_k,\ k \in \mathbb{Z}\}$ with the generating function b. Then, for all $t \in \mathbb{R}$,*

[3] We distinguish between a generating function of an adjoint partition (in this paper, denoted by b) and a generating function of a fuzzy partition (in this paper, denoted by a). The latter is characterized in Definition 1, while the former is associated with an adjoint partition and can have values outside the interval $[0, 1]$.

$$\sum_{k=-\infty}^{\infty} b\left(\frac{t}{H}-k\right)=1, \tag{16}$$

$$\sum_{k=-\infty}^{\infty} b_k(t)=\frac{1}{H^2}, \tag{17}$$

$$\sum_{k=-\infty}^{\infty} b^2\left(\frac{t}{H}-k\right)=\|b\|^2<\infty, \tag{18}$$

where $\|\cdot\|$ *is the norm in* $L_2(\mathbb{R})$.

At the end of this subsection, we give a particular example of an h-uniform partition of $\mathbb{R}$ and its adjoint where the latter has an analytic representation.

Example 1. We consider an h-uniform partition $\{\delta_k,\ k\in\mathbb{Z}\}$ of $\mathbb{R}$, where $\delta_k(t)=\delta(t-t_k)$, $t_k=k\cdot h$ and δ is the Dirac's delta function[4]. Although this partition is not fuzzy (it is generated by the non-bounded delta function), it fulfills all the assumptions of Lemma 2, including the main condition (11). The latter is because for all $\omega\in\mathbb{R}$, $\widehat{\delta}(\omega)=1$, so that we can choose an arbitrary bounded interval $[-\Omega,\Omega]$ where this condition is fulfilled. We choose $\Omega=\pi$ and apply the proof of Lemma 2 to the partition $\{\delta_k,\ k\in\mathbb{Z}\}$. After substitution into (12), we easily obtain the generating function sinc of the adjoint to $\{\delta_k,\ k\in\mathbb{Z}\}$ partition, so that

$$b(t)=\frac{1}{2\pi}\int_{-\pi}^{\pi} e^{i\omega t}d\omega=\frac{1}{\pi t}\sin(\pi t)=\operatorname{sinc}(t). \tag{19}$$

The resulting adjoint h-uniform partition is given by the set of functions $\{\operatorname{sinc}_k,\ k\in\mathbb{Z}\}$, where $\operatorname{sinc}_k(t)=\operatorname{sinc}(t-t_k)$, so that sinc is its generating function.

In Fig. 1, we demonstrate graphs of generating functions of the two adjoint uniform partitions of $\mathbb{R}$ with respect to two uniform partitions with the following generating functions: δ (Dirac's delta) and a^{cos} (raised cosine). The latter is given by (3), and it is of the fuzzy type.

In almost all cases, a computation of a generating function b of an adjoint partition cannot be performed analytically. It is a matter of a numeric computation on the basis of the expression (12). The example given Fig. 1, has been numerically computed as well.

4.2 Main Result

In this subsection, we show that a function that fulfills the same conditions as in the Nyquist-Shannon-Kotelnikov theorem (also known as a sampling theorem)

[4] Strictly speaking, the Dirac's delta is not a function, but a generalized function or a linear functional. Therefore, it makes sense to use it only if it appears inside an integral. In our paper, we always follow this restriction.

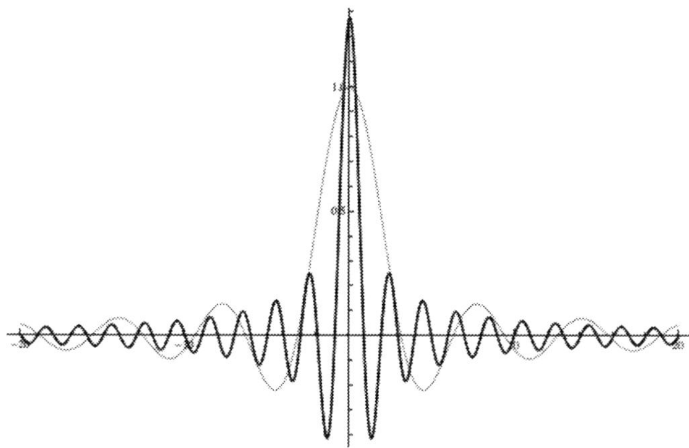

Fig. 1. Generating functions of the two adjoint uniform partitions of $\mathbb{R}$ with respect to uniform partitions with generating functions δ (in gray) and the raised cosine a^{cos} (in black).

can be reconstructed from a countable set of its F-transform components. Moreover, we obtain the sampling theorem as a particular case. The following theorem explicitly describes all conditions required for the successful reconstruction. The proof can be found in [11].

Theorem 2 (Reconstruction from the direct F-transform). *Let function $x \in L_2(\mathbb{R})$ be continuous and band-limited, i.e., $\widehat{x}(\omega) = 0$ for $|\omega| > \Omega$, where Ω is some positive constant. Let $h = \frac{\pi}{\Omega}$, $H > h/2$ and a_H an H-rescaled version of the generating function a, such that for all $\omega \in [-\Omega, \Omega]$, $\widehat{a_H}(\omega) \neq 0$.*

Let $\{b_k,\, k \in \mathbb{Z}\}$ be the adjoint (h, H)-uniform partition of $\mathbb{R}$ with respect to that given by $\{a_k,\, k \in \mathbb{Z}\}$, where $a_k(s) = a_H(t_k - s)$ and $t_k = k \cdot h$, $k \in \mathbb{Z}$.

Finally, let the sequence $\{X_k,\, k \in \mathbb{Z}\}$ consist of the F-transform components of x with respect to the fuzzy partition $\{a_k,\, k \in \mathbb{Z}\}$.

Then, x can be uniquely determined by its F-transform components, so that the following representation holds:

$$x(t) = \frac{H\pi}{\Omega} \sum_{k=-\infty}^{\infty} X_k \cdot b_k(t). \tag{20}$$

Below, we give another expression for reconstruction formula (20) in terms of generating function b of partition $\{b_k,\, k \in \mathbb{Z}\}$.

Corollary 1. *Let function x fulfill the assumptions of Theorem 2. Then, x can be reconstructed from its F-transform components so that*

$$x(t) = \frac{h}{H} \sum_{k=-\infty}^{\infty} X_k \cdot b\left(\frac{t - t_k}{H}\right), \tag{21}$$

where $b \in L_2(\mathbb{R})$ is a generating function of the adjoint (h, H)-uniform partition $\{b_k,\, k \in \mathbb{Z}\}$.

Remark 2. If in (21), we assume that $H = h$ (in other words, $\{a_k,\, k \in \mathbb{Z}\}$ is an h-uniform fuzzy partition of $\mathbb{R}$), then the reconstruction from the F-transform components takes the form

$$x(t) = \sum_{k=-\infty}^{\infty} X_k \cdot b\left(\frac{t - t_k}{h}\right) = \sum_{k=-\infty}^{\infty} X_k \cdot b\left(\frac{t}{h} - k\right), \tag{22}$$

where $b \in L_2(\mathbb{R})$ is the function whose Fourier transform is equal to

$$\widehat{b}(\omega) = \frac{\mathbf{1}_{[-\pi,\pi]}}{\widehat{a}(\omega)}. \tag{23}$$

Reconstruction (22) is similar to the Nyquist-Shannon-Kotelnikov formula (2).

In the below given corollary, we extend the range of applicability of Theorem 2 to the h-uniform partition $\{\delta_k,\, k \in \mathbb{Z}\}$ introduced in the Example 1. By this, we obtain the Nyquist-Shannon-Kotelnikov reconstruction in the form of (2), see the proof in [11].

Corollary 2. *Let the assumptions of Theorem 2 be fulfilled and the Dirac's delta δ and* sinc *be chosen as generating function of an h-uniform partition $\{\delta_k,\, k \in \mathbb{Z}\}$ and the corresponding adjoint h-uniform partition $\{\mathrm{sinc}_k,\, k \in \mathbb{Z}\}$. Then, after respective substitutions the reconstruction formula (22) becomes equivalent with the Nyquist-Shannon-Kotelnikov reconstruction in the form of (2).*

Remark 3. The principal difference between the Nyquist-Shannon-Kotelnikov and the proposed reconstruction is that the former one works as an interpolating technique, while the latter one is able to perform reconstruction even from averaged values of a given function.

5 Conclusion

We discussed the problem of reconstruction from a set of F-transform components. We introduced the adjoint fuzzy partition and the inversion formula and proved that a function can be reconstructed from its F-transform components. Moreover, we showed that if the Dirac's delta δ is chosen as generating function of an h-uniform partition, then the reconstruction from the F-transform components becomes equivalent with the Nyquist-Shannon-Kotelnikov reconstruction.

Acknowledgement. This work was partially supported by the project LQ1602 IT4Innovations excellence in science.

References

1. Crouzet, J.-F.: Fuzzy projection versus inverse fuzzy transform as ampling/interpolation schemes. Fuzzy Sets Syst. **193**, 108–121 (2012)
2. Holčapek, M., Tichý, T.: A smoothing filter based on fuzzy transform. Fuzzy Sets Syst. **180**, 69–97 (2011)
3. Holčapek, M., Perfilieva, I., Novák, V., Kreinovich, V.: Necessary and sufficient conditions for generalized uniform fuzzy partitions. Fuzzy Sets Syst. **277**, 97–121 (2015)
4. Kotelnikov, V.A.: On the carrying capacity of the ether and wire in telecommunications. In: Material for the First All-Union Conference on Questions of Communication, Izd. Red. Upr. Svyazi RKKA, Moscow (1933). (in Russian)
5. Novák, V., Štěpnička, M., Dvořák, A., Perfilieva, I., Pavliska, V., Vavříčková, L.: Analysis of seasonal time series using fuzzy approach. Int. J. General Syst. **39**(3), 305–328 (2010)
6. Nyquist, H.: Certain topics in telegraph transmission theory. Trans. AIEE **47**, 617–644 (1928)
7. Patanè, G.: Fuzzy transform and least-squares approximation: analogies differences and generalizations. Fuzzy Sets Syst. **180**, 41–54 (2011)
8. Perfilieva, I.: Fuzzy transforms: theory and applications. Fuzzy Sets Syst. **157**, 993–1023 (2006)
9. Perfilieva, I.: Fuzzy transforms: a challenge to conventional transforms. In: Hawkes, P.W. (ed.) Advances in Images and Electron Physics, vol. 147, pp. 137–196. Elsevier Academic Press, San Diego (2007)
10. Perfilieva, I.: F-transform. In: Kacprzyk, J., Pedrycz, W. (eds.) Springer Handbook of Computational Intelligence, pp. 113–130. Springer, Heidelberg (2015)
11. Perfilieva, I., Holčapek, M., Kreinovich, V.: A new reconstruction from the F-Transform components. Fuzzy Sets Syst. **288**, 3–25 (2016)
12. Shannon, C.E.: Communication in the presence of noise. Proc. Inst. Radio Eng. **37**(1), 10–21 (1949)
13. Stefanini, L.: F-transform with parametric generalized fuzzy partitions. Fuzzy Sets Syst. **180**, 98–120 (2011)
14. Takagi, T., Sugeno, M.: Fuzzy identification of systems and its application to modeling and control. IEEE Trans. Syst. Man Cybern. **15**, 116–132 (1985)
15. Zadeh, L.A.: Fuzzy sets. Inf. Control **8**, 338–353 (1965)

How to Incorporate Excluding Features in Fuzzy Relational Compositions and What for

Nhung Cao(✉) and Martin Štěpnička

Centre of Excellence IT4Innovations, Institute for Research and Applications of Fuzzy Modeling, University of Ostrava, 30. dubna 22, 701 03 Ostrava 1, Czech Republic
{nhung.cao,martin.stepnicka}@osu.cz
http://irafm.osu.cz/

Abstract. The aim of this paper is, first, to recall fuzzy relational compositions (products) and, to introduce an idea, how the so-called excluding features could be incorporated into the theoretical background. Apart from rather natural definitions, we provide readers with a theoretical investigation that provides and answer to a rather natural question, under which conditions, in terms of the underlying algebraic structures, the proposed incorporation of excluding features preserves the same properties as the incorporation in the classical relational compositions. The positive impact of the incorporation on reducing the suspicions provided by the basic "circlet" composition without losing the possibly correct suspicion, as in the case of the use of the Bandler-Kohout products, is demonstrated on an example.

Keywords: Fuzzy relational compositions · Fuzzy relational products · Bandler-Kohout products · Residuated algebraic structures · Medical diagnosis · Classification

1 Introduction and Preliminaries

1.1 Introduction

Fuzzy relational compositions[1] have an important role in many areas of fuzzy mathematics, including the formal constructions of fuzzy inference systems [1,2], medical diagnosis [3], architectures of information processing [4] or in flexible queries to relational databases [5]. They provide an extension of classical relational compositions and have been firstly studied by Willis Bandler and Ladislav Kohout

M. Štěpnička—This research was partially supported by the NPU II project LQ1602 "IT4Innovations excellence in science" provided by the MŠMT.

[1] Instead of the term composition, one may often encounter the term "product" denoting the same mappings or objects. This terminology naturally comes from the product-like matrix calculation of the compositions.

J.P. Carvalho et al. (Eds.): IPMU 2016, Part II, CCIS 611, pp. 470–481, 2016.
DOI: 10.1007/978-3-319-40581-0_38

between late 70's and early 80's. There are numerous studies deeply elaborating various aspects of fuzzy relation compositions, see e.g. [6,7].

Thinking about the medical diagnosis problem, we show, how the "excluding symptoms" may be incorporated into the fuzzy relational products in order to improve and give precision to the suspicion provided the standard fuzzy relational products. The contribution of this extension will be demonstrated on an illustrative classification example.

1.2 Background Algebraic Structure of Truth Values

Recall some basic definitions of underlying algebraic structures and some properties, that hold in these structures and that will be used in the sequel.

Definition 1. An algebra $\mathcal{L} = \langle L, \wedge, \vee, \otimes, \rightarrow, 0, 1\rangle$ is a *residuated lattice* if

1. $\langle L, \wedge, \vee, 0, 1\rangle$ is a lattice with the least and the greatest element
2. $\langle L, \otimes, 0, 1\rangle$ is a commutative monoid such that $\otimes$ is isotone in both arguments
3. the operation $\rightarrow$ is a residuation with respect to $\otimes$, i.e.

$$a \otimes b \leq c \quad \text{iff} \quad a \rightarrow c \geq b. \tag{1}$$

Let us list some of the useful properties that are immediately available to us for any $a, b, c \in \mathcal{L}$ [8]:

$$a \rightarrow b = 1 \text{ whenever } a \leq b, \tag{2}$$

$$a \rightarrow c \geq b \rightarrow c \text{ whenever } a \leq b, \tag{3}$$

$$a \rightarrow b \leq a \rightarrow c \text{ whenever } b \leq c, \tag{4}$$

$$a \otimes (a \rightarrow b) \leq b, \tag{5}$$

$$a \rightarrow (b \rightarrow c) = (a \otimes b) \rightarrow c = (b \otimes a) \rightarrow c, \tag{6}$$

We can define additional operations for all $a, b \in L$, namely: biresiduation (biimplication, residual equivalence), negation, and addition, respectively:

$$a \leftrightarrow b = (a \rightarrow b) \wedge (b \rightarrow a),$$

$$\neg a = a \rightarrow 0,$$

$$a \oplus b = \neg(\neg a \otimes \neg b).$$

Lemma 1. *Let $\langle L, \wedge, \vee, \otimes, \rightarrow, 0, 1\rangle$ be a residuated lattice. Then for all $a, b \in L$:*

$$\neg(a \otimes b) \geq \neg a \oplus \neg b. \tag{7}$$

Proof. Due to (5) and the adjunction property, we get $a \leq (a \rightarrow 0) \rightarrow 0$. Then using the definition of the addition, negation and the antitonicity of residual implication $\rightarrow$ in the first argument (3), we get

$$\begin{aligned}\neg a \oplus \neg b &= (((a \rightarrow 0) \rightarrow 0) \otimes ((b \rightarrow 0) \rightarrow 0)) \rightarrow 0\\ &\leq (a \otimes b) \rightarrow 0 = \neg(a \otimes b).\end{aligned}$$

□

Let us recall two more notions, that will be used in the sequel in order to narrow the class of residuated lattices where some desirable properties will be preserved.

Definition 2. Let $\langle L, \wedge, \vee, \otimes, \rightarrow, 0, 1\rangle$ be a residuated lattice. We say that the negation $\neg$ is *strict*, if the following holds:

$$\neg a = \begin{cases} 0, & a > 0 \\ 1, & a = 0. \end{cases}$$

Furthermore, we say that an element $a \in L \smallsetminus \{0, 1\}$ is a *zero divisor* of $\otimes$ if there exists some $b \in L \smallsetminus \{0, 1\}$ such that $a \otimes b = 0$.

Finally, let us recall the MV-algebra.

Definition 3. An MV-algebra is an algebra $\mathcal{L} = \langle L, \oplus, \otimes, \neg, 0, 1\rangle$ with two binary operations $\oplus, \otimes$, a unary operation $\neg$ and two constants such that $\langle L, \oplus, 0\rangle$ and $\langle L, \otimes, 1\rangle$ are commutative monoids and the following identities hold:

$$\begin{aligned} a \oplus \neg a &= 1, & a \otimes \neg a &= 0, \\ \neg(a \oplus b) &= \neg a \otimes \neg b, & \neg(a \otimes b) &= \neg a \oplus \neg b, \\ a &= \neg\neg a, & \neg 0 &= 1, \\ \neg(\neg a \oplus b) \oplus b &= \neg(\neg b \oplus a) \oplus a. \end{aligned}$$

Every MV-algebra $\langle L, \oplus, \otimes, \neg, 0, 1\rangle$ is a residuated lattice $\langle L, \wedge, \vee, \otimes, \rightarrow, 0, 1\rangle$ by putting

$$\begin{aligned} a \vee b &= \neg(\neg a \oplus b) \oplus b = (a \otimes \neg b) \oplus b, \\ a \wedge b &= \neg(\neg a \vee \neg b) = (a \oplus \neg b) \otimes b, \\ a \rightarrow b &= \neg a \oplus b, \end{aligned}$$

where $\rightarrow$ is a residuation operation with respect to $\otimes$.

A Heyting algebra is a residuated lattice with $\otimes = \wedge$, i.e., $\langle L, \vee, \wedge, \rightarrow, 0, 1\rangle$. An MV-algebra or residuated lattice is complete if the underlying lattice is complete.

Let us recall [8], that in a complete residuated lattice, the following holds for any $a, b, c \in \mathcal{L}$ and for any index set I:

$$\bigvee_{i \in \mathcal{I}} a_i \rightarrow b = \bigwedge_{i \in \mathcal{I}} (a_i \rightarrow b). \tag{8}$$

The following equality we immediately get as a consequence of (8) and the definition of the negation:

$$\neg \bigvee_{i \in I} a_i = \bigwedge_{i \in I} \neg a_i.$$

The well known examples of residuated lattice is the Łukasiewicz algebra, the Gödel algebra or the Goguen algebra. The two latter ones, however, are not MV-algebras and thus, the double negation law does not hold in these algebras.

1.3 Relational and Fuzzy Relational Compositions

Let us recall some basic facts about relational compositions, from which the fuzzy relational compositions naturally stem. Following the original work of Willis Bandler and Ladislav Kohout (cf. [9]), and for the sake of illustrative nature explaining the semantic of the compositions, we assume that X be a finite set of patients, Y be a finite set of symptoms and Z be a set of diseases. Let R be a binary relation on $X \times Y$ and S be a binary relation on $Y \times Z$. Then four fundamental compositions are given by

$$R \circ S = \{(x, z) \in X \times Z \mid \exists\, y \in Y : (x, y) \in R \,\&\, (y, z) \in S\}, \tag{9}$$
$$R \lhd S = \{(x, z) \in X \times Z \mid \forall\, y \in Y : (x, y) \in R \Rightarrow (y, z) \in S\}, \tag{10}$$
$$R \rhd S = \{(x, z) \in X \times Z \mid \forall\, y \in Y : (x, y) \in R \Leftarrow (y, z) \in S\}, \tag{11}$$
$$R \,\square\, S = \{(x, z) \in X \times Z \mid \forall\, y \in Y : (x, y) \in R \Leftrightarrow (y, z) \in S\}. \tag{12}$$

and are called *basic (direct/circlet)* composition, *Bandler-Kohout (abbr. BK) subproduct*, *Bandler-Kohout superproduct* and *Bandler-Kohout square product*, respectively.

The idea behind these compositions stems from the assumption that the meaning of $(x, y) \in R$ is that patient x has a symptom y and that $(y, z) \in S$ expresses the fact that symptom y belongs to diseases z. Then the semantic of $(x, z) \in R \circ S$ is that patient x has at least one symptom belonging to disease z and therefore, it expresses a suspicion of having this disease. The "triangle" and square compositions (10)–(12) provide a sort of more accurate specification or a strengthening of the initial suspicion. The fact that $(x, z) \in R \lhd S$ means that all symptoms of patient x belong to disease z; the fact that $(x, z) \in R \rhd S$ means that patient x has all symptoms belonging to disease z; and finally, the meaning of $(x, z) \in R \,\square\, S$ is that patient x has all symptoms of the disease z and all symptoms of the patient belong to disease z.

As mentioned above, the relational compositions have been extended for fuzzy relations $R \mathrel{\underset{\sim}{\subset}} X \times Y$ and $S \mathrel{\underset{\sim}{\subset}} Y \times Z$ in order to deal with partial connection between elements from distinct universes X, Y, and Z. Below we recall a standard definition of the extension of all the four above given compositions of classical compositions.

Definition 4. Let X, Y, Z be non-empty universes, let $R \mathrel{\underset{\sim}{\subset}} X \times Y, S \mathrel{\underset{\sim}{\subset}} Y \times Z$. Then the *compositions* $\circ, \lhd, \rhd, \square$ *of fuzzy relations* R and S are fuzzy relations on $X \times Z$ defined as follows:

$$(R \circ S)(x,z) = \bigvee_{y \in Y} (R(x,y) \otimes S(y,z)),$$
$$(R \lhd S)(x,z) = \bigwedge_{y \in Y} (R(x,y) \to S(y,z)),$$
$$(R \rhd S)(x,z) = \bigwedge_{y \in Y} (R(x,y) \leftarrow S(y,z)),$$
$$(R \square S)(x,z) = \bigwedge_{y \in Y} (R(x,y) \leftrightarrow S(y,z)),$$

for all $x \in X$ and $z \in Z$.

For the sake of completeness, let us also mention so-called inf-S compositions [10,11] that were neither motivated by the medical diagnosis problem nor studied by Bandler and Kohout in their original studies in 70's and 80's, however, also fit into the theory of fuzzy relational compositions and will be used in the sequel.

Definition 5. Let X, Y, Z be non-empty universes, let $R \mathrel{\underset{\sim}{\subset}} X \times Y, S \mathrel{\underset{\sim}{\subset}} Y \times Z$. Then the *composition* $\triangledown$ *of fuzzy relations* R and S is a fuzzy relation on $X \times Z$ defined as follows:

$$(R \triangledown S)(x,z) = \bigwedge_{y \in Y} (R(x,y) \oplus S(y,z))$$

for all $x \in X$ and $z \in Z$.

However, as it has been noted e.g. in [12,13] one may often encounter a situation, when the basic suspicion $\circ$ is high for many pairs (x, z), i.e. nearly each patient is suspicious of having many diseases. This becomes when too many common symptoms are shared by many diseases and thus, many patients have at least one symptom belonging to nearly any disease. The problem mentioned [12,13] is that the BK products $\lhd, \rhd$ or $\square$ do not have to be helpful in this case because patients may have most but not all symptoms of a given disease and, analogously, patients may have not only symptoms of a single disease. In such cases, the BK products give only low values and do not help to strengthen the suspicion by excluding some of the diseases originally viewed as potential ones by the direct product $\circ$. Therefore, the authors proposed to replace the existential and universal quantifiers, represented in the Definition 4 by suprema and infima, by generalized intermediate quantifiers [14] determined by fuzzy measure [15] in order to build "softer" compositions searching, e.g., for pairs of patients and diseases such that the patients has "most" of the symptoms belonging to the given disease.

Remark 1. Besides [12,13], there were also other approaches to improve the precision of the suspicions determined by the basic compositions. We may recall e.g. [16] where the authors introduce a sort of weight parameter in order to emphasize distinct influence of features. However, this approach does not use fuzzy sets but their extensions – interval-valued fuzzy sets. Moreover, they focus on the BK-subproduct and not the direct product and thus, their approach is not compatible with our approach introduced below.

1.4 Excluding Symptoms

This investigation is motivated by the existence of excluding symptoms for particular diseases, i.e., symptoms, that directly exclude possibility having a particular disease, if being carried by a given patient, no matter how many other symptoms linking the given patient with the given diseases exist. Below, we will show how this idea can be incorporated using the fuzzy relational compositions and how it can be useful. The demonstrative example will also show, that although the implementation of the excluding symptoms does something totally different than the use of the generalized quantifiers [13], it can be similarly helpful in reducing the number of suspicions and thus, potential, if being combined with the generalized quantifiers, it may become a very efficient tool.

So, let us assume that $E \subset Y \times Z$ is a relation with the semantic that $(y, z) \in E$ means that y is an excluding symptom for disease z. Then we may introduce relation $R \circ S^{\backslash} E \subseteq X \times Z$ as follows

$$\begin{aligned} R \circ S^{\backslash} E = & \{(x, z) \in X \times Z \mid (\exists\, y \in Y : (x, y) \in R \;\&\; (y, z) \in S) \;\& \\ & (\nexists y \in Y : (x, y) \in R \;\&\; (y, z) \in E)\}. \end{aligned} \tag{13}$$

The proposed composition provides the desirable meanings, i.e., the fact that $(x, z) \in R \circ S^{\backslash} E$ means that patient x has at least one symptom belonging to a disease z and there is no excluding symptoms related to disease z carried by this patient.

In the case of classical binary relations, the above introduced incorporation of excluding symptoms may be rewritten in two alternative ways.

Lemma 2. *Let X, Y, Z be non-empty universes, let $R \subseteq X \times Y$, $S \subseteq Y \times Z$ and $E \subseteq Y \times Z$ and let $R \circ S^{\backslash} E$ be given by (13). Then it holds*

$$\begin{aligned} R \circ S^{\backslash} E = & \{(x, z) \in X \times Z \mid (\exists\, y \in Y : (x, y) \in R \;\&\; (y, z) \in S) \;\& \\ & (\forall y \in Y : (x, y) \notin R \;\vee\; (y, z) \notin E)\}, \end{aligned} \tag{14}$$

$$\begin{aligned} R \circ S^{\backslash} E = & \{(x, z) \in X \times Z \mid (\exists\, y \in Y : (x, y) \in R \;\&\; (y, z) \in S) \;\& \\ & (\forall y \in Y : (x, y) \in R \;\Rightarrow\; (y, z) \notin E)\}. \end{aligned} \tag{15}$$

Proof. It follows from the fact that, in classical case:

$$\begin{aligned} \nexists y : (R(x, y) \;\&\; E(y, z)) & \equiv \forall y : \neg(R(x, y) \;\&\; E(y, z)) \\ & \equiv \forall y : (\neg R(x, y) \vee \neg E(y, z)) \\ & \equiv \forall y : (R(x, y) \Rightarrow \neg E(y, z)). \end{aligned}$$

□

If we introduce the notation naturally symbolizing the used notations, i.e., let $R \circ S^{\backprime}E^{\triangledown}$ denotes the right hand side of (14) and let $R \circ S^{\backprime}E^{\triangleleft}$ denotes the right hand side of (15). Then we can easily write:

$$R \circ S^{\backprime}E = (R \circ S^{\backprime}E)^{\triangledown} = (R \circ S^{\backprime}E)^{\triangleleft}.$$

This equivalence is somehow consistent with a human intuition and thus, it is a desirable property. However, in case of the incorporation of excluding symptoms in the fuzzy relational compositions, the equality does not come automatically and some additional properties have to be assumed. Investigation focusing on this will be provided in the next section.

We present a short lemma that demonstrates that the concept of the excluding features cannot be easily avoided by using only the $\circ$ composition and an appropriate choice of the fuzzy relation S.

Lemma 3. *Let $T = S \cap \neg E$. Then $R \circ S^{\backprime}E \neq R \circ T$.*

2 Fuzzy Relational Compositions and Excluding Features

2.1 Residuated Lattice as the Background Algebraic Structure

In this Section, we assume that the background algebraic structure is the complete residuated lattice $\langle L, \wedge, \vee, \rightarrow, \otimes, 0, 1\rangle$ and the R, S, E are binary fuzzy relations on $X \times Y$ an $Y \times Z$, respectively

Definition 6. Let X, Y, Z be non-empty universes, let $R \mathrel{\underset{\sim}{\subset}} X \times Y, S, E \mathrel{\underset{\sim}{\subset}} Y \times Z$. Then $R \circ S^{\backprime}E$ is a fuzzy relations on $X \times Z$ defined as follows:

$$(R \circ S^{\backprime}E)(x,z) = \bigvee_{y \in Y} (R(x,y) \otimes S(y,z)) \otimes \neg \bigvee_{y \in Y} (R(x,y) \otimes E(y,z)), \tag{16}$$

$$(R \circ S^{\backprime}E)^{\triangleleft}(x,z) = \bigvee_{y \in Y} (R(x,y) \otimes S(y,z)) \otimes \bigwedge_{y \in Y} (R(x,y) \rightarrow \neg E(y,z)), \tag{17}$$

$$(R \circ S^{\backprime}E)^{\triangledown}(x,z) = \bigvee_{y \in Y} (R(x,y) \otimes S(y,z)) \otimes \bigwedge_{y \in Y} (\neg R(x,y) \oplus \neg E(y,z)). \tag{18}$$

By the initial definition of fuzzy relational composition, expressions (16)–(18) can be rewritten as follows:

$$(R \circ S^{\backprime}E)(x,z) = (R \circ S)(x,z) \otimes \neg(R \circ E)(x,z), \tag{19}$$

$$(R \circ S^{\backprime}E)^{\triangleleft}(x,z) = (R \circ S)(x,z) \otimes (R \triangleleft \neg E)(x,z), \tag{20}$$

$$(R \circ S^{\backprime}E)^{\triangledown}(x,z) = (R \circ S)(x,z) \otimes (\neg R \triangledown \neg E)(x,z), \tag{21}$$

respectively.

Remark 2. Note, that (19)–(21) actually present that the fuzzy relational composition incorporating excluding features always as a combination of two fundamental compositions, either $\circ$ and $\triangleleft$, or $\circ$ and $\triangledown$. This is not a new approach and even the Bandler-Kohout products were re-defined by B. De Baets and E. Kerre using such a combination in order to avoid trivial suspicions, see [6].

Lemma 4. *Let the underlying algebraic structure* $\langle L, \wedge, \vee, \rightarrow, \otimes, 0, 1\rangle$ *be a complete residuated lattice. Then*

$$(R \circ S^{\backprime}E)(x, z) = (R \circ S^{\backprime}E)^{\lhd}(x, z), \tag{22}$$

$$(R \circ S^{\backprime}E)(x, z) \supseteq (R \circ S^{\backprime}E)^{\triangledown}(x, z). \tag{23}$$

Proof. In order to prove (22) it is sufficient to prove that $\neg(R \circ E) = (R \lhd \neg E)$. Using the fact $\neg \bigvee_{i\in I} a = \bigwedge_{i\in I} \neg a$ for an arbitrary index set I and using the property $(a \otimes b) \rightarrow c = a \rightarrow (b \rightarrow c)$, we may proceed as follows:

$$\begin{aligned}\neg \bigvee_{y\in Y} (R(x, y) \otimes E(y, z)) &= \bigwedge_{y\in Y} \neg(R(x, y) \otimes E(y, z)) \\ &= \bigwedge_{y\in Y} ((R(x, y) \otimes E(y, z)) \rightarrow 0) \\ &= \bigwedge_{y\in Y} (R(x, y) \rightarrow (E(y, z) \rightarrow 0)) \\ &= \bigwedge (R(x, y) \rightarrow \neg E(y, z)) .\end{aligned}$$

The proof of (23) uses property (7):

$$\begin{aligned}\neg \bigvee_{y\in Y} (R(x, y) \otimes E(y, z)) &= \bigwedge_{y\in Y} \neg(R(x, y) \otimes E(y, z)) \\ &\geq \bigwedge_{y\in Y} (\neg R(x, y) \oplus \neg E(y, z)).\end{aligned}$$

□

Note again, that the concept of the excluding features cannot be easily avoided and that Lemma 3.

2.2 The Case of Particular Classes of Residuated Lattices

As we can see, the equality of $(R \circ S^{\backprime}E)$ and $(R \circ S^{\backprime}E)^{\triangledown}$ is not generally preserved in the residuated lattice. The question is, whether adding some additional properties narrowing the class of residuated lattices as the underlying structures would be sufficient in order to obtain the equality. The answer is given by the following lemma.

Lemma 5. *Let the underlying algebraic structure* $\langle L, \wedge, \vee, \rightarrow, \otimes, 0, 1\rangle$ *be a complete residuated lattice such that the negation* $\neg a = a \rightarrow 0$ *is strict and* $\otimes$ *has no zero divisors. Then*

$$(R \circ S^{\backprime}E)(x, z) = (R \circ S^{\backprime}E)^{\triangledown}(x, z). \tag{24}$$

Proof. First, let us prove that $\neg a \oplus \neg b = \neg(a \otimes b)$ holds in the given structure. We know, that in any residuated lattice, the following holds

$$\neg a \oplus \neg b = \neg(\neg(\neg a) \otimes \neg(\neg b)) = (((a \rightarrow 0) \rightarrow 0) \otimes ((b \rightarrow 0) \rightarrow 0)) \rightarrow 0.$$

Now, let us assume that a or b equals to 0 (w.l.o.g $a = 0$). Then

$$\begin{aligned}\neg a \oplus \neg b &= (((0 \to 0) \to 0) \otimes ((b \to 0) \to 0)) \to 0 \\ &= ((1 \to 0) \otimes ((b \to 0) \to 0)) \to 0 \\ &= (0 \otimes ((b \to 0) \to 0)) \to 0 \\ &= 0 \to 1 = 1 \geq \neg(a \otimes b)\end{aligned}$$

and as the opposite inequality holds generally in any residuated lattice, we get $\neg a \oplus \neg b = \neg(a \otimes b)$. Now, assume that $a, b > 0$. Then, as $\otimes$ has no zero divisors, $a \otimes b > 0$ and as $\neg$ is strict, we get

$$\begin{aligned}\neg(a \otimes b) &= (a \otimes b) \to 0 \\ &= 0 \leq \neg a \oplus \neg b\end{aligned}$$

from which, we again get the equality $\neg a \oplus \neg b = \neg(a \otimes b)$.

Then the proof proceeds easily as follows:

$$\begin{aligned}\neg \bigvee_{y \in Y} (R(x,y) \otimes E(y,z)) &= \bigwedge_{y \in Y} \neg(R(x,y) \otimes E(y,z)) \\ &= \bigwedge_{y \in Y} (\neg R(x,y) \oplus \neg E(y,z)).\end{aligned}$$

□

The following corollaries demonstrate the importance of the above results for some of the most usual algebraic structures.

Corollary 1. *Let the underlying algebraic structure $\langle L, \wedge, \vee, \to, \otimes, 0, 1\rangle$ be the complete Gödel algebra or the complete Goguen algebra. Then*

$$(R \circ S^{\backprime} E)(x,z) = (R \circ S^{\backprime} E)^{\triangledown}(x,z).$$

Corollary 2. *Let the underlying algebraic structure $\langle L, \wedge, \vee, \to, 0, 1\rangle$ be the complete Heyting algebra. Then*

$$(R \circ S^{\backprime} E)(x,z) = (R \circ S^{\backprime} E)^{\triangledown}(x,z).$$

One may also naturally ask, whether the equalities hold for the MV-algebra $\langle L, \oplus, \otimes, \neg, 0, 1\rangle$ which is the most natural generalization of Boolean algebra. MV-algebra is a residuated lattice, but $\otimes$ has zero divisors and the negation $\neg$ is not strict, yet it is involutive. Typical example of the MV-algebra is the Łukasiewicz algebra. The answer is positive.

Lemma 6. *Let the underlying algebraic structure $\langle L, \oplus, \otimes, \neg, 0, 1\rangle$ be a complete MV-algebra. Then*

$$\begin{aligned}(R \circ S^{\backprime} E)(x,z) &= (R \circ S^{\backprime} E)^{\triangleleft}(x,z), \\ (R \circ S^{\backprime} E)(x,z) &= (R \circ S^{\backprime} E)^{\triangledown}(x,z).\end{aligned}$$

Proof. The first equality holds because MV-algebra is a residuated lattice, where the inequality holds generally. The second equality is proved analogously as in the proof of Lemma 5 using the fact, that $\neg a \oplus \neg b = \neg(a \otimes b)$ holds in any MV-algebra. □

3 Demonstrative Example

Let us demonstrate the influence of the incorporation of the excluding features into the fuzzy relational compositions on a short yet illustrative example. Let Z be a set of families of animals (z_1 - Bird, z_2 - Fish, z_3 - Dog, z_4 - Equidae, z_5 - Mosquito, z_6 - Monotreme, z_7 - Reptile), Y be a set of animal features (y_1 - flies, y_2 - feathers, y_3 - fins, y_4 - claws, y_5 - hair, y_6 - teeth, y_7 - beak, y_8 - scales, y_9 - swims) and let X be a set of particular animals (Platypus, Emu, Hairless dog, Aligator, Parrotfish, Puffin) we want to classify to their families. Furthermore, let $S, E \subsetsim Y \times Z$ be given as follows

S	z_1	z_2	z_3	z_4	z_5	z_6	z_7
y_1	0.8	0	0	0	1	0	0
y_2	1	0	0	0	0	0	0
y_3	0	1	0	0	0	0.5	0
y_4	0.9	0	1	0	0	0.8	0.3
y_5	0	0	0.8	1	0	0.9	0
y_6	0	0.6	1	1	0	0	0.7
y_7	1	0.1	0	0	0	0.5	0
y_8	0.7	0.9	0	0	0	0	1
y_9	0.5	1	0.8	0.6	0.1	0.7	0.8

E	z_1	z_2	z_3	z_4	z_5	z_6	z_7
y_1	0	1	1	1	0	1	1
y_2	0	1	1	1	1	1	1
y_3	1	0	1	1	1	0	1
y_4	0	1	0	1	1	0	0
y_5	0.8	1	0	0	1	0	1
y_6	1	0	0	0	1	1	0
y_7	0	0.1	1	1	1	0	1
y_8	0	0	1	0	1	1	0
y_9	0	0	0	0	0.8	0	0

and let $R \subsetsim X \times Y$ be given as follows

R	y_1	y_2	y_3	y_4	y_5	y_6	y_7	y_8	y_9
Platypus	0	0	0	1	1	0	1	0	0.9
Emu	0	1	0	1	0	0	1	0.5	0.4
Hairless dog	0	0	0	1	0.2	1	0	0	0.7
Aligator	0	0	0	1	0	1	0	1	0.9
Parrotfish	0	0	1	0	0	0.9	0.8	1	1
Puffin	1	1	0	1	0	0	1	0.4	0.9

If we use the Łukasiweicz algebra as the underlying algebraic structure then we get the following compositions

$R \circ S$	z_1	z_2	z_3	z_4	z_5	z_6	z_7
Platypus	1	0.9	1	1	0	0.9	0.7
Emu	1	0.4	1	0	0	0.8	0.5
Hairless dog	0.9	0.7	1	1	0	0.8	0.7
Aligator	0.9	0.9	1	1	0	0.8	1
Parrotfish	0.8	1	0.9	0.9	0.1	0.7	1
Puffin	1	0.9	1	0.5	1	0.8	0.7

$R \lhd S$	z_1	z_2	z_3	z_4	z_5	z_6	z_7
Platypus	0	0	0	0	0	0.5	0
Emu	0.9	0	0	0	0	0	0
Hairless dog	0	0	1	0	0	0	0.3
Aligator	0	0	0	0	0	0	0.3
Parrotfish	0	0.3	0	0	0	0	0
Puffin	0.6	0	0	0	0	0	0

$R \square S$	z_1	z_2	z_3	z_4	z_5	z_6	z_7
Platypus	0	0	0	0	0	0.5	0
Emu	0.2	0	0	0	0	0	0
Hairless dog	0	0	0.4	0	0	0	0
Aligator	0	0	0	0	0	0	0.3
Parrotfish	0	0.3	0	0	0	0	0
Puffin	0.6	0	0	0	0	0	0

$R \circ S^{\backslash} E$	z_1	z_2	z_3	z_4	z_5	z_6	z_7
Platypus	0.2	0	0	0	0	0.9	0
Emu	1	0	0	0	0	0	0
Hairless dog	0	0	1	0	0	0	0.5
Aligator	0	0	0	0	0	0	1
Parrotfish	0	1	0	0	0	0	0
Puffin	1	0	0	0	0	0	0

As one may see, the initial suspicion given by $R \circ S$ is too high for too many pairs and does not help to classify a given animal into a correct family. The strengthening of the suspicion given by $R \lhd S$ is, on the other hand, at some cases too much restrictive and if we add also the other triangle product $\rhd$ and finally calculate the square product $R \square S$, we get nearly no more suspicion. This is a consequence of the use of the universal quantifier and the frequent occurrence of animals that do not carry all the typical features fully (e.g. Hairless dog has nearly no hair, Emu does not fly) and also the fact, that some animals carry features, do not belong so much to the given family of animals (e.g. Parrotfish has a beak). However, if the concept of excluding features is incorporated appropriately, it can help to eliminate fake initial suspicion without lowering the membership degrees to the correct families. This requires an appropriate choice of values in E which should not exclude the family of fishes from the suspicions ones if a given animal has a beak or, it should not exclude the family of Birds, if a given animal does not fly. On the other hand, if an animal has hair, it should exclude Birds and vice-versa, if an animal has feathers, Monotreme should be excluded.

4 Conclusions and Future Work

We have shortly recalled the definitions of relational and fuzzy relational compositions and motivated the incorporation of so-called excluding features into the basic "circlet" composition. We have theoretically investigated, under which conditions, particularly in which algebras, the incorporation of the excluding features preserves the same property as in the classical relational compositions. The investigated property was the equivalence of three distinct types how this idea could be incorporated, which is undoubtedly a desirable property preventing non-uniqueness and confusing of results. We have proved that in applications the most often used algebras, i.e. Gödel and Goguen BL-algebras and the Łukasiewcz

MV-algebras are among the appropriate ones. Finally the use of the excluding features was demonstrated on an illustrative example, which was by purpose chosen out of the medical diagnosis practice in order to demonstrate also other potential areas of application. This choice was not arbitrary as this investigation is motivated by an on-going interdisciplinary research leading to building a mobile application automatically classifying a given odonata (dragonfly) based on features inserted into the application by a volunteer biologist. This project coincides with the goal of the city of Ostrava to support preservation biology and technological services including citizen sciences.

References

1. Pedrycz, W.: Applications of fuzzy relational equations for methods of reasoning in presence of fuzzy data. Fuzzy Sets Syst. **16**, 163–175 (1985)
2. Štěpnička, M., Jayaram, B.: On the suitability of the Bandler-Kohout subproduct as an inference mechanism. IEEE Trans. Fuzzy Syst. **18**(2), 285–298 (2010)
3. Bandler, W., Kohout, L.: Semantics of implication operators and fuzzy relational products. Int. J. Man-Mach. Stud. **12**(1), 89–116 (1980)
4. Bandler, W., Kohout, L.: Relational-product architectures for information processing. Inf. Sci. **37**, 25–37 (1985)
5. Dubois, D., Prade, H.: Semantics of quotient operators in fuzzy relational databases. Fuzzy Sets Syst. **78**, 89–93 (1996)
6. De Baets, B., Kerre, E.: Fuzzy relational compositions. Fuzzy Sets Syst. **60**, 109–120 (1993)
7. Běhounek, L., Daňková, M.: Relational compositions in fuzzy class theory. Fuzzy Sets Syst. **160**(8), 1005–1036 (2009)
8. Novák, V., Perfilieva, I., Močkoř, J.: Mathematical Principles of Fuzzy Logic. Kluwer Academic Publishers, Boston (1999)
9. Bandler, W., Kohout, L.: Fuzzy power sets and fuzzy implication operators. Fuzzy Sets Syst. **4**, 183–190 (1980)
10. Pedrycz, W.: Fuzzy relational equations with generalized connectives and their applications. Fuzzy Sets Syst. **10**, 185–201 (1983)
11. Chung, F., Lee, T.: Analytical resolution and numerical identification of fuzzy relational systems. IEEE Trans. Syst. Man Cybern. **28**, 919–924 (1998)
12. Štěpnička, M., Holčapek, M.: Fuzzy relational compositions based on generalized quantifiers. In: Laurent, A., Strauss, O., Bouchon-Meunier, B., Yager, R.R. (eds.) IPMU 2014, Part II. CCIS, vol. 443, pp. 224–233. Springer, Heidelberg (2014)
13. Cao, N., Štěpnička, M.: An existence of fuzzy relational compositions using generalized quantifiers. In: Proceedings of the 16th World Congress of the International Fuzzy Systems Association (IFSA), 9th Conference of the European Society for Fuzzy-Logic, Technology (EUSFLAT). Advances in Intelligent Systems Research, vol. 89, pp. 49–58. Atlantis press, Gijón (2015)
14. Murinová, P., Novák, V.: A formal theory of generalized intermediate syllogisms. Fuzzy Sets Syst. **186**(1), 47–80 (2012)
15. Dvořák, A., Holčapek, M.: L-fuzzy quantifiers of type $\langle 1 \rangle$ determined by fuzzy measures. Fuzzy Sets Syst. **160**(23), 3425–3452 (2009)
16. Lim, C., Chan, C.: A weighted inference engine based on interval-valued fuzzy relational theory. Expert Syst. with Appl. **42**, 3410–3419 (2015)

Towards Fuzzy Partial Set Theory

Libor Běhounek and Martina Daňková(✉)

Institute for Research and Applications of Fuzzy Modeling, NSC IT4Innovations, Division University of Ostrava, 30. dubna 22, 701 03 Ostrava 1, Czech Republic
{libor.behounek,martina.dankova}@osu.cz
http://irafm.osu.cz/

Abstract. We sketch a simple theory of fuzzy partial sets, i.e., fuzzy sets that can have undefined membership degrees. The theory is developed in the semantic framework of a first-order extension of the recently proposed fuzzy partial propositional logic. We introduce a selection of basic notions of fuzzy partial set theory, discuss their variants, and present a few initial results on the properties of fuzzy partial class operations and relations.

Keywords: Fuzzy partial logic · Fuzzy partial set · Fuzzy set theory

1 Introduction

A simple system of fuzzy partial propositional logic, i.e., a fuzzy propositional logic which admits undefined truth degrees, has recently been proposed in [3]. A natural next step is to develop its first- or higher-order variants and formalize the theory of fuzzy partial sets (i.e., fuzzy sets that can have undefined membership degrees) within this framework. This paper presents first steps toward this goal, in a manner similar to [6, Chap. 18] or [1]. We define the semantics of a simple first-order extension of fuzzy partial propositional logic and a simple theory of fuzzy partial sets of the first order; introduce a selection of basic fuzzy partial set–theoretic notions within its framework; and present a few results about these notions. Because of space limitation, we omit all proofs; they will be given in the upcoming full paper. In the present paper we only develop the semantic aspects of the proposed theory; its deductive apparatus is left for future research, as is its extension by fuzzy partial relations and classes of higher orders.

2 Fuzzy Partial Propositional Logic

Fuzzy partial logic proposed in [3] is based on (any implicative expansion of) the well-known fuzzy logic $\mathrm{MTL}_\triangle$of left-continuous t-norms. We assume the reader's familiarity with $\mathrm{MTL}_\triangle$ (or some of its axiomatic extensions, such as Łukasiewicz logic with the $\triangle$-operator; see, e.g., [2,6–8]).

Furhter on, let the logic L be any implicative expansion of $\mathrm{MTL}_\triangle$, i.e., a logic expanding $\mathrm{MTL}_\triangle$ by any set of axioms, derivation rules, and connectives

J.P. Carvalho et al. (Eds.): IPMU 2016, Part II, CCIS 611, pp. 482–494, 2016.
DOI: 10.1007/978-3-319-40581-0_39

of arbitrary arities, such that each added connective c is congruent w.r.t. biimplication: $\varphi \leftrightarrow \psi \vdash_{\mathrm{L}} c(\chi_1, \ldots, \varphi, \ldots, \chi_n) \leftrightarrow c(\chi_1, \ldots, \psi, \ldots, \chi_n)$. Since L is assumed to expand $\mathrm{MTL}_\triangle$, the language $\mathscr{S}$ of L contains at least the connectives $\wedge$, $\vee$, $\&$, $\rightarrow$, $\leftrightarrow$, 0, 1, and $\triangle$.

The fuzzy partial propositional logic L^* based on L is defined as follows (for details see [3]):

- The *language* (or signature) $\mathscr{S}^*$ of L^* extends the language $\mathscr{S}$ of L by the truth constant $*$ (representing the undefined truth degree of propositions), the unary connective ! (for the crisp modality "is defined"), and the binary connective $\bar{\wedge}$ (for Kleene-style min-conjunction).
- Intended *algebras* (of truth values) for L^* are defined by expanding the algebras for L by a dummy element $*$ (to be assigned to propositions with undefined truth). In the intended L^*-algebra $\boldsymbol{L}_* = \boldsymbol{L} \cup \{*\}$ (where $\boldsymbol{L}$ is an L-algebra), the connectives of L^* are interpreted as described by the following truth tables, for all unary connectives $u \in \mathscr{S}$, binary connectives $c \in \mathscr{S}$ (and similarly for higher arities), $\alpha, \beta \in \boldsymbol{L}$ and $\gamma, \delta \in \boldsymbol{L} \setminus \{0\}$:

$$
\begin{array}{c|c} & ! \\ \hline \alpha & 1 \\ * & 0 \end{array}
\qquad
\begin{array}{c|c} & u \\ \hline \alpha & u\alpha \\ * & * \end{array}
\qquad
\begin{array}{c|cc} c & \beta & * \\ \hline \alpha & \alpha\, c\, \beta & * \\ * & * & * \end{array}
\qquad
\begin{array}{c|ccc} \bar{\wedge} & 0 & \delta & * \\ \hline 0 & 0 & 0 & 0 \\ \gamma & 0 & \gamma \wedge \delta & * \\ * & 0 & * & * \end{array}
\tag{1}
$$

- *Tautologies* of L^* are defined as formulae that are evaluated to 1 under all evaluations in all intended L^*-algebras. *Entailment* in L^* is defined as transmission of the value 1 under all evaluations in all intended L^*-algebras. As usual, we write $\models \varphi$ to indicate the tautologicity of φ in L^* and $\Gamma \models \varphi$ to denote the fact that the set Γ of formulae entails the formula φ in L^*.
- An *axiomatic system* for L^* has been proposed in [3], which extends the (suitably modified) axioms and rules of L by 4 additional derivation rules and 10 additional axiom schemata. The general, linear, and (if enjoyed by L) standard *completeness theorems* (respectively w.r.t. L^*-algebras $\boldsymbol{L}_*$ over all, linear, or standard L-algebras $\boldsymbol{L}$) can be proved for this axiomatic system. However, since in this paper we only deal with the semantics of fuzzy partial logic and fuzzy partial set theory, we leave the axiomatic system for L^* aside.

The connectives of $\mathscr{S}^*$ make a broad class of derived connectives available in L^*. The class includes all connectives determined in intended L^*-algebras by truth-tables similar to those of (1) above, with fields delimited and filled by $\mathscr{S}^*$-formulae (for more details see [3]). This includes several useful families of connectives well-known from three-valued logic (see, e.g., [4]), such as:

- The *Bochvar-style* connectives, which treat $*$ as the annihilator. Recall that in L^*, the connectives of the original language $\mathscr{S}$ of the underlying fuzzy logic L are actually interpreted Bochvar-style: see the truth tables (1) above.

- The *Sobociński-style* connectives $\hat{c} \in \{\hat{\wedge}, \hat{\vee}, \hat{\&}\}$, which treat $*$ as the neutral element; and the Sobociński-style implication $\hat{\rightarrow}$ residuated with $\hat{\&}$:

$\hat{c}$	β	$*$
α	$\alpha\, c\, \beta$	α
$*$	β	$*$

$\hat{\rightarrow}$	β	$*$
α	$\alpha \rightarrow \beta$	$\neg\alpha$
$*$	β	$*$

(2)

- The *Kleene-style* connectives $\bar{\wedge}$, $\bar{\vee}$, $\bar{\&}$, and $\bar{\rightarrow}$, which keep the annihilators of the corresponding connectives of L and are evaluated Bochvar-style otherwise; i.e., for $c \in \{\wedge, \&\}$:

$\bar{c}$	0	β	$*$
0	0	0	0
α	0	$\alpha\, c\, \beta$	$*$
$*$	0	$*$	$*$

$\bar{\vee}$	δ	1	$*$
γ	$\gamma \vee \delta$	1	$*$
1	1	1	1
$*$	$*$	1	$*$

$\bar{\rightarrow}$	δ	1	$*$
0	1	1	1
α	$\alpha \rightarrow \delta$	1	$*$
$*$	$*$	1	$*$

(3)

- The *McCarthy-style* sequential binary connectives (evaluated Kleene-style if their first argument is defined and Bochvar-style otherwise); the *Bochvar-external* style connectives (which treat $*$ as 0); the *best-case* (or *worst-case*) connectives (which yield the largest or lowest truth value when $*$ is replaced by any defined truth degree); etc.

Moreover, the following useful unary and binary connectives are L*-definable:

x	$?x$	$\downarrow x$	$\uparrow x$
α	0	α	α
$*$	1	0	1

x	$\circledast x$
γ	0
1	$*$
$*$	0

(4)

$\equiv$	β	$*$
α	$\triangle(\alpha \leftrightarrow \beta)$	0
$*$	0	1

$\sqsubseteq$	β	$*$
α	$\triangle(\alpha \leftrightarrow \beta)$	0
$*$	1	1

$\trianglelefteq$	β	$*$
α	$\triangle(\alpha \rightarrow \beta)$	0
$*$	1	1

(5)

for $\alpha, \beta \neq *$ and $\gamma \notin \{1, *\}$. For more details on fuzzy partial propositional logic (including examples of valid laws and several metamathematical results) see [3].

3 Fuzzy Partial First-Order Logic

A simple first-order variant L$\forall^*$ of L* can be defined in a manner analogous to other fuzzy first-order logics (cf., e.g., [7] or [2]). Like in propositional L*, the (standardly defined) first-order formulae of L$\forall_*$ are evaluated in L*-algebras of truth values. The interpretation of a unary (and analogously higher-arity) predicate symbol P in a given model $\mathbf{M}$ is thus a *total* function $P_{\mathbf{M}} \colon D_{\mathbf{M}} \rightarrow \boldsymbol{L}_*$, where $D_{\mathbf{M}}$ is the domain of $\mathbf{M}$ and $\boldsymbol{L}_*$ is an intended L*-algebra of truth values. The assignment $P_{\mathbf{M}}(a) = *$ then represents the undefined truth of Px in $\mathbf{M}$ under the evaluation $x \mapsto a$. Function symbols F of arity $n \geq 0$ are interpreted in $\mathbf{M}$ as usual, by functions $F_{\mathbf{M}} \colon \left(D_{\mathbf{M}}\right)^n \rightarrow D_{\mathbf{M}}$. In this paper we do not consider undefined individuals, since the language of partial fuzzy class theory, which is the main focus of the present paper, only contains class terms with always well-defined values; a fuller account of fuzzy partial first-order logic that accommodates undefined individuals is left for future work.

The Tarski conditions for terms, atomic formulae, and propositional connectives are defined as usual; because of space limitations, we omit them here and refer the reader to [7, Chap. 5] or [2, Sect. 5]. The clauses for quantifiers are introduced in (6)–(7) below. Let the truth value (in $\boldsymbol{L}_*$) of the formula φ in a model $\mathbf{M}$ under an evaluation e of individual variables be denoted by $\|\varphi\|^{\mathbf{M}}_e$; if $\mathbf{M}$ and e are fixed or arbitrary, we shall write simply $\|\varphi\|$. Similarly, the value of the term t in $\mathbf{M}$ under e will be denoted by $\|t\|^{\mathbf{M}}_e$ (or simply $\|t\|$). The evaluation that assigns $a \in D_{\mathbf{M}}$ to x and coincides with e on all other individual variables will be denoted by $e[x \mapsto a]$.

The primitive quantifiers $\forall$, $\exists$ of L$\forall^*$ will be interpreted Bochvar-style, yielding $*$ whenever there is an undefined instance of the quantified formula:

$$\|(\forall x)\varphi\|^{\mathbf{M}}_e = \begin{cases} * & \text{if } \|\varphi\|^{\mathbf{M}}_{e[x\mapsto a]} = * \text{ for some } a \in D_{\mathbf{M}} \\ \inf_{a\in D_{\mathbf{M}}} \|\varphi\|^{\mathbf{M}}_{e[x\mapsto a]} & \text{otherwise} \end{cases} \tag{6}$$

$$\|(\exists x)\varphi\|^{\mathbf{M}}_e = \begin{cases} * & \text{if } \|\varphi\|^{\mathbf{M}}_{e[x\mapsto a]} = * \text{ for some } a \in D_{\mathbf{M}} \\ \sup_{a\in D_{\mathbf{M}}} \|\varphi\|^{\mathbf{M}}_{e[x\mapsto a]} & \text{otherwise} \end{cases} \tag{7}$$

While Bochvar quantifiers $\forall, \exists$ themselves are of limited utility, they are sufficient for the definability of further useful quantifiers in L$\forall^*$ by means of the connectives of L*. For instance, consider the Sobociński-style quantifiers with the following Tarski conditions:

$$\|(\hat\forall x)\varphi\|^{\mathbf{M}}_e = \begin{cases} * & \text{if } \|\varphi\|^{\mathbf{M}}_{e[x\mapsto a]} = * \text{ for all } a \in D_{\mathbf{M}} \\ \inf_{a\in D_{\mathbf{M}}} \|{\uparrow}\varphi\|^{\mathbf{M}}_{e[x\mapsto a]} & \text{otherwise} \end{cases} \tag{8}$$

$$\|(\hat\exists x)\varphi\|^{\mathbf{M}}_e = \begin{cases} * & \text{if } \|\varphi\|^{\mathbf{M}}_{e[x\mapsto a]} = * \text{ for all } a \in D_{\mathbf{M}} \\ \sup_{a\in D_{\mathbf{M}}} \|{\downarrow}\varphi\|^{\mathbf{M}}_{e[x\mapsto a]} & \text{otherwise} \end{cases} \tag{9}$$

They can be defined by means of $\forall, \exists$ and the connectives of L* as follows:

$$(\hat\forall x)\varphi \equiv_{\mathrm{df}} (\forall x){\uparrow}\varphi \vee \circledast(\forall x)?\varphi \tag{10}$$

$$(\hat\exists x)\varphi \equiv_{\mathrm{df}} (\exists x){\downarrow}\varphi \vee \circledast(\forall x)?\varphi \tag{11}$$

Similarly, the Kleene-style quantifiers can be defined as follows:

$$(\bar\forall x)\varphi \equiv_{\mathrm{df}} (\forall x)\varphi \mathbin{\bar\wedge} (\hat\forall x)\varphi \tag{12}$$

$$(\bar\exists x)\varphi \equiv_{\mathrm{df}} (\exists x)\varphi \mathbin{\bar\vee} (\hat\exists x)\varphi \tag{13}$$

Further useful quantifiers arise by combining these quantifiers with the connectives $\uparrow, \downarrow$ of (4); among them, the most meaningful are the following ones:

$$(\forall x)\uparrow\varphi \qquad \text{Sette}\,\forall \tag{14}$$
$$(\forall x)\downarrow\varphi \qquad \text{Bochvar-external}\,\forall \tag{15}$$
$$(\exists x)\downarrow\varphi \qquad \text{Bochvar-external}\,\exists \tag{16}$$

The names of all the quantifiers are derived from the three-valued connectives (see, e.g., [4]) to which they reduce in a single-element domain over the three-valued L*-algebra. All of the quantifiers reduce to the usual quantifiers $\forall, \exists$ of the first-order fuzzy logic L$\forall$ if all instances of the quantified formula are defined.

As usual, validity in a model of L$\forall^*$ is defined as 1-truth under all evaluations; tautologicity as validity in all models; and entailment as transmission of validity in each model, analogously to other first-order fuzzy logics (with 1 the only designated truth value); again we refer the reader to [7, Chap. 5] or [2, Sect. 2]. Since the focus of this paper is on the semantics, we leave the axiomatization of L$\forall^*$ aside. Let us just hint that L$\forall^*$ turns out to be implicative (see [9]) w.r.t. the connective $\trianglelefteq$ of (5), so it can be axiomatized straightforwardly by adding Rasiowa's axioms for quantifiers (and, optionally, Hájek's axiom ($\forall$3) of [7] to ensure completeness w.r.t. safe models over *linear* intended L*-algebras) to the axiomatic system for L*: this axiomatizes the quantifiers $\forall$ and $\hat{\exists}$, which correspond to inf/sup w.r.t. the order generated by $\trianglelefteq$; the Bochvar-style quantifier $\exists$ is definable from $\forall, \hat{\exists}$ and the connectives of L*.

4 A Simple Theory of Fuzzy Partial Sets

Fuzzy partial logic of the first order provides the means for formalization of the theory of fuzzy partial sets, in a similar manner as fuzzy set theory can be formalized in fuzzy first-order logic. A simple representation of fuzzy sets in fuzzy first-order logic L treats membership to a fuzzy set as a binary predicate $\in$ between two sorts of objects, representing elements of a fixed domain and fuzzy subsets of this domain. The membership predicate $\in$ is required to satisfy the axioms of extensionality and comprehension (suitably adapted for L, see [1, Sect. 3.1]) or the corresponding semantic conditions on the models of $\in$ [6, Chap. 18.1].

A simple theory of *partial* fuzzy sets can be obtained by mimicking this approach in fuzzy partial first-order logic L$\forall^*$, i.e., by evaluating $\in$ in L*-algebras $\boldsymbol{L}_*$ instead of L-algebras $\boldsymbol{L}$. In this section we briefly introduce such a simple theory of fuzzy sets, called PFCT$_1$(L*). The construction is analogous to the theory of L-valued first-order fuzzy classes introduced in [1, Sect. 3], which we shall denote here by FCT$_1$(L). Further extension of PFCT$_1$(L*) by tuples (needed for fuzzy partial relations) and fuzzy partial sets of higher orders (yielding a fully fledged fuzzy partial higher-order logic L*) can then be done analogously as in FCT$_1$(L): cf. [1, Sect. 4–5].

Similarly to $\mathrm{FCT}_1(\mathrm{L})$, the theory $\mathrm{PFCT}_1(\mathrm{L}^*)$ of fuzzy partial first-order classes is a theory in multi-sorted first-order logic $\mathrm{L}\forall^*$ (the multi-sorted generalization of $\mathrm{L}\forall^*$ is defined analogously to multi-sorted $\mathrm{L}\forall$, cf. [1]). The language of $\mathrm{PFCT}(\mathrm{L}^*)$ consists of:

- Variables for *elements*, denoted by lowercase letters $x, y, \ldots$.
- Variables for (fuzzy partial) *classes*, denoted by uppercase letters $A, B, \ldots$.
- The (crisp total) *equality predicate* $=$ on each sort.
- The (fuzzy partial) *membership predicate* $\in$ between elements and classes. By convention, the formula $x \in A$ can be abbreviated by Ax.
- *Class terms* $\{x \mid \varphi(x)\}$ for each variable x and each $\mathrm{PFCT}_1(\mathrm{L}^*)$-formula φ.

Analogously to the intended models of $\mathrm{FCT}_1(\mathrm{L})$ (see [1, Sect. 3.1] for details), the intended models of $\mathrm{PFCT}_1(\mathrm{L}^*)$ consist of all $\boldsymbol{L}_*$-valued membership functions (representing fuzzy partial classes) over a given domain of elements, with the obvious interpretation of the predicates $=$ (as crisp identity), $\in$, and class terms:

$$\|x \in A\| = \|A\|(\|x\|) \tag{17}$$

$$\|\{x \mid \varphi(x)\}\| \colon \|x\| \mapsto \|\varphi(x)\| \tag{18}$$

(i.e., the truth value of membership is the degree assigned by the membership function, and class terms denote the membership functions evaluated according to their formulae).

Even though the theory of the intended models described above is not axiomatizable (as it contains classical second-order logic), we conjecture that similarly to classical second-order logic and $\mathrm{FCT}_1(\mathrm{L})$, the following Henkin-style axioms (for all $\mathrm{PFCT}_1(\mathrm{L}^*)$-formulae φ) are sufficient for most practical purposes:

- *Equality axioms:* $x = x$ and $x = y \to (\varphi(x) \equiv \varphi(y))$, and the same for classes
- *Extensionality:* $(\forall x)(Ax \equiv Bx) \to A = B$
- *Class comprehension:* $y \in \{x \mid \varphi(x)\} \equiv \varphi(y)$

In the subsequent sections, the framework of $\mathrm{PFCT}_1(\mathrm{L}^*)$ is only used as a language for expressing statements about partial fuzzy sets. The validity of all theorems given here can be proved semantically by considering the truth values in all intended models of $\mathrm{PFCT}_1(\mathrm{L}^*)$, without employing its deductive apparatus (which is left for future research).

5 Partial Variants of Fuzzy Set–Theoretic Notions

Since the fuzzy partial logic $\mathrm{L}\forall^*$ has a richer set of meaningful connectives and quantifiers than the fuzzy logic $\mathrm{L}\forall$, fuzzy set-theoretic notions defined in $\mathrm{FCT}_1(\mathrm{L})$ can usually be generalized to $\mathrm{PFCT}_1(\mathrm{L}^*)$ in several ways, each with its own uses. In this section we shall discuss such variations of partial fuzzy set–theoretic notions, obtained by varying the quantifiers and connectives in their definitions. Due to space limitations, only a few examples will be given to illustrate the general point. Some of the laws governing these notions will be given in Sect. 7.

5.1 Domains and Equality

The main difference between $\mathrm{FCT}_1(\mathrm{L})$ and $\mathrm{PFCT}_1(\mathrm{L}^*)$ is that membership functions representing fuzzy classes in $\mathrm{FCT}_1(\mathrm{L})$ are total, while in $\mathrm{PFCT}_1(\mathrm{L}^*)$ they can be partial (with the undefined membership degrees represented by the truth value $*$). Several notions specific to fuzzy partial set theory thus come from the theory of partial functions. For instance, we define the domain of a class as the domain of its membership function:

$$\operatorname{dom} A =_{\mathrm{df}} \{x \mid !Ax\} \qquad \text{domain} \qquad (19)$$

A class is total if it is defined for all elements:

$$\operatorname{Tot} A \equiv_{\mathrm{df}} (\forall x)!Ax \qquad \text{totalness} \qquad (20)$$

Clearly domains are total (and crisp) classes: $\mathrm{PFCT}_1(\mathrm{L}^*) \models \operatorname{Tot}(\operatorname{dom} A)$.

Furthermore, like in the theory of partial functions, we can introduce several notions of (crisp) equality for partial fuzzy classes, based on the various notions of equality of their partial membership functions:

$$A = B \equiv_{\mathrm{df}} (\forall x)(Ax \equiv Bx) \qquad \text{strong equality} \qquad (21)$$
$$A \doteqdot B \equiv_{\mathrm{df}} (\forall x)((Ax \equiv Bx) \vee ?Ax \vee ?Bx) \qquad \text{weak equality} \qquad (22)$$
$$A \sqsubseteq B \equiv_{\mathrm{df}} (\forall x)(Ax \sqsubseteq Bx) \qquad \text{subfunction} \qquad (23)$$

Note that the strong equality predicate $=$ is actually the primitive equality predicate of $\mathrm{PFCT}_1(\mathrm{L}^*)$ governed by the axioms equality of extensionality. By the semantics of $\equiv$, $?$, and $\sqsubseteq$ (see (4)–(5)), all of these notions are crisp and defined (i.e., non-$*$) for all pairs of classes; $=$ is a (total crisp) equivalence relation; and $\sqsubseteq$ is a (total crisp) partial order on fuzzy classes. Moreover,

$$\mathrm{PFCT}_1(\mathrm{L}^*) \models (A = B) \leftrightarrow ((A \doteqdot B) \wedge (\operatorname{dom} A = \operatorname{dom} B)) \qquad (24)$$
$$\mathrm{PFCT}_1(\mathrm{L}^*) \models (A = B) \leftrightarrow ((A \sqsubseteq B) \wedge (B \sqsubseteq A)) \qquad (25)$$
$$\mathrm{PFCT}_1(\mathrm{L}^*) \models (A \sqsubseteq B) \rightarrow (A \doteqdot B). \qquad (26)$$

(Further on we omit the "$\mathrm{PFCT}_1(\mathrm{L}^*) \models$" prefix when the claim of validity in $\mathrm{PFCT}_1(\mathrm{L}^*)$ is clearly intended.)

5.2 Class Constants and Shifts

Like in $\mathrm{FCT}_1(\mathrm{L})$, the *empty class* $\emptyset$ can be defined in $\mathrm{PFCT}_1(\mathrm{L}^*)$ as the class to which all elements belong to degree 0:

$$\emptyset =_{\mathrm{df}} \{x \mid 0\} \qquad \text{empty class} \qquad (27)$$

In $\mathrm{PFCT}_1(\mathrm{L}^*)$, we can define another class which also contains no elements to any defined positive degree (so, in this sense, is also 'empty'), namely:

$$\lambda =_{\mathrm{df}} \{x \mid *\} \qquad \text{totally undefined class} \qquad (28)$$

Clearly $\lambda \neq \emptyset$ (though $\lambda \sqsubseteq \emptyset$, and so by (26), $\lambda \doteqdot \emptyset$; in general, $\lambda \sqsubseteq A$ and $\lambda \doteqdot A$ for any class A). The property of having no elements to any defined positive degree is shared by all subfunctions of $\emptyset$; we can call these classes *uninhabited;* the class λ is the smallest and $\emptyset$ the largest uninhabited class w.r.t. $\sqsubseteq$.

The *universal class* is the class which contains all elements to degree 1:

$$\mathrm{V} =_{\mathrm{df}} \{x \mid 1\} \qquad \text{universal class} \qquad (29)$$

Clearly, $\mathrm{dom}\,\mathrm{V} = \mathrm{dom}\,\emptyset = \mathrm{V}$, while $\mathrm{dom}\,\lambda = \emptyset$. Observe that λ is the unique class with empty domain.

Recall the connectives $\downarrow$ and $\uparrow$ of (4), which reinterpret $*$ respectively as 0 and 1. The corresponding class operations turn out to be quite useful:

$$\Uparrow A =_{\mathrm{df}} \{x \mid {\uparrow} Ax\} \qquad \text{up-shift} \qquad (30)$$

$$\Downarrow A =_{\mathrm{df}} \{x \mid {\downarrow} Ax\} \qquad \text{down-shift} \qquad (31)$$

The up- and down-shifts of all classes are total: $\mathrm{dom} \Uparrow A = \mathrm{dom} \Downarrow A = \mathrm{V}$. Obviously $\Uparrow\lambda = \mathrm{V}$ and $\Downarrow\lambda = \emptyset$. For total classes, $\Uparrow A = \Downarrow A = A$.

5.3 Height and Plinth

For totally defined fuzzy classes, the fuzzy set–theoretic notions of height and plinth are defined in $\mathrm{FCT}_1(\mathrm{L})$ as follows:

$$\mathrm{Hgt}\, A \equiv_{\mathrm{df}} (\exists x) Ax \qquad \text{height} \qquad \|\mathrm{Hgt}\, A\| = \sup_x \|A\|(x) \qquad (32)$$

$$\mathrm{Plt}\, A \equiv_{\mathrm{df}} (\forall x) Ax \qquad \text{plinth} \qquad \|\mathrm{Plt}\, A\| = \inf_x \|A\|(x) \qquad (33)$$

As we have seen in Sect. 3, in $\mathrm{L}\forall^*$ we have a choice of several meaningful existential and universal quantifiers, yielding several different notions of height and plinth for partial fuzzy classes:

$$\mathrm{Hgt}\, A \equiv_{\mathrm{df}} (\exists x) Ax \quad \mathrm{Plt}\, A \equiv_{\mathrm{df}} (\forall x) Ax \quad \text{Bochvar height and plinth} \qquad (34)$$

$$\hat{\mathrm{H}}\mathrm{gt}\, A \equiv_{\mathrm{df}} (\hat{\exists} x) Ax \quad \hat{\mathrm{P}}\mathrm{lt}\, A \equiv_{\mathrm{df}} (\hat{\forall} x) Ax \quad \text{Sobociński height and plinth} \qquad (35)$$

$$\bar{\mathrm{H}}\mathrm{gt}\, A \equiv_{\mathrm{df}} (\bar{\exists} x) Ax \quad \bar{\mathrm{P}}\mathrm{lt}\, A \equiv_{\mathrm{df}} (\bar{\forall} x) Ax \quad \text{Kleene height and plinth} \qquad (36)$$

Further meaningful notions of height and plinth can be obtained by combining the quantifiers with shifts, cf. (14)–(16): Bochvar-external height $\mathrm{Hgt} \Downarrow A$, Bochvar-external plinth $\mathrm{Plt} \Downarrow A$, and Sette plinth $\mathrm{Plt} \Uparrow A$.

For total fuzzy classes, all of these notions coincide with the usual notions of height (32) and plinth (33) known from fuzzy set theory; however, they differ for non-total classes: Bochvar height and plinth yield $*$ for any non-total class, while Sobociński heights and plinths ignore the undefined values and only yield $*$ for the totally undefined class λ (see Fig. 1). Kleene height coincides with Bochvar heihgt except for classes with Sobociński height 1 (and dually for Kleene plinth). Bochvar-external height and Sette plinth, respectively, coincide with Sobociński height and plinth, save for the class λ (where $\mathrm{Hgt} \Downarrow\lambda \equiv 0$ and $\mathrm{Plt} \Uparrow\lambda \equiv 1$, while $\widehat{\mathrm{Hgt}}\, \lambda \equiv \widehat{\mathrm{Plt}}\, \lambda \equiv *$). Bochvar-external plinth is 0 for any non-total class.

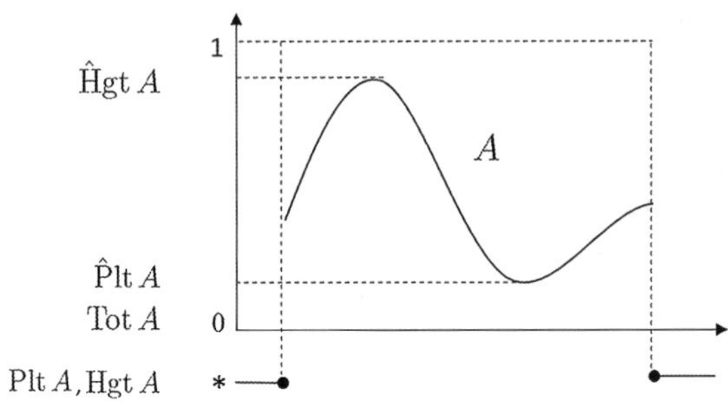

Fig. 1. Bochvar and Sobociński heights and plinths

5.4 Kernels and Supports

There are several meaningful ways of defining kernels and supports of partial fuzzy sets. The Bochvar-style kernel $\ker A =_{\mathrm{df}} \{x \mid \triangle Ax\}$ and support $\mathrm{supp}\, A =_{\mathrm{df}} \{x \mid \neg\triangle\neg Ax\}$ are undefined whenever Ax is undefined, and coincide with the usual notions of kernel and support on the domain of A. Consequently, $\mathrm{dom}(\ker A) = \mathrm{dom}(\mathrm{supp}\, A) = \mathrm{dom}\, A$.

Another meaningful option is to regard elements outside $\mathrm{dom}\, A$ as *not* belonging to the kernel (nor support) of A. This variant is definable as $\ker \Downarrow A$ and $\mathrm{supp} \Downarrow A$ (or equivalently, $\Downarrow \ker A$ and $\Downarrow \mathrm{supp}\, A$). Since these notions treat $*$ as 0, a fitting name for them is *Bochvar-external* kernel and support.

Another useful notion of support is the following:

$$\|(\mathrm{supp}^* A)x\| = \begin{cases} 1 & \text{if } \|Ax\| \notin \{0, *\} \\ * & \text{otherwise} \end{cases}$$

It can be defined in $\mathrm{PFCT}_1(\mathrm{L}^*)$ as $\mathrm{supp}^* A =_{\mathrm{df}} \{x \mid (\neg\triangle\neg Ax) \vee \circledast(\triangle\neg Ax)\}$. An analogous notion of kernel is $\ker^* A =_{\mathrm{df}} \mathrm{supp}^* \ker A$. The domains of supp^* and $\ker^*$ are Bochvar-external kernels and supports: $\mathrm{dom}(\ker^* A) = \Downarrow \ker A$ and $\mathrm{dom}(\mathrm{supp}^* A) = \Downarrow \mathrm{supp}\, A$. The three notions of kernel and support are illustrated in Fig. 2.

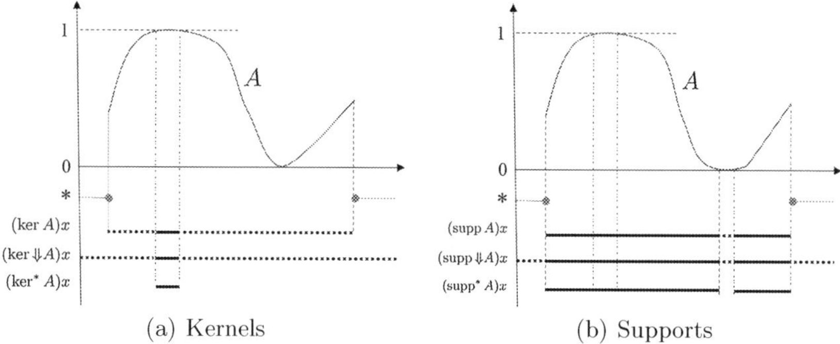

(a) Kernels (b) Supports

Fig. 2. Various notions of kernel and support. Bold lines indicate the degree 1, dotted lines indicate the degree 0, and empty lines indicate the undefined degree $*$.

5.5 Basic First-Order Class Operations

Like in the theory of total fuzzy or crisp sets, basic partial fuzzy class operations result from combining the membership degrees by propositional connectives of L^*. Again, different families of propositional connectives give rise to different families of class operations. Thus, e.g., we have the following variants of min-intersection of partial fuzzy classes:

$$A \cap B =_{\text{df}} \{x \mid Ax \wedge Bx\} \qquad \text{Bochvar min-intersection} \tag{37}$$

$$A \mathbin{\hat{\cap}} B =_{\text{df}} \{x \mid Ax \mathbin{\hat{\wedge}} Bx\} \qquad \text{Sobociński min-intersection} \tag{38}$$

$$A \mathbin{\bar{\cap}} B =_{\text{df}} \{x \mid Ax \mathbin{\bar{\wedge}} Bx\} \qquad \text{Kleene min-intersection} \tag{39}$$

Similar class operations can be derived from all propositional connectives of all families of L^*-definable connectives introduced in Sect. 2: for example, Bochvar strong intersection $A \mathbin{\dot{\cap}} B =_{\text{df}} \{x \mid Ax \,\&\, Bx\}$, Sobociński max-union $A \mathbin{\hat{\cup}} B =_{\text{df}} \{x \mid Ax \mathbin{\hat{\vee}} Bx\}$, etc. Again, for total fuzzy classes they coincide with the corresponding operations known from fuzzy set theory, but their behaviors differ for non-total operands: for instance, observe that

$$\operatorname{dom}(A \cap B) = \operatorname{dom} A \cap \operatorname{dom} B \tag{40}$$

$$\operatorname{dom}(A \mathbin{\hat{\cap}} B) = \operatorname{dom} A \cup \operatorname{dom} B \tag{41}$$

(and analogously for the domains of all Bochvar and Sobociński operations).

An even greater variation is found in $\text{PFCT}_1(\text{L}^*)$-notions defined by more complex formulae, such as graded inclusion, fuzzy equality, etc., in which the variants of connectives and quantifiers occurring in the definition can be freely combined (as, e.g., in Bochvar–Sobociński graded inclusion $(\forall x)(Ax \mathbin{\hat{\rightarrow}} Bx)$). Many of these combinations express meaningful ways of handling undefined membership degrees. However, due to the number of possible variants, we leave a more detailed discussion of such combined notions aside here.

6 Representation of Fuzzy Partial Classes

Like ordinary fuzzy sets, a fuzzy partial set A can be represented by the system of crisp horizontal level sets $A_\alpha = \{x \mid \|Ax\| = \alpha\}$ for each $\alpha \in \boldsymbol{L}_*$, which partition the universe of elements; the only difference from ordinary fuzzy sets is the presence of an extra level set A_* of elements x such that $\|Ax\| = *$. Obviously A_* is the complement of $\operatorname{dom} A$. A fuzzy partial class A is thus determined by a crisp domain $X = \operatorname{dom} A$ and a fuzzy subset F of X (represented in a model of $\mathrm{PFCT}_1(\mathrm{L}^*)$ by a membership function $\|F\|$). This agrees very well with Zadeh's original conception of fuzzy set [10], specified as a membership function on a crisp domain. Partial fuzzy set theory such as $\mathrm{PFCT}_1(\mathrm{L}^*)$ thus renders Zadeh's notion of fuzzy set more faithfully than common fuzzy set theories formalized in fuzzy first- or higher-order logic (incl. FCT of [1]): in the models of the latter theories, all fuzzy sets are total, and so restricted to a *single* fixed crisp domain.

In models of $\mathrm{PFCT}_1(\mathrm{L}^*)$, partial fuzzy classes can thus be represented by pairs $\langle X, F\rangle$ of *total* fuzzy classes (where X is moreover crisp). Given X and F, the represented partial fuzzy class can be reconstructed as $A = F \cap \operatorname{supp}^* X$. Conversely, given a partial fuzzy class A, the representing pair of total classes can be defined as $\langle \operatorname{dom} A, \Downarrow\! A\rangle$. This correspondence makes it possible to emulate partial fuzzy set theory within the framework of ordinary fuzzy set theory (such as FCT: a formal syntactic interpretation of $\mathrm{PFCT}_1(\mathrm{L}^*)$ in FCT is left for future work). An advantage of $\mathrm{PFCT}_1(\mathrm{L}^*)$ over such emulation is that its relations and operations handle both components $\langle X, F\rangle$ simultaneously, in a variety of predefined manners. For example, Bochvar and Sobociński max-unions perform the following $\mathrm{FCT}_1(\mathrm{L})$-operations on the representing pairs (cf. (40)–(41)):

$$\langle X_1, F_1\rangle \cup \langle X_2, F_2\rangle =_{\mathrm{df}} \langle X_1 \cap X_2, F_1 \cup F_2\rangle \tag{42}$$

$$\langle X_1, F_1\rangle \,\hat{\cup}\, \langle X_2, F_2\rangle =_{\mathrm{df}} \langle X_1 \cup X_2, F_1 \cup F_2\rangle \tag{43}$$

7 Theorems on Elementary Relations and Operations

Like in the case of FCT, the formal apparatus of $\mathrm{PFCT}_1(\mathrm{L}^*)$ enables proving schematic theorems on the laws governing its defined notions. As an example, we will give an analogue of the metatheorem of FCT [1, Theorem 33], which provides a correspondence between propositional laws of L^* and the laws (of certain forms) governing fuzzy partial class relations and operations in $\mathrm{PFCT}_1(\mathrm{L}^*)$.

In this section, the expression $\varphi(p_1, \ldots, p_n)$ will imply that all variables of the propositional formula φ are among $p_1, \ldots, p_n$. The formula resulting from $\varphi(p_1, \ldots, p_n)$ by substituting ψ_i for p_i (for each $i \leq n$) is denoted by $\varphi(\psi_1, \ldots, \psi_n)$.

Definition 1. *Let $\varphi(p_1, \ldots, p_k)$ be a propositional formula of L^*. The elementary class operation generated by φ is defined in* $\mathrm{PFCT}(\mathrm{L}^*)$ *as:*

$$\mathrm{Op}_\varphi(A_1, \ldots, A_k) =_{\mathrm{df}} \{x \mid \varphi(A_1 x, \ldots, A_k x)\}.$$

Furthermore, the k-ary elementary class relation generated by φ *and* $Q \in \{\forall, \hat{\forall}, \bar{\forall}, \exists, \hat{\exists}, \bar{\exists}\}$ *is defined as:*

$$\mathrm{Rel}^{Q}_{\varphi}(A_1, \ldots, A_k) \equiv_{\mathrm{df}} (Qx)\varphi(A_1x, \ldots, A_kx).$$

We may write just Op_c or Rel^{Q}_{c} if φ is $c(p_1, \ldots, p_k)$ for c a (primitive or defined) connective of L*. Thus, e.g., $\mathrm{Op}_{\triangle}$ is ker, $\mathrm{Op}_{\hat{\wedge}}$ is $\hat{\cap}$, $\mathrm{Rel}^{\forall}_{\equiv}$ is $=$, etc.

Theorem 1. *Let* $\varphi(p_1, \ldots, p_n)$ *and* $\psi_i(p_{i,1}, \ldots, p_{i,k_i})$, $1 \leq i \leq n$, *be propositional formulae of* L^* *and* $Q \in \{\forall, \bar{\forall}, \hat{\forall}, \exists, \bar{\exists}, \hat{\exists}\}$. *Then the following conditions are equivalent:*

1. $\mathrm{L}^* \models \varphi(\psi_1, \ldots, \psi_n)$
2. $\mathrm{PFCT}_1(\mathrm{L}^*) \models \mathrm{Rel}^{Q}_{\varphi}(\mathrm{Op}_{\psi_1}(A_{1,1}, \ldots, A_{1,k_1}), \ldots, \mathrm{Op}_{\psi_n}(A_{n,1}, \ldots, A_{n,k_n}))$

Theorem 1 effectively reduces elementary class relations between class operations to propositional logic. We give just a few examples of its corollaries:

$\mathrm{PFCT}_1(\mathrm{L}^*)$		MTL*
$\models A \dot{\cap} (B \dot{\cap} C) = (A \dot{\cap} B) \dot{\cap} C$	by	$\models p \,\&\, (q \,\&\, r) \equiv (p \,\&\, q) \,\&\, r$
$\not\models \emptyset \subseteq A$	by	$\not\models 0 \to p$
$\models \lambda \sqsubseteq A$	by	$\models (* \equiv p) \vee ?*$
$\models \mathrm{dom}(A \mathbin{\hat{\cap}} B) = \mathrm{dom}\, A \cup \mathrm{dom}\, B$	by	$\models !(p \mathbin{\hat{\vee}} q) \equiv (!p \vee !q)$, etc.

8 Conclusion

In this paper we proposed a (simplistic) system of first-order fuzzy partial logic and fuzzy partial set theory, extending usual fuzzy logics (e.g., $\mathrm{MTL}_{\triangle}$, $\mathrm{BL}_{\triangle}$, $\mathrm{Ł}_{\triangle}$, ŁΠ, etc.) by an extra truth value $*$ for *undefined* truth (not *unknown* truth, though: cf. [5]). This makes it suitable, e.g., for handling non-denoting linguistic terms, non-applicable fuzzy conditions, or various error states in fuzzy contexts (for motivation see [3, Sect. 1]; envisaged applications are left for future work).

Acknowledgments. The work was supported by grant No. 16–191705 "Fuzzy partial logic" of GA ČR and project LQ1602 "IT4I XS" of MŠMT ČR.

References

1. Běhounek, L., Cintula, P.: Fuzzy class theory. Fuzzy Sets Syst. **154**(1), 34–55 (2005)
2. Běhounek, L., Cintula, P., Hájek, P.: Introduction to mathematical fuzzy logic. In: Cintula, P., Hájek, P., Noguera, C. (eds.) Handbook of Mathematical Fuzzy Logic, pp. 1–101. College Publications (2011)

3. Běhounek, L., Novák, V.: Towards fuzzy partial logic. In: Proceedings of the IEEE 45th International Symposium on Multiple-Valued Logics (ISMVL 2015), pp. 139–144. IEEE (2015)
4. Ciucci, D., Dubois, D.: A map of dependencies among three-valued logics. Inf. Sci. **250**, 162–177 (2013)
5. Dubois, D.: Reasoning about ignorance and contradiction: many-valued logics versus epistemic logic. Soft Comput. **16**, 1817–1831 (2012)
6. Gottwald, S.: A Treatise on Many-Valued Logics. Research Studies Press, Baldock (2001)
7. Hájek, P.: Metamathematics of Fuzzy Logic. Kluwer, Dordrecht (1998)
8. Novák, V., Perfilieva, I., Močkoř, J.: Mathematical Principles of Fuzzy Logic. Kluwer, Boston (1999)
9. Rasiowa, H.: An Algebraic Approach to Non-Classical Logics. North-Holland, Amsterdam (1974)
10. Zadeh, L.A.: Fuzzy sets. Inf. Control **8**(3), 338–353 (1965)

On Perception-based Logical Deduction with Fuzzy Inputs

Antonín Dvořák and Martin Štěpnička[(✉)]

Centre of Excellence IT4Innovations, Institute for Research and Applications of Fuzzy Modeling, University of Ostrava, 30. dubna 22, 701 03 Ostrava 1, Czech Republic
{antonin.dvorak,martin.stepnicka}@osu.cz
http://irafm.osu.cz/

Abstract. We present and analyze inference method called Perception-based Logical Deduction (PbLD) aimed at the treatment of fuzzy IF-THEN rules as linguistically expressed genuine logical implications. We analyze two variants of PbLD (original and balancing) that differ in the selection of fired IF-THEN rules. We concentrate on a situation when inputs into inference are fuzzy sets (fuzzy inputs). We study the conditions under which both variants fulfill the interpolativity property.

Keywords: Perception-based Logical Deduction · Fuzzy interpolation · Inference mechanism · Implicative fuzzy rules · Łukasiewicz algebra

1 Introduction

1.1 Fuzzy Rules and Inference Mechanisms

Consider a finite set of *fuzzy IF-THEN rules* (a *fuzzy rule base*):

$$\mathcal{R}_i := \mathsf{IF}\ X \text{ is } \mathcal{A}_i\ \mathsf{THEN}\ Y \text{ is } \mathcal{B}_i, \quad i = 1, \ldots, n, \tag{1}$$

where X, Y are linguistic variables and $\mathcal{A}_i, \mathcal{B}_i$ are linguistic labels (e.g., "small", "around ten" etc.). The interpretation of expressions $\mathcal{A}_i, \mathcal{B}_i$ is modeled by appropriate antecedent and consequent fuzzy sets A_i, B_i defined on some non-empty universes U, V, respectively. For our purposes, we can restrict our choice of U, V to closed real intervals. There exist two standard approaches to modelling a given fuzzy rule base by an appropriate fuzzy relation, namely by $\check{R}, \hat{R} \in \mathcal{F}(U \times V)$:

$$\check{R}(u,v) = \bigvee_{i=1}^{n} \left(A_i(u) * B_i(v)\right), \quad \hat{R}(u,v) = \bigwedge_{i=1}^{n} \left(A_i(u) \to B_i(v)\right) \tag{2}$$

where $*$ is a t-norm and $\to$ is a fuzzy implication, jointly usually forming an adjoint pair. For further details we refer to relevant sources [1–3].

M. Štěpnička—This research was partially supported by the NPU II project LQ1602 "IT4Innovations excellence in science" provided by the MŠMT.

J.P. Carvalho et al. (Eds.): IPMU 2016, Part II, CCIS 611, pp. 495–509, 2016.
DOI: 10.1007/978-3-319-40581-0_40

There are many inference mechanisms that, with the help of fuzzy rules, deduce an appropriate output $B_0 \in \mathcal{F}(V)$ based on a given observation (input) $A_0 \in \mathcal{F}(U)$. Let us restrict our focus on the fuzzy relational inference systems that directly use an image of a fuzzy set under the fuzzy relation as a model of the inference mechanism. Most often, the *direct image* $\circ$ (also CRI [4]) is used, however, the *Bandler-Kohout subproduct* $\lhd$ was proposed in [5] and later on, in [6], it was shown that both inference mechanisms are equally good. Recall that

$$(A_0 \circ R)(v) = \bigvee_{u \in U} (A_0(u) * R(u,v)), \quad (A_0 \lhd R)(v) = \bigwedge_{u \in U} (A_0(u) \to R(u,v)). \quad (3)$$

1.2 Motivation

Models of fuzzy rule bases and inference mechanisms described in the previous subsection based on CRI or BK-subproduct are well suited for the purpose of approximation of an unknown function characterized imprecisely by fuzzy rules. The fuzzy sets A_i, B_i, which interpret linguistic labels $\mathcal{A}_i$, $\mathcal{B}_i$, are usually of one of standard shapes, e.g., triangular, trapezoidal etc. Further, they are usually uniformly distributed along the intervals U and V, forming a fuzzy partition.

If we are interested in capturing the meaning of linguistic labels, which are used by humans most often, we have to search for another model of these labels. We argue that, according to intuition shared by humans, if something is extremely small, it is, at the same time, small. Therefore, the interpretation of "extremely small" should be a fuzzy subset of the interpretation of "small". Typical shapes of interpretations of these linguistic expressions are depicted in Fig. 1. If we accept that a model of these linguistic expressions should possess this property, we find that inference mechanisms and fuzzy rule base models described above are not well suited for it.

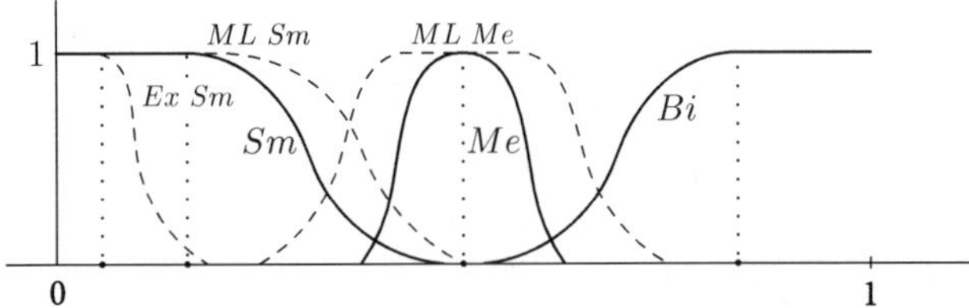

Fig. 1. Graphical representations of fuzzy sets that interpret linguistic expressions *extremely small*, *small*, *more or less small*, *more or less medium*, *medium* and *big*.

A fuzzy relational model is interpolative (correct) if and only if it models a certain type of continuous behavior [7]. In general, this can be viewed as a feature rather than a disadvantage, but there are situations where even a smooth fuzzy model should be able to provide a discontinuous jump [7]. Consider the following fuzzy rule base $\mathcal{RB}_1$:

$$\mathcal{R}_1 := \mathsf{IF}\ X \text{ is } \textit{small}\ \mathsf{THEN}\ Y \text{ is } -\textit{big},$$
$$\mathcal{R}_2 := \mathsf{IF}\ X \text{ is } \textit{extremely small}\ \mathsf{THEN}\ Y \text{ is } +\textit{big},$$

where X is a distance of a vehicle from a traffic signal when the yellow light appears on it and Y is a desired change of speed of that vehicle. Obviously, the change from braking to speeding up (big negative and positive changes in speed) is something that cannot be provided by a correct (consistent) model using a fuzzy IF-THEN rule base within a fuzzy relational inference [7]. Moreover, if interpretations of *extremely small* and *small* are as in Fig. 1 and the observation is $u_0 = 0$, we expect that the rule $\mathcal{R}_2$ has to be used, and the result has to correspond to linguistic expression $+big$. However, results of both fuzzy relational interpretations (2) are unsatisfactory. The reason is that both rules $\mathcal{R}_1$ and $\mathcal{R}_2$ are fired and consequents of these rules are very different. Thus, the inference mechanism either put both contradictory consequents into the conclusion, or it annihilates both consequents into the empty fuzzy set.

Based on these considerations, a method called *Perception-based Logical Deduction* (abbr. PbLD) has been developed [8,9]. *Perception-based*, because to an input we assign only the most fitting linguistic expression(s) from antecedents of a fuzzy rule base, call them *perceptions* and fire the corresponding rules. *Logical*, because it understands IF-THEN rules as logical implications and it has been developed within a formal logical theory. It can be shown that for the PbLD method it is no problem to provide conclusions based on the fuzzy rule base $\mathcal{RB}_1$ in accordance with human intuition.

As we mentioned above, PbLD fires the rules whose antecedents fit the observation best. But what does it mean? Consider again the fuzzy rule base $\mathcal{RB}_1$. There is a substantial difference between consequents corresponding to antecedents *small* (rule $\mathcal{R}_1$) and *extremely small* (rule $\mathcal{R}_2$). Intuitively, if an observation is extremely small, only $\mathcal{R}_2$ should be used. Similarly, if an observation is small (but not extremely small), only $\mathcal{R}_1$ should be used. So, we use the rule with the maximal membership degree of the observation in fuzzy sets interpreting antecedent linguistic expressions.[1] If there are more such rules, then we use the rule whose antecedent is most *specific*.[2]

The study and applications of Perception-based Logical Deduction have been conducted mainly for the case of crisp observations and in [10] extended for the case of fuzzy observations that occur, for instance, in hierarchical fuzzy systems. However, as we will show below, the generalized firing degree proposed in [10] had a certain drawback that requires to revisit the topic and to investigate theoretical properties for fuzzy inputs based on the newly axiomatically defined generalized firing degree. By the theoretical properties, we mean, mainly, the

[1] We will call this degree a *firing degree* of the observation u_o in fuzzy rule $\mathcal{R}_i$.

[2] E.g., *extremely small* is more specific than *small*.

preservation of modus ponens that directly leads to the *interpolativity*. Briefly, it says that if a fuzzy input (observation) is equal to the antecedent of fuzzy rule $\mathcal{R}_i$, then the inferred conclusion should be equal to the consequent of $\mathcal{R}_i$.

2 PbLD Revisited

2.1 Mathematical Background

We fix the Łukasiewicz algebra $\mathcal{L} = \langle [0,1], \wedge, \vee, *, \rightarrow, 0, 1\rangle$ as the background algebraic structure. Let U be a universe and $A, B \in \mathcal{F}(U)$. We say that A is a *subset* of B and denote it by $A \subseteq B$ if for all $u \in U$, $A(u) \leq B(u)$. By $\emptyset$ we denote the empty fuzzy set on U, that is, $\emptyset(u) = 0$ for all $u \in U$. Let $u_0 \in U$. By χ_{u_0} we denote the fuzzy set (*fuzzy singleton*) at point u_0 defined as $\chi_{u_0}(u_0) = 1$ and $\chi_{u_0}(u) = 0$ for $u \neq u_0$.

Major components of systems of fuzzy IF-THEN rules are evaluative linguistic expressions, *evaluative expressions* for short [11]. A simple form of an evaluative expression has the following structure:

$$\langle\text{linguistic hedge}\rangle\langle\text{atomic evaluative expression}\rangle.$$

An *atomic evaluative expression* is one of the *canonical* adjectives: *small, medium* and *big*, which we abbreviate in the following as Sm, Me and Bi, respectively.

Linguistic hedges are specific adverbs that make the interpretations of atomic expressions more or less precise. We may distinguish between hedges with a *narrowing effect* and with a *widening effect*, (a special case is the empty hedge). Of course, the number of hedges is limited in practical applications. Without loss of generality, we use the hedges introduced in Table 1. Note that our hedges are of the so-called inclusive type, which means that the interpretations (fuzzy sets) of more specific evaluative expressions are included in those of less specific ones, as shown in Fig. 1. Note also that we always suppose the interpretations of evaluative expressions to be *normal* fuzzy sets, i.e., their kernels are non-empty.

Table 1. Linguistic hedges and their abbreviations.

Narrowing effect	Widening effect
very (Ve)	more or less (ML)
significantly (Si)	roughly (Ro)
extremely (Ex)	quite roughly (QR)

Whenever we use the above recalled theory of evaluative linguistic expressions with linguistic hedges of an inclusive type, the use of a single fuzzy relation – either $\hat{R}$ or $\check{R}$ – is not appropriate anymore, as we argued in Sect. 1.2. In order to distinguish the situation from the, say, "standard" fuzzy rule base, which deal with fuzzy partitions and may use a single fuzzy relation as a model, the set of rules (1) will be called *linguistic description* and denoted by $LD = \{\mathcal{R}_1, \ldots, \mathcal{R}_n\}$.[3] The following conventions will be kept in this paper: N_n will denote the set $\{1, \ldots, n\}$ of natural numbers; if LD is given, then $\mathcal{A}_i$ and $\mathcal{B}_i$, $i \in N_n$, will denote antecedent and consequent evaluative expressions from the i-th fuzzy IF-THEN rule $\mathcal{R}_i$, respectively; $A_i \in \mathcal{F}(U)$ and $B_i \in \mathcal{F}(V)$ will denote their interpretations (fuzzy sets on the closed real intervals U, V).

A specificity ordering relation on the set of evaluative expressions is defined in order to allow us to state the relationships (inclusions) among evaluative expressions (or their models). First, let us define the ordering $\leq_{\mathrm{H}}$ on the set of hedges that can be defined on the hedges from Table 1 as follows:

$$\mathrm{Ex} \leq_{\mathrm{H}} \mathrm{Si} \leq_{\mathrm{H}} \mathrm{Ve} \leq_{\mathrm{H}} \langle \mathrm{empty} \rangle \leq_{\mathrm{H}} \mathrm{ML} \leq_{\mathrm{H}} \mathrm{Ro} \leq_{\mathrm{H}} \mathrm{QR}.$$

Based on $\leq_{\mathrm{H}}$, we define the ordering $\leq_{\mathrm{LE}}$ of evaluative expressions. Let $\mathcal{A}_1 := \langle \mathrm{hedge} \rangle_1 \mathcal{A}$ and $\mathcal{A}_2 := \langle \mathrm{hedge} \rangle_2 \mathcal{A}$, where $\mathcal{A}$ is an atomic expression. Then,

$$\mathcal{A}_1 \leq_{\mathrm{LE}} \mathcal{A}_2$$

if $\langle \mathrm{hedge} \rangle_1 \leq_{\mathrm{H}} \langle \mathrm{hedge} \rangle_2$. So, evaluative expressions of the same type (with identical atomic expressions) are ordered according to their specificity (i.e., hedges), $\mathcal{A}_1$ and $\mathcal{A}_2$ with different atomic expressions cannot be ordered by $\leq_{\mathrm{LE}}$.

Further, we adopt the extension of the theory of evaluative expressions by the following two axioms, namely the *partition axiom* and *inclusion axiom* [9], and newly we add the *uniqueness axiom.*

Axiom 1 (Partition Axiom). *Let $\mathcal{A}_1, \mathcal{A}_2$ be evaluative expressions with different atomic expressions that are modeled by $A_1, A_2 \in \mathcal{F}(U)$. Then, for all $u \in U$:*

$$A_1(u) + A_2(u) < 2.$$

Axiom 2 (Inclusion Axiom). *Let $\mathcal{A}_1, \mathcal{A}_2$ be two nonequal evaluative expressions ordered as $\mathcal{A}_1 \leq_{\mathrm{LE}} \mathcal{A}_2$ and modeled by fuzzy sets A_1, A_2, respectively. Then,*

$$A_1 \subseteq A_2 \quad \text{and} \quad \mathrm{Ker}(A_1) \subset \mathrm{Ker}(A_2),$$

where $\mathrm{Ker}(A)$ *denotes the kernel of a fuzzy set A.*

Axiom 3 (Uniqueness Axiom). *Let LD be given. Then*

$$(\forall i \in N_n)(\forall j \in N_n)((i \neq j) \Rightarrow (\mathcal{A}_i \neq \mathcal{A}_j)). \tag{4}$$

It should be noted that antecedents of a linguistic description usually contain more than one variable. In this case, antecedent variables are compounded by conjunction and the ordering $\leq_{\mathrm{LE}}$ of compound evaluative expressions is extended in a straightforward way.

---

[3] LD is viewed as a set, hence we omit multiple occurrences and each rule can be contained in LD only once.

2.2 PbLD with Fuzzy Inputs

In this section we describe the PbLD with inputs formed by fuzzy sets (we call it *PbLD with fuzzy inputs*). For formal treatment of PbLD with crisp inputs, see [10, Sects. 2.2–2.3].

We start with a generalization of the *firing degree*. Recall that the firing degree of a crisp input u_0 associated with the rule $\mathcal{R}_i$ from LD is simply $A_i(u_0)$. In [10, Sect. 2.4], we proposed to use as a generalized fired degree operations $\circ$ and $\lhd$ derived from (3) and defined as follows:

$$A_1 \circ A_2 = \bigvee_{u \in U} (A_1(u) * A_2(u)) \,, \quad A_1 \lhd A_2 = \bigwedge_{u \in U} (A_1(u) \to A_2(u)) \,. \quad (5)$$

However, we showed in [10, Sect. 3] that using $\circ$ as the generalized firing degree does not provide satisfactory results from the point of view of interpolativity. Results with $\lhd$ were much more promising, but later we noticed a certain drawback of this operation for our purposes, namely that if there is an $u \in U$ such that $A_1(u) = 1$ and $A_2(u) = 0$, then $A_1 \lhd A_2 = 0$. This is a problem for balancing PbLD. Consider, for example, fuzzy sets A_1, A_2 being interpretations of evaluative expression *small* and *more or less medium*, respectively (see Fig. 1). Then $A_1 \lhd A_2 = A_2 \lhd A_1 = 0$. But this is not in accordance with the idea behind balancing PbLD (see the discussion below). Therefore, we searched for less strict operations similar to $\lhd$. For this purpose, let us introduce an *inclusion measure* by the following axiomatic definition.

Definition 1. Let U be a universe. We say that the operation $Im : \mathcal{F}(U) \times \mathcal{F}(U) \to [0, 1]$ is an *inclusion measure* if it fulfills the following axioms for all $A_1, A_2, A_3 \in \mathcal{F}(U)$ and all $u_0 \in U$:

(Ax1) $Im(A_1, A_2) = 1$ iff $A_1 \subseteq A_2$,
(Ax2) if $A_1 \neq \emptyset$, then $Im(A_1, A_2) = 0$ iff $A_1 \cap A_2 = \emptyset$,
(Ax3) if $A_1 = \chi_{u_0}$, then $Im(A_1, A_2) = A_2(u_0)$,
(Ax4) if $A_2 \subseteq A_3$, then $Im(A_1, A_2) \leq Im(A_1, A_3)$.

Remark 1. 1. The first axiom (Ax1) says that A_1 being a subset of A_2 is a necessary and sufficient condition for an inclusion measure to attain the highest possible degree. The second axiom (Ax2) states that an inclusion degree is equal to zero if and only its arguments are disjoint (the exception is the empty fuzzy set as the first argument, because the empty fuzzy set is a subset of any fuzzy set, hence the axiom (Ax1) applies). The correct behavior of inclusion measures with respect to fuzzy singletons is guaranteed by (Ax3). Finally, by (Ax4) we require the monotonicity of Im in the second argument. Naturally, if we expand A_2, but A_1 does not change (hence, the size of the intersection of A_1 and A_2 may increase), we do not want the decrease of $Im(A_1, A_2)$.

2. The class of operations that fulfill axioms (Ax1)–(Ax4) can be also understood as a subclass of the class of *generalized fuzzy quantifiers of type* $\langle 1, 1\rangle$. Recall that this class is formed, given a universe U, just by all mappings

$\mathcal{F}(U) \times \mathcal{F}(U) \to L$, where L is a structure of truth values, see [12]. The operation $\lhd$ applied on fuzzy sets A_1, A_2 corresponds to the fuzzy quantifier *all* A_1 *are* A_2. From the point of view of the theory of generalized fuzzy quantifiers, inclusion measures are related to fuzzy quantifier *many* [13].

Example 1. Let U be finite. Define $RC : \mathcal{F}(U) \times \mathcal{F}(U) \to [0,1]$ as follows. If $A_1 = \emptyset$, then $RC(A_1, A_2) := 1$. Otherwise,

$$RC(A_1, A_2) := \frac{\sum_{u \in U} (A_1 \cap A_2)(u)}{\sum_{u \in U} A_1(u)}.$$

It is easy to see that RC (relative cardinality) is an inclusion measure according to Definition 1.

Example 2. $\lhd$ given in (5) is not an inclusion measure with respect to Definition 1. It fulfills axioms (Ax1), (Ax3) and (Ax4), but it does not fulfill (Ax2).

Definition 2. Let $A_i, A_0 \in \mathcal{F}(U)$ and let $Im : \mathcal{F}(U) \times \mathcal{F}(U) \to [0,1]$ be an inclusion measure. The *generalized firing degree* of the observation A_0 with respect to the IF-THEN rule $\mathcal{R}_i$ is defined as $Im(A_0, A_i)$. If the operation Im is fixed, we write $A_i(A_0)$ instead of $Im(A_0, A_i)$.

In the rest of this paper, we suppose that Im is a fixed inclusion measure for a given universe U and write $A_i(A_0)$ instead of $Im(A_0, A_i)$.

Now we present two variants of the PbLD with fuzzy inputs, namely the *original* PbLD and the *balancing* PbLD. They differ in the selection of fired IF-THEN rules. The original PbLD chooses the rule with the maximal firing degree among all rules from a given LD. The balancing PbLD chooses the rule with the maximal firing degree *in any group of rules with the same atomic evaluative expression*. The reason for introducing this variant is an effort to use all available information in situations when the fuzzy input is, for example, placed in-between antecedent fuzzy sets modeling evaluative expressions *small* and *medium*, see Example 3. The only formal difference between two variants of PbLD is in the definition of the ordering of antecedent fuzzy sets with respect to a given fuzzy input.

Definition 3. Let LD be a linguistic description. Let $A_0 \in \mathcal{F}(U)$.

1. We write (the O in $\leq^O_{A_0}$ stand for *original* PbLD)

$$A_i \leq^O_{A_0} A_j$$

$$\begin{aligned} \text{either if} \quad & A_i(A_0) > A_j(A_0), \\ \text{or if} \quad & A_i(A_0) = A_j(A_0) \text{ and } \mathcal{A}_i \leq_{\mathrm{LE}} \mathcal{A}_j. \end{aligned}$$

2. Let A_i, A_j be such that $\mathcal{A}_i := \langle\text{hedge}\rangle_i \mathcal{A}$ and $\mathcal{A}_j := \langle\text{hedge}\rangle_j \mathcal{A}$, where $\mathcal{A}$ is an atomic expression. We write (the B in $\leq^B_{A_0}$ stand for *balancing* PbLD)

$$A_i \leq^B_{A_0} A_j$$

$$\begin{aligned} \text{either if} \quad & A_i(A_0) > A_j(A_0), \\ \text{or if} \quad & A_i(A_0) = A_j(A_0) \text{ and } \mathcal{A}_i \leq_{\mathrm{LE}} \mathcal{A}_j. \end{aligned}$$

The local perception with respect to a given linguistic description LD and a given fuzzy input A_0 will then be given as follows.

Definition 4. Let LD be a linguistic description. Let $T \in \{O, B\}$. The *local perception* is a mapping $P^{LD}_T : \mathcal{F}(U) \to \mathcal{P}(N_n)$ that assigns to each fuzzy set $A_0 \in \mathcal{F}(U)$ the set

$$P^{LD}_T(A_0) = \{i \in N_n \mid A_i(A_0) > 0 \,\&\, (\forall j \in N_n)((A_j \leq^T_{A_0} A_i) \Rightarrow (\mathcal{A}_j = \mathcal{A}_i))\}.$$

Now we can define the deduction rule of PbLD.

Definition 5. Let LD be a linguistic description. Let $T \in \{O, B\}$. Let $@ \in \{\circ, \triangleleft\}$ (cf. formulas (3)). Let us be provided with an observation $A_0 \in \mathcal{F}(U)$. Then, the *rule of PbLD* (r^T_{PbLD}) is given as follows:

$$r^T_{PbLD} : \frac{P^{LD}_T(A_0), LD}{A_0 @ \hat{R}_{A_0}}, \tag{6}$$

where for all $u \in V$ and all $v \in V$,

$$\hat{R}_{A_0}(u, v) = \bigwedge_{i \in P^{LD}_T(A_0)} (A_i(u) \to B_i(v)).$$

One can easily check that, in case of a crisp (singleton) input, the above definitions are equivalent to Definitions 1–4 from [10].

Example 3. Let us demonstrate the behavior of the PbLD with fuzzy inputs on the following linguistic description $\mathcal{RB}_2$:

$$\begin{aligned} \mathcal{R}_1 &:= \mathsf{IF}\ X \text{ is } \textit{Ve Sm}\ \mathsf{THEN}\ Y \text{ is } \textit{Bi}, \\ \mathcal{R}_2 &:= \mathsf{IF}\ X \text{ is } \textit{Ro Sm}\ \mathsf{THEN}\ Y \text{ is } \textit{ML Me}, \\ \mathcal{R}_3 &:= \mathsf{IF}\ X \text{ is } \textit{Ro Me}\ \mathsf{THEN}\ Y \text{ is } \textit{QR Sm}, \\ \mathcal{R}_4 &:= \mathsf{IF}\ X \text{ is } \textit{Me}\ \mathsf{THEN}\ Y \text{ is } \textit{Ro Bi}. \end{aligned}$$

Let us consider two fuzzy inputs A_{01} and A_{02} (see Fig. 2(a)). The fuzzy sets interpreting the consequent evaluative expressions from $\mathcal{RB}_2$ are on Fig. 2(b). In Table 2, we can see generalized firing degrees $A_i(A_{0j})$, $i = 1, \ldots, 4$, $j = 1, 2$, computed using the inclusion measure RC from Example 1. The greatest values in both rows are marked by bold font, the second greatest by italics. For the

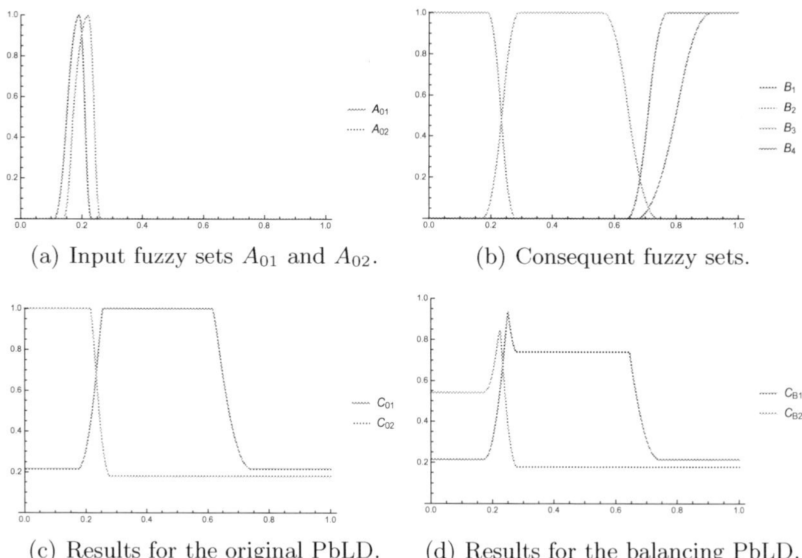

(a) Input fuzzy sets A_{01} and A_{02}.

(b) Consequent fuzzy sets.

(c) Results for the original PbLD.

(d) Results for the balancing PbLD.

Fig. 2. Example of PbLD for the linguistic description $\mathcal{RB}_2$. (Color figure online)

Table 2. Generalized firing degrees of fuzzy inputs A_{01} and A_{01}.

$A_i(A_{0j})$	A_1	A_2	A_3	A_4
A_{01}	0.006	**0.764**	*0.219*	0.031
A_{02}	0	*0.414*	**0.536**	0.115

original PbLD, the rule with the greatest generalized firing degree is used, that is, for the fuzzy inputs A_{01} and A_{02}, the rules $\mathcal{R}_2$ and $\mathcal{R}_3$ are used, respectively. Results for the original PbLD, marked as C_{01} and C_{02}, are depicted in Fig. 2(c). For the balancing PbLD, we use rules with the greatest non-zero generalized firing degrees *among rules whose antecedents have the same atomic evaluative expression*. In our case, for both fuzzy inputs A_{01} and A_{02}, rules $\mathcal{R}_2$ and $\mathcal{R}_3$ are used. Results for the balancing PbLD are depicted in Fig. 2(d).

We can see that for the original PbLD, the resulting fuzzy sets that correspond to fuzzy inputs A_{01} and A_{02} are, in fact, modified consequent fuzzy sets B_2 and B_3, respectively. For the balancing PbLD, both consequent fuzzy sets B_2 and B_3 are "mixed", but for the fuzzy input A_{01}, the influence of the rule $\mathcal{R}_2$ is stronger (the result is "more like medium"). Analogously, for the fuzzy input A_{02}, the influence of $\mathcal{R}_3$ is stronger (the result is "more like small").

3 Fuzzy Interpolation

Each inference mechanism should possess a fundamental property—preservation of modus ponens. If an input fuzzy set $A_0 \in \mathcal{F}(U)$ is equal to one of the antecedents, say to the A_i, then the inferred output $B_0 \in \mathcal{F}(V)$ should be equal to the respective consequent B_i. This requirement leads to the following *system of fuzzy relational equations*

$$A_i @ R = B_i\,, \qquad i = 1, \ldots, n, \tag{7}$$

where $R \in \mathcal{F}(U \times V)$ is unknown. An R satisfying (7) is called a *solution* of the system. Solution R *interpolates* pairs (A_i, B_i) and can be seen as a correct model of the given fuzzy rule base in the given fuzzy inference system.

Obviously, not all systems (7) are solvable, i.e., not for all sets of pairs $(A_i, B_i)_{i=1}^n$ there exists a fuzzy relation that interpolates them. The question of solvability of such systems was addressed by many researchers. In this section, we recall only the most fundamental results [14–16].

Theorem 1. *System* (7) *with* $@ = \circ$ ($@ = \lhd$) *is solvable if and only if* $\hat{R}$ ($\check{R}$) *is a solution of this system. In case of solvability,* $\hat{R}$ ($\check{R}$) *is the greatest (the least) solution of* (7) *with* $@ = \circ$ ($@ = \lhd$)*.*

Theorem 1 actually states that the implicative model $\hat{R}$ (Mamdani-Assilian model $\check{R}$) should be the first choice whenever dealing with the inference modelled by $\circ$ ($\lhd$). If there exist some reasons (e.g., robustness, low computational complexity or the existence of hierarchical inference that is identical to the non-hierarchical one [6]), why the combination of Mamdani-Assilian model $\check{R}$ and the CRI inference $\circ$ (or the combinations of the implicative model $\hat{R}$ and the BK-subproduct inference $\lhd$) should be preferred, one should first of all check whether the interpolativity is preserved also in this case. The answer to this question is provided by the following theorems that collect results from [15,17].

Theorem 2. *Let all* A_i, $i \in N_n$, *be normal. Then* $\check{R}$ ($\hat{R}$) *is a solution of* (7) *with* $@ = \circ$ ($@ = \lhd$) *if and only if the condition*

$$\bigvee_{u \in U} (A_i(u) * A_j(u)) \leq \bigwedge_{v \in V} (B_i(v) \leftrightarrow B_j(v)) \tag{8}$$

holds for any $i, j \in N_n$.

Due to the monotonicity of images, namely:

$$A @ R_1 \subseteq A @ R_2, \quad @ \in \{\circ, \lhd\}$$

for any $R_1, R_2 \in \mathcal{F}(U \times V)$ such that $R_1 \subseteq R_2$, we can state the following corollary.

Corollary 1. *Let* $R_1, R_2 \in \mathcal{F}(U \times V)$ *be two solutions of system* (7) *with* $@ = \circ$ ($@ = \lhd$)*. Then any* $R \in \mathcal{F}(U \times V)$*, such that* $R_1 \subseteq R \subseteq R_2$*, is a solution of this system, too.*

3.1 Interpolativity in Case of the Original PbLD

The interpolativity of the PbLD turns into an investigation of the following problem: *Given LD and the antecedent* $A_i \in \mathcal{F}(U)$ *of the i-th rule,* $i = 1, \ldots, n$, *it should hold that after the application of* r_{PbLD}:

$$r^T_{PbLD} : \frac{P^{LD}_T(A_i), LD}{A_i @ \hat{R}_{A_i}},$$

the conclusion $A_i @ \hat{R}_{A_i}$ *is equal to the consequent* $B_i \in \mathcal{F}(V)$ *of this i-th rule.*[4]

Lemma 1. *Let LD be a linguistic description, let* $i \in N_n$*. Then* $P^{LD}_O(A_i) = \{i\}$.

Sketch of the Proof: We search for $P^{LD}_O(A_0)$ for $A_0 = A_i$. Due to (Ax1), $A_i(A_0) = 1$. If there does not exist any other $j \in N_n$ such that $A_0 \subseteq A_j$, then i is the only index for which $A_i(A_0) = 1$ and thus $P^{LD}_O(A_0) = 1$.

Now, let us assume that there exists $j \in N_n$ such that $A_0 \subseteq A_j$ and thus $A_j(A_0) = 1$. But as $A_i = A_0$, we also have $A_i \subseteq A_j$, which consequently means that $\mathcal{A}_i \leq_{\mathrm{LE}} \mathcal{A}_j$ and together with the *Uniqueness axiom* gives $\mathcal{A}_i \neq \mathcal{A}_j$ and therefore, $j \notin P^{LD}_O(A_0)$. As this holds for any $j \neq i$, $P^{LD}_O(A_i) = \{i\}$. □

Corollary 2. *Let LD be a linguistic description and let* r^O_{PbLD} *be used. Then the interpolativity is preserved for both* $@ = \circ$ *and* $@ = \lhd$.

One may see that with the original PbLD, the interpolativity is obtained automatically and thus, the method is somehow safe from the logical point of view (preservation of modus ponens). However, the question is for what price we get this safety. In particular, whether this approach is not too restrictive in situations, when more[5] rules should be fired. Therefore, naturally we continue with investigating the same problem for the balancing PbLD.

3.2 Interpolativity in Case of the Balancing PbLD

For the case of the balancing PbLD, there is no such a strong result as the one for the original PbLD provided by Lemma 1. However, at least a weaker result of the similar nature may be provided also here.

Lemma 2. *Let LD be a linguistic description, let* $i \in N_n$*. Then* $i \in P^{LD}_B(A_i)$.

Sketch of the Proof: The proof is analogous to the proof of Lemma 1. □

The interpolativity problem in the setting of PbLD differs from the fuzzy relational one by solving each of the n fuzzy relational equations separately. Hence, we may focus on a single equation

$$A_i @ R = B_i \tag{9}$$

for a fixed yet arbitrarily chosen $i \in N_n$.

---

[4] Compared to (7), where a single fuzzy relation has to be a solution of the whole system, here not all rules are fired and each equation is solved separately.

[5] When $A_0 \neq A_i$, not necessarily a single rule is fired. However, the original PbLD, compared to the fuzzy relational approach, notably reduces the number of fired rules.

Lemma 3. *Let $i \in N_n$ and r^B_{PbLD} is used. The sufficient condition for $\hat{R}_{A_i}$ being the solution of (9) is the following one*

$$\bigvee_{u \in U} (A_j(u) * A_i(u)) \leq \bigwedge_{v \in V} (B_i(v) \to B_j(v)) \tag{10}$$

for all $j \in P^{LD}_B(A_i)$, regardless of $@ = \lhd$ or $@ = \circ$.

Sketch of the Proof: Let us fix any $i \in N_n$. The corresponding fuzzy relational equation

$$A_i \lhd R = B_i \tag{11}$$

is always solvable and (8) holds. Thus, according to Theorem 2, $\hat{R}_i(u, v) = A_i(u) \to B_i(v)$ is a solution of the equation (11) and, according to Theorem 1, the least solution of this equation is $\check{R}_i(u, v) = A_i(u) * B_i(v)$.

In order to prove that $\hat{R}_{A_i}$ defined as

$$\hat{R}_{A_i}(u, v) = \bigwedge_{j \in P^{LD}_B(A_i)} (A_j(u) \to B_j(v))$$

is a solution of the equation (11), it is sufficient to prove that it lies between $\check{R}_i$ and $\hat{R}_i$ (cf. Corollary 1).

Since $i \in P^{LD}(A_i)$, obviously $\hat{R}_{A_i} \subseteq \hat{R}_i$ holds. As $\check{R}_i$ is the least solution, $\hat{R}_{A_i}$ will become a solution if and only if the second inclusion $\check{R}_i \subseteq \hat{R}_{A_i}$ is preserved. The inclusion can be expanded as follows

$$A_i(u) * B_i(v) \leq \bigwedge_{j \in P^{LD}_B(A_i)} (A_j(u) \to B_j(v)) \quad \forall u \forall v$$

and will be preserved if and only if for all $j \in P^{LD}_B(A_i)$, the following inequality will be preserved

$$A_i(u) * B_i(v) \leq A_j(u) \to B_j(v) \quad \forall u \forall v$$

which, using adjunction, associativity of $*$ and again adjunction, turns to be equivalent to the preservation of inequality (10). The proof for $@ = \circ$ proceeds analogously, the fact that the role of $\hat{R}_i$ and $\check{R}_i$ is switched is irrelevant, again, both fuzzy relations are solutions of the given equation, and we prove that $\hat{R}_{A_i}$ lies in between of them. □

Despite the positive result given by Lemma 3, one may expect to get a more comfortable condition at least for the case of $\circ$, which is more appropriate for the implicative interpretation of rules. The problem of (10) comes from the fact that it is not so straightforward and easy to check. The so-called *finitary* condition [18,19], that firstly appeared in [20] under the name *boundary* condition, is such a type of comfortable condition that is easy to check or even to ensure the solvability in advance, because it focuses on the antecedent fuzzy sets only.

But first of all, let us present a result that we could obtain due to Lemma 2.

Lemma 4. *Let LD be a linguistic description, let $i \in N_n$. If r^B_{PbLD} is used then*

$$A_i \circ \hat{R}_{A_i} \subseteq B_i. \tag{12}$$

Sketch of the Proof: Due to Lemma 2 we get $\hat{R}_{A_i} \subseteq \hat{R}_i$, where $\hat{R}_i(u,v) = A_i(u) \to B_i(v)$ solves the equation. Then the monotonicity of $*$ proves (12). □

Now, we introduce a modified finitary condition. Recall that in [18–20] it was always defined as the existence of exclusive points, that is, points, that fully belong to a single antecedent fuzzy set and do not belong at all to any other antecedent fuzzy set. For the evaluative linguistic expressions, which are modelled by fully overlapping fuzzy sets, this condition does not hold. However, as we will show below, due to the use of the PbLD inference, it is sufficient if such points exist only for all the antecedent fuzzy sets of the fired rules.

Definition 6. Let LD be given. We say that fired rules fulfill the *modified finitary condition* if for all $i \in N_n$ there exists $u_i \in U$ such that $A_i(u_i) = 1$ and $A_j(u_i) = 0$ for all $j \in P^{LD}_B(A_i)$, $j \neq i$.

Proposition 1. *Let LD be given and the antecedent fuzzy sets fulfill the modified finitary condition. Then the following holds for any $i \in N_n$*

$$A_i \circ \hat{R}_{A_i} = B_i. \tag{13}$$

Sketch of the Proof: Let us fix $i \in N_n$ and choose the $u_i \in U$ such that $A_i(u_i) = 1$ and $A_j(u_i) = 0$ for any $j \in P^{LD}_B(A_i)$, $j \neq i$. Using the equality $a * (a \to b) = b$ and the normality of antecedents, we get

$$(A_i \circ \hat{R}_{A_i})(v) = \bigvee_{u \in U} \left(A_i(u) * \bigwedge_{j \in P^{LD}_B(A_i)} (A_j(u) \to B_j(v)) \right)$$

$$= \cdots = B_i(v) \wedge \left(1 * \bigwedge_{\substack{j \in P^{LD}_B(A_i) \\ j \neq i}} (0 \to B_j(v)) \right) = B_i(v).$$

□

It is important to note that using fuzzy sets modeling the evaluative linguistic expressions ensures the fulfillment of the modified finitary condition and thus, the interpolativity property (preservation of the modus ponens), whenever we deal with the balancing PbLD with $\circ$.

4 Concluding Remarks

We revisited the PbLD inference method for fuzzy inputs firstly defined in [10] and redefined the generalized firing degree in order to capture the intuitive behaviour form the crisp input case. This led to a new investigation of mathematical properties of the PbLD. We showed that using the redefined generalized firing degree, the preservation of modus pones, as the most fundamental property of any inference system, is preserved under much weaker conditions than before and moreover, under weaker conditions than in the case of the fuzzy relational inference systems.

References

1. Mamdani, E.H., Assilian, S.: An experiment in linguistic synthesis with a fuzzy logic controller. Int. J. Man Mach. Stud. **7**, 1–13 (1975)
2. Dubois, D., Prade, H.: What are fuzzy rules and how to use them. Fuzzy Sets Syst. **84**, 169–185 (1996)
3. Štěpnička, M., Bodenhofer, U., Daňková, M., Novák, V.: Continuity issues of the implicational interpretation of fuzzy rules. Fuzzy Sets Syst. **161**, 1959–1972 (2010)
4. Zadeh, L.A.: Outline of a new approach to the analysis of complex systems and decision processes. IEEE Trans. Syst. Man Cybern. **3**(1), 28–44 (1973)
5. Pedrycz, W.: Applications of fuzzy relational equations for methods of reasoning in presence of fuzzy data. Fuzzy Sets Syst. **16**, 163–175 (1985)
6. Štěpnička, M., Jayaram, B.: On the suitability of the Bandler-Kohout subproduct as an inference mechanism. IEEE Trans. Fuzzy Syst. **18**(2), 285–298 (2010)
7. Perfilieva, I., Novák, V.: System of fuzzy relation equations as a continuous model of IF-THEN rules. Inf. Sci. **177**, 3218–3227 (2007)
8. Novák, V., Perfilieva, I.: On the semantics of perception-based fuzzy logic deduction. Int. J. Intell. Syst. **19**, 1007–1031 (2004)
9. Dvořák, A., Štěpnička, M., Štěpničková, L.: On redundancies in systems of fuzzy/linguistic IF-THEN rules under perception-based logical deduction inference. Fuzzy Sets Syst. **277**, 22–43 (2015)
10. Štěpnička, M., Dvořák, A.: On perception-based logical deduction and its variants. In: Proceedings of 16th IFSA World Congress/9th EUSFLAT Conference, pp. 341–350. Atlantis Press, Gijón (2015)
11. Novák, V.: A comprehensive theory of trichotomous evaluative linguistic expressions. Fuzzy Sets Syst. **159**(22), 2939–2969 (2008)
12. Holčapek, M.: Monadic **L**-fuzzy quantifiers of the type $\langle 1^n, 1 \rangle$. Fuzzy Sets Syst. **159**(14), 1811–1835 (2008)
13. Murinová, P., Novák, V.: Analysis of the intermediate quantifier "many" in fuzzy natural logic. In: Proceedings of 16th IFSA World Congress/9th EUSFLAT Conference, pp. 1147–1153. Atlantis Press, Gijón (2015)
14. De Baets, B.: Analytical solution methods for fuzzy relational equations. In: Dubois, D., Prade, H. (eds.) The Handbook of Fuzzy Set Series, vol. 1, pp. 291–340. Academic Kluwer Publ, Boston (2000)
15. Klawonn, F.: Fuzzy points, fuzzy relations and fuzzy functions. In: Novák, V., Perfilieva, I. (eds.) Discovering the World with Fuzzy Logic, pp. 431–453. Springer, Berlin (2000)
16. Perfilieva, I., Nosková, L.: System of fuzzy relation equations with inf-$\rightarrow$ composition: complete set of solutions. Fuzzy Sets Syst. **159**, 2256–2271 (2008)
17. Nosková, L.: Systems of fuzzy relation equation with inf-$\rightarrow$ composition: solvability and solutions. J. Electr. Eng. **12**(s), 69–72 (2005)
18. Perfilieva, I.: Finitary solvability conditions for systems of fuzzy relation equations. Inf. Sci. **234**, 29–43 (2013)

19. Štěpnička, M., Jayaram, B.: Interpolativity of at-least and at-most models of monotone fuzzy rule bases with multiple antecedent variables. Fuzzy Sets Syst. (in press). doi:10.1016/j.fss.2015.09.013
20. Chung, F.-L., Lee, T.: A new look at solving a system of fuzzy relational equations. Fuzzy Sets Syst. **88**(3), 343–353 (1997)

Graded Dominance and Cantor-Bernstein Equipollence of Fuzzy Sets

Michal Holčapek(✉)

Institute for Research and Applications of Fuzzy Modelling, NSC IT4Innovations, University of Ostrava, 30. dubna 22, 701 03 Ostrava 1, Czech Republic
michal.holcapek@osu.cz
http://irafm.osu.cz

Abstract. The aim of the paper is to propose a graded dominance for fuzzy sets that assigns to each pair of fuzzy sets a degree in which one fuzzy set has less cardinality than another one or the cardinalities of both fuzzy sets are approximately equal. The graded dominance for fuzzy sets is a natural generalization of the dominance relation for sets. The graded dominance is then used for the introduction of a fuzzy class equivalence that satisfies a graded version of the Cantor-Bernstein theorem.

Keywords: Fuzzy set theory · Fuzzy sets · Fuzzy classes · Graded dominance · Graded equipollence · Cantor-Bernstein theorem

1 Introduction

In classical set theory, we say that a set x is dominated by a set y (symbolically, $x \precsim y$) if there exists a one-to-one function of x to y. Further, we say that x and y are equipollent (symbolically, $x \sim y$) if there exists a one-to-one correspondence (bijection) between x and y. The Cantor-Bernstein theorem (CBT for short) states the famous relationship between the dominance relation and the equipollence of sets, namely,

$$x \precsim y \,\&\, y \preceq x \Rightarrow x \sim y.$$

The concept of dominance and equipollence of sets has been generalized for fuzzy sets by Wygralak in [9][1]. He proved, among others, that the Cantor-Bernstein theorem is satisfied for the proposed concepts. A more complex task appears in the case when the both concepts should be designed in a graded style. It means that a degree of truth to which one fuzzy set is dominated by another one or two fuzzy sets are approximately equipollent should be specified in a natural way. The first definition of graded (many-valued) equipollence of fuzzy sets has been also introduced by Wygralak in [9] by means of the Łukasiewicz biresiduum and generalized (fuzzy) cardinals. Unfortunately, this concept has

[1] More precisely, Wygralak studied in [9] more general objects called "vaguely defined objects", where fuzzy sets form a special case.

J.P. Carvalho et al. (Eds.): IPMU 2016, Part II, CCIS 611, pp. 510–521, 2016.
DOI: 10.1007/978-3-319-40581-0_41

not been fully elaborated to obtain a serious graded cardinal theory of fuzzy sets. A different approach to the graded equipollence of fuzzy sets based on one-to-one correspondences between fuzzy sets has been proposed by Holčapek and Turčan in [7] (see also [6]). Since this approach follows the Cantor's idea of comparing sizes of sets, we will refer to this equipollence as the graded *Cantor euipollence.* In [5], a graded generalization of one version of the CBT[2] has been proved for fuzzy sets with finite supports and membership degrees interpreted in a linearly ordered residuated lattice. Unfortunately, this graded version of the CBT does not remain true for fuzzy sets with infinite supports, because one-to-one correspondences between fuzzy sets are too "sensitive" to neglect certain membership degrees.[3]

The failure of the graded Cantor's equipollence in proving one of the essential theorems of set theory motivates us to introduce another type of graded equipollence for fuzzy sets that satisfies the graded generalization of the CBT. We will refer to this equipollence as the graded *Cantor-Bernstein equipollence.* To introduce the graded Cantor-Bernstein equipollence, it is reasonable to begin with a simpler concept of graded dominance for fuzzy sets that expresses degrees of truth to which one fuzzy set can be embedded into another one. The graded Cantor-Bernstein equipollence is then defined using the conjunction of degrees of truth to which one fuzzy set is dominated by another one and vice versa. As a simple consequence of this definition, we obtain the satisfaction of the graded generalization of the CBT.

The paper is organized as follows. Section 2 is devoted to preliminaries, where we introduce algebraic structures of truth values and basic elements of fuzzy set theory in a universe of sets. Sections 3 and 4 are devoted to the graded dominance and the graded Cantor-Bernstein equipollence, respectively, and their properties. The last section is a conclusion.

2 Preliminaries

2.1 Algebraic Structures of Truth Values

In this paper, the truth values are interpreted in a complete linearly ordered residuated lattice. Recall that an algebra $\mathbf{L} = \langle L, \wedge, \vee, \otimes, \rightarrow \bot, \top\rangle$ with four binary operations and two constants is said to be a *residuated lattice* provided that

(i) $\langle L, \wedge, \vee, \bot, \top\rangle$ is a bounded lattice, where $\bot$ is the least element and $\top$ is the greatest element of L, respectively,
(ii) $\langle L, \otimes, \top\rangle$ is a commutative monoid,
(iii) the pair $\langle \otimes, \rightarrow\rangle$ forms an adjoint pair, i.e.,

$$a \leq b \rightarrow c \quad \text{if and only if} \quad a \otimes b \leq c \tag{1}$$

holds for each $a, b, c \in L$ ($\leq$ denotes the corresponding lattice ordering).

[2] See, Theorem 4 on p. 518 in this paper.
[3] See, Example 2 on p. 518 in this paper.

We say that a residuated lattice is *complete* (*linearly ordered*) if $\langle L, \wedge, \vee, \bot, \top \rangle$ is a complete (linearly ordered) lattice. Details and examples of residuated lattices can be found in [1] or [8].

2.2 Fuzzy Sets and Fuzzy Classes in a Universe of Sets

Fuzzy Sets. A fuzzy set is standardly defined as a function from a fixed non-empty universe of discourse to a set (lattice) of truth values. Nevertheless, a theory of fuzzy sets that admits the constructions like power fuzzy sets or exponentiation of fuzzy sets naturally requires a system of universes of discourse rather than one fixed universe (cf., [3]). Therefore, we proposed in [7] (see also [6]) a universe of sets over a complete residuated lattice and introduce the fuzzy sets and the fuzzy classes in such a way that their universes of discourse are elements and subclasses of the universe of sets, respectively. The both definitions were introduced to imitate the position of sets and classes (auxiliary objects) in the Zermelo-Fraenkel axiomatic theory with the axiom of choice (ZFC for short). In what follows, we use $x \in y$ to denote that the set x is a member of set y, further, we use $P(x)$, $\mathscr{D}(f)$ and $\mathscr{R}(f)$ to denote the power set of a set x, the domain and the range of a function f, respectively.

Definition 1. *Let* $\mathbf{L}$ *be a complete linearly ordered residuated lattice. A universe of sets over* $\mathbf{L}$ *is a non-empty class* $\mathfrak{U}$ *of sets in ZFC satisfying the following properties:*

(U1) $x \in y$ *and* $y \in \mathfrak{U}$, *then* $x \in \mathfrak{U}$,
(U2) $x, y \in \mathfrak{U}$, *then* $\{x, y\} \in \mathfrak{U}$,
(U3) $x \in \mathfrak{U}$, *then* $\mathrm{P}(x) \in \mathfrak{U}$,
(U4) $x \in \mathfrak{U}$ *and* $y_i \in \mathfrak{U}$ *for any* $i \in x$, *then* $\bigcup_{i \in x} y_i \in \mathfrak{U}$,
(U5) $x \in \mathfrak{U}$ *and* $f : x \longrightarrow L$, *then* $\mathscr{R}(f) \in \mathfrak{U}$,

where L *denotes the support of* $\mathbf{L}$.

Basic examples of the universes of sets are the classes of all or finite sets. If the ZFC is extended by the axiom admitting the existence of strongly inaccessible cardinals, one can introduce a universe of sets over $\mathbf{L}$ to be a Grothendieck universe.

Definition 2. *Let* $\mathfrak{U}$ *be a universe of sets over* $\mathbf{L}$. *A function* $A : z \longrightarrow L$ *(in ZFC) is called a* fuzzy set *in* $\mathfrak{U}$ *if* $z \in \mathfrak{U}$.[4]

A consequence of (U5) is that any fuzzy set in $\mathfrak{U}$ belongs to $\mathfrak{U}$. Let $A : z \longrightarrow L$ be a fuzzy set in $\mathfrak{U}$. The set $\mathscr{D}(A)$ is called the *universe of discourse of* A (*universe of* A for short). The set $\mathscr{S}(A) = \{x \in z \mid A(x) > \bot\}$ is called the *support of fuzzy set* A. An element $x \in z$ such that $x \notin \mathscr{S}(A)$ is said to be *negligible in* A. A fuzzy set A is said to be *crisp* and referred to a *crisp set* if $A(x) \in \{\bot, \top\}$ for

[4] For simplicity, we use in the definition the term of "fuzzy sets", although, a more convenient denotation should be $\mathbf{L}$-fuzzy sets with reference to the lattice $\mathbf{L}$.

any $x \in z$. One can see that the empty function as a vacuous fuzzy set is crisp, since the assumption on a crisp set is trivially satisfied. If $x \subseteq y$ are sets in $\mathfrak{U}$, we use χ_x to denote the *characteristic function of* x *on* y, i.e., $\chi_x : y \longrightarrow L$, which is defined by $\chi_x(z) = \top$ if $z \in x$, and $\chi_x(z) = \bot$, otherwise. Obviously, a fuzzy set is crisp if and only if it is a characteristic function of a set. We write $A \subseteq B$ provided that $\mathscr{D}(A) \subseteq \mathscr{D}(B)$ and $A(a) \leq B(a)$ for any $a \in \mathscr{D}(A)$. One can check that $\subseteq$ is a partial ordering on the class $\mathfrak{F}(\mathfrak{U})$ of all fuzzy sets in $\mathfrak{U}$.

In literature on fuzzy set theory, a function that assigns $\bot$ to each element of its domain is usually referred to the empty fuzzy set. In our theory, we use a different interpretation of the empty fuzzy set as follows.

Definition 3. *The empty function* $\emptyset : \emptyset \longrightarrow L$ *is called the empty fuzzy set.*

Identity up to Negligibility. We say that two fuzzy sets A and B are *identical* if their domains are identical, i.e., $\mathscr{D}(A) = \mathscr{D}(B)$, and A and B coincide as the functions, i.e., $A(a) = B(a)$ for any $a \in \mathscr{D}(A)$. Another essential predicate in our theory of fuzzy sets that generalizes the concept of identity for fuzzy sets is a binary relation saying that two fuzzy sets are identical up to their negligible elements.

Definition 4. *We say that fuzzy sets* A *and* B *are identical up to negligibility (symbolically,* $A \equiv B$*) if* $\mathscr{S}(A) = \mathscr{S}(B)$ *and* $A(a) = B(a)$ *for any* $a \in \mathscr{S}(A)$.

Obviously, if $\mathscr{S}(A) = \emptyset$, then the previous definition states that a fuzzy set B is identical with A up to negligibility if and only if $\mathscr{S}(B) = \emptyset$. It is easy to see that the relation "to be identical up to negligibility" is an equivalence on $\mathfrak{F}(\mathfrak{U})$. We use $\mathrm{cls}(A)$ to denote the equivalence class of all fuzzy sets from $\mathfrak{F}(\mathfrak{U})$ that are identical with A up to negligibility.

Example 1. One can simply verify that $\emptyset \equiv \{0/a, 0/b\}$ and $\{0/a, 0/b\} \in \mathrm{cls}(\emptyset)$, or $\{0.9/a\} \equiv \{0.9/a, 0/b\}$ and $\{0.9/a\} \in \mathrm{cls}(\{0.9/a, 0/b\})$.

Fuzzy Set Relations. Let I be a non-empty set of indices, and let $\prod_{i \in I} z_i$ be the Cartesian product of sets z_i from $\mathfrak{U}$. A fuzzy set $R : \prod_{i \in I} z_i \to L$ is called a *fuzzy set relation* (or *fuzzy relation* for short). In fuzzy set theory, the concept of fuzzy equivalence belongs among the most important concepts (see, e.g., [1,4]).

Definition 5. *A fuzzy relation* $R : z \times z \longrightarrow L$ *is called a fuzzy equivalence provided that the following axioms hold for any* $a, b, c \in z$:

(FE1) $R(a, a) = \top$,
(FE2) $R(a, b) = R(b, a)$,
(FE3) $R(a, b) \otimes R(b, c) \leq R(a, c)$.

Axioms (FE1) and (FE2) straightforwardly generalize the definition of reflexivity and symmetry of relations, axiom (FE3) generalizes the transitivity where the operation of multiplication $\otimes$ of the residuated lattice $\mathbf{L}$ is used for the interpretation of the conjunction. A fuzzy partial ordering has been established and studied by Bodenhofer in [2]. The following definition is a slight modification of Bodenhofer's definition.

Definition 6. *Let R be a fuzzy equivalence on z. A fuzzy relation $S : z \times z \to L$ is called an R-fuzzy partial ordering provided that the following axioms hold for any $a, b, c \in z$:*

(FPO1) $S(a,a) = \top$,
(FPO2) $S(a,b) \otimes S(b,a) \leq R(a,b)$,
(FPO3) $S(a,b) \otimes S(b,c) \leq S(a,c)$.

Functions in Certain Degrees of Truth. We use $\mathfrak{Func}$ to denote the class of all functions in $\mathfrak{U}$ and $\mathfrak{Func}(x,y)$, $\mathfrak{Func}_{1\text{-}1}(x,y)$ and $\mathfrak{Func}^{1\text{-}1}_{\text{corr}}(x,y)$ to denote the set of all functions, one-to-one functions and one-to-one correspondences of x onto y, respectively, provided that $x, y \in \mathfrak{U}$. Further, we use $[f \in \mathfrak{Func}(x,y)]$ to denote the truth value (i.e., $\bot$ or $\top$) expressing that f is a member of $\mathfrak{Func}(x,y)$, and similarly for $[f \in \mathfrak{Func}_{1\text{-}1}(x,y)]$ and $[f \in \mathfrak{Func}^{1\text{-}1}_{\text{corr}}(x,y)]$.

Definition 7. *Let $A, B \in \mathfrak{F}(\mathfrak{U})$, and let $f \in \mathfrak{Func}$. We say that f is approximately a function of A to B in the degree α provided that*

$$\alpha = [f \in \mathfrak{Func}(\mathscr{D}(A), \mathscr{D}(B))] \otimes \bigwedge_{(a,f(a)) \in \mathscr{D}(A) \times \mathscr{D}(B)} (A(a) \to B(f(a)). \qquad (2)$$

We use $[f : A \longrightarrow B]$ to denote the degree of truth in which the function f is approximately a function of A to B. Let us emphasize that if f is not a function of $\mathscr{D}(A)$ to $\mathscr{D}(B)$, then $[f : A \longrightarrow B] = \bot$ even if the infimum in (2) is greater than $\bot$. Similarly we define degrees of truth for one-to-one functions and correspondences.

Definition 8. *Let $A, B \in \mathfrak{F}(\mathfrak{U})$, and let $f \in \mathfrak{Func}$. We say that f is approximately a 1-1 function of A to B in the degree α provided that*

$$\alpha = [f \in \mathfrak{Func}_{1\text{-}1}(\mathscr{D}(A), \mathscr{D}(B))] \otimes [f : A \longrightarrow B]. \qquad (3)$$

We say that f is approximately a one-to-one correspondence between A and B in the degree α provided that

$$\alpha = [f \in \mathfrak{Func}^{1\text{-}1}_{\text{corr}}(\mathscr{D}(A), \mathscr{D}(B))] \otimes \bigwedge_{(a,f(a)) \in \mathscr{D}(A) \times \mathscr{D}(B)} (A(a) \leftrightarrow B(f(a)). \qquad (4)$$

We use $[f : A \xrightarrow{1\text{-}1} B]$ and $[f : A \xrightarrow[\text{corr}]{1\text{-}1} B]$ to denote the degree of truth to which f is a one-to-one function of A to B and a one-to-one correspondence between A and B, respectively.

Fuzzy Classes. Although the fuzzy sets in $\mathfrak{U}$ are the major objects in our theory, it seems to be useful, analogously to set theory, to introduce the concept of fuzzy class in $\mathfrak{U}$.

Definition 9. *Let $\mathfrak{U}$ be a universe of sets over* $\mathbf{L}$. *A class function $\mathcal{A} : \mathfrak{Z} \longrightarrow L$ (in ZFC) is called a fuzzy class in $\mathfrak{U}$ if $\mathfrak{Z} \subseteq \mathfrak{U}$.*

Note that each fuzzy set is a fuzzy class because of (U1), but not vice versa. Hence, a fuzzy class $\mathcal{A}$ is said to be *proper* if there is no fuzzy set which is identical to $\mathcal{A}$ up to negligibility (the relation $\equiv$ is extended here to fuzzy classes).

Fuzzy Class Relations. Fuzzy class relations are defined similarly to fuzzy set relations, only fuzzy sets are replaced by fuzzy classes. For the purpose of this paper, we introduce the fuzzy class equivalence and fuzzy class partial ordering.

Definition 10. *A fuzzy class relation $\mathcal{R} : \mathfrak{Z} \times \mathfrak{Z} \longrightarrow L$ is called a fuzzy class equivalence if it satisfies (FE1)-(FE3) of Definition 5.*

Definition 11. *Let $\mathcal{R}$ be a fuzzy class equivalence. A fuzzy class relation $\mathcal{S} : \mathfrak{Z} \times \mathfrak{Z} \longrightarrow L$ is called an $\mathcal{R}$-fuzzy class partial ordering if it satisfies (FPO1)-(FPO3) of Definition 6.*

3 Graded Dominance of Fuzzy Sets

As we have mentioned in Introduction, a set x is dominated by a set y if there exists a one-to-one function of x to y. A straightforward generalization of the dominance relation for sets to a graded dominance for fuzzy sets (symbolically, $\precapprox$) should consist in seeking of a one-to-one function which is approximately a one-to-one function of one fuzzy set to another fuzzy set with the highest degree of truth. Formally, it means that

$$A \precapprox B \Leftrightarrow (\exists f \in \mathfrak{Func})(f : A \xrightarrow{1\text{-}1} B). \tag{5}$$

Thus, the fuzzy set A is approximately dominated by the fuzzy set B in the degree of truth, which is equal to the supremum of all degrees of truth in which functions are approximately one-to-one functions of A to B. Nevertheless, this generalization is insufficient to well describe the graded dominance. Indeed, considering fuzzy sets with different cardinality of their domains, we can find the problem that no one-to-one function of one fuzzy set to another one can be constructed even if they are identical up to negligibility. This motivates us to rewrite (5) into the following more complex formula:

$$A \precapprox B \Leftrightarrow (\exists A' \in \mathrm{cls}(A)\, \exists B' \in \mathrm{cls}(B)\, \exists f \in \mathfrak{Func})(f : A' \xrightarrow{1\text{-}1} B'). \tag{6}$$

One can see that the previous definition is free of the choice of fuzzy sets that are identical up to negligibility, which seems to be a natural requirement. Since the all quantifications used in (6) are made over proper classes, which means to compute the supremum over these classes, first, we prove a useful lemma that enables a restriction of the evaluation to sets.

Lemma 1. *Let $A, B \in \mathfrak{F}(\mathfrak{U})$ be such that*

(i) $|\mathscr{D}(A)| \leq |\mathscr{D}(B)|$,
(ii) $|\mathscr{S}(A)| \leq |\mathscr{D}(B) \backslash \mathscr{S}(B)|$

and let

$$\alpha = \bigvee_{f \in \mathfrak{Func}_{1\text{-}1}(\mathscr{D}(A), \mathscr{D}(B))} [f : A \xrightarrow{1\text{-}1} B]. \tag{7}$$

Then, it holds $[f : C \xrightarrow{1\text{-}1} D] \leq \alpha$ *for any* $C \in \mathrm{cls}(A)$, $D \in \mathrm{cls}(B)$ *and* $f \in \mathfrak{Func}$.

Proof. Let $A, B \in \mathfrak{F}(\mathfrak{U})$ satisfy (i) and (ii). If $\mathscr{D}(A) = \emptyset$, the statement is trivially satisfied, since $\alpha = \top$. Let $\mathscr{D}(A) \neq \emptyset$. Since $\bot \to \beta = \top$ holds for any $\beta \in L$, we may assume, without loss of generality, that $\mathscr{D}(A) = \mathscr{S}(A)$. Let $C \in \mathrm{cls}(A)$, $D \in \mathrm{cls}(B)$ and $f \in \mathfrak{Func}$. If $f \notin \mathfrak{Func}_{1\text{-}1}(\mathscr{D}(C), \mathscr{D}(D))$, then $[f : C \xrightarrow{1\text{-}1} D] = \bot \leq \alpha$. Let $f \in \mathfrak{Func}_{1\text{-}1}(\mathscr{D}(C), \mathscr{D}(D))$. It is easy to see that

$$[f : C \xrightarrow{1\text{-}1} D] = \bigwedge_{a \in \mathscr{D}(C)} (C(a) \to D(f(a))) = \bigwedge_{a \in \mathscr{S}(C)} (C(a) \to D(f(a))). \tag{8}$$

Let g be the restriction of f to $x = \mathscr{S}(C) = \mathscr{D}(A)$. Further, let $y = g(x)$ (the image of x under g), $y_1 = y \cap \mathscr{S}(D)$ and $y_2 = y \backslash y_1$. By assumption (ii), we find that $|y_2| \leq |\mathscr{S}(A)| \leq |\mathscr{D}(B) \backslash \mathscr{S}(B)|$. Hence, there exists a one-to-one function $h : y \to \mathscr{D}(B)$ such that $h(a) = a$ for any $a \in y_1$ and $h(a) \in \mathscr{D}(B) \backslash \mathscr{S}(B)$ for any $a \in y_2$. Obviously, $h \circ g : \mathscr{D}(A) \longrightarrow \mathscr{D}(B)$ is a one-to-one function such that $D(f(a)) = B(h \circ g(a))$ holds for any $a \in \mathscr{S}(A)$. Due to the equality in (8), we obtain

$$[f : C \xrightarrow{1\text{-}1} D] = \bigwedge_{a \in \mathscr{S}(C)} (C(a) \to D(f(a))) =$$

$$\bigwedge_{a \in \mathscr{D}(A)} (A(a) \to B(h \circ g(a))) = [h \circ g : A \xrightarrow{1\text{-}1} B] \leq \alpha,$$

which concludes the proof. □

Remark 1. An open problem is whether the statement in Lemma 1 remains valid without the assumption (ii). We cannot find an example showing the importance of (ii), but also we do not know how to prove the statement without (ii).

If $A, B \in \mathfrak{F}(\mathfrak{U})$ are fuzzy sets that satisfy (i) and (ii) from the previous lemma, we say that A is *cardinal separable* in B. It is easy to see that to each fuzzy set A there exists a fuzzy set B such that A is cardinal separable in B. Now, we can proceed to the definition that specifies the degree in which one fuzzy set is approximately dominated by another fuzzy set.

Definition 12. *Let $A, B \in \mathfrak{F}(\mathfrak{U})$. We say that A is approximately dominated by B to the degree α provided that*

$$\alpha = \bigvee_{f \in \mathfrak{Func}(\mathscr{D}(A), \mathscr{D}(C))} [f : A \xrightarrow{1\text{-}1} C], \tag{9}$$

for $C \in \mathrm{cls}(B)$ such that A is cardinal separable in C.

From Lemma 1, one can see that the definition does not depend on the choice of $C \in \mathrm{cls}(B)$, it is sufficient to consider that A is cardinal separable in C. We use $\precapprox$ to denote the fuzzy class relation to be approximately dominated in a certain degree and $[A \precapprox B]$ to denote the degree to which A is approximately dominated by B.

Definition 13. *The fuzzy class relation* $\precapprox$ *is called the graded dominance.*

Theorem 1. *If* $A, B, C, D \in \mathfrak{F}(\mathfrak{U})$ *such that* $A \equiv C$ *and* $B \equiv D$*, then*

$$[A \precapprox B] = [C \precapprox D].$$

Proof. Without loss of generality, we assume that A and C are cardinal separable in B and D, respectively. Let $X \in \mathrm{cls}(A)$ such that $\mathscr{D}(X) = \mathscr{S}(X)$. Since $X \subseteq A$, X has to be cardinal separable in B. From the rule $\bot \to \beta = \top$, one can see that $[X \precapprox B] = [A \precapprox B]$. Similarly $X \subseteq C$ implies that X is cardinal separable in D, whence $[X \precapprox D] = [C \precapprox D]$. According to the definition of the graded dominance, we find that $[X \precapprox B] = [X \precapprox C]$, which implies the desirable equality. $\square$

Theorem 2. *The graded dominance is a fuzzy class preordering on* $\mathfrak{F}(\mathfrak{U})$*, i.e.,* $\precapprox$ *satisfies (FE1) and (FE3) of Definition 6.*

Proof. Obviously, (FE1) is true. To prove the transitivity, let $A, B, C \in \mathfrak{F}(\mathfrak{U})$. Without loss of generality, we assume that A is cardinal separable in B and simultaneously B is cardinal separable in C. One can simply prove that for arbitrary $f, g \in \mathfrak{Func}$ such that $g \circ f \in \mathfrak{Func}$, it holds

$$[f : A \xrightarrow{1\text{-}1} B] \otimes [g : C] \leq [g \circ f : A \xrightarrow{1\text{-}1} C].$$

Then,

$$\begin{aligned}
&[A \precapprox B] \otimes [B \precapprox C] = \\
&\bigvee_{f \in \mathfrak{Func}(\mathscr{D}(A), \mathscr{D}(B))} [f : A \xrightarrow{1\text{-}1} B] \otimes \bigvee_{g \in \mathfrak{Func}(\mathscr{D}(B), \mathscr{D}(C))} [g : B \xrightarrow{1\text{-}1} C] = \\
&\bigvee_{f \in \mathfrak{Func}(\mathscr{D}(A), \mathscr{D}(B))} \bigvee_{g \in \mathfrak{Func}(\mathscr{D}(B), \mathscr{D}(C))} [f : A \xrightarrow{1\text{-}1} B] \otimes [g : B \xrightarrow{1\text{-}1} C] \leq \\
&\bigvee_{f \in \mathfrak{Func}(\mathscr{D}(A), \mathscr{D}(B))} \bigvee_{g \in \mathfrak{Func}(\mathscr{D}(B), \mathscr{D}(C))} [g \circ f : A \xrightarrow{1\text{-}1} C] \leq \\
&\bigvee_{h \in \mathfrak{Func}(\mathscr{D}(A), \mathscr{D}(C))} [h : A \xrightarrow{1\text{-}1} C] = [A \precapprox C],
\end{aligned}$$

where we used the distributivity of $\otimes$ over $\bigvee$, which holds in each complete residuated lattice. $\square$

4 Graded Cantor-Bernstein Equipollence

In [7], we introduced the concept of graded (Cantor) equipollence. Let us recall its definition. We say that fuzzy sets A, B are *cardinal separable* if A is cardinal separable in B and vice versa. Obviously, to each pair of fuzzy sets, there are cardinal separable fuzzy sets that are identical to original ones up to negligibility.

Definition 14. *Let $A, B \in \mathfrak{F}(\mathfrak{U})$, and let $C \in \mathrm{cls}(A)$ and $D \in \mathrm{cls}(B)$ be fuzzy sets that are cardinal separable. We say that A and B are approximately Cantor equipollent to the degree α provided that*

$$\alpha = \bigvee_{f \in \mathfrak{Func}(\mathscr{D}(C), \mathscr{D}(D))} [f : C \xrightarrow[\text{corr}]{\text{1-1}} D]. \tag{10}$$

We use $\overset{c}{\approx}$ to denote the fuzzy class relation to be approximately Cantor equipollent to a certain degree and $[A \overset{c}{\approx} B]$ to denote the degree to which the fuzzy sets A and B are approximately Cantor equipollent.

Definition 15. *The fuzzy class relation $\overset{c}{\approx}$ is called the graded Cantor equipollence (or the graded C-equipollence for short) of fuzzy sets.*

A simple consequence of the definition of C-equipollence is the following fact.

Theorem 3. *If $A, B, C, D \in \mathfrak{F}(\mathfrak{U})$ such that $A \equiv C$ and $B \equiv D$, then*

$$[A \overset{c}{\approx} B] = [C \overset{c}{\approx} D].$$

One can show that the C-equipollence is a fuzzy class equivalence.[5] Moreover, the following graded version of the Cantor-Bernstein theorem (GCBT) has been proved in [5] for fuzzy sets with finite supports. Note that the linearity of residuated lattice is here an essential assumption.

Theorem 4. *Let $A, B, C, D \in \mathfrak{F}(\mathfrak{U})$ be fuzzy sets with finite supports such that $C \subseteq A$ and $D \subseteq B$. Then,*

$$[A \overset{c}{\approx} D] \wedge [C \overset{c}{\approx} B] \leq [A \overset{c}{\approx} B]. \tag{11}$$

Note that a special case of the previous theorem is the following statement: *if $C \subseteq A \subseteq B$, then* $[C \overset{c}{\approx} B] \leq [A \overset{c}{\approx} B]$. Nevertheless, the GCBT is false for arbitrary fuzzy sets, as the following example demonstrates.

Example 2. Let ω denote the set of natural numbers. Let o and e be the sets of all odd and even numbers, respectively. Define $A : \omega \longrightarrow L$, where L is the support of a residuated lattice $\mathbf{L}$ with an element $\alpha \notin \{\bot, \top\}$, by

$$A(n) = \begin{cases} \top, & n \in o; \\ \alpha, & n \in e. \end{cases}$$

[5] The proof can be designed similarly to the proof of Theorem 5.6 in [5] for finite fuzzy sets.

It is easy to see that $[\chi_o \overset{c}{\approx} \chi_\omega] = \top$, where χ_o and χ_ω are the characteristic functions of o and ω on ω, respectively, and $\chi_o \subset A \subset \chi_\omega$. But, one can simply show that $[\chi_o \overset{c}{\approx} A] = [A \overset{c}{\approx} \chi_\omega] = \alpha \leftrightarrow \top = \alpha < \top$.

A relationship between the graded dominance and the graded C-equipollence of fuzzy sets is described in the following statement.

Theorem 5. *Let $A, B \in \mathfrak{F}(\mathfrak{U})$. Then,*

$$[A \precapprox B] = \bigvee_{\substack{C \in \mathfrak{F}(\mathfrak{U}) \\ C \subseteq B}} [A \overset{c}{\approx} C].$$

Proof. Without loss of generality, assume that $\mathscr{D}(A) \neq \emptyset$ and A is cardinal separable in B.[6] Let $f \in \mathfrak{Func}(\mathscr{D}(A), \mathscr{D}(B))$. Obviously, it is sufficient to restrict ourselves to the cases when $[f : A \xrightarrow{1\text{-}1} B] > \bot$. In what follows, we will define $C \subseteq B$ such that

$$[f : A \xrightarrow{1\text{-}1} B] = [f : A \xrightarrow[\text{corr}]{1\text{-}1} C]. \tag{12}$$

Recall that $\mathbf{L}$ is linearly ordered. Let $x = f(\mathscr{D}(A))$, and let $C : x \longrightarrow L$ be defined by

$$C(b) = \begin{cases} B(b), & A(f^{-1}(b)) > B(b), \\ A(f^{-1}(b)), & \text{otherwise}, \end{cases} \tag{13}$$

for any $b \in x$. Obviously, $C \subseteq B$. Moreover, it holds

$$A(a) \to B(f(a)) = (A(a) \to C(f(a)) \wedge (C(f(a)) \to A(a)) = A(a) \leftrightarrow C(f(a))$$

for any $a \in \mathscr{D}(A)$; therefore, (12) is satisfied. Since $[f : A \xrightarrow[\text{corr}]{1\text{-}1} C] \leq [A \overset{c}{\approx} C]$, we simply find that

$$[A \precapprox B] \leq \bigvee_{\substack{C \in \mathfrak{F}(\mathfrak{U}) \\ C \subseteq B}} [A \overset{c}{\approx} C].$$

Let $C \subseteq B$. If $[A \overset{c}{\approx} C] = \bot$, we obtain trivially $[A \overset{c}{\approx} C] \leq [A \precapprox B]$. Assume that $[A \overset{c}{\approx} C] > \bot$. Let $A' \in \text{cls}(A)$ and $C' \in \text{cls}(C)$ be fuzzy sets that are cardinal separable. Moreover, let $B' \in \text{cls}(B)$ such that $C' \subseteq B'$. From the assumption on A' and C', we obtain that A' is cardinal separable in B'. Moreover, for any $a \in \mathscr{D}(A')$, we have

$$A'(a) \leftrightarrow C'(f(a)) \leq A'(a) \to C'(f(a)) \leq A'(a) \to B'(f(a));$$

therefore, $[f : A' \xrightarrow[\text{corr}]{1\text{-}1} C'] \leq [f : A' \xrightarrow{1\text{-}1} B']$. Using Theorems 1 and 3 and the previous inequality, we find that

$$[A \overset{c}{\approx} C] = [A' \overset{c}{\approx} C'] = \bigvee_{f \in \mathfrak{Func}(\mathscr{D}(A'), \mathscr{D}(C'))} [f : A' \xrightarrow[\text{corr}]{1\text{-}1} C'] \leq$$
$$\bigvee_{f \in \mathfrak{Func}(\mathscr{D}(A'), \mathscr{D}(B'))} [A' \xrightarrow{1\text{-}1} B'] = [A' \precapprox B'] = [A \precapprox B].$$

Since $[A \overset{c}{\approx} C] \leq [A \precapprox B]$ holds for any $C \subseteq B$, the inequality remains valid also after the application of the supremum over all fuzzy subsets of B. □

[6] We assume $\mathscr{D}(A) \neq \emptyset$ to avoid the empty function.

As a simple consequence of the previous theorem, we obtain

$$[A \stackrel{c}{\approx} B] \leq [A \precapprox B] \wedge [B \precapprox A], \tag{14}$$

which is another demonstration that the GCBT fails in a full generality. To ensure the satisfaction of the GCBT, we need a fuzzy class equivalence on $\mathfrak{F}(\mathfrak{U})$ such that arbitrary two fuzzy sets are equivalent in a degree that is not less than the conjunction of degrees to which fuzzy sets dominate each other. On the other hand, no fuzzy set can be dominated by another fuzzy set to a degree, which is less than the degree in which these fuzzy sets are equivalent. Hence, we obtain a unique fuzzy class relation satisfying the both requirements.

Definition 16. *Let $A, B \in \mathfrak{F}(\mathfrak{U})$. We say that A and B are approximately Cantor-Bernstein equipollent to the degree α provided that*

$$\alpha = [A \precapprox B] \wedge [B \precapprox A]. \tag{15}$$

We use $\stackrel{cb}{\approx}$ to denote the fuzzy class relation to be approximately Cantor-Bernstein equipollent to a certain degree and $[A \stackrel{cb}{\approx} B]$ to denote the degree in which the fuzzy sets A and B are approximately Cantor-Bernstein equipollent.

Definition 17. *The fuzzy class relation $\stackrel{cb}{\approx}$ is called the graded Cantor-Bernstein equipollence (or the graded CB-equipollence for short) of fuzzy sets.*

From the inequality in (14), one can see that $[A \stackrel{c}{\approx} B] \leq [A \stackrel{cb}{\approx} B]$. Moreover, there are fuzzy sets (see Example 2) for which we get the strict inequality. The both types of equipollence coincide on fuzzy sets with finite supports.

Theorem 6. *Let $A, B \in \mathfrak{F}(\mathfrak{U})$ be fuzzy sets with finite supports. Then,*

$$[A \stackrel{c}{\approx} B] = [A \stackrel{cb}{\approx} B]. \tag{16}$$

Proof. Let $A, B \in \mathfrak{F}(\mathfrak{U})$. Without loss of generality, assume that A, B are cardinal separable, and let $|\mathscr{D}(A)| = |\mathscr{D}(B)| = n$ for a natural number n. Since the both domains are finite and the lattice is linearly ordered, we find that

$$[A \precapprox B] = [f : A \xrightarrow{1\text{-}1} B] \quad \text{and} \quad [B \precapprox A] = [g : B \xrightarrow{1\text{-}1} A]$$

for suitable one-to-one functions. Let $x = \{g(b) \mid b \in \mathscr{D}(B) \,\&\, B(b) > A(g(b))\}$ and $y = \{f(a) \mid a \in \mathscr{D}(A) \,\&\, A(a) > B(f(a))\}$. Define C as the restriction of A to x and D as the restriction of B to y. Obviously, we have $C \subseteq A$ and $D \subseteq B$. Moreover, we find that

$$[A \precapprox B] = [f : A \xrightarrow{1\text{-}1} B] = [f : A \xrightarrow[\text{corr}]{1\text{-}1} D] \leq [A \stackrel{c}{\approx} D]$$

and similarly $[B \precapprox A] \leq [C \stackrel{c}{\approx} B]$. Using the graded version of the Cantor-Bernstein theorem (Theorem 4), we obtain

$$[A \stackrel{cb}{\approx} B] = [A \precapprox B] \wedge [B \precapprox A] \leq [A \stackrel{c}{\approx} D] \wedge [C \stackrel{c}{\approx} B] \leq [A \stackrel{c}{\approx} B].$$

Since the opposite inequality is true by (14), we obtain the desirable equality. □

Theorem 7. *The graded CB-equipollence is a fuzzy class equivalence on* $\mathfrak{F}(\mathfrak{U})$.

Proof. Obviously, (FE1) and (FE2) are satisfied. The transitivity of $\overset{\text{cb}}{\approx}$ is a simple consequence of the transitivity of the graded dominance. □

Theorem 8. *The graded dominance is a* $\overset{\text{cb}}{\approx}$*-fuzzy class partial ordering on* $\mathfrak{F}(\mathfrak{U})$.

Proof. According to Theorem 2, the graded dominance is a fuzzy class preordering on $\mathfrak{F}(\mathfrak{U})$, i.e., (FPO1) and (FPO3) are satisfied. (FPO2) follows immediately from the definition of $\overset{\text{cb}}{\approx}$ and the fact that $\alpha \otimes \beta \leq \alpha \wedge \beta$, where $\alpha, \beta \in L$. □

5 Conclusion

In this paper, we introduced the concept of graded dominance for fuzzy sets and showed some of its basic properties. Motivated by the satisfaction of the Cantor-Bernstein theorem, we introduced a new type of graded equipollence based on the graded dominance of fuzzy sets. We proved that the original definition of graded equipollence based on one-to-one correspondences between fuzzy sets coincides with the new graded equipollence assuming fuzzy sets with finite supports.

Acknowledgments. This work was supported by the project LQ1602 IT4Innovations excellence in science.

References

1. Bělohlávek, R.: Fuzzy Relational Systems: Foundations and Principles. Kluwer Academic Publisher, New York (2002)
2. Bodenhofer, U.: A similarity-based generalization of fuzzy orderings preserving the classical axioms. Int. J. Uncertain. Fuzziness Knowl. Based Syst. **08**(05), 593–610 (2000)
3. Gottwald, S.: Set theory for fuzzy sets of higher level. Fuzzy Sets Syst. **2**, 125–151 (1979)
4. Höhle, U.: M-valued sets and sheaves over integral commutative CL-monoid. In: Rodabaugh, S.E., Klement, E.P., Höhle, U. (eds.) Applications to Category Theory to Fuzzy Subsets, pp. 33–72. Kluwer Academic Publishers, Dordrecht (1992)
5. Holčapek, M.: A graded approach to cardinal theory of finite fuzzy sets, part I: Graded equipollence. Fuzzy Sets Syst. (2015). http://dx.doi.org/10.1016/j.fss.2015.08.010
6. Holčapek, M.: Graded equipollence and fuzzy c-measures of finite fuzzy sets. In: Proceedings of 2011 IEEE International Conference on Fuzzy Systems, pp. 2375–2382, DnE Taiwan (2011)
7. Holčapek, M., Turčan, M.: Graded equipollence of fuzzy sets. In: Carvalho, J.P., Kaymak, D.U., Sousa, J.M.C. (eds.) Proceedings of IFSA/EUSFLAT 2009, pp. 1565–1570. European Soc Fuzzy logic & Technology, Johannes Kepler university, Austria (2009)
8. Novák, V., Perfilieva, I., Močkoř, J.: Mathematical Principles of Fuzzy Logic. Kluwer Academic Publisher, Boston (1999)
9. Wygralak, M.: Vaguely defined objects. Representations, fuzzy sets and nonclassical cardinality theory. Theory and Decision Library. Mathematical and Statistical Methods, vol. 33. Kluwer Academic Publisher, Dordrecht (1996)

Uninorms on Interval-Valued Fuzzy Sets

Martin Kalina[1(✉)] and Pavol Král[2]

[1] Department of Mathematics, Faculty of Civil Engineering, Slovak University of Technology in Bratislava, Radlinského 11, 810 05 Bratislava, Slovakia
kalina@math.sk

[2] Department of Quantitative Methods and Information Systems, Faculty of Economics, Matej Bel University, Tajovského 10, 975 90 Banská Bystrica, Slovakia
pavol.kral@umb.sk

Abstract. This paper is a kind of continuation of the paper by G. Deschrijver 'Uninorms which are neither conjunctive nor disjunctive in interval-valued fuzzy set theory', which was published in Information Sciences in 2013. In that paper he constructed uninorms whose neutral element is arbitrary of the type $\mathbf{e} = (e, e)$ and annihilator, $\mathbf{a}$, is arbitrary point that is incomparable with $\mathbf{e}$. In the present paper we intend to show what are all possibilities of the position of the pair $(\mathbf{e}, \mathbf{a})$.

Keywords: Interval-valued fuzzy set · Uninorm · Uninorm neither conjunctive nor disjunctive

1 Introduction and Preliminaries

Uninorms, or more general, associative commutative and monotone (binary) operations on the unit interval, due to their associativity, can be straightforwardly extended to n-ary operations for arbitrary $n \in \mathbb{N}$. This means that they are special types of aggregation functions. As such they have proven their importance in various fields of applications. It is important to have many families of such operations to be at hand for researchers. Studying their behaviour is interesting also from the theoretical point of view. Associative commutative and monotone operations are recently studied also on bounded lattices (see, e.g., [2,11,18,19]). This paper is a contribution to the theory of uninorms on bounded lattices showing what are possible positions of the neutral element and annihilator of a uninorm within the framework of interval-valued fuzzy sets.

1.1 Uninorms

Uninorms appeared for the first time in the paper by Dombi [9] under the name 'aggregative operators'. Dombi's aggregative operators were constructed with the aim to fuzzify evaluation of objects in the theory of multicriteria decision-making. With the help of aggregative operators, objects can be divided into two

J.P. Carvalho et al. (Eds.): IPMU 2016, Part II, CCIS 611, pp. 522–531, 2016.
DOI: 10.1007/978-3-319-40581-0_42

classes – those which satisfy given criteria at least at a threshold level α, and those which do not. Aggregative operators introduced by Dombi are nowadays known under the name representable uninorms.

Independently of Dombi, also in the paper by Czogała and Drewniak [4] uninorms (but also other associative operations) were studied. In 1996 Yager and Rybalov [24] re-introduced uninorms as a natural generalisation of both t-norms and t-conorms (for details on t-norms and their duals, t-conorms, see, e.g., [20]). Since that time researchers study properties of several distinguished families of uninorms.

Definition 1. *A uninorm U is a function $U : [0,1]^2 \to [0,1]$ that is increasing, commutative, associative and has a neutral element $e \in [0,1]$.*

Remark 1. If $U : [0,1]^2 \to [0,1]$ is a uninorm and $e = 0$ is its neutral element then U is a t-conorm. If $e = 1$ is its neutral element then U is a t-norm. For more information on t-norms and t-conorms see, e.g., [20].

Example 1. Basic t-norms which will be important for our purposes, are

- minimum t-norm, $T_M(x,y) = \min\{x,y\}$,
- drastic product,

$$T_D(x,y) = \begin{cases} 0, & \text{if } \max\{x,y\} < 1, \\ \min\{x,y\}, & \text{if } \max\{x,y\} = 1. \end{cases}$$

The basic t-conorms (dual to the above t-norms) are:

- maximum t-conorm, $S_M(x,y) = \max\{x,y\}$,
- drastic sum,

$$S_D(x,y) = \begin{cases} 1, & \text{if } \min\{x,y\} > 0, \\ \max\{x,y\}, & \text{if } \min\{x,y\} = 0. \end{cases}$$

Remark 2. If $e \notin \{0,1\}$ is the neutral element of U, we say that U is a proper uninorm.

Every uninorm U has a distinguished element a called annihilator, for which the following holds $U(a,x) = U(0,1) = a \in \{0,1\}$. A uninorm U is said to be *conjunctive* if $U(x,0) = 0$, and U is said to be disjunctive if $U(1,x) = 1$, for all $x \in [0,1]$.

There are only conjunctive and disjunctive uninorms on $[0,1]$.

Conjunctive and disjunctive uninorms are dual in the following way

$$U_d(x,y) = 1 - U_c(1-x, 1-y),$$

where U_c is an arbitrary conjunctive uninorm and U_d its dual disjunctive uninorm. Assuming U_c has a neutral element e, the neutral element of U_d is $1-e$.

Some special classes of uninorms were studied, e.g., in papers [11–13,17,21–23]. An overview of basic properties of uninorms is in [3]. Because of lack of space we provide only a very brief introduction to uninorms.

For an arbitrary uninorm U with neutral element $e \in]0,1[$ and arbitrary $(x,y) \in]0,e[\times]e,1] \cup]e,1] \times]0,e[$ we have

$$\min\{x,y\} \le U(x,y) \le \max\{x,y\}. \quad (1)$$

1.2 Interval-Valued Fuzzy Sets and Uninorms on Interval-Valued Fuzzy Sets

Interval-valued fuzzy set theory [16,25] is an extension of the fuzzy set theory such that to each element of the universe a closed subinterval of the unit interval is assigned as its membership grade (or, is an approximation of the unknown membership grade). Another extension of the fuzzy set theory is the so-called IF-set theory [1] where, instead of a closed interval, a membership grade and non-membership grade are assigned. In [7] it is shown that IF-sets and interval-valued fuzzy sets are equivalent structures and they both are equivalent structures to L-valued fuzzy sets in the sense of Goguen [14] with respect to a special lattice $\mathcal{L}$. In our considerations we will use pairs $(u_1, u_2) \in [0,1]^2$ such that $u_1 \le u_2$ instead of closed intervals.

Definition 2. *Denote $L = \{(u_1,u_2) \in [0,1]^2; u_1 \le u_2\}$ and for $(u_1,u_2) \in L, (v_1,v_2) \in L$ define $(u_1,u_2) \le_{\mathcal{L}} (v_1,v_2)$ if $u_1 \le v_1$ and $u_2 \le v_2$. Then $\mathcal{L} = (L, \le_{\mathcal{L}})$ is a lattice. If X is a universal set, then any mapping $A\colon X \to L$ is an interval-valued fuzzy set.*

Let $\tilde{\mathcal{L}} = (\tilde{L}, \le_{\tilde{\mathcal{L}}}, 0_{\tilde{\mathcal{L}}}, 1_{\tilde{\mathcal{L}}})$ be arbitrary bounded lattice. We will write $u \parallel v$ for elements $u, v \in \tilde{L}$ which are incomparable with respect to $\le_{\tilde{\mathcal{L}}}$.

Similarly as on the unit square, also uninorms on L^2 can be defined. Namely, a mapping $\mathcal{U}\colon L^2 \to L$ is a uninorm if it is associative, commutative, increasing and has a neutral element. To distinguish uninorms (t-norms, t-conorms) defined on L^2 from those defined on $[0,1]^2$, we will call the former ones interval-valued uninorms (interval-valued t-norms, interval-valued t-conorms, respectively) and the latter ones uninorms (t-norms, t-conorms) on $[0,1]$. Similarly, for $\eta \in L$ we will call a t-norm with its domain $[\mathbf{0}, \eta]^2$ a $[\mathbf{0}, \eta]$-valued t-norm, and a t-conorm with its domain $[\eta, \mathbf{1}]^2$ an $[\eta, \mathbf{1}]$-valued t-conorm.

In [19] the authors have shown the following theorem.

Theorem 1 [19]. *Let $\tilde{\mathcal{L}} = (\tilde{L}, \le_{\tilde{\mathcal{L}}}, 0_{\tilde{\mathcal{L}}}, 1_{\tilde{\mathcal{L}}})$ be a bounded lattice and $e \in \tilde{L} \setminus \{0_{\tilde{\mathcal{L}}}, 1_{\tilde{\mathcal{L}}}\}$ is arbitrary element. If $T_e\colon [0_{\tilde{\mathcal{L}}}, e]^2 \to [0_{\tilde{\mathcal{L}}}, e]$ is a lattice-valued t-norm and $S_e\colon [e, 1_{\tilde{\mathcal{L}}}]^2 \to [e, 1_{\tilde{\mathcal{L}}}]$ a lattice-valued t-conorm, then the functions defined by*

$$\mathcal{U}(u,v) = \begin{cases} T_e(u,v) & \text{if } (u,v) \in [0_{\tilde{\mathcal{L}}}, e]^2, \\ u \vee_{\tilde{L}} v & \text{if } (u,v) \in [0_{\tilde{\mathcal{L}}}, e] \times [e, 1_{\tilde{\mathcal{L}}}] \cup [e, 1_{\tilde{\mathcal{L}}}] \times [0_{\tilde{\mathcal{L}}}, e], \\ u & \text{if } v \in [0_{\tilde{\mathcal{L}}}, e] \text{ and } u \parallel e, \\ v & \text{if } u \in [0_{\tilde{\mathcal{L}}}, e] \text{ and } v \parallel e, \\ 1_{\tilde{\mathcal{L}}} & \text{otherwise} \end{cases}$$

and

$$\mathcal{U}(u,v) = \begin{cases} S_e(u,v) & \text{if } (u,v) \in [e, 1_{\tilde{\mathcal{L}}}]^2, \\ u \wedge_{\tilde{L}} v & \text{if } (u,v) \in [0_{\tilde{\mathcal{L}}}, e[\times [e, 1_{\tilde{\mathcal{L}}}] \cup [e, 1_{\tilde{\mathcal{L}}}] \times]0_{\tilde{\mathcal{L}}}, e], \\ u & \text{if } v \in [e, 1_{\tilde{\mathcal{L}}}] \text{ and } u \parallel e, \\ v & \text{if } u \in [e, 1_{\tilde{\mathcal{L}}}] \text{ and } v \parallel e, \\ 0_{\tilde{\mathcal{L}}} & \text{otherwise} \end{cases}$$

are uninorms on L.

As Theorem 1 shows, we can choose arbitrary element $e \in \tilde{L} \setminus \{0_{\tilde{\mathcal{L}}}, 1_{\tilde{\mathcal{L}}}\}$ and we find a conjunctive and a disjunctive proper uninorm $\mathcal{U}$ whose neutral element is e.

From now on, by L we will always denote the set $L = \{(u_1, u_2) \in [0,1]^2; u_1 \leq u_2\}$, $\mathbf{0} = (0,0)$ and $\mathbf{1} = (1,1)$.

In [6] Deschrijver has proven the following theorem for interval-valued uninorms.

Theorem 2 [6]. *Denote $D = \{(u,u); u \in [0,1]\}$. Let $\varepsilon = (e,e) \in D \setminus \{\mathbf{0}, \mathbf{1}\}$, $\alpha \in L$, T_1, T_2 be t-norms and S_1, S_2 t-conorms on $[0,1]$ such that*

- $\alpha \parallel \varepsilon$,
- *there exist t-norms T_{1a} and T_{1b} on $[0,1]$ such that $T_1 = (\langle T_{1a}, 0, \frac{a_1}{e}\rangle, \langle T_{1b}, \frac{a_1}{e}, 1\rangle)$,*
- *there exist t-conorms S_{2a} and S_{2b} on $[0,1]$ such that $S_2 = (\langle S_{2a}, 0, \frac{a_2-e}{1-e}\rangle, \langle S_{2b}, \frac{a_2-e}{1-e}, 1\rangle)$,*
- $T_1(u_1,u_2) \leq T_2(u_1,u_2)$ *and* $S_1(u_1,u_2) \leq S_2(u_1,u_2)$ *for all* $(u_1,u_2) \in [0,1]^2$.

For all $\mathbf{u} = (u_1,u_2) \in L$ and $\mathbf{v} = (v_1,v_2) \in L$ define the following mappings $\mathcal{U}_x \colon L^2 \to L$ and $\mathcal{U}_y \colon L^2 \to L$

$$\mathcal{U}_x(\mathbf{u},\mathbf{v}) = \begin{cases} a_1 & \text{if } u_1 < a_1 \text{ and } v_1 \geq a_1 \text{ and } v_2 > e \\ & \text{ or } v_1 < a_1 \text{ and } u_1 \geq a_1 \text{ and } u_2 > e, \\ U_1(u_1,v_1) & \text{otherwise,} \end{cases}$$

$$\mathcal{U}_y(\mathbf{u},\mathbf{v}) = \begin{cases} a_2 & \text{if } u_2 > a_2 \text{ and } v_2 \leq a_2 \text{ and } v_1 < e \\ & \text{ or } v_2 > a_2 \text{ and } u_2 \leq a_2 \text{ and } u_1 < e, \\ U_2(u_2,v_2) & \text{otherwise,} \end{cases}$$

where for all $u_1, u_2, v_1, v_2 \in [0,1]$

$$U_1(u_1,v_1) = \begin{cases} e \cdot T_1\left(\frac{u_1}{e}, \frac{v_1}{e}\right) & \text{if } \max(u_1,v_1) \leq e, \\ e + (1-e)S_1\left(\frac{u_1-e}{1-e}, \frac{v_1-e}{1-e}\right) & \text{if } \min(u_1,v_1) \geq e, \\ \min(u_1,v_1) & \text{otherwise,} \end{cases}$$

$$U_2(u_2,v_2) = \begin{cases} e \cdot T_2\left(\frac{u_2}{e}, \frac{v_2}{e}\right) & \text{if } \max(u_2,v_2) \leq e, \\ e + (1-e)S_2\left(\frac{u_2-e}{1-e}, \frac{v_2-e}{1-e}\right) & \text{if } \min(u_2,v_2) \geq e, \\ \max(u_2,v_2) & \text{otherwise.} \end{cases}$$

Then $\mathcal{U}(\mathbf{u},\mathbf{v}) = (\mathcal{U}_x(\mathbf{u},\mathbf{v}), \mathcal{U}_y(\mathbf{u},\mathbf{v}))$ is an interval-valued uninorm with neutral element ε and annihilator α.

2 Construction of Interval-Valued Uninorms

First, we consider conjunctive uninorms with arbitrary neutral element $\varepsilon \neq \mathbf{0}$. We provide here a construction which is slightly different from that in Theorem 1.

Proposition 1. *Let $\varepsilon = (e_1, e_2) \neq \mathbf{0}$ be any arbitrary element of L. Assume that $\mathcal{S}\colon [\varepsilon, \mathbf{1}]^2 \to [\varepsilon, \mathbf{1}]$ is an $[\varepsilon, \mathbf{1}]$-valued t-conorm. Choose t-norms T_1, T_2, T_3, T_4 and T_5 on $[0,1]$ such that $T_1 \le T_4$ and $T_2 \le T_5$. We will consider the following ordinal sums of t-norms*

$$T_l = (\langle T_1, 0, e_1\rangle, \langle T_2, e_1, e_2\rangle, \langle T_3, e_2, 1\rangle)\,,\ T_r = (\langle T_4, 0, e_1\rangle, \langle T_5, e_1, e_2\rangle, \langle T_3, e_2, 1\rangle)\,. \tag{2}$$

Then

$$\mathcal{U}(\mathbf{x}, \mathbf{y}) = \begin{cases} \mathcal{S}(\mathbf{x}, \mathbf{y}) & \text{if } \mathbf{x} \ge \varepsilon \text{ and } \mathbf{y} \ge \varepsilon. \\ \mathbf{x} & \text{if } \mathbf{y} \ge \varepsilon \text{ and } \mathbf{x} \not\ge \varepsilon, \\ \mathbf{y} & \text{if } \mathbf{x} \ge \varepsilon \text{ and } \mathbf{y} \not\ge \varepsilon, \\ (T_l(x_1, y_1), T_r(x_2, y_2)) & \text{otherwise,} \end{cases} \tag{3}$$

is a conjunctive interval-valued uninorm whose neutral element is ε.

Dually we have the following assertion for disjunctive interval-valued uninorms.

Proposition 2. *Let $\varepsilon = (e_1, e_2) \neq \mathbf{1}$ be any arbitrary element of L. Assume that $\mathcal{T}\colon [\mathbf{0}, \varepsilon]^2 \to [\mathbf{0}, \varepsilon]$ is a $[\mathbf{0}, \varepsilon]$-valued t-norm. Choose t-conorms S_1, S_2, S_3, S_4 and S_5 on $[0,1]$ such that $S_2 \le S_4$ and $S_3 \le S_5$. We will consider the following ordinal sums of t-conorms*

$$S_l = (\langle S_1, 0, e_1\rangle, \langle S_2, e_1, e_2\rangle, \langle S_3, e_2, 1\rangle)\,,\ S_r = (\langle S_1, 0, e_1\rangle, \langle S_4, e_1, e_2\rangle, \langle S_5, e_2, 1\rangle)\,. \tag{4}$$

Then

$$\mathcal{U}(\mathbf{x}, \mathbf{y}) = \begin{cases} \mathcal{T}(\mathbf{x}, \mathbf{y}) & \text{if } \mathbf{x} \le \varepsilon \text{ and } \mathbf{y} \le \varepsilon. \\ \mathbf{x} & \text{if } \mathbf{y} \le \varepsilon \text{ and } \mathbf{x} \not\le \varepsilon, \\ \mathbf{y} & \text{if } \mathbf{x} \le \varepsilon \text{ and } \mathbf{y} \not\le \varepsilon, \\ (S_l(x_1, y_1), S_r(x_2, y_2)) & \text{otherwise,} \end{cases} \tag{5}$$

is a disjunctive interval-valued uninorm whose neutral element is ε.

Now, we present a conjunctive and a disjunctive interval-valued uninorm with their neutral element equal to $(0,1)$. The two uninorms defined in the following lemma are different from those introduced in Propositions 1 and 2, respectively.

Lemma 1. *Let $T\colon [0,1]^2 \to [0,1]$ and $S\colon [0,1]^2 \to [0,1]$ be a t-norm and a t-conorm, respectively. Then*

(a)

$$\hat{\mathcal{U}}((x_1, x_2), (y_1, y_2)) = (\min\{S(x_1, y_1), T(x_2, y_2)\}, T(x_2, y_2))$$

is a conjunctive interval-valued uninorm with the neutral element $\varepsilon = (0,1)$,

(b)

$$\tilde{\mathcal{U}}((x_1, x_2), (y_1, y_2)) = (S(x_1, y_1), \max\{S(x_1, y_1), T(x_2, y_2)\}) \tag{6}$$

is a disjunctive interval-valued uninorm with the neutral element $\varepsilon = (0,1)$.

Notation. Let $\mathcal{U}_d$ be the interval-valued uninorm defined by (6) using the maximum t-conorm S_M and the minimum t-norm T_M, and $(a,c) \leq_L (b,d)$ be elements of L. Then we denote

$$\tilde{\mathcal{U}}_a^b(\mathbf{x},\mathbf{y}) = \mathcal{U}_d\left(\left(\frac{x_1-a}{b-a},\frac{y_1-a}{b-a}\right),\left(\frac{x_2-c}{d-c},\frac{y_2-c}{d-c}\right)\right) + (a,c). \tag{7}$$

Next we present two constructions of interval-valued uninorms with a neutral element ε and an annihilator α such that $\varepsilon \parallel \alpha$. As we can see on Fig. 1, there are two different possibilities for the positions of $\varepsilon = (e_1, e_2)$ and $\alpha = (a_1, a_2)$. Namely, $a_1 < e_1 \leq e_2 < a_2$ (treated in Theorem 3) and $e_1 < a_1 \leq a_2 < e_2$ (treated in Theorem 4).

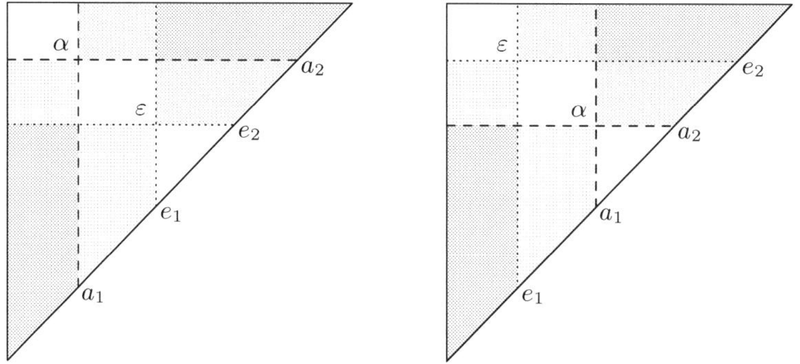

Fig. 1. Possible positions of annihilator α and neutral element ε for interval-valued uninorms

Theorem 3. *Let $\alpha = (a_1, a_2) \in L$ and $\varepsilon = (e_1, e_2) \in L$ be incomparable elements such that $a_1 < e_1 \leq e_2 < a_2$. Let T_1, T_2, T_3, T_4 be t-norms on $[0,1]$ such that $T_1 \leq T_3$ and $T_2 \leq T_4$. Further, let S_1, S_2, S_3, S_4 be t-conorms on $[0,1]$ such that $S_1 \leq S_3$ and $S_2 \leq S_4$. Denote by $T_x : [0,e_1]^2 \to [0,e_1]$, $T^y\colon [0,e_2]^2 \to [0,e_2]$, $S_x\colon [e_1,1]^2 \to [e_1,1]$ and $S^y\colon [e_2,1]^2 \to [e_2,1]$ the respective ordinal sums*

$$\begin{aligned}
T_x &= (\langle T_1, 0, a_1\rangle, \langle T_2, a_1, e_1\rangle),\\
T^y &= (\langle T_3, 0, a_1\rangle, \langle T_4, a_1, e_1\rangle, \langle T_D, e_1, e_2\rangle),\\
S_x &= (\langle S_D, e_1, e_2\rangle, \langle S_1, e_2, a_2\rangle, \langle S_2, a_2, 1\rangle),\\
S^y &= (\langle S_3, e_2, a_2\rangle, \langle S_4, a_2, 1\rangle).
\end{aligned}$$

Denote by $\mathcal{O} = \{(u_1, u_2) \in L; u_1 \geq e_1,\ u_2 \leq e_2\}$. *Define the following functions*

$$\mathcal{U}_x(\mathbf{u},\mathbf{v}) = \begin{cases} T_x(u_1,v_1) & \text{if } \mathbf{u} \leq \varepsilon,\ \mathbf{v} \leq \varepsilon, \\ & \text{or } u_1 < a_1 \text{ and } \mathbf{v} \leq \varepsilon, \\ & \text{or } v_1 < a_1 \text{ and } \mathbf{u} \leq \varepsilon, \\ & \text{or } a_1 \leq u_1 \leq e_1 \text{ and } a_1 \leq v_1 < e_1, \\ a_1 & \text{if } \min(u_2,v_2) > e_2,\ \min(u_1,v_1) \leq a_1, \\ & \text{ and } \max(u_1,v_1) \geq a_1, \\ & \text{or } (\mathbf{u} \leq \alpha \text{ and } \mathbf{v} \geq \alpha) \text{ or } (\mathbf{v} \leq \alpha \text{ and } \mathbf{u} \geq \alpha), \\ \min(u_1,v_1) & \text{if } \mathbf{u} \in \mathcal{O} \text{ and } v_1 < e_1, \\ & \text{or } \mathbf{v} \in \mathcal{O} \text{ and } u_1 < e_1, \\ & \text{or } \min(u_1,v_1) \in [a_1, e_1[\text{ and } (\mathbf{u} \geq \varepsilon \text{ or } \mathbf{v} \geq \varepsilon), \\ S_x(u_1,v_1) & \text{if } \mathbf{u} \geq \varepsilon,\ \mathbf{v} \geq \varepsilon,\ \min(u_2,v_2) > e_2, \\ & \text{or } \mathbf{u} \in \mathcal{O} \text{ and } \mathbf{v} > \varepsilon,\ v_2 > e_2, \\ & \text{or } \mathbf{v} \in \mathcal{O} \text{ and } \mathbf{u} > \varepsilon,\ u_2 > e_2, \\ \tilde{\mathcal{U}}_{e_1}^{e_2}(\mathbf{u},\mathbf{v})_x & \text{if } (\mathbf{u},\mathbf{v}) \in \mathcal{O}^2, \end{cases} \tag{8}$$

$$\mathcal{U}^y(\mathbf{u},\mathbf{v}) = \begin{cases} T^y(u_2,v_2) & \text{if } \mathbf{u} \leq \varepsilon,\ \mathbf{v} \leq \varepsilon,\ \max(u_1,v_1) < e_1, \\ & \text{or } \mathbf{u} \leq \varepsilon,\ u_1 < e_1 \text{ and } \mathbf{v} \in \mathcal{O}, \\ & \text{or } \mathbf{v} \leq \varepsilon,\ v_1 < e_1 \text{ and } \mathbf{u} \in \mathcal{O}, \\ a_2 & \text{if } \max(u_1,v_1) < e_1,\ \min(u_2,v_2) \leq a_2, \\ & \text{ and } \max(u_2,v_2) \geq a_2, \\ & \text{or } (\mathbf{u} \leq \alpha \text{ and } \mathbf{v} \geq \alpha) \text{ or } (\mathbf{v} \leq \alpha \text{ and } \mathbf{u} \geq \alpha), \\ \max(u_2,v_2) & \text{if } \max(u_2,v_2) \in]e_2, a_2] \text{ and } (\mathbf{u} \leq \varepsilon \text{ or } \mathbf{v} \leq \varepsilon), \\ & \text{if } \mathbf{u} \in \mathcal{O} \text{ and } v_2 > e_2, \\ & \text{or } \mathbf{v} \in \mathcal{O} \text{ and } u_2 > e_2, \\ S^y(u_2,v_2) & \text{if } \mathbf{u} \geq \varepsilon \text{ and } \mathbf{v} \geq \varepsilon, \\ & \text{or } u_2 > a_2 \text{ and } \mathbf{v} \geq \varepsilon, \\ & \text{or } v_2 > a_2 \text{ and } \mathbf{u} \geq \varepsilon, \\ & \text{or } (u_2,v_2) \in [e_2, a_2]^2, \\ \tilde{\mathcal{U}}_{e_1}^{e_2}(\mathbf{u},\mathbf{v})_y & \text{if } (\mathbf{u},\mathbf{v}) \in \mathcal{O}^2. \end{cases} \tag{9}$$

Then $\mathcal{U} : L^2 \to L$ *given by*

$$\mathcal{U}(\mathbf{u},\mathbf{v}) = (\mathcal{U}_x(\mathbf{u},\mathbf{v}), \mathcal{U}^y(\mathbf{u},\mathbf{v})),$$

is an interval-valued uninorm whose neutral element is $\varepsilon = (e_1, e_2)$ *and annihilator is* $\alpha = (a_1, a_2)$.

Now, we are going to treat the other possibility of general positions of annihilator and neutral element of an interval-valued uninorm.

Theorem 4. *Let* $\alpha = (a_1, a_2) \in L$ *and* $\varepsilon = (e_1, e_2) \in L$ *be incomparable elements such that* $e_1 < a_1 \leq a_2 < e_2$. *Let* T_1, T_2 *and* T_3 *be t-norms on* $[0,1]$ *such that* $T_1 \leq T_2$. *Further, let* S_1, S_2 *and* S_3 *be t-conorms on* $[0,1]$ *such that* $S_2 \leq S_3$. *Denote by* $T_x : [0, e_1]^2 \to [0, e_1]$ *and* $S^y : [e_2, 1]^2 \to [e_2, 1]$ *the following affine transformations of* T_1 *and* S_3, *respectively*

$$T_x = (\langle T_1, 0, e_1 \rangle),$$
$$S^y = (\langle S_3, e_2, 1 \rangle).$$

Further denote by $T^y\colon [0,e_2]^2 \to [0,e_2]$ *and* $S_x\colon [e_1,1]^2 \to [e_1,1]$ *the respective ordinal sums*

$$T^y = (\langle T_2, 0, e_1\rangle, \langle T_D, e_1, a_1\rangle, \langle T_3, a_1, a_2\rangle, \langle T_D, a_2, e_2\rangle),$$
$$S_x = (\langle S_D, e_1, a_1\rangle, \langle S_1, a_1, a_2\rangle, \langle S_D, a_2, e_2\rangle, \langle S_2, e_2, 1\rangle).$$

Denote $\mathcal{O} = \{(u_1,u_2) \in L; u_1 \geq a_1,\ u_2 \leq a_2\}$. *Define the following functions*

$$\mathcal{U}_x(\mathbf{u},\mathbf{v}) = \begin{cases} T_x(u_1,v_1) & \textit{if } u_1 \leq e_1 \textit{ and } v_1 \leq e_1, \\ S_x(u_1,v_1) & \textit{if } \mathbf{u} \geq \varepsilon,\ \mathbf{v} \geq \varepsilon, \\ v_1 & \textit{if } \mathbf{u} \parallel \varepsilon,\ u_1 \leq e_1,\ v_1 > e_1, \\ & \textit{or } u_1 < a_1,\ u_2 > a_2,\ v_1 \geq a_1, \\ & \textit{or } \mathbf{u} \geq (\varepsilon \vee \alpha),\ v_1 > a_1,\ v_2 < e_2, \\ u_1 & \textit{if } \mathbf{v} \parallel \varepsilon,\ v_1 \leq e_1,\ u_1 > e_1, \\ & \textit{or } v_1 < a_1,\ v_2 > a_2,\ u_1 \geq a_1, \\ & \textit{or } \mathbf{v} \geq (\varepsilon \vee \alpha),\ u_1 > a_1,\ u_2 < e_2, \\ a_1 & \textit{if } (\mathbf{u} \leq \alpha \textit{ and } \mathbf{v} \geq \alpha) \textit{ or } (\mathbf{v} \leq \alpha \textit{ and } \mathbf{u} \geq \alpha), \\ & \textit{or } \mathbf{u} \leq \alpha \textit{ and } \mathbf{v} \in \mathcal{O}, \\ & \textit{or } \mathbf{v} \leq \alpha \textit{ and } \mathbf{u} \in \mathcal{O} \\ \max(u_1,v_1) & \textit{if } u_1 \leq a_1,\ u_2 < e_2,\ v_1 \in]e_1,a_1], \\ & \textit{or } v_1 \leq a_1,\ v_2 < e_2,\ u_1 \in]e_1,a_1], \\ \min(u_1,v_1) & \textit{if } \min(u_1,v_1) \geq a_1,\ \max(u_2,v_2) \leq e_2, \end{cases} \tag{10}$$

$$\mathcal{U}^y(\mathbf{u},\mathbf{v}) = \begin{cases} S^y(u_2,v_2) & \textit{if } \min(u_2,v_2) \geq e_2, \\ T^y(u_2,v_2) & \textit{if } \mathbf{u} \leq \varepsilon,\ \mathbf{v} \leq \varepsilon, \\ v_2 & \textit{if } \mathbf{u} \parallel \varepsilon,\ u_1 \leq e_1,\ v_2 < e_2, \\ & \textit{or } u_2 > a_2,\ u_1 < a_1,\ v_2 \leq a_2, \\ & \textit{or } \mathbf{u} \leq (\varepsilon \wedge \alpha),\ v_1 > e_1,\ v_2 < a_2, \\ u_2 & \textit{if } \mathbf{v} \parallel \varepsilon,\ v_1 \leq e_1,\ u_2 < e_2, \\ & \textit{or } v_2 > a_2,\ v_1 < a_1,\ u_2 \leq a_2, \\ & \textit{or } \mathbf{v} \leq (\varepsilon \wedge \alpha),\ u_1 > e_1,\ u_2 < a_2, \\ a_2 & \textit{or } (\mathbf{u} \leq \alpha \textit{ and } \mathbf{v} \geq \alpha) \textit{ or } (\mathbf{v} \leq \alpha \textit{ and } \mathbf{u} \geq \alpha), \\ & \textit{or } \mathbf{u} \geq \alpha \textit{ and } \mathbf{v} \in \mathcal{O}, \\ & \textit{or } \mathbf{v} \geq \alpha \textit{ and } \mathbf{u} \in \mathcal{O} \\ \min(u_2,v_2) & \textit{if } u_1 > e_1,\ u_2 \geq a_2,\ v_2 \in [a_2,e_2[, \\ & \textit{or } v_1 > e_1,\ v_2 \geq a_2,\ u_2 \in [a_2,e_2[, \\ \max(u_2,v_2) & \textit{if } \min(u_1,v_1) \geq e_1,\ \max(u_2,v_2) \leq a_2, \end{cases} \tag{11}$$

Then $\mathcal{U} : L^2 \to L$ *given by*

$$\mathcal{U}(\mathbf{u},\mathbf{v}) = (\mathcal{U}_x(\mathbf{u},\mathbf{v}), \mathcal{U}^y(\mathbf{u},\mathbf{v}))\,,$$

is an interval-valued uninorm whose neutral element is $\varepsilon = (e_1, e_2)$ *and annihilator is* $\alpha = (a_1, a_2)$.

Remark 3. The main idea of the construction in Theorem 3 is the following: we use the t-norm T_x for aggregating the first coordinates if they are below e_1 with some exceptions when the result is a_1, and the t-conorm S_x (or max) if the first coordinates are above e_1. Similarly the second coordinates are aggregating via S^y if the second coordinates are above e_2 with some exceptions when the result is a_2, and the t-norm T_y if the second coordinates are below e_2. As a problematic area is that one denoted as $\mathcal{O}$ where we use the formula (7).

The main idea of the construction in Theorem 4 is that we use min to aggregate the first coordinates if they are above a_1, but outside of the area which is above ε. We use max to aggregate the first coordinates if they are from $]e_1, a_1]$, but also only outside of the area which is above ε. If the first coordinate of one element is below e_1, of the other element is above e_1, the result is aggregated either as a_1 or via max. And similarly (dually) we aggregate also the second coordinates.

3 Conclusions

In this paper we have presented constructions of interval-valued uninorms. In general there are three possibilities

- conjunctive uninorms with annihilator $\alpha = \mathbf{0}$ and arbitrary neutral element $\varepsilon \neq \mathbf{0}$,
- disjunctive uninorms with annihilator $\alpha = \mathbf{1}$ and arbitrary neutral element $\varepsilon \neq \mathbf{1}$,
- uninorms which are neither conjunctive nor disjunctive whose neutral element and annihilator are arbitrary incomparable elements of the lattice of interval-valued fuzzy sets.

Acknowledgments. The work of Martin Kalina has been supported from the Science and Technology Assistance Agency under contract No. APVV-14-0013, and from the VEGA grant agency, grant number 2/0069/16.

Pavol Král has been supported from the project VEGA 1/0647/14.

References

1. Atanasov, K.T.: Intuitionistic Fuzzy Sets. Springer, Heidelberg (1999)
2. Bodjanova, S., Kalina, M.: Construction of uninorms on bounded lattices. In: IEEE 12th International Symposium on Intelligent Systems and Informatics, SISY 2014, Subotica, pp. 61–66 (2014)
3. Calvo, T., Kolesárová, A., Komorníková, M., Mesiar, R.: Aggregation operators: properties, classes and construction methods. In: Calvo, T., Mayor, G., Mesiar, R. (eds.) Aggregation Operators. Studies in Fuzziness and Soft Computing, vol. 97, pp. 3–104. Springer, Heidelberg (2002)
4. Czogała, E., Drewniak, J.: Associative monotonic operations in fuzzy set theory. Fuzzy Sets Syst. **12**, 249–269 (1984)
5. Deschrijver, G.: A representation of t-norms in interval valued L-fuzzy set theory. Fuzzy Sets Syst. **159**, 1597–1618 (2008)

6. Deschrijver, G.: Uninorms which are neither conjunctive nor disjunctive in interval-valued fuzzy set theory. Inf. Sci. **244**, 48–59 (2013)
7. Deschrijver, G., Kerre, E.E.: On the relationship between some extensions of fuzzy set theory. Fuzzy Sets Syst. **133**(2), 227–235 (2003)
8. Deschrijver, G., Kerre, E.E.: Uninorms in L^*-fuzzy set theory. Fuzzy Sets Syst. **148**, 243–262 (2004)
9. Dombi, J.: Basic concepts for a theory of evaluation: the aggregative operator. Eur. J. Oper. Res. **10**, 282–293 (1982)
10. Dombi, J.: A general class of fuzzy operators, the DeMorgan class of fuzzy operators and fuzziness measures induced by fuzzy operators. Fuzzy Sets Syst. **8**, 149–163 (1982)
11. Drygaś, P.: On monotonic operations which are locally internal on some subset of their domain. In: Štepnicka, M., et al. (eds.) New Dimensions in Fuzzy Logic and Related Technologies, Proceedings of the 5th EUSFLAT Conference 2007, vol. 2, pp. 359–364. Universitas Ostraviensis, Ostrava (2007)
12. Fodor, J., De Baets, B.: A single-point characterization of representable uninorms. Fuzzy Sets Syst. **202**, 89–99 (2012)
13. Fodor, J., Yager, R.R., Rybalov, A.: Structure of uninorms. Int. J. Uncertain. Fuzziness Knowl Based Syst. **5**, 411–422 (1997)
14. Goguen, J.A.: L-fuzzy sets. J. Math. Anal. Appl. **18**(1), 145–174 (1967)
15. González-Hidalgo, M., Massanet, S., Mir, A., Ruiz-Aguilera, D.: Information processing and management of uncertainty in knowledge-based systems. In: Laurent, A., Strauss, O., Bouchon-Meunier, B., Yager, R.R. (eds.) IPMU 2014. Communications in Computer and Information Science, vol. 443, pp. 184–193. Springer, Switzerland (2014)
16. Gorzałczany, M.B.: A method of inference in approximate reasoning based on interval-valued fuzzy sets. Fuzzy Sets Syst. **21**(1), 1–17 (1987)
17. Hu, S., Li, Z.: The structure of continuous uninorms. Fuzzy Sets Syst. **124**, 43–52 (2001)
18. Karaçal, F., Ince, M.A., Mesiar, R.: Nullnorms on bounded lattices. Inf. Sci. **325**, 227–236 (2015)
19. Karaçal, F., Mesiar, R.: Uninorms on bounded lattices. Fuzzy Sets Syst. **261**, 33–43 (2015)
20. Klement, E.P., Mesiar, R., Pap, E.: Triangular Norms. Springer, Heidelberg (2000)
21. Mas, M., Massanet, S., Ruiz-Aguilera, D., Torrens, J.: A survey on the existing classes of uninorms. J. Intell. Fuzzy Syst. **29**(3), 1021–1037 (2015)
22. Petrík, M., Mesiar, R.: On the structure of special classes of uninorms. Fuzzy Sets Syst. **240**, 22–38 (2014)
23. Ruiz-Aguilera, D., Torrens, J., De Baets, B., Fodor, J.: Some remarks on the characterization of idempotent uninorms. In: Hüllermeier, E., Kruse, R., Hoffmann, F. (eds.) IPMU 2010. LNCS, vol. 6178, pp. 425–434. Springer, Heidelberg (2010)
24. Yager, R.R., Rybalov, A.: Uninorm aggregation operators. Fuzzy Sets Syst. **80**, 111–120 (1996)
25. Sambuc, R.: Fonctions Φ-floues. application à l'aide au diagnostic en pathologie thyroidienne, Ph.D. thesis, Université de Marseille, France (1975)

Algorithm for Generating Finite Totally Ordered Monoids

Milan Petrík[1,2(✉)] and Thomas Vetterlein[3]

[1] Department of Mathematics, Faculty of Engineering, Czech University of Life Sciences, Prague, Czech Republic
petrikm@tf.czu.cz, petrik@cs.cas.cz
[2] Institute of Computer Science, Academy of Sciences, Prague, Czech Republic
[3] Department of Knowledge-Based Mathematical Systems, Johannes Kepler University, Linz, Austria
Thomas.Vetterlein@jku.at

Abstract. The semantics of fuzzy logic is typically based on negative totally ordered monoids. This contribution describes an algorithm generating in a step-wise fashion all finite structures of this kind.

Keywords: Discrete triangular norm · Finite negative totally ordered monoid · Rees coextension · Rees congruence · Reidemeister closure condition · Tomonoid partition

1 Introduction

Partially ordered monoids are structures occurring in several fields of mathematics and computer sciences, in particular in logic. In non-classical logic, the canonical set of truth values is often endowed with a binary operation making this set into a partially ordered monoid. The monoidal operation then corresponds to the conjunction.

The algebraic semantics of the fuzzy logic MTL is the variety of MTL-algebras—commutative residuated lattices where the top element is the monoidal identity—and every MTL algebra is a subdirect product of MTL chains (i.e., totally ordered MTL algebras). This makes negative totally ordered monoids (which are, in fact, monoidal reducts of MTL chains) important structures worth of studying.

In this contribution, we focus on finite structures as they may be used, e.g., in finite-valued fuzzy logics. We note that, under the additional assumption of commutativity, the structures that we consider can be identified with linearly ordered finite MTL-algebras; MTL-algebras are in turn the algebraic counterpart of the fuzzy logic MTL [8].

M. Petrík—supported by the Czech Science Foundation (GAČR) under Project 15-07724Y (Totally ordered monoids).

T. Vetterlein—supported by the Austrian Science Fund (FWF) under Project I 1923-N25 (New perspectives on residuated posets).

J.P. Carvalho et al. (Eds.): IPMU 2016, Part II, CCIS 611, pp. 532–543, 2016.
DOI: 10.1007/978-3-319-40581-0_43

Further, this contribution can be seen as a practical appendix to our previous paper [13] which has yielded a method to describe all the one-element Rees coextensions of a given finite, negative, totally ordered monoid (shortly a f.n. tomonoid) S, that is, all the f.n. tomonoids greater by one element such that S is their common Rees quotient. This way, starting from the trivial monoid, one can generate all the possible f.n. tomonoids up to a given finite size. While the cited paper has been focused on describing the coextensions and giving a proof that all the existing f.n. tomonoids are necessarily obtained this way, this paper intends to give a practical description of the algorithm that produces the tomonoids. This paper can be also viewed as a continuation of our previous result [12] where we have, however, dealt with the Archimedean case only.

2 Basic Notions

We begin with an introduction of the basic notions of the paper. A *monoid* is an algebra $(S; \odot, 1)$ of type $\langle 2, 0\rangle$ such that $(a \odot b) \odot c = a \odot (b \odot c)$ and $a \odot 1 = 1 \odot a = a$ for every $a, b, c \in S$. A total (linear) order $\leqslant$ on a monoid S is called *compatible* if $a \leqslant b$ implies both $a \odot c \leqslant b \odot c$ and $c \odot a \leqslant c \odot b$ for every $a, b, c \in S$. In such a case, we call $(S; \leqslant, \odot, 1)$ a *totally ordered monoid* or a *tomonoid*, for short. We also say that $\odot$ is *monotone* with respect to $\leqslant$. Further, S is called *commutative* if $a \odot b = b \odot a$ for every $a, b \in S$. Finally, S is

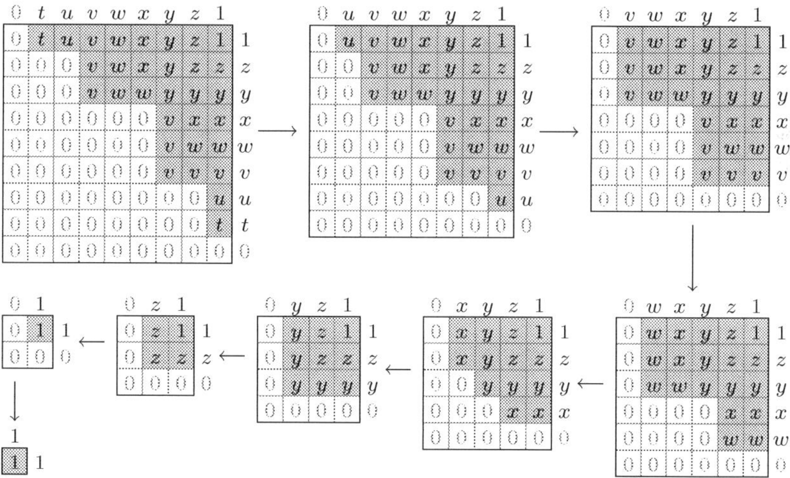

Fig. 1. Examples of f.n. tomonoids depicted by their multiplication tables. Seeing the cells of a table as ordered pairs from S^2, the level equivalence classes correspond with the maximal sets of the cells with the same symbol. The depicted f.n. tomonoids are actually created as one-element Rees quotients starting with the first f.n. tomonoid of size 9. As we may observe, the one-element Rees quotient arises by "cutting off" the column and the row indexed by the zero and by merging the zero and the atom classes into one. Finally, we reach the trivial monoid.

called *negative* if 1 is the top element. We note that in the context of residuated lattices, usually the notion "integral" is used instead.

This paper is focused mainly on finite, negative, totally ordered monoids which we abbreviate by "*f.n. tomonoids*". In general, we do not assume the monoids to be commutative [9], although, we deal with the commutativity, as well. Let us remark that commutative f.n. tomonoids correspond to *discrete triangular norms* [6]. The smallest monoid that consists of the monoidal identity 1 alone, is called the *trivial tomonoid.*

The illustrations in this paper depict f.n. tomonoids by their multiplication tables, see Fig. 1.

3 Level Set Representation of Tomonoids

We do not work with f.n. tomonoids directly but we rather work with their level set representations. In the the following text, by S^2 we denote the Cartesian product of the set S with itself, i.e., $S^2 = S \times S$.

For a tomonoid $(S; \leqslant, \odot, 1)$ and two pairs $(a, b), (c, d) \in S^2$ we define $(a, b) \sim (c, d)$ iff $a \odot b = c \odot d$ and we call $\sim$ the *level equivalence* associated with S.

Let $(S; \leqslant)$ be a totally ordered set. By $\trianglelefteqslant$ we denote the componentwise order on S^2, i.e., for every $a, b, c, d \in S$, we put $(a, b) \trianglelefteqslant (c, d)$ iff $a \leqslant b$ and $c \leqslant d$. Let $1 \in S$ and let $\sim$ be an equivalence on S^2 such that the following holds:

(P1) For every $a, b, c, d, e \in S$, $(a, b) \sim (1, d)$ and $(b, c) \sim (1, e)$ imply $(d, c) \sim (a, e)$. (See an illustration in Fig. 2-left.)

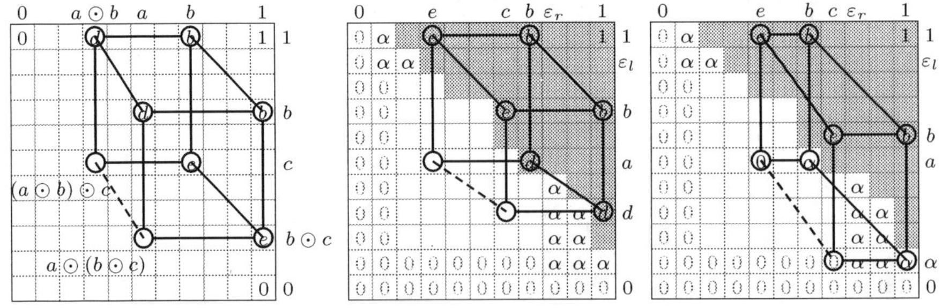

Fig. 2. Left: Illustration of Property (P1). Consider two rectangles such that the first one hits the upper edge and the second one hits the right edge of the multiplication table. Assume that the upper left, upper right, and lower right vertices are in the same level equivalence classes, respectively. Then also the remaining lower left vertices are elements of the same level equivalence class. This property is directly related to the associativity of the tomonoid and corresponds to the *Reidemeister condition* known from web geometry [1,5]. **Middle: Illustration of (E2).** For every two pairs $(a, b), (b, c) \in \mathcal{P}$ we relate $(a, e) \dot{\sim} (d, c)$. **Right: Illustration of (E3'a).** Let $(a, b) \in \mathcal{Q}$, let $c < \varepsilon_r$, and let $(b, c) \sim e$. If $(a, b) \dot{\sim} 0$ then also $(a, e) \dot{\sim} 0$ according to the monotonicity. If $(a, b) \dot{\sim} \alpha$ then $(a, e) \dot{\sim} 0$ according to (P1).

(P2) For every $a, b \in S$ there is exactly one $c \in S$ such that $(a, b) \sim (1, c) \sim (c, 1)$.
(P3) For every $a, b, c, d, a', b', c', d' \in S$, $(a, b) \sim (a', b') \trianglelefteq (c, d) \sim (c', d') \trianglelefteq (a, b)$ implies $(a, b) \sim (c, d)$.

Then we call $(S^2; \trianglelefteq, \sim, (1{,}1))$ a *tomonoid partition*. The following two propositions show that tomonoids and tomonoid partitions are in a one-to-one correspondence.

Proposition 1 [13]. *Let $(S; \leqslant, \odot, 1)$ be a tomonoid and let $\sim$ be its level equivalence. Then $(S^2; \trianglelefteq, \sim, (1{,}1))$ is a tomonoid partition.*

Proposition 2 [13]. *Let $(S^2; \trianglelefteq, \sim, (1{,}1))$ be a tomonoid partition. For every $a, b \in S$, let $a \odot b$ be given as the unique c such that $(a, b) \sim (1, c) \sim (c, 1)$. Then $(S; \leqslant, \odot, 1)$ is the unique tomonoid such that $(S^2; \trianglelefteq, \sim, (1{,}1))$ is its associated tomonoid partition.*

In the following text, we will write $(a, b) \sim c$ instead of $(a, b) \sim (1, c) \sim (c, 1)$.

4 Rees Quotients and Coextensions

In this section we introduce the notion of a one-element coextension of a f.n. tomonoid.

Let $(S; \leqslant, \odot, 1)$ be a f.n. tomonoid. We call its least element the *zero* (and we denote it by 0), we call its second smallest element the *atom* (and we denote it by α), and we call its second greatest element the *coatom* (and we denote it by κ). Recall that 1 is the greatest element of S.

A *tomonoid congruence* on S is an equivalence relation $\approx$ on S such that

1. $\approx$ is a congruence [10] of S as a monoid and
2. each equivalence class is convex.

The operation induced by $\odot$ on the quotient $\langle S \rangle_{\approx}$ we denote again by $\odot$. For $a, b \in S$, we define $\langle a \rangle_{\approx} \leqslant \langle b \rangle_{\approx}$ if $a \approx b$ or $a < b$. We may observe that $(\langle S \rangle_{\approx}; \leqslant, \odot, \langle 1 \rangle_{\approx})$ is a tomonoid again and we call $\langle S \rangle_{\approx}$ the *tomonoid quotient* with respect to $\approx$. This procedure preserves the properties of finiteness, negativity, and commutativity.

We proceed with the notion of the Rees congruence which is commonly used for semigroups [11]. Let $q \in S$. For $a, b \in S$ we define $a \approx_q b$ if $a = b$ or $a, b \leqslant q$. Then $\approx_q$ is a tomonoid congruence and we call it the *Rees congruence* with respect to q. We denote the corresponding quotient by S/q and we call it the *Rees quotient* of S with respect to q. Furthermore, we call S a *Rees coextension* of S/q [13]. If moreover $q = \alpha$, we call S/q the *one-element Rees quotient* of S and we call S the *one-element Rees coextension* (or, shortly, the *one-element coextension*) of S/q. See an illustration in Fig. 1.

5 One-Element Rees Coextensions of f.n. Tomonoids

The algorithm we are going to present is based on a theorem [13] which we briefly describe here. Let $(S; \leqslant, \odot, 1)$ be a f.n. tomonoid. We denote $S^\star = S \setminus \{0\}$. A *zero doubling extension* of S is a totally ordered set $\bar{S} = S^\star \dot{\cup} \{0, \alpha\}$ such that $0 < \alpha < a$ for every $a \in S^\star$. We call $a \in S$ an *idempotent* if $a \odot a = a$. Obviously, 0 and 1 are idempotents of every f.n. tomonoid.

Let $\sim_1$ and $\sim_2$ be two equivalence relations on S^2. We say that $\sim_2$ is a *coarsening* of $\sim_1$ if $\sim_1 \subseteq \sim_2$, that is, if each equivalence class of $\sim_2$ is a union of some equivalence classes of $\sim_1$.

Let $(S; \trianglelefteq, \sim, (1,1))$ be a f.n. tomonoid partition. Let $\bar{S}$ be the zero doubling extension of S. Define

$$\mathcal{P} = \{(a,b) \in \bar{S}^2 \mid a, b \in S^\star \text{ and there is } c \in S^\star \text{ s.t. } (a,b) \sim c\}, \tag{1}$$

$$\mathcal{Q} = \bar{S}^2 \setminus \mathcal{P}. \tag{2}$$

Let $(\varepsilon_l, \varepsilon_r)$ be a pair of non-zero idempotents of S and let $\dot{\sim}$ be the smallest equivalence relation on $\bar{S}^2$ such that the following conditions hold:

- (E1) We have $(a,b) \dot{\sim} (c,d)$ for every $(a,b), (c,d) \in \mathcal{P}$ such that $(a,b) \sim (c,d)$.
- (E2) We have $(d,c) \dot{\sim} (a,e)$ for every $(a,b), (b,c) \in \mathcal{P}$ and $(d,c), (a,e) \in \mathcal{Q}$ such that $(a,b) \sim d$ and $(b,c) \sim e$. (See an illustration in Fig. 2-middle.)
- (E3'a) We have $(a,e) \dot{\sim} 0$ for every $a, b, c, e \in S^\star$ such that $(a,b) \in \mathcal{Q}$, $(b,c) \sim e$, and $c < \varepsilon_r$.

  Furthermore, we have $(d,c) \dot{\sim} 0$ for any $a, b, c, d \in S^\star$ such that $(b,c) \in \mathcal{Q}$, $(a,b) \sim d$, and $a < \varepsilon_l$. (See an illustration in Fig. 2-right.)
- (E3'b) We have $(a,e) \dot{\sim} (a,b)$ for every $a, b, c, e \in S^\star$ such that $(a,b) \in \mathcal{Q}$, $(b,c) \sim e$, and $c \geq \varepsilon_r$.

  Furthermore, we have $(d,c) \dot{\sim} (b,c)$ for every $a, b, c, d \in S^\star$ such that $(b,c) \in \mathcal{Q}$, $(a,b) \sim d$, and $a \geq \varepsilon_l$. (See illustrations in Fig. 3-left and middle.)
- (E3'c) We have $(a,b) \dot{\sim} 0$ for every $a, b, c > 0$ such that $(a,b), (b,c) \in \mathcal{Q}$, $a < \varepsilon_l$, and $c \geq \varepsilon_r$.

  Furthermore, we have $(b,c) \dot{\sim} 0$ for every $a, b, c > 0$ such that $(a,b), (b,c) \in \mathcal{Q}$, $a \geq \varepsilon_l$, and $c < \varepsilon_r$. (See an illustration in Fig. 3-right.)
- (E4'a) We have $(1,0) \dot{\sim} (0,1) \dot{\sim} (a,\alpha) \dot{\sim} (\alpha,b)$ for every $a < \varepsilon_l$ and $b < \varepsilon_r$.

  Furthermore, we have $(a,b) \dot{\sim} 0$ for every $(a,b), (c,d) \in \mathcal{Q}$ such that $(a,b) \trianglelefteq (c,d) \dot{\sim} 0$.
- (E4'b) We have $(1,\alpha) \dot{\sim} (\alpha,1) \dot{\sim} (\varepsilon_l,\alpha) \dot{\sim} (\alpha,\varepsilon_r)$.

  Furthermore, we have $(a,b) \dot{\sim} \alpha$ for every $(a,b), (c,d) \in \mathcal{Q}$ such that $(a,b) \trianglerighteq (c,d) \dot{\sim} \alpha$.

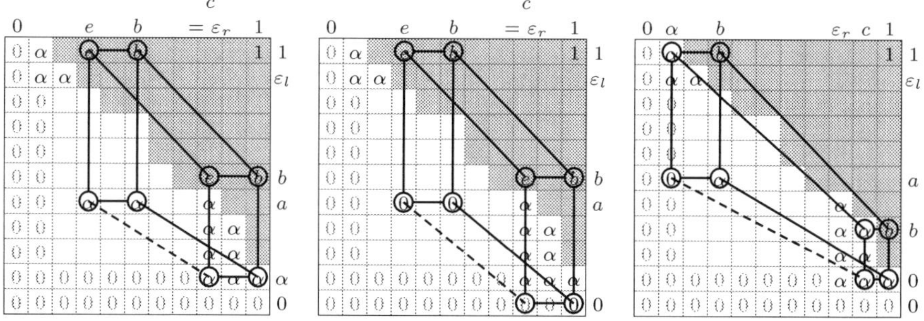

Fig. 3. Left and middle: Illustration of (E3'b). Let $(a,b) \in \mathcal{Q}$, let $c \geqslant \varepsilon_r$, and let $(b,c) \sim e$. If $(a,b) \dot{\sim} \alpha$ then, according to (P1), $(a,e) \dot{\sim} \alpha$, as well (left figure). If $(a,b) \dot{\sim} 0$ then, according to (P1) or monotonicity, $(a,e) \dot{\sim} 0$, as well (middle figure). **Right: Illustration of (E3'c).** Let $(a,b),(b,c) \in \mathcal{Q}$, let $c \geq \varepsilon_r$, and let $a < \varepsilon_l$. Then $(a,b) \dot{\sim} 0$. Indeed, if we had $(a,b) \dot{\sim} \alpha$ then, according to (P1), we would also have $(a,\alpha) \dot{\sim} \alpha$ which is a contradiction.

We call the structure $(\bar{S}^2; \trianglelefteqslant, \dot{\sim}, (1,1))$ the $(\varepsilon_l, \varepsilon_r)$*-ramification* of $(S^2; \trianglelefteqslant, \sim, (1,1))$.

Theorem 1 [13]. *Let $(S; \trianglelefteqslant, \sim, (1,1))$ be a f.n. tomonoid partition and let $(\varepsilon_l, \varepsilon_r)$ be a pair of its non-zero idempotents. Let $(\bar{S}^2; \trianglelefteqslant, \dot{\sim}, (1,1))$ be the $(\varepsilon_l, \varepsilon_r)$-ramification of $(S^2; \trianglelefteqslant, \sim, (1,1))$.*

If $(1,0) \dot{\sim} (1,\alpha)$ then there is no one-element coextension of S^2 with respect to $(\varepsilon_l, \varepsilon_r)$. Otherwise, let $\bar{\sim}$ be a coarsening of $\dot{\sim}$ such that the following holds: the $\bar{\sim}$-class of each $c \in S^\star$ coincides with the $\dot{\sim}$-class of c, the $\bar{\sim}$-class of 0 is downward closed, and each $\bar{\sim}$-class contains exactly one element of the form $(1,c)$ for some $c \in \bar{S}$. Then $(\bar{S}^2; \trianglelefteqslant, \bar{\sim}, (1,1))$ is a one-element coextension of S^2 with respect to $(\varepsilon_l, \varepsilon_r)$.

Moreover, all one-element coextensions of S^2 with respect to $(\varepsilon_l, \varepsilon_r)$, if there are any, arise in this way.

6 Representation of f.n. Tomonoids

The aim of this paper is to describe an algorithmic implementation of Theorem 1. The crucial part is to choose a suitable representation of the f.n. tomonoids (and the corresponding tomonoid partitions). F.n. tomonoids can be naturally represented by two-dimensional arrays representing their multiplication tables (see, e.g., Fig. 1). However, this approach has shown as unsuitable for the implementation. Performing the algorithm, we mainly need to work with the level equivalence classes; we need, for example, to add pairs to this classes or we need to merge two classes into one.

Therefore we have decided to represent a f.n. tomonoid $(S; \leqslant, \odot, 1)$ as a collection of level equivalence classes of pairs from S^2. Such a collection forms a partition of S^2, i.e., every pair belongs to an (exactly one) equivalence class. Each level equivalence class is either assigned to a unique value of the f.n. tomonoid (which means that it must contain two pairs of the form $(a, 1)$ and $(1, a)$ where $a \in S$) or it is "unassigned". An "unassigned" class can be a singleton.

7 Methods

Two methods, that recursively call each other, create the core of the implemented algorithm:

- a method that adds a pair (a, b) to a z-level equivalence class, we denote it by $(a, b) \dot\sim z$,
- a method that relates a pair (a, b) with a pair (c, d), we denote it by $(a, b) \dot\sim (c, d)$.

When implementing these two methods, it is first crucial that the transitivity of $\dot\sim$ is preserved. That is, when we add a pair (a, b) to a certain level equivalence class, we consequently need to add to the same class also all the pairs that are already related to (a, b).

Second, it is important that the monotonicity of the constructed tomonoid is not violated. This task is easier by the fact that (except for Part (E1), see below) we work only with pairs that are assigned to 0, α, or unassigned. Thus, when performing $(a, b) \dot\sim z$, z is either 0 or α. If $z = 0$ then we need to be sure that also all the pairs lower that (a, b) are assigned to 0. If $z = \alpha$ we proceed analogously for the pairs greater that (a, b). The details are described in the next two subsections.

Method Implementing $(a, b) \dot\sim z$

Recall that z is either 0 or α. We delete the whole level equivalence class containing (a, b) and we add all the deleted pairs to the z-level equivalence class. If $z = 0$ then for every pair (x, y) in the deleted class:

- for every pair $(u, v) \in \mathcal{Q}$ such that $(u, v) \trianglelefteqslant (x, y)$:
  - perform $(u, v) \dot\sim 0$.

If $z = \alpha$ then for every pair (x, y) in the deleted class:

- for every pair $(u, v) \in \mathcal{Q}$ such that $(u, v) \trianglerighteqslant (x, y)$:
  - perform $(u, v) \dot\sim \alpha$.

If (a, b) is already contained in a y-level equivalence class and $y \neq z$ an error is emitted signalizing that the constructed coextension is not possible.

Method Implementing $(a, b) \dot\sim (c, d)$

If both the pairs (a, b) and (c, d) belong to unassigned level equivalence classes, we simply delete one of the classes and add all its pairs to the second one.

If one of the pairs, say (a, b), belongs to a z-level equivalence class (z is either 0 or α), we perform $(c, d) \dot\sim z$.

If (a, b) belongs to a z-level equivalence class and (c, d) belongs to a y-level equivalence class then either $y = z$ which means that both (a, b) and (c, d) belong to the same level equivalence class and thus we do not perform anything, or $y \neq z$ which means that it is not possible to construct such a coextension. In the latter case an error is emitted stopping the process.

8 Algorithm

Input:

- $(S^2; \trianglelefteq, \sim, (1{,}1))$ ... tomonoid partition of a f.n. tomonoid $(S; \leqslant, \odot, 1)$
- $(\varepsilon_l, \varepsilon_r)$ ... pair of its non-zero idempotents

Output:

- $(\bar{S}^2; \trianglelefteq, \bar{\sim}, (1{,}1))$ ... a one-element coextension of $(S^2; \trianglelefteq, \sim, (1{,}1))$ with respect to $(\varepsilon_l, \varepsilon_r)$

Algorithm:

Initialization:

1. Let $\bar{S}$ be the zero doubling extension of S.
2. Let 0, α, and κ be the zero, the atom, and the coatom of $\bar{S}$, respectively. Let $\mathcal{P}$ and $\mathcal{Q}$ be given by (1) and (2), respectively.
3. Let $\dot{\sim}$ be an equivalence relation on $\bar{S}^2$. (The following steps are going to define this relation.)

Part (E1):

4. For every $(a, b), (c, d) \in \mathcal{P}$:
   - define $(a, b) \dot{\sim} (c, d)$
     if $(a, b) \sim (c, d) \sim e$ for some $e \in \bar{S} \setminus \{0, \alpha\}$.

Part (E2):

5. For every $(a, b), (b, c) \in \mathcal{P}$:
   - let $d \in \bar{S}$ be such that $(a, b) \sim d$,
   - let $e \in \bar{S}$ be such that $(b, c) \sim e$,
   - perform $(a, e) \dot{\sim} (d, c)$.

Part (E4'):

6. Perform $(1, 0) \dot{\sim} (0, 1) \dot{\sim} 0$.
7. Perform $(a, \alpha) \dot{\sim} (\alpha, b) \dot{\sim} 0$ for $a < \varepsilon_l$ and $b < \varepsilon_r$.
8. Perform $(\varepsilon_l, \alpha) \dot{\sim} (\alpha, \varepsilon_r) \dot{\sim} \alpha$.

Part (E3'a):

9. For every $a \in \bar{S}$ such that $\alpha < a < \varepsilon_l$:
   - let $b \in \bar{S}$ be the highest element such that $(a, b) \in \mathcal{Q}$,
   - let $c \in \bar{S}$ be the highest element such that $c < \varepsilon_r$,
   - let $e \in \bar{S}$ be such that $(b, c) \sim e$,
   - if $e > \alpha$ then perform $(a, e) \dot{\sim} 0$.
10. For every $c \in \bar{S}$ such that $\alpha < c < \varepsilon_r$:
   - let $b \in \bar{S}$ be the highest element such that $(b, c) \in \mathcal{Q}$,
   - let $a \in \bar{S}$ be the highest element such that $a < \varepsilon_l$,
   - let $d \in \bar{S}$ be such that $(a, b) \sim d$,
   - if $d > \alpha$ then perform $(d, c) \dot{\sim} 0$.

Part (E3'c):

11. For every $a \in \bar{S}$ such that $\varepsilon_l \leqslant a < 1$:
   - let $b \in \bar{S}$ be the highest element such that $(a, b) \in \mathcal{Q}$,
   - let $c \in \bar{S}$ be the highest element such that $(b, c) \in \mathcal{Q}$ and $c < \varepsilon_r$,
   - perform $(b, c) \dot{\sim} 0$.
12. For every $c \in \bar{S}$ such that $\varepsilon_r \leqslant c < 1$:
   - let $b \in \bar{S}$ be the highest element such that $(b, c) \in \mathcal{Q}$,
   - let $a \in \bar{S}$ be the highest element such that $(a, b) \in \mathcal{Q}$ and $a < \varepsilon_l$,
   - perform $(a, b) \dot{\sim} 0$.

Part (E3'b):

13. For every $b \in \bar{S}$ such that $\alpha < b < 1$:
   - let $e \in \bar{S}$ be such that $(b, \varepsilon_r) \sim e$,
   - if $e < b$ then:
     - for every $a \in \bar{S}$ s.t. $\alpha < a < \varepsilon_l$ and $(a, b) \in \mathcal{Q}$:
       * perform $(a, e) \dot{\sim} (a, b)$.
14. For every $b \in \bar{S}$ such that $\alpha < b < 1$:
   - let $d \in \bar{S}$ be such that $(\varepsilon_l, b) \sim d$,
   - if $d < b$ then:
     - for every $c \in \bar{S}$ s.t. $\alpha < c < \varepsilon_r$ and $(b, c) \in \mathcal{Q}$:
       * perform $(d, c) \dot{\sim} (b, c)$.

Coarsening:

15. Let $\bar{\sim} := \dot{\sim}$.
16. For every pair $(a, b) \in \bar{S}^2$, that belongs to an unassigned level equivalence class, perform arbitrarily either $(a, b) \bar{\sim} 0$ or $(a, b) \bar{\sim} \alpha$.

Remark 1. Let $\varphi \in S$ be the lowest non-zero idempotent of S. In Step 5 we may omit those pairs $(a, b), (b, c) \in \mathcal{P}$ where $(a, b), (b, c) \trianglerighteq (\varphi, \varphi)$ since, in such a case, $(a, e), (d, c) \in \mathcal{P}$.

Remark 2. In order to obtain all the one-element coextensions of S we simply repeat the procedure for every possible pair of its non-zero idempotents including $(1, 1)$. Furthermore, we create an additional coextension in the following way:

- Perform Steps 1 and 2.
- Perform $(1, 0) \bar{\sim} (0, 1) \bar{\sim} 0$.
- Perform $(\alpha, \alpha) \bar{\sim} \alpha$.

9 Example

Let us perform the algorithm taking the first f.n. tomonoid of size 9 in Fig. 1. As we can see, it has three non-zero idempotents: y, z, and 1. We are going to construct all the one-element coextensions with respect to (z, y).

- Initialization, Part (E1), and Part (E4'):
  - We obtain the values depicted in Fig. 4a.
- Part (E3'a) (see Fig. 4b):
  - Step 9:
    * For $(b, c) = (t, z)$ we perform $(y, t) \dot{\sim} 0$.
    * For $(b, c) = (u, z)$ we perform $(y, u) \dot{\sim} 0$.
    * For $(b, c) = (v, x)$ we perform $(v, v) \dot{\sim} 0$.
    * For $(b, c) = (w, x)$ we perform $(v, w) \dot{\sim} 0$.
    * For $(b, c) = (x, x)$ we perform $(v, x) \dot{\sim} 0$.
    * For $(b, c) \in \{(y, u), (u, z)\}$ we do not perform anything.
  - Step 10:
    * for $(a, b) = (z, t)$ we perform $(x, t) \dot{\sim} 0$.
    * for $(a, b) = (z, u)$ we perform $(x, u) \dot{\sim} 0$.
    * for $(a, b) \in \{(x, v), (x, w), (x, x), (u, y), (x, z)\}$ we do not perform anything.
- Part (E3'c) (see Fig. 4c):
  - Step 11:
    * For $a = y$ we obtain $b = u$ and $c = y$. Thus we perform $(u, y) \dot{\sim} 0$.
    * For $a = z$ we obtain $b = u$ and $c = y$. Thus we perform $(u, y) \dot{\sim} 0$.

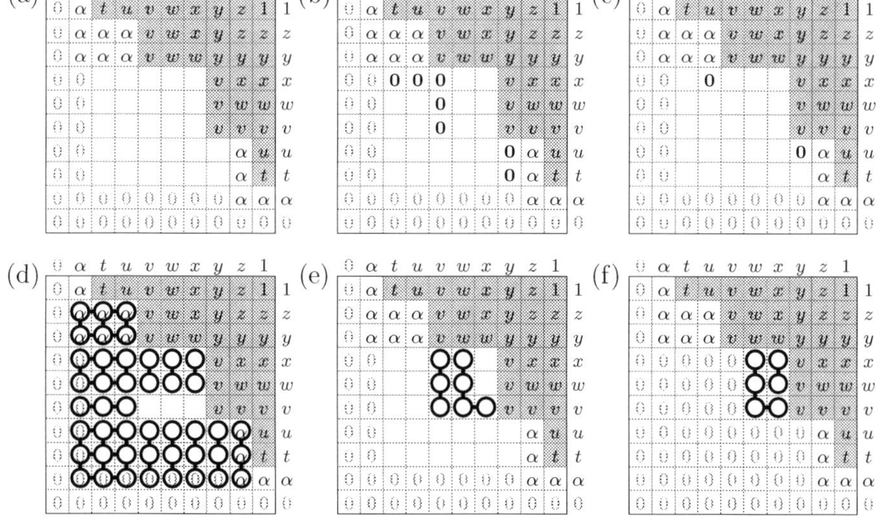

Fig. 4. Illustration of the algorithm.

- Step 12:
    * For $c = z$ we obtain $b = u$ and $a = x$. Thus we perform $(x, u) \dot{\sim} 0$.

– Part (E3'b) (see Fig. 4d):
- Step 13:
    * For $b = t$ we obtain $e = \alpha$ and we perform $(a, \alpha) \dot{\sim} (a, t)$ for every a from α to x.
    * For $b = u$ we obtain $e = \alpha$ and we perform $(a, \alpha) \dot{\sim} (a, u)$ for every a from α to x.
- Step 14:
    * For $b = t$ we obtain $d = \alpha$ and we perform $(\alpha, c) \dot{\sim} (t, c)$ for every c from α to y.
    * For $b = u$ we obtain $d = \alpha$ and we perform $(\alpha, c) \dot{\sim} (u, c)$ for every c from α to y.
    * For $b = x$ we obtain $d = w$ and we perform $(w, c) \dot{\sim} (x, c)$ for every c from α to x.
    * For $b = z$ we obtain $d = y$ and we perform $(y, c) \dot{\sim} (z, c)$ for every c from α to u.

– Part (E2) (see Fig. 4e):
- For $a = x$, $b = y$, and $c = x$ perform $(x, w) \dot{\sim} (v, x)$.
- For $a = w$, $b = y$, and $c = x$ perform $(w, w) \dot{\sim} (v, x)$.
- For $a = v$, $b = y$, and $c = x$ perform $(v, w) \dot{\sim} (v, x)$.
- For $a = x$, $b = y$, and $c = w$ perform $(x, w) \dot{\sim} (v, w)$.
- For $a = w$, $b = y$, and $c = w$ perform $(w, w) \dot{\sim} (v, w)$.
- For $a = x$, $b = y$, and $c = v$ perform $(x, v) \dot{\sim} (v, v)$.
- For $a = w$, $b = y$, and $c = v$ perform $(w, v) \dot{\sim} (v, v)$.

– Finally, we obtain the situation depicted in Fig. 4f. As we can see, there are three distinct one-element coextensions of the tomonoid.

10 Comparison with Existing Algorithms

We are aware that there already exists a number of results with similar goals [2–4,7]. We have, however, decided to introduce a new algorithm; firstly, since we wanted to have a practical support for our theoretical results [12], and secondly, since we believe that our approach is more effective. Let us make a short comparison.

The algorithm by Bartušek and Navara [2,3] is based on one-element Rees coextensions, as well, although this notion is not used. In this approach, all the existing coextensions of a given commutative f.n. tomonoid are obtained by checking the associativity for every newly defined $(x, y) \in \mathcal{Q}$.

The algorithm by Bělohlávek and Vychodil [4] uses a recursive backtracking procedure to test every possible tomonoid multiplication table on associativity.

As these two algorithms are both based on variants of the brute-force approach, our algorithm promises to give a better performance.

The comparison with the algorithm by De Baets and Mesiar [7] is planned to be made later when we have its description.

11 Conclusion

All the steps of the algorithm run in a polynomial time (with respect to the size of S) except for Step 16 where we, actually, obtain all the possible one-element coextensions with respect to the given pair of idempotents $(\varepsilon_l, \varepsilon_r)$. This step runs in exponential time since also the number of the coextensions is bounded from below by an exponential function depending on the size of S [12].

The validity of the algorithm's output is assured by Theorem 1. Remark that, since every f.n. tomonoid is a one-element Rees quotient of another f.n. tomonoid, it follows from Theorem 1 that every existing f.n. tomonoid can be obtained by the described procedure.

If we wish to obtain all the commutative one-element coextensions of a commutative f.n. tomonoid S, we simply perform $(a, b) \dot{\sim} (b, a)$ for every $(a, b) \in \mathcal{Q}$ right after Initialization.

The algorithm has been implemented and tested in the programming language Python; see http://cmp.felk.cvut.cz/~petrikm/extensions.php.

References

1. Aczél, J.: Quasigroups, nets and nomograms. Adv. Math. **1**, 383–450 (1965)
2. Bartušek, T.: Fuzzy operations on finite sets of truth values (in Czech). Master's thesis, Czech Technical University, Prague (2001)
3. Bartušek, T., Navara, M.: Program for generating fuzzy logical operations and its use in mathematical proofs. Kybernetika **38**, 235–244 (2002)
4. Bělohlávek, R., Vychodil, V.: Residuated lattices of size $\leq$ 12. Order **27**, 147–161 (2010)
5. Blaschke, W., Bol, G.: Geometrie der Gewebe, topologische Fragen der Differentialgeometrie (in German). Springer, Berlin (1939)
6. De Baets, B., Mesiar, R.: Discrete triangular norms. In: Rodabaugh, S.E., Klement, E.P. (eds.) Topological and Algebraic Structures in Fuzzy Sets. Trends in Logic, vol. 20, pp. 389–400. Springer, Netherlands (2003)
7. De Baets, B., Mesiar, R.: Triangular norms on product lattices. Fuzzy Sets Syst. **104**, 61–75 (1999)
8. Esteva, F., Godo, L.: Monoidal t-norm based logic: towards a logic for left-continuous t-norms. Fuzzy Sets Syst. **124**, 271–288 (2001)
9. Evans, K., Konikoff, M., Madden, J.J., Mathis, R., Whipple, G.: Totally ordered commutative monoids. Semigroup Forum **62**, 249–278 (2001)
10. Grillet, P.A.: Semigroups. An Introduction to the Structure Theory. Marcel Dekker, New York (1995)
11. Howie, J.M.: An Introduction to Semigroup Theory. Academic Press, London (1976)
12. Petrík, M., Vetterlein, T.: Algorithm to generate the Archimedean, finite, negative tomonoids. In: Joint 7th International Conference on Soft Computing and Intelligent Systems and 15th International Symposium on Advanced Intelligent Systems, pp. 42–47. Kitakyushu, December 2014
13. Petrík, M., Vetterlein, T.: Rees coextensions of finite, negative tomonoids. J. Log. Comput. (2015). doi:10.1093/logcom/exv047

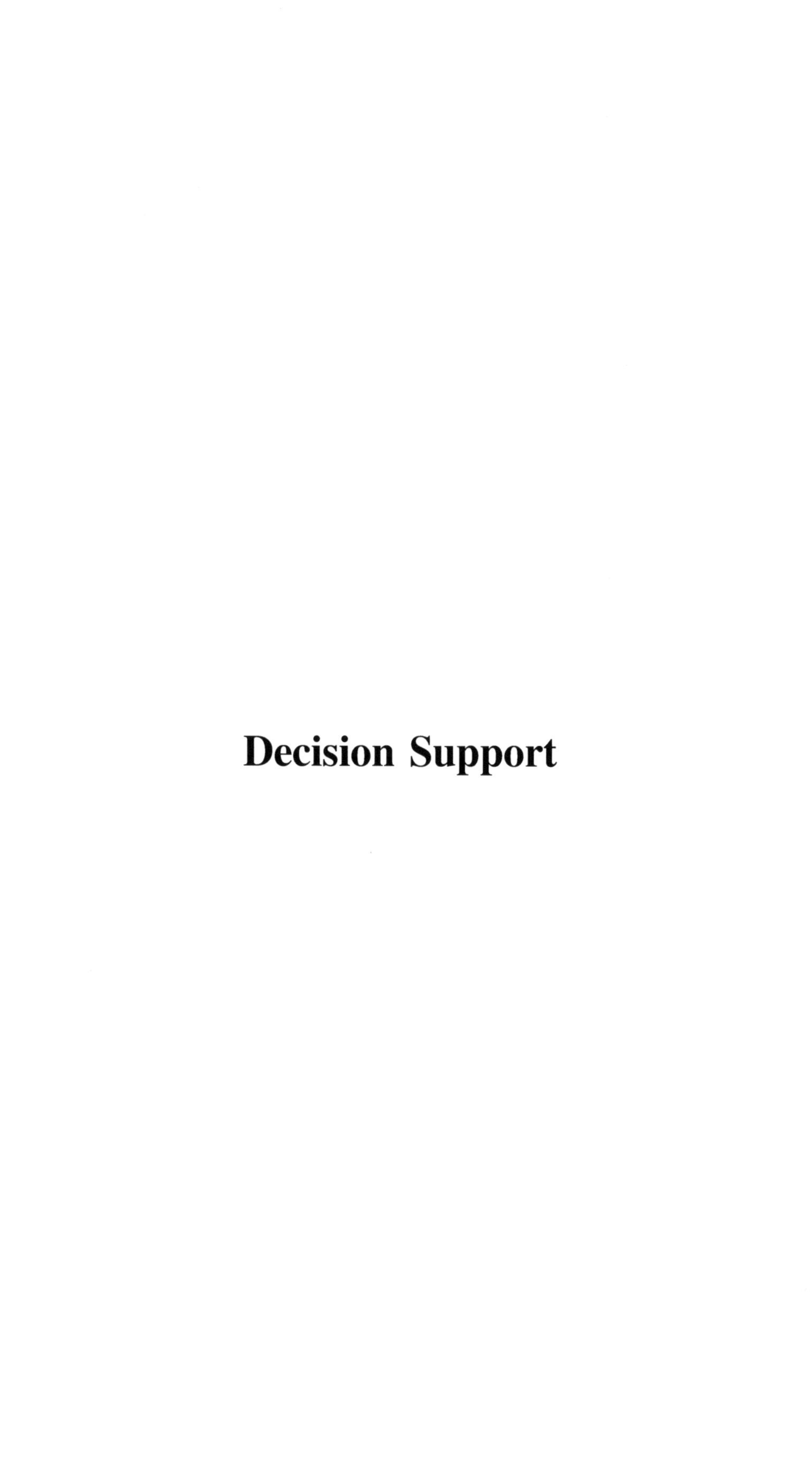

Decision Support

Constructing Preference Relations from Utilities and Vice Versa

Thomas A. Runkler(✉)

Siemens AG, Corporate Technology, Otto–Hahn–Ring 6, 81739 Munich, Germany
thomas.runkler@siemens.com

Abstract. In decision making, the suitability of decision options may be specified by utility values for each option or by preference values for each pair of options. This paper introduces a new approach to construct a matrix of pairwise preference values from a vector of utility values, which is called the U2P transformation. The U2P transformation satisfies reciprocity, triangle condition, weak transitivity, restricted max–min transitivity, and restricted max–max transitivity, but it violates (unrestricted) max–min transitivity, (unrestricted) max–max transitivity, multiplicative transitivity, and additive transitivity. Inversion of the U2P transformation yields the P2U transformation which can be used to construct a vector of utility values from a matrix of pairwise preference values. Numerical experiments with movie ratings illustrate the practical use of the U2P and P2U transformations.

Keywords: Preference relations · Utility theory · Rating · Ranking · Decision making · Movie ranking

1 Introduction

Consider a (finite) set O of $n \in \{2, 3, \ldots\}$ options, for example in a decision making process. The utility [4] of each option $i \in \{1, \ldots, n\}$, is specified by $u_i \in [0, 1]$, where the best utility is $u_i = 1$, the worst utility is $u_i = 0$, and $u_i > u_j$ indicates that the utility of option i is higher than the utility of option $j \in \{1, \ldots, n\}$. The vector $u = (u_1, \ldots, u_n) \subset [0, 1]^n$ of utility values can be interpreted as a vector of membership values of a fuzzy set of options, for example ratings of movies.

A (pairwise) *preference relation* [5] on $O \times O$ can be defined by an $n \times n$ preference matrix P with elements $p_{ij} \in [0, 1]$, where the highest preference of option i to option j is $p_{ij} = 1$, and the lowest preference is $p_{ij} = 0$. For a *reciprocal* preference relation we require

$$p_{ij} + p_{ji} = 1 \tag{1}$$

for all $i, j = 1, \ldots, n$. This implies that

$$p_{ii} = 0.5 \tag{2}$$

J.P. Carvalho et al. (Eds.): IPMU 2016, Part II, CCIS 611, pp. 547–558, 2016.
DOI: 10.1007/978-3-319-40581-0_44

for all $i = 1, \ldots, n$. Hence, $p_{ij} = 0.5$ indicates that none of the options i or j is preferred to the other one.

Preferences for decision options may be specified by a utility vector u or by a (pairwise) preference matrix P, and many methods of decision making based on utility values or preferences have been proposed in the literature, for example fuzzy decision making [2], multi–attribute fuzzy decision making [12], decision making based on fuzzy preference relations [11], group decision making based on fuzzy preference relations [15], based on incomplete preference relations [6], or based on interval fuzzy preference relations [17]. Successful applications of these approaches include direct marketing [13], optimization of logistic processes [14], movie rating [8], or selection of propulsion and manoeuvring systems [10].

In this paper, we consider the problem of generating preference matrices from utility vectors and the problem of generating utility vectors from preference matrices. Based on five assumptions we develop a so–called *U2P transformation* that generates preference matrices from utility vectors (Sect. 2). We show that the proposed U2P transformation satisfies some properties for fuzzy preference relations taken from the literature but violates some other properties (Sect. 3). By inversion of the U2P transformation we obtain the the *P2U transformation* that generates utility vectors from preference matrices (Sect. 4). Conclusions and possible directions for future research are summarized in Sect. 5.

2 The U2P Transformation

Our approach to generate preference matrices from utility vectors is based on the following five assumptions:

1. If two options have the same utility, then none of these options is preferred to the other one:
$$u_i = u_j \quad \Rightarrow \quad p_{ij} = 0.5. \tag{3}$$
2. If one option has zero utility and another option has nonzero utility, then the degree of preference of the first option to the second option is zero:
$$u_i = 0 \quad \wedge \quad u_j > 0 \quad \Rightarrow \quad p_{ij} = 0. \tag{4}$$
3. If one option has nonzero utility and another option has zero utility, then the degree of preference of the first option to the second option is one:
$$u_i > 0 \quad \wedge \quad u_j = 0 \quad \Rightarrow \quad p_{ij} = 1. \tag{5}$$
4. If one option has utility one and another option has utility less than one, then the degree of preference of the first option to the second option is one:
$$u_i = 1 \quad \wedge \quad u_j < 1 \quad \Rightarrow \quad p_{ij} = 1. \tag{6}$$
5. If one option has utility less than one and another option has utility one, then the degree of preference of the first option to the second option is zero:
$$u_i < 1 \quad \wedge \quad u_j = 1 \quad \Rightarrow \quad p_{ij} = 0. \tag{7}$$

There are infinitely many mappings from u to P and vice versa that satisfy these five assumptions. In this paper we propose a mapping that appears mathematically elegant and, as we will see in the next section, has some interesting mathematical properties. To construct a mapping from u to P we first consider assumptions 2 and 5, i.e. Eqs. (4) and (7). Equation (4) can be satisfied by

$$p_{ij} = \frac{u_i}{u_j} \tag{8}$$

and Eq. (7) can be satisfied by

$$p_{ij} = \frac{1-u_j}{1-u_i} \tag{9}$$

and hence both (4) and (7) can be satisfied by

$$\begin{aligned} p_{ij} &= \frac{A}{1/\left(\frac{u_i}{u_j}\right) + 1/\left(\frac{1-u_j}{1-u_i}\right)} \\ &= \frac{A \cdot u_i \cdot (1-u_j)}{u_j \cdot (1-u_j) + u_i \cdot (1-u_i)} \quad \text{for } (u_i, u_j) \in [0,1]^2 \setminus \{0,1\}^2 \end{aligned} \tag{10}$$

with a suitable constant A, for example $A = 1$, and

$$(3) \Rightarrow p_{ij} = 0.5 \text{ for } (u_i, u_j) = (0,0) \tag{11}$$

$$(4) \Rightarrow p_{ij} = 0 \quad \text{for } (u_i, u_j) = (0,1) \tag{12}$$

$$(5) \Rightarrow p_{ij} = 1 \quad \text{for } (u_i, u_j) = (1,0) \tag{13}$$

$$(3) \Rightarrow p_{ij} = 0.5 \text{ for } (u_i, u_j) = (1,1) \tag{14}$$

We want to choose A, so that for $(u_i, u_j) \in [0,1]^2 \setminus \{0,1\}^2$ not only assumptions 2 and 5 but also assumptions 1, 3, and 4 are satisfied, i.e. Eqs. (3), (5), and (6). For $u_i = u_j$ (3), Eq. (10) yields

$$p_{ij} = \frac{A}{2} \tag{15}$$

so we can achieve $p_{ij} = 0.5$ as required by (3) if $A = 1$. For $u_i > 0$ and $u_j = 0$ (5), Eq. (10) yields

$$p_{ij} = \frac{A}{1-u_i} \tag{16}$$

so we can achieve $p_{ij} = 1$ as required by (5) if $A = 1 - u_i$. For $u_i = 1$ and $u_j < 1$ (6), Eq. (10) yields

$$p_{ij} = \frac{A}{u_j} \tag{17}$$

so we can achieve $p_{ij} = 1$ as required by (6) if $A = u_j$. To summarize, we want to choose A so that the following three conditions hold:

$$A = 1 \qquad \text{if } u_i = u_j \tag{18}$$

$$A = 1 - u_i \qquad \text{if } u_i > 0 \wedge u_j = 0 \tag{19}$$

$$A = u_j \qquad \text{if } u_i = 1 \wedge u_j < 1 \tag{20}$$

Again, there are infinitely many choices for A that satisfy these three conditions. Here we choose

$$A = 1 - u_i + u_j \tag{21}$$

and the reader may easily verify that (21) satisfies (18)–(20). Finally, we insert (21) into (10), take (11)–(14), and obtain what we call the *U2P transformation*

$$p_{ij} = \begin{cases} \frac{u_i \cdot (1-u_j) \cdot (1-u_i+u_j)}{u_j \cdot (1-u_j) + u_i \cdot (1-u_i)} & \text{for } (u_i, u_j) \in [0,1]^2 \setminus \{0,1\}^2 \\ 0 & \text{for } (u_i, u_j) = (0,1) \\ 0.5 & \text{for } (u_i, u_j) \in \{(0,0),(1,1)\} \\ 1 & \text{for } (u_i, u_j) = (1,0) \end{cases} \tag{22}$$

The limits at the four edge points $(u_i, u_j) \in \{0,1\}^2$ are

$$\lim_{\substack{u_i=0 \\ u_j \to 0}} p_{ij} = \lim_{u_j \to 0} \frac{0}{u_j \cdot (1-u_j)} = 0 \neq 0.5 \tag{23}$$

$$\lim_{\substack{u_i \to 0 \\ u_j=0}} p_{ij} = \lim_{u_i \to 0} \frac{u_i \cdot (1-u_i)}{u_i \cdot (1-u_i)} = 1 \neq 0.5 \tag{24}$$

so p_{ij} is not continuous at $(u_i, u_j) = (0,0)$, and

$$\lim_{\substack{u_i=0 \\ u_j \to 1}} p_{ij} = \lim_{u_j \to 1} \frac{0}{u_j \cdot (1-u_j)} = 0 \tag{25}$$

$$\lim_{\substack{u_i \to 0 \\ u_j=1}} p_{ij} = \lim_{u_i \to 0} \frac{0)}{u_i \cdot (1-u_i)} = 0 \tag{26}$$

so p_{ij} is continuous at $(u_i, u_j) = (0,1)$, and

$$\lim_{\substack{u_i=1 \\ u_j \to 0}} p_{ij} = \lim_{u_j \to 0} \frac{(1-u_j) \cdot u_j}{u_j \cdot (1-u_j)} = 1 \tag{27}$$

$$\lim_{\substack{u_i \to 1 \\ u_j=0}} p_{ij} = \lim_{u_i \to 1} \frac{u_i \cdot (1-u_i)}{u_i \cdot (1-u_i)} = 1 \tag{28}$$

so p_{ij} is continuous at $(u_i, u_j) = (1,0)$, and

$$\lim_{\substack{u_i=1 \\ u_j \to 1}} p_{ij} = \lim_{u_j \to 1} \frac{(1-u_j) \cdot u_j}{u_j \cdot (1-u_j)} = 1 \neq 0.5 \tag{29}$$

$$\lim_{\substack{u_i \to 1 \\ u_j=1}} p_{ij} = \lim_{u_i \to 1} \frac{0)}{u_i \cdot (1-u_i)} = 0 \neq 0.5 \tag{30}$$

so p_{ij} is not continuous at $(u_i, u_j) = (1,1)$. Figure 1 shows a plot of the U2P transformation function. Obviously, p_{ij} is (not strictly) monotonically increasing with u_i and (not strictly) monotonically decreasing with u_j. Moreover, p_{ij} is symmetric with respect to reflection at the plane $u_i = 1 - u_j$ and symmetric with respect to rotation by 180° around the line $u_i = u_j$ and $p_{ij} = 0.5$. All preference values p_{ij} are in the interval $[0,1]$.

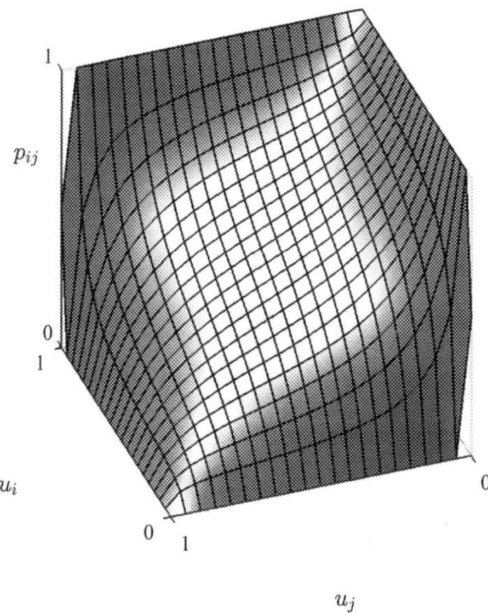

Fig. 1. 3D plot of the U2P transformation function.

3 Properties of the U2P Transformation

In this section we analyze some important mathematical properties of the U2P transformation.

3.1 Reciprocity

Preference relations generated by the U2P transformation are *reciprocal* (1), since for $(u_i, u_j) \in [0,1]^2 \setminus \{0,1\}^2$ Eq. (22) yields

$$\begin{aligned} p_{ij}+p_{ji} &= \frac{u_i \cdot (1-u_j) \cdot (1-u_i+u_j)}{u_j \cdot (1-u_j) + u_i \cdot (1-u_i)} + \frac{u_j \cdot (1-u_i) \cdot (1-u_j+u_i)}{u_i \cdot (1-u_i) + u_j \cdot (1-u_j)} \\ &= \frac{(1-u_j) \cdot u_i \cdot (1-u_i) + u_i \cdot u_j \cdot (1-u_j) + (1-u_i) \cdot u_j \cdot (1-u_j) + u_j \cdot u_i \cdot (1-u_i)}{u_j \cdot (1-u_j) + u_i \cdot (1-u_i)} \\ &= \frac{u_i \cdot (1-u_i) + u_j \cdot (1-u_j)}{u_j \cdot (1-u_j) + u_i \cdot (1-u_i)} = 1 \end{aligned} \tag{31}$$

and for the four edge points $(u_i, u_j) \in \{0,1\}^2$ we obtain

$$(u_i, u_j) = (0,0) \Rightarrow p_{ij} + p_{ji} = 0.5 + 0.5 = 1 \tag{32}$$
$$(u_i, u_j) = (0,1) \Rightarrow p_{ij} + p_{ji} = 0 + 1 = 1 \tag{33}$$
$$(u_i, u_j) = (1,0) \Rightarrow p_{ij} + p_{ji} = 1 + 0 = 1 \tag{34}$$
$$(u_i, u_j) = (1,1) \Rightarrow p_{ij} + p_{ji} = 0.5 + 0.5 = 1 \tag{35}$$

In the following subsections we examine the U2P transformation with respect to the eight properties of fuzzy preference relations that are listed and discussed in [7]: triangle condition [9], weak transitivity [16], max–min transitivity [3,18], max–max transitivity [3,18], restricted max–min transitivity [16], restricted max–max transitivity [16], multiplicative transitivity [16], and additive transitivity [15,16].

3.2 Triangle Condition

The triangle condition [9] is defined as

$$p_{ij} + p_{jk} \geq p_{ik} \quad \forall i, j, k \tag{36}$$

The U2P transformation has the monotonicity properties mentioned above, and all preference values are in $[0, 1]$, so

$$u_j \leq u_k \quad \Rightarrow \quad p_{ij} \geq p_{ik} \quad \Rightarrow \quad p_{ij} + p_{jk} \geq p_{ik} \tag{37}$$

and

$$u_i \leq u_j \quad \Rightarrow \quad p_{jk} \geq p_{ik} \quad \Rightarrow \quad p_{ij} + p_{jk} \geq p_{ik} \tag{38}$$

There are six possible situations for the order of u_i, u_j and u_k:

$$u_i \leq u_j \leq u_k \tag{39}$$

$$u_i \leq u_k \leq u_j \tag{40}$$

$$u_j \leq u_i \leq u_k \tag{41}$$

$$u_j \leq u_k \leq u_i \tag{42}$$

$$u_k \leq u_i \leq u_j \tag{43}$$

$$u_k \leq u_j \leq u_i \tag{44}$$

For (39), (41), and (42) we have (37), and for (40) and (43) we have (38). In the only remaining case (44) we have

$$p_{ij} \geq 0.5 \quad \wedge \quad p_{jk} \geq 0.5 \quad \Rightarrow \quad p_{ij} + p_{ik} \geq 1 \quad \Rightarrow \quad p_{ij} + p_{jk} \geq p_{ik} \tag{45}$$

This means that the triangle condition (36) holds for all six possible orders, so the U2P transformation satisfies the triangle condition.

3.3 Weak Transitivity

Weak transitivity [16] is defined as

$$p_{ij} \geq 0.5 \quad \wedge \quad p_{jk} \geq 0.5 \quad \Rightarrow \quad p_{ik} \geq 0.5 \quad \forall i, j, k \tag{46}$$

For the U2P transformation we have

$$p_{ij} \geq 0.5 \quad \Leftrightarrow \quad u_i \geq u_j \tag{47}$$

and so

$$p_{ij} \geq 0.5 \ \wedge \ p_{jk} \geq 0.5 \ \Rightarrow \ u_i \geq u_j \geq u_k \ \Rightarrow \ u_i \geq u_k \ \Rightarrow \ p_{ik} \geq 0.5 \tag{48}$$

so the U2P transformation satisfies the weak transitivity criterion.

3.4 Max–min Transitivity

Max–min transitivity [3,18] is defined as

$$p_{ik} \geq \min(p_{ij}, p_{ik}) \quad \forall i, j, k \tag{49}$$

Consider the case

$$u_i = 0.3, \quad u_j = 0.6, \quad u_k = 0.8 \tag{50}$$

With the U2P transformation we obtain

$$p_{ij} \approx 0.35, \quad p_{jk} \approx 0.36, \quad p_{ik} \approx 0.24 \tag{51}$$

and so

$$p_{ik} \approx 0.24 \ngeq \min(p_{ij}, p_{ik}) \approx 0.35 \tag{52}$$

which violates (49), and so the U2P transformation violates max–min transitivity.

3.5 Max–max Transitivity

Max–max transitivity [3,18] is defined as

$$p_{ik} \geq \max(p_{ij}, p_{ik}) \quad \forall i, j, k \tag{53}$$

For (50) the U2P transformation yields (51), hence

$$p_{ik} \approx 0.24 \ngeq \max(p_{ij}, p_{ik}) \approx 0.36 \tag{54}$$

which violates (53), and so the U2P transformation violates max–max transitivity.

3.6 Restricted Max–min Transitivity

Restricted max–min transitivity [16] is defined as

$$p_{ij} \geq 0.5 \quad \wedge \quad p_{jk} \geq 0.5 \quad \Rightarrow \quad p_{ik} \geq \min(p_{ij}, p_{ik}) \quad \forall i, j, k \tag{55}$$

For the U2P transformation we have (47), so

$$p_{ij} \geq 0.5 \quad \Rightarrow \quad u_i \geq u_j \quad \Rightarrow \quad p_{ik} \geq p_{jk} \tag{56}$$

$$p_{jk} \geq 0.5 \quad \Rightarrow \quad u_j \geq u_k \quad \Rightarrow \quad p_{ij} \leq p_{ik} \tag{57}$$

and further

$$p_{ik} \geq p_{ij} \quad \wedge \quad p_{ik} \geq p_{jk} \quad \Rightarrow \quad p_{ik} \geq \min(p_{ij}, p_{ik}) \tag{58}$$

so the U2P transformation satisfies restricted max–min transitivity.

3.7 Restricted Max–max Transitivity

Restricted max–max transitivity [16] is defined as

$$p_{ij} \geq 0.5 \quad \wedge \quad p_{jk} \geq 0.5 \quad \Rightarrow \quad p_{ik} \geq \max(p_{ij}, p_{ik}) \quad \forall i,j,k \tag{59}$$

For the U2P transformation we have (56), (57), and so

$$p_{ik} \geq p_{ij} \quad \wedge \quad p_{ik} \geq p_{jk} \quad \Rightarrow \quad p_{ik} \geq \max(p_{ij}, p_{ik}) \tag{60}$$

so the U2P transformation satisfies restricted max–max transitivity.

3.8 Multiplicative Transitivity

Multiplicative transitivity [16] is defined as

$$\frac{p_{ji}}{p_{ij}} \cdot \frac{p_{kj}}{p_{jk}} = \frac{p_{ki}}{p_{ik}} \quad \forall i,j,k \tag{61}$$

For (50) the U2P transformation yields (51) and

$$\begin{aligned} \frac{p_{ji}}{p_{ij}} \cdot \frac{p_{kj}}{p_{jk}} &= \frac{1-p_{ij}}{p_{ij}} \cdot \frac{1-p_{jk}}{p_{jk}} \approx \frac{1-0.35}{0.35} \cdot \frac{1-0.36}{0.36} \approx 3.6 \\ &\neq \frac{p_{ki}}{p_{ik}} = \frac{1-p_{ik}}{p_{ik}} \approx \frac{1-0.24}{0.24} \approx 3.2 \end{aligned} \tag{62}$$

which violates (61), and so the U2P transformation violates multiplicative transitivity.

3.9 Additive Transitivity

Additive transitivity [15, 16] is defined as

$$(p_{ij} - 0.5) + (p_{jk} - 0.5) = (p_{ik} - 0.5) \quad \forall i,j,k \tag{63}$$

For (50) the U2P transformation yields (51) and

$$\begin{aligned} (p_{ij} - 0.5) + (p_{jk} - 0.5) &\approx (0.35 - 0.5) - (0.36 - 0.5) = -0.39 \\ &\neq (p_{ik} - 0.5) \approx (0.24 - 0.5) = -0.26 \end{aligned} \tag{64}$$

which violates (63). In this example all preferences p_{ij}, p_{jk}, and p_{ik} are < 0.5. However, the U2P transformation also violates additive transitivity for the following case

$$u_i = 0.9, \quad u_j = 0.3, \quad u_k = 0.2 \tag{65}$$

where

$$p_{ij} \approx 0.84, \quad p_{jk} \approx 0.58, \quad p_{ik} \approx 0.86 \tag{66}$$

so all preferences p_{ij}, p_{jk}, and p_{ik} are >0.5, but still

$$\begin{aligned} (p_{ij} - 0.5) + (p_{jk} - 0.5) &\approx (0.84 - 0.5) - (0.58 - 0.5) = 0.42 \\ &\neq (p_{ik} - 0.5) \approx (0.86 - 0.5) = 0.36 \end{aligned} \tag{67}$$

which violates (63), and so the U2P transformation violates additive transitivity.

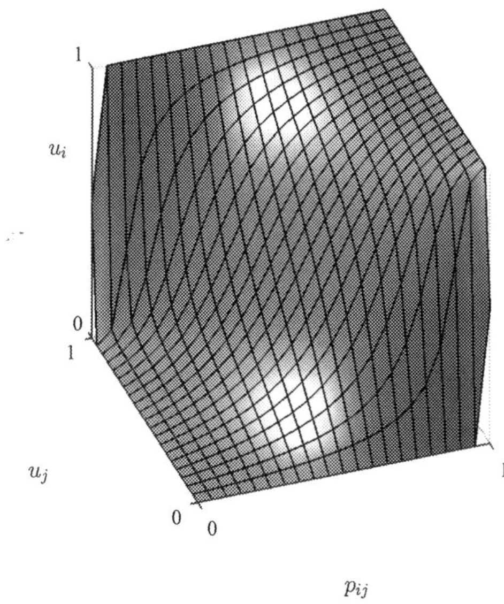

Fig. 2. 3D plot of the P2U transformation function.

4 The P2U Transformation

In the previous sections we have considered the problem of generating preference matrices from utility vectors. In this section we will consider the reverse problem: generating utility vectors from preference matrices. The basic idea is to initially specify a utility value for one of the options, say u_j, $j \in \{1, \ldots, n\}$, and then compute the utility values of all the other options u_i, $i = 1, \ldots, n$, $i \neq j$, using a function of u_j and p_{ij}. This function can be obtained by solving (the first case of) Eq. (22) for u_i and defining the values of u_i for the four edges $(p_{ij}, u_j) \in \{0, 1\}^2$, which after several conversions (omitted here due to limited space) yields

$$u_i = \begin{cases} \frac{\sqrt{s_{ij}}+p_{ij}+u_j^2-1}{2\cdot(p_{ij}+u_j-1)} & \text{for } p_{ij}, u_j \in [0,1],\ p_{ij} \neq 1-u_j \\ p_{ij} & \text{for } u_j \in (0,1),\ p_{ij} = 1-u_j \\ 0.5 & \text{for } (p_{ij}, u_j) \in \{(0,1),(1,0)\} \end{cases} \tag{68}$$

with

$$\begin{aligned} s_{ij} = {} & p_{ij}^2 \cdot \left(-4 \cdot u_j^2 + 4 \cdot u_j + 1\right) \\ & + p_{ij} \cdot \left(-4 \cdot u_j^3 + 10 \cdot u_j^2 - 4 \cdot u_j - 2\right) \\ & + u_j^4 - 2 \cdot u_j^2 + 1 \end{aligned} \tag{69}$$

In a similar way as in (23)–(30) we can show that (68) is continuous at $(p_{ij}, u_j) \in [0,1]^2 \setminus \{(0,1),(1,0)\}$ and not continuous at $(p_{ij}, u_j) \in \{(0,1),(1,0)\}$. Figure 2 shows a plot of the P2U transformation function.

5 Numerical Experiments

To briefly illustrate the use of the U2P and P2U transformations we consider two artificial data set motivated by the Netflix challenge [1] where a data set containing movie ratings was analyzed. Here, we consider three movies with the ratings 0, 50 %, and 100 %, so we have the utility vector

$$U = (0, 0.5, 1) \tag{70}$$

Using the U2P transformation (22) we obtain the preference matrix

$$P = \begin{pmatrix} 0.5 & 0 & 1 \\ 1 & 0.5 & 0 \\ 0 & 1 & 0.5 \end{pmatrix} \tag{71}$$

which indicates that the second movie is completely preferred to the first movie ($p_{21} = 1$), and the third movie is completely preferred to the second movie ($p_{32} = 1$). If we pick the second movie as a reference, $j = 2$, and set the utility of the second movie to $u_2 = 0.5$, then the P2U transformation (68) of (71) recovers the original utility vector from (70). Next, we consider three other movies with the ratings 25 %, 50 %, and 75 %, so

$$U = (0.25, 0.5, 0.75) \tag{72}$$

and the P2U transformation yields the preference matrix

$$P \approx \begin{pmatrix} 0.5 & 0.3571 & 0.25 \\ 0.6429 & 0.5 & 0.3571 \\ 0.75 & 0.6429 & 0.5 \end{pmatrix} \tag{73}$$

If we pick the reference $u_2 = 0.5$ again, then the P2U transformation of (73) will recover the original utility vector (72). However, if we choose a different utility value for the second movie, for example 75 % ($u_2 = 0.75$), then the P2U transformation of (73) will yield

$$U \approx (0.5, 0.75, 0.8738) \tag{74}$$

which is different from the original utility vector at (72) but yields the same preference matrix (73). This example shows that the U2P transformation induces an equivalence relation, and we call a pair of utility vectors U_1 and U_2 *preference equivalent* with respect to the U2P transformation, if and only if the corresponding preference matrices are equal, $P_1 = P_2$.

6 Conclusions

Based on five assumptions about utilities and preferences we have developed the U2P transformation (22) that maps utility (or membership) vectors to (fuzzy) preference matrices. The U2P transformation satisfies reciprocity, triangle condition, weak transitivity, restricted max–min transitivity, and restricted max–max transitivity, but it violates (unrestricted) max–min transitivity, (unrestricted) max–max transitivity, multiplicative transitivity, and additive transitivity. By inversion of the U2P transformation we have also developed the P2U transformation (68) that maps (fuzzy) preference matrices to utility (or membership) vectors. Numerical experiments with movie ratings have illustrated the practical use of the U2P and P2U transformations, and have shown that the U2P transformation induces an equivalence relation.

Some questions are left open that may be considered interesting for future research: What are the properties of the equivalence relation induced by the U2P transformation? Is it possible to satisfy our five assumptions and also satisfy (unrestricted) max–min transitivity, (unrestricted) max–max transitivity, multiplicative transitivity, or additive transitivity? Which modification will yield a transformation that is continuous on the whole unit square? Multiplicative transitivity is considered a very important property by some authors. How can our five assumptions be relaxed, so that preferences are obtained which satisfy multiplicative transitivity?

References

1. Bell, R.M., Koren, Y.: Lessons from the Netflix prize challenge. ACM SIGKDD Explor. Newsl. **9**(2), 75–79 (2007)
2. Bellman, R., Zadeh, L.: Decision making in a fuzzy environment. Manage. Sci. **17**(4), 141–164 (1970)
3. Dubois, D., Prade, H.: Fuzzy Sets and Systems. Academic Press, London (1980)
4. Fishburn, P.C.: Utility Theory. Wiley Online Library, New York (1988)
5. Fodor, J.C., Roubens, M.R.: Fuzzy Preference Modelling and Multicriteria Decision Support, vol. 14. Springer, New York (1994)
6. Herrera-Viedma, E., Alonso, S., Chiclana, F., Herrera, F.: A consensus model for group decision making with incomplete fuzzy preference relations. IEEE Trans. Fuzzy Syst. **15**(5), 863–877 (2007)
7. Herrera-Viedma, E., Herrera, F., Chiclana, F., Luque, M.: Some issues on consistency of fuzzy preference relations. Eur. J. Oper. Res. **154**(1), 98–109 (2004)
8. Jeon, T., Cho, J., Lee, S., Baek, G., Kim, S.: A movie rating prediction system of user propensity analysis based on collaborative filtering and fuzzy system. In: IEEE International Conference on Fuzzy Systems, Jeju, Korea, pp. 507–511 (2009)
9. Luce, R.D., Suppes, P.: Preferences utility and subject probability. In: Luce, R.D., Bush, R.R., Eugene, G.E. (eds.) Handbook of Mathematical Psychology, vol. III, pp. 249–410. Wiley, New York (1963)
10. Ölçer, A.I., Odabaşi, A.Y.: A new fuzzy multiple attributive group decision making methodology and its application to propulsion/manoeuvring system selection problem. Eur. J. Oper. Res. **166**(1), 93–114 (2005)
11. Orlovsky, S.A.: Decision-making with a fuzzy preference relation. Fuzzy Sets Syst. **1**(3), 155–167 (1978)
12. Ribeiro, R.A.: Fuzzy multiple attribute decision making: a review and new preference elicitation techniques. Fuzzy Sets Syst. **78**(2), 155–181 (1996)

13. Setnes, M., Kaymak, U.: Fuzzy modeling of client preference from large data sets: an application to target selection in direct marketing. IEEE Trans. Fuzzy Syst. **9**(1), 153–163 (2001)
14. Sousa, J.M., Palm, R.H., Silva, C.A., Runkler, T.A.: Optimizing logistic processes using a fuzzy decision making approach. IEEE Trans. Syst. Man Cybern. A **33**(2), 245–256 (2003)
15. Tanino, T.: Fuzzy preference orderings in group decision making. Fuzzy Sets Syst. **12**(2), 117–131 (1984)
16. Tanino, T.: Fuzzy preference relations in group decision making. In: Non-conventional Preference Relations in Decision Making, pp. 54–71. Springer, New York (1988)
17. Türkşen, İ.B., Bilgiç, T.: Interval valued strict preference with Zadeh triples. Fuzzy Sets Syst. **78**(2), 183–195 (1996)
18. Zimmermann, H.J.: Fuzzy Set Theory and Its Applications. Kluwer Academic Publishers, Boston (1985)

A Characterization of the Performance of Ordering Methods in TTRP with Fuzzy Coefficients in the Capacity Constraints

Isis Torres-Pérez[1], Carlos Cruz[2], Alejandro Rosete-Suárez[1], and José Luis Verdegay[2(✉)]

[1] Instituto Superior Politécnico José Antonio Echeverría, Havana, Cuba
{itorres,rosete}@ceis.cujae.edu.cu
[2] University of Granada, Granada, Spain
{carloscruz,verdegay}@decsai.ugr.es

Abstract. Recently, the Truck and Trailer Routing Problem (TTRP) has been tackled with uncertainty in the coefficients of constrains. In order to solve this problem it is necessary to use methods for comparison fuzzy numbers. The problem of ordering fuzzy quantities has been addressed by many authors and there are many indices to perform this task. However, it is impossible to give a final answer to the question on what ranking method is the best in this problem. In this paper we focus our attention on a model to characterize TTRP instances. We use a data mining algorithm to derive a decision tree that determined the best method for comparison based on the characteristics of the TTRP problem to be solved.

Keywords: Fuzzy optimization · Truck and Trailer Routing Problem (TTRP) · Fuzzy coefficients · Fuzzy constraints · Ranking function · Decision tree

1 Introduction

Fuzzy Optimization models and methods has been one of the most and well-studied topics inside the broad area of Soft Computing. Particularly relevant is the field of Fuzzy Linear Programming (FLP) that constitutes the basis for solving fuzzy optimization problems. FLP models are classified according to the way the fuzziness is introduced. In the last past years several kinds of FLP models have appeared in the literature [1], but one main is: FLP models in which coefficients of the constraints and right hand values are defined as fuzzy numbers. This type of model can be stated in the following form:

$$\begin{aligned} & max/min\ f(x_j, c_j) \\ & \quad s.t.\ h(x_j, a^f_{ij})\{\leq^f, \geq^f\}b^f_i \\ & \quad x_j \geq 0 \end{aligned} \tag{1}$$

J.P. Carvalho et al. (Eds.): IPMU 2016, Part II, CCIS 611, pp. 559–568, 2016.
DOI: 10.1007/978-3-319-40581-0_45

where x_j are the decision variables, c_j are the coefficients of the objective function, a_{ij}^f are the fuzzy coefficients of the constraints and b_i^f are fuzzy right hand values of the constraints. The functions $f(x_j, c_j)$ and $h(x_j, a_{ij}^f)$ can be linear or nonlinear functions. The first version of this problem appeared in [2] (although supposing imprecision in the objective as well). In [3] is developed a general solution strategy that manage the imprecision in the comparison by introducing a fuzzy number τ_i for each single constraint, given by the decision maker. This value represents the allowed maximum violation in the $i-th$ constraint. The solution approach transforms the fuzzy model (1) in an equivalent auxiliary traditional model that is expressed as follows.

$$\begin{aligned} &max/min\ f(x_j, c_j) \\ &\qquad s.t.\ h(x_j, a_{ij}^f)\{\leq_g, \geq_g\}b_i^f + \tau_i^f(1-\alpha) \\ &\qquad x_j \geq 0, \alpha \in [0,1] \end{aligned} \tag{2}$$

where the symbols $\leq_g$ and $\geq_g$, stands for a comparison relation between fuzzy numbers using a ranking function g for the constraints, and α is a satisfaction level defined by the decision maker.

The ranking methods for fuzzy numbers has been used by many researchers working in the area of ordering fuzzy quantities. Although such methods are required in other applications. The methods for ranking fuzzy numbers in many cases can induce different rankings. In consequence, a long list of different auxiliary models and set of possible different solutions are obtained according to the comparison relation between fuzzy numbers used. In the study carried out by [4] these methods can be classified into three classes:

- Methods based on the definition of an ordering function: is constructed a mapping function to transform fuzzy quantities into a positive real number and then simply ranks based on the comparison of the obtained real numbers. In this group are the following methods: [5–9].
- Methods based on the comparison of alternatives: is defined some reference set(s) and evaluates each fuzzy quantity by calculating and comparing the closeness of fuzzy quantities to the reference set(s). The definition of the reference set can be done in two forms. Several methods in this class are: [10–12].
- Methods based on a relationship of preference: is constructed a fuzzy binary preference relation to manipulate pairwise comparisons. The result of all pairwise comparisons are used to obtain a order relation among fuzzy quantities. Examples of these methods are: [3,13,14].

Before such variety of methods the following question emerges: which comparison method is more convenient to use or which is the one that gets the most adequate results? Studies about this topic where exist comparative analysis are very few. More recent research is presented in [15] that investigates differences/similarities between ranking methods. However, most of the time choosing a method rather than another is a matter of preference or is context dependent.

The comparison methods have multiple applications in the fuzzy models of transport problems. One of the most important kind of problems nowadays is the intermodal freight transport which has great relevance in the research lines of the Europe Union Research and Innovation Programme Horizon 2020. Concretely, a real-world application is in the Truck and Trailer Routing Problem, TTRP, when the decision-maker is willing to allow some violations in the accomplishment of the capacity constraints and their coefficients i.e., the so-called Fuzzy TTRP. This last problem will be focused on this paper.

Consequently the rest of the paper is organized as follows. In Sect. 2 one presents the basic elements of the TTRP. Also, it is proposes a fuzzy model for dealing with the imprecision in the set of constraints and a general approach for solving this model is described. Section 3 provides a experimental study to illustrate the usefulness of characterizing the performance of the methods of comparison. Finally, Sect. 4 summarizes the work presented.

2 TTRP with Fuzzy Demands and Capacities

The Truck and Trailer Routing Problem (TTRP) is an optimization problem which by its nature favors the presence of vagueness, imprecision and uncertainty in the information handled. This problem consists of a heterogeneous fleet composed of m_c trucks and m_r trailers to serve a set of customers dispersed $V = \{v_1, v_2, \ldots, v_n\}$ from a central depot denoted as node v_0. Each customer $v_i \in V$ has a non-negative demand $q_i > 0$. The capacities of the trucks and the trailers are Q_c and Q_r, respectively; and the distance $c_{(ij)}$ between any two nodes $v_i, v_j \in V \cup \{v_0\}$ is known. Some customers with accessibility constraints must be served just by truck, while others can be served either by truck or by a complete vehicle (a truck pulling a trailer). These access constraints partition the customers into two subsets: the subset of truck customers V_c accessible only by truck (TC), and the subset of vehicle customers V_v, accessible either by a truck or by a complete vehicle (VC). A solution of the TTRP is generally composed of three types of routes: Pure Vehicle Route performed by a complete vehicle and contains only vehicle customer. Pure Truck Route performed by a truck alone and may visit both customer type and Complete Vehicle Route consisting of a main tour traveled by a complete vehicle, and at least one sub-tour traveled by the truck alone [16]. One example is shown in Fig. 1 to illustrate these kinds of route.

Also, each route is limited by capacity of vehicle used. In general, the goal for this NP-hard problem [17] is to find a set of least cost vehicle routes that start and end at the central depot such that each customer is serviced exactly once; the total demand of any vehicle route does not exceed the total capacity of the allocated vehicles used in that route; and the number of required trucks and trailers is not greater than m_c and m_r, respectively.

The solution approaches published in the literature about this topic can be divided into three groups: exact approaches [18], approximated approaches (including heuristic and metaheuristics) [16, 19–23], or a combination of these

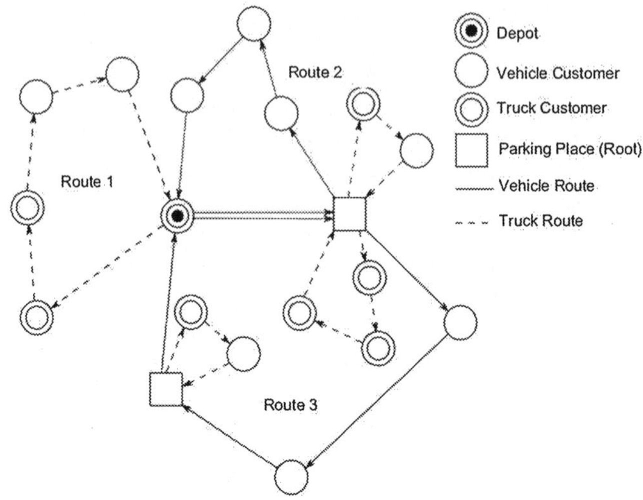

Fig. 1. Routes TTRP.

approaches (matheuristics) [24]. However, to the best of our knowledge, point out that most of models and approaches used for the TTRP in the literature assume that the data available are accurate, still when in many practical problems the available knowledge about some data and parameters of the model involving uncertainty.

In the reality, this problem is very complex and the information is not always available with sufficient precision and completeness as desired for adequate planning and management. In previous works [25,26] the authors deal with a Fuzzy TTRP modeled as a fuzzy linear programming where the decision maker tolerates violations in the accomplishment of the constraints; i.e. the decision maker permits the constraints to be satisfied" as well as possible. However, there are other situations in which the decision maker permits some violations in the constraints and also the data that define them ($a_i x$ and b_i), have a vague nature. For example, the information regarding customer demand typically is not established with any level of precision or is not available to the decision maker. A similar situation occurs with truck loads. These loads depend on demand and truck and trailers capacities as well as the numbers of units that are used in the transport. These parameters are vague and can be expressed by means of fuzzy numbers. In this case the capacity constraints for TTRP with fuzzy demands and capacities can be described as:

$$\sum_{i=0}^{n}\sum_{j=1}^{n} q_j^f x_{ij}^{k1} + \sum_{i=0}^{n}\sum_{j=1}^{n} q_j^f x_{ij}^{k0} \leq_f Q_c^f + Q_r^f \tag{3}$$

$$\sum_{i=0}^{n}\sum_{j=1}^{n} q_j^f x_{ij}^{k1} \leq_f Q_c^f \tag{4}$$

where x_{ij}^{kl} is a binary variable equal to 1 if and only if the vehicle k with or without trailer ($l = 0$ or $l = 1$) is used from i to j, and 0 otherwise. The customers demand (q_j^f) and vehicle capacities (Q_c^f and Q_r^f) are fuzzy numbers. Also, both constraints are considered fuzzy and the symbol $\leq_f$ is used to indicate it. According to (2), these constraints can be replaced by the following constraints:

$$\sum_{i=0}^{n}\sum_{j=1}^{n} q_j^f x_{ij}^{k1} + \sum_{i=0}^{n}\sum_{j=i}^{n} q_j^f x_{ij}^{k0} \leq_g (Q_c^f + Q_r^f) + \tau_1^f (1-\alpha) \tag{5}$$

$$\sum_{i=0}^{n}\sum_{j=1}^{n} q_j^f x_{ij}^{k1} \leq_g Q_c^f + \tau_2^f (1-\alpha) \tag{6}$$

where τ_1^f and τ_2^f are fuzzy numbers that represents the tolerance levels of each capacity constraint. Last year this model was introduced in [27]. That paper was the first approach where demands and capacities in TTRP were modeled making use of the concept of fuzzy number. Unlike previous contributions [25,26] is tackled the TTRP with the set of constraints totally fuzzy. In this case, the solution to the model is obtained by particularization of the different comparison methods of fuzzy numbers. However, it is important to know which of these methods is more convenient to use or which is the one that gets the most adequate results. These and other questions can arise when solving our model. A strategy to face this problem may be to have a model that indicates the best comparison method based on the characteristics of the problem to be solved. This model would be able to offer knowledge on the types of instances where each method works better or worse. This knowledge can be of great utility for the final users.

3 Experiments

In order to generate our model, we used 21 TTRP benchmark problems reported by [16]. These public test instances were derived from seven classical vehicle routing problem. Table 1 shows the characteristics of problems.

The experiments were performed on a computer with an Intel Xeon running at 2.40 GHz under Linux Ubuntu with 23 GB of RAM. We decided to use an algorithm based on local search (Hill Climbing), which is available from the BiCIAM library [28]. The results were obtained with 30 independent runs with 100000 fitness evaluations for each problem. The instances of TTRP were solved for $\alpha = \{0.0, 0.1, 0.2, 0.3, 0.4, 0.5, 0.6, 0.7, 0.8, 0.9, 1.0\}$. The amount demand of each customer and the limited vehicle capacities are considered triangular fuzzy numbers. In each case were obtained in the form of 10 % variation in the modal value. Tolerance levels τ_1 and τ_2 are considered fuzzy numbers.

Also, we used six ranking function to obtain a particular order relation between fuzzy numbers. In the following, we will briefly describe each function will be used to test our proposal.

Table 1. Instances TTRP.

Problem	Customers			Trucks		Trailers	
	Total	VC	TC	Number	Capacity	Number	Capacity
1	50	38	12	5	100	3	100
2		25	25				
3		13	37				
4	75	57	18	9	100	5	100
5		38	37				
6		19	56				
7	100	75	25	8	150	4	100
8		50	50				
9		25	75				
10	150	113	37	12	150	6	100
11		75	75				
12		38	112				
13	199	150	49	17	150	9	100
14		100	99				
15		50	149				
16	120	90	30	7	150	4	100
17		60	60				
18		30	90				
19	100	75	25	10	150	5	100
20		50	50				
21		25	75				

1. Chang in [9] proposed a ranking method based on the following index

$$C_I(\tilde{u}) = \int_{z \in \sup \mu_{\tilde{u}}} z\mu_{\tilde{u}}(z)dz \tag{7}$$

2. Dubois and Prade propose a set of four indices able to completely describe the relative location of two fuzzy numbers [13]. In particular we use:

$$PD(\tilde{u_i}, \tilde{u_j}) = \sup \min(\mu_{\tilde{u_i}}, \mu_{\tilde{u_j}}) \tag{8}$$

$$ND(\tilde{u_i}, \tilde{u_j}) = \inf \sup \min(1 - \mu_{\tilde{u_i}}, \mu_{\tilde{u_j}}) \tag{9}$$

3. In [5,7,8], Yager proposed four ranking methods, where he does not assume any hypothesis of normality or convexity. In this study we use the following three methods proposed by Yager:

$$Y_1(\tilde{u}) = \frac{\int_0^1 g(z)\mu_{\tilde{u}}(z)dz}{\int_0^1 \mu_{\tilde{u}}(z)dz} \tag{10}$$

$$Y_2(\tilde{u}) = \int_0^{\alpha-max} M(U_\alpha)d\alpha \tag{11}$$

$$Y_4(\tilde{u}) = \sup_{z\in[0,1]} \min(z, \mu_{\tilde{u}}(z)) \tag{12}$$

With the results we performed Friedman test [29] with $\alpha = 0.05$ as the level of confidence. The results point out that Y_4 dominates the other ranking functions and achieved the highest rankings.

Also, we can raise the 15 hypotheses of equality among the 6 methods of our study, and apply the post-hoc Shaffer [30] and Holm [31] to contrast them. Nine of these hypotheses confirm the improvement of Y_4 over the rest of the comparison methods. Furthermore, the C_I method was overcome by all methods considered. Finally, only 6 hypotheses can be rejected using these procedures. Each one does not find any significant difference between ND and PD, Y_1 and Y_2. Clearly, this is visible in the graphic of the Fig. 2.

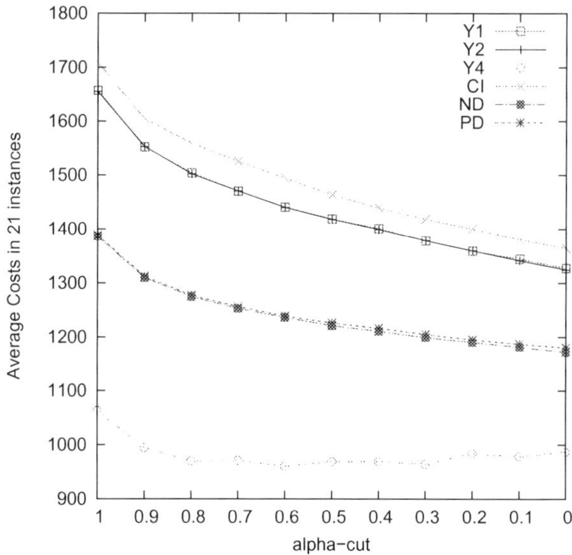

Fig. 2. Behavior average of the comparison methods in 21 TTRP instances.

In this point, we decide to generate a decision tree model that indicates the best comparison method based on the characteristics of the problem to be solved. The following figure shows the obtained model using J48 algorithm of tool KNIME on a minable view of 231 tuples (21 instances × 11 α-cuts) (Fig. 3).

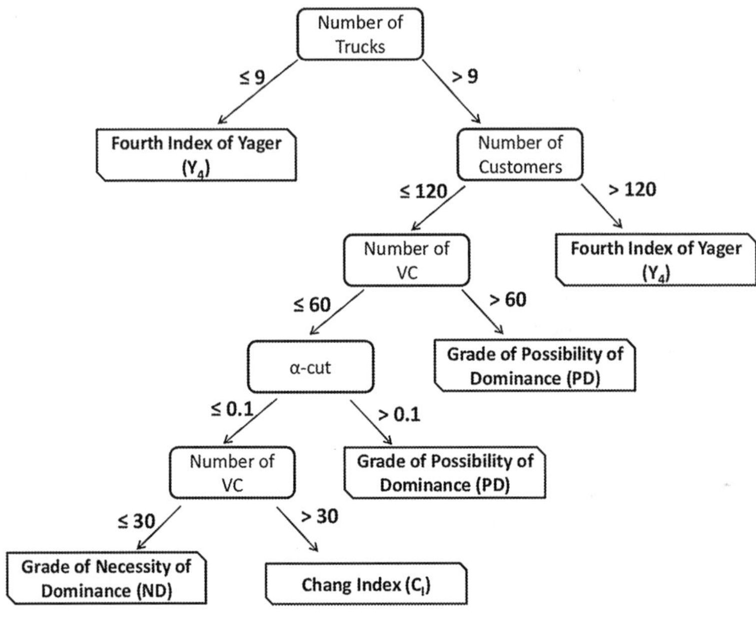

Fig. 3. Decision tree model.

This model comprise a set of rules to determine the best method for comparison based on the characteristics of the problem to be solved. For example, this model suggests using the method Grade of Necessity of Dominance (ND) when the parameter number of trucks is strictly greater than (>) 9, the number of customers is less than or equal to ($\leq$) 120 with no more than 30 customers of type VC and α-cut equal to 0.0 or 0.1. Another conclusion is that the methods First Index of Yager (Y_1) and Second Index of Yager (Y_2) are not adequate to solve any of the instances. Also, it is important to note that the most important parameters to decide the best method are number of trucks, number of customers, number of VC and α-cuts.

4 Summary

In this paper we introduce a decision model useful for users as it allows to define strategies for selecting comparison methods in solving the TTRP with fuzzy demands and capacities. This knowledge can be generalized into a learning mechanism to determine which methods to use depending on the characteristics of a problem. Furthermore, the model can be improved if new TTRP problems are incorporated or comparison methods.

Acknowledgments. This work was supported by the projects TIN2014-55024-P from the Spanish Ministry of Economy and Competitiveness, and P11-TIC-8001 from the Andalusian Government (including FEDER funds).

References

1. Baykasoglu, A., Tolunay, G.: A review and classification of fuzzy mathematical programs. J. Intell. Fuzzy Syst. **19**(3), 205–229 (2008)
2. Tanaka, H., Ichihashi, H., Asai, K.: A formulation of fuzzy linear programming problems based on comparison of fuzzy numbers. Control Cybernet. **13**, 185–194 (1984)
3. Delgado, M., Verdegay, J.L., Vila, M.A.: A general model for fuzzy linear programming. Fuzzy Sets Syst. **29**(1), 21–29 (1989)
4. Wang, X., Kerre, E.: Fuzzy Logic Foundations and Industrial Applications. Part I, Springer, US, International Series in Intelligent Technologies, cap. On the Classification and the Dependencies of the Ordering Methods Advances in Intelligent Systems Research, vol. 8, pp. 73–90 (1996)
5. Yager, R.R.: Ranking fuzzy subsets over the unit interval. In: Proceeding 1978 CDC, pp. 1435–1437. IEEE Conference, New York (1978)
6. Adamo, J.M.: Fuzzy decision trees. Fuzzy Sets Syst. **4**, 207–219 (1980)
7. Yager, R.R.: On choosing between fuzzy subsets. Kybernetes **9**, 151–154 (1980)
8. Yager, R.R.: A procedure for ordering fuzzy subsets of the unit interval. Inf. Sci. **24**, 143–161 (1981)
9. Chang, W.: Ranking of fuzzy utilities with triangular membership functions. In: Proceeding of International Conference on Policy Analysis and Systems, pp. 263–272 (1981)
10. Jain, R.: A procedure for multiple-aspect decision making using fuzzy set. Int. J. Syst. Sci. **8**(1), 1–7 (1977)
11. Chen, S.: Ranking fuzzy numbers with maximizing set and minimizing set. Fuzzy Sets Syst. **17**, 113–129 (1985)
12. Kim, K., Park, K.S.: Ranking fuzzy numbers with index of optimism. Fuzzy Sets Syst. **35**(2), 143–150 (1990)
13. Dubois, D., Prade, H.: Ranking fuzzy numbers in the setting of possibility theory. Inf. Sci. **30**(3), 183–224 (1983)
14. Nakahara, Y.: User oriented ranking criteria and its application to fuzzy mathematical programming problems. Fuzzy Sets Syst. **94**, 275–286 (1998)
15. Brunelli, M., Mezeib, J.: How different are ranking methods for fuzzy numbers? A numerical study. Int. J. Approximate Reasoning **54**(4), 627–639 (2013)
16. Chao, I.-M.: A tabu search method for the truck and trailer routing problem. Comput. Oper. Res. **29**(1), 33–51 (2002)
17. Garey, M., Johnson, D.: Computers and Intractability: A Guide to the Theory of NP-completeness. W. H. Freeman, San Francisco (1979)
18. Drexl, M.: A Branch-and-price algorithm for the truck-and-trailer routing problem. Technical report, Deutsche Post Endowed Chair of Optimization of Distribution Networks (2006)

19. Scheuerer, S.: A tabu search heuristic for the truck and trailer routing problem. Comput. Oper. Res. **33**(4), 894–909 (2006)
20. Lin, S.-W., Yu, V.F., Chou, S.Y.: Solving the truck and trailer routing problem based on a simulated annealing heuristic. Computers & Operation Research. **36**(5), 1683–1692 (2009)
21. Villegas, J.G., Prins, C., Prodhon, C., Medaglia, A.L., Velasco, N.: A grasp with evolutionary path relinking for the truck and trailer routing problem. Comput. Oper. Res. **38**(9), 1319–1334 (2011)
22. Derigs, U., Pullmann, M., Vogel, U.: Truck and trailer routing - problems, heuristics and computational experience. Comput. Oper. Res. **40**(2), 536–546 (2013)
23. Mirmohammadsadeghi, S., Ahmed, S., Nadirah, E.: Application of memetic algorithm to solve truck and trailer routing problems. In: Proceedings of the 2014 International Conference on Industrial Engineering and Operations Management, Bali, pp. 747–755 (2014)
24. Villegas, J.G., Prins, C., Prodhon, C., Medaglia, A.L., Velasco, N.: A matheuristic for the truck and trailer routing problem. Eur. J. Oper. Res. **230**(2), 231–244 (2013)
25. Torres, I., Rosete, A., Cruz, C., Verdegay, J.: Fuzzy constraints in the truck and trailer routing problem. In: Proceedings of the Fourth International Workshop on Knowledge Discovery, Management and Decision Support (EUREKA-2013), Mazatln, pp. 71–78 (2013)
26. Torres, I., Cruz, C., Verdegay, J.: Solving the truck and trailer routing problem with fuzzy constraints. Int. J. Comput. Intell. Syst. **8**(4), 713–724 (2015)
27. Torres, I., Rosete, A., Cruz, C., Verdegay, J.: Truck and Trailer Routing Problem under fuzzy environment. In: Proceedings of the 16th World Congress of the International Fuzzy Systems Association (IFSA) and 9th Conference of the European Society for Fuzzy Logic and Technology (EUSFLAT), Gijón, pp. 1189–1194 (2015)
28. Fajardo, J., Masegosa, A., Pelta, D.: Algorithm portfolio based scheme for dynamic optimization problems. Int. J. Comput. Intell. Syst. **8**(4), 667–689 (2015)
29. Friedman, M.: A comparison of alternative tests of significance for the problem of m rankings. Ann. Math. Stat. **11**(40), 86–92 (1940)
30. Shaffer, J.: Modified sequentially rejective multiple test procedures. J. Am. Stat. Assoc. **81**(395), 826–831 (1986)
31. Holm, S.: A simple sequentially rejective multiple test procedure. Scand. J. Stat. **6**(2), 65–70 (1979)

Preferences on Gambles Representable by a Choquet Expected Value with Respect to Conditional Belief and Plausibility Functions

Letizia Caldari[1], Giulianella Coletti[1(✉)], Davide Petturiti[1], and Barbara Vantaggi[2]

[1] Dip. Matematica e Informatica, University of Perugia, Perugia, Italy
letizia.caldari@studenti.unipg.it,
{giulianella.coletti,davide.petturiti}@unipg.it
[2] Dip. S.B.A.I., "La Sapienza", University of Rome, Roma, Italy
barbara.vantaggi@sbai.uniroma1.it

Abstract. We deal with preference relations on finite sets of (conditional) gambles, providing necessary and sufficient conditions for their representability by a (conditional) Choquet expected value with respect to a (conditional) belief function or a (conditional) plausibility function.

Keywords: Choquet expected value · Conditional belief function · Conditional plausibility function · Preference relation · Representation

1 Introduction

The aim of the paper is to provide axioms assuring the representability of a preference relation on an arbitrary finite set of (conditional) gambles through a specific functional involving a (conditional) non-additive measure of uncertainty. The problem of dealing with preferences on a finite set of "objects" has been studied in [5,6,8,9], in the context of decisions under risk: in [9] these "objects" are lotteries, i.e., random quantities equipped with a probability distribution and the decision model of reference is the expected utility, while in [5,6,8] such "objects" are generalized lotteries [21], i.e., random quantities equipped with a belief function or a convex capacity and the decision model of reference is the Choquet expected utility [2,19,20].

In this paper we consider, in the setting of decisions under uncertainty [12,24,28], situations in which a preference relation $\precsim$ is given on a finite set $\mathcal{F}$ of gambles and the decision model of reference is the Choquet expected value with respect to a belief or a plausibility function. For this aim we introduce two rationality principles which are necessary and sufficient conditions for the existence of a belief function or a plausibility function φ such that the corresponding Choquet integral represents $\precsim$, i.e., for every $f, g \in \mathcal{F}$,

$$f \precsim g \Longleftrightarrow \oint f \,\mathrm{d}\varphi \leq \oint g \,\mathrm{d}\varphi.$$

J.P. Carvalho et al. (Eds.): IPMU 2016, Part II, CCIS 611, pp. 569–580, 2016.
DOI: 10.1007/978-3-319-40581-0_46

Nevertheless, a decision maker could be either not able or not interested in giving preferences between gambles, but he could only express his preferences under the hypothesis that a particular event happens. In other words, he could not be able to express his preference relation under a generic scenario, but just under specific scenarios, which are taken into account at the same time.

To manage the above situation, the first problem to solve is the choice of the most suited notion of conditioning to which refer among those present in the literature [4,11,13,14,16,21,25]. Among the different proposals, the paper adopts the axiomatic definition of conditional belief and plausibility function given in [10,13], which generalizes the one introduced in [14], allowing conditioning to events of null plausibility (for a discussion about the different axiomatic definitions see [7]). The main reason for adopting this definition is that, in analogy with Savage's probabilistic framework [24], if we consider a single scenario (i.e., a conditioning event H) for all the gambles f's, the conditional preference relation between gambles conditioned to H agrees with the unconditional relation on gambles which are null outside of H. Moreover, since this definition of conditioning satisfies a form of chain rule, it allows an easy computation of the "updated" degrees of belief.

Inside the above conditional framework, we propose two conditional rationality principles, which are necessary and sufficient conditions for the existence of a conditional belief function or a conditional plausibility function whose related Choquet conditional expected values represent all the preferences between the conditional acts $f|H$'s.

2 Preliminaries

Let $S = \{s_1, \ldots, s_n\}$ be a finite set of states of nature and denote by $\wp(S)$ the power set of S, whose elements are interpreted as the events of interest.

We recall that a *belief function* Bel [14,25] on $\wp(S)$ is a function such that $Bel(\emptyset) = 0$, $Bel(S) = 1$ and satisfying the n-monotonicity property for every $n \geq 2$, i.e., for every $A_1, \ldots, A_n \in \wp(S)$,

$$Bel\left(\bigcup_{i=1}^{n} A_i\right) \geq \sum_{\emptyset \neq I \subseteq \{1,\ldots,n\}} (-1)^{|I|+1} Bel\left(\bigcap_{i \in I} A_i\right).$$

The previous property implies the monotonicity of Bel with respect to set inclusion $\subseteq$, hence belief functions are particular *normalized capacities* [15]. The *dual* function Pl defined, for every $A \in \wp(S)$, as $Pl(A) = 1 - Bel(A^c)$, is called *plausibility function*.

A belief function Bel on $\wp(S)$ is completely singled out by its *Möbius inverse* [3], called *basic (probability) assignment* [25], defined for every $A \in \wp(S)$ as

$$m(A) = \sum_{B \subseteq A} (-1)^{|A \setminus B|} Bel(B).$$

Such a function $m : \wp(S) \to [0,1]$ is such that $m(\emptyset) = 0$, $\sum_{A \in \wp(S)} m(A) = 1$, and, for every $A \in \wp(S)$,

$$Bel(A) = \sum_{B \subseteq A} m(B) \quad \text{and} \quad Pl(A) = \sum_{B \cap A \neq \emptyset} m(B).$$

A set A in $\wp(S)$ is a *focal element* for m (and so also for the corresponding Bel) whenever $m(A) > 0$. In particular, a belief function (and so its dual plausibility function) is a probability measure if all its focal elements are singletons.

For a function $f : S \to \mathbb{R}$, if φ is a capacity on $\wp(S)$ and σ is a permutation of $\{1, \ldots, n\}$ such that $f(s_{\sigma(1)}) \leq \ldots \leq f(s_{\sigma(n)})$ (see [15]), the *Choquet integral* of f w.r.t. φ is defined, denoting $E_i^\sigma = \{s_{\sigma(i)}, \ldots, s_{\sigma(n)}\}$ for $i = 1, \ldots, n$ and $E_{n+1}^\sigma = \emptyset$, as

$$\oint f \, \mathrm{d}\varphi = \sum_{i=1}^{n} f(s_{\sigma(i)})(\varphi(E_i^\sigma) - \varphi(E_{i+1}^\sigma)).$$

In particular, when φ reduces, respectively, to a belief function Bel or to a plausibility function Pl, we have

$$\oint f \, \mathrm{d}Bel = \sum_{i=1}^{n} f(s_{\sigma(i)}) \left(\sum_{\{s_{\sigma(i)}\} \subseteq B \subseteq E_i^\sigma} m(B) \right),$$

$$\oint f \, \mathrm{d}Pl = \sum_{i=1}^{n} f(s_{\sigma(i)}) \left(\sum_{\{s_{\sigma(i)}\} \subseteq B \subseteq (E_{i+1}^\sigma)^c} m(B) \right),$$

to which a lower/upper prevision interpretation can be given (see, e.g., [23,27]).

Let $\mathcal{H} \subseteq \wp(S) \setminus \{\emptyset\}$ be an additive class (i.e., a set of events closed under finite unions). A function $Pl : \wp(S) \times \mathcal{H} \to [0,1]$ is a *conditional plausibility function* if it satisfies the following conditions:

(i) $Pl(E|H) = Pl(E \cap H|H)$, for every $E \in \wp(S)$ and $H \in \mathcal{H}$;
(ii) $Pl(\cdot|H)$ is a plausibility function on $\wp(S)$, for every $H \in \mathcal{H}$;
(iii) $Pl(E \cap F|H) = Pl(E|H) \cdot Pl(F|E \cap H)$, for every $E \cap H, H \in \mathcal{H}$ and $E, F \in \wp(S)$.

Moreover, given a conditional plausibility function, the dual *conditional belief function* $Bel(\cdot|\cdot)$ is defined for every event $E|H \in \wp(S) \times \mathcal{H}$ as

$$Bel(E|H) = 1 - Pl(E^c|H).$$

The function $Bel(\cdot|\cdot)$ satisfies the following conditions *(i')–(iii')*, where S_P is the probabilistic t-conorm (i.e., $S_P(x,y) = x + y - xy$, for $x, y \in [0,1]$):

(i') $Bel(E|H) = Bel(E \cap H|H)$, for every $E \in \wp(S)$ and $H \in \mathcal{H}$;
(ii') $Bel(\cdot|H)$ is a belief function on $\wp(S)$, for every $H \in \mathcal{H}$;
(iii') $Bel(E \cup F|H) = S_P(Bel(E|H), Bel(F|E^c \cap H))$, for every $H, E^c \cap H \in \mathcal{H}$ and $E, F \in \wp(S)$.

As follows by the results in [1], every conditional plausibility function $Pl(\cdot|\cdot)$ on $\wp(S) \times \mathcal{H}$ is completely determined by a linearly ordered class of plausibility functions on $\wp(S)$ with disjoint sets of focal elements, which is called *agreeing class* of plausibility functions. In general, if $\mathcal{H} \subset \wp(S) \setminus \{\emptyset\}$ such class is not unique, but uniqueness is obtained in case $\mathcal{H} = \wp(S) \setminus \{\emptyset\}$. Among the agreeing classes giving rise to a $Pl(\cdot|\cdot)$ on $\wp(S) \times \mathcal{H}$ there is a unique agreeing class $\{Pl_0, \dots, Pl_k\}$ of plausibility functions on $\wp(S)$, called *minimal agreeing class,* such that

- $Pl_0(\cdot) = Pl(\cdot|H_0^0)$ with $H_0^0 = \bigcup_{H \in \mathcal{H}} H$;
- for $\alpha > 0$, $Pl_\alpha(\cdot) = Pl(\cdot|H_0^\alpha)$ with $H_0^\alpha = \bigcup\{H \in \mathcal{H} : Pl_\beta(H) = 0, \beta = 0, \dots, \alpha - 1\} \neq \emptyset$.

The class $\{Pl_0, \dots, Pl_k\}$ is such that for every $H \in \mathcal{H}$ there is $\alpha \in \{0, \dots, k\}$ such that $Pl_\alpha(H) > 0$. Moreover, $\{Pl_0, \dots, Pl_k\}$ agrees with the conditional plausibility $Pl(\cdot|\cdot)$ on $\wp(S) \times \mathcal{H}$ in the sense that, for every $E|H \in \wp(S) \times \mathcal{H}$, denoting with α_H the minimum index in $\{0, \dots, k\}$ such that $Pl_{\alpha_H}(H) > 0$, it holds that

$$Pl(E|H) = \frac{Pl_{\alpha_H}(E \cap H)}{Pl_{\alpha_H}(H)}.$$

3 Preferences on Gambles

Throughout this section we consider the following decision theoretic setting:

- $S = \{s_1, \dots, s_n\}$ is a finite set of states of nature;
- $\wp(S)^0 = \wp(S) \setminus \{\emptyset\} = \{A_1, \dots, A_{2^n-1}\}$ is the set of not impossible events;
- $X = \{x_1, \dots, x_m\} \subseteq [0, +\infty)$ is a finite set of outcomes (money payoffs);
- $\mathcal{F} \subseteq X^S$ is a finite set of gambles, where a gamble $f : S \rightarrow X$ is a state-contingent payoff (for instance a financial asset);
- $\precsim$ is a complete binary relation on $\mathcal{F}$, expressing the preferences of the decision maker on the considered gambles, whose asymmetric and symmetric parts are denoted as $\prec$ and $\sim$, respectively.

Recall that $\precsim$ is non-trivial if $\prec$ is not empty.

Usually the decision maker is uncertain about which state will be true: sometimes he/she possesses a measure of uncertainty φ on $\wp(S)$, which can be, e.g., a probability (in this case we have lotteries [29]) or a convex capacity (in this case we have generalized lotteries [5,6,8,21]). In both cases the aim is to find a utility function $u : X \rightarrow \mathbb{R}$ such that a functional $\Phi_{(u,\varphi)}(\cdot)$ represents $\precsim$, that is,

$$f \precsim g \Longleftrightarrow \Phi_{(u,\varphi)}(f) \leq \Phi_{(u,\varphi)}(g).$$

The situation just described connotes a decision problem under risk [17,18]. In this paper, we cope with decision problems under uncertainty where the utility function u is tacitly assumed to be the identity function and the issue is to determine the uncertainty measure φ, which is asked to be either a belief or a plausibility function, where the Choquet integral is the reference decision functional (in analogy with, e.g., [20,26]).

Definition 1. *The* **generalized lower and upper gambles** *corresponding to a gamble $f : S \to X$ are the functions $f^{\mathbf{L}}, f^{\mathbf{U}} : \wp(S)^0 \to X$ defined as*

$$f^{\mathbf{L}}(A_i) = \min_{s \in A_i} f(s), \quad \text{for } i = 1, \dots, 2^n - 1,$$
$$f^{\mathbf{U}}(A_i) = \max_{s \in A_i} f(s), \quad \text{for } i = 1, \dots, 2^n - 1,$$

also denoted as the row vectors

$$f^{\mathbf{L}} = (f^{\mathbf{L}}(A_1), \dots, f^{\mathbf{L}}(A_{2^n-1})), f^{\mathbf{U}} = (f^{\mathbf{U}}(A_1), \dots, f^{\mathbf{U}}(A_{2^n-1})).$$

The following proposition is an immediate consequence of the definition of $\oint f \,\mathrm{d}Bel$ and $\oint f \,\mathrm{d}Pl$ as proved in [20].

Proposition 1. *Let $Bel : \wp(S) \to [0,1]$ be a belief function with associated plausibility function Pl and basic probability assignment m. For every gamble $f : S \to X$, it holds*

$$\oint f \,\mathrm{d}Bel = \sum_{i=1}^{2^n-1} f^{\mathbf{L}}(A_i) m(A_i) \quad \text{and} \quad \oint f \,\mathrm{d}Pl = \sum_{i=1}^{2^n-1} f^{\mathbf{U}}(A_i) m(A_i).$$

Definition 2. *A complete binary relation $\precsim$ on $\mathcal{F}$ is* **Bel-rational** *if it satisfies the following condition:*

(B-R) *for every $f_i, g_i \in \mathcal{F}$ with $f_i \precsim g_i$ and for every $\lambda_i > 0$, for $i = 1, \dots, h$, with $h \in \mathbb{N}$, it holds*

$$\sum_{i=1}^{h} \lambda_i g_i^{\mathbf{L}} \leq \sum_{i=1}^{h} \lambda_i f_i^{\mathbf{L}} \Longrightarrow f_i \sim g_i \quad \text{for } i = 1, \dots, h.$$

Theorem 1. *For a non-trivial complete binary relation $\precsim$ on $\mathcal{F}$ the following statements are equivalent:*

(i) $\precsim$ is Bel-rational (i.e., satisfies **(B-R)***);*
(ii) there exists a belief function $Bel : \wp(S) \to [0,1]$ such that, for every $f, g \in \mathcal{F}$,

$$f \precsim g \Longleftrightarrow \oint f \,\mathrm{d}Bel \leq \oint g \,\mathrm{d}Bel.$$

Proof. Denote $\mathcal{R}^{\prec} = \{(f_i, g_i) \in \mathcal{F}^2 : f_i \prec g_i\}$ and $\mathcal{R}^{\precsim} = \{(f_j, g_j) \in \mathcal{F}^2 : f_j \precsim g_j \text{ and } \neg(f_j \prec g_j)\}$ with $n_1 = \operatorname{card} \mathcal{R}^{\prec}$ and $n_2 = \operatorname{card} \mathcal{R}^{\precsim}$. Note that $\mathcal{R}^{\prec}$ is not empty due to non-triviality of $\precsim$, but it could be $\mathcal{R}^{\precsim} = \emptyset$.

We show that *(ii)* is equivalent to *(i)*. Let $A = [a^i]$ and $B = [b^j]$ be the $(n_1 \times (2^n - 1))$ and $(n_2 \times (2^n - 1))$ real matrices with rows $a^i = g_i^{\mathbf{L}} - f_i^{\mathbf{L}}$, for $i = 1, \dots, n_1$, and $b^j = g_j^{\mathbf{L}} - f_j^{\mathbf{L}}$, for $j = 1, \dots, n_2$. By Proposition 1, condition *(ii)* is equivalent to the existence of a $((2^n - 1) \times 1)$ column vector $\mathbf{w}$ which is a solution of the following system

$$\mathcal{S} : \begin{cases} A\mathbf{w} > \mathbf{0}, \\ B\mathbf{w} \geq \mathbf{0}, \\ \mathbf{w} \geq \mathbf{0}, \\ \mathbf{w} \neq \mathbf{0}. \end{cases}$$

Then setting $m(\emptyset) = 0$ and $m(A_i) = \frac{w_i}{\sum_{j=1}^{2^n-1} w_j}$, for $i = 1, \ldots, 2^n - 1$, we obtain a basic probability assignment on $\wp(S)$ whose corresponding Bel is such that the functional $\oint f \, \mathrm{d}Bel$ defined for every $f \in \mathcal{F}$ represents $\precsim$.

By a well-known alternative theorem (see, e.g., [22]) the solvability of $\mathcal{S}$ is equivalent to the non-solvability of the following system

$$\mathcal{S}' : \begin{cases} \mathbf{y}A + \mathbf{z}B \leq \mathbf{0}, \\ \mathbf{y}, \mathbf{z} \geq \mathbf{0}, \\ \mathbf{y} \neq \mathbf{0}, \end{cases}$$

where $\mathbf{y}$ and $\mathbf{z}$ are, respectively, $(1 \times n_1)$ and $(1 \times n_2)$ unknown row vectors. In turn, the non-solvability of $\mathcal{S}'$ is equivalent to condition **(B-R)**. Indeed, $\mathcal{S}'$ has solution $(\mathbf{y}, \mathbf{z})$ if and only if

$$\sum_{i=1}^{n_1} y_i g_i^{\mathbf{L}} + \sum_{j=1}^{n_2} z_j g_j^{\mathbf{L}} \leq \sum_{i=1}^{n_1} y_i f_i^{\mathbf{L}} + \sum_{j=1}^{n_2} z_j f_j^{\mathbf{L}},$$

with at least an index $i \in \{1, \ldots, n_1\}$ such that $y_i > 0$, for which $f_i \prec g_i$, i.e., if and only if condition **(B-R)** does not hold. □

Definition 3. *A complete binary relation* $\precsim$ *on* $\mathcal{F}$ *is* **Pl-rational** *if it satisfies the following condition:*

(P-R) *for every* $f_i, g_i \in \mathcal{F}$ *with* $f_i \precsim g_i$ *and for every* $\lambda_i > 0$, *for* $i = 1, \ldots, h$, *with* $h \in \mathbb{N}$, *it holds*

$$\sum_{i=1}^{h} \lambda_i g_i^{\mathbf{U}} \leq \sum_{i=1}^{h} \lambda_i f_i^{\mathbf{U}} \Longrightarrow f_i \sim g_i \quad \text{for } i = 1, \ldots, h.$$

Theorem 2. *For a non-trivial complete binary relation* $\precsim$ *on* $\mathcal{F}$ *the following statements are equivalent:*

(i) $\precsim$ *is Pl-rational (i.e., satisfies* **(P-R)***);*
(ii) there exists a plausibility function $Pl : \wp(S) \to [0,1]$ *such that, for every* $f, g \in \mathcal{F}$,

$$f \precsim g \Longleftrightarrow \oint f \, \mathrm{d}Pl \leq \oint g \, \mathrm{d}Pl.$$

Proof. The proof goes along the same line of the proof of Theorem 1 considering the $(n_1 \times (2^n - 1))$ and $(n_2 \times (2^n - 1))$ real matrices $A = [a^i]$ and $B = [b^j]$ with rows $a^i = g_i^{\mathbf{U}} - f_i^{\mathbf{U}}$, for $i = 1, \ldots, n_1$, and $b^j = g_j^{\mathbf{U}} - f_j^{\mathbf{U}}$, for $j = 1, \ldots, n_2$. □

The following example shows an application of conditions **(B-R)** and **(P-R)**.

Example 1. Let $S = \{s_1, s_2, s_3\}$ and $X = \{1, 2, 3\}$ (in millions of €) with the gambles $\mathcal{F} = \{f, g, h\}$ reported below and the complete preference relation $\precsim$ on $\mathcal{F}$ such that $f \sim h \prec g$.

S	s_1	s_2	s_3
f	1	2	2
g	2	3	1
h	1	1	2

$\wp(S)^0$	$\{s_1\}$	$\{s_2\}$	$\{s_3\}$	$\{s_1,s_2\}$	$\{s_1,s_3\}$	$\{s_2,s_3\}$	S
$f^{\mathbf{L}}$	1	2	2	1	1	2	1
$g^{\mathbf{L}}$	2	3	1	2	1	1	1
$h^{\mathbf{L}}$	1	1	2	1	1	1	1
$f^{\mathbf{U}}$	1	2	2	2	2	2	2
$g^{\mathbf{U}}$	2	3	1	3	2	3	3
$h^{\mathbf{U}}$	1	1	2	1	2	2	2

We show that $\precsim$ is Bel-rational. For that, consider a linear combination with weights $\lambda_i \geq 0$ for $i = 1, \ldots, 4$ related to pairs in $\precsim$,

$$\lambda_1 g^{\mathbf{L}} + \lambda_2 g^{\mathbf{L}} + \lambda_3 f^{\mathbf{L}} + \lambda_4 h^{\mathbf{L}} \leq \lambda_1 f^{\mathbf{L}} + \lambda_2 h^{\mathbf{L}} + \lambda_3 h^{\mathbf{L}} + \lambda_4 f^{\mathbf{L}},$$

where reflexive comparisons are omitted since they cancel out. Simple computations show that for no choice of $\lambda_1 > 0$ or $\lambda_2 > 0$ the inequality above can hold. In turn, this implies that 'for every finite subset of comparisons with positive coefficients λ_i's condition **(B-R)** is satisfied.

A *Bel* whose corresponding Choquet expected value on $\mathcal{F}$ represents $\precsim$ can be found solving the system $\mathcal{S}$ in the proof of Theorem 1 with

$$A = \begin{bmatrix} g^{\mathbf{L}} - f^{\mathbf{L}} \\ g^{\mathbf{L}} - h^{\mathbf{L}} \end{bmatrix} \quad \text{and} \quad B = \begin{bmatrix} f^{\mathbf{L}} - h^{\mathbf{L}} \\ h^{\mathbf{L}} - f^{\mathbf{L}} \end{bmatrix},$$

for which a solution is $\mathbf{w} = (0,0,0,2,2,0,0)^T$. The solution $\mathbf{w}$ determines the basic probability assignment m on $\wp(S)$ such that $m(\{s_3\}) = m(\{s_1,s_3\}) = \frac{1}{2}$ and 0 otherwise. Then, the corresponding belief function *Bel* is such that $\oint f \,\mathrm{d}Bel = \oint h \,\mathrm{d}Bel = 1 < \frac{3}{2} = \oint g \,\mathrm{d}Bel$.

With an analogous procedure it is possible to show that $\precsim$ is also Pl-rational, moreover, the basic probability assignment m' on $\wp(S)$ such that $m'(\{s_1,s_3\}) = m'(\{s_2,s_3\}) = \frac{1}{2}$ and 0 otherwise, gives rise to the plausibility function Pl' such that $\oint f \,\mathrm{d}Pl' = \oint h \,\mathrm{d}Pl' = 2 < \frac{5}{2} = \oint g \,\mathrm{d}Pl'$.

4 Preferences on Conditional Gambles

Let $\mathcal{L} \subseteq \wp(S)^0$ be a finite set of possible scenarios and consider the set of conditional gambles $\mathcal{C} = \mathcal{F} \times \mathcal{L}$, denoted as $f|H$'s, together with the family of complete binary relations $\{\precsim_H\}_{H \in \mathcal{L}}$, each one defined on $\mathcal{F} \times \{H\}$, for $H \in \mathcal{L}$. As usual, we denote with $\prec_H$ and $\sim_H$ the asymmetric and symmetric parts of $\precsim_H$, respectively.

Definition 4. *For a gamble $f : S \to X$ and an event $H \in \wp(S)^0$, the H-***cut lower generalized gamble** *is the function $f^{\mathbf{L},H} : \wp(S)^0 \to X$ defined as*

$$f^{\mathbf{L},H}(A_i) = \begin{cases} 0 & \text{if } A_i \subseteq H^c, \\ f^{\mathbf{L}}(A_i \cap H) & \text{otherwise,} \end{cases} \quad \text{for } i = 1, \ldots, 2^n - 1,$$

also denoted as the row vector

$$f^{\mathbf{L},H} = (f^{\mathbf{L},H}(A_1), \ldots, f^{\mathbf{L},H}(A_{2^n-1})).$$

The following proposition gives a characterization of the Choquet integral w.r.t. a conditional plausibility and belief function in terms of the generalized acts introduced in Definitions 1 and 4.

Proposition 2. *Let* $Pl : \wp(S) \times \mathcal{H} \to [0,1]$ *be a conditional plausibility function generated by the minimal agreeing class of plausibility functions* $\{Pl_0, \dots, Pl_k\}$ *on* $\wp(S)$ *with basic probability assignments* $\{m_0, \dots, m_k\}$*, and* $Bel(\cdot|\cdot)$ *the dual conditional belief function. For every gamble* $f : S \to X$ *and every* $H \in \mathcal{H}$ *with* $\alpha_H \in \{0, \dots, k\}$ *the minimum index such that* $Pl_{\alpha_H}(H) > 0$*, it holds*

$$\oint f \,\mathrm{d}Pl(\cdot|H) = \frac{1}{Pl_{\alpha_H}(H)} \sum_{i=1}^{2^n-1} (fI_H)^{\mathbf{U}}(A_i) m_{\alpha_H}(A_i),$$

$$\oint f \,\mathrm{d}Bel(\cdot|H) = \frac{1}{Pl_{\alpha_H}(H)} \sum_{i=1}^{2^n-1} f^{\mathbf{L},H}(A_i) m_{\alpha_H}(A_i).$$

Proof. By the representation of $Pl(\cdot|H)$ and $Bel(\cdot|H)$ through $\{Pl_0, \dots, Pl_k\}$, we have that

$$\begin{aligned}
\oint f \,\mathrm{d}Pl(\cdot|H) &= \frac{1}{Pl_{\alpha_H}(H)} \oint f \,\mathrm{d}Pl_{\alpha_H}(\cdot \cap H) \\
&= \frac{1}{Pl_{\alpha_H}(H)} \oint fI_H \,\mathrm{d}Pl_{\alpha_H}(\cdot) \\
&= \frac{1}{Pl_{\alpha_H}(H)} \sum_{i=1}^{2^n-1} (fI_H)^{\mathbf{U}}(A_i) m_{\alpha_H}(A_i), \\
\oint f \,\mathrm{d}Bel(\cdot|H) &= \frac{1}{Pl_{\alpha_H}(H)} \oint f \,\mathrm{d}[Pl_{\alpha_H}(H) - Pl_{\alpha_H}((\cdot)^c \cap H)] \\
&= \frac{1}{Pl_{\alpha_H}(H)} \sum_{i=1}^{n} f(s_{\sigma(i)})(Pl((E^{\sigma}_{i+1})^c \cap H) - Pl((E^{\sigma}_{i})^c \cap H)) \\
&= \frac{1}{Pl_{\alpha_H}(H)} \sum_{i=1}^{2^n-1} f^{\mathbf{L},H}(A_i) m_{\alpha_H}(A_i).
\end{aligned}$$

□

We first cope with the representation of a family of complete binary relations $\{\precsim_H\}_{H \in \mathcal{L}}$ by means of a Choquet integral w.r.t. a conditional plausibility function.

Definition 5. *A family of complete binary relations* $\{\precsim_H\}_{H \in \mathcal{L}}$ *each one defined on* $\mathcal{F} \times \{H\}$*, for* $H \in \mathcal{L}$*, is* **CPl-rational** *if it satisfies the following condition:*

(CP-R) *for every* $f_j|H_j, g_j|H_j \in \mathcal{C}$ *with* $f_j|H_j \precsim_{H_j} g_j|H_j$ *there exists* $\delta_j \geq 0$ *such that* $\delta_j > 0$ *if and only if* $f_j|H_j \prec_{H_j} g_j|H_j$*, and for every* $f_i|H_i, g_i|H_i \in \mathcal{C}$

with $f_i|H_i \precsim_{H_i} g_i|H_i$ *and for every* $\lambda_i > 0$, *for* $i = 1, \ldots, h$, *with* $h \in \mathbb{N}$, *denoting* $H_0^0 = \bigcup_{i=1}^h H_i$, *it holds*

$$\max_{A_k \cap H_0^0 \neq \emptyset} \left[\sum_{i=1}^h \lambda_i \left((g_i I_{H_i})^{\mathbf{U}} - (f_i I_{H_i})^{\mathbf{U}} - \delta_i I_{H_i}^{\mathbf{U}} \right) \right] \geq 0$$

Theorem 3. *For a family of non-trivial complete binary relations* $\{\precsim_H\}_{H\in\mathcal{L}}$ *each one defined on* $\mathcal{F} \times \{H\}$, *for* $H \in \mathcal{L}$, *the following statements are equivalent:*

(i) $\{\precsim_H\}_{H\in\mathcal{L}}$ *is CPl-rational (i.e., satisfies* **(CP-R)***);*
(ii) there exists a conditional plausibility function $Pl : \wp(S) \times \mathcal{H} \to [0,1]$, *where* $\mathcal{H}$ *is the additive class obtained closing* $\mathcal{L}$ *w.r.t. finite unions, such that, for every* $f|H, g|H \in \mathcal{C}$,

$$f|H \precsim_H g|H \Longleftrightarrow \oint f \,\mathrm{d}Pl(\cdot|H) \leq \oint g \,\mathrm{d}Pl(\cdot|H).$$

Proof. Denote $\mathcal{R}_0^{\precsim_H} = \{(f_i|H_i, g_i|H_i) \in \mathcal{C}^2 : f_i|H_i \precsim_{H_i} g_i|H_i\}$, with $n_0 = \operatorname{card} \mathcal{R}_0^{\precsim_H}$.

We prove that *(ii)* is equivalent to *(i)*. Define $H_0^0 = \bigcup_{H\in\mathcal{L}} H$ and let $\mathcal{H}$ be the additive class generated by $\mathcal{L}$. By Proposition 1 and the bijection between conditional plausibility functions on $\wp(S) \times \mathcal{H}$ and minimal agreeing classes, condition *(ii)* is equivalent to the existence of a minimal agreeing class $\{Pl_0, \ldots, Pl_k\}$ of plausibility functions on $\wp(S)$ generating a conditional plausibility $Pl(\cdot|\cdot)$ on $\wp(S) \times \mathcal{H}$, such that $\oint f \,\mathrm{d}Pl(\cdot|H)$ for every $f|H \in \mathcal{C}$ represents $\precsim_H$.

In turn, this is equivalent to the compatibility of the following sequence of systems $\mathcal{S}_0, \ldots, \mathcal{S}_k$, where for every $f_i|H_i, g_i|H_i \in \mathcal{C}$ with $f_i|H_i \precsim_{H_i} g_i|H_i$ we have a parameter $\delta_i \geq 0$ such that $\delta_i > 0$ if and only if $f_i|H_i \prec_{H_i} g_i|H_i$. The first system has the form

$$\mathcal{S}_0 : \begin{cases} A^0\mathbf{w}^0 > \mathbf{0}, \\ B^0\mathbf{w}^0 \geq \mathbf{0}, \\ \mathbf{w}^0 \geq \mathbf{0}, \\ \mathbf{w}^0 \neq \mathbf{0}, \end{cases}$$

with $A^0 = [I_{H_0^0}^{\mathbf{U}}]$ and $B^0 = [b^i]$ the $(1 \times (2^n - 1))$ and $(n_0 \times (2^n - 1))$ real matrices with rows $b^i = (g_i I_{H_i})^{\mathbf{U}} - (f_i I_{H_i})^{\mathbf{U}} - \delta_i I_{H_i}^{\mathbf{U}}$, for $i = 1, \ldots, n_0$, and $\mathbf{w}^0$ a $((2^n - 1) \times 1)$ unknown column vector.

For $\alpha > 0$, define $H_0^\alpha = \bigcup \left\{ H \in \mathcal{L} : \sum_{A_i \cap H \neq \emptyset} w_i^\beta = 0, \beta = 0, \ldots, \alpha - 1 \right\}$, $\mathcal{R}_\alpha^{\precsim_H} = \left\{ (f_j|H_j, g_j|H_j) \in \mathcal{R}_{\alpha-1}^{\precsim_H} : \sum_{A_i \cap H_j \neq \emptyset} w_i^\beta = 0, \beta = 0, \ldots, \alpha - 1 \right\}$, and let $n_\alpha = \operatorname{card} \mathcal{R}_\alpha^{\precsim}$. The system $\mathcal{S}_\alpha$ has the form

$$\mathcal{S}_\alpha : \begin{cases} A^\alpha\mathbf{w}^\alpha > \mathbf{0}, \\ B^\alpha\mathbf{w}^\alpha \geq \mathbf{0}, \\ \mathbf{w}^\alpha \geq \mathbf{0}, \\ \mathbf{w}^\alpha \neq \mathbf{0}, \end{cases}$$

with $A^\alpha = [I^{\mathbf{U}}_{H_0^\alpha}]$ and $B^\alpha = [b^i]$ the $(1 \times (2^n - 1))$ and $(n_\alpha \times (2^n - 1))$ real matrices with rows and $b^j = (g_j I_{H_j})^{\mathbf{U}} - (f_j I_{H_j})^{\mathbf{U}} - \delta_j I^{\mathbf{U}}_{H_j}$, for $j = 1, \dots, n_\alpha$, and $\mathbf{w}^\alpha$ a $((2^n - 1) \times 1)$ unknown column vector.

Given a sequence of solutions $\mathbf{w}^0, \dots, \mathbf{w}^k$, setting $m_\alpha(\emptyset) = 0$, $m_\alpha(A_i) = 0$ for $A_i \subseteq (H_0^\alpha)^c$, and otherwise $m_\alpha(A_i) = \frac{w_i^\alpha}{\sum_{A_j \cap H_0^\alpha \neq \emptyset} w_j^\alpha}$, for $i = 1, \dots, 2^n - 1$ and $\alpha = 0, \dots, k$, we obtain a class of basic probability assignments $\{m_0, \dots, m_k\}$ on $\wp(S)$ giving rise to the searched minimal agreeing class of plausibility functions $\{Pl_0, \dots, Pl_k\}$ on $\wp(S)$.

For $\alpha = 0, \dots, k$, by the same alternative theorem quoted in the proof of Theorem 1, the solvability of $\mathcal{S}_\alpha$ is equivalent to the non-solvability of the following system

$$\mathcal{S}'_\alpha : \begin{cases} \mathbf{y}^\alpha A^\alpha + \mathbf{z}^\alpha B^\alpha \leq \mathbf{0}, \\ \mathbf{y}^\alpha, \mathbf{z}^\alpha \geq \mathbf{0}, \\ \mathbf{y}^\alpha \neq \mathbf{0}, \end{cases}$$

where $\mathbf{y}^\alpha$ and $\mathbf{z}^\alpha$ are, respectively, (1×1) and $(1 \times n_\alpha)$ unknown row vectors. In turn, the non-solvability of $\mathcal{S}'_\alpha$ is equivalent to condition **(CP-R)**. Indeed, $\mathcal{S}'_\alpha$ has solution $(\mathbf{y}^\alpha, \mathbf{z}^\alpha)$ if and only if

$$y_1^\alpha I^{\mathbf{U}}_{H_0^\alpha} + \sum_{j=1}^{n_\alpha} z_j^\alpha ((g_j I_{H_j})^{\mathbf{U}} - (f_j I_{H_j})^{\mathbf{U}} - \delta_j I^{\mathbf{U}}_{H_j}) \leq \mathbf{0},$$

with $y_1^\alpha > 0$ and $z_j^\alpha \geq 0$, for $j = 1, \dots, n_\alpha$. That is, the solvability of $\mathcal{S}_\alpha$ is equivalent to

$$\max_{A_i \cap H_0^\alpha = \emptyset} \sum_{j=1}^{n_\alpha} z_j^\alpha ((g_j I_{H_j})^{\mathbf{U}} - (f_j I_{H_j})^{\mathbf{U}} - \delta_j I^{\mathbf{U}}_{H_j}) \geq 0,$$

which is equivalent to condition **(CP-R)**. □

We turn then to the representation of a family of complete binary relations $\{\precsim_H\}_{H \in \mathcal{L}}$ by means of a Choquet integral w.r.t. a conditional belief function.

Definition 6. *A family of complete binary relations $\{\precsim_H\}_{H \in \mathcal{L}}$ each one defined on $\mathcal{F} \times \{H\}$, for $H \in \mathcal{L}$, is* **CBel-rational** *if it satisfies the following condition:*

(CB-R) *for every $f_j|H_j, g_j|H_j \in \mathcal{C}$ with $f_j|H_j \precsim_{H_j} g_j|H_j$ there exists $\delta_j \geq 0$ such that $\delta_j > 0$ if and only if $f_j|H_j \prec_{H_j} g_j|H_j$, and for every $f_i|H_i, g_i|H_i \in \mathcal{C}$ with $f_i|H_i \precsim g_i|H_i$ and for every $\lambda_i > 0$, for $i = 1, \dots, h$, with $h \in \mathbb{N}$, denoting $H_0^0 = \bigcup_{i=1}^h H_i$, it holds*

$$\max_{A_k \cap H_0^0 \neq \emptyset} \left[\sum_{i=1}^{h} \lambda_i \left(g_i^{\mathbf{L}, H_i} - f_i^{\mathbf{L}, H_i} - \delta_i I^{\mathbf{U}}_{H_i} \right) \right] \geq 0$$

Theorem 4. *For a family of non-trivial complete binary relations $\{\precsim_H\}_{H \in \mathcal{L}}$ each one defined on $\mathcal{F} \times \{H\}$, for $H \in \mathcal{L}$, the following statements are equivalent:*

(i) $\{\precsim_H\}_{H\in\mathcal{L}}$ *is CBel-rational (i.e., satisfies* **(CB-R)***);*

(ii) there exists a conditional belief function $Bel : \wp(S) \times \mathcal{H} \to [0,1]$*, where* $\mathcal{H}$ *is the additive class obtained closing* $\mathcal{L}$ *w.r.t. finite unions, such that, for every* $f|H, g|H \in \mathcal{C}$*,*

$$f|H \precsim_H g|H \iff \oint f \,\mathrm{d}Bel(\cdot|H) \leq \oint g \,\mathrm{d}Bel(\cdot|H).$$

Proof. The proof goes along the same line of the proof of Theorem 3 considering, for $\alpha = 0,\ldots,k$, the $(1 \times (2^n - 1))$ and $(n_\alpha \times (2^n - 1))$ real matrices $A = [I^{\mathbf{U}}_{H_0^\alpha}]$ and $B = [b^i]$ with rows $b^j = g_j^{\mathbf{L},H_j} - f_j^{\mathbf{L},H_j} - \delta_j I^{\mathbf{U}}_{H_j}$, for $j = 1,\ldots,n_\alpha$. □

Let us stress that all conditions **(B-R)**, **(P-R)**, **(CB-R)** and **(CP-R)** imply that all the considered preference relations are transitive and so are weak orders.

In [28] a rationality condition assuring the representability of a preference relation on conditional gambles (also with different conditioning events) through a conditional expected value has been provided. Limiting to the decision theoretic setting of this paper, the condition in [28] implies both **(CB-R)** and **(CP-R)**.

5 Conclusions

We introduce rationality conditions which are necessary and sufficient for the existence of a (conditional) belief or plausibility function, whose (conditional) Choquet expected value represents a preference relation on a set of (conditional) gambles. The main difference with the existing literature is that we focus on a finite setting, which is the most common situation in real decision problems. An open issue is to extend the present results to the infinite case, but excluding restrictive requirements both on the set of gambles and on the preference relation.

Acknowledgements. Work partially supported by the INdAM-GNAMPA Project 2015 U2015/000418 and by the Italian MIUR PRIN 2010-11 2010FP79LR_003.

References

1. Capotorti, A., Coletti, G., Vantaggi, B.: Standard and nonstandard representability of positive uncertainty orderings. Kybernetika **50**(2), 189–215 (2014)
2. Chateauneuf, A., Cohen, M.: Choquet expected utility model: a new approach to individual behavior under uncertainty and social choice welfare. Fuzzy Meas. and Int. Th. App., pp. 289–314. Physica, Heidelberg (2000)
3. Chateauneuf, A., Jaffray, J.-Y.: Some characterizations of lower probabilities and other monotone capacities through the use of Möbous inversion. Math. Soc. Sci. **17**, 263–283 (1989)
4. Chateauneuf, A., Kast, R., Lapied, A.: Conditioning capacacities and Choquet integrals: the role of comonotony. Th. Dec. **51**, 367–386 (2001)
5. Coletti, G., Petturiti, D., Vantaggi, B.: Choquet expected utility representation of preferences on generalized lotteries. In: Laurent, A., Strauss, O., Bouchon-Meunier, B., Yager, R.R. (eds.) IPMU 2014, Part II. CCIS, vol. 443, pp. 444–453. Springer, Heidelberg (2014)

6. Coletti, G., Petturiti, D., Vantaggi, B.: Rationality principles for preferences on belief functions. Kybernetika **51**(3), 486–507 (2015)
7. Coletti, G., Petturiti, D., Vantaggi, B.: Conditional belief functions as lower envelopes of conditional probabilities in a finite setting. Inf. Sci. **339**, 64–84 (2016)
8. Coletti, G., Petturiti, D., Vantaggi, B.: Decisions under risk and partial knowledge modelling uncertainty and risk aversion. In: Proceedings of the 9th International Symposium on Imprecise Probability: Theories and Applications, pp. 77–86 (2015)
9. Coletti, G., Regoli, G.: How can an expert system help in choosing the optimal decision? Th. Dec. **33**(3), 253–264 (1992)
10. Coletti, G., Scozzafava, R.: From conditional events to conditional measures: a new axiomatic approach Ann. Math. Art. Int. **32**(1–4), 373–392 (2001)
11. Coletti, G., Scozzafava, R.: Toward a general theory of conditional beliefs. Int. J. Int. Sys. **21**, 229–259 (2006)
12. Coletti, G., Vantaggi, B.: Representability of ordinal relations on a set of conditional events. Th. Dec. **60**(2), 137–174 (2006)
13. Coletti, G., Vantaggi, B.: A view on conditional measures through local representability of binary relations. Int. J. Approx. Reas. **47**, 268–283 (2008)
14. Dempster, A.P.: Upper and lower probabilities induced by a multivalued mapping. Ann. Math. Stat. **38**(2), 325–339 (1967)
15. Denneberg, D.: Non-additive Measure and Integral. Theory and Decision Library: Series B, vol. 27. Kluwer Academic, Dordrecht (1994)
16. Fagin, R., Halpern, J.Y.: Uncertainty, belief and probability. Comput. Int. **7**(3), 160–173 (1991)
17. Gajdos, T., Tallon, J.M., Vergnaud, J.C.: Decision making with imprecise probabilistic information. J. Math. Econ. **40**(6), 647–681 (2004)
18. Gajdos, T., Hayashi, T., Tallon, J.M., Vergnaud, J.C.: Attitude toward imprecise information. J. Econ. Th. **140**(1), 27–65 (2008)
19. Gilboa, I., Schmeidler, D.: Maxmin expected utility with non-unique prior. J. Math. Econ. **18**(2), 141–153 (1989)
20. Gilboa, I., Schmeidler, D.: Additive representations of non-additive measures and the Choquet integral. Ann. Op. Res. **52**, 43–65 (1994)
21. Jaffray, J.-Y.: Linear utility theory for belief functions. Op. Res. Let. **8**(2), 107–112 (1989)
22. Mangasarian, O.L.: Nonlinear Programming. Classics in Applied Mathematics, vol. 10. SIAM, Philadelphia (1994)
23. Miranda, E., de Cooman, G., Couso, I.: Lower previsions induced by multi-valued mappings. J. Stat. Plan. and Inf. **133**, 173–197 (2005)
24. Savage, L.: The Foundations of Statistics. Wiley, New York (1954)
25. Shafer, G.: A Mathematical Theory of Evidence. Princeton University Press, Princeton (1976)
26. Schmeidler, D.: Subjective probability and expected utility without additivity. Econometrica **57**(3), 571–587 (1989)
27. Schmeidler, D.: Integral representation without additivity. Proc. Am. Math. Soc. **97**(2), 255–261 (1986)
28. Vantaggi, B.: Incomplete preferences on conditional random variables. Representability by coherent conditional previsions. Math. Soc. Sci. **60**, 104–112 (2010)
29. von Neumann, J., Morgenstern, O.: Theory of Games and Economic Behavior. Princeton University Press, Princeton (1944)

A New Vision of Zadeh's Z-numbers

Sebastia Massanet(✉), Juan Vicente Riera, and Joan Torrens

University of the Balearic Islands, Crta. Valldemossa km 7.5, 07122 Palma, Spain
{s.massanet,jvicente.riera,jts224}@uib.es

Abstract. From their introduction Z-numbers have been deeply studied and many investigations have appeared trying to reduce the inherent complexity in their computation. In this line, this paper presents a new vision of Z-numbers based on discrete fuzzy numbers with support in a finite chain L_n. In this new approach, a Z-number associated with a variable, X, is a pair (A, B) of discrete fuzzy numbers, where A is interpreted as a fuzzy restriction on X, while the estimation of the reliability of A is interpreted as a linguistic valuation based on the discrete fuzzy number B. In this non-probabilistic approach an aggregation method is proposed with the aim of applying it in group decision making problems.

Keywords: Z-numbers · Discrete fuzzy number · Aggregation function · Decision making · Computing with words

1 Introduction

The human brain is characterized by its capability to do many tasks without the necessity to make any specific measurement or calculation. Indeed, it is able to reach precise decisions just from its own perceptions and by using quite inaccurate data. It was from this fact that L. Zadeh introduced the idea of Computing With Words (CWW) [25], as a computation based on *words*, or *perceptions*, or even sentences of the natural language, instead of the traditional computation based on numbers. There are cases where CWW is specially adequate and even necessary because of two reasons. Namely, computing with words becomes a necessity when the available information is too imprecise to justify the use of numbers, and also in situations where there is a tolerance for imprecision which can be exploited to achieve tractability, robustness, low solution cost, and better rapport with reality.

On the other hand, uncertainty is a common factor in a wide range of real-world decision-making problems and the task of handling it properly us a tough task. This uncertainty often comes from the vagueness of meanings that are used by experts in problems where qualitative information is used. This fact has motivated that computation with words have turned into a usual resource in the field of decision making. For this reason, several different linguistic models have been presented in the literature with the purpose of modelling experts' opinions. In [13] it is presented a systematic review process about multi-granular fuzzy linguistic model approaches (FLM) considering six different categories: Traditional

J.P. Carvalho et al. (Eds.): IPMU 2016, Part II, CCIS 611, pp. 581–592, 2016.
DOI: 10.1007/978-3-319-40581-0_47

multi-granular FLM based on fuzzy membership functions [9], Ordinal multi-granular FLM based on a basic Linguistic Term Set [5], Ordinal multi-granular FLM based on 2-tuple FLM [6], Ordinal multi-granular FLM based on hierarchical trees [8], Multi-granular FLM based on qualitative description spaces [21], and Ordinal multi-granular FLM based on discrete fuzzy numbers, or dfn for short [12,16,17].

Following with the previous ideas, that is, the accurate modelling of natural language, Zadeh [26] in 2011, introduced the concept of Z-number as an ordered pair of fuzzy numbers (A, B). Thus, when a Z-number is associated with a real-valued uncertain variable X, the ordered triple (X, A, B) is referred to as a Z-valuation, where the first component A is interpreted not as a value of X, but as a restriction on the values which X can take; and the second one, B is referred to as certainty (sureness, confidence, reliability, probability, possibility...) about the value of A. For instance, the opinion *it is very likely that the investment risk is very low*can be modelled as the Z-valuation (investment risk, very low, very likely). Since then, many researchers have focused their studies on Z-numbers from different aspects (theoretical knowledge or practical applications). In this way, Yager [24] assumes that X is a random variable and considers a particular kind of parametrized distributions (normal and uniform) in order to simplify the computations. Pal et al. [14] propose an algorithm for CWW using Z-numbers. Patel et al. [15] provide an applied model of Z-numbers and implement this model into an expert system shell for CWW. Aliev et al. [1,3] introduce the idea of discrete Z-numbers (Z-numbers whose components are discrete fuzzy numbers) and present an approach to decision making based on Z-information. Kang et al. [10] give a method of converting Z-numbers to classical fuzzy numbers according to the fuzzy expectation of fuzzy set, and finally, we wish to highlight also the monograph [2] that presents a comprehensive and self-contained theory of Z-arithmetic and its applications.

Zadeh [26] pointed out "*Problems involving computation with Z-numbers are easy to state but far from easy to solve*". This complexity has led to the proposal of many approaches in the literature (see [1,14,24]). Moreover, it should be noted that when we consider a Z-valuation (X, A, B) the underlying probability distributions are not known in general or it is necessary to fix them previously to simplify computations. Thus, the modelling of an expert opinion in a decision making problem becomes more rigid, since probability distributions previously determined by the expert system must be used.

Therefore, we propose in this article another vision of Zadeh's Z-numbers in which similarly to [1] we consider Z-information expressed as couples of discrete fuzzy numbers. However, we do not regard the second component from a probabilistic point of view but as a dfn-evaluation [12,16] that represents the sureness or confidence of the first component. This approach not only increases the flexibility of the expert opinions, but it also eases the management and the operations between Z-valuations by using aggregation operators in the set of dfns [4,17–19]. These advantages allow this theory to be considered for decision making problems.

2 Preliminaries

In this section we will present the main concepts related to discrete fuzzy numbers that will be used later.

By a fuzzy subset of $\mathbb{R}$, we mean a function $A : \mathbb{R} \to [0, 1]$. For each fuzzy subset A, let $A^\alpha = \{x \in \mathbb{R} : A(x) \geq \alpha\}$ for any $\alpha \in (0, 1]$ be its α-level set (or α-cut). By $supp(A)$, we mean the support of A, i.e., the set $\{x \in \mathbb{R} : A(x) > 0\}$.

Definition 1 [23]. *A fuzzy subset A of $\mathbb{R}$ with membership mapping $A : \mathbb{R} \to [0, 1]$ is called a* discrete fuzzy number, *or* dfn *for short, if its support is finite, i.e., there exist $x_1, ..., x_n \in \mathbb{R}$ with $x_1 < x_2 < ... < x_n$ such that $supp(A) = \{x_1, ..., x_n\}$, and there are natural numbers s, t with $1 \leq s \leq t \leq n$ such that:*

1. *$A(x_i) = 1$ for all i with $s \leq i \leq t$.* (core)
2. *$A(x_i) \leq A(x_j)$ for all i, j with $1 \leq i \leq j \leq s$.*
3. *$A(x_i) \geq A(x_j)$ for all i, j with $t \leq i \leq j \leq n$.*

In Fig. 1, a graphical representation of a general discrete fuzzy number is displayed.

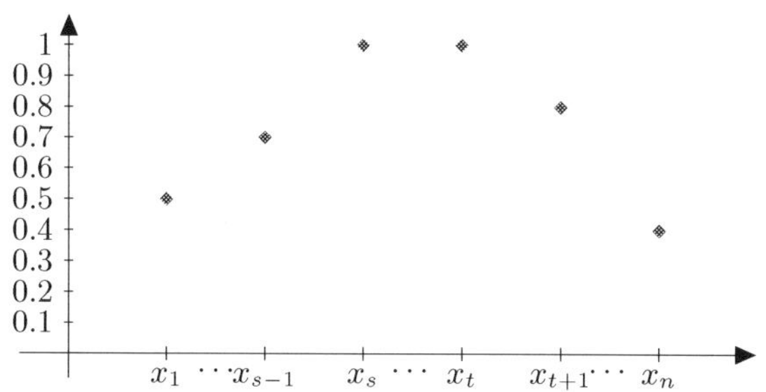

Fig. 1. Graphical representation of a general discrete fuzzy number with support $\{x_1, \ldots, x_n\}$ and core $\{x_s, \ldots, x_t\}$.

From now on, we will denote by L_n the finite chain $L_n = \{0, 1, \ldots, n\}$ and by $\mathcal{A}_1^{L_n}$ the set of discrete fuzzy numbers whose support is a subinterval of the finite chain L_n. Note that in this case, any α-cut is also a subinterval of L_n that will be denoted by A^α. Aggregation functions defined on L_n have been extended to $\mathcal{A}_1^{L_n}$ (see for instance [4,17]) according to the next result.

Theorem 1 [4,17]. *Consider a binary aggregation function F on the finite chain L_n. The binary operation on $\mathcal{A}_1^{L_n}$ defined as follows*

$$\begin{aligned} \mathcal{F} : \mathcal{A}_1^{L_n} \times \mathcal{A}_1^{L_n} &\longrightarrow \mathcal{A}_1^{L_n} \\ (A, B) &\longmapsto \mathcal{F}(A, B) \end{aligned}$$

being $\mathcal{F}(A, B)$ the discrete fuzzy number whose α-cuts are the sets

$$\{z \in L_n \mid F(\min A^\alpha, \min B^\alpha) \leq z \leq F(\max A^\alpha, \max B^\alpha)\}$$

for each $\alpha \in [0, 1]$ is an aggregation function on $\mathcal{A}_1^{L_n}$.

2.1 Linguistic Model Based on Discrete Fuzzy Numbers

In this section we recall the fuzzy linguistic model based on discrete fuzzy numbers in $\mathcal{A}_1^{L_n}$ that was presented in [12].

First of all, note that we can consider a bijective mapping between the ordinal scale $\mathfrak{L} = \{s_0, \ldots, s_n\}$ and the finite chain L_n which keeps the original order. Furthermore, each normal discrete convex fuzzy subset defined on the ordinal scale $\mathfrak{L}$ can be considered like a discrete fuzzy number belonging to $\mathcal{A}_1^{L_n}$, and vice-versa.

For example, consider the linguistic hedge

$$\mathfrak{L} = \{N, VL, L, M, H, VH, T\} \tag{1}$$

where the letters refer to the linguistic terms None, Very Low, Low, Medium, High, Very High and Total and they are listed in an increasing order:

$$N \prec VL \prec L \prec M \prec H \prec VH \prec T$$

and the finite chain L_6. Thus, $A = \{0.5/2, 0.75/3, 1/4, 0.75/5\} \in \mathcal{A}_1^{L_6}$ can be also expressed as $A = \{0.5/L, 0.75/M, 1/H, 0.75/VH\}$. Note that this discrete fuzzy number, A, can be interpreted as a possible flexibilization of the linguistic label High H (see Fig. 2(a)). The previous discrete fuzzy number would be suitable for an expert who hesitates about his opinion. He thinks that the best grade would be the linguistic label H but he cannot discard other grades around it in some degree. On the other hand, the discrete fuzzy number $A = \{1/4\}$, or equivalently $A = \{1/H\}$, would be used by an expert who is completely sure of his opinion. This fact shows that the model based on discrete fuzzy numbers generalizes any linguistic model where the experts' evaluations are limited to choose a single linguistic label.

Furthermore, in [7,16] it was shown that the evaluations based on discrete fuzzy numbers generalize also the concept of Interval-valued evaluations or Hesitant Fuzzy Linguistic Term Sets (HFLTS) (see [20,22] for details). In this way, it is possible to define different *flexibilizations* of a linguistic expression through the following subjective evaluations:

$$\begin{aligned} \text{Between } s_i \text{ and } s_j &= \{A \in \mathcal{A}_1^{L_n} \mid core(A) = [s_i, s_j]\} \\ \text{Worse than } s_i &= \{A \in \mathcal{A}_1^{L_n} \mid core(A) = [s_0, s_{i-1}]\} \\ \text{Better than } s_i &= \{A \in \mathcal{A}_1^{L_n} \mid core(A) = [s_{i+1}, s_n]\} \end{aligned} \tag{2}$$

for all $0 \leq i, j \leq n$. Thus, discrete fuzzy numbers $A \in \mathcal{A}_1^{L_n}$ with $core(A) = [s_i, s_j]$, but with a different support, can be interpreted as flexibilizations of the subjective evaluation "between s_i and s_j", and similarly with the other expressions. For instance, the discrete fuzzy number $B = \{0.5/1, 1/2, 1/3, 1/4, 0.25/5\}$, that can be also expressed as $B = \{0.5/VL, 1/L, 1/M, 1/H, 0.25/VH\}$, is a possible flexibilization of the HFLTS "between Low and High" (see Fig. 2(c)). On the other hand, if in these evaluations the core coincides with the support, we retrieve HFLTS or interval-valued evaluations. For instance, $B = \{1/L, 1/M, 1/H\}$, or

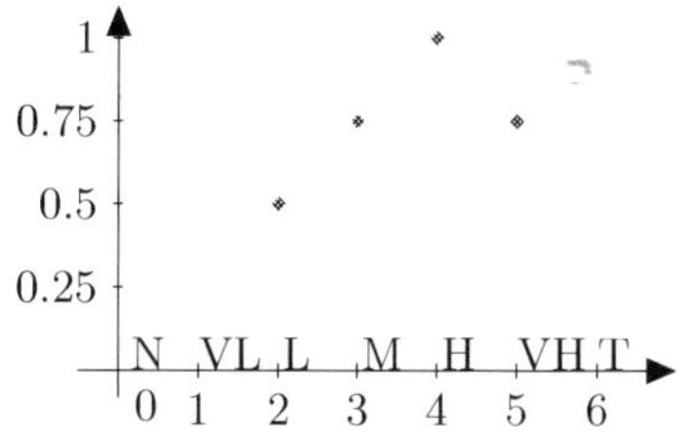

(a) A possible flexibilization of H

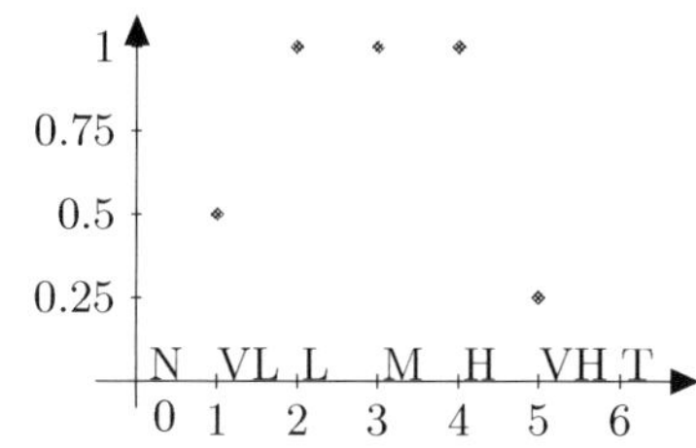

(b) A possible flexibilization of "between Low and High"

Fig. 2. Different types of experts' evaluations using discrete fuzzy numbers.

equivalently $B = \{1/2, 1/3, 1/4\}$, corresponds to the interval-valued evaluation or HFLTS given by $[L, H]$.

From the above discussion, we introduce the following definition.

Definition 2 [12]. *Let $L_n = \{0, \ldots, n\}$ be a finite chain. We call a dfn-evaluation to each discrete fuzzy number A belonging to $\mathcal{A}_1^{L_n}$.*

3 A Review on Zadeh's Z-numbers

In this section we recall the main concepts about this topic and we also analyse some of the most interesting ideas published in this framework.

Definition 3 [26]. *An ordered pair of fuzzy numbers (A, B) is a Z-number. A Z-number is associated with a real-valued uncertain variable, X, with the first component, A, playing the role of fuzzy restriction on X, while the fuzzy number, B is an imprecise estimation of reliability of A. The ordered triple (X, A, B) is referred as a Z-valuation and it is equivalent to an assignment statement, X is (A, B).*

Remark 1. When X is a random variable, X is (A, B) can be interpreted as Prob(X is A) is B where Prob(X is A) is the probability measure of the fuzzy event A in the sense of [26].

Operations with Z-numbers: Let $Z_1 = (A_1, B_1)$ and $Z_2 = (A_2, B_2)$ be Z-numbers describing values of uncertain real-valued variables X_1 and X_2 respectively. The inference rule is represented as follows:

$$\frac{\begin{array}{c} Z_1 \ is \ (A_1, B_1) \\ Z_2 \ is \ (A_2, B_2) \end{array}}{Z_1 * Z_2 \ is \ (A_1 * A_2, B_1 \circ B_2)} \tag{3}$$

where $*$ represents an arithmetical operation and $A_1 * A_2$ is computed according to Zadeh's extension principle and $B_1 \circ B_2$ is computed applying the version of the extension principle which relates to probabilistic restrictions (for more details see [26]).

Note that the complexity of this operation is well known and usually yields a very complex non-linear variational problem (see [1,15]). To avoid or relax this complexity some different perspectives in the field of Z-numbers have been proposed to simplify the operations as well as the computational cost of their implementation. Let us highlight below some of them.

Yager [24] proposes to consider a Z-valuation (X, A, B) in terms of a possibility distribution G over the space P of all probability distributions on X, and the relation between Z-numbers and linguistic summaries. Moreover, he suggests an alternative formulation of Z-information in terms of a Dempster-Shafer belief structures which involves type-2 fuzzy sets. In this approach only typical distributions are considered (exponential and uniform). However, if the second component models the degree of credibility (certainty, security, sureness) about the values on the first component, it is always difficult to think that such certainty can be modelled by a known probability distribution function or a previously fixed one.

In [15] the authors present an applied model of Z-numbers and implement this model into an expert system shell for CWW called CWShell. The main idea of this model is to limit the number of probability distributions deemed appropriate by the domain expert (so-called pool of distributions) and convert them into a discrete form for an easier calculation. Again the number of distribution functions and the number of parameters to be chosen by the experts are reduced in order to simplify the process and the computational cost.

Pal et al. [14] propose an algorithm for CWW and describe simulation experiments of CWW using Z-numbers. Another interesting aspect of this work is an analysis of the strengths and the challenge about this topic suggesting some possible solutions. In this sense, the identification of appropriate fuzzy set models for the perceptions of words in both components are analysed.

Aliev et al. [1] present a new vision of Z-numbers, the discrete Z-numbers, where the two components of a Z-valuation are discrete fuzzy numbers. In this approach basic arithmetic operations and a ranking method are proposed. Furthermore, the authors justify this model on three main aspects. Firstly, the computation with discrete fuzzy numbers and discrete probability distributions has a significantly lower computational complexity than that with continuous fuzzy numbers and density functions. The second consideration is due to the fact that linguistic information is always described by a set of meaningful linguistic term sets which can be represented by ordinal linguistic scales. Finally, in this case it is not necessary to assume a type of distribution constraining the modelling ability, but one can consider a general case.

From the previous analysis we highlight the following:

(i) Z-numbers can be conceptualized as a formidable tool in the design of discourse-oriented decision-making systems, risk assessments, etc.

(ii) It is necessary to find new linguistic models based on fuzzy sets to collect the main ideas established by Zadeh [26] in order to reduce the computational cost of the inference process.

It is clear that this second point is the central idea that generates all the above mentioned papers. In the same direction, we propose in the next section a new interpretation of Z-numbers based on discrete fuzzy numbers in $\mathcal{A}_1^{L_n}$.

4 A New Look of Zadeh's Z-numbers

In the previous section we have seen that Zadeh's original concept can be a very appropriate tool to model the reasoning with words. However this idea presents some problems when we want to compute with Z-valuations. The different proposals previously analysed consider the second component from a probabilistic point of view according to the original idea of Z-numbers. However, in the seminal paper [26], Zadeh also states that the second component, B, is a measure of reliability (certain) of the first component, and closely related to certainty there are many concepts as: sureness, confidence, reliability, probability. That is, B can be interpreted from different points of view.

From this idea we present in this section a new approach to Z-numbers based on discrete fuzzy numbers in $\mathcal{A}_1^{L_n}$, where the second component is also interpreted as a discrete fuzzy number, avoiding in this way the probabilistic aspect considered in the previous approaches. This second component represents the sureness or confidence of the first component. Thus, we present Z-numbers as a couple of discrete fuzzy numbers where each component takes values in a different finite chain. In this sense, each one of these chains will refer to the linguistic terms set used by the experts in their valuations. Formally,

Definition 4. *Let us consider L_n and L_m two finite scales. An ordered pair of discrete fuzzy numbers (A, B) with $A \in \mathcal{A}_1^{L_n}$, $B \in \mathcal{A}_1^{L_m}$ is a (L_n, L_m)-discrete Z-number. An (L_n, L_m)-discrete Z-number is associated with an uncertain variable, X, with the first component, A, playing the role of fuzzy restriction on X, while the discrete fuzzy number, B is an imprecise estimation of reliability of A. The ordered triple (X, A, B) is referred as a Z-valuation and it is equivalent to an assignment statement, X is (A, B).*

Example 1. Unlikely, the investment risk in this country is low can be interpreted as the Z-valuation $Z =$ (investment risk, low, unlikely). For instance if we consider the linguistic term sets

$$S = \{\text{Very High, High, Neutral, Low, Very Low}\},$$
$$S' = \{\text{Impossible, Very Unlikely, Unlikely, Maybe or Maybe Not, Likely, Very Likely, Sure}\},$$

to express the first and second components of Z respectively, the Z-valuation can be expressed by the couple $Z = (A, B)$ where

$$A = \{0.5/H, 0.8/N, 1/L, 0.7/VL\} = \{0.5/1, 0.8/2, 1/3, 0.7/4\} \in \mathcal{A}_1^{L_4},$$
$$B = \{0.3/VU, 1/U, 0.8/M, 0.5/L\} = \{0.3/1, 1/2, 0.8/3, 0.5/4\} \in \mathcal{A}_1^{L_6}.$$

Remark 2. In our approximation we have supposed a linguistic interpretation of each component. However, discrete fuzzy numbers whose support is a finite chain L_n can also be understood from a numerical point of view. Next sentence shows an example of a Z-valuation where the first component is read as a numerical value (*about 30 millions*) but the second one (*not sure*) as a linguistic interpretation:

It is very unlikely that the anticipated budget deficit will be about 30 millions, can be interpreted as the Z-valuation $Z=$ (anticipated budget deficit, about 30 millions, very unlikely).

Now, if $Z_i = (A_i, B_i)$ for $i = 1, \ldots, k$ are (L_n, L_m)-discrete Z-numbers describing values of an uncertain variable X and we want to aggregate them, the inference rule represented as in (3) can be used similarly. Specifically, if $\mathcal{G}_1, \mathcal{G}_2$ are aggregation functions on $\mathcal{A}_1^{L_n}$ and $\mathcal{A}_1^{L_m}$ respectively (obtained according to Theorem 1), we will denote by $\mathbb{G} = (\mathcal{G}_1, \mathcal{G}_2)$ the resulting operation on Z-valuation given by $\mathbb{G}(Z_1, \ldots, Z_k) = (\mathcal{G}_1(A_1, \ldots, A_k), \mathcal{G}_2(B_1, \ldots, B_k))$. Note that the operational complexity pointed out in previous sections is considerably reduced in contrast to the classical approaches.

Example 2. Following with the Z-valuation $Z = (A, B)$ given in Example 1, if we consider another valuation $Z' =$ (Investment risk, High, Likely) showed by the pair $Z' = (C, D)$ where

$$C = \{0.5/0, 1/1, 0.9/2, 0.7/3\} \in \mathcal{A}_1^{L_4}, \quad D = \{0.8/2, 0.9/3, 1/4, 0.5/5\} \in \mathcal{A}_1^{L_6},$$

respectively, these two Z-valuations can be aggregated using for instance the extension of the kernel aggregation function [11] with parameter $k = 3$ in L_4 and L_6, respectively, obtaining

$$\frac{\begin{array}{c} Investment\ Risk\ is\ \ (Low,\ UnLikely) \\ Investment\ Risk\ is\ \ (High,\ Likely) \end{array}}{Z = (High, Unlikely)} \tag{4}$$

where

$$\text{High} = \{0.5/0, 1/1, 0.9/2, 0.7/3\}, \quad \text{Unlikely} = \{0.3/1, 1/2, 0.8/3, 0.5/4\}.$$

From the previous considerations, let us propose now a method of getting a final decision on a specific problem by aggregating the opinions given by some experts when these opinions are expressed through Z-valuations based on (L_n, L_m)-discrete Z-numbers. Suppose that we have k experts that give their opinions on r different variables X_j for $j = 1, \ldots, r$.

1. Each expert chooses the linguistic scales that he will use to make his/her Z-valuations for each variable, that we will denote by (L^1_{ij}, L^2_{ij}) for $1 \leq i \leq k$ and $1 \leq j \leq r$. Let us also denote by (A_{ij}, B_{ij}) the Z-valuation given by expert i with respect to the variable X_j for each $1 \leq i \leq k$ and $1 \leq j \leq r$ (see Table 1).

Table 1. Expert opinions expressed as Z-valuations.

	Variables			
	X_1	X_2	...	X_r
E_1	$(A_{1,1}, B_{1,1})$	$(A_{1,2}, B_{1,2})$	...	$(A_{1,r}, B_{1,r})$
E_2	$(A_{2,1}, B_{2,1})$	$(A_{2,2}, B_{2,2})$	...	$(A_{2,r}, B_{2,r})$
$\vdots$	$\vdots$	$\vdots$	$\vdots$	$\vdots$
E_k	$(A_{k,1}, B_{k,1})$	$(A_{k,2}, B_{k,2})$	...	$(A_{k,r}, B_{k,r})$

Table 2. Valuation for each variable and Final Valuation (FV).

	Variables		
	X_1	...	X_r
E_1	$Z_{11} = (\tilde{A}_{1,1}, \tilde{B}_{1,1})$	...	$Z_{1r} = (\tilde{A}_{1,r}, \tilde{B}_{1,r})$
E_2	$Z_{21} = (\tilde{A}_{2,1}, \tilde{B}_{2,1})$	...	$Z_{2r} = (\tilde{A}_{2,r}, \tilde{B}_{2,r})$
$\vdots$	$\vdots$	$\vdots$	$\vdots$
E_k	$Z_{k1} = (\tilde{A}_{k,1}, \tilde{B}_{k,1})$	...	$Z_{kr} = (\tilde{A}_{k,r}, \tilde{B}_{k,r})$
	$\mathbb{G}_1(Z_{11}, \cdots, Z_{k1})$	...	$\mathbb{G}_r(Z_{1r}, \cdots, Z_{kr})$
FV	$\mathbb{G}(\mathbb{G}_1(Z_{11}, \cdots, Z_{k1}), ..., \mathbb{G}_r(Z_{1r}, \cdots, Z_{kr}))$		

2. In order to reduce to a common linguistic scale the valuations of all experts with respect to all variables, we choose two linguistic scales (L_1, L_2) and we convert all Z-valuations (A_{ij}, B_{ij}) for $1 \leq i \leq k$ and $1 \leq j \leq r$ to Z-valuations $(\tilde{A}_{i,j}, \tilde{B}_{i,j})$ expressed all of them into the common linguistic scale (L_1, L_2), according to the process stated in [12].
3. To obtain a global valuation for each variable X_j with $1 \leq j \leq r$ according to the experts' opinions, we construct the operations $\mathbb{G}_j$, where each $\mathbb{G}_j$ is obtained by $\mathbb{G}_j = (\mathcal{G}_{j1}, \mathcal{G}_{j2})$ with $\mathcal{G}_{j1}, \mathcal{G}_{j2}$ aggregation functions on $\mathcal{A}_1^{L_n}$ and $\mathcal{A}_1^{L_m}$, respectively (obtained from Theorem 1). Finally, the final valuation about the considered problem will be obtained using a (possibly new) operation $\mathbb{G}$ (see Table 2).

Example 3. Let us suppose that a company considers to invest in a foreign country and it decides to hire three experts $E = \{E_1, E_2, E_3\}$ who are specialists in assessing the following variables $\{X_1, X_2\}$, where X_1=*Laboral unrest* and X_2=*Political instability in the medium and long term*, respectively. The company will take a final decision using the global variable $X = \{\text{Invesment risk}\}$, that will be obtained via the aggregation method of X_1 and X_2 explained in this section. To simplify the example we will suppose that all valuations have already been reduced to the linguistic scale (L_1, L_2) given by

$$L_1 = \{N,\ VL,\ L,\ N,\ H,\ VH,\ T\}, \qquad L_2 = \{I,\ VU,\ U,\ MN,\ L,\ VL,\ S\},$$

where items in L_1 stand for *Null, Very Low, Low, Neutral, High, Very High, Total* and items in L_2 stand for the labels used in the scale S' in Example 1.

Thus, all the Z-valuations given by the experts in both components can be interpreted as discrete fuzzy numbers in $\mathcal{A}_1^{L_6}$. Let us suppose that they are given by

$$Z_{11} = (L = \{0.5/1, 1/2, 0.8/3, 0.7/4\}, L = \{0.3/2, 0.6/3, 1/4, 0.8/5\})$$
$$Z_{21} = (VL = \{0.6/0, 1/1, 0.7/2, 0.7/3\}, VL = \{0.4/3, 0.6/4, 1/5, 0.8/6\})$$
$$Z_{31} = (N = \{0.6/1, 0.7/2, 1/3, 0.6/4\}, L = \{0.5/3, 0.7/4, 1/5, 0.6/6\})$$

for the variable X_1 and

$$Z_{12} = (H = \{0.5/2, 0.8/3, 1/4, 0.7/5\}, VL = \{0.8/4, 1/5\})$$
$$Z_{22} = (L = \{0.6/0, 0.7/1, 1/2, 0.6/3\}, L = \{0.4/3, 1/4, 0.7/5, 0.5/6\})$$
$$Z_{32} = (VL = \{0.6/0, 1/1, 0.7/2, 0.6/3\}, L = \{0.5/2, 0.8/3, 1/4, 0.6/5\})$$

for the variable X_2. These valuations can be viewed in Table 3. Now, according to point 3 in the proposed method, let us choose the same aggregation function in all cases. In particular, we will consider $\mathcal{G}$ the extension to $\mathcal{A}_1^{L_6}$ of the kernel aggregation function on L_6 with parameter $k = 3$ (see [11]) and the operations $\mathbb{G}, \mathbb{G}_1$ and $\mathbb{G}_2$ are obtained as $\mathbb{G} = \mathbb{G}_1 = \mathbb{G}_2 = (\mathcal{G}, \mathcal{G})$.

Thus, we get the global evaluations for the variables X_1 and X_2

$$\mathbb{G}_1(Z_{11}, Z_{21}, Z_{31}) = (\{0.6/0, 1/1, 0.7/2, 0.7/3\}, \{0.3/2, 0.6/3, 1/4, 0.8/5\}),$$
$$\mathbb{G}_2(Z_{12}, Z_{22}, A_{32}) = (\{0.6/0, 1/1, 0.7/2, 0.6/3\}\{0.5/2, 0.8/3, 1/4, 0.6/5\}),$$

respectively. Finally, the final valuation is obtained as:

$$\mathbb{G}(\mathbb{G}_1(Z_{11}, Z_{21}, Z_{31}), \mathbb{G}_2(Z_{12}, Z_{22}, A_{32})) =$$
$$(\{0.6/0, 1/1, 0.7/2, 0.6/3\}, \{0.3/2.0.6/3.1/4, 0.8/5\}) = (Very\ Low,\ Likely).$$

Then it is likely that the investment risk is very low.

Table 3. Z-valuations expressed by the experts on $L = (L_1, L_2)$.

	Variables	
	X_1	X_2
E_1	$Z_{11} = (Low, Likely)$	$Z_{12} = (High, Very\ Likely)$
E_2	$Z_{21} = (Very\ Low, Very\ Likely)$	$Z_{22} = (Low, Likely)$
E_3	$Z_{31} = (Neutral, Very\ Likely)$	$Z_{32} = (Very\ Low, Likely)$

5 Conclusions and Future Work

In this paper we have presented a new approach on Z-numbers based on discrete fuzzy numbers with support in a finite chain L_n. In this approach a Z-number associated with a variable, X, is interpreted as a pair (A, B) of discrete fuzzy numbers, where A is considered as a fuzzy restriction on X, while the estimation of the reliability of A is modelled as a linguistic valuation based on the discrete fuzzy number B. Thus, the known arithmetic on discrete fuzzy numbers can be used for avoiding or reducing the natural computational complexity of usual Z-numbers. In addition, an aggregation method of these Z-numbers based on aggregation functions defined on the set of discrete fuzzy numbers $\mathcal{A}_1^{L_n}$ has been proposed. An example illustrating this method has also been included.

As a future work, we want to use this new interpretation of Z-numbers and the proposed aggregation method in the solution of group decision making problems. As it is well known, decision making problems usually consist in two phases: *Aggregation phase of linguistic information* and *Exploitation phase*. The first one could be managed using the aggregation method described in this paper. The second one will require the study of a rank ordering method among the alternatives according to the collective Z-valuation obtained through the aggregation method in order to choose the best alternatives. This will be precisely our next goal in the topic.

Acknowledgments. This paper has been partially supported by the Spanish Grant TIN2013-42795-P.

References

1. Aliev, R., Alizadeh, A., Huseynov, O.: The arithmetic of discrete Z-numbers. Inf. Sci. **290**, 134–155 (2015)
2. Aliev, R.A., Huseynov, O.H., Aliyev, R.R., Alizadeh, A.A.: The Arithmetic of Z-Numbers: Theory and Applications. World Scientific Publishing, Singapore (2015)
3. Aliev, R.A., Mraiziq, D., Huseynov, O.H.: Expected utility based decision making under Z-information and its application. Comput. Intell. Neurosci. (2015)
4. Casasnovas, J., Riera, J.V.: Extension of discrete t-norms and t-conorms to discrete fuzzy numbers. Fuzzy Sets Syst. **167**(1), 65–81 (2011)
5. Herrera, F., Herrera-Viedma, E., Martínez, L.: A fusion approach for managing multi-granularity linguistic term sets in decision making. Fuzzy Sets Syst. **114**, 43–58 (2000)
6. Herrera, F., Martínez, L.: A 2-tuple fuzzy linguistic representation model for computing withwords. IEEE Trans. Fuzzy Syst. **8**(6), 746–752 (2000)
7. Herrera-Viedma, E., Riera, J.V., Massanet, S., Torrens, J.: Some remarks on the fuzzy linguistic model based on discrete fuzzy numbers. In: Angelov, P., et al. (eds.) Intelligent Systems 2014. AISC, vol. 322, pp. 319–330. Springer, Heidelberg (2015)
8. Huynh, V., Nakamori, Y.: A satisfactory-oriented approach to multiexpert decision-making withlinguistic assessments. IEEE Trans. Syst. Man Cybernet. **35**, 184–196 (2005)

9. Jiang, Y., Fan, Z., Ma, J.: A method for group decision making with multigranularity linguistic assessment information. Inf. Sci. **178**, 1098–1109 (2008)
10. Kang, B., Wei, D., Li, Y., Deng, Y.: A method of converting Z-number to classical fuzzy number. J. Inf. Comput. Sci. **9**(3), 202–209 (2012)
11. Mas, M., Monserrat, M., Torrens, J.: Kernel aggregation functions on finite scales. Constructions from their marginals. Fuzzy Sets Syst. **241**, 27–40 (2014)
12. Massanet, S., Riera, J.V., Torrens, J., Herrera-Viedma, E.: A new linguistic computational model based on discrete fuzzy numbers for computing with words. Inf. Sci. **258**, 277–290 (2014)
13. Morente-Molinera, J., Pérez, I., Ureña, M., Herrera-Viedma, E.: On multi-granular fuzzy linguistic modeling in group decision making problems: a systematic review and future trends. Knowl. Based Syst. **74**, 49–60 (2015)
14. Pal, S.K., Banerjee, R., Dutta, S., Sarma, S.: An insight into the Z-number approach to CWW. Fundamentae Informaticae **124**, 197–229 (2013)
15. Patel, P., Khorasani, E.S., Rahimi, S.: Modeling and implimentation of Z-numbers. Soft Comput. **20**(4), 1341–1364 (2016)
16. Riera, J.V., Massanet, S., Herrera-Viedma, E., Torrens, J.: Some interesting properties of the fuzzy linguistic model based on discrete fuzzy numbers to manage hesitant fuzzy linguistic information. Appl. Soft Comput. **36**, 383–391 (2015)
17. Riera, J.V., Torrens, J.: Aggregation of subjective evaluations based on discrete fuzzy numbers. Fuzzy Sets Syst. **191**, 21–40 (2012)
18. Riera, J.V., Torrens, J.: Aggregation functions on the set of discrete fuzzy numbers defined from a pair of discrete aggregations. Fuzzy Sets Syst. **241**, 76–93 (2014)
19. Riera, J.V., Torrens, J.: Using discrete fuzzy numbers in the aggregation of incomplete qualitative information. Fuzzy Sets Syst. **264**, 121–137 (2015)
20. Rodríguez, R., Martínez, L., Herrera, F.: Hesitant fuzzy linguistic term sets for decision making. IEEE Trans. Fuzzy Syst. **20**(1), 109–119 (2012)
21. Roselló, L., Sánchez, M., Agell, N., Prats, F., Mazaira, F.: Using consensus and distances between generalized multi-attributelinguistic assessments for group decision-making. Inf. Fusion **32**, 65–75 (2011)
22. Torra, V.: Hesitant fuzzy sets. Int. J. Intell. Syst. **25**(6), 529–539 (2010)
23. Voxman, W.: Canonical representations of discrete fuzzy numbers. Fuzzy Sets Syst. **118**(3), 457–466 (2001)
24. Yager, R.: On Z-valuations using Zadeh's Z-numbers. Int. J. Intell. Syst. **27**, 259–278 (2012)
25. Zadeh, L.: Fuzzy logic = computing with words. IEEE Trans. Fuzzy Syst. **4**, 103–111 (1996)
26. Zadeh, L.: A note on Z-numbers. Inf. Sci. **9**(1), 43–80 (2011)

Comparison Measures

Comparing Interval-Valued Estimations with Point-Valued Estimations

Hugo Saulnier[1,2(✉)], Olivier Strauss[1], and Ines Couso[2]

[1] LIRMM, 161 Rue Ada, Montpellier, France
{hsaulnier,strauss}@lirmm.fr
[2] Universidad de Oviedo, Oviedo, Spain
couso@uniovi.es

Abstract. In the last decade, numerous proposals have been made to deal with imprecision in estimation problems. Those approaches, many of which involve dealing with interval-valued outputs, deal with the subtle difference between uncertainty and imprecision. One of the crucial points – which to our knowledge has never been addressed – is "how to compare an interval-valued method with a precise valued method?"

The usual way to compare two estimation methods is to use benchmark data with ground truths and to compute a distance between the estimates of each method and the ground truth. However, most of the mathematical available extensions of distances are either biased in favor of a precise approach or in favor of an imprecise approach.

This paper proposes a new tool, the weighted variation of the mid-point distance (WVD), that is more suitable to achieve this kind of comparison, dealing with imprecision with a particular semantic. After reviewing existing distances, we introduce the WVD, first from an intuitive perspective, then from a more mathematical point of view. Its very satisfactory properties are highlighted through an experiment.

Keywords: Interval-valued data · Imprecise probabilities · Engineering

1 Introduction

Scientists willing to consider imprecise data and methods in their analysis (such as [1,2] or [3]) face the problem of comparing methods with interval-valued output with methods with precisely valued outputs. In this paper, we consider imprecise valued regression methods producing interval valued estimates of a precise reference. One of the most usual ways for assessing a preference between one method and another is to use a set of benchmark data to compute a distance between the output of each method and the known ground truth, the current fashion in engineering problems being to rather use the $\mathcal{L}_1$ distance (see [4] for example). However, if comparing two interval-valued methods or two precise valued methods is straightforward, a comparison between an interval-valued method and a precise valued method is more intricate since any existing extension of the $\mathcal{L}_1$ distance is either biased towards or against imprecision: the

J.P. Carvalho et al. (Eds.): IPMU 2016, Part II, CCIS 611, pp. 595–604, 2016.
DOI: 10.1007/978-3-319-40581-0_48

supremum (or Hausdorff[1]) distance tends to disfavor imprecise valued estimates while the *infimum* distance promotes immoderately imprecision.

In order to establish a preference between methods we need to compute the distance between the reference (precise) and the estimate (precise or imprecise). This preference must express as wisely as possible the intricate pros and cons of using an imprecise estimation versus using a precise estimation. Figure 1 illustrates the kind of situation we could fall into. A reference (plain line) is estimated (dotted line) in four different settings.

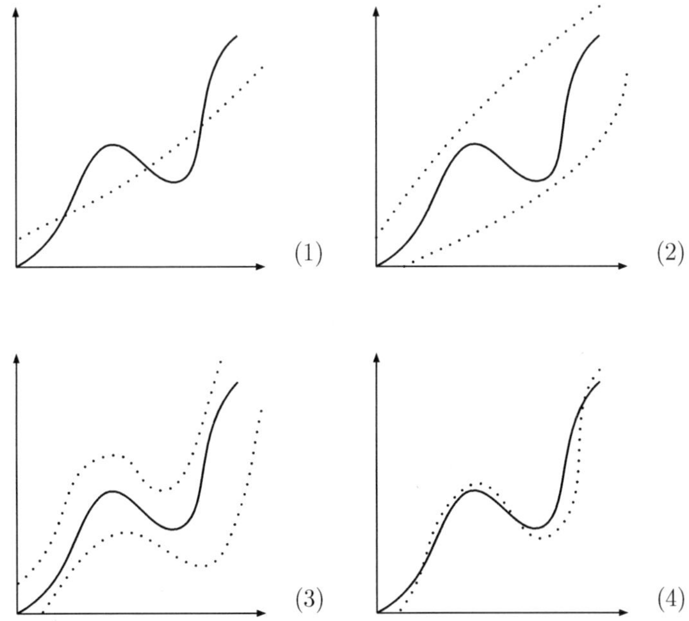

Fig. 1. Precise estimations (1,4) and imprecise estimation (2,3). Plain line is the reference, dotted lines are estimations.

Probably (2) would be preferred to (1). The estimation of (2) is imprecise but informative. It describes the possible variations of the reference. On the other hand the precise estimation of (1) gives an inaccurate description of the data. Again (3) is better than (2), because it has narrower bounds while still being informative. But (4) should be preferred among all, because it is both precise and accurate.

In this paper, we propose a nice candidate to achieve this kind of ordering. The rest of the paper will be organized in three main sections: after reviewing existing metrics, we introduce the WVD, first from an intuitive perspective, then

[1] The Hausdorff distance is also called the Pompeiu-Hausdorff distance. Here we simply refer to it as the Hausdorff distance.

from a more mathematical point of view. Finally some experiments highlight its very nice properties.

2 A Review of Existing Distances Between Points and Intervals

2.1 Notations

Let $\mathbb{R}$ be the real line and $\mathbb{IR}$ be the set of closed intervals of $\mathbb{R}$. X will denote a closed interval of $\mathbb{IR}$, $\underline{x} \in \mathbb{R}$ its lower element, $\overline{x} \in \mathbb{R}$ its upper element, $\tilde{x} = \frac{\underline{x}+\overline{x}}{2}$ its center, and $r = \frac{\overline{x}+\underline{x}}{2}$ its radius.

Here we are interested in defining distances between a finite sequence of points (scalar vector) and a finite sequence of intervals (interval vector). In the following we will consider distances from a reference vector of scalars $\mathbf{s} = (s_1, \ldots, s_N)$ to an indexed collection (vector) of intervals $\mathbf{X} = (X_1, \ldots, X_N)$ where $\forall i \in \{1, \ldots, N\}, X_i = [\underline{x}_i, \overline{x}_i]$. We also denote by $\tilde{\mathbf{x}}$ the vector of the center values of $\mathbf{X}$. The literature includes different alternatives to compute the distance between pairs of intervals of the real line. Based on each of those proposals, one can define the distance from a point $s \in \mathbb{R}$ to an interval $X \in \mathbb{IR}$ as the distance between the singleton $\{s\}$ (which is in turn an element in $\mathbb{IR}$) and X. We can therefore define the distance from $\mathbf{s} = (s_1, \ldots, s_N)$ to $\mathbf{X}$ as the arithmetic mean of the distances between their components. We next review different proposals of distances between pairs of real intervals from the literature, and construct their associated distances from vectors of points to vectors of intervals.

2.2 Hausdorff Distance

Let (U, d) be a complete metric space. Let $\mathcal{K}(U)$ denote the family of non-empty compact subsets of U. The Hausdorff distance between two non-empty compact subsets $A, B \in \mathcal{K}(U)$ is

$$d_H(A, B) = \max\{\sup_{a \in A} \inf_{b \in B} d(a, b), \sup_{b \in B} \inf_{a \in A} d(a, b)\}.$$

In particular, we can define the *Hausdorff distance from* $s \in \mathbb{R}$ *to* $X \in \mathbb{IR}$ as the Hausdorff distance[2] between $\{s\}$ and X, i.e.:

$$d_H(s, I) = \max\{\sup_{x \in I} d(s, x), \inf_{x \in I} d(s, x)\} = \sup_{x \in I} d(s, x), \quad (1)$$

The Hausdorff distance is one the most widely used distances from points to sets [5]. If d is the $\mathcal{L}_1$ distance, Eq. (1) can be simplified in:

$$d_H(s, I) = |s - \tilde{x}| + r, \quad (2)$$

[2] We will use the same notation d_H in order to denote the mapping defined on $\mathbb{R} \times \mathbb{IR}$ derived from $d_H : \mathcal{K}(\mathbb{R}) \times \mathcal{K}(\mathbb{R}) \to \mathbb{R}$. Obviously this new mapping does not satisfy metric properties. It is not even applied to objects of the same kind.

where $\tilde{x}$ and r respectively denote the mid-point and the radius of the closed interval $X = [\tilde{x} - r, \tilde{x} + r]$. Computation of the distance from $\mathbf{s} \in \mathbb{R}^N$ to $\mathbf{X} \in \mathbb{IR}^N$ is obtained by averaging Expression (2) on their components:

$$d_H(\mathbf{s}, \mathbf{X}) = \frac{1}{N} \sum_{i=1}^{N} (|s_i - \tilde{x}_i| + r_i). \tag{3}$$

2.3 Mid-Spread Distance

Expression (2) considers the interval X as a pair composed of its centre $\tilde{x}$ and its radius r like in the mid-spread distance approach [6], where the distance is computed as a linear combination of the distance between the two centers and the distance between the two radii (the radius of s being equal to 0). The mid-spread distance between two intervals $X = [\tilde{x} - r, \tilde{x} + r]$ and $Y = [\tilde{y} - r', \tilde{y} + r']$ is defined as:

$$d_{ms}^{\gamma}(X, Y) = |\tilde{x} - \tilde{y}| + \gamma |r - r'|.$$

According to it, we can define the *mid-spread distance from s to $X = [\tilde{x} + r, \tilde{x} + r]$ as:*

$$d_{ms}^{\gamma}(s, X) = |s - \tilde{x}| + \gamma r.$$

Based on it we can calculate the mid-spread distance from $\mathbf{s}$ to $\mathbf{X}$ as:

$$d_{ms}^{\gamma}(\mathbf{s}, \mathbf{X}) = \frac{1}{N} \sum_{i=1}^{N} (|s_i - \tilde{x}_i| + \gamma r_i), \tag{4}$$

where $\gamma \in \mathbb{R}^+$ is a weight given to the radius w.r.t. the mid. This approach can be seen as a generalization of the Hausdorff distance since $d_H = d_{ms}^1$.

2.4 Mid-Point Distance

The mid-point distance leads to a less technical distance. It is just the $\mathcal{L}_1$ distance between $\mathbf{s}$ and $\tilde{\mathbf{x}}$. Within this approach the intervals are reduced to points and the imprecision is simply ignored. This approach is easy both conceptually and computationally. For this reason it is used a lot in practice [7]. It is defined as:

$$d_m(\mathbf{s}, \mathbf{X}) = \frac{1}{N} \sum_{i=1}^{N} |s_i - \tilde{x}_i|. \tag{5}$$

It can also be considered either as a special case of the mid-spread distance, where $\gamma = 0$ or as a two step Hausdorff distance. In this last interpretation, the method consists of choosing, as a representative single point of the interval $\mathbf{X}$, the precise value that minimizes the Hausdorff distance of the $\mathcal{L}_1$ distance, i.e. its center $\tilde{\mathbf{x}}$, and then compute the usual distance between $\mathbf{s}$ and this representative.

2.5 Infimum Distance

The classical way to define the distance from a point s to a set X in topology is to consider the shortest distance from s to any point in X, i.e.:

$$d_{inf}(s, X) = \inf_{x \in X} d(s, x). \tag{6}$$

When, in particular, d denotes the $\mathcal{L}_1$ distance from a point s to a closed interval $X = [\tilde{x} - r, \tilde{x} + r]$, Expression (7) reduces to:

$$d_{inf}(s, X) = \begin{cases} 0 & \text{if} \quad x \in X \\ |\tilde{x} - s| - r & \text{if} \quad x \notin X \end{cases} \tag{7}$$

Based on this definition, we can compute the distance from $\mathbf{s}$ to $\mathbf{X}$ by averaging expression (7) on the N components:

$$d_{inf}(\mathbf{s}, \mathbf{X}) = \frac{1}{N} \sum_{i=1}^{N} d_{inf}(s_i, X_i). \tag{8}$$

This definition captures the notion of imprecision, but does not have enough separating power. As a long as a point is included in the interval, the distance will be zero. It means that an estimation consisting of intervals of the form $(-\infty, +\infty)$, i.e. a completely vacuous estimation, would always have a distance of zero to any precise estimation, because it contains them all. This distance has exactly the opposite flaw than the Hausdorff distance: the wider the interval, the lower the distance. Thus an interval-valued estimation will always be considered as less distant from a reference value than an equivalent precise-valued estimation: $\forall \mathbf{y} \in \mathbf{X}$, $d(\mathbf{s}, \mathbf{y}) \geq d_{inf}(\mathbf{s}, \mathbf{X})$.

3 Weighted Variation of the Mid-Point Distance

3.1 Definition

A perfect candidate would be an extension which would not penalize imprecision when it conveys information, for example when it reflects coherently the variability of the quantities under consideration. This is in line with the guaranteed approach of [8] and the confidence interval interpretation of imprecision [9].

The weighted variation of the mid-point distance is simply defined as the mean of deviations from the center of the intervals weighted by the inverse of the radius of the interval:

$$d_w(\mathbf{s}, \mathbf{X}) = \frac{1}{N} \sum_{i=1}^{N} \frac{\overline{r}}{r_i} |s_i - \tilde{x}_i|, \tag{9}$$

with

$$\overline{r} = \frac{1}{N}\sum_{i=1}^{N} r_i.$$

It can be seen as a mid-point distance computed on a space where the imprecision is uniform (usually unitary). It is then back-projected to the original space by multiplying the result by the mean radius of the intervals. The weights of the weighted extension are proportional to the inverses of the radii of the intervals. Thus for a certain index, the wider the radius, the lower will be the impact of an important deviation from the center of the interval. It is a very straightforward way of translating the idea that if an interval estimation suitably describes the variability of the reference, then it should be less penalized. Still, having wide radii will be penalized by increasing the mean radius $\overline{r}$.

Obviously the WVD does not satisfy metric properties. The main problem is that if the vector $\mathbf{X}$ contains both intervals of strictly positive radii and degenerated intervals (i.e. singletons) then the WVD we propose is not defined. It thus does not formally stand as an extension of the $\mathcal{L}_1$ distance. However we shall insist that the aim of this tool is to propose a practical solution to the unsolved problem of computing distances between points and intervals. When we restrict its use to cases where $\mathbf{X}$ contains only non-degenerated intervals – which happens most of the time in practical cases – then the WVD fulfills its role in a way that is both very simple computationally and that makes sense from a formal point of view.

3.2 Formal Interpretation

Let us consider a regression problem where $\mathcal{X} : \Omega \rightarrow \mathbb{R}^n$ denotes the vector of attributes and $\mathcal{Y} : \Omega \rightarrow \mathbb{R}$ represents the response variable. Let us consider a sample of size N $((\mathbf{x}_1, y_1), \ldots, (\mathbf{x}_N, y_N))$. Let us consider a regression model $f : \mathbb{R}^N \rightarrow \mathbb{R}$ and let $\hat{y}_i = f(\mathbf{x}_i)$ be the punctual estimation of y_i based on it. The average distance $\overline{d} = \frac{1}{N}|y_i - \hat{y}_i|$ can be used to estimate the degree of goodness of our model, in terms of $\mathcal{L}_1$ loss function. Let now $X_{i,\alpha} = [\hat{y}_i - c_\alpha r_i, \hat{y}_i + c_\alpha r_i]$ represent a prediction interval, with (exact) confidence level $1 - \alpha$, based on $\mathbf{x}_i$ for the value of the response variable of another individual, randomly picked in the subpopulation of those whose vector of attributes coincides with the vector of attributes of the ith individual of our initial sample, $\mathbf{x}_i$. Due to the variability of this subpopulation, such a response value does not necessarily coincide with the observed y_i. The length of prediction intervals in linear regression problems, under the usual normality assumptions, takes this form of $c_\alpha r_i$. The average radius of the N prediction intervals calculated from the sample is $c_\alpha \overline{r}$. Let us now consider another test sample of size N of pairs of the form $(\mathbf{x}_i, y_i')$. An unbiased estimation of $1 - \alpha$ based on this test sample is:

$$1 - \hat{\alpha} = \frac{1}{N}\#\{i|\, |\hat{y}_i - y_i'| \leq c_\alpha r_i\} = \frac{1}{N}\#\{i|\, \frac{|\hat{y}_i - y_i'|}{r_i} \leq c_\alpha\}.$$

Let the random variable $\overline{\mathcal{D}}$ denote the average $\mathcal{L}_1$ distance between the true values y_i' and their estimations $\hat{y}_i$ based on the $\mathbf{x}_i$'s. Let us prove that the WVD from the vector of responses y_i' to the sequence of confidence intervals $X_{i,\alpha}$ is an unbiased estimation of the expectation of $\overline{\mathcal{D}}$. According to above assumptions, for every $j = 1, \ldots, N$, the probability that D_j is less than $c_\alpha \cdot r_j$ is $1-\alpha$. Furthermore, as we have shown, the relative cumulative frequency $\frac{1}{N}\#\{i|\frac{|y_i'-\hat{y}_i|}{r_i} \leq c_\alpha\}$ is an unbiased estimation of $1-\alpha$. In other words, $\hat{F}(c_\alpha) = \frac{1}{N}\#\{i|\frac{|y_i'-\hat{y}_i|}{r_i} \leq c_\alpha\}$ is an unbiased estimation of the cumulative probability $F_{\frac{D_j}{r_j}}(c_\alpha)$. Therefore the relative frequency $\hat{p}(x) = \frac{1}{N}\sum_{i=1}^{N} \mathbb{1}_{\frac{|y_i'-\hat{y}_i|}{r_i}}(x)$ is an unbiased estimation of the probability $P\left(\frac{D_j}{r_j} = x\right) = P(D_j = r_j \cdot x)$. Thus, the expectation of D_j is unbiasedly estimated by $\hat{E}(D_j) = \frac{1}{N}\sum_{i=1}^{N} \frac{|y_i'-\hat{y}_i|}{r_i} r_j$. Therefore, the expectation of $\overline{\mathcal{D}}$, $E(\overline{\mathcal{D}}) = \frac{1}{N}\sum_{j=1}^{N} E(D_j)$ can be unbiasedly estimated by:

$$\hat{E}(\overline{\mathcal{D}}) = \frac{1}{N}\sum_{i=1}^{N} \frac{|y_i' - \hat{y}_i|}{r_i}\overline{r}.$$

4 Experiment

4.1 Experimental Procedure

This section aims at comparing the behavior of the proposed WVD of the $\mathcal{L}_1$ distance to the four other extensions we have presented in Sect. 2. The experiment is based on a data set composed of $K = 3900$ 8×8 subsampled patches extracted from a 2117×3006 high resolution image of the painting *La Joconde* by Leonardo Da Vinci. The reduction factor is set to 5. The subsampling procedures aims at simulating acquisitions of images of this painting using different imagers having the same numerical resolution but different point spread functions.

To do so, we used the so-called imprecise filtering sub-sampling method [10]. It consists of replacing the smoothing anti-aliasing kernel used to transform a high resolution image into a low resolution image by a capacity that represents a convex set of bell-shaped smoothing kernels. Filtering a patch with this method leads to an interval-valued subsampled patch that represent the convex set of all the patches that would have been obtained by subsampling the original patch with all the kernels belonging to the so-defined convex set (here a set of unimodal centered smooth kernels whose support is lower than 8). Practically, each interval-valued patch is composed of an upper patch $\overline{P}_k$ and a lower patch $\underline{P}_k$ (for the kth patch).

4.2 How Consistent Is the Weighted Variation of the Mid-Point Distance?

This experiment aims at illustrating the fact that the extension we propose behaves consistently. Ideally an imprecise patch contains the information of several precise patches. Therefore a distance between imprecise and precise patches which wisely uses the information of imprecision would behave the same way as the $\mathcal{L}_1$ distance would when used to compare precise patches drawn randomly inside the imprecise patches. The comparison is achieved by computing a ratio of imprecise-to-precise $\mathcal{L}_1$ distances on various configuration of patches. If the distribution of this ratio is centered around 1 for a large number of samples, it means that the considered extension is consistent, because it reflects the information conveyed by the imprecise patches in a way that is consistent with how the $\mathcal{L}_1$ distance would behave on a set of precise patches. Moreover, the lower the variance, the more consistent is the extension.

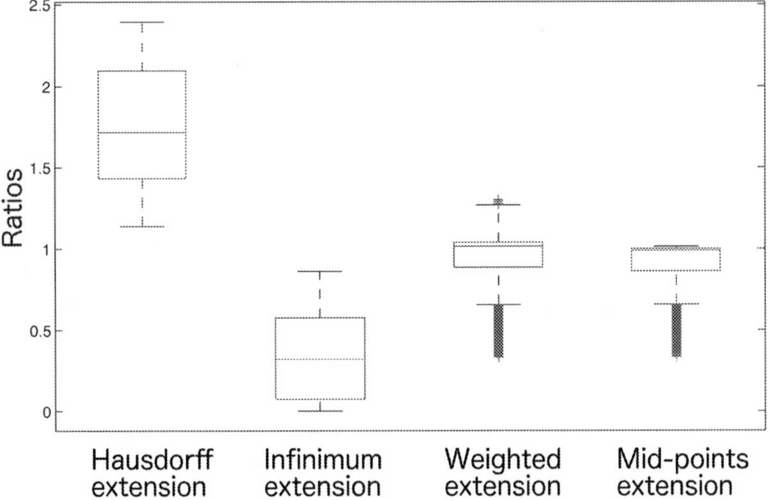

Fig. 2. Box-plot results of the first experiment computed on the four extensions. Weighted extension is WVD.

Table 1. Moments of the ratios.

	Hausdorff	Infinimum	Mid-points	Weighted
mean	1.54	0.503	0.95	0.99
median	1.71	0.32	0.99	1.01
standard deviation	0.36	0.27	0.14	0.16

The experiment is carried as follows. We divide the set of 3900 patches into two subsets of 1950 patches. The motivation for separating our data into two subsets is to compare patches having non-null distances (even if some patches bear some similarities).

We take the first subset ($k = 1, \ldots, 1950$) to define references by considering the mid patches $\tilde{P}_k = \frac{1}{2}(\overline{P}_k + \underline{P}_k)$. We take the second subset ($j = 1951, \ldots, 3900$) to define the imprecise estimations defined by their upper ($\overline{P}_j$) and lower ($\underline{P}_j$) patches. For each imprecise patch of the second subset, we draw 300 precise patches P_j^n ($n = 1, \ldots, 300$) included in the imprecise patch $[\underline{P}_j, \overline{P}_j]$ (300 offers a good tradeoff between a statistically meaningful sampling of the interval and a reasonable computation time). We then compute, for each $(k, j, n) \in [1, 1950] \times [1951, 3900] \times [1, 300]$, a ratio-distance which is defined as:

$$r_e = \frac{d_e(\tilde{P}_k, [\underline{P}_j, \overline{P}_j])}{d_(P_j^n, \tilde{P}_k)}, \qquad (10)$$

where d is the (precise) $\mathcal{L}_1$ distance between two patches and d_e is one of the extensions of the $\mathcal{L}_1$ distance ($e \in \{$Hausdorff, infinimum, mid, weighted $\}$. The distribution of the ratios for the different extensions are presented as a box-plot in Fig. 2. Table 1 shows the mean, median and standard deviation of the ratios distributions for each extension.

As might have been expected, the Hausdorff distance always leads to an over-evaluated distance, while, on contrary, the infimum distance leads to an under-evaluated distance. WVD and mid-point distance seem to provide distances that are consistent in that the means of their ratios are close to 1. For these extensions, comparing the imprecise patch with a precise one is statistically equivalent to comparing this precise patch with any patch contained in the imprecise patch. Although the ratios computed for the WVD have a slightly higher standard-deviation (which can be explained by the divergent behavior it has when some intervals tend to points) a Wilcoxon test applied on all the extensions showed the distribution of ratios of the WVD to be the closest to 1.

5 Conclusion

In this article we have introduced a new mathematical tool, the weighted variation of the mid-point distance, that allows to compare the performances of an interval-valued method with those of a conventional precise-valued method. We have presented its behavior through an experiment, where we compared it with other possible alternatives, namely the Hausdorff, mid-point and infimum distances. The WVD has some problems, such as its divergent behavior when some of the intervals radii tend to zero. However when considering a case of imprecise estimations where the imprecision has to be informative, meaning that the radius of the intervals should reflect the quality of the information provided, the new tool we proposed proved itself to have the best tradeoff between informativeness and consistency with the $\mathcal{L}_1$ distance.

Acknowledgements. This work is partially supported by TIN2014-56967-R (Spanish Ministry of Science and Innovation), FC-15-GRUPIN14-073 (Regional Ministry of the Principality of Asturias) and Merimee exchange program.

References

1. Beer, M., Ferson, S., Kreinovich, V.: Imprecise probabilities in engineering analyses. Mech. Syst. Signal Process. **37**, 4–29 (2013)
2. Aughenbaugh, J.M., Paredis, C.J.: The value of imprecise probabilities in engineering design. In: 2005 ASME Design Engineering Technical Conference and Computers and Information in Engineering Conference, American Society of Mechanical Engineers (ASME) (2005)
3. Oberguggenberger, M., King, J., Schmelzer, B.: Classical and imprecise probability methods for sensitivity analysis in engineering: a case study. Int. J. Approximate Reasoning **50**(4), 680–693 (2009). Imprecise Probability Models and their Applications (Issues in Imprecise Probability)
4. Afonso, M., Sanches, J.M.: Image reconstruction under multiplicative speckle noise using total variation. Neurocomputing **150**, 200–213 (2015)
5. de Souza, R.M., de Carvalho, F.: Clustering of interval data based on city-block distances. Pattern Recognit. Lett. **25**(3), 353–365 (2004)
6. Trutschnig, W., González-Rodrìguez, G., Colubi, A., Gil, M.A.: A new family of metrics for compact, convex (fuzzy) sets based on a generalized concept of mid and spread. Inf. Sci. **179**, 3964–3972 (2009)
7. Augustin, T., Walter, G., Coolen, F.P.A.: Statistical inference, pp. 135–189 (2014)
8. Jaulin, L., Kieffer, M., Didrit, O., Walter, E.: Applied Interval Analysis. Springer, London (2001)
9. Mauris, G.: Possibility distributions: a unified representation of usual direct-probability-based parameter estimation methods. Int. J. Approximate Reasoning **52**, 1232–1242 (2011)
10. Strauss, O.: Non-additive interval-valued f-transform. Fuzzy Sets Syst. **270**, 1–24 (2015)

On Different Ways to be (dis)similar to Elements in a Set. Boolean Analysis and Graded Extension

Henri Prade[1,2(✉)] and Gilles Richard[1]

[1] IRIT, Université Paul Sabatier, 31062 Toulouse Cedex 9, France
{prade,richard}@irit.fr
[2] QCIS, University of Technology, Sydney, Australia

Abstract. We investigate here two questions, first in a Boolean setting and then in a gradual setting: Can we give a formal meaning to "being at odds" (in the sense of being an outlayer) with regard to a subset and, as a dual problem, can we give a meaning to "being even" (in the sense of conforming to a given set of values). Is there a relation between oddness and evenness? Such questions emerge from recent proposals for using oddness or evenness measures in classification problems. This paper is dedicated to a formal study of the oddness and evenness indices in the case of subsets with three or four elements, which are at the basis of the associated measures. Triples are indeed the only subsets such that adding an item that conforms to the triple minority, if any, destroys the majority. It appears that the notions of oddness and evenness are not simple dual of each other; a third notion of being "balanced" interplays with the two others. This is discussed in the setting of squares and hexagons of opposition. The notions of oddness and evenness are related to the study of homogeneous and heterogeneous logical proportions that link four Boolean variables through the conjunction of two equivalences between similarity or dissimilarity indicators pertaining to pairs of these variables. Although elementary, the analysis provides an organized view of new notions that appear to be meaningful when revisiting the old ideas of similarity and dissimilarity in a new perspective. As a side result, it is also mentioned that the logical proportion underlying the idea of being balanced corresponds to the logical encoding of Bongard problems.

Keywords: Similarity · Dissimilarity · Logical proportion

1 Introduction

The similarity, or the dissimilarity of two items is often a matter of degree, when items are described in terms of graded features, or when one counts the number of features having identical values. The idea of dissimilarity is then closely related to the idea of distance [8].

For two items, judging the similarity or dissimilarity between the values of a given Boolean feature reduces to check identity or difference. For three or more

J.P. Carvalho et al. (Eds.): IPMU 2016, Part II, CCIS 611, pp. 605–618, 2016.
DOI: 10.1007/978-3-319-40581-0_49

items, the set of possible situations becomes richer: all values may be identical, all values except one may be identical, values may be balanced, etc.

The idea of classifying an object into a class in such a way that the newcomer is as little as possible an outlayer with respect to the items in the class [7], or such that the class augmented with the newcomer remains as homogeneous as possible [5,6], has motivated the introduction of oddness and evenness measures based on Boolean-valued indices. The oddness index indicates if a new value is at odds (or not) among a multiset of three or more values. The evenness index says if a multiset of three values remains homogeneous (or not), when introducing a fourth value. Oddness and evenness indices constitute the basis for defining oddness and evenness measures of an item with respect to a multiset.

The special interest in subsets of four elements has its roots in the recent study of logical proportions [14], and more particularly of homogeneous ones (which includes analogical proportion) and heterogeneous ones [15]. Logical proportions connect four ordered Boolean variables a, b, c, d through a conjunction of two equivalences between similarity or dissimilarity indicators pertaining respectively to the pairs (a, b) and (c, d). The eight existing homogeneous or heterogeneous logical proportions are the only ones that are independent from the way the features are encoded in terms of what is true and what is false.

The aim of this paper is to provide a logical study of oddness and evenness indices, of their relations, of how they relate to logical proportions, and of how they can be generalized to graded features. Section 2 offers a short background on homogeneous and heterogeneous proportions and then, an informal discussion of the different situations encountered when a new item is added to a singleton, a pair or a triple. This leads to a new reading of logical proportion. Section 3 discusses the logical expressions of oddness and evenness in the Boolean case. Section 4 investigates the logical relation between oddness and evenness indices which does not reduce to a simple binary opposition. This relation, which involves the third notion of being balanced, is discussed in the setting of square and hexagonal opposition. A logical proportion, known as inverse paralogy, encodes balancedness and turns out to be at the basis of the logical encoding of Bongard problems [4] (which are a particular type of intelligence puzzle). Section 5 studies the extension of oddness and evenness indices in the case of graded features.

2 Logical Proportions and Conformity

Considering n Boolean features, a single item is described as an element of $\mathbb{B}^n$. We focus on a particular feature. We consider the process where a new item with feature value x is added to a singleton $\{a\}$, or a pair $\{a, b\}$, or a triple $\{a, b, c\}$ of values of the same feature for another, two other, three other items respectively. In case of triples, what is obtained can be related to the notion of (homogeneous and heterogeneous) logical proportions, that we first briefly recall.

2.1 Logical Proportions: Brief Background

Logical proportions [14,15] connect four Boolean variables through a conjunction of two equivalences between similarity or dissimilarity indicators pertaining respectively to two pairs (a, b) and (c, d). There are two similarity indicators, namely $a \wedge b$ and $\overline{a} \wedge \overline{b}$, and two dissimilarity indicators $\overline{a} \wedge b$ and $a \wedge \overline{b}$. A prototypical logical proportion is the analogical proportion, which expresses that "a is to b as c is to d", or more formally that "a differs from b as c differs from d (and b differs from a as d differs from c)", which is logically expressed as [12] by the quaternary connective (where $\overline{x} = 1$ if $x = 0$ and $\overline{x} = 0$ if $x = 1$):

$$\text{Ana}(a, b, c, d) = ((a \wedge \overline{b}) \equiv (c \wedge \overline{d})) \wedge ((\overline{a} \wedge b) \equiv (\overline{c} \wedge d))$$

Beyond distinctive properties ($\text{Ana}(a, b, a, b) = 1$; $\text{Ana}(a, b, c, d) = \text{Ana}(c, d, a, b)$; $\text{Ana}(a, b, c, d) = \text{Ana}(a, c, b, d)$), Ana also satisfies $\text{Ana}(a, b, c, d) = \text{Ana}(\overline{a}, \overline{b}, \overline{c}, \overline{d})$ which is a remarkable property, namely "code independency", expressing independency with respect to the way the considered feature is encoded in terms of what is true or what is false. It has been established [14] that there exist eight logical proportions satisfying "code independency", which split into four homogeneous logical proportions (they include the analogical proportion) and four heterogeneous ones. The names and expressions of the three other homogeneous logical proportions are given below:

- *reverse analogy* : $\text{Rev}(a, b, c, d) = ((a \wedge \overline{b}) \equiv (\overline{c} \wedge d)) \wedge ((\overline{a} \wedge b) \equiv (c \wedge \overline{d}))$
- *paralogy* : $\text{Par}(a, b, c, d) = ((a \wedge b) \equiv (c \wedge d)) \wedge ((\overline{a} \wedge \overline{b}) \equiv (\overline{c} \wedge \overline{d}))$
- *inverseparalogy* : $\text{Inv}(a, b, c, d) = ((a \wedge b) \equiv (\overline{c} \wedge \overline{d})) \wedge ((\overline{a} \wedge \overline{b}) \equiv (c \wedge d))$

$\text{Rev}(a, b, c, d)$ reverses the analogy into "a is to b as d is to c". $\text{Par}(a, b, c, d)$ expresses that what a and b have in common (positively or negatively), c and d have it also, and conversely; $\text{Inv}(a, b, c, d)$ states that what a and b have in common, c and d do not have it. All logical proportions (not only the code independent ones) have the property to be true for exactly 6 patterns among 2^4 possible ones. In Table 1 below, we give the 6 patterns that make true the four homogeneous logical proportions. The four heterogeneous logical proportions that are code independent have a quite different semantics. They express that there is an intruder among $\{a, b, c, d\}$, which is not a (H_1), which is not b (H_2), which is not c (H_3), and which is not d (H_4) respectively. Their logical expressions are given below:

- $H_1(a, b, c, d) = (\overline{a} \wedge b \equiv \overline{c} \wedge \overline{d}) \wedge (a \wedge \overline{b} \equiv c \wedge d)$
- $H_2(a, b, c, d) = (\overline{a} \wedge b \equiv c \wedge d) \wedge (a \wedge \overline{b} \equiv \overline{c} \wedge \overline{d})$
- $H_3(a, b, c, d) = (\overline{a} \wedge \overline{b} \equiv \overline{c} \wedge d) \wedge (a \wedge b \equiv c \wedge \overline{d})$
- $H_4(a, b, c, d) = (\overline{a} \wedge \overline{b} \equiv c \wedge \overline{d}) \wedge (a \wedge b \equiv \overline{c} \wedge d)$

together to the six patterns that make them true in Table 2.

Table 1. Analogy, Reverse analogy, Paralogy, Inverse Paralogy truth tables

Ana				Rev				Par				Inv			
0	0	0	0	0	0	0	0	0	0	0	0	1	1	0	0
1	1	1	1	1	1	1	1	1	1	1	1	0	0	1	1
0	0	1	1	0	0	1	1	1	0	0	1	1	0	0	1
1	1	0	0	1	1	0	0	0	1	1	0	0	1	1	0
0	1	0	1	0	1	1	0	0	1	0	1	0	1	0	1
1	0	1	0	1	0	0	1	1	0	1	0	1	0	1	0

Table 2. H_1, H_2, H_3, H_4 Boolean truth tables

H_1				H_2				H_3				H_4			
1	1	1	0	1	1	1	0	1	1	1	0	1	1	0	1
0	0	0	1	0	0	0	1	0	0	0	1	0	0	1	0
1	1	0	1	1	1	0	1	1	0	1	1	1	0	1	1
0	0	1	0	0	0	1	0	0	1	0	0	0	1	0	0
1	0	1	1	0	1	1	1	0	1	1	1	0	1	1	1
0	1	0	0	1	0	0	0	1	0	0	0	1	0	0	0

2.2 Conformity Analysis Within 2, 3 and 4-Elements Multisets

Considering a Boolean item, a and another one x to be added to build a pair (a, x), there are only two situations. Either they are identical namely $(a, x) = (1, 1)$ or $(0, 0)$ and $a = x$, or they are different $(a, x) = (1, 0)$ or $(0, 1)$ and $a \neq x$. Now, dealing with two Boolean items a, b, and a third one x to be added to build a triple (a, b, x), there are still two situations. Either they are all identical, i.e. $(a, b, x) = (1, 1, 1)$ or $(0, 0, 0)$ and $a = b = x$, or not, i.e. $(a, b, x) = (1, 1, 0), (1, 0, 1)$, $(1, 0, 0), (0, 1, 1), (0, 1, 0)$ or $(0, 0, 1)$. Since (a, b, x) results from the process of adding x to (a, b), one should make a difference between the multisets $(1, 1, 0), (0, 0, 1)$ where $a = b \neq x$ and the four other multisets $(1, 0, 1), (1, 0, 0)$, $(0, 1, 1), (0, 1, 0)$ where $a \neq b$.

If we now consider three Boolean items a, b, c, and then x is added, the situations become richer. Apart from the situation where all are identical, namely $(a, b, c, x) = (1, 1, 1, 1)$ or $(0, 0, 0, 0)$ and $a = b = c = x$, there are two very distinct situations, the one where truth and false values are equally balanced between a, b, c, x, i.e. $(a, b, c, x) = (1, 1, 0, 0), (0, 0, 1, 1), (1, 0, 1, 0), (0, 1, 0, 1), (1, 0, 0, 1)$, or $(0, 1, 1, 0)$, and the situation where there is an intruder among a, b, c, x, i.e. $(a, b, c, x) = (1, 0, 0, 0)$, $(0, 1, 1, 1)$, $(1, 0, 1, 1)$, $(0, 1, 0, 0)$, $(1, 1, 0, 1)$, $(0, 0, 1, 0)$, $(1, 1, 1, 0)$, or $(0, 0, 0, 1)$. Moreover, as in the previous ternary case, considering that (a, b, c, x) is obtained by the addition of x to (a, b, c), the last set can be split into two groups, namely the last two 4-tuples $(1, 1, 1, 0)$, $(0, 0, 0, 1)$ where x does not respect the preexisting identity $a = b = c$, and the 6 others, where there is an intruder but which is not the latest entrant, which means that x agrees with the majority in $\{a, b, c\}$ (note that majority makes sense as soon as there are three elements a, b, c).

Note that for larger subsets S, having 4 elements only (rather than 3), such as $\{a, b, c, d\}$, it becomes possible that the newcomer x increases the minority (e.g., going from one '0' to two '0'), without changing the majority (e.g., made of three '1'). Indeed, the majority value that may be shared by 3 elements in the 4-elements subset will then remain unchanged in the 5-elements subset $\{a, b, c, d, x\}$ resulting from the arrival of a fifth element whatever its value. A similar phenomenon takes place if we start with larger subsets S having 5 elements or more. So we are losing a distinctive property of 3-elements subsets which have a different behavior depending if x conforms or not to the minority in the 3-elements subset $\{a, b, c\}$ when the triple is not fully homogeneous. Triples are the only subsets such that adding an item that conforms to the triple minority, if any, destroys the majority. Thus, 3-elements subsets are able to clearly discriminate, among different x those that conform to the majority of the triple. This discussion highlights the particular place of 3-elements multisets regarding the idea of conformity. When adding a fourth element, we are back to the pattern of a logical proportion, but with a new reading.

3 Oddness and Evenness Measures

A recent series of papers has reported promising experiments on benchmarks with classifiers based on the simple ideas that a new item should go in the class where this item is the less at odds with the members of the class, or where the class augmented with the new item should remain as even as possible. These classifiers are making use of an oddness measure for an item with respect to a class [7], or of an evenness measure [5,6] of a set of items. The global oddness measure is built, up to a normalization factor, by first adding the values of a Boolean oddness index over the n features describing the items, and then by cumulating the results obtained for all the subsets of items of a given size in the class. Similarly, the global evenness measure of a class is computed by a double cumulation of the values of a Boolean evenness index over features and over subsets with three items (up to normalization). In this section, we restate the expressions of the oddness index and of the evenness index. As we shall see, both of them have expressions that are based on heterogeneous proportions.

3.1 An Oddness Index for Boolean Data

The *oddness* index can be defined from heterogeneous proportions. Let us remember the meaning of H_i: $H_i(a, b, c, d)$ holds if and only if there is an intruder among a, b, c, d and the value in position i is not this intruder. Thus, $H_1(a, b, c, d) = H_2(a, b, c, d) = H_3(a, b, c, d) = 1$ means that there is an intruder which is out of the multiset $\{a, b, c\}$. This has led to the definition of the oddness index of x with respect to $\{a, b, c\}$ as follows:

$$Odd(\{a, b, c\}, x) =_{def} H_1(a, b, c, x) \wedge H_2(a, b, c, x) \wedge H_3(a, b, c, x) \qquad (1)$$

Table 3. H_1, H_2, H_3 and *Odd* truth values

a	b	c	d	H_1	H_2	H_3	Odd
0	0	0	0	0	0	0	0
0	0	0	1	**1**	**1**	**1**	**1**
0	0	1	0	1	1	0	0
0	0	1	1	0	0	0	0
0	1	0	0	1	0	1	0
0	1	0	1	0	0	0	0
0	1	1	0	0	0	0	0
0	1	1	1	0	1	1	0
1	0	0	0	0	1	1	0
1	0	0	1	0	0	0	0
1	0	1	0	0	0	0	0
1	0	1	1	1	0	1	0
1	1	0	0	0	0	0	0
1	1	0	1	1	1	0	0
1	1	1	0	**1**	**1**	**1**	**1**
1	1	1	1	0	0	0	0

Since $H_1(a,b,c,d) \wedge H_2(a,b,c,d) \wedge H_3(a,b,c,d) \wedge H_4(a,b,c,d)$ is a contradiction (as can be seen on the truth tables), we have $Odd(\{a,b,c\},x) \to \neg H_4(a,b,c,x)$. It can be checked that the right hand side of this definition is stable with respect to any permutation of a, b, c which means that the multiset notation on the left hand side is justified. The truth table of *Odd* is given in Table 3. It is clear that $Odd(\{a,b,c\},x)$ holds only when the value of x is seen at odds among the other values, i.e. when x is the intruder. Moreover $Odd(\{a,b,c\},x)$ does not hold in the opposite situation where there is a majority among values in a, b, c, x and x belongs to this majority (e.g., $Odd(\{0,1,0\},0) = 0$), or when there is no majority at all (e.g., $Odd(\{0,1,1\},0) = 0$).

A simple examination of Table 3 shows that $Odd(\{a,b,c\},x)$ is equivalent to

$$Odd(\{a,b,c\},d) \equiv ((a \wedge b \wedge c) \not\equiv d)) \wedge ((a \vee b \vee c) \not\equiv d) \quad (2)$$

More precisely, given a multiset of three identical Boolean values a, b, c, $Odd(\{a,b,c\},x)$ can act as a flag indicating if the 4th value x is different from the common value of a, b, c. Then the value x is at odds with respect to the other values. Moreover, such a definition can be easily generalized to the expression of the oddness of x with respect to any subset S (instead of just $\{a,b,c\}$) by replacing $a \wedge b \wedge c$ (resp. $a \vee b \vee c$) by the conjunction (resp. disjunction) of all elements in S in (2).

3.2 An Evenness Index for Boolean Data

Contrary to the oddness definition, where all $H_i, i = 1, 2, 3$, are involved in defining the measure, only H_4 is useful for defining $Even_4(\{a, b, c, x\})$. Adopting an opposite viewpoint, we want to know if adding a new element to a given subset of items, keeps this subset as homogeneous, as even as it was before (i.e., x does not appear as an intruder in subset $\{a, b, c\}$).

Let us consider three Boolean variables a, b, c. Then, there are two possibilities, either $a = b = c$, or only two of the three are equal. In both cases, a strict majority takes place. Let m denote the majority value.

Now consider the newcomer x, either $x = m$, and m remains the majority value in $\{a, b, c, x\}$, or $x \neq m$, and *there is no longer any majority in* $\{a, b, c, x\}$ (i.e., two values are equal to 1 and two values to 0) if a, b, c are not identical. Only if $x = m$, x conforms to the majority. This idea of majority helps us to define an *evenness* index via an heterogeneous proportion. Since $H_i(a, b, c, d) = 1$ holds if and only if there is an intruder among a, b, c, d and the value in position i is not this intruder, $H_i(a, b, c, d) = 1$ implies that there is a majority of values among (a, b, c, d) and the value in position i conforms to the majority of values appearing among the three other positions (i.e. the multiset of values $\{a, b, c, d\}$ is more or less even). But the reverse implication does not hold since when the four parameters have identical value, $\forall i \in [1, 4], H_i(a, b, c, d) = 0$. In order to have a Boolean definition for "there is a majority of values among the parameters a, b, c, x and the parameter in position i belongs to this majority of values", we need to consider the case where all the values are identical, which leads to:

$$Even_i(\{a, b, c, d\}) =_{def} H_i(a, b, c, d) \vee Eq(a, b, c, d) \tag{3}$$

where $Eq(a, b, c, d) =_{def} (a \equiv b) \wedge (b \equiv c) \wedge (c \equiv d)$. Thus, with $Even_i$ we take into account the special case where all the values are equal. The truth table of $Even_4$ is given in Table 4.

It is clear that $Even_4(\{a, b, c, x\})$ holds only when the value of x belongs to a majority of the feature values. $Even_4(\{a, b, c, x\})$ does not hold in an opposite situation where there is no majority among values as it is the case for $Even_4(\{0, 0, 1, 1\})$ or $Even_4(\{0, 1, 1, 0\})$.

The situations where $Even_4(\{a, b, c, x\}) = 1$ exactly cover the two cases already mentioned where x is identical to the majority value in the triple $\{a, b, c\}$ (x is not the intruder), namely either $a = b = c = x$, or two of the three a, b, c are equal to x. So the fact that x joins the subset $\{a, b, c\}$, when $Even(a, b, c, d) = 1$, leaves the resulting subset at least as *even* as it was, hence the name, and in fact the majority is reinforced by the arrival of x. Note also that $Even_4(a, b, c, d)$ is left unchanged by any permutation of $\{a, b, c\}$. This means that the ordering inside triples does not matter. Besides, $Even_4(a, b, c, d) = 1$ entails $Even_4(\overline{a}, \overline{b}, \overline{c}, \overline{d}) = 1$, expressing that $Even_4(a, b, c, d)$ does not depend on the information encoding: both *Odd* and $Even_4$ are code-independent.

Table 4. H_4, Eq and $Even_4$ truth values

a	b	c	d	H_4	Eq	$Even_4$
0	0	0	0	0	1	1
0	0	0	1	0	0	0
0	0	1	0	1	0	1
0	0	1	1	0	0	0
0	1	0	0	1	0	1
0	1	0	1	0	0	0
0	1	1	0	0	0	0
0	1	1	1	1	0	1
1	0	0	0	1	0	1
1	0	0	1	0	0	0
1	0	1	0	0	0	0
1	0	1	1	1	0	1
1	1	0	0	0	0	0
1	1	0	1	1	0	1
1	1	1	0	0	0	0
1	1	1	1	0	1	1

4 Relation Between Oddness and Evenness Indices

The oddness and evenness Boolean indices have been proposed independently. Despite the fact that their name might suggest that oddness and evenness capture dual concepts, it is not the case that $Even_4(a, b, c, d) \equiv \neg Odd(\{a, b, c\}, d)$, as it can be easily seen on truth tables. However, these two indices exhibit noticeable links that we are going to lay bare in the setting of structures of opposition. Indeed $Odd(\{a, b, c\}, d)$ and $Even(a, b, c, d)$ (from now on we omit the subscript 4) interplay with a third entity whose truth table coincides with the one of inverse paralogy $Inv(a, b, c, d)$, as shown below. The last part of the section shows that $Inv(a, b, c, d)$ makes sense by itself and is the cornerstone of Bongard problems, logically speaking.

4.1 Oddness, Evenness and Balancedness: A Hexagon of Opposition

First, let us notice that Odd and $Even$ are mutually exclusive. It can be easily checked on truth tables that the complement of their disjunction is nothing but the inverse paralogy introduced in Sect. 2, namely

$$\neg Even(a, b, c, d) \wedge \neg Odd(\{a, b, c\}, d) \equiv Inv(a, b, c, d).$$

This reflects the fact that Odd and $Even$ and Inv are mutually exclusive and cover the 16 possible Boolean 4-tuples of values for (a, b, c, d) (i.e., the disjunction $Even(a, b, c, d) \vee Odd(\{a, b, c\}, d) \vee Inv(a, b, c, d)$ is a tautology). Thus, we have

$$Even(a,b,c,d) \equiv \neg Odd(\{a,b,c\},d) \wedge \neg Inv(a,b,c,d)$$
$$Odd(\{a,b,c\},d) \equiv \neg Even(a,b,c,d) \wedge \neg Inv(a,b,c,d)$$

Let us also note that $Even(a,b,c,d) \rightarrow \neg Odd(\{a,b,c\},d)$,

$$\neg Even(a,b,c,d) \rightarrow \neg H_4(a,b,c,d),$$
$$Odd(\{a,b,c\},d) \rightarrow \neg H_4(a,b,c,d).$$

which exhibits the opposite nature of *Odd* and *Even*. Their structure of opposition may be better understood using a hexagon of opposition which generalizes the classical square of opposition as explained now.

The square of opposition is an old graphical device which exhibits three different forms of opposition between pairs of logical statements among four ones (contradiction, not being true together, not being false together). The reader is referred to [1,13] for details. Starting in the middle of the XXth century, a revival of interest has led to the introduction of a hexagon of opposition which involves three squares of opposition [2,3]. It has been noticed that such a structure can be displayed as soon as we have at hand a partition with three elements [9].

In Fig. 1, we display such a hexagon based on the partition made of the three mutually exclusive quaternary connectives whose patterns that make them true are exhibited. We recognize $Eq(a,b,c,d)$ expressing that a,b,c,d are identical, the disjunction $H_1(a,b,c,d) \vee H_2(a,b,c,d) \vee H_3(a,b,c,d) \vee H_4(a,b,c,d)$ which expresses that there is an intruder among a,b,c,d, and the inverse paralogy $Inv(a,b,c,d)$ which corresponds to the six patterns where true and false are equally balanced. In a hexagon of opposition, diagonals feature contradiction via negation. Thus $\neg(H_1(a,b,c,d) \vee H_2(a,b,c,d) \vee H_3(a,b,c,d) \vee H_4(a,b,c,d))$ is nothing but $Ana(a,b,c,d) \vee Rev(a,b,c,d) \vee Par(a,b,c,d) \vee Inv(a,b,c,d))$, known as Klein's operator [14]. Arrows $\longrightarrow\!\!\!\gg$ define logical entailments, the operators that are at vertices where two arrows arrive are the disjunction of the operators associated with the other extremities of the two arrows, while the operators that are at vertices from which two arrows start are the conjunction of the operators at the other extremities of the two arrows.

In Fig. 2 below, we display a hexagon based on the partition made by $Odd(\{a,b,c\},d)$, $Inv(a,b,c,d)$ and $Even(a,b,c,d)$. Diagonals and arrows have the same respective logical meaning as in the previous hexagon, namely contradiction and entailment. While in Fig. 1 the underlying partition was reflecting the number of truth values that are equal in the patterns (4, 3, and 2 respectively), Fig. 2 is based on a partition with a more subtle meaning: either the fourth element is at odds with the three others, or the set $\{a,b,c\}$ remains even after the addition of d, or the addition of d makes $\{a,b,c,d\}$ equally balanced with regard to truth or falsity.

4.2 Inverse Paralogy and Bongard Problems

The inverse paralogy operator $Inv(a,b,c,d)$, which can be viewed as expressing that the addition of d makes $\{a,b,c,d\}$ equally balanced between truth and falsity, has another more static reading expressing that what a and b have in common, c and d do not have it, and conversely as said in subsect. 2.2. This

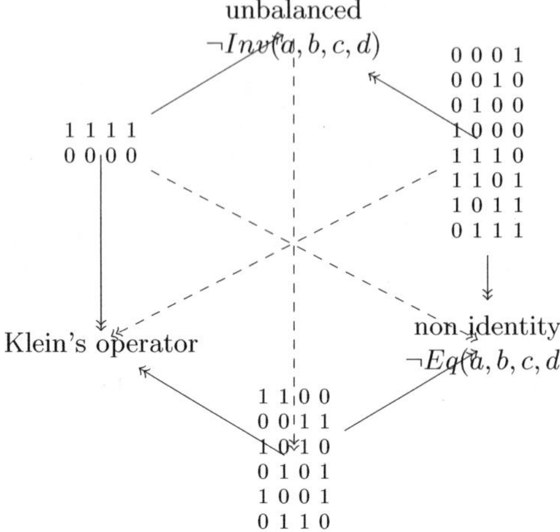

Fig. 1. Hexagon induced by $Eq(a,b,c,d)$, $Inv(a,b,c,d)$ and $\bigvee_i H_i(a,b,c,d)$

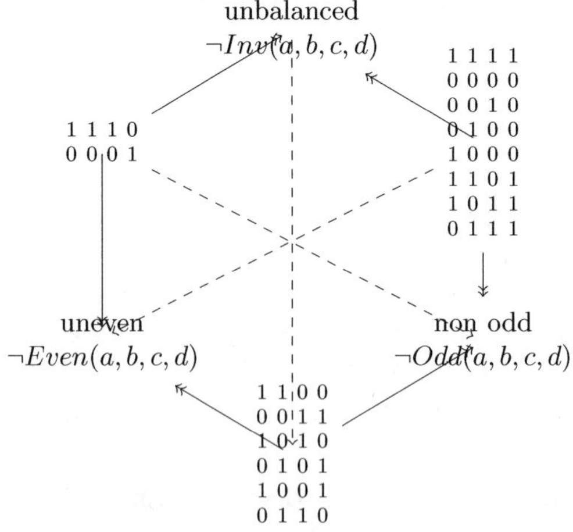

Fig. 2. Hexagon induced by $Odd(\{a,b,c\},d)$, $Inv(a,b,c,d)$ and $Even(a,b,c,d)$

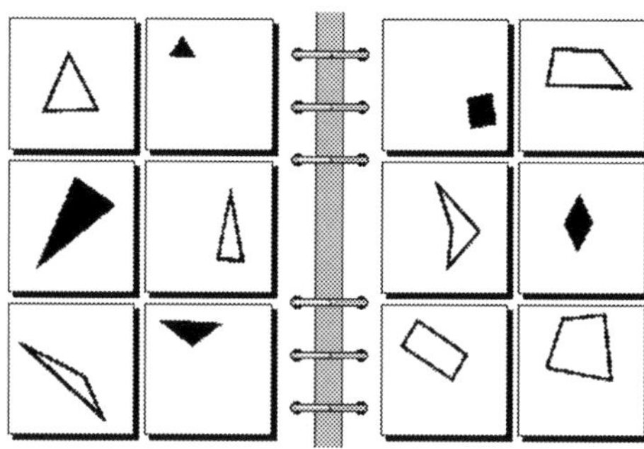

Fig. 3. A Bongard problem

second understanding is the one we should have in mind for understanding and showing that $Inv(a, b, c, d)$ is at work in Bongard problems.

In such visual puzzles, which first appeared in the appendix of a book published by Bongard [4] (see also [10,11]), you are in presence of two sets of simple diagrams A and B. All the diagrams from set A have a common feature, which is lacking in all the diagrams of set B. The problem is to identify this common factor in A. Figure 3 shows an example of such a problem [10], where A (on the left) and B (on the right) are both made of six boxes with simple drawings. As usual with visual problems, a part of the difficulty is to identify the relevant features. Having non relevant features is not a problem, but missing a key feature will forbid you to find the solution. In the example of Fig. 3, features are pretty obvious (being a triangle (t), being a quadrangle (q), being black (b), being white (w), and maybe we may add information about being small, in a corner, including the center of the box (c), having some orientation, and so on. Difficult problems come from the fact that some relevant feature may be difficult to identify (e.g., some cardinality condition).

Assume we use the Boolean features (t, q, b, w, c). Let us denote the left boxes Bl_i and right boxes Br_i numbered from 1 to 6 in the reading order. Then Br_1 is encoded by $(1, 0, 0, 1, 1)$, Br_2 by $(1, 0, 1, 0, 0)$, while Bl_1 is $(0, 1, 1, 0, 0)$, and Bl_2 $(0, 1, 0, 1, 0)$. If we consider two left boxes and two right boxes, we may expect that they satisfy an inverse paralogy relation on a number of features. Table 5 shows the patterns linking the above boxes for five features.

As can be seen, for features t, q, b, w we have patterns that validate an inverse paralogy. In particularly, the first two patterns are characteristic of the inverse paralogy. The first one identifies t (being a triangle) as corresponding to a property that is possessed by Bl_1 and Bl_2 and by none of Br_1 and Br_2. This is a feature which is common to Bl_1 and Bl_2 and lacking in both Br_1 and Br_2. So the procedure is to check it for any two pairs of left boxes and any two pairs of right boxes, we have always this pattern for a common feature (here t is the only candidate). Observe also that feature c introduces some noise, since the corresponding pattern in Table 5 is of the intruder type and is satisfied by

Table 5. Example of partial encoding of the Bongard problem

	Br_1	Br_2	Bl_1	Bl_2
t	1	1	0	0
q	0	0	1	1
b	0	1	1	0
w	1	0	0	1
c	1	0	0	0

H_2, H_3, and H_4, but not by *Inv*. This is unavoidable when it happens in such problem for some "ill-chosen feature".

5 Extension to Numerical Features

We briefly discuss the extension of oddness and evenness indices to graded values. Feature values are now assumed to belong to $[0, 1]$ after suitable normalization.

Oddness A direct translation of formula (2), taking min for $\wedge$, max for $\vee$, and $1 - |\cdot - \cdot|$ for $\equiv$ as in Łukasiewicz logic, leads to:

$$Odd(\{a, b, c\}, d) = \min(|\max(a, b, c) - d|, |\min(a, b, c) - d|) \tag{4}$$

First, it is easy to check that *Odd* remains code independent, changing graded values into their complement to 1. Let us examine some situations to get a precise understanding of the formula for numerical data and to check that this *oddness* measure fits with the intuition. We have

- $Odd(\{u, u, u\}, v) = |u - v|$
  Indeed the larger $|u - v|$, the more v is at odds with regard to $\{u, u, u\}$.
- $Odd(\{v, u, u\}, v) = 0$ which is consistent with the expected semantics of *Odd*.
- $Odd(\{u, v, w\}, \max(u, v, w)) = Odd(\{u, v, w\}, \min(u, v, w)) = 0$, and in any case, $Odd(\{u, v, w\}, u) \leq 0.5$.

This suggests that the proposed definition fits with the initial intuition and provides high truth values when d appears *Odd* with regard to the multiset $\{a, b, c\}$ and low truth values when d is not very different from the other values.

It is also worth noticing that this oddness index is not limited to multisets $\{a, b, c\}$ with 3 elements, and can be extended to the oddness $Odd(\mathcal{S}, x)$ of an item x with regard to a multiset S of any size. as follows: $Odd(\mathcal{S}, x) = \min(|\max(S) - x|, |\min(S) - x|)$. As can be seen, we only compare x to the upper and lower values in S, which may be considered as a meaningful summary of S only if S is very small (when we have no additional information about the distribution of values in S), i.e. $|S| = 1, 2, 3$, or may be 4.

Evenness A translation of formula (3), taking min for $\wedge$, max for $\vee$, and $1-|\cdot-\cdot|$ for $\equiv$, leads to the following expression for $Even_4(a,b,c,d)$:

$$max(min(1-|min(a,b)-min(1-c,d)|, 1-|min(1-a,1-b)-min(c,1-d)|),\\ 1-|max(a,b,c,d)-min(a,b,c,d)|)$$

Let us examine the behavior of this definition. Let us consider $f(x) = Even(0,x,x,x)$: we would expect f to get the constant value 1, since, whatever its value, the last element x, cannot be considered as an intruder in the multiset $\{0,x,x\}$; and $g(x) = Even(0,x,x,0)$: here we expect a function decreasing from 1 to 0 when x goes from 0 to 1. Indeed, the smaller x, the closer to 1 $Even(0,x,x,0)$ should be, while the larger x the more 0 appears to be equal to the minority value in the multiset $\{0,x,x\}$. As can be checked, f is not a constant function and g is not monotonically decreasing. This contrasts with $Odd(\{0,x,x\},x) = 0$ and $Odd(\{0,x,x\},0) = 0$. Since a direct translation of the Boolean definition (3) does not fit with the expected meaning of evenness in the case of graded truth values, we may try to start from the property $Even \equiv \neg Odd \wedge \neg I$ to get another translation as: $Even(a,b,c,d) = min(1-Odd(\{a,b,c\},d), 1-I(a,b,c,d))$. This new definition leads to $Even(0,x,x,x) = 1-x$ if $x \le 0.5$ and $1-min(x,2-2x)$ when $x \ge 0.5$. Then $g(x) = Even(0,x,x,0)$ is satisfactory since we get the decreasing function $1-x$. However $f(x) = Even(0,x,x,x)$ may be far from 1 (in particular, $Even(0,\frac{2}{3},\frac{2}{3},\frac{2}{3}) = \frac{1}{3}$). Such a behavior for f is not satisfactory. It is an open question to find a better definition for $Even$ in the graded case, which would coincide with the Boolean case when $a,b,c,d \in \{0,1\}$.

6 Concluding Remarks

This paper has provided an organized view of two logical indices referring to the ideas of evenness or of oddness of an item with respect to a multiset, and their relationship in the setting of logical proportions. In particular, we have shown that a third notion, expressing that a set of truth values is balanced between truth and falsity, complements them in the Boolean case. While analogical proportion has been shown to be useful in solving analogical puzzles and Raven progressive matrix tests, heterogeneous logical proportions for solving "pick up the one which does not fit" quizzes, it is of interest to notice that the third logical operator underlies another type of puzzle, Bongard problems.

References

1. Béziau, J.-Y.: New light on the square of oppositions and its nameless corner. Logical Invest. **10**, 218–233 (2003)
2. Béziau, J.-Y.: The power of the hexagon. Log. Univers. **6**(1–2), 1–43 (2012)
3. Blanché, R.: Structures Intellectuelles. Essai sur l'Organisation Systématique des Concepts. Vrin, Paris (1966)

4. Bongard, M.M.: Pattern Recognition. Hayden Book Co., Spartan Books, Rochelle Park (1970). Original publication in Russian (1967)
5. Bounhas, M., Prade, H., Richard, G.: Evenness-based reasoning with logical proportions applied to classification. In: Beierle, C., Dekhtyar, A. (eds.) SUM 2015. LNCS, vol. 9310, pp. 139–154. Springer, Heidelberg (2015)
6. Bounhas, M., Prade, H., Richard, G.: A new view of conformity and its application to classification. In: van der Gaag, L.C. (ed.) ECSQARU 2015. LNAI, vol. 9161, pp. 221–232. Springer, Switzerland (2015)
7. Bounhas, M., Prade, H., Richard, G.: Oddness-based classifiers for boolean or numerical data. In: Hölldobler, S., et al. (eds.) KI 2015. LNCS, vol. 9324, pp. 32–44. Springer, Heidelberg (2015). doi:10.1007/978-3-319-24489-1_3
8. Deza, M., Deza, E.: Encyclopedia of Distances, 3rd edn. Springer, Heidelberg (2014)
9. Dubois, D., Prade, H.: From Blanché's hexagonal organization of concepts to formal concept analysis and possibility theory. Log. Univers. **6**, 149–169 (2012)
10. Foundalis, H.: Phaeaco: A Cognitive Architecture Inspired by Bongards Problems. Doctoral dissertation, Indiana University, Center for Research on Concepts and Cognition (CRCC), Bloomington (2006)
11. Hofstadter, D.R.: Gödel, Escher, Bach: an Eternal Golden Braid. Basic Books, New York (1979)
12. Miclet, L., Prade, H.: Handling analogical proportions in classical logic and fuzzy logics settings. In: Sossai, C., Chemello, G. (eds.) ECSQARU 2009. LNCS, vol. 5590, pp. 638–650. Springer, Heidelberg (2009)
13. Parsons, T.: The traditional square of opposition. In: Zalta, E.N. (ed.) The Stanford Encyclopedia of Philosophy (2008)
14. Prade, H., Richard, G.: From analogical proportion to logical proportions. Log. Univers. **7**(4), 441–505 (2013)
15. Prade, H., Richard, G.: Homogenous and heterogeneous logical proportions. IfCoLog J. Logics Appl. **1**(1), 1–51 (2014)

Comparing System Reliabilities with Ill-Known Probabilities

Lanting Yu, Sébastien Destercke(✉), Mohamed Sallak, and Walter Schon

Sorbonne Universités, Université de Technologie de Compiègne, CNRS, Heudiasyc, 57 Avenue de Landshut, 60203 Compiègne, France
{lanting.yu,sebastien.destercke,mohamed.sallak,walter.schon}@hds.utc.fr

Abstract. In reliability analysis, comparing system reliability is an essential task when designing safe systems. When the failure probabilities of the system components (assumed to be independent) are precisely known, this task is relatively simple to achieve, as system reliabilities are precise numbers. When failure probabilities are ill-known (known to lie in an interval) and we want to have guaranteed comparisons (i.e., declare a system more reliable than another when it is for any possible probability value), there are different ways to compare system reliabilities. We explore the computational problems posed by such extensions, providing first insights about their pros and cons.

Keywords: System design · Reliability analysis · Imprecise probability · Comparison

1 Introduction

Being able to compare system reliabilities is essential when designing systems. Provided the structure function mapping single component reliabilities to the overall system reliability is known, this step poses no particular problem (at least from a theoretical standpoint) when failure probabilities are precisely known.

However, in practice, it may be difficult to provide precise assessments of such probabilities, for example because little data exist for the components (they may be issued from new technologies), or because they are given by expert opinions. This typically happens in early-stage phase design of new systems. In such a case, the problem of comparing system reliabilities become much more difficult, both conceptually and computationally speaking.

In this paper, we explore what happens when the component probabilities of functioning are ill-known, that is are only known to lie in an interval. Several aspects of reliability analysis have been extended to the case of ill-known probabilities, such as importance indices [8], multi-state systems [4], common cause failure problems [9], ... Yet, to our knowledge the problem of system reliability comparison remain to be formally studied within this setting.

In Sect. 3, we extend usual system comparisons (recalled in Sect. 2) to interval-valued probabilities in two different ways, discussing the theoretical and

J.P. Carvalho et al. (Eds.): IPMU 2016, Part II, CCIS 611, pp. 619–629, 2016.
DOI: 10.1007/978-3-319-40581-0_50

practical pros and cons of each extension. Section 4 provides a more complex examples than the very simple, illustrative ones provided along the paper. The necessary basics of reliability as well as notations are briefly recalled in Sect. 2.

2 System Modelling and Comparison: Basics

In this paper, we assume that we want to compare the designs of K systems $S_1, \ldots, S_K$ in terms of reliability, in order to choose (one of) the safest among them. The kth system will be composed of a set of r^k components, and a given component can belong to one of T populations (types) of components, all components of a population being assumed to have the same stochastic behaviour (i.e., same failure rate).

We will denote by $p_j \in [0,1]$ the possibly ill-known probability that a component of type j is functioning, and $1 - p_j$ the probability that it is inoperative or malfunctioning. We will also denote by $x^k_{i,j_{ik}} \in \{0,1\}$ both the ith component of kth system, which is of type j_{ik}, as well as its state ($x^k_{i,j_{ik}} = 0$ if malfunctioning, 1 if working). $p_{j_{ik}}$ is then the probability of $x^k_{i,j_{ik}} = 1$. Table 1 summarises these notations.

Table 1. Notation summary

Variable	Domain	Meaning
K	$\mathbb{Z}$	Number of systems
r^k	$\mathbb{Z}$	Number of components in the kth system
T	$\mathbb{Z}$	Number of component types (of possible stochastic behaviors)
p_j	$[0,1], j \in \{1,\ldots,T\}$	Probability that a component of type j will be working
$x^k_{i,j_{ik}}$	$\{0,1\}$	ith component of kth system, of type j_{ik}, and its state

In this paper, we will assume that we know the structure function $\phi^k : \{0,1\}^{r^k} \to \{0,1\}$ of the kth system and that it is written in the "simple" following way:

$$\phi^k(x^k_{1,j_{1k}}, \ldots, x^k_{r_k,j_{r_k k}}) = \sum_{A \subseteq \{1,\ldots,r_k\}} d^k_A \prod_{i \in A} x^k_{i,j_{ik}} \tag{1}$$

with d^k_A real-valued coefficients (some subsets A can receive $d_A = 0$) that can either be positive or negative. At least in principle, every system and structure function can be put in the form of Eq. (1), that is a multi-linear form [2]. We also make the classical assumption in reliability that each system is coherent, meaning that ϕ^k is increasing[1] and that we have the boundary conditions[2] $\phi^k(0,\ldots,0) = 0$, $\phi^k(1,\ldots,1) = 1$. Going from the structure function to the reliability R^k of the system is then quite simple, as it simply consists in replacing $x_{i,j_{ik}}$ by the corresponding probability $p_{j_{ik}}$, that is

$$R^k(p_{j_{1k}},\ldots,p_{j_{r_k k}}) = \sum_{A\subseteq\{1,\ldots,r_k\}} d_A^k \prod_{i\in A} p_{j_{ik}} \tag{2}$$

To simplify notations, we will simply note $R^k(p_{j_{1k}},\ldots,p_{j_{r_k k}})$ by R^k. Note that R^k is a function of the probabilities p_j, that can appear multiple times for one subset A. Note that being a coherent system means that the functions R^j are increasing in every variable p_j.

Example 1. Assume we have two (very) simple series system ($K = 2$), a first with two components, a second with three, and three different component types ($T = 3$). The two first components of each system are of the same type (1 and 2, respectively). The systems are illustrated in Fig. 1, and we do have

$$R^1 = p_1 \cdot p_2$$

$$R^2 = p_1 \cdot p_2 \cdot p_3$$

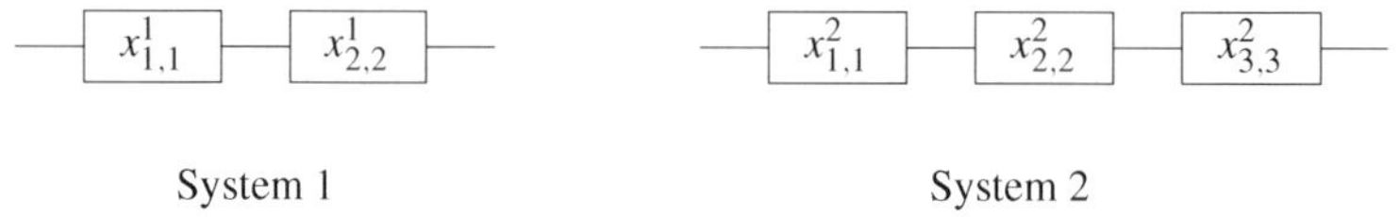

Fig. 1. Two simple series systems

Comparing two systems (say, the kth and ℓth) then comes down to compare their reliabilities R^k and R^ℓ. System S_k is then said to be preferred to system S_ℓ, denoted by $S_k \succ S_\ell$, if and only if

$$R^k > R^\ell \tag{3}$$

or, equivalently when probabilities p_j are precisely known, if and only if

$$R^k - R^\ell > 0. \tag{4}$$

Example 2. Let us continue Example 1 by using the precisely valued probabilities $p_1 = 0.8, p_2 = 0.9$ and $p_3 = 0.8$. We then have

$$R^1 = 0.72 \text{ and } R^2 = 0.576$$

[1] If one component goes from failing to working, then the system state can only improve.

[2] The system works (fails) if all components work (fail).

meaning that system S_2 should be discarded. We also have

$$R^1 - R^2 = p_1 \cdot p_2 - p_1 \cdot p_2 \cdot p_3 = p_1 \cdot p_2 \cdot (1 - p_3) = 0.144.$$

We can also notice that whatever the values of p_1, p_2, p_3, we will always have $S_1 \succ S_2$ (since $R^1 - R^2$ is a product of positive terms).

Let us now investigate what becomes of such a comparison when probabilities $p_j \in [\underline{p}_j, \overline{p}_j]$ are only known to lie in intervals.

3 Comparing Systems with Interval Probabilities

In this section, we investigate the most natural extensions of Eqs. (3) and (4) to an imprecise setting. We will see that in the imprecise case, they do no longer coincide, and the first extension only provides an approximation of the second one, but is computationally more tractable.

Note that in this paper, we are interested in guaranteed comparisons, that is we want to assess that S^k is more reliable than S^ℓ when this is true for any values of p_j within $[\underline{p}_j, \overline{p}_j]$ and for $j = 1, \dots, T$. For convenience, we will denote by $\mathscr{P} = \times_{j=1}^{T} [\underline{p}_j, \overline{p}_j]$ the Cartesian product of those intervals.

3.1 Interval Comparison: Definition

A first way to extend the comparison is to compute bounds over R^k, obtaining the interval $[\underline{R}^k, \overline{R}^k]$ such that

$$\underline{R}^k = \inf_{p_{j_{ik}} \in [\underline{p}_{j_{ik}}, \overline{p}_{j_{ik}}]} R^k = \sum_{A \subseteq \{1, \dots, r_k\}} d_A^k \prod_{i \in A} \underline{p}_{j_{ik}} \tag{5}$$

and

$$\overline{R}^k = \inf_{p_{j_{ik}} \in [\underline{p}_{j_{ik}}, \overline{p}_{j_{ik}}]} R^k = \sum_{A \subseteq \{1, \dots, r_k\}} d_A^k \prod_{i \in A} \overline{p}_{j_{ik}}. \tag{6}$$

where the fact that probability values can be replaced by their corresponding bounds follows from the increasing monotonicity of reliability functions. We can then straightforwardly extend Eq. (3) by saying that system S^k is interval-preferred to system S^ℓ, denoted $S^k \succ_{IC} S^\ell$, if and only if

$$\underline{R}^k > \overline{R}^\ell, \tag{7}$$

that is we are absolutely certain that S^k is more reliable than S^ℓ. In this case, comparing two systems just come down to make four computations instead of two to get the corresponding intervals. If the two intervals overlap, then systems S^k and S^ℓ are incomparable according to this criterion.

However, comparison (7) is very rough, in the sense that it will often result in incomparability of systems, even if it is obvious that one system is preferable to another, as Example 3 shows.

Example 3. Let us consider the systems of Example 2 with the following bounds

$$p_1 \in [0.7, 0.9], p_2 \in [0.8, 1] \text{ and } p_3 \in [0.7, 0.9].$$

We then obtain the intervals

$$R^1 \in [0.56, 0.9] \text{ and } R^2 \in [0.392, 0.81]$$

meaning that the system are not comparable according to $\succ_{IC}$.

3.2 Difference Comparison:definition

Interval comparison somehow extends Eq. (3), but a second way to extend the precise comparison is to extend Eq. (4). Before doing so, let us simplify notations by adopting the convention that $R^{k-\ell} := R^k - R^\ell$. We can then say that system S^k is difference-preferred to system S^ℓ, denoted $S^k \succ_{DC} S^\ell$, if and only if the value

$$\underline{R}^{k-\ell} = \inf_{\substack{p_{j_{ik}} \in [\underline{p}_{j_{ik}}, \overline{p}_{j_{ik}}] \\ p_{j_i\ell} \in [\underline{p}_{j_{i\ell}}, \overline{p}_{j_{i\ell}}]}} R^k - R^\ell \tag{8}$$

$$= \inf_{\substack{p_{j_{ik}} \in [\underline{p}_{j_{ik}}, \overline{p}_{j_{ik}}] \\ p_{j_{i\ell}} \in [\underline{p}_{j_{i\ell}}, \overline{p}_{j_{i\ell}}]}} \sum_{A \subseteq \{1,\dots,r_k\}} d_A^k \prod_{i \in A} p_{j_{ik}} - \sum_{A \subseteq \{1,\dots,r_\ell\}} d_A^\ell \prod_{i \in A} p_{j_{i\ell}} \tag{9}$$

is positive, i.e., $\underline{R}^{k-\ell} > 0$. In practice, this comes down to ask R^k to be higher than R^ℓ for all possible values of p_j, hence it also gives a guaranteed comparison. Example 4 and Corollary 1 show that this way of comparing systems is actually better than the previous, in the sense that it still gives guarantee but is less conservative. Yet, computing $\underline{R}^{k-\ell}$ can be far from straightforward (in contrast with the case of interval comparison), and we try to characterize in the next section when this task will be easy.

Example 4. Let us apply Eq. (7) to Example 3. In this case we have from Example 2 that $R^{1-2} = p_1 \cdot p_2 \cdot (1 - p_3)$ and so

$$\underline{R}^{1-2} = \inf_{\substack{p_1 \in [0.7, 0.9], \\ p_2 \in [0.8, 1], \\ p_3 \in [0.7, 0.9]}} p_1 \cdot p_2 \cdot (1 - p_3) = 0.7 \cdot 0.8 \cdot 0.1 = 0.056$$

which is indeed quite low, but still higher than zero, hence $S^1 \succ_{DC} S^2$, allowing us to reach a decision where we could not before.

And indeed, we always have the following relation between the two notions:

Proposition 1. $\underline{R}^{k-\ell} \geq \underline{R}^k - \overline{R}^\ell$

Proof. The inequality $\inf_{x \in D} f(x) + g(x) \geq \inf_{x \in D} f(x) + \inf_{x \in D} g(x)$ with x a vector of values and D a convex set is known to be true. If we define x as the vector of probability values $p_1, \ldots, p_T$, and take $f = R^k$, $g = -R^\ell$, $D = \mathscr{P}$, we get

$$\inf_{x \in \mathscr{P}} R^k(x) - R^\ell(x) \geq \inf_{x \in \mathscr{P}} R^k(x) + \inf_{x \in \mathscr{P}} -R^\ell(x) \geq \inf_{x \in \mathscr{P}} R^k(x) - \sup_{x \in \mathscr{P}} R^\ell(x)$$

We then get the following corollary, showing that if $S^k \succ_{IC} S^\ell$, then $S^k \succ_{DC} S^\ell$, but not the reverse. Actually, a similar problem is known under the name "dependency problem" in interval arithmetic, for which many solutions have been proposed [3].

Corollary 1. *If* $\underline{R}^k - \overline{R}^\ell > 0$, *then* $\underline{R}^{k-\ell} > 0$

So $\underline{R}^{k-\ell}$ is definitely a more accurate way of comparing systems. Let us now study a bit the problem of actually computing it.

Remark 1. In imprecise probability theory, a similar relation exists between the maximality decision rule and the interval dominance decision rule [10]. However, two main differences, in terms of optimization problems, between imprecise probabilities and our study are that $\mathscr{P}$ is here an hypercube and that optimization has to be done over non-linear functions in general, while imprecise probabilities is concerned with bounds of expectations over a subset of the unit simplex. Note that we could also search to adapt other imprecise probability decision rules: maximin and maximax extend directly by using Eqs. (5) and (6), while the notion of E-admissibility may require more involved investigation, especially as it is not based on a pairwise comparison scheme.

3.3 Computing $\underline{R}^{k-\ell}$

In general, $R^k - R^\ell$ will be a polynomial in variables p_j that is neither decreasing nor increasing in those variables. Computing bounds over such polynomials when variables lie in a hyper-cube (which is our case) is known to be NP-hard [6], hence infeasible in practice. Two solutions are then to look for approximations that remain close to $\underline{R}^{k-\ell}$ but are more tractable (using interval bounds provides a crude approximation), or to identify those sub-cases for which the solution will be easier to find. In this paper, we explore the second alternative, and leave the first for future works.

Before studying in detail how $\underline{R}^{k-\ell}$ can be computed, we have to recall the notions of global monotonicity and of local monotonicity of a function [5] $f(x_1, \ldots, x_n)$ in a variable x_i

Definition 1 (Global monotonicity). *Function* $f(x_1, \ldots, x_n)$ *is globally increasing (decreasing) in* x_i *if it is always increasing (decreasing) in* x_i*, irrespectively of the other variable values.*

If f is globally increasing in x_i, then its lower and upper bounds are known to be obtained for $x_i = \underline{x}_i$ and $x_i = \overline{x}_i$ when $x_i \in [\underline{x}_i, \overline{x}_i]$, respectively.

Definition 2 (Local monotonicity). *Function $f(x_1, \ldots, x_n)$ is locally increasing (decreasing) in x_i if it is either increasing or decreasing in x_i when the other variables $x_1, \ldots, x_{i-1}, x_{i+1}, \ldots, x_n$ values are fixed.*

If f is locally monotonic in x_i, then its bounds are known to be obtained for $x_i = \underline{x}_i$ or $x_i = \overline{x}_i$, but which value to take between these two ones depends on the other variable values, in contrast with global monotonicity (where the value to consider is fixed, whatever the other variable values). A function will be said to be non-monotone in x_i if it is not locally or globally monotone in it.

Example 5. Consider the following functions of x_1, x_2 with $x_i \in [-2, 1]$, then the functions

$$f_1(x_1, x_2) = x_1 - x_2, \quad f_2(x_1, x_2) = -(x_1 \cdot x_2), \quad f_3(x_1, x_2) = x_1^2 \cdot x_2^2$$

are respectively globally, locally, and not monotone in each of their variables. x_1 (x_2) is globally increasing (decreasing) in f_1. f_2 is decreasing in x_1 (x_2) when x_2 (x_1) is positive, and increasing when x_2 (x_1) is negative (hence the monotonicity depends on the value of the other variables). f_3 is neither locally nor globally monotone in both variables (i.e., $\underline{f}_3 = f_3(0, 0)$).

Given two systems S^k and S^ℓ, we now define the following subsets of component types:

- The subsets

$$T_k = \{j \in \{1, \ldots, T\} | \forall p_{j_{i\ell}}, i = 1, \ldots, r^\ell, j_{i\ell} \neq j\}$$

$$T_\ell = \{j \in \{1, \ldots, T\} | \forall p_{j_{ik}}, i = 1, \ldots, r^k, j_{ik} \neq j\}$$

  that denote the types of components that are encountered only in system S^k (T_k) or S^ℓ (T_ℓ).
- The subset

$$\begin{aligned} T_{\ell \cap k,1} = \{j \in \{1, \ldots, T\} | \exists p_{j_{i\ell}}, p_{j_{i'k}} \ s.t. \ j_{i\ell} = j_{i'k} = j \quad & \wedge \\ \exists! i \ s.t. \ j_{i\ell} = j \quad & \wedge \\ \exists! i' \ s.t. \ j_{i'k} = j\} \quad & \end{aligned}$$

  that includes all component types that are in both systems, but only once in each of them.
- The subset

$$\begin{aligned} T_{\ell \cap k,+} = \{j \in \{1, \ldots, T\} | \exists p_{j_{i\ell}}, p_{j_{i'k}} \ s.t. \ j_{i\ell} = j_{i'k} = j \quad & \wedge \\ (\exists i, i' \ s.t. \ j_{i\ell} = j_{i'\ell} = j \quad & \vee \\ \exists i, i' \ s.t. \ j_{ik} = j_{i'k} = j)\} \quad & \end{aligned}$$

  that includes all component types that are in both systems and appear more than once in at least one of the two systems.

The subsets $T_k, T_\ell, T_{\ell\cap k,1}, T_{\ell\cap k,+}$ form a well-defined partition of the component types in systems S^k and S^ℓ. We can then show a first property

Proposition 2. *$R^k - R^\ell$ is a globally monotonic function in variables p_j, $j \in T_k \cup T_\ell$. It is increasing (decreasing) in variables $p_j, j \in T_k$ ($p_j, j \in T_\ell$)*

Proof. Without loss of generality, let us assume that $p_1, \dots, p_i \in T_k$ and $p_{i+1}, \dots, p_j \in T_\ell$. By assumption, we have

$$R^k - R^\ell = R^k(p_1, \dots, p_i, p_{j+1}, \dots, p_T) - R^\ell(p_{i+1}, \dots, p_T),$$

therefore the monotonicity with respect to $p_1, \dots, p_i$ ($p_{i+1}, \dots, p_j$) depends only of their monotonicity with respect to R^k (R^ℓ), which are known to both be increasing in those variables.

This means that if $p_i \in T_k$ or T_ℓ, we know for which value of p_i the lower bound is obtained ($\underline{p}_i$ if $p_i \in T_k$, else $\overline{p}_i$) . Also note that in the particular case where $T_{k\cap\ell,1} = T_{k\cap\ell,+} = \emptyset$, the following result follows:

Lemma 1. *if $T_{k\cap\ell,1} = T_{k\cap\ell,+} = \emptyset$, then $\underline{R}^k - \overline{R}^\ell = \underline{R}^{k-\ell}$*

Proof. When $T_{k\cap\ell,1} = T_{k\cap\ell,+} = \emptyset$, there are no shared variables between R^k and R^ℓ, meaning that $j = T$ in proof of Proposition 2 and that

$$\inf_{p_1,\dots,p_i} \inf_{p_{i+1},\dots,T} R^k - R^\ell = R^k(\underline{p}_1, \dots, \underline{p}_i) - R^\ell(\overline{p}_{i+1}, \dots, \underline{p}_T)$$

Proposition 3. *$R^k - R^\ell$ is a locally monotonic function in variables p_j, $j \in T_{k\cap\ell,1}$.*

Proof. (sketch) We know that both R^k and R^ℓ are equivalent to replacing the $x^k_{i,j_{ik}}$ in Eq. (1) by their probability types. If a type p_i of component is present once (and exactly once) in each system, this means that for every subset A, p_j power will be either zero or one in the products $\prod_{i\in A^k} p_{j_{ik}}$ and $\prod_{i\in A^\ell} p_{j_{i\ell}}$ of Eq. (2). Therefore, $R^k - R^\ell$ will be a sum of products where p_j has power zero or one, meaning that if the other variables $p_1, \dots, p_{j-1}, p_{j+1}, \dots, p_T$ are fixed, the derivative of $R^k - R^\ell$ with respect to p_j will be a constant (whose positivity or negativity will depend of $p_1, \dots, p_{j-1}, p_{j+1}, \dots, p_T$ values), hence that $R^k - R^\ell$ is either decreasing or increasing in p_j.

Fig. 2. Two simple series systems with common component

Example 6. Let us consider the series systems of Fig. 2 with three types of components, where $p_2 \in T_{1 \cap 2,1}$. We have

$$R^1 - R^2 = p_1 \cdot p_2 - p_3 \cdot p_2 = p_2 \cdot (p_1 - p_3)$$

which is indeed locally, but not globally, monotone in p_2 (it is increasing if $p_1 > p_3$, decreasing if $p_1 < p_3$).

This means that, if we have $N = |T_{k \cap \ell,1}|$ variables p_j for which we are locally monotone, we know that the lower bound is obtained for one of the 2^N vertices of the corresponding hypercube $\times_{i \in T_{k \cap \ell,1}} [\underline{p}_i, \overline{p}_i]$. If N is not too high, then we can think of simply enumerating the set of possible values.

Finally, we cannot guarantee any kind of monotonicity for the variables $p_j, j \in T_{\ell \cap k,+}$. However, if the cardinality of $T_{\ell \cap k,+}$ is not too high, it is always possible to make a random search within its defined area.

Example 7. Let us consider the very simple case depicted in Fig. 3, where we have

$$R^1 = p_1^2 \text{ and } R^2 = p_1$$

hence $R^{1-2} = p_1^2 - p_1 = p_1(p_1 - 1)$, which will always be negative. However, if $p_1 \in [0.4, 0.6]$, the bound $\underline{R}^{1-2} = -0.25$ is obtained for $p_1 = 0.5$, which does not correspond to one of the bounds $\underline{p}_1, \overline{p}_1$.

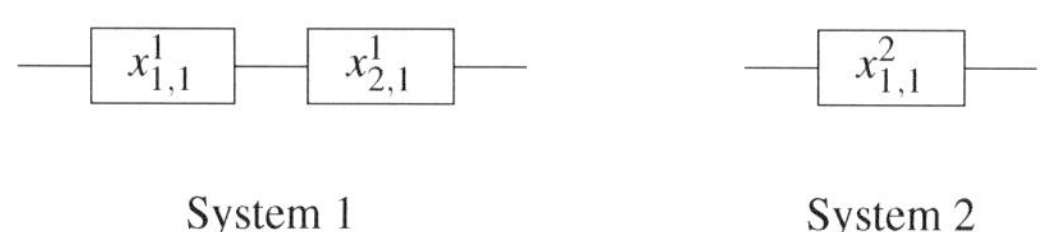

Fig. 3. Two simple series systems with redundancy

It means that when confronted with too much components present in both systems and multiple times in at least one of them, computing the bound may quickly become intractable in practice. This becomes even truer if the monotonicity of other variables (those in $T_{\ell \cap k,1}$) depends on those variables in $T_{\ell \cap k,+}$.

An easy solution is to "duplicate" each variable p_j in $T_{\ell \cap k,+}$ with variables having the same interval bound, so that each variable is present at most once in each system. In the case of Fig. 3, this means considering a variable p'_1 for the second component of System 1. Such a straightforward approach has two potential drawbacks: the increase of the number of component types in $T_{\ell \cap k,1}$, and the fact that the approximation can be quite loose. Such a strategy is therefore likely to be useful only when the number of component types.

4 A More Complex Example

Let us now consider two slightly more complex systems, where we want to chose the most reliable design. The systems are depicted by Fig. 4, and consider three types of components, with $p_1 \in [0.9, 1]$, $p_2 \in [0.8, 0.9]$ and $p_3 \in [0.85, 0.95]$, where one hesitates between choosing a 2 out of 3 architecture with slightly less reliable components, and a parallel architecture with potentially more reliable components. The reliabilities of the systems are

$$R^1 = p_1 \cdot p_2^2 \cdot (3 - 2p_2)$$

and

$$R^2 = p_1 \cdot p_3 \cdot (2 - p_3).$$

Intervals $[\underline{R}^1, \overline{R}^1]$ and $[\underline{R}^2, \overline{R}^2]$ intersect, hence interval comparison is not sufficient to tell us whether S_1 is better than S_2, or the reverse. We have $T_{1\cap 2,1} = \{1\}$, $T_1 = \{2\}$ and $T_2 = \{3\}$, therefore if we want to compute $\underline{R}^{1-2}$, our previous results tell us that

$$\underline{R}^{1-2} = p_1^* \cdot \underline{p}_2^2 \cdot (3 - 2\underline{p}_2) - p_1^* \cdot \overline{p}_3 \cdot (2 - \overline{p}_3)$$

with $p_1^* \in \{\underline{p}_1, \overline{p}_1\}$. The result is obtained for $\overline{p}_1$ (the function is decreasing in p_1 for $p_2 = \underline{p}_2$ and $p_3 = \overline{p}_3$), and $\underline{R}^{1-2} = -0.1015$, meaning that we cannot conclude that $S_1 \succ_{DC} S_2$. Following a similar line of reasoning for $\underline{R}^{2-1}$ (which is increasing in p_1), we get $\underline{R}^{2-1} = 0.00495$ and are able to tell that $S_2 \succ_{DC} S_1$.

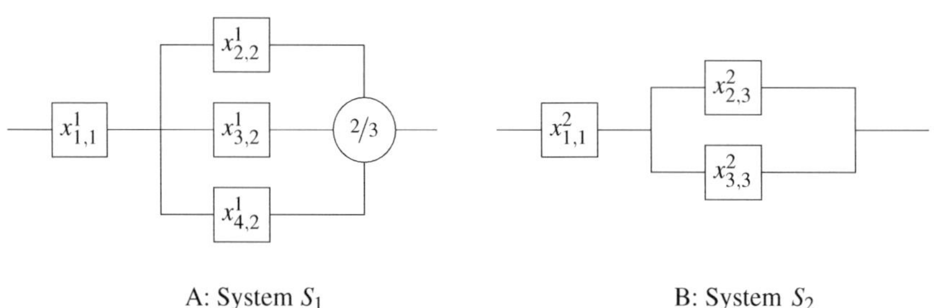

Fig. 4. Two system designs to compare.

5 Conclusion

In this paper, we have studied how comparisons of system reliabilities can be extended when probabilities are ill-known, or interval-valued. In particular, we have focused on comparison notions that allows for incomparability when the information is too weak to be certain that one system is more reliable than another.

We have seen that computing the lower bound over the difference of reliabilities is less conservative, but more computationally demanding than just comparing reliability bounds of each systems taken individually. While we have pointed out ways to reduce the complexity of such computations (by focusing on global and local comonotonicity), it remains to investigate how to approximate $\underline{R}^{k-\ell}$ with a lower bound better than $\underline{R}^k - \overline{R}^\ell$, but computationally more tractable than computing $\underline{R}^{k-\ell}$. A first way to do so is to exploit bounds used when the reliability probabilities are precisely known, but when computing the output probability is computationally prohibitive, see e.g., [7].

An additional interesting problem to explore is to formalize which information we should query to make two incomparable systems comparable. For instance, we may formulate it as an expert elicitation problem [1].

References

1. Abdallah, N.B., Destercke, S.: Optimal expert elicitation to reduce intervaluncertainty. In: Proceedings of the Thirty-First Conference on Uncertainty in Artificial Intelligence, UAI 2015, pp. 12–21, 12–16 July, 2015, Amsterdam, The Netherlands (2015)
2. Borgonovo, E.: The reliability importance of components and prime implicants in coherent and non-coherent systems including total-order interactions. Eur. J. Oper. Res. **204**(3), 485–495 (2010)
3. De Figueiredo, L.H., Stolfi, J.: Affine arithmetic: concepts and applications. Numer. Algorithms **37**(1–4), 147–158 (2004)
4. Ding, Y., Lisnianski, A.: Fuzzy universal generating functions for multi-state system reliability assessment. Fuzzy Sets Syst. **159**(3), 307–324 (2008)
5. Fortin, J., Dubois, D., Fargier, H.: Gradual numbers and their application to fuzzy interval analysis. IEEE Trans. Fuzzy Syst. **16**(2), 388–402 (2008)
6. Kreinovich, V., Lakeyev, A., Rohn, J.: Computational complexity of interval algebraic problems: some are feasible and some are computationally intractable-a survey. Math. Res. **90**, 293–306 (1996)
7. Mteza, P.Y.: Bounds for the reliability of binary coherent systems. Ph.D. thesis (2014)
8. Sallak, M., Schon, W., Aguirre, F.: Extended component importance measures considering aleatory and epistemic uncertainties. IEEE Trans. Reliab. **62**(1), 49–65 (2013)
9. Troffaes, M.C., Walter, G., Kelly, D.: A robust bayesian approach to modeling epistemic uncertainty in common-cause failure models. Reliab. Eng. Syst. Saf. **125**, 13–21 (2014)
10. Troffaes, M.: Decision making under uncertainty using imprecise probabilities. Int. J. Approximate Reasoning **45**, 17–29 (2007)

Machine Learning

Visualization of Individual Ensemble Classifier Contributions

Catarina Silva[1,2](✉) and Bernardete Ribeiro[1]

[1] Center for Informatics and Systems, University of Coimbra, Coimbra, Portugal
{catarina,bribeiro}@dei.uc.pt
[2] School of Technology and Management, Polytechnic Institute of Leiria, Leiria, Portugal

Abstract. Ensembles of classifiers are usually considered a valuable approach in different scenarios. A broad range of methods to deal with the construction, diversity and combination of multiple predictive models have been extensively studied. While the focus is often to obtain more accurate and robust predictions than single models seldom the individual contribution of classifiers which could contribute to a better understanding of the uncertainty associated with ensembles' outputs is taken into account. In this work we look into this issue and focus on evaluating the individual ensemble classifier contributions using several scenarios. We propose a visual web model that allows for the evaluation of both individual contributions as well as their interactions. We apply the proposed approach on a benchmark dataset and show how it can visually be used to better understand the uncertainty associated with the construction of ensembles, presenting some insight on the individual contributions and interactions.

Keywords: Ensembles · Classification · Visualization

1 Introduction

In the last decades the need to construct classifiers that include some degree of transparency has increased exponentially, due to the increased demand of users to understand the decisions automatic classifiers propose, specially in decision support systems [1]. As a consequence, there is an ever-increasing need for frameworks that despite the use of black-box algorithms exhibit some degree of transparency, in applications such as, medical support systems, data processing and visualization, text/web mining, digital information search, and patent analysis.

The task in visualization approaches can be described as bridging information from the algorithm feature space to the user feature space. This task is usually non-trivial, since there seldom exists a direct transform/interpretation of results. However, the potential advantages of visual data exploration/results interpretation is that the user is directly involved in the learning/classification [2].

J.P. Carvalho et al. (Eds.): IPMU 2016, Part II, CCIS 611, pp. 633–642, 2016.
DOI: 10.1007/978-3-319-40581-0_51

Different methods have been proposed and selected according to experiments on a particular problem instance [3], but interesting methods should be sufficiently generic to be applied to at least a set of problems.

In classification scenarios, on a particular problem different algorithms may obtain different results, but over all problems, they are indistinguishable [4]. It follows that if an algorithm achieves superior results on some problems, it must pay with inferiority on other problems, i.e. the probability distribution on problem instances is such that all problem solvers have similarly distributed results [5]. Hence, we propose the combination of different and heterogeneous models in an ensemble, e.g. Support Vector Machines (SVM) [6], Fuzzy Rules [7], K-Nearest neighbors (KNN) [8] or Naive Bayes [9]. Such an ensemble approach circumvents the problems associated with making a specific algorithm transparent, moving the research issue to a more generic one of making the ensemble more transparent (or less opaque).

In this work we propose a general framework for visualization of individual ensemble classifier contributions. We explore the different and similar error patterns to correlate the baseline models presenting a visual interpretation of results. The proposed framework can be adapted to any ensemble, despite the number and nature of the underlying models and combination algorithms.

The rest of the paper is organized as follows. In Sect. 2 we present the background that supports our approach and describe current approaches. In Sect. 3 we detail the proposed framework for visualization of individual ensemble classifier contributions. In Sect. 4, we describe the experimental setup used, including performance metrics and benchmark datasets, including also the main results and analysis. Finally, Sect. 5 addresses the conclusions and future research lines.

2 Background

In this section we will introduce the background on ensemble systems and on visualization, including current approaches.

2.1 Ensembles

Ensemble based systems (also known under various other names, such as multiple classifier systems, committee of classifiers, or mixture of experts) have shown favorable results compared to those of single-expert systems for a broad range of applications requiring automated decision making under a variety of scenarios.

In matters of great importance that have financial, medical, or other implications, we often seek a second opinion before making a decision, sometimes more [10]. In doing so, we analyze each one, and combine them using some implicit process to reach a final decision that is apparently the best informed one. This process of consulting several experts before making a final decision is perhaps second nature to us; yet, the extensive benefits of such a process in classification systems is still being discovered by the computational intelligence community [11]. In

[12], a sample of the vast literature on classifier combination can be found, on both the theory and implementation of ensemble based classifiers.

According to [13] there are three main reasons to use ensembles: statistical, representational and computational. Regarding statistical ones, either there is not sufficient data to find the optimal hypothesis or there are many different hypothesis with limited data. In the representational set of reasons, unknown functions may not be present in the hypotheses space or even a combination of present hypotheses may expand it. For the computational issue, the learning algorithms may stuck in local minima, therefore using an ensemble might be useful.

In summary, an ensemble rationale is that, given a task that requires specific knowledge, k experts may perform better than one, given that their individual responses are duly combined. A classifier committee is then characterized by (i) a choice of k classifiers, and (ii) a choice of a combination function [14,15], usually denominated voting algorithm. The classifiers should be as independent as possible to guarantee a large number of inductions on the data [16].

A common voting algorithm is majority voting, where each base classifier (expert) votes on the class the example should belong to and the majority wins (in two-class problems an odd number of classifiers should be used).

2.2 Visualization

As previously referred, recently there has been a growing demand for more transparent classifiers that offer users with some degree of information and involvement in the classification process. Progress in this area has been accelerated when users can readily access visualization techniques relevant to the given problem. To achieve this goal different approaches can, and have been, followed, usually focusing on explaining a specific model or algorithm, like in [17] where an approach to visualize high-dimensional fuzzy classification rules is presented.

Current research results are usually mapped to a myriad of different possibilities, given that the human interaction and understanding is under-rated. Amongst those methodologies, one can find interactive visualization by using multi-level pie charts, multi bar charts, histograms, scatter plots, tree maps and dataflow diagrams. The different visualization techniques help in understanding different levels of information hidden in very large data sets [21].

In [23], a classification of information visualization and visual data mining techniques based on the data type to be visualized, on the visualization technique, and on the interaction and distortion technique is proposed.

In [21], visual data mining applications for enhancing business decisions are discussed. The visual data mining concept is implemented by presenting results in the form of visual interpretation.

In [18], methods for visualizing multidimensional data are shown to augment clinical disease risk assessment by providing reduced-dimensional displays which stratify patient data points according to risk level while providing additional insight into clinically important individual risk factor variables.

Visualization is currently one of the key research trends. Data are continuously acquired for a variety of purposes. The ability to make timely decisions based on available data is crucial in various scenarios, e.g., business success, clinical treatments, cyber and national security, and disaster management [24]. Visualization frameworks are beginning to provide decision-makers powerful tools that will certainly evolve in the next years.

3 Proposed Approach

Figure 1 generically depicts the proposed ensemble approach that constitutes the base of the framework for visualization of individual ensemble classifier contributions.

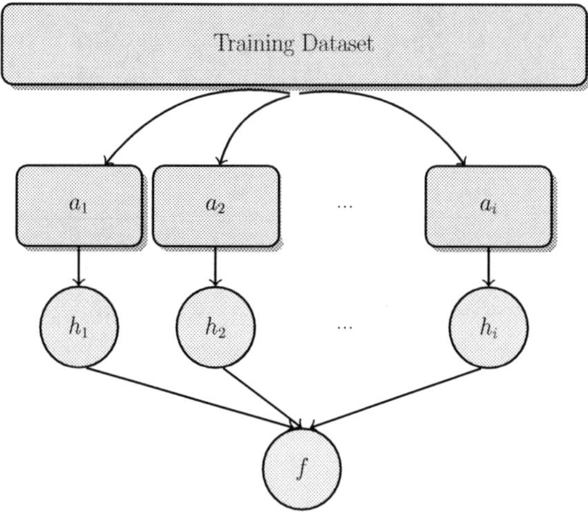

Fig. 1. Ensemble approach.

Considering a D training dataset:

$$D = \{\mathbf{x}_n, \mathbf{y}_n\}_{n=1}^{N}, \tag{1}$$

with N examples, where $\mathbf{x}_n$ are the input feature vectors and $\mathbf{y}_n$ is the label vector, the ensemble is constructed by a set of T inducers:

$$A_T = \{\mathbf{a}_i(.)\}_{i=1}^{T} \tag{2}$$

This set of inducers will generate a set of T hypothesis (models or classifiers):

$$H_T = \{\mathbf{h}_i(.)\}_{i=1}^{T} \tag{3}$$

These models will then produce output classifications:

$$h_i : \mathcal{X} \mapsto \mathcal{Y}, \mathcal{Y} = \{0, 1\}, \tag{4}$$

assuming two classes $\{0, 1\}$ with no loss of generality. Having these T classifiers, then an aggregation function f must be defined, for instance a majority voting function.

These T baseline classifiers should convey the diversity needed to assure that different views on the training data exist, and hence different error patterns that can be exploited to improve the final classification. An error pattern is characterized by two sets: for a given example, the set of classifiers that correctly classify it and the set of classifiers that do not correctly classify it. Such heterogeneity can be achieved with different data splits for each classifier and/or with different algorithms in the creation of the models.

The ensemble can result of any combination and, as referred in Sect. 2.1, the majority voting is a common and simple choice for f.

Using both the individual classifications of each baseline classifier and the final majority voted ensemble classification, our framework constructs a radial web with the q most frequent co-occurring error patterns, given insight on the individual contribution of baseline classifiers, as we will show in the following section.

An error pattern is defined as a subset of test examples $\mathcal{D}_{test}$ where the same subset of baseline classifiers, $\mathcal{H}_{fail}$, fails to deliver the correct classification (and hence the same subset of classifiers delivers the right classification).

The resulting visual representation is a radial web with one radial section for each q selected error patterns, depicting: (i) the number of examples in the set $\mathcal{D}_{test}$ by the radius of the graph; and (ii) which classifiers are part of $\mathcal{H}_{fail}$ by presenting a representation in their radial slot.

Figure 2 presents an example of the visual representation. In this example the radial web includes 8 sections ($q = 8$) and the ensemble is composed by 5 baseline classifiers ($T = 5$). Here, we only show hypothetical results for one of the error patterns, represented by the radial sections in quadrant 1. In this scenario, we can see that baseline classifiers h_1, h_2 and h_5 have failed in the same $\mathcal{D}_{test}$. Moreover, by the radius of the radial colored sections one can visually infer that the number of testing examples in $\mathcal{D}_{test}$ is relevant (it is in fact around 80 % of the errors).

In the next section we are going to present the deployment of the proposed approach to a specific scenario, showing its effectiveness in providing users with helpful aid in visualization of individual ensemble classifier contributions.

4 Experimental Setup and Results

To test the proposed visualization of individual ensemble classifier contributions approach we first define the performance metrics and introduce a binary case study: the Pima Diabetes dataset, available at https://archive.ics.uci.edu/ml/datasets/Pima+Indians+Diabetes and described in following. The experimental

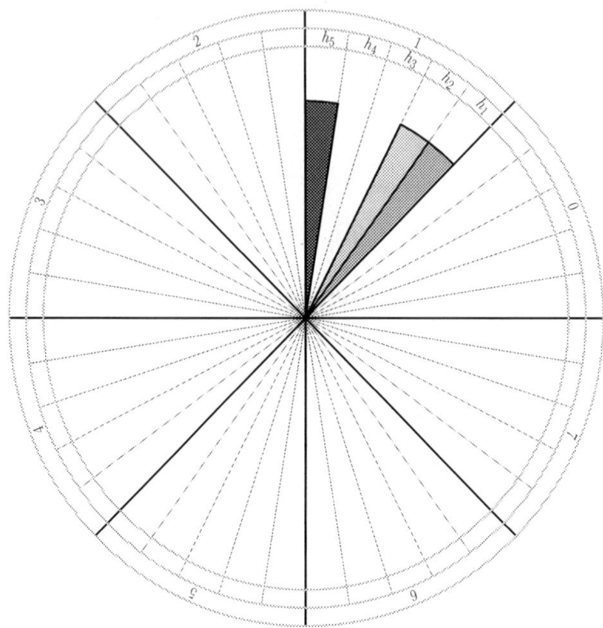

Fig. 2. Example of the visualization using the proposed framework (Color figure online).

setup also includes the construction of a majority voting ensemble with its baseline classifiers and the depiction and analysis of the visual result of the framework.

4.1 Performance Metrics

In order to evaluate a binary decision task we first define a contingency matrix representing the possible outcomes of the classification, as shown in Table 1.

Table 1. Contingency table for binary classification.

	Class Positive	Class Negative
Assigned Positive	a	b
	(True Positives)	(False Positives)
Assigned Negative	c	d
	(False Negatives)	(True Negatives)

4.2 Diabetes Dataset

The *Pima Diabetes* dataset corresponds to a binary problem of the diagnostic of diabetes, i.e. given a set of attributes about a patient the ground truth is the definition of whether that person has (or does not have) diabetes. The dataset includes patient records with 8 attributes and the target class. The dataset includes 768 examples, of which 500 are negative and 268 are positive. The numeric attributes, including the class, are:

1. Number of times pregnant
2. Plasma glucose concentration a 2 h in an oral glucose tolerance test
3. Diastolic blood pressure (mm Hg)
4. Triceps skin fold thickness (mm)
5. 2-Hour serum insulin (mu U/ml)
6. Body mass index (weight in kg/$(height\ in\ m)^2$)
7. Diabetes pedigree function
8. Age (years)
9. Class variable (0 or 1)

The division in train and test sets was carried out as defined in Table 2, using a random 66/33 split.

Table 2. Split train/test defined for the Diabetes dataset.

	Class Positive	Class Negative
Train	179	333
Test	89	167

4.3 Ensemble Classifiers and Performance Results

The heterogeneous baseline classifiers defined to test the framework were:

- Support Vector Machine (SVM)
- Fuzzy Rules
- K-Nearest Neighbor (KNN)
- Naive Bayes (NB)

These baseline classifiers were constructed using Weka 3.7.13 (http://www.cs.waikato.ac.nz/ml/weka/) with default parameters. The baseline individual performance is shown in Table 3 together with the final ensemble classification perfomance.

As can be gleaned from Table 3, the ensemble result outperforms the best of the classifiers as usually expected, and analyzing the table, one can perceive that the gain is specially significant in Recall values, i.e., there are less examples classified as False Negatives.

However, regarding evaluation of the individual contribution this type of quantitative results alone does not offer much help. In the following section we will show how more information can be extracted.

Table 3. Performance results on Diabetes dataset.

	Accuracy	Precision	Recall	F1
SVM	78.52 %	81.48 %	49.44 %	61.54 %
Fuzzy	79.69 %	74.19 %	51.69 %	60.93 %
KNN	79.69 %	54.26 %	57.30 %	55.74 %
Naive Bayes	79.69 %	81.48 %	49.44 %	61.54 %
Ensemble	79.69 %	75.34 %	61.80 %	**67.90%**

4.4 Visualization of Individual Contributions

Following the case study introduced in the last section, the settings for deployment of the proposed framework were defined as:

- number of baseline classifiers: $T = 5$
- number of frequent error patterns: $q = 8$

These heuristically defined parameters should be tuned for each application. In the future we will pursue the semi-automatic definition of such parameters. The resulting radial web is presented in Fig. 3. Analyzing this web several aspects

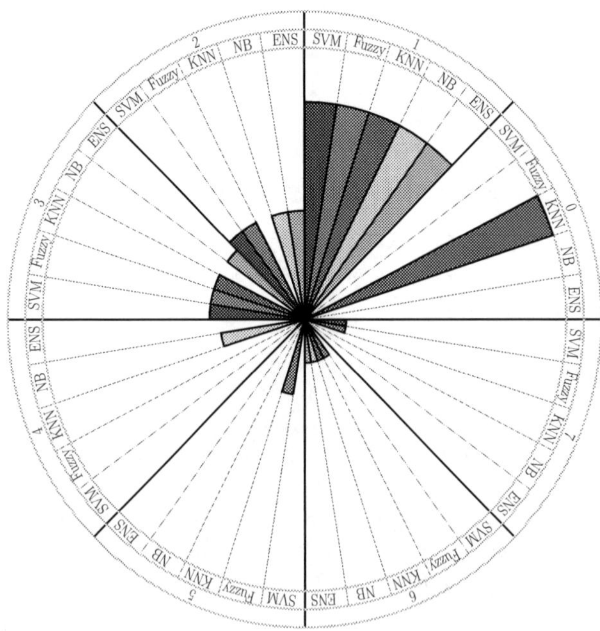

Fig. 3. Combination of error patterns in classifiers in the Diabetes dataset (Color figure online).

of both individual contribution of classifiers, as well as, as of correlation between error patterns are evident.

Quadrants 0, 4, and 5 show the distinctiveness in error patterns of KNN, Fuzzy and NB. In these examples, these classifiers did not predict well the output class, but all others were correct. On the opposite, quadrants 2, 3, and 6 show particularity in accuracy for KNN, NB and SVM. In each set of examples in these quadrants only these specific classifiers had the correct prediction, determining that their distinction could be better explored.

Finally, quadrant 1 shows that in the hard to classify examples all classifiers make mistakes, and in fact these represent a considerable part of the errors. For a researcher this information can be extremely valuable, stating that probably the baseline classifiers should be further enriched.

5 Conclusions and Future Work

In this paper we presented a framework for visualization of individual ensemble classifier contributions.

A radial web model was proposed to both evaluate individual contributions and interactions between classifiers. We applied the proposed approach on a benchmark dataset and the results show it can visually be used to better understand the uncertainty associated with the construction of ensembles, presenting conclusions on the radial webs that were constructed, namely on the evaluation of the heterogeneity of baseline classifiers.

Future lines of research will consist in automatically determining the threshold for the number of q and generalizing the framework for any classification problem, namely multi-label, multi-classification scenarios.

References

1. Liu, L., Wang, L.: What has my classifier learned? visualizing the classification rules of bag-of-feature model by support region detection. In: IEEE Conference on Computer Vision and Pattern Recognition (CVPR), pp. 3586–3593 (2012)
2. Keim, D.: Information visualization and visual data mining. IEEE Trans. Visual Comput. Graphics **8**(1), 1–8 (2002)
3. Wehrend, S., Lewis, C.: A problem-oriented classification of visualization techniques. In: IEEE Visualization, pp. 139–143 (1990)
4. Wolpert, D.H., Macready, W.G.: No free lunch theorems for optimization. IEEE Trans. Evol. Comput. **1**, 67 (1997)
5. Silva, C., Lotrič, U., Ribeiro, B., Dobnikar, A.: Distributed text classification with an ensemble kernel-based learning approach. IEEE Trans. Syst. Man Cybern. Part C Appl. Rev. **40**, 287–297 (2010)
6. Vapnik, V.: The Nature of Statistical Learning Theory. Springer, New York (1998)
7. Huehn, J.C., Huellermeier, E.: FURIA: an algorithm for unordered fuzzy rule induction. In: Hühn, J., Hüllermeier, E. (eds.) Data Mining and Knowledge Discovery, vol. 19, 3rd edn, pp. 293–319. Springer, Heidelberg (2009)

8. Altman, N.S.: An introduction to kernel and nearest-neighbor nonparametric regression. Am. Stat. **46**(3), 175–185 (1992)
9. Russell, S., Norvig, P.: Artificial Intelligence: A Modern Approach, 2nd edn. Prentice Hall, Upper Saddle River (2003)
10. Salgado, C., Vieira, S.M., Mendonça, L.F., Finkelstein, S., Sousa, J.M.C.: Ensemble fuzzy models in personalized medicine: application to vasopressors administration. Eng. Appl. Artif. Intell. **49**, 141–148 (2016)
11. Polikar, R.: Ensemble based systems in decision making. IEEE Circuits Syst. Mag. **6**(3), 21–45 (2006)
12. Kuncheva, L.: Combining Pattern Classifiers - Methods and Algorithms. Wiley, Hoboken (2004)
13. Dietterich, T.G.: Ensemble methods in machine learning. In: Kittler, J., Roli, F. (eds.) MCS 2000. LNCS, vol. 1857, pp. 1–15. Springer, Heidelberg (2000)
14. Silva, C., Ribeiro, B. (eds.): Inductive Inference for Large Scale Text Classification, vol. 225. Springer, Heidelberg (2010)
15. Silva, C., Ribeiro, B.: Background on text classification. In: Inductive Inference for Large Scale Text Classification. Studies in Computational Intelligence, vol. 255. Springer (2010)
16. Gacquer D., Piechowiak S., Delmotte, F., Delcroix, V.: A genetic approach for training diverse classifier ensembles. In: International Conference on Information Processing and Management of Uncertainty in Knowledge-Based Systems, IPMU, pp. 798–805 (2008)
17. Rehm, F., Klawonn, F., Kruse, R.: Rule classification visualization of high-dimensional data. In: International Conference on Information Processing and Management of Uncertainty in Knowledge-Based Systems, IPMU, pp. 1944–1948 (2006)
18. Harle, C., Neill, D., Padman, R.: An information visualization approach to classification and assessment of diabetes risk in primary care. In: Li, J., Aleman, D., Sikora, R. (eds.) Proceedings of the 3rd INFORMS Workshop on Data Mining and Health Informatics (DM-HI 2008) (2008)
19. Talbot, J., Lee, B., Kapoor, A., Tan, D.: EnsembleMatrix: interactive visualization to support machine learning with multiple classifiers. In: ACM CHI (2009)
20. Wang, J., Yu, B., Gasser, L.: Classification Visualization with Shaded Similarity Matrix. University of Illinois at Urbana-Campaign, Technical report GSLIS (2002)
21. Velu, C.M., Kashwan, K.R.: Performance Analysis for visual data mining classification techniques of decision tree, ensemble and SOM. Int. J. Comput. Appl. **57**(22) (2012)
22. Holten, D.: Hierarchical edge bundles: visualization of adjacencyrelations in hierarchical data. IEEE Trans. Visualizationand Comput. Graphics **12**(5), 741–748 (2006)
23. Keim, D.A., Mansmann, F., Schneidewind, J., Thomas, J., Ziegler, H.: Visual analytics: scope and challenges. In: Simoff, S.J., Böhlen, M.H., Mazeika, A. (eds.) Visual Data Mining. LNCS, vol. 4404, pp. 76–90. Springer, Heidelberg (2008)
24. Keim, D., Qu, H., Ma, K.-L.: Big-data visualization. IEEE Comput. Graphics Appl. **33**(4), 20–21 (2013)
25. van Rijsbergen, C.: Information Retrieval, Butterworths (ed.) (1979)

Feature Selection from Partially Uncertain Data Within the Belief Function Framework

Asma Trabelsi[1(✉)], Zied Elouedi[1], and Eric Lefevre[2]

[1] Université de Tunis, Institut Supérieur de Gestion de Tunis, LARODEC, Le Bardo, Tunisia
trabelsyasma@gmail.com, zied.elouedi@gmx.fr

[2] Univ. Artois, EA 3926, Laboratoire de Génie Informatique et d'Automatique de l'Artois (LGI2A), 62400 Béthune, France
eric.lefevre@univ-artois.fr

Abstract. With the rapid growth of high dimensional data, feature selection has become a substantial task for several machine learning problems. In fact, it is regarded as an important process for classification performance owing to its ability to remove redundant and inconsistent features. The rough set theory is regarded as a well known tool allowing relevant feature selection. As the task of attribute selection using rough sets is an NP-hard problem to solve it, several heuristic algorithms have been introduced. The Johnson's algorithm, handling data characterized by certain and precise attribute values, is one of the most known ones. In this paper, we propose to extend this latter algorithm to an uncertain context, precisely where data contain uncertain condition attribute values represented within the belief function framework. We test the performance of our belief Johnson's algorithm through several experiments on synthetic databases.

Keywords: Classification · Feature selection · Rough set theory · Heuristic algorithms · Belief function theory

1 Introduction

Classification is regarded as a substantial problem in the fileds of machine learning and data mining and it has become increasingly challenging owing to the exponential data growth in both sample size and dimensionality [21]. Dimensionality reduction techniques such as feature selection are widely used to deal with high-dimensionality [15,20]. In fact, they allow to exclude as much as possible irrelevant and redundant attributes from the original set of attributes for the purpose of reducing the computational cost and the dimensionality space of huge data sets as well to improve the classification accuracy. Mainly, there exist three feature selection approaches: wrapper, filter and embedded. Wrapper methods incorporate classification algorithms to search and select a subset of attributes, while filter methods select a subset of attributes independently of any classification algorithm. In contrast to filter and wrapper, embedded methods performed

J.P. Carvalho et al. (Eds.): IPMU 2016, Part II, CCIS 611, pp. 643–655, 2016.
DOI: 10.1007/978-3-319-40581-0_52

the learning and the feature selection levels together. As the classification of each attribute or subsets of attributes is costly in term of computation time, we resort, in this investigation, to filter approaches. Rough set theory is considered as one of the most efficient filter methods allowing to find out the minimal set of relevant attributes also called reduct [14]. The main advantages of the reduct is its ability to predict the decision concepts as well as the whole set of attributes. Basically, finding the set of all reducts or finding the optimal reduct is regarded as an NP-hard problem which has led to the introduction of several heuristic approaches such as the QuickReduct algorithm [3], the Johnson's algorithm [8], etc. In this paper, we propose to adapt this latter heuristic algorithm thanks to its capacity to discover only one reduct with the minimal number of attributes generally close to the optimal from a given data. It is substantial to note that Johnson's algorithm handles only the case of perfect data. However, real world databases may be susceptible to imprecision, incompleteness and uncertainty. Such cases require to adapt the concept of rough sets heuristic algorithms to an uncertain environment. In several domains, uncertainty may exist either in decision attributes or in condition attributes or in both decision and condition attributes. For instance, in medicine, patients' symptoms (condition attributes) or patients' diseases (decision attribute) can be uncertain. Therefore, in this paper, we propose to adapt the Johnson's heuristic algorithm to an uncertain environment. Several theories have been discussed in the literature to handle uncertainty such as the bayesian theory [1], the fuzzy theory [6], the belief function theory [5], etc. As this latter formalism has the advantage to deal with partial or even total ignorance, we propose a belief Johnson algorithm to find reducts from a partially uncertain decision table. More precisely, we tackle the problem where uncertainty exists only in the condition attributes. This paper is organized as follows. Section 2 is dedicated to hightailing the basic concepts of the rough set theory. We detail Johnson's algorithm in Sect. 3. Section 4 provides an overview of the fundamental concepts of the belief function theory. Our novel approach for feature selection based on the belief Johnson's algorithms has been presented in Sect. 5. Section 6 describes the experimental results yielded from several uncertain databases under the classifier fusion framework, in order to evaluate the performance of our novel approach. In Sect. 7, we draw conclusion and we highlight some future works.

2 Rough Set Theory

Rough Sets (RS), introduced by Pawlak [14], is a valid mathematical tool for dealing with imperfect knowledge (vague, imprecise and uncertain) in variety of applications related to machine learning area which mainly includes the problems of knowledge discovery, clustering [13], classification [7,10], feature selection [2,11], etc. This latter paradigm consists of extracting the smallest subsets of relevant features, also called reducts, from the original set of features of a given data. In a practical point of view, information and knowledge are represented by a decision information system which is defined as a pair $A = (U, R)$, where

$U = \{O_1, \ldots, O_n\}$ is a non-empty, finite set of objects called the universe and $R = C \cup D$ is a finite set of attribute, $C = \{c_1, \ldots, c_K\}$ is a non-empty, finite set of K condition attributes, v_{c_k} is a non-empty set of values of $c_k \in C$, $D = \{\text{d}\}$ is the decision attribute set and v_d is the decision attribute value [23].

Given a subset of condition attributes $B \subseteq C$, an indiscernibility relation, denoted by $IND(B)$, is defined as follows:

$$IND(B) = U/B = \{(O_i, O_j) \in U \times U | \forall c_k \in B, v_{c_k}(O_i) = v_{c_k}(O_j)\} \quad (1)$$

The indiscernibility relation based on the decision attribute $\{d\}$, denoted by $IND(\{d\})$, is set to:

$$IND(\{d\}) = U/\{d\} = \{O_j \in U | [O_j]_{\{d\}}\} \quad (2)$$

Let $B \subseteq C$ and $X \subseteq U$. We can approximate X by using only the information contained by constructing the B-lower and B-upper approximations of X, denoted respectively by $\underline{B}(X)$ and $\overline{B}(X)$ and defined by:

$$\underline{B}(X) = \{O_j | [O_j]_B \subseteq X\} \quad (3)$$

and

$$\overline{B}(X) = \{O_j | [O_j]_B \cap X = \emptyset\} \quad (4)$$

where

$$[O_j]_B = \{O_i | \forall c_k \in B, v_{c_k}(O_i) = v_{c_k}(O_j)\} \quad (5)$$

The positive region, embracing all objects of U that can be classified to blocks of $U/\{d\}$ by means of the condition attributes B, is defined as :

$$Pos_B(\{d\}) = \bigcup_{X \in U/\{d\}} \underline{B}(X) \quad (6)$$

Keeping only attributes that preserve the positive region is regarded as a practical way for feature reduction. It is noteworthy that there exist several subsets of condition attributes and those which are minimal are called reducts. A subset $B \subseteq C$ is a reduct of C with respect to D, if B is minimal and:

$$Pos_B(D) = Pos_C(D) \quad (7)$$

In other terms, the attributes that do not belong to any reduct are unnecessary for the classification of the universe elements. Authors in [17], have introduced the notation of discernibility matrix and function as other ways for finding reducts for a decision table DT. The discernibility matrix of DT, denoted by M, is a $|U| \times |U|$ matrix, in which the element $M(O_i, O_j)$ for an object pair (O_i, O_j) is defined by:

$$M(O_i, O_j) = \{v_c \in C | v_c(O_i) \neq v_c(O_j) \text{ and } v_d(O_i) \neq v_d(O_j)\} \forall i, j = \{1, \ldots, n\}$$

The matrix element $M(O_i, O_j)$ represents the set of all condition attributes discerning objects O_i and O_j that do not have the same value of the decision attribute d. The notion of discernibility function can be defined from the discernibility matrix as follows:

$$f(M) = \wedge\{\vee(M(O_i, O_j))|\forall O_i, O_j \in U, M(O_i, O_j) = \emptyset\} \tag{8}$$

Reducts may be yielded by transforming the discernibility function from conjunctive normal form into disjunctive normal form. The major shortcoming of this solution is its costly operation which makes it impractical for medium sized or large sized data sets. Therefore, several heuristic algorithms have been discussed to overcome this drawback. The best known of them is the Johnson's heuristic algorithm [8].

3 Johnson's Heuristic Algorithm

Johnson's algorithm proposed in [8] is an heuristic algorithm that uses a greedy search technique which consists of picking out attributes having the most frequency appearing in the discernibility matrix. Algorithm 1 below underlines the main steps of the Johnson algorithm.

Algorithm 1. Johnson's Algorithm(U,C ∪ d)

```
input:U: a finite set of instances, C: a set of conditional attributes, d: a set of
  decision attributes
Output: R:reduct, R ⊆ C
R ← ∅
M ← DiscernibilityMatrix (U,C ∪ d)
repeat
  c ← SelectAttributeWithMaxWeight(M)
  R ← R ∪ {c}
  for i=1 to |U| do
    for j=1 to |U| do
      if c ∈ M(O_i, O_j) then
        M(O_i, O_j) = ∅
      end if
    end for
  end for
until (M(O_i, O_j) =∅ ∀ i,j)
```

Johnson's algorithm begins by setting the reduct candidate, denoted by R, to an emptyset. Subsequently, it computes the number of occurrences of each attribute in the discernibility matrix. The attribute that has the highest count of appearances will be added to R and all cells containing this attribute will be removed from the discernibility matrix. This process should be repeated until all non empty cells are removed. Then, the algorithm returns R as a final reduct.

Though Johnson's algorithm guarantees to uncover a single reduct, it is unuseful in the case where data sets are characterized by uncertain attributes. Thus, we propose to extend this algorithm to an uncertain context, more particulary to the context of the belief function theory.

4 Belief Function Theory

The belief function theory, also known as Dempster-Shafer Theory (DST) or theory of evidence [16], is considered as a useful theory for representing and managing uncertain knowledge. In what follows, we briefly introduce the main concepts of the belief function theory as interpreted in the Transferable belief Model (TBM) [19].

Let Θ be a finite non-empty set of N elementary events related to a given problem, these events are assumed to be exhaustive and mutually exclusive. Such Θ is called the frame of discernment. The power set of Θ, denoted by 2^{Θ}, is composed of all the subsets of Θ.

The impact of evidence assigned to each subsets of the frame of discernment Θ is named basic belief assignment (bba) and is defined as:

$$m : 2^{\Theta} \rightarrow [0, 1]$$
$$\sum_{A \subseteq \Theta} m(A) = 1 \qquad (9)$$

The amount $m(A)$, known as basic belief mass (bbm), expresses the degree of belief committed exactly to the event A.

To make decision within the belief function framework, we must transform the bba into a probability measure called pignistic probability denoted $BetP$ and defined as follows [18]:

$$BetP(A) = \sum_{B \subseteq \Theta} \frac{|A \cap B|}{|B|} \frac{m(B)}{1 - m(\emptyset)} \forall A \in \Theta \qquad (10)$$

5 Belief Johnson's Algorithm for Partially Uncertain Data

This Section is devoted to describing our heuristic approach for feature selection form partially uncertain decision table. Our proposed solution, namely belief Johnson's algorithm, aims to extract the subset of relevant attributes which enables the same classification ability as the entire set of attributes. In what follows, we provide firstly a brief description of a partially uncertain decision table under the belief function framework and then we detail our heuristic approach.

5.1 Partially Uncertain Decision Table

Our partially uncertain decision table will be defined as a pair UDT=(U, uC ∪ d) where U is a finite set of n objects U={O_1,. . .,O_n} described by a set of K uncertain condition attributes denoted by uC={c_1, . . ., c_K} and a certain decision attribute denoted by {d}. In this work, we suggest to represent the uncertainty of each condition attribute within the belief function framework. Thus, a basic belief assignment $m_i^{\Theta_k}$, defined on the frame of discernment Θ_k which represents all possible values of a condition attribute $c_k \in uC$, will be assigned to each condition attribute value v_{c_k} of an instance O_i. These bbas can be induced by one or several agents and they may express the case of total certainty ($m_i^{\Theta_k}(\{v_{c_k}\}) = 1$ and $m_i^{\Theta_k}(\Theta_k) = 0$) or even the case of total ignorance ($m_i^{\Theta_k}(\{v_{c_k}\}) = 0$ and $m_i^{\Theta_k}(\Theta_k) = 1$).

Example: Let Table 1 be our uncertain decision table composed with eight instances characterized by three uncertain categorical condition attributes $uC = \{Hair, Eye, Height\}$ and a certain decision attribute d with possible values {d_1, d_2}. To simplify the notations, we will use 1, 2 and 3 instead of $Hair$, Eye and $Height$. The basic belief assignments, which are randomly affected to the condition attribute values, will be defined on the frame of discernments $\Theta_1 = \{Blond, Dark\}$, $\Theta_2 = \{Brown, Blue\}$ and $\Theta_3 = \{Short, Middle, Tall\}$.

Table 1. Uncertain decision table

	$Hair$	Eye	$Height$	d
O_1	$m_1^{\Theta_1}(\{Dark\})$=0.5 $m_1^{\Theta_1}(\Theta_1)$=0.5	$m_1^{\Theta_2}(\{Brown\})$=1 $m_1^{\Theta_2}(\Theta_2)$=0	$m_1^{\Theta_3}(\{Middle\})$=0.95 $m_1^{\Theta_3}(\Theta_3)$=0.05	d_1
O_2	$m_2^{\Theta_1}(\{Blond\})$=0.1 $m_2^{\Theta_1}(\Theta_1) = 0.9$	$m_2^{\Theta_2}(\{Blue\})$=0.82 $m_2^{\Theta_2}(\Theta_2)$=0.18	$m_2^{\Theta_3}(\{Middle\})$=1 $m_2^{\Theta_3}(\Theta_3)$=0	d_1
O_3	$m_3^{\Theta_1}(\{Blond\})$=0.6 $m_3^{\Theta_1}(\Theta_1) = 0.4$	$m_3^{\Theta_2}(\{Brown\})$=0.2 $m_3^{\Theta_2}(\Theta_2)$=0.8	$m_3^{\Theta_3}(\{Tall\})$=0.55 $m_3^{\Theta_3}(\Theta_3)$=0.45	d_2
O_4	$m_4^{\Theta_1}(\{Dark\})$=0.7 $m_4^{\Theta_1}(\Theta_1) = 0.3$	$m_4^{\Theta_2}(\{Brown\})$=0 $m_4^{\Theta_2}(\Theta_2)$=1	$m_4^{\Theta_3}(\{Short\})$=1 $m_4^{\Theta_3}(\Theta_3)$=0	d_1
O_5	$m_5^{\Theta_1}(\{Blond\})$=1 $m_5^{\Theta_1}(\Theta_1) = 0$	$m_5^{\Theta_2}(\{Blue\})$=0.18 $m_5^{\Theta_2}(\Theta_2)$=0.82	$m_5^{\Theta_3}(\{Middle\})$=0.15 $m_5^{\Theta_3}(\Theta_3)$=0.85	d_2
O_6	$m_6^{\Theta_1}(\{Blond\})$=0.3 $m_6^{\Theta_1}(\Theta_1) = 0.7$	$m_6^{\Theta_2}(\{Brown\})$=0.13 $m_6^{\Theta_2}(\Theta_2)$=0.87	$m_6^{\Theta_3}(\{Tall\})$=0.8 $m_6^{\Theta_3}(\Theta_3)$=0.2	d_2
O_7	$m_7^{\Theta_1}(\{Dark\})$=1 $m_7^{\Theta_1}(\Theta_1) = 0$	$m_7^{\Theta_2}(\{Brown\})$=0.8 $m_7^{\Theta_2}(\Theta_2)$=0.2	$m_7^{\Theta_3}(\{Tall\})$=0.25 $m_7^{\Theta_3}(\Theta_3)$=0.75	d_1
O_8	$m_8^{\Theta_1}(\{Dark\})$=0.5 $m_8^{\Theta_1}(\Theta_1) = 0.5$	$m_8^{\Theta_2}(\{Blue\})$=0.22 $m_8^{\Theta_2}(\Theta_2)$=0.78	$m_8^{\Theta_3}(\{Short\})$=0.1 $m_8^{\Theta_3}(\Theta_3)$=0.9	d_1

5.2 Reducts for Partially Uncertain Decision Table

Let us remind that the reduct, using Johnson's algorithm, is constructed by sequentially adding the most discernable attributes for a given decision attribute. Therefore, the computation of the discernibility matrix M will be a preliminary step in Johnson's algorithm. However, computing M from partially uncertain decision table UDT=(U, $uC \cup d$) remains really a challenging task which has not attracted great attention yet. To cope with this problem, we propose to adapt Johnson's heuristic algorithm to an uncertain environment, precisely to the belief function framework. Our belief Jonson's algorithm tackles mainly the problem where the uncertainty exists only in the condition attributes and represented within the framework of belief functions. In such cases, dissimilarity metrics must be used to discern all pairs of objects with different decision values. Consequently, entries of the discernibility matrix should be set as follows $\forall\ i,j \in \{1,\ldots,n\}$ and $k \in \{1,\ldots,K\}$:

$$M'(O_i,O_j) = \{c_k \in C|dist(m_i^{\Theta_k}, m_j^{\Theta_k}) > S \text{ and } v_d(O_i) \neq v_d(O_j)\} \tag{11}$$

where S denotes a tolerance threshold and $dist$ corresponds to a distance measure between two bbas. Different distance metrics have been investigated in the literature such as the Tessems distance [22], the Euclidean distance [4], the Jousselme distance [9], etc. This latter is one of the most commonly used distances. Given two bbas m_1 and m_2, the Jousselme distance measure is computed as follows:

$$dist(m_1,m_2) = \sqrt{\frac{1}{2}(m_1 - m_2)^T D(m_1 - m_2)} \tag{12}$$

with D is the Jaccard index matrix, the elements of which are calculated as follows:

$$D(A,B) = \begin{cases} 1 & \text{if } A{=}B{=}\ \emptyset \\ \dfrac{|A \cap B|}{|A \cup B|} & \forall\ A,B \in 2^{\Theta} \end{cases} \tag{13}$$

Once the discernibility matrix is computed, the reduct will be incrementally composed by adding the condition attribute that occurs with the most frequency and then removing any cells contain this attribute. This procedure must be repeated until all non-empty cells will be eliminated.

Example: In order to extract the reduct relative to our partially uncertain decision table (see Table 1), we start by computing the discernibility matrix M' where the threshold S is setting to 0.1 (see Table 2). To simplify, we use the notations Ha, E and He respectively for $Hair$, Eye and $Height$. For instance, $M'(O_1,O_2)$=$\emptyset$ due to the fact that $v_d(O_1) = v_d(O_2)$. Another example, $M'(O_1,O_5) = \{Ha, E, He\}$ because $dist(m_1^{\Theta_1}, m_5^{\Theta_1}) = 0.5 > 0.1$, $dist(m_1^{\Theta_2}, m_5^{\Theta_2}) = 0.7185 > 0.1$, $dist(m_1^{\Theta_3}, m_5^{\Theta_3}) = 0.6532 > 0.1$ and $v_d(O_1) \neq v_d(O_5)$.

Table 2. Discernibility matrix M'

	O_1	O_2	O_3	O_4	O_5	O_6	O_7	O_8
O_1	-							
O_2	-	-						
O_3	E, He	Ha, E, He						
O_4	-	-	Ha, E, He					
O_5	Ha, E, He	Ha, E, He	-	Ha, E, He	-			
O_6	Ha, E, He	Ha, E	-	Ha, He	-	-		
O_7	-	-	Ha, E, He	-	Ha, E, He	Ha, E, He		
O_8	-	-	E, He	-	Ha, E, He	Ha, E, He	-	-

Table 3. Description of databases

Databases	#Instances	#Attributes
Tic-Tac-Toe	958	9
SPECT Heart	267	22
Lymphography	148	18
Voting Records	435	16
Zoo	101	17

Let us now compute the reduct using our belief Johnson's algorithm. Firstly, we count the number of occurrences relative to each condition attribute and the feature with the highest frequency will be added to the reduct. In our discernibility matrix (Table 2), the attributes *Eye* and *Height* appear 14 times, while the attribute *Hair* appears 13 times. As attributes *Eye* and *Height* have equal weights, we randomly add one among them to the reduct R. If the attribute *Eye* is chosen then we remove all cells containing *Eye* from M' and the next best feature will be selected. By removing *Eye*, we still have *Hair* and *Height* with weights equal to 1. As *Hair* and *Height* have equal weights, we add either *Hair* or *Height* to R and then we remove the chosen attribute from M': if we remove the attribute *Hair*, R will be set to $R = \{Eye, Hair\}$ and M' will be empty. By against, if we remove the attribute *Height*, R will be equal to $R = \{Eye, Height\}$ and M' will be empty.

6 Experimentations

In order to evaluate the performance of our heuristic feature selection approach, we propose to carry out several experimental tests on real world databases obtained from the U.C.I. repository [12]. Table 3 gives a brief description of the databases where #Instances and #Attributes denote respectively the total number of instances and the total number of condition attributes.

As all these databases do not contain uncertain condition attributes represented within the belief function framework, we propose to generate synthetic databases by taking into account the original database D and a degree of uncertainty P to transform actual condition attribute value v_{c_k} of each object O_i, where $c_k \in uC$, into a basic belief assignment as follows:

$$m_i^{\Theta_k}(\{v_{c_k}\}) = 1 - P$$
$$m_i^{\Theta_k}(\Theta_k) = P \qquad (14)$$

The degree of uncertainty P takes value in the interval [0,1]: Certain Case (P=0), Low Uncertainty ($0 \leq P < 0.4$), Middle Uncertainty ($0.4 \leq P < 0.7$) and High Uncertainty ($0.7 \leq P \leq 1$).

To check the validity of our proposed heuristic approach, we try to perform an empirical comparison in terms of dimensionality space and classification accuracy criterion (PCC) between results yielded by our initial databases and those obtained by our belief Johnson's algorithm in both certain and uncertain cases. In order to compare PCCs, we resort to three well known classification algorithms, namely the Decision Tree classifier (DT), the Naive Bayes classifier (NB) and the k-Nearest Neighbor classifier (k-NN) with k equals to 1. As these classification algorithms cannot handle data characterized by uncertain condition attributes represented within the framework of belief functions, we perform the pignistic transformation, using Eq. 9, to make decision about condition attribute values which should be chosen. Once computing the pignistic probability of all condition attribute beliefs, we run the three mentioned classifiers using the leave one out cross validation approach which divides a data set with N instances into N-1 instances for training and the remaining instance for testing. This procedure will be repeated N times where each existing instance is used once as a test set. Experimental results are given from Table 4–6 where #F denotes the number of selected attributes. Note that, for the sake of simplification, we have replaced the attribute names in the reduct by numbers according to their order in the databases.

We remind that our ultimate objective is to reduce dimensionality space as well as the computational time and keep or increase the classification accuracy. Let us note that in certain case our belief Johnson's algorithm gives exactly

Table 4. Classification accuracy (%) without dimensionality reduction

Databases	NB	DT	1-NN
Tic-Tac-Toe	82.04	69.41	99.16
SPECT Heart	84.64	79.40	82.39
Lymphography	79.05	83.78	82.43
Voting Records	96.55	91.37	90.84
Zoo	92.07	95.04	96.03

Table 5. Belief Johnson's algorithm: certain case

Databases	Reduct	#F	PCC (%)		
			NB	DT	1-NN
Tic-Tac-Toe	R={1, 2, 3, 4, 5, 6, 8, 9}	8	80.58	71.71	81.41
SPECT Heart	R={1, 3, 4, 5, 6, 7, 8, 9, 10, 13, 14, 16, 19, 20, 21, 22}	16	79.40	79.77	80.52
Lymphography	R={2, 13, 14, 15, 16, 18}	6	77.02	80.40	81.75
Voting Records	R=1, 2, 3, 4, 11, 13, 15, 16	8	96.12	93.96	94.82
Zoo	R={4, 7, 9, 11, 14}	5	94.05	90.09	96.03

the same results as the original Johnson's algorithm. From the results given in Tables 4, 5 and 6, we have deduced that in both certain and uncertain cases, our belief Johnson's algorithm allows a significant dimensionality reduction. For instance, applying our belief Johnson's algorithm to the Lymphography database containing 18 condition attributes we obtain 6 selected features in certain case, while applying this proposed algorithm to synthetic Lymphography database we obtain 5 selected features for both low and middle uncertainty cases and 6 selected feature for high uncertainty case.

In terms of the PCC criterion, we emphasize that for our certain case feature reduction allows the improvement of the PCC criterion compared to those yielded by initial databases, though not always. However, the PCCs yielded following to the feature reduction process are often close to those obtained with the initial databases. For example, for the initial Spect-Heart database, we have 84.64 %, 79.40 % and 82.39 % as PCCs relative to respectively DT, NB and 1-NN classifiers, while applying our belief Johnson's algorithm in certain case, we get 79.40 %, 79.77 % and 80.52 % as PCCs relative to respectively DT, NB and 1-NN classifiers. Consequently, we can admit that feature reduction allows not only to reduce dimensionality space and computational time, but also to provide significant classification accuracies and thus, it is worth applying belief Johnson's algorithm to partially uncertain databases. Concretely, we have tackled three levels of uncertainty: Low, Middle and High. From Table 6, we can deduce that the Decision Tree, the Naive Bayes and the 1-Nearest Neighbors classifiers have yielded interesting PCC values for the different synthetic databases obtained by using the three levels of uncertainty. For instance, for Voting Records database, we have gotten 95.68 %, 94.39 % and 96.12 % as PCCs obtained respectively by the DT, the NB and the 1-NN classifiers in low uncertainty case, for the middle uncertainty case, we have obtained 95.68 %, 95.25 % and 96.12 % as PCCs relative respectively to the DT, the NB and the 1-NN classifiers. Also, we have reported 96.55 %, 94.39 % and 95.68 % as PCCs obtained respectively by the DT, the NB and the 1-NN classifiers in high uncertainty case.

Table 6. Belief Johnson's algorithm: uncertain case

Databases	Low					Middle					High				
	Reduct	#F	PCC (%)			Reduct	#F	PCC (%)			Reduct	#F	PCC (%)		
			NB	DT	1-NN			NB	DT	1-NN			NB	DT	1-NN
Tic-Tac-Toe	R={1, 2, 3, 4, 5, 6, 7, 9}	8	82.56	73.48	92.90	R={1, 2, 3, 5, 7, 8, 9}	7	78.81	69.83	89.53	R={1, 2, 3, 4, 5, 6, 7, 8, 9}	9	82.04	69.41	99.16
SPECT Heart	R={1, 3, 4, 5, 6, 7, 8, 9, 10, 13, 14, 15, 20, 22 }	14	82.77	79.77	80.52	R={1, 3, 7, 8, 9, 10, 11, 13, 14, 16, 17, 20, 21, 22 }	14	84.26	80.14	81.64	R={1, 5, 6, 7, 8, 9, 10, 13, 16, 17, 20, 22 }	12	83.52	81.64	82.02
Lymphography	R={2, 13, 14, 15, 18}	5	75.67	76.35	79.72	R={2, 13, 14, 15, 18}	5	75.67	76.35	79.72	R={2, 13, 14, 15, 16, 18}	6	79.05	78.37	70.05
Voting Records	R={2, 3, 4, 11,13, 16}	6	95.68	94.39	96.12	R={1, 2, 3, 4, 11, 13}	6	95.68	95.25	96.12	R={2, 3, 4, 5, 6, 9, 11, 13, 15}	9	96.55	94.39	95.68
Zoo	R={2, 4, 7, 9, 14}	5	93.06	87.12	95.04	R={4, 5, 7, 9, 14}	5	94.05	85.14	95.04	R={2, 5, 6, 7, 8, 9, 13, 14}	8	95.04	91.08	91.08

7 Conclusion

In this paper, we have proposed a new heuristic approach for relevant feature selection from partially uncertain decision table, precisely where uncertainty exists only in the condition attributes and represented within the belief function framework. Our experimental tests have shown the efficiency of our proposed method in terms of dimensionality reduction and classification accuracy. The major limitation of our proposed approach consists on the one hand of its inability to give optimal reducts and, on the other hand of the adaptation of classical machine learning to handle what is called uncertain databases. So, as a future work, we look forward to improving our proposed method by allowing the optimal reduct. We also regard to use learning algorithms adapted to uncertain data in order to check the validity of our approach. We further intend to introduce uncertainty in both condition and decision attributes.

References

1. Bernardo, J.M., Smith, A.F.: Bayesian theory, vol. 405. John Wiley & Sons, New York (2009)
2. Bhatt, R.B., Gopal, M.: On fuzzy-rough sets approach to feature selection. Pattern Recogn. Lett. **26**(7), 965–975 (2005)
3. Chouchoulas, A., Shen, Q.: Rough set-aided keyword reduction for text categorization. Appl. Artif. Intell. **15**(9), 843–873 (2001)
4. Cuzzolin, F.: A geometric approach to the theory of evidence. IEEE Trans. Syst. Man Cybern. Part C Appl. Rev. **38**(4), 522–534 (2008)
5. Denœux, T.: Reasoning with imprecise belief structures. Int. J. Approximate Reasoning **20**(1), 79–111 (1999)
6. Haag, T., Hanss, M.: Comprehensive modeling of uncertain systems using fuzzy set theory. In: Elishakoff, I., Soize, C. (eds.) Nondeterministic Mechanics, vol. 539, pp. 193–226. Springer, Vienna (2012)
7. Jensen, R., Cornelis, C.: Fuzzy-rough nearest neighbour classification and prediction. Theoret. Comput. Sci. **412**(42), 5871–5884 (2011)
8. Johnson, D.S.: Approximation algorithms for combinatorial problems. In: The 5th Annual ACM Symposium on Theory of Computing, pp. 38–49. ACM (1973)
9. Jousselme, A., Grenier, D., Bossé, E.: A new distance between two bodies of evidence. Inf. fusion **2**(2), 91–101 (2001)
10. Khoo, L., Tor, S., Zhai, L.: A rough-set-based approach for classification and rule induction. Int. J. Adv. Manuf. Technol. **15**(6), 438–444 (1999)
11. Kumar, P., Vadakkepat, P., Poh, L.A.: Fuzzy-rough discriminative feature selection and classification algorithm, with application to microarray and image datasets. Appl. Soft Comput. **11**(4), 3429–3440 (2011)
12. Murphy, P., Aha, D.: UCI repository databases (1996). http://www.ics.uci.edu/mlearn
13. Parmar, D., Wu, T., Blackhurst, J.: MMR: an algorithm for clustering categorical data using rough set theory. Data Knowl. Eng. **63**(3), 879–893 (2007)
14. Pawlak, Z.: Rough sets. Int. J. Comput. Inform. Sci. **11**(5), 341–356 (1982)
15. Pizzi, N.J., Pedrycz, W.: Aggregating multiple classification results using fuzzy integration and stochastic feature selection. Int. J. Approximate Reasoning **51**(8), 883–894 (2010)

16. Shafer, G.: A mathematical theory of evidence, vol. 1. Princeton University Press, Princeton (1976)
17. Skowron, A., Rauszer, C.: The discernibility matrices and functions in information systems. In: Słowiński, R. (ed.) Intelligent Decision Support, pp. 331–362. Springer, Netherlands (1992)
18. Smets, P.: The transferable belief model for quantified belief representation. Handb. Defeasible Reasoning Uncertainty Manag. Syst. **1**, 267–301 (1998)
19. Smets, P., Kennes, R.: The transferable belief model. Artif. Intell. **66**(2), 191–234 (1994)
20. Sorzano, C.O.S., Vargas, J., Montano, A.P.: A survey of dimensionality reduction techniques. arXiv preprint arXiv:1403.2877
21. Sun, C., Rampalli, N., Yang, F., Doan, A.: Chimera: large-scale classification using machine learning, rules, and crowdsourcing. Proc. VLDB Endowment **7**(13), 1529–1540 (2014)
22. Tessem, B.: Approximations for efficient computation in the theory of evidence. Artif. Intell. **61**(2), 315–329 (1993)
23. Yao, Y., Zhao, Y.: Discernibility matrix simplification for constructing attribute reducts. Inf. sci. **179**(7), 867–882 (2009)

On the Suitability of Type-1 Fuzzy Regression Tree Forests for Complex Datasets

Fathi Gasir[1(✉)] and Keeley Crockett[2]

[1] Computer Science Department, Faculty of Information Technology, Misurata University, Misurata, Libya
F.Gasir@it.misuratau.edu.ly

[2] The Intelligent Systems Group, School of Computing, Mathematics and Digital Technology, Manchester Metropolitan University, Manchester, Chester Street M1 5GD, UK
K.Crockett@mmu.ac.uk

Abstract. One of the challenges in data mining practices is that the datasets vary in complexity and often have different characteristics such as number of attributes, dependent variables characteristics etc. In terms of regression problems, the features that describe the dataset will vary in their complexity, sparseness verses coverage in relation to the decision space, and the number of outcome classes. Fuzzy Decision trees are well-established classifiers in terms of building robust, representative models of the domain. In order to represent different perspectives of the same domain, fuzzy trees can be used to construct fuzzy decision forests to enhance the predictive ability of singular trees. This paper describes an empirical study which examines the applicability of fuzzy tree regression forests to seven different datasets which have complex properties. The relationship between dataset characteristics and the performance of fuzzy regression tree forests is debated.

Keywords: Fuzzy decision trees · Fuzzy regression forests

1 Introduction

It is a known problem that the complexity of data is becoming increasing challenging for traditional machine learning algorithms to deal with, especially in the Big Data arena where data variety, veracity and volume have to be taken into consideration. However, the debate continues on whether the focus should be on developing better algorithms or to generate models using more data [1]. In the context of Big Data, Kwona and Simb [2] performed a comprehensive study on the performance of classification algorithms in relation to a datasets features. The experimental study found that legacy classification algorithms performed differently depending on how the data was structured, its content and context in which it was applied [2]. For example, the number of features in any data set not only affects the time to produce an optimal model, but also influences the performance when using classification algorithms [3–5].

Fuzzy decision trees allow data instances to simultaneously fire multiple branches of a node with different degrees of membership whereby allowing all information to

J.P. Carvalho et al. (Eds.): IPMU 2016, Part II, CCIS 611, pp. 656–663, 2016.
DOI: 10.1007/978-3-319-40581-0_53

contribute towards the final classification [6, 7]. More specifically, fuzzy regression trees are used where there is a non-linear relationship between input and output variables. Fuzzy decision tree forests have been shown to improve the predictive power of a singular fuzzy trees by allowing numerous insights and interpretations of the datasets that are being modelled [8–13]. Fuzzy Forests designed for classification problems have also be shown to be tolerant to noisy data [9].

The study presented in this paper investigates the relationship between the datasets characteristics, number of features, and datasets sizes and the performance of fuzzy regression tree forests in the context of regression tree problems. An empirical study on seven known datasets and generates for each fuzzy decision forests comprising each of five fuzzy trees using the Elgasir algorithm [13]. Fuzzification is optimised in each case by using the adapted version of an artificial immune network model (opt-aiNet [14]). A series of experiments is conducted to determine whether the characteristics of the data affects the performance of the fuzzy regression forests. This is determined through a comparison with singular crisp regression trees. This paper is structured as follows: Sect. 2 provides an overview of related work in the field of fuzzy regression trees and forests. Section 3 describes the algorithm for constructing type-1 fuzzy forests using the Elgasir algorithm. The characteristics of the dataset are described in 4, with the experimental methodology and results in Sect. 5. Finally, conclusions are presented in Sect. 6.

2 Related Work

Regression tree induction algorithms [15] are a technical approach which are used to construct a set of rules that will predict events in a given domain. Regression tree induction algorithms induce rules from the knowledge of a set of examples, known as a training dataset, whose predicted outcome is already known. The process of regression tree induction involves selecting [15]. CHAID provides a set of rules that can be applied to a new (unseen) dataset to predict the target or outcome. The CHAID algorithm stops growing a tree before overfitting occurs, as a result of using its unique dynamic branching strategy for determining the optimal number of branches. This strategy merges together attribute values that are shown to be statistically homogenous (similar), retaining the values that are heterogeneous (dissimilar). Trees generated from traditional tree induction algorithms are often referred to as "crisp" and suffer from sharp decision boundaries which results of using the strict partitioning for regression trees induction [7] and values are restricted to a limited number of discrete values as a result of using a discrete function to generate the tree output.

Fuzzy decision tree rule induction algorithms overcome such problems by allowing gradual transitions to exist between continuous attributes at tree nodes and utilizing fuzzy inference to combine information throughout the tree rather than following a single root to leaf node path. Early methods of fuzzy decision tree development replied on experts in the domain to pre-fuzzify the data prior to induction – a task that introduced a further uncertainty through subjectivity. Specific to this paper is attempts to fuzzily the CHAID algorithm. First achieved by Fowdar et al. [6], the Fuzzy CHAID Induction Algorithm produced robust fuzzy trees with significantly higher accuracies than its crisp counterpart. The fuzzy regression tree algorithm known as Elgasir, also based on CHAID, incorporated degrees of uncertainty typical in data through the use of trapezoidal membership functions and Takagi-Sugeno fuzzy inference is applied to aggregate a final continuous output value. Elgasir alleviates Fowdar's defuzzification problem as a result of using Takagi-Sugeno fuzzy inference to aggregate fuzzy regression tree output as a single numeric value [16].

Fuzzy decision tree forests or assembles allow the concepts of fuzzy decision trees to be applied to allow different models of the same domain to be combined. Of significance in the field was Bonissone et al. [12] approach which used a fuzzy learning algorithm to create singular fuzzy trees using Breiman's metholdogy and then applied different configurations of combining leaf information. A different approach described in Crockett et al. [7] involved the use of creating multiple fuzzy C4.5 decision trees from non-fuzzy really world data by selecting as the root attributes with high to low information content. Cadenas et al. [11] showed used a fuzzy random forest assemble method to select features for classification problems thus reducing dimensionality and improving classification accuracy. Work in this field has focused on classification and little work has reported on regression problems.

3 An Algorirthm for Constructing Type-1 Fuzzy Decision Tree Forests

This section outlines the Elgasir fuzzy regression tree rule induction algorithm and describes how it is used to generate fuzzy regression tree forests.

3.1 The Elgasir Algorithm

The aim of the fuzzy regression tree algorithm Elgasir [13] was to apply appropriate membership functions to all branch split points in order to master the weakness of crisp decision trees, by allowing all the information used throughout the tree to contribute towards the outcome. Elgasir's foundations were based on Kass's CHAID Algorithm [15]. CHAID is a highly efficient statistical technique used to induce standard regression trees that are easy for humans to interpret. In order to reduce the strict partitioning at nodes and represent uncertainty, Elgasir combined principles of fuzzy theory and Takagi-Sugeno fuzzy inference technique to produce type-1 fuzzy regression trees. [16]. In order to optimise fuzzy set boundaries throughout the tree, an adapted version of an artificial immune network model (opt-aiNet [13]) was applied. A brief overview of the algorithm is provided below and a full description can be found in [13].

1. Randomization and Partition the dataset into training and test data using multi-fold cross validation.
2. Apply CHAID Crisp Regression Trees rule induction algorithm for the first subset data.
   a. Generate optimal crisp CHAID regression tree from the training dataset by empirically applying various values to CHAID regression parameters.
   b. Evaluate performance of crisp tree using test dataset.

3. Convert near-optimal crisp tree into a set of IF-THEN rules.
4. Fuzzification of crisp CHAID regression tree.
   a. Repeat until the near-optimal performance of the fuzzy regression tree is reached.
      (i) Apply adapted opt-aiNet to determine optimal amount of fuzzification to membership functions in all branch split points within the antecedent rules and Repeat
      (ii) Parameterize consequent part of IF-THEN rules, where n is the total number of IF-THEN rules converted from the near-optimal crisp tree (step 2.a).
      (iii) Identification of consequent parameters using the training dataset.
      (iv) Evaluation of grades of membership.

   .
5. Repeat steps 2 to 4, until each subset has been used once as a test dataset.

   Step 6. Report on overall average error rate.

3.2 Constructing Forests

The Elgasir algorithm described in Sect. 3.1 can be used to create fuzzy decision forests comprising of *n* fuzzy regression trees from one training sample where each tree represents a different perspective of the training sample. This allows better coverage of the domain which is less sensitive to noise in the data. The methodology reported in [13] comprises of three stages. Stage 1 generates *n* crisp regression trees using the CHAID algorithm and converts in to a fuzzy rule base; Stage 2 involves determination of fuzzy sets around each tree node and associated membership functions; Stage 3 requires optimization of fuzzy membership functions using the immune network opt-aiNet. In this work optimal forests are conducted for all datasets in this study.

4 Characteristics of Data

Seven known datasets are used in this study. Based in three criterion: the number of instances, number of attributes and number of unique values. They were selected the Boston Housing dataset is used to predict the median value of owner occupied homes, in $1,000's, as collected by the U.S Census Service concerning housing in the area of Boston, Massachusetts [17]. The Abalone dataset is concerned with predicting the age of abalone from physical measurements and has 28 unique outcome values [17]. The Compactiv dataset [18] is a collection of computer systems activity measures where the prediction task is to predict the variable *usr*, the portion of time that CPUs run in user mode. The Elevators dataset [17] is obtained from the task of controlling an F16 aircraft, and the goal variable is related to an action taken on the elevators of the aircraft. The Stock prices dataset [17] contains daily stock prices, from January 1988 through to October 1991, for ten aerospace companies. The task is to approximate the price of the 10th company, given the price of the others. The Concrete Compressive Strength Dataset comprising of 938 attributes is used to predict the concrete compressive strength [18]. Finally, the Communities and Crime dataset (120 attributes) is used to predict the per capita violent crime, 121 instances were left after the instances with missing attributes were remove [18]. This dataset has 120 attributes describing various social, economic and criminal factors. Table 1 presents a summary of the characteristics of these datasets.

Table 1. Dataset Characteristics

Name	Number of instances	Number of attributes	Unique Values
Boston housing	506	14	229
Abalone	4177	9	28
Compactiv	8192	21	56
Elevators	16599	18	61
Stock prices	950	9	203
Crime	121	120	115
Concrete	1030	8	938

5 Experimental Methodology

For each dataset in Table 1, stratified 10-fold cross validation was applied. The training cases were partitioned into 10 equal-sized blocks with similar class distributions. Each block in turn is used to evaluate singular CHAID decision trees and the optimised fuzzy trees which were incrementally added the fuzzy forest. The singular CHAID trees were first optimised through parameter tuning to prevent any bias occurring. To create the second and subsequent trees in the forest, the attribute having the lowest p-value (the highest ranking) was constrained from formulating the root of the second tree. Five fuzzy trees were induced and compiled into each forest as it has been shown that increasing the number of trees further would result in an increase in the error rate [13]. Fuzzification was optimised across each forest using opt-aiNet.

6 Results and Discussion

Table 2 present the result the average error rate of five Crisp CHAID regression trees for seven datasets and Table 3 shows the results the average of each of five fuzzy regression tree forests for all datasets. The best result was obtained from the Concrete dataset where the error rate of fuzzy regression tree forests was reduced by 42 % compared to Crisp CHAID regression trees which obtained a P-Value 0.0203. The Abalone dataset results show that fuzzy regression tree forests reduced the error rate by 41 % compared to Crisp CHAID regression trees with P-Value 0.0213. The reduction of the error rate was 34 % on the Stock Price dataset by fuzzy regression tree forests compared to Crisp CHAID regression trees obtaining a P-Value of 0.0265. For the Crime dataset, the error was reduced by 27 % by fuzzy regression tree forests compared to Crisp CHAID regression trees (P-Value 0.0422). The reduction of the error rate was 27 % on the Compactiv dataset by fuzzy regression tree forests compared to Crisp CHAID regression trees with P-Value 0.0393. Whilst a 26 % reduction in the error rate was achieved for the Boston housing dataset by fuzzy regression tree forests compared to Crisp CHAID regression trees which obtained a P-Value 0.0395. The Elevators dataset results show that fuzzy regression tree forests reduced the error rate by 24 % compared to Crisp CHAID regression trees obtaining a P-Value of 0.0412. Results of applying a paired t-test between results obtained from the singular crisp CHAID tree and Fuzzy regression tree forests can be found in Table 4. These results of all datasets show a statistically significant ($P < 0.05$) in performance of fuzzy regression tree forests comparing with Crisp CHAID regression trees.

Table 2. Result the average of the five Crisp CHAID regression trees

Dataset	Training dataset (error value)	Test dataset (error value)
Boston housing	21.0576	21.4086
Abalone	4.4833	4.4982
Compactiv	25.4945	25.7666
Elevators	0.0000140389	0.0000140794
Stock prices	7.6938	7.849
Crime	0.3419	0.3451
Concrete	0.1401	0.1492

Table 3. Result the average of the five fuzzy regression tree forests

Dataset	Training dataset (error value)	Test dataset (error value)
Boston housing	13.4618	15.8973
Abalone	2.41916	2.6545
Compactiv	17.9268	18.84
Elevators	0.0000106117	0.0000106593
Stock prices	5.1245	5.1959
Crime	0.2489	0.2515
Concrete	0.0802	0.0863

Table 4. Results of paired t-test and Test Dataset of Crisp regression tree and Fuzzy regression tree forests

Dataset	Crisp regression tree (error value)	FRTF (error value)	P-Value
Boston housing	21.4086	15.8973	0.0395
Abalone	4.4982	2.6545	0.0213
Compactiv	25.7666	18.84	0.0393
Elevators	0.0000140794	0.0000106593	0.0412
Stock Prices	7.849	5.1959	0.0265
Crime	0.3451	0.2515	0.0422
Concrete	0.1492	0.0863	0.0203

According to Tables 1 and 4, the number of attributes of dataset have been found to be significantly correlated to the performance of fuzzy regression tree forests. The biggest improvement in performance was obtained on the Concrete dataset, Abalone dataset and Stock Price dataset which have the smallest number of attributes 8,9 and 9 respectively compared with the rest of datasets. Based on these results, the number of attributes have inverse proportional relationship with the performance accuracy of the proposed method. On the other hand, the other dataset characteristics such as dataset size and unique outcome value been found not to be significantly correlated to the performance of fuzzy regression tree forests.

7 Conclusion

This empirical study has shown that fuzzy regression tree forests, once optimized, can outperform singular crisp regression trees regardless of the number of instances, number of attributes and number of unique values. Optimization of each individual forest was domain dependent. As Elgasir is based on CHAID, the Chi-Square test of significance is used to evaluate all values of the predictor variable to select at each tree node the most significant attribute based on significance. Therefore, insignificant attributes are removed prior to the crisp trees fuzzification which typically reduces the number of attributes in the dataset that are modelled. In this study the relationship between dataset characteristics and the performance of fuzzy regression tree forests have been highlighted. The empirical results of seven datasets have shown that the number of attributes in a dataset have been found to be significantly correlated to the performance of fuzzy regression tree forests.

References

1. Amatriain, X.: In Machine Learning, What is Better: More Data or better Algorithms. https://www.quora.com/In-machine-learning-is-more-data-always-better-than-better-algorithms/answer/Xavier-Amatriain. Accessed 21 Feb 2016
2. Kwona, O., Simb, J.M.: Effects of data set features on the performances of classification algorithms. Expert Syst. Appl. **40**(5), 1847–1857 (2013)
3. Cadenas, J.M., Garrido, M.C., Martinez, R., Bonissone, P.P.: Towards the learning from low quality data in a fuzzy random forest ensemble. In: 2011 IEEE International Conference on Fuzzy Systems (FUZZ), pp. 2897–2904, 27-30 June, 2011
4. Cadenas, J.M., Garrido, M.C., Martínez, R.: Selecting features from low quality datasets by a fuzzy ensemble. In: Madani, K., Correia, A.D., Rosa, A., Filipe, J. (eds.) CI. SCI, vol. 577, pp. 229–244. Springer, Heidelberg (2015)
5. Bhatt, N., Thakkar, A., Ganatra, A., Bhatt, N.: Ranking of Classifiers based on Dataset Characteristics using Active Meta Learning. Int. J. Comput. Appl. **69**(20), 0975–8887 (2013)
6. Crockett, K., Bandar, Z., O'Shea, J., Fowdar, J.: A fuzzy numeric inference strategy for classification and regression problems. Int. J. Knowl. Based Intell. Eng. Syst. **12**(4), 255–269 (2008)
7. Crockett, K., Bandar, Z., McLean, D., O'Shea, J.: On Constructing a Fuzzy Inference Framework using Crisp Decision Trees. Fuzzy Sets Syst. **157**(21), 2809–2832 (2006)

8. Marsala, C., Rifqi, M., Summarizing fuzzy decision forest by subclass discovery. In: 2013 IEEE International Conference on Fuzzy Systems (FUZZ), pp. 1–6, 7–10 July 2013
9. De Matteis, A.D.; Marcelloni, F.; Segatori, A.: A new approach to fuzzy random forest generation. In: 2015 IEEE International Conference on Fuzzy Systems (FUZZ-IEEE), pp. 1–8, 2–5 August, 2015
10. Jiang, R., Bouridane, A., Crookes, D., Celebi, M.E., Wei, H.-L.: Privacy-protected facial biometric verification via fuzzy forest learning. In: IEEE Transactions on Fuzzy Systems, vol. PP, no. 99, p. 1 (2015)
11. Cadenas, J.M., Garrido, M.C., Martinez, R.: Learning in a fuzzy random forest ensemble from imperfect data. In: 2011 IEEE International Conference on Systems, Man, and Cybernetics (SMC), pp. 277–282, 9–12 October 2011. doi:10.1109/ICSMC.2011.6083678
12. Bonissone, P., Cadenas, J.M., Garrido, M.C., Díaz-Valladares, R.A.: A fuzzy random forest. Int. J. Approximate Reasoning **51**(7), 729–747 (2010)
13. Gasir, F., Crockett, K., Bandar, Z.: Inducing fuzzy regression tree forests using artificial immune systems. Int. J. Uncertainty Fuzziness Knowl. Based Syst. **20**, 133–157 (2012). doi: 10.1142/S0218488512400181
14. De Castro, L.N., Timmis, J.: An artificial immune network for multimodal function optimisation. In: Proceedings of IEEE World Congress on Evolutionary Computation, pp. 669–674 (2002)
15. Kass, G.V.: An exploratory technique for investigating large quantities of categorical data. Appl. Stat. **29**(2), 119–127 (1979)
16. Takagi, T., Sugeno, M.: Fuzzy identification of a system and its application to modeling and control. IEEE Trans. Syst. Man Cybern. **15**, 116–132 (1985)
17. Lichman, M.: UCI Machine Learning Repository http://archive.ics.uci.edu/ml. University of California, School of Information and Computer Science, Irvine, CA (2013)
18. KEEL repository Knowledge Extraction based on Evolutionary Learning 2016. http://sci2s.ugr.es/keel/datasets.php (2016)

Social Data Processing

Dynamic Analysis of Participatory Learning in Linked Open Data: Certainty and Adaptation

Marek Z. Reformat[1(✉)], Ronald R. Yager[2], and Jesse Xi Chen[1]

[1] University of Alberta, Edmonton, Canada
{Marek.Reformat,Jesse.Chen}@ualberta.ca

[2] Iona College, New Rochelle, NY, USA
yager@panix.com

Abstract. Graph-based data formats are popular ways of representing information, while graph-processing engines and graph databases become preferable tools for handling data of different size. World Wide Web Consortium has introduced a graph-based data format called Resource Description Framework (RDF) as the part of its Semantic Web initiative. The intrinsic features of RDF, i.e., its interconnectivity and simplicity of expressing information as triples containing two entities connected by a property, provide new possibilities of analyzing and absorbing information.

The participatory learning of propositional knowledge is an attractive way of integrating and updating knowledge bases built based on symbolic data equipped with uncertainty. In such context, an idea of considering RDF triples as propositions allowed us to use the principles of participatory learning for assimilating RDF triples and handling different levels of uncertainty associated with them.

The paper examines the RDF-based participatory learning process from the perspective of its dynamics. The emphasis is put on aspects related to handling certainty, accepting new pieces of information, and dealing with contradicting information. The learning process is presented, and the results of analysis are provided.

1 Introduction

The web's role as data repository is fully established. The users access it on regular basis, process the encountered data, and extract valuable information from it. More and more often the users anticipate that such tasks should be done in an automatic way resulting in 'ready-to-use' knowledge repositories. At the same time, the web data is stored in multiple locations, is associated with different levels of certainty, and could contain contradicting pieces of information.

In the recent years, a number of graph-based data formats have emerged as a very attractive and promising way of representing data and express its semantics. One of the most popular methods of representing web data as graphs is Resource Description Framework (RDF) [10]. RDF has been proposed and standardized by W3C [11]. It denotes information as triples, where each triple is a single piece of information. Multiple triples are connected between each other constituting graphs. RDF is a fundamental data format used in Semantic Web [1] and Linked Open Data [2, 6].

J.P. Carvalho et al. (Eds.): IPMU 2016, Part II, CCIS 611, pp. 667–677, 2016.
DOI: 10.1007/978-3-319-40581-0_54

A paradigm of participatory learning has been introduced in [7], and successfully applied in multiple areas, for example [3, 4]. Its fundamental principle is based on a simple, yet very realistic idea of 'conditional' learning where a process of accepting new information depends on the already known information. The 'original' version of the participatory learning has been adopted to deal with propositions [8], and further to handle RDF triples [5].

The RDF version of participatory learning [5] brings quite a different view at the process of collecting data and information. The process of collecting new pieces of information, i.e., RDF triples, and assimilating them resembles a realistic learning activity. Especially attractive are methods used to determine a degree of compatibility between the new and known data, as well as procedures governing acceptance of the new data.

We foresee a scenario where software agents can collect a new data, and then depending on the already known data they can accept the new 'findings' to a degree. It is important to understand how this acceptation process looks like and what factors influence it. This paper addresses these issues. It analyzes the RDF-based participatory learning from the point of view of its dynamics, i.e., from the perspective of its ability to accept a new data via the process of updating levels of certainty associated with the known data. Additionally, we investigate a mechanism of handling contradicting information.

The paper includes a very short introduction to RDF data format and its characteristics (Sect. 2), as well as a brief description of basic steps of participatory learning (Sect. 3). In this case the emphasis is put on the explanation of fundamental ideas of participatory learning with RDF triples. The main contributions of this paper are related to the analysis of behavior of RDF-based learning (Sect. 4). In particular, we examine:

- important features of a human-like learning process that are inherently present in participatory learning, i.e., learning without previous knowledge (Sect. 4.1), and learning via exposure to facts equipped with different levels of confidence (Sects. 4.2 and 4.3);
- 'quickness' of a learning process and its influence on the ability to assimilate new facts and to determine confidence in them (Sect. 4).

Overall, this paper shows how the RDF-based participatory leaning mimics a human-like learning process, and how this process can be perceived as an experience-based learning. Assimilation of data is governed by a degree of difference in confidence (possibility values) between known and new pieces of information.

2 Background

2.1 Introduction to RDF

The most important, and yet simple, idea introduced with RDF is to represent any piece of information as a triple < `subject-property-object` > where the `subject` is an entity that is being described/defined, the `property` represents a relation that exists between the `subject` and the `object`, and the `object` is another entity or a literal (number or string) that is ‘linked’ via the `property` to the `subject` [10]. An example of a set of triples that share the same object – `dbpedia.org/page/The_Beatles` – is shown in Fig. 1. All these triples constitute a description of the entity ‘Beatles’. Some of these features are: < `-genre-Pop_Band` > , < `-genre-Rock_Band` > , < `-name-TheBeatles` > , < `-activeYearStart-Year-1960` > , < `-homeTown-Liverpool` > , < `-artistOf-Help` > , < `-bandMember-John_Lennon` > , and < `-bandMember-Paul_McCartney` > . As we can see, each RDF triple is perceived as a single feature of the entity composed of two elements: a property and a object. In other words, a property ‘defines’ a type of relation between a given subject and a object. A set of triples that share the same subject represents a set of features of this subject – a definition of the entity.

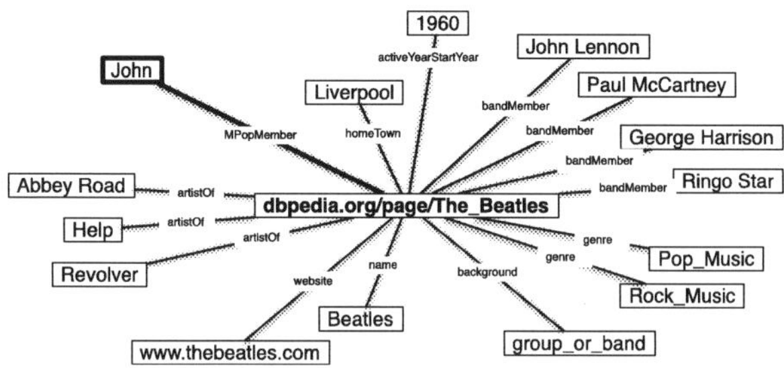

Fig. 1. RDF-based description of Berkeley

Quite often subjects and objects of one triple can be involved in multiple other triples, i.e., they can be objects or subjects of other triples. In such a case, multiple definitions – RDF-stars – can share features, and some features can be centers of another RDF-stars. Such interconnected triples constitute a network of interleaving definitions of entities.

2.2 RDF Triples as Propositions

The ability to perceive a single feature of an entity as an RDF triple leads to a very important observation that becomes a basic idea of the application of participatory learning to RDF data. We state that each RDF triple is a single proposition defined on the domain of values that the RDF triple’s property can assume. For example, the highlighted RDF triple in Fig. 1:

<dbpedia.org/page/The_Beatles – MPMember – John>

can be expressed into a proposition with a fuzzy subset *S*

$$P: \quad V_{MPMember} \text{ is } S: \quad \{\frac{1.0}{John}, \quad \frac{0.0}{other\ values\ of\ V_{MPMember}}\}$$

where $V_{MPMember}$ is a variable representing the most popular member of a band, and *John* as its value is associated with the possibility of *1.0*, while all other possible values of $V_{MPMember}$ (alternatives) have the possibility of *0.0* [8]. This means that the *most popular member* of the British band *The Beatles* is *John*. At this stage it is very important to contrast this with the proposition

$$P:\ V_{MPMember} \text{ is } S:\ \{\frac{1.0}{John},\ \frac{1.0}{other\ values\ of\ V_{MPMember}}\}$$

that represents a statement 'do not know'. The value of *1.0* is associated with *John* as well as with *other values of* $V_{MPMember}$ – this means that everything is possible and there is no indication that some values are more or less possible then others. With such an approach, a set of RDF triples that defines a given entity can be treated as a knowledge base with propositions denoting features of the entity.

3 Participatory Learning with RDF

3.1 Introduction

A description of the same entity can exist in multiple places on the web. We can state that collecting such sets of RDFs can be perceived as a learning process. In such a case, accumulating descriptions of one and the same entity is equivalent to a repetitive process of acquiring information and eventually gaining confidence in gathered descriptions, i.e., RDF triples. This idea is a pivotal aspect of the learning approach described here. As the result, the information about an entity is composed of triples associated with different levels of confidence (possibility values assigned to different alternatives) [8].

Before we analyze dynamics of the RDF participatory learning, let us take a look at a formal description of this learning involving RDF triples. Here, we present only the most essential aspects; we recommend [5, 8] for more detailed descriptions. The presented equations show a process of determining consistencies and compatibilities of known and new information represented as RDF triples, and a mechanism of combining both knowledge bases. Let the known knowledge base, *KB*, be a set of propositions P_i built based on RDF triples (NOTE: we will use the subscript *i* for propositions)

$$P_i{:}V_i \text{ is } S_i \text{ is } \alpha_i - certain \tag{1}$$

They can be represented as equivalent propositions:

$$P_i{:}V_i \text{ is } [F_i(x_i) = Max(S_i(x_i), (1-\alpha_i))] \tag{2}$$

where $x_i \in X_i$, and X_i is a domain of the variable V_i.

3.2 Consistency and Compatibility

The consistency *Con* of the *KB* is determined based on propositions, in the form of Eq. 2, in the following way:

$$Con(KB) = Con(KB) = Min\{\underset{X_1}{Max}[F_1(x_1)],\ \ldots, \underset{X_i}{Max}[F_i(x_i)], \ldots\} \tag{3}$$

The same approach is used for representing propositions of a new knowledge base (Eq. 4) and consistency of this base (Eq. 5):

$$ {}^{N}P_p{:}V_p \text{ is } {}^{N}\mathrm{F}_p(x_p) = Max({}^{N}S_p(x_p), (1-\alpha_p)) \tag{4} $$

(NOTE: for the new knowledge base we will use the subscript p propositions).

$$ Con({}^{N}KB) = Min\{\underset{X_1}{Max}[{}^{N}F_1(x_1)], \ldots, \underset{X_p}{Max}[{}^{N}F_p(x_p)], \ldots\} \tag{5} $$

Right now, we can determine a compatibility level ρ between the known and new knowledge bases:

$$ Comp(KB) = \frac{Con(KB \cup {}^{N}KB)}{Con(KB)} = \rho \tag{6} $$

where

$$ \begin{aligned} Con(KB \cup {}^{N}KB) = Min\{ \quad & \text{if } X_i = X_p\text{:} \quad \underset{X_i}{Max}[F_i(x_i) \wedge {}^{N}F_p(x_p)]\ , \ldots\ , \\ & \text{otherwise:} \quad \underset{X_i}{Max}[F_i(x_i)], \underset{X_p}{Max}[{}^{N}F_p(x_p)]\ , \ldots\} \end{aligned} \tag{7} $$

3.3 Learning Coefficients

As we can see, the consistency of a combined sets of propositions, i.e., $KB \cup {}^{N}KB$, is determined by finding the maximum of intersection of fuzzy sets on the same domain. Now we are ready to adjust values of two parameters: a compatibility ratio ρ and an arousal level δ. An adjusted value of δ is:

$$ \delta^* = \delta + Comt({}^{N}KB) \cdot Con({}^{N}KB) \cdot \beta \cdot ((1-\rho) - \delta) \tag{8} $$

where *Comt* is the commitment of the observation [8]:

$$ Comt({}^{N}KB) = 1 - Max(\underset{X_i}{Min}[F_i(x_i), \ldots]), \underset{X_p}{Min}[F_p(x_p), \ldots]), \ldots) \tag{8a} $$

and β is a learning rate set up by a user/individual. The modified value of ρ is:

$$ \rho^+ = \rho \vee (\delta^* \wedge Con({}^{N}KB)) \tag{9} $$

Once we determine the values of δ^* and ρ^+ we can combine the known and new knowledge bases. The updated knowledge base KB^* is built using the following approach:

$$ KB^* = (KB, (1-\delta^*)) \cup ({}^{N}KB, \rho^+) = (F_i(x) \vee \delta^*) \wedge ({}^{N}F_p(x) \vee (1-\rho^+)) \tag{10} $$

4 Dynamics of Learning Process

The dynamics of participatory learning process – the main attention of this paper – is governed by three parameters: the compatibility ratio ρ between the known and new knowledge bases, the arousal level δ indicating a repetitive degree of mismatch between the known and new knowledge bases, and the learning rate β that determines a 'speed' of a learning process. Among these three parameters, the user is controls only the learning rate β. Based on the Eqs. (8) and (9), we can say that β also influences the rate of change of the arousal δ^* and impacts, indirectly, the value of the compatibility ρ^+.

In our analysis of the learning process we focus on effects the value of learning rate β has on the agent's ability to learn new facts (propositions), as well as on the reaction to changes of these facts. A set of experiments have been conducted in order to learn how the known knowledge base changes regarding confidence in facts (possibility values associated with alternatives) – due to encountering new pieces of information.

4.1 Learning Without Prior Knowledge

The first of the presented experiments (*e1*) illustrates the scenario when an agent does not know anything about popularity of members of the band *The Beatles*. It means, the known knowledge base KB_{known} contains the following proposition:

$$P^{e1}_{KBknown}{:}V_{MPMember} \text{ is } \left\{ \frac{1}{V_{MPMember} = Paul}, \frac{1}{V_{MPMember} = John}, \frac{1}{V_{MPMember} = other} \right\} \alpha^{e1}_{KBknown} = 0.0 \quad (11)$$

The agent, in its process of collecting data, encounters different facts related to the popularity of the band's members. In this paper, we consider two propositions P_{KBnew1} and P_{KBnew2} related to the property *MPMember* that are a new knowledge base KB_{new}:

$$P_{KBnew1}{:}V_{MPMember} \text{ is } \left\{ \frac{0}{V_{MPMember} = Paul}, \frac{1}{V_{MPMember} = John}, \frac{0}{V_{MPMember} = other} \right\} \alpha_{KBnew1} = 1.0 \quad (12)$$

$$P_{KBnew2}{:}V_{MPMember} \text{ is } \left\{ \frac{1}{V_{MPMember} = Paul}, \frac{0}{V_{MPMember} = John}, \frac{0}{V_{MPMember} = other} \right\} \alpha_{KBnew2} = 1.0 \quad (13)$$

The agent is 'exposed' to these two propositions in a temporal fashion, i.e., is sees them in a sequence presented in Fig. 2. The P_{KBnew1} is seen in time steps 1-5 and 15-25, while P_{KBnew2} is seen in time steps 6-14.

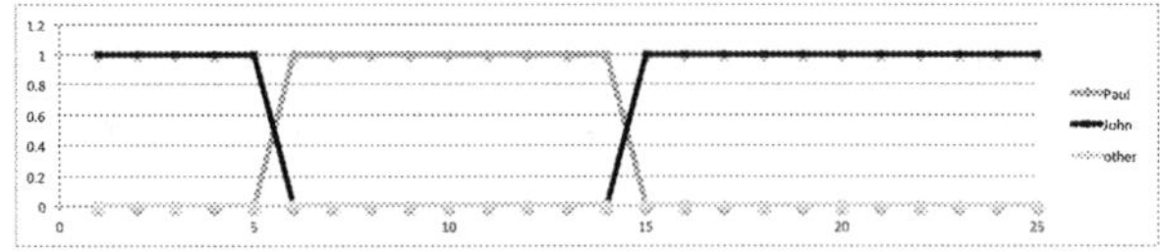

Fig. 2. The agent's exposure to the propositions P_{KBnew1} and P_{KBnew2} at different moments of time in the experiment no. 1 (x: time steps, y: possibility value of alternatives)

The changes in possibility values in the proposition $P^{e1}_{KBknown}$ are examined for two values of the learning rate β: 0.95, Fig. 3; and 0.25, Fig. 4. The figures illustrate changes, at each time step, in the values of possibilities associated with different alternatives of this proposition after assimilation of new proposition. The updated (U) proposition is of the form:

$$^{U}P^{e1}_{KBknown}:V_{MPMember} \text{ is } \left\{ \frac{A}{V_{MPMember}=Paul}, \frac{B}{V_{MPMember}=John}, \frac{C}{V_{MPMember}=other} \right\} \quad (14)$$

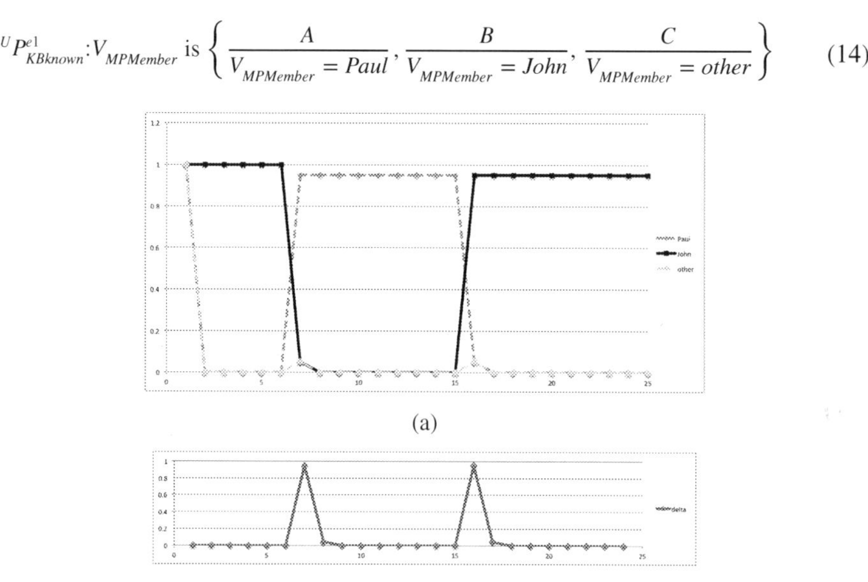

(a)

(b)

Fig. 3. Learnt possibility values for three alternatives of the proposition $^{U}P^{e1}_{KBknown}$(a) and the values of δ (b) for $\beta = 0.95$ (x: time steps, y: possibility values)

The values of A, B, and C change depending on the encounter propositions. If we take a look at Fig. 3 (a) we can observe that the agent – which has a high value of the learning rate of 0.95 – very quickly adopts the changes. We can see that the value of arousal δ 'jumps' very quickly to 1.0, and once the 'change' in the possibility values associated with alternatives occurs, it quickly goes back to 0.0. The values of A, B and C at 25th time step are 0.00, 0.95 and 0.00, respectively. The situation looks quite differently for a 'slow' learner – a small value of the learning rate: 0.25, Fig. 4. The possibility values, i.e., the values of A, B, and C are quite different now: 0.00, 0.31 and 0.00. Additionally, the value of arousal changes differently, Fig. 4 (b). All this is affected by the learning rate: a lower rate leads to a slower assimilation process and lower possibility values. We can say that the agent with a lower learning rate reacts to changes with delay, and possibility levels of alternatives are lower after the new information is assimilated.

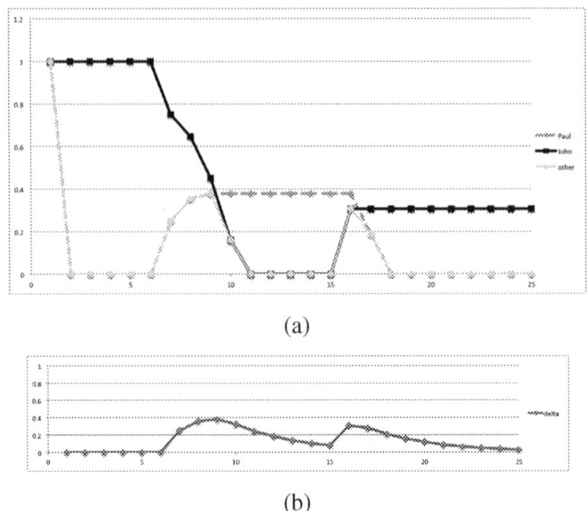

(a)

(b)

Fig. 4. Learnt possibility values for three alternatives of the proposition ${}^{U}P^{e1}_{KBknown}$(a) and the values of δ (b) for $\beta = 0.25$ (x: time steps, y: possibility values)

4.2 Learning Based on High Confidence Knowledge

In the second experiment (*e2*), the known knowledge base contains the proposition:

$$P^{e2}_{KBknown}:V_{MPMember} \text{ is } \left\{ \frac{0}{V_{MPMember} = Paul}, \frac{1}{V_{MPMmember} = John}, \frac{0}{V_{MPMember} = other} \right\} \alpha_0 = 1.0 \tag{15}$$

It means that the agent knows who is the most popular member of the band, and is fully confident (possibility value = 1.0 for *John*, 0.0 for others) in this knowledge. The agent encounters the same facts as in the previous experiment (Eqs. 12 and 13) but in a different temporal sequence, Fig. 5.

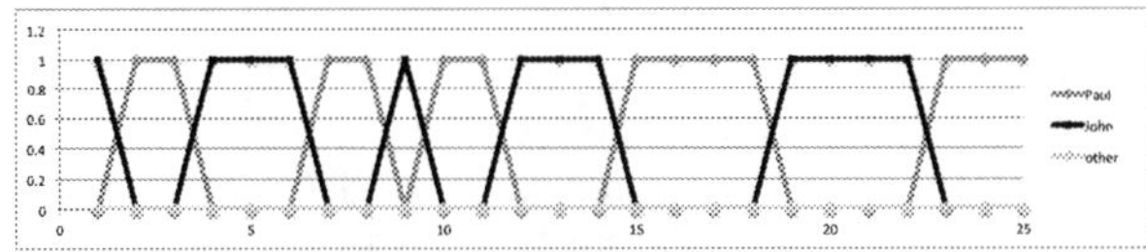

Fig. 5. The agent's exposure to the propositions P_{KBnew1} and P_{KBnew2} at different moments of time in the experiment no. 2 (x: time steps, y: possibility value of alternatives)

The results for $\beta = 0.95$ are shown in Fig. 6 (left side). There is some delay in the agent's reaction to different propositions, but overall the agent follows the changes encountered during an information collection process. The changes happen despite the agent's high confidence in the proposition $P^{e2}_{KBknown}$ of the known knowledge base.

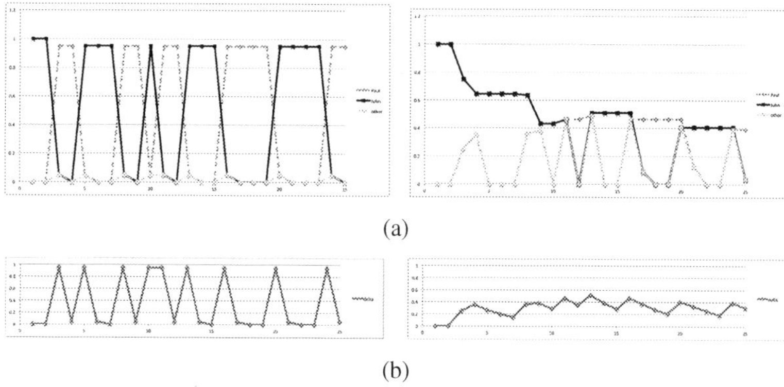

(a)

(b)

Fig. 6. Learnt possibility values for three alternatives of the proposition ${}^{U}P^{e2}_{KBknown}$ (a) and the values of δ (b) for $\beta = 0.25$ (left side), and for $\beta = 0.25$ (right side); (x: time steps, y: possibility values)

The learning process leads to quite different results when the agent's learning rate is just 0.25, Fig. 6 (right side). The low learning rate makes the possibility values associated with alternatives much smaller. The new knowledge base proposition 'seen' in the last three time steps:

$$P_{KBnew2}\!:\!V_{MPMember} \text{ is } \left\{ \frac{1}{V_{MPMember} = Paul}, \frac{0}{V_{MPMember} = John}, \frac{0}{V_{MPMember} = other} \right\} \tag{16}$$

leads to the following possibility values associated with alternatives from the proposition of the known knowledge base:

$${}^{U}P^{e2}_{KBknown}\!:\!V_{MPMember} \text{ is } \left\{ \frac{0.39}{V_{MPMember} = Paul}, \frac{0.04}{V_{MPMember} = John}, \frac{0.04}{V_{MPMember} = other} \right\} \tag{17}$$

For the case of $\beta = 0.95$, these values have been 0.95, 0.00, and 0.00, respectively. We could postulate that a low learning rate makes the agent more sensitive to changes in the propositions' possibility values. Eventually, this leads to lower confidence in the propositions after the assimilation process. Overall, the knowledge base becomes also less consistent.

4.3 Learning Based on Low Confidence Knowledge

The third experiment (*e3*) illustrates the scenario when the agent's known knowledge about popularity of the band's most popular member consists of the proposition:

$$P^{e3}_{KBknown}\!:\!V_{MPMember} \text{ is } \left\{ \frac{1}{V_{MPMember} = Paul}, \frac{0}{V_{MPMember} = John}, \frac{0}{V_{MPMember} = other} \right\} \alpha_0 = 1.0 \tag{18}$$

However, in the process of collecting data, the agent encounters propositions representing ‘do not know’ condition, i.e., possibility values associated with different alternatives are quite high and comparable, Sect. 2.2. The two propositions are:

$$P_{KBnew3}{:}V_{MPMember} \text{ is } \left\{ \frac{0.8}{V_{MPMember} = Paul}, \frac{1}{V_{MPMember} = John}, \frac{0.8}{V_{MPMember} = other} \right\} \tag{19}$$

$$P_{NBnew4}{:}V_{MPMember} \text{ is } \left\{ \frac{0.5}{V_{MPMember} = Paul}, \frac{1}{V_{MPMember} = John}, \frac{0.5}{V_{MPMember} = other} \right\} \tag{20}$$

These encountered propositions are ‘seen’ by the agent according to the temporal sequence presented in Fig. 7.

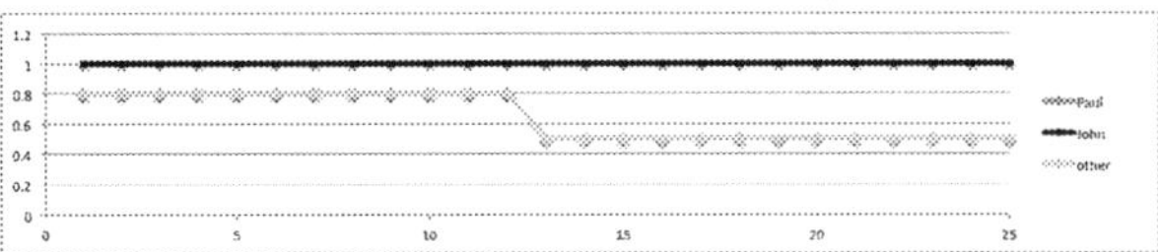

Fig. 7. The agent’s exposure to the propositions P_{KBnew3} and P_{KBnew4} at different moments of time in the experiment no. 3 (x: time steps, y: possibility value of alternatives)

As in the case of the previous two experiments, also here we perform learning with two values of the learning rate: $\beta = 0.95$ and $\beta = 0.25$. The obtained results are presented in Fig. 8 left and right side, respectively.

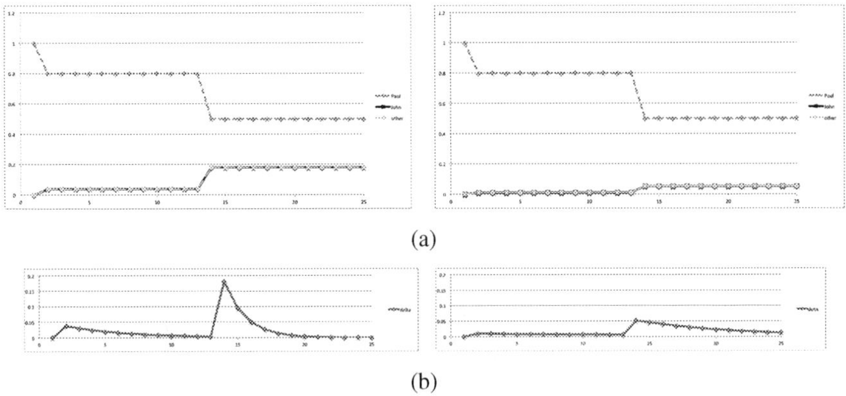

Fig. 8. Learnt possibility values for three alternatives of the proposition ${}^{U}P^{e3}_{KBknown}$ and the values of δ (b) for $\beta = 0.25$ (left side), and for $\beta = 0.25$ (right side); (x: time steps, y: possibility values)

An interesting observation is that the possibility value for the alternative $V_{MPMember} = Paul$ (the initial state of the known knowledge of the agent) decreases with the agent’s exposure to different information. This decrease is bigger when the possibility value for $V_{MPMember} = John$ is higher, i.e., 0.5 versus 0.8 (Eqs. 19 and 20).

This decrease is the same no matter what is the value of the learning rate. However, the increase in the possibility value for the alternative $V_{MPMember} = John$ depends on the learning rate. As we can see in Fig. 8 (a, left side) the learning rate of 0.95 leads to the possibility value of 0.36 for the $V_{MPMember} = John$. For the learning rate of 0.25, Fig. 8 (a, right side), the possibility value for $V_{MPMember} = John$ increases only to the value of 0.10. As we can see in Fig. 8 (b) the values of arousal are influenced by the learning rate, and this influences the increase in the possibility value for $V_{MPMember} = John$.

5 Conclusion

The paper focuses on presentation and analysis of a participatory learning process involving RDF data. The emphasis is put on evaluation of dynamics of the learning. We show how the learning activity depends on the identified rate of learning (learning rate β). We compare the learning process of 'quick learner' with the learning behavior of 'slow learner'. We investigate how the learning rates influence confidence levels in the assimilated new pieces of data. The RDF-based participatory learning is an example of a learning activity we could label 'not everything is learned at once'. This is a process of discovering new information, and accepting it in a piece-by-piece manner. This process can be a bit faulty at the beginning but eventually it leads to determining an adequate level of certainty in the known data based on a newly collected data. We would like to postulate that the observed dynamic behavior of the participatory learning is comparable with elements of a human-like experience-based learning process.

References

1. Berners-Lee, T., Hendler, J., Lassila, O.: The semantic web. Sci. Am. **284**, 34–43 (2001)
2. Bizer, C., Heath, T., Berners-Lee, T.: Linked data-the story so far. Int. J. Semantic Web and Inf. Syst. **4**, 1–22 (2009)
3. Hell, M., Ballini, R, Costa, P., Gomide, F.: Training Neurofuzzy Networks with Participatory Learning. In: FUZZ-IEEE Conference, pp. 1–6 (2007)
4. de Oliveira, J.V., Pedrycz, W.: Advances in Fuzzy Clustering and its Application. Wiley, New York (2007)
5. Reformat, M.Z., Yager, R.R.: Participatory Learning in Linked Open Data, IFSA-EUSFLAT (2015)
6. Shadbolt, N., Hall, W., Berners-Lee, T.: The semantic web revisited. Intell. Syst. **21**, 96–101 (2006)
7. Yager, R.R.: A model of participatory learning. IEEE Trans. Syst. Man Cybern. **20**, 1229–1234 (1990)
8. Yager, R.R.: Participatory learning of propositional knowledge. IEEE Trans. Fuzzy Sets Syst. **20**, 715–727 (2012)
9. Zadeh, L.A.: A theory of approximate reasoning. Mach. Intell. **9**, 149–194 (1979)
10. RDF Primer. http://www.w3.org/TR/2014/NOTE-rdf11-primer-20140225/. Accessed 31 March, 2016
11. World Wide Web Consortium (W3c). https://www.w3.org. Accessed 31 March, 2016

Online Fuzzy Community Detection by Using Nearest Hubs

Pascal Held(✉) and Rudolf Kruse

Faculty of Computer Science, Otto von Guericke University of Magdeburg, Universitätsplatz 2, 39106 Magdeburg, Germany
{pascal.held,rudolf.kruse}@ovgu.de
http://fuzzy.cs.uni-magdeburg.de

Abstract. Community and cluster detection is a popular field of social network analysis. Most algorithms focus on static graphs or series of snapshots.

In this paper we present an algorithm, which detects communities in dynamic graphs. The method is based on the shortest paths to high-connected nodes, so called hubs. Due to local message passing, we can update the clustering results with low computational effort.

The presented algorithm is compared with the Louvain method on large-scale real-world datasets with given community structure. The detected community structure is compared to the given with NMI scores. The advantage of the algorithm is the good performance in dynamic scenarios.

1 Introduction

Social network analysis has become very popular in recent years, e.g. for friendship analysis, disease transmission or detection of interesting entities. One part of this scientific field is cluster or community detection. A cluster, or synonymous community, in a social network is a group of entities, which have a higher connection inside the group than to other entities. As a popular example, we consider the famous Karate club example from [23], which will be used to illustrate the algorithm results. The Zachary karate club contains 34 members. One day there was a conflict between the leader and the trainer of the club, so the club split up into two sub-clubs. The community analysis based on the relations can predict which subgroup each member chose.

Most algorithms in community detection focus on a single analysis of a static graph, e.g. [4,15]. In real-world, most of these graphs are dynamic [10]. So nodes appear or disappear and relations evolve over time. To handle these dynamics, the developed algorithms were used on several snapshots of the same graph. Changes in the clustering could be tracked with different algorithms, e.g. the FOCUS [9], DEMON [8] or MONIC [19] framework. There is also work done on dynamic graphs, e.g. the DENGRAPH algorithm [5,6].

In this paper we present an online algorithm to find communities based on high-connected hubs. It is based on the assumption that there exists some highly

J.P. Carvalho et al. (Eds.): IPMU 2016, Part II, CCIS 611, pp. 678–689, 2016.
DOI: 10.1007/978-3-319-40581-0_55

connected nodes, called hubs, which will group people around them. The proposed algorithm detects these hubs and assigns to all non-hub elements the closest hub as a cluster label.

The rest of the paper is structured as follows. First, in Sect. 2 we give a brief overview about cluster and community structure and related algorithms. Next, we present the proposed algorithm in Sect. 3 and experiments in Sect. 4. The paper will end with a conclusion in Sect. 5.

2 Related Work

In this section, we introduce the term of cluster and community structure. Also, we present two properties of social networks which are important for this work. First, the typical community structure of these networks and second, the scale-free property which is used to determine hubs. In the second part of this section, we give a brief introduction into related algorithms. These algorithms are used to compare the results of our algorithm.

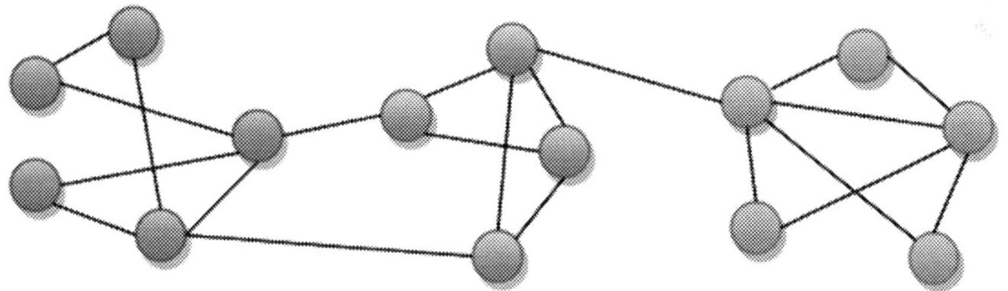

Fig. 1. Cluster structure of a network with three clusters

2.1 Cluster- and Community Structure

Social networks have the property that there are subgroups of nodes which are more densely connected to each other than to the rest of the network. For example, this could be a group of friends, see also Fig. 1. These subgroups are called communities or clusters, which should be detected by community detection algorithms. These entities will be modeled as nodes in graphs. The relationship between them will represent edges between the nodes.

In social network analysis we can distinguish between a graph partition and covering. The partition is related to the cluster structure. Every node is assigned to exactly one cluster. Covering is more advanced, communities (or fuzzy cluster [12,13]) can overlap, and their nodes can be assigned to more than one community at a time. The evaluation of the quality of the overlapping community structure is difficult, so we focus on crisp assignments.

Two simple measures for evaluating the quality of a partition are the intra-cluster density and the inter-cluster sparseness. The intra-cluster density describes the ratio of existing and possible edges within the clusters. If this measure is high, clusters are strongly connected. The inter-cluster sparseness describes the ratio of the number of existing edges and the number of all possible pairwise edges between nodes of different clusters. This value should be small, to get good separations between the clusters [7].

Another very popular measurement is the q-modularity proposed by Newman and Girvan [16]. The basic idea is that graphs with the same degree distribution, but randomly connected nodes, should have no cluster structure. The observed density within a cluster should be higher than the density in such a random graph. A maximization of the q-modularity should yield into a good clustering, because densely connected nodes are grouped together. This maximization is the main approach of the Louvain method.

In our experiments we use datasets with a known community structure. There are several measures to compare the quality of an artificial partition with the true community structure of a graph. First we use the Normalized Mutual Information score [20], which is a normalized form of the Mutual Information created by Cover and Thomas [3].

In 1999, Barabási et al. [1] introduced the idea of scale free networks. This means that the node degree distribution of all nodes follow a power-law function.

$$P(k) \sim k^{-\gamma},$$

where $P(k)$ is the probability of the degree k of a node. For most observed social networks, γ is between 2 and 3.

From this distribution we get a lot of nodes with a low node degree and only a few nodes with higher degrees. These highly connected nodes are called hubs.

A plausible assumption is that these hubs are distributed equally over the whole network. As they have a lot of connections, they group a lot of entities around each other which will result in local clusters. In our algorithm, we use these hubs as starting points for the clusters and all surrounding nodes will be assigned to the closest hub.

2.2 Related Algorithms

In this section two common graph partition algorithms will be briefly introduced. Spectral Clustering is a common graph-based clustering algorithm, while the Louvain method is one of the best community detection algorithms.

Spectral Clustering. Spectral clustering [18] uses an eigenvalue analysis of the normalized Laplacian matrix. As input, we use the connectivity matrix of our graphs. The first k eigenvectors are used for a dimension reduction.

On the lower-space data, we use either k-means [17] as partition algorithm or the discretization proposed by Yu and Shi [22].

Due to the eigenvalue decomposition, this algorithm has a runtime complexity of $O(n^3)$. If the discretization approach is used, the results are deterministic, otherwise not.

Louvain Method. In 2008 Blandel et al. [2] proposed an algorithm to extract community structure from large datasets. The method is based on heuristics and q-modularity optimization. The main idea is, that firstly each node is assigned an individual community. Then iteratively, for every node, a q-modularity gain for switching to adjacent communities is calculated. If there is a positive q-modularity gain, the node will be put into the corresponding community with the highest q-modularity gain. This process is done until no further increment is possible.

Finally, there is a new graph built on the community structure of the input graph. Each community becomes a node. The edges in the new graph represent the sum of edges between two communities.

The procedure will be repeated until there is no q-modularity gain after merging.

This algorithm is a greedy approach with almost linear runtime. The results are not deterministic, so the algorithm should be run multiple times.

3 An Online Algorithm for Dynamic Community Detection

Our dynamic nearest hub clustering algorithm (NHC) is based on two steps. First, we determine all hubs in the network, which will be used as cluster centers. Starting from them, we can propagate the shortest paths through the whole network.

For the dynamic scenario, we start with the first graph, then iteratively change the resultant clustering by applying changes. Allowed modifications are adding or removing nodes or edges.

In the following, we start with a formal definition and the selection of hub nodes. Afterwards, we present the basic algorithm.

3.1 Formal Algorithm Results

The algorithm is based on two major values for each node. First we define the distance to the closest hub $dist_{hub}$ as follow:

$$dist_{hub}(x) = \begin{cases} 0 & \text{if } x \in hubs \\ min\left(\{\omega(x,y) + dist_{hub}(y) : y \in neighbors(x)\}\right) & \text{otherwise} \end{cases}$$

where $\omega(x, y)$ is the weight of the (x, y)-edge.

The second value is the fuzzy membership $\mu_h(x)$ of the node x to the hub h which is based on $P(x) = \{y : y \in neighbors(x) \wedge \omega(x, y) + dist_{hub}(y) = dist_{hub}(x)\}$, as follow:

$$\mu_h(x) = \begin{cases} 1 & \text{if } x \in hubs \wedge x = h \\ 0 & \text{if } x \in hubs \wedge x \neq h \\ \frac{1}{|P(x)|} \sum_{p \in P(x)} \mu_h(p) & \text{otherwise} \end{cases}$$

The partition representing membership is the result algorithm.

3.2 Determine Nodes as Hubs

As mentioned in Sect. 2.1, hubs are highly connected nodes. The simplest method to detect hubs is to determine the n most connected nodes. The main drawback of this method is, that all hubs have to be tracked and a decision if another node gets a hub could be done only with these tracked hubs.

To decide whether a node is a suitable hub individually, we introduce a threshold d_{min} as minimal node degree. Every node with $\deg(node) \geq d_{min}$ will be marked as a hub node. A good threshold is dependent from the underlying graph structure as well as from the expected cluster size. This has the advantage that not all hubs have to be known for the decision, so the analysis can be performed without having the full graph. This enables the algorithm to be run in parallel.

3.3 Basic Algorithm

The algorithm is based on passing hub information through the network. Important changes in the network structure are propagated to all relevant nodes.

Each node stores a hub information table T with the tuple entries (h, p, α), where h represents the corresponding hub and p the parent node, with the shortest path to the hub. α represents a weight of this information. Additionally, we store the hub distance d.

The Message. $M_{x \to y}(T', d')$ sent from a node to the neighbor nodes contains the basic hub information table

$$T' = \left\{ \left(h, \frac{\alpha}{\sum_{i=1}^{|T|} \alpha_i} \right) : (h, p, \alpha) \in T \right\}.$$

where α_i is the α-value of entry i. The distance is set to

$$d' = d + \omega(x, y),$$

where $\omega(x, y)$ is the weight of the (x, y)-edge. For unweighted graphs $\omega(x, y) = 1$ if the edge exists, 0 otherwise. For weighted graphs, the weight must not be lower then 0.

Processing Messages. $M_{x \to y}(T', d')$ in the target node. We focus on three different cases. First, if $d' < d$ the new node distance in lower than the current distance. The hub information table is set to the table from the message, where p is set to the sending node. Also the distance is updated to the new distance.

The second case is, that the message has the same distance value than the current one $d' = d$. In this case we remove all tuples concerning the sending node and append the new information.

$$T_{new} = \left(\bigcup_{(h,p,\alpha)\in T \wedge p\neq x} (h,p,\alpha) \right) \cup \{(h,x,\alpha) : (h,\alpha) \in T'\}$$

In both cases, the new hub information table will be propagated to all other neighbors.

If $d' \geq d + \omega(y,x)$, the sender has a worse connection than the receiver. In this case, the table is not updated, but the current table is propagated to the sender, so the sending node can update its distances.

Otherwise the message is dropped, with no further steps.

Altering the Graph. Changes in the graph structure are handled as follows: If a new edge (x,y) is added, node x sends an information message to y. Due to message processing, either y will update its hub information or send its information to x. Additionally, the new structure is propagated through the network.

Whenever an edge is removed, we have to check which of the nodes was the connection to the parent node. Associated hub information tuples have to be removed. If this clears the table, the distance is set to $d = \infty$. Changes have to be propagated to all neighbors.

If a node is removed, then all connected edges also have to be removed and processed. Pure node creation does not influence the structure and does not have to be handled, because they have no connections.

Defining New Hubs. If a node n gets higher connected and becomes a hub, the distance value is set to $d = 0$ and the hub information table is set to $T = (n, nil, 1)$. The information has to be propagated to all neighbors.

Removing Hubs. If a node n loses its hub state, the distance is set to $d = \infty$ and the information table is cleared. After propagating this information, all neighbor nodes send alternative hub information.

Calculate Membership Values μ. Based on the α value in the hub information table of each node, we can calculate the membership for each node and hub as a fraction of the sum of membership-assignments to a given hub and the total membership assignment as follows:

$$\mu_h(x) = \frac{\sum_{(h^*,p,\alpha)\in T \wedge h^*=h} \alpha}{\sum_{(h^*,p,\alpha)\in T} \alpha}$$

If a crisp partition is needed, we use $\text{argmax}_h \mu_h(x)$ as defuzzification of the cluster assignment. On tie, we use the hub with the lower index.

The algorithm runs almost linear as a function of the number of edges. The resulting membership values are deterministic and the results are optimal, with respect to the hub distance.

4 Experiments

We will do a two step evaluation of our presented algorithm, first, a static and second, a dynamic case. In the static experiment, we compare the clustering results with other algorithms. The dynamic experiment checks the performance for dynamic graphs.

4.1 Static Experiment

First we check the algorithm in a static experiment setting. We take well-known datasets to evaluate and compare clustering results with the spectral clustering and Louvain method.

We start with a deeper view on the karate dataset from Zachary [23]. He observed the relationship of the members of the club, after the club has split up into two groups.

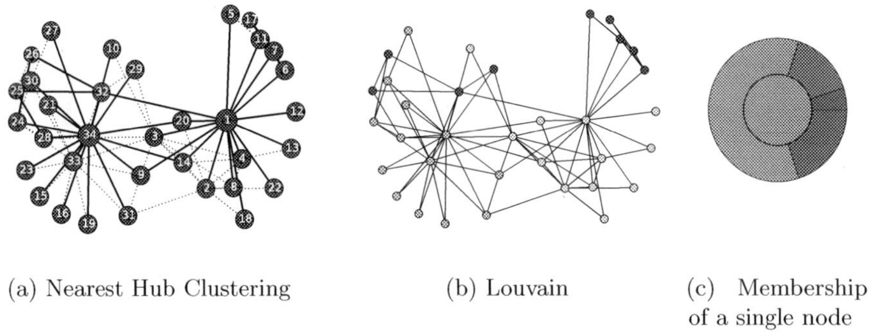

(a) Nearest Hub Clustering (b) Louvain (c) Membership of a single node

Fig. 2. Clustering of karate club dataset

Figure 2a shows the results of the clustering with our Nearest Hub Clustering. The outer circle of a node describes the community membership distribution of the different communities. The inner circle is the crisp cluster association described in Sect. 3. In Fig. 2c, a single node is shown where the membership distribution is more visible. The solid lines describe connections, which are used for next hub propagation. The dotted lines do not carry nearest hub information, but they are still in the dataset.

Figure 2a shows the two cluster center when using NHC. The nodes 9, 14, 20, and 32 are exactly in-between the two center nodes. Due to crisp partition, they are associated to cluster 34, but they could also be assigned to cluster 1.

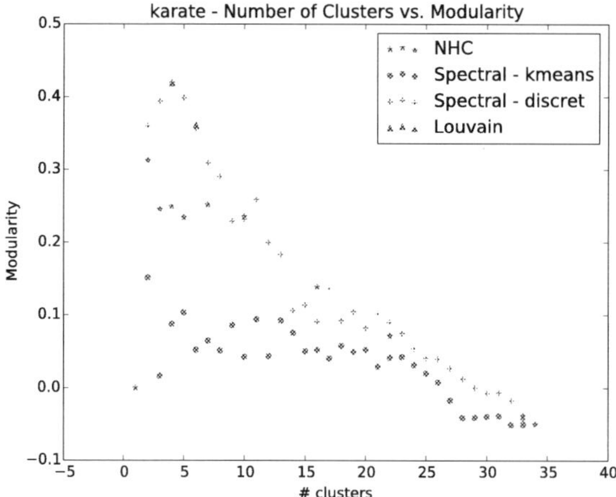

Fig. 3. Modularity comparison for different number of clusters (karate dataset)

Figure 2b shows the same graph clustered with the Louvain method. This method offers four clusters, where the two main clusters are similar to the NHC results.

In Fig. 3 we show the reached q-modularity for NHC, spectral clustering, and the Louvain method for different number of clusters. The Louvain method only offers results for four and six clusters. A more detailed structure is not possible. NHC outperforms the spectral clustering with k-means for every number of clusters. Especially for a small number of clusters, the spectral clustering with discretization and the Louvain method produce good results for the q-modularity. For 10 and more clusters, NHC produces similar results to discretized spectral clustering. The lower values for lower number of clusters might be caused by the assignment method for equal distributed memberships.

In Fig. 4 we present the q-modularity for the dolphins network [14]. The dataset represents the social structure of 62 dolphins. NHC does not reach the q-modularity of the other methods for lower number of clusters, but for higher numbers, the values are similar to spectral clustering with discretization. Again, the Louvain method could not produce results for finer structures.

Large Scale Networks. Yang and Leskovec [21] provide a set of large scale online communities including a ground truth. We used the DBLP, YouTube, and Amazon dataset to check the cluster prediction performance of our algorithm. These datasets contain from 300,000 to 1,100,000 nodes and from 925,000 to almost 3,000,000 edges. The dataset contains from 8,300 to 75,000 non-overlapping communities.

We use the NMI-score to evaluate the cluster performance. Due to the computational complexity of this measure, especially for a large amount of clusters, we use the top 5000 communities provided by Yang and Leskovec. The selection was done by different community measures.

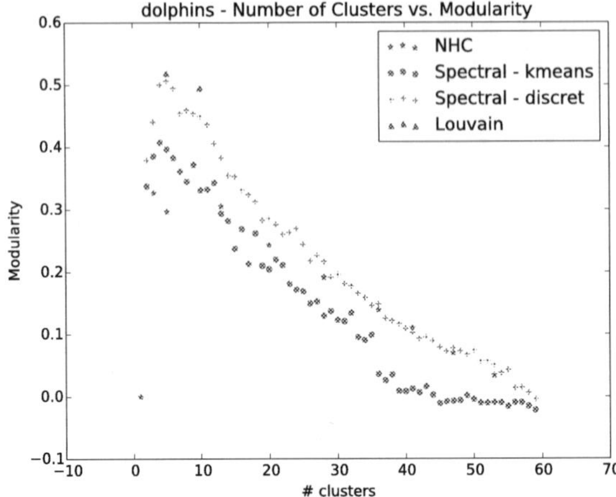

Fig. 4. Modularity comparison for different number of clusters (dolphins dataset)

Due to the complexity of spectral clustering and the fact we have to check a lot of different number of communities, we skip this algorithm for this experiment. We will compare Louvain results with our proposed algorithm.

Table 1. Large scale network comparison for Louvain and Nearest Hub Clustering

	DBLP	Amazon	YouTube
Number of nodes	317, 080	334, 863	1, 134, 890
Number of edges	1, 049, 866	925, 872	2, 987, 624
Number of communities	13, 477	75, 149	8, 385
Top 5000 communities - nodes	112, 228	67, 462	72, 959
Louvain - time	98,7s	81.7s	252,2s
Louvain - NMI-score	0.530	0.871	0.510
Louvain - q-modularity	0.818	0.926	0.710
NHC - time	135.35s	168.31s	154,8s
NHC - min-degree	12	16	4
NHC - NMI-score	0.746	0.945	0.842
NHC - q-modularity	0.432	0.613	0.297

Table 1 shows detailed experiment results. The Louvain method is a fast greedy approach to optimize the modularity. So it produces good cluster results under the assumption that the q-modularity is a good cluster evaluation measure. If, within a given cluster, there exists more edges than expected from the node degree distribution, the q-modularity yields into larger values.

As expected, the Louvain method gets much higher values for q-modularity. But a comparison to the given ground truth shows that this is not correlated to better cluster similarity scores. NHC get NMI-scores from 0.746 to 0.945, while Louvain get NMI-scores from 0.510 to 0.871. This indicates that a large amount of community structure could be found by our algorithm.

The required runtime shows that NHC requires a similar amount of time as the Louvain method.

4.2 Dynamic Experiment

In this section we check the dynamic behavior of the presented algorithm. We generate random clustered powerlaw-graphs with the algorithm proposed by Holme and Kim [11]. As parameters we chose $n = 1000$ nodes, which are connected each to $m = 10$ other nodes. With a probability of $p = 0.7$ the model will create a triangle to increase the cluster coefficient of the resulting network.

The analysis will focus on the amount of messages which have to be processed during adding and removing nodes. The quality of the resulting partition is not checked because the algorithm produces optimal results with respect to the target function of finding closest hubs. The quality of this objective function is already checked in the static case.

Adding Edges. To investigate the adding edges behavior, we generated 100 graphs and added 100 random edges to each graph. This yields into 10000 adding edged events. In Fig. 5a, we show a logarithmic histogram of the processed messages distribution. If a new edge does not create a new shorter path to the hubs, which happens in 63.3% of the cases, only two messages between the two new connected nodes have to be processed. On average 7.73 messages (standard deviation of 10.88) and maximal 549 messages have been sent.

Due to the low amount of messages needed to update the clustering during adding edges to the network, the presented algorithm performs well when dynamically adding nodes and edges.

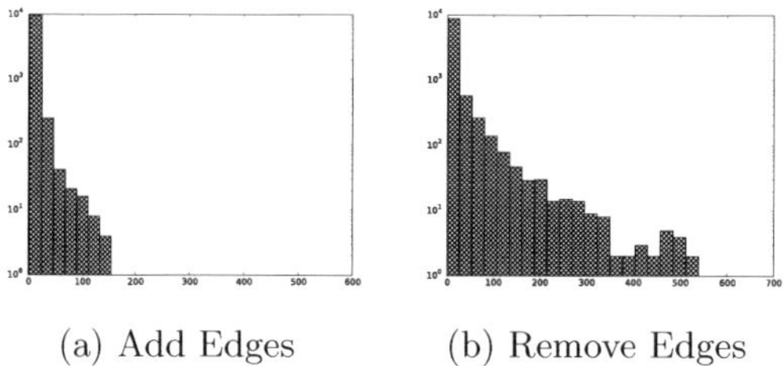

(a) Add Edges (b) Remove Edges

Fig. 5. Logarithmic histogram of messages sent during 10000 events.

Removing Edges. We performed similar experiments for removing edges. Again we created 100 graphs and removed randomly 100 existing edges from each graph, so we get 10000 removing events. Figure 5b presents a logarithmic histogram of the distribution of sent messages. If the removed edge does not destroy the shortest path to a hub, we do not have to sent any messages at all, which happens in 60.1% of the events. On average we need 14.2 messages (standard deviation of 42.15). Maximally 674 messages have been sent. This is a very rare case where the structure near the hubs is changed.

Also the removing process works with a low amount of sent messages. This shows that the presented algorithm performs well on dynamic graph structures.

5 Conclusion

We presented an algorithm for graph clustering. In static experiments it produces high q-modularity results for a finer structure. In contrast to the Louvain method or the spectral clustering with discretization, the algorithm did not reach the global optimum of q-modularity.

Promising experiments on large-scale real-world datasets show that large parts of the underlying community structure can be found by our algorithm.

The main advantage of the algorithm is the dynamic behavior. If the graph changes over time, only a small amount of processing steps have to be done to update the clustering. This enables it for online cluster and community analysis.

The algorithm itself generates overlapping communities and provides fuzzy membership degrees. These results are deterministic. Randomness influences only the crisp cluster assignment.

In the future we will extend our experiments, e.g. directly compare the results of different cluster algorithms or the investigation of more datasets. A larger range of datasets could also provide more insights in the effective runtime behavior. Furthermore, a hierarchical version of the algorithm is in progress. The main idea is that all possible hub thresholds are processed at the same time, so the user can adjust the value afterwards, without recalculating the partition.

References

1. Barabási, A.L., Albert, R., Jeong, H.: Mean-field theory for scale-free random networks. Physica A **272**, 173–187 (1999)
2. Blondel, V.D., Guillaume, J.L., Lambiotte, R., Lefebvre, E.: Fast unfolding of communities in large networks. J. Stat. Mech: Theory Exp. **2008**(10), P10008 (2008)
3. Cover, T.M., Thomas, J.A.: Elements of Information Theory. Wiley, New York (1991)
4. Donath, W.E., Hoffman, A.J.: Lower bounds for the partitioning of graphs. IBM J. Res. Dev. **17**(5), 420–425 (1973)
5. Falkowski, T., Barth, A., Spiliopoulou, M.: DENGRAPH: a density-based community detection algorithm. In: IEEE/WIC/ACM International Conference of Web Intell, pp. 112–115. IEEE (November 2007)

6. Falkowski, T., Barth, A., Spiliopoulou, M.: Studying community dynamics with an incremental graph mining algorithm. Am., 1–11 (2008)
7. Fortunato, S.: Community detection in graphs. Phys. Rep. **486**(3–5), 75–174 (2010)
8. Ganti, V., Gehrke, J., Ramakrishnan, R.: DEMON: mining and monitoring evolving data. IEEE Trans. Knowl. Data Eng. **13**(1), 50–63 (2001)
9. Gehrke, J., Ganti, V., Gehrke, J., Ramakrishnan, R.: A framework for measuring changes in data characteristics. Focus (Madison) **578**, 126–137 (1999)
10. Held, P., Dockhorn, A., Kruse, R.: Generating events for dynamic social network simulations. In: Laurent, A., Strauss, O., Bouchon-Meunier, B., Yager, R.R. (eds.) IPMU 2014, Part II. CCIS, vol. 443, pp. 46–55. Springer, Heidelberg (2014)
11. Holme, P., Kim, B.J.: Growing scale-free networks with tunable clustering. Phys. Rev. E: Stat., Nonlin, Soft Matter Phys. **65**, 2–5 (2002)
12. Höppner, F., Klawonn, F., Kruse, R., Runkler, T.: Fuzzy Clustering. Wiley, Chichester (1999)
13. Kruse, R., Borgelt, C., Klawonn, F., Moewes, C., Steinbrecher, M., Held, P.: Computational Intelligence: A Methodological Introduction. Texts in Computer Science. Springer, New York (2013)
14. Lusseau, D., Schneider, K., Boisseau, O.J., Haase, P., Slooten, E., Dawson, S.M.: The bottlenose dolphin community of doubtful sound features a large proportion of long-lasting associations: can geographic isolation explain this unique trait? Behav. Ecol. Sociobiol. **54**, 396–405 (2003)
15. MacQueen, J., et al.: Some methods for classification and analysis of multivariate observations. In: Proceedings of Fifth Berkeley Symposium on Mathematical Statistical and Probability, vol. 1, pp. 281–297, Oakland, CA, USA (1967)
16. Newman, M.E.J., Girvan, M.: Finding and evaluating community structure in networks. Phys. Rev. E: Stat., Nonlin, Soft Matter Phys. **69**, 1–15 (2004)
17. Ng, A.Y., Jordan, M.I., Weiss, Y.: On Spectral Clustering: Analysis and an algorithm. Adv. Neural Inf. Process. Syst. **2**, 849–856 (2002)
18. Shi, J., Malik, J.: Normalized Cuts and Image Segmentation. IEEE Trans. Pattern Anal. Mach. Intell. **22**(8), 888–905 (2000)
19. Spiliopoulou, M., Ntoutsi, I., Theodoridis, Y., Schult, R.: MONIC - Modeling and Monitoring Cluster Transitions. In: Proceedings of KDD-06, pp. 706–711 (2006)
20. Strehl, A., Ghosh, J.: Cluster ensembles a knowledge reuse framework for combining multiple partitions. J. Mach. Learn. Res. **3**, 583–617 (2002)
21. Yang, J., Leskovec, J.: Defining and evaluating network communities based on ground-truth. Knowl. Inf. Syst., pp. 1–33 (2013)
22. Yu, S., Shi, J.: Multiclass spectral clustering. In: Proceedings of Ninth IEEE International Conference on Computer Vision, vol. 1 (2003)
23. Zachary, W.: An information flow modelfor conflict and fission in small groups1. J. Anthropol. Res. **33**(4), 452–473 (1977)

Creating Extended Gender Labelled Datasets of Twitter Users

Marco Vicente[1,2], Fernando Batista[1,2], and Joao Paulo Carvalho[1,3](✉)

[1] L^{2}F–Spoken Language Systems Laboratory, INESC-ID, Lisbon, Portugal
m.vicente.pt@gmail.com,
{marco.vicente,fernando.batista,joao.carvalho}@inesc-id.pt
[2] Instituto Universitário de Lisboa (ISCTE-IUL), Lisbon, Portugal
[3] Instituto Superior Técnico, Universidade de Lisboa, Lisbon, Portugal

Abstract. The gender information of a Twitter user is not known *a priori* when analysing Twitter data, because user registration does not include gender information. This paper proposes an approach for creating extended gender labelled datasets of Twitter users. The process involves creating a smaller database of active Twitter users and to manually label the gender. The process follows by extracting features from unstructured information found on each user profile and by creating a gender classification model. The model is then applied to a larger dataset, thus providing automatic labels and corresponding confidence scores, which can be used to estimate the most accurately labeled users. The resulting databases can be further enriched with additional information extracted, for example, from the profile picture and from the user location. The proposed approach was successfully applied to English and Portuguese users, leading to two large datasets containing more than 57 K labeled users each.

Keywords: Gender classification · Twitter users · Gender database · Text mining

1 Introduction

Existing social networking services provide means for people to communicate and express their feelings in a easy way. Such user generated content contains clues of user's behaviors and preferences, as well as other metadata information that is now available for scientific research. Twitter, in particular, has become a relevant source for social networking studies, mainly because: it provides a simple way for users to express their feelings, ideas, and opinions; makes the user generated content and associated metadata available to the community; and furthermore provides easy-to-use web interfaces and application programming interfaces (API) to access data. For many studies, the different attributes about a user may be relevant. However, Twitter registration does not explicitly include relevant information such as, for example, gender (not even optionally). For that reason, many previous studies involving Twitter had to rely on small manually labelled datasets of users. Manual labelling represents a labor-intensive task and

J.P. Carvalho et al. (Eds.): IPMU 2016, Part II, CCIS 611, pp. 690–702, 2016.
DOI: 10.1007/978-3-319-40581-0_56

Table 1. Twitter labelled datasets reported in the literature.

Study	Users	Tweets	Languages	Geography
Rao et al. (2010) [18]	1000	405 k	English	India
Burger et al. (2011) [5]	183729	4.1M	Several	
Bergsma et al. (2013) [3]				
Liu et al. (2012) [11]	400	N/A	English	Canada
Bamman et al. (2012) [2]	14464	9.2M	English	United States
Deitrick et al. (2012) [7]	N/A	3031	English	
Fink et al. (2012) [8]	11155	18.5M	English	Nigerian
Miller et al. (2012) [14]	3000	N/A	English	
Al Zamal et al. (2012) [1]	400	N/A	English	Canada
Liu and Ruths (2013) [12]	8000	8M	English	
Ciot et al. (2013) [6]	8118	N/A	Several	
Kokkos and Tzouramanis (2014) [10]	N/A	10000	English	
Ugheoke (2014) [19]	1000	N/A	English	
van Helteren and Speerstra (2014) [9]	600	N/A	Dutch	
Nguyen et al. (2014) [15]	3000	N/A	Dutch	
Van Zegbroeck (2014) [20]	8791	N/A	Flemish	
Vicente et al. (2015) [22]	1464		English, Portuguese	

is a very demanding challenge when analysing social media given the usual huge number of users.

The creation of Twitter datasets is commonly reported in the literature, and researchers have built databases of Twitter users for many geographic regions and languages, including English [13,17] and Portuguese [4]. Due to the above reason, most of the reported databases are not labelled with user attributes like gender or age (user age was also not available on Twitter until late 2015). Previous studies reported the task of labelling users with their gender to be demanding, labor-intensive and in many cases not reusable. Table 1 presents a summarised list of labelled datasets reported in the literature, revealing that most of the studies use small labelled datasets when compared to the number of existing users. These studies involve several languages but in the reported literature, English represents 66.7 % of the users, Portuguese represents 14.4 % and Spanish represents about 6 %. In [18], 1000 profiles were manually annotated through the gender/name association using the Twitter profile information (*user name* and *screen name*). Burger et al. [5] report the most extensive database, which was created by following the blogging website links available in the profile of Twitter users, and extracting the gender from the corresponding profiles. To evaluate the accuracy of their method, the authors randomly selected 1000 Twitter users and manually validated them. Only 15 % of the sample had explicit gender information. In this case, filtering only Twitter users with blogs may bias the dataset, but also filters bots and spammers. Liu and Ruths [12] labelled their data using the Amazon Mechanical Turk platform, a platform developed for the distribution of tasks to human workers where each human intelligence task (HIT) is performed by an individual in exchange for a small fee. The reliability of such method is uncertain, even when the same task is performed by more than one person. In [5], the accuracy of Amazon Mechanical Turk human gender classification was only of 68.7 %, when averaged across workers. In [1,6,7,10,11, 14–16,19,21], Liu et al. manually labelled users to produce datasets, observing either *user name*, *screen name*, profile picture, tweets or a combination of those attributes. Information available in social media profiles such as Facebook and LinkedIn and associated blogging websites, when provided by Twitter users was also used by [15,21,22].

This paper describes a method for creating extended gender labelled datasets in a semi-automatic fashion. The proposed methodology is language independent, but the focus was currently given to Portuguese and English users. Based on the proposed methodology, two extended labelled datasets of English and Portuguese Twitter users have been created, which can be used, not only to provide additional information for further processing stages, but also to create gender models based on the users' generated content and profile information. This paper is structured as follows: Section summarises the whole process of creating gender labelled datasets, while Sects. 2 and 3 provide details about two major stages: the creation of core extended labelled datasets, and the enrichment the datasets previously created, respectively. Section 5 describes the data validation. Finally, Sect. 6 summarises our work and presents the conclusions.

2 Proposed Approach

The approach to create extended labelled datasets is depicted in Fig. 1. The first step in the pipeline consists of extracting data from the Streaming API. It involves restricting tweets to a given geolocation, to the set of languages under consideration, and also involves filtering the data in order to remove undesirable users, such as companies and chat bots.

The second step uses a small dataset of labelled users in order to create a gender classification model. The model is based on a set of features extracted

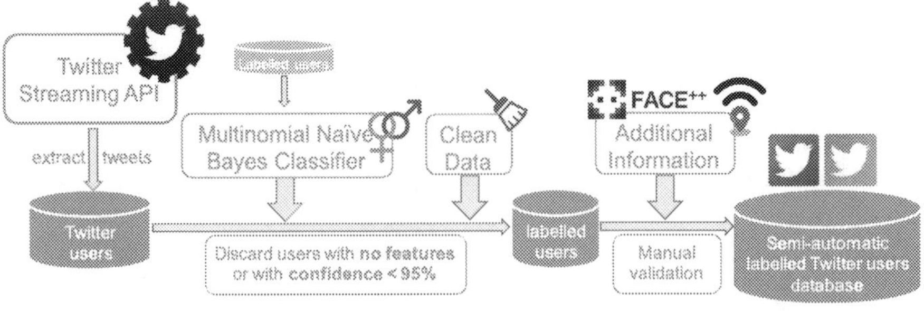

Fig. 1. Semi-automatic gender labelled dataset creation diagram.

from unstructured information found on two profile attributes: *user name*(up to 15 characters), *screen name* (up to 20 characters). The feature extraction process, detailed in [21,22], considers a number of normalisation steps, such as: *repeated vowels* (e.g.: "eriiiiiiiic" → "eric"), and *leet speak* (e.g.: "3ric" →"eric"). After finding one or more names, the following elements are addressed in each feature: "case", "boundaries", "separation" and "position". E.g.: Considering the *screen name* "johnGaines", three names can be extracted: "john", "aine" and "ines". The name "aine" has no valid boundaries, since is preceded and succeeded by alphabetic characters. The feature found is weak and the size of the name is lower than the previously defined threshold. Consequently, the name is discarded. The name "ines" has a valid end boundary, as it is not succeeded by alphabetic characters. Finally, the name "john" has a valid end boundary and starts at the beginning of the screen name. Different thresholds have been defined for specific features, e.g. names with such type of boundary (valid end boundary) and such position (start of the screen name) must contain at least 3 characters. At the end of this process 192 features are extracted, including examples such as: "male_name_correct_beginning_separation_and_case", "female_name_beginning_no_separation" [21]. The classification model is then applied to the large dataset, where users having a classification accuracy below a given threshold may be discarded in order to minimise the number of classification errors and improve data quality. The procedure may include optional manual steps to remove other undesirable users that were not detected previously.

Finally, the dataset created can be enriched with additional information that includes, for example, attributes derived from the profile picture and from the user location.

In the scope of this paper, two datasets of Twitter users have been created: a dataset of English users, and a dataset of Portuguese users. The English dataset was extracted from one year of tweets, collected from January 2014 to December 2014 using the Twitter *streaming/sample* API. The data was restricted to active users that have produced at least 100 tweets in English language, either in the United Kingdom or in the United States. From those we kept around 100 K users. The Portuguese dataset is the full dataset of the data described in [4], and corresponds to a database of Portuguese users, restricted by users that have tweeted at least 100 tweets in Portuguese language, geolocated in the Portuguese mainland. After creating the datasets we have partially validated the data in order to assess its quality, and split it into *train*, *development* and *evaluation* subsets. The Twitter *streaming/sample* API is technically limited to about 1 % of the actual public tweets, but since the amount of geolocated tweets filtered for the Portuguese language within the country geographical area is under 1 % of the total number of tweets, the limitations imposed by Twitter are not relevant.

3 Dataset with Core Gender Labels

As previously stated, we have automatically produced two core datasets containing gender labels, based on 192 features extracted from the *screen name*

Table 2. Number of users that have triggered a given number of gender features.

Dataset	Number of users	No features		1 to 10 features		> 10 features	
English	100000	27110	27 %	65559	66 %	7331	7%
Portuguese	105450	44559	42 %	57440	55 %	3451	3 %

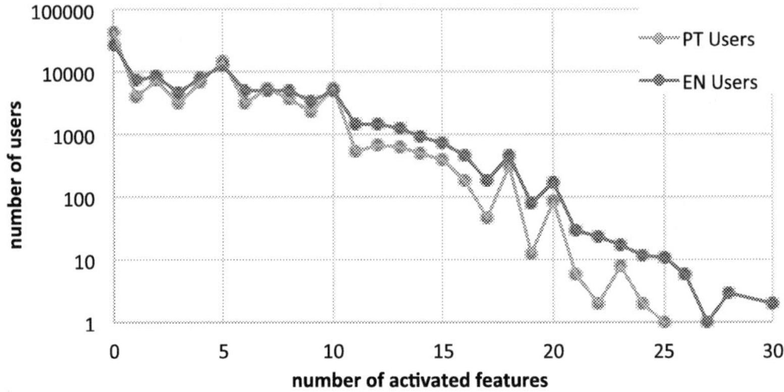

Fig. 2. Automatic gender classification-features per users. (Color figure online)

and from the *user name*. This approach was applied to both English and Portuguese users leading to two datasets. Table 2 and Fig. 2 present the distribution of the users according to the number of features they positively trigger, revealing a higher portion of English users that did not trigger any of the features (42 %) and that were, therefore, discarded from the data. The considerably high proportion of discarded users signal two situations: (i) the profile of such users information provide little or no clues for gender detection; (ii) the feature set can be improved in order to take into account other possible gender detection clues.

In order find the profile attribute that mostly contributes with features to the gender classification task, we have analysed the distribution of features that were extracted either from the *screen name* or from the *user name*. Table 3 shows the obtained results, where columns *user name* and *screen name* represent users activating only features extracted from the corresponding attribute. Results reveal that the two attributes are equally relevant for gender detection.

Based on the extracted features, we have applied supervised machine learning in order to automatically guess the gender label of each user. Different methods have been tested, including: Naive Bayes variants, Logistic Regression, Support Vector Machines, Fuzzy c-Means clustering and k-means, but Multinomial Naive Bayes (MNB) turned out to achieve the best performance [21]. We have used two existing MNB models for English and Portuguese gender classification, previously created from the smaller datasets manually labelled with gender [22]. In the absence of any previously annotated datasets, an alternative approach

Table 3. Number of users that have triggered gender features per profile attribute.

Dataset	None		User name		Screen name		Both	
English	27110	27 %	20845	21 %	20580	21 %	31465	31 %
Portuguese	44599	42 %	17776	18 %	18443	17 %	24672	23 %

Table 4. Some of the gender indicative words.

English		Portuguese	
Male	Female	Male	Female
Father	Mother	pai	mãe
Boy	Girl	rapaz	rapariga
Boyfriend	Girlfriend	Namorado	namorada
Grandfather	Grandmother	avô	avó

would be to use an unsupervised gender classification procedure based on Fuzzy C-means clustering [21], which performed almost as well as MNB (96 % classification accuracy). After the automatic gender classification stage, users with no features or with features, but classified with a confidence score lower than 95 % were discarded in order to minimise the number of classification errors. All the remaining users were added to a dataset, as well as the text of their 100 most recent tweets.

In order to further improve the quality of the data, we have manually validated a subset of the data as follows: (i) we have randomly selected a sample of the labeled dataset to manually validate and correct data; and (ii) we selected a sample of the labelled dataset by searching for gender related words in the users' descriptions. Concerning the second task, Table 4 describes some of the words more informative about the gender. Some of these words are associated to the opposite gender when preceded by possessive determiners (e.g.: "my husband" is considered female[1], while "husband" is male). This second task may be considered as biased, since the probability of finding wrong classification is higher, but it improves the quality of the dataset.

Finally, each one of the resulting datasets were randomly partitioned into 3 subsets: (i) *train* – includes 60 % of the users and can be used to train models; (ii) *development* – includes 20 % of the users and can be used to train or to tune models, minimising problems, such as overfitting; (iii) *test* – includes 20 % of the users and can be used to assess the performance of the final models. Table 5 shows the number of tweets and users included in each one of the subsets.

[1] Gay and transsexual users, as profiles from companies, are not in the scope of this study.

4 Enriching the Datasets with Additional Information

In order to further enhance the datasets, we have added information about two new features for each user: gender recognition from profile picture, and detailed geographical information based on the last known location. The first attribute provides useful information for improving or confirming the gender classification performed previously, while the second attribute may be relevant for tackling region specific phenomena in further automatic processing.

4.1 Gender Based on the Profile Picture

To the best of our knowledge, the use of the gender attribute extracted from the profile picture has not been reported in previous work. However, the profile picture might contain clues regarding the gender of the user.

Face++ [2] is a recent facial recognition API that is publicly available and can be used to analyse the users' profile picture. We have used it through its API to extract the gender and the corresponding confidence, and the resulting info was stored in the datasets. The API was invoked with the profile picture URL extracted from the last stored tweet of each user. Figure 3 illustrates the usage of Face++, where the first picture was correctly classified. Many of the users were correctly classified using this method, but it still presents the following limitations: (i) our datasets contains data back from 2014, and some of the users

Table 5. Split of the obtained semi-automatic gender labelled datasets.

Dataset	#tweets	Train	Development	Test	Total
English	6.5M	39043	13015	13015	65073
Portuguese	5.8M	34625	11540	11540	57705

Fig. 3. Face++ gender detection examples.

[2] http://www.faceplusplus.com/.

Table 6. Number of users involved when Face++ was applied to guess the user gender.

	English		Portuguese	
Image unavailable	31076	54 %	28605	44 %
No face detected	9777	17 %	12995	20 %
Male	9156	16 %	10805	17 %
Female	7857	14 %	12649	19 %

Table 7. Examples of geolocation information.

United States	United Kingdom	Portugal
New York, **NY**	*North East*, United Kingdom	**Lisboa**, Portugal
St. James, **NY**	Westminster, *London*	Paços de Ferreira, **Porto**
New York, US	Cardiff, **Wales**	*Vila Nova de Gaia*, Portugal
New Jersey, USA		

have changed their profile picture in the meanwhile; (ii) some of the pictures do not contain faces; (iii) Face++ is sometimes unable to correctly detect the face in the picture, as exemplified in the second picture presented in the figure.

Table 6 summaries the gender data retrieved from the Face++ API, showing that 54 % of the English users and 44 % of the Portuguese users do not have a profile picture or have removed it since 2014. From the users with an existing profile picture, no face was detected for 36 % in both datasets. In the English dataset, more male than female users have a profile picture with a face, but the opposite occurs in the Portuguese dataset. The gender information provided based on the profile picture could be combined with our previous labels in order to enhance the classification prediction of the whole system. However, it was not used for that purpose in the scope of this paper.

4.2 Geographical Location

People may write differently according to their location, and Twitter provides geolocation within each tweet as long as the user allows it. Despite not providing additional clues about the user gender, in order to better characterise our data and provide extended usage, we have added geographical information to our datasets based on the existing metadata.

We took different approaches depending on the dataset. The English dataset contains tweets in English from more than 200 countries. Adding state or district information for each country would be almost impossible and in most cases unnecessary, since for more than 100 countries the dataset contains only a few number of users, sometimes less than 10. From the entire labelled dataset, 78 % users' last geographical location was the United States and 11 % the United Kingdom. For the United States' users, we added the information regarding the

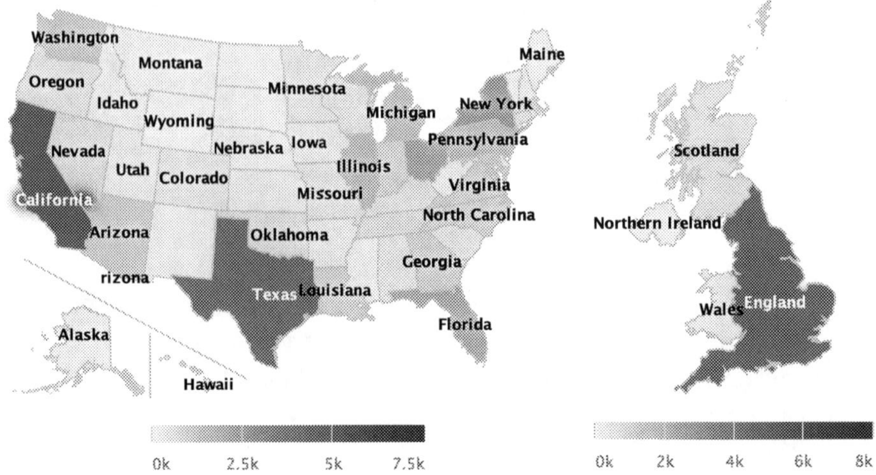

Fig. 4. Labelled users in the United States and in the United Kingdom.

Table 8. Manual validation of the automatic gender classification.

Dataset	Users					Incorrect classification					
	Total	Female		Male		Total		Female		Male	
English	3030	1883	62.2 %	1147	37.9 %	274	9.0 %	187	68.3 %	87	31.8 %
Portuguese	3028	1754	57.9 %	1274	42.1 %	93	3.1 %	76	81.7 %	17	18.3 %

location's state. We extracted the last location from the users and searched for a city or state. The first and second columns of Table 7 shows examples of possible values for geolocation for the United States and United Kingdom, where bold represent states and countries. Twitter usually provides the state code for tweets geolocated in the United States (from the standard INCITS 38[3]). When the code was not found, we extracted the location and mapped it to the corresponding state code. For the United Kingdom labelled users, the distinction added was the country: Scotland, Northern Ireland, England and Wales. We extracted the last location from the users and searched for a city, a state or a country. Figure 4 shows the distribution of the labelled users in the United States and in the United Kingdom.

In the Portuguese dataset, we added a feature with the district of the location. We extracted the last location from the user and searched for a city or district. After finding the cities, they were mapped to the corresponding district. The third column of Table 7 shows possible values for geolocation for the Portuguese territory, where districts are represented in bold, and cities or locations and represented in italics. In the case of the Portuguese archipelagos, we aggregated each location in its archipelago, Madeira and Azores. Finally, we added the

[3] http://geonames.usgs.gov/domestic/download_data.htm.

District	Female	Male	Total
Açores	515	469	984
Aveiro	4110	2712	6822
Beja	400	292	692
Braga	2202	1427	3629
Bragança	517	323	840
Castelo Branco	1600	1324	2924
Coimbra	1715	1189	2904
Évora	440	251	691
Faro	2377	1749	4126
Guarda	66	68	134
Leiria	1384	962	2346
Lisboa	9743	8387	18130
Madeira	340	254	594
Portalegre	307	175	482
Porto	2680	1883	4563
Santarém	1179	796	1975
Setúbal	1764	1279	3043
Viana do Castelo	454	331	785
Vila Real	347	214	561
Viseu	626	431	1057

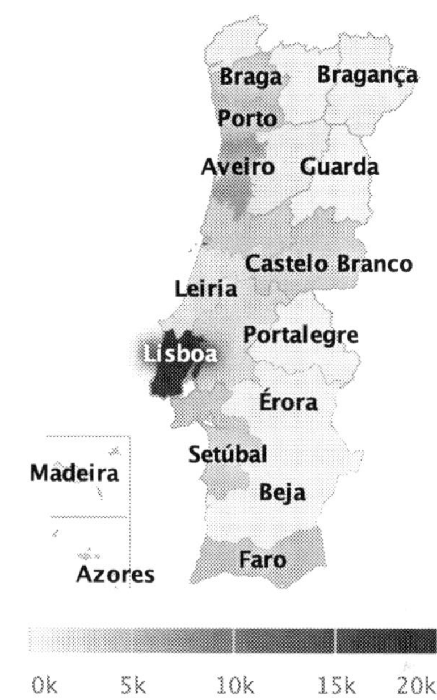

Fig. 5. Labelled Portuguese users per district and gender.

district information to each user. Figure 5 shows the geographical distribution of Portuguese labelled users by district.

5 Data Validation

In order to independently assess the quality of the data, a manual validation was performed. About 3000 users were randomly selected from each labelled dataset, and the gender label was validated using both the Twitter profile content and the blogging sites (when available). We looked for names both in the *user name* and in the *screen name* of the profile, analysed the profile picture of the user and, if the user had blogging sites associated to their profile, we followed those URLs and cross validated the data found with their gender classification. Table 8 summaries the results obtained. We were expecting classification accuracy around 97.3 % for the English dataset [22], which was the accuracy achieved for the test set of the smaller dataset, but we have detected around 9 % of incorrectly classified users. In the Portuguese dataset only 3 % of the users were considered incorrectly classified. That was an expected results because the Portuguese language has a construction of names with more clues to gender than English. The difference in the accuracy may be also related to the higher number of features triggered for English users, probably due to noise found in the attributes. Most of the incorrect classifications in the datasets were due to the four unavoidable reasons: (i) Twitter profile was not of a person; (ii) user was transsexual; (iii) profile was removed and the manual validation was impossible to perform; (iv) gender was incorrectly assigned. By looking at the profiles that were incorrectly classified, we noticed that female names represent a higher percentage in both datasets. In the English dataset the percentage (68 %) is in accordance with variation of the sample. In the Portuguese dataset the difference is noticeable. Female users represent 82 % of the errors, even though the random sample contained only 58 % of female users. The overall result is very satisfactory and, to the best of our knowledge, these results are much better than any other reported non-manual gender dataset.

6 Conclusion

This paper presents an approach for creating extended gender labelled datasets of Twitter users in a semi-automatic fashion. The proposed approach was successfully applied to English and Portuguese users, and two large datasets of labeled users were created. The creation of datasets of Twitter users in commonly reported in the literature. However, most of the datasets are either not labeled with user gender or they are rather small in size. Labelled datasets reported in this paper are only surpassed in size by the work reported by Burger et al. [5], but we have employed less effort and more limited resources. The datasets obtained using the presented procedure constitute a valuable resource that can be used either for creating gender models or to perform gender dependent analyses of Twitter content. As ongoing work, we have already explored supervised and unsupervised gender classification models using the developed datasets and obtained an accuracy of 93.2 % with English users and an accuracy of 96.9 % with Portuguese users in our test sets.

Despite the encouraging results, the proposed approach has still several limitations that will be addressed in the near future: (i) Twitter users might not use their real names and for that reason the reliability of self-declared names is uncertain (e.g.: a male user can have a female gender associated *user name*); (ii) The proposed approach is not robust when facing profiles of companies and other organisations; (iii) Twitter metadata might be incorrect. For example, a tweet identified by Twitter as being written in Portuguese may be written in a different language.

Acknowledgments. This work was supported by national funds through Fundação para a Ciência e a Tecnologia (FCT) with reference UID/CEC/50021/2013.

References

1. Al Zamal, F., Liu, W., Ruths, D.: Homophily and latent attribute inference: inferring latent attributes of twitter users from neighbors. In: ICWSM 270 (2012)
2. Bamman, D., Eisenstein, J., Schnoebelen, T.: Gender in twitter: styles, stances, and social networks. CoRR abs/1210.4567 (2012)

3. Bergsma, S., Dredze, M., Van Durme, B., Wilson, T., Yarowsky, D.: Broadly improving user classification via communication-based name and location clustering on twitter. In: HLT-NAACL, pp. 1010–1019 (2013)
4. Brogueira, G., Batista, F., Carvalho, J.P., Moniz, H.: Expanding a database of portuguese tweets. In: Pereira, M.J.V., Leal, J.P., Simões, A. (eds.) 3rd Symposium on Languages, Applications and Technologies. OpenAccess Series in Informatics (OASIcs), vol. 38, pp. 275–282. Schloss Dagstuhl-Leibniz-Zentrum fuer Informatik, Dagstuhl, Germany (2014). http://drops.dagstuhl.de/opus/volltexte/2014/4576
5. Burger, J.D., Henderson, J., Kim, G., Zarrella, G.: Discriminating gender on twitter. In: Proceedings of the Conference on Empirical Methods in Natural Language Processing (EMNLP 2011), pp. 1301–1309. Association for Computational Linguistics, Stroudsburg, PA, USA (2011). http://dl.acm.org/citation.cfm?id=2145432.2145568
6. Ciot, M., Sonderegger, M., Ruths, D.: Gender inference of twitter users in non-english contexts. In: EMNLP, pp. 1136–1145 (2013)
7. Deitrick, W., Miller, Z., Valyou, B., Dickinson, B., Munson, T., Hu, W.: Gender identification on twitter using the modified balanced winnow (2012)
8. Fink, C., Kopecky, J., Morawski, M.: Inferring gender from the content of tweets: a region specific example. In: ICWSM (2012)
9. van Halteren, H., Speerstra, N.: Gender recognition on dutch tweets. Comput. Linguist. Neth. J. **4**, 171–190 (2014)
10. Kokkos, A., Tzouramanis, T.: A robust gender inference model for online social networks and its application to linkedin and twitter. First Monday 19(9) (2014)
11. Liu, W., Al Zamal, F., Ruths, D.: Using social media to infer gender composition of commuter populations. In: Proceedings of the When the City Meets the Citizen Workshop, The International Conference on Weblogs and Social Media (2012)
12. Liu, W., Ruths, D.: What's in a name? using first names as features for gender inference in twitter. In: AAAI Spring Symposium: Analyzing Microtext (2013)
13. McCreadie, R., Soboroff, I., Lin, J., Macdonald, C., Ounis, I., McCullough, D.: On building a reusable twitter corpus. In: Proceedings of the 35th International ACM SIGIR Conference on Research and Development in Information Retrieval, pp. 1113–1114. ACM (2012)
14. Miller, Z., Dickinson, B., Hu, W.: Gender prediction on twitter using stream algorithms with N-gram character features. Int. J. Intell. Sci. **2**(24), 143–148 (2012)
15. Nguyen, D., Trieschnigg, D., Dogruöz, A.S., Gravel, R., Theune, M., Meder, T., de Jong, F.: Why gender and age prediction from tweets is hard: lessons from a crowdsourcing experiment (2014)
16. Pennacchiotti, M., Popescu, A.M.: A machine learning approach to twitter user classification. In: ICWS, vol. 11, pp. 281–288 (2011)
17. Petrović, S., Osborne, M., Lavrenko, V.: The edinburgh twitter corpus. In: Proceedings of the NAACL HLT 2010 Workshop on Computational Linguistics in a World of Social Media (WSA 2010), pp. 25–26. Association for Computational Linguistics, Stroudsburg, PA, USA (2010). http://dl.acm.org/citation.cfm?id=1860667.1860680
18. Rao, D., Yarowsky, D., Shreevats, A., Gupta, M.: Classifying latent user attributes in twitter. In: Proceedings of the 2nd International Workshop on Search and Mining User-Generated Contents (SMUC 2010), pp. 37–44. ACM, New York, NY, USA (2010). http://doi.acm.org/10.1145/1871985.1871993
19. Ugheoke, T.O.: Detecting the Gender of a Tweet Sender. Master's thesis (2014)
20. Van Zegbroeck, E.: Predicting the Gender of Flemish Twitter Users Using an Ensemble of Classifiers. Master's thesis (2014)

21. Vicente, M., Batista, F., Carvalho, J.P.: Twitter gender classification using user unstructured information. In: Proceedings of IEEE International Conference on Fuzzy Systems (FUZZ-IEEE), Istambul, Turkey, August 2015. http://fuzzieee2015.org
22. Vicente, M., Carvalho, J.P., Batista, F.: Using unstructured profile information for gender classification of portuguese and english twitter users. In: Sierra-Rodríguez, J.-L., Leal, J.-P., Simões, A. (eds.) SLATE 2015. CCIS, vol. 563, pp. 57–64. Springer, Heidelberg (2015). doi:10.1007/978-3-319-27653-3_6

Temporal Data Processing

Suppression of High Frequencies in Time Series Using Fuzzy Transform of Higher Degree

Michal Holčapek and Linh Nguyen(✉)

Institute for Research and Applications of Fuzzy Modelling, NSC IT4Innovations, University of Ostrava, 30. dubna 22, 701 03 Ostrava 1, Czech Republic
{michal.holcapek,linh.nguyen}@osu.cz

Abstract. In this paper, we provide a theoretical justification for the application of higher degree fuzzy transform in time series analysis. We demonstrate that the higher degree fuzzy transform technique can be used for the suppression of high frequencies in time series, which belongs among the essential assumptions for a successful extraction of the trend (trend-cycle) of time series. More precisely, if a time series can be additively decomposed into a trend-cycle, a seasonal component and a noise, we show that high frequencies appearing in the seasonal component can be arbitrarily suppressed using the fuzzy transform of higher degree with a reasonable adjustment of parameters of a generalized uniform fuzzy partition.

Keywords: Fuzzy transform · High frequency · Seasonal component · Time series

1 Introduction

The fuzzy transform (F-transform) theory has been introduced by Perfilieva in [2] and then generalized to a higher degree in [3]. An application of fuzzy transform to the time series analysis was published in [5,6]. The trend (trend-cycle) extraction is one of the major tasks in the time series analysis. In literature, for a review some modern methods, we refer to [7], we can find different approaches for the trend extraction like a model-based approach (an ARIMA model or a state space model), nonparametric linear filtering (the Henderson, LOESS, and Hodrick-Prescott filters), or singular spectrum analysis. A successful extraction of the trend-cycle requires a significant suppression or better elimination of high (positive or negative) frequencies and random fluctuations. In the case of the higher degree F-transform technique, the problem of trend-cycle extraction has been partially solved in [1,9], where a suppression of high frequencies in seasonal components has been demonstrated for the F^0- and F^1-transform with respect to the triangle and raised cosine uniform fuzzy partitions. This paper is a continuation of our research of the suppression of high frequencies in time series using the higher degree F-transform technique, where the particular cases are now replaced by more general ones. The results should bring us a unified view on the trend-cycle extraction possibilities using the higher degree F-transform technique.

J.P. Carvalho et al. (Eds.): IPMU 2016, Part II, CCIS 611, pp. 705–716, 2016.
DOI: 10.1007/978-3-319-40581-0_57

2 Preliminaries

Let $\mathbb{N}$, $\mathbb{Z}$, $\mathbb{R}$ and $\mathbb{C}$ denote the set of natural numbers, integers, reals and complex numbers, respectively. For any complex number $c \in \mathbb{C}$, we use $|c|$ to denote the absolute value of c. i.e., $|c| = (c \cdot \bar{c})^{\frac{1}{2}}$, where $\bar{c}$ is the complex conjugate of c.

2.1 Generalized Uniform Fuzzy Partition

A fuzzy partition of an interval or the real line is a core of the (higher degree) F-transform. In this paper, we restrict ourselves to fuzzy partitions that are uniformly spread along the real line and are determined by a generating function, which definition is as follows.

Definition 1. *A real–valued function $a : \mathbb{R} \to [0,1]$ is said to be a* generating function *provided that a is a continuous and even function that is non-increasing in $[0,1]$ and vanishing outside of $[-1,1]$.*

Basic examples of generating functions frequently appearing in applications of the F-transform technique are the triangle a and raised cosine functions.

Example 1. Functions $a^{tr}, a^{rc} : \mathbb{R} \to [0,1]$ defined by

$$a^{tr}(t) = \max(1 - |t|, 0) \tag{1}$$

$$a^{rc}(t) = \begin{cases} \frac{1}{2}(1 + \cos(\pi t)), & -1 \leq t \leq 1; \\ 0, & \text{otherwise}, \end{cases} \tag{2}$$

for any $t \in \mathbb{R}$, are called the *triangle* and *raised cosine* generating functions, respectively.

The triangle generating function is a special case of B-spline generating functions that were used in a little modified form in [10, 11].

Example 2. Let us define a rectangular pulse β^0 as follows:

$$\beta^0(t) = \begin{cases} 1, & -\frac{1}{2} < x < \frac{1}{2}, \\ \frac{1}{2}, & |x| = \frac{1}{2}, \\ 0, & \text{otherwise}. \end{cases} \tag{3}$$

A central B–spline of degree n denoted by β^n is constructed from the $(n+1)$-fold convolution of the rectangular pulse β^0:

$$\beta^n(t) = \underbrace{\beta^0 \star \beta^0 \star \cdots \star \beta^0(t)}_{(n+1) \text{ times}}. \tag{4}$$

A B–spline generating function of degree n is denoted by $a^{bs,n}(t)$ and defined by rescaling the support of $\beta^n(t)$, precisely,

$$a^{bs,n}(t) = \beta^n\left(\frac{(n+1) \cdot t}{2}\right). \tag{5}$$

Obviously, $a^{bs,1}(t) = a^{tr}(t)$.

Let a be a generating function, and let h and r be positive constants. The parameters h and r are called the *bandwidth* and the *shift*, respectively. Define by $t_k = k \cdot r$, $k \in \mathbb{N}$, the k-th node of the real line. The function $a_{h,r,k} : \mathbb{R} \to [0,1]$ given by

$$a_{h,r,k}(t) = a\left(\frac{t - t_k}{h}\right) \tag{6}$$

is said to be a *scaled generating function placed at the k-th node of the real line.* Now, we can proceed to the definition of a generalized uniform fuzzy partition of $\mathbb{R}$ as has been proposed in [4].

Definition 2. *Let a be a generating function, and let h and r be positive constants.* A generalized uniform fuzzy partition of the real line determined by the triplet (a, h, r) *is the collection $\{a_{h,r,k}\}_{k\in\mathbb{Z}}$ of scaled generating functions placed at all nodes $t_k = k \cdot r$, $k \in \mathbb{Z}$, which satisfies* a la Ruspini condition*:*

$$\sum_{k\in\mathbb{Z}} a_{h,r,k}(t) = 1 \tag{7}$$

for any $t \in \mathbb{R}$.

In what follows, we provide a sufficient condition for a generalized uniform fuzzy partition to be determined from a generating function, bandwidth and shift.

Theorem 1. *Let a be a generating function, and let $\gamma = \int_{-1}^{1} a(t)dt$. If a satisfies the γ-symmetry condition, i.e.,*

$$\sum_{k\in\mathbb{Z}} a(t - k\gamma) = 1, \quad t \in [0,1], \tag{8}$$

and $\frac{\gamma h}{r} \in \mathbb{N}$ for $h, r > 0$, then the triplet $\left(\frac{r}{\gamma h} \cdot a, h, r\right)$ determines a generalized uniform fuzzy partition of $\mathbb{R}$.

Proof. Let us assume that $\frac{\gamma h}{r} = m \in \mathbb{N}$. We show that a la Ruspini condition is satisfied for the triplet $\left(\frac{r}{\gamma h} \cdot a, h, r\right)$. For any $t \in \mathbb{R}$, we have

$$\sum_{k\in\mathbb{Z}} \frac{r}{\gamma h} \cdot a\left(\frac{t - kr}{h}\right) = \frac{1}{m}\sum_{n\in\mathbb{Z}} a\left(\frac{t}{h} - \frac{k\gamma}{m}\right) = \frac{1}{m}\sum_{i=0}^{m-1}\sum_{\ell\in\mathbb{Z}} a\left(\frac{t}{h} - \frac{i\gamma}{m} - \ell\gamma\right),$$

where we used the fact that any integer n can be uniquely expressed as $\ell m + i$ for a certain integer ℓ and $i = 0, \dots, m-1$. Moreover, for any $i = 0, 1, \dots, m-1$, there exists $\ell^i \in \mathbb{Z}$ such that

$$\frac{t}{h} - \frac{i\gamma}{m} - \ell^i\gamma \in [0,1]. \tag{9}$$

Indeed, let $s = \frac{t}{h} - \frac{i\gamma}{m}$. If $s - \ell\gamma \notin [0,1]$ for any $\ell \in \mathbb{Z}$, then $\gamma > 1$, but this is a contradiction with the definition of the generating function. By the γ–symmetry condition, for any $i = 0, \ldots, m-1$, we find that

$$\sum_{\ell \in \mathbb{Z}} a\left(\frac{t}{h} - \frac{i\gamma}{m} - \ell\gamma\right) = \sum_{n \in \mathbb{Z}} a\left(\frac{t}{h} - \frac{i\gamma}{m} - \ell^i\gamma - n\gamma\right) = 1.$$

Hence, we obtain

$$\sum_{k \in \mathbb{Z}} \frac{r}{\gamma h} \cdot a\left(\frac{t - kr}{h}\right) = \frac{1}{m} \sum_{i=0}^{m-1} \sum_{\ell \in \mathbb{Z}} a\left(\frac{t}{h} - \frac{i\gamma}{m} - \ell\gamma\right) = 1.$$

□

As a straightforward consequence of this theorem we obtain a sufficient condition for the determination of the raised cosine and B–spline generalized uniform fuzzy partitions.

Corollary 1. *If $\frac{h}{r} \in \mathbb{N}$ for $h, r > 0$, then the triplet $\left(\frac{r}{h} \cdot a^{rc}, h, r\right)$ determines a raised cosine generalized uniform fuzzy partition.*

Corollary 2. *If $\frac{2h}{r(n+1)} \in \mathbb{N}$ for $h, r > 0$, n is a positive natural number, then the triplet $\left(\frac{r(n+1)}{2h} \cdot a^{bs,n}, h, r\right)$ determines a B–spline generalized uniform fuzzy partition of degree n.*

In the sequel, when we consider a raised cosine or B–spline generalized uniform fuzzy partition, we assume that it is determined by the respective triplet of parameters stated in the mentioned above corollaries.

2.2 Higher Degree Fuzzy Transform

In the original paper [3] on the higher degree F-transform, an orthogonal basis of polynomials derived by the Gram-Schmidt orthogonalization process with respect to the basic functions of a given fuzzy partition were used to find the components of its direct phase. In this paper, we consider another approach based on the monomial basis and special matrices.

Definition 3. *Let $\{a_{h,r,k}\}_{k \in \mathbb{Z}}$ be a generalized uniform fuzzy partition of the real line determined by the triplet (a, h, r), and f be a piecewise continuous function. The* direct fuzzy transform of m-th degree *(F^m–transform) of f with respect to $\{a_{h,r,k}\}_{k \in \mathbb{Z}}$ is the collection $F_m^{\rightarrow}[f] = \{Q_k \mid k \in \mathbb{Z}\}$ where*

$$Q_k(t) = \beta_{k,0} + \beta_{k,1}(t - t_k) + \ldots + \beta_{k,m}(t - t_k)^m \tag{10}$$

determined by $(\beta_{k,0}, \beta_{k,1}, \ldots, \beta_{k,m})^T = H^{-1} \cdot C^{-1} \cdot Y$ *where* $H = \mathrm{diag}(1, \mathrm{h}, \ldots, \mathrm{h}^{\mathrm{m}})$, *and* $C = (C_{ij})$ *and* $Y = (Y_i)$ *are matrices of the sizes* $(m+1) \times (m+1)$ *and* $(m+1) \times 1$, *respectively determined as follows*

$$C_{ij} = \int_{-1}^{1} t^{i+j-2} a(t) dt, \tag{11}$$

$$Y_i = \int_{-1}^{1} f(th + t_k) \cdot t^{i-1} a(t) dt. \tag{12}$$

The polynomial $F^{\rightarrow}_{m,k}[f] = Q_k$ *is called the* k-th component of direct F^m–transform of f with respect to $\{a_{h,r,k}\}_{k\in\mathbb{Z}}$.

The inverse phase of the higher degree F-transform is linear like combination of basic functions with coefficients represented by the F-transform components (polynomials).

Definition 4. *Let* $F^{\rightarrow}_m[f] = \{Q_k \mid k \in \mathbb{Z}\}$ *be a direct fuzzy transform of* m*-th degree of a function* f *with respect to. The* inverse fuzzy transform of f of m-th degree *with respect to a generalized uniform fuzzy partition* $\{a_{h,r,k}\}_{k\in\mathbf{Z}}$ *is defined as follows*

$$\hat{f}(t) = \sum_{k\in\mathbb{Z}} Q_k(t) a_{h,r,k}(t), \quad t \in \mathbb{R}. \tag{13}$$

In what follows, we omit the reference to the degree of the direct F-transform in all cases when no confusion can appear.

3 Suppression of High Frequencies in Time Series

In this section, the k-th component of direct F^m-transform of a function f with respect to the generalized uniform fuzzy partition $\{a_{h,r,k}\}_{k\in\mathbb{Z}}$ determined by a triplet (a, h, r) will be denoted by $F^{\rightarrow}_{m,k,a}[f]$ to stress the type of generating function.

3.1 Assumptions on Time Series

Let $X(t), t \in \mathbb{R}$, be a time series, i.e., a realization of a random process.[1] Similarly to the papers [1,9], we assume that $X(t)$ can be additively decomposed and written in the following form

$$X(t) = TC(t) + S(t) + R(t), \quad t \in \mathbb{R}, \tag{14}$$

where $TC(t)$, $S(t)$, $R(t)$ denote the trend-cycle, seasonal and noise component, respectively. Further, we assume that the seasonal component $S(t)$ is the sum of a finite number of waves as follows

$$S(t) = \sum_{k=0}^{s} p_k e^{i\omega_k t}, \tag{15}$$

[1] The set of real numbers $\mathbb{R}$ can be replaced here by any real interval, e.g., $[0, T]$.

where p_k is a complex number, ω_k is the k–th frequency, and i denotes the imaginary unit.

In this paper, we show that each summand in (15) can be arbitrarily suppressed by the application of the higher degree fuzzy transform with respect to a generalized uniform fuzzy partition, where the bandwidth h and a number N of derivatives of respective generating function are major parameters. A consequence of this fact is that the higher degree fuzzy transform technique can arbitrarily suppressed the seasonal component.

3.2 General Justifications

Let $p(\omega, t) = e^{i\omega t}$, where t, ω and i denotes the time, frequency and imaginary unit, respectively. The complex-valued function $p(\omega, t)$ represents a summand in (15). The following theorem states that the size of each component $F^{\rightarrow}_{m,k,a}[p(\omega, t)]$ of direct higher degree F-transform of $p(\omega, t)$ at each point $t \in [t_k - h, t_k + h]$ can be arbitrarily small. Before, let us recall the well known Riemann-Lebesgue lemma in the Fourier analysis (see, e.g., [8]).

Lemma 1 (Riemann–Lebesgue). *Let f be a function of $L^1[a, b]$. Then, the following integrals*

$$\int_a^b f(x) \cos \lambda x dx \quad \textit{and} \quad \int_a^b f(x) \sin \lambda x dx$$

tend to zero as $|\lambda| \to \infty$.

Theorem 2. *Let $\{a_{h,r,k}\}_{k\in\mathbb{Z}}$ be a generalized uniform fuzzy partition determined by the triplet (a, h, r), and let $\omega \in \mathbb{R}$ and $d = \omega h$. Then, for any $m \in \mathbb{N}$ and $t \in [t_k - h, t_k + h]$, it holds*

$$\lim_{|d|\to\infty} \left|F^{\rightarrow}_{m,k,a}[p(\omega, t)](t)\right| = 0. \tag{16}$$

Proof. From the definition of the direct F^m-transform; (Definition 3), we have

$$F^{\rightarrow}_{m,k,a}[p(\omega, t)](t) = \beta_{k,0} + \beta_{k,1}(t - t_k) + \cdots + \beta_{k,m}(t - t_k)^m,$$

determined by

$$\boldsymbol{\beta}_k = (\beta_{k,0}, \cdots, \beta_{k,m})^T = H^{-1} \cdot C^{-1} \cdot Y,$$

where $H = \text{diag}(1, h, \ldots, h^m)$, and $C = (C_{ij})$ and $Y = (Y_i)$ are matrices of the sizes $(m + 1) \times (m + 1)$ and $(m + 1) \times 1$, which are determined by

$$C_{ij} = \int_{-1}^{1} t^{i+j-2} a(t) dt \quad \text{and} \quad Y_i = \int_{-1}^{1} p(\omega, th + t_k) \cdot t^{i-1} a(t) dt,$$

respectively. Then, one can simply find that

$$\left|F^{\rightarrow}_{m,k,a}[p(\omega, t)](t)\right| \le \sum_{j=0}^{m} |t - t_k|^j |\beta_{k,j}|.$$

Moreover, putting $T = C^{-1}$, we obtain

$$|\beta_{k,j}| \le \frac{\|T\|}{h^j} \sum_{\ell=0}^{m} |Y_{\ell+1}|$$

for any $j = 0, 1, \ldots, m$, where $\|T\| = \max\{|T_{ij}||i, j = 1, 2, \ldots, m+1\}$. Hence

$$\left|F^{\rightarrow}_{m,k,a}[p(\omega,t)](t)\right| \le \|T\| \sum_{j,\ell=0}^{m} \left|\frac{t-t_k}{h}\right|^j |Y_{\ell+1}| \le (m+1)\|T\| \sum_{\ell=0}^{m} |Y_{\ell+1}|.$$

In addition, for any $\ell = 0, 1, \ldots, m$, we have

$$\begin{aligned} Y_{\ell+1} &= \int_{-1}^{1} p(\omega, ht+t_k) t^\ell a(t) dt = \int_{-1}^{1} e^{i\omega(ht+t_k)} t^\ell a(t) dt \\ &= e^{i\omega t_k} \int_{-1}^{1} t^\ell a(t) e^{i\omega ht} dt = e^{i\omega t_k} \int_{-1}^{1} t^\ell a(t) e^{itd} dt \\ &= e^{i\omega t_k} \left(\int_{-1}^{1} t^\ell a(t) \cos(td) dt + i \int_{-1}^{1} t^\ell a(t) \sin(td) dt \right) \end{aligned}$$

Using Lemma 1 and the previous equality, we obtain the desirable results. □

From this theorem, one can see that if a generalized uniform fuzzy partition is fixed, i.e., we do not change the generating function and its bandwidth h, then the absolute value of $F^{\rightarrow}_{m,k}[p(\omega,t)](t)$ converges to zero for large (positive or negative) frequencies, more precisely, for the frequencies that go in the absolute value to infinity. On the other hand, if a frequency ω is fixed, then the same effect is obtained by considering large bandwidths. Hence, we can assert that the high frequencies, which appear in the seasonal component of a time series, can be significantly suppressed by a reasonable setting of the bandwidth of a generating function from which a generalized uniform fuzzy partition is constructed.

Although, the previous theorem has a major importance for the high frequency suppression, it says nothing about the speed of the convergence. The following theorem shows that the speed of convergence depends on the differentiability of generating functions.

Theorem 3. *Let $\{a_{h,r,k}\}_{k\in\mathbb{Z}}$ be a generalized uniform fuzzy partition determined by the triplet (a, h, r), and let $\omega \in \mathbb{R}$ and $d = \omega h$. If $a \in C^N(\mathbb{R})$, i.e., a is N-times differentiable and its N derivatives are continuous on $\mathbb{R}$, then*

$$\left|F^{\rightarrow}_{m,k,a}[p(\omega,t)](t)\right| = O(d^{-N}) \quad \text{as} \quad |d| \to \infty, \tag{17}$$

for any $m \in \mathbb{N}$ and $t \in [t_k - h, t_k + h]$.

Proof. By the same arguments used in the proof of Theorem 2, we find that

$$\left|F^{\rightarrow}_{m,k,a}[p(\omega,t)](t)\right| \le (m+1)\|T\| \sum_{\ell=0}^{m} |Y_{\ell+1}|, \tag{18}$$

where

$$|Y_{\ell+1}| = \left| \int_{-1}^{1} t^\ell a(t) e^{itd} dt \right|.$$

Since $a \in C^N(\mathbb{R})$, we have $t^\ell a \in C^N(\mathbb{R})$. Using the integration by parts, we obtain

$$\begin{aligned} \int_{-1}^{1} t^\ell a(t) e^{itd} dt &= \frac{(-1)^N}{(i \cdot d)^N} \cdot \int_{-1}^{1} (t^\ell a(t))^{(N)} \cdot e^{itd} dt \\ &\quad + e^{itd} \sum_{j=0}^{N-1} \frac{(-1)^j}{(i \cdot d)^{j+1}} \cdot (t^\ell a(t))^{(j)} \Bigg|_{t=-1}^{t=1}. \end{aligned}$$

Form the continuity of $a^{(j)}$ for any $j = 0, 1, \ldots, N$, we have $(t^\ell a(t))^{(j)}\big|_{t=\pm 1} = 0$ for any $j = 0, 1, \ldots, N-1$. Therefore,

$$\int_{-1}^{1} t^\ell a(t) e^{itd} dt = \frac{(-1)^N}{(i \cdot d)^N} \cdot \int_{-1}^{1} (t^\ell a(t))^{(N)} \cdot e^{itd} dt.$$

Furthermore,

$$|Y_{\ell+1}| = \left| \int_{-1}^{1} t^\ell a(t) e^{itd} dt \right| \leq \frac{1}{|d|^N} \cdot \int_{-1}^{1} \left| (t^\ell a(t))^{(N)} \right| dt.$$

From this inequality and (18), we obtain the desirable statement.

□

3.3 Particular Cases of Generating Functions

In Theorem 3, we showed that the speed of the convergence is d^{-N} provided that the fuzzy partition is based on a generating function a belonging to C^N. The following theorems state that in particular cases, the speed of the convergence can be even higher.

Theorem 4. *Let $\{\tilde{a}^{rc}_{h,r,k}\}_{k \in \mathbb{Z}}$ be a raised cosine generalized uniform fuzzy partition determined by a triplet $(\tilde{a}^{rc}, h, r)$, where $\tilde{a}^{rc} = \frac{r}{h} \cdot a^{rc}$, and let $\omega \in \mathbb{R}$ and $d = \omega h$. Then, for any $m \in \mathbb{N}$ and $t \in [t_k - h, t_k + h]$, it holds*

$$\left| F^{\rightarrow}_{m,k,\tilde{a}^{rc}}[p(\omega, t)](t) \right| = O(d^{-3}) \quad as \quad |d| \to \infty. \tag{19}$$

Proof. Using the integration by parts, for any $\ell = 0, 1, \ldots, m$, we obtain

$$\begin{aligned} \int_{-1}^{1} t^\ell \tilde{a}^{rc}(t) e^{itd} dt &= \frac{(-1)^3}{(i \cdot d)^3} \int_{-1}^{1} (t^\ell \tilde{a}^{rc}(t))^{(3)} e^{itd} dt \\ &\quad + e^{itd} \cdot \sum_{j=0}^{2} \frac{(-1)^j}{(i \cdot d)^{j+1}} \cdot (t^\ell \tilde{a}^{rc}(t))^{(j)} \Bigg|_{t=-1}^{t=1} \end{aligned}$$

By a straightforward computation, we find that $(t^\ell\tilde{a}^{rc}(t))^{(j)}\big|_{t=\pm1} = 0$ for any $j = 0, 1$. Hence, we obtain

$$\int_{-1}^{1} t^\ell\tilde{a}^{rc}(t)e^{itd}dt = \frac{1}{i\cdot d^3}\int_{-1}^{1}(t^\ell\tilde{a}^{rc}(t))^{(3)}e^{itd}dt \\ -\frac{e^{id}}{i\cdot d^3}\cdot(t^\ell\tilde{a}^{rc}(t))''(1)+\frac{e^{-id}}{i\cdot d^3}\cdot(t^\ell\tilde{a}^{rc}(t))''(-1).$$

Therefore,

$$\left|\int_{-1}^{1} t^\ell\tilde{a}^{rc}(t)e^{itd}dt\right| \leq \\ \frac{1}{|d|^3}\left(\int_{-1}^{1}\left|(t^\ell\tilde{a}^{rc}(t))^{(3)}\right|dt + |(t^\ell\tilde{a}^{rc}(t))''(1)| + |(t^\ell\tilde{a}^{rc}(t))''(-1)|\right).$$

By similar arguments used in the proof of Theorem 3, we obtain the desirable statement.

□

Theorem 5. *Let* $\left\{\tilde{a}^{bs,n}_{h,r,k}\right\}_{k\in\mathbb{Z}}$ *be a B-spline generalized uniform fuzzy partition of degree* n *determined by the triplet* $(\tilde{a}^{bs,n}, h, r)$*, where* $\tilde{a}^{bs,n} = \frac{r(n+1)}{2h}\cdot a^{bs,n}$*,* $\omega\in\mathbb{R}$ *and* $d = \omega h$*. Then,*

$$\left|F^{\rightarrow}_{m,k,\tilde{a}^{bs,n}}[p(\omega,t)](t)\right| = O(d^{-(n+1)}) \quad as \quad |d|\to\infty, \tag{20}$$

for any $m\in\mathbb{N}$ *and* $t\in[t_k-h, t_k+h]$*.*

Proof. Let $f = -\frac{d}{2\pi}$. Recall that the Fourier transform (unitary with ordinary frequency) of $\beta^n(t)$ is the function $\mathrm{sinc}^{n+1}(f)$, where f is the frequency, and $\mathrm{sinc}(f) = \frac{sin(\pi f)}{\pi f}$. By the definition of $\tilde{a}^{bs,n}$ and the basic properties of the Fourier transform, for any $\ell = 0, 1, \ldots, m$, we obtain

$$\int_{-1}^{1} t^\ell\tilde{a}^{bs,n}(t)e^{idt}dt = \frac{r(n+1)}{2h}\int_{-1}^{1}t^\ell\beta^n\left(\frac{(n+1)t}{2}\right)e^{-i2\pi tf}dt \\ = \frac{r(n+1)}{2h}\cdot\mathcal{F}\left[t^\ell\beta^n\left(\frac{(n+1)t}{2}\right)\right](f) = \frac{r}{h}\cdot\left(\frac{i}{2\pi}\right)^\ell\cdot\frac{d^\ell}{df^\ell}\mathrm{sinc}^{n+1}\left(\frac{2f}{n+1}\right).$$

Moreover, one can prove that

$$\frac{d^\ell}{df^\ell}\mathrm{sinc}^{n+1}\left(\frac{2f}{n+1}\right) = O(f^{-(n+1)}) \quad \text{as} \quad |f|\to\infty,$$

for any $\ell = 0, 1, \ldots, m$. Therefore,

$$\left|\int_{-1}^{1} t^\ell\tilde{a}^{bs,n}(t)e^{idt}dt\right| = O(d^{-(n+1)}) \quad \text{as} \quad |d|\to\infty.$$

By similar arguments used in the proof of Theorem 3, we obtain the desirable statement. □

As we have mentioned in Example 2, the triangular generalized uniform fuzzy partition is a B-spline generalized uniform fuzzy partition of degree $n = 1$. The following statement is a consequence of the previous theorem.

Corollary 3. *Let $\{\tilde{a}^{tr}_{h,r,k}\}_{k\in\mathbb{Z}}$ be a generalized uniform fuzzy partition determined by a triplet $(\tilde{a}^{tr}, h, r)$, where $\tilde{a}^{tr} = \frac{r}{h} \cdot a^{tr}$, and let $\omega \in \mathbb{R}$ and $d = \omega h$. Then,*

$$\left|F^{\rightarrow}_{m,k,\tilde{a}^{tr}}[p(\omega,t)](t)\right| = O(d^{-2}) \qquad \text{as} \qquad |d| \to \infty. \tag{21}$$

for any $m \in \mathbb{N}$ and $t \in [t_k - h, t_k + h]$.

From a comparison of Theorems 3 and 4, one can see that the use of B-spline generalized uniform fuzzy partitions determined by B-splines of higher degrees ($n \geq 3$) should be favorable in the high frequencies suppression against the application of the raised cosine uniform fuzzy partition.

3.4 Illustrative Examples

In this subsection, we provide an example demonstrating how the higher degree F-transform can suppress the high frequencies in time series.

We consider an artificial time series $X(t)$, $t \in \{0, 1, \ldots, 100\}$ defined as

$$X(t) = 20\sin(0.063t) + 5\sin(0.63t + 1.5) + 5\sin(1.26t + 0.35) + 15\sin(2.7t + 1.12) + 7\cos(0.41t + 0.79).$$

One can see that the functions $20\sin(0.063t)$ and $7\cos(0.41t + 0.79)$ are two waves in $X(t)$ corresponding to the lowest and the second lowest frequency in the time series, respectively. Firstly, we apply the F^2-transform that is considered with respect to three types of fuzzy partitions, namely, the triangle, raised cosine and B-spline (of degree 3) uniform fuzzy partition with the same bandwidth $h = 10$. In Fig. 1, the dark curves depict the results of the inverse F^2-transforms of $X(t)$ with respect to the all considered fuzzy partitions. In Fig. 1c, one can see that the F^2-transform with respect to the B-spline uniform fuzzy partition of degree 3 filters out almost all of high frequencies in $X(t)$ except the lowest frequency in the wave expressed by the function $20\sin(0.063t)$. On the other hand, the lowest and the second lowest frequencies are not filtered out by the F^2-transform with respect to the remaining uniform fuzzy partitions as one can see in Fig. 1a, b. More precisely, in these figures, the inverse F^2-transforms of $X(t)$ coincide approximately with the sum of waves expressed by the functions $20\sin(0.063t)$ and $7\cos(0.41t+0.79)$ and depicted by the gray curves. If we want to suppress also the wave with the second lowest frequency in $X(t)$, i.e., the wave expressed by the function $7\cos(0.41t+0.79)$, we have to increase the value of the bandwidth h. In Fig. 2, one can see the effect when the bandwidth h is increased for 10 to 16, namely, the F^2-transform with respect to the triangle and raised cosine uniform fuzzy partitions significantly suppressed the wave with the second lowest frequency.

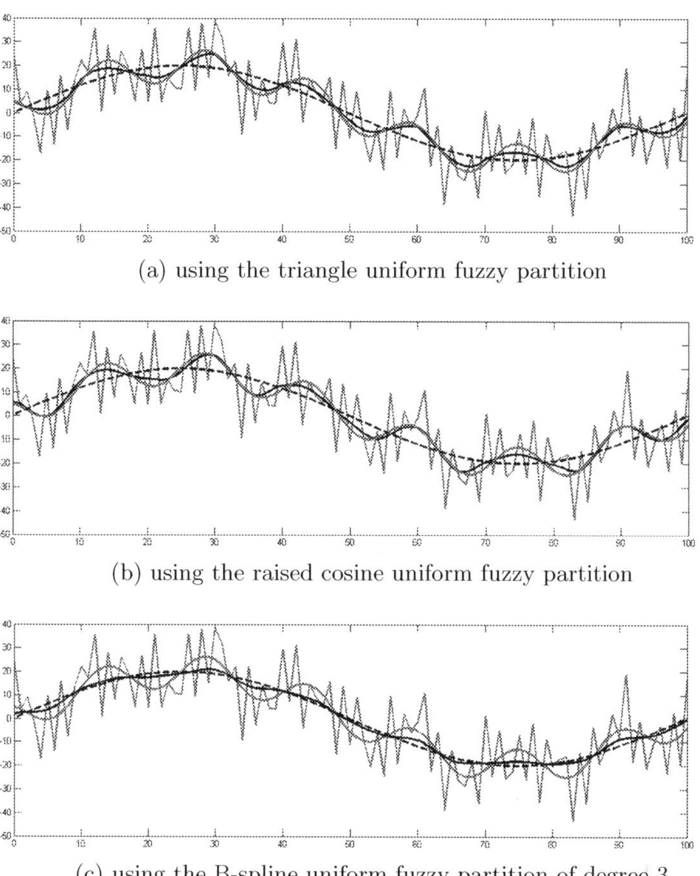

(a) using the triangle uniform fuzzy partition

(b) using the raised cosine uniform fuzzy partition

(c) using the B-spline uniform fuzzy partition of degree 3

Fig. 1. Suppression of high frequencies in $X(t)$ (broken line) by the F^2-transform with respect to various fuzzy partitions. The dashed curve and the gray curve are the plots of functions $20\sin(0.063t)$ and $20\sin(0.063t) + 7\cos(0.41t + 0.79)$, respectively. The inverse F^2-transform of $X(t)$ is depicted by the dark curve.

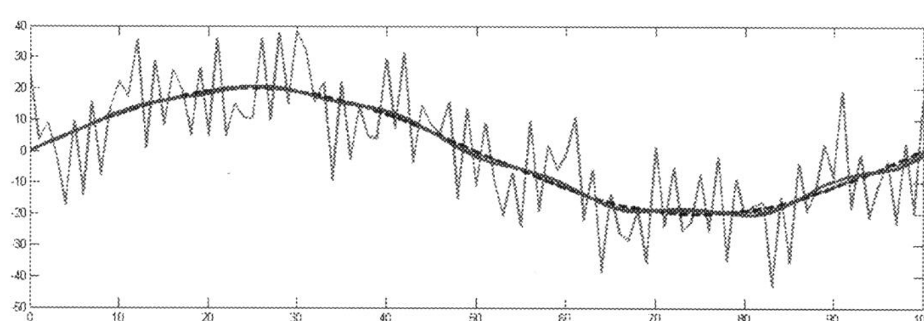

Fig. 2. Suppression of the second lowest frequencies in $X(t)$. The dashed curve is the plot of the wave corresponding to the lowest frequency $20\sin(0.063t)$. The gray and dark curves depict the inverse F^2-transform of $X(t)$ with respect to the triangle and raised cosine uniform fuzzy partitions, respectively.

4 Conclusions

In this paper, we proved that the high frequencies in time series (presented in the seasonal component) can be successfully suppressed by setting of parameters of higher degree F-transform. Namely, the bandwidth and the choice of the generating function with respect to its differentiability are the most important parameters. The results provide a theoretical justification of the application of higher degree fuzzy transform in time series analysis focusing on the trend (trend-cycle) extraction.

Acknowledgment. This work was supported by the project LQ1602 IT4Innovations excellence in science. The additional support was also provided by the Czech Science Foundation through the project of No.16-09541S.

References

1. Novák, V., Perfilieva, I., Holčapek, M., Kreinovich, V.: Filtering out high frequencies using F-transform. Inf. Sci. **274**, 192–209 (2014)
2. Perfilieva, I.: Fuzzy transforms: theory and applications. Fuzzy Sets Syst. **157**, 993–1023 (2006)
3. Perfilieva, I., Daňková, M., Bede, B.: Towards a higher degree F-transform. Fuzzy Sets Syst. **180**, 3–19 (2011)
4. Holčapek, M., Perfilieva, I., Novák, V., Kreinovich, V.: Necessary and sufficient conditions for generalized uniform fuzzy partition. Fuzzy Sets Syst. **277**, 97–121 (2015)
5. Novák, V., Štěpnička, M., Dvořák, A., Perfilieva, I., Pavliska, V., Vavříčková, L.: Analysis of seasonal time series using fuzzy approach. Int. J. Gen. Syst. **39**, 305–328 (2010)
6. Novák, V., Štěpnička, M., Perfilieva, I., Pavliska, V.: Analysis of periodical time series using soft computing methods. In: Ruan, D., Montero, J., Lu, J., Martnez, L., D'hondt, P., Kerre, E. (eds.) Computational Intelligence in Decision and Control. World Scientific, New Jersey (2008)
7. Alexandrov, T., Bianconcini, S., Bee Dagum, E., Maass, P., McElroy, T.: A Review of Some Modern Approaches to the Problem of Trend Extraction. Research Report Series, Statistics 2008–3, U.S. Census Bureau, Washington (2009)
8. Titchmarsh, E.C.: Introduction to The Theory of Fourier Integrals, 2nd edn. Oxford University Press, Oxford (1948)
9. Nguyen, L., Novák, V.: Filtering out high frequencies in time series using F-transform with respect to raised cosine generalized uniform fuzzy. In: Proceedings of IEEE International Conference on Fuzzy Systems, Istanbul (2015)
10. Kodorane, I., Asmuss, S.: On approximation properties of spline based F-transform with respect to fuzzy m-partition. In: Pasi, G., Montero, J., Ciucci, D. (eds.) Proceedings of the 8th Conference of the European Society for Fuzzy Logic and Technology (EUSFLAT 2013), pp. 772–779. Atlantis Press (2013)
11. Kokainis, M., Asmuss, S.: Approximation properties of higher degree F-transforms based on B-splines. In: Proceedings of the IEEE International Conference on Fuzzy Systems, Istanbul (2015)

A Modular Fuzzy Expert System Architecture for Data and Event Streams Processing

Jean-Philippe Poli(✉) and Laurence Boudet

CEA, LIST, Data Analysis and System Intelligence Laboratory,
91191 Gif-sur-Yvette Cedex, France
{jean-philippe.poli,laurence.boudet}@cea.fr

Abstract. In many decision making scenarios, fuzzy expert systems have been useful to deduce a more conceptual knowledge from data. With the emergence of the Internet of Things and the growing presence of cloud-based architectures, it is necessary to improve fuzzy expert systems to support higher level operators, large rule bases and an abundant flow of inputs.

In this paper, we present a modular fuzzy expert system which takes data or event streams in input and which outputs decisions on the fly. Its architecture relies on both a graph-based representation of the rule base and the cooperation of four customizable modules. Stress tests regarding the number of rules have been carried out to characterize its efficiency.

Keywords: Fuzzy expert system · Complex event processing · Data stream processing · Rule base representation · Policies

1 Introduction

The emergence of connected objects and of the Internet of Things is leading towards a continuous data acquisition from different devices and sensors. Before this recent phenomenon, the data were stored in data warehouse, queried at once and manipulated by algorithms as a whole. With such data in motion, the use cases have changed: for instance, new database management paradigms are introduced, special efforts are made on data compression to avoid networks overload, and supervised or unsupervised learning algorithms are rethought.

Cugola and Margara [8] define the *Information Flow Processing* (IFP) domain as the domain of tools capable of processing information as it flows. Usually, the flow is coming from multiple sources and processed in order to extract relevant knowledge. They also distinguish two subdomains: *Complex Event Processing* (CEP) and *Data Stream Processing* (DSP). On the one hand, DSP consists in processing data flows and in producing a new data flow as output. The Federal Standard defines a data stream as a "sequence of digitally encoded signals used to represent information in transmission". Algorithms for processing such data have to be fast and incremental [21]. In [12], the authors are revealing the open challenges which must be addressed in the domain of data

J.P. Carvalho et al. (Eds.): IPMU 2016, Part II, CCIS 611, pp. 717–728, 2016.
DOI: 10.1007/978-3-319-40581-0_58

stream mining, including privacy issues, developing a methodology for stream preprocessing, developing online monitoring systems and balancing resources. On the other hand, CEP differs by the type of data items it considers: an item is a notification of event [15]. CEP aims at managing thousands of events per second [23], for instance, up to 125000 events for a financial software [22]. In this domain, processing mainly consists in filtering, gathering and combining those events in order to build a higher level information [2]. In many real world cases, DSP and CEP have became usual considerations. We can cite for example: system monitoring and fault detection, home automation, security and finance [8].

Whatever the type of items in the stream, i.e. either data or events, the information may be incomplete and imprecise by nature [3]. For instance, sensors may be out of order or inaccurate, and data may be noisy. Fuzzy logic [24] has been specifically designed to mathematically represent uncertainty and vagueness and is a popular tool for dealing with imprecision in many real world problems. Taking advantage of fuzzy logic, fuzzy expert systems allow to easily represent human knowledge about data and phenomena and have been successfully applied to many domains [9,20].

Comparing with boolean logic expert systems, fuzzy expert systems are often associated with a higher computational cost. Indeed, the whole rule base has to be evaluated in order to compute the outputs, whereas in classical expert systems, a subset of the rules are applied one by one to produce the inference. Moreover, it has been showed that fuzzy rule bases need to be more complicated if only piecewise-linear functions (e.g. trapezoids...) are used instead of non-linear membership functions (e.g. sigmoids...) [5]. Consequently, expensive functions in terms of computation are needed to evaluate the aggregation and the defuzzification [13]. Moreover, in real-world applications, it is possible to have very large rule bases which require a great amount of processor time [1]. Fuzzy controllers have been introduced to overcome these drawbacks and are able to process inputs in real-time [18]. Other papers address the acceleration of fuzzy computation either with dedicated hardware [4] or with the help of Graphics Processing Units (GPU) [10].

To our experience, fuzzy expert softwares which run on CPU platforms are more convenient for many reasons. Firstly, they are easier to interface with an existing system than electronic chipsets. Then, DSP and CEP both rely on software intensive architectures. In terms of scalability, it is possible to use from a single core of a machine to several machines and it can all be done transparently for the user; for instance, it can take advantage of the virtualization as in cloud-based services. To the best of our knowledge, current fuzzy expert systems, both open source or commercial, rely on straightforward architectures which only manage the classical fuzzy operators (and, or, not). To describe the relations between the data or the events, more sophisticated operators are needed for temporal [7,16], space [6,19] or even spatio-temporal [14] reasoning. These operators imply a higher computation cost. Moreover, traditional fuzzy expert systems compute output values only when input values have changed. This is not compliant with event streams whose events are potentially arriving in an irregular manner: in such a case, expressions may change before the next event arrival (see Sect. 2.3).

Using the terminology of Cugola and Margara, we aim at developing a fuzzy expert system to process information flows, handling the imprecision brought by noisy data, sensors or network problems with fuzzy logic. The motivation of our work is to provide an efficient fuzzy expert system in operational contexts. To enable human experts to author more complex rules, our system is able to efficiently evaluate complex fuzzy relations [16]. To ensure it can interface easily with the various information systems of our partners, we chose to avoid specific architectures (like GPU) and to develop a software for data and event streams processing on regular CPU platforms. Finally, in industrial applications, the efficiency is important not only because there must be a lot of rules, but also because the rules can be applied to a huge number of objects or events per second.

The paper is structured as follows: Sect. 2 describes the architecture of our fuzzy expert system. Section 3 presents the protocol and the results of experiments on synthetic data. Finally, Sect. 4 points out the conclusions.

2 Architecture Description

During fuzzy inference, when a group of inputs change at a time t, all the rules containing at least one of those inputs have to be reevaluated. In information streams, inputs may change several times per second, or rules must be applied on thousands of incoming events per second; the evaluation of the whole rule base may thus need a huge computation time. We introduce an architecture which tends to avoid the system saturation.

2.1 Architecture Overview

Figure 1 presents the overview of the proposed architecture. The modularity is ensured by a separation of the tasks and a customization provided by the use of policies. A policy is a set of parameters which customize the behavior of each module. The combination of the behaviors of all the modules enable to address a lot of applications and issues: regular or irregular data rate, delay before inference, etc. The architecture is composed of several modules:

- the **active input queue** gathers the input and group them by timestamp,
- the **scheduler** is able to monitor the system (via the operating system) and to decide which inputs group has to be processed,
- the **evaluator** is in charge of the evaluation of the rules,
- the **output change broadcaster** informs the user about outputs changes.

The different modules are supposed to avoid an overload of the system (for instance, the active input queue selects the inputs which should be treated) or user overfeeding (for instance, the output change broadcaster displays only the

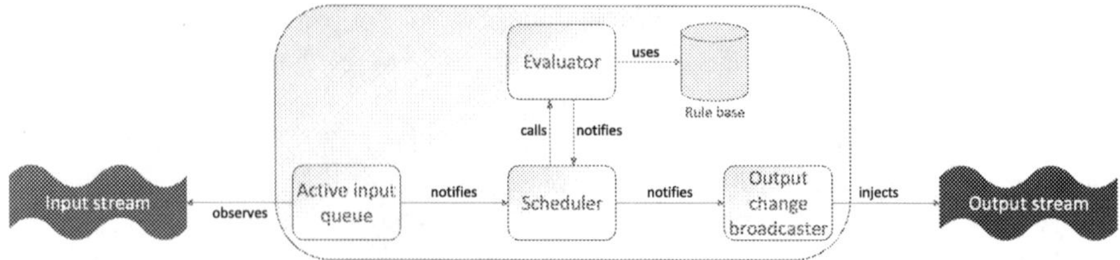

Fig. 1. Architecture overview.

relevant information). We first introduce how we optimize the rule base representation by common subexpression elimination and the concept of expiration of expressions. We then describe each module of the architecture and give some examples of policies.

2.2 Rule Base Representation

The rule base in-memory model plays a major role in the efficiency of the fuzzy expert system. Expressions are usually modeled with a tree [17], as in Fig. 2(a). However, some expressions can be included in several rules or other expressions: thus, in a tree representation, it is difficult to check the redundancy of such expressions, and it is necessary to evaluate them several times when a group of inputs changed. This problem is known as common subexpression elimination (CSE).

To address the CSE problem in our architecture, we chose to represent each expression by a unique node: thus, the rule base is not represented by a tree anymore but by a graph (Fig. 2(b)). More precisely, we use an acyclic directed graph to avoid loops during the evaluation. In the graph, an edge $A \longrightarrow B$ means that if the value of the node A changes, it affects the node B and B has to be evaluated again. A node can represent fuzzy expressions (including fuzzy propositions) or rules, and we consider particular nodes for defuzzification and aggregation. Thus, the changes propagate from input nodes to output nodes. The propagation stops if there are no changes during the evaluation of the current node.

The propagation is achieved as particular breadth-first traversal of the graph. However, for a fuzzy n-ary expression, it is necessary to evaluate its n predecessors before its own evaluation, otherwise it would be evaluated n times, and worst, at a certain time, its value would be inconsistent. To avoid this effect, we added a priority information to the nodes. Before starting the fuzzy inference engine, the graph is traversed and a recursive function $priority : Node \rightarrow integer$ is applied. Let N be the current node to be treated, the function $priority$ is defined as follow:

- if N is an input node, then $priority(N) = 0$,
- otherwise, let s_i be the successors of N, $priority(N) = \max_i(priority(s_i)) + 1$.

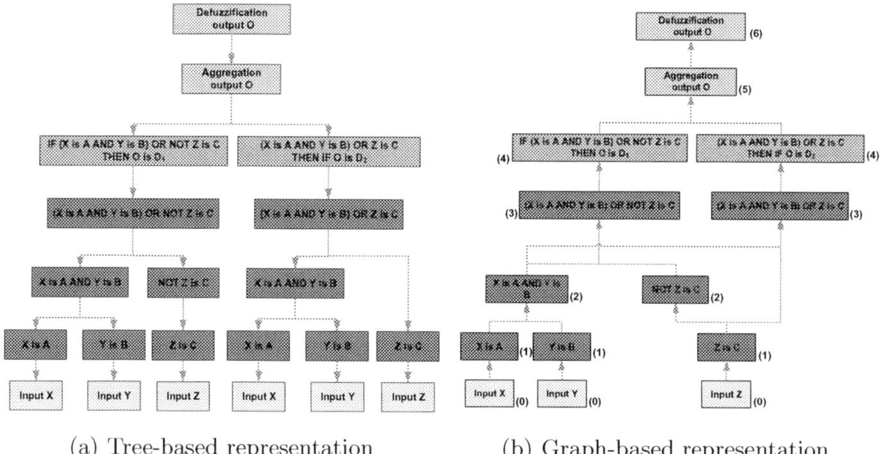

(a) Tree-based representation (b) Graph-based representation

Fig. 2. Representations of a base of two rules.

Let X, Y and Z be three input linguistic variables, and A, B, C a term from respectively X, Y, Z. Let D_1 and D_2 be two terms of an output linguistic variable O. Then, the rule base is composed of two rules:

- IF (X is A AND Y is B) OR NOT Z is C THEN O is D_1,
- IF (X is A AND Y is B) OR Z is C THEN O is D_2.

In Fig. 2(b), numbers in brackets represent the evaluation priority of each node; the three inputs are at the bottom of the figure and have a null priority, which means they need to be evaluated first. We will develop in Sect. 2.6 the use of the priority during evaluation.

To the best of our knowledge, current fuzzy expert system does not implement CSE. This is due to the fact that they only use classical fuzzy logic operators which are really fast to evaluate.

2.3 Expiration

Among the sophisticated relations we have implemented, temporal operators [16] and those which depend from them need a special attention when applied on event streams. The particularity of event streams is that the system is noticed of events irregularly. For instance, let consider the fact "the temperature was too hot on Monday from 2 am to 3 am". The system has received two events: at 2 am, a temperature high enough to activate the fuzzy proposition "the temperature is too hot", and at 3 am, a lower temperature such as "the temperature is too hot" is false. Now, we consider the temporal operator "occurrence" from [7] which indicates that a phenomenon has occurred on a certain scope in the past: for instance, it can express that "the temperature was too hot during the last 24 h". Until the next Tuesday 3 am, the degree of truth of this occurrence is strictly greater than 0. After 24 h, its degree of truth equals 0, whereas the system inputs have not changed since Monday 3 am.

Classical fuzzy expert systems cannot perform this trick since they need that inputs change to compute the outputs. We thus introduce in our system the notion of expiration. Some expressions in the rule base are marked as "expirable" and signal to the scheduler (see Sect. 2.5) that they need to be evaluated again. Expirable components must provide an expiration frequency and a set of criteria to stop the expiration. The expiration frequency is a parameter which depends on the application and the set of criteria depend only on the definition of the operator. To implement expiration, the values of all the expressions are stored in memory to allow partial recalculation.

2.4 Active Input Queue

Sensor networks are a particular case of information stream. Some sensors measure (data stream) and some others detect (event stream), but they usually work in an asynchronous way. Moreover, some delays can appear in such networks. The active input queue is thus in charge of:

- listening to the information stream,
- fetching the interesting values inside,
- grouping those values by timestamp,
- enqueueing those groups of inputs,
- signaling the scheduler that a new group has been enqueued.

Different policies can be conceived for this component. For instance, in some applications, it is necessary to wait for delayed sensors or delayed network packets before signaling the scheduler. Conversely, it can ignore delays and late arrivals, and thus filter these data. It may also be seen as a firewall which protects the scheduler from irrelevant inputs.

2.5 Scheduler

The scheduler has an important role to play in order to limit the delay between the arrival of the data and the decision making. When a new input set is announced, it decides, regarding its own policy, whether it is important to evaluate it immediately, later or not at all.

In the simplest implementation, the scheduler just gets the first element in the active input queue, asks the evaluator to evaluate this group of inputs and gives the results to the broadcaster. With the use of policies, his behavior can be more sophisticated. For instance, one particular implementation can monitor the system to determine how busy the CPU cores are and to decide whether a group of inputs can be skipped. Moreover, the scheduler implements the expiration. All the expirable components of the rule base whose evaluation has changed are placed in another queue, waiting to expire.

Another implementation may consist in evaluating on different processor cores of the machine. Each core receives a sub-part of the input set. A simple algorithm based on the graph representation of the rule base is used to separate independent inputs on different sub-parts: this is simply achieved by finding connected components of graph with well-known algorithms of graph theory [11].

2.6 Evaluator

The evaluator is the component which evaluates the different expressions and rules in the rule base. For a set of inputs, it gives a particular set of outputs. It also takes advantage of the rule base representation to perform the computation only when necessary.

In order to evaluate the different nodes of the graph representing the rule base, the evaluator traverses the graph in a certain order. To ensure the right order, we use a priority queue Q. The priority queue Q is implemented such as the nodes with the lowest priority are placed first and such as it contains only one occurrence of each node. The general evaluation algorithm is given below:

```
Q ← changed inputs
while Q ≠ ∅ do
  current ← first(Q)
  Q ← dequeue(Q)
  Evaluate(current)
  if current has changed then
    for all child of current do
      Q ← enqueue(Q, child)
    end for
  end if
end while
return all the values
```

The priority is important in some cases. In Fig. 2(b), the priority queue ensures the node "(X is A AND Y is B) OR Z is C" is evaluated at the right time. It ensures that if several paths lead to the same node N, all nodes on the paths are assessed before N.

In fuzzy logic, different functions can be used for operators (conjunction, disjunction, negation, implication) evaluation. The policies of the evaluator indicate which version of the operators must be used.

2.7 Output Change Broadcast

The broadcaster is also an important module because it is in charge of building the output stream. The last step is indeed to inform on the fly the calling system or the user that some outputs have changed. The policies are used to determine when and how the outputs have to be broadcast, for instance:

- the changes can be gathered and sent at regular time intervals,
- only outputs which have changed are broadcast with their new value,
- the changes can be sent with a trace of the activated rules.

It may gather information from the graph and the evaluation of its node to build justifications (to explain why the decision has been made).

3 Experiments

The experiments aim at comparing the performances of the evaluation of different rule bases regarding two different policies of the evaluator module:

- full recalculation mode: all the expressions and nodes are reassessed each time an input changes,
- partial recalculation mode: last values of the nodes are kept in memory and are reassessed only when needed.

The second mode is the one on which our architecture relies. Its modularity, through the use of policies, allows to easily switch between the two modes. In both cases, we only consider the graph-based representation of a rule base, whose evaluation is indeed faster than with a tree-based representation.

3.1 Protocol

These experiments have been carried out on artificial rule bases and data sets whose generation is described hereafter. Let $\{v_i\}_{1 \leq i \leq n}$ be n input linguistic variables, each defined by p terms $T_i^1, \ldots, T_i^p$. Let w be an unique output linguistic variable whose terms are $W_1, ..., W_K$. Those input variables combine into rules by the full conjunctive combination principle:

$$\texttt{IF}\, v_1, \texttt{is}\, T_1^{l_1}\, \texttt{and} \ldots \texttt{and}\, v_n\, \texttt{is}\, T_n^{l_n}\, \texttt{THEN}\, w\, \texttt{is}\, W_k$$

where $T_i^{l_i}$ refers to a term of v_i with $1 \leq l_i \leq p$ and $k = \sum_{i=1}^{n} l_i - n + 1$. Thus, for a given couple (n, p), there are p^n possible combinations of those inputs (i.e. rules) and w has $K = n(p-1) + 1$ terms.

For the sake of simplicity, the terms $T_i^{l_i}$ of each variable v_i are defined by triangular membership functions on the domain $[0, p+1]$. By construction, the support of each term $T_i^{l_i}$ is $[l_i - 1; l_i + 1]$ and its kernel is $\{l_i\}$. The same construction is used for the terms W_k of w. Figure 3 shows an example of a linguistic variable characterized by 3 terms.

Each input variable v_i receives a data stream of 20 values, which have been generated following an uniform distribution $\mathcal{U}([0, p+1])$.

The architecture has been configured as follows: the active input queue is set in DSP mode, i.e. it waits to receive a value for each input. The scheduler evaluates this group as soon as possible, then the new value of the output is broadcast. This is the most simple configuration of these modules. The two modes of evaluation of the architecture have been obtained by configuring the policy of the evaluator: in one case, it uses its memory functionality; in the other case, it has to compute all the values of the nodes again. The same input data streams have been used for both cases.

Finally, by varying both the number of inputs n and the number of terms p from 2 to 10, we are able to assess the performance of the architecture on large rule bases and to draw some conclusions. Due to the computational cost, the

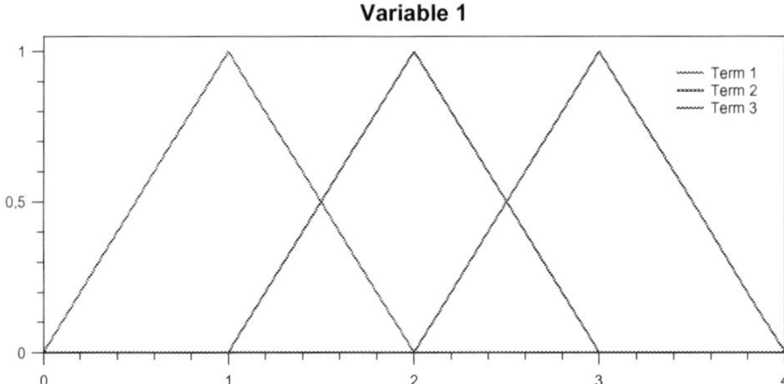

Fig. 3. Linguistic variable with 3 terms defined on the domain [0, 4]. (Color figure online)

largest configuration was obtained with 6 input variables and 9 linguistic terms. This represents a set of 531441 rules to compute the value of the output w. Even if this is not a realistic case, it is useful to benchmark the proposed system.

3.2 Results

In this section, we first compare the average number of nodes being reevaluated in each mode and then compare the average evaluation time of the rule base. The averages are computed over the 20 values of the data stream in order to decrease the possible biases.

Figure 4 shows the number of evaluated nodes regarding the number of rules, in both modes (full and partial recalculation); the two axes are shown in log-scale. Point clouds confirm the intuition: storing the value of each node allows to stop propagating the changes, and strongly decreases the number of nodes to evaluate. For a rule base with 16807 rules, from $n = 5$ linguistic variables and $p = 7$ terms, 36449 nodes may be evaluated in the full recalculation mode, whereas in the partial one, only 120 nodes in average are evaluated.

The drastic reduction of the number of nodes to be evaluated can be explained by a theoretical analysis. Indeed, the number of nodes N_g of the graph-based representation can be evaluated by the following equation:

$$N_g(n,p) = \underbrace{n}_{inputs} + \underbrace{n \times p}_{propositions} + \underbrace{\sum_{i=2}^{n} p^i}_{conjunctions} + \underbrace{p^n}_{implications} + \underbrace{1}_{aggregation} + \underbrace{1}_{defuzzification}$$

The fuzzy partitions used to create the terms of the linguistic variables explain why, at each time, for each variable, at most 2 terms out of p are activated. Thus, at most $N_g(n, 2)$ nodes have to be evaluated: for $n = 5$, at most 109 nodes will be activated. But a large number of them are null because of the conjunctive combination of the inputs. Now, in order to count the number of needed reevaluations, we should consider the worst case: all the active

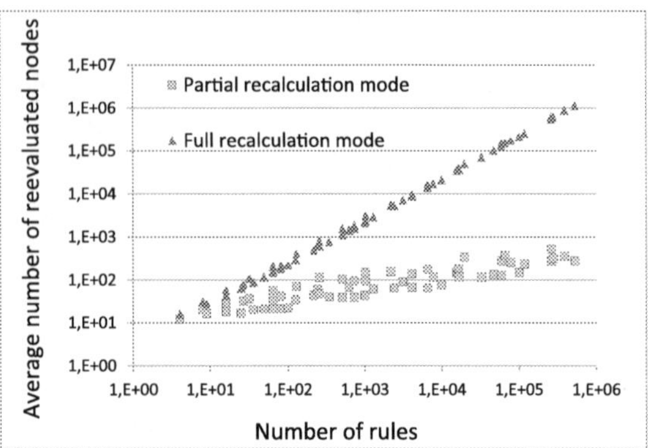

Fig. 4. Average number of reevaluated nodes in function of the number of rules (log-scale) for both modes.

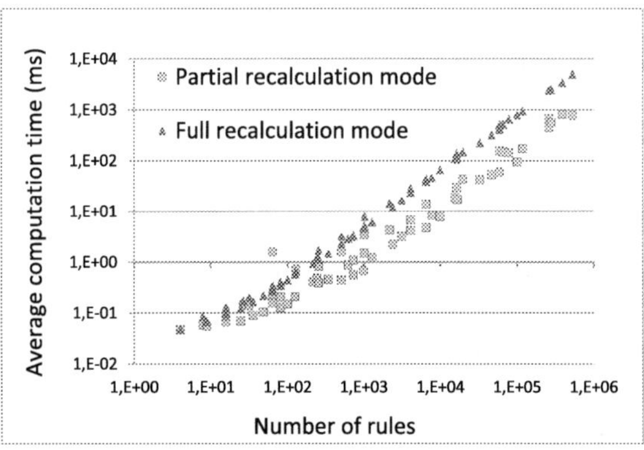

Fig. 5. Computation time (in ms) in function of the number of rules (in log scale).

elementary propositions become null, and the same number of propositions get a non-null value. This gives $2 \times N_g(n, 2)$ as a pessimistic upper bound of the number of nodes that need to be reevaluated.

Figure 5 shows the duration of the evaluation of the rule bases in both modes. These tests have been processed on only one core of an Intel Xeon X5650 at 2.67 GHz on a Windows server. The system is implemented in C#. With the same example as before, to evaluate 16807 rules, full recalculation mode needs approximately 106.8 ms whereas the partial one needs only 17.1 ms, i.e. the latter one is more than 6 times faster than the former one on this rule base structure. It seems that saved computational time is not as high as we could expect considering the saved computations shown just before. But saved computations correspond to the evaluation of a null value by a quite simple function (mainly either by the membership function evaluation or by a conjunctive combination of two expressions) and to affect it to nodes that were already null. Considering this remark, a gain of approximately half an order of magnitude is a good result by only avoiding the computation of null values of graph nodes.

These tests are good stress tests because all the inputs change at the same time. For rule bases of conventional sizes, for instance 300 rules, the engine needs only 0.67 ms in the partial recalculation mode. Thus, we can handle inputs which change more than 1000 times per second on only one core.

4 Conclusion

In this paper, we have presented a modern architecture for a fuzzy expert system designed to handle information streams (data streams or event streams). The architecture relies on two aspects. Firstly, the graph representation of the rule base indicates the dependency between inputs, expressions, rules and outputs. Secondly, the use of four cooperating modules permits to filter and to decide when it is possible to process a set of inputs. The introduction of policies in the four modules allows to customize their behaviors regarding the addressed projects or issues.

The described architecture has been implemented and used in several industrial projects in different domains: home automation, decision making in industry and home care services. All projects needed to process either data stream or event stream, sometimes both of them at the same time.

Uncertainty and imprecision are real-world challenges, but others emerge. Users need more fuzzy relations to be able to describe their scenarios or to characterize what they want to extract from the streams. Considering CEP and several thousands of inputs per second, we should parallelize the computations. Finally, online rule base optimization will allow users to sketch first rules and then let the system evolve.

References

1. Acampora, G., Loia, V.: Fuzzy control interoperability and scalability for adaptive domotic framework. IEEE Trans. Industr. Inf. **1**(2), 97–111 (2005)
2. Alevizos, E., Skarlatidis, A., Artikis, A., Paliouras, G.: Complex event processing under uncertainty: a short survey. In: Fischer, P.M., Alonso, G., Arenas, M., Geerts, F. (eds.) Proceedings of the Workshops of the EDBT/ICDT 2015 Joint Conference. CEUR Workshop Proceedings, vol. 1330, pp. 97–103. CEUR-WS.org (2015)
3. Artikis, A., Baber, C., Bizarro, P.: Canudas-de Wit, C., Etzion, O., Fournier, F., Goulart, P., Howes, A., Lygeros, J., Paliouras, G., Schuster, A., Sharfman, I.: Scalable proactive event-driven decision making. IEEE Technol. Soc. Mag. **33**(3), 35–41 (2014)
4. Basterretxea, K., Del Campo, I.: Electronic hardware for fuzzy computation. In: Scalable Fuzzy Algorithms for Data Management and Analysis: Methods and Design, pp. 1–30. Information Science Reference (2010)
5. Basterretxea, K., Tarela, J.M., de Campo, I., Bosque, G.: An experimental study on nonlinear function computation for neural/fuzzy hardware design. IEEE Trans. Neural Netw. **18**(1), 266–283 (2007)

6. Bloch, I.: Fuzzy spatial relationships for image processing and interpretation: a review. Image Vis. Comput. **23**(2), 89–110 (2005)
7. Cariñena, P., Bugarín, A., Mucientes, M., Barro, S.: A language for expressing fuzzy temporal rules. Mathware Soft Comput. **7**(2–3), 213–227 (2000)
8. Cugola, G., Margara, A.: Processing flows of information: from data stream to complex event processing. ACM Comput. Surv. **44**(3), 15:1–15:62 (2012)
9. Garibaldi, J.M.: Fuzzy expert systems. In: Gabrys, B., Leiviskä, K., Strackeljan, J. (eds.) Do Smart Adaptive Systems Exist? Best Practice for Selection and Combination of Intelligent Methods. STUDFUZZ, vol. 173, pp. 105–132. Springer, Heidelberg (2005)
10. Harvey III., N.R.H.L., Keller, J.M., Anderson, D.: Speedup of fuzzy logic through stream processing on graphics processing units. IEEE Congress on Evolutionary Computation, pp. 3809–3815 (2008)
11. Hopcroft, J., Tarjan, R.: Algorithm 447: efficient algorithms for graph manipulation. Commun. ACM **16**(6), 372–378 (1973)
12. Krempl, G., Žliobaite, I., Brzeziński, D., Hüllermeier, E., Last, M., Lemaire, V., Noack, T., Shaker, A., Sievi, S., Spiliopoulou, M., Stefanowski, J.: Open challenges for data stream mining research. SIGKDD Explor. Newsl. **16**(1), 1–10 (2014)
13. Laurent, A., Lesot, M.J. (eds.): Scalable Fuzzy Algorithms for Data Management and Analysis: Methods and Design. Information Science Reference (2010)
14. Le Yaouanc, J.-M., Poli, J.-P.: A fuzzy spatio-temporal-based approach for activity recognition. In: Castano, S., Vassiliadis, P., Lakshmanan, L.V.S., Lee, M.L. (eds.) ER 2012 Workshops 2012. LNCS, vol. 7518, pp. 314–323. Springer, Heidelberg (2012)
15. Luckham, D.C.: The Power of Events: An Introduction to Complex Event Processing in Distributed Enterprise Systems. Addison-Wesley Longman Publishing Co. Inc., Boston (2001)
16. Poli, J.P., Boudet, L.: Online temporal reasoning for event and data streams processing. In: 2016 IEEE Conference on Fuzzy Systems, FUZZ-IEEE (2016) (to appear)
17. Preiss, B.R.: Data Structures and Algorithms with Object-Oriented Design Patterns in Java. Worldwide Series in Computer Science. Wiley, New York (2000)
18. Reznik, L.: Fuzzy Controllers Handbook. Newnes, Oxford (1997)
19. Schockaert, S., Cock, M.D., Kerre, E.: Reasoning About Fuzzy Temporal and Spatial Information from the Web. World Scientific, Singapore (2010)
20. Siler, W., Buckley, J.: Fuzzy Expert Systems and Fuzzy Reasoning. Wiley-Interscience, Hoboken (2005)
21. Silva, J.A., Faria, E.R., Barros, R.C., Hruschka, E.R., de Carvalho, A.C.P.L.F., Gama, J.: Data stream clustering: a survey. ACM Comput. Surv. **46**(1), 13:1–13:31 (2013)
22. Stonebraker, M., Çetintemel, U., Zdonik, S.: The 8 requirements of real-time stream processing. SIGMOD Rec. **34**(4), 42–47 (2005)
23. Wu, E., Diao, Y., Rizvi, S.: High-performance complex event processing over streams. In: Proceedings of the 2006 ACM SIGMOD International Conference on Management of Data, pp. 407–418. ACM, New York (2006)
24. Zadeh, L.: Fuzzy sets. Inf. Control **8**(3), 338–353 (1965)

Estimation and Characterization of Activity Duration in Business Processes

Rodrigo M. T. Gonçalves[1(✉)], Rui Jorge Almeida[1], João M. C. Sousa[2], and Remco M. Dijkman[1]

[1] Information Systems Group, Eindhoven University of Technology, De Lismortel 2, 5612AR Eindhoven, The Netherlands
{r.m.t.goncalves,rjalmeida,r.m.dijkman}@tue.nl
[2] IDMEC, Instituto Superior Técnico, Universidade de Lisboa, Lisboa, Portugal
jmsousa@tecnico.ulisboa.pt

Abstract. Process-aware information systems are typically used to log events in a variety of domains (e.g. commercial, logistics, healthcare) describing the execution of business processes. The analysis of such logs can provide meaningful knowledge for organizations to improve the quality of their services as well as their efficiency. The prediction of activity durations, based on historic data from execution logs, allows the creation of feasible plans for business processes. However, a problem arises when there are discrepancies between execution logs and the actual execution. When event logs are partially human-generated there is an underlying uncertainty related to the time at which events (recorded by means of user interaction) are logged. If not taken into account, this uncertainty can lead to wrong predictions of activity durations. In this paper, we focus on creating assumptions to estimate activity durations and analyse their impact in the stochastic characterization. A partially human-generated logistics database is used as example.

Keywords: Event logs · Stochastic characterization · Business processes

1 Introduction

As information systems are becoming more intertwined with the operational processes they support, multitudes of events (e.g., transaction logs or audit trails) are recorded. These execution logs can be extracted from almost every process aware information system (PAIS) and provide a chronological record of events referring to the business activities that have been carried out. This gives a detailed overview about the process history [1,2]. However, despite the increasing amount of sensors and respective data, many PAIS still lack of a completely automated log process. Instead, they rely on the interaction of users to log activities that the system can not keep track of. As an example, in transportation processes, drive activities can be easily detected by analysing GPS coordinates but load and unload tasks, which represent the core of logistic processes, are

J.P. Carvalho et al. (Eds.): IPMU 2016, Part II, CCIS 611, pp. 729–740, 2016.
DOI: 10.1007/978-3-319-40581-0_59

commonly manually introduced since they are performed outside of the PAIS environment. When event logs are partially human-generated there is an underlying uncertainty related to the time at which events (introduced by means of user interaction) are logged [3]. In other words, users are able to log events before, or after, the execution of such occurrences. This behaviour may lead to wrong estimations while assessing the duration of activities that were manually logged [4]. As a consequence, process planning often leads to violated time windows, unnecessary delays and underutilized resources. The real-world complexity of planning is caused by the high level of detail that is required to get executable plans and the large volumes of data that must be collected and processed to gather the information required to create the planning. Nonetheless, the value of such analysis is strongly dependent on the quality and suitability of the input event log data [5]. This paper provides an analysis of the temporal uncertainty inherent to partially human-generated event logs and its impact on stochastic activity durations. Moreover, we present a technique to correct temporal inaccuracies that arise from inaccurate log of events by users.

Activity logs can be derived from lower-level data modification logs by means of event log schema transformation techniques such as trace segmentation [6]. Users may provide different pieces of data that, when clustered, constitute a single activity [7]. A lot of work have been done in conformance checking by measuring the alignment between event logs and process models, [8–12], since it is common to find that the execution of a certain process does not conform to the plan that was made in advance [8]. However, to our best knowledge, there is no similar approach to quantify the discrepancies between execution logs and the actual execution of business processes. We have developed techniques for detecting and characterizing such discrepancies by using activity timelines to describe the execution logs and estimate the duration of activities that are executed out of the environment context. Estimation hypothesis are created based on common sense assumptions and there impact on the stochastic results is analysed using a partially human-generated logistics database as example.

2 Preliminary Considerations

A business process is a collection of related and structured activities that produce a specific service, or product, for a particular customer. A complex business process may be decomposed into several sub-processes, which have their own attributes, but also contribute to achieving the goal of the super-process. The analysis of a business process typically includes the mapping of processes and sub-processes down to the activity level [13]. Depending on which activities are meant to be characterized, we divide the overall process into *sub-processes of interest* delimited by $[t_j, t_n]$, where t_j and t_n are the start time and end time of a *sub-process*, respectively. Activities performed outside the interval $[t_j, t_n]$ are not considered during the estimation process.

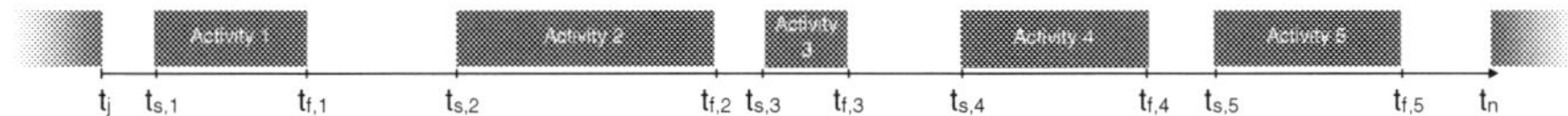

Fig. 1. Activity timeline

2.1 Activity Timelines

Each *sub-process* is described by an activity timeline. A graphical way of showing the performed activities in chronological order. Each bar represents a performed activity and its duration is given by the length of the corresponding bar. However, in the majority of the cases, information systems do not support the explicit notion of activities/tasks. Instead, they only log and support low-level events. Based on the work of [14], we assume events are logged in a way such that (i) each event refers to a specific occurrence, (ii) each event has an associated performer ID (the entity executing or initiating the occurrence), (iii) events have a time-stamp and are totally ordered, (iv) all activities performed during the process have an associated event(s) that makes the logging possible, (v) all activities performed during the process are logged and (vi) there is no activity concurrency. Table 1 shows an example of a low-level event log from a logistic process. To derive activity logs from low-level event logs, we define an activity as follows:

Definition 1. ***Activity*** - *an activity, a_i, is defined as a finite sequence of events, e, in chronological order over a finite period of time, where each event in the activity is a documented occurrence at a given time stamp t, defined as:*

$$a_i = \left\langle e_j^{t_j = t_{s,i}}, e_{j+1}^{t_{j+1}}, \ldots, e_f^{t_{f,i}} \right\rangle \tag{1}$$

The duration of an activity, Δt_{a_i}, is given by the elapsed time between its first and last event.

$$\Delta t_{a_i} = t_{f,i} - t_{s,i} \tag{2}$$

If the event descriptions in the event log are known to be correct (as shown in Table 1), the activity recognition process is straightforward: simply identify the events that indicate the start and end of an activity and group in between events into one activity. Otherwise, more complex approaches such as [6] or [7] are needed. In order to clarify the setting in which these approaches can be of interest, one needs to reflect on how discrepancies between the execution and its log may emerge. The lack of alignment between them is assumed to be caused by users when logging activities. In that sense, we define two sets of activities depending on how they are logged. The definitions are given bellow:

Definition 2. ***System activities,*** S - *activities which log is automatically done by the system. They are assumed to be correctly logged, from a temporal perspective, since the system is fully aware of the current state: performed id, time and knowledge about the activity taking place.*

Table 1. Example of a low-level execution log from a logistic process

PID	Timestamp	Description	Events	Activities
A1	14:57:58	Start of Break	$e_1^{t_1=t_{s,1}}$	
A1	15:21:45	End of Break	$e_2^{t_2=t_{f,1}}$	$a_1 \in U$
A1	15:21:46	Cancellation of	$e_3^{t_3}$	-
A1	15:22:21	Start of Drive	$e_4^{t_4=t_{s,2}}$	
A1	15:23:15	End of Drive	$e_5^{t_5=t_{f,2}}$	$a_2 \in S$
	...			-
A1	18:24:56	Start of Unload	$e_6^{t_6=t_{s,3}}$	
A1	18:26:29	Contact OFF	$e_7^{t_7}$	
A1	18:28:52	Contact ON	$e_8^{t_8}$	
A1	18:28:53	Task Finished	$e_9^{t_9}$	$a_3 \in U$
A1	18:30:46	End of Unload	$e_{10}^{t_{10}=t_{f,3}}$	
	...			-

System crashes, errors, skewed system clocks and timezone problems are not considered.

Definition 3. ***User activities,*** *U - activities which log is dependent on the interaction between user and system. Only the user has knowledge about which activity is taking place. The time at which the activity took place is uncertain since users are able to log activities before, while or after the actual activity take place.*

3 Stochastic Characterization of Process Activities

The estimation process for the activities duration (i.e. *user activities*) is done based on the *dead* time (timespan where no activity was logged) available on the neighbourhood of such activities. However, it is not reasonable to assume that all sub-processes have their user activities wrongly logged. In order to quantify how good is a log of a sub-process, we make use of the logged ratio. The logged ratio, ϕ, represents the amount sub-process time that was logged. If the *logged ratio*, ϕ, is bigger than a threshold value, θ, the *user activities* of the sub-process are assumed to have been correctly logged and the estimation process skips the activity timeline. If the logged ratio does not meet the threshold value, the start or end times of the user activities must be corrected. In that case sub-processes are divided into time slots formed by the system activities. The amount of *dead* time is calculated and assigned to the user activities according to the formulated hypothesis. These steps are described in the following sections.

3.1 Logged Ratio

The existence of *dead* time in between the log of two activities does not strictly indicate that such *user activities* were incorrectly logged. Referring to the assumption made in the previous section - that the system supports the log of all activities performed during the process - activity timelines should be fulfilled up to at least a certain degree, since an activity is always being performed. Thus, in order to consider if a sub-process should have the duration of its *user activities* estimated, we evaluate the amount of *dead* time through the logged ratio. If the logged ratio is 1, there is no *dead* time in between activities and we can assume they were correctly logged. The likelihood of activities being correctly logged diminishes as the amount of *dead* time increases (and the logged ratio decreases). The *logged ratio* of a *sub-process* is obtained as follows:

$$\phi = \frac{\sum_{i=1}^{i=h}(t_{f,i} - t_{s,i})}{t_n - t_j}, \tag{3}$$

where t_j and t_n represent the start time and end time of a *sub-process*, respectively, h is the total number of activities in the sub-process and $t_{s,i}$ and $t_{f,i}$ indicate the start and end time of a performed activity a_i.

3.2 Time Slots

System activities are used as anchor activities. They are the starting point for the estimation process since their log is assumed to be correct. The time-span which they occupy have to be preserved. A time slot can be defined in three ways: (i) as the gap between every two consecutive *system activities*, (ii) as the gap between an estimation boundary (t_j or t_n) and the closest *system activity* or (iii) as the gap between t_j and t_n, if there are no *system activities* in the sub-process. In other words, time slots are portions of activity timelines where no activity was logged. They are characterized by a start and end date, in addition with the *user activities* that were performed on such gap. Each sub-process contains $m+1$ time slots, where m is the total number of *system activities* in the sub-process. Equation 4 shows how to defined the start and end dates of the time slots. t_s^* and t_f^* represent the beginning and end of a *system activity*, respectively.

$$time\_slots^T = \begin{bmatrix} t_j & t_{f,1}^* & \dots & t_{f,m}^* \\ t_{s,1}^* & t_{s,2}^* & \dots & t_n \end{bmatrix} \tag{4}$$

Figure 2 shows an example of the time slots for an activity timeline with two *systems activities*. In this case, $t_{s,1}^* = t_{s,2}$ and $t_{f,1}^* = t_{f,2}$ since the first *system activity* is the second overall activity in the timeline. The same principle applies to the second *system activity* which is the fifth overall activity, $t_{s,2}^* = t_{s,5}$ and $t_{f,2}^* = t_{f,5}$. The time slot lengths (i.e. duration), Δt_s, are given by Eq. 5.

$$\Delta t_{s_i} = time\_slots(i,2) - time\_slots(i,1) \tag{5}$$

The information of each *sub-process* is summarized in a table. Time slots are described by their length and the duration of the activities performed on it, as seen in Table 2.

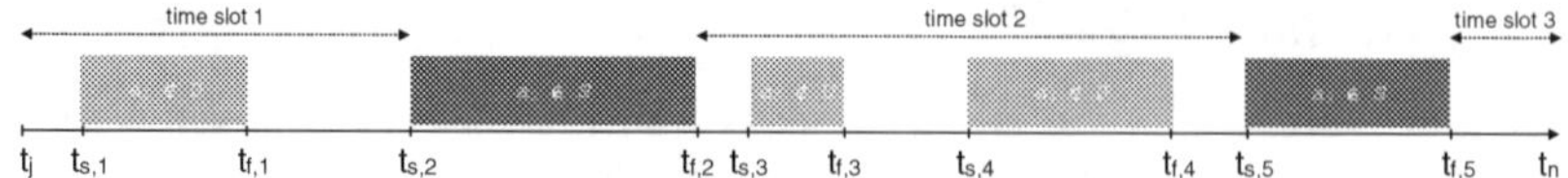

Fig. 2. Time slots

Table 2. Time slot description

Time slot	Length Δt_s	Activities	
		#	Dur Δt_a
1	$t^*_{s,1} - t_j$	1	$t_{f,1} - t_{s,1}$
2	$t^*_{s,2} - t^*_{f,1}$	2	$t_{f,3} - t_{s,3}$
			$t_{f,4} - t_{s,4}$
3	$t^*_n - t^*_{f,2}$	0	-

3.3 Estimation Hypothesis

In order to estimate *user activity* durations we make use of three different hypothesis which differ from each other in the assumptions made. All the available information about the process execution comes directly from the execution logs, in the form of a low-level event logs. As previously explained, events belonging to automatically logged activities are assumed to have correct timestamps. The underlying uncertainty of the execution logs comes from the events logged by users. Therefore, the only information about *users activities* that can be used is the number of activities performed, their sequential order and the timespan where they took place, given by the length of the time slot to which they belong.

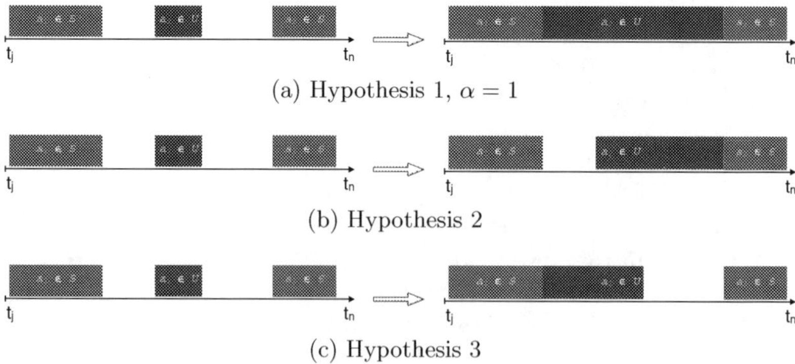

(a) Hypothesis 1, $\alpha = 1$

(b) Hypothesis 2

(c) Hypothesis 3

Fig. 3. Estimation hypothesis

Hypothesis 1 - The first hypothesis consist in assuming the duration of all *user activities* performed inside the time slot to have the same duration. Hence the estimated duration of the *user activities* is given by the ratio between time slot length, Δt_s, and the number of performed activities, $\#a$.

$$\Delta t_a^* = \Delta t_s / \#a \tag{6}$$

It is implicit to this formulation that, after the estimation, there are no dead times in between activities since the whole time slot length Δt_s is assigned to *user activities*. However, such definition can be relax by using only a portion of the time slot length for estimation, $\alpha \Delta t_s$. Figure 3a shows an estimation example of a *user activity* with $\alpha = 1$. The estimated duration of the *user activity* a_2 is given by the timespan from the end of activity a_1 to the start of activity a_3.

Hypothesis 2 - Another possibility is to assume that the starting events of *user activities* were correctly logged and that a *user activity* ends as soon as the next *user activity* starts. The original end dates of *user activities* are set to the starting dates of the following activity. If the activity being estimated, a_i, is the last in the time slot (i.e. $i = h$), its new end date is set to t_n. The amount of dead time in each time slot, after estimation, is given by the difference between the start time of the first *user activity* and the end time of the first *system activity*. If there is non (i.e. $a_1 \in U$), the dead time is given by $t_{s,1} - t_j$ instead.

$$\Delta t_{a_i}^* = \begin{cases} t_{s,i+1} - t_{s,i} & \text{if } i < h \\ t_n - t_{s,i} & \text{if } i = h \end{cases}, \tag{7}$$

where $\Delta t_{a_i}^*$ is the estimated duration of activity a_i and h is the total number of activities in the time slot. This hypothesis is consistent with a real scenario where users log both start and end events of an activity right before accomplishing it. An example of this estimation hypothesis can be seen in Fig. 3b. The end event of activity a_2 is shifted forwards so that it coincides with the start event of activity a_3.

Hypothesis 3 - Lastly, opposing to hypothesis 2, we assume that end events of *user activities* were correctly logged and that *user activities* start as soon as the prior *user activity* ends. This means that users are assumed to log both start and end events of an activity right after accomplishing it. The original start dates of *user activities* are shifted to the end dates of previously performed activities or, if the activity being estimated is the first in the time slot (i.e. $i = 1$), to t_j. In this case, the amount of dead time in each time slot is given by the difference between the start time of the last *system activity* and the end time of the prior *user activity*. If the last activity is a *user activity* (i.e. $a_h \in U$), the dead time is given by $t_n - t_{f,h}$ instead. An example of this estimation hypothesis can be seen in Fig. 3c. The start event of activity a_2 is shifted backwards so that it coincides with the end event of activity a_1.

$$\Delta t^*_{a_i} = \begin{cases} t_{f,i} - t_{f,i-1} & \text{if } i > 1 \\ t_{f,i} - t_j & \text{if } i = 1 \end{cases} \tag{8}$$

4 Real Example: Logistic Business Process

The event log used is part of logistic data-base from a fleet of international road transport trucks (TIR). The fleet has an integrated management system and each truck has a terminal where users are able to log activity related events, such as the start and end of an activity. To evaluate the performance of the developed methodology, a comparison is made between the original and the estimated durations of *user activities* for each hypothesis. Firstly, as described in Sect. 2, the activity log is inferred from the low-level event log. Using domain knowledge, the activity types found are divided into two sets, depending on the logging process of the correspondent events. In this case, the activities are divided as follows:

$$\begin{aligned} \textbf{\textit{System activites}} &= \{\text{Sign Up, Log Out, Log In, Drive}\} \\ \textbf{\textit{User activites}} &= \{\text{Load, Unload, Rest, Break, Wait, Arrive, Refueling}\} \end{aligned}$$

4.1 Defining Sub-processes

A logistic process is divided in two main sub-processes: the transportation process and the package handling process. The first, consists essentially of drive activities to transport cargo between locations and, the second, of activities such as load and unload. The start time, t_j, and end time, t_n, of the sub-processes of interest, in this case the handling process, can be identified by the end time and start time of a drive activity, respectively. Figure 4 shows the activity timeline of

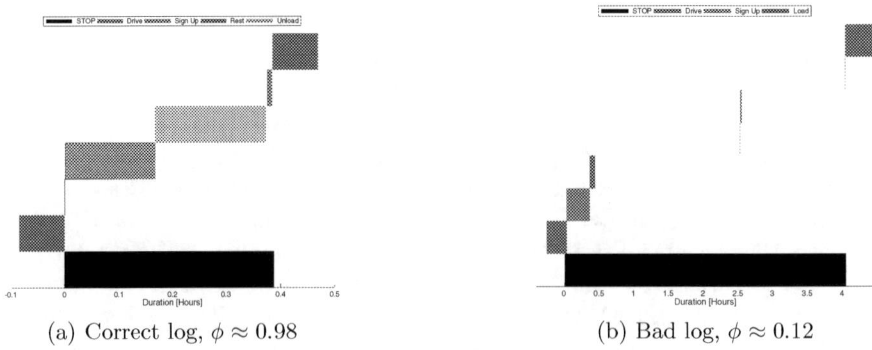

(a) Correct log, $\phi \approx 0.98$ (b) Bad log, $\phi \approx 0.12$

Fig. 4. Examples of activity timelines from two truck stops (Color figure online)

two sub-processes. The black bar, delimited by t_j and t_n, represents the timespan where the sub-process took place. Each additional bar represents a performed activity. Load and unload activities are highlighted with a dotted line. The main difference between examples, is the amount of *dead* time. Considering the *dead* time between the load activities and the prior drive activities, the most probable scenario is that the driver, after parking at the load site, executed the actual load activity and, only after completion, logged the "start of activity" and "end of activity" events, within a seconds time period.

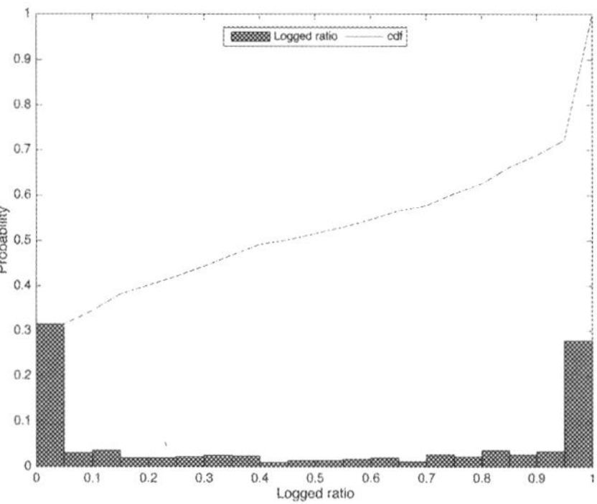

Fig. 5. Logged ratio

To check how common this behaviour is, we evaluated the ratio between logged time and sub-process duration. The logged ratio, ϕ, represents the amount sub-process time that was logged. If $\phi = 0$, no *user activities* were logged in the sub-process. If $\phi = 1$, all sub-process timespan was logged with *user activities*. As it can be seen in Fig. 5, about 50 % of the *sub-processes* have a logged ratio smaller than 0.4. In addition, 32 % of the analysed sub-processes show a logged ratio smaller than 0.05. These numbers would be even higher if we had accounted for sub-processes were no activity was logged, since for those the logged ratio is equal to zero. These proves that the behaviour described by Fig. 4b is common among the users.

4.2 Comparing Original Durations with Estimated Durations

In order to evaluate the impact of the estimation hypothesis, a random customer was selected from the logistic data-base. A frequency analysis is made to the original and estimated durations of load activities. The threshold value Θ was set to zero so that all *user activities* are estimated independently of the logged ratio ϕ. The results are shown in the form of histograms where each bar has a width of five minutes and its height represents the percentage of load activities whose duration is on a given duration interval. As it can be seen in Fig. 6a, 45 % of load activities have an original duration of less than five minutes. For

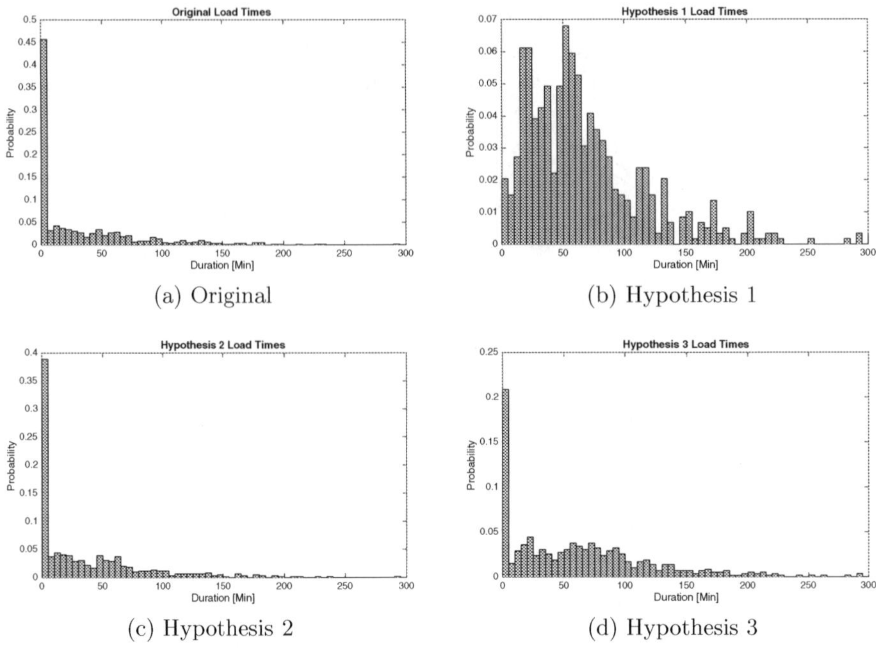

(a) Original (b) Hypothesis 1

(c) Hypothesis 2 (d) Hypothesis 3

Fig. 6. Load activity durations

this particular logistic process, load activities are not expected to be so quick due to the largo amount of cargo. The problem of human influence in event logs becomes even more clear by analysing the quantiles of original load durations, in Table 3. The probability of a load activity to be smaller than 0.167 min (≈10 s), which is a very improbable fact, is 10 %. The quantiles of original *user activity* durations shows that this problem is not only associated with load activities, but with every activity whose log is human influenced. Both original histograms, for load activities, Fig. 6a, and for all *user activities*, reveal similar results. Even when accounting for rest activities, whose duration is expected to be up to eight hours, there is a 80 % confidence level of an activity to have their duration between 0.133 min (≈8 s) and 117 min.

Since hypothesis 1 estimates all *user activities* in the time slot equally (i.e. same duration) the histogram of estimated durations is less skewed than the others. The amount of load activities which duration was originally shorter than five minutes drops from 45 % to about 2 %. Estimations are higher in all quantiles, when comparing with the originals, especially in the last one. In cases with large time-slots and a single *user activity* logged, it is possible for activities to be overestimated leading to higher durations in the 0.99 quantile. In any case, there is a 80 % confidence level for an activity to the duration in the interval [17.1, 171.2] minutes. When using hypothesis 2 for estimation, the starting events are assumed to be correctly logged and end events are shifted to the start event

Table 3. Quantiles of activities durations [Min]

	1 %	2.5 %	5 %	10 %	50 %	90 %	95 %	97.5 %	99 %
Original Load	0.1	0.117	0.133	0.167	10.93	97.17	133.15	164.83	195.51
H1 Load	1.383	7.56	11.975	17.808	57.896	154.1	202.18	346.56	1374.2
H2 Load	0.2	0.233	0.282	0.317	18.492	110.05	149.8	197.95	594.75
H3 Load	0.233	0.267	0.3	0.333	56.167	152.68	203.87	293.3	584.16
Original User Act.	0.05	0.067	0.083	0.133	12.142	117.45	180.19	550.93	789.4
H1 User Act.	1.372	6.041	10.164	17.137	56.183	171.28	361.89	886.13	2124.8
H2 User Act.	0.146	0.2	0.233	0.3	16.483	111.26	178.53	558.01	1279.4
H3 User Act.	0.217	0.253	0.3	0.367	52.20	153.95	229.34	591	775.41

of the following activity. The obtained histogram is very similar to the original one. Apart from the 0.99 quantile, all of them reveal estimations close to the original ones including the mean value. Hypothesis 3, that consists in shifting the start events to the end event of the prior activity, show similar results for the lower quantiles. However, in the higher ones, the estimations are closer to the ones obtained with hypothesis 1.

The obtained results clearly indicate the likelihood of users when manually logging events. Since there is no big difference between the original durations and the ones estimated using hypothesis 2 we can conclude that the wrongly logged event is usually the starting one. Users log the start and end events after accomplishing the activities, resulting in activity durations of scarce seconds in the execution logs.

4.3 Conclusions and Future Work

When event logs are partially human-generated there is an underlying uncertainty related to the time at which events (recorded by means of user interaction) are logged. This leads to discrepancies between the execution logs and the actual execution of business processes. The analysis of these logs can provide meaningful information for companies to improve the quality and efficiency of their services, if they are aligned with the processes they represent. Therefore, it is essential to take into account the uncertainty of execution logs originated by users while logging events. In this paper, we demonstrate the feasibility of estimating activity durations based on low-level event logs and hypothesis from real-world scenarios. Even though these type of estimation can applied to a broad range of business processes, we focused on a particular usage scenario for a given logistic process. We demonstrate that a simple analysis to the original execution logs, when partially human-generated, is not enough to assess activity durations for planning purposes. In addition, we formulate different hypothesis for estimation and discuss their impact on the stochastic results. As future work, we suggest the formulation of different hypothesis including not only the duration and number of the original activity but also the type. The implementation of fuzzy intervals is off interest too since it can provide a comparison study.

Acknowledgments. This work was supported by FCT, through IDMEC, under LAETA, project UID/EMS/50022/2013 and SusCity (MITPTB/CS/0026/2013). This work was supported by the DAIPEX project grand funded by Dinalog.

References

1. Van der Aalst, W.M.P., Weijters, T., Maruster, L.: Workflow mining: discovering process models from event logs. IEEE Trans. Knowl. Data Eng. **16**, 1128–1142 (2004)
2. Muehlen, M.Z., Rosemann, M.: Workflow-based process monitoring and controlling-technical and organizational issues. In: Proceedings of the 33rd Annual Hawaii International Conference on System Sciences, 10 pp. IEEE (2000)
3. Cugola, G., Di Nitto, E., Fuggetta, A., Ghezzi, C.: A framework for formalizing inconsistencies and deviations in human-centered systems. ACM Trans. Softw. Eng. Methodol. (TOSEM) **5**, 191–230 (1996)
4. He, C., Ma, C.: Measuring behavioral correspondence to a timed concurrent model. In: Proceedings of the IEEE International Conference on Software Maintenance (ICSM 2001), p. 332. IEEE Computer Society (2001)
5. Luckham, D.C., Frasca, B.: Complex event processing in distributed systems. Computer Systems Laboratory Technical report CSL-TR-98-754. Stanford University, Stanford 28 (1998)
6. Günther, C.W., Rozinat, A., van der Aalst, W.M.P.: Activity mining by global trace segmentation. In: Rinderle-Ma, S., Sadiq, S., Leymann, F. (eds.) BPM 2009. LNBIP, vol. 43, pp. 128–139. Springer, Heidelberg (2010)
7. Günther, C.W., van der Aalst, W.M.P.: Mining activity clusters from low-level event logs. Beta, Research School for Operations Management and Logistics (2006)
8. Cook, J.E., Wolf, A.L.: Software process validation: quantitatively measuring the correspondence of a process to a model. ACM Trans. Softw. Eng. Methodol. (TOSEM) **8**, 147–176 (1999)
9. Van der Aalst, W.M.P., Rubin, V., Verbeek, H., van Dongen, B.F., Kindler, E., Günther, C.W.: Process mining: a two-step approach to balance between underfitting and overfitting. Softw. Syst. Model. **9**, 87–111 (2010)
10. Van Der Aalst, W.M.P.: Process Mining: Discovery, Conformance and Enhancement of Business Processes. Springer, Heidelberg (2011)
11. Van der Aalst, W.M.P., Dumas, M., Ouyang, C., Rozinat, A., Verbeek, E.: Conformance checking of service behavior. ACM Trans. Internet Technol. (TOIT) **8**, 13 (2008)
12. Rozinat, A., van der Aalst, W.M.P.: Conformance checking of processes based on monitoring real behavior. Inf. Syst. **33**, 64–95 (2008)
13. Belhajjame, K., Brambilla, M.: Ontology-based description and discovery of business processes. In: Halpin, T., Krogstie, J., Nurcan, S., Proper, E., Schmidt, R., Soffer, P., Ukor, R. (eds.) Enterprise, Business-Process and Information Systems Modeling. LNBIP, vol. 29, pp. 85–98. Springer, Heidelberg (2009)
14. van der Aalst, W.M.P., Reijers, H.A., Weijters, A.J., Dongen, B.F., Medeiros, A.A., Song, M., Verbeek, H.: Business process mining: an industrial application. Inf. Syst. **32**, 713–732 (2007)

Fuzzy Modeling Based on Mixed Fuzzy Clustering for Multivariate Time Series of Unequal Lengths

Cátia M. Salgado(✉), Susana M. Vieira, and João M. C. Sousa

IDMEC, Instituto Superior Técnico, Universidade de Lisboa, Lisboa, Portugal
catiasalgado@live.com, {susana.vieira,jmsousa}@tecnico.ulisboa.pt

Abstract. The sampling rate of variables collected in the hospital setting is dependent on several factors. Patients have different lengths of stay in the hospital, during which distinct physiological parameters are measured. The frequency of measurements depends ultimately in the type of variable and in the patient condition. Hence, when performing data based modeling for knowledge discovery in medical databases, one should have in consideration the heterogeneity of variables. This paper proposes an extension of a mixed fuzzy clustering algorithm in order to handle time invariant and time variant features of unequal lengths. Additionally, a novel approach for deriving Takagi-Sugeno fuzzy models, based on feature transformation using fuzzy c-means is implemented and compared with approaches based on mixed fuzzy clustering. The proposed approaches are tested on real data for mortality prediction in intensive care units of patients diagnosed with acute kidney injury and for ICU readmission prediction. Overall, mixed fuzzy clustering yields better results than fuzzy c-means. Moreover, the proposed extension for time series of unequal lengths improves previous results. Mortality is classified with an AUC of 0.73 and readmissions with an AUC of 0.64.

Keywords: Fuzzy modeling · Classification · Mixed fuzzy clustering · Multivariate time series · Time variant · Time invariant

1 Introduction

Electronic Health Records (EHR) are systematic collections of longitudinal electronic patient health information generated by one or more encounters in any care delivery setting. This information includes patient demographics, progress notes, medications, vital signs, laboratory results, radiology reports, procedures and diagnoses. Most of the observations present in these different sources have an indication of the time over which the measurement was made, which is uniquely important, as the sequentially of events may indicate some impending outcomes. Depending on the type of variable, varying degrees of sparsity and irregularities exist. Whilst weight (e.g.) can be registered every time the patient has a hospital encounter, vital signs are measured continuously during his/her stay.

J.P. Carvalho et al. (Eds.): IPMU 2016, Part II, CCIS 611, pp. 741–751, 2016.
DOI: 10.1007/978-3-319-40581-0_60

Medical data, if properly analyzed and interpreted, could greatly improve the development of best clinical practices [1]. In order to make sense of data and do data-based classification modeling, machine learning methods able to handle heterogeneity in data are required. In fact, in any machine learning task, data needs to be processed in order to be usable for model construction and the quality of data processing highly influences the quality of the models created. In EHRs, there are a lot of issues concerning the structure of data that need to be properly investigated, in particular, the existence of varying sampling rates of distinct variables, varying sampling rates for the same variable across different periods of the patient stay (uneven time series), varying sampling rates for the same variable across different patients (misaligned time series), missing data, noise and outliers.

This paper is focused on the problem of clustering heterogeneous data extracted from EHRs. Regarding data that has a time component, distinctions can be made as to whether the data are discrete or real, uniformly or non-uniformly sampled, univariate or multivariate and of equal or unequal length; clustering is ultimately dependent on these factors [2]. Mixed Fuzzy Clustering (MFC) [3] is a clustering algorithm that allows identification of patterns in time variant and time invariant data. MFC can be utilized in various scenarios, as long as there is an interest in mining time variant and time invariant data simultaneously. There are some challenges associated with the method, in particular the restriction to use time series of equal lengths over different variables, which renders the method unsuitable for a wide range of applications. This challenge is particularly pronounced in the medical domain due to the heterogeneity of data present in EHRs. Hence, in this paper we study MFC for modeling EHR data, and propose a new MFC scheme to account for heterogeneity. Additionally, a novel approach for deriving models is implemented and compared with previous approaches.

The structure of this paper is as follows. MFC algorithm for variables with unequal lengths is presented in Sect. 2 and Takagi-Sugeno fuzzy models in Sect. 3. Details of the data, processing methodology and results are provided in Sect. 4. Conclusions are presented in Sect. 5.

2 Mixed Fuzzy Clustering for Time Series with Unequal Lengths

Mixed fuzzy clustering (MFC) is a clustering method based on Fuzzy c-means [4] that allows the clustering of time variant and time invariant features simultaneously [3].

In order to extend the spatiotemporal clustering method proposed in [5] which only deals with one time-series to the case of multiple time-series, [3] introduced a new dimension to handle P time variant features with fixed length Q. In this work we extend this formulation to allow clustering of multivariate time series of different lengths, taking in consideration the distinct sampling rate of variables.

Input data x is characterized by features whose value is constant in time, $\mathbf{x}^s$, and by features that change over time, X^t:

$$x = (\mathbf{x}^s, X^t), \tag{1}$$

where:

$$\mathbf{x}^s = \begin{pmatrix} x^s_{11} & x^s_{12} & \cdots & x^s_{1R} \\ x^s_{21} & x^s_{22} & \cdots & x^s_{2R} \\ \vdots & \vdots & \ddots & \vdots \\ x^s_{N1} & x^s_{N2} & \cdots & x^s_{NR} \end{pmatrix}, \tag{2}$$

with N equal to the number of samples and R equal to the number of time invariant features, and:

$$X^t = \begin{Bmatrix} \mathbf{x}^t_{11} & \mathbf{x}^t_{12} & \cdots & \mathbf{x}^t_{1P} \\ \mathbf{x}^t_{21} & \mathbf{x}^t_{22} & \cdots & \mathbf{x}^t_{2P} \\ \vdots & \vdots & \ddots & \vdots \\ \mathbf{x}^t_{N1} & \mathbf{x}^t_{N2} & \cdots & \mathbf{x}^t_{NP} \end{Bmatrix}, \tag{3}$$

with number of columns equal to the number of time variant features P, and rows equal to the number of samples.

Each element of X^t, $\mathbf{x}^t_{ip}$, where $i = \{1, ..., N\}$ and $p = \{1, ..., P\}$, is composed of an array of values of length Q dependent on p:

$$\mathbf{x}^t_{ip} = \left(x^t_{i1}, x^t_{i2}, \cdots, x^t_{iQ(p)}\right), \tag{4}$$

The main difference between MFC and the classical FCM relies on the distance function [3]. In MFC, a new pondering element λ is included, factoring the importance to be given to the time variant component. The distance is also calculated separately for each time-series.

The static prototypes $\mathbf{v}^s_l$ for each cluster l and the temporal prototypes $\mathbf{v}^t_{l,p}$ for each cluster l and feature p are computed following Eqs. (5) and (6), respectively. The matrix of temporal prototypes for cluster l is represented by V^t_l.

$$\mathbf{v}^s_l = \frac{\sum_{i=1}^{N} u^m_{li}\mathbf{x}^s_i}{\sum_{i=1}^{N} u^m_{li}} \tag{5}$$

$$\mathbf{v}^t_{lp} = \frac{\sum_{i=1}^{N} u^m_{li}\mathbf{x}^t_{ip}}{\sum_{i=1}^{N} u^m_{li}} \tag{6}$$

The distance function between a sample and the static and temporal prototype of a cluster is computed following Eq. (7), where δ^2 represents the squared euclidean distance.

$$d^2_\lambda(\mathbf{v}^s_l, V^t_l, x_i) = ||\mathbf{v}^s_l - \mathbf{x}^s_i||^2 + \lambda \sum_{p=1}^{P} \delta^2(\mathbf{v}^t_{lp}, \mathbf{x}^t_{ip}) \tag{7}$$

The degree of membership of an entity i to cluster l is given by Eq. (8).

$$u_{li} = \frac{1}{\sum_{o=1}^{C} \left(\frac{d_\lambda(\mathbf{v}_l^s, V_l^t, x_i)}{d_\lambda(\mathbf{v}_o^s, V_o^t, x_i)} \right)^{\frac{2}{m-1}}}, \tag{8}$$

where m is the degree of fuzziness. The $C \times N$ matrix $U = [u_{li}]$ satisfies the following conditions:

$$u_{li} \in [0,1] \quad \forall l; \;\; 0 < \sum_{i=1}^{N} u_{li} < N \quad \forall l, i; \;\; \sum_{l=1}^{C} u_{li} = 1 \quad \forall l.$$

Equation (9) presents the augmented FCM objective function.

$$J = \sum_{l=1}^{C} \sum_{i=1}^{N} u_{li}^m d_\lambda^2(\mathbf{v}_l^s, V_l^t, x_i) \tag{9}$$

3 Modeling

3.1 Takagi-Sugeno

Fuzzy models are "grey box" and transparent models that allow the approximation of non linear systems with no previous knowledge of the system to be modeled. Fuzzy models have the advantage of not only providing transparency, but also linguistic interpretation in the form of rules.

In this work, Takagi-Sugeno (TS) fuzzy models (FM) [6] are derived from data. These consist of fuzzy rules where each rule describes a local input-output relation. With TS-FM, each discriminant function consists of rules of the type:

$$\begin{aligned} R_i: \text{ If } x_1 \text{is } A_{i1} \text{ and ... and } x_M \text{ is } A_{iM} \\ \text{then } y(\mathbf{x}) = f_i(\mathbf{x}), i = 1, 2, ..., K \end{aligned} \tag{10}$$

where f_i is the consequent function of rule R_i and y is the output. The degree of activation of the ith rule is given by $\beta_i = \prod_{j=1}^{M} \mu_{A_{ij}}(\mathbf{x})$, where $\mu_{A_{ij}}(\mathbf{x}) : \mathbb{R} \rightarrow [0,1]$. The output is computed by aggregating the individual rules contributions:

$$y(\mathbf{x}) = \frac{\sum_{i=1}^{K} \beta_i f_i(\mathbf{x})}{\sum_{i=1}^{K} \beta_i} \tag{11}$$

Given that this is a classification problem, a threshold γ is required to turn the continuous output $y \in [0,1]$ into the binary output $y \in \{0,1\}$. This way, a sample $\mathbf{x}$ is labeled as 1 if $y(\mathbf{x}) \geqslant \gamma$.

The number of rules K and the antecedent fuzzy sets A_{ij} are determined by fuzzy clustering in the product space of the input and output variables. The consequent functions $f_i(\mathbf{x})$ are linear functions determined by ordinary-least squares in the space of the input and output variables.

3.2 Model Design

Distinct TS-FM approaches based on [3] were considered for this study. The strategies differ in the type of input data and in the methodology used to determine the antecedent fuzzy sets. In particular, the antecedent fuzzy sets and the number of rules of the TS-FM are determined based either on the partition matrix generated by the FCM algorithm (FCM FM), or in the partition matrix generated by MFC (MFC FM). The input variables consist of (i) time variant and time invariant features or (ii) transpose of the partition matrix generated by MFC (U^{MFC}).

When time variant and invariant data are used as input for the fuzzy models, each time stamp of the time series is treated as one feature, i.e., the input of the model consists of a $N \times (R + Q_1 + ... + Q_P)$ matrix. When using the partition matrix, each feature corresponds to the degree of membership of the entities to the clusters such that the number of features equals the number of clusters determined in the clustering step. In particular, the input becomes the $N \times C$ matrix U^T, which corresponds to the transpose of the partition matrix $U = [u_{li}]$ in (8). This approach can be seen as a type of feature transformation method.

Considering the method based on feature transformation, proposed in [3], we wanted to test if the reported improvement in results could be attributed to using a new type of clustering scheme, to the transformation of the input feature space, or both. For this reason, we propose a new approach based on feature transformation using fuzzy c-means (U^{FCM}), which is a more standard fuzzy clustering method.

The modeling strategies are summarized in the following:

- FCM FM: Antecedent fuzzy sets determined by FCM in the product space of the input and output variables.
- MFC FM: Antecedent fuzzy sets determined by MFC in the product space of the input and output variables.
- FCM–U^{MFC} FM: Antecedent fuzzy sets determined by FCM in the product space of the partition matrix generated by MFC and output variable.
- FCM–U^{FCM} FM: Antecedent fuzzy sets determined by FCM in the product space of the partition matrix generated by FCM and output variable.

4 Experimental Results

4.1 Data Description

The current study made use of the Multiparameter Intelligent Monitoring in Intensive Care (MIMIC) II and III databases. MIMIC-III is a large, openly-available database comprising deidentified health-related data associated with patients who stayed in intensive critical care units of the Beth Israel Deaconess Medical Center, in Boston, between 2001 and 2012 [7,8]. MIMIC-III is publicly available on the PhysioNet website (http://www.physionet.org/) [8]. The database includes information such as demographics, vital sign measurements, laboratory test results, procedures, medications, caregiver notes, imaging reports, and other third party clinical information systems. The MIMIC-III relational database (version 1.3) used in this work contains data from 46,476 subjects, corresponding to 61,532 ICU admissions [9]. This database is used to extract information regarding patients diagnosed with acute kidney injury.

We also use MIMIC II, which is a previous version of MIMIC III that contains 32,535 patients. This database is used to assess the performance of fuzzy models in the classification of readmissions.

Acute Kidney Injury. The cohort included adult patients (age $\geq$ 15 years old) with ICU length of stay of more than 24 h, who had at least two serum creatinine measurements and at least one interval of six hours' urine output observation. In addition, in order to be included in this study, the patients had a primary ICD-9 diagnosis (code *584.9*) of acute renal injury at the time of their hospital admission [10,11]. Patients who underwent renal replacement therapy (RRT) on the day of or prior to their hospital admission were excluded. Patients were also excluded if they had end-stage renal disease (ESRD), given by ICD-9 code *585.6*, or first serum creatinine level of > 4 mg/dL [11,12].

Time invariant variables selected were weight and age on admission, and time variant variables were heart rate (beats/min), respiratory rate (breaths/min), oxygen saturation in the blood (%), non-invasive blood pressure (NBP) mean (mmHg) [12–14]. The output consists in the patient classification regarding mortality: 1 if the patient died within one year after discharge and 0 if not. Table 1 shows the mean sampling rate of each variable in the dataset.

Readmissions. In order to exploit the advantage of using features of unequal lengths and the advantage of feature transformation based on fuzzy c-means, we evaluate the dataset providing worse results in [3]. This dataset was built for the prediction of early readmissions, i.e., readmissions to the ICU within 24 to 72 h after discharge. The same set of features was used and for comparison purposes we select again the same number of measurements collected before discharge, regardless of the day they were taken. We also investigate performance results using smaller lengths of the time series per day and unequal lengths between variables.

Table 1. Average sampling rate (samples/day) for AKI dataset.

	Class 0	Class 1
Creatinine	1.3	2.1
Heart rate	35.0	29.6
NBP	27.5	26.6
Urine output	17.2	16.2
SpO2	33.3	28.6
Respiratory rate	37.3	32.4

Table 2. Average sampling rate (samples/day) for readmissions dataset.

	Class 0	Class 1
Heart rate	28.2	27.8
NBP	16.7	14.7
Temperature	6.9	6.6
Creatinine	2.1	2.0
SpO2	28.1	27.6
Platelets	2.2	2.0
Lactic acid	8.2	6.2

The selected time invariant features were age, weight, Simplified Acute Physiology Score (SAPS II) and Sequential Organ Failure Assessment (SOFA) score on admission. Gender was excluded because since it is a binary variable, it could highly influence the finding of clusters. Time variant features were heart rate (beats/min), temperature (°C), platelets (cells $\times 10^3/\mu L$), non-invasive blood pressure mean (mmHg), oxygen saturation in the blood (%), lactic acid (mg/dL) and creatinine (mg/dL). Table 2 shows the mean sampling rate of each variable in the dataset.

4.2 Data Processing

In this work, all measurements regarding the chosen set of time variant features are extracted from the database. The mean sampling rate of each variable is calculated, and the number of samples is selected according to it. This way, each feature is treated independently, and more information can be used.

Consider that Q measurements per day are required for variable p, and that the observation window of variable p in a certain patient stay is given by the time elapsed between the first and last measurements of p in one day. Measurements are distributed among Q bins, according to the observation window. If the number of measurements available exceeds the required, the median among bins containing more than one measurement is used as the value for that time point. On the other hand, if the number of measurements is smaller than the required Q, there are empty bins. In this case, a zero order hold procedure is applied so that previous values are used.

According to expert medical knowledge, considering the full stay of a patient stay should be more informative of adverse events and outcomes, than merely considering short time windows. Contrarily to [3], where the last 10 measurements of each patient were used, this data extraction strategy allows that information spanning a longer period can be used, and that key time periods of the day (e.g. morning, afternoon and night) are incorporated in the models. Moreover, less missing data needs to be imputed for those variables with smaller sampling rates.

Table 3. AUC performance of different fuzzy models for AKI and readmissions datasets of varying time series lengths and observation windows. $Q(p)$ gives the number of samples per day for each time variant feature (see order of feature vector in Tables 1 and 2). The number of patients and percentage of patients from class 1 in each dataset is also given. Results in bold highlight the best performer in each dataset. Results are presented as mean (standard deviation).

AKI						
$Q(p)$	Observation window	Patients	FCM–U^{MFC}	FCM–U^{FCM}	MFC	FCM
{10,10,10,10,10,10}	last 10	2192 (40 %)	0.70±0.03	0.70±0.02	**0.73±0.01**	0.69±0.02
{10,10,10,10,10,10}	1	1843 (39 %)	0.69±0.03	0.68±0.04	**0.72±0.02**	0.67±0.04
{1,15,13,8,14,16}	1	1843 (39 %)	0.69±0.03	0.69±0.03	**0.71±0.02**	0.66±0.05
{1,30,27,16,29,32}	1	1843 (39 %)	0.67±0.01	0.68±0.02	**0.70±0.05**	0.64±0.03
{1,3,3,2,3,3}	2	1280 (43 %)	0.70±0.04	0.68±0.05	**0.72±0.04**	0.69±0.04
{1,3,3,2,3,3}	3	910 (46 %)	**0.71±0.02**	0.69±0.02	0.68±0.04	0.63±0.07
Readmissions						
{10,10,10,10,10,10}	last 10	2660 (7.6 %)	0.56±0.04	0.56±0.04	0.57±0.02	**0.58±0.07**
{10,10,10,10,10,10}	1	1389 (9.5 %)	**0.58±0.06**	0.53±0.08	0.53±0.07	0.49±0.07
{14,7,3,1,14,1,3}	1	1389 (9.5 %)	**0.64±0.06**	0.61±0.04	0.56±0.07	0.54±0.08
{28,15,7,2,28,2,6}	1	1389 (9.5 %)	**0.59±0.04**	0.57±0.04	0.54±0.06	0.52±0.09
{3,2,2,1,3,1,2}	2	978 (10 %)	**0.58±0.07**	0.57±0.06	0.53±0.06	0.53±0.06
{3,2,2,1,3,1,2}	3	741 (11 %)	0.56±0.05	**0.58±0.05**	0.55±0.07	0.56±0.06

In this work, values outside acceptable physiological ranges were deleted. Values were normalized between 0 and 1 for clustering and modeling purposes.

4.3 Model Performance

This section presents the experimental evaluation of the fuzzy models described in Sect. 3.2 using real EHRs data. The performance of the models is evaluated in terms of area under the receiver operating characteristic curve (AUC) [15], accuracy (correct classification rate), sensitivity (true positive classification rate, also called recall) and specificity (true negative classification rate).

The dataset is initially partitioned into train and test folds, such that 80 % of data is reserved for train and 20 % for test, using 5 fold cross validation. For each fixed number of clusters a grid search is performed to find λ and threshold γ by creating models using 50 % of the train data and testing them with 50 % of the train set, i.e., new train set partitions are created from the train set to select the model parameter λ and γ. For FCM–U^{MFC} and FCM–U^{FCM}, the input U used for testing the models is obtained by calculating the distance from the test set to the initially created cluster prototypes and updating the partition matrix using the calculated distance. The γ giving the smaller difference between sensitivity and specificity and the λ giving the best performance in terms of AUC, for the test set, are selected. This model is then used to test with the 20 % test set. In each round of the cross validation, the model giving the best performance for

Table 4. Performance of different fuzzy models for the best AKI and readmissions datasets performers in Table 3. The number of clusters C and average weighting parameter λ selected by grid search are also given. Results are presented as mean (standard deviation).

AKI						
FM	C	λ	AUC	ACC	Sensitivity	Specificity
FCM–U^{MFC}	4	2.0±0.7	0.70±0.03	**0.66±0.03**	**0.68±0.03**	0.65±0.06
FCM–U^{FCM}	3	-	0.70±0.02	0.65±0.03	0.64±0.09	**0.67±0.09**
MFC	2	3.4±0.5	**0.73±0.01**	**0.66±0.01**	0.67±0.04	0.65±0.03
FCM	2	-	0.69±0.02	0.63±0.02	0.63±0.03	0.63±0.06
Readmissions						
FCM–U^{MFC}	3	3.4±3.8	**0.64±0.06**	**0.57±0.04**	**0.66±0.08**	**0.56±0.05**
FCM–U^{FCM}	4	-	0.61±0.04	0.56±0.06	0.58±0.06	0.55±0.07
MFC	3	1.6±1.3	0.56±0.07	0.49±0.06	0.57±0.08	0.49±0.06
FCM	3	-	0.54±0.08	0.51±0.06	0.57±0.07	0.50±0.06

each number of clusters is selected and the results are averaged over the rounds. In the end, results are shown for C giving the best average.

The four modeling approaches were applied to each dataset, with the model parameters - C, λ and γ - being selected by grid search for each method, such that $\gamma = 0, 0.01, ..., 1$, $\lambda = 0, 1, ..., 10$ and $m = 2$. Cross validation is performed separately for $C = 2, 3, 4$. Results in terms of AUC are shown in Table 3.

In a previous study using fuzzy modeling based on mixed fuzzy clustering [3], the classification of readmissions yield results of maximum AUC of 0.58. In this work, the proposed feature extraction approach is able to increase the performance of fuzzy models to AUC = 0.64 for the method using MFC feature transformation, when approximately half of the average sampling rate in class 1 is considered for an observation window of 1 day. In readmissions, MFC feature transformation method performs better than FCM in 4 out of 6 datasets, equal in 1 and worse in 1. MFC FMs perform better than FCM in 3, equal in 1 and worse in 1. Considering AKI, MFC performs better than FCM in 6 out of 6 datasets. The best performance overall, in terms of AUC, is achieved when the last 10 measurements per variable are used. These results are comparable with the achieved in Celi et al. [16], where the first 72 h of ICU admission were used (AUC of 0.74).

Note that as the observation window increases so does the ratio of patients from the class of interest, suggesting that these patients tend to stay longer in the ICU than patients from class 0. Also, when the last 10 measurements are considered, measurements from any part of the patient stay can be selected, i.e., data might include measurements spanning 10 days of observation. On the other hand, when a an observation period of 1, 2 or 3 days are chosen, measurements are restricted to one day period, always starting from the last. If measurements in the chosen day are not available the patient is discarded, which explains the difference in size of datasets using the same $Q(p)$.

In both classification datasets, methods using MFC yield the best AUC. In the case of readmissions, the best performer is FCM–U^{MFC} with an AUC of 0.64 and in AKI the best performer is MFC with an AUC of 0.73. In Table 4, C, AUC, accuracy, sensitivity and specificity are presented for these two datasets. Note that readmissions is very imbalanced in terms of sensitivity and specificity, which can be explained by the high class imbalance. Values of λ greater than 1 are able to increase the performance of fuzzy c-means rule-based models, by increasing the relevance of the time variant component of data.

5 Conclusions

This work proposes an extension of MFC for time series of unequal lengths and a new feature transformation approach based on FCM. In summary, this work allowed the investigation of different observation windows (1–3 days and last 10 measurements); unequal lengths of time series across different features; different Takagi-Sugeno fuzzy models (including feature transformation and MFC for deriving the antecedents fuzzy sets). The proposed MFC fuzzy models are able to improve the performance of fuzzy c-means based models.

Future work should focus in testing the algorithm in different application domains, e.g. energy consumption, spatiotemporal applications, where different data problems exist. In order to improve the AUC and accuracy of the models, wrapper feature selection should be performed to find an optimum subset of time variant and invariant features.

Acknowledgments. This work was supported by FCT, through IDMEC, under project iDecision4Care (IF/00833/2014/CP1238/CT0002). S. Vieira acknowledges support by Program Investigador FCT (IF/00833/ 2014) from FCT, co-funded by the European Social Fund (ESF) through the Operational Program Human Potential (POPH). This work was supported by FCT, through IDMEC, under LAETA, project UID/EMS/50022/2013.

References

1. Celi, L.A., Mark, R.G., Stone, D.J., Montgomery, R.A.: Big data in the intensive care unit closing the data loop. Am. J. Respir. Crit. Med. **187**(11), 1157–1160 (2013)
2. Liao, T.W.: Clustering of time series data - a survey. Pattern Recogn. **38**, 1857–1874 (2005)
3. Ferreira, M.C., Salgado, C.M., Viegas, J.L., Schafer, H., Azevedo, C.S., Vieira, S.M., Sousa, J.: Fuzzy modeling based on mixed fuzzy clustering for health care applications. In: 2015 IEEE International Conference on Fuzzy Systems (FUZZ-IEEE), pp. 1–5. IEEE (2015)
4. Bezdek, J.C., Ehrlich, R., Full, W.: FCM: the fuzzy c-means clustering algorithm. Comput. Geosci. **10**(2), 191–203 (1984)

5. Izakian, H., Pedrycz, W., Jamal, I.: Clustering spatiotemporal data: an augmented fuzzy C-means. IEEE Trans. Fuzzy Syst. **21**(5), 855–868 (2013)
6. Takagi, T., Sugeno, M.: Fuzzy Identification of Systems and Its Application to Modeling and Control (1985)
7. Saeed, M., Villarroel, M., Reisner, A.T., Clifford, G., Lehman, L., Moody, G., Heldt, T., Kyaw, T.H., Moody, B., Mark, R.G.: Multiparameter intelligent monitoring in intensive care ii (mimic-ii): a public-access intensive care unit database. Crit. Care Med. **39**(5), 952 (2011)
8. Goldberger, A.L., Amaral, L.A.N., Glass, L., Hausdorff, J.M., Ivanov, P.C., Mark, R.G., Mietus, J.E., Moody, G.B., Peng, C., Stanley, H.E.: Physiobank, physiotoolkit, and physionet components of a new research resource for complex physiologic signals. Circulation **101**(23), 215–220 (2000)
9. Overview of the mimic-iii data (2015). http://mimic.physionet.org/gettingstarted/overview/
10. Fuchs, L., Lee, J., Novack, V., Baumfeld, Y., Scott, D., Celi, L., Mandelbaum, T., Howell, M., Talmor, D.: Severity of acute kidney injury and two-year outcomes in critically ill patients. Chest **144**(3), 866–875 (2013)
11. Mandelbaum, T., Scott, D.J., Lee, J., Mark, R.G., Malhotra, A., Waikar, S.S., Howell, M.D., Talmor, D.: Outcome of critically ill patients with acute kidney injury using the akin criteria. Crit. Care Med. **39**(12), 2659–2664 (2011)
12. Celi, L.A.G., Tang, R.J., Villarroel, M.C., Davidzon, G.A., Lester, W.T., Chueh, H.C.: A clinical database-driven approach to decision support: predicting mortality among patients with acute kidney injury. J. Healthc. Eng. **2**(1), 97–110 (2011). 22844575[pmid]
13. Fialho, A.S., Cismondi, F., Vieira, S.M., Reti, S.R., Sousa, J.M.C., Finkelstein, S.N.: Data mining using clinical physiology at discharge to predict ICU readmissions. Expert Syst. Appl. **39**(18), 13158–13165 (2012)
14. Pereira, R.D.M.A., Salgado, C.M., Dejam, A., Reti, S.R., Vieira, S.M., Sousa, J.M.C., Celi, L.A., Finkelstein, S.N.: Fuzzy modeling to predict severely depressed left ventricular ejection fraction following admission to the intensive care unit using clinical physiology. Sci. World J. **2015**, 1–9 (2015). 26345130[pmid]
15. Hanley, J.A., McNeil, B.J.: The meaning and use of the area under a receiver operating characteristic (ROC) curve. Radiology **143**(4), 29–36 (1982)
16. Celi, L., Tang, R., Villarroel, M., Davidzon, G., Lester, W., Chueh, H.: A clinical database-driven approach to decision support: predicting mortality among patients with acute kidney injury. J. Healthc. Eng. **2**(1), 97–110 (2011)

Time Varying Correlation Estimation Using Probabilistic Fuzzy Systems

Nalan Baştürk[1(✉)] and Rui Jorge Almeida[2(✉)]

[1] Department of Quantitative Economics, Maastricht University, P.O. Box 616, 6200 MD Maastricht, The Netherlands
n.basturk@maastrichtuniversity.nl

[2] School of Industrial Engineering, Eindhoven University of Technology, P.O. Box 513, 5600 MB Eindhoven, The Netherlands
rjalmeida@tue.nl

Abstract. Accurate financial risk analysis has drawn considerable attention after the recent financial crisis. Several regulatory agencies recently documented the need for proper assessment and reporting of financial risk for banks and other financial institutions. It is stressed that risk analysis should take into account changing risk properties over time. For a set of financial assets, risk analysis relies on the correlation and covariance structure among these returns from these assets. Therefore analyzing changes in the correlations and covariances of assets is essential to document changing risk properties. In this paper we show that a PFS can be used to model unobserved time-varying correlation between financial returns. The method is applied to simulated data and real data of daily NASDAQ and HSI stock returns. We show that the PFS application improves over the conventional moving window approximation of time-varying correlation by decreasing the sensitivity of the results to the selection of the window length.

Keywords: Probabilistic fuzzy systems · Time-varying correlations · Risk analysis

1 Introduction

Assessing measures of risk for financial returns has an important role in investment decisions, portfolio analysis and for regulatory purposes [1,2]. The importance of accurate estimation of risk of asset returns have been discussed extensively, particularly after the recent financial crisis [2,3]. Measures of risk often need to be analyzed for more than one financial asset since most investment decisions are based on a selected portfolio of multiple assets, where the investor aims to diversify of risk [4].

For a single series of financial returns, it is well documented that the associated risk changes substantially over time, which is denoted by time-varying conditional volatility in asset returns [5]. Methods which avoid strong distributional assumptions prove to be useful for estimating such time-varying volatility

J.P. Carvalho et al. (Eds.): IPMU 2016, Part II, CCIS 611, pp. 752–763, 2016.
DOI: 10.1007/978-3-319-40581-0_61

[6–10]. For multiple financial returns, an additional important feature in portfolio management is the correlation between the returns of different assets, where the portfolio risk depends on the correlation of the returns from each asset forming the portfolio [11]. For example, if the portfolio is based on two negatively correlated stocks, the portfolio is said to be 'well-diversified' with small risk [4]. Thus an accurate risk calculation, *e.g.* for a portfolio of financial assets, requires the accurate calculation of assets' correlation at the given decision time.

Earlier research has shown that correlations between financial returns, particularly across international markets change over time, e.g. during major financial crises [12]. Different methodologies, with different distributional assumptions for correlation, have been proposed to calculate time-varying correlations for returns. Parametric models are proposed to estimate correlations between returns as well as other model parameters [13]. Alternative methodologies are based on moving window correlation estimates, where time-varying correlation at a given time is approximated by a proxy, namely the sample correlation at a selected time window. Moving window estimates have the advantage of avoiding strong distributional assumptions and are shown to perform well particularly in forecasting [14]. However, these estimates are also shown to be sensitive to the selection of a window size and there is a natural trade-off between capturing time variation in correlations and obtaining an accurate proxy for correlation at a given time. If the selected window size is too large, proxies of correlation, *i.e.* sample correlation, approaches the long run mean of correlation. Hence time variation in correlation cannot be captured. On the contrary, if the selected window size is too small, time variation in correlation is captured, but each proxy of correlation has high uncertainty and correlation estimation is not accurate [15].

In this paper, we propose the first PFS for modeling time-varying correlations of financial returns and to improve moving window correlation estimates. The method captures time varying correlation and conditional volatility without an underlying restricted statistical model for the correlations. PFS has previously been shown to perform well for conditional volatility and risk estimation [7,9,16]. The proposed model is different from the earlier PFS models. In the current paper, the antecedents and the consequents of the system are based on proxies, i.e. approximations of time-varying correlations instead of observed data. The use of approximations of correlations instead of the actual values leads to measurement errors. We show that the PFS model takes into account the imprecision resulting from these measurement errors through the use of fuzzy sets. In addition, the trade-off between capturing time-variation in correlation and obtaining accurate correlation estimates is mitigated using the proposed PFS model. These features are illustrated using simulated data and a real data application.

2 Probabilistic Fuzzy Systems

Probabilistic fuzzy systems combine two different types of uncertainty, namely fuzziness or linguistic vagueness, and probabilistic uncertainty. A probabilistic fuzzy system follows an idea similar to [17,18] where the different concepts

[19–22] of fuzzy sets and probabilities are complementary [20]. In this work we consider that the probabilistic uncertainty relate to aleatoric variability, while fuzzy sets are used to represent gradualness, epistemic uncertainty or bipolarity [21,23].

The PFS consists of a set of rules whose antecedents are fuzzy conditions and whose consequents are probability distributions. Assuming that the input space is a subset of $\mathbb{R}^n$ and that the rule consequents are defined on a finite domain $Y \subseteq \mathbb{R}$, a probabilistic fuzzy system consists of a system of rules R_q, $q = 1, \ldots, Q$, of the type

$$R_q : \text{If } \mathbf{x} \text{ is } A_q \text{ then } f(y) \text{ is } f(y|A_q)\,, \tag{1}$$

where $\mathbf{x} \in \mathbb{R}^n$ is an input vector, $A_q : X \longrightarrow [0, 1]$ is a fuzzy set defined on X and $f(y|A_q)$ is the conditional pdf of the stochastic output variable $\underline{y}$ given the fuzzy event A_q. The interpretation is as follows: if fuzzy antecedent A_q is fully valid ($x \in \text{core}(A_q)$), then y is a sample value from the probability distribution with conditional pdf $f(y|A_q)$.

A PFS has been described with two possible and equivalent reasoning mechanisms, namely the fuzzy histogram approach and the probabilistic fuzzy output approach [24]. In this work we focus on the fuzzy histogram approach since the pdf obtained from the approach can be used to assess the precision in correlation estimates. We replace in each rule of (1) the true pdf $f(y|A_q)$ by its fuzzy approximation (fuzzy histogram) $\hat{f}(y|A_q)$ yielding the rule set $\hat{R}_q$, $q = 1, \ldots, Q$ defined as

$$\hat{R}_q : \text{If } \mathbf{x} \text{ is } A_q \text{ then } f(y) \text{ is } \hat{f}(y|A_q)\,. \tag{2}$$

The fuzzy histogram $\hat{f}(y|A_q)$ for each rule is obtained from a fuzzy partition of the compact output space Y with $j = 1, \ldots, J$ fuzzy classes $C_j|A_q$ with probability estimates $\hat{\Pr}(C_j|A_q)$ and the corresponding membership function $u_{C_j}(y)$ [25]

$$\hat{f}(y|A_q) = \sum_{j=1}^{J} \frac{\hat{\Pr}(C_j|A_q) u_{C_j}(y)}{\int_{-\infty}^{\infty} u_{C_j}(y) dy}\,, \tag{3}$$

where the probability estimates $\hat{\Pr}(C_j|A_q)$ satisfy the conditions $\hat{\Pr}(C_j|A_q) \geq 0$ and $\sum_{j=1}^{J} \hat{\Pr}(C_j|A_q) = 1$, and they can be calculated using the maximum likelihood method [7]. In this paper we do not assume any particular algebraic structure for the conditional probability of fuzzy events. There are several examples of definitions of conditional probabilities of fuzzy events that satisfy the classical axioms of conditional probabilities, such as [26].

The interpretation of this type of reasoning is as follows. Given the occurrence of a (multidimensional) antecedent fuzzy event A_q, which is a conjunction of the fuzzy conditions defined on input variables, an estimate of the conditional probability density function based on a fuzzy histogram $\hat{f}(y|A_q)$ is calculated.

Given an input vector $\mathbf{x}$, the output of a probabilistic fuzzy system is a conditional density function which can be computed as

$$\hat{f}(y|\mathbf{x}) = \sum_{j=1}^{J}\sum_{q=1}^{Q} \beta_q(\mathbf{x})\hat{\Pr}(C_j|A_q)\frac{u_{C_j}(y)}{\int_{-\infty}^{\infty} u_{C_j}(y)dy}, \tag{4}$$

where $\beta_q(\mathbf{x}) = u_{A_q}(\mathbf{x})/\sum_{q'=1}^{Q} u_{A_{q'}}(\mathbf{x})$ is the normalised degree of fulfillment of rule R_q and u_{A_q} is the degree of fulfillment of rule R_q. When $\mathbf{x}$ is n-dimensional, u_{A_q} is determined as a conjunction of the individual memberships in the antecedents computed by a suitable t-norm, *i.e.*, $u_{A_q}(\mathbf{x}) = u_{A_{q1}}(x_1) \circ \cdots \circ u_{A_{qn}}(x_n)$, where $x_i, i = 1, \ldots, n$ is the i-th component of $\mathbf{x}$ and $\circ$ denotes a t-norm.

It can be shown [24] that the conditional density output $\hat{f}(y|\mathbf{x})$ of a PFS is a proper probability density function *i.e.* $\int_{-\infty}^{\infty} \hat{f}(y|\mathbf{x})dy = 1$ and the expected value $\hat{\mathrm{E}}(\underline{y}|\mathbf{x})$ and the second moment $\hat{\mathrm{E}}(\underline{y}^2|\mathbf{x})$, exist if the given the partitioning of the output space, since the output membership values satisfy $\sum_{j=1}^{J} u_{C_j}(y) = 1, \forall y \in Y, y < \infty$. Under these conditions, a crisp output using the expected value can be calculated as

$$\hat{\mu}_{y|\mathbf{x}} = \hat{E}(y|\mathbf{x}) = \int_{-\infty}^{\infty} y\hat{f}(y|\mathbf{x})dy = \sum_{q=1}^{Q}\sum_{j=1}^{J} \beta_q(\mathbf{x})\hat{\Pr}(C_j|A_q)z_{1,j}, \tag{5}$$

where $z_{1,j} = \int_{-\infty}^{\infty} yu_{C_j}(y)dy/\int_{-\infty}^{\infty} u_{C_j}(y)dy$ is the centroid of the jth output fuzzy set.

3 Correlation Estimation Using PFS

In this paper we consider a model for two returns $y_t = (y_{1,t}, y_{2,t})'$:

$$y_t = H_t^{1/2} z_t \tag{6}$$

where $t = 1, \ldots, T$ indicates the time period, $z_t = (z_{1,t}, z_{2,t})'$ is such that $z_{i,t}$ for $i = 1, 2$ are random variables with mean 0 and variance 1, H_t is a 2×2 positive definite matrix and $H_t^{1/2}$ denotes the Choleski decomposition of H_t. In most models, e.g. in multivariate GARCH models, the distribution of $z_{i,t}$ is defined as a standard normal distribution. We focus on two assets for illustration purposes, but the model and the applications can be generalized to any number of assets.

The covariance of two returns in (6) is $\mathrm{Var}(y_t) = H_t^{1/2}H_t'^{1/2} = H_t$, i.e. the matrix H_t represents the time-varying variance-covariances of y_t, which by construction are not observable. Different models have been proposed to model the time-varying conditional variance-covariance matrix H_t in (6) [13]. A common feature of these models is the dependency of the current covariances H_t and past covariances $H_{t-1}, \ldots, H_{t-p}$. Similar to univariate GARCH models, such a dependency on past values ensure smooth changes in the variance-covariance structure over time. In addition, any modeling approach for H_t should ensure that this matrix is a positive definite matrix at each time period. This necessary condition may lead to additional parameter restrictions in models [13].

The following decomposition of the variance-covariance matrix is often used to identify variances and correlation coefficients [13]:

$$H_t = D_t R_t D_t = \begin{pmatrix} h_{1,1,t}^{1/2} & 0 \\ 0 & h_{2,2,t}^{1/2} \end{pmatrix} \begin{pmatrix} 1 & \rho_t \\ \rho_t & 1 \end{pmatrix} \begin{pmatrix} h_{1,1,t}^{1/2} & 0 \\ 0 & h_{2,2,t}^{1/2} \end{pmatrix} \tag{7}$$

where D_t is the diagonal matrix with variances of each series in diagonals and R_t matrix includes the correlations of the two series ρ_t. Using this decomposition, H_t is a positive definite matrix as long as the diagonal elements of D_t are positive and $\rho_t \in (-1, 1)$ for all t. The advantage of the decomposition in (7) is that the diagonal elements of the matrix D_t can be estimated using a given conditional volatility model, for example using [7] or [10], for each series $y_1 = (y_{1,t}, \dots, y_{1,t})$ and $y_2 = (y_{2,t}, \dots, y_{2,t})$. This estimation can be performed independent of the estimation of correlation coefficients in R_t since D_t defines the unconditional variance of each series at time t, which is by definition independent of correlations R_t.

For the two series in (6), moving–window (MW) correlation estimates $\hat{\rho}_t$ using window length m can be calculated using Pearson's linear correlation coefficient:

$$\hat{\mu}_{i,t} = \frac{\sum_{t'=t-m+1}^{t} y_{i,t'}}{m}, \quad \text{for } i = 1, 2 \tag{8}$$

$$\hat{\sigma}_{i,t}^2 = \frac{\sum_{t'=t-m+1}^{t} (y_{i,t'} - \hat{\mu}_{i,t})^2}{m-1}, \quad \text{for } i = 1, 2 \tag{9}$$

$$\hat{\rho}_t^{(m)} = \sum_{t'=t-m+1}^{t} \frac{(y_{1,t'} - \hat{\mu}_{1,t})(y_{2,t'} - \hat{\mu}_{2,t})}{(m-1)\sigma_{1,t}\sigma_{2,t}}. \tag{10}$$

The correlation estimate in (10) has an asymptotic normal distribution with variance $(1 - (\hat{\rho}_t^{(m)})^2)/(m - 2)$. However, for small m, this asymptotic property does not necessarily hold, hence asymptotic variances do not reflect the actual variance of the correlation estimate. In all following examples we report the estimation uncertainty in $\hat{\rho}$ in (10) using bootstrap [27] results based on 1000 bootstrap samples of size $m/2$. When the purpose is to forecast future correlations between assets, the common method is to use past information as follows [15]:

$$E\left(\hat{\rho}_{t+1}^{(m)} | y_1, \dots, y_t\right) = \hat{\rho}_t^{(m)}, \tag{11}$$

where $\hat{\rho}_t^{(m)}$ is obtained from (10).

The PFS for correlation modeling has the following rules for $q = 1, \ldots, Q$:

$$\hat{R}_q : \text{If } \hat{\rho}_{t-1}^{(m)} \text{ is } A_q \text{ then } f(\hat{\rho}_t^{(m)}) \text{ is } \hat{f}(\hat{\rho}_t^{(m)}|A_q), \tag{12}$$

where both the antecedent (past correlation) and the consequent (current correlation) are estimated using (10) with a pre-selected window size m, and $\hat{f}(\hat{\rho}_t^{(m)}|A_q)$ is a fuzzy histogram described as [25]

$$\hat{f}(\hat{\rho}^{(m)}|A_q) = \sum_{j=1}^{J} \frac{\hat{\Pr}(C_j|A_q) u_{C_j}(\hat{\rho}_t^{(m)})}{\int_{-\infty}^{\infty} u_{C_j}(\hat{\rho}_t^{(m)}) d\hat{\rho}_t^{(m)}}, \tag{13}$$

i.e. both the antecedent and the consequent variables are only approximations of the real variable of interest, correlations.

The parameters of the probabilistic fuzzy systems are estimated using a procedure similar to [7], here briefly summarized. Following the distinction between input and output present in the rule structure of (2), the optimization problem is divided in two parts. First we obtain the input membership parameters by using a fuzzy clustering heuristic, that uses the fuzzy c-means algorithm, set the output membership parameters as Gaussian, shouldered at the edges and finally optimize the probability parameters $\hat{\Pr}(C_j|A_q)$ using maximum likelihood estimation.

4 Simulated Data with Time-Varying Correlation

In this section we illustrate the performance of the PFS model using simulated data and compare the results with MW estimates of correlation, which are often used as proxies for correlation [15]. In the described PFS, both the input and the output of PFS are approximations of actual (unobserved) correlation. We use simulation experiments to study the effect of these approximated inputs and outputs in PFS on the approximation capability of PFS, particularly in comparison to MW approximation. In addition, for the simulation studies, actual correlation is known. We can therefore compare obtained results from the two methods, MW and PFS, with actual correlation values. Such a comparison is not possible using real data, unless a loss function is defined [15].

As an example, we simulate $T = 500$ observations $y_t = (y_{1,t}, y_{2,t})'$ for $t = 1, \ldots, T$ from a model with highly persistent time-varying correlations following an auto-regressive process, described by:

$$\begin{aligned} y_t &\sim N\left(\begin{pmatrix}0\\0\end{pmatrix}, \begin{pmatrix}1 & \rho_t\\ \rho_t & 1\end{pmatrix}\right) \\ \rho_t &= \max\left(-1+\epsilon, \min\left(1-\epsilon, 0.1 + 0.8\rho_{t-1} + \eta_t\right)\right) \end{aligned} \tag{14}$$

where $\epsilon = 10^{-5}$, $\eta_t \sim NID(0, 0.005)$ and the restriction $\rho_t \in (-1, 1)$ of the covariance decomposition (7) is satisfied. As shown in Fig. 1a, the model presented in (14) has time-varying correlations between times series $y_{1,t}$ and $y_{2,t}$, and furthermore, in periods of high correlation, series $y_{1,t}$ and $y_{2,t}$ have common upward or downward movements.

4.1 PFS Application with True Consequents and Proxies for Antecedents

We first consider a PFS model of the form (12) where the inputs are correlation estimates calculated from (10) and the output of PFS are the true observed correlation (given by the simulated parameters). We again note that this application is not realistic since correlations are unobserved in reality, but it serves the purpose of documenting the effect of using approximations of correlation as the input variables on the PFS results, isolating this the effect from the approximation of the output variable.

Each PFS rule in (12) are adjusted such that the output of PFS is the actual correlation at time t, ρ_t as the output:

$$\hat{R}_q : \text{If } \hat{\rho}_{t-1}^{(m)} \text{ is } A_q \text{ then } f(\rho_t) \text{ is } \hat{f}(\rho_t|A_q)\,. \tag{15}$$

We first consider a window length of $m = 10$ for obtaining correlation approximations in (13), for moving window forecasts and for the input variable of PFS. Results of the MW correlation estimates and those from PFS with 4 antecedents and 9 consequents are shown in Fig. 1b and c. MW estimates of correlation in Fig. 1b change substantially over time, capturing changing correlation levels. In addition, the obtained 99 % intervals around these estimates are often too wide, covering all values within $(-1, 1)$, indicating that the uncertainty around the estimated values are high. Figure 1b also shows that the peaks of MW estimates are mostly after the peaks in correlation. I.e MW estimates are often late and inaccurate in capturing correlation changes. PFS estimates in Fig. 1c, however, follow the increases and decreases of the actual correlation smoothly, with tighter confidence intervals.

We next compare MW and PFS estimates for time-varying correlation by using mean absolute error (MAE) between estimated and actual correlations to compare the accuracy of the methods. We emphasize that such a comparison is only possible if actual correlation is known, i.e. in a simulation setting. MAE for the two methods are calculated as follows:

$$\text{MAE}^{(m)} = \frac{1}{T}\sum_{t=1}^{T}\left|\hat{\rho}_t^{(m)} - \rho_t\right| \tag{16}$$

where PFS estimates of $\hat{\rho}_t^{(m)}$ are obtained from (5), MW estimates of $\hat{\rho}_t^{(m)}$ are obtained from (11), and ρ_t is the simulated value of correlation at time t. MAE from MW and PFS using different window sizes are shown in Fig. 2a. According to MAE, PFS results are smaller than those of MW estimation for all window sizes. Regardless of the window size, MAE from PFS is around 0.3 %, while MA estimates lead to very high MAE, especially with small window sizes. It is particularly interesting that the PFS estimates with a small window size are still accurate. This result follows from the addition of probability parameters in the model in PFS, which are estimated using the full sample information. Even though the antecedent is calculated inaccurately, e.g. with a too small

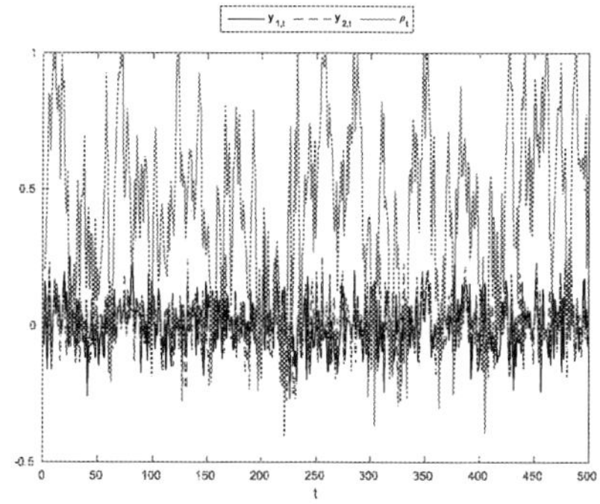

(a) Simulated data and time-varying correlations from (14).

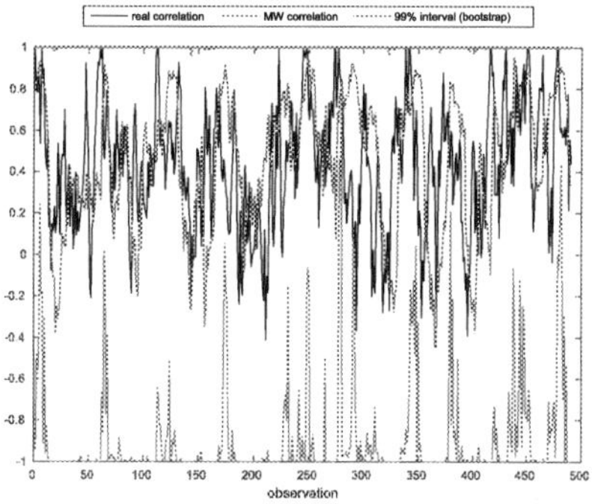

(b) MW estimates with window size 10

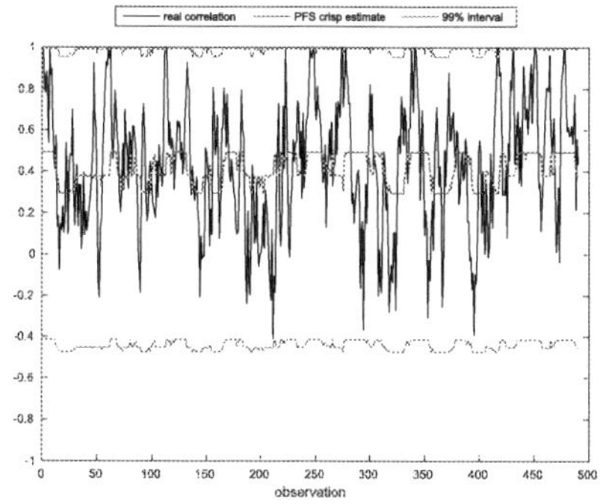

(c) PFS estimates with antecedents from MW, known consequents

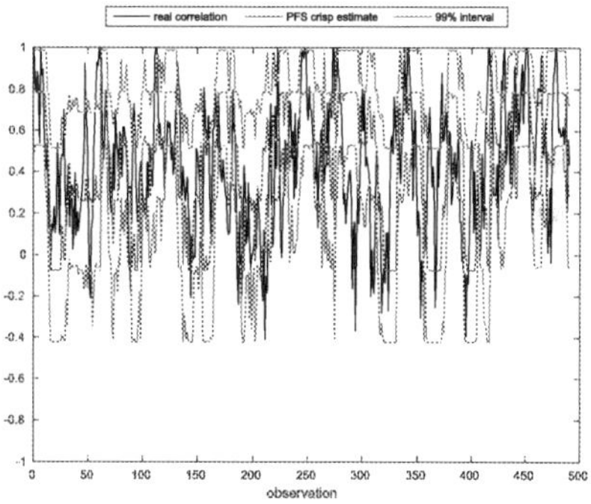

(d) PFS estimation with antecedents and consequents from MW

Fig. 1. MW and PFS estimates of time-varying correlation for simulated data. (Color figure online)

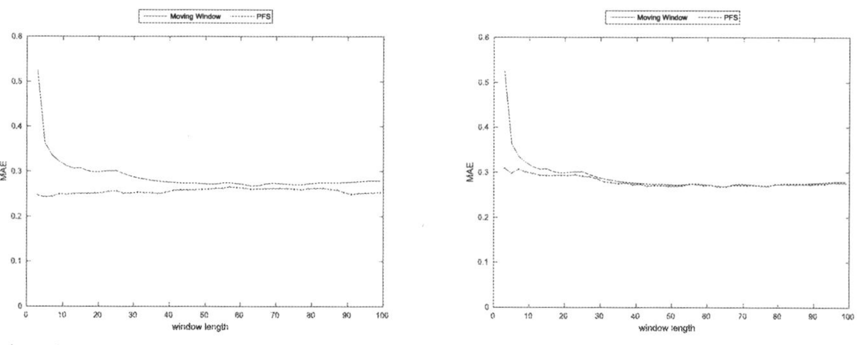

(a) MAE for MW and PFS with known consequents

(b) MAE for MW and PFS with consequents from MW

Fig. 2. MAE comparisons of MW and PFS for simulated data. (Color figure online)

window size, PFS parameters incorporate full sample information and regulate the correlation estimates through the use of fuzzy sets. Finally, the variance of mean absolute errors from MW estimates in Fig. 2a is 0.02 while that of PFS is approximately 1.7e-5, i.e. PFS estimates are clearly less sensitive to the window size selection compared to MW estimates.

4.2 PFS Application with Proxies for Antecedents and Consequents

In this section, we consider a more realistic PFS set-up compared to Sect. 4.1, where both the antecedents and consequents in PFS are obtained from MW estimation. Specifically, both the antecedents and the consequents of PFS are obtained for an MW estimation in (10) with a pre-selected window size m, as defined in (12)–(13).

The estimated time-varying correlations and the 99 % intervals for MW and PFS applications for a single simulation study and window size $m = 10$ are shown in Fig. 1d. The PFS application is based on 4 antecedents and 9 consequents. We compare the results to those obtained by MW estimation, reported in Fig. 1b. Correlation estimates and the 99 % intervals are smoother when PFS model is used compared the MW results, even though both the antecedent and consequent of PFS are based on the approximations of correlation instead of actual correlation. Similarly, the uncertainty in the time-varying correlation ρ_t, illustrated by the 99 % interval is much smaller using PFS. Figure 2b presents MAE obtained from MW estimation and PFS using different window sizes. MAE from PFS are between 0.2 and 0.3, regardless of the window size, while MAE from MW estimation varies substantially with the window size. In addition, the variance of the MAE from MW estimates in Fig. 2b is 0.04 while that of PFS is approximately 0.02. Hence PFS estimates are less sensitive to the choice of the window length used for correlation estimates, while MW performs particularly poorly when the window length is small. In other words, PFS decreases the sensitivity of the results to the choice of the window length. Furthermore, we note that the number of antecedents and consequents in PFS has a small effect on the obtained MAE values; for all cases PFS models provide good approximations to actual correlations.

5 Real Data Application

For the real data illustration, we use 1463 daily percentage returns for the Hong Kong Hang Seng Index (HSI) and NASDAQ index between 04 January 2006 and 15 December 2011, where returns r_t at time t are calculated as $r_t = 100 \times (\ln(p_t) - \ln(p_{t-1}))$ where p_t is the closing price of the index at time t. We select this period of stock returns to ensure that the recent financial crisis is included in the data period, and thus we can analyze potential changes in the correlation between the two indexes during the crisis. Daily returns between 04/01/2006 and 18/11/2009 are used as the estimation sample, and the remaining 500 returns after 18/11/2009 are taken as the forecast sample. We note that

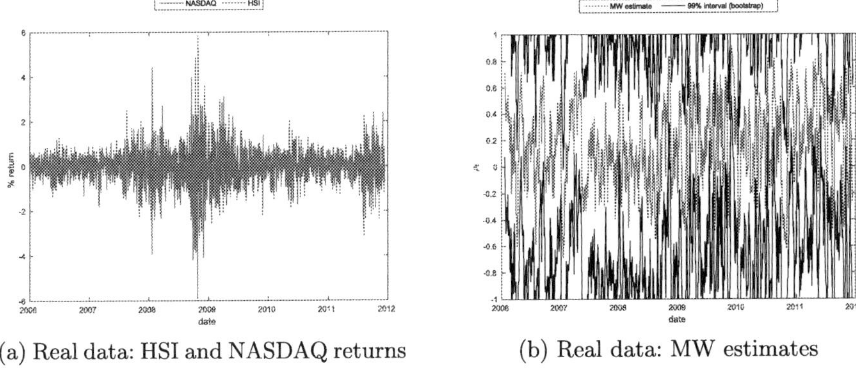

(a) Real data: HSI and NASDAQ returns

(b) Real data: MW estimates

Fig. 3. HSI returns, NASDAQ returns and MW correlation estimates. (Color figure online)

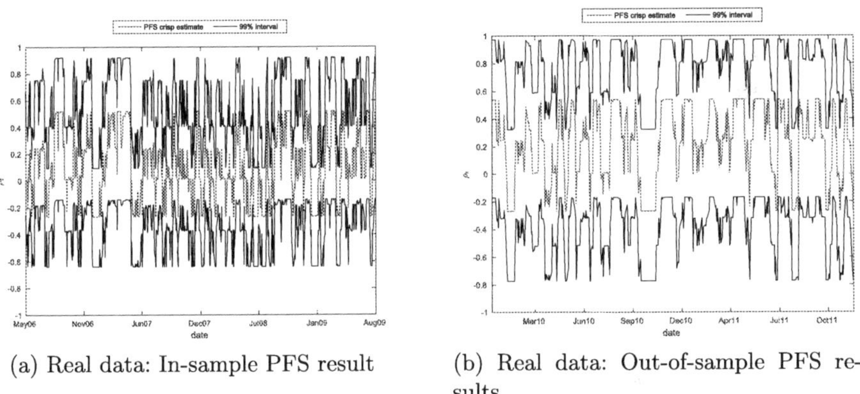

(a) Real data: In-sample PFS result

(b) Real data: Out-of-sample PFS results

Fig. 4. PFS results for HSI and NASDAQ returns. (Color figure online)

the days for which the stock market is closed are slightly different for the two stock markets. We analyze the data during days where both stock markets were open. Percentage returns for both indexes are shown in Fig. 3a.

We first employ the covariance decomposition in (7) on the returns, and obtain the MW estimates of correlation using (10) with a window size of $m = 10$. Since the purpose is to obtain correlation estimates of these stock returns, we do not estimate or report the variance matrix D_t in (7). Given the MW estimates of correlation, we apply the PFS model in (12) with 4 antecedents and 9 consequents, using MW estimates of the previous day as the input variable, and MW estimates of the current day as the output variable for PFS.

MW estimates of correlation are provided in Fig. 3b for the estimation and forecast sample. PFS estimates, on the other hand are reported in Fig. 4a for the estimation sample and in Fig. 4b for the forecast sample. The reported PFS estimate corresponds to the expected value of the PFS output (5). The general findings confirm the simulation study in Sect. 4. MW estimates of correlation are very volatile, and such extreme variation in correlation is counter-intuitive for daily stock prices. In addition, the uncertainty around the MW estimates, represented by the 99 % intervals in Fig. 3b, are very high in all periods. We conclude that the mowing window estimation is unlikely to provide accurate estimates of correlation. PFS estimates of correlation in Fig. 4a, on the other hand, are more stable with smaller 99 % intervals compared to MW estimation. We conclude that, even with this relatively small window length, the obtained results from PFS capture time-varying correlation with substantial accuracy in estimates. Hence the trade-off between capturing time-variation in correlation and obtaining accurate correlation estimates is mitigated using PFS.

6 Conclusions

In this paper we show that a PFS can be used to model unobserved time-varying correlation between financial returns. The proposed method avoids strong distributional assumptions on the correlation process and uses the conventional approximation of time-varying correlation, namely sample correlations from moving windows, as antecedents and consequents. The method is applied to simulated and real data where we show that the PFS application improves over the conventional moving window approximation of time-varying correlation in terms of decreasing the sensitivity to the selection of the window length. In future work, we plan to apply the PFS to intra-day correlation between different stock prices where accurate estimation depends heavily on moving window estimates and the sizes of the moving windows. In addition, we plan to analyze the theoretical foundations and study the interpretability of the rules of the proposed methodology.

References

1. Basel Committee on Banking Supervision: principles for effective risk data aggregation and risk reporting (2013) Bank for International Settlements report
2. Basel Committee on Banking Supervision: Regulatory consistency assessment programme (RCAP)-analysis of risk-weighted assets for credit risk in the banking book (2013) Bank for International Settlements report
3. Embrechts, P., Puccetti, G., Rüschendorf, L., Wang, R., Beleraj, A.: An academic response to Basel 3.5. Risks **2**, 25–48 (2014)
4. Markowitz, H.M.: Portfolio Selection: Efficient Diversification of Investments, vol. 16. Yale University Press, Connecticut (1968)
5. Bollerslev, T.: Generalized autoregressive conditional heteroskedasticity. J. Econometrics **31**, 307–327 (1986)
6. Andersen, T.G., Bollerslev, T., Diebold, F.X.: Parametric and nonparametric volatility measurement. In: Ait-Sahalia, Y., Hansen, L.P. (eds.) Handbook of Financial Econometrics. Tools and Techniques, vol. 1, pp. 67–138. Elsevier (2010)
7. Almeida, R.J., Baştürk, N., Kaymak, U., Milea, V.: A multi-covariate semi-parametric conditional volatility model using probabilistic fuzzy systems. In: 2012 IEEE Conference on Computational Intelligence for Financial Engineering Economics (CIFEr), pp. 1–8 (2012)

8. Almeida, R.J., Baştürk, N., Kaymak, U., da Costa Sousa, J.M.: Conditional density estimation using fuzzy GARCH models. In: Kruse, R., Berthold, M., Moewes, C., Gil, M.A., Grzegorzewski, P., Hryniewicz, O. (eds.) Synergies of Soft Computing and Statistics. AISC, vol. 190, pp. 173–182. Springer, Heidelberg (2013)
9. Almeida, R.J., Baştürk, N., Kaymak, U.: Probabilistic fuzzy systems for seasonality analysis and multiple horizon forecasts. In: 2014 IEEE Conference on Computational Intelligence for Financial Engineering Economics (CIFEr), pp. 497–504 (2014)
10. Almeida, R.J., Basturk, N., Kaymak, U., da Costa Sousa, J.M.: Estimation of flexible fuzzy GARCH models for conditional density estimation. Inf. Sci. **267**, 252–266 (2014)
11. Krishnan, C., Petkova, R., Ritchken, P.: Correlation risk. J. Empir. Finan. **16**, 353–367 (2009)
12. Hamao, Y., Masulis, R.W., Ng, V.: Correlations in price changes and volatility across international stock markets. Rev. Finan. Stud. **3**, 281–307 (1990)
13. Bauwens, L., Laurent, S., Rombouts, J.V.: Multivariate GARCH models: a survey. J. Appl. Econometrics **21**, 79–109 (2006)
14. Becker, R., Clements, A.E., Doolan, M.B., Hurn, A.S.: Selecting volatility forecasting models for portfolio allocation purposes. Int. J. Forecast. **31**, 849–861 (2015)
15. Jacobs, M., Karagozoglu, A.K.: On the characteristics of dynamic correlations between asset pairs. Res. Int. Bus. Finan. **32**, 60–82 (2014)
16. Xu, D., Kaymak, U.: Value-at-risk estimation by using probabilistic fuzzy systems. In: Proceedings of the 2008 World Congress on Computational Intelligence, Hong-Kong, pp. 2109–2116 (2008)
17. Liu, Z., Li, H.X.: A probabilistic fuzzy logic system for modeling and control. IEEE Trans. Fuzzy Syst. **13**, 848–859 (2005)
18. Zhang, G., Li, H.X.: An efficient configuration for probabilistic fuzzy logic system. IEEE Trans. Fuzzy Syst. **20**, 898–909 (2012)
19. Zadeh, L.A.: Probability measures of fuzzy events. J. Math. Anal. Appl. **23**, 421–427 (1968)
20. Zadeh, L.A.: Discussion: probability theory and fuzzy logic are complementary rather than competitive. Technometrics **37**, 271–276 (1995)
21. Dubois, D., Prade, H.: The three semantics of fuzzy sets. Fuzzy Sets Syst. **90**, 141–150 (1997). Fuzzy Sets: Where Do We Stand? Where Do We Go?
22. Bertoluzza, C., Gil, M.A., Ralescu, D.A. (eds.): Statistical Modeling, Analysis and Management of Fuzzy Data. Studies in Fuzziness and Soft Computing. Physica Verlag, Heidelberg (2002)
23. Dubois, D., Prade, H.: Gradualness, uncertainty and bipolarity: making sense of fuzzy sets. Fuzzy Sets Syst. **192**, 3–24 (2012). Fuzzy Set Theory Where Do We Stand and Where Do We Go?
24. van den Berg, J., Kaymak, U., Almeida, R.J.: Conditional density estimation using probabilistic fuzzy systems. IEEE Trans. Fuzzy Syst. **21**, 869–882 (2013)
25. van den Berg, J., van den Bergh, W.M., Kaymak, U.: Probabilistic and statistical fuzzy set foundations of competitive exception learning. In: Proceedings of the Tenth IEEE International Conference on Fuzzy Systems, Melbourne, Australia, vol. 2, pp. 1035–1038 (2001)
26. Coletti, G., Scozzafava, R.: Conditional probability and fuzzy information. Comput. Stat. Data Anal. **51**, 115–132 (2006)
27. Wu, C.F.J.: Jackknife, bootstrap and other resampling methods in regression analysis. Ann. Statist. **14**, 1261–1295 (1986)

Aggregation

Fitting Aggregation Functions to Data: Part I - Linearization and Regularization

Maciej Bartoszuk[1], Gleb Beliakov[2], Marek Gagolewski[1,3(✉)], and Simon James[2]

[1] Faculty of Mathematics and Information Science, Warsaw University of Technology, ul. Koszykowa 75, 00-662 Warsaw, Poland
bartoszukm@mini.pw.edu.pl, gagolews@ibspan.waw.pl
[2] School of Information Technology, Deakin University, 221 Burwood Hwy, Burwood, VIC 3125, Australia
{gleb,sjames}@deakin.edu.au
[3] Systems Research Institute, Polish Academy of Sciences, ul. Newelska 6, 01-447 Warsaw, Poland

Abstract. The use of supervised learning techniques for fitting weights and/or generator functions of weighted quasi-arithmetic means – a special class of idempotent and nondecreasing aggregation functions – to empirical data has already been considered in a number of papers. Nevertheless, there are still some important issues that have not been discussed in the literature yet. In the first part of this two-part contribution we deal with the concept of regularization, a quite standard technique from machine learning applied so as to increase the fit quality on test and validation data samples. Due to the constraints on the weighting vector, it turns out that quite different methods can be used in the current framework, as compared to regression models. Moreover, it is worth noting that so far fitting weighted quasi-arithmetic means to empirical data has only been performed approximately, via the so-called linearization technique. In this paper we consider exact solutions to such special optimization tasks and indicate cases where linearization leads to much worse solutions.

Keywords: Aggregation functions · Weighted quasi-arithmetic means · Least squares fitting · Regularization · Linearization

1 Introduction

In various situations, one is faced with a need to combine $n \geq 2$ numeric values in the unit interval, so that a single representative output is produced. Usually, some idempotent aggregation function, see, e.g., [7,13], is the required data fusion tool.

Definition 1. *We say that* $\mathsf{F} : [0,1]^n \to [0,1]$ *is an idempotent aggregation function, whenever it is nondecreasing in each variable, and for all* $x \in [0,1]$ *it holds* $\mathsf{F}(x,\ldots,x) = x$.

J.P. Carvalho et al. (Eds.): IPMU 2016, Part II, CCIS 611, pp. 767–779, 2016.
DOI: 10.1007/978-3-319-40581-0_62

Among useful idempotent aggregation functions we find weighted quasi-arithmetic means.

Definition 2. *Let* $\varphi : [0,1] \to \bar{\mathbb{R}}$ *be a continuous and strictly monotonic function and* $\mathbf{w}$ *be a weighting vector of length* n*, i.e., one such that for all* i *it holds* $w_i \geq 0$ *and* $\sum_{i=1}^n w_i = 1$*. Then a* weighted quasi-arithmetic mean *generated by* φ *and* $\mathbf{w}$ *is an idempotent aggregation function* $\mathsf{WQAMean}_{\varphi,\mathbf{w}} : [0,1]^n \to [0,1]$ *given for* $\mathbf{x} \in [0,1]^n$ *by:*

$$\mathsf{WQAMean}_{\varphi,\mathbf{w}}(\mathbf{x}) = \varphi^{-1}\left(\sum_{i=1}^{n} w_i \varphi(x_i)\right) = \varphi^{-1}\left(\mathbf{w}^T \varphi(\mathbf{x})\right).$$

Here are a few interesting cases in the above class:

- $\mathsf{WAMean}_{\mathbf{w}}(\mathbf{x}) = \sum_{i=1}^n w_i x_i = \mathbf{w}^T \mathbf{x}$,
  (weighted arithmetic mean, convex combination of inputs, $\varphi(x) = x$)
- $\mathsf{WHMean}_{\mathbf{w}}(\mathbf{x}) = \frac{1}{\sum_{i=1}^n w_i / x_i}$, (weighted harmonic mean, $\varphi(x) = 1/x$)
- $\mathsf{WGMean}_{\mathbf{w}}(\mathbf{x}) = \prod_{i=1}^n x_i^{w_i}$, (weighted geometric mean, $\varphi(x) = \log x$)
- $\mathsf{PMean}_{r,\mathbf{w}}(\mathbf{x}) = \left(\sum_{i=1}^n w_i x_i^r\right)^{1/r}$ for some $r \neq 0$,
  (weighted power mean, $\varphi(x) = x^r$)
- $\mathsf{EMean}_{\gamma,\mathbf{w}}(\mathbf{x}) = \frac{1}{\gamma} \log\left(\sum_{i=1}^n w_i e^{\gamma x_i}\right)$ for some $\gamma \neq 0$.
  (weighted exponential mean, $\varphi(x) = e^{\gamma x}$)

Let us presume that we observe $m \geq n$ input vectors $\mathbf{X} = [\mathbf{x}^{(1)}, \ldots, \mathbf{x}^{(m)}] \in [0,1]^{n \times m}$ together with m desired output values $\mathbf{Y} = [y^{(1)}, \ldots, y^{(m)}] \in [0,1]^{1 \times m}$ and that we would like to fit a *model* that determines the *best* functional relationship between the input values and the desired outputs. Generally, such a task is referred to as regression in machine learning. Nevertheless, classical regression models do not guarantee any preservation of important algebraic properties like the mentioned nondecreasingness or idempotence. Therefore, in our case, for a fixed generator function φ, we shall focus on the task concerning fitting a weighted quasi-arithmetic mean $\mathsf{WQAMean}_{\varphi,\mathbf{w}}$, compare, e.g., [4,7,12,21], to an empirical data set.

Given a *loss function* $E : \mathbb{R}^m \to [0,\infty)$ that is strictly decreasing towards $\mathbf{0}$, $E(0,\ldots,0) = 0$, the task of our interest may be expressed as an optimization problem:

$$\text{minimize } E\left(\varphi^{-1}\left(\sum_{i=1}^{n} w_i \varphi(x_i^{(1)})\right) - y^{(1)}, \ldots, \varphi^{-1}\left(\sum_{i=1}^{n} w_i \varphi(x_i^{(m)})\right) - y^{(m)}\right)$$

with respect to $\mathbf{w}$, under the constraints that $\mathbf{1}^T \mathbf{w} = 1$ and $\mathbf{w} \geq \mathbf{0}$. Typically, E is an L_p norm, in particular: $E(e_1, \ldots, e_m) = \sqrt{\sum_{i=1}^m e_i^2}$ (least squared error fitting, LSE), $E(e_1, \ldots, e_m) = \sum_{i=1}^m |e_i|$ (least absolute deviation fitting, LAD), or $E(e_1, \ldots, e_m) = \bigvee_{i=1}^m |e_i|$ (least maximum absolute deviation fitting, LMD).

In the weighted arithmetic mean case ($\varphi(x) = x$), it is well-known that an LSE fit can be expressed as a quadratic programming task, and both LAD and

LMD fits may be solved by introducing a few auxiliary variables and then by applying some linear programming solvers, see [6,9,12].

More generally, for arbitrary but fixed φ (note that if φ is unknown one may rely on a notion of spline functions to model a generator function, see [2,3,5,6,9,10]), the weight fitting task has up to now been solved approximately via a technique called *linearization*, compare [5,6,8,9,19]. Observing that if $y^{(j)}$ is not subject to any measurement error, i.e., we have $\varphi^{-1}\left(\sum_{i=1}^{n} w_i \varphi(x_i^{(j)})\right) = y^{(j)}$, instead of minimizing a function of $\varphi^{-1}\left(\sum_{i=1}^{n} w_i \varphi(x_i^{(j)})\right) - y^{(j)}$ we may consider a function of $\sum_{i=1}^{n} w_i \varphi(x_i^{(j)}) - \varphi(y^{(j)})$. Thus, the input and output values can be transformed prior to applying a weight fit procedure and then we may proceed in the same manner as when $\varphi(x) = x$ (and in fact deal with a linear interpolation problem). However, in practice this is rarely the case.

What is more, as noted recently in [12], we usually observe that a model may be overfit to a training data set and thus perform weakly on test or validation samples. Also, sometimes we would like to fit a function which is nondecreasing and idempotent but the input values need to be properly normalized prior to aggregate them so that these two important properties are meaningful.

The aim of this two-part paper is to complement the mentioned results (and extend the preliminary outcomes listed in [12]) concerning fitting of weighted quasi-arithmetic means (weighted arithmetic means in particular). In Sect. 2 we discuss possible ways to fit a weighted quasi-arithmetic mean without relying on the linearization technique. Moreover, we perform various numerical experiments that enable us to indicate in which cases linearization leads to a significant decrease in the fit quality. In Sect. 3 we discuss different ways of regularizing a model so as to prevent overfitting. One of the possible approaches consists of adding a special penalty, which is a common procedure in machine learning. However, as parameters of our model fulfill specific constraints (non-negative weights adding up to 1), other ways are possible in our setting too. Section 4 concludes this part of the contribution.

Moreover, in the second part [1] we deal with the problem of properly normalizing (transforming) discordant input values in such a way that idempotent aggregation functions may be fit. We present an application of such a procedure in a classification task dealing with the identification of pairs of similar R [17] source code chunks.

2 Linearization

From now on let us assume that $E(e_1, \ldots, e_m) = \sqrt{\sum_{i=1}^{m} e_i^2}$, i.e., we would like to find a least squares error fit. Such an approach is perhaps most common in machine learning literature [14]. What is more, most of the ideas presented in this paper can be quite easily applied in other settings.

Torra in [19,20] (compare also, e.g., [6,8,9]) already discussed weighted quasi-arithmetic mean fitting tasks. Nevertheless, it was noted that the problem is difficult in general, so one may simplify the problem assuming that the desired outputs are not subject to errors. In such a case, noting that a fixed generator function φ is surely invertible, we have for all j:

$$\sum_{i=1}^{n} w_i \varphi(x_i^{(j)}) = \varphi(y^{(j)}).$$

Using this assumption, instead of minimizing:

$$\|\varphi^{-1}\left(\mathbf{w}^T \varphi(\mathbf{X})\right) - \mathbf{Y}\|_2^2$$

one may decide to minimize a quite different (in general) goodness of fit measure:

$$\|\mathbf{w}^T \varphi(\mathbf{X}) - \varphi(\mathbf{Y})\|_2^2.$$

Such an approach is often called *linearization* of inputs.

Let us suppose, however, that we would like to solve the original weight fit problem and not the simplified (approximate) one.

Example 1 ([12]). Suppose that $n = 5$ and we are given $m = 9$ toy data points given as below. Here $\mathbf{Y}$ was generated using $\mathbf{w} = (0.33, 0.43, 0.10, 0.08, 0.06)$ and $\varphi(x) = x^2$ with white noise was added ($\sigma = 0.05$).

j	1	2	3	4	5	6	7	8	9
$x_1^{(j)}$	0.12	0.48	0.65	0.07	0.37	0.22	0.29	0.57	0.84
$x_2^{(j)}$	0.73	0.41	0.45	0.79	0.92	0.23	0.90	0.40	0.57
$x_3^{(j)}$	0.43	0.84	0.70	0.96	0.81	0.86	0.72	0.53	0.42
$x_4^{(j)}$	0.52	0.75	0.48	0.40	0.62	0.28	0.80	0.92	0.79
$x_5^{(j)}$	0.69	0.70	0.24	0.22	0.92	0.34	0.15	0.50	0.50
$y^{(j)}$	0.65	0.58	0.70	0.51	0.82	0.56	0.70	0.64	0.75

Here are the true $\eth_1$, $\eth_2$, and $\eth_\infty$ error measures in the case of the linearized and the exact LSE and LAD minimization tasks.

E	$\eth_1$	$\eth_2$	$\eth_\infty$
LAD – linearization	0.7385	0.4120	0.2798
LSE – linearization	0.7423	0.2859	0.1626
LAD – optimal	0.7157	0.3170	0.2044
LSE – optimal	0.7587	0.2817	0.1501

In the above example, the differences are relatively small, but not negligible. While the use of the linearization technique for least-squared error fitting of quasi-arithmetic means will often lead to reliable results, there are clearly some situations where such a technique may not be justified. Fitting to the transformed dataset $\varphi(\mathbf{X}), \varphi(\mathbf{Y})$ essentially stretches the space along which the residuals are distributed and for some functions this will have a larger impact than others.

As an example, consider the geometric mean with generator $\varphi(x) = \log x$. For lower values of y, differences in the transformed residuals can become disproportionately large and pull the weights towards these data points. Further on we shall perform a few numerical experiments to indicate the generator functions that lead to much greater discrepancies.

2.1 Algorithms

We aim to:

$$\text{minimize} \sum_{j=1}^{m} \left(\varphi^{-1} \left(\sum_{i=1}^{n} w_i \varphi\left(x_i^{(j)}\right) \right) - y^{(j)} \right)^2 \quad \text{w.r.t. } \mathbf{w}$$

subject to $\mathbf{w} \geq \mathbf{0}$ and $\mathbf{1}^T\mathbf{w} = 1$. By homogeneity and triangle inequality of $\|\cdot\|_2$ we have that this is a convex optimization problem.

A Solution Based on a Nonlinear Solver. First of all, we may consider the above as a generic nonlinear optimization task. To drop the constraints on $\mathbf{w}$, we can use an approach considered (in a different context) by Filev and Yager [11], see also [20]. We take a different parameter space, $\boldsymbol{\lambda} \in \mathbb{R}^n$, such that:

$$w_i = \frac{\exp(\lambda_i)}{\sum_{k=1}^{n} \exp(\lambda_k)}.$$

Assuming that φ^{-1} is differentiable, let us determine the gradient $\nabla E(\boldsymbol{\lambda})$. For any $k = 1, \ldots, n$ it holds:

$$\frac{\partial}{\partial \lambda_k} E(\boldsymbol{\lambda}) = 2 \frac{\exp(\lambda_k)}{\sum_{i=1}^{n} \exp(\lambda_i)} \sum_{j=1}^{m} \left(\varphi^{-1} \left(\frac{\sum_{i=1}^{n} \exp(\lambda_i) \varphi\left(x_i^{(j)}\right)}{\sum_{i=1}^{n} \exp(\lambda_i)} \right) - y^{(j)} \right)$$
$$\cdot (\varphi^{-1})' \left(\frac{\sum_{i=1}^{n} \exp(\lambda_i) \varphi\left(x_i^{(j)}\right)}{\sum_{i=1}^{n} \exp(\lambda_i)} \right) \cdot \left(\varphi\left(x_k^{(j)}\right) - \frac{\sum_{i=1}^{n} \exp(\lambda_i) \varphi\left(x_i^{(j)}\right)}{\sum_{i=1}^{n} \exp(\lambda_i)} \right).$$

Assuming that $\mathbf{Z} = \mathbf{w}^T \varphi(\mathbf{X})$ and $\mathbf{w} = \exp(\boldsymbol{\lambda}) / \mathbf{1}^T \exp(\boldsymbol{\lambda})$, we have:

$$\nabla E(\boldsymbol{\lambda}) = 2 \cdot \mathbf{w} \cdot \left(\left((\phi^{-1}(\mathbf{Z}) - Y) \cdot (\varphi^{-1})'(\mathbf{Z}) \right) \times \left(\varphi(\mathbf{X})^T - \mathbf{Z} \right) \right),$$

where $\cdot, -$ stand for elementwise vectorized multiplication and subtraction, respectively, $\times$ denotes matrix multiplication, and $\varphi(\mathbf{X})^T - \mathbf{Z}$ means that we subtract $\mathbf{Z}$ from each column in $\varphi(\mathbf{X})^T$. The solution may be computed using, e.g., a quasi-Newton nonlinear optimization method by Broyden, Fletcher, Goldfarb and Shanno (the BFGS algorithm, see [16]). However, let us note that while using the mentioned reparametrization, the BFGS algorithm may occasionally fail to converge as now the solution is not unique.

Another possible way of solving the above problem would be to rewrite the objective as a function of $n-1$ variables $v_1, \ldots, v_{n-1}$ such that $w_i = v_i$ for $i = 1, \ldots, n-1$ and $w_n = 1 - \sum_{i=1}^{n-1} v_i$. This is a constrained optimization problem, but the constraints on $\mathbf{v}$ are linear: $\mathbf{v} \geq \mathbf{0}$ and $\mathbf{1}^T\mathbf{v} \leq 1$. With generic nonlinear solvers, such an optimization task is usually determined by adding an appropriate barrier function (e.g., logarithmic barrier, see [16]) term to the objective function.

Compensation Factors in Linearization. Another possible way is to apply a compensation factor such that for any residual in the linearization technique $v^{(j)} = \left(\sum_{i=1}^{n} \varphi(x_i^{(j)})\right) - \varphi(y^{(j)})$, we estimate the true residual $r^{(j)} = \varphi^{-1}\left(\sum_{i=1}^{n} \varphi(x_i^{(j)})\right) - y^{(j)}$.

We begin with an estimate of our true residual, which we denote by r_{est}. The estimated residual for any known $v^{(j)}$ is then calculated as:

$$\text{est}(r^{(j)}) = \frac{v^{(j)} r_{\text{est}}}{\varphi(y^{(j)} + r_{\text{est}}) - \varphi(y^{(j)})}.$$

In other words, we calculate the average rate of change between $y^{(j)}$ and $y^{(j)} + r_{\text{est}}$ and then use the reciprocal of this as our compensation factor. A visual illustration of this process is shown in Fig. 1.

Where $y^{(j)} + r_{\text{est}}$ is outside our domain $[0, 1]$, we can instead use the average rate of change between $y^{(j)}$ and the boundary point (or very close to the boundary if φ is infinite).

Obviously the average rate of change will differ depending on whether $v^{(j)}$ is positive or negative, and so we can split $v^{(j)}$ into its positive and negative components so that a different compensation factor can be applied.

We let $v^{(j)} = v_+^{(j)} + v_-^{(j)}$ where $v_+^{(j)} \geq 0, v_-^{(j)} \leq 0$ which we then use as decision variables in our quadratic programming task. We optimize with respect to these,

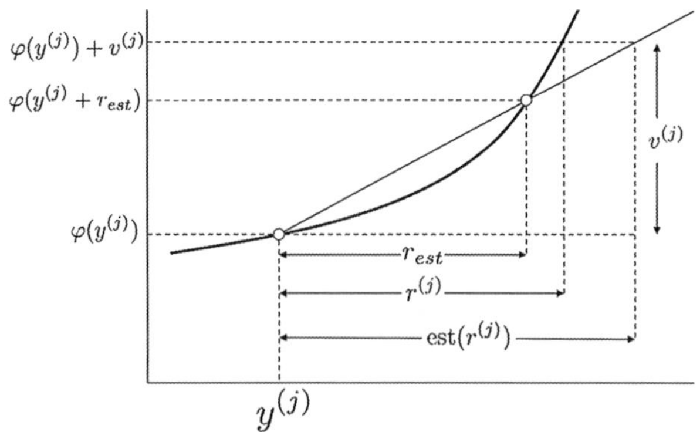

Fig. 1. Illustration of how compensation factor is calculated. From a generating function φ we estimate $r^{(j)}$ using the average rate of change between $y^{(j)}$ and $y^{(j)} + r_{\text{est}}$.

and add constraints:

$$\left(\sum_{i=1}^{n} \varphi(x_i^{(j)})\right) + v_+^{(j)} + v_-^{(j)} = \varphi(y^{(j)}).$$

In summary, we have the following quadratic programming task.

$$\text{minimize} \sum_{j=1}^{m} \left(\frac{v_+^{(j)} r_{\text{est}}}{\varphi(y^{(j)}+r_{\text{est}})-\varphi(y^{(j)})}\right)^2 + \left(\frac{v_-^{(j)} r_{\text{est}}}{\varphi(y^{(j)}-r_{\text{est}})-\varphi(y^{(j)})}\right)^2 \quad \text{w.r.t. } \mathbf{w}, \mathbf{v}_-, \mathbf{v}_+$$

$$\text{such that} \quad \sum_{i=1}^{n} w_i = 1, w_i \geq 0, i = 1, \ldots, n,$$

$$\left(\sum_{i=1}^{n} \varphi(x_i^{(j)})\right) + v_+^{(j)} + v_-^{(j)} = \varphi(y^{(j)}),$$

$$v_+^{(j)} \geq 0, -v_-^{(j)} \geq 0, j = 1, \ldots, m.$$

Since the usefulness of this method will depend on r_{est}, we can also set up a bilevel optimization problem such that it is optimized for the given training set. Nevertheless, note that this time we deal with an approximate method. The following experiments show that the method is useful for compensating for the stretching effect of the generating function and also help us identify some specific instances of where linearization by itself has poor performance.

2.2 Experiments

For each of the generating functions $\varphi(x) = x^2, x^3, x^{1/2}, \log x$ we created data sets with $m = 20$ and $m = 100$ test points. For each trial, we generated $\mathbf{w}$ randomly and then after calculating the desired output values we added Gaussian noise with $\sigma = 0.05$ and 0.1. We then measured the LSE using:

(i) the linearization technique where the data are transformed using $\varphi(\mathbf{X}), \varphi(\mathbf{Y})$, i.e., the sum of $\left(\left(\sum_{i=1}^{n} \varphi(x_i^{(j)})\right) - \varphi(y^{(j)})\right)^2$,
(ii) the method proposed here with $r_{\text{est}} = 0.1$,
(iii) the method proposed here with r_{est} optimized,
(iv) a general non-linear optimization solver.

The $\mathbf{x}^{(j)}$ vectors were generated both from the uniform distribution on $[0,1]^n$ as well as an exponential distribution scaled to the interval $[0,1]$. The uniform data would be expected to result in y values distributed around the middle of the interval, while exponentially distributed data would often result in outputs closer to the lower end of the interval. We expect the latter case to result in worse performance for linearization.

After obtaining the fitted weighted vectors, we calculated the total LSE and normalized these values by expressing them as a proportion of the optimal LSE from the weighting vector obtained from the generalized solver. The results are shown in Table 1.

Table 1. Relative LSE calculated as proportion of total LSE obtained using a general nonlinear solver *(iv)*. Results represent averages over 10 trials

Uniformly distributed $\mathbf{X}$						
$m = 20$	$\sigma = 0.05$			$\sigma = 0.1$		
$\varphi(x)$	*(i)*	*(ii)*	*(iii)*	*(i)*	*(ii)*	*(iii)*
x^2	0.0623	0.0011	0.0005	0.0270	0.0159	0.0026
x^3	0.0936	0.0114	0.0025	0.1329	0.1038	0.0131
$x^{1/2}$	0.0233	0.0006	0.0004	0.0725	0.0099	0.0008
$\log x$	0.1492	0.0042	0.0025	0.4763	0.0358	0.0055
$m = 100$	$\sigma = 0.05$			$\sigma = 0.1$		
$\varphi(x)$	*(i)*	*(ii)*	*(iii)*	*(i)*	*(ii)*	*(iii)*
x^2	0.0072	0.0004	0.0003	0.0106	0.0070	0.0015
x^3	0.0266	0.0017	0.0014	0.0356	0.0499	0.0080
$x^{1/2}$	0.0080	0.0002	0.0001	0.0061	0.0014	0.0005
$\log x$	0.1921	0.0013	0.0011	0.2550	0.0064	0.0017
Exponentially distributed $\mathbf{X}$						
$m = 20$	$\sigma = 0.05$			$\sigma = 0.1$		
$\varphi(x)$	*(i)*	*(ii)*	*(iii)*	*(i)*	*(ii)*	*(iii)*
x^2	0.0479	0.0066	0.0035	0.0795	0.0946	0.0132
x^3	0.1560	0.0234	0.0145	0.2345	0.3030	0.0266
$x^{1/2}$	0.1048	0.0061	0.0036	0.0522	0.0204	0.0047
$\log x$	1.0970	0.0133	0.0076	2.7748	0.0558	0.0194
$m = 100$	$\sigma = 0.05$			$\sigma = 0.1$		
$\varphi(x)$	*(i)*	*(ii)*	*(iii)*	*(i)*	*(ii)*	*(iii)*
x^2	0.0198	0.0042	0.0041	0.0273	0.0520	0.0055
x^3	0.0579	0.0100	0.0040	0.0820	0.1641	0.0117
$x^{1/2}$	0.0167	0.0015	0.0012	0.0204	0.0098	0.0045
$\log x$	1.8401	0.0095	0.0049	0.7770	0.0460	0.0228

A number of observations can be made from this data. Firstly, we note that linearization for $\varphi(x) = x^2$ tended only to produce increases in LSE of about 2–7 %, regardless of the data distribution. On the other hand, $\varphi(x) = x^3$ showed increases in LSE of between 9–24 % when there were only $m = 20$ data instances.

The most dramatic results were obtained when fitting the geometric mean. For exponentially distributed input vectors, the method of linearization increased the error by up to 277 % when the noise added to y was generated using $\sigma = 0.1$. There was large variability in these trials – in the best case using linearization increased the error by 45 %, while at other times the difference was 10 fold (LSE was 0.8752 compared with 0.0940 when $r_{\text{est}} = 0.1$ was used, about a 3.2 % increase on the optimal LSE). The worst results seemed to occur where the data set included y values equal to zero. The weight associated with the lowest input for that instance is pulled up to try and reduce the very large error. The compensation factor however did seem to obtain decent improvements even when the data was exponentially distributed.

While a setting of $r_{\text{est}} = 0.1$ tended to result in significant improvements for small errors ($\sigma = 0.05$), increased error of $\sigma = 0.1$ often had optimal values that were closer to 0.2.

3 Regularization

In this section we discuss a few possible ways to prevent a model being overfit to a training sample. In other words, we would like that for other samples of the same kind (e.g., following the same statistical distribution) the model performance does not decrease drastically.

3.1 Tikhonov Regularization

The Tikhonov regularization [18] is a basis for the ridge regression [15] method. It has a form of an additional penalty term dependent on a scaled squared L_2 norm of the weighting vector.

In our case we may consider, for some λ, an optimization task:

$$\text{minimize} \sum_{j=1}^{m} \left(\varphi^{-1} \left(\sum_{i=1}^{n} w_i \varphi(x_i^{(j)}) \right) - y^{(j)} \right)^2 + \lambda \sum_{i=1}^{n} w_i^2 \quad \text{w.r.t. } \mathbf{w}$$

subject to $\mathbf{1}^T \mathbf{w} = 1$ and $\mathbf{w} \geq \mathbf{0}$. Note that due to the usual constraints on $\mathbf{w}$, the use of the L_1 norm instead of squared L_2 (like, e.g., in Lasso regression) does not make much sense at this point: we always have $\|\mathbf{w}\|_1 = 1$.

In the simplest case ($\varphi(x) = x$), the above optimization problem can be written in terms of the following quadratic programming task:

$$\text{minimize } 0.5\, \mathbf{w}^T (\mathbf{X}\mathbf{X}^T + \lambda \mathbf{I}) \mathbf{w} - (\mathbf{X}\mathbf{Y}^T)^T \mathbf{w} \quad \text{w.r.t. } \mathbf{w}$$

subject to $\mathbf{w} \geq \mathbf{0}$, $\mathbf{1}^T \mathbf{w} = 1$, which minimizes the squared error plus a $\lambda \|\mathbf{w}\|_2^2$ penalty term. Note that for other generator functions φ we can easily incorporate an appropriate penalty to an optimization task considered in the previous section.

Remark 1. Unlike in regression problems, where we always presuppose that $\lambda \geq 0$, in our framework we are bounded with additional constraints on $\mathbf{w}$ which, for large λ, tend to generate weighting vectors such that $w_i \to 1/n$. On the other hand, in the current framework the case of $\lambda < 0$ may also lead to useful outcomes. Yet, we should note that for $\lambda \to -\infty$ we observe that $w_j \to 1$ for some j.

Example 2. Let us consider a data set generated randomly with R as follows:

```
set.seed(321)
n <- 10; m <- 100
realw <- rbeta(n, 1, 5)
realw <- realw/sum(realw) # real weights ~ beta distribution
X <- t(matrix(runif(n*m), nrow=m))
Y <- t(realw) %*% X + rnorm (m, 0, 0.1)
Y[,] <- pmax(0, pmin(1, Y))
X <- round(X, 2) # uniform distribution, rounded
Y <- round(Y, 2) # sigma=0.1, truncated to [0,1], rounded
train <- sample(1:m, m*0.8)
X_test <- X[,-train,drop=FALSE] # test sample
Y_test <- Y[,-train,drop=FALSE]
X <- X[,train,drop=FALSE]        # training sample
Y <- Y[,train,drop=FALSE]
```

The data points are divided into two groups: a training sample (80 % of the observations, used to estimate the weights) and a test sample (20 %, used to compute the error). Figure 2 depicts squared error measures as a function of Tikhonov regularization coefficient λ. We see we were able to improve the error measure by ca. 9 % by using $\lambda \simeq 4.83$.

E	$\eth_1$	$\eth_2$	$\eth_\infty$
— (using `realw`)	1.291	0.3637	0.1915
LAD	1.488	0.4027	0.1937
LSE	1.436	0.4036	0.1931
LMD	1.475	0.4025	0.1699
LSE + regularization, $\lambda \simeq 4.83$	1.371	0.3705	0.1891

3.2 Weights Dispersion Entropy

Similar to minimizing sum of squared weights, maximizing weights dispersion entropy can also have benefits, measuring the degree to which the function takes into account all the inputs, compare [9] for its use for a different purpose. It is given by:

$$\mathsf{Disp}(\mathbf{w}) = -\sum_{i=1}^{n} w_i \log w_i,$$

with the convention $0 \cdot \log 0 = 0$.

In cases where fitting results in multiple solutions, for example when there is too few data for there to be a singular minimizer, Torra proposed the use of weights dispersion as an additional criterion to determine the best solution [19]. It is implemented as a second level of the optimization. After obtaining a

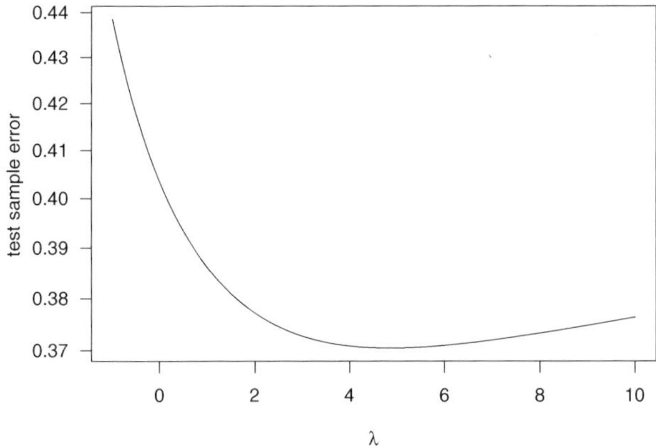

Fig. 2. Three error measures on a test data set from Example 2 as a function of regularization penalty λ.

minimum A to the objective in the standard least squares fitting problem, one then solves:

$$\text{minimize} \sum_{i=1}^{n} w_i \log w_i + \lambda \left(\sum_{j=1}^{m} \left(\varphi^{-1} \left(\sum_{i=1}^{n} w_i \varphi(x_i^{(j)}) \right) - y^{(j)} \right)^2 - A \right)^2 \quad \text{w.r.t. } \mathbf{w}$$

$$\text{such that} \quad \sum_{i=1}^{n} w_i = 1, w_i \geq 0, i = 1, \ldots, n,$$

for some $\lambda > 0$.

One may alternatively consider a one-level task like:

$$\text{minimize} \sum_{j=1}^{m} \left(\varphi^{-1} \left(\sum_{i=1}^{n} w_i \varphi(x_i^{(j)}) \right) - y^{(j)} \right)^2 + \lambda \sum_{i=1}^{n} w_i \log w_i \quad \text{w.r.t. } \mathbf{w}$$

subject to the standard constraints.

Remark 2. In fact, weights dispersion and sum of squared weights are both examples of functions used to model income inequality in economics and evenness in ecology. There are numerous other functions used in these fields that could also be used as secondary objectives to achieve the task of weight regularization (e.g., the Gini index).

4 Conclusion

We have considered some practical issues concerning fitting weighted quasi arithmetic means to empirical data using supervised learning-like approaches. First of all, we pointed out the drawbacks of the commonly applied linearization technique, which may lead to far-from-optimal solutions. Moreover, we analyzed some ways to prevent model overfitting.

Note that the discussion can be easily generalized to other error measures and fitting other aggregation functions parametrized via a weighting vector, e.g., weighted Bonferroni means and generalized OWA operators. In future applications, the compensation and regularization techniques proposed here can be used to learn more useful and informative models.

Acknowledgments. This study was supported by the National Science Center, Poland, research project 2014/13/D/HS4/01700.

References

1. Bartoszuk, M., Beliakov, G., Gagolewski, M., James, S.: Fitting aggregation functions to data: Part II - idempotization. In: Carvalho, J.P., Lesot, M.-J., Kaymak, U., Vieira, S., Bouchon-Meunier, B., (eds.) IPMU 2016, Part II, CCIS, vol. 611, pp. 780–789. Springer, Heidelberg (2016)
2. Beliakov, G.: Shape preserving approximation using least squares splines. Approximation Theory Appl. **16**(4), 80–98 (2000)
3. Beliakov, G.: Monotone approximation of aggregation operators using least squares splines. Int. J. Uncertainty Fuzziness Knowl. Based Syst. **10**, 659–676 (2002)
4. Beliakov, G.: How to build aggregation operators from data. Int. J. Intell. Syst. **18**, 903–923 (2003)
5. Beliakov, G.: Learning weights in the generalized OWA operators. Fuzzy Optim. Decis. Making **4**, 119–130 (2005)
6. Beliakov, G.: Construction of aggregation functions from data using linear programming. Fuzzy Sets Syst. **160**, 65–75 (2009)
7. Beliakov, G., Bustince, H., Calvo, T.: A Practical Guide to Averaging Functions. STUDFUZZ. Springer, Switzerland (2016)
8. Beliakov, G., James, S.: Using linear programming for weights identification of generalized bonferroni means in R. In: Narukawa, Y., López, B., Villaret, M., Torra, V. (eds.) MDAI 2012. LNCS, vol. 7647, pp. 35–44. Springer, Heidelberg (2012)
9. Beliakov, G., Pradera, A., Calvo, T.: Aggregation Functions: A Guide for Practitioners. Springer, Heidelberg (2007)
10. Beliakov, G., Warren, J.: Appropriate choice of aggregation operators in fuzzy decision support systems. IEEE Trans. Fuzzy Syst. **9**(6), 773–784 (2001)
11. Filev, D., Yager, R.R.: On the issue of obtaining OWA operator weights. Fuzzy Sets Syst. **94**, 157–169 (1998)
12. Gagolewski, M.: Data Fusion: Theory, Methods, and Applications. Institute of Computer Science, Polish Academy of Sciences, Warsaw (2015)
13. Grabisch, M., Marichal, J.L., Mesiar, R., Pap, E.: Aggregation Functions. Cambridge University Press, New York (2009)
14. Hastie, T., Tibshirani, R., Friedman, J.: The Elements of Statistical Learning: Data Mining, Inference, and Prediction. Springer, New York (2013)
15. Hoerl, A., Kennard, R.: Ridge regression: biased estimation for nonorthogonal problems. Technometrics **12**(1), 55–67 (1970)

16. Nocedal, J., Wright, S.: Numerical Optimization. Springer, New York (2006)
17. R Development Core Team: R: a language and environment for statistical computing. R Foundation for Statistical Computing, Vienna, Austria (2016). http://www.R-project.org
18. Tikhonov, A., Arsenin, V.: Solution of Ill-posed Problems. Winston & Sons, Washington (1977)
19. Torra, V.: Learning weights for the quasi-weighted means. IEEE Trans. Fuzzy Syst. **10**(5), 653–666 (2002)
20. Torra, V.: OWA operators in data modeling and reidentification. IEEE Trans. Fuzzy Syst. **12**(5), 652–660 (2004)
21. Torra, V.: Aggregation operators and models. Fuzzy Sets Syst. **156**, 407–410 (2005)

Fitting Aggregation Functions to Data: Part II - Idempotization

Maciej Bartoszuk[1], Gleb Beliakov[2], Marek Gagolewski[1,3(✉)], and Simon James[2]

[1] Faculty of Mathematics and Information Science, Warsaw University of Technology, ul. Koszykowa 75, 00-662 Warsaw, Poland
bartoszukm@mini.pw.edu.pl

[2] School of Information Technology, Deakin University, 221 Burwood Hwy, Burwood, VIC 3125, Australia
{gleb,sjames}@deakin.edu.au

[3] Systems Research Institute, Polish Academy of Sciences, ul. Newelska 6, 01-447 Warsaw, Poland
gagolews@ibspan.waw.pl

Abstract. The use of supervised learning techniques for fitting weights and/or generator functions of weighted quasi-arithmetic means – a special class of idempotent and nondecreasing aggregation functions – to empirical data has already been considered in a number of papers. Nevertheless, there are still some important issues that have not been discussed in the literature yet. In the second part of this two-part contribution we deal with a quite common situation in which we have inputs coming from different sources, describing a similar phenomenon, but which have not been properly normalized. In such a case, idempotent and nondecreasing functions cannot be used to aggregate them unless proper preprocessing is performed. The proposed idempotization method, based on the notion of B-splines, allows for an automatic calibration of independent variables. The introduced technique is applied in an R source code plagiarism detection system.

Keywords: Aggregation functions · Weighted quasi-arithmetic means · Least squares fitting · Idempotence

1 Introduction

Idempotent aggregation functions – mappings like $\mathsf{F} : [0,1]^n \to [0,1]$ being nondecreasing in each variable and fulfilling $\mathsf{F}(x,\ldots,x) = x$ for all $x \in [0,1]$ – have numerous applications, including areas like decision making, pattern recognition, and data analysis, compare, e.g., [8,11].

For a fixed $n \geq 2$, let $\mathbf{w} \in [0,1]^n$ be a weighting vector, i.e., one with $\sum_{i=1}^{n} w_i = 1$. In the first unit [1] of this two-part contribution we dealt with two important practical issues concerning supervised learning of weights of weighted

J.P. Carvalho et al. (Eds.): IPMU 2016, Part II, CCIS 611, pp. 780–789, 2016.
DOI: 10.1007/978-3-319-40581-0_63

quasi-arithmetic means with a known continuous and strictly monotone generator $\varphi : [0,1] \to \bar{\mathbb{R}}$, that is idempotent aggregation functions given for arbitrary $\mathbf{x} \in [0,1]^n$ by the formula:

$$\mathsf{WQAMean}_{\varphi,\mathbf{w}}(\mathbf{x}) = \varphi^{-1}\left(\sum_{i=1}^{n} w_i \varphi(x_i)\right).$$

First of all, we observed that most often researchers considered an approximate version of weight learning tasks and relied on a linearization of input variables, compare, e.g., [7]. Therefore, we discussed possible implementations of the exact fitting procedure and identified some cases where linearization leads to solutions of significantly worse quality in terms of the squared error between the desired and generated outputs. Secondly, we noted that the computed models may overfit a training data set and perform weakly on test and validation samples. Thus, some regularization methods were proposed to overcome this limitation. We indicated that due to the typical constraints on the weighting vector (nonnegative coefficients summing up to 1), not all the regularization techniques known from machine learning [13] can be applied, but – on the other hand – we may consider new, quite different ones instead.

Assume that we are given $m \geq n$ input vectors in a form $\mathbf{X} = [\mathbf{x}^{(1)}, \ldots, \mathbf{x}^{(m)}] \in [0,1]^{n \times m}$ together with m desired output values $\mathbf{Y} = [y^{(1)}, \ldots, y^{(m)}] \in [0,1]^{1 \times m}$. For simplicity, we shall focus only on fitting weighted arithmetic means to $(\mathbf{X}, \mathbf{Y})$ using the least squared error criterion, noting that the key ideas presented further on can be extrapolated to other settings. And so, we aim to:

$$\text{minimize} \sum_{j=1}^{m} \left(\sum_{i=1}^{n} w_i x_i^{(j)} - y^{(j)}\right)^2 \quad \text{w.r.t. } \mathbf{w},$$

under the constraints that $\mathbf{1}^T\mathbf{w} = 1$ and $\mathbf{w} \geq \mathbf{0}$, compare, e.g., [5].

However, let us presume that the input values represent m measurements of the same phenomenon done via n different methods which output numeric values that cannot be directly compared: each of them is defined up to a strictly increasing and continuous transformation and a proper input data idempotization scheme has to be applied prior to fitting a model.

Example 1. In the R [15] language plagiarism detection system described in [2,3], the similarity of a source code chunk pair is assessed via $n = 4$ diverse methods. Each of them reflects quite different ideas behind what plagiarism really is in its nature:

- $x_1^{(j)}$ – is based on the so-called program dependence graphs (PDGs),
- $x_2^{(j)}$ – simply computes the Levenshtein distance between source texts,
- $x_3^{(j)}$ – determines the longest common subsequence of two corresponding token strings,
- $x_4^{(j)}$ – compares the number of common R function calls.

Each of the four variables is a real number in $[0,1]$, but the way they have been mapped to the unit interval is quite arbitrary – in fact, initially, we should treat them as values on an ordinal – and not interval – scale. Most common machine learning methods (e.g., regression and classification) should work quite well on such data, but the construction of aggregation functions does not make much sense on raw inputs of this kind. However, if appropriate strictly monotone transformations $\varphi_1,\dots,\varphi_4:[0,1]\to[0,1]$ were determined, we would have the values normalized in such a way that they can be compared to each other ($x_1^{(j)} = 0.5$ would denote the same similarity level as $x_2^{(j)} = 0.5$, hence we could expect – by idempotence – the aggregated similarity to be equal to 0.5 too). Moreover – by nondecreasingness – we could be sure that any increase of a similarity level never leads to a decrease in the aggregated similarity – a constraint not guaranteed by any classical machine learning method.

Therefore, in this part of the contribution, we deal with the problem which aims to construct an *idempotized* model for a given data set, that is we are going to:

$$\text{minimize} \sum_{j=1}^{m}\left(\sum_{i=1}^{n} w_i\tilde{x}_i^{(j)} - y^{(j)}\right)^2 \quad \text{w.r.t. } \mathbf{w},$$

under the standard constraints on a weighting vector $\mathbf{w}$, where $\tilde{x}_i^{(j)} = \varphi_i(x_i^{(j)})$ for some automatically generated monotone and continuous $\varphi_1,\dots,\varphi_n:[0,1]^n\to[0,1]$. This enables us to develop idempotent aggregation functions-based regression (and, as a by-product, binary classification) models [13], which – by construction – fulfill some desired algebraic properties, and hence posses a better, more intuitive interpretation than classical approaches on data sets similar to the one in Example 1.

The paper is set out as follows. In the next section we recall the notion of B-splines, which we shall use for modeling the φ_i functions. Section 3 discusses the proposed idempotization and aggregation function fitting procedure, together with some key implementation details. Note that in order to increase its performance on test samples, the model employs a regularization term which we discussed in the first part of this contribution [1]. Section 4 discusses the results of an experimental study conducted on the aforementioned plagiarism detection system data. Finally, Sect. 5 summarizes the paper and indicates future research directions.

2 B-splines

In a quasi-arithmetic mean fitting task, Beliakov et al. [4,6,9,10] rely on the notion of B-splines to model an unknown generator function.

Let $p\ge 1$ and $\mathbf{t}=(t_1,\dots,t_k)$ be an increasingly ordered *knot vector* of length k for some $k\ge 0$ such that $0<t_i<t_{i+1}<1$ for all $i=1,\dots,k$. For simplicity, we presume that $t_i=0$ for $i<1$ and $t_i=1$ whenever $i>k$.

Definition 1. B-spline basis functions *for* $j = 0, \ldots, p$ *and* $x \in [0, 1]$ *are defined recursively as:*

$$N^{\mathbf{t}}_{i,j}(x) = \begin{cases} 1 \text{ if } x \in [t_{i-1}, t_i], \\ 0 \text{ otherwise}, \end{cases} \qquad (j = 0)$$

$$N^{\mathbf{t}}_{i,j}(x) = \frac{x - t_{i-1}}{t_{i+j-1} - t_{i-1}} N^{\mathbf{t}}_{i,j-1}(x) + \frac{t_{i+j} - x}{t_{i+j} - t_i} N^{\mathbf{t}}_{i+1,j-1}(x), \quad (j > 0)$$

with convention $\cdot/0 = 0$.

Note that a vector of equidistant knots is a quite common setting. Figure 1 depicts exemplary B-spline basis functions.

Additionally, let $\mathbf{v} \in [0, 1]^\eta$ be a vector of *control points*, where $\eta = p + k + 1$.

Definition 2. *A function* $B^{\mathbf{t}}_{\mathbf{v}} : [0, 1] \to [0, 1]$ *given by:*

$$B^{\mathbf{t}}_{\mathbf{v}}(x) = \sum_{i=1}^{\eta} v_i N^{\mathbf{t}}_{i-p,p}(x) \tag{1}$$

is called a nonperiodic B-spline of degree p *based on a knot vector* $\mathbf{t}$ *and generated by a control points vector* $\mathbf{v}$, *see, e.g., [16].*

In particular, for $p = 1$ we get a piecewise linear function interpolating $(0, v_1), (t_1, v_2), \ldots, (t_k, v_{\eta-1}), (1, v_\eta)$. On the other hand, for $p = 3$ we get a cubic B-spline.

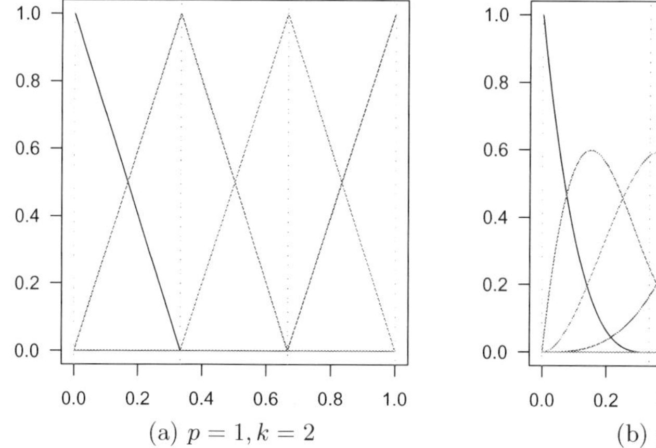

(a) $p = 1, k = 2$

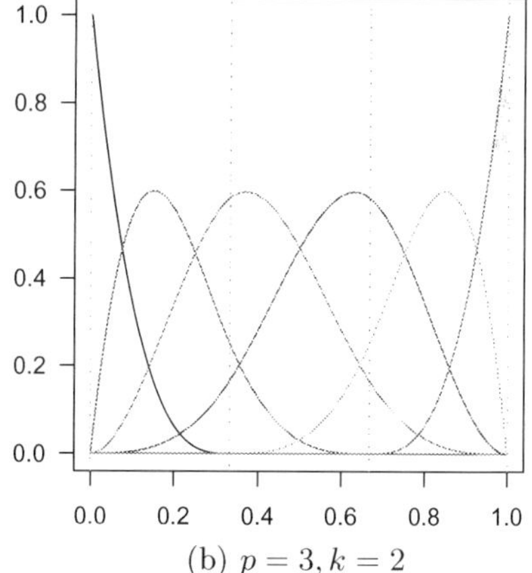

(b) $p = 3, k = 2$

Fig. 1. Exemplary B-spline basis functions $N^{\mathbf{t}}_{j-p,p}(x)$ as a function of x, $j = 1, \ldots\ldots, p + k + 1$; $\mathbf{t}$ is a vector of equidistant knots, k is the number of the internal knots, while p is the polynomial degree (Color figure online).

3 Proposed Idempotization Method

The proposed idempotization and weighted arithmetic mean fitting task seeks an OMA operator-like [14] function:

$$\mathsf{F}_{\varphi,\mathbf{w}}(\mathbf{x}) = \sum_{i=1}^{n} w_i \varphi_i(x_i),$$

that minimizes the total squared differences between $\mathbf{Y}$ and $\mathsf{F}_{\varphi,\mathbf{w}}(\mathbf{X})$. Here, $\mathbf{w}$ stands for a weighting vector and $\varphi_1, \ldots, \varphi_n : [0,1] \to [0,1]$ are some strictly increasing continuous bijections.

Of course, as there are uncountably many functions that can be used in the model we are looking for, we should restrict the feature space in order to make the task solvable on a computer. In our case, for a fixed p and a knot vector $\mathbf{t}$ of length k, we assume that φ_i – used to normalize the i-th variable, $i = 1, \ldots, n$ – is a nonperiodic B-spline:

$$\varphi_i(x) = \sum_{j=1}^{\eta} c_i N^{\mathbf{t}}_{j-p,p}(x)$$

for some vector of $\eta = p + k + 1$ control points $\mathbf{c}$ ordered increasingly. Note that the condition $0 = c_1 < c_2 < \cdots < c_{\eta-1} < c_\eta = 1$ guarantees that φ_i is strictly increasing, continuous, and onto $[0,1]$.

Therefore, the feasible set consists of functions:

$$\mathsf{F}_{\mathbf{c},\mathbf{w}}(\mathbf{x}) = \sum_{i=1}^{n} w_i \tilde{x}_i = \sum_{i=1}^{n} w_i \sum_{j=1}^{\eta} c_j^{(i)} N^{\mathbf{t}}_{j-p,p}(x_i),$$

where $w_1, \ldots, w_n \geq 0$, $\sum_{i=1}^{n} w_i = 1$, $c_1^{(i)} = 0$, $c_\eta^{(i)} = 1$, $c_2^{(i)} - c_1^{(i)} > 0, \ldots,$ $c_\eta^{(i)} - c_{\eta-1}^{(i)} > 0$ for all $i = 1, \ldots, n$. Please observe that $\mathsf{F}_{\mathbf{c},\mathbf{w}}$ is an idempotent and nondecreasing in each variable function of each $\tilde{\mathbf{x}}^{(j)} = (\varphi_1(x_1^{(j)}), \ldots, \varphi_n(x_n^{(j)}))$.

Also, as in the first part of our contribution [1], we would like to prevent overfitting to the training data set, so we should consider some form of the model regularization.

To sum up, for some fixed Tiknohov regularization coefficient $\lambda_w \in \mathbb{R}$, in this paper we are interested in the following optimization task:

$$\text{minimize} \sum_{l=1}^{m} \left(\left(\sum_{i=1}^{n} w_i \sum_{j=1}^{\eta} c_j^{(i)} N^{\mathbf{t}}_{j-p,p}\left(x_i^{(l)}\right) \right) - y^{(l)} \right)^2 + \lambda_w \sum_{i=1}^{n} w_i^2 \quad \text{w.r.t. } \mathbf{w}, \mathbf{c},$$

under the above-mentioned constraints.

As far as computer implementation is concerned, we can rewrite the above equation in terms of a bi-level minimization procedure. The inner-level part, for a fixed $\mathbf{w}$, optimizes for $\mathbf{c}$ and in fact can be written in the form of a standard quadratic programming task (with linear constraints, note that we may precompute the values of B-spline basis functions for each element in $\mathbf{X}$ and store them for further reference). The outer-level component, optimizing for $\mathbf{w}$, can be solved via some non-linear solver – in our case we propose to rely on the CMA-ES [12] algorithm and logarithmic barrier functions that enable us to ensure that the constraints on $\mathbf{w}$ are met.

4 Experimental Results

In this section we apply the proposed method on the data set from Example 1, that is, four different similarity measures for each (unordered) pair of R functions in the benchmark data set discussed in [2,3]. The number of unique observations equals to $m = 30628$. The benchmark data set is of the following form:

j	1	2	3	4	5	6	7	8	...
$x_1^{(j)}$	0.82	0.58	0.15	0.37	0.17	0.22	0.69	0.87	...
$x_2^{(j)}$	0.73	0.41	0.25	0.26	0.02	0.13	0.90	0.70	...
$x_3^{(j)}$	0.63	0.84	0.38	0.40	0.11	0.46	0.72	0.83	...
$x_4^{(j)}$	0.92	0.75	0.48	0.39	0.12	0.28	0.80	0.92	...
$y^{(j)}$	1.00	0.75	0.50	0.25	0.00	0.25	0.75	1.00	...

The meaning of the four variables has been explained in Example 1. The output variable, y, is a value in the set $\{0.0, 0.25, 0.5, 0.75, 1.0\}$ and reflects an expert's assessment of a similarity degree originally provided on a linguistic scale, one of "totally dissimilar", "dissimilar", "hard to say", "similar", or "very similar". We can conceive the y variable as a kind of censored data – it would of course be impossible for an expert to provide a precise similarity degree assessment in a form of a real number in the $[0, 1]$ interval. At a design stage, an (ordered) linguistic scale seemed a much more user-friendly choice.

We may observe that, as far as raw data are concerned, there is no weighted arithmetic mean which returns the value of 1.00 for input values like $(0.82, 0.73, 0.63, 0.92)$ or $(0.87, 0.70, 0.83, 0.92)$.

We split the data set into two parts: 80 % randomly chosen observations are used for training, while the remaining 20 % is left for testing purposes. Let us verify the usefulness of the proposed method in the two following scenarios:

- we treat the fitted function as a regression model which describes the relationship between the explanatory variables and the dependent variable,
- we partition the range of the predicted y into intervals and label the values in y according to which interval they fall; as a result, we obtain a binary classifier.

For simplicity, we assume that the B-splines' knots are equidistant.

In the first scenario, we are simply interested in the sum of squared differences (denoted with $\mathfrak{d}_2^2$) between the predicted and desired ys. For the sake of comparison, we chose a classical linear regression model (to recall, there are no constraints on the form of coefficients in its case).

In the second case, we marked the y values greater or equal to 0.5 as cases of plagiarism (class 1) and the other ones as non-suspicious ones (class 2). In order to be able to use the considered regression models in such a task (i.e., the proposed method as well as linear regression), after finding the optimal fit we also seek for the parameter α that splits the predicted y range into two subintervals in such a way that the two naturally obtained classes maximize the F-measure on the training data set. To recall, the *F-measure* is the harmonic mean of precision and recall. These two classifiers are compared with logistic regression and the random forest algorithm.

The sum-of-squares error and the F-measure are negatively correlated, although the correspondence between them is not a monotone function. There are cases, where increasing $\mathfrak{d}_2^2$ leads to an increase in the F-measure and oppositely.

Firstly, let us study the influence of the B-splines degrees and the number of internal knots used on the model performance. We examined the polynomials with corresponding parameters ranging from $p = 1$ and $k = 1$ up to $p = 5$ and $k = 5$ (25 cases in total). Surprisingly, it turns out that the impact is relatively small. What is more, we observe that higher degree polynomials may also lead to a decreased model performance. The difference between the minimal and the maximal F-measure value is equal to ca. 0.02. For the $\mathfrak{d}_2^2$ error, the model of the lowest quality has been obtained for $p = 1$, $k = 1$. As can be seen in Fig. 2, fitted polynomials of higher degrees of course have shapes similar to those of lower degrees.

Moreover, let us consider the effect of using the λ_w regularization coefficient. The F-measure and the $\mathfrak{d}_2^2$ error as a function of λ_w is depicted in Fig. 3

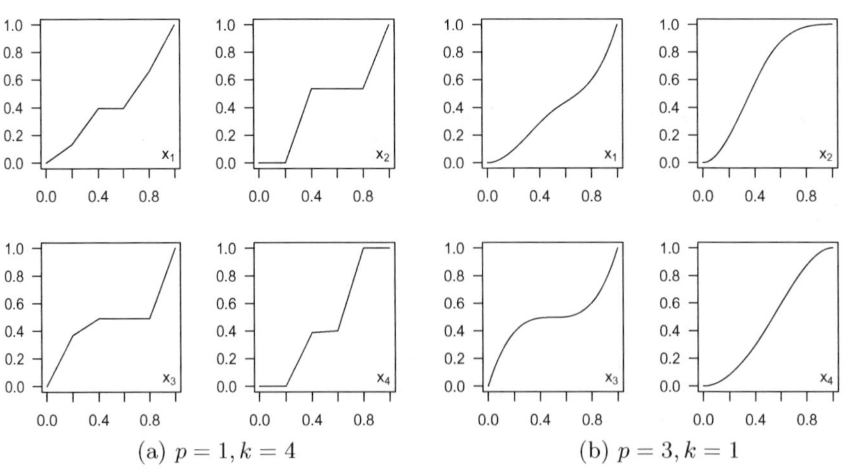

(a) $p = 1, k = 4$ (b) $p = 3, k = 1$

Fig. 2. Best B-splines of different degrees fit to the training sample.

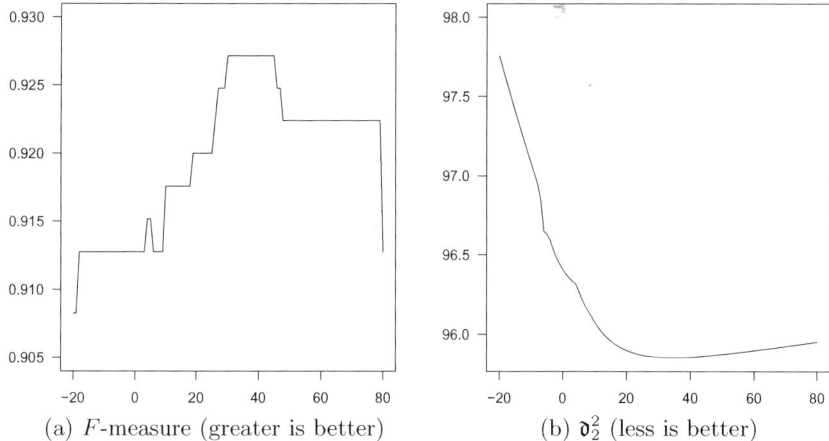

(a) F-measure (greater is better)

(b) $\mathfrak{d}_2^2$ (less is better)

Fig. 3. F-measure and squared error as a function of the λ_w regularization coefficient.

($p = 3$, $k = 1$ was used for F-measure and $p = 1$, $k = 4$ for $\mathfrak{d}_2^2$ error). The highest value of the F-measure was obtained for $p = 3$, $k = 1$, $\lambda_w = 33$, while the smallest $\mathfrak{d}_2^2$ error – for $p = 1$, $k = 4$, $\lambda_w = 30$.

Table 1 summarizes the performance measures of the four considered algorithms. The proposed method gives a higher F-measure than linear and logistic regression as well as a lower $\mathfrak{d}_2^2$ error than linear regression. Even though we get a lower F-measure than in the random forest case, please note that our method comes with important algebraic properties of the resulting aggregation function (such as idempotence and nondecreasingness in each variable), as well as a nice model interpretability. The random forest algorithm does not posses such advantages.

Finally, let us study the impact of introducing idempotization to a weighted arithmetic mean-based model, by comparing the performance of the model on

Table 1. Performance of the fitted models (accuracy, precision, recall, F-measures, squared L_2 error). The proposed method is based on $\lambda_w = 33$, $w_1 = 0.35$, $w_2 = 0.15$, $w_3 = 0.15$, $w_4 = 0.35$, $p = 3$, $k = 1$ for optimizing F-measure (a) and $\lambda_w = 30$, $w_1 = 0.30$, $w_2 = 0.16$, $w_3 = 0.15$, $w_4 = 0.39$, $p = 1$, $k = 4$ for optimizing $\mathfrak{d}_2^2$ (b).

Method	Accuracy	Precision	Recall	F	$\mathfrak{d}_2^2$
Proposed method (a)	0.997	0.921	0.933	0.927	106.62
Proposed method (b)	0.997	0.900	0.920	0.910	95.85
Linear regression	0.995	0.810	0.969	0.883	103.53
Logistic regression	0.997	0.885	0.960	0.921	—
Random forest	0.998	0.927	0.956	0.941	—

Table 2. Performance measures as functions of different weighting vectors; $p = 3$, $k = 1$, $\lambda_w = 0$, with and without idempotization.

w_1	w_2	w_3	w_4	Accuracy	Precision	Recall	F	$\mathfrak{d}_2^2$	Idempot.
1	0	0	0	0.992	0.848	0.693	0.763	186.30	Yes
0	1	0	0	0.995	0.927	0.787	0.851	208.74	Yes
0	0	1	0	0.994	0.803	0.853	0.828	316.04	Yes
0	0	0	1	0.996	0.904	0.840	0.871	136.67	Yes
0.27	0.06	0.38	0.29	0.996	0.952	0.800	0.870	137.34	No
0.41	0.12	0.07	0.40	0.997	0.919	0.907	0.913	107.69	Yes

the raw data set and on the version transformed with optimal B-splines. Table 2 summarizes the model quality measures for a fixed λ_w equal to 0. We observe that relying on a single feature does not lead to particularly good performance. The same happens if the weighting vector is optimized for but the idempotization scheme is not applied at all. Therefore, there is a positive impact of both factors. Similar observations can be done for other λ_w coefficients.

5 Conclusion

We have discussed a supervised learning method for weights of weighted arithmetic means in cases where data come from an ordinal scale and they have to be properly mapped to the $[0, 1]$ interval prior to computing an optimal fit.

The introduced method has been applied on a real-world data set consisting of data on similarity degrees of pairs of R functions. It has many advantages:

- determining the φ_i functions enables us to normalize the input values so that they become mutually comparable and easily interpretable to the plagiarism detection system's users; other machine learning methods can be applied on the transformed sample too;
- the fitted weighted arithmetic mean serves as a regression model for our data and explains the relationship between the individual similarity degrees and the aggregated similarity assessment; as the weights are by definition nonnegative and they sum up to one, we have a clear intuition of which of the four methods has the highest impact;
- the fitted model fulfills two important properties: it is idempotent and nondecreasing in each variable; thus, its behavior is much more natural and understandable to end-users;
- the obtained regression model can be easily transformed in such a way that it is suitable for using in, e.g., binary classification tasks.

Let us note that the introduced method can be easily extended to the case of fitting arbitrary weighted quasi-arithmetic means, with or without known generator functions.

For future work, inspecting different regularization schemes could lead to models of increased quality. In particular, one may think of introducing a regularization component for the vector of control points, e.g., for some $\lambda_c \in \mathbb{R}$, of the form $\lambda_c \sum_{i=1}^{n} \sum_{j=2}^{\eta} \left(c_j^{(i)} - c_{j-1}^{(i)}\right)^2$.

Acknowledgments. This study was supported by the National Science Center, Poland, research project 2014/13/D/HS4/01700.

References

1. Bartoszuk, M., Beliakov, G., Gagolewski, M., James, S.: Fitting aggregation functions to data: part I - linearization and regularization. In: Carvalho, J.P., Lesot, M.-J., Kaymak, U., Vieira, S., Bouchon-Meunier, B., Yager, R.R. (eds.) IPMU 2016, Part II. CCIS, vol. 611, pp. 767–779. Springer, Heidelberg (2016)
2. Bartoszuk, M., Gagolewski, M.: A fuzzy R code similarity detection algorithm. In: Laurent, A., Strauss, O., Bouchon-Meunier, B., Yager, R.R. (eds.) IPMU 2014, Part III. CCIS, vol. 444, pp. 21–30. Springer, Heidelberg (2014)
3. Bartoszuk, M., Gagolewski, M.: Detecting similarity of R functions via a fusion of multiple heuristic methods. In: Alonso, J., Bustince, H., Reformat, M. (eds.) Proceedings of IFSA/Eusflat 2015, pp. 419–426. Atlantic Press (2015)
4. Beliakov, G.: Monotone approximation of aggregation operators using least squares splines. Int. J. Uncertainty Fuzziness Knowl. Based Syst. **10**, 659–676 (2002)
5. Beliakov, G.: How to build aggregation operators from data. Int. J. Intell. Syst. **18**, 903–923 (2003)
6. Beliakov, G.: Learning weights in the generalized OWA operators. Fuzzy Optim. Decis. Making **4**, 119–130 (2005)
7. Beliakov, G.: Construction of aggregation functions from data using linear programming. Fuzzy Sets Syst. **160**, 65–75 (2009)
8. Beliakov, G., Bustince, H., Calvo, T.: A Practical Guide to Averaging Functions. STUDFUZZ, vol. 329. Springer, Heidelberg (2016)
9. Beliakov, G., Pradera, A., Calvo, T.: Aggregation Functions: A Guide for Practitioners. STUDFUZZ, vol. 221. Springer, Heidelberg (2007)
10. Beliakov, G., Warren, J.: Appropriate choice of aggregation operators in fuzzy decision support systems. IEEE Trans. Fuzzy Syst. **9**(6), 773–784 (2001)
11. Gagolewski, M.: Data Fusion: Theory, Methods, and Applications. Institute of Computer Science, Polish Academy of Sciences, Warsaw, Poland (2015)
12. Hansen, N.: The CMA evolution strategy: a comparing review. In: Lozano, J., Larranga, P., Inza, I., Bengoetxea, E. (eds.) Towards a New Evolutionary Computation. STUDFUZZ, vol. 192, pp. 75–102. Springer, Heidelberg (2006)
13. Hastie, T., Tibshirani, R., Friedman, J.: The Elements of Statistical Learning: Data Mining, Inference, and Prediction. Springer, New York (2013)
14. Mesiar, R., Mesiarová-Zemánková, A.: The ordered modular averages. IEEE Trans. Fuzzy Syst. **19**(1), 42–50 (2011)
15. R Development Core Team: R: A language and environment for statistical computing. R Foundation for Statistical Computing, Vienna, Austria (2016). http://www.R-project.org
16. Schumaker, L.: Spline Functions: Basic Theory. Cambridge University Press, Cambridge (2007)

Mean Estimation Based on FWA Using Ranked Set Sampling with Single and Multiple Rankers

Bekir Cetintav[1,2], Gozde Ulutagay[1,2], Selma Gurler[1,2(✉)], and Neslihan Demirel[1,2]

[1] Department of Statistics, Dokuz Eylul University, 35170 Izmir, Turkey
{bekir.cetintav,selma.erdogan,neslihan.ortabas}@deu.edu.tr
[2] Department of Industrial Engineering, Izmir Univerity, 35170 Izmir, Turkey
gozde.ulutagay@izmir.edu.tr

Abstract. The Ranked Set Sampling (RSS) is an advanced sampling method which improves the precision and accuracy of the mean estimator. In RSS, the units in the random sets which are drawn from a population are ranked by a ranking mechanism, and one of these ranked units is sampled from each set with a specific scheme. Ranking the units (visually or by a concomitant variable) could not be perfect because there is an uncertainty in decision making about the rank of a unit. In this study, we propose a fuzzy set perspective for RSS and an estimator for the population mean based on Fuzzy Weighted Average (FWA) operator. A real data application is given to illustrate the new approach for the single and multiple rankers.

Keywords: Ranked set sampling · Uncertainty · Fuzzy sets · Fuzzy weighted average · Multiple rankers

1 Introduction

Ranked set sampling (RSS) is a useful and alternative sampling method where the knowledge about the ranks (orders) of the units is used. Because of the additional information on the ranks of the units, RSS is more representative than the simple random sampling (SRS) counterpart with equal sample size. Takahasi and Wakimoto [1], and Dell and Clutter [2] construct the statistical background of this method, introduced earlier by McIntyre [3], and show that the efficiency of the mean estimator based on RSS is greater than or equal to the efficiency of the mean estimator based on SRS. For detailed information about RSS, see the book of Chen et al. [4] and the review of Wolfe [5]. In simple terms, RSS procedure consists of three parts. Random sets are drawn from a specific population,

B. Cetintav—This study is supported by the Scientific and Technological Research Council of Turkey (TUBITAK-COST Grant No. 115F300) under ISCH COST Action IS1304.

J.P. Carvalho et al. (Eds.): IPMU 2016, Part II, CCIS 611, pp. 790–797, 2016.
DOI: 10.1007/978-3-319-40581-0_64

the units in the sets are ranked by a mechanism, and one of these ranked units is sampled from each set with a specific scheme. In the ranking mechanism, the ranker could be an expert-researcher or a highly-correlated concomitant variable and makes the decisions about the rank of the units. In practice, the decisions of ranks should be made without actual measurement of the concerning variable. Thus, these decisions could not be always perfect even if ranking is done by a powerful criterion. This unavoidable uncertainty in the ranking mechanism is mentioned as imprecise/imperfect ranking in the literature. Several studies, Bohn and Wolfe [6], MacEachern et al. [7], Frey [8], Oztrk [9–11] focused on the modeling uncertainty in a probabilistic way. According to us, fuzzy sets could be an alternative way to dealing with the uncertainty occurs in the ranking mechanism. For example, the uncertain knowledge about the experience/correlation level of the ranker could be described with linguistic terms such as fairly experienced or poor, easily. Then the uncertainty could be included in the inference process by constructing the fuzzy sets of accuracy levels. In this study, we propose a new perspective for RSS and a Fuzzy Weighted Average (FWA) operator for the estimation of the population mean. The basic properties of RSS and our motivation are given in Sect. 2. General information about FWA and our FWA operator are defined in Sect. 3. We also give a real data example for the new method in Sect. 4. Finally, the conclusions are given in Sect. 5.

2 The Motivation and Fundamentals of RSS

Before starting the RSS procedure, the size of the sets (m) and sample size (N) should be decided. After that there are six-steps to obtain the sample.

1. Select m units at random from a specified population.
2. Rank these m units with some expert judgment without measuring them.
3. Retain the smallest judged unit and return the others.
4. Select the second m units, rank them, retain the second smallest judged unit and return the others. Continue to the process until m ordered units are measured.
5. First five steps are repeated for n times to get n cycle and mn observations.

These steps are illustrated by Fig. 1 for $m = 3$. These mn observations are called a standard ranked set sample. $X_{((h)j)}$ means hth ordered unit in a specific set in the jth cycle for $h = 1, 2, \ldots, m$ and $j = 1, 2, \ldots, n$. In the probabilistic perspective, the hth ordered units $X_{((h))}$ in the random sets are the representatives of hth order statistics. Suppose that $X_{((h)1)}, X_{((h)2)}, \ldots, X_{((h)n)}$ are the hth ordered observations of n random sets from the same population. Then the expected values and variances of these observations are given as follows:

$$E(X_{(h)1}) = E(X_{(h)2}) = \ldots = E(X_{(h)n}) = E(X_{(h)}. \tag{1}$$

$$Var(X_{(h)1}) = Var(X_{(h)2}) = \ldots = Var(X_{(h)n}) = Var(X_{(h)}. \tag{2}$$

Cycle 1	Cycle 2		Cycle n
$X_{(1)1} \leq X_{(2)1} \leq X_{(3)1} \rightarrow X_{(1)1}$	$X_{(1)2} \leq X_{(2)2} \leq X_{(3)2} \rightarrow X_{(1)2}$		$X_{(1)n} \leq X_{(2)n} \leq X_{(3)n} \rightarrow X_{(1)n}$
$X_{(1)1} \leq X_{(2)1} \leq X_{(3)1} \rightarrow X_{(2)1}$	$X_{(1)2} \leq X_{(2)2} \leq X_{(3)2} \rightarrow X_{(2)2}$	...	$X_{(1)n} \leq X_{(2)n} \leq X_{(3)n} \rightarrow X_{(2)n}$
$X_{(1)1} \leq X_{(2)1} \leq X_{(3)1} \rightarrow X_{(3)1}$	$X_{(1)2} \leq X_{(2)2} \leq X_{(3)2} \rightarrow X_{(3)2}$		$X_{(1)n} \leq X_{(2)n} \leq X_{(3)n} \rightarrow X_{(3)n}$

Fig. 1. RSS Procedure for n cycle with set size is 3

Suppose that X_h consists of $X_{((h)j)}$, and denotes the set of hth ranked units for cycle size $j = 1, 2, ..., n$. Since these ordered units in the random sets are drawn without actual measurement, it is expected that there is an uncertainty in ranking of the units. In practice, we can define a fuzzy set by using observations as the uncertain representatives of hth ranked unit obtained in RSS. If $\widetilde{X}_h$ represents the fuzzy set of hth ranked unit with membership function $\mu_{(\widetilde{X}_h)}(X_{(h)j})$, then it can be defined as follows for $h = 1, 2, ..., m$, and $j = 1, 2, ..., n$.

$$\widetilde{X}_h = \{X_{(h)j}, \mu_{(\widetilde{X}_h)}(X_{(h)j}) \mid X_{(h)j} \in X_h\}. \tag{3}$$

On the other side, the ranking criteria, which might be the personal judgment of an expert or the knowledge from the concomitant variable, also has an uncertainty in nature. In practice, the knowledge about the accuracy level of the ranker is uncertain and hardly measurable. However, it can be easily described with linguistic terms and these linguistic terms can be defined as fuzzy sets with a reasonable way. Suppose that, we have a ranker with accuracy level on the ranking the units in one of the five level; poor, fairly poor, fair, fairly good and good. Let W_r represent the crisp universal set of the ranking accuracy level of a human expert r or concomitant variable r, for $r = 1, 2, ..., K$. W_r contains $w_{(r,v)}$ values, for the sake of the simplicity $v = 1, 2, 3$, which represents the minimum, most possible and maximum values of accuracy level of the ranker (see Fig. 2). An accuracy value $w_{(r,v)}$ is between 0 and 1. Let $\widetilde{W}_r$ represent the fuzzy set for accuracy level of rth ranker with membership function $\mu_{(\widetilde{W}_r)}(w_{(r,v)})$. Then we can define the fuzzy sets of rankers experience level as follows:

$$\widetilde{W}_h = \{w_{r,v}, \mu_{(\widetilde{W}_h)}(w_{r,v}) \mid w_{r,v} \in W_h\}. \tag{4}$$

for $r = 1, 2, \ldots, K$ where K is the number of ranker. For the case of single ranker, there is only one fuzzy set of the rankers accuracy level as:

$$\widetilde{W} = \{w_v, \mu_{(\widetilde{W})}(w_v) \mid w_v \in W\}. \tag{5}$$

Both $\widetilde{X}_h$ and $\widetilde{W}_r$ are convex fuzzy sets naturally arising in the ranking part of RSS where X_h and W_r are the crisp universal sets.

3 FWA Operator Based on RSS

With a basic definition, FWA is a fuzzy extension of classical weighted average which is commonly used in estimation and decision making. Especially in

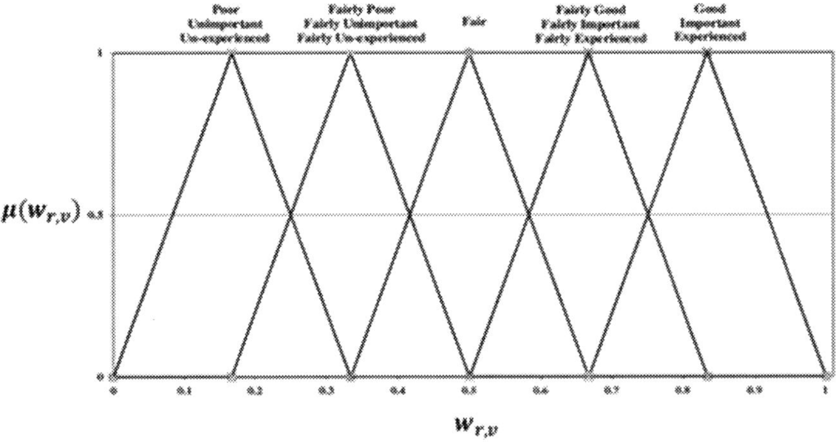

Fig. 2. Fuzzy sets of rankers accuracy levels (Alhumaidi, 2015)

multiple criteria decision making, the nature of the criteria are usually uncertain and difficult to determine numerically. Let $\widetilde{X}_i$ be the fuzzy set of decision i and $\widetilde{W}_i$ be the fuzzy set of the weight of decision i, then the fuzzy weighted average $\widetilde{Y}$, is denoted as follows (see, Kao and Liu [16]).

$$\widetilde{Y} = \frac{\sum_{i=1}^{N} \widetilde{W}_i \widetilde{X}_i}{\sum_{i=1}^{N} \widetilde{W}_i} \tag{6}$$

where $\widetilde{W}_i = \{w_i, \mu_{(\widetilde{W}i)}(w_i) \mid w_i \in W_i\}$ and $\widetilde{X}_i = \{X_i, \mu_{(\widetilde{X}i)}(X_i) \mid X_i \in X_i\}$ for $i = 1, 2, \ldots, N$.

Generally, the membership function of the FWA formulated as a nonlinear programming problem in literature. Nonlinearity of the model and non-differentiable nature of the memberships makes the problem hard to solve. Dong and Wong [12] introduce an α-cut approach to find an exact solution to fuzzy weighted averages. It gives discrete solutions for each-cut level within 2^{2n} permutations for n criteria. Liou and Wang [13], Guh et al. [14] and Lee and Park [15] propose new methods to reduce the complexity and increase the efficiency in the solution of the problem. For different aspects see also Borek and Noppen [17], Mokhtarian [18]. In this study, we use Kao and Liu [16] fractional programming approach to construct the membership function for FWA based on the α-cut levels of fuzzy sets and the extension principle because of its easiness to apply. In our motivation, each ranker has specific criteria on ranking the units. For that reason, there are different fuzzy sets of hth ranked units, $\widetilde{X}_{(h,r)}$, for each ranker r. Similar to FWA given by Kao and Liu (2001), we can define a formula for FWA of each hth ranked units as follows:

$$FWA_{RSS}^{(h)} = \frac{\sum_{r=1}^{K} \widetilde{W}_r \widetilde{X}_{h,r}}{\sum_{r=1}^{K} \widetilde{W}_r}. \tag{7}$$

where K and m are the number of rankers and the set size, respectively, for $r = 1, 2, \ldots, K$ and $h = 1, 2, \ldots, m$. Then FWA estimation of the population mean based on RSS is given as follows:

$$FWA_{RSS} = \frac{1}{m} \sum_{h=1}^{m} FWA_{RSS}^{(h)}. \tag{8}$$

Since $\widetilde{X}_h$ and $\widetilde{W}_r$ are convex fuzzy sets, also FWA_{RSS} will be a fuzzy set. If the researcher wants a crisp value for the estimation of the population mean, some well known defuzzification methods, such as center of area, bisector of area, should be used (see for detail, Näaz et al. [20]).

4 Real Data Application

In this section, we illustrate the new FWA operator with real data application for two cases, single and multiple rankers. We use a real data set from a biometrical study from Nish et al. [21] which was constructed to predict the age of abalone from physical measurements. Generally, the age of abalone is determined by cutting the shell through the cone, staining it, and counting the number of rings through a microscope. However it is a time-consuming task. In the original study, the researchers suggest to use other physical measurements, which are easier to obtain, to predict the age of abalones. The data set contains 4177 units with the variables, Sex, Length (mm), Diameter (mm), Height (mm), Whole weight (grams), Shucked weight (grams), Viscera weight (grams), Shell weight (grams), Rings (integer +1.5 gives the age in years).

4.1 Single Ranker Case

Let our objective be to obtain the mean estimation of the viscera weight, X, which has also more time-consuming measurement process comparing with the other physical measurements. Diameter is chosen as the concomitant variable Y for ranking the abalones in the random sets without actual measurement of their X values. Suppose that, we randomly select 3 units for each set and repeat the cycle 4 times. The ranked set sample is given as Table 1.

Table 1. Ranked set sample of viscera weight of abalones based on diameter as the concomitant variable Y

	Cycle			
	j = 1	j = 2	j = 3	j = 4
h = 1	0.1350	0.1020	0.0455	0.0585
h = 2	0.1135	0.2615	0.1635	0.3110
h = 3	0.0730	0.4015	0.2860	0.5145

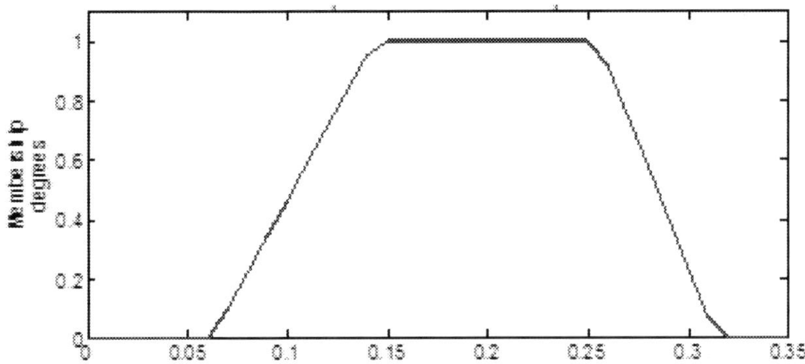

Fig. 3. RSS Procedure for n cycle with set size is 3

It is clear that the FWA formula defined in Eq. (8) degenerates and transforms the arithmetic mean of h fuzzy numbers. For $r = 1$, then

$$FWA_{RSS} = \frac{1}{m} \sum_{h=1}^{m} \widetilde{X}_h. \tag{9}$$

By using Eq. (9), trapezoidal fuzzy number of the FWA_{RSS} is determined as [0.0773, 0.1693, 0.255, 0.3201] (see, Fig. 3). We can defuzzify FWA_{RSS} to obtain a crisp estimation for population mean by using the methods center of area and bisector of area. The results are 0.2042 and 0.2100, respectively.

4.2 Multiple Rankers Case

We can extend the design given above under the same conditions, $setsize = 3$ and $cyclesize = 4$ for multiple rankers. Viscera weight is chosen as the concerning variable X again. Y_1,Y_2 and Y_3 are chosen as the concomitant variables diameter, height and whole weight, respectively. It is expected that whole weight variable is a good ranker and diameter and height variables are fairly good rankers (for the fuzzy set representation, see Fig. 2).

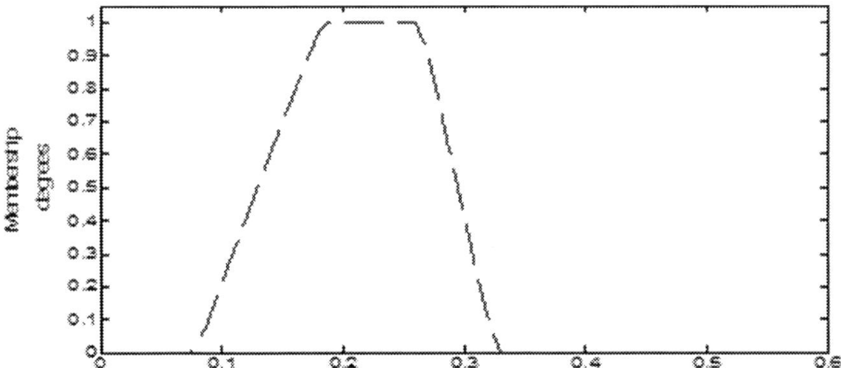

Fig. 4. RSS Procedure for n cycle with set size is 3

Table 2. Ranked set samples of viscera weight of abalones based on *diameter*, *height* and *wholeweight* as the concomitant variables Y_1,Y_2 and Y_3

		Cycle			
		j = 1	j = 2	j = 3	j = 4
h = 1	Y_1	0.1350	0.1020	0.0455	0.0585
	Y_2	0.1350	0.1020	0.0755	0.0585
	Y_3	0.0880	0.1020	0.0455	0.0585
h = 2	Y_1	0.1135	0.2615	0.1635	0.3110
	Y_2	0.1135	0.3460	0.1990	0.3110
	Y_3	0.1135	0.3460	0.2135	0.3110
h = 3	Y_1	0.0730	0.4015	0.2860	0.5145
	Y_2	0.0730	0.4015	0.2860	0.5145
	Y_3	0.0730	0.4015	0.2860	0.5145

There are slight differences among the ranked set samples based on each concomitant variable. This means different rankers (in this study concomitant variables) can make different rank decisions for same units. We obtain $FWA_{RSS}^{(h)}$ trapezoidal fuzzy numbers for $h = 1, 2, 3$ by using Kao and Lius (2001) fractional programming approach and they are $[0.0491, 0.0638, 0.097, 0.1244]$ and $[0.1134, 0.1929, 0.2952, 0.3361]$ and $[0.073, 0.286, 0.4015, 0.5145]$, respectively. Finally, we calculate the FWA average $FWA_{RSS} = [0.0785, 0.1809, 0.2646, 0.325]$ and crisp estimation for population mean by using the methods center of area and bisector of area as 0.2106 and 0.2100, respectively (see, Fig. 4). The standard ranked set sample means based on the concomitant variables Y_1,Y_2 and Y_3 are 0.2055, 0.2180 and 0.2128, respectively. When we compare mean estimation results of our new approach with the standard ranked set samples results given above, they seem acceptable for both cases, single and multiple rankers (Table 2).

5 Conclusions

In this study, we introduce a new approach for modeling uncertainty in ranked set sampling. Also, a Fuzzy Weighted Average (FWA) operator is proposed for the estimation of population mean. After some theoretical background and motivation, real data application is given for single ranker and multiple rankers. In further work, a well-rounded simulation study will be designed to reach decisive conclusions.

Acknowledgments. This study is supported by the Scientific and Technological Research Council of Turkey (TUBITAK-COST Grant No. 115F300) under ISCH COST Action IS1304.

References

1. Takahasi, K., Wakimoto, K.: On unbiased estimates of the population mean based on the sample stratified by means of ordering. Ann. Inst. Stat. Math. **20**, 131 (1968)

2. Dell, T.R., Clutter, J.L.: Ranked set sampling theory with order statistics background. Biometrika **28**, 545–555 (1972)
3. McIntyre, G.A.: A method of unbiased selective sampling using ranked sets. Aust. J. Agric. Res. **3**, 385–390 (1952)
4. Chen, Z., Bai, Z., Sinha, B.K.: Ranked Set Sampling: Theory and Application. Springer, New York (2004)
5. Wolfe, D.A.: Ranked Set Sampling: Its Relevance and Impact on Statistical Inference. ISRN Probability and Statistics (2012)
6. Bohn, L.L., Wolfe, D.A.: The effect of imperfect judgment rankings on properties of procedures based on the ranked-set samples analog of the Mann-Whitney-Wilcoxon statistic. J. Am. Stat. Assoc. **89**, 168–176 (1994)
7. MacEachern, S.N., Stasny, E.A., Wolfe, D.A.: Judgement post-stratification with imprecise rankings. Biometrics **60**, 207–215 (2004)
8. Frey, J.: New imperfect ranking models for ranked set sampling. J. Stat. Plan Infer. **137**, 1433–1445 (2007)
9. Ozturk, O.: Statistical inference in presence of ranking error in ranked set sampling. Can. J. Stat. **35**, 577–594 (2008)
10. Ozturk, O.: Sampling from partially rank-ordered sets. Environ. Ecol. Stat. **18**, 757–779 (2011)
11. Ozturk, O.: Combining multi-ranker information in judgment post stratified and ranked set samples when sets are partially ordered. Can. J. Stat. **41**, 304–324 (2013)
12. Dong, W.M., Wong, F.S.: Fuzzy weighted averages and implementation of the extension principle. Fuzzy Sets Syst. **21**(2), 183–199 (1987)
13. Liou, T.S., Wang, M.J.: Fuzzy weighted average: an improved algorithm. Fuzzy Sets Syst. **49**(3), 307–315 (1992)
14. Guh, Y.Y., Hong, C.C., Wang, K.M., Lee, E.S.: Fuzzy weighted average: a max-min paired elimination method. Comput. Math. Appl. **32**, 115–123 (1996)
15. Lee, D.H., Park, D.: An efficient algorithm for fuzzy weighted average. Fuzzy Sets Syst. **87**, 39–45 (1997)
16. Kao, C., Liu, S.T.: Fractional programming approach to fuzzy weighted average. Fuzzy Sets Syst. **120**, 435–444 (2001)
17. Van den Broek, P., Noppen, J.: Fuzzy Weighted Average: Alternative approach Fuzzy Information Processing Society, NAFIPS, Annual meeting of the North American, pp. 126–130. IEEE (2006)
18. Mokhtarian, M.N.: A new fuzzy weighted average (FWA) method based on left and right scores: an application for determining a suitable location for a gas oil station. Comput. Math. Appl. **61**, 3136–3145 (2011)
19. Alhumaidi, H.M.: Construction contractors ranking method using multiple decision-makers and multi-attribute fuzzy weighted average. J. Constr. Eng. Manage. **141**(4) (2015)
20. Naaz, S., Alam, A., Biswas, R.: Effect of different defuzzification methods in a fuzzy based load balancing application. IJCSI Int. J. Comput. Sci. **8**, 261–267 (2011)
21. Nash, J.W., Sellers, T.L., Talbot, S.R., Cawthorn, A.J., Ford, W.B.: The Population Biology of Abalone (Haliotis species) in Tasmania. I. Blacklip Abalone (H. rubra) from the North Coast and Islands of Bass Strait, Sea Fisheries Division, Technical report No. 48 (1994). ISSN: 1034–3288

On the Sensitivity of the Weighted Relevance Aggregation Operator and Its Application to Fuzzy Signatures

István Á. Harmati[1(✉)] and László T. Kóczy[2,3]

[1] Department of Mathematics and Computational Sciences, Széchenyi István University, Győr, Hungary
harmati@sze.hu

[2] Department of Automation, Széchenyi István University, Győr, Hungary
koczy@sze.hu

[3] Department of Telecommunications and Media informatics, Budapest University of Technology and Economics, Budapest, Hungary

Abstract. The weighted relevance aggregation operator is a modified, flexible version of the general power mean. In this paper we discuss the sensitivity of this operator, namely we give bounds on the change of the output in terms of vector norms of the change of the input variables. We apply these results to characterize to sensitivity of fuzzy signatures which are equipped with these operators in its nodes.

Keywords: Sensitivity · Aggregation operator · Weighted relevance · Fuzzy signature

1 Introduction

Modeling of complex systems always a difficult task, particularly when human activities, lack of data, irreproducibility are also features of the system. In these problems the exact mathematical model is not known or too difficult.

The fuzzy signature based approach offers a users friendly solution. In this modeling technique the complex systems is described by a set of qualitative measures, which are also arranged into a hierarchical framework expressing interconnections and dependencies, and modeling the human approach to the problem. There is a wide variety of applications, for example in economy, in the medical field [1], and in several fields of engineering and informatics, for example robotics [2], data mining [3] and civil engineering [4,5].

In mathematical point of view, fuzzy signatures are hierarchical representations of data structuring into vectors of fuzzy values [6]. A fuzzy signature is defined as a special multidimensional fuzzy data structure, which is a generalization of vector valued fuzzy sets [7]. Vector valued fuzzy sets are special cases of L-fuzzy sets which were introduced in [8]. A fuzzy signature is defined by

$$A\colon X \to S^{(n)}, \tag{1}$$

J.P. Carvalho et al. (Eds.): IPMU 2016, Part II, CCIS 611, pp. 798–808, 2016.
DOI: 10.1007/978-3-319-40581-0_65

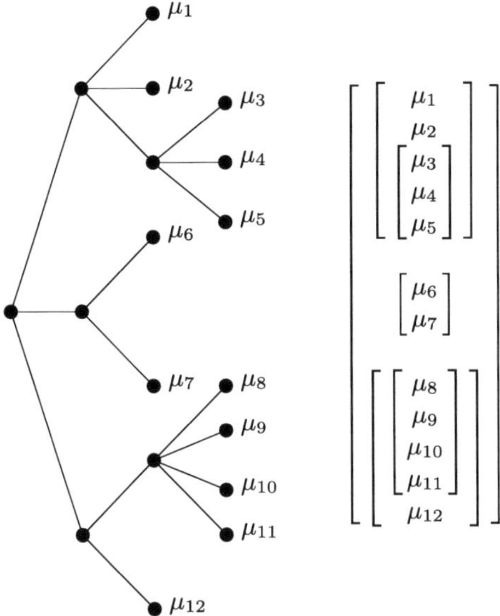

Fig. 1. A fuzzy signature graph and the corresponding nested vectors.

where X is the universe of discourse, $1 \leq n$ and

$$S^{(n)} = \times_{i=1}^{n} S_i \qquad S_i = \begin{cases} [0,1] \\ S^{(m)} \end{cases} \tag{2}$$

A fuzzy signature can be represented by a nested vector value fuzzy sets and by a tree graph also (see Fig. 1), the latter one is more expressive [7].

Values at the leaves or input values (μ-s) usually depend on the opinion of human experts or determined by estimation methods. The final conclusion, the output of the fuzzy signature is computed from the inputs applying suitable aggregation functions, this is the membership value of the whole fuzzy signature. Due to the built-in uncertainty or lack of detailed information of the complex system that we are going to model, different human experts or different kind of estimation methods may give different scores to the same situation. In real applications a fuzzy signature based model should have some robustness, so the output should not change *too* much if the input values change a *little*.

In the following we discuss the issue how the membership value of the whole fuzzy signature changes if the membership values in the nested vectors change. In other words, if we think of the tree graph representation, how the membership value of the root changes if the membership values of leaves change. For answering this question we have to know the structure of the signature tree and the applied aggregation operators. The case when the fuzzy signature has weighted general mean aggregation operators in the nodes was discussed in [9,10] and in [11]. The weighted relevance aggregation operator, which we analyse here, is a kind of modification of the aggregation operator mentioned previously. With less strict conditions on the parameters we get an operator whose parameters we can easily compute (more exactly, we can determine them by learning algorithms), and on the other hand this operator is an especially useful tool. The applied methodology is similar to the way we followed in [9], the main difference is the investigated function.

In the remaining part of the paper in Sect. 2 we review some mathematical definitions and theorems, in Sect. 3 the sensitivity of the weighted relevance aggregation operator is discussed, in Sect. 4 we examine the sensitivity of fuzzy signatures equipped with WRAOs in their nodes.

2 Basic Definitions and Theorems

First we recall the definition of the p-norm (see for example [14])

Definition 1 (p-norm). *Let $p \geq 1$ a real number and $\underline{x} = (x_1, \ldots, x_n) \in \mathbb{R}^n$. Then the p-norm of $\underline{x}$*

$$\|\underline{x}\|_p = \left(\sum_{k=1}^{n} |x_k|^p \right)^{\frac{1}{p}} \tag{3}$$

Some widely used p-norms:

- $p = 1$ (taxicab norm) $\|\underline{x}\|_1 = |x_1| + \ldots + |x_n|$
- $p = 2$ (euclidean norm) $\|\underline{x}\|_2 = \sqrt{x_1^2 + \ldots + x_n^2}$
- $p = \infty$ (maximum norm) $\|\underline{x}\|_\infty = \max(|x_1|, \ldots, |x_n|)$

Two important properties of the p-norm:

- If $1 \leq p \leq q \leq \infty$ then $\|\underline{x}\|_q \leq \|\underline{x}\|_p$.
- If $1 \leq p \leq q \leq \infty$ then $\|\underline{x}\|_p \leq \|\underline{x}\|_q \cdot n^{1/p - 1/q}$.

We will use the generalization of the triangular inequality, the so called Minkowski's inequality.

Theorem 1 (Minkowski's inequality). *(see for example [12] or [13]). Let $\underline{a}$, $\underline{b} \in \mathbb{R}^n$, $p \geq 1$, then the following inequality holds:*

$$\|\underline{a} + \underline{b}\|_p \leq \|\underline{a}\|_p + \|\underline{b}\|_p \tag{4}$$

The generalization of the reverse triangular inequality also holds:

Corollary 1. *If $\underline{a}$, $\underline{b} \in \mathbb{R}^n$, $p \geq 1$, then*

$$\left| \|\underline{a}\|_p - \|\underline{b}\|_p \right| \leq \|\underline{a} - \underline{b}\|_p \tag{5}$$

Theorem 2 (Hölder's inequality). *(see for example [12] or [13]). Let $\underline{a}$, $\underline{b} \in \mathbb{R}^n$, $r, s, t \geq 1$ and $1/r = 1/t + 1/s$. Then the following inequality holds:*

$$\left[\sum_{i=1}^{n} |a_i \cdot b_i|^r \right]^{1/r} \leq \left[\sum_{i=1}^{n} |a_i|^t \right]^{1/t} \cdot \left[\sum_{i=1}^{n} |b_i|^s \right]^{1/s} \tag{6}$$

or in terms of p-norms:

$$\|\underline{a} \circ \underline{b}\|_r \leq \|\underline{a}\|_t \cdot \|\underline{b}\|_s \tag{7}$$

where '∘' denotes the element-wise product (also known as Hadamard- or Schur-product).

The multivariate case of Lagrange's mean value theorem:

Theorem 3. *Let G be an open subset of $\mathbb{R}^n$ and let $f\colon G \subset \mathbb{R}^n \to \mathbb{R}$. If $\underline{x}$, $\underline{y} \in G$ and f is differentiable at each point of the line segment $\underline{xy}$, then there exists on that line segment a point $\underline{\xi}$ between $\underline{x}$ and $\underline{y}$ such that*

$$f(\underline{y}) - f(\underline{x}) = \nabla f(\underline{\xi}) \cdot (\underline{y} - \underline{x})$$

or in other form:

$$f(\underline{y}) - f(\underline{x}) = \sum_{i=1}^{n} \frac{\partial f(\underline{\xi})}{\partial x_i} \cdot (y_i - x_i)$$

Corollary 2. *Applying Lagrange's mean value theorem and Hölder's inequality we get an upper estimation of the change of f, where $1/s + 1/t = 1$:*

$$\left|f(\underline{y}) - f(\underline{x})\right| \leq \|\nabla f(\underline{\xi})\|_t \cdot \|\underline{y} - \underline{x}\|_s \tag{8}$$

3 Sensitivity of the Weighted Relevance Aggregation Operator

The weighted relevance aggregation operator was introduced by B. S. U. Mendis (see [15,17]) as a flexible modification of the weighted general mean, with less constrains on the weights, but very suitable for machine learning [16].

Definition 2 (WRAO). *Let $x_1, \ldots, x_n \in [0, 1]$ and $a_1, \ldots, a_n$ be nonnegative real numbers, $0 < \max\{a_i\} \leq 1$, and $\alpha \in \mathbb{R}$ ($\alpha \neq 0$). Then the weighted relevance aggregation of $x_1, \ldots, x_n$ with weights $a_1, \ldots, a_n$ and with parameter α is the following:*

$$H_\alpha(x_1, \ldots, x_n) = \left[\frac{1}{n}\sum_{k=1}^{n}(a_k \cdot x_k)^\alpha\right]^{\frac{1}{\alpha}} \tag{9}$$

As we mentioned above, the examined weighted relevance aggregation operator is the modification of the weighted power mean (or in general, the weighted quasi arithmetic mean [18]) aggregation operator. The input sensitivity or lipschitzian property of quasi arithmetic mean operators was discussed in details in [19]. Although the WRAO is closely related to the class of quasi arithmetic means, it not belongs to this.

We note that the usual property of the weights (they sum equals 1) is omitted here. This modification yields a more flexible operator, which is in fact not an aggregation operator if the a_i-s are already fixed (except for the case when $a_i = 1$ for all i). The usefulness was demonstrated by several examples, compared to other, more conventional operators. The comparison with OWA was discussed in [20], the comparison with Choquet integral can be found in [17]. Moreover, practical applications also appear in [17] with different values of the exponent (α).

More properties of the WRAO:

The limits at $\pm\infty$ regardless to the weights are the minimum and maximum:

$$\lim_{\alpha\to\infty}\left[\frac{1}{n}\sum_{k=1}^{n}(a_k\cdot x_k)^\alpha\right]^{\frac{1}{\alpha}}=\max(a_i\cdot x_i) \tag{10}$$

$$\lim_{\alpha\to-\infty}\left[\frac{1}{n}\sum_{k=1}^{n}(a_k\cdot x_k)^\alpha\right]^{\frac{1}{\alpha}}=\min(a_i\cdot x_i) \tag{11}$$

The limit is the geometric mean if $\alpha\to 0$:

$$\lim_{\alpha\to 0}\left[\frac{1}{n}\sum_{k=1}^{n}(a_k\cdot x_k)^\alpha\right]^{\frac{1}{\alpha}}=\left(\prod_{i=1}^{n}(a_i\cdot x_i)\right)^{1/n} \tag{12}$$

The sensitivity of the weighted relevance aggregation operator can be examined in two stages. Namely, in the learning (or training) stage, when we search for good a_i-s for our model; and in the application stage, when a_i-s are constant values and x_i-s are the inputs. We consider the second case and we discuss the following question. If the input values $x_1,\ldots,x_n$ change a *little*, then how large will be the change of the output H? In mathematical point of view, we search for a C_γ, for which

$$|\Delta H|\le C_\gamma\cdot\|\Delta\underline{x}\|_\gamma \tag{13}$$

The answer naturally depends on the change of the inputs ($\Delta\underline{x}$) and on the parameter α. In the following we discuss this question on the whole range of α.

Let us introduce the following notations:

$\underline{x}=(x_1,\ldots,x_n)$	the input vector
$\underline{x}^*=(x_1^*,\ldots,x_n^*)$	the perturbed input vector
$\Delta\underline{x}=\underline{x}^*-\underline{x}$	the change of the input vector
$H^*=\left[\frac{1}{n}\sum_{k=1}^{n}(a_k\cdot x_k^*)^\alpha\right]^{\frac{1}{\alpha}}$	the new output
$\Delta H=H^*-H$	the change of the output
$\underline{a}=(a_1,\ldots,a_n)$	the weighting vector
$\underline{a}'=n^{-1/\alpha}\cdot\underline{a}$	the weighting vector including n

Moreover, we use the '$\circ$' symbol for element-wise product of vectors (a.k.a. Schur-product or Hadamard-product).

3.1 Case $\alpha \geq 1$

Since $\alpha \geq 1$ in this case the WRAO behaves like a p-norm. Because of this reason we can apply vector norm related inequalities to estimate $|\Delta H|$.

$$|\Delta H| = \left| \left[\frac{1}{n} \sum_{k=1}^{n} (a_k \cdot x_k^*)^\alpha \right]^{\frac{1}{\alpha}} - \left[\frac{1}{n} \sum_{k=1}^{n} (a_k \cdot x_k)^\alpha \right]^{\frac{1}{\alpha}} \right| \tag{14}$$

$$= \left| \|\underline{a}' \circ \underline{x}^*\|_\alpha - \|\underline{a}' \circ \underline{x}\|_\alpha \right| \leq \|\underline{a}' \circ \underline{x}^* - \underline{a}' \circ \underline{x}\|_\alpha = \|\underline{a}' \circ \Delta \underline{x}\|_\alpha \tag{15}$$

We got an upper estimation of $|\Delta H|$. If all of the Δx_i-s are less than $\varepsilon > 0$, then we have

$$|\Delta H| \leq \|\underline{a}' \circ \Delta \underline{x}\|_\alpha \leq \varepsilon \cdot \|\underline{a}'\|_\alpha = \varepsilon \cdot \frac{1}{n} \cdot \|\underline{a}\|_\alpha \tag{16}$$

Since $a_i \leq 1$ for all $i = 1, \ldots, n$, it follows that $1/n \cdot \|\underline{a}\|_\alpha \leq 1$ so $|\Delta H| \leq \varepsilon$ (moreover, except for extreme cases $a_i < 1$ holds, and the upper bound is less than ε).

If we want to measure the change of the input vector $\underline{x}$ in a vector norm, namely in a p-norm, then by Hölder's inequality we can give a further upper estimation of $|\Delta H|$:

$$|\Delta H| \leq \|\underline{a}' \circ \Delta \underline{x}\|_\alpha \leq \|\underline{a}'\|_\beta \cdot \|\Delta \underline{x}\|_\gamma \tag{17}$$

where $1/\alpha = 1/\beta + 1/\gamma$ and $\beta, \gamma \geq 1$, with the convention that $1/\infty = 0$. From this equality we get that

$$\beta = \frac{\gamma \cdot \alpha}{\gamma - \alpha}$$

This must be greater or equal to 1, so

$$\gamma \geq \frac{\alpha}{1 - \alpha}$$

This inequality is fulfilled if $\gamma > \alpha$, so we can conclude that if $\gamma > \alpha$ then the upper estimation is

$$|\Delta H| \leq \|\underline{a}'\|_\beta \cdot \|\Delta \underline{x}\|_\gamma \tag{18}$$

where $\beta = \gamma \cdot \alpha / (\gamma - \alpha)$.

If $\gamma \leq \alpha$ then there is no such β. Using the monotonicity property of p-norm we get

$$|\Delta H| \leq \|\underline{a}' \circ \Delta \underline{x}\|_\alpha \leq \|\underline{a}' \circ \Delta \underline{x}\|_\gamma \leq \|\underline{a}'\|_\infty \cdot \|\Delta \underline{x}\|_\gamma \tag{19}$$

The most widely used p-norms are the taxicab ($p = 1$), the euclidean ($p = 2$) and the maximum ($p = \infty$) norms. In these cases the upper bounds are the following:

– if $\gamma = 1$:

$$|\Delta H| \leq \|\underline{a}'\|_\infty \cdot \|\Delta \underline{x}\|_1$$

- if $\gamma = 2$:
  - if $1 \leq \alpha < 2$:

$$|\Delta H| \leq \|\underline{a}'\|_\beta \cdot \|\Delta \underline{x}\|_2$$

where $\beta = 2\alpha/(2-\alpha)$
  - if $\alpha \geq 2$:

$$|\Delta H| \leq \|\underline{a}'\|_\infty \cdot \|\Delta \underline{x}\|_2$$

- if $\gamma = \infty$

$$|\Delta H| \leq \|\underline{a}'\|_\infty \cdot \|\Delta \underline{x}\|_\infty$$

3.2 Case $0 < \alpha < 1$

We will show that in this case we cannot give a general C_γ for $|\Delta H| \leq C_\gamma \cdot \|\Delta \underline{x}\|_\gamma$. Let us apply Corollary 2:

$$|\Delta H| = |\nabla H_{\underline{\xi}} \cdot \Delta \underline{x}| \leq \|\nabla H_{\underline{\xi}}\|_s \cdot \|\Delta \underline{x}\|_t \tag{20}$$

The partial derivatives of H:

$$\frac{\partial H}{\partial x_i} = \left[\sum_{k=1}^{n} \frac{a_k^\alpha}{n} \cdot x_k^\alpha\right]^{1/\alpha - 1} \cdot \frac{a_i^\alpha}{n} \cdot x_i^{\alpha-1} \tag{21}$$

with this

$$\|\nabla H\|_s = \left[\sum_{i=1}^{n} \left|\frac{\partial H}{\partial x_i}\right|^s\right]^{1/s} = \left[\sum_{i=1}^{n} \left(\frac{a_i^\alpha}{n}\right)^s \left(\frac{1}{x_i}\right)^{s(1-\alpha)} \cdot \left[\frac{1}{n}\sum_{k=1}^{n}(a_k \cdot x_k)^\alpha\right]^{1/\alpha \cdot s(1-\alpha)}\right]^{1/s} \tag{22}$$

by the monotonicity of the WRAO this is greater or equal to the following expression:

$$\left[\sum_{i=1}^{n} \left(\frac{a_i^\alpha}{n}\right)^s \left(\frac{1}{x_i}\right)^{s(1-\alpha)} \cdot \left[\prod_{k=1}^{n}(a_k \cdot x_k)^{1/n}\right]^{s(1-\alpha)}\right]^{1/s} \tag{23}$$

If $x_1 = 1/K$ and $x_2 = \ldots = x_n = 1$ then this expression can be arbitrarily large if K is large enough. We conclude that if $0 < \alpha < 1$ then we cannot give a *general* upper bound for $|\Delta H|$ in terms of $\|\Delta \underline{x}\|_\gamma$.

3.3 Case $\alpha = 0$

We define the weighted relevance aggregation operator for $\alpha = 0$ by the limit. Similarly to the previous case, will show that we cannot give a general C_γ for $|\Delta H| \leq C_\gamma \cdot \|\Delta \underline{x}\|_\gamma$.

$$H = \lim_{\alpha \to 0} \left[\frac{1}{n}\sum_{k=1}^{n}(a_k \cdot x_k)^\alpha\right]^{\frac{1}{\alpha}} = \left(\prod_{i=1}^{n}(a_i \cdot x_i)\right)^{1/n} \tag{24}$$

The partial derivatives of H:

$$\frac{\partial H}{\partial x_i} = \frac{1}{x_i} \cdot \frac{1}{n} \cdot \prod_{k=1}^{n} (a_k \cdot x_k)^{1/n} \tag{25}$$

The norm of ∇H:

$$\|\nabla H\|_s = \left[\sum_{i=1}^{n} \left| \frac{\partial H}{\partial x_i} \right|^s \right]^{1/s} = \prod_{k=1}^{n} (a_k \cdot x_k)^{1/n} \cdot \left[\sum_{i=1}^{n} \left(\frac{1}{x_i \cdot n} \right)^s \right]^{1/s} \tag{26}$$

If $x_1 = 1/K$ and $x_2 = \ldots = x_n = 1$ then this expression can be arbitrarily large if K is large enough, so if $\alpha = 0$ then we cannot give a *general* upper bound for $|\Delta H|$ in terms of $\|\Delta \underline{x}\|_\gamma$.

3.4 Case $\alpha < 0$

In practical application this case not occurs very often, but from mathematical point of view it is necessary to discuss the whole range of α. By series of transformation and using the fact that the WRAO is monotone increasing we get:

$$\|\nabla H\|_s \leq \|\underline{a}'\|_\infty \tag{27}$$

So the upper bound is

$$|\Delta H| \leq \|\underline{a}'\|_\infty \cdot \|\Delta \underline{x}\|_\gamma \tag{28}$$

Table 1. Values of the coefficient C_γ for $|\Delta H| \leq C_\gamma \cdot \|\Delta \underline{x}\|_\gamma$.

Value of α	$\alpha < 0$	$0 \leq \alpha < 1$	$\alpha = 1$	$\alpha \leq \gamma$	$\alpha > \gamma$
C_γ	$\|\underline{a}'\|_\infty$	-	$\|\underline{a}'\|_\infty$	$\|\underline{a}'\|_{\alpha \cdot \gamma/(\gamma-\alpha)}$	$\|\underline{a}'\|_\infty$

4 Sensitivity of Fuzzy Signatures

Applying the results of the previous section we can analyse the sensitivity of fuzzy signatures in which the values are determined by a WRAO operator in every nodes. The sensitivity bound of the whole fuzzy signature can be derived from the bounds of the WRAOs (Table 1), according to the graph structure of the signature. The whole computation can be carried out from the leaves of the signature to the root.

Let us denote by C_{11} the bound for the WRAO applied in the root of the signature and by $\Delta \underline{x}_{11}$ of the changing of its input vector; the bounds for their WRAO operators are $C_{21}, \ldots, C_{2n_2}$ (n_2 is the number of vertices to the root), the changing of their inputs are $\Delta \underline{x}_{21}, \ldots, \Delta \underline{x}_{2n_2}$ etc., till the end of the graph. Let us denote the output of the fuzzy signature (the membership value) by h.

Then the change of the output value can be estimated by the following way:

- In $\|\cdot\|_\gamma$ norm of the input vector:

  Now it is more convenient to deal with $|\Delta h|^\gamma$ instead of $|\Delta h|$. The estimation:

$$\begin{aligned}
|\Delta h|^\gamma &\leq C_{11}^\gamma \cdot \|\Delta \underline{x}_{11}\|_\gamma^\gamma \\
&\leq C_{11}^\gamma \cdot \left(C_{21}^\gamma \cdot \|\Delta \underline{x}_{21}\|_\gamma^\gamma + \cdots + C_{2n_2}^\gamma \cdot \|\Delta \underline{x}_{2n_2}\|_\gamma^\gamma\right) \\
&\vdots \\
&\leq \sum_{i=1}^{N} C_i^\gamma \cdot |\Delta x_i|^\gamma \leq \max(C_i^\gamma) \cdot \sum_{i=1}^{N} |\Delta x_i|^\gamma \\
&= \max(C_i^\gamma) \cdot \|\Delta \underline{x}\|_\gamma^\gamma
\end{aligned}$$

  where C_i is the product of the bounds from the root to the i-th leaf.
- In $\|\cdot\|_\infty$ norm of the input vector: Because of the max operator this case behaves in a different way. The D_{**}-s are the bounds for the WRAO operators for $\gamma = \infty$.

$$\begin{aligned}
|\Delta h| &\leq D_{11} \cdot \|\Delta \underline{x}_{11}\|_\infty \\
&\leq D_{11} \cdot \max\left(D_{21} \cdot \|\Delta \underline{x}_{21}\|_\infty, \ldots, D_{2n_2} \cdot \|\Delta \underline{x}_{2n_2}\|_\infty\right) \\
&\vdots \\
&\leq \\
&= \max(D_i) \cdot \|\Delta \underline{x}\|_\infty
\end{aligned}$$

  where D_i is the product of the greatest bounds at every level.

5 Summary

We discussed the sensitivity of the weighted relevance aggregation operator. The sensitivity highly depends on the value of the parameter α, and in some cases (if $0 \leq \alpha < 1$) the operator not behaves well, a very small change of the input vector can cause a large deviation in the output value, so the practitioners should avoid WRAO with $0 \leq \alpha < 1$.

The sensitivity of fuzzy signatures with WRAOs in the nodes was also discussed. Unfortunately an elegant formula cannot be given, in general the recursive method is the only choice.

Acknowledgments. This research was supported by National Research, Development and Innovation Office (NKFIH) K105529 and K108405.

References

1. Wong, K.W., Gedeon, T.D., Kóczy, L.T.: Construction of fuzzy signature from data: an example of SARS pre-clinical diagnosis system. In: Proceedings of the IEEE International Conference on Fuzzy Systems (FUZZ-IEEE2004), Budapest, pp. 1649–1654 (2004)
2. Ballagi, Á., Kóczy, L.T., Gedeon, T.D.: Robot cooperation without explicit communication by fuzzy signatures and decision trees. In: Proceedings of the Joint International Fuzzy Systems Association World Congress and European Society of Fuzzy Logic and Technology Conference (IFSA-EUSFLAT2009), Lisbon, Portugal, pp. 1468–1473 (2009)
3. Vámos, T., Kóczy, L.T., Biró, G.: Fuzzy signatures in datamining. In: Proceedings of the Joint 9th IFSA World Congress and 20th NAFIPS International Conference, Vancouver, BC, Canada, pp. 2842–2846 (2001)
4. Molnárka, G., Kóczy, L.T.: Decision support system for evaluating existing apartment buildings based on fuzzy signatures. Int. J. Comput. Commun. Control, **VI**(3), 442–457 (2011)
5. Bukovics, Á., Kóczy, L.T.: Fuzzy Signature-based model for qualification and ranking of residential buildings. XXXVIII. IAHS World Congress on Housing, Istanbul, Turkey, pp. 290–297 (2012)
6. Pozna, C., Minculete, N., Precup, R.E., Kóczy, L.T., Ballagi, Á.: Signatures: definitions, operators and applications to fuzzy modelling. Fuzzy Sets Syst. **201**, 86–104 (2012)
7. Kóczy, L.T., Vámos, T., Biró, G.: Fuzzy signatures. In: Proceedings of the 4th Meeting of the Euro Working Group on Fuzzy Sets and the 2nd International Conference on Soft and Intelligent Computing (EUROPUSE-SIC99), Budapest, Hungary, pp. 210–217 (1999)
8. Goguen, J.A.: L-fuzzy sets. J. Math. Anal. Appl. **18**(1), 145–174 (1967)
9. Harmati, I.Á., Bukovics, Á., Kóczy, L.T.: Sensitivity analysis of the weighted generalized mean aggregation operator and its application to fuzzy signatures. In: IEEE World Congress on Computational Intelligence (WCCI 2014 - FUZZ-IEEE 2014), Peking, China, pp. 1327–1332 (2014)
10. Harmati, I.Á., Bukovics, Á., Kóczy, L.T.: Sensitivity analysis of fuzzy signatures using Minkowski's inequality. In: Proceedings in Adaptation, Learning and Optimization: Proceedings of the 18th Asia Pacific Symposium on Intelligent and Evolutionary Systems, Singapore, pp. 587–596. Springer, Switzerland (2014)
11. Harmati, I.Á., Bukovics, Á., Kóczy, L.T.: Minkowski's inequality based sensitivity analysis of fuzzy signatures. J. Artif. Intell. Soft Comput. Res. (accepted article, in press)
12. Hardy, G.H., Littlewood, J.E., Pólya, G.: Inequalities. Cambridge University Press, Cambridge (1952)
13. Bullen, P.S.: Handbook of Means and Their Inequalities. Kluwer Academic Publishers, Dordrecht (2003)
14. Golub, G.H., van Loane, C.F.: Matrix Computations. John Hopkins University Press, Baltimore (1996)
15. Mendis, B.S.U., Gedeon, T.D., Botzheim, J., Kóczy, L.T.: Generalised weighted relevance aggregation operators for hierarchical fuzzy signatures. In: International Conference on Computational Intelligence for Modelling, Control and Automation, and International Conference on Intelligent Agents, Web Technologies and Internet Commerce, pp. 198–198. IEEE (2006)

16. Mendis, B.S.U., Gedeon, T.D., Kóczy, L.T.: On the issue of learning weights from observations for fuzzy signatures. In: World Automation Congress, WAC 2006. pp. 1–6. IEEE (2006)
17. Mendis, B.S.U., Gedeon, T.D.: Complex structured decision making model: a hierarchical frame work for complex structured data. Inf. Sci. **194**, 85–106 (2012)
18. Beliakov, G., Pradera, A., Calvo, T.: Aggregation Functions: A Guide for Practitioners, vol. 221. Springer, Heidelberg (2007)
19. Beliakov, G., Calvo, T., James, S.: On Lipschitz properties of generated aggregation functions. Fuzzy Sets Syst. **161**(10), 1437–1447 (2010)
20. Mendis, B.S.U., Gedeon, T.D.: Aggregation selection for hierarchical fuzzy signatures: a comparison of hierarchical OWA and WRAO. In: Proceedings of IPMU, pp. 1376–1383 (2008)

Some Results on Extension of Lattice-Valued XOR, XOR-Implications and E-Implications

Eduardo Palmeira[1(✉)] and Benjamín Bedregal[2]

[1] Programa de Pós-Graduação em Modelagem Computacional em Ciência e Tecnologia - PPGMC, Departamento de Ciências Exatas e Tecnológicas - DCET, Universidade Estadual de Santa Cruz - UESC, Ilhéus, BA 45662-900, Brazil
espalmeira@uesc.br

[2] Departamento de Informtica e Matemtica Aplicada - DIMAp, Universidade Federal do Rio Grande do Norte - UFRN, Natal, RN 59078-970, Brazil
bedregal@dimap.ufrn.br

Abstract. The extension problem is an important and interesting issue that be addressed for many different classes of operator. For instance, one can thing who to extend a fuzzy operator from a lattice to a bigger one preserving its algebraic properties. In this paper we attempt to the extension of lattice-valued version of Xor (exclusive) operator using a special method based on retractions. Also we discuss about Xor-implications end E-implications.

Keywords: Extension · Xor-implications · E-implications

1 Introduction

It is well known that Xor connective has many application in computer science. It is used as a primitive operation in many encryption algorithms, as in sample threshold activated neural network, in the identification of elemental emission spectra, in the construction of conflict-free hash functions, etc. [4].

In order to gain a more realistic framework many researchers have been turned their attention to work on lattice structures and a very interesting issue for lattice is state a generic way to extend operators from a lattice (sublattice) to a bigger one.

In this paper we apply an extension method proposed in [18] for lattice-valued fuzzy xor connective. We start presenting some preliminaries definitions in Sect. 2. In the following section we discuss about the extension method via retractions for t-norms, t-conorms and fuzzy negations and present the main results on extension of xor operators, xor-implications and E-implications.

2 Some Results on Lattice Theory

In this paper we are considering definitions and properties known from the lattice theory in the algebraic sense. For a detailed review of them we strongly recommend the following references [6,9–15,20]. In the whole paper we write L for a lattice and M for its sublattice.

J.P. Carvalho et al. (Eds.): IPMU 2016, Part II, CCIS 611, pp. 809–820, 2016.
DOI: 10.1007/978-3-319-40581-0_66

2.1 (r,s)-Sublattices

In general given a bounded lattice L and a nonempty subset $M \subseteq L$ it is said that M is a sublattice of L if for all $x, y \in M$ the following conditions hold:

$$x \wedge_L y \in M \quad and \quad x \vee_L y \in M$$

In other words, M equipped with the restriction of the operations $\wedge_L$ and $\vee_L$ inherits the lattice structure of L.

We would like to work with a relaxed notion of sublattice in which the condition $M \subseteq L$ is somewhat weakened.

Definition 1. *[6] A homomorphism r of a lattice L onto a lattice M is said to be a* **retraction** *if there exists a homomorphism s of M into L which satisfies $r \circ s = id_M$. A lattice M is called a* **retract** *of a lattice L if there is a retraction r of L onto M and s is then called a* **pseudo-inverse** *of r.*

It is important to point out here that a retraction r as in Definition 1 is surjective (onto) and hence s is naturally injective since $r \circ s = id_M$. These properties are used in some proofs throughout this paper.

Definition 2. *Let L and M be arbitrary bounded lattices. We say that M is a (r, s)-***sublattice** *of L if M is a retract of L (i.e. M is a sublattice of L up to isomorphisms). In other words, M is a (r, s)-sublattice of L if there is a retraction r of L onto M with pseudo-inverse $s : M \rightarrow L$.*

Remark 1. Throughout this paper we consider the notion of (r, s)-sublattice as in Definition 2 instead of the ordinary notion of sublattice. Whenever the usual definition of sublattice is used and this is not clear from the context, this sublattice will be called ordinary sublattice.

Definition 3. *Every retraction $r : L \longrightarrow M$ (with pseudo-inverse s) which satisfies $s \circ r \leqslant id_L$*[1] *($id_L \leqslant s \circ r$) is called a* **lower** *(an upper)* **retraction**. *In this case M is called a* **lower** *(an upper)* **retract** *of L.*

Notice that both in Definitions 2 and 3 the pseudo-inverse s of a retraction r needs not be unique. This is an advantage of our notion of sublattice since if there exist more than one pseudo-inverse for the same retraction it is possible to identify M with a subset of L in different ways. This gives us the possibility of choosing the best one for our proposes. But it must be clear that when we say that M is a (lower, upper or neither) (r, s)-sublattice of L we are considering the existence of at least one pseudo-inverse s and fixing it. In any case, no matter which pseudo-inverse is taken every result presented here holds.

Definition 4. *Let M be a (r_1, s)-sublattice of L. We say that*

[1] If f and g are functions on a lattice L it is said that $f \leqslant g$ if and only if $f(x) \leqslant_L g(x)$ for all $x \in L$.

1. *M is a* **lower** *(r_1, s)*-**sublattice** *of L if r_1 is a lower retraction. Notation: $M < L$ with respect to (r_1, s);*
2. *M is an* **upper** *(r_1, s)*-**sublattice** *of L if r_1 is an upper retraction. Notation: $M > L$ with respect to (r_1, s);*
3. *If r_1 is a lower retraction and there is an upper retraction $r_2 : L \longrightarrow M$ such that its pseudo-inverse is also s then M is called a* **full** *(r_1, r_2, s)*-**sublattice** *of L. Notation: $M \trianglelefteq L$ with respect to (r_1, r_2, s).*

2.2 T-norms and T-conorms on L

A short formalization for the notion of t-norm and t-conorm on bounded lattices is presented here. Moreover, some results are demonstrated as well.

Definition 5. *[2] Let L be a bounded lattice. A binary operation $T : L \times L \longrightarrow L$ is a* **t-norm** *if for all $x, y, z \in L$ it satisfies:*

1. *$T(x, y) = T(y, x)$ (commutativity);*
2. *$T(x, T(y, z)) = T(T(x, y), z)$ (associativity);*
3. *If $x \leqslant_L y$ then $T(x, z) \leqslant_L T(y, z)$, $\forall\, z \in L$ (monotonicity);*
4. *$T(x, 1_L) = x$ (boundary condition).*

Dually, a function $S : L \times L \longrightarrow L$ that is commutative, associative, monotone and has 0_L as the boundary element is called a **t-conorm** on L.

Notice that $T(x, y) \leqslant_L x$ (or $T(x, y) \leqslant_L y$) and $x \leqslant_L S(x, y)$ (or $y \leqslant_L S(x, y)$) for all $x, y \in L$. In fact, $T(x, y) \leqslant_L x \wedge_L y \leqslant_L x$ and $x \leqslant_L x \vee_L y \leqslant_L S(x, y)$.

2.3 Negations on L

There are several approaches to the notion of fuzzy negation in order to have a generalization of the classical one [3, 5, 8, 17, 18].

Definition 6. *A function $N : L \longrightarrow L$ is called a* **fuzzy negation** *if it satisfies:*

(N1) $N(0_L) = 1_L$ and $N(1_L) = 0_L$;
(N2) *If* $x \leqslant_L y$ then $N(y) \leqslant_L N(x)$, for all $x, y \in L$.
Moreover, if a fuzzy negation N on L satisfies the involute property, namely
(N3) $N(N(x)) = x$, for all $x \in L$
it is called a **strong** fuzzy negation.
Negations satisfying Property (N4) are called **frontier**.
(N4) $N(x) \in \{0_L, 1_L\}$ if and only if $x = 0_L$ or $x = 1_L$
Finally, if $x \in L$ is such that $N(x) = x$ then x is said to be an **equilibrium point** of N.

2.4 Fuzzy Implications

This section is devoted to present the notion of fuzzy implication on lattices and some properties. It is well-known that there are some different ways to interpret fuzzy implications (see [1,5,7,16,21]) but here we consider the notion considered in [1].

Definition 7. *A* **fuzzy implication** *on bounded lattice L is a function $I : L \times L \longrightarrow L$ such that for each $x, y, z \in L$ the following properties hold:*

(CC1) $I(0_L, 0_L) = 1_L$ (Corner condition 1);
(CC2) $I(1_L, 1_L) = 1_L$ (Corner condition 2);
(CC3) $I(0_L, 1_L) = 1_L$ (Corner condition 3);
(CC4) $I(1_L, 0_L) = 0_L$ (Corner condition 4).

In which follows we consider some properties for a fuzzy implication I on L:

(FPA) if $x \leqslant_L y$ then $I(y, z) \leqslant_L I(x, z)$ (First variable antitonicity);
(SPI) if $y \leqslant_L z$ then $I(x, y) \leqslant_L I(x, z)$ (Second variable isotonicity);
(LB) $I(0_L, y) = 1_L$, for all $y \in L$;
(RB) $I(x, 1_L) = 1_L$, for all $x \in L$;
(NP) $I(1_L, y) = y$ for each $y \in L$ (left neutrality principle);
(L-NP) $I(1_L, y) \leqslant_L y$ for each $y \in L$;
(EP) $I(x, I(y, z)) = I(y, I(x, z))$ for all $x, y, z \in L$ (exchange principle);
(IP) $I(x, x) = 1_L$ for each $x \in L$ (identity principle);
(IBL) $I(x, I(x, y)) = I(x, y)$ for all $x, y, z \in L$ (iterative Boolean law);
(CP) $I(x, y) = I(N(y), N(x))$ for each $x, y \in L$ with N a fuzzy negation on L (law of contraposition);
(P) $I(x, y) = 0_L$ if and only if $x = 1_L$ and $y = 0_L$ (Positive);
(LEM) $S(N(x), x) = 1_L$ for each $x \in L$ (law of excluded middle).

2.5 Fuzzy Xor Connective

Usually Xor connective is considered in order to evaluate the value with which one and only one of its immediate antecedents is true. An usual definition of this connective for the unit interval $[0, 1]$ is given by

Definition 8. *[4] A binary operation $W : [0, 1] \times [0, 1] \longrightarrow [0, 1]$ is called a* **Xor operator** *if, for all $x, y, z \in [0, 1]$ it follows that:*

1. $W(x, y) = W(y, x)$ (commutativity);
2. $W(x, W(y, z)) = W(W(x, y), z)$ (associativity);
3. $W(0, x) = x$ (0-identity);
4. $W(1, 1) = 0$ (boundary condition).

It follows from items 3. and 4. of Definition 8 that fuzzy Xor connective generalizes the classical one.

Let's see some properties of fuzzy Xor (see [4]):

(W1) $W(x, x) = 0$;
(W2) $W(W(x, y), x) = y$;
(W3) $N_W(x) = W(x, 1)$ is a strong fuzzy negation;
(W4) If $W(x, y) = 0$ then $x = y$;
(W5) $W(N_W(x), x) = 1$.

3 Extension Method via Retractions (EMR)

Let M be a complete ordinary sublattice of L and T^M be a t-norm on M. Under these conditions, Saminger-Platz et al. in [19] have proposed a method for extending the t-norm T^M from M to L. They start by describing a way to extend this t-norm for 0_L and 1_L using the function

$$T^{M\cup\{0_L,1_L\}}(x,y) = \begin{cases} x \wedge_L y, \, if \; 1_L \in \{x,y\}; \\ 0_L, \, if \; 0_L \in \{x,y\}; \\ T^M(x,y), \; otherwise. \end{cases} \tag{1}$$

Then, considering a function that takes an element x belonging to L and assigns to it $x^* = \sup_M\{z \mid z \leqslant_L x, z \in M \cup \{0_L, 1_L\}\}$ it is proved that the t-norm given by

$$J^L_{T^M}(x,y) = \begin{cases} x \wedge_L y, \, if \; 1_L \in \{x,y\}; \\ T^{M\cup\{0_L,1_L\}}(x^*,y^*), \; otherwise. \end{cases} \tag{2}$$

is an extension of T^M from M to L, i.e. the restriction of $J^L_{T^M}$ to M is equal to T^M.

Note that this is a natural but drastic and particular way to extend t-norms since it collapses all elements of $L\backslash M$ on M and only considers complete sublattices.

Looking for a more general and flexible extension method, Palmeira and Bedregal presented in [18] another way to extend t-norms, t-conorms and fuzzy negations considering (r, s)-sublattices. This method generalizes (2).

Proposition 1. *[18] Let $M < L$ with respect to (r, s). If T is a t-norm on M then $T^E : L \times L \longrightarrow L$ defined by*

$$T^E(x,y) = \begin{cases} x \wedge_L y, \; if \; 1_L \in \{x,y\} \\ s(T(r(x),r(y))), \; otherwise. \end{cases} \tag{3}$$

is a t-norm which extends T from M to L.

In a similar way, it is possible to extend t-conorms as follows

Proposition 2. *[18] Let $M > L$ with respect to (r, s). If S is a t-conorm on M then $S^E : L \times L \longrightarrow L$ defined by*

$$S^E(x,y) = \begin{cases} x \vee_L y, \; if \; 0_L \in \{x,y\} \\ s(S(r(x),r(y))), \; otherwise. \end{cases} \tag{4}$$

is a t-conorm which extends S from M to L.

For negations we have

Proposition 3. *[18] Let M be a (r, s)-sublattice of L and $N : M \longrightarrow M$ be a fuzzy negation. Then $N^E(x) = s(N(r(x)))$ for each $x \in L$ is a fuzzy negation that extends N from M to L.*

3.1 Lattice-Valued Xor Operator

Now we aim to present a definition of Xor operator on a lattice and then apply the EMR method shown above to extend this operator from M to L.

Definition 9. *A binary operation $W : L \times L \longrightarrow L$ is called a* **Xor operator** *if for all $x, y, z \in L$ it follows that:*

1. $W(x, y) = W(y, x)$ *(commutativity);*
2. $W(x, W(y, z)) = W(W(x, y), z)$ *(associativity);*
3. $W(0_L, x) = x$ *(0-identity);*
4. $W(1_L, 1_L) = 0_L$ *(boundary condition).*

So naturally the following theorem holds.

Theorem 1. *Let M be a (r, s)-sublattice of L. If W is a xor operator on M then $W^E : L \times L \longrightarrow L$ defined by*

$$W^E(x,y) = \begin{cases} x \vee_L y, & \text{if } 0_L \in \{x,y\} \\ s(W(r(x),r(y))), & \text{otherwise.} \end{cases} \tag{5}$$

is a xor operator on L.

Proof. It is clear from the proof of Proposition 2 that W^E is associative, commutative and satisfies the 0-identity property. We shall only prove that boundary condition holds. Thus

$$\begin{aligned} W^E(1_L, 1_L) &= s(W(r(1_L), r(1_L))) \\ &= s(W(1_M, 1_M)) = s(0_M) = 0_L \end{aligned}$$

□

Notice that in Proposition 2 one of the hypothesis is that $M > L$ what is not necessary for extending xor operators as one can see in Theorem 8. This fact occur just because of assumption $M > L$ is required to prove the monotonicity property of t-conorms.

The following theorem shows that properties (W1) and (W5) are preserved by the extension method of retractions.

Theorem 2. *Let M be a (r, s)-sublattice of L. If W is a xor operator on M satisfying properties (W1) and (W2) then its extension W^E as defined in Theorem 1 satisfies properties (W1) and (W5) by considering $N_{W^E} = W^E(x, 1_L)$ for all $x \in L$.*

Proof. Suppose that W satisfies property (W1) i.e. $W(x, x) = 0_M$ for all $x \in M$. Thus, if $x = 0_L$ then $W^E(x, x) = x \vee x = 0_L$. Otherwise, $W^E(x, x) = s(W(r(x), r(x))) = s(0_M) = 0_L$. Therefore, W^E satisfies (W1).

Now assume that W satisfies (W2). In this case, we have three possibles:

(i) If $x = 0_L$ then $W^E(N_{W^E}(x), x) = W^E(1_L, 0_L) = 1_L$;

(ii) When $x = 1_L$ it follows that $W^E(N_{W^E}(x), x) = W^E(0_L, 1_L) = 1_L$ since $N_{W^E}(1_L) = 0_L$;

(iii) For other cases we have

$$\begin{aligned} & W^E(N_{W^E}(x), x) = \\ & = s(W(r(W^E(x, 1_L), r(x)))) \\ & = s(W(W(r(x), r(1_L)), r(x))) \quad by\ Def.2.1 \\ & = s(W(W(r(x), 1_M), r(x))) \\ & = s(1_M) \quad by\ (W2) \\ & = 1_L \end{aligned}$$

Then, by (i), (ii) and (iii) it can be concluded that W^E satisfies (W5). □

Proposition 4. *Let M be a (r, s)-sublattice of L and W be a xor operator on M. Thus*

1. *If W satisfies (W2) and r is a lower (upper) retraction then $W^E(W^E(x, y), x) \leqslant_L y$ ($W^E(W^E(x, y), x) \geqslant_L y$);*
2. *$W^E(x, y) = 0_L$ implies $r(x) = r(y)$ whenever (W4) holds for W. Moreover, if r is a lower (upper) retraction then $W^E(x, y) = 0_L$ implies $x \leqslant y$ ($x \geqslant y$).*

Proof. 1. It is clear that $W^E(W^E(x, y), x) = y$ for all $y \in L$ if $x = 0_L$. Now suppose that $x \neq 0_L$ and r is a lower retraction, that is,

$$r \circ s = Id_M \quad and \quad s \circ r \leqslant Id_L \tag{6}$$

Thus

$$\begin{aligned} W^E(W^E(x, y), x) &= s(W(r(W^E(x, y)), r(x))) \\ &= s(W(W(r(x), r(y)), r(x))) \\ &= s(W(r(y), W(r(x), r(x)))) \\ &= s(W(r(y), 0_M)) \\ &= s(r(y)) \leqslant_L y \end{aligned}$$

In case r is an upper retraction we have that $s \circ r \geqslant Id_L$ and hence we have that $W^E(W^E(x, y), x) \geqslant_L y$.

2. Suppose that $0_L \in \{x, y\}$. Without loss of generality take $x = 0_L$. Thus, if $x \vee_L y = W^E(x, y) = 0_L$ then $x = y = 0_L$.

On the other hand, if $0_L \in \{x, y\}$ then $W^E(x, y) = 0_L$ which implies that $s(W(r(x), r(y))) = 0_M$, that is, $W(r(x), r(y)) = 0_M$. Since W satisfies (W4) it follows that $r(x) = r(y)$. Moreover, if r is a lower retraction, by Identities (6) we can conclude that $x \leqslant_L y$. Dually it can be proof that $x \geqslant_L y$ if r is an upper retraction. □

The Proposition 4 presents a weaker version of properties (W2) and (W4) which shows that the extension method via retractions fails in preserving those properties.

It is possible to obtain fuzzy xor operators from t-norms, t-conorms and fuzzy negations. Bedregal states in Proposition 3.4 of [4] that if T is a t-norm, S is a t-conorm and N is a fuzzy negation then the function defined as in Eq. (7) below is a fuzzy xor connective.

$$W_T(x,y) = T(S(x,y), N(T(x,y))) \quad x,y \in [0,1] \tag{7}$$

It is clear that W still being a xor operator if we take $x, y \in L$. Taking this fact into account, we can prove the following proposition.

Proposition 5. *Let M be a (r,s)-sublattice of L and W be a xor operator on M. If T is a t-norm, S is a t-conorm and N is a fuzzy negation, all defined on M, then for all $x, y \in L$*

$$W_T{}^E(x,y) = T^E(S^E(x,y), N^E(T^E(x,y))) \tag{8}$$

is a fuzzy xor operator on L.

Proof. From commutativity and associativity of T^E, S^E and N^E it follows that W^E is a commutative and associative operator. Notice also that

$$\begin{aligned} W_T{}^E(0_L, y) &= T^E(S^E(0_L, y), N^E(T^E(0_L, y))) \\ &= T^E(y, N^E(0_L)) \\ &= T^E(y, 1_L) = y \end{aligned}$$

Thus 0_L is the neutral element of W^E. It remains to prove that $W^E(1_L, 1_L) = 0_L$. Indeed

$$\begin{aligned} W_T{}^E(1_L, 1_L) &= T^E(S^E(1_L, 1_L), N^E(T^E(1_L, 1_L))) \\ &= T^E(1_L, N^E(1_L)) \\ &= T^E(1_L, 0_L) = 0_L \end{aligned}$$

□

Another way to build a fuzzy xor from the triplet S, T, N is as follows:

$$W_S(x,y) = S(T(N(x), y), T(x, N(y))) \tag{9}$$

for all $x, y \in L$ (see [4]).

Proposition 6. *Let M be a (r,s)-sublattice of L and W be a xor operator on M. If T is a t-norm, S is a t-conorm and N is a fuzzy negation, all defined on M, then for all $x, y \in L$*

$$W_S{}^E(x,y) = S^E(T^E(N^E(x), y), T^E(x, N^E(y))) \tag{10}$$

is a fuzzy xor operator on L.

Similar to the proof of Proposition 5.

3.2 Lattice-Valued Xor-Implications

Proposition 7. *Given a t-conorm S, a fuzzy negation N and a fuzzy Xor operator W, all of them defined on a lattice L, thus the function $I_{W,S,N} : L \times L \to L$ taken as*

$$I_{W,S,N}(x, y) = W(x, S(N(x), N(y))) \tag{11}$$

for all $x, y \in L$, is a fuzzy implication.

Proof. Notice that
(CC1) $I_{W,S,N}(0_L, 0_L) = W(0_L, S(N(0_L), N(0_L))) = W(0_L, S(1_L, 1_L)) = 1_L$;
(CC2) $I_{W,S,N}(1_L, 1_L) = W(1_L, S(N(1_L), N(1_L))) = W(1_L, S(0_L, 0_L)) = 1_L$;
(CC3) $I_{W,S,N}(0_L, 1_L) = W(0_L, S(N(0_L), N(1_L))) = W(0_L, S(1_L, 0_L)) = 1_L$;
(CC4) $I_{W,S,N}(1_L, 0_L) = W(1_L, S(N(1_L), N(0_L))) = W(1_L, S(0_L, 1_L)) = 0_L$.

Then $I_{W,S,N}$ is a fuzzy implication. □

The fuzzy operator defined in Eq. (11) is called a lattice-valued **fuzzy Xor-implication**.

Proposition 8. *Let W be a xor operator on lattice L and S be a t-conorm on L. Then*

1. *I_{W,S,N_W} satisfies property (NP);*
2. *If W satisfies $(W3)$ then $I_{W,S,N_W}(x, 0_L)$ is a strong negation;*
3. *If W satisfies $(W1)$ and there exists an $e \in L$ such that $N_W(e) = e$ then $I_{W,S,N}(e, 1_L) = 0_L$.*

Proof. (NP): For all $x \in L$ it follows that

$$\begin{aligned} I_{W,S,N_W}(1_L, x) &= W(1_L, S(N_W(1_L), N_W(x))) \\ &= W(1_L, S(0_L, N_W(x))) \\ &= W(S(1_L, 0_L), N_W(x)) \\ &= W(1_L, N_W(x)) = N_W(N_W(x)) = x \end{aligned}$$

Let be $N(x) = I_{W,S,N_W}(x, 0_L)$. Then

$$I_{W,S,N_W}(x, 0_L) = W(x, S(N_W(x), N_W(0_L))) = W(x, S(N_W(x), 1_L)) = W(x, 1_L)$$

which is a strong negation by $(W3)$.

Finally, suppose W satisfies $(W1)$ and there exists an $e \in L$ such that $N_W(e) = e$. Hence, $I_{W,S,N}(e, 1_L) = W(e, S(N(e), N(1_L))) = W(e, S(e, 0_L)) = W(e, e) = 0_L$ □

Proposition 9. *Let M be a (r, s)-sublattice of L. If W is a xor operator on lattice M and S is a t-conorm on M then*

1. *$I_{W^E, S^E, N_{W^E}}$ is a fuzzy implication on L;*
2. *If W satisfies $(W3)$ then $I_{W^E, S^E, N_{W^E}}$ satisfies properties (NP) and the negation $I_{W^E, S^E, N_{W^E}}(x, 0_L)$ is strong;*

Proof. 1. Note that $I_{W^E,S^E,N_W{}^E}(x,y) = W^E(x, S^E(N(x), N_W{}^E(y)))$ for all $x, y \in L$. Since W^E is a xor operator on L (by Theorem 1), S^E is a t-conorm on L (by Proposition 2) and $N_W{}^E$ is a fuzzy negation on L (by Proposition 3) then $I_{W^E,S^E,N_W{}^E}$ is a fuzzy implication on L.
2. Similar to proof of Proposition 8.

□

3.3 Lattice-Valued E-implications

Proposition 10. *Let S, N and W be a t-conorm, a fuzzy negation and a fuzzy xor operator on L, respectively. Then, the function $I_{W,S,N} : L \times L \to L$ taken, for all $x, y \in L$, as*

$$I_{S,N,W}(x,y) = S(N(x), W(N(x), y)) \tag{12}$$

is a fuzzy implication, called lattice-valued E-implication.

Proof. It follows that,
(CC1) $I_{S,N,W}(0_L, 0_L) = S(N(0_L), W(N(0_L), 0_L)) = S(1_L, W(1_L, 0_L)) = 1_L$;
(CC2) $I_{S,N,W}(0_L, 1_L) = S(N(0_L), W(N(0_L), 1_L)) = S(1_L, W(1_L, 1_L)) = 1_L$;
(CC3) $I_{S,N,W}(1_L, 0_L) = S(N(1_L), W(N(1_L), 1_L)) = S(0_L, W(0_L, 1_L)) = 1_L$;
(CC4) $I_{S,N,W}(1_L, 0_L) = S(N(1_L), W(N(1_L), 0_L)) = S(0_L, W(0_L, 0_L)) = 0_L$;

□

Proposition 11. *Let W be a xor operator on lattice L and S be a t-conorm on L. Then,*

1. *$I_{S,N,W}$ satisfies property (NP);*
2. *$N(x) \leqslant_L I_{S,N,W}(x, 0_L)$, for all $x \in L$.*

Proof. For all $y \in L$,

$$I_{S,N,W}(1_L, y) = S(N(1_L), W(N(1_L), y)) = S(0_L, W(0_L, y)) = S(0_L, y) = y$$

i.e. property (NP) holds.

Notice also that, for all $x \in L$, we have

$$I_{S,N,W}(x, 0_L) = S(N(x), W(N(x), 0_L)) = S(N(x), N(x)) \geqslant_L N(x)$$

□

Proposition 12. *Let M be a (r,s)-sublattice of L. If W is a xor operator on lattice M and S is a t-conorm on M then*

1. *$I_{S^E,N_W{}^E,W^E}$ is a fuzzy implication on L;*
2. *If W satisfies $(W3)$ then $I_{S^E,N_W{}^E,W^E}$ satisfies property (NP);*
3. *$N_W{}^E(x) \leqslant_L I_{S^E,N_W{}^E,W^E}(x, 0_L)$ for all $x \in L$;*
4. *$I_{S^E,N_W{}^E,W^E}(x,x) = 1_L$ whenever W satisfies (W2).*

Proof. Items 1, 2 and 3 are straightforward from the above results. In order to show item 4, suppose that W satisfies (W2) and hence, by Theorem 2 W^E satisfies (W5). Thus

$$I_{S^E,{N_W}^E,W^E}(x,x) = S^E({N_W}^E(x), W({N_W}^E(x), x)) = S^E({N_W}^E(x), 1_L) = 1_L$$

for all $x \in L$. □

4 Final Remarks

The main contribution of this is paper is related to the study of extension of Xor operator, Xor-implications and E-implications taking into account the method EMR. Results shown that some properties of those operators can be preserved by the extension method and there are some other that fails. In order to fix the weakening of the extension method, for future works we wish apply the extension method via e-operator [17] for xor operators and make a comparison of results.

References

1. Baczyński, M., Jayaram, B.: Fuzzy Implications. Studies in Fuzziness and Soft Computing. Springer, Heidelberg (2008)
2. Bedregal, B.C., Santos, H.S., Callejas-Bedregal, R.: T-norms on bounded lattices: T-norm morphisms and operators. IEEE International Conference on Fuzzy Systems, pp. 22–28 (2006)
3. Bedregal, B.C.: On interval fuzzy negations. Fuzzy Sets Syst. **161**, 2290–2313 (2010)
4. Bedregal, B.C.: Xor-implications and E-implications: classes of fuzzy implications based on fuzzy xor. Eletron. Notes Theoret. Comput. Sci. **274**, 5–18 (2009)
5. Bedregal, B.C., Beliakov, G., Bustince, H., Fernandez, J., Pradera, A., Reiser, R.: (S, N)-implications on bounded lattices. In: Baczyński, M., et al. (eds) Advances in Fuzzy Implication Functions. Studies in Fuzziness and Soft Computing. Springer, Berlin (2013)
6. Birkhoff, G.: Lattice Theory. American Mathematical Society, Providence (1973)
7. Bustince, H., Burillo, P., Soria, F.: Automorphisms, negations and implication operators. Fuzzy Sets Syst. **134**(2), 209–229 (2003)
8. Calvo, T.: On mixed de morgan triplets. Fuzzy Sets Syst. **50**, 47–50 (1992)
9. Chen, G., Pham, T.T.: Fuzzy Sets, Fuzzy Logic and Fuzzy Control Systems. CRC Press, Boca Raton (2001)
10. De Cooman, G., Kerre, E.E.: Order norms on bounded partially ordered sets. J. Fuzzy Math. **2**, 281–310 (1994)
11. Davey, B.A., Priestley, H.A.: Introduction to Lattices and Order, 2nd edn. Cambridge University Press, Cambridge (2002)
12. Hajek, P.: Metamathematics of Fuzzy Logic. Kluwer Academic Publishers, Dordrecht (1998)
13. Klement, E.P., Mesiar, R.: Logical, Algebraic, Analytic, and Probabilistic Aspects of Triangular Norms. Elsevier B.V., The Netherlands (2005)
14. Klement, E.P., Mesiar, R., Pap, E.: Triangular Norms. Kluwer Academic Publishers, Dordrecht (2000)

15. Klir, G.J., Yuan, B.: Fuzzy Sets and Fuzzy Logic, Theory and Applications. Prentice Hall PTR, NJ (1995)
16. Mas, M., Monserrat, M., Torrens, J., Trillas, E.: A survey on fuzzy implications functions. IEEE Trans. Fuzzy Syst. **15**(6), 1107–1121 (2007)
17. Palmeira, E.S., Bedregal, B.C., Mesiar, R., Fernandez, J.: A new way to extend T-Norms, T-conorms and negations. Fuzzy Sets Syst. **240**, 1–21 (2014)
18. Palmeira, E.S., Bedregal, B.C.: Extension of fuzzy logic operators defined on bounded lattices via retractions. Comput. Math. Appl. **63**, 1026–1038 (2012)
19. Saminger-Platz, S., Klement, E.P., Mesiar, R.: On extensions of triangular norms on bounded lattices. Indagationes Mathematicae **19**(1), 135–150 (2008)
20. Takano, M.: Strong completeness of lattice-valued logic. Arch. Math. Logic **41**, 497–505 (2002)
21. Yager, R.R.: On the implication operator in fuzzy logic. Inf. Sci. **31**(2), 141–164 (1983)

Fuzzy Block-Pulse Functions and Its Application to Solve Linear Fuzzy Fredholm Integral Equations of the Second Kind

Shokrollah Ziari[1(✉)] and Reza Ezzati[2]

[1] Department of Mathematics, Firoozkooh Branch, Islamic Azad University, Firoozkooh, Iran
shok_ziari@yahoo.com

[2] Department of Mathematics, Karaj Branch, Islamic Azad University, Karaj, Iran

Abstract. In this paper, a method of successive approximations based on the fuzzy block-pulse functions is proposed to solve linear fuzzy Fredholm integral equations of the second kind. Moreover, the error estimation of the approximation solution is given. Finally, illustrative example is included to show the accuracy and the efficiency of the proposed method.

Keywords: Fuzzy Fredholm integral equations · Fuzzy block-pulse functions · Modulus of continuity · Lipschitz condition

1 Introduction

Block pulse functions are a set of orthogonal functions with piecewise constant value and usually applied as a useful tool in the analysis, synthesis identification and other problems of control and systems sciences [18]. Many authors used the block pulse functions to obtain numerical solutions of differential equations, integral and integro-differential equations in the crisp case.

One of the most interesting research matters in fuzzy sets and systems is to study fuzzy integral equations. Some authors have studied on fuzzy integral equations from the theoretical and practical point of view. The investigation of the existence of a unique solution for fuzzy integral equations had been carried out in [5,19–21]. The Banach fixed point theorem is the important tool in studying the existence and uniqueness of the solution for fuzzy integral equations which can be appear in numerical procedures for solving fuzzy integral equations, based on the iterative techniques. The iterative numerical methods for solving fuzzy integral equations can be found in [5–7] and [12–14]. The Adomian decomposition method, Nystorm technique, fuzzy Bernstein polynomials and fuzzy Haar wavelet were applied to solve the fuzzy integral equations of the second kind in [1,3,10,23]. Bica and Popescu in [7], applied the method of successive approximations for solving the fuzzy Hammerstein integral equation. Ezzati and Ziari in [11], proved the convergence of the method of successive approximations for solving nonlinear fuzzy Fredholm integral equations of the

J.P. Carvalho et al. (Eds.): IPMU 2016, Part II, CCIS 611, pp. 821–832, 2016.
DOI: 10.1007/978-3-319-40581-0_67

second kind, and they proposed an iterative procedure based on the trapezoidal quadrature. Recently, Baghmisheh and Ezzati in [4], approximated the fuzzy function by the hybrid Taylor and block-pulse functions and estimated the error approximation. Also, an iterative procedure is constructed based on the hybrid Taylor and block-pulse functions for solving nonlinear Fredholm fuzzy integral equations by them. Recently, in [24], Ziari and Bica obtained the error estimation of the iterative method based on trapezoidal formula to solve nonlinear fuzzy Hammerstein-Fredholm integral equations of the second kind given in terms of uniform and partial modulus of continuity. In the present paper, we propose a iterative numerical method based on fuzzy block pulse functions for solving fuzzy Fredholm integral equations with Lipschitzian fuzzy-number-valued function as a starting point. Also, We prove the convergence theorem of presented successive approximation method, and to demonstrate the accuracy of the proposed method we present one illustrative example. This paper is organized as follows:

In Sect. 2, we review some elementary concepts of the fuzzy set theory. In Sect. 3, we review of bloc pulse functions. Section 4 presents the fuzzy function approximation based on fuzzy block pulse functions. The fuzzy Fredholm integral equations of the second kind based on fuzzy block pulse functions proposed in Sect. 5. In Sect. 6, the error estimate of the numerical solution obtained by using fuzzy block pulse functions. One numerical example for proposed method is given in Sect. 7. Finally, Sect. 8 gives our concluding remarks.

2 Preliminaries

Definition 1. [9] A fuzzy number is a function $u : R \rightarrow [0, 1]$ having the properties:

(1) u is normal, that is $\exists x_0 \in R$ such that $u(x_0) = 1$,
(2) u is fuzzy convex set

$$(i.e.\ u(\lambda x + (1-\lambda)y) \geq \min\{u(x), u(y)\} \quad \forall x, y \in R, \lambda \in [0,1]),$$

(3) u is upper semicontinuous on R,
(4) the $\overline{\{\, x \in R : u(x) > 0\}}$ is compact set.

The set of all fuzzy numbers is denoted by R_F.

Definition 2. [13] An arbitrary fuzzy number is represented, in parametric form, by an ordered pair of functions $(\underline{u}(r), \overline{u}(r)), 0 \leq r \leq 1$, which satisfy the following requirements:

(1) $\underline{u}(r)$ is a bounded left continuous nondecreasing function over [0,1],
(2) $\overline{u}(r)$ is a bounded left continuous nonincreasing function over [0,1],
(3) $\underline{u}(r) \leq \overline{u}(r), \quad 0 \leq r \leq 1$.

The addition and scaler multiplication of fuzzy numbers in R_F are defined as follows:

(1) $u \oplus v = (\underline{u}(r) + \underline{v}(r), \overline{u}(r) + \overline{v}(r))$,

(2) $(\lambda \odot u) = \begin{cases} (\lambda \underline{u}(r), \lambda \overline{u}(r)) & \lambda \geq 0, \\ (\lambda \overline{u}(r), \lambda \underline{u}(r)) & \lambda < 0. \end{cases}$

Definition 3. [17] For arbitrary fuzzy numbers $u = (\underline{u}(r), \overline{u}(r))$ and $v = (\underline{v}(r), \overline{v}(r))$, the quantity

$$D(u, v) = \sup_{r \in [0,1]} \max\{|\underline{u}(r) - \underline{v}(r)|, |\overline{u}(r) - \overline{v}(r)|\}$$

is the distance between u and v.

Lemma 1. [22] The following properties are hold [11]:

(1) $D(u \oplus w, v \oplus w) = D(u, v) \quad \forall\, u, v, w \in R_F$,
(2) $D(k \odot u, k \odot v) = |k|\, D(u, v) \quad \forall\, u, v \in R_F \quad \forall k \in R$,
(3) $D(u \oplus v, w \oplus e) \leq D(u, w) + D(v, e) \quad \forall\, u, v, w, e \in R_F$,
(4) $D(u \oplus v, \tilde{0}) \leq D(u, \tilde{0}) + D(v, \tilde{0}), \quad \forall\ u, v \in R_F$,
(5) $D(k \odot u, k \odot v) = |k| D(u, v), \quad \forall\ u, v \in R_F, \quad \forall\ k \in R$.

In [22], it is proved that (R_F, D) is a complete metric space.

Remark 1. The properties (4) in Lemma 1 introduce the definition of a function $\|.\| : R_F \to R^+$ by $\|u\| = D(u, \tilde{0})$, which has the properties of the usual norms. In [4] the properties of this function are presented as follows:
(i) $\|u\| \geq 0, \quad \forall\ u \in R_F$, and $\|u\| = 0$ iff $u = \tilde{0}$,
(ii) $\|\lambda.u\| = |\lambda|\|u\|$ and $\|u \oplus v\| \leq \|u\| + \|v\|, \quad \forall\ u, v \in R_F, \quad \forall\ \lambda \in R$,
(iii) $|\|u\| - \|v\|| \leq D(u, v)$ and $D(u, v) \leq \|u\| + \|v\| \quad \forall\ u, v \in R_F$.

Definition 4. [8] Let $f, g : [a, b] \to R_F$, be fuzzy real number valued functions. The uniform distance between f, g is defined by

$$D^*(f, g) = \sup\{D(f(x), g(x) \mid x \in [a, b]\}$$

Definition 5. [2] Let $f : [a, b] \to R_F$ be a bounded function, then function $\omega_{[a,b]}(f, .) : R_+ \cup \{0\} \to R_+$,

$$\omega_{[a,b]}(f, \delta) = \sup\{D(f(x), f(y))| \; x, y \in [a, b], |x - y| \leq \delta\},$$

where R_+ is the set of positive real numbers, is called the modulus of continuity of f on $[a, b]$.

The following properties of the modulus of continuity are introduced in [13] as follows:

Theorem 1. The following properties holds:

(1) $D(f(x), f(y)) \leq \omega_{[a,b]}(f, |x-y|)$ *for any* $x, y \in [a,b]$,
(2) $\omega_{[a,b]}(f, |x-y|)$ *is increasing function of* δ,
(3) $\omega_{[a,b]}(f, 0) = 0$,
(4) $\omega_{[a,b]}(f, \delta_1 + \delta_2) \leq \omega_{[a,b]}(f, \delta_1) + \omega_{[a,b]}(f, \delta_2)$ *for any* $\delta_1, \delta_2 \geq 0$,
(5) $\omega_{[a,b]}(f, n\delta) \leq n\omega\, [a,b]\, (f, \delta)$ *for any* $\delta \geq 0$ $n \in N$,
(6) $\omega_{[a,b]}(f, \lambda\delta) \leq (\lambda + 1)\omega_{[a,b]}(f, \delta)$ *for any* $\delta, \lambda \geq 0$,
(7) *If* $[c,d] \subseteq [a,b]$ *then* $\omega_{[c,d]}(f, \delta) \leq \omega_{[a,b]}(f, \delta)$.

Definition 6. [15] A fuzzy real number valued function $f : [a,b] \to R_F$ is said to be continuous in $x_0 \in [a,b]$, if for each $\varepsilon > 0$ there is $\delta > 0$ such that $D(f(x), f(x_0)) < \varepsilon$, whenever $x \in [a,b]$ and $|x - x_0| < \delta$. We say that f is fuzzy continuous on $[a,b]$ if f is continuous at each $x_0 \in [a,b]$, and denote the space of all such functions by $C_F[a,b]$.

Definition 7. [5] Let $f : [a,b] \to R_F$. f is fuzzy-Riemann integrable to $I \in R_F$ if for any $\varepsilon > 0$, there exists $\delta > 0$ such that for any division $P = \{[u,v]\,;\xi\}$ of $[a,b]$ with the norms $\Delta(p) < \delta$, we have

$$D\left(\sum_P (v-u) \odot f(\xi), I\right) < \varepsilon,$$

where $\sum$ denotes the fuzzy summation. In this case it is denoted by $I = (FR)\int_a^b f(x)dx$.

Lemma 2. [5] If $f,\ g : [a,b] \subseteq R \to R_F$ are fuzzy continuous functions, then the function $F : [a,b] \to R_+$ by $F(x) = D(f(x), g(x))$ is continuous on $[a,b]$ and

$$D\left((FR)\int_a^b f(x)dx, (FR)\int_a^b g(x)dx\right) \leq \int_a^b D(f(x), g(x))dx.$$

Theorem 2. [16] If $f, g : [a,b] \to R_F$ are (FR) integrable fuzzy functions, and α, β are real numbers, then

$$(FR)\int_0^1 (\alpha \odot f(t) \oplus \beta \odot g(t))dt = \alpha \odot (FR)\int_0^1 f(t)dt \oplus \beta \odot (FR)\int_0^1 g(t)dt.$$

Definition 8. [5] A function $f : [a,b] \to R_F$ is said to be L-Lipschitz if $D(f(x), f(y)) \leq L|x-y|$ for any $x, y \in [a,b]$.

3 Review of Block-Pulse Functions

Firstly, we recall definition of the crisp block pulse functions in the following definition.

Definition 9. (see [18]). Block-pulse functions on the unit interval $[0, 1)$ is defined as follows:

$$\phi_i(t) = \begin{cases} 1 & t \in [\frac{i-1}{n}, \frac{i}{n}), \\ 0 & otherwise, \end{cases} \tag{1}$$

where $i = 1, 2, ..., n$ with a positive integer value for m. Also, ϕ_i is called i^{th} block-pulse function (BPF).

The BPFs satisfy in the properties of disjointness, orthogonality and completeness [18].

The disjointness property can be directly obtained from the definition of BPFs:

$$\phi_i(t)\phi_j(t) = \begin{cases} \phi_i(t) & i = j, \\ 0 & i \neq j, \end{cases}$$

where, $i, j = 1, 2, ..., n$.

The orthogonality property of BPF is given by

$$< \phi_i(t), \phi_j(t) >= \int_0^1 \phi_i(t)\phi_j(t)dt = \begin{cases} \frac{1}{n} & i = j, \\ 0 & i \neq j, \end{cases}$$

where $< ., . >$ denotes the inner product form and $i, j = 1, 2, ..., n$.

The completeness property is as follows: For every real bounded function $f(t)$ which is square integrable in the interval $t \in [0, 1)$, when n approaches to the infinity, Parseval's identity holds:

$$\int_0^1 f^2(t)dt = \sum_{i=0}^{\infty} f_i^2 \parallel \phi_i(t) \parallel^2,$$

where

$$f_i = n \int_0^1 f(t)\phi_i(t)dt.$$

Now, we defined the fuzzy block-pulse functions as follows:

Definition 10. For $f \in C_F[0, 1)$, the fuzzy block-pulse function like operator, $\Phi_m^{(F)}(f)(t)$, relative to the crisp knot sequence $(t_1, \cdots, t_n)$ is a function from the real line to the set of fuzzy numbers as the following form:

$$\Phi_n^{(F)}(f)(t) = \sum_{i=1}^{n} f(t_i) \odot \phi_i(t), \quad n \in N, \quad t \in [0, 1). \tag{2}$$

where $\phi_i(t)$'s are the crisp block-pulse basis functions defined by (1) and $t_i = \frac{i-0.5}{n}$ $\forall\ i = 1,...,n$. It is obvious that $\phi_i(t) \geq 0$, for all $t \in [0,1)$, $\phi_1(t), \phi_2(t), ..., \phi_m(t)$ are linearly independent, and

$$\sum_{i=1}^{n} \phi_i(t) = 1.$$

4 Function Approximation by Fuzzy Block-Pulse Functions

For f in $C_F([0,1))$, let us consider a fuzzy block pulse functions operator,

$$\Phi_n^{(F)}(f)(t) = \sum_{i=1}^{n} f(t_i) \odot \phi_i(t). \tag{3}$$

where $B_i(t)$ is defined by (1) and $t_i = \frac{i-0.5}{n}$ $\forall\ i = 1,...,n$.

Theorem 3. If $f \in C_F([0,1))$ and f satisfies in Lipschitz condition then

$$D^*(\Phi_n^{(F)}(f), f) \leq \frac{L_f}{n}. \tag{4}$$

Proof. Let $t \in [0,1)$, so there exists $j \in \{1,...,n\}$ such that $t \in [t_{j-1,}t_j]$. Then from Eq. (3), we observe that

$$D(\Phi_n^{(F)}(f)(t), f(t)) = D\left(\sum_{i=1}^{n} f(t_i) \odot \phi_i(t), f(t)\right)$$

$$= D\left(\sum_{i=1}^{n} f(t_i) \odot \phi_i(t), \sum_{i=1}^{n} \phi_i(t) \odot f(t)\right).$$

By the properties of Lemma 2.1 we have:

$$D(\Phi_n^{(F)}(f)(t), f(t)) \leq \sum_{i=1}^{n} \phi_i(t) D\left(f(t_i), f(t)\right).$$

It is obviously that $\text{supp}(\phi_j) = [t_{j-1}, t_j)$. For $t \in \text{supp}(\phi_j)$ by properties of the modulus of continuity we obtain:

$$D(\Phi_n^{(F)}(f)(t), f(t)) \leq \omega(f, |t_j - t_{j-1}|) \leq \omega(f, \frac{1}{n}).$$

Finally we get:

$$D^*(\Phi_n^{(F)}(f), f) \leq \frac{L_f}{n}.$$

□

Remark 2. It can be easily shown that $\lim_{n\to\infty} D^*(\Phi_n^{(F)}(f), f) = 0$.

5 Fuzzy Integral Equations

Here, we consider the linear fuzzy integral equation of Fredholm type

$$F(t) = f(t) \oplus \lambda \odot (FR) \int_a^b k(s,t) \odot F(s)ds, t \in [a,b]. \tag{5}$$

where $\lambda > 0$, $K(s,t)$ is an arbitrary kernel function on the square $a \leq s, t \leq b$, f, F are continuous fuzzy-number-valued functions.
In the following theorem are given sufficient conditions for the existence of an unique solution of Eq. (5) (see e.g. [14]).

Theorem 4. Let $k(s,t)$ be continuous for $a \leq s, t \leq b, \lambda > 0$, and $f(t)$ a fuzzy continuous of $t, a \leq t \leq b$. If

$$\lambda < \frac{1}{M(b-a)},$$

where

$$M = \max_{a \leq s,t \leq b} |k(s,t)|,$$

then the iterative procedure

$$F_0(t) = f(t),$$

$$F_k(t) = f(t) \oplus \lambda \odot (FR) \int_a^b k(s,t) \odot F_{k-1}(s)ds, \quad k \geq 1,$$

converges to the unique solution F of (5). Specially,

$$D^*(F, F_k) \leq \frac{L^k}{1-L} D^*(F_0, F_1),$$

where $L = \lambda M(b-a)$.

Remark 3. The upper bound of $D^*(F_0, F_1)$ can be obtained as follows:

$$D(F_0(t), F_1(t)) = D(f(t), f(t) \oplus \lambda \odot (FR) \int_a^b k(s,t) \odot f(s)ds \leq \lambda(b-a)MD(\tilde{0}, f(t)).$$

Thus, we have:

$$D^*(F_0, F_1) \leq L\|f\|_F,$$

and also, we have:

$$D^*(F, F_k) \leq \frac{L^{k+1}}{1-L}\|f\|_F. \tag{6}$$

Throughout this paper, we consider fuzzy Fredholm integral Eq. (5) with $a = 0$ and $b = 1$.

Now, we introduce the numerical method to find the approximate solution of Eq. (5). In this way, we consider the following uniform partitions of the interval $[0, 1)$:

$$\Delta : \ t_1 < ... < t_{n-1} < t_n \tag{7}$$

with $t_i = \frac{i-0.5}{n}, \ \ 1 \leq i \leq n$. Then the following iterative procedure gives the approximate solution of Eq. (5) in the point $t \in [0, 1)$ using fuzzy block-pulse functions:

$$y_0(t) = f(t),$$

$$y_m(t) = f(t) \oplus \lambda \odot (FR) \int_0^1 K(s,t) \odot \Phi_n^F(y_{m-1})(s)ds, \quad \forall t \in [0,1), \quad m \geq 1. \tag{8}$$

The above recursive relation can be written in the following form:

$$y_0(t) = f(t),$$

$$y_m(t) = f(t) \oplus \lambda \odot \sum_{i=1}^{n} G_i(t) \odot y_{m-1}(t_i), \quad \forall \ \ t \in [0,1), \quad m \geq 1, \tag{9}$$

where

$$G_i(t) = \int_0^1 K(s,t) \phi_i(s) ds.$$

6 Error Estimation

In this section, we investigate the convergence of the iterative proposed method to the solution of Eq. (4).

Theorem 5. Consider the Fredholm Eq. (5) with continuous kernel $K(s, t)$ having constant sign on $[a, b] \times [a, b]$ and f continuous on $[0, 1)$. Under the hypotheses of Theorem 4, iterative procedure Eq. (8) converges to the unique solution F of Eq. (5) and the following error estimate holds true:

$$D^*(F, y_m) \leq \frac{1}{1-L} \left(L^{m+1} \|f\|_F + \frac{L'}{n} \right),$$

where $L' = \max\{L_f, L_{y_1}, \cdots, L_{y_{m-1}}\}$.

Proof. Since

$$F_1(t) = f(t) \oplus \lambda \odot (FR) \int_0^1 K(s,t) \odot F_0(s) ds,$$

we have

$$D(F_1(t), y_1(t)) = D(f(t), f(t))$$

$$+\lambda D\left((FR)\int_0^1 K(s,t)\odot F_0(s)ds, (FR)\int_0^1 K(s,t)\odot \Phi_n^F(y_0)(s)ds\right)$$

$$\leq \lambda \int_0^1 |K(s,t)| D\left(f(s), \Phi_n^{(F)}(f)(s)\right) ds$$

$$\leq \lambda M \int_0^1 D\left(f(s), \Phi_n^{(F)}(f)(s)\right) ds.$$

From Theorem 4.1, we obtain:

$$D^*(F_1, y_1) \leq L.\frac{L_f}{n}. \tag{10}$$

Now, since

$$F_2(t) = f(t) \oplus \lambda \odot (FR)\int_0^1 K(s,t)\odot F_1(s)ds,$$

we conclude:

$$D(F_2(t), y_2(t)) = \lambda D\left((FR)\int_0^1 K(s,t)\odot F_1(s)ds, (FR)\int_0^1 K(s,t)\odot \Phi_n^F(y_1)(s)ds\right)$$

$$\leq \lambda \int_0^1 |K(s,t)| D\left(F_1(s), \Phi_n^{(F)}(y_1)(s)\right) ds$$

$$\leq \lambda M D^*\left(F_1, \Phi_n^{(F)}(y_1)\right)$$

$$\leq LD^*(F_1, y_1) + LD^*\left(y_1, \Phi_n^F(y_1)\right).$$

By using Eq. (10) and Theorem 4.1, we obtain:

$$D^*(F_2, y_2) \leq L^2.\frac{L_f}{n} + L.\frac{L_{y_1}}{n}.$$

By induction, we get:

$$D(F_m(t), y_m(t)) \leq L^m.\frac{L_f}{n} + L^{m-1}.\frac{L_{y_1}}{n} + \cdots + L.\frac{L_{y_{m-1}}}{n}.$$

Let us consider $L' = \max\{L_f, L_{y_1}, \cdots, L_{y_{m-1}}\}$ therefore we have:

$$D^*(F_m, y_m) \leq \frac{L'(1-L^m)}{n(1-L)}.$$

Since $L < 1$, according to $\frac{1-L^m}{n(1-L)} \leq \frac{1}{n(1-L)}$ for each $m \in N$, we get:

$$D^*(F_m, y_m) \leq \frac{L'}{n(1-L)}.$$

Using Eq. (6), we obtain:

$$D^*(F, F_m) \leq \frac{L^{m+1}}{1-L}\|f\|_F, \quad \forall \ t \in [0,1), \quad m \geq 1.$$

So we have

$$\begin{aligned} D^*(F, y_m) &\leq D^*(F, F_m) + D^*(F_m, y_m) \\ &\leq \frac{L^{m+1}}{1-L}\|f\|_F + \frac{L'}{n(1-L)} \\ &\leq \frac{1}{1-L}\left(L^{m+1}\|f\|_F + \frac{L'}{n}\right). \end{aligned}$$

Remark 4. Since $L < 1$, it can be easily proved that

$$\lim_{m\to\infty,\ n\to\infty} D^*(F, y_m) = 0,$$

that shows the convergence of the method.

7 Numerical Examples

In this section, we apply the presented method in Sect. 5 for solving the fuzzy integral Eq. (5) in one example. The approximate solution is calculated for different values of m and n. Also, we compare the numerical solution obtained by using the proposed method with the exact solution. The computations associated with the example was performed using Mathematica software.

Example 1. Consider the following linear fuzzy Fredholm integral equation:

$$F(t) = f(t) \oplus (FR)\int_0^1 k(s,t) \odot F(s)ds, \qquad t \in [0,1]$$

$$\underline{f}(t,r) = rt - \frac{1}{6}rt^2 - \frac{1}{12}r, \qquad t \in [0,1], \quad r \in [0,1]$$

$$\overline{f}(t,r) = 2t - rt - \frac{1}{3}t^2 - \frac{1}{6} + \frac{1}{6}rt^2 + \frac{1}{12}r, \qquad t \in [0,1], \quad r \in [0,1]$$

and kernel

$$k(s,t) = \frac{s^2+t^2}{3}, \qquad s,t \in [0,1].$$

The exact solution in this case is given by

$$(\underline{F}(t,r), \overline{F}(t,r)) = \big(rt, (2-r)t\big).$$

To compare the error with $m = 10,\ n = 10$ and $m = 12,\ n = 50$, see Table 1.

Table 1. The accuracy on the level sets for Example 2 in $t = 0.5$

r-level	m = 10	n = 10	m = 12	n = 50
	$\lvert \underline{F} - \underline{y}_{21} \rvert$	$\lvert \overline{F} - \overline{y}_{22} \rvert$	$\lvert \underline{F} - \underline{y}_{21} \rvert$	$\lvert \overline{F} - \overline{y}_{22} \rvert$
0.00	0.0000E+0	1.0548E-3	0.0000E+0	4.2225E-5
0.25	1.3183E-4	9.2292E-4	5.2782E-6	3.6947E-5
0.50	2.6369E-4	7.9108E-4	1.0556E-5	3.1669E-5
0.75	3.9554E-4	6.5923E-4	1.5835E-5	2.6391E-5
1.00	5.2739E-4	5.2739E-4	2.1113E-5	2.1113E-5

8 Conclusions

In this paper, we have presented an iterative procedure by using fuzzy block pulse functions to solve the linear Ferdholm fuzzy integral (5). The error estimate of the approximated function was obtained by using the fuzzy block pulse functions for the Lipschitzian function. The error estimate of the present method is proved; for obtaining the best approximating solution of the equation, the numbers m and n must be chosen sufficiently large. The analyzed example illustrates the ability and reliability of the fuzzy block pulse functions method for (5).

References

1. Abbasbandy, S., Babolian, E., Alavi, M.: Numerical method for solving linear Fredholm fuzzy integral equations of the second kind. Chaos Solitons Fractals **31**(1), 138–146 (2007)
2. Anastassiou, G.A.: Fuzzy Mathematics: Approximation Theory. Springer, Berlin (2010)
3. Babolian, E., Goghary, H.S., Abbasbandy, S.: Numerical solution of linear Fredholm fuzzy integral equations of the second kind by Adomian method. Appl. Math. Comput. **161**, 733–744 (2005)
4. Baghmisheh, M., Ezzati, R.: Numerical solution of nonlinear fuzzy Fredholm integral equations of the second kind using hybrid of block-pulse functions and Taylor series. Adv. Differ. Equat. **51** (2015). doi:10.1186/s13662-015-0389-7
5. Bede, B., Gal, S.G.: Quadrature rules for integrals of fuzzy-number-valued functions. Fuzzy Sets Syst. **145**, 359–380 (2004)
6. Bica, A.M.: Error estimation in the approximation of the solution of nonlinear fuzzy Fredholm integral equations. Inf. Sci. **178**, 1279–1292 (2008)
7. Bica, A.M., Popescu, C.: Approximating the solution of nonlinear Hammerstein fuzzy integral equations. Fuzzy Sets Syst. **245**, 1–17 (2014)
8. Congxin, W., Cong, W.: The supremum and infimum of the set of fuzzy numbers and its applications. J. Math. Anal. Appl. **210**, 499–511 (1997)
9. Dubois, D., Prade, H.: Fuzzy numbers: an overview. In: Analysis of Fuzzy Information, vol. 1, pp. 3–39. CRC Press, Boca Raton (1987)
10. Ezzati, R., Ziari, S.: Numerical solution and error estimation of fuzzy Fredholm integral equation using fuzzy Bernstein polynomials. Aust. J. Basic Appl. Sci. **5**(9), 2072–2082 (2011)

11. Ezzati, R., Ziari, S.: Numerical solution of nonlinear fuzzy Fredholm integral equations using iterative method. Appl. Math. Comput. **225**, 33–42 (2013)
12. Fariborzi Araghi, M.A., Parandin, N.: Numerical solution of fuzzy Fredholm integral equations by the Lagrange interpolation based on the extension principle. Soft. Comput. **15**, 2449–2456 (2011)
13. Friedman, M., Ma, M., Kandel, A.: Numerical solutions of fuzzy differential and integral equations. Fuzzy Sets Syst. **106**, 35–48 (1999)
14. Friedman, M., Ma, M., Kandel, A.: Solutions to fuzzy integral equations with arbitrary kernels. Int. J. Approx. Reason. **20**, 249–262 (1999)
15. Fang, J.-X., Xue, Q.-Y.: Some properties of the space fuzzy-valued continuous functions on a compact set. Fuzzy Sets Syst. **160**, 1620–1631 (2009)
16. Goetschel, R., Voxman, W.: Elementary fuzzy calculus. Fuzzy Sets Syst. **18**, 31–43 (1986)
17. Gong, Z., Wu, C.: Bounded variation absolute continuity and absolute integrability for fuzzy-number-valued functions. Fuzzy Sets Syst. **129**, 83–94 (2002)
18. Jiang, Z.H., Schanfelberger, W.: Block-Pulse Functions and Their Applications in Control Systems. Springer, Berlin (1992)
19. Park, J.Y., Han, H.K.: Existence and uniqueness theorem for a solution of fuzzy Volterra integral equations. Fuzzy Sets Syst. **105**, 481–488 (1999)
20. Park, J.Y., Jeong, J.U.: On the existence and uniqueness of solutions of fuzzy Volttera-Fredholm integral equations. Fuzzy Sets Syst. **115**, 425–431 (2000)
21. Subrahmanyam, P.V., Sudarsanam, S.K.: A note on fuzzy Volterra integral equations. Fuzzy Sets Syst. **81**, 237–240 (1996)
22. Wu, C., Gong, Z.: On Henstock integral of fuzzy-number-valued functions (I). Fuzzy Sets Syst. **120**, 523–532 (2001)
23. Ziari, S., Ezzati, R., Abbasbandy, S.: Numerical solution of linear fuzzy Fredholm integral equations of the second kind using fuzzy haar wavelet. In: Greco, S., Bouchon-Meunier, B., Coletti, G., Fedrizzi, M., Matarazzo, B., Yager, R.R. (eds.) IPMU 2012, Part III. CCIS, vol. 299, pp. 79–89. Springer, Heidelberg (2012)
24. Ziari, S., Bica, A.M.: New error estimate in the iterative numerical method for nonlinear fuzzy Hammerstein-Fredholm integral equations. Fuzzy Sets Syst. **295**, 136–152 (2015)

Author Index

Printed in the United States
By Bookmasters